NSF
GP 35602-X
UW 144·D444
PURCHASED DEC, 1974

PRINCIPAL INVESTIGATOR
John P. Walters

THE HYDROGEN MOLECULE WAVELENGTH
TABLES OF GERHARD HEINRICH DIEKE

GERHARD HEINRICH DIEKE, *Spectra and Energy Levels of Rare Earth Ions in Crystals,* edited by H. M. Crosswhite and Hannah Crosswhite, 1968

THE HYDROGEN MOLECULE
WAVELENGTH TABLES OF
GERHARD HEINRICH DIEKE

EDITED BY

H. M. CROSSWHITE
The Johns Hopkins University

WILEY-INTERSCIENCE, a Division of John Wiley & Sons, Inc.
New York • London • Sydney • Toronto

Library of Congress Catalog Card Number: 77-38780

ISBN 0-471-18890-5

Printed in the United States of America.

10 9 8 7 6 5 4 3 2 1

Preface

The death of Gerhard Heinrich Dieke on August 25, 1965, brought to an end over forty years of research on the spectra of molecular hydrogen and its isotopes, which he had begun in 1924 as a student of P. Ehrenfest at Leiden and had continued in 1926-1927 with J. J. Hopfield at Berkeley. After the discovery of deuterium he began work on D_2 and DH at The Johns Hopkins University, where he had come in 1930. When the third isotope, tritium, became available after World War II work was also begun (in cooperation with F. S. Tomkins of the Argonne National Laboratory and E. S. Robinson of the Los Alamos Scientific Laboratory) on the other three molecular isotopes T_2, TH, and TD. As experimental techniques were improved and the analytical work proceeded, it became more and more apparent that the older literature on H_2 was no longer adequate. Beginning in 1952, remeasurement of the H_2 spectrum itself (exclusive of the vacuum ultraviolet) was added to the program. This work was extended to nearly 3 microns in the PbS region of the infrared when a new type of spectrometer designed and built by W. G. Fastie was put into operation. Papers were published from time to time as the analysis progressed. The present volume makes available the accumulated experimental information on H_2 for use by theorists in extending the structural analysis and by experimental spectroscopists requiring detailed information in particular spectral regions. This description is essentially as he left it, except that the process of entering the data on punched cards, which he had begun, has now been completed and additional spectrum line identifications involving existing levels have been entered in the tables.

H. M. CROSSWHITE

Baltimore, Maryland
November 1971

v

Contents

Introduction

THE MEASUREMENTS

The tables in this book contain all of the significant experimental information between 2800 Å and 2.9 microns that was available from G. H. Dieke's records and notebooks at the time of his death. Many of the analytical results have already been reported,[1] including a detailed description of the experimental conditions used as well as literature references to previous work. The present table contains the experimental data on which the 1958 paper was based, together with some more recent unpublished extensions of the analysis. Except where noted in the tables, the information is from research at The Johns Hopkins University, principally since 1952. Because transitions involving the lowest electronic state $(1s^2\ ^1\Sigma_g^+)$ all occur in the vacuum ultraviolet,[2] they are not covered here.

The main spectrograph used was a 21-ft. Paschen with a 1200-groove/mm grating having a first-order dispersion of 1.25 Å/mm and a resolution limit of about 0.1 cm^{-1}. In most exposures of the H$_2$ spectrum the widths of the hydrogen lines themselves limited the attainable accuracy, which was a few hundredths of a cm^{-1}. This spectrograph was used in the second order below 5000 Å and in the first order between 5000 and 9000 Å. Above this, to the photographic limit near 12000 Å, exposures were taken on a 21-ft. Wadsworth spectrograph. With this spectrograph the limiting resolution was approximately 0.2 cm^{-1}, with a corresponding deterioration in accuracy.

These photographic wavelength measurements were calibrated by lines from an iron-neon hollow-cathode discharge tube.[3] Small residual corrections for individual plates were made when called for by comparison with H$_2$ interferometric measurements such as those of Foster and Richardson[4] and several hundred additional ones made in this laboratory.[5]

Above 1 micron, to the sensitivity limit of PbS detectors near 3 microns, additional measurements were made[6] on a 72-in. plane grating spectrometer of the Fastie-Ebert type capable of resolving about 0.04 cm^{-1}, but again the accuracy attained was limited by the light source broadening of the lines. This series of measurements was particularly valuable in that it permitted the linking together of two previously unconnected groups of triplet levels. These wavelength data (in Å) and associated wavenumbers (in cm^{-1}) are compiled in columns 1, 2, and 8 of the long T table that makes up the bulk of this book.

Three source conditions were chosen to assist in distinguishing the rotational and vibrational quantum assignments: (1) a low-temperature (77 K), low-pressure (0.1–0.2 mm) electrodeless discharge of low to medium power for minimum line broadening; (2) a high-power microwave discharge at 1 mm for maximum light intensity; and (3) a high-power electrodeless discharge with 80 mm of helium to suppress high vibrational quantum states. Columns 4, 5, and 6 of Table T give the corresponding estimated intensities (I_1, I_2, I_3) as seen in these three sources. Since these are eye estimates (on an approximately logarithmic scale), they give only a qualitative indication of the true intensity. The code letters occasionally appearing in these columns have the following significance:

b   Broad line with sharp sides (usually due to overexposure)
d   Diffuse line
r   Line shaded to red
v   Line shaded to violet

The values I_4 in column 7 represent an attempt to make the intensity values more quantitative. Photoelectric measurements in selected spectral regions for the low-pressure, low-temperature condition were made[7] to serve as a calibration for densitometric work on the photographic plates. Many entries have been made for the stronger lines, but this phase of the work was interrupted by Dieke's death.

Column 3 (Ref.) contains code letters referring to earlier literature:

F   W. Finkelnburg, *Z. Physik* **52,** 27 (1928)
G   H. G. Gale, G. S. Monk, and K. O. Lee, *Astrophys. J.* **67,** 89 (1928)
K   N. A. Kent, *Astrophys. J.* **84,** 585 (1936)
      N. A. Kent and R. G. Lacount, *Astrophys. J.* **86,** 311 (1937)
      N. A. Kent and R. G. Lacount, *Phys. Rev.* **51,** 241 (1937)
      L. A. Combes, R. H. Fry, and N. A. Kent, *Phys. Rev.* **56,** 678 (1939)
R   E.W. Foster and O.W. Richardson, *Proc. Roy. Soc.* **A189,** 149 (1947)

When this column contains an i, an impurity line is indicated. This is usually due to a small argon impurity in the helium gas used in condition (3), but lines of He, Si, O, Ag (from the discharge tube wall coating used to promote recombination of hydrogen atoms), Na, and Ne have also been identified.

Columns 9 and 10 identify the upper and lower electronic and vibrational states, using a notation explained in the next section. Column 11 gives the branch type and rotational quantum number of the lower state, thus completing the identification of the two levels involved. When more than one classification can be made for an individual line, all pertinent information of columns 1 to 8 is repeated.

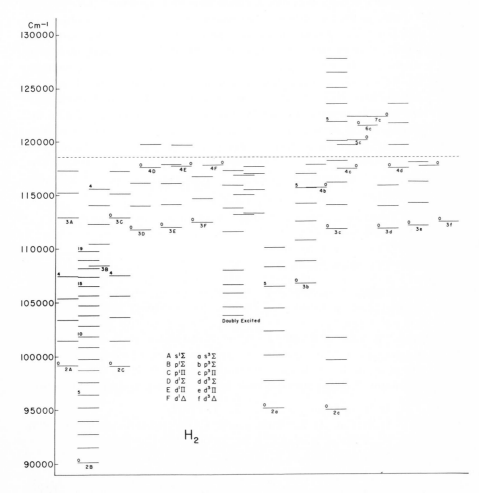

FIGURE 1. The electronic and vibrational states of the H_2 molecule. The energy scale is with reference to zero for the ground state $1s^2 \ ^1\Sigma$. The code letters (capital for singlets, lowercase for triplets) refer to singly-excited electronic states, with one electron remaining in a 1s orbital. The positions of some doubly-excited singlet states are also indicated. The dotted line gives the location of the dissociation energy.

THE ENERGY LEVELS

The energy levels given in the E tables on pages E1 to E13 are essentially those of the 1958 paper, with a few additions and corrections. The electronic and vibrational structure of the most important of these is shown in Figure 1, which also gives the code used in identifying them in the T tables. As was done in the 1958 paper, levels in the E tables that belong to orthohydrogen are given in *italics* and those for para-hydrogen in ordinary type. Some of these must be regarded as tentative because of the unfinished nature of the analysis. In particular, levels of 2B ($2p\ {}^1\Sigma_u^+$) having vibrational quantum numbers higher than 9 are not reproduced here, as they are not firmly established. However, the classifications (columns 9 to 11 of T tables) have been retained in aid of further work on this point. Some of the higher levels involving a 3d orbital have also been questioned as being too far out of line with *ab initio* calculations of their position,[8] but these have been retained.

ACKNOWLEDGMENTS

The extensive experimental investigations and the accumulation of the prodigious quantity of data that has gone into this compilation was made possible by the continuing interest of the Division of Research of the Atomic Energy Commission, covering a period of more than twenty years, and to a grant from the Rockefeller Foundation for assistance with editorial expenses. A large share of the credit must also be given to a long series of students working on various phases of the problem, R. W. Blue, Margaret N. Lewis, N. Ginsburg, C. S. Rainwater, S. P. Cunningham, F. T. Byrne, S. P. S. Porto, C. L. Beckel, A. M. Wittenberg, D. L. Dimock, P. Gloersen, and R. J. Boye; and to Mrs. D. Wittenberg and Mrs. A. T. Young for long hours spent with comparator and notebooks.

REFERENCES

1. G. H. Dieke, *J. Molecular Spectrosc.* **2,** 494 (1958); S. P. S. Porto and G. H. Dieke, *J. Opt. Soc. Am.* **45,** 447 (1955); P. Gloersen and G. H. Dieke, *J. Molecular Spectrosc.* **16,** 191 (1965); G. H. Dieke and S. P. Cunningham, *J. Molecular Spectrosc.* **18,** 288 (1965); and references contained there.

2. P. G. Wilkinson, *Can. J. Phys.* **46,** 1225 (1968); G. Herzberg, *Phys. Rev. Letters* **23,** 1081 (1969); G. Herzberg, *J. Molecular Spectrosc.* **33,** 147 (1970); and references contained there.

3. H. M. Crosswhite, *The Iron-Neon Hollow Cathode Spectrum,* NBS Monograph, in press.

4. E. W. Foster and O. W. Richardson, *Proc. Roy. Soc.* **A189,** 149 (1947).

5. H. M. Crosswhite, unpublished measurements.

6. P. Gloersen and G. H. Dieke, *J. Molecular Spectrosc.* **16,** 191 (1965); P. Gloersen, dissertation, The Johns Hopkins University, 1956.

7. D. L. Dimock, dissertation, The Johns Hopkins University, 1957.

8. C. B. Wakefield and E. R. Davidson, *J. Chem. Phys.* **43,** 834 (1965); W. M. Wright and E. R. Davidson, *J. Chem. Phys.* **43,** 840 (1965); and M. L. Ginter, *J. Chem. Phys.* **46,** 3687 (1967).

THE HYDROGEN MOLECULE WAVELENGTH
TABLES OF GERHARD HEINRICH DIEKE

SINGLET LEVELS

(singly excited)

2A (2s ¹Σ)

K	v = 0	1	2	3	4
0	99 156.76	101 486.81	103 551.59	105 376.90	107 417.87
1	220.19	545.98	597.58	407.23	441.65
2	346.54	663.65	682.13	465.97	488.55
3	534.77	841.41	987.21	548.84	557.33
4	783.30	102 072.98	104 151.79	650.55	646.30
5	100 090.29	359.22	378.90		
6	453.18				
7	869.42				
8	101 335.90				
9	849.26				

3A (3s ¹Σ)

K	v = 0	1	2
0	112 949.57	115 243.52	117 289.09
1	113 008.73	288.88	330.50
2	126.08	385.84	447.50
3	295.44	536.94	582.25
4			751.31
5			991.56

SINGLET LEVELS

(singly excited)

2B (2p ¹Σ)

K	v = 0	1	2	3	4
0	90 195.46	91 513.80	92 795.28	94 041.99	95 255.00
1	234.31	550.72	830.56	075.81	287.57
2	311.61	624.16	900.76	143.14	352.20
3	426.67	733.61	93 005.43	243.55	448.72
4	578.35	878.04	143.63	376.24	576.34
5	765.31	92 056.30	314.34	540.23	734.09
6	986.06	267.00	516.32	734.34	921.09
7	91 238.44	508.37	747.91	957.17	96 135.82
8	520.62	778.59	94 007.72	95 207.30	376.97
9	830.19	93 075.86	293.45	482.60	642.93
10	92 165.44	398.02	604.05	781.40	

K	5	6	7	8	9
0	96 434.95	97 582.45	98 697.94	99 781.27	100 834.02
1	466.34	612.66	726.98	807.12	857.6
2	528.53	672.53	784.57	865.86	915.2
3	621.46	762.10	870.90	946.15	999.0
4	744.37	880.49	984.80	100 060.18	101 098.2
5	896.34	98 027.00	99 126.38	187.60	132.9
6	97 076.61	201.00			
7	283.69	401.00			
8	516.49	625.66			

3B (3p ¹Σ)

K	v = 0	1	2	3	4
0	108 435.66	110 470.51	112 350.52	114 067.17	115 598.39
1		521.41	396.47	110.35	639.06
2	602.61	622.82	488.17	196.40	720.02
3		774.08	625.70	324.99	841.47
4		974.16	808.50	495.35	116 003.56
5		111 221.60	113 035.60	706.83	
6		514.93	305.75	956.70	

SINGLET LEVELS
(singly excited)
2C (2p ¹Π)

K	v = 0	1	2	3	4
1–	99 142.86	101 448.37	103 619.91	105 660.16	107 573.7
+	144.03	449.60	620.79	652.80	
2	264.37	563.61	729.04	766.96	766.1
	267.84	567.00	730.46	775.81	
3	445.46	735.27	891.51	916.66	812.2
	451.97	741.20	887.97		
4	684.73	962.08	104 106.17	106 121.4	108 000.5
	694.68	968.66	133.70		
5	980.33	102 242.33	371.37	371.4	239.9
	993.23	251.02			
6	100 330.16	573.72	634.94		
7	731.71	954.24	105 044.76		

3C (3p ¹Π)

K	v = 0	1	2
1–	112 923.63	115 146.27	117 239.86
+	927.27	147.87	243.69
2	113 041.20	257.89	345.38
	051.70	261.65	355.32
3	216.59	424.00	502.54
	235.04	429.94	518.88
4	448.20	643.36	710.14
	478.04	649.51	730.43
5	734.40	914.53	966.60
6	114 072.92		
7	461.52		
8			
9	115 377.72		

3D (3d ¹Σ)

K	v = 0	1	2	3
0	111 804.63	114 036.66	116 156.81	
1	797.11	022.86	095.65	117 839.36
2	819.78	038.36	140.36	865.75
3	885.07	085.58	206.23	911.56
4	997.49	172.25	297.38	979.71
5	112 162.01	310.58	416.46	118 098.46
6	379.49	500.14	565.33	
7	632.99	737.01	761.06	
8	978.63	115 027.47		
9	113 341.26	383.23		
10	759.56			
11	114 193.89			

4D (4d ¹Σ)

v = 0
117 428.82
403.08
383.59
445.99
559.98
.730.03
955.20
118 233.58

SINGLET LEVELS
(singly excited)

3E (3d $^1\Pi$)

K	v = 0	1	2	3
1-	112 064.91	114 164.13	116 106.42	117 873.49
+	127.23	215.95	225.76	932.18
2	139.61	244.86	189.52	956.15
	274.24	345.75	341.14	118 019.29
3	264.09	371.11	316.01	080.35
	463.04	494.49	487.05	131.30
4	441.12	544.96	486.72	246.04
	695.70	669.65	666.92	249.58
5	671.10	767.09	701.73	452.91
	958.43		875.64	
6	953.21	115 037.29	960.88	700.33
	113 283.62			
7	285.57	353.78	117 262.90	995.71
	693.53			
8	666.18	715.34	606.40	119 312.35
9	114 093.34	116 119.13	989.30	
10	561.25	563.70	118 409.71	

4E (4d $^1\Pi$)

K	v = 0	1	2
1-	117 581.25	119 724.05	
+	682.54		
2	595.37	759.61	
3	702.18	863.35	
	872.79		
4	870.14	120 023.55	121 778.90

3F (3d $^1\Delta$)

K	v = 0	1	2	3
2-	112 517.95	114 710.24	116 779.20	118 727.12
+	528.75	713.42	779.75	
3	735.56	906.55	955.16	882.35
	766.60	915.50	952.16	868.44
4	113 010.39	115 158.62	117 183.62	
	070.18	173.56		
5	338.57	462.26		
	407.49			
6	716.85	816.17		

4F (4d $^1\Delta$)

v = 0
117 826.75
928.32
118 063.35
194.41
348.84
523.89

SINGLET LEVELS

(doubly excited)
2K $((2p\sigma)^2\ ^1\Sigma)$

K	v = 2	3	4	5	6
0	103 830.54	104 722.61	105 958.16	106 705.07	108 090.56
1	*849.83*	*739.33*	*983.22*	*726.19*	*114.14*
2	895.00	772.20	106 034.58	768.46	161.59
3	781.97	820.38	*114.35*	*831.66*	*232.64*
4	858.84	883.67	224.94	916.53	
5		*964.04*	*366.20*	107 025.08	

V

K	v = 0	1
0		
1	*111 379.13*	*112 118.13*
2	539.02	271.12
3	*731.36*	*455.12*
4	776.25	
5		

Z

	v = 0	1	2
0	111 620.81	113 853.40	
1	642.27	*871.34*	116 023.65
2	685.73	912.31	039.33
3	751.92	*979.54*	080.60
4	837.27		143.54
5	*953.96*		224.41

Y

K	v = 1	2
0	113 250.24	116 907.41
1	*269.69*	*923.86*
2	308.60	956.40
3	*370.41*	117 008.19
4		
5		

X

	v = 0	1	2
0	113 385.50		117 706.66
1	*410.81*	115 569.68	706.27
2	462.58	598.77	724.89
3	*542.35*	*645.92*	*760.98*
4	654.45		828.26
5	*764.52*		

U

K	v = 1
0	
1	*114 520.53*
2	558.16
3	*628.18*
4	777.41
5	*996.91*

W

	v = 1	2
0	115 091.84	117 073.43
1	*128.70*	*098.45*
2	199.34	148.38
3	*302.01*	*224.03*
4	442.26	332.37
5	*638.21*	*512.58*

T

	v = 0
0	117 325.78
1	376.40
2	446.26
3	*560.73*

TRIPLET LEVELS

(singly excited)

2a (2s $^3\Sigma$)

K	v = 0	1	2	3	4
0	95 226.00	97 750.32	100 138.60	102 394.73	104 521.63
1	292.67	813.74	198.86	451.90	575.78
2	425.50	940.06	318.89	565.81	683.65
3	623.44	98 128.30	497.73	735.49	844.33
4	884.93	376.95	733.92	959.52	105 056.20
5	96 208.04	684.14	101 025.74	103 236.33	
6	590.38	99 047.64	370.86	563.61	
7	97 029.25	464.76			
8	521.55	932.45			
9	98 064.32	100 448.25			
10	653.79	101 008.13			
11	99 286.82				
12	959.54				

K	5	6	7	8
0	106 521.22	108 394.05		
1	572.35	442.16	110 184.36	111 784.59
2	674.27	538.15		
3	825.69	680.73		
4	107 026.21			
5				
6				
7				
8				

3a (3s $^3\Sigma$)

K	v = 0	1	2
0	112 021.48	114 290.21	
1	082.71	347.65	
2	199.72	461.71	116 597.85
3	372.94	631.88	761.03
4	607.42	856.66	976.65
5		115 134.10	
6		461.80	
7		837.06	

(singly excited)

3b (3p ³Σ)

K	v = 0	1	2	3	4
0	106 831.61	108 894.31	110 821.59	112 610.75	114 254.32
1	884.73	944.36	868.52	654.68	296.58
2	990.50	109 044.09	962.06	742.22	380.09
3	107 148.24	192.74	111 101.53	872.77	507.52
4	356.79	389.34	286.01	113 045.36	688.58
5	614.77	632.46	514.04	258.73	
6	920.35	920.42	784.15	511.46	
7	108 271.48	110 251.14	112 094.47	821.62	
8	666.02	621.30	442.87		
9	109 101.30		826.87		
10	574.96		113 244.54		
11	110 083.91				
12	625.62				
13	111 196.90				

K	5	6
0	115 731.42	
1	773.29	117 066.87
2	849.14	139.80
3	959.33	266.67

4b (4p ³Σ)

K	v = 0	1	2
0	115 751.92	117 895.50	
1	801.64	948.62	
2	908.43	118 054.46	120 114.80
3	116 070.76	210.98	
4	286.32		
5	553.56		

TRIPLET LEVELS

(singly excited)

2c (2p ³Π)

K	v = 0	1	2	3	4
1–	*95 091.32*	*97 430.14*	*99 646.91*	*101 744.94*	*103 725.88*
+	091.32	430.10	646.88	744.94	860.34
2	212.11	545.17	756.28	848.88	964.08
	212.06	*545.24*	*756.68*	*849.42*	
3	*392.10*	*716.62*	*919.48*	*102 004.12*	
	392.01	716.78	919.97	004.70	104 119.80
4	629.95	943.16	100 135.04	208.92	
	629.77	*943.49*	*135.78*	*210.16*	
5	*924.01*	*98 223.30*	*401.55*	*462.15*	
	923.67	223.52	402.56	462.98	
6	96 272.09	554.76	716.85		
	271.61	*555.01*	*718.09*		
7	*672.22*	*935.40*	*101 078.81*		
	653.08	935.57			
8	97 105.93				

3c (3p ³Π)

K	v = 0	1	2	3	4
1–	*111 904.09*	*114 144.33*	*116 259.15*	*118 252.64*	*120 128.14*
+	904.57	143.97	260.11	254.64	
2	112 021.86	256.09	365.16	353.03	233.04
	023.22	*254.44*	*367.80*	*357.44*	
3	*197.36*	*422.67*	*523.08*	*502.56*	*364.39*
	199.91	416.14	527.83	508.99	
4	429.22	642.70	731.65	700.01	550.73
	433.09	*629.98*	*738.52*		
5	*715.66*	*914.57*	*989.38*	*943.86*	
	720.88	866.98	997.79		
6	113 054.81	115 236.34	117 294.22		
	060.95		*303.09*		
7		*605.43*			
	450.46		650.34		
8	886.13	116 019.48			

K	5	6	7	8	9
1–	*121 888.26*	*123 534.98*	*125 069.49*	*126 492.09*	*127 802.15*
+					
2	977.78	619.18	148.39		
3	*122 110.79*	*744.19*	*265.54*		
4	287.19	909.51			

(singly excited)

4c (4p ³Π)

K	v = 0	1	2	3	4
1−	117 555.90	119 765.54	121 849.41	123 811.50	125 654.54
+	555.76				
2	672.68	876.28	954.30	910.75	748.22
	672.09				
3	846.72	120 041.29	122 110.52	124 058.58	887.84
	845.10				
4	118 076.65	259.23	316.96	253.84	126 072.06
	073.11				
5	360.69	528.56	571.94	495.38	299.52
	353.74				
6	686.79				
	684.62				
7	119 082.91				

K	5	6	7
1−	127 382.24	128 996.08	130 496.91
2	470.43	129 078.63	
3	602.00	202.54	

5c (5p ³Π)

K	v = 0	1	2	3
1−	120 131.94	122 327.64	124 397.63	126 347.00
+	133.05	343.64		
2	248.49	438.91		444.11
	255.30	454.65		
3	421.80	604.17		588.15
	434.59	618.91		
4	650.89	821.98		779.18
		844.50		
5	933.95	123 090.82		

TRIPLET LEVELS

(singly excited)

6c (6p $^3\Pi$)

K	v = 0	1
1–	121 516.93	123 705.64
2	633.26	
3	806.37	
4	122 034.41	
5	315.95	

7c (7p $^3\Pi$)

K	v = 0
1–	122 353.40
+	
2	
+	468.81
3	642.35

3d (3d $^3\Sigma$)

K	v = 0	1	2	3
0	111 948.18	114 035.99	115 934.57	117 658.34
1	946.19	033.41	944.14	676.91
2	976.23	060.56	979.00	721.10
3	112 046.32	130.02	116 051.59	799.17
4	166.96	246.94	167.12	915.52
5	340.44	413.43	328.09	118 073.04
6	566.95	628.94		272.67
7	844.84	893.08		
8	113 154.75	115 204.13		

4d (4d $^3\Sigma$)

K	v = 0	1	2	3
0	117 616.61	119 765.91	121 761.42	123 621.46
1	567.41	715.06	720.26	588.45
2	569.32	716.37	726.56	597.99
3	622.39	770.79	781.06	653.04
4	735.79	879.46	883.61	
5	915.84	120 040.91	122 032.66	
6	118 161.38			
7	469.91			

(singly excited)

3e (3d ³Π)

K	v = 0	1	2	3
1–	112 216.42	114 329.78	116 295.29	118 108.43
+	*303.30*	*418.76*	*331.82*	*133.26*
2	*290.57*	*408.61*	*377.51*	*190.77*
	460.60	*568.99*	*468.89*	*256.46*
3	414.57	533.93	502.97	314.44
	653.33	*763.75*	*653.79*	*428.48*
4	*591.33*	*707.32*	*672.93*	*480.31*
	890.64	999.84	881.06	634.25
5	821.43	929.20	887.68	
	113 176.66	*115 278.75*	*117 149.32*	*888.85*
6	*103.52*	*200.41*	*147.01*	
		601.33	456.83	
7	434.34		449.70	
8			*794.31*	

4e (4d ³Π)

K	v = 0	1	2	3
1–	117 713.82	119 878.23	121 926.72	123 516.17
+		*911.92*	*940.43*	*818.42*
2	*749.15*	*919.74*	*963.45*	*953.08*
	868.68	120 034.99	122 067.88	565.37
3	855.84	022.65	061.18	666.64
	118 039.83	*211.40*	*246.33*	*124 137.29*
4	*023.79*	*182.58*	*209.03*	*371.45*
	267.43	441.72	475.05	123 818.36
5	248.91	394.23	407.40	
	548.96	*722.10*	*752.21*	*124 650.39*
6	*528.54*	*658.23*	*609.64*	
	880.68			
7		975.70		
	119 256.23			

T R I P L E T   L E V E L S

(singly excited)

3f (3d ³Δ)

K	v = 0	1	2	3
2–	*112 663.62*	*114 856.34*	*116 925.90*	*118 874.80*
+	679.25	861.68	927.52	875.22
3	881.92	115 054.16	117 104.40	119 034.83
	929.09	*074.30*	*111.78*	*037.28*
4	*113 157.24*	*307.81*	*335.95*	*243.85*
	242.56	350.64	353.50	249.61
5	485.86	613.01	616.70	497.88
	610.19	*683.00*	*648.23*	*514.15*
6	*864.58*	*966.27*	*942.96*	
	114 026.32	116 064.26	990.63	
7		364.24	118 311.18	
		488.83		

4f (4d ³Δ)

K	v = 0	1	2	3
2–	*117 979.95*	*120 146.74*	*122 188.65*	*124 106.36*
+	118 108.08		325.61	245.72
3	216.86	373.40	307.70	309.62
	393.42		*587.49*	*499.58*
4	*502.24*	*643.90*	*654.92*	*530.46*
	760.31		920.32	813.64
5				
	119 190.80		*123 304.97*	

(doubly excited)

u

K	v = 0	1	2
0			
1	113 144.44	115 529.20	117 701.91
2	*242.57*	*626.00*	*793.15*
3	407.00	777.38	947.69
4		*995.22*	*118 169.19*

t

K	v = 0	1
0	*113 501.97*	*115 681.69*
1	455.79	
2	*476.94*	*898.28*
3	571.10	
4	*739.61*	*116 017.20*

g

K	v = 0	1
0	*118 520.96*	*120 978.10*
1	396.36	891.49
2	*339.47*	*859.08*
3	348.78	886.02
4	*423.80*	*981.79*
5		121 140.00

w

K	v = 0
0	120 355.67
1	*312.85*
2	304.14

λ_{air}	λ_{vac}	Ref.	I_1	I_2	I_3	I_4	ν	Upper	Lower	Br.
2805.251	2806.077		0.2				35636.94			
2828.238	2829.070		0.1d				35347.30			
2834.638	2835.472		0.2d				35267.50			
2836.772	2837.606		0.2d				35240.97			
2837.013	2837.847		0.1				35237.98			
2837.625	2838.459		0.1d				35230.38			
2838.341	2839.176		0.3				35221.49			
2838.963	2839.798		0.1d				35213.77			
2881.579	2882.424		0.5				34693.02			
2888.229	2889.076		0.3				34613.14			
2945.105	2945.966	1	0.7				33944.72			
2977.563	2978.432		0.2b				33574.71			
2987.723	2988.595		0.2				33460.54			
2998.216	2999.090		0.2d				33343.45			
3002.415	3003.290		0.3				33296.82			
3009.142	3010.019		0.2				33222.38			
3012.919	3013.797		0.3				33180.74			
3041.421	3042.306		0.2				32869.80			
3057.910	3058.799		0.2d				32692.57			
3059.689	3060.579		0.2				32673.56			
3060.892	3061.782		0.2b				32660.72			
3079.855	3080.750		0.1d				32459.63			
3079.959	3080.853		0.2d				32458.54			
3091.653	3092.550		0.2				32335.77			
3107.711	3108.613		0.3				32168.69			
3114.928	3115.832		0.3				32094.16			
3119.631	3120.536		0.3				32045.78			
3121.949	3122.854		0.2	0.3			32021.99			
3124.617	3125.523			0.1			31994.65			
3127.736	3128.643			0.1			31962.74			
3127.833	3128.740			0.1			31961.75			
3128.350	3129.257			0.2			31956.47			
3128.416	3129.322			0.1			31955.80			
3128.618	3129.525		0.6	0.4			31953.73			
3132.007	3132.915		0.2	0.1			31919.16			
3132.246	3133.154			0.1			31916.72			
3132.332	3133.239			0.2			31915.85			
3132.689	3133.597			0.1			31912.21			
3132.867	3133.775			0.1			31910.40			
3133.509	3134.417			0.1			31903.86			
3133.670	3134.578			0.1			31902.22			
3133.780	3134.688			0.1			31901.10			
3135.350	3136.258			0.2			31885.13			
3135.432	3136.341			0.1			31884.29			
3135.642	3136.550			0.2			31882.16			
3136.287	3137.196			0.1			31875.60			
3137.275	3138.184			0.1			31865.56			
3140.219	3141.129			0.2			31835.69			
3140.722	3141.632			0.1			31830.59			
3141.086	3141.996			0.1			31826.90			
3141.657	3142.567			0.2			31821.12			
3142.280	3143.190			0.2			31814.81			
3142.448	3143.358			0.2			31813.11			
3142.693	3143.603			0.2			31810.63			
3142.869	3143.779			0.2			31808.85			
3143.612	3144.523			0.2d			31801.33			
3146.542	3147.453			0.1			31771.72			
3150.022	3150.934			0.1			31736.62			
3151.207	3152.120			0.2			31724.68			
3151.408	3152.321		0.3	0.3			31722.66			

λ_{air}	λ_{vac}	Ref.	I_1	I_2	I_3	I_4	ν	Upper	Lower	Br.
3151.860	3152.773			0.2			31718.11			
3152.418	3153.331			0.1			31712.50			
3152.854	3153.767			0.2			31708.11			
3153.001	3153.914			0.1			31706.63			
3153.306	3154.219			0.2d			31703.57			
3153.747	3154.660			0.2d			31699.14			
3154.496	3155.409			0.2b			31691.61			
3154.960	3155.873			0.2d			31686.95			
3155.232	3156.145			0.1			31684.22			
3156.815	3157.729			0.1			31668.33			
3158.567	3159.482			0.2			31650.76			
3158.884	3159.798			0.1			31647.59			
3158.980	3159.894			0.1			31646.63			
3159.732	3160.647			0.2			31639.09			
3164.790	3165.706			0.1			31588.53			
3165.832	3166.749			0.2			31578.13			
3167.270	3168.186			0.1			31563.80			
3169.342	3170.259			0.1			31543.16			
3171.180	3172.098			0.1			31524.88			
3174.489	3175.408			0.2			31492.02			
3175.638	3176.556			0.1			31480.63			
3175.721	3176.639			0.2			31479.81			
3176.146	3177.065			0.1			31475.59			
3176.986	3177.905		0.3	0.4			31467.27	4E 2	2B 0	Q 2
3179.334	3180.254			0.1			31444.03			
3179.703	3180.623			0.1			31440.38			
3181.685	3182.605			0.2			31420.80			
3182.300	3183.220			0.1			31414.73			
3183.430	3184.350			0.1			31403.58			
3184.620	3185.541			0.2			31391.84			
3186.481	3187.402			0.1			31373.51			
3187.108	3188.029		0.1				31367.34			
3187.252	3188.174		0.2	0.2			31365.92			
3187.375	3188.297			0.1			31364.71			
3187.745	3188.667	i	1.0	0.4			31361.07			
3188.758	3189.680			0.1			31351.11			
3189.119	3190.041			0.2d			31347.56			
3189.509	3190.431			0.2			31343.73			
3191.285	3192.208			0.1			31326.28			
3192.546	3193.469			0.1			31313.91			
3194.110	3195.033			0.1			31298.58			
3194.837	3195.760			0.1			31291.46			
3198.189	3199.113			0.1			31258.66			
3198.389	3199.313		0.2	0.2			31256.71			
3198.521	3199.445			0.1			31255.42			
3198.694	3199.618			0.1			31253.73			
3198.972	3199.897			0.2d			31251.01			
3199.113	3200.038			0.2			31249.63			
3199.512	3200.436			0.1			31245.74			
3199.958	3200.883			0.1			31241.38			
3206.456	3207.383		0.2b				31178.07			
3208.980	3209.907			0.2			31153.55			
3209.572	3210.500		0.1d				31147.80			
3210.954	3211.881			0.2			31134.40			
3211.457	3212.385			0.2			31129.52			
3211.639	3212.567		0.2d				31127.76			
3213.171	3214.099			0.2d			31112.92			
3213.331	3214.259		0.3	0.3			31111.37			
3213.441	3214.370			0.2			31110.30			
3213.725	3214.654			0.2			31107.55			

λ_{air}	λ_{vac}	Ref.	I_1	I_2	I_3	I_4	ν	Upper	Lower	Br.
3217.984	3218.914			0.1			31066.38			
3218.730	3219.660		0.2	0.3			31059.18			
3220.397	3221.327			0.1			31043.11			
3220.837	3221.767			0.1			31038.87			
3221.597	3222.528			0.1			31031.54			
3222.964	3223.895			0.1			31018.38			
3224.568	3225.500		0.2	0.2			31002.95			
3225.012	3225.943			0.2d			30998.69			
3227.270	3228.202			0.1			30977.00			
3228.774	3229.706			0.2			30962.57			
3229.170	3230.103			0.1			30958.77			
3231.207	3232.139		0.2	0.1d			30939.26			
3234.233	3235.167			0.1			30910.31			
3234.858	3235.791			0.1			30904.34			
3235.554	3236.488			0.1			30897.69			
3238.139	3239.073			0.1			30873.03			
3238.663	3239.598			0.2d			30868.03			
3239.132	3240.067			0.1			30863.56			
3239.464	3240.399			0.1			30860.40			
3241.571	3242.506		0.2	0.0d			30840.34			
3242.070	3243.006			0.1			30835.59			
3244.422	3245.358		0.1	0.1			30813.24			
3245.294	3246.230			0.2			30804.96			
3247.292	3248.229			0.3			30786.01			
3248.088	3249.025			0.1			30778.46			
3248.957	3249.894		0.1				30770.23			
3249.110	3250.048			0.1			30768.78			
3249.263	3250.201			0.1			30767.33			
3251.414	3252.352			0.1			30746.98			
3251.579	3252.517		0.2	0.3			30745.42			
3251.922	3252.860		0.3	0.2			30742.18			
3253.496	3254.434		0.3				30727.31			
3255.580	3256.519		0.2				30707.64			
3257.591	3258.530			0.1			30688.68			
3260.115	3261.055			0.1			30664.92			
3260.363	3261.303		0.3				30662.59			
3261.029	3261.969		0.5				30656.33			
3262.742	3263.683			0.1			30640.23			
3263.374	3264.315			0.2			30634.30			
3265.885	3266.826			0.2			30610.75			
3266.333	3267.275		0.1				30606.55			
3266.426	3267.367		0.1	0.2d			30605.68			
3267.014	3267.956			0.2d			30600.17			
3267.076	3268.018			0.3			30599.59			
3268.041	3268.983		0.2d				30590.55			
3268.195	3269.137			0.0			30589.11			
3268.661	3269.603			0.1			30584.75			
3269.831	3270.773		0.4				30573.81			
3269.964	3270.907		0.1				30572.56			
3270.648	3271.591			0.2			30566.17			
3270.794	3271.736			0.1			30564.81			
3271.572	3272.515			0.1			30557.54			
3271.982	3272.925			0.0			30553.71			
3274.035	3274.979			0.1			30534.55			
3275.018	3275.961		0.2				30525.39			
3275.177	3276.121			0.2b			30523.90			
3276.243	3277.187			0.2			30513.97			
3276.731	3277.675			0.2			30509.43			
3276.817	3277.761			0.1			30508.63			
3277.598	3278.542		0.1	0.1			30501.36			

λ_{air}	λ_{vac}	Ref.	I_1	I_2	I_3	I_4	ν	Upper	Lower	Br.
3277.682	3278.626			0.2b			30500.58			
3278.522	3279.467			0.1			30492.76			
3278.587	3279.532			0.0			30492.16			
3278.729	3279.673			0.1			30490.84			
3279.129	3280.074			0.3			30487.12			
3279.454	3280.399		0.1				30484.10			
3280.121	3281.066			0.0			30477.90			
3280.409	3281.354			0.0			30475.22			
3280.535	3281.480			0.2			30474.05			
3280.619	3281.564			0.4			30473.27			
3280.658	3281.603			0.8v			30472.91			
3280.672	3281.617	1	0.2	1.5			30472.78			
3280.697	3281.642			1.0			30472.55			
3280.763	3281.709			0.5			30471.93			
3280.893	3281.838			0.2			30470.73			
3281.352	3282.298		0.2b				30466.46			
3281.399	3282.344			0.5d			30466.03			
3281.426	3282.371		0.2b				30465.78			
3281.568	3282.513			0.0			30464.46			
3281.626	3282.572			0.2b			30463.92			
3282.385	3283.330			0.1			30456.88			
3283.558	3284.504		0.1	0.1d			30446.00			
3283.801	3284.748		0.2				30443.74			
3284.319	3285.266			0.2b			30438.94			
3285.427	3286.373			0.2			30428.68			
3285.632	3286.578		0.2				30426.78			
3286.532	3287.478		0.3				30418.45			
3287.072	3288.019			0.1			30413.45			
3287.333	3288.280			0.1			30411.04			
3287.430	3288.377		0.6	0.6			30410.14			
3288.296	3289.243			0.1			30402.13			
3288.777	3289.725			0.1			30397.68			
3288.937	3289.885			0.2			30396.20			
3288.976	3289.924			0.1			30395.84			
3289.090	3290.038			0.0			30394.79			
3289.336	3290.283			0.0			30392.52			
3289.439	3290.386			0.1			30391.57			
3289.587	3290.534			0.1			30390.20			
3289.708	3290.656			0.1			30389.08			
3289.807	3290.754			0.1			30388.17			
3289.888	3290.835			0.1			30387.42			
3290.569	3291.517			0.1d			30381.13			
3290.697	3291.645			0.1			30379.95			
3291.383	3292.331			0.3			30373.62			
3291.687	3292.635			0.2			30370.81			
3291.944	3292.892			0.2			30368.44			
3292.096	3293.044			0.2			30367.04			
3292.269	3293.218			0.3			30365.44			
3292.499	3293.447			0.0			30363.32			
3292.818	3293.766			0.0			30360.38			
3292.948	3293.897			0.1			30359.18			
3293.179	3294.128			0.1			30357.05			
3293.256	3294.205			0.1			30356.34			
3293.473	3294.422			0.2			30354.34			
3293.559	3294.508			0.2			30353.55			
3294.069	3295.018			0.2d			30348.85			
3294.551	3295.500			0.2			30344.41			
3294.686	3295.635			0.0			30343.17			
3294.805	3295.754			0.2			30342.07			
3294.921	3295.870			0.1			30341.00			

λ_{air}	λ_{vac}	Ref.	I_1	I_2	I_3	I_4	ν	Upper	Lower	Br.
3295.196	3296.145			0.1			30338.47			
3295.363	3296.312			0.1			30336.93			
3295.526	3296.475			0.2			30335.43			
3295.940	3296.890			0.1			30331.62			
3296.015	3296.965			0.2			30330.93			
3296.461	3297.410			0.1			30326.83			
3296.781	3297.730			0.1			30323.89			
3297.007	3297.956			0.1			30321.81			
3297.281	3298.230			0.1			30319.29			
3297.868	3298.818		0.2				30313.89			
3298.201	3299.151			0.0			30310.83			
3298.336	3299.286			0.0			30309.59			
3298.470	3299.420			0.2			30308.36			
3298.676	3299.625			0.0			30306.47			
3298.885	3299.835			0.0			30304.55			
3299.333	3300.283			0.0			30300.43			
3299.702	3300.652			0.1			30297.04			
3300.013	3300.963			0.0			30294.19			
3300.237	3301.187			0.2			30292.13			
3300.407	3301.357			0.1			30290.57			
3300.532	3301.483			0.2			30289.42			
3300.713	3301.664			0.1			30287.76			
3301.052	3302.003			0.1			30284.65			
3301.156	3302.106			0.1			30283.70			
3301.559	3302.510			0.0			30280.00			
3301.758	3302.708			0.1			30278.18			
3301.891	3302.841			0.2			30276.96			
3302.159	3303.110			0.1			30274.50			
3302.306	3303.257			0.1			30273.15			
3302.365	3303.316	1		0.7			30272.61			
3302.978	3303.929	1		0.5			30266.99			
3303.511	3304.462			0.2			30262.11			
3304.334	3305.286			0.1			30254.57			
3304.753	3305.704			0.1			30250.74			
3305.453	3306.405			0.2d			30244.33			
3309.951	3310.904		0.0d	0.1			30203.23			
3312.260	3313.213		0.2	0.1			30182.18			
3312.793	3313.747		0.0	0.1			30177.32			
3313.007	3313.961		0.3	0.3			30175.37			
3313.103	3314.057		0.1	0.0			30174.50			
3313.868	3314.822			0.2d			30167.53			
3314.219	3315.173			0.2			30164.34			
3314.814	3315.767		0.2	0.5			30158.93			
3315.189	3316.144		0.1	0.0			30155.51			
3315.275	3316.230		0.9	0.7			30154.72	4E 2	2B 1	Q 2
3315.763	3316.718			0.1			30150.29			
3315.908	3316.862		0.2	0.3			30148.98			
3316.918	3317.872			0.1			30139.80			
3318.147	3319.102			0.2			30128.63			
3318.351	3319.306			0.1			30126.78			
3319.108	3320.063			0.0			30119.91			
3320.040	3320.996		0.1				30111.45			
3326.456	3327.413		0.3	0.1			30053.38			
3327.493	3328.451		0.0				30044.01			
3329.503	3330.460		0.4	0.6			30025.88			
3329.606	3330.563			0.0			30024.95			
3331.271	3332.229		0.1				30009.94			
3331.397	3332.355		0.6	0.6			30008.81			
3331.816	3332.775			0.1			30005.03			
3332.766	3333.724			0.2			29996.48			

λ_{air}	λ_{vac}	Ref.	I_1	I_2	I_3	I_4	ν	Upper	Lower	Br.
3333.199	3334.158		0.2b	0.0			29992.58			
3333.346	3334.305		0.1b	0.0			29991.26			
3336.095	3337.054			0.1			29966.55			
3336.249	3337.209			0.1			29965.16			
3336.469	3337.428			0.1			29963.19			
3338.769	3339.729		0.1				29942.55			
3339.013	3339.973		0.3	0.1			29940.36			
3339.135	3340.095			0.0			29939.27			
3339.320	3340.280			0.1			29937.61			
3340.564	3341.525			0.2			29926.46			
3340.595	3341.556			0.1			29926.18			
3341.194	3342.154		0.2d	0.1d			29920.82			
3341.417	3342.378			0.2			29918.82			
3341.463	3342.424	1		0.6v			29918.41			
3341.480	3342.440	1		0.9			29918.26			
3341.501	3342.462	1		0.8			29918.07			
3342.883	3343.844		0.3	0.4			29905.70			
3343.934	3344.896		0.1				29896.30			
3344.266	3345.228			0.1			29893.33			
3345.040	3346.001		0.6	0.5			29886.42			
3345.296	3346.258			0.2			29884.13			
3345.993	3346.955		0.2	0.2			29877.90			
3347.108	3348.070			0.1			29867.95			
3347.800	3348.762			0.2			29861.78			
3348.409	3349.371			0.1			29856.35			
3349.571	3350.534	F					29845.99			
3349.747	3350.710		0.1d				29844.42			
3349.948	3350.911		0.2	0.2d			29842.63			
3351.613	3352.576			0.0			29827.81			
3351.744	3352.708			0.2			29826.64			
3352.725	3353.689		0.1b	0.1			29817.91			
3352.850	3353.814		0.3	0.3			29816.80			
3353.440	3354.403		0.0	0.2			29811.56			
3354.277	3355.241		0.1d	0.1			29804.12			
3354.551	3355.516		0.3				29801.68			
3354.665	3355.629		0.1				29800.67			
3354.908	3355.872		0.5	0.2			29798.51			
3355.163	3356.127			0.1			29796.25			
3355.298	3356.262		0.1				29795.05			
3355.436	3356.401		0.2	0.2			29793.82			
3355.786	3356.750		0.1				29790.72			
3356.276	3357.240		0.1b				29786.37			
3356.429	3357.394		0.1d				29785.01			
3356.601	3357.566			0.1			29783.48			
3356.739	3357.704		0.5	0.4			29782.26			
3357.206	3358.170		0.5	0.3			29778.12			
3357.463	3358.428		0.0	0.0			29775.84			
3357.579	3358.544			0.1			29774.81			
3357.626	3358.591		0.1				29774.39			
3357.737	3358.702		0.2	0.1d			29773.41			
3357.921	3358.885		0.1				29771.78			
3358.008	3358.973		0.9	0.4			29771.00			
3358.142	3359.107		0.1				29769.82			
3358.241	3359.206			0.1			29768.94			
3359.592	3360.557			0.1			29756.97			
3359.724	3360.689		0.6	0.3			29755.80			
3360.861	3361.827		0.1				29745.73			
3362.965	3363.932		0.1d				29727.12			
3365.042	3366.008		0.5	0.2			29708.78			
3366.792	3367.760			0.2			29693.33			

λ_{air}	λ_{vac}	Ref.	I_1	I_2	I_3	I_4	ν	Upper	Lower	Br.
3367.086	3368.053		0.1				29690.74			
3367.709	3368.676		0.2				29685.25			
3368.246	3369.213		0.1	0.1d			29680.52			
3368.326	3369.294		0.1				29679.81			
3368.408	3369.376		1.5	0.7			29679.09			
3368.503	3369.471		0.0	0.2			29678.25			
3369.011	3369.978		0.1				29673.78			
3369.112	3370.080		0.2				29672.89			
3369.482	3370.450		0.1				29669.63			
3369.617	3370.585			0.2			29668.44			
3369.770	3370.738			0.1			29667.09			
3369.907	3370.875			0.2			29665.89			
3370.309	3371.277		0.2				29662.35			
3371.085	3372.053		0.1d				29655.52			
3374.897	3375.866			0.1			29622.03			
3375.193	3376.162		0.2b				29619.43			
3376.365	3377.334		0.1				29609.15			
3377.143	3378.112			0.2			29602.33			
3377.248	3378.217		0.1				29601.41			
3377.400	3378.370		0.3				29600.07			
3377.509	3378.479		0.1	0.1			29599.12			
3377.836	3378.806		0.1d				29596.25			
3379.171	3380.142		0.1				29584.56			
3379.395	3380.365		0.2				29582.60			
3379.602	3380.572		0.1				29580.79			
3379.747	3380.717		0.3				29579.52			
3380.123	3381.094			0.3			29576.23			
3380.502	3381.473			0.2			29572.91			
3381.016	3381.987		0.2				29568.42	4c 5	2a 1	Q 1
3381.939	3382.910		0.1d	0.0			29560.35			
3382.383	3383.354			0.1			29556.47			
3382.630	3383.601			0.1			29554.31			
3382.748	3383.719		0.1d	0.2			29553.28			
3382.820	3383.791			0.4			29552.65			
3382.878	3383.850	1		1.0			29552.14			
3382.888	3383.859		0.3	0.1			29552.06			
3382.907	3383.878			0.8			29551.89			
3382.971	3383.942			0.4			29551.33			
3383.812	3384.783			0.1			29543.99			
3384.042	3385.013			0.1			29541.98			
3384.280	3385.252			0.0			29539.90			
3384.608	3385.580			0.1			29537.04			
3385.443	3386.415			0.2			29529.75	4c 5	2a 1	Q 2
3385.549	3386.521		0.1				29528.83			
3385.732	3386.704		0.1	0.0			29527.23			
3387.017	3387.990		0.2	0.2			29516.03			
3387.504	3388.476			0.1			29511.79			
3387.821	3388.793			0.2			29509.03			
3387.931	3388.903		0.2				29508.07			
3388.236	3389.209			0.2			29505.41			
3388.507	3389.480		0.5	0.3			29503.05			
3388.657	3389.629		0.1	0.0			29501.75			
3389.960	3390.933		0.3	0.1			29490.41			
3390.043	3391.016		1.0	0.5			29489.69	4E 1	2B 0	Q 1
3390.148	3391.121		0.1	0.0			29488.77			
3390.467	3391.440		0.1				29486.00			
3391.801	3392.775		0.1				29474.40			
3391.983	3392.957		0.1				29472.82			
3392.103	3393.076		0.3	0.3			29471.78			
3392.198	3393.172			0.2b			29470.95			

λ_{air}	λ_{vac}	Ref.	I_1	I_2	I_3	I_4	ν	Upper	Lower	Br.
3392.301	3393.274			0.0			29470.06			
3392.469	3393.443		0.2d	0.2			29468.60			
3392.722	3393.696		0.1				29466.40			
3392.854	3393.828		0.4				29465.25			
3393.431	3394.405		0.2				29460.24			
3394.139	3395.113		0.5	0.1d			29454.10			
3394.490	3395.465		0.1	0.2			29451.05			
3394.598	3395.572		0.1				29450.12			
3394.662	3395.636		0.4	0.2			29449.56			
3394.752	3395.726		0.2	0.1			29448.78			
3394.844	3395.818	G	1.5b	0.8			29447.99	4E 1	2B 0	Q 2
3394.934	3395.909			0.3			29447.20			
3395.093	3396.068		0.2d				29445.82			
3395.171	3396.146	G	0.9	0.5			29445.14	4E 1	2B 0	Q 4
3395.551	3396.526		0.4				29441.85			
3395.753	3396.728		0.1				29440.10			
3396.066	3397.040		0.2				29437.39			
3396.151	3397.125		0.9	0.5			29436.66	4E 1	2B 0	Q 3
3396.207	3397.182	F					29436.16			
3396.265	3397.240		0.1				29435.66			
3396.351	3397.325			0.1			29434.92			
3397.284	3398.259		0.1				29426.83			
3397.702	3398.677			0.1			29423.21			
3397.858	3398.833		0.2				29421.86			
3398.222	3399.197		0.2				29418.71			
3400.059	3401.034		0.2				29402.82			
3401.043	3402.019		0.2				29394.31			
3401.274	3402.250		0.3				29392.31			
3401.905	3402.881		0.2	0.1			29386.86			
3402.881	3403.858		0.1				29378.43			
3403.614	3404.590		0.2				29372.11			
3404.403	3405.380		0.1	0.1			29365.30			
3404.483	3405.460		0.9	0.4			29364.61			
3404.577	3405.554		0.2				29363.80			
3404.723	3405.700			0.2d			29362.54			
3404.841	3405.818		0.1	0.0			29361.52			
3406.177	3407.154			0.2			29350.01			
3406.308	3407.285		0.2				29348.88			
3406.799	3407.776			0.1			29344.65			
3406.849	3407.826			0.1			29344.22			
3407.389	3408.366			0.1			29339.57			
3407.983	3408.961		0.1				29334.45			
3408.690	3409.668			0.1			29328.37			
3408.872	3409.850		0.1				29326.80			
3408.982	3409.960		0.4	0.2			29325.86			
3409.441	3410.419		0.1	0.0			29321.91			
3409.643	3410.621			0.0			29320.17			
3410.100	3411.079			0.2b			29316.24			
3410.411	3411.389		0.2	0.0			29313.57			
3410.476	3411.455			0.1			29313.01			
3410.653	3411.631			0.1			29311.49			
3410.811	3411.790		0.1	0.1			29310.13			
3410.993	3411.971		0.1	0.0			29308.57			
3411.209	3412.188		0.1	0.2			29306.71			
3411.582	3412.560		0.1				29303.51			
3412.953	3413.932		0.1d				29291.74			
3413.213	3414.192			0.0			29289.51			
3414.297	3415.276		0.1				29280.21			
3416.276	3417.255		0.1d				29263.25			
3416.900	3417.880			0.2			29257.90			

λ_{air}	λ_{vac}	Ref.	I_1	I_2	I_3	I_4	ν	Upper	Lower	Br.
3417.299	3418.279			0.2b			29254.49			
3418.274	3419.255			0.2			29246.14			
3418.609	3419.589		0.1				29243.28			
3418.673	3419.653		0.1				29242.73			
3418.849	3419.830		0.4	0.2b			29241.22			
3418.943	3419.924			0.1			29240.42			
3419.108	3420.088			0.1			29239.01			
3419.266	3420.246			0.1			29237.66			
3419.432	3420.412		0.1d				29236.24			
3420.062	3421.043		0.1				29230.85			
3421.035	3422.016		0.1d				29222.54			
3421.981	3422.962			0.2			29214.46			
3422.109	3423.090		0.2d				29213.37			
3422.676	3423.657		0.4	0.1			29208.53			
3422.956	3423.938		0.1				29206.14			
3423.920	3424.901		0.1	0.2			29197.92			
3424.189	3425.171		0.2				29195.62			
3425.595	3426.577		0.1	0.1d			29183.64	4f 3	2c 0	Q 4
3425.671	3426.654		0.2				29182.99			
3425.851	3426.833			0.0			29181.46			
3426.044	3427.026			0.2			29179.82			
3426.135	3427.118			0.2			29179.04			
3426.267	3427.249		0.1	0.2d			29177.92			
3427.360	3428.343		0.1d				29168.61			
3427.544	3428.526		0.3				29167.05			
3428.804	3429.787			0.1			29156.33			
3431.014	3431.998		0.3	0.1			29137.55			
3431.484	3432.468		0.4	0.2			29133.56			
3431.622	3432.605			0.1			29132.39			
3431.830	3432.814			0.2			29130.62			
3432.316	3433.300		0.2				29126.50			
3432.886	3433.870		0.2d				29121.66			
3433.569	3434.553			0.2			29115.87			
3433.794	3434.778		0.1				29113.96			
3433.907	3434.892		0.1				29113.00			
3433.999	3434.984		0.3				29112.22			
3436.038	3437.023		0.2				29094.95			
3436.300	3437.285		0.1				29092.73			
3436.477	3437.462		0.2b				29091.23			
3436.588	3437.573			0.2d			29090.29			
3436.726	3437.711		0.2				29089.12			
3436.882	3437.867			0.1			29087.80			
3436.990	3437.975			0.2			29086.89			
3437.042	3438.027			0.2			29086.45			
3437.462	3438.448			0.1			29082.89			
3437.570	3438.555			0.1			29081.98			
3438.271	3439.257		0.2	0.1d			29076.05			
3438.404	3439.389			0.2			29074.93			
3439.430	3440.416		0.1				29066.25			
3439.533	3440.519			0.1			29065.38			
3439.822	3440.808			0.2			29062.94			
3440.941	3441.927			0.2d			29053.49			
3441.224	3442.210			0.1			29051.10			
3441.345	3442.331		0.1d				29050.08			
3441.550	3442.536			0.2b			29048.35			
3441.760	3442.746		0.1				29046.58			
3441.904	3442.891			0.1			29045.36			
3442.638	3443.625			0.2			29039.17			
3443.561	3444.547			0.2			29031.39			
3444.240	3445.227		0.2				29025.66			

λ_{air}	λ_{vac}	Ref.	I$_1$	I$_2$	I$_3$	I$_4$	ν	Upper	Lower	Br.
3445.075	3446.062			0.1			29018.63			
3445.704	3446.692			0.1			29013.33			
3446.499	3447.487			0.1			29006.64			
3447.034	3448.021			0.1			29002.14			
3447.464	3448.452			0.1			28998.52			
3447.586	3448.574	1	0.6	0.0			28997.49			
3447.664	3448.652		0.1				28996.84			
3448.726	3449.714		0.1				28987.91			
3450.522	3451.511			0.2			28972.82			
3450.709	3451.698		0.2	0.0			28971.25			
3451.398	3452.387		0.2	0.1			28965.47			
3451.499	3452.488		0.3	0.2			28964.62			
3451.612	3452.601		0.2	0.1			28963.67			
3451.778	3452.767		0.2	0.1			28962.28			
3451.975	3452.964			0.2			28960.63			
3453.070	3454.060			0.1			28951.44			
3454.077	3455.067			0.1			28943.00			
3454.699	3455.689		0.1	0.1			28937.79			
3456.205	3457.195		0.1				28925.18	4e 3	2c 0	R 2
3456.577	3457.567			0.2			28922.07			
3457.370	3458.360			0.1			28915.44			
3458.363	3459.354		0.3				28907.13			
3458.433	3459.424		0.2				28906.55			
3458.671	3459.662		0.2d				28904.56			
3458.986	3459.977		0.2d	0.0			28901.93			
3459.236	3460.227		0.1	0.1			28899.84			
3459.456	3460.447		0.1	0.1			28898.00			
3460.022	3461.014		0.2				28893.27			
3461.644	3462.635			0.2			28879.74			
3461.737	3462.729		0.1	0.2d			28878.96			
3461.833	3462.825		0.6	0.4			28878.16	4E 2	2B 2	Q 2
3463.298	3464.290		0.3	0.2			28865.94			
3463.841	3464.833		0.1	0.0			28861.42			
3464.265	3465.257			0.0			28857.89			
3464.925	3465.917			0.1			28852.39			
3465.031	3466.023			0.2			28851.51			
3465.105	3466.098			0.1			28850.89			
3466.411	3467.404		0.2d				28840.02			
3466.571	3467.564			0.1			28838.69			
3467.751	3468.744			0.1			28828.88			
3468.313	3469.306			0.2			28824.21			
3468.606	3469.600		0.2				28821.77			
3468.730	3469.724		0.1				28820.74			
3469.294	3470.287			0.1			28816.06			
3471.209	3472.203		0.2				28800.16			
3471.342	3472.336			0.2			28799.06			
3471.569	3472.563		0.2	0.1			28797.17	4c 6	2a 2	Q 1
3474.029	3475.024		0.2				28776.78			
3474.604	3475.599			0.1			28772.02			
3475.070	3476.065			0.1			28768.16			
3475.300	3476.295			0.1			28766.26			
3478.231	3479.227			0.0			28742.02			
3478.699	3479.695			0.1			28738.15			
3478.933	3479.929			0.1			28736.22			
3479.409	3480.405			0.1			28732.29			
3479.801	3480.797			0.0			28729.05			
3479.934	3480.931			0.0			28727.95			
3480.022	3481.018			0.0			28727.23	4e 3	2c 0	Q 1
3480.126	3481.122			0.0			28726.37	4e 3	2c 0	Q 5
3480.429	3481.425			0.1			28723.87			

λ_{air}	λ_{vac}	Ref.	I_1	I_2	I_3	I_4	ν	Upper	Lower	Br.
3480.545	3481.541			0.1			28722.91			
3480.859	3481.855			0.0			28720.32			
3481.449	3482.446			0.0			28715.45			
3481.647	3482.644		0.1	0.1			28713.82	4f 3	2c 0	P 3
3481.794	3482.790			0.1			28712.61			
3482.019	3483.016			0.1			28710.75			
3482.137	3483.134		0.1	0.1			28709.78			
3482.506	3483.502			0.2d			28706.74			
3482.706	3483.703			0.1			28705.09	4c 6	2a 2	Q 3
3482.963	3483.960		0.2d	0.0			28702.97			
3483.632	3484.629			0.1			28697.46			
3483.717	3484.714		0.2	0.2			28696.76			
3483.782	3484.780		0.7	0.3			28696.22			
3484.036	3485.033		0.4	0.2			28694.13			
3484.189	3485.186			0.1d			28692.87			
3484.314	3485.311		0.1				28691.84			
3485.028	3486.026			0.2			28685.96			
3485.120	3486.117			0.2			28685.21			
3485.364	3486.361		0.2				28683.20			
3486.109	3487.107			0.2b			28677.07			
3486.425	3487.423		0.2				28674.47			
3487.269	3488.267			0.1			28667.53			
3487.656	3488.654			0.2			28664.35			
3488.130	3489.129		0.3				28660.45			
3488.610	3489.608			0.1			28656.51			
3488.852	3489.851		0.1	0.2d			28654.52			
3489.208	3490.206		0.2d				28651.60			
3489.689	3490.688		0.2				28647.65			
3489.992	3490.991			0.0			28645.16			
3490.626	3491.625		0.2	0.1			28639.96			
3491.282	3492.281			0.1			28634.58			
3492.022	3493.021		0.2				28628.51			
3492.137	3493.136		0.3				28627.57			
3492.310	3493.309			0.2			28626.15			
3493.325	3494.325		0.3d				28617.83			
3493.940	3494.939			0.2d			28612.80			
3494.190	3495.190		0.1				28610.75			
3494.393	3495.393		0.1				28609.09			
3494.740	3495.740		0.2				28606.25	4e 3	2c 0	P 2
3494.740	3495.740		0.2				28606.25	4f 3	2c 0	P 5
3495.071	3496.071			0.1			28603.54			
3495.575	3496.576		0.3	0.2			28599.41			
3495.700	3496.700			0.1			28598.39			
3495.887	3496.887			0.1			28596.86			
3496.189	3497.189			0.0			28594.39			
3496.319	3497.319			0.1			28593.33			
3496.748	3497.749		0.1				28589.82			
3497.103	3498.103		0.2d				28586.92			
3497.606	3498.606		0.0				28582.81			
3498.069	3499.070		0.2				28579.02			
3498.564	3499.565		0.1				28574.98			
3499.069	3500.070		0.2d				28570.86			
3499.165	3500.166		0.7	0.2			28570.07			
3499.545	3500.546		0.2	0.0			28566.97			
3499.676	3500.677		0.2				28565.90			
3499.815	3500.816		0.7	0.2			28564.77			
3500.072	3501.073			0.0			28562.67			
3500.269	3501.271		0.2				28561.06	4e 3	2c 0	P 3
3500.534	3501.535		0.2				28558.90			
3500.752	3501.754			0.1			28557.12	3F 3	2B 0	R 2

λ_{air}	λ_{vac}	Ref.	I_1	I_2	I_3	I_4	ν	Upper	Lower	Br.
3501.828	3502.829		0.1				28548.35			
3501.942	3502.944		0.2d				28547.42			
3502.084	3503.086		0.1				28546.26			
3502.167	3503.169			0.0			28545.58			
3502.505	3503.507		0.5				28542.83			
3502.979	3503.981			0.1			28538.97			
3504.229	3505.231		0.1				28528.79			
3504.462	3505.464		0.1				28526.89			
3504.720	3505.723		0.2				28524.79			
3504.945	3505.947		0.1				28522.96			
3505.069	3506.072		0.4	0.1			28521.95			
3505.354	3506.357		0.1				28519.63			
3505.454	3506.456		0.3				28518.82	4c 3	2a 0	Q 1
3505.561	3506.563		0.1				28517.95			
3505.893	3506.895		0.2				28515.25			
3506.662	3507.665		0.3	0.0			28508.99			
3507.462	3508.465			0.0			28502.49			
3507.883	3508.886		0.2				28499.07			
3508.815	3509.819		0.1				28491.50			
3509.385	3510.389		0.2				28486.87			
3510.209	3511.213		0.1				28480.19			
3510.319	3511.324			0.1			28479.29			
3510.419	3511.423		0.5	0.2			28478.48			
3510.984	3511.988			0.0			28473.90	4e 3	2c 0	R 1
3511.306	3512.310		0.2	0.0			28471.29			
3511.461	3512.466			0.0			28470.03			
3511.972	3512.976		0.1				28465.89			
3512.213	3513.217		0.5	0.2			28463.94			
3512.521	3513.526		0.2d				28461.44			
3512.612	3513.617			0.0			28460.70			
3512.720	3513.724		0.2b				28459.83	5c 3	2a 1	Q 3
3512.891	3513.896		1.0	0.3			28458.44			
3513.035	3514.039		0.2				28457.28			
3513.142	3514.147			0.0			28456.41			
3513.561	3514.565		0.1				28453.02			
3513.908	3514.913		0.1				28450.21			
3514.088	3515.093		0.4	0.2			28448.75			
3514.641	3515.647		0.1				28444.27			
3515.317	3516.323		0.2d				28438.80			
3515.561	3516.566		0.2				28436.83			
3515.676	3516.681		0.3				28435.90			
3516.213	3517.218		0.1				28431.56			
3516.938	3517.943		0.2d	0.0			28425.70	4e 3	2c 0	R 3
3517.102	3518.108		0.2d				28424.37			
3517.217	3518.223		0.1				28423.44			
3518.210	3519.216		0.2				28415.42	3F 3	2B 0	Q 2
3518.367	3519.373		0.1				28414.15			
3518.454	3519.460		0.4				28413.45	6c 1	2a 0	Q 1
3518.683	3519.689		0.1				28411.60			
3519.151	3520.157		0.2				28407.82			
3519.366	3520.372		0.1				28406.09			
3519.613	3520.620		0.2				28404.09			
3519.752	3520.759			0.1			28402.97			
3519.912	3520.918		0.2				28401.68			
3520.186	3521.192			0.1			28399.47			
3520.390	3521.397		0.2				28397.82			
3520.580	3521.587		0.1d				28396.29			
3520.751	3521.758		0.2				28394.91			
3520.925	3521.932		0.2	0.0			28393.51			
3522.234	3523.241		0.1				28382.96			

λ_air	λ_vac	Ref.	I₁	I₂	I₃	I₄	ν	Upper	Lower	Br.
3522.740	3523.747		0.1				28378.88	4e 3	2c 0	P 6
3522.894	3523.901			0.0			28377.64			
3523.152	3524.160			0.0			28375.56			
3523.398	3524.405		0.3				28373.58			
3524.270	3525.278		0.4	0.1			28366.56			
3524.514	3525.521		0.1				28364.60			
3524.949	3525.956		0.3	0.1			28361.10			
3525.246	3526.253		0.1				28358.71			
3525.359	3526.367		0.2				28357.80			
3525.558	3526.566		0.1				28356.20			
3525.819	3526.827		0.3				28354.10			
3526.362	3527.370			0.1			28349.73			
3526.464	3527.472			0.1			28348.91			
3526.865	3527.873		0.2				28345.69			
3527.074	3528.082			0.0			28344.01			
3527.268	3528.276			0.0			28342.45			
3527.670	3528.679		0.2				28339.22			
3527.920	3528.929		0.4				28337.21			
3528.692	3529.701			0.0			28331.01			
3528.856	3529.864		0.2				28329.70			
3529.266	3530.274			0.1			28326.41			
3529.425	3530.434			0.1			28325.13			
3530.179	3531.188			0.0			28319.08			
3530.650	3531.660		0.2				28315.30			
3531.117	3532.126			0.1			28311.56			
3531.386	3532.396		0.1d				28309.40			
3531.748	3532.757		0.3				28306.50			
3531.832	3532.841	F					28305.83			
3532.142	3533.152			0.0			28303.34			
3532.316	3533.325			0.0			28301.95			
3533.424	3534.434			0.1			28293.07			
3533.527	3534.537			0.1			28292.25			
3533.730	3534.740		0.1				28290.62			
3534.446	3535.457			0.0			28284.89			
3534.881	3535.892			0.1			28281.41			
3535.013	3536.023		0.4	0.0			28280.36			
3535.514	3536.524		0.1				28276.35			
3536.571	3537.581		0.2				28267.90			
3536.638	3537.649		0.2				28267.36			
3537.756	3538.767		0.2				28258.43			
3538.033	3539.044		0.3				28256.22			
3539.224	3540.235			0.0			28246.71			
3539.414	3540.426		0.2				28245.19			
3539.610	3540.621		0.2				28243.63			
3539.810	3540.822		0.1				28242.03	3c 6	2a 0	Q 1
3541.049	3542.061		0.3				28232.15			
3541.364	3542.376			0.0			28229.64			
3541.724	3542.736		0.1d	0.1			28226.77			
3541.802	3542.814		0.2d				28226.15			
3542.079	3543.091		0.2				28223.94			
3542.237	3543.250		0.3				28222.68			
3543.316	3544.328		0.1				28214.09			
3543.452	3544.464		0.2				28213.01			
3543.732	3544.744		0.2d	0.1			28210.78			
3543.877	3544.890		0.2				28209.62			
3544.276	3545.288			0.0			28206.45			
3544.375	3545.388		0.2d				28205.66			
3544.462	3545.474		0.2				28204.97			
3544.541	3545.554		0.5	0.2			28204.34			
3544.741	3545.754		0.1	0.1			28202.75			

λ_{air}	λ_{vac}	Ref.	I_1	I_2	I_3	I_4	ν	Upper	Lower	Br.
3544.899	3545.912		0.2				28201.49			
3545.158	3546.171			0.1			28199.43			
3545.741	3546.755		0.1				28194.79			
3545.974	3546.987			0.1			28192.94			
3546.548	3547.561		0.1				28188.38	4e 3	2c 0	Q 4
3546.703	3547.716		0.2				28187.15			
3546.815	3547.828			0.0			28186.26			
3547.478	3548.491		0.5d	0.2			28180.99			
3547.759	3548.772		0.2				28178.76			
3547.913	3548.927		0.1	0.1			28177.53			
3548.158	3549.171		0.2				28175.59			
3548.236	3549.250		0.2				28174.97			
3548.447	3549.461		0.4	0.2			28173.29	4E 1	2B 1	Q 1
3548.447	3549.461		0.4	0.2			28173.29	4e 3	2c 0	P 3
3548.677	3549.691		0.2				28171.47			
3548.956	3549.970		0.1				28169.25			
3549.289	3550.303		0.1				28166.61			
3549.418	3550.432		0.3	0.2			28165.59			
3549.579	3550.593		0.1				28164.31			
3549.668	3550.682		0.2				28163.60			
3549.770	3550.785		0.3				28162.79			
3549.918	3550.932		0.2				28161.62			
3550.121	3551.135		0.4				28160.01			
3550.141	3551.155		0.3b	0.3			28159.85			
3550.165	3551.179		0.4				28159.66			
3550.286	3551.300		0.2				28158.70			
3550.328	3551.342		0.1				28158.37			
3550.464	3551.478		0.2d				28157.29			
3550.831	3551.845		0.2				28154.38			
3551.242	3552.257		0.2b				28151.12			
3551.304	3552.318		0.1				28150.63			
3551.796	3552.811		0.1				28146.73			
3551.954	3552.968		0.3	0.1			28145.48	4E 1	2B 1	Q 4
3552.100	3553.115		0.2b				28144.32			
3552.135	3553.150		0.2				28144.04			
3552.508	3553.523		0.1				28141.09			
3552.671	3553.686		0.2				28139.80			
3552.895	3553.910			0.1			28138.02			
3552.998	3554.013		0.1	0.0			28137.21			
3553.221	3554.236		0.9	0.3			28135.44	4E 1	2B 1	Q 2
3553.335	3554.350			0.0			28134.54			
3553.658	3554.673			0.1			28131.98			
3553.831	3554.846		0.2d				28130.61			
3553.950	3554.965		0.2				28129.67	4E 1	2B 1	Q 3
3554.170	3555.185				0.1		28127.93			
3554.295	3555.310	1			0.3		28126.94			
3554.390	3555.405		0.2				28126.19			
3554.457	3555.472		0.3	0.2			28125.66			
3554.692	3555.707		0.1	0.1			28123.80			
3554.816	3555.831		0.2				28122.82			
3554.892	3555.907		1.0	0.3			28122.22			
3554.988	3556.003		0.1				28121.46			
3555.046	3556.061		0.1				28121.00	3c 6	2a 0	Q 3
3555.199	3556.214		0.1				28119.79			
3555.425	3556.441		0.1				28118.00			
3555.624	3556.639		0.2				28116.43			
3555.983	3556.999		0.5	0.1			28113.59			
3556.756	3557.772			0.0			28107.48			
3556.952	3557.968		0.1				28105.93			
3556.646	3557.662		0.1				28108.35			

λ_{air}	λ_{vac}	Ref.	I_1	I_2	I_3	I_4	ν	Upper	Lower	Br.
3557.066	3558.082		0.1	0.0			28105.03			
3557.341	3558.357		0.1				28102.86			
3557.847	3558.863		0.1				28098.86			
3557.944	3558.961		0.3				28098.09	4F 0	2B 0	R 3
3558.687	3559.703		0.1				28092.23			
3558.972	3559.988		0.1				28089.98			
3559.575	3560.592		0.2				28085.22			
3559.941	3560.958		0.2				28082.33			
3560.039	3561.056		0.4				28081.56			
3560.161	3561.177		0.4	0.0			28080.60			
3560.300	3561.317		0.2				28079.50			
3561.084	3562.101		0.2d	0.1			28073.32			
3561.538	3562.555		0.1				28069.74			
3561.806	3562.823		0.2d	0.1			28067.63			
3562.037	3563.054			0.0			28065.81			
3562.122	3563.139		0.2b				28065.14			
3562.198	3563.215		0.2				28064.54			
3562.263	3563.280		0.7	0.2			28064.03			
3562.349	3563.366		0.1				28063.35			
3562.461	3563.478			0.0			28062.47			
3562.748	3563.765		0.2				28060.21			
3563.282	3564.300	i			0.2		28056.00			
3563.807	3564.825		0.1				28051.87			
3564.014	3565.032		0.6	0.2			28050.24			
3564.203	3565.221		0.1				28048.75			
3564.283	3565.301				0.1		28048.12			
3564.541	3565.559			0.1			28046.09			
3564.625	3565.643		0.2				28045.43			
3564.685	3565.703		0.3	0.1	0.1		28044.96	4c 7	2a 3	Q 1
3564.830	3565.848		0.1				28043.82			
3565.817	3566.835		0.2				28036.06	4e 3	2c 0	P 4
3566.156	3567.175		0.1				28033.39			
3566.408	3567.427		0.1				28031.41			
3566.617	3567.635		0.1				28029.77			
3566.700	3567.718		0.4	0.1			28029.12			
3567.135	3568.154		0.2				28025.70			
3567.241	3568.259		0.1				28024.87	3c 6	2a 0	Q 4
3567.526	3568.544				0.1		28022.63			
3567.650	3568.669	i		0.1	0.3		28021.65			
3567.746	3568.765		0.2				28020.90			
3568.057	3569.076			0.0			28018.46			
3568.199	3569.218		0.1				28017.34			
3570.061	3571.080		0.2d				28002.73			
3571.062	3572.082			0.0	0.1		27994.88			
3571.241	3572.260		0.2				27993.48			
3571.447	3572.467		0.1				27991.86			
3571.590	3572.610		0.2	0.1			27990.74			
3571.658	3572.678		0.4				27990.21			
3571.775	3572.795		0.1				27989.29			
3571.992	3573.012		0.1				27987.59			
3572.300	3573.320		0.1		0.2		27985.18			
3572.417	3573.437		0.1d				27984.26			
3572.494	3573.514		0.2				27983.66			
3572.918	3573.938		0.2				27980.34			
3573.827	3574.848		0.2				27973.22			
3573.959	3574.979		0.1				27972.19			
3574.273	3575.294		0.1				27969.73			
3574.740	3575.760		0.4	0.1			27966.08			
3575.086	3576.107		0.1				27963.37			
3575.191	3576.212		0.1				27962.55			

λ_{air}	λ_{vac}	Ref.	I_1	I_2	I_3	I_4	ν	Upper	Lower	Br.
3575.333	3576.354		0.2				27961.44			
3575.453	3576.474		0.4				27960.50			
3575.839	3576.860		0.2d	0.0			27957.48			
3575.983	3577.004		0.2				27956.36			
3576.088	3577.109		0.2				27955.54			
3576.706	3577.727			0.0			27950.71			
3577.069	3578.090		0.2				27947.87			
3577.600	3578.622		0.2				27943.72			
3577.877	3578.898		0.2				27941.56			
3577.988	3579.010		0.1				27940.69			
3578.319	3579.340		0.2				27938.11			
3578.395	3579.417		0.6	0.1			27937.51			
3578.822	3579.844		0.2d				27934.18			
3578.994	3580.015		0.2				27932.84			
3579.139	3580.160			0.1			27931.71			
3579.440	3580.462		0.2d				27929.36			
3579.644	3580.665		0.2				27927.77			
3580.077	3581.099		0.1				27924.39			
3580.186	3581.208		0.2				27923.54			
3580.540	3581.562		0.1				27920.78			
3580.631	3581.653		0.3	0.1			27920.07			
3580.708	3581.730		0.1				27919.47			
3580.783	3581.806		0.6	0.2	0.1		27918.88			
3580.880	3581.902		0.2				27918.13			
3581.016	3582.038		0.3b	0.1			27917.07			
3581.679	3582.701		0.1				27911.90			
3581.755	3582.777		0.2				27911.31			
3582.568	3583.591		0.1b				27904.97			
3583.029	3584.052		0.2	0.1			27901.38			
3583.146	3584.169		0.2				27900.47			
3583.288	3584.310		0.2	0.0			27899.37			
3583.462	3584.485		0.2				27898.01			
3583.742	3584.765		0.1				27895.83			
3583.932	3584.955		0.1				27894.35	4e 3	2c 0	P 5
3584.336	3585.359		0.5	0.2			27891.21			
3584.448	3585.471		0.1				27890.34			
3584.639	3585.662		0.1				27888.85			
3584.787	3585.810		0.2b				27887.70			
3585.160	3586.183		0.1d	0.1			27884.80			
3585.422	3586.445		0.6	0.4			27882.76	4F 0	2B 0	R 2
3585.522	3586.546		0.1				27881.98			
3585.733	3586.757			0.1			27880.34			
3585.827	3586.851		0.1				27879.61			
3585.953	3586.977		0.2d				27878.63			
3586.235	3587.259		0.1				27876.44			
3586.433	3587.457		0.1d	0.0			27874.90			
3586.763	3587.786		0.1				27872.34			
3586.864	3587.888		0.2d				27871.55			
3587.082	3588.106		0.1				27869.86			
3587.164	3588.188		0.1				27869.22			
3587.267	3588.291		0.3				27868.42			
3587.391	3588.415			0.1			27867.46			
3587.774	3588.798		0.1d	0.1			27864.48			
3588.014	3589.038		0.1	0.0			27862.62			
3588.099	3589.123			0.0			27861.96			
3588.291	3589.315		0.1				27860.47			
3588.425	3589.449		0.3				27859.43			
3588.622	3589.646		0.2				27857.90			
3589.404	3590.428		0.2	0.0			27851.83			
3589.813	3590.837		0.0				27848.66			

λ_{air}	λ_{vac}	Ref.	I_1	I_2	I_3	I_4	ν	Upper	Lower	Br.
3589.878	3590.903			0.0			27848.15			
3590.028	3591.052		0.3	0.0			27846.99			
3590.292	3591.317			0.1			27844.94			
3590.462	3591.487			0.1			27843.62			
3590.822	3591.847		0.7	0.2			27840.83	4c 4	2a 1	Q 1
3591.012	3592.037		0.2				27839.36			
3592.484	3593.509		0.2	0.1			27827.95			
3592.798	3593.823		0.1				27825.52			
3592.875	3593.901		0.1	0.0			27824.92			
3593.017	3594.043		0.1				27823.82			
3593.129	3594.154		0.5	0.3			27822.96	3E 3	2B 0	R 3
3593.272	3594.297		0.2				27821.85			
3593.367	3594.393	G	0.6	0.2			27821.11			
3593.444	3594.469		0.3d	0.1			27820.52			
3593.556	3594.582	G	1.0	0.4			27819.65	3E 3	2B 0	R 2
3593.662	3594.687		0.2				27818.83			
3593.823	3594.849		0.2b	0.1			27817.58			
3593.931	3594.956		0.1d				27816.75			
3594.141	3595.167		0.3	0.1			27815.12			
3594.569	3595.595		0.3	0.2			27811.81			
3594.709	3595.734		0.1				27810.73			
3594.834	3595.860		0.0				27809.76			
3594.981	3596.007		0.2b	0.0			27808.62	4c 4	2a 1	Q 2
3595.261	3596.287		0.0	0.1			27806.46			
3595.487	3596.513		0.2				27804.71			
3595.623	3596.649		0.3	0.1			27803.66			
3595.892	3596.918		0.2	0.0			27801.58			
3596.171	3597.197			0.1			27799.42			
3596.652	3597.679		0.4b	0.2			27795.70			
3596.972	3597.998		0.2	0.1			27793.23			
3597.047	3598.074		0.1				27792.65			
3597.127	3598.154		0.2				27792.03	3E 3	2B 0	Q 8
3597.247	3598.273		0.5	0.2			27791.11			
3597.505	3598.532		0.1				27789.11			
3597.615	3598.642		0.3	0.1			27788.26			
3597.720	3598.747		0.1				27787.45			
3597.865	3598.892		0.1	0.0			27786.33			
3598.053	3599.080		0.6	0.3			27784.88	3E 3	2B 0	R 1
3598.330	3599.357		0.2				27782.74			
3598.487	3599.514		0.1b	0.1			27781.53			
3598.933	3599.960			0.1			27778.09			
3599.632	3600.659		0.1				27772.69			
3599.701	3600.728		0.1				27772.16			
3599.911	3600.938		0.5	0.4	0.3		27770.54	4F 0	2B 0	Q 4
3600.239	3601.266		0.1	0.0			27768.01			
3600.351	3601.378		0.1				27767.15			
3601.338	3602.365		0.2				27759.54	4F 0	2B 0	P 5
3601.338	3602.365		0.2				27759.54	4c 4	2a 1	Q 3
3601.971	3602.999		0.1	0.0			27754.66			
3602.519	3603.546		0.1				27750.44			
3602.597	3603.624		0.5	0.2			27749.84			
3602.808	3603.836		0.1				27748.21			
3602.860	3603.888			0.0			27747.81			
3602.973	3604.001		0.1				27746.94			
3603.102	3604.130		0.1d	0.1			27745.95			
3603.391	3604.419		0.1	0.1			27743.72			
3603.441	3604.469		0.2				27743.34			
3603.516	3604.544		0.1	0.0			27742.76			
3603.591	3604.619		0.7	0.3			27742.18			
3603.813	3604.842		0.2				27740.47			

λ_{air}	λ_{vac}	Ref.	I_1	I_2	I_3	I_4	ν	Upper	Lower	Br.
3603.951	3604.979		0.1	0.1			27739.41			
3604.177	3605.205		0.1d	0.1			27737.67			
3604.307	3605.335		0.5	0.3			27736.67	3E 3	2B 0	R 0
3604.431	3605.459			0.0			27735.72			
3604.614	3605.642			0.1			27734.31			
3604.745	3605.774		0.2				27733.30			
3605.109	3606.138			0.1			27730.50			
3605.883	3606.912			0.1			27724.55			
3606.120	3607.148			0.1			27722.73			
3606.242	3607.271		0.2d	0.2			27721.79			
3606.406	3607.435			0.1			27720.53			
3606.507	3607.536	i	0.0		0.3		27719.75			
3607.045	3608.074		0.1				27715.62			
3607.493	3608.522		0.1				27712.18			
3607.615	3608.644			0.1			27711.24			
3607.719	3608.748			0.0			27710.44			
3608.360	3609.389		0.1				27705.52			
3609.031	3610.060		0.3				27700.37			
3609.362	3610.391		0.1				27697.83			
3609.480	3610.510			0.0			27696.92			
3609.861	3610.890		0.4	0.3	0.1b		27694.00	4F 0	2B 0	R 1
3610.720	3611.750		0.0				27687.41	3E 3	2B 0	Q 5
3610.797	3611.827		0.2				27686.82			
3610.908	3611.938		0.0				27685.97			
3611.034	3612.064		0.6	0.3	0.1d		27685.00			
3611.350	3612.380			0.1			27682.58			
3611.696	3612.726		0.0				27679.93			
3611.796	3612.826		0.2				27679.16			
3612.044	3613.074		0.2	0.1			27677.26			
3612.146	3613.176		0.1				27676.48			
3612.361	3613.392		0.1				27674.83	4f 2	2c 0	R 4
3612.514	3613.544	F					27673.66			
3612.806	3613.837		0.2	0.2			27671.42			
3612.901	3613.931	GK	1.0	0.6	0.3d		27670.70			
3613.018	3614.049			0.1			27669.80			
3613.100	3614.131		0.2				27669.17			
3613.210	3614.241		0.1	0.1			27668.33			
3613.303	3614.333		0.3	0.3	0.1		27667.62	3E 3	2B 0	Q 4
3613.405	3614.435		0.1				27666.84			
3613.640	3614.670	i	0.8	0.1	0.3		27665.04			
3613.941	3614.972				0.1		27662.73			
3614.171	3615.202		0.1	0.2			27660.97			
3614.563	3615.594		0.2				27657.97			
3614.699	3615.730		0.2				27656.93			
3615.046	3616.077		0.1				27654.28			
3615.127	3616.158		0.3	0.3	0.1		27653.66	3E 3	2B 0	Q 3
3615.201	3616.232	F					27653.09			
3615.346	3616.378		0.2	0.1			27651.98			
3615.574	3616.605		0.2				27650.24			
3615.796	3616.827		0.2				27648.54			
3615.941	3616.973		0.1				27647.43			
3616.102	3617.134		0.1	0.0			27646.20			
3616.220	3617.251		0.2				27645.30			
3616.321	3617.352	G	0.9	0.5			27644.53	3E 3	2B 0	Q 2
3616.410	3617.441	G	0.9	0.4			27643.85	3D 3	2B 0	R 0
3616.533	3617.564		0.3				27642.91			
3616.682	3617.713			0.1			27641.77			
3616.862	3617.894		0.1				27640.39			
3617.023	3618.055		0.4	0.3			27639.16	3E 3	2B 0	Q 1
3617.179	3618.211		0.1	0.1			27637.97			

λ_{air}	λ_{vac}	Ref.	I_1	I_2	I_3	I_4	ν	Upper	Lower	Br.
3617.249	3618.280		0.1	0.1			27637.44			
3617.349	3618.381		0.6	0.4	0.2		27636.67	4F 0	2B 0	Q 3
3617.461	3618.492		0.4	0.2			27635.82	4E 2	2B 3	Q 2
3617.584	3618.615		0.2				27634.88			
3617.665	3618.696		0.1				27634.26			
3617.779	3618.810		0.2	0.1			27633.39			
3617.882	3618.914		0.2				27632.60			
3617.965	3618.996		0.2				27631.97			
3618.039	3619.071		0.7	0.3			27631.40	3D 3	2B 0	R 1
3618.123	3619.155		0.7	0.3			27630.76			
3618.259	3619.291		0.1	0.1			27629.72			
3618.436	3619.468		0.2				27628.37			
3618.564	3619.596		0.2	0.1			27627.39			
3618.669	3619.701		0.2				27626.59			
3618.740	3619.772	G	1.5	0.4			27626.05			
3618.861	3619.892		0.2d				27625.13			
3619.298	3620.330		0.2				27621.79			
3619.403	3620.435		0.3	0.1			27620.99			
3619.491	3620.523		0.1				27620.32	3E 3	2B 0	P 2
3619.635	3620.667		0.1				27619.22			
3619.765	3620.797		0.1				27618.23			
3619.973	3621.005		0.2				27616.64			
3620.049	3621.081	GK	0.9	0.5	0.3d		27616.06	4F 0	2B 0	P 4
3620.184	3621.216		0.1				27615.03	4c 4	2a 1	Q 5
3620.328	3621.361		0.2				27613.93			
3620.512	3621.544		0.1	0.1			27612.53			
3621.002	3622.035		0.1				27608.79			
3621.258	3622.291		0.2				27606.84			
3621.660	3622.692			0.1			27603.78			
3621.778	3622.810		0.2	0.1			27602.88	3c 9	2a 2	Q 1
3621.925	3622.957		0.1				27601.76			
3622.077	3623.110		0.2				27600.60			
3622.165	3623.198	GK	0.9	0.4			27599.93	3D 3	2B 0	R 2
3622.493	3623.526	F					27597.43			
3622.952	3623.985	F					27593.93			
3623.039	3624.072	K	0.5	0.3			27593.27			
3623.122	3624.155		0.2				27592.64	3E 3	2B 0	P 3
3623.287	3624.320		0.1	0.1			27591.38			
3624.325	3625.358		0.1	0.0			27583.48			
3624.600	3625.633		0.1				27581.39			
3624.669	3625.703	F					27580.86			
3625.233	3626.267		0.3	0.2			27576.57			
3625.558	3626.592		0.3	0.1			27574.10			
3625.789	3626.823			0.0			27572.34			
3626.283	3627.316		0.3	0.2			27568.59			
3626.461	3627.495		0.1				27567.23			
3626.527	3627.561		0.3		0.2d		27566.73			
3626.626	3627.660		0.1				27565.98			
3626.906	3627.940		0.2d	0.0			27563.85			
3627.118	3628.152		0.1	0.1			27562.24			
3627.252	3628.286	K	0.6	0.4	0.1d		27561.22	4E 0	2B 0	R 2
3627.762	3628.796		0.1	0.0			27557.35			
3627.851	3628.885		0.2				27556.67			
3628.050	3629.084				0.1b		27555.16			
3628.191	3629.225		0.1				27554.09			
3628.345	3629.379		0.4	0.3			27552.92	3D 3	2B 0	R 3
3628.345	3629.379		0.4	0.3			27552.92	3E 3	2B 0	P 4
3628.419	3629.453		0.1				27552.36			
3628.515	3629.549		0.3	0.2			27551.63			
3629.260	3630.295			0.1			27545.97			

λ_{air}	λ_{vac}	Ref.	I_1	I_2	I_3	I_4	ν	Upper	Lower	Br.
3629.723	3630.758		0.3	0.1			27542.46			
3629.864	3630.899		0.1	0.0			27541.39			
3630.095	3631.130		0.0		0.1		27539.64			
3630.701	3631.736		0.3d	0.1			27535.04			
3630.872	3631.908		0.2				27533.74			
3631.111	3632.146		0.2				27531.93			
3631.667	3632.702		0.4	0.1			27527.72	3D 3	2B 0	P 2
3632.057	3633.092		0.2				27524.76			
3632.197	3633.232		0.6	0.3			27523.70			
3632.305	3633.341		0.2d				27522.88			
3632.448	3633.483		0.2d	0.0			27521.80			
3632.503	3633.539		0.2				27521.38			
3632.568	3633.603		0.2				27520.89			
3632.680	3633.716	GK	0.7	0.5	0.2		27520.04	3D 3	2B 0	R 4
3632.885	3633.920		0.2b	0.1			27518.49			
3633.013	3634.048		0.1	0.1			27517.52			
3633.096	3634.132		0.2				27516.89			
3633.163	3634.199		0.3	0.1			27516.38			
3633.248	3634.283		0.2				27515.74			
3633.322	3634.357	GK	1.5	0.7	0.3		27515.18	4F 0	2B 0	Q 2
3633.425	3634.460		0.1				27514.40			
3633.505	3634.541			0.0			27513.79			
3633.820	3634.855		0.1				27511.41			
3633.891	3634.927		0.4	0.2	0.1		27510.87	X 2	2B 0	R 0
3634.075	3635.111		0.2				27509.48			
3634.232	3635.268		0.4		0.1		27508.29			
3634.384	3635.420		0.2				27507.14			
3634.448	3635.484				0.2d		27506.65			
3634.519	3635.555		0.2b				27506.12			
3634.737	3635.773		0.1				27504.47			
3634.841	3635.877		0.4		0.1d		27503.68			
3634.931	3635.967		0.2				27503.00			
3635.005	3636.041		0.2				27502.44			
3635.105	3636.141	GK	0.8	0.5	0.2		27501.68	4F 0	2B 0	P 3
3635.212	3636.249		0.3b				27500.87			
3635.450	3636.487		0.1				27499.07			
3635.655	3636.692		0.1				27497.52			
3635.838	3636.874		0.1				27496.14			
3635.938	3636.975			0.0			27495.38			
3636.123	3637.160		0.1d				27493.98			
3636.491	3637.528		0.1				27491.20			
3636.574	3637.611		0.2	0.0			27490.57	X 2	2B 0	R 1
3636.712	3637.749		0.1				27489.53			
3636.879	3637.915		0.2b				27488.27			
3637.034	3638.070	GK	0.8	0.4	0.2		27487.10	4E 0	2B 0	R 0
3637.139	3638.176		0.1				27486.30			
3637.298	3638.335		0.3		0.1		27485.10			
3638.033	3639.070		0.3				27479.55			
3638.163	3639.200		0.2d				27478.57			
3638.266	3639.303		0.3d				27477.79			
3638.400	3639.437		0.2d				27476.78			
3638.966	3640.004	F					27472.50	X 2	2B 0	P 1
3639.219	3640.257		0.4	0.2			27470.59			
3639.437	3640.474				0.1		27468.95			
3639.573	3640.611		0.2	0.2	0.2		27467.92			
3639.716	3640.754			0.0			27466.84			
3640.009	3641.047		0.3				27464.63			
3640.090	3641.128	F					27464.02			
3640.406	3641.443		0.2				27461.64			
3640.818	3641.856	F					27458.53			

λ_{air}	λ_{vac}	Ref.	I_1	I_2	I_3	I_4	ν	Upper	Lower	Br.
3641.206	3642.244		0.2	0.0			27455.60			
3641.277	3642.314		0.2				27455.07			
3641.577	3642.614		0.6	0.2			27452.81			
3641.720	3642.758		0.1	0.1			27451.73			
3641.890	3642.927			0.0			27450.45			
3642.032	3643.070		0.6	0.3			27449.38	X 2	2B 0	R 2
3642.561	3643.599		0.4	0.2			27445.39			
3642.991	3644.029		0.2	0.0			27442.15			
3643.115	3644.153				0.1		27441.22			
3643.267	3644.306			0.1			27440.07			
3643.502	3644.541			0.0			27438.30	3D 3	2B 0	P 3
3643.825	3644.863			0.1			27435.87			
3644.407	3645.445		0.3	0.1			27431.49			
3644.496	3645.534			0.0			27430.82			
3644.553	3645.592		0.1				27430.39			
3644.859	3645.897		0.4	0.2			27428.09			
3645.070	3646.109		0.1d				27426.50			
3645.267	3646.305		0.1	0.1			27425.02			
3645.466	3646.505		0.2d				27423.52			
3645.974	3647.013		0.4	0.2			27419.70			
3646.094	3647.133		0.2				27418.80			
3646.485	3647.524		0.3	0.1			27415.86			
3646.683	3647.722		0.1		0.1		27414.37			
3646.841	3647.880	GK	0.8	0.4	0.2		27413.18	3A 2	2B 0	R 4
3646.908	3647.947		0.1				27412.68			
3646.960	3647.999		0.0	0.0			27412.29			
3647.132	3648.172			0.1			27410.99			
3647.400	3648.439			0.0			27408.98			
3648.159	3649.198		0.1				27403.28			
3648.336	3649.375	F					27401.95			
3648.381	3649.421	K	0.6	0.3			27401.61	X 2	2B 0	P 3
3648.627	3649.667		0.2	0.1			27399.76			
3648.830	3649.869			0.0			27398.24			
3649.040	3650.080		0.1d				27396.66			
3649.136	3650.176		0.1	0.0			27395.94			
3649.303	3650.342		0.7	0.3	0.1		27394.69	X 2	2B 0	P 2
3649.398	3650.438			0.1			27393.97			
3649.498	3650.538		0.2				27393.22			
3649.638	3650.678		0.2d				27392.17			
3649.753	3650.793			0.0			27391.31			
3649.826	3650.866			0.0	0.2		27390.76			
3649.926	3650.966			0.0	0.1		27390.01			
3649.995	3651.035		0.1				27389.49			
3650.090	3651.130		0.6	0.4	0.3		27388.78			
3650.158	3651.198	K		0.1	0.2		27388.27			
3650.275	3651.315		0.2				27387.39			
3650.398	3651.438		0.4	0.2			27386.47			
3650.533	3651.573		0.1				27385.46			
3650.645	3651.685			0.1			27384.62			
3650.767	3651.807	F					27383.70			
3650.853	3651.893		0.0d	0.0			27383.06			
3651.173	3652.213	F					27380.66	4f 2	2c 0	Q 5
3651.363	3652.404				0.1b		27379.23			
3651.635	3652.676		0.3	0.1			27377.19			
3651.703	3652.744		0.3				27376.68			
3652.072	3653.112		0.0	0.0			27373.92			
3652.305	3653.346		0.1	0.1			27372.17			
3652.373	3653.414		0.0	0.2			27371.66			
3652.468	3653.509	GK	1.5b	0.8	0.4	1.29	27370.95	4E 0	2B 0	P 2
3652.565	3653.606		0.1				27370.22			

λ_{air}	λ_{vac}	Ref.	I_1	I_2	I_3	I_4	ν	Upper	Lower	Br.
3652.780	3653.821		0.0d	0.0			27368.61			
3653.048	3654.089			0.0			27366.60			
3653.227	3654.268		0.3				27365.26			
3653.724	3654.765		0.1d				27361.54			
3654.289	3655.330			0.1			27357.31			
3654.556	3655.597		0.0	0.0			27355.31			
3654.842	3655.883	K					27353.17			
3654.914	3655.956		0.2		0.1		27352.63			
3655.314	3656.355				0.1		27349.64			
3655.513	3656.554			0.1			27348.15			
3655.672	3656.714	GK	1.5b	0.7	0.3	1.09	27346.96	4E 0	2B 0	Q 1
3655.772	3656.814		0.1				27346.21			
3655.855	3656.897			0.1			27345.59			
3656.004	3657.045		0.0				27344.48			
3656.112	3657.154			0.0			27343.67			
3656.432	3657.473		0.2				27341.28			
3656.529	3657.571		0.3	0.2			27340.55			
3656.778	3657.820			0.1			27338.69			
3657.297	3658.339			0.0			27334.81			
3657.516	3658.558			0.2	0.1		27333.17	3D 3	2B 0	P 4
3657.573	3658.615		0.4	0.3	0.2		27332.75			
3657.800	3658.842		0.1	0.1			27331.05			
3657.874	3658.916		0.3d				27330.50			
3657.954	3658.996	K	0.6	0.4	0.3		27329.90	4E 0	2B 0	P 5
3658.281	3659.323	F					27327.46			
3658.441	3659.484				0.1		27326.26			
3658.547	3659.589			0.1			27325.47			
3658.662	3659.705	GK	0.7	0.4	0.2		27324.61	3A 2	2B 0	R 3
3658.759	3659.801		0.1				27323.89			
3658.905	3659.947		0.0d	0.0			27322.80			
3659.057	3660.100				0.1		27321.66			
3659.285	3660.327		0.1		0.0		27319.96			
3659.407	3660.449		0.2				27319.05			
36 9.494	3660.536		0.2		0.2		27318.40			
3659.615	3660.657		0.0d				27317.50			
3659.806	3660.849		0.0				27316.07			
3660.184	3661.227		0.2b				27313.25			
3660.294	3661.337				0.1		27312.43			
3660.491	3661.534		0.1	0.1			27310.96			
3660.877	3661.920		0.0d				27308.08			
3661.066	3662.109			0.1			27306.67			
3661.200	3662.243			0.1			27305.67			
3661.286	3662.329			0.0			27305.03			
3661.632	3662.675		0.1		0.1		27302.45			
3661.718	3662.761		0.1				27301.81			
3661.837	3662.880		0.1				27300.92			
3661.934	3662.977		0.2				27300.20			
3662.079	3663.122		0.2	0.0			27299.12			
3662.195	3663.238		0.2				27298.25	X 2	2B 0	P 3
3662.297	3663.340		0.2	0.2	0.1		27297.49			
3662.383	3663.426		0.2				27296.85			
3662.468	3663.511		0.9				27296.22			
3662.480	3663.523	GK	0.5	0.4	0.2		27296.13	4E 0	2B 0	P 3
3662.503	3663.546		0.8				27295.96			
3662.602	3663.645		0.2	0.1			27295.22			
3662.704	3663.747	GK	0.9	0.5	0.3		27294.46	4E 0	2B 0	P 4
3662.803	3663.846		0.1	0.1			27293.72			
3662.894	3663.938		0.1	0.2			27293.04			
3662.971	3664.014			0.1			27292.47			
3663.060	3664.103	GK	1.5	0.8	0.5	1.22	27291.81	4E 0	2B 0	Q 4

λ_{air}	λ_{vac}	Ref.	I_1	I_2	I_3	I_4	ν	Upper	Lower	Br.
3663.180	3664.224		0.2				27290.91			
3663.264	3664.307		0.4	0.2			27290.29	4f 2	2c 0	Q 4
3663.371	3664.414		0.4	0.0			27289.49			
3663.464	3664.507		0.4				27288.80			
3663.480	3664.523		0.4	0.3	0.1		27288.68			
3663.501	3664.545		0.4				27288.52			
3663.576	3664.620		0.1				27287.96			
3663.701	3664.745		0.1				27287.03			
3663.758	3664.801		0.0	0.2			27286.61			
3663.967	3665.011		0.3	0.2			27285.05			
3664.045	3665.089		0.2	0.2			27284.47			
3664.138	3665.181	GK	1.5b	0.9	0.5	1.66	27283.78	4E 0	2B 0	Q 2
3664.234	3665.278		0.3	0.1	0.1		27283.06			
3664.362	3665.406		0.1d				27282.11			
3664.476	3665.520		0.2	0.2			27281.26			
3664.574	3665.618	G	0.5	0.3	0.2b		27280.53			
3664.696	3665.740		0.2				27279.62			
3664.770	3665.814		0.2b	0.1			27279.07			
3664.882	3665.926	GK	0.9	0.4	0.2d		27278.24			
3664.968	3666.012		0.1				27277.60			
3665.079	3666.123		0.2	0.0			27276.77			
3665.138	3666.182		0.4				27276.33			
3665.247	3666.291	GK	1.0	0.6	0.3	1.09	27275.52	4E 0	2B 0	Q 3
3665.304	3666.348		0.3				27275.10			
3665.418	3666.462		0.2				27274.25			
3665.501	3666.545			0.1			27273.63			
3665.609	3666.653		0.3	0.2	0.1		27272.83			
3665.739	3666.783		0.3	0.2			27271.86			
3665.810	3666.855		0.2				27271.33			
3665.905	3666.949	GK	1.5b	1.0	0.3	1.69	27270.63	3A 2	2B 0	R 2
3665.997	3667.041		0.3	0.1			27269.94			
3666.052	3667.097		0.1d				27269.53			
3666.159	3667.203		0.1	0.1			27268.74			
3666.254	3667.298		0.5r	0.2			27268.03			
3666.429	3667.473		0.2b				27266.73			
3666.491	3667.535		0.0d				27266.27			
3666.605	3667.649		0.4	0.2			27265.42			
3666.670	3667.714		0.3				27264.94			
3666.757	3667.801		0.2				27264.29			
3666.827	3667.871		0.1d				27263.77			
3667.019	3668.064			0.0			27262.34			
3667.229	3668.274		0.3	0.2	0.2b		27260.78			
3667.430	3668.474		0.2	0.0			27259.29			
3667.508	3668.552		0.7				27258.71			
3667.520	3668.564	GK	0.5	0.4			27258.62			
3667.547	3668.591		0.6				27258.42			
3667.766	3668.811		0.1	0.2			27256.79			
3667.902	3668.947		0.2d				27255.78	3c 7	2a 1	Q 1
3668.102	3669.147				0.1		27254.29			
3668.288	3669.333		0.1				27252.91			
3668.373	3669.418		0.1d				27252.28			
3668.738	3669.783		0.2d	0.1			27249.57			
3668.850	3669.894	GK	0.6	0.6	0.3		27248.74	T 0	2B 0	R 2
3668.933	3669.978			0.0			27248.12			
3669.011	3670.056	GK	0.7	0.6	0.3		27247.54	4D 0	2B 0	R 6
3669.092	3670.137		0.2	0.1			27246.94			
3669.363	3670.408		0.2	0.1	0.0		27244.93			
3669.567	3670.612			0.0			27243.41			
3669.706	3670.751		0.1				27242.38			
3669.915	3670.960		0.2				27240.83			

λ_{air}	λ_{vac}	Ref.	I_1	I_2	I_3	I_4	ν	Upper	Lower	Br.
3669.967	3671.013				0.1		27240.44			
3670.082	3671.127		0.1b				27239.59			
3670.151	3671.196		0.1d				27239.08			
3670.365	3671.410		0.0d		0.1		27237.49			
3670.571	3671.616			0.1			27235.96			
3670.659	3671.704				0.2		27235.31			
3670.719	3671.765		0.2	0.2			27234.86			
3671.000	3672.045				0.1		27232.78			
3671.276	3672.322			0.0			27230.73			
3671.442	3672.488		0.1				27229.50			
3671.551	3672.597		0.2	0.0			27228.69			
3671.969	3673.015	G					27225.59			
3672.104	3673.150		0.2				27224.59			
3672.182	3673.228		0.1d				27224.01			
3672.265	3673.310		0.2				27223.40			
3672.343	3673.389	GK	0.8	0.3			27222.82			
3672.397	3673.443		0.3	0.1			27222.42			
3672.563	3673.609		0.3	0.2			27221.19	5c 1	2a 0	R 3
3672.688	3673.734			0.1			27220.26			
3672.807	3673.853		0.1				27219.38			
3672.872	3673.918		0.4				27218.90			
3672.966	3674.012		0.0d				27218.20			
3673.054	3674.100		0.2				27217.55			
3673.317	3674.363		0.1				27215.60			
3673.389	3674.435		0.3	0.2			27215.07			
3673.478	3674.524		0.2	0.1			27214.41	3D 3	2B 0	P 5
3673.548	3674.594		0.2d	0.2			27213.89			
3673.644	3674.690	GK	1.5b	0.9	0.2	1.67	27213.18	3A 2	2B 0	R 1
3673.733	3674.779		0.2				27212.52			
3673.810	3674.856	GK	1.5	0.6	0.2		27211.95	T 0	2B 0	R 1
3673.918	3674.964		0.0				27211.15			
3673.981	3675.028		0.3	0.2			27210.68			
3674.034	3675.080		0.1				27210.29			
3674.141	3675.187		0.1	0.0			27209.50			
3674.239	3675.286		0.3b	0.3			27208.77			
3674.309	3675.356		0.3	0.2			27208.25	3c 7	2a 1	Q 2
3674.396	3675.442	GK	1.5b	1.0	0.6	2.04	27207.61	4D 0	2B 0	R 0
3674.488	3675.534		0.5	0.1			27206.93			
3674.569	3675.615		0.2	0.2			27206.33			
3674.651	3675.698		0.2				27205.72			
3674.754	3675.800		0.3	0.1			27204.96			
3674.862	3675.908	F					27204.16			
3674.979	3676.026		0.3	0.0	0.2		27203.29			
3675.159	3676.206		0.6				27201.96			
3675.235	3676.281				0.2		27201.40			
3675.374	3676.421		0.1				27200.37			
3675.542	3676.588		0.3	0.1			27199.13			
3675.633	3676.680		0.1				27198.45			
3675.721	3676.768		0.4	0.1			27197.80			
3675.929	3676.976		0.3				27196.26			
3676.225	3677.272		0.4	0.3			27194.07	4D 0	2B 0	P 1
3676.382	3677.429		0.3	0.1			27192.91	5c 1	2a 0	R 2
3676.620	3677.667		0.1	0.1			27191.15			
3676.685	3677.732						27190.67			
3676.796	3677.843	GK	0.8	0.7	0.4		27189.85	4D 0	2B 0	R 5
3676.89	3677.942		0.1				27189.12			
3677.104	3678.151		0.2d				27187.57			
3677.240	3678.287		0.1				27186.57			
3677.392	3678.440		0.0				27185.44			
3677.506	3678.553			0.1			27184.60			

λ_{air}	λ_{vac}	Ref.	I$_1$	I$_2$	I$_3$	I$_4$	ν	Upper	Lower	Br.
3677.670	3678.717		0.5	0.2			27183.39	4c 5	2a 2	Q 1
3677.793	3678.840		0.4	0.1			27182.48	X 2	2B 0	P 4
3677.874	3678.921		0.8	0.2			27181.88			
3678.000	3679.047	GK	1.0	0.4	0.1		27180.95	T 0	2B 0	R 0
3678.107	3679.154		0.1	0.1			27180.16			
3678.287	3679.334			0.0			27178.83			
3678.365	3679.413		0.1d				27178.25			
3678.525	3679.573		0.1				27177.07			
3678.623	3679.670		0.0				27176.35	7c 0	2a 0	R 1
3678.753	3679.800		0.6	0.3			27175.39			
3679.012	3680.060	F					27173.47			
3679.098	3680.145		0.2	0.2			27172.84			
3679.411	3680.458				0.1		27170.53			
3679.471	3680.519			0.1			27170.08			
3679.700	3680.748		0.1d		0.1		27168.39			
3679.771	3680.819		0.3				27167.87			
3679.859	3680.907				0.2d		27167.22			
3680.127	3681.175		0.2	0.2			27165.24			
3680.382	3681.430		0.1				27163.36			
3680.441	3681.489		0.2				27162.92			
3680.540	3681.588	GK	1.5	0.6	0.2	1.00	27162.19	5c 1	2a 0	R 1
3680.641	3681.689		0.3	0.1			27161.45			
3680.719	3681.767		0.2				27160.87			
3680.791	3681.839		0.2	0.0			27160.34			
3680.939	3681.987		0.1		0.1		27159.25			
3681.050	3682.098		0.0				27158.43			
3681.199	3682.247		0.2	0.0	0.1		27157.33			
3681.271	3682.319		0.1				27156.80			
3681.345	3682.394	GK	0.9	0.5	0.2		27156.25			
3681.444	3682.493		0.2				27155.52			
3681.524	3682.573		0.3	0.1			27154.93			
3681.637	3682.685		0.3	0.2			27154.10			
3681.801	3682.849		0.2	0.2	0.2		27152.89			
3681.878	3682.927		0.1	0.2			27152.32			
3681.968	3683.016	GK	1.5b	1.0	0.7	1.51	27151.66	4D 0	2B 0	R 4
3681.968	3683.016	GK	1.5b	1.0	0.7	1.51	27151.66	4c 5	2a 2	Q 2
3682.065	3683.114		0.2				27150.94			
3682.185	3683.233		0.1	0.1			27150.06			
3682.292	3683.340	GK	1.5b	0.8	0.3	1.28	27149.27	4D 0	2B 0	R 1
3682.365	3683.414		0.2	0.1			27148.73	3F 3	2B 1	Q 3
3682.474	3683.522		0.3	0.2			27147.93			
3682.559	3683.608		0.3	0.1			27147.30			
3682.630	3683.678		0.1		0.2		27146.78			
3682.748	3683.796		0.2b				27145.91			
3683.060	3684.108		0.3				27143.61			
3683.197	3684.245		0.0				27142.60			
3683.468	3684.517			0.2	0.1		27140.60			
3683.546	3684.594		0.6				27140.03			
3683.692	3684.741		0.0		0.1		27138.95			
3683.789	3684.837		0.2	0.2			27138.24			
3683.897	3684.946		0.3	0.2	0.2d		27137.44	3c 7	2a 1	Q 3
3683.962	3685.011		0.2	0.1			27136.96			
3684.061	3685.110		0.2				27136.23			
3684.150	3685.199		0.4	0.3	0.2		27135.58			
3684.226	3685.275		0.4	0.1			27135.02	3A 2	2B 0	R 0
3684.313	3685.362	GK	1.5b	1.0	0.6	2.08	27134.38	4D 0	2B 0	R 2
3684.406	3685.455		0.3				27133.69			
3684.457	3685.505	GK	1.5	0.8	0.4		27133.32	4D 0	2B 0	R 3
3684.565	3685.614		0.3	0.1			27132.52			
3684.671	3685.720		0.2				27131.74			

λ_{air}	λ_{vac}	Ref.	I_1	I_2	I_3	I_4	ν	Upper	Lower	Br.
3684.829	3685.878		0.1				27130.58			
3684.895	3685.944		0.1		0.1		27130.09			
3684.996	3686.045		0.4	0.2			27129.35			
3685.077	3686.126		0.5				27128.75			
3685.161	3686.211		0.2	0.2			27128.13			
3685.273	3686.322		0.0d				27127.31			
3685.383	3686.432			0.1			27126.50			
3685.546	3686.595		0.1				27125.30			
3685.641	3686.690		0.3	0.2			27124.60			
3685.848	3686.897		0.2	0.1			27123.08			
3686.049	3687.098		0.1				27121.60			
3686.147	3687.196		0.4	0.3			27120.88			
3686.228	3687.278		0.2				27120.28			
3686.386	3687.435				0.2		27119.12			
3686.586	3687.635		0.3	0.2			27117.65	5c 1	2a 0	R 0
3686.830	3687.880				0.2		27115.85			
3686.911	3687.960		0.3				27115.26			
3687.106	3688.156		0.0d	0.1			27113.82	4f 2	2c 0	Q 2
3687.300	3688.349	GK	0.7	0.5	0.2		27112.40	3D 3	2B 0	P 6
3687.418	3688.468		0.1				27111.53			
3687.502	3688.552			0.1			27110.91			
3687.553	3688.602		0.4	0.2			27110.54			
3687.672	3688.722			0.2	0.1		27109.66			
3687.786	3688.836		0.3	0.2			27108.82			
3688.001	3689.051			0.1			27107.24			
3688.230	3689.280		0.5	0.2			27105.56			
3688.316	3689.366			0.1			27104.93			
3688.400	3689.450		0.2	0.2			27104.31	4c 5	2a 2	Q 3
3688.483	3689.533			0.1			27103.70			
3688.596	3689.646	GK	0.9	0.4			27102.87	3F 3	2B 1	Q 2
3688.705	3689.755		0.2				27102.07			
3688.814	3689.864		0.3				27101.27			
3689.260	3690.311		0.1				27097.99			
3689.375	3690.425		0.3				27097.15	4f 2	2c 0	R 1
3689.375	3690.425		0.3				27097.15	4f 3	2c 1	R 3
3689.609	3690.659			0.1			27095.43	4f 2	2c 0	R 2
3689.704	3690.755		0.1d				27094.73			
3689.812	3690.862		0.0				27093.94			
3689.999	3691.049			0.1			27092.57			
3690.057	3691.107			0.0			27092.14			
3690.148	3691.199	GK	1.5b	0.7	0.3		27091.47	4D 0	2B 0	P 2
3690.148	3691.199	GK	1.5b	0.7	0.3		27091.47	T 0	2B 0	P 1
3690.274	3691.324		0.4	0.2	0.2		27090.55			
3690.349	3691.399		1.0	0.4			27090.00			
3690.547	3691.598		0.1				27088.54			
3690.913	3691.963	1	0.3	0.1	0.2		27085.86			
3691.131	3692.181		0.0				27084.26			
3691.227	3692.278			0.0			27083.55	4e 2	2c 0	R 3
3691.724	3692.774		0.3				27079.91			
3691.983	3693.034			0.1			27078.01			
3692.131	3693.182		0.2				27076.92			
3692.322	3693.373		0.2d				27075.52			
3692.786	3693.837		0.2				27072.12			
3692.883	3693.934		0.2				27071.41			
3693.004	3694.055			0.1			27070.52			
3693.172	3694.223				0.1		27069.29			
3693.319	3694.371		0.1	0.0			27068.21			
3693.467	3694.518				0.1		27067.13			
3693.692	3694.743	F					27065.48			
3693.793	3694.844		0.1				27064.74	T 0	2B 0	P 2

λ_air	λ_vac	Ref.	I₁	I₂	I₃	I₄	ν	Upper	Lower	Br.
3694.030	3695.082		0.4	0.3	0.2		27063.00	X 2	2B 0	P 5
3694.340	3695.392	GK	0.7	0.3			27060.73	7c 0	2a 0	Q 1
3694.923	3695.975		0.0				27056.46			
3695.009	3696.061		0.2	0.2			27055.83			
3695.080	3696.132		0.2				27055.31			
3695.156	3696.207	GK	1.5	0.4			27054.76	3A 2	2B 0	P 1
3695.272	3696.323		0.2				27053.91			
3695.930	3696.982		0.2				27049.09			
3696.229	3697.281		0.3	0.1			27046.90			
3696.548	3697.600		0.0				27044.57			
3696.690	3697.742		0.1				27043.53			
3696.739	3697.791		0.3				27043.17			
3696.802	3697.854		0.3	0.3			27042.71			
3697.082	3698.135		0.2				27040.66			
3697.256	3698.308				0.2		27039.39			
3697.524	3698.576		0.2	0.0	0.2		27037.43			
3697.713	3698.765				0.2		27036.05			
3697.772	3698.824		0.1				27035.62			
3697.859	3698.912	GK	1.5	0.5	0.2d		27034.98	5c 1	2a 0	Q 1
3697.859	3698.912	GK	1.5	0.5	0.2d		27034.98	4e 2	2c 0	R 2
3698.014	3699.066			0.1			27033.85	4f 2	2c 0	P 6
3698.260	3699.312			0.0			27032.05			
3698.618	3699.671		0.2				27029.43			
3698.800	3699.853				0.2		27028.10			
3699.015	3700.068		0.2		0.1		27026.53			
3699.076	3700.128		0.2				27026.09			
3699.290	3700.343		0.3				27024.52	4f 2	2c 0	Q 4
3699.388	3700.440				0.1		27023.81			
3699.538	3700.591		0.5	0.4	0.2		27022.71			
3699.698	3700.751		0.1	0.1			27021.54			
3699.792	3700.844	GK	1.5	0.8	0.2	1.25	27020.86	3A 2	2B 0	P 3
3699.890	3700.943		0.3	0.0			27020.14			
3699.964	3701.017		0.2				27019.60	T 0	2B 0	P 3
3700.059	3701.112	GK	1.5b	0.8	0.2	1.30	27018.91	3A 2	2B 0	P 2
3700.059	3701.112	GK	1.5b	0.8	0.2	1.30	27018.91	7c 0	2a 0	Q 3
3700.200	3701.253		0.2	0.1			27017.88			
3700.270	3701.323		0.5	0.3	0.2		27017.37			
3700.409	3701.462		0.1	0.1			27016.35			
3700.565	3701.618		0.3				27015.21			
3700.701	3701.754		0.2	0.0	0.1		27014.22			
3700.813	3701.866		0.6	0.4			27013.40	5c 1	2a 0	Q 2
3700.949	3702.002				0.2		27012.41			
3701.038	3702.091			0.1			27011.76			
3701.346	3702.400		0.4	0.1	0.2		27009.51			
3701.455	3702.508			0.0			27008.72			
3701.572	3702.626				0.1		27007.86			
3701.703	3702.756		0.0d	0.2	0.1		27006.91			
3701.782	3702.836		0.2	0.1			27006.33			
3701.887	3702.941	GK	0.9	0.7	0.3		27005.56	3A 2	2B 0	P 6
3701.966	3703.019		0.1				27004.99			
3702.019	3703.073		0.2	0.1			27004.60			
3702.116	3703.170	GK	1.5b	0.9	0.3	1.63	27003.89	3A 2	2B 0	P 4
3702.224	3703.277		0.2	0.2			27003.11			
3702.366	3703.420		0.2d				27002.07			
3702.483	3703.536		0.2	0.0			27001.22			
3702.549	3703.602	F					27000.74			
3702.629	3703.683		0.1				27000.15			
3702.827	3703.881		0.2				26998.71			
3702.890	3703.944		0.3				26998.25			
3703.078	3704.132				0.2		26996.88	4f 2	2c 0	P 5

λ_{air}	λ_{vac}	Ref.	I_1	I_2	I_3	I_4	ν	Upper	Lower	Br.
3703.313	3704.366		0.1				26995.17			
3703.474	3704.528		0.2		0.2d		26993.99			
3703.658	3704.712		0.3				26992.65			
3703.734	3704.788		0.3	0.2			26992.10			
3703.940	3704.994		0.2	0.2			26990.60	3F 3	2B 1	P 4
3704.225	3705.279		0.2		0.1		26988.52			
3704.421	3705.475			0.1			26987.09			
3704.583	3705.637	GK	0.8	0.6	0.2		26985.91	3A 2	2B 0	P 5
3704.806	3705.860		0.1		0.1		26984.29			
3704.881	3705.935		0.2				26983.74			
3704.999	3706.054	i	0.7		0.3		26982.88			
3705.138	3706.192		0.3	0.0			26981.87	T 0	2B 0	P 4
3705.296	3706.350		0.6	0.4			26980.72	5c 1	2a 0	Q 3
3705.477	3706.532		0.1				26979.40			
3705.763	3706.817		0.2d				26977.32	4f 2	2c 0	Q 2
3705.763	3706.817		0.2d				26977.32	4e 2	2c 0	R 1
3706.123	3707.177		0.0d				26974.70			
3706.263	3707.318		0.0d				26973.68			
3706.451	3707.506		0.1				26972.31			
3706.520	3707.575				0.1		26971.81			
3706.703	3707.758		0.1				26970.48			
3706.788	3707.843		0.5				26969.86			
3707.004	3708.059		0.1				26968.29			
3707.183	3708.237		0.2				26966.99			
3707.591	3708.646				0.1		26964.02			
3707.739	3708.794		0.2				26962.94			
3707.816	3708.871		0.1				26962.38			
3708.103	3709.158			0.1			26960.30			
3708.200	3709.255		0.3	0.2			26959.59			
3708.437	3709.492		0.2				26957.87	4f 2	2c 0	P 4
3708.514	3709.569		0.2	0.2			26957.31			
3708.570	3709.625		0.4				26956.90	4D 0	2B 0	P 3
3708.650	3709.705		0.0				26956.32			
3708.749	3709.804		0.1				26955.60			
3708.844	3709.899			0.1			26954.91			
3708.949	3710.004		0.1				26954.15	4f 3	2c 1	R 2
3709.090	3710.146		0.1				26953.12			
3709.834	3710.889		0.2				26947.72			
3710.033	3711.089		0.1	0.1			26946.27			
3710.153	3711.209			0.0			26945.40			
3710.256	3711.312	F					26944.65			
3710.353	3711.408						26943.95			
3710.540	3711.596		0.3	0.1	0.2		26942.59			
3710.938	3711.994		0.1				26939.70			
3710.992	3712.048				0.2		26939.31			
3711.091	3712.147		0.0				26938.59			
3711.229	3712.285		0.2				26937.59			
3711.303	3712.359		0.2	0.1			26937.05	5c 1	2a 0	Q 4
3711.355	3712.411		0.2				26936.67			
3711.440	3712.495		0.4	0.1			26936.06			
3711.551	3712.607	F					26935.25			
3711.692	3712.748		0.2	0.2			26934.23	W 2	2B 0	R 4
3711.814	3712.870		0.1				26933.34	4f 2	2c 0	P 3
3711.931	3712.988			0.0			26932.49			
3712.102	3713.158			0.1			26931.25			
3712.304	3713.360				0.2		26929.79			
3712.493	3713.549						26928.42			
3712.742	3713.798		0.0		0.1		26926.61			
3712.941	3713.997		0.2		0.1		26925.17			
3713.261	3714.317		0.3	0.2			26922.85			

λ_{air}	λ_{vac}	Ref.	I_1	I_2	I_3	I_4	ν	Upper	Lower	Br.
3713.422	3714.478		0.2				26921.68			
3713.585	3714.641		0.3	0.2			26920.50			
3713.825	3714.881		0.2d	0.1			26918.76			
3713.985	3715.041		0.4	0.3			26917.60	5c 1	2a 0	P 2
3714.108	3715.164		0.1				26916.71			
3714.211	3715.268		0.2				26915.96	4f 2	2c 0	Q 3
3714.337	3715.393		0.2				26915.05			
3714.407	3715.464		0.1				26914.54			
3714.472	3715.529	K	0.8	0.4	0.2		26914.07	W 2	2B 0	R 1
3714.596	3715.653		0.8	0.3			26913.17			
3714.700	3715.757	GK	0.8	0.5	0.2		26912.42	W 2	2B 0	R 2
3714.805	3715.861		0.2				26911.66			
3714.892	3715.948		0.1		0.1		26911.03			
3715.228	3716.285		0.2				26908.59			
3715.286	3716.343			0.0			26908.17			
3715.344	3716.401				0.1		26907.75			
3715.576	3716.633		0.4	0.3	0.2		26906.07	W 2	2B 0	R 3
3715.724	3716.781		0.1				26905.00			
3715.844	3716.901		0.2d	0.2			26904.13			
3715.927	3716.984		0.2				26903.53			
3716.000	3717.058	GK	1.5	0.7	0.2	1.00	26903.00	W 2	2B 0	R 0
3716.099	3717.156		0.3				26902.29			
3716.191	3717.248		0.2				26901.62			
3716.281	3717.338		0.4	0.3			26900.97			
3716.380	3717.438		0.1		0.1		26900.25			
3716.516	3717.573		0.0				26899.27			
3716.597	3717.655			0.1			26898.68			
3716.774	3717.831		0.2				26897.40			
3716.846	3717.903				0.1		26896.88			
3716.976	3718.033			0.1			26895.94			
3717.117	3718.174			0.0			26894.92			
3717.237	3718.295		0.3	0.2	0.2		26894.05			
3717.315	3718.372	GK	0.8	0.5			26893.49	4E 1	2B 2	Q 1
3717.409	3718.466		0.1				26892.81			
3717.497	3718.555			0.0			26892.17			
3718.017	3719.074			0.1			26888.41			
3718.130	3719.188				0.2		26887.59			
3718.378	3719.436		0.0d				26885.80			
3718.728	3719.786		0.1				26883.27			
3718.796	3719.853		0.2	0.1			26882.78	5c 1	2a 0	Q 5
3718.888	3719.946		0.3	0.3			26882.11			
3719.051	3720.109		0.2				26880.93			
3719.193	3720.251	GK	0.7	0.6	0.2		26879.91	4E 1	2B 2	Q 4
3719.295	3720.353		0.1				26879.17			
3719.444	3720.502		0.2				26878.09			
3719.609	3720.667		0.1				26876.90			
3719.696	3720.754		0.2				26876.27			
3719.927	3720.986	G	0.2				26874.60			
3720.247	3721.305		0.2d				26872.29	4e 2	2c 0	R 1
3720.611	3721.670		0.2				26869.66			
3720.711	3721.769		0.2				26868.94			
3720.787	3721.846		0.5	0.3			26868.39			
3720.891	3721.950		0.2				26867.64	4D 0	2B 0	P 4
3721.057	3722.116			0.1			26866.44			
3721.149	3722.207		0.2				26865.78			
3721.225	3722.283	K	0.5	0.4			26865.23			
3721.330	3722.389		0.3				26864.47			
3721.416	3722.475		0.2				26863.85			
3721.524	3722.583		0.1				26863.07			
3721.660	3722.719		0.2				26862.09			

λ_{air}	λ_{vac}	Ref.	I_1	I_2	I_3	I_4	ν	Upper	Lower	Br.
3721.744	3722.803		0.2				26861.48			
3721.814	3722.872		0.2		0.2		26860.98			
3721.900	3722.958		0.2				26860.36			
3721.966	3723.025		0.2				26859.88			
3722.026	3723.084		0.2	0.1			26859.45			
3722.106	3723.165	GK	1.5b	0.7	0.2	1.45	26858.87	4E 1	2B 2	Q 2
3722.234	3723.292		0.7	0.4			26857.95	4E 1	2B 2	Q 3
3722.346	3723.405		0.2				26857.14			
3722.446	3723.504		0.2				26856.42			
3722.681	3723.740			0.1			26854.72			
3722.882	3723.941		0.3				26853.27			
3723.185	3724.244		0.1				26851.09			
3723.294	3724.353	F					26850.30			
3723.368	3724.427				0.2		26849.77			
3723.656	3724.715		0.2				26847.69			
3723.792	3724.851		0.2				26846.71			
3723.975	3725.034			0.0	0.2		26845.39	7c 0	2a 0	P 3
3724.076	3725.136		0.2				26844.66			
3724.269	3725.329		0.4	0.1			26843.27			
3724.387	3725.447			0.2			26842.42			
3724.537	3725.596		0.0				26841.34			
3724.690	3725.749		0.2	0.1			26840.24			
3724.851	3725.910	GK	1.5b	0.6	0.2		26839.08	W 2	2B 0	P 1
3724.942	3726.002		0.2	0.1			26838.42			
3725.016	3726.075		0.2		0.2		26837.89			
3725.121	3726.181		0.1				26837.13			
3725.213	3726.272		0.1				26836.47			
3725.295	3726.354		0.1		0.2		26835.88			
3725.432	3726.492		0.1	0.0			26834.89			
3725.625	3726.685		0.2				26833.50			
3725.727	3726.786		0.1				26832.77			
3725.791	3726.850			0.1			26832.31			
3725.884	3726.943						26831.64	5c 1	2a 0	P 3
3726.343	3727.403			0.2			26828.33			
3726.459	3727.518		0.3	0.4	0.2d		26827.50			
3726.648	3727.707			0.1			26826.14			
3726.732	3727.792			0.0			26825.53			
3726.866	3727.926		0.1	0.0			26824.57			
3727.035	3728.095		0.4	0.3			26823.35			
3727.348	3728.408		0.0				26821.10			
3727.428	3728.488		0.1d				26820.52			
3727.783	3728.843			0.1			26817.97			
3728.160	3729.220		0.0d				26815.26	4f 3	2c 1	R 1
3728.364	3729.424			0.0			26813.79	4f 3	2c 1	R 3
3728.720	3729.780		0.1d				26811.23			
3729.012	3730.073		0.3	0.3	0.1		26809.13			
3729.226	3730.287		0.0d				26807.59			
3729.634	3730.695				0.1		26804.66			
3729.715	3730.775		0.1	0.0			26804.08			
3729.780	3730.841		0.2				26803.61			
3729.854	3730.915		0.2	0.1			26803.08			
3730.118	3731.179		0.3				26801.18			
3730.191	3731.251		0.3		0.2		26800.66			
3730.440	3731.501		0.2				26798.87			
3730.539	3731.599		0.2				26798.16			
3730.681	3731.742		0.3	0.4	0.2		26797.14			
3730.735	3731.796		0.4				26796.75			
3730.848	3731.909			0.1			26795.94			
3731.047	3732.108		0.2	0.2			26794.51	4D 0	2B 0	P 5
3731.207	3732.268		0.1				26793.36			

λ_air	λ_vac	Ref.	I₁	I₂	I₃	I₄	ν	Upper	Lower	Br.
3731.332	3732.393		0.4		0.2		26792.46			
3731.401	3732.462		0.2	0.1			26791.97			
3731.519	3732.580		0.2				26791.12	4F 0	2B 1	R 3
3731.552	3732.613		0.1	0.1			26790.88			
3731.703	3732.764		0.2d				26789.80			
3731.793	3732.855				0.1		26789.15			
3731.947	3733.008		0.3	0.3			26788.05			
3732.020	3733.082		0.3	0.2			26787.52			
3732.112	3733.174	GK	1.5b	0.9	0.2	1.46	26786.86	W 2	2B 0	P 2
3732.214	3733.275		0.3				26786.13			
3732.397	3733.458			0.1			26784.82			
3732.666	3733.727						26782.89	4f 3	2c 1	Q 3
3732.717	3733.779		0.1	0.1			26782.52			
3732.879	3733.940		0.4		0.2		26781.36			
3733.017	3734.078	F			0.1		26780.37			
3733.082	3734.144				0.1		26779.90			
3733.184	3734.246			0.1			26779.17			
3733.470	3734.532		0.1	0.1			26777.12	4e 2	2c 0	R 4
3733.549	3734.611			0.2			26776.55			
3733.608	3734.670		0.1d				26776.13			
3733.681	3734.742		0.2				26775.61			
3733.969	3735.031		0.2b	0.2	0.2		26773.54			
3734.202	3735.264				0.2		26771.87			
3734.263	3735.325		0.1				26771.43			
3734.371	3735.433	K	0.3b				26770.66			
3734.482	3735.544			0.1			26769.86			
3734.541	3735.603	F					26769.44			
3734.646	3735.708			0.1			26768.69			
3735.066	3736.128			0.0			26765.68			
3735.332	3736.394		0.2				26763.77			
3735.444	3736.506		0.2	0.1			26762.97			
3735.618	3736.681		0.2	0.2			26761.72			
3735.814	3736.876		0.2				26760.32			
3735.882	3736.944			0.1			26759.83			
3735.987	3737.049		0.1				26759.08			
3736.064	3737.126		0.3	0.3			26758.53			
3736.128	3737.190		0.4				26758.07			
3736.173	3737.235		0.2	0.2			26757.75			
3736.439	3737.502		0.2				26755.84			
3736.506	3737.569		0.1				26755.36			
3736.599	3737.661		0.1				26754.70			
3736.840	3737.903				0.1		26752.97			
3737.459	3738.522		0.0d				26748.54			
3737.592	3738.655			0.2			26747.59			
3737.860	3738.923		0.2	0.1			26745.67			
3737.983	3739.046		0.2				26744.79			
3738.034	3739.096		0.1				26744.43			
3738.087	3739.149	G	0.3	0.2			26744.05	4D 0	2B 0	P 6
3738.499	3739.562		0.1				26741.10			
3738.576	3739.639		0.2d		0.1		26740.55			
3738.765	3739.828				0.1		26739.20			
3738.867	3739.930		0.2	0.1			26738.47			
3739.001	3740.064				0.2		26737.51			
3739.088	3740.151			0.1			26736.89			
3739.323	3740.386	F					26735.21			
3739.439	3740.502			0.1			26734.38			
3739.530	3740.593			0.0			26733.73	5c 1	2a 0	P 4
3739.647	3740.710	GK	0.7	0.3			26732.89			
3739.791	3740.855			0.1			26731.86			
3739.870	3740.933		0.2d				26731.30			

λ_{air}	λ_{vac}	Ref.	I_1	I_2	I_3	I_4	ν	Upper	Lower	Br.
3739.937	3741.000						26730.82	4f 2	2c 0	P 5
3740.070	3741.133		0.1	0.1			26729.87			
3740.184	3741.248		0.1				26729.05			
3740.277	3741.340	GK	0.7	0.3			26728.39	Y 2	2B 0	R 0
3740.428	3741.491		0.2				26727.31			
3740.487	3741.550		0.2				26726.89			
3740.746	3741.809		0.2b				26725.04			
3740.996	3742.060			0.2	0.1		26723.25			
3741.156	3742.219		0.3				26722.11	Y 2	2B 0	R 1
3741.276	3742.280	GK	1.0	0.4	0.2	1.00	26721.68	W 2	2B 0	P 3
3741.331	3742.395		0.3	0.1	0.1		26720.86			
3741.523	3742.586	F					26719.49			
3741.678	3742.742		0.1				26718.38			
3741.783	3742.847	F					26717.63			
3741.885	3742.949		0.2d				26716.90	4D 0	2B 0	P 7
3741.990	3743.054		0.2				26716.15			
3742.132	3743.196		0.1	0.1			26715.14			
3742.268	3743.332		0.4	0.2	0.2		26714.17	4e 2	2c 0	P 2
3742.642	3743.706		0.4	0.1			26711.50			
3742.758	3743.822		0.1		0.1		26710.67			
3742.928	3743.992				0.2		26709.46			
3743.040	3744.104		0.2				26708.66			
3743.110	3744.174	F					26708.16			
3743.187	3744.251				0.2		26707.61			
3743.280	3744.344		0.1				26706.95	4e 3	2c 1	R 4
3743.456	3744.520		0.2	0.1			26705.69			
3743.574	3744.638		0.1		0.2		26704.85			
3743.657	3744.721		0.2				26704.26			
3743.842	3744.906			0.0			26702.94			
3744.160	3745.224		0.2	0.2			26700.67	4f 3	2c 1	Q 2
3744.247	3745.311				0.1		26700.05			
3744.407	3745.471		0.4				26698.91			
3744.488	3745.553				0.2		26698.33			
3744.735	3745.800		0.2	0.2	0.2		26696.57	Y 2	2B 0	R 2
3744.892	3745.957	G	0.3	0.3	0.2b		26695.45			
3745.143	3746.208	F					26693.66			
3745.282	3746.347			0.2			26692.67			
3745.665	3746.730		0.2				26689.94			
3745.822	3746.887		0.5	0.3	0.2		26688.82			
3746.039	3747.103		0.2	0.2			26687.28			
3746.123	3747.188		0.2				26686.68			
3746.229	3747.294				0.1		26685.92			
3746.283	3747.348		0.2d				26685.54	4e 2	2c 0	R 5
3746.336	3747.401		0.2				26685.16			
3746.764	3747.830		0.5	0.3			26682.11		2B 4	Q 2
3747.106	3748.171				0.2		26679.68			
3747.264	3748.330		0.1				26678.55			
3747.361	3748.427			0.1			26677.86	4f 2	2c 0	P 4
3747.422	3748.487	F					26677.43			
3747.495	3748.560		0.3	0.0			26676.91			
3747.741	3748.806		0.2	0.0	0.2		26675.16			
3747.866	3748.931		0.2				26674.27			
3748.039	3749.104	K	0.7	0.3	0.2		26673.04	Y 2	2B 0	P 1
3748.502	3749.568		0.1				26669.74	4d 2	2c 0	P 1
3748.740	3749.805		0.3	0.0			26668.05			
3748.956	3750.022			0.0			26666.51			
3749.059	3750.125			0.0			26665.78			
3749.762	3750.828		0.2b				26660.78			
3750.135	3751.201	K	0.3b				26658.13			
3750.433	3751.499		0.2				26656.01			

λ_{air}	λ_{vac}	Ref.	I_1	I_2	I_3	I_4	ν	Upper	Lower	Br.
3750.629	3751.695				0.2		26654.62	·4e 3	2c 1	R 3
3750.824	3751.890						26653.23			
3751.111	3752.178		0.3				26651.19			
3751.248	3752.314		0.1				26650.22			
3751.492	3752.558		0.1d				26648.49			
3751.527	3752.593		0.2				26648.24			
3751.615	3752.682		0.2				26647.61			
3751.717	3752.783		0.1	0.1			26646.89			
3751.887	3752.954	GK	1.5b	0.9	0.3	1.40	26645.68	W 2	2B 0	P 4
3751.980	3753.047		0.5	0.0			26645.02			
3752.089	3753.155		0.3				26644.25			
3752.536	3753.603			0.0			26641.07			
3752.622	3753.689	GK	0.8	0.5	0.2		26640.46	3F 2	2B 0	R 2
3752.721	3753.788		0.2				26639.76			
3752.862	3753.928		0.0				26638.76			
3753.052	3754.119				0.2		26637.41			
3753.173	3754.240			0.1			26636.55	5c 1	2a 0	P 5
3753.256	3754.323			0.1			26635.96			
3753.344	3754.410		0.6	0.3			26635.34			
3753.518	3754.585		0.1		0.2		26634.10			
3753.761	3754.828		0.2d	0.2			26632.38			
3753.884	3754.950		0.2	0.1			26631.51			
3754.102	3755.169			0.1			26629.96			
3754.191	3755.258		0.2				26629.33			
3754.387	3755.454				0.2		26627.94			
3754.525	3755.592			0.0			26626.96			
3754.614	3755.681				0.1		26626.33			
3754.759	3755.826		0.2				26625.30			
3754.947	3756.014			0.0	0.1		26623.97			
3755.109	3756.176	K	0.6	0.3	0.2		26622.82			
3755.209	3756.276		0.2				26622.11			
3755.316	3756.383				0.1		26621.35			
3755.364	3756.431		0.2				26621.01			
3755.483	3756.550		0.1		0.1		26620.17			
3755.677	3756.745		0.0d	0.1			26618.79			
3755.816	3756.883		0.2	0.0	0.1		26617.81			
3755.888	3756.955	K	0.3	0.3			26617.30			
3755.931	3756.999				0.2		26616.99			
3756.120	3757.188		0.0d				26615.65			
3756.408	3757.476		0.2				26613.61			
3756.516	3757.583		0.2	0.2			26612.85			
3756.599	3757.667	GK	1.0	0.6	0.2		26612.26	Y 2	2B 0	P 2
3756.665	3757.733		0.2				26611.79			
3756.787	3757.854		0.1b	0.0			26610.93			
3756.925	3757.993		0.4	0.2			26609.95			
3757.172	3758.240		0.1	0.0			26608.20			
3757.377	3758.445		0.3	0.1			26606.75			
3757.457	3758.525				0.1		26606.18			
3757.593	3758.661	GK	0.6	0.5	0.2		26605.22	3F 2	2B 0	Q 4
3757.727	3758.795		0.0d				26604.27			
3757.929	3758.997				0.2		26602.84			
3758.003	3759.071			0.0			26602.32			
3758.104	3759.172			0.0			26601.60			
3758.381	3759.449				0.1		26599.64			
3758.497	3759.565		0.1				26598.82			
3758.654	3759.722	F					26597.71			
3758.896	3759.964			0.0			26596.00	3c 5	2a 0	Q 1
3759.149	3760.217		0.1				26594.21			
3759.219	3760.288		0.1				26593.71			
3759.321	3760.389		0.6	0.3			26592.99	4f 3	2c 1	Q 3

λ_{air}	λ_{vac}	Ref.	I_1	I_2	I_3	I_4	ν	Upper	Lower	Br.
3759.461	3760.530		0.2	0.0			26592.00	4e 3	2c 1	R 2
3759.499	3760.568		0.1				26591.73			
3759.576	3760.644	GK	0.9	0.3			26591.19			
3759.655	3760.723		0.2				26590.63			
3759.738	3760.807		0.5	0.2			26590.04	4f 3	2c 1	P 5
3759.800	3760.869		0.1				26589.60			
3759.921	3760.989		0.2				26588.75			
3760.038	3761.107		0.3				26587.92			
3760.175	3761.244		0.1		0.2		26586.95	4f 3	2c 1	Q 4
3760.377	3761.446		0.2d				26585.52			
3760.476	3761.545		0.1				26584.82			
3760.608	3761.677		0.2				26583.89	5c 2	2a 1	Q 1
3760.741	3761.810	K	0.4	0.4	0.2		26582.95			
3761.233	3762.302				0.1d		26579.47			
3761.743	3762.812	F					26575.87			
3761.883	3762.952		0.2				26574.88			
3761.992	3763.061		0.2				26574.11			
3762.084	3763.153	F					26573.46			
3762.206	3763.275			0.0			26572.60			
3762.546	3763.615		0.4	0.3			26570.20	4F 0	2B 1	R 2
3762.669	3763.738		0.3	0.1			26569.33			
3762.806	3763.875		0.0				26568.36			
3762.917	3763.986		0.0				26567.58			
3762.996	3764.065	GK	0.8	0.6			26567.02	W 2	2B 0	P 5
3763.097	3764.166		0.2	0.0			26566.31			
3763.182	3764.251			0.1			26565.71			
3763.745	3764.815		0.2				26561.73	4f 3	2c 1	Q 2
3763.957	3765.026		0.2				26560.24			
3764.329	3765.399		0.4	0.2			26557.61			
3764.453	3765.522	GK	0.8	0.4			26556.74	4c 2	2a 0	Q 1
3764.514	3765.583		0.2d				26556.31	4f 3	2c 1	P 4
3764.853	3765.922		0.0d				26553.92			
3765.084	3766.154			0.0			26552.29	3c 5	2a 0	Q 2
3765.669	3766.739			0.1			26548.16			
3765.833	3766.903		0.2				26547.01			
3766.069	3767.140		0.6	0.3			26545.34	3F 2	2B 0	R 1
3766.237	3767.307		0.6	0.3			26544.16	4c 6	2a 3	Q 1
3766.409	3767.479		0.2				26542.95			
3766.511	3767.581		0.3	0.1			26542.23			
3766.875	3767.946			0.1			26539.66			
3767.093	3768.163		0.1	0.0			26538.13			
3767.734	3768.805		0.0	0.1			26533.61	3E 3	2B 1	Q 8
3768.193	3769.264		0.3	0.1			26530.38			
3768.287	3769.357	GK	0.8	0.4			26529.72	Y 2	2B 0	P 3
3768.412	3769.483		0.3				26528.84	4c 2	2a 0	Q 2
3768.412	3769.483		0.3				26528.84	4f 3	2c 1	Q 3
3768.466	3769.537	K	0.3	0.3			26528.46	3F 2	2B 0	Q 3
3768.632	3769.703		0.1	0.1			26527.29			
3768.736	3769.807	GK	0.5	0.5	0.2		26526.56	W 2	2B 0	P 6
3768.825	3769.896		0.2				26525.93			
3769.119	3770.190		0.2d				26523.86			
3769.279	3770.349	F					26522.74	4e 3	2c 1	R 1
3769.436	3770.507		0.1				26521.63			
3769.633	3770.704		0.2d				26520.25			
3770.069	3771.140			0.0			26517.18			
3770.156	3771.227		0.1	0.1			26516.57			
3770.244	3771.315	GK	0.9	0.7	0.1d	1.07	26515.95	3E 3	2B 1	R 3
3770.601	3771.672		0.3		0.1		26513.44			
3770.655	3771.726	K	0.3				26513.06			
3770.719	3771.790		0.3				26512.61	4c 6	2a 3	Q 2

λ_{air}	λ_{vac}	Ref.	I_1	I_2	I_3	I_4	ν	Upper	Lower	Br.
3771.348	3772.419		0.3				26508.19	4d 2	2c 0	P 2
3771.407	3772.479		0.2	0.0			26507.77			
3771.507	3772.578	GK	1.5b	0.9	0.2	1.48	26507.07	3E 3	2B 1	R 2
3771.607	3772.678		0.3	0.1			26506.37			
3771.757	3772.829			0.0			26505.31			
3772.083	3773.155		0.4	0.2			26503.02			
3772.187	3773.259		0.2				26502.29			
3772.290	3773.361		0.4				26501.57			
3772.720	3773.791		0.2				26498.55			
3772.895	3773.967		0.3				26497.32			
3773.377	3774.449		0.2	0.2			26493.93			
3773.531	3774.603		0.2				26492.85			
3773.885	3774.957		0.1d				26490.37			
3774.168	3775.240		0.2d	0.1			26488.38			
3774.318	3775.390	GK	1.5b	0.7	0.3	1.20	26487.33	3E 3	2B 1	Q 7
3774.318	3775.390	GK	1.5b	0.7	0.3	1.20	26487.33	3c 5	2a 0	Q 3
3774.318	3775.390	GK	1.5b	0.7	0.3	1.20	26487.33	4c 2	2a 0	Q 3
3774.473	3775.545		0.2	0.1			26486.24			
3774.607	3775.679		0.2b	0.1			26485.30			
3774.664	3775.736	F					26484.90			
3774.717	3775.789		0.2	0.2			26484.53			
3775.090	3776.163		0.2	0.0			26481.91			
3775.196	3776.268		0.2				26481.17			
3775.415	3776.488		0.3				26479.63			
3775.576	3776.649	GK	1.0	0.6		1.11	26478.50			
3775.636	3776.709		0.5	0.4	0.2b		26478.08			
3775.917	3776.990		0.1				26476.11			
3776.041	3777.114			0.0			26475.24			
3776.221	3777.294		0.2				26473.98			
3776.348	3777.421		0.2				26473.09			
3776.670	3777.743	K	0.4	0.3	0.2		26470.83	4F 0	2B 1	Q 4
3776.759	3777.832		0.1	0.1			26470.21			
3776.893	3777.966		0.5	0.1			26469.27			
3776.947	3778.020		0.5	0.4	0.3b		26468.89			
3777.009	3778.082	GK	1.5	0.7		1.19	26468.46	3E 3	2B 1	R 1
3777.009	3778.082	GK	1.5	0.7		1.19	26468.46	4F 0	2B 1	P 5
3777.133	3778.206	GK	1.0	0.5			26467.59	3F 2	2B 0	Q 2
3777.217	3778.290		0.1				26467.00	4c 6	2a 3	Q 3
3777.311	3778.384		0.1				26466.34			
3777.398	3778.471		0.3	0.2			26465.73			
3777.508	3778.581	GK	0.7	0.5	0.2		26464.96			
3777.601	3778.674		0.0	0.1			26464.31			
3777.794	3778.867		0.2				26462.96			
3778.125	3779.198		0.3	0.3			26460.64			
3778.226	3779.299			0.0			26459.93			
3778.319	3779.392		0.3				26459.28			
3778.499	3779.572		0.3	0.2	0.2		26458.02			
3778.583	3779.657		0.2				26457.43			
3778.753	3779.827		0.3	0.2			26456.24			
3779.073	3780.147		0.2d				26454.00			
3779.443	3780.517		0.3	0.1			26451.41			
3779.746	3780.820		0.1				26449.29			
3779.845	3780.918		0.2				26448.60			
3779.931	3781.004			0.0			26448.00			
3780.038	3781.111			0.1			26447.25			
3780.134	3781.207		0.4	0.2			26446.58			
3780.308	3781.382	K	0.4	0.3	0.2		26445.36			
3780.475	3781.549		0.4	0.3			26444.19			
3780.567	3781.641		0.1				26443.55			
3780.675	3781.749		0.6	0.4			26442.79			

λ_{air}	λ_{vac}	Ref.	I_1	I_2	I_3	I_4	ν	Upper	Lower	Br.
3780.777	3781.851		0.1				26442.08			
3780.894	3781.968		0.2b	0.0			26441.26			
3781.026	3782.100		0.5	0.3			26440.34			
3781.212	3782.286		0.2				26439.04			
3781.277	3782.351		0.2	0.2			26438.58			
3781.400	3782.474		0.1				26437.72			
3781.528	3782.602		0.4	0.2			26436.83			
3781.616	3782.690		0.2	0.1			26436.21			
3781.732	3782.806		0.2				26435.40			
3781.810	3782.884		0.2d				26434.86			
3781.945	3783.020		0.2	0.2			26433.91			
3782.024	3783.098	K	0.5	0.6	0.2		26433.36	3E 3	2B 1	Q 6
3782.269	3783.343		0.2d				26431.65	4c 2	2a 0	Q 4
3782.353	3783.428		0.2				26431.06	4e 2	2c 0	P 4
3782.442	3783.516		0.2	0.1			26430.44			
3782.525	3783.599	GK	1.0	0.6			26429.86	Y 2	2B 0	P 4
3782.597	3783.671		0.2				26429.36			
3782.664	3783.738		0.2	0.1			26428.89			
3782.757	3783.831		0.9		0.2d		26428.24			
3782.774	3783.848	GK	0.7	0.5			26428.12			
3782.798	3783.873		0.8				26427.95	4e 3	2c 1	Q 4
3782.894	3783.969		0.2	0.₀2			26427.28	4e 3	2c 1	Q 5
3782.973	3784.047	GK	0.8	0.5	0.2d		26426.73	4E 2	2B 4	Q 2
3783.125	3784.199		0.1				26425.67			
3783.384	3784.458		0.3	0.2			26423.86			
3783.507	3784.582		0.2	0.1			26423.00			
3783.619	3784.693		0.2				26422.22			
3783.755	3784.829		0.2b	0.1	0.1		26421.27			
3783.832	3784.907			0.0			26420.73	4e 3	2c 1	Q 3
3783.948	3785.023		0.4	0.4	0.2b		26419.92			
3784.000	3785.074		0.1				26419.56			
3784.093	3785.168		0.2	0.1			26418.91			
3784.173	3785.248	GK	0.8	0.8	0.2		26418.35	3E 3	2B 1	R 0
3784.262	3785.337		0.2				26417.73			
3784.530	3785.605		0.1				26415.86			
3784.699	3785.774		0.0				26414.68			
3784.833	3785.908		0.1d		0.1		26413.74			
3785.397	3786.472		0.3b	0.2			26409.81			
3785.699	3786.774			0.0			26407.70	4e 3	2c 1	Q 2
3786.489	3787.565		0.1d				26402.19	3c 5	2a 0	Q 4
3786.984	3788.060		0.2				26398.74			
3787.076	3788.151		0.1				26398.10			
3787.165	3788.240		0.1				26397.48			
3787.297	3788.372	K	0.3b	0.5	0.1		26396.56	3E 3	2B 1	Q 5
3787.420	3788.496	GK	0.7	0.5	0.2d		26395.70			
3787.508	3788.583			0.0			26395.09			
3787.598	3788.674		0.2				26394.46			
3787.720	3788.796		0.1				26393.61			
3787.788	3788.863		0.2				26393.14			
3787.871	3788.947		0.2b	0.1			26392.56			
3788.421	3789.496		0.0d	0.0d			26388.73	4d 2	2c 0	Q 3
3788.421	3789.496		0.0d	0.0d			26388.73	4e 3	2c 1	Q 1
3788.785	3789.861		0.1				26386.19			
3789.090	3790.166			0.0d			26384.07			
3789.256	3790.332		0.5	0.4	0.2d		26382.91			
3789.416	3790.492		0.0				26381.80			
3789.521	3790.597		0.1	0.1			26381.07			
3789.762	3790.838		0.4	0.2b			26379.39			
3789.809	3790.886		0.2				26379.06			
3789.911	3790.988		0.1d				26378.35			

λ_{air}	λ_{vac}	Ref.	I_1	I_2	I_3	I_4	ν	Upper	Lower	Br.
3789.996	3791.072		0.1	0.1d			26377.76	4F 0	2B 1	R 1
3790.434	3791.511		0.0				26374.71			
3790.541	3791.617			0.1			26373.97	3F 2	2B 0	P 4
3790.774	3791.850			0.0			26372.35			
3790.884	3791.961		0.3	0.2			26371.58			
3790.971	3792.047	GK	0.6	0.6	0.3	1.08	26370.98			
3791.061	3792.138		0.1	0.1			26370.35			
3791.245	3792.322		0.1d	0.2			26369.07			
3791.329	3792.405		0.1	0.1			26368.49			
3791.411	3792.487	GK	0.7	0.9	0.4	1.54	26367.92	3E 3	2B 1	Q 4
3791.501	3792.578		0.2	0.1			26367.29			
3791.595	3792.671		0.4	0.3			26366.64	4f 3	2c 1	P 4
3791.675	3792.752		0.1				26366.08			
3791.741	3792.818		0.0	0.0			26365.62			
3791.836	3792.913		0.1	0.0			26364.96			
3791.905	3792.982		0.2	0.2			26364.48			
3792.010	3793.087		0.4	0.3			26363.75	4c 2	2a 0	Q 5
3792.125	3793.202		0.1d				26362.95			
3792.246	3793.323		0.2				26362.11			
3792.448	3793.525			0.1			26360.71			
3792.813	3793.890		0.2	0.2			26358.17			
3792.920	3793.997		0.1				26357.43			
3793.061	3794.138		0.1				26356.45			
3793.138	3794.215		0.2				26355.91			
3793.222	3794.299			0.0			26355.33			
3793.340	3794.417		0.2	0.2			26354.51			
3793.452	3794.529		0.0				26353.73			
3793.534	3794.611		0.3				26353.16	3F 2	2B 0	P 3
3793.569	3794.646			0.2			26352.92			
3793.592	3794.669		0.2				26352.76			
3793.651	3794.728		0.2				26352.35			
3793.704	3794.781		0.0	0.2			26351.98			
3794.038	3795.115			0.1			26349.66			
3794.195	3795.272		0.0				26348.57			
3794.289	3795.366			0.2			26347.92			
3794.388	3795.465		0.0	0.0			26347.23			
3794.467	3795.545	GK	0.7	0.9	0.3	1.41	26346.68	3E 3	2B 1	Q 3
3794.569	3795.647		0.3	0.2			26345.97			
3794.709	3795.787		0.3	0.3			26345.00			
3794.830	3795.908		0.2		0.1		26344.16			
3794.917	3795.994		0.1				26343.56			
3794.989	3796.066		0.2	0.2d	0.1d		26343.06			
3795.072	3796.150		0.1				26342.48			
3795.134	3796.212		0.0				26342.05			
3795.238	3796.316		0.2	0.0d			26341.33			
3795.334	3796.412		0.2				26340.66			
3795.466	3796.543		0.1d	0.0			26339.75			
3795.564	3796.641		0.1				26339.07			
3795.627	3796.705		0.3	0.2			26338.63			
3795.757	3796.834		0.2d	0.0			26337.73			
3795.911	3796.989		0.2d	0.1			26336.66			
3796.062	3797.140	GK	0.6	0.5			26335.61			
3796.177	3797.255		0.0	0.1			26334.81			
3796.227	3797.304		0.2				26334.47	4d 2	2c 0	P 3
3796.327	3797.405		0.1				26333.77			
3796.431	3797.509		0.4	0.4			26333.05			
3796.515	3797.593		0.2	0.1			26332.47			
3796.594	3797.672	GK	1.5b	1.0	0.5	2.01	26331.92	3E 3	2B 1	Q 2
3796.694	3797.772		0.4	0.3			26331.23			
3796.809	3797.887		0.1	0.2			26330.43			

λ_{air}	λ_{vac}	Ref.	I_1	I_2	I_3	I_4	ν	Upper	Lower	Br.
3796.904	3797.982	K	0.3	0.4	0.1		26329.77	4F 0	2B 1	Q 3
3796.965	3798.043		0.2				26329.35			
3797.047	3798.125		0.1				26328.78			
3797.141	3798.219	GK	0.8b	0.7	0.3		26328.13			
3797.247	3798.326		0.2				26327.39			
3797.360	3798.438		0.5	0.3			26326.61			
3797.435	3798.513		0.2	0.1			26326.09			
3797.521	3798.600	GK	1.0b	1.0	0.3	1.86	26325.49	3D 3	2B 1	R 0
3797.627	3798.705		0.4	0.2			26324.76			
3797.760	3798.838		0.3	0.3			26323.84			
3797.832	3798.910		0.0				26323.34			
3797.917	3798.995	GK	0.8	0.9	0.3	1.48	26322.75	3E 3	2B 1	Q 1
3798.015	3799.093		0.3				26322.07			
3798.177	3799.255			0.0			26320.95			
3798.308	3799.386		0.1	0.1			26320.04			
3798.381	3799.460		0.1				26319.53			
3798.490	3799.568		0.4	0.3			26318.78			
3798.594	3799.672		0.2				26318.06			
3798.676	3799.754		0.1	0.2			26317.49			
3798.777	3799.856		0.3	0.2			26316.79			
3798.840	3799.919	GK	0.6	0.6	0.3		26316.35	4F 0	2B 1	P 4
3798.947	3800.026		0.3				26315.61			
3799.041	3800.120	GK	0.9b	0.8	0.3	1.54	26314.96	3D 3	2B 1	R 1
3799.138	3800.217		0.3	0.1			26314.29			
3799.252	3800.331		0.4	0.3			26313.50			
3799.311	3800.390		0.2	0.2			26313.09			
3799.406	3800.485		0.4	0.4			26312.43			
3799.539	3800.618		0.1				26311.51			
3799.726	3800.804		0.0				26310.22			
3799.805	3800.884		0.1				26309.67			
3799.892	3800.971		0.0				26309.07			
3799.958	3801.037		0.2	0.3	0.2b		26308.61			
3800.053	3801.132	GK	0.7	0.7		1.01	26307.95	3E 3	2B 1	P 2
3800.150	3801.229		0.1	0.1d			26307.28	4f 3	2c 1	P 5
3800.196	3801.275		0.1				26306.96			
3800.409	3801.488		0.1d	0.0d			26305.49			
3800.494	3801.573		0.0				26304.90			
3800.617	3801.696		0.4	0.4	0.2b		26304.05			
3800.719	3801.799		0.1				26303.34			
3800.860	3801.939	K	0.2	0.3			26302.37			
3800.942	3802.021		0.4b	0.3			26301.80			
3801.052	3802.131		0.2				26301.04			
3801.124	3802.203		0.3	0.3			26300.54			
3801.205	3802.284		0.0				26299.98			
3801.260	3802.339		0.1	0.1d			26299.60			
3801.329	3802.409		0.1				26299.12			
3801.444	3802.523		0.3	0.3			26298.33			
3801.535	3802.614		0.1	0.2			26297.70			
3801.611	3802.691	GK	0.6	0.6	0.2		26297.17	3E 2	2B 0	R 4
3801.837	3802.916		0.2	0.2d			26295.61			
3801.989	3803.068		0.0	0.1			26294.56			
3802.084	3803.163	GK	0.7	0.6	0.1	0.93	26293.90			
3802.179	3803.259		0.2				26293.24	3c 8	2a 2	Q 1
3802.268	3803.347		0.0	0.0			26292.63			
3802.370	3803.450		0.2				26291.92			
3802.480	3803.560		0.0				26291.16			
3802.609	3803.689		0.2	0.2			26290.27			
3802.668	3803.748		0.2				26289.86			
3802.767	3803.846		0.2				26289.18			
3802.864	3803.943		0.4	0.3			26288.51			

λ_{air}	λ_{vac}	Ref.	I_1	I_2	I_3	I_4	ν	Upper	Lower	Br.
3802.947	3804.027		0.3	0.1			26287.93			
3803.033	3804.113	GK	1.5b	1.0	0.4	1.98	26287.34	3D 3	2B 1	R 2
3803.133	3804.212		0.4	0.2			26286.65			
3803.202	3804.282		0.2				26286.17			
3803.288	3804.367	GK	0.6	0.6		1.16	26285.58	3E 3	2B 1	P 3
3803.386	3804.466		0.2				26284.90	4e 2	2c 0	P 5
3803.468	3804.548		0.1				26284.33			
3803.681	3804.761		0.2	0.0d			26282.86			
3804.091	3805.171		0.1d				26280.03			
3804.505	3805.585			0.0d			26277.17			
3804.754	3805.834		0.0				26275.45			
3804.825	3805.905		0.0				26274.96			
3804.923	3806.003		0.3	0.2			26274.28			
3805.829	3806.909		0.3	0.3			26268.03			
3805.911	3806.992		0.2d				26267.46			
3806.108	3807.189		0.2	0.1d			26266.10			
3806.239	3807.319		0.1				26265.20			
3806.339	3807.419		0.1				26264.51			
3806.453	3807.534		0.4	0.3			26263.72			
3806.626	3807.706		0.2	0.1			26262.53			
3806.710	3807.790		0.3				26261.95			
3806.740	3807.821	GK	0.6	0.5	0.2		26261.74			
3806.862	3807.943		0.2				26260.90			
3807.490	3808.571		0.4	0.4	0.1		26256.57			
3807.599	3808.679		0.1				26255.82			
3807.730	3808.811			0.0			26254.91			
3807.855	3808.936			0.0d			26254.05	4d 2	2c 0	Q 4
3807.970	3809.051	GK	0.6	0.5			26253.26	3E 3	2B 1	P 4
3808.155	3809.236		0.1	0.1			26251.98			
3808.301	3809.382			0.0			26250.98			
3808.415	3809.496			0.0			26250.19			
3808.486	3809.567		0.0d				26249.70			
3808.565	3809.646			0.1			26249.16			
3808.641	3809.723		0.4	0.4	0.1		26248.63	4E 0	2B 1	R 2
3808.776	3809.858		0.1				26247.70			
3808.897	3809.978		0.1	0.1			26246.87			
3809.026	3810.107	GK	0.8	0.8	0.2	1.28	26245.98	3D 3	2B 1	R 3
3809.120	3810.202		0.2b				26245.33			
3809.274	3810.356	F	0.3b	0.3	0.2d		26244.27	3E 2	2B 0	Q10
3809.672	3810.753		0.1d				26241.53			
3809.774	3810.855		0.0	0.1			26240.83			
3809.878	3810.960	G	0.5	0.3			26240.11	3E 2	2B 0	R 3
3809.977	3811.058		0.0				26239.43			
3810.853	3811.934	1		0.0d			26233.40			
3811.040	3812.122		0.0	0.0d			26232.11			
3811.180	3812.261	K	0.4	0.4	0.1		26231.15	4d 2	2c 0	Q 1
3811.562	3812.644		0.1d	0.1d			26228.52			
3811.761	3812.843		0.5	0.4	0.2		26227.15			
3812.014	3813.096		0.2	0.1			26225.41			
3812.102	3813.184		0.0				26224.80			
3812.179	3813.262	GK	0.7	0.6	0.2	0.92	26224.27	6c 0	2a 0	Q 1
3812.271	3813.353		0.2				26223.64			
3812.342	3813.424		0.1				26223.15			
3812.469	3813.551		0.0d	0.0			26222.28			
3812.584	3813.666		0.2	0.2			26221.49			
3812.669	3813.752			0.2			26220.90			
3812.752	3813.835	GK	0.8	0.9	0.4	1.37	26220.33	3D 3	2B 1	R 4
3812.854	3813.936		0.2	0.2			26219.63			
3813.132	3814.214			0.1d			26217.72			
3813.434	3814.517		0.2	0.0			26215.64			

λ_air	λ_vac	Ref.	I₁	I₂	I₃	I₄	ν	Upper	Lower	Br.
3813.510	3814.592	GK	0.6	0.6			26215.12	3D 3	2B 1	P 2
3813.610	3814.693		0.1				26214.43			
3813.789	3814.872		0.1				26213.20			
3813.993	3815.076		0.2				26211.80			
3814.399	3815.482			0.1			26209.01			
3814.590	3815.672	GK	0.4	0.4	0.2b		26207.70	6c 0	2a 0	Q 2
3814.680	3815.763		0.1				26207.08			
3814.904	3815.987			0.1			26205.54			
3815.021	3816.103			0.1			26204.74			
3815.142	3816.224		0.4	0.4			26203.91			
3815.334	3816.417	GK	0.6	0.7	0.3	1.12	26202.59	4F 0	2B 1	Q 2
3815.436	3816.519		0.2	0.1			26201.89			
3815.529	3816.612		0.1	0.0			26201.25			
3815.945	3817.028			0.0			26198.39			
3816.157	3817.240		0.1d	0.0			26196.94			
3816.264	3817.348		0.0				26196.20			
3816.356	3817.439	GK	0.5	0.6	0.3		26195.57			
3816.481	3817.565	GK	0.4	0.6	0.2		26194.71	4F 0	2B 1	P 3
3816.684	3817.767		0.2b	0.2			26193.32	3E 3	2B 1	P 5
3816.799	3817.882	GK	0.6	0.6	0.1	0.99	26192.53	X 2	2B 1	R 0
3816.911	3817.995		0.1				26191.76	4d 3	2c 1	P 1
3817.059	3818.142			0.1			26190.75			
3817.245	3818.328		0.2b	0.0			26189.47			
3817.681	3818.764		0.2				26186.48			
3818.107	3819.190		0.2	0.1			26183.56			
3818.196	3819.279	GK	0.5	0.6	0.3		26182.95	6c 0	2a 0	Q 3
3818.324	3819.408		0.1				26182.07			
3818.880	3819.964	GK	0.5	0.5			26178.26			
3819.011	3820.095		0.2d				26177.36			
3819.125	3820.209		0.1d				26176.58			
3819.247	3820.331		0.3				26175.74			
3819.294	3820.378	GK	0.6	0.6	0.1		26175.42	3E 2	2B 0	R 2
3819.388	3820.471		0.0	0.1			26174.78			
3819.472	3820.556	GK	0.5	0.6			26174.20	X 2	2B 1	R 1
3819.596	3820.680	1	0.3d	0.2	0.5		26173.35			
3819.741	3820.825				0.2		26172.36	4E 0	2B 1	R 1
3819.922	3821.006		0.2				26171.12			
3820.041	3821.125		0.2				26170.30			
3820.177	3821.261		0.0				26169.37			
3820.263	3821.347		0.4	0.4	0.2		26168.78	4E 0	2B 1	R 0
3820.463	3821.547		0.1				26167.41	4d 3	2c 1	R 1
3820.697	3821.781		0.0d		0.1		26165.81			
3820.923	3822.008		0.1	0.0	0.1		26164.26			
3821.173	3822.257				0.1		26162.55			
3821.264	3822.348		0.1d	0.1			26161.93			
3821.694	3822.779	K	0.3	0.3	0.1		26158.98	3E 2	2B 0	Q 9
3821.877	3822.962	GK	0.6	0.4			26157.73			
3822.165	3823.250		0.3	0.3			26155.76	X 2	2B 1	P 1
3822.316	3823.400		0.3	0.3			26154.73			
3822.409	3823.494		0.2	0.2			26154.09			
3822.848	3823.932			0.0			26151.09	4d 2	2c 0	P 4
3822.962	3824.046		0.2	0.2	0.1		26150.31			
3823.013	3824.098			0.1			26149.96			
3823.089	3824.174	GK	0.6	0.5			26149.44	6c 0	2a 0	Q 4
3823.263	3824.348		0.2	0.2			26148.25	5c 3	2a 2	Q 1
3823.263	3824.348		0.2	0.2			26148.25	4e 3	2c 1	P 5
3823.410	3824.495		0.2	0.3			26147.24			
3823.615	3824.700			0.2			26145.84			
3823.797	3824.882		0.1				26144.60			
3823.899	3824.984		0.0				26143.90			

λ_{air}	λ_{vac}	Ref.	I_1	I_2	I_3	I_4	ν	Upper	Lower	Br.
3824.009	3825.094		0.1d				26143.15			
3824.282	3825.367		0.0				26141.28			
3824.777	3825.862		0.0				26137.90			
3824.939	3826.025	GK	0.7	0.8	0.2	1.30	26136.79	X 2	2B 1	R 2
3825.040	3826.126		0.2				26136.10			
3825.444	3826.530		0.0				26133.34			
3825.629	3826.714	GK	0.4	0.3			26132.08	3D 3	2B 1	P 3
3826.150	3827.235		0.2	0.2	0.2		26128.52			
3827.410	3828.496		0.3	0.3			26119.92			
3827.697	3828.783		0.1				26117.96			
3827.920	3829.006		0.1				26116.44			
3828.354	3829.440		0.2	0.3			26113.48	3A 2	2B 1	R 4
3828.500	3829.586		0.2b				26112.48			
3828.595	3829.682		0.2				26111.83			
3828.681	3829.767		0.1				26111.25			
3828.786	3829.872		0.2	0.1			26110.53			
3828.890	3829.977	K	0.3	0.4	0.2b		26109.82			
3829.007	3830.094		0.0d				26109.02	4d 2	2c 0	Q 5
3829.167	3830.254	K	0.3	0.5	0.3		26107.93	6c 0	2a 0	Q 5
3829.167	3830.254	K	0.3	0.5	0.3		26107.93	4d 3	2c 1	R 2
3829.335	3830.421		0.3	0.4			26106.79	3E 2	2B 0	R 1
3829.415	3830.502				0.1		26106.24			
3829.505	3830.591			0.1			26105.63			
3829.624	3830.710		0.1d	0.2			26104.82			
3829.729	3830.816		0.0				26104.10			
3829.813	3830.900		0.1				26103.53			
3829.992	3831.079	G	0.4	0.4			26102.31			
3830.238	3831.325		0.2				26100.63			
3830.344	3831.431		0.2	0.1			26099.91			
3830.461	3831.548		0.2				26099.11			
3830.523	3831.610			0.1			26098.69			
3830.665	3831.752		0.2d	0.2			26097.72			
3830.806	3831.893		0.1	0.1			26096.76			
3830.930	3832.017		0.1	0.0			26095.92			
3831.015	3832.102		0.2	0.1			26095.34	4e 3	2c 1	P 6
3831.112	3832.199	GK	0.5	0.6	0.1		26094.68	X 2	2B 1	R 3
3831.204	3832.291		0.1				26094.05			
3831.526	3832.613		0.1				26091.86			
3831.658	3832.745		0.1				26090.96			
3831.783	3832.870		0.2				26090.11	5c 3	2a 2	Q 3
3831.918	3833.005		0.1				26089.19			
3831.987	3833.074		0.2	0.2			26088.72			
3832.103	3833.190		0.2	0.2			26087.93			
3832.234	3833.321		0.2				26087.04			
3832.301	3833.389		0.2				26086.58	4e 3	2c 1	Q 1
3832.428	3833.515	GK	0.6	0.6	0.4		26085.72	3E 2	2B 0	Q 8
3832.550	3833.637		0.0				26084.89	4c 3	2a 1	R 1
3832.739	3833.827		0.1	0.1			26083.60			
3832.842	3833.930			0.0			26082.90			
3832.949	3834.037		0.4	0.3			26082.17	X 2	2B 1	P 2
3833.374	3834.462		0.1d	0.0d			26079.28			
3833.493	3834.581	GK	0.5	0.5	0.2		26078.47			
3833.590	3834.678		0.1				26077.81			
3833.667	3834.754			0.1			26077.29			
3833.812	3834.900		0.2				26076.30			
3833.934	3835.022		0.1	0.1d			26075.47			
3834.109	3835.197			0.0			26074.28			
3834.184	3835.272		0.4	0.3			26073.77			
3834.334	3835.422		0.1				26072.75			
3834.406	3835.494		0.2				26072.26			

λ_{air}	λ_{vac}	Ref.	I_1	I_2	I_3	I_4	ν	Upper	Lower	Br.
3834.528	3835.616		0.2	0.1d			26071.43			
3834.668	3835.756	1	0.0d		0.3		26070.48			
3834.962	3836.050		0.3	0.2			26068.48			
3835.084	3836.172			0.1			26067.65			
3835.208	3836.296		0.0				26066.81			
3835.312	3836.401			0.3			26066.10			
3835.402	3836.490	GK	0.5b	0.3	0.4b		26065.49			
3835.589	3836.677		0.2	0.3			26064.22			
3835.798	3836.886		0.2	0.3			26062.80			
3835.883	3836.972	G	0.4	0.4	0.3b		26062.22			
3836.184	3837.272		0.1				26060.18			
3836.265	3837.353		0.2	0.2			26059.63			
3836.449	3837.537	GK	0.8	0.8	0.4	1.35	26058.38	4E 0	2B 1	P 2
3836.549	3837.637		0.2				26057.70			
3836.877	3837.966		0.0				26055.47			
3837.269	3838.358		0.0				26052.81	4d 3	2c 1	Q 2
3837.453	3838.542		0.1	0.1			26051.56			
3837.521	3838.609		0.2				26051.10			
3837.628	3838.717	GK	0.5	0.5	0.3		26050.37			
3837.761	3838.850		0.0				26049.47			
3837.892	3838.981		0.1	0.0			26048.58			
3838.387	3839.476		0.1	0.1			26045.22	5c 3	2a 2	Q 4
3838.629	3839.718		0.3	0.3			26043.58	4d 3	2c 1	P 2
3838.754	3839.843			0.1			26042.73			
3838.887	3839.976		0.3	0.4	0.2		26041.83			
3839.018	3840.107			0.0d			26040.94			
3839.148	3840.237			0.1			26040.06			
3839.313	3840.402	GK	0.4	0.5	0.3		26038.94	4E 0	2B 1	P 5
3839.546	3840.635		0.2d				26037.36			
3839.670	3840.759			0.0d			26036.52			
3839.789	3840.879		0.1d				26035.71			
3839.888	3840.977		0.1	0.0			26035.04			
3840.002	3841.091		0.2				26034.27			
3840.046	3841.135		0.2	0.0			26033.97			
3840.111	3841.200	K	0.3	0.2			26033.53	3D 3	2B 1	P 4
3840.193	3841.283	GK	0.5	0.5	0.2d		26032.97			
3840.387	3841.476			0.1d			26031.66			
3840.555	3841.644	GK	0.6b	0.6	0.3	1.09	26030.52	4E 0	2B 1	Q 1
3840.555	3841.644	GK	0.6b	0.6	0.3	1.09	26030.52	3E 2	2B 0	R 0
3840.664	3841.754		0.1				26029.78			
3840.798	3841.888			0.0			26028.87			
3841.387	3842.477			0.0			26024.88			
3841.460	3842.549	G	0.5	0.5	0.3		26024.39	3E 2	2B 0	Q 7
3841.570	3842.660		0.1	0.1			26023.64			
3841.654	3842.744			0.2			26023.07			
3841.812	3842.902			0.0			26022.00			
3842.301	3843.391		0.1d				26018.69			
3842.371	3843.460		0.1				26018.22			
3842.450	3843.540	GK	0.5	0.5	0.2		26017.68	3A 2	2B 1	R 3
3842.542	3843.632			0.0			26017.06			
3842.982	3844.072	K					26014.08			
3843.068	3844.158			0.1d			26013.50			
3843.168	3844.258		0.4	0.3			26012.82			
3843.372	3844.462		0.2	0.0			26011.44			
3843.570	3844.660			0.0d			26010.10			
3843.866	3844.956		0.1				26008.10			
3844.033	3845.123	K	0.2	0.2	0.2		26006.97			
3844.226	3845.317	K					26005.66			
3844.311	3845.401		0.1d				26005.09			
3844.623	3845.713			0.1			26002.98			

λ_{air}	λ_{vac}	Ref.	I$_1$	I$_2$	I$_3$	I$_4$	ν	Upper	Lower	Br.
3844.912	3846.003			0.0			26001.02			
3845.006	3846.096	K					26000.39			
3845.290	3846.380		0.1				25998.47			
3845.392	3846.482	GK	0.5	0.4			25997.78	4c 3	2a 1	Q 1
3845.489	3846.580		0.2				25997.12			
3845.556	3846.646		0.1	0.1			25996.67			
3845.602	3846.692			0.0			25996.36			
3845.671	3846.762		0.2	0.2	0.2d		25995.89			
3845.744	3846.834		0.1	0.1			25995.40			
3845.837	3846.928	GK	0.6	0.6	0.3	1.02	25994.77	4E 0	2B 1	P 4
3845.948	3847.039		0.1	0.0			25994.02			
3846.060	3847.151			0.2			25993.26			
3846.131	3847.222			0.0			25992.78			
3846.227	3847.318	GK	0.7	0.8	0.5	1.29	25992.13	4E 0	2B 1	Q 4
3846.333	3847.423		0.1	0.1			25991.42	X 2	2B 1	P 3
3846.452	3847.543			0.1			25990.61			
3846.529	3847.620			0.0			25990.09			
3846.645	3847.736	K					25989.31			
3846.692	3847.783		0.0				25988.99	4E 0	2B 1	P 3
3846.982	3848.073			0.0			25987.03			
3847.397	3848.488		0.2d	0.1			25984.23			
3847.758	3848.850		0.2				25981.79			
3847.892	3848.983	K					25980.89			
3847.979	3849.070		0.1d	0.1			25980.30			
3848.118	3849.210			0.0			25979.36			
3848.274	3849.365		0.1d	0.0			25978.31			
3848.474	3849.565		0.3	0.2			25976.96			
3848.619	3849.710		0.1	0.1			25975.98			
3848.712	3849.804		0.2				25975.35			
3848.800	3849.891	GK	0.7	0.7	0.5		25974.76	3E 2	2B 0	Q 6
3848.887	3849.979		0.1	0.0			25974.17			
3848.963	3850.054	G	0.3	0.1			25973.66			
3849.041	3850.133	K	0.1				25973.13			
3849.166	3850.257		0.3	0.1			25972.29			
3849.235	3850.327		0.1	0.0			25971.82			
3849.321	3850.413	GK	0.9	0.8	0.5	1.72	25971.24	4E 0	2B 1	Q 2
3849.321	3850.413	GK	0.9	0.8	0.5	1.72	25971.24	4e 3	2c 1	P 2
3849.427	3850.518		0.3	0.1			25970.53	4c 3	2a 1	Q 2
3849.563	3850.655		0.1	0.1			25969.61			
3849.634	3850.726		0.1	0.0			25969.13			
3849.717	3850.809	GK	0.6	0.6	0.3	1.18	25968.57	4E 0	2B 1	Q 3
3849.830	3850.922		0.3	0.2			25967.81	3F 3	2B 2	R 2
3850.006	3851.098	GK	0.6	0.7	0.5	0.93	25966.62	4D 0	2B 1	R 6
3850.101	3851.193		0.2	0.0			25965.98			
3850.147	3851.239		0.1				25965.67			
3850.362	3851.454		0.1b				25964.22			
3850.641	3851.733	K					25962.34			
3851.107	3852.199		0.2	0.0			25959.20			
3851.179	3852.272		0.1				25958.71			
3851.273	3852.365	GK	0.8	0.9	0.3	1.58	25958.08	3A 2	2B 1	R 2
3851.377	3852.469		0.2	0.1	0.2		25957.38			
3851.513	3852.605	K					25956.46			
3852.191	3853.284			0.1			25951.89			
3852.392	3853.484	K					25950.54			
3852.567	3853.660		0.2				25949.36	4e 3	2c 1	Q 3
3852.913	3854.006		0.1				25947.03			
3853.197	3854.289	K	0.2	0.2			25945.12			
3853.330	3854.423		0.1				25944.22			
3853.482	3854.575	K					25943.20			
3853.705	3854.797		0.1	0.1			25941.70			

λ_{air}	λ_{vac}	Ref.	I_1	I_2	I_3	I_4	ν	Upper	Lower	Br.
3853.828	3854.921		0.3	0.4			25940.87			
3854.029	3855.121	K	0.3	0.3			25939.52			
3854.306	3855.399		0.1				25937.65	4e 2	2c 0	P 7
3854.503	3855.596	GK	0.5	0.7	0.3		25936.33	3E 2	2B 0	Q 5
3854.503	3855.596	GK	0.5	0.7	0.3		25936.33	T 0	2B 1	R 2
3854.503	3855.596	GK	0.5	0.7	0.3		25936.33	4d 3	2c 1	Q 3
3854.596	3855.689		0.1				25935.70			
3854.809	3855.902		0.2b	0.1d			25934.27			
3855.100	3856.193		0.2				25932.31			
3855.173	3856.266		0.1	0.1			25931.82			
3855.287	3856.381		0.2b				25931.05			
3855.399	3856.492	GK	0.4	0.4	0.2		25930.30	4c 3	2a 1	Q 3
3855.540	3856.634		0.4	0.4			25929.35			
3855.661	3856.754	K	0.2				25928.54			
3855.777	3856.870		0.1d				25927.76			
3855.927	3857.020		0.1				25926.75			
3856.015	3857.108		0.1d	0.0d	0.2		25926.16			
3856.235	3857.328			0.0			25924.68			
3856.340	3857.434	GK	0.5	0.5	0.3		25923.97			
3856.468	3857.562		0.0	0.1			25923.11	3D 3	2B 1	P 5
3856.546	3857.639	G					25922.59	3D 2	2B 0	P 1
3856.584	3857.678		0.5	0.4			25922.33			
38 6.638	38 7.731	K					25921.97			
3856.758	3857.852		0.4	0.3			25921.16	4c 7	2a 4	Q 1
3856.866	3857.959		0.1	0.1d			25920.44			
3856.946	3858.040		0.0				25919.90			
3857.069	3858.163			0.1d			25919.07			
3857.376	3858.470			0.1d			25917.01			
3857.452	3858.546			0.1			25916.50			
3857.593	3858.687			0.1			25915.55			
3857.642	3858.736		0.4	0.4			25915.22			
3857.745	3858.839		0.3				25914.53	3E 2	2B 0	P 3
3857.791	3858.885	GK	0.7	0.8	0.3	1.32	25914.22	3E 2	2B 0	P 2
3857.900	3858.994		0.2				25913.49			
3858.077	3859.171		0.1				25912.30			
3858.153	3859.247		0.1	0.1			25911.79			
3858.259	3859.353		0.2				25911.08			
3858.320	3859.414			0.1d			25910.67			
3858.524	3859.618		0.3	0.2	0.2		25909.30			
3858.613	3859.707		0.7	0.5			25908.70	3E 2	2B 0	P 4
3858.683	3859.777	GK	0.7	0.9	0.5	1.57	25908.23	3E 2	2B 0	Q 4
3858.784	3859.879		0.2				25907.55			
3858.863	3859.958		0.4	0.3	0.1		25907.02			
3858.935	3860.029		0.1				25906.54			
3859.015	3860.110	GK	0.8	0.7	0.1	1.24	25906.00	3D 2	2B 0	R 1
3859.127	3860.221		0.1	0.1			25905.25			
3859.209	3860.303			0.1			25904.70			
3859.380	3860.475		0.2	0.1d			25903.55			
3859.586	3860.680		0.2	0.1			25902.17			
3859.710	3860.804		0.3	0.2			25901.34	3F 2	2B 0	P 5
3859.801	3860.895		0.3	0.0			25900.73			
3859.889	3860.983	GK	1.0	0.9	0.3	1.53	25900.14	3D 2	2B 0	R 0
3859.988	3861.083		0.2	0.0			25899.47			
3860.064	3861.159	GK	0.6	0.6	0.4		25898.96	4D 0	2B 1	R 5
3860.383	3861.478	GK	0.4	0.4	0.1		25896.82	3A 2	2B 1	R 1
3860.479	3861.573		0.1				25896.18			
3860.567	3861.661	GK	0.7	0.6	0.2		25895.59	T 0	2B 1	R 1
3860.716	3861.810	GK	1.0b	1.0	0.4	1.80	25894.59	3D 2	2B 0	R 2
3860.828	3861.922		0.3	0.2			25893.84			
3860.931	3862.025		0.1	0.1			25893.15			

λ_{air}	λ_{vac}	Ref.	I_1	I_2	I_3	I_4	ν	Upper	Lower	Br.
3861.013	3862.107		0.1d	0.0			25892.60			
3861.111	3862.206		0.5	0.5	0.2		25891.94	6c 1	2a 1	Q 1
3861.212	3862.307		0.1				25891.26			
3861.262	3862.356		0.2				25890.93			
3861.327	3862.422		0.3	0.2			25890.49			
3861.502	3862.597	GK	1.0b	1.0	0.6	1.94	25889.32	3E 2	2B 0	Q 3
3861.502	3862.597	GK	1.0b	1.0	0.6	1.94	25889.32	3E 2	2B 0	P 6
3861.502	3862.597	GK	1.0b	1.0	0.6	1.94	25889.32	4D 0	2B 1	R 0
3861.605	3862.700		0.3				25888.63			
3861.800	3862.895		0.1				25887.32			
3861.903	3862.998		0.2				25886.63			
3861.984	3863.079		0.1				25886.09			
3862.183	3863.279		0.2				25884.75			
3862.473	3863.568	K	0.1				25882.81	X 2	2B 1	P 4
3862.533	3863.628		0.2				25882.41			
3862.637	3863.732		0.3	0.1	0.1		25881.71			
3862.746	3863.841		0.2	0.1d			25880.98	4d 3	2c 1	P 3
3862.834	3863.929		0.1				25880.39			
3862.943	3864.038		0.2	0.0			25879.66			
3863.031	3864.126		0.4	0.3d			25879.07			
3863.118	3864.213		0.2	0.2			25878.49			
3863.206	3864.301	GK	1.5b	1.0	0.5	1.88	25877.90	4D 0	2B 1	P 1
3863.206	3864.301	GK	1.5b	1.0	0.5	1.88	25877.90	3E 2	2B 0	Q 2
3863.304	3864.400		0.4				25877.24			
3863.346	3864.442	K	0.3	0.3	0.2		25876.96	4c 3	2a 1	Q 4
3863.346	3864.442	K	0.3	0.3	0.2		25876.96	3F 3	2B 2	Q 3
3863.434	3864.530		0.1d	0.0			25876.37			
3863.567	3864.663		0.0				25875.48	4e 3	2c 1	Q 4
3863.809	3864.905				0.2d		25873.86			
3863.911	3865.006			0.1			25873.18			
3863.993	3865.088			0.0			25872.63			
3864.087	3865.182	GK	0.7	0.8	0.3	1.23	25872.00	3E 2	2B 0	Q 1
3864.199	3865.294		0.2	0.0			25871.25			
3864.287	3865.383	GK	0.6	0.7	0.3	1.19	25870.66	3D 2	2B 0	R 3
3864.381	3865.477		0.2				25870.03			
3864.559	3865.655		0.1d	0.0			25868.84			
3864.825	3865.921		0.0d	0.1			25867.06			
3865.131	3866.227	K	0.4d	0.3			25865.01			
3865.308	3866.403		0.2d	0.2b			25863.83			
3865.388	3866.484		0.1				25863.29			
3865.485	3866.581	GK	0.8	0.8	0.2	1.44	25862.64	T 0	2B 1	R 0
3865.583	3866.678		0.2	0.1			25861.99			
3865.757	3866.853			0.0			25860.82			
3865.855	3866.951		0.1d	0.1			25860.17			
3866.006	3867.102	GK	0.4	0.3			25859.16			
3866.104	3867.200		0.0	0.0			25858.50			
3866.260	3867.356		0.1	0.2b			25857.46			
3866.433	3867.529	G	0.4	0.3	0.2		25856.30			
3866.620	3867.716			0.1			25855.05			
3866.768	3867.864			0.0			25854.06			
3866.898	3867.995		0.3	0.2			25853.19			
3866.991	3868.087		0.2				25852.57			
3867.076	3868.173	GK	1.0b	1.0	0.9b	1.71	25852.00	4D 0	2B 1	R 4
3867.171	3868.267		0.3	0.1			25851.37			
3867.311	3868.408		0.1	0.2			25850.43			
3867.452	3868.548		0.1	0.1	0.2		25849.49			
3867.563	3868.659			0.2			25848.75	4e 3	2c 1	P 3
3867.615	3868.711	K	0.1				25848.40			
3867.732	3868.828	K	0.6	0.4			25847.62			
3867.839	3868.936		0.2	0.0d			25846.90			

λ_{air}	λ_{vac}	Ref.	I_1	I_2	I_3	I_4	ν	Upper	Lower	Br.
3867.925	3869.021		0.1				25846.33			
3868.001	3869.098	GK	0.7	0.5	0.2d	1.01	25845.82			
3868.104	3869.201		0.1				25845.13			
3868.257	3869.354		0.1	0.1d			25844.11			
3868.380	3869.476	K	0.2				25843.29			
3868.483	3869.580		0.2	0.2			25842.60			
3868.706	3869.803		0.2	0.2			25841.11			
3868.814	3869.911	K		0.1			25840.39			
3868.856	3869.953		0.2				25840.11			
3868.931	3870.027		0.1				25839.61			
3868.991	3870.087		0.3	0.3			25839.21			
3869.074	3870.171		0.1	0.0			25838.65			
3869.161	3870.258	GK	0.8	0.9	0.4	1.43	25838.07	3D 2	2B 0	R 4
3869.266	3870.363		0.2	0.1			25837.37			
3869.399	3870.496	K			0.1		25836.48			
3869.524	3870.621		0.1	0.1d	0.1		25835.65			
3869.661	3870.758		0.3	0.3			25834.73			
3869.766	3870.863		0.3	0.2			25834.03			
3869.865	3870.962		0.2				25833.37			
3869.939	3871.036	GK	1.0b	0.9	0.4	1.63	25832.88	4D 0	2B 1	R 1
3870.039	3871.136		0.3	0.1			25832.21			
3870.144	3871.241	GK	0.5	0.6	0.2		25831.51	3D 3	2B 1	P 6
3870.261	3871.358		0.2	0.1			25830.73			
3870.435	3871.532		0.3				25829.57			
3870.490	3871.587		0.2	0.2b			25829.20			
3870.541	3871.638		0.2				25828.86			
3870.638	3871.736	GK	0.9	0.9	0.3	1.47	25828.21	Z 2	2B 0	R 0
3870.743	3871.840		0.4	0.2			25827.51			
3870.839	3871.936		0.2	0.1			25826.87			
3870.914	3872.011	GK	0.9	0.9	0.6	1.65	25826.37	4D 0	2B 1	R 3
3870.914	3872.011	GK	0.9	0.9	0.6	1.65	25826.37	3F 3	2B 2	Q 2
3871.015	3872.112		0.3	0.1d			25825.70			
3871.100	3872.197		0.1				25825.13			
3871.149	3872.247		0.2	0.2			25824.80			
3871.206	3872.304		0.2				25824.42			
3871.245	3872.343		0.1	0.1			25824.16			
3871.325	3872.422		0.2	0.1			25823.63			
3871.379	3872.476	K			0.3		25823.27			
3871.425	3872.523		0.4	0.3			25822.96			
3871.512	3872.610		0.3	0.1			25822.38			
3871.596	3872.694	GK	1.5b	1.0	0.8b	2.19	25821.82	4D 0	2B 1	R 2
3871.698	3872.796		0.4	0.3			25821.14			
3871.782	3872.880		0.1		0.2		25820.58			
3871.860	3872.958		0.2	0.1			25820.06			
3871.938	3873.036		0.0	0.2	0.3		25819.54			
3871.977	3873.075		0.3				25819.28			
3872.090	3873.187		0.3	0.0	0.3		25818.53			
3872.190	3873.288		0.4	0.3			25817.86			
3872.280	3873.378		0.3	0.0	0.4		25817.26			
3872.358	3873.456	GK	1.5b	1.0	0.4	2.07	25816.74	3A 2	2B 1	R 0
3872.358	3873.456	GK	1.5b	1.0	0.4	2.07	25816.74	Z 2	2B 0	R 3
3872.460	3873.558		0.4	0.2			25816.06			
3872.564	3873.661		0.2	0.1			25815.37			
3872.630	3873.727		0.2				25814.93			
3872.702	3873.799		0.0		0.3		25814.45			
3872.783	3873.880		0.1				25813.91			
3872.846	3873.943		0.2	0.2	0.2		25813.49			
3872.936	3874.033		0.2	0.0			25812.89			
3872.999	3874.096		0.1				25812.47			
3873.068	3874.166	K	0.4	0.3	0.3		25812.01			

λ_{air}	λ_{vac}	Ref.	I_1	I_2	I_3	I_4	ν	Upper	Lower	Br.
3873.180	3874.278			0.0			25811.26	4c 3	2a 1	Q 5
3873.251	3874.349		0.1	0.1	0.1		25810.79			
3873.396	3874.494	K	0.2				25809.82			
3873.522	3874.620		0.4	0.3	0.4		25808.98			
3873.639	3874.737		0.1				25808.20			
3873.799	3874.897		0.2				25807.14			
3873.944	3875.042		0.2b	0.1			25806.17			
3874.061	3875.159		0.1	0.0	0.4		25805.39			
3874.118	3875.216	GK	0.9	0.8	0.2	1.27	25805.01	Z 2	2B 0	R 1
3874.228	3875.326		0.2				25804.28			
3874.285	3875.383		0.3	0.2			25803.90			
3874.374	3875.472		0.1				25803.31			
3874.438	3875.536				0.1		25802.88			
3874.545	3875.643			0.0	0.2		25802.17			
3874.663	3875.762		0.1	0.0			25801.38			
3874.788	3875.886				0.2		25800.55			
3874.863	3875.961	G	0.5	0.6	0.2		25800.05	3D 2	2B 0	R 5
3874.979	3876.077		0.3	0.2			25799.28			
3875.320	3876.418				0.3		25797.01			
3875.467	3876.565		0.2	0.2			25796.03			
3875.646	3876.744			0.0			25794.84			
3875.966	3877.064				0.4		25792.71			
3876.053	3877.152		0.2				25792.13			
3876.145	3877.243		0.2	0.1			25791.52			
3876.211	3877.310		0.2	0.0			25791.08			
3876.260	3877.359		0.1				25790.75			
3876.408	3877.506		0.1				25789.77			
3876.572	3877.670			0.0			25788.68			
3876.722	3877.821			0.1			25787.68			
3876.886	3877.985			0.0			25786.59			
3876.977	3878.076				0.2		25785.98			
3877.066	3878.165		0.1d	0.1d			25785.39			
3877.183	3878.282		0.1	0.0	0.1		25784.61			
3877.272	3878.371	G	0.6	0.6	0.2		25784.02	3D 2	2B 0	P 2
3877.383	3878.482		0.2d				25783.28			
3877.472	3878.571		0.0		0.3		25782.69			
3877.583	3878.683		0.2	0.1			25781.95			
3878.038	3879.137		0.4	0.3			25778.93	4D 0	2B 1	P 2
3878.188	3879.287		0.2	0.1			25777.93			
3878.272	3879.372		0.2				25777.37			
3878.418	3879.518		0.2d		0.3		25776.40			
3878.640	3879.739	G	0.6	0.7	0.4		25774.93	T 0	2B 1	P 1
3878.640	3879.739	G	0.6	0.7	0.4		25774.93	3D 2	2B 0	R 6
3878.751	3879.850		0.2				25774.19			
3878.923	3880.022			0.1d	0.2		25773.05			
3878.996	3880.096			0.1			25772.56			
3879.076	3880.175		0.4	0.4			25772.03	X 2	2B 1	P 5
3879.187	3880.287		0.2d	0.1d			25771.29			
3879.359	3880.459		0.3	0.2			25770.15	4c 3	2a 1	P 3
3879.458	3880.558		0.2	0.0			25769.49			
3879.531	3880.630	GK	1.0	0.9	0.3	1.68	25769.01	Z 2	2B 0	R 2
3879.639	3880.739		0.4	0.2			25768.29			
3879.741	3880.841	G	0.6	0.6	0.4		25767.61			
3879.872	3880.972		0.1	0.2			25766.74			
3879.993	3881.093		0.2	0.1	0.2		25765.94			
3880.037	3881.136			0.0			25765.65			
3880.145	3881.245	G	0.4	0.4			25764.93			
3880.282	3881.382		0.1				25764.02			
3880.374	3881.474		0.0		0.2		25763.41			
3880.635	3881.734				0.1d		25761.68			

λ_air	λ_vac	Ref.	I₁	I₂	I₃	I₄	ν	Upper	Lower	Br.
3880.719	3881.819		0.0				25761.12	4d 2	2c 0	P 6
3880.824	3881.924				0.1		25760.42			
3881.074	3882.174		0.1				25758.76			
3881.136	3882.236			0.0			25758.35			
3881.201	3882.301		0.2				25757.92			
3881.462	3882.562		0.3	0.2	0.3		25756.19			
3881.653	3882.753			0.0d			25754.92			
3881.890	3882.990		0.1	0.0d			25753.35			
3882.054	3883.154	GK	0.7	0.6	0.2d	1.22	25752.26	T 0	2B 1	P 2
3882.142	3883.242		0.1	0.0			25751.68			
3882.200	3883.301		0.2				25751.29			
3882.268	3883.368		0.0				25750.84			
3882.369	3883.469		0.2				25750.17			
3882.707	3883.807		0.0d	0.1			25747.93			
3883.378	3884.479		0.1	0.1			25743.48			
3883.490	3884.590		0.2b	0.0			25742.74			
3883.627	3884.728		0.4	0.3			25741.83			
3883.718	3884.818		0.1				25741.23			
3883.798	3884.898						25740.70			
3883.985	3885.085		0.3	0.1			25739.46			
3884.072	3885.173		0.2	0.0			25738.88			
3884.149	3885.250	GK	0.9b	0.8	0.2	1.60	25738.37	3A 2	2B 1	P 1
3884.247	3885.348		0.2				25737.72			
3884.317	3885.417		0.3	0.3			25737.26			
3884.369	3885.470		0.2	0.1	0.2		25736.91			
3884.529	3885.630		0.2b	0.2			25735.85			
3884.643	3885.744		0.1				25735.10			
3884.768	3885.869		0.2d		0.3		25734.27			
3884.859	3885.960		0.1				25733.67			
3884.972	3886.073		0.2				25732.92			
3885.097	3886.198		0.2	0.2	0.2		25732.09			
3885.431	3886.532		0.1	0.2			25729.88			
3885.595	3886.697		0.1d	0.1			25728.79			
3885.914	3887.015		0.2	0.2			25726.68			
3886.098	3887.200		0.2	0.1	0.2		25725.46			
3886.219	3887.321	GK	0.7	0.8	0.4	1.22	25724.66	3A 2	2B 1	P 6
3886.219	3887.321	GK	0.7	0.8	0.4	1.22	25724.66	3F 3	2B 2	P 4
3886.313	3887.414		0.2				25724.04			
3886.408	3887.509		0.1	0.0	0.3		25723.41	4e 3	2c 1	P 4
3886.515	3887.617			0.1d			25722.70			
3886.600	3887.701			0.0			25722.14			
3886.701	3887.803				0.6		25721.47			
3886.750	3887.851			0.1			25721.15	3c 6	2a 1	Q 1
3886.904	3888.005			0.0			25720.13			
3886.975	3888.076				0.2		25719.66			
3887.049	3888.150		0.3	0.2			25719.17			
3887.164	3888.265			0.1	0.2		25718.41			
3887.282	3888.383			0.1			25717.63			
3887.390	3888.492	GK	0.6	0.6	0.3	0.99	25716.91	Z 2	2B 0	R 3
3887.425	3888.527				0.3		25716.68			
3887.498	3888.599		0.0	0.0			25716.20			
3887.563	3888.664				0.2		25715.77			
3887.685	3888.787		0.2	0.2d			25714.96			
3887.844	3888.946	GK	0.7	0.8	0.4	1.51	25713.91	3A 2	2B 1	P 3
3887.844	3888.946	GK	0.7	0.8	0.4	1.51	25713.91	3D 2	2B 0	P 3
3887.953	3889.055		0.0d		0.4		25713.19			
3888.006	3889.108				0.4		25712.84	T 0	2B 1	P 3
3888.006	3889.108				0.4		25712.84	4D 0	2B 0	P 8
3888.142	3889.244		0.1d	0.0d			25711.94	Z 2	2B 0	P 2
3888.260	3889.362				0.4		25711.16			

λ_{air}	λ_{vac}	Ref.	I_1	I_2	I_3	I_4	ν	Upper	Lower	Br.
3888.585	3889.687		0.2				25709.01			
3888.642	3889.744	1	0.6	0.7	1.5b		25708.63			
3888.747	3889.849		0.2				25707.94			
3888.813	3889.915		0.4	0.2			25707.50			
3888.904	3890.006		0.2	0.0			25706.90			
3888.987	3890.089	GK	0.8b	1.0	1.0b	2.23	25706.35	3A 2	2B 1	P 2
3889.089	3890.191		0.5b	0.2			25705.68			
3889.142	3890.244		0.2	0.1			25705.33			
3889.317	3890.419	GK	0.7b	1.0	0.5	2.05	25704.17	3A 2	2B 1	P 4
3889.408	3890.510		0.3	0.2	0.2		25703.57			
3889.500	3890.602			0.0			25702.96			
3889.577	3890.680			0.2			25702.45			
3889.679	3890.781			0.2	0.3		25701.78			
3889.850	3890.952			0.1	0.6		25700.65			
3890.019	3891.122		0.2	0.0			25699.53			
3890.095	3891.197				0.1		25699.03			
3890.210	3891.312				0.3		25698.27			
3890.272	3891.375		0.3	0.3	0.2		25697.86			
3890.393	3891.496		0.0	0.3			25697.06			
3890.575	3891.677		0.0		0.9		25695.86			
3890.707	3891.809	GK	0.5	0.7	0.2	1.12	25694.99	3A 2	2B 1	P 5
3890.831	3891.933			0.1			25694.17			
3890.922	3892.024		0.3	0.2			25693.57			
3891.034	3892.136		0.2		0.2		25692.83			
3891.190	3892.292				0.7		25691.80			
3891.252	3892.355		0.2	0.2			25691.39			
3891.340	3892.442		0.2	0.2	0.2		25690.81			
3891.490	3892.592				0.1		25689.82			
3891.891	3892.994		0.5b	0.1			25687.17			
3892.100	3893.203			0.3			25685.79	4g 1	2a 0	P 1
3892.513	3893.616				0.2		25683.07			
3892.578	3893.681			0.1d	0.6		25682.64	T 0	2B 1	P 4
3892.867	3893.970			0.2d			25680.73			
3892.919	3894.022				0.4		25680.39			
3893.058	3894.161				0.6		25679.47	3c 6	2a 1	Q 2
3893.207	3894.310			0.1d			25678.49			
3893.592	3894.695						25675.95			
3893.900	3895.003		0.3				25673.92			
3893.994	3895.097		0.3	0.3			25673.30			
3894.076	3895.179		0.1d				25672.76			
3894.129	3895.232		0.1d				25672.41			
3894.208	3895.311		0.0d				25671.89			
3894.417	3895.521			0.1			25670.51			
3894.488	3895.592				0.2		25670.04			
3894.642	3895.745	1			0.2		25669.03			
3894.736	3895.839		0.0d				25668.41			
3894.850	3895.953		0.1d	0.0d			25667.66			
3895.071	3896.175				0.6		25666.20			
3895.303	3896.407		0.1d		0.2		25664.67			
3895.457	3896.560				0.6		25663.66			
3895.545	3896.648			0.2			25663.08			
3896.049	3897.153			0.2b	0.0d		25659.76			
3896.524	3897.628				0.4		25656.63			
3896.728	3897.832				0.1		25655.29			
3897.116	3898.221			0.2b	0.0d		25652.73			
3897.274	3898.379				0.1		25651.69			
3897.458	3898.562		0.2		0.4		25650.48			
3897.534	3898.638	G	0.6	0.6			25649.98	4D 0	2B 1	P 3
3897.636	3898.740		0.2		0.2		25649.31			
3897.735	3898.839		0.2d				25648.66			

λ_{air}	λ_{vac}	Ref.	I_1	I_2	I_3	I_4	ν	Upper	Lower	Br.
3897.797	3898.901		0.4d	0.4			25648.25	4E 1	2B 3	Q 1
3897.937	3899.041	G	0.4	0.5	0.2		25647.33	4E 1	2B 3	Q 4
3898.068	3899.172		0.0	0.0	0.5		25646.47			
3898.130	3899.234	GK	0.6	0.6			25646.06	Z 2	2B 0	R 4
3898.236	3899.341			0.0			25645.36			
3898.384	3899.488				0.6		25644.39			
3898.524	3899.628		0.3	0.2			25643.47			
3899.056	3900.161			0.0			25639.97			
3899.205	3900.310			0.1			25638.99			
3899.330	3900.434		0.3d	0.3	0.2b		25638.17			
3899.888	3900.993	G	0.5	0.5	0.3		25634.50	W 2	2B 1	R 4
3899.984	3901.089		0.2d				25633.87			
3900.074	3901.178			0.0			25633.28			
3900.822	3901.927				0.2b		25628.36			
3900.901	3902.007	G	0.4	0.4			25627.84	3D 2	2B 0	P 4
3900.990	3902.095		0.1				25627.26			
3901.587	3902.692		0.1	0.1	0.5		25623.34			
3901.856	3902.961				0.5		25621.57			
3901.957	3903.062		0.2d		0.3		25620.91			
3902.014	3903.120			0.0	0.2		25620.53			
3902.124	3903.230	G	0.4	0.5			25619.81	4E 1	2B 3	Q 3
3902.281	3903.386			0.0			25618.78			
3902.461	3903.566		0.1d	0.1d			25617.60			
3902.523	3903.629			0.1			25617.19			
3902.633	3903.738	GK	0.9	0.7	0.4	1.47	25616.47	4E 1	2B 3	Q 2
3902.735	3903.841		0.2	0.1			25615.80	3c 6	2a 1	Q 3
3902.907	3904.013			0.0			25614.67			
3903.088	3904.194			0.1			25613.48			
3903.273	3904.379		0.3d	0.4	0.1d		25612.27	Z 2	2B 0	P 3
3903.534	3904.639			0.0			25610.56			
3904.192	3905.298				0.2d		25606.24			
3904.518	3905.625			0.1d			25604.10			
3904.884	3905.991		0.1d		0.1d		25601.70			
3905.051	3906.157		0.1d				25600.61			
3905.165	3906.271		0.3	0.3	0.1d		25599.86	W 2	2B 1	R 2
3905.264	3906.371		0.1				25599.21			
3905.330	3906.436			0.0d			25598.78	W 2	2B 1	R 3
3905.382	3906.488		0.1				25598.44			
3905.507	3906.613	G	0.5	0.4	1.0		25597.62	W 2	2B 1	R 1
3905.655	3906.761			0.1	0.4		25596.65			
3905.792	3906.899		0.4	0.4	0.1		25595.75			
3905.891	3906.998				0.4		25595.10	4e 3	2c 1	P 5
3906.171	3907.277			0.0d			25593.27			
3906.656	3907.763			0.0			25590.09			
3906.757	3907.863			0.0			25589.43			
3907.300	3908.407		0.1d				25585.87			
3907.397	3908.503		0.1				25585.24			
3907.481	3908.587	GK	0.8	0.7	0.2	1.16	25584.69	W 2	2B 1	R 0
3907.566	3908.673		0.1d				25584.13			
3907.759	3908.866		0.2b	0.2	0.4		25582.87			
3908.041	3909.148	G	0.4	0.3			25581.02			
3908.147	3909.254		0.0				25580.33			
3908.258	3909.365			0.0	0.4		25579.60			
3908.457	3909.564		0.1	0.2			25578.30			
3908.545	3909.653		0.1	0.1			25577.72			
3909.001	3910.108		0.1	0.1			25574.74			
3909.091	3910.198		0.4	0.3	0.1		25574.15			
3909.207	3910.315		0.0d	0.0			25573.39			
3909.305	3910.412			0.1			25572.75			
3909.770	3910.877		0.1d				25569.71			

λ_{air}	λ_{vac}	Ref.	I₁	I₂	I₃	I₄	ν	Upper	Lower	Br.
3909.877	3910.984			0.1			25569.01			
3910.038	3911.145	GK	0.8	0.7	0.2	0.91	25567.96	4D 0	2B 1	P 4
3910.316	3911.423			0.2	0.4		25566.14	4g 1	2a 0	R 1
3910.446	3911.553		0.2d	0.2			25565.29			
3910.522	3911.630				0.4		25564.79			
3910.694	3911.801						25563.67			
3910.839	3911.947			0.0			25562.72			
3910.995	3912.103			0.0			25561.70			
3911.443	3912.551			0.1			25558.77			
3911.537	3912.645			0.2			25558.16			
3911.967	3913.075		0.1d	0.0			25555.35			
3912.604	3913.712		0.3	0.4	0.2		25551.19			
3912.843	3913.9 1	G					25549.63			
3913.204	3914.313			0.0			25547.27			
3913.334	3914.443				0.1		25546.42			
3913.731	3914.840			0.1			25543.83			
3913.858	3914.967		0.5	0.4	0.1		25543.00			
3914.009	3915.117	G	0.3b	0.4	0.5		25542.02			
3914.413	3915.522				0.5		25539.38			
3915.450	3916.558				0.2		25532.62	3c 6	2a 1	Q 4
3915.601	3916.710				0.4		25531.63	3D 2	2B 0	P 5
3915.778	3916.887				0.4		25530.48			
3916.074	3917.183			0.1			25528.55			
3916.189	3917.298	G	0.4	0.4	0.2		25527.80			
3916.350	3917.459			0.1			25526.75			
3916.612	3917.722						25525.04			
3916.795	3917.904		0.2d	0.2d			25523.85			
3916.885	3917.995		0.1				25523.26			
3916.971	3918.081	GK	0.8	0.6	0.2	0.93	25522.70	W 2	2B 1	P 1
3917.067	3918.176		0.1				25522.08			
3917.152	3918.262			0.1d	0.1		25521.52			
3917.424	3918.534			0.1			25519.75	4F 0	2B 2	R 3
3917.536	3918.646		0.1d				25519.02			
3917.857	3918.967		0.1d				25516.93			
3917.972	3919.082	GK	0.5	0.6	0.3		25516.18			
3918.124	3919.234			0.0			25515.19			
3918.286	3919.395			0.0			25514.14			
3919.405	3920.515			0.1			25506.85			
3919.565	3920.675		0.3	0.3			25505.81			
3919.716	3920.826		0.2d	0.0			25504.83			
3919.890	3921.000		0.3d	0.3	0.1		25503.70	4D 0	2B 1	P 5
3920.119	3921.229		0.2d	0.1			25502.21	Z 2	2B 0	P 4
3920.485	3921.595		0.0d	0.0			25499.83			
3920.714	3921.824				0.4		25498.34			
3920.860	3921.970		0.0d	0.0			25497.39			
3921.067	3922.178				0.5		25496.04			
3921.538	3922.648				0.3		25492.98			
3921.698	3922.809				0.3		25491.94			
3922.466	3923.577		0.3d	0.2d			25486.95			
3922.623	3923.734		0.1				25485.93			
3923.063	3924.174						25483.07			
3923.745	3924.856		0.0d				25478.64			
3923.987	3925.098		0.2b	0.0			25477.07			
3924.069	3925.180						25476.54			
3924.150	3925.261		0.2				25476.01			
3924.255	3925.366		0.3	0.2			25475.33			
3924.330	3925.442		0.2	0.0			25474.84			
3924.415	3925.526	GK	1.5	1.0	0.3	1.68	25474.29	W 2	2B 1	P 2
3924.517	3925.628		0.3	0.1			25473.63			
3924.631	3925.742			0.1			25472.89			

λ_{air}	λ_{vac}	Ref.	I_1	I_2	I_3	I_4	ν	Upper	Lower	Br.
3924.831	3925.943		0.1d				25471.59			
3924.907	3926.018			0.1			25471.10			
3925.021	3926.132			0.0			25470.36			
3925.213	3926.325			0.2			25469.11			
3925.327	3926.439			0.1			25468.37			
3925.449	3926.561			0.1			25467.58			
3925.620	3926.732			0.1			25466.47	4g 1	2a 0	P 2
3925.784	3926.895						25465.41			
3925.972	3927.083	G	0.5	0.5	0.3		25464.19			
3926.132	3927.244		0.3	0.4	0.1		25463.15	4D 0	2B 1	P 6
3926.371	3927.483		0.3b	0.2			25461.60			
3926.552	3927.663		0.1				25460.43	4g 1	2a 0	R 2
3927.110	3928.222		0.1				25456.81			
3927.195	3928.307	G	0.5	0.6	0.3		25456.26			
3927.283	3928.395	G	0.6	0.5			25455.69	4c 4	2a 2	Q 1
3927.383	3928.495		0.2	0.2	0.1		25455.04	4D 0	2B 1	P 8
3927.914	3929.026	G	0.4	0.5	0.5		25451.60			
3928.216	3929.329			0.2d	0.5		25449.64			
3928.315	3929.427		0.1d				25449.00			
3928.392	3929.505	GK	0.7	0.8	0.3b	0.98	25448.50			
3928.500	3929.613		0.1d				25447.80			
3928.612	3929.724	1		0.1			25447.08			
3928.775	3929.888			0.0			25446.02			
3929.191	3930.303		0.2b	0.2d	0.2		25443.33			
3929.333	3930.445				0.1		25442.41			
3929.643	3930.756			0.1			25440.40			
3929.825	3930.938		0.3	0.3d	0.2b		25439.22			
3929.895	3931.008		0.1				25438.77			
3930.230	3931.343			0.1			25436.60			
3930.366	3931.479			0.2	0.2		25435.72			
3930.694	3931.807	GK	0.6	0.6	0.3		25433.60			
3930.790	3931.903		0.0				25432.98			
3930.895	3932.008			0.0			25432.30			
3931.207	3932.320			0.0d			25430.28	3D 2	2B 0	P 6
3931.352	3932.465		0.3	0.3			25429.34	4c 4	2a 2	Q 2
3931.685	3932.798			0.0			25427.19			
3932.091	3933.205			0.0			25424.56			
3932.637	3933.751						25421.03			
3932.936	3934.050		0.2	0.2			25419.10			
3933.191	3934.305		0.2d	0.2d			25417.45			
3933.445	3934.559		0.1d	0.1d			25415.81			
3933.608	3934.721	GK	0.8	0.8	0.2	1.18	25414.76	W 2	2B 1	P 3
3933.705	3934.819		0.2				25414.13			
3934.258	3935.372		0.1d				25410.56			
3934.332	3935.446	G	0.4	0.4			25410.08	Y 2	2B 1	R 0
3934.518	3935.632			0.1			25408.88			
3934.676	3935.790		0.1	0.2d	0.2		25407.86			
3935.011	3936.125		0.2	0.2			25405.70	Y 2	2B 1	R 1
3935.088	3936.202			0.1			25405.20			
3935.187	3936.301	G	0.5	0.6	0.3		25404.56			
3935.638	3936.752				0.4b		25401.65			
3935.921	3937.036		0.1d		0.4		25399.82			
3936.014	3937.129			0.2			25399.22			
3936.104	3937.219		0.0	0.0d			25398.64			
3936.214	3937.329		0.3	0.2			25397.93			
3936.424	3937.538		0.1	0.1d			25396.58			
3936.611	3937.726		0.2	0.3d	0.2		25395.37			
3936.895	3938.009			0.0d			25393.54			
3937.121	3938.236				0.1		25392.08			
3937.261	3938.375		0.2d	0.1d			25391.18			

λ_{air}	λ_{vac}	Ref.	I_1	I_2	I_3	I_4	ν	Upper	Lower	Br.
3937.423	3938.538	F	0.4	0.4			25390.13	4c 4	2a 2	Q 3
3937.591	3938.706		0.2d				25389.05			
3937.845	3938.960		0.3	0.2	0.2		25387.41			
3938.191	3939.306		0.2	0.2			25385.18			
3938.401	3939.516			0.0			25383.83	Y 2	2B 1	R 2
3938.683	3939.798		0.1	0.0			25382.01			
3938.845	3939.960		0.0	0.0			25380.97			
3938.946	3940.061		0.3	0.3			25380.32			
3939.242	3940.357			0.1			25378.41	Z 2	2B 0	P 5
3939.722	3940.837		0.0d	0.2			25375.32			
3939.779	3940.894			0.1			25374.95			
3939.893	3941.008	G	0.4	0.3			25374.22	X 1	2B 0	R 0
3939.986	3941.101		0.1				25373.62			
3940.332	3941.447		0.1d	0.1			25371.39			
3940.874	3941.990			0.1			25367.90			
3941.081	3942.196		0.3	0.3	0.3		25366.57			
3941.256	3942.372			0.0			25365.44			
3941.421	3942.537		0.3	0.3			25364.38	X 1	2B 0	R 1
3941.738	3942.854			0.1			25362.34			
3942.161	3943.277			0.1d			25359.62			
3942.296	3943.412		0.4	0.3			25358.75	4g 1	2a 0	R 3
3942.489	3943.605						25357.51			
3942.615	3943.731	G	0.5	0.4			25356.70	Y 2	2B 1	P 1
3942.747	3943.863		0.1				25355.85			
3942.944	3944.061			0.0			25354.58			
3943.153	3944.269		0.2d	0.2			25353.24			
3943.419	3944.535			0.1			25351.53			
3943.620	3944.736		0.2	0.3			25350.24	3c 9	2a 3	Q 1
3943.812	3944.929		0.1d	0.1			25349.00			
3943.924	3945.041			0.2	0.4		25348.28			
3944.114	3945.231		0.2	0.2d			25347.06			
3944.181	3945.298		0.2		0.4		25346.63			
3944.285	3945.402	GK	1.0	1.0	0.4	1.65	25345.96	W 2	2B 1	P 4
3944.384	3945.500		0.2	0.1			25345.33			
3944.469	3945.586						25344.78			
3944.573	3945.690		0.1	0.1			25344.11			
3944.657	3945.774			0.1			25343.57			
3944.752	3945.869		0.1				25342.96			
3944.835	3945.951	GK	0.7	0.7	0.2	0.68	25342.43			
3944.953	3946.070		0.2	0.1			25341.67			
3945.012	3946.129		0.1				25341.29			
3945.339	3946.456		0.1b				25339.19			
3945.500	3946.616		0.2	0.2			25338.16	4c 4	2a 2	Q 4
3946.109	3947.226	GK	0.6	0.5	0.1	0.91	25334.25	X 1	2B 0	R 2
3946.224	3947.341		0.2				25333.51			
3946.274	3947.391		0.4	0.3	0.2		25333.19			
3946.453	3947.570		0.1				25332.04			
3946.598	3947.715		0.3	0.2			25331.11			
3946.814	3947.931		0.2d	0.0			25329.72			
3946.923	3948.041		0.2				25329.02			
3947.000	3948.117			0.1			25328.53			
3947.098	3948.215	GK	0.7	0.7	0.2		25327.90	3F 2	2B 1	R 2
3947.207	3948.324	K	0.2				25327.20			
3947.274	3948.391						25326.77	3D 2	2B 0	P 7
3947.377	3948.494		0.3	0.0d			25326.11			
3947.499	3948.616	K 1	0.1d		0.4		25325.33			
3947.605	3948.722	K					25324.65			
3947.824	3948.942		0.0	0.1			25323.24			
3948.008	3949.126						25322.06			
3948.050	3949.168		0.3d	0.2b			25321.79			

λ_{air}	λ_{vac}	Ref.	I_1	I_2	I_3	I_4	ν	Upper	Lower	Br.
3948.097	3949.215						25321.49			
3948.195	3949.313		0.0				25320.86			
3948.266	3949.383		0.1	0.1			25320.41			
3948.422	3949.539		0.2d	0.3d	0.2		25319.41			
3948.624	3949.742			0.0d			25318.11			
3948.893	3950.010			0.2			25316.39			
3948.978	3950.096	1	0.0		1.0		25315.84			
3949.091	3950.208	G	0.5	0.5			25315.12			
3949.195	3950.313		0.1d				25314.45			
3949.376	3950.494		0.2	0.3			25313.29			
3949.434	3950.552	GK	0.5	0.5	0.3b		25312.92			
3949.518	3950.636		0.1d	0.0			25312.38			
3950.152	3951.270	G	0.4	0.4	0.2d		25308.32			
3950.309	3951.427		0.2				25307.31			
3950.416	3951.534		0.2	0.1			25306.63			
3950.586	3951.704	GK	0.8	0.9	0.4	1.03	25305.54	3F 2	2B 1	Q 4
3950.723	3951.841		0.2	0.1			25304.66	3E 3	2B 2	Q 8
3950.789	3951.907			0.0d			25304.24			
3950.957	3952.076			0.1d			25303.16			
3951.020	3952.138			0.0			25302.76			
3951.129	3952.247	G	0.5	0.6	0.2		25302.06			
3951.331	3952.449		0.2d	0.1			25300.77			
3951.498	3952.616	GK	0.8	0.7	0.2	1.00	25299.70	Y 2	2B 1	P 2
3951.596	3952.714		0.1				25299.07			
3952.438	3953.557		0.1d				25293.68	4F 0	2B 2	R 2
3952.599	3953.718		0.3	0.3	0.2		25292.65			
3952.821	3953.940	GK	0.4	0.4	0.3b		25291.23			
3952.971	3954.090				0.3		25290.27			
3953.485	3954.604		0.0				25286.98			
3953.756	3954.875				0.1		25285.25			
3954.014	3955.133			0.0			25283.60			
3954.158	3955.277			0.0			25282.68			
3954.263	3955.382		0.0				25282.01			
3954.378	3955.497			0.1			25281.27			
3955.014	3956.133		0.2				25277.21			
3955.109	3956.228		0.2	0.1			25276.60			
3955.190	3956.310	GK	0.7	0.7	0.3	1.03	25276.08	W 2	2B 1	P 5
3955.298	3956.418		0.2d	0.1			25275.39			
3955.555	3956.674						25273.75	4c 4	2a 2	Q 5
3955.799	3956.919			0.1d			25272.19			
3955.956	3957.075		0.1d	0.0			25271.19			
3956.272	3957.392			0.0d			25269.17			
3956.585	3957.705		0.3b	0.3d	0.2d		25267.17			
3956.729	3957.849		0.2				25266.25			
3956.831	3957.951		0.1d				25265.60			
3956.925	3958.045		0.2				25265.00			
3957.022	3958.142	GK	0.6	0.5	0.3		25264.38	w 0	2c 0	P 1
3957.248	3958.367						25262.94			
3957.320	3958.440			0.2d			25262.48			
3957.530	3958.650		0.2	0.3d			25261.14			
3957.735	3958.855		0.3	0.2d			25259.83			
3957.909	3959.029		0.1d				25258.72			
3958.003	3959.123		0.3d	0.2d			25258.12	X 1	2B 0	P 2
3958.244	3959.364		0.1d				25256.58			
3958.373	3959.493			0.2			25255.76			
3958.506	3959.626		0.1d				25254.91	4g 1	2a 0	R 4
3958.755	3959.875		0.1d				25253.32			
3959.114	3960.234			0.2			25251.03			
3959.215	3960.335	G	0.5	0.4			25250.39	4E 2	2B 5	Q 2
3959.315	3960.435		0.1				25249.75			

λ_{air}	λ_{vac}	Ref.	I_1	I_2	I_3	I_4	ν	Upper	Lower	Br.
3959.492	3960.612			0.0			25248.62			
3959.619	3960.740	GK	0.4	0.4	0.1		25247.81	3E 3	2B 2	Q 7
3959.790	3960.911			0.0			25246.72			
3959.959	3961.080	GK	0.7	0.8	0.4	1.05	25245.64	W 2	2B 1	P 6
3960.055	3961.176		0.3	0.2			25245.03			
3960.190	3961.311	GK	0.6	0.7	0.2d	0.98	25244.17	3E 3	2B 2	R 3
3960.316	3961.436		0.2	0.2			25243.37			
3960.372	3961.493		0.1				25243.01			
3960.488	3961.609		0.2				25242.27			
3960.604	3961.725		0.1	0.1			25241.53			
3960.785	3961.905	G	0.4	0.4	0.2		25240.38	3D 2	2B 0	P 8
3960.965	3962.086		0.1	0.2			25239.23			
3961.155	3962.276	K		0.2d			25238.02	Z 2	2B 0	P 6
3961.256	3962.376				0.2		25237.38			
3961.474	3962.595			0.1			25235.99	4g 1	2a 0	P 3
3961.659	3962.780			0.1			25234.81			
3961.895	3963.016		0.2				25233.31			
3961.970	3963.091		0.2	0.2d			25232.83			
3962.066	3963.187		0.1				25232.22			
3962.165	3963.286		0.2	0.2	0.2		25231.59			
3962.242	3963.363			0.0			25231.10			
3962.339	3963.460	GK	0.9	0.9	0.3	1.45	25230.48	3E 3	2B 2	R 2
3962.446	3963.567		0.3	0.1			25229.80			
3962.574	3963.696	G	0.5	0.5			25228.98	3F 2	2B 1	R 1
3962.705	3963.826		0.2	0.1			25228.15			
3962.794	3963.916						25227.58			
3962.981	3964.103		0.2	0.2			25226.39			
3963.065	3964.186		0.1				25225.86			
3963.153	3964.274	GK	0.9	1.0	0.3	1.56	25225.30	3A 1	2B 0	R 2
3963.263	3964.384		0.3	0.0			25224.60			
3963.413	3964.535		0.1d	0.0d			25223.64			
3963.544	3964.665	GK	0.6	0.6			25222.81	Y 2	2B 1	P 3
3963.651	3964.772		0.1	0.0			25222.13			
3963.740	3964.862	GK	0.6	0.7	0.2		25221.56	w 0	2c 0	Q 1
3963.740	3964.862	GK	0.6	0.7	0.2		25221.56	3F 2	2B 1	Q 3
3963.842	3964.964		0.2	0.0			25220.91			
3963.985	3965.107			0.1			25220.00			
3964.111	3965.233			0.0			25219.20			
3964.312	3965.434		0.2	0.2			25217.92			
3964.438	3965.560			0.1			25217.12			
3964.548	3965.670			0.1			25216.42			
3964.647	3965.769		0.3	0.3			25215.79			
3964.726	3965.848	1	0.2	0.3	1.0b		25215.29			
3965.108	3966.230	GK	0.5	0.5	0.2		25212.86	w 0	2c 0	R 1
3965.221	3966.343		0.1				25212.14			
3965.346	3966.467		0.1d				25211.35			
3965.486	3966.607		0.1d				25210.46	4F 0	2B 2	P 5
3965.627	3966.749	F	0.2d	0.2			25209.56			
3965.707	3966.829			0.1			25209.05			
3965.808	3966.930		0.2d				25208.41			
3966.307	3967.429	F	0.3d	0.3			25205.24	4F 0	2B 2	Q 4
3966.716	3967.838			0.0			25202.64			
3966.820	3967.942	GK	0.6	0.5		0.86	25201.98			
3966.928	3968.051		0.3				25201.29			
3966.998	3968.120		0.1	0.0			25200.85			
3967.098	3968.221		0.2	0.2d			25200.21			
3967.412	3968.534				0.1		25198.22			
3967.705	3968.827		0.1d	0.1			25196.36			
3967.846	3968.969		0.1d	0.0			25195.46			
3968.083	3969.205		0.1d	0.0			25193.96			

λ_{air}	λ_{vac}	Ref.	I_1	I_2	I_3	I_4	ν	Upper	Lower	Br.
3968.239	3969.361		0.1d	0.1			25192.97			
3968.565	3969.687		0.3	0.3	0.2		25190.90			
3968.752	3969.875		0.1d				25189.71			
3968.818	3969.941		0.1	0.2	0.2		25189.29			
3968.919	3970.042	GK	0.8	0.7	0.2	0.86	25188.65	3E 3	2B 2	R 1
3969.029	3970.152		0.2				25187.95			
3969.130	3970.253		0.1		0.2		25187.31			
3969.275	3970.398		0.3	0.3	0.2d		25186.39			
3969.488	3970.611		0.1d	0.0			25185.04			
3969.650	3970.773	F	0.4	0.6	0.2		25184.01	3E 3	2B 2	Q 6
3969.764	3970.887		0.0	0.0			25183.29			
3969.868	3970.991		0.0				25182.63			
3970.071	3971.195	K 1	0.9	0.6	1.5b		25181.34			
3970.240	3971.363		0.1				25180.27			
3970.694	3971.818			0.1			25177.39			
3970.876	3971.999			0.0			25176.24			
3971.018	3972.141		0.2	0.2d			25175.34			
3971.289	3972.412			0.1d			25173.62			
3971.462	3972.586	GK	0.7	0.7	0.3		25172.52			
3971.579	3972.703		0.2				25171.78	X 1	2B 0	P 3
3971.649	3972.772			0.0			25171.34			
3971.716	3972.840		0.1				25170.91			
3971.918	3973.042			0.1			25169.63			
3972.030	3973.154		0.2d	0.1			25168.92			
3972.229	3973.353				0.2		25167.66			
3972.319	3973.443		0.1	0.0			25167.09			
3972.442	3973.566			0.0	0.2		25166.31			
3972.622	3973.746		0.1				25165.17			
3972.720	3973.844		0.2				25164.55			
3972.878	3974.002		0.2d	0.3			25163.55			
3972.974	3974.098		0.1				25162.94			
3973.124	3974.248		0.1d	0.1			25161.99			
3973.358	3974.482		0.2	0.2			25160.51			
3973.511	3974.635			0.1			25159.54			
3973.696	3974.820		0.0d				25158.37			
3973.835	3974.959		0.2	0.2			25157.49			
3974.069	3975.193		0.3	0.2			25156.01			
3974.132	3975.256		0.0				25155.61			
3974.224	3975.348	GK	0.9	0.8	0.2b	1.16	25155.03	3F 2	2B 1	Q 2
3974.325	3975.449		0.2	0.1			25154.39			
3974.409	3975.533						25153.86			
3974.527	3975.651		0.0				25153.11			
3974.605	3975.729		0.2	0.0d			25152.62			
3974.679	3975.803		0.0				25152.15			
3974.782	3975.906	GK	0.8	0.7	0.2b	1.20	25151.50	3A 1	2B 0	R 1
3974.881	3976.006		0.2				25150.87			
3975.248	3976.372		0.2	0.0			25148.55			
3975.338	3976.462			0.0			25147.98			
3975.471	3976.595		0.2d	0.2			25147.14			
3975.583	3976.708		0.0				25146.43			
3975.681	3976.806	GK	0.9	0.8	0.4b	1.37	25145.81			
3975.796	3976.921		0.1				25145.08			
3975.863	3976.988		0.2	0.2			25144.66			
3975.934	3977.059		0.1				25144.21			
3976.173	3977.298		0.2d	0.1			25142.70			
3976.666	3977.791			0.2			25139.58			
3976.752	3977.877		0.3	0.4			25139.04			
3976.832	3977.957	GK	0.3	0.6	0.2b		25138.53	3E 3	2B 2	Q 5
3976.907	3978.032		0.2				25138.06			
3977.019	3978.144		0.2				25137.35			

λ_{air}	λ_{vac}	Ref.	I_1	I_2	I_3	I_4	ν	Upper	Lower	Br.
3977.095	3978.220	GK	0.7	0.7	0.2	0.89	25136.87	3E 3	2B 2	R 0
3977.207	3978.332		0.2	0.1			25136.16			
3977.339	3978.464		0.3	0.3	0.2		25135.33			
3977.486	3978.611		0.1d	0.0			25134.40			
3977.663	3978.788		0.1d	0.0			25133.28			
3977.823	3978.948			0.0			25132.27			
3977.908	3979.034			0.0			25131.73			
3977.965	3979.091		0.2	0.1			25131.37			
3978.070	3979.195		0.1				25130.71			
3978.155	3979.281	GK	0.8	0.7	0.3	1.26	25130.17	Y 2	2B 1	P 4
3978.265	3979.390		0.2	0.0			25129.48			
3978.347	3979.472		0.3	0.3			25128.96			
3978.459	3979.585	K	0.5	0.5	0.3		25128.25			
3978.562	3979.688		0.1	0.0			25127.60			
3978.649	3979.775	GK	0.6	0.6	0.5b	0.98	25127.05			
3978.724	3979.849		0.4	0.5			25126.58			
3978.831	3979.957		0.2	0.2	0.1		25125.90			
3979.012	3980.137						25124.76			
3979.140	3980.266		0.0d				25123.95			
3979.332	3980.458	GK	0.7	0.7	0.5		25122.74			
3979.424	3980.549		0.2				25122.16			
3979.546	3980.671		0.0				25121.39			
3979.758	3980.884	GK	0.7	0.7		1.02	25120.05			
3979.880	3981.006			0.0			25119.28			
3979.950	3981.076			0.0	0.5		25118.84			
3980.623	3981.749		0.0	0.2d	0.1		25114.59			
3980.783	3981.909		0.3d	0.2			25113.58			
3980.898	3982.024			0.0			25112.86			
3980.999	3982.125	K	0.3	0.3	0.2		25112.22			
3981.186	3982.312	K	0.3	0.3	0.2		25111.04			
3981.327	3982.453		0.0d				25110.15			
3981.438	3982.564			0.0			25109.45			
3981.546	3982.672	K	0.3	0.3	0.3		25108.77			
3981.728	3982.855			0.1			25107.62			
3981.889	3983.015		0.1d	0.0			25106.61			
3981.993	3983.120			0.1			25105.95			
3982.079	3983.205	K	0.4	0.4	0.2		25105.41			
3982.212	3983.339			0.0			25104.57			
3982.371	3983.497		0.2b	0.2b			25103.57			
3982.477	3983.603		0.1				25102.90			
3982.564	3983.691	GK	0.9	1.0d	0.5		25102.35	3E 3	2B 2	Q 4
3982.647	3983.773	G	0.1d			1.69	25101.83			
3982.758	3983.884	K	0.3	0.3	0.3		25101.13	w 0	2c 0	P 2
3982.937	3984.064			0.1			25100.00			
3982.986	3984.113				0.2		25099.69			
3983.051	3984.178						25099.28			
3983.148	3984.275			0.1d			25098.67			
3983.283	3984.410			0.1			25097.82	4F 0	2B 2	R 1
3983.378	3984.505		0.1d	0.2			25097.22			
3983.539	3984.665			0.1			25096.21			
3983.664	3984.791	K	0.3	0.4	0.4		25095.42			
3983.820	3984.946	K	0.4	0.4			25094.44			
3983.980	3985.107	GK	1.0	0.9	0.2	1.56	25093.43	3A 1	2B 0	R 0
3984.104	3985.231		0.3	0.2d			25092.65	4e 1	2c 0	R 4
3984.364	3985.491		0.1d	0.0			25091.01			
3984.552	3985.679		0.1d	0.2b			25089.83			
3984.709	3985.836	K	0.4	0.5	0.3		25088.84			
3984.973	3986.100		0.2d	0.3	0.2		25087.18			
3985.182	3986.309			0.0			25085.86			
3985.292	3986.419			0.0			25085.17			

λ_{air}	λ_{vac}	Ref.	I_1	I_2	I_3	I_4	ν	Upper	Lower	Br.
3985.422	3986.549	K	0.6	0.6	0.3		25084.35			
3985.624	3986.751		0.2d	0.2			25083.08			
3985.737	3986.864	GK	0.4	0.4	0.2		25082.37			
3985.889	3987.017		0.4	0.4	0.3		25081.41	4f 2	2c 1	Q 5
3986.048	3987.176		0.2	0.2			25080.41			
3986.129	3987.257			0.2	0.3		25079.90			
3986.538	3987.665			0.1			25077.33			
3986.690	3987.818		0.1				25076.37			
3986.738	3987.866		0.1	0.2			25076.07			
3986.838	3987.966		0.0	0.1			25075.44			
3986.931	3988.058	GK	0.9	0.9	0.3	1.49	25074.86	3E 3	2B 2	Q 3
3987.056	3988.184		0.5	0.3			25074.07	3F 2	2B 1	P 4
3987.174	3988.302		0.2d	0.0			25073.33			
3987.338	3988.465	G	0.4b	0.3	0.3		25072.30			
3987.489	3988.616		0.1				25071.35			
3987.619	3988.747		0.3	0.2	0.3d		25070.53			
3987.745	3988.873		0.1d	0.0			25069.74			
3987.853	3988.981		0.2d	0.2	0.2		25069.06			
3987.947	3989.075		0.1				25068.47			
3988.055	3989.183		0.3	0.3			25067.79	X 1	2B 0	P 4
3988.184	3989.312		0.0				25066.98			
3988.270	3989.398		0.2	0.3	0.2		25066.44			
3988.348	3989.476		0.0d				25065.95			
3988.410	3989.538		0.4	0.3	0.2		25065.56			
3988.547	3989.675		0.1d		0.1		25064.70			
3988.723	3989.851		0.1d	0.1			25063.59			
3988.835	3989.963		0.0				25062.89			
3988.943	3990.071		0.3d	0.3	0.3		25062.21			
3989.048	3990.176		0.1d				25061.55			
3989.153	3990.281	G	0.4d	0.4	0.2		25060.89			
3989.320	3990.448		0.3d	0.3			25059.84	W 1	2B 0	R 4
3989.390	3990.518				0.2		25059.40			
3989.451	3990.579		0.1	0.1			25059.02			
3989.616	3990.745		0.4b	0.3d	0.1		25057.98	4F 0	2B 2	Q 3
3989.683	3990.812						25057.56			
3989.753	3990.882		0.2				25057.12			
3989.820	3990.948						25056.70			
3989.868	3990.996		0.6	0.4			25056.40			
3989.946	3991.074		0.3	0.2			25055.91			
3990.038	3991.167	GK	1.5b	1.5	0.8	2.15	25055.33	3E 3	2B 2	Q 2
3990.038	3991.167	GK	1.5b	1.5	0.8	2.15	25055.33	4f 1	2c 0	R 1
3990.145	3991.273		0.5	0.3	0.2		25054.66			
3990.242	3991.371						25054.05			
3990.325	3991.453		0.2				25053.53			
3990.413	3991.541		0.4	0.3	0.3		25052.98			
3990.572	3991.700			0.1			25051.98			
3990.682	3991.810		0.0d				25051.29			
3990.765	3991.893	K	0.3	0.3			25050.77	4F 0	2B 2	P 4
3990.870	3991.998		0.1				25050.11			
3990.954	3992.083	K	0.3	0.3			25049.58	4e 1	2c 0	R 3
3991.050	3992.178		0.1	0.1			25048.98			
3991.149	3992.277	GK	0.8	0.6	0.3	1.06	25048.36			
3991.254	3992.383		0.2	0.1	0.2		25047.70			
3991.415	3992.544		0.3	0.2			25046.69			
3991.514	3992.642	G	0.4	0.3			25046.07	3F 2	2B 1	P 3
3991.665	3992.794		0.4	0.2			25045.12			
3991.757	3992.886		0.3	0.0			25044.54			
3991.842	3992.971	GK	1.5	0.8	0.4	1.95	25044.01	3D 3	2B 2	R 0
3991.946	3993.074		0.3				25043.36			
3992.012	3993.141	K	0.9	0.7	0.4		25042.94	3E 3	2B 2	Q 1

λ_{air}	λ_{vac}	Ref.	I_1	I_2	I_3	I_4	ν	Upper	Lower	Br.
3992.122	3993.251		0.3	0.0			25042.25	4f 2	2c 1	R 2
3992.252	3993.381		0.1d	0.0			25041.44			
3992.338	3993.467	K	0.8	0.6	0.4		25040.90			
3992.444	3993.574		0.1d	0.1			25040.23			
3992.582	3993.711		0.2				25039.37			
3992.681	3993.810		0.3	0.4	0.2		25038.75			
3992.811	3993.940		0.2d	0.2			25037.93			
3992.894	3994.023		0.0	0.2			25037.41			
3992.980	3994.109		0.3	0.2			25036.87			
3993.079	3994.208		0.3	0.2			25036.25			
3993.162	3994.291		0.3	0.1			25035.73			
3993.259	3994.389	GK	1.5	0.5	0.3	1.52	25035.12	3D 3	2B 2	R 1
3993.366	3994.496		0.3				25034.45			
3993.446	3994.575						25033.95			
3993.527	3994.657		0.0				25033.44			
3993.583	3994.713		0.1				25033.09			
3993.684	3994.813		0.2d				25032.46			
3993.770	3994.899		0.1				25031.92			
3993.859	3994.989	GK	1.0	0.6	0.3	1.16	25031.36	3E 3	2B 2	P 2
3994.003	3995.132	K	0.6	0.2			25030.46			
3994.113	3995.242		0.1d				25029.77			
3994.277	3995.407		0.0d	0.1			25028.74			
3994.416	3995.546	K	0.8	0.3	0.3d		25027.87			
3994.514	3995.643		0.0				25027.26			
3994.678	3995.808		0.0		0.2		25026.23			
3994.753	3995.883		0.0				25025.76			
3994.854	3995.983	K	0.6	0.4	0.2		25025.13			
3994.988	3996.117		0.0				25024.29			
3995.077	3996.207		0.0				25023.73			
3995.143	3996.272		0.3	0.2d	0.2		25023.32			
3995.245	3996.374		0.2d				25022.68			
3995.380	3996.510			0.1d			25021.83			
3995.532	3996.662		0.0				25020.88			
3995.725	3996.855		0.2d		0.2		25019.67			
3995.829	3996.959		0.3	0.2d			25019.02			
3996.113	3997.243		0.1d				25017.24			
3996.241	3997.371		0.2d	0.1			25016.44			
3996.471	3997.601		0.2d	0.0			25015.00	W 1	2B 0	R 3
3996.567	3997.697		0.1				25014.40	4f 1	2c 0	Q 4
3996.658	3997.788	GK	0.8	0.7	0.2	1.06	25013.83	3E 3	2B 2	P 3
3996.770	3997.900		0.4	0.4	0.2		25013.13			
3996.872	3998.003		0.2				25012.49			
3996.933	3998.063						25012.11			
3996.989	3998.119	K	0.5b	0.5d	0.3		25011.76	3E 2	2B 1	Q10
3997.056	3998.186		0.2				25011.34			
3997.147	3998.278	GK	1.5	1.5	0.4	1.96	25010.77	3D 3	2B 2	R 2
3997.248	3998.378		0.3	0.3			25010.14			
3997.305	3998.436		0.2	0.0			25009.78			
3997.392	3998.522	K	0.8	0.6	0.2		25009.24	3A 1	2B 0	P 1
3997.392	3998.522	K	0.8	0.6	0.2		25009.24	5c 0	2a 0	R 2
3997.502	3998.632		0.2	0.0			25008.55			
3997.582	3998.712						25008.05			
3997.660	3998.791			0.1			25007.56			
3997.836	3998.967		0.1	0.1d			25006.46			
3997.988	3999.119			0.0			25005.51			
3998.146	3999.277		0.3	0.3			25004.52			
3998.199	3999.330		0.0		0.2		25004.19			
3998.263	3999.394	GK	0.5	0.6	0.2	0.94	25003.79			
3998.375	3999.506		0.2				25003.09			
3998.487	3999.618		0.0				25002.39			

λ_{air}	λ_{vac}	Ref.	I_1	I_2	I_3	I_4	ν	Upper	Lower	Br.
3998.583	3999.714		0.2				25001.79			
3998.717	3999.848				0.2		25000.95	4g 1	2a 0	P 4
3998.842	3999.973			0.0			25000.17			
3998.956	4000.086		0.0				24999.46	4e 1	2c 0	R 2
3999.008	4000.139			0.1			24999.13			
3999.136	4000.267		0.3	0.2d			24998.33			
3999.268	4000.398	GK	0.6	0.7	0.3	0.69	24997.51	3E 2	2B 1	R 4
3999.367	4000.498		0.1d				24996.89			
3999.536	4000.667			0.1			24995.83			
3999.724	4000.855			0.0d	0.1		24994.66			
3999.813	4000.944	G					24994.10			
3999.895	4001.026			0.0			24993.59	5c 0	2a 0	R 2
4000.037	4001.168		0.1				24992.70			
4000.114	4001.245		0.0	0.0			24992.22			
4000.215	4001.346	GK	0.9	0.9	0.3	1.26	24991.59			
4000.391	4001.522		0.2b	0.2			24990.49	W 1	2B 0	R 2
4000.513	4001.644		0.1	0.1d			24989.73			
4000.582	4001.713		0.0				24989.30			
4000.673	4001.804		0.4	0.3d	0.2		24988.73			
4000.756	4001.887		0.0				24988.21			
4000.843	4001.974	GK	0.9	0.9	0.2	1.45	24987.67	3E 3	2B 2	P 4
4000.955	4002.086		0.3	0.2			24986.97			
4001.121	4002.252		0.2	0.0			24985.93			
4001.254	4002.385		0.1d				24985.10			
4001.334	4002.465		0.2				24984.60			
4001.366	4002.498		0.1	0.1			24984.40			
4001.522	4002.653		0.1d				24983.43			
4001.637	4002.768	K	0.1				24982.71			
4001.863	4002.994		0.1	0.1d			24981.30	4f 1	2c 0	Q 3
4001.916	4003.047		0.0				24980.97			
4002.058	4003.190	GK	0.6	0.5	0.2b	0.81	24980.08			
4002.185	4003.316		0.4	0.3			24979.29			
4002.335	4003.467		0.3	0.2d			24978.35			
4002.417	4003.549		0.2	0.0			24977.84			
4002.504	4003.635	GK	1.0	0.9	0.3	1.68	24977.30	3A 1	2B 0	P 2
4002.504	4003.635	GK	1.0	0.9	0.3	1.68	24977.30	4f 2	2c 1	Q 4
4002.614	4003.746		0.3	0.0			24976.61			
4002.662	4003.794			0.0			24976.31			
4002.771	4003.903	G	0.4	0.4	0.2		24975.63			
4002.861	4003.993		0.1	0.1			24975.07			
4002.936	4004.068		0.2	0.2			24974.60			
4002.997	4004.129	GK	0.8	0.8	0.3	1.30	24974.22	3D 3	2B 2	R 3
4003.125	4004.257		0.1d	0.1			24973.42			
4003.236	4004.368			0.1d	0.2		24972.73			
4003.345	4004.477		0.2	0.1			24972.05	4E 0	2B 2	R 2
4003.403	4004.535	G	0.4	0.4	0.3		24971.69			
4003.523	4004.655		0.2	0.2			24970.94			
4003.635	4004.767			0.1			24970.24			
4003.730	4004.862		0.3	0.2			24969.65			
4003.802	4004.934	GK	0.5	0.5	0.3		24969.20	4E 0	2B 2	R 2
4003.962	4005.094		0.1	0.1			24968.20			
4004.033	4005.165		0.2	0.1d			24967.76			
4004.081	4005.213						24967.46			
4004.307	4005.439		0.0				24966.05			
4004.354	4005.486			0.1			24965.76			
4004.476	4005.608	G	0.5	0.4			24965.00	W 1	2B 0	R 1
4004.641	4005.773	GK	0.5	0.5	0.2		24963.97			
4004.744	4005.876		0.1	0.1			24963.33			
4004.851	4005.983	G	0.5	0.4	0.3		24962.66	5c 0	2a 0	R 1
4004.986	4006.118		0.2	0.1			24961.82			

λ_{air}	λ_{vac}	Ref.	I_1	I_2	I_3	I_4	ν	Upper	Lower	Br.
4005.109	4006.242	G	0.5	0.4			24961.05			
4005.296	4006.428		0.2	0.2d			24959.89			
4005.406	4006.539	GK	0.8	0.7			24959.20	3A 1	2B 0	P 3
4005.507	4006.640	GK	0.9	0.9	0.4	1.81	24958.57	3A 1	2B 0	P 4
4005.616	4006.749		0.2	0.2			24957.89			
4005.685	4006.818		0.1	0.1			24957.46			
4005.787	4006.919		0.3	0.1d			24956.83			
4005.860	4006.993		0.2	0.0			24956.37			
4005.958	4007.091	GK	1.0	0.9	0.5	1.37	24955.76			
4006.116	4007.248	GK	0.7	0.7	0.2		24954.78	3D 3	2B 2	R 4
4006.215	4007.348		0.2	0.1			24954.16			
4006.337	4007.470			0.0			24953.40			
4006.418	4007.550			0.1d			24952.90			
4006.612	4007.745			0.1d			24951.69			
4006.729	4007.862		0.2				24950.96			
4006.800	4007.932		0.3	0.3	0.2b		24950.52			
4006.951	4008.083		0.3	0.2	0.1		24949.58			
4007.029	4008.162		0.1				24949.09			
4007.142	4008.275			0.1			24948.39			
4007.253	4008.386		0.2	0.3	0.2		24947.70			
4007.344	4008.477	G	0.4	0.4			24947.13			
4007.503	4008.636		0.1d				24946.14			
4007.577	4008.710		0.1				24945.68			
4007.665	4008.798		0.1				24945.13			
4007.743	4008.876		0.2	0.2	0.1		24944.65			
4007.955	4009.088		0.1	0.1d			24943.33	4e 1	2c 0	R 1
4008.215	4009.348						24941.71			
4008.397	4009.530		0.2d	0.2d			24940.58			
4008.549	4009.683		0.2	0.1			24939.63			
4008.635	4009.768		0.2	0.0			24939.10			
4008.726	4009.859	GK	0.9	0.8	0.1	1.00	24938.53	3D 3	2B 2	P 2
4008.829	4009.962		0.0				24937.89			
4008.901	4010.035		0.3	0.3	0.2		24937.44			
4009.106	4010.239			0.1			24936.17			
4009.253	4010.387	GK	0.5	0.5	0.2	0.83	24935.25	3E 3	2B 2	P 5
4009.337	4010.471		0.2	0.1			24934.73	4f 1	2c 0	Q 2
4009.395	4010.528		0.1	0.0			24934.37			
4009.469	4010.602		0.1				24933.91			
4009.572	4010.705	GK	0.9	0.7	0.2	1.10	24933.27	W 1	2B 0	R 0
4009.572	4010.705	GK	0.9	0.7	0.2	1.10	24933.27	3E 2	2B 1	R 3
4009.668	4010.802		0.2				24932.67			
4009.707	4010.840		0.1				24932.43			
4009.771	4010.905			0.1d			24932.03			
4009.847	4010.980		0.1				24931.56			
4009.948	4011.082		0.4	0.3	0.3		24930.93			
4010.040	4011.174	GK	0.6	0.4			24930.36	4c 5	2a 3	Q 1
4010.157	4011.291		0.1	0.0d			24929.63			
4010.217	4011.350		0.2				24929.26			
4010.579	4011.713		0.1d				24927.01			
4010.738	4011.872	GK	0.6	0.5	0.3	0.80	24926.02	4F 0	2B 2	Q 2
4010.833	4011.967		0.3	0.2			24925.43			
4010.971	4012.105		0.4	0.3			24924.57			
4011.103	4012.237		0.1	0.1			24923.75			
4011.224	4012.358	GK	0.6	0.5	0.2		24923.00	4F 0	2B 2	P 3
4011.319	4012.453		0.0				24922.41			
4011.404	4012.538		0.0d	0.0			24921.88			
4011.506	4012.640		0.1	0.0			24921.25			
4011.575	4012.709		0.3	0.3			24920.82			
4011.646	4012.780		0.1				24920.38			
4011.702	4012.836		0.3	0.2			24920.03			

λ_{air}	λ_{vac}	Ref.	I$_1$	I$_2$	I$_3$	I$_4$	ν	Upper	Lower	Br.
4011.784	4012.918	GK	0.4	0.4			24919.52			
4011.937	4013.071		0.1	0.1			24918.57			
4012.201	4013.335			0.0			24916.93			
4012.473	4013.608			0.0			24915.24			
4012.673	4013.807		0.2		0.2		24914.00			
4012.759	4013.893	GK	0.5	0.5	0.2		24913.47	3E 2	2B 1	Q 9
4012.870	4014.004		0.1	0.1			24912.78			
4012.976	4014.110		0.0	0.1			24912.12	w 0	2c 0	P 3
4013.066	4014.201			0.0			24911.56			
4013.150	4014.284	GK	0.8	0.7		1.14	24911.04	X 2	2B 2	R 0
4013.253	4014.388		0.2	0.1d			24910.40			
4013.356	4014.491			0.1d			24909.76			
4013.461	4014.595			0.1d			24909.11			
4013.577	4014.711	G	0.4	0.3			24908.39			
4013.678	4014.813		0.0	0.1			24907.76			
4013.793	4014.927	F	0.4	0.2			24907.05	5c 0	2a 0	R 0
4013.894	4015.029	1	0.1	0.0			24906.42			
4014.012	4015.147		0.2d	0.2			24905.69			
4014.073	4015.208			0.1			24905.31			
4014.175	4015.310	GK	0.5	0.5			24904.68	4c 5	2a 3	Q 2
4014.307	4015.442		0.3	0.2			24903.86			
4014.915	4016.050			0.1			24900.09			
4015.070	4016.205		0.2d				24899.13			
4015.142	4016.277			0.2			24898.68			
4015.407	4016.542		0.3	0.3			24897.04			
4015.655	4016.790		0.2	0.2d			24895.50	4f 2	2c 1	R 1
4015.752	4016.887		0.1	0.1			24894.90			
4015.839	4016.974	GK	0.7	0.5		0.88	24894.36	X 2	2B 2	R 1
4015.945	4017.081		0.1d				24893.70	4f 3	2c 2	R 3
4016.033	4017.168						24893.16			
4016.126	4017.261		0.1d				24892.58			
4016.228	4017.363		0.2	0.2			24891.95			
4016.321	4017.457	GK	0.7	0.4		0.72	24891.37			
4016.444	4017.579		0.0	0.1			24890.61			
4016.557	4017.692	GK	0.7	0.6	0.3	0.75	24889.91			
4016.668	4017.804		0.0	0.1			24889.22			
4016.759	4017.894		0.1				24888.66			
4016.981	4018.117		0.2b	0.2			24887.28	4E 0	2B 2	R 0
4017.175	4018.311			0.0			24886.08			
4017.358	4018.493		0.1d	0.0			24884.95			
4017.454	4018.590	G					24884.35			
4017.524	4018.659		0.4	0.3			24883.92			
4017.653	4018.789		0.0				24883.12			
4017.779	4018.915	GK	0.5	0.5	0.2		24882.34			
4018.147	4019.283			0.2			24880.06			
4018.276	4019.412		0.1d	0.2			24879.26			
4018.406	4019.541	G	0.5	0.5	0.2		24878.46			
4018.536	4019.672		0.2	0.1			24877.65			
4018.580	4019.716						24877.38			
4018.638	4019.774		0.2	0.0			24877.02			
4018.733	4019.869	G	0.5	0.4			24876.43			
4018.816	4019.952			0.0			24875.92	X 2	2B 2	P 1
4018.908	4020.044	GK	0.8	0.6		0.82	24875.35			
4019.047	4020.183		0.1	0.1			24874.49			
4019.152	4020.288	G	0.5	0.4	0.2		24873.84			
4019.236	4020.372			0.2			24873.32			
4019.331	4020.467		0.3	0.2			24872.73			
4019.480	4020.616		0.1				24871.81			
4019.611	4020.747		0.2d	0.0			24871.00	4f 2	2c 1	Q 3
4019.611	4020.747		0.2d	0.0			24871.00	3c 7	2a 2	Q 1

λ_{air}	λ_{vac}	Ref.	I_1	I_2	I_3	I_4	ν	Upper	Lower	Br.
4019.774	4020.910		0.0d				24869.99			
4019.870	4021.006		0.4	0.3			24869.40			
4020.057	4021.193		0.1				24868.24			
4020.209	4021.345		0.1d				24867.30			
4020.335	4021.471	G	0.4	0.4			24866.52	4c 5	2a 3	Q 3
4020.474	4021.611		0.1				24865.66			
4020.701	4021.837		0.2d				24864.26			
4020.760	4021.897			0.1d			24863.89			
4020.833	4021.970						24863.44			
4020.930	4022.067	GK	0.8	0.7	0.2b	1.03	24862.84	3E 2	2B 1	R 2
4021.032	4022.169		0.1				24862.21			
4021.137	4022.274		0.2d	0.2			24861.56			
4021.263	4022.400			0.0			24860.78			
4021.351	4022.487	GK	0.8	0.7	0.2	1.34	24860.24	3D 3	2B 2	P 3
4021.351	4022.487	GK	0.8	0.7	0.2	1.34	24860.24	X 2	2B 2	R 2
4021.464	4022.601		0.1	0.1			24859.54			
4021.561	4022.698	GK		0.4			24858.94			
4021.618	4022.754			0.2	0.2		24858.59			
4021.718	4022.855		0.1	0.1			24857.97			
4021.797	4022.934	GK	0.9	0.7	0.2		24857.48	W 1	2B 0	P 1
4021.904	4023.041		0.2				24856.82			
4022.008	4023.144			0.1			24856.18			
4022.169	4023.306		0.1d	0.1d			24855.18			
4022.289	4023.426			0.0d			24854.44			
4022.585	4023.722			0.0d			24852.61			
4022.750	4023.887			0.1d			24851.59			
4023.171	4024.308			0.1d			24848.99			
4023.345	4024.482	GK	0.5	0.5	0.2	0.56	24847.92	3A 2	2B 2	R 4
4023.575	4024.712		0.0	0.0			24846.50			
4023.805	4024.942		0.3	0.1			24845.08			
4023.983	4025.120						24843.98			
4024.059	4025.196		0.0d				24843.51			
4024.310	4025.447		0.3d	0.2			24841.96			
4024.571	4025.708		0.3	0.2			24840.35			
4024.655	4025.792		0.2	0.0			24839.83			
4024.744	4025.882	GK	1.0	0.8	0.3	1.26	24839.28	5c 0	2a 0	Q 1
4024.848	4025.985		0.3	0.0			24838.64			
4024.937	4026.074						24838.09	4f 2	2c 1	R 3
4025.044	4026.181		0.0d				24837.43			
4025.117	4026.254			0.1			24836.98			
4025.355	4026.493			0.2			24835.51	3c 4	2a 0	Q 1
4025.417	4026.554		0.3b	0.2			24835.13			
4025.738	4026.875			0.0d			24833.15			
4025.880	4027.018		0.1d				24832.27			
4026.031	4027.169		0.2d	0.1			24831.34			
4026.104	4027.242			0.1			24830.89			
4026.187	4027.325	1			1.5b		24830.38			
4026.214	4027.352	GK	0.8d	0.6		0.82	24830.21	3F 1	2B 0	Q 6
4026.349	4027.487		0.3	0.1	0.4		24829.38	3c 7	2a 2	Q 2
4026.484	4027.621		0.3	0.2			24828.55			
4026.521	4027.659						24828.32	4e 1	2c 0	R 1
4026.613	4027.751	GK	0.8	0.8	1.0b	1.01	24827.75	3E 2	2B 1	Q 8
4026.724	4027.862		0.2	0.2			24827.07			
4026.998	4028.136		0.1d	0.1d			24825.38			
4027.174	4028.313		0.1d	0.2			24824.29			
4027.254	4028.392			0.1			24823.80			
4027.385	4028.523	GK	0.8b	0.8	0.7b	1.22	24822.99	5c 0	2a 0	Q 2
4027.385	4028.523	GK	0.8b	0.8	0.7b	1.22	24822.99	X 2	2B 2	R 3
4027.385	4028.523	GK	0.8b	0.8	0.7b	1.22	24822.99	4e 1	2c 0	Q 2
4027.587	4028.725			0.0d			24821.75	4b 2	2a 0	R 1

λ_{air}	λ_{vac}	Ref.	I_1	I_2	I_3	I_4	ν	Upper	Lower	Br.
4027.652	4028.790			0.0d			24821.35			
4027.789	4028.928		0.3	0.2			24820.50	4e 1	2c 0	Q 1
4027.875	4029.014		0.1				24819.97			
4027.970	4029.108		0.2	0.3			24819.39	4e 1	2c 0	Q 3
4028.065	4029.204		0.2				24818.80			
4028.161	4029.299		0.3	0.2			24818.21			
4028.247	4029.385		0.5	0.4	0.3		24817.68			
4028.338	4029.476	GK	1.5	1.0	0.4	1.92	24817.12	W 1	2B 0	P 2
4028.447	4029.585		0.4	0.2			24816.45			
4028.541	4029.679			0.0			24815.87			
4028.666	4029.804		0.1	0.2			24815.10			
4028.713	4029.852		0.1d		0.2d		24814.81			
4028.812	4029.951		0.0	0.0			24814.20			
4028.864	4030.003		0.1				24813.88			
4028.935	4030.074		0.2	0.3	0.1d		24813.44			
4029.020	4030.158	GK	0.4	0.3			24812.92			
4029.218	4030.357			0.0			24811.70	4e 1	2c 0	Q 4
4029.382	4030.521		0.3d	0.2			24810.69	4e 1	2c 0	R 2
4029.582	4030.721		0.3	0.2			24809.46			
4029.754	4030.893		0.3	0.3			24808.40	4e 2	2c 1	R 4
4029.905	4031.044		0.3	0.3			24807.47	3c 4	2a 0	Q 2
4029.905	4031.044		0.3	0.3			24807.47	3E 1	2B 0	Q13
4030.007	4031.146	G	0.4	0.4			24806.84			
4030.115	4031.254		0.3	0.3			24806.18			
4030.225	4031.364	K	0.1d	0.1			24805.50	X 2	2B 2	P 2
4030.347	4031.486		0.4	0.2			24804.75			
4030.625	4031.764			0.0			24803.04			
4030.756	4031.896			0.1			24802.23			
4030.901	4032.040			0.1			24801.34			
4031.099	4032.239		0.0				24800.12			
4031.200	4032.339			0.2			24799.50			
4031.316	4032.455			0.2			24798.79			
4031.385	4032.525	GK	0.7	0.7	0.6b	1.06	24798.36	5c 0	2a 0	Q 3
4031.385	4032.525	GK	0.7	0.7	0.6b	1.06	24798.36	4e 1	2c 0	Q 5
4031.501	4032.640						24797.65			
4031.587	4032.726			0.2			24797.12			
4031.668	4032.808			0.1			24796.62			
4031.764	4032.904	GK	0.8	0.7	0.3	0.99	24796.03			
4031.878	4033.018		0.1	0.1d			24795.33			
4032.241	4033.380			0.2d			24793.10			
4032.364	4033.504		0.0d	0.1			24792.34			
4032.680	4033.820	GK	0.5	0.5		0.49	24790.40	4e 1	2c 0	R 3
4032.680	4033.820	GK	0.5	0.5		0.49	24790.40	3E 2	2B 1	R 1
4032.809	4033.948		0.2	0.1d			24789.61			
4033.385	4034.524		0.1d	0.0d			24786.07			
4033.577	4034.716		0.2	0.2			24784.89			
4033.676	4034.816			0.1			24784.28			
4033.777	4034.917	GK	0.6	0.6	0.3	0.65	24783.66			
4033.956	4035.096		0.1	0.2			24782.56			
4034.083	4035.223	GK	0.7	0.7	0.3		24781.78	4E 0	2B 2	P 2
4034.219	4035.359		0.2	0.4			24780.94			
4034.242	4035.382						24780.80	4f 2	2c 1	Q 2
4034.379	4035.519						24779.96			
4034.465	4035.605						24779.43			
4034.649	4035.789		0.0				24778.30			
4034.957	4036.097		0.1				24776.41			
4035.078	4036.218		0.3b	0.3			24775.67			
4035.190	4036.330		0.1				24774.98			
4035.309	4036.449		0.2				24774.25			
4035.381	4036.521		0.2	0.1	0.1		24773.81	4g 1	2a 0	P 5

λ_{air}	λ_{vac}	Ref.	I_1	I_2	I_3	I_4	ν	Upper	Lower	Br.
4035.493	4036.633		0.0				24773.12			
4035.571	4036.711	GK	0.9	0.9	0.3	1.35	24772.64	W 1	2B 0	P 3
4035.679	4036.819	K	0.2	0.2			24771.98			
4035.773	4036.914						24771.40			
4035.943	4037.083		0.0				24770.36			
4036.179	4037.319		0.1				24768.91			
4036.242	4037.383		0.1	0.1			24768.52			
4036.344	4037.484	GK	0.6	0.5			24767.90	3D 3	2B 2	P 4
4036.344	4037.484	GK	0.6	0.5			24767.90	3c 7	2a 2	Q 3
4036.445	4037.585		0.0				24767.28			
4036.534	4037.675		0.1				24766.73			
4036.661	4037.802	K	0.3	0.4	0.3		24765.95	5c 0	2a 0	Q 4
4036.710	4037.851	G					24765.65			
4037.105	4038.245			0.0d			24763.23			
4037.291	4038.431		0.2				24762.09	4f 2	2c 1	R 2
4037.361	4038.501		0.3	0.3			24761.66			
4037.545	4038.686		0.2b	0.2			24760.53			
4037.914	4039.054		0.1d				24758.27	4e 2	2c 1	R 3
4037.914	4039.054		0.1d				24758.27	4e 2	2c 1	R 1
4038.026	4039.167		0.0	0.2d			24757.58			
4038.204	4039.345			0.0			24756.49			
4038.318	4039.459	K	0.3	0.4	0.3		24755.79			
4038.414	4039.555			0.1			24755.20			
4038.525	4039.666	GK	0.7	0.8	0.5	0.65	24754.52	3E 2	2B 1	Q 7
4038.525	4039.666	GK	0.7	0.8	0.5	0.65	24754.52	4f 1	2c 0	P 3
4038.701	4039.842			0.1			24753.44			
4038.984	4040.125			0.1			24751.71			
4039.069	4040.210						24751.19			
4039.150	4040.291	GK	0.4	0.4	0.2	0.56	24750.69	4E 0	2B 2	Q 1
4039.317	4040.458		0.1	0.0			24749.67	4f 2	2c 1	P 6
4039.524	4040.665			0.1			24748.40			
4039.612	4040.753						24747.86			
4039.654	4040.796	K	0.3b	0.3	0.3		24747.60	3F 1	2B 0	R 3
4039.986	4041.127	G	0.3b	0.3			24745.57	3A 2	2B 2	R 3
4040.121	4041.263						24744.74			
4040.272	4041.413	GK	0.4	0.3d			24743.82	4f 1	2c 0	P 4
4040.548	4041.689		0.2	0.1			24742.13			
4040.659	4041.800						24741.45	3c 4	2a 0	Q 3
4041.077	4042.219			0.1			24738.89			
4041.257	4042.398						24737.79			
4041.382	4042.524			0.2			24737.02			
4041.554	4042.696		0.2	0.2			24735.97			
4041.799	4042.941			0.1			24734.47			
4041.876	4043.018			0.0			24734.00	4e 1	2c 0	R 5
4041.971	4043.112		0.4	0.4			24733.42			
4042.101	4043.243		0.1				24732.62			
4042.221	4043.363		0.2	0.2			24731.89			
4042.484	4043.626		0.4	0.3			24730.28			
4042.590	4043.732						24729.63			
4042.665	4043.807	K	0.5	0.4	0.2		24729.17	4E 0	2B 2	P 4
4042.794	4043.937		0.3	0.3			24728.38			
4042.922	4044.064			0.1			24727.60			
4043.002	4044.144			0.0			24727.11			
4043.100	4044.242	GK	0.6	0.5		0.92	24726.51	4E 0	2B 2	Q 4
4043.194	4044.337	K	0.3	0.4	0.4b		24725.93	5c 0	2a 0	Q 5
4043.280	4044.422		0.2				24725.41	3F 3	2B 3	R 2
4043.393	4044.535		0.3	0.3			24724.72			
4043.478	4044.620		0.2	0.2			24724.20			
4043.568	4044.710	GK	1.0	1.0	0.5	1.79	24723.65	W 1	2B 0	P 4
4043.674	4044.817		0.3	0.1			24723.00			

λ_{air}	λ_{vac}	Ref.	I_1	I_2	I_3	I_4	ν	Upper	Lower	Br.
4043.780	4044.923						24722.35			
4043.934	4045.077						24721.41			
4044.111	4045.253		0.3	0.3			24720.33	4f 1	2c 0	P 5
4044.263	4045.406		0.1	0.1			24719.40	X 2	2B 2	P 3
4044.320	4045.463			0.1			24719.05			
4044.420	4045.563	1	0.6	0.7	1.5b		24718.44			
4044.464	4045.607	GK				1.00	24718.17			
4044.494	4045.636		0.6	0.7			24717.99			
4044.610	4045.753		0.2	0.3			24717.28	4D 0	2B 2	R 6
4044.661	4045.803						24716.97			
4044.774	4045.916	G	0.5	0.6	0.3	0.46	24716.28	5c 1	2a 1	R 3
4044.867	4046.010			0.0			24715.71			
4044.944	4046.087						24715.24			
4045.283	4046.425		0.0	0.0			24713.17			
4045.400	4046.543						24712.45			
4045.479	4046.622	GK	0.7	0.6		0.70	24711.97	3E 2	2B 1	R 0
4045.592	4046.735		0.2	0.1			24711.28	4f 2	2c 1	Q 4
4045.795	4046.938			0.3			24710.04			
4045.980	4047.123	1	0.4	0.4	0.4		24708.91			
4046.106	4047.249						24708.14			
4046.152	4047.295						24707.86			
4046.201	4047.344	K	0.3	0.4			24707.56	5c 0	2a 0	P 2
4046.201	4047.344	K	0.3	0.4			24707.56	4e 1	2c 0	Q 2
4046.453	4047.596		0.1				24706.02			
4046.555	4047.698	K					24705.40			
4046.612	4047.755		0.2	0.2			24705.05			
4046.701	4047.844						24704.51			
4046.807	4047.950						24703.86	4e 1	2c 0	R 6
4047.161	4048.304		0.0	0.0			24701.70			
4047.358	4048.501						24700.50	4e 2	2c 1	R 2
4047.525	4048.668						24699.48	4e 1	2c 0	P 2
4047.581	4048.724	GK	0.7	0.4b	0.2d	0.53	24699.14			
4047.717	4048.860			0.1			24698.31			
4047.861	4049.004						24697.43			
4047.962	4049.106	GK	0.7	0.7	0.3	0.90	24696.81	4E 0	2B 2	Q 3
4047.962	4049.106	GK	0.7	0.7	0.3	0.90	24696.81	3F 1	2B 0	Q 5
4047.962	4049.106	GK	0.7	0.7	0.3	0.90	24696.81	4f 2	2c 1	P 5
4048.164	4049.308						24695.58			
4048.248	4049.391						24695.07			
4048.323	4049.467	GK	0.8	0.7	0.3	1.37	24694.61	4E 0	2B 2	Q 2
4048.453	4049.596	GK	0.9	1.0	1.0b	1.35	24693.82	3E 2	2B 1	Q 6
4048.561	4049.704		0.2	0.2			24693.16			
4048.774	4049.918						24691.86			
4049.005	4050.149			0.0			24690.45			
4049.095	4050.239						24689.90			
4049.189	4050.333	G	0.7	0.5		1.06	24689.33			
4049.354	4050.498	K	0.3	0.5	0.5		24688.32			
4049.520	4050.664		0.2	0.1			24687.31			
4049.599	4050.743			0.0			24686.83			
4049.820	4050.964						24685.48			
4050.334	4051.478		0.1	0.0			24682.35			
4050.477	4051.621		0.3	0.4			24681.48	3A 2	2B 2	R 2
4050.583	4051.728		0.0	0.2d			24680.83			
4050.662	4051.806						24680.35			
4050.751	4051.895			0.1			24679.81			
4050.835	4051.979		0.1	0.2			24679.30			
4050.938	4052.082		0.1	0.2	0.2		24678.67	5c 1	2a 1	R 2
4050.938	4052.082		0.1	0.2	0.2		24678.67	4f 3	2c 2	Q 4
4051.040	4052.184		0.2				24678.05			
4051.155	4052.299						24677.35			

λ_{air}	λ_{vac}	Ref.	I_1	I_2	I_3	I_4	ν	Upper	Lower	Br.
4051.227	4052.371	GK	0.8	0.8	0.3	1.16	24676.91	W 1	2B 0	P 5
4051.330	4052.475			0.1			24676.28			
4051.434	4052.578		0.1	0.2			24675.65			
4052.013	4053.158						24672.12			
4052.214	4053.358			0.1			24670.90			
4052.344	4053.488			0.1			24670.11			
4052.465	4053.610						24669.37			
4052.518	4053.662			0.1			24669.05			
4052.619	4053.764						24668.43			
4052.808	4053.953			0.2			24667.28			
4053.142	4054.287			0.2			24665.25	3D 3	2B 2	P 5
4053.246	4054.390		0.0	0.2			24664.62			
4053.390	4054.535		0.3	0.3	0.2		24663.74			
4053.523	4054.668	GK	0.6	0.6	0.3	0.72	24662.93			
4053.628	4054.773			0.0			24662.29			
4053.699	4054.844			0.0			24661.86			
4053.813	4054.958		0.0	0.1d			24661.17			
4053.895	4055.040			0.1			24660.67			
4054.003	4055.148	GK	0.7	0.6	0.2	0.88	24660.01	T 0	2B 2	R 2
4054.125	4055.270						24659.27			
4054.212	4055.357		0.1	0.2d			24658.74			
4054.383	4055.528	GK	0.6	0.6	0.6	0.49	24657.70	3E 1	2B 0	Q12
4054.531	4055.676	i		0.3			24656.80			
4054.725	4055.870						24655.62			
4054.801	4055.946			0.1d	0.1		24655.16	7c 0	2a 1	R 1
4055.087	4056.232		0.4	0.3			24653.42			
4055.191	4056.336			0.1			24652.79			
4055.288	4056.433	GK	0.8	0.9	0.7	1.31	24652.20	W 1	2B 0	P 6
4055.396	4056.542		0.1	0.0			24651.54			
4055.465	4056.611						24651.12			
4055.544	4056.690						24650.64			
4055.627	4056.772		0.1	0.2			24650.14			
4055.753	4056.899		0.4	0.4	0.2		24649.37			
4055.914	4057.060		0.3	0.3			24648.39			
4056.048	4057.193		0.1				24647.58			
4056.173	4057.319		0.3	0.2b			24646.82			
4056.341	4057.486						24645.80			
4056.416	4057.562	GK	0.8	0.9	0.6	1.16	24645.34	3E 2	2B 1	Q 5
4056.522	4057.668		0.2	0.1			24644.70			
4056.703	4057.849			0.0			24643.60	4f 2	2c 1	Q 2
4056.703	4057.849			0.0			24643.60	4f 2	2c 1	P 4
4056.979	4058.125			0.1			24641.92			
4057.067	4058.213			0.0			24641.39			
4057.156	4058.302	GK	0.5	0.6	0.4	0.53	24640.85	5c 1	2a 1	R 1
4057.156	4058.302	GK	0.5	0.6	0.4	0.53	24640.85	4D 0	2B 2	R 5
4057.261	4058.407			0.1			24640.21			
4057.345	4058.491			0.1			24639.70			
4057.436	4058.582		0.3	0.3			24639.15	3F 3	2B 3	Q 3
4057.556	4058.702	GK	0.7	0.7	0.3	1.19	24638.42	4e 2	2c 1	R 1
4057.676	4058.822						24637.69			
4057.776	4058.923		0.3	0.4	0.2		24637.08			
4057.992	4059.138						24635.77			
4058.340	4059.486			0.0			24633.66			
4058.442	4059.588	G	0.5	0.5	0.2	0.45	24633.04			
4058.542	4059.689		0.1	0.0			24632.43			
4058.598	4059.745	G					24632.09			
4058.636	4059.783		0.5	0.5	0.3		24631.86	5c 0	2a 0	P 3
4058.743	4059.890			0.0			24631.21			
4058.276	4059.422						24634.05			
4058.795	4059.941	G					24630.90			

λ_{air}	λ_{vac}	Ref.	I_1	I_2	I_3	I_4	ν	Upper	Lower	Br.
4058.854	4060.000	G	0.0	0.0			24630.54	4e 1	2c 0	Q 3
4058.996	4060.142						24629.68			
4059.078	4060.224		0.2	0.1			24629.18			
4059.169	4060.315			0.1			24628.63			
4059.264	4060.411	GK	0.8	0.6	0.2b	0.89	24628.05			
4059.390	4060.536		0.0				24627.29			
4059.469	4060.615			0.1			24626.81			
4059.564	4060.711		0.2	0.2			24626.23			
4059.816	4060.963	G	0.3	0.2			24624.70	4d 1	2c 0	R 1
4059.912	4061.059				0.3		24624.12			
4059.960	4061.107		0.3	0.2			24623.83	4d 1	2c 0	Q 1
4060.130	4061.277			0.1			24622.80			
4060.392	4061.539		0.2	0.1			24621.21			
4060.463	4061.610		0.3	0.3	0.3d		24620.78			
4060.572	4061.719						24620.12	4c 1	2a 0	R 2
4060.902	4062.049	G					24618.12			
4061.006	4062.153			0.0			24617.49	X 2	2B 2	P 4
4061.096	4062.243	GK	0.6	0.5	0.2d	0.81	24616.94	3A 2	2B 2	R 1
4061.205	4062.352						24616.28			
4061.303	4062.450	GK	0.6	0.4	0.2	0.61	24615.69	T 0	2B 2	R 1
4061.459	4062.606	G	0.0	0.1		0.57	24614.74			
4061.534	4062.681				0.2		24614.29			
4061.605	4062.752		0.3	0.2			24613.86			
4061.723	4062.870						24613.14			
4061.809	4062.956			0.1			24612.62			
4061.940	4063.087						24611.83			
4062.034	4063.181		0.3b	0.3d			24611.26	4f 3	2c 3	R 3
4062.133	4063.280		0.2				24610.66	3E 2	2B 1	P 5
4062.242	4063.389		0.2	0.2			24610.00			
4062.331	4063.478		0.1	0.3			24609.46			
4062.413	4063.561		1.5				24608.96	4f 2	2c 1	P 3
4062.413	4063.561		1.5				24608.96	3E 2	2B 1	P 4
4062.450	4063.597	GK		1.5b	1.5b	2.29	24608.74	3E 2	2B 1	P 6
4062.450	4063.597	GK		1.5b	1.5b	2.29	24608.74	3E 2	2B 1	Q 4
4062.491	4063.638		1.5				24608.49			
4062.600	4063.747		1.0				24607.83	4D 0	2B 2	R 0
4062.630	4063.777	K		0.9b	0.5		24607.65			
4062.658	4063.805		1.0				24607.48	3E 2	2B 1	P 3
4062.768	4063.916		0.2				24606.81			
4062.831	4063.978						24606.43			
4062.913	4064.061	GK	0.8	0.6			24605.93	3D 2	2B 1	P 1
4063.021	4064.168		0.2	0.2			24605.28			
4063.085	4064.233			0.1			24604.89			
4063.166	4064.314			0.2			24604.40			
4063.264	4064.411	GK	0.8	0.7	0.3	1.01	24603.81	3F 1	2B 0	R 2
4063.313	4064.461	K					24603.51			
4063.356	4064.504						24603.25			
4063.450	4064.598		0.3	0.3			24602.68			
4063.553	4064.700		0.2				24602.06			
4063.635	4064.783	GK	1.5	1.5	0.6	1.81	24601.56	3E 2	2B 1	P 2
4063.739	4064.887		0.3	0.2			24600.93			
4063.835	4064.983		0.2	0.3			24600.35			
4064.007	4065.155			0.1			24599.31			
4064.048	4065.196						24599.06	4f 3	2c 2	R 1
4064.126	4065.274			0.0			24598.59			
4064.184	4065.332	G					24598.24	4D 0	2B 2	P 1
4064.303	4065.451			0.1			24597.52			
4064.430	4065.578			0.2			24596.75			
4064.496	4065.644		0.3b	0.3			24596.35			
4064.577	4065.725			0.3			24595.86			

λ_{air}	λ_{vac}	Ref.	I_1	I_2	I_3	I_4	ν	Upper	Lower	Br.
4064.696	4065.844		0.1	0.1			24595.14			
4064.787	4065.935			0.2			24594.59			
4064.861	4066.009		0.4	0.3			24594.14			
4064.987	4066.135			0.2			24593.38	5c 1	2a 1	R 0
4065.083	4066.231						24592.80			
4065.151	4066.299	G	0.3	0.3			24592.39			
4065.265	4066.413						24591.70			
4065.306	4066.454		0.2	0.2			24591.45			
4065.443	4066.591						24590.62	4f 2	2c 1	Q 3
4065.476	4066.624		0.4	0.5	0.3		24590.42			
4065.541	4066.689						24590.03			
4065.618	4066.767	GK	1.0	1.0	0.3	1.67	24589.56	3D 2	2B 1	R 1
4065.723	4066.871		0.3	0.2			24588.93			
4065.828	4066.977						24588.29			
4065.908	4067.056						24587.81			
4065.980	4067.129		0.1	0.1			24587.37			
4066.033	4067.182			0.2			24587.05			
4066.141	4067.289	GK	0.8	0.9	0.7		24586.40	4D 0	2B 2	R 4
4066.291	4067.440			0.1			24585.49			
4066.329	4067.478					1.32	24585.26			
4066.435	4067.584		0.4	0.3			24584.62			
4066.498	4067.647			0.2			24584.24	3F 3	2B 3	Q 2
4066.626	4067.774			0.1			24583.47			
4066.690	4067.839						24583.08			
4066.713	4067.862		0.5	0.3			24582.94			
4066.723	4067.872						24582.88			
4066.794	4067.943						24582.45	3E 2	2B 1	Q 3
4066.826	4067.974	K					24582.26	3D 3	2B 2	P 6
4066.867	4068.016	GK	1.5b	1.5b	0.9b	2.50	24582.01	3D 2	2B 1	R 0
4066.867	4068.016	GK	1.5b	1.5b	0.9b	2.50	24582.01	4e 1	2c 0	P 4
4066.867	4068.016	GK	1.5b	1.5b	0.9b	2.50	24582.01	3D 2	2B 1	R 2
4067.003	4068.151		0.5	0.3			24581.19	T 0	2B 2	R 0
4067.085	4068.234						24580.69			
4067.172	4068.320	GK	0.8	0.9	0.7	1.53	24580.17	3F 1	2B 0	Q 4
4067.172	4068.320	GK	0.8	0.9	0.7	1.53	24580.17	4f 3	2c 2	Q 3
4067.272	4068.421		0.1				24579.56			
4067.349	4068.497						24579.10			
4067.398	4068.547			0.1			24578.80			
4067.484	4068.633		0.1	0.2			24578.28			
4067.532	4068.681						24577.99			
4067.605	4068.754		0.4	0.4	0.2		24577.55			
4067.683	4068.832						24577.08			
4067.724	4068.873		0.2	0.2			24576.83			
4067.850	4068.999						24576.07			
4067.885	4069.034		0.2	0.2			24575.86			
4067.979	4069.128		0.2	0.1			24575.29			
4068.151	4069.300		0.2	0.2			24574.25			
4068.282	4069.431		0.2	0.2			24573.46			
4068.461	4069.610			0.1			24572.38			
4068.580	4069.729		0.1	0.2			24571.66			
4068.747	4069.896		0.2				24570.65			
4068.908	4070.057		0.4	0.4			24569.68			
4068.988	4070.137	GK	0.3	0.4			24569.20			
4069.165	4070.314		0.3				24568.13			
4069.218	4070.367		0.3	0.4b			24567.81			
4069.271	4070.420						24567.49			
4069.355	4070.504	GK	0.3	0.2			24566.98			
4069.451	4070.600		0.5	0.4			24566.40			
4069.541	4070.690		0.7	0.3			24565.86			
4069.637	4070.786	GK	1.5b	1.5	1.0b	2.60	24565.28	3E 2	2B 1	Q 2

λ_{air}	λ_{vac}	Ref.	I_1	I_2	I_3	I_4	ν	Upper	Lower	Br.
4069.740	4070.889		0.7	0.4			24564.66			
4069.816	4070.965						24564.20			
4069.900	4071.050	GK	0.9	0.9	0.4	1.72	24563.69	3D 2	2B 1	R 3
4070.015	4071.164		0.3	0.3			24563.00			
4070.094	4071.243						24562.52			
4070.155	4071.305		0.2	0.2			24562.15			
4070.223	4071.373						24561.74			
4070.321	4071.471	G	0.4	0.5	0.2		24561.15			
4070.371	4071.520						24560.85			
4070.442	4071.592		0.1				24560.42			
4070.557	4071.706		0.2				24559.73			
4070.641	4071.791		0.0				24559.22			
4070.737	4071.887		0.3	0.2			24558.64	4d 1	2c 0	R 2
4070.772	4071.922	K					24558.43			
4070.941	4072.091		0.3	0.2			24557.41			
4071.065	4072.215		0.3	0.3			24556.66			
4071.150	4072.300		0.2	0.2			24556.15			
4071.248	4072.397	GK	1.0	1.5	0.5	1.78	24555.56	3E 2	2B 1	Q 1
4071.346	4072.495	GK	0.2				24554.97			
4071.414	4072.563	GK	0.7	0.6	0.3		24554.56	4D 0	2B 2	R 3
4071.565	4072.714		0.1	0.2d			24553.65	4f 3	2c 2	R 2
4071.666	4072.815	GK	0.7	0.7		1.07	24553.04	4e 1	2c 0	Q 4
4071.666	4072.815	GK	0.7	0.7		1.07	24553.04	4D 0	2B 2	R 1
4071.773	4072.923		0.2	0.2			24552.39			
4071.936	4073.086		0.2	0.2			24551.41			
4072.069	4073.219		0.2	0.2			24550.61			
4072.193	4073.343		0.2	0.3	0.1		24549.86	5c 0	2a 0	P 4
4072.193	4073.343		0.2	0.3	0.1		24549.86	4g 1	2a 0	P 6
4072.351	4073.501	G	0.2	0.2d			24548.91			
4072.445	4073.595		0.0	0.0			24548.34			
4072.543	4073.693	G	0.4	0.4	0.2		24547.75			
4072.658	4073.808		0.2				24547.06			
4072.785	4073.935		0.3	0.2d			24546.29			
4072.882	4074.032		0.1	0.0			24545.71			
4072.963	4074.113	GK	1.0	1.0	0.6	1.71	24545.22	4D 0	2B 2	R 2
4073.067	4074.218		0.3	0.2			24544.59			
4073.170	4074.320						24543.97			
4073.265	4074.415		0.1	0.2			24543.40			
4073.378	4074.528		0.2	0.2			24542.72			
4073.419	4074.569						24542.47			
4073.481	4074.631			0.1			24542.10			
4073.560	4074.711						24541.62			
4073.635	4074.785			0.2			24541.17			
4073.713	4074.863		0.3b	0.3			24540.70			
4073.796	4074.946						24540.20			
4073.892	4075.043	K	0.4	0.4			24539.62	7c 0	2a 1	Q 1
4074.002	4075.152			0.3			24538.96			
4074.102	4075.252	GK	1.0	1.5	0.9	1.80	24538.36	3D 2	2B 1	R 4
4074.209	4075.360		0.3	0.3			24537.71			
4074.294	4075.445						24537.20			
4074.369	4075.519		0.1	0.1d			24536.75			
4074.462	4075.612		0.4	0.4	0.3		24536.19			
4074.523	4075.674						24535.82			
4074.616	4075.767	G	0.4	0.4		0.68	24535.26	3A 2	2B 2	R 0
4074.749	4075.900		0.4	0.4	0.3		24534.46			
4074.892	4076.043		0.2	0.3			24533.60	5c 0	2a 0	P 4
4074.892	4076.043		0.2	0.3			24533.60	4e 2	2c 1	R 1
4075.010	4076.161	G	0.3	0.3		0.46	24532.89	4e 2	2c 1	Q 4
4075.131	4076.282						24532.16			
4075.327	4076.478		0.2	0.2			24530.98			

λ_{air}	λ_{vac}	Ref.	I_1	I_2	I_3	I_4	ν	Upper		Lower		Br.	
4075.547	4076.697		0.0				24529.66	4e	2	2c	1	Q	3
4075.593	4076.744						24529.38						
4075.731	4076.882		0.3	0.3			24528.55	4e	2	2c	1	Q	5
4075.889	4077.040	G	0.4	0.5	0.3	0.20	24527.60	4e	1	2c	0	P	3
4076.000	4077.151						24526.93						
4076.126	4077.277	G	0.2	0.3			24526.17						
4076.233	4077.384			0.1d			24525.53						
4076.376	4077.527			0.2d			24524.67						
4076.512	4077.663						24523.85						
4076.627	4077.778			0.2			24523.16	4e	2	2c	1	Q	2
4076.796	4077.948	G	0.3	0.3			24522.14						
4076.893	4078.044			0.0			24521.56						
4076.991	4078.142	GK	0.4	0.4	0.4	0.11	24520.97	3E	1	2B	0	Q11	
4077.294	4078.445						24519.15						
4077.348	4078.500			0.2d			24518.82						
4077.472	4078.623		0.2	0.0d			24518.08	4e	1	2c	0	P	5
4077.608	4078.759						24517.26						
4077.696	4078.847						24516.73						
4077.784	4078.936		0.2	0.1			24516.20						
4077.824	4078.975	K		0.1			24515.96	4e	2	2c	1	R	2
4077.976	4079.127		0.2	0.1			24515.05						
4078.075	4079.227			0.0			24514.45	4e	3	2c	2	R	4
4078.168	4079.320	GK	0.8	0.7	0.2d	0.89	24513.89	5c	1	2a	1	Q	1
4078.168	4079.320	GK	0.8	0.7	0.2d	0.89	24513.89	X	2	2B	2	P	5
4078.168	4079.320	GK	0.8	0.7	0.2d	0.89	24513.89	7c	0	2a	1	Q	3
4078.282	4079.433		0.2	0.1			24513.21						
4078.391	4079.543		0.2	0.2			24512.55						
4078.495	4079.646	G	0.6	0.5			24511.93						
4078.619	4079.771						24511.18						
4078.658	4079.809		0.4	0.3			24510.95						
4078.747	4079.899		0.4	0.2			24510.41	4e	2	2c	1	Q	1
4078.842	4079.994	GK	1.0b	1.0b	1.0v	2.18	24509.84	Z	2	2B	1	R	0
4078.957	4080.109	GK	0.5	0.8			24509.15	3D	2	2B	1	R	5
4079.030	4080.182		0.3				24508.71						
4079.129	4080.280						24508.12						
4079.195	4080.347			0.2			24507.72						
4079.340	4080.492		0.0	0.2			24506.85						
4079.435	4080.587	G					24506.28						
4079.548	4080.700		0.1	0.2b			24505.60						
4079.661	4080.813			0.1			24504.92						
4079.829	4080.981	K		0.1			24503.91	4d	1	2c	0	Q	2
4079.971	4081.123			0.2			24503.06	4d	1	2c	0	P	2
4080.093	4081.245	GK	0.5	0.5		0.66	24502.33	4D	0	2B	2	P	2
4080.207	4081.359		0.2	0.2			24501.64						
4080.339	4081.491		0.3	0.2			24500.85						
4080.416	4081.568	G	0.1	0.3		0.11	24500.39						
4080.491	4081.643			0.1			24499.94						
4080.579	4081.731			0.0			24499.41						
4080.674	4081.826	GK	0.4	0.5		0.46	24498.84	5c	1	2a	1	Q	2
4080.774	4081.926			0.2			24498.24						
4080.869	4082.021		0.3	0.3			24497.67						
4081.019	4082.171			0.2			24496.77	4e	2	2c	1	Q	1
4081.135	4082.288			0.1			24496.07						
4081.234	4082.386			0.2	0.2		24495.48	T	0	2B	2	P	1
4081.340	4082.492			0.2			24494.84						
4081.402	4082.554			0.1			24494.47						
4081.482	4082.634	GK	0.8	0.9	0.8	0.88	24493.99	3D	2	2B	1	R	6
4081.583	4082.736	G	0.2	0.1			24493.38						
4081.707	4082.859		0.1	0.1d			24492.64	3F	3	2B	3	P	4
4081.707	4082.859		0.1	0.1d			24492.64	4e	2	2c	1	R	3

λ_{air}	λ_{vac}	Ref.	I_1	I_2	I_3	I_4	ν	Upper	Lower	Br.
4081.923	4083.076			0.1d			24491.34	4b 2	2a 0	P 3
4082.000	4083.153		0.2b				24490.88			
4082.098	4083.251		0.1				24490.29			
4082.165	4083.318						24489.89			
4082.203	4083.356		0.4	0.3			24489.66	4f 3	2c 2	Q 2
4082.292	4083.444		0.2	0.2			24489.13			
4082.385	4083.538	GK	1.0	1.0	0.4	1.88	24488.57	Z 2	2B 1	R 1
4082.487	4083.639		0.4	0.2			24487.96			
4082.594	4083.746						24487.32	4d 1	2c 0	R 3
4082.689	4083.841		0.2	0.1d	0.1		24486.75			
4082.732	4083.885						24486.49			
4082.782	4083.935						24486.19			
4082.914	4084.066		0.4	0.4	0.2		24485.40			
4082.972	4084.125						24485.05			
4083.054	4084.207						24484.56			
4083.202	4084.355		0.1	0.2			24483.67			
4083.254	4084.407						24483.36			
4083.339	4084.492	GK	0.5	0.4		0.30	24482.85			
4083.454	4084.607						24482.16			
4083.532	4084.685		0.1	0.2			24481.69			
4083.661	4084.814		0.3	0.3			24480.92			
4083.756	4084.909			0.1			24480.35			
4083.846	4084.999	GK	0.6	0.7	0.3	1.04	24479.81	3F 1	2B 0	Q 3
4083.970	4085.123	K	0.6	0.6			24479.07	3F 1	2B 0	R 1
4084.038	4085.191	G	0.4	0.6	0.4		24478.66			
4084.143	4085.296						24478.03			
4084.213	4085.366						24477.61			
4084.317	4085.470			0.0			24476.99			
4084.407	4085.560		0.2	0.2			24476.45			
4084.502	4085.655	GK	0.5	0.7	0.2	0.73	24475.88	T 0	2B 2	P 2
4084.502	4085.655	GK	0.5	0.7	0.2	0.73	24475.88	5c 1	2a 1	Q 3
4084.604	4085.757		0.3	0.3			24475.27	3A 2	2B 2	P 6
4084.694	4085.847						24474.73			
4084.827	4085.980		0.2	0.2			24473.93			
4084.922	4086.076						24473.36			
4085.008	4086.161	GK	0.9	0.9	0.3d	1.30	24472.85	4c 1	2a 0	Q 1
4085.114	4086.268	1		0.1			24472.21			
4085.161	4086.314						24471.93	4f 2	2c 1	P 3
4085.241	4086.395	GK	0.9	0.9	0.4	1.68	24471.45	3D 2	2B 1	P 2
4085.336	4086.490	G	0.6	0.6	0.7		24470.88	4e 1	2c 0	Q 5
4085.448	4086.602						24470.21			
4085.528	4086.682						24469.73			
4085.629	4086.782						24469.13			
4085.710	4086.864		0.3	0.3			24468.64			
4085.832	4086.986		0.1	0.1			24467.91			
4085.978	4087.131		0.1				24467.04			
4086.096	4087.250		0.2	0.1			24466.33			
4086.288	4087.442	G	0.4	0.3			24465.18			
4086.390	4087.544						24464.57			
4086.474	4087.627		0.3	0.4	0.2		24464.07	4e 2	2c 1	R 4
4086.584	4087.738		0.1				24463.41			
4086.711	4087.865		0.3	0.3	0.3		24462.65			
4086.840	4087.993						24461.88			
4086.895	4088.048		0.3	0.2			24461.55			
4087.007	4088.160		0.4	0.3	0.2d		24460.88			
4087.142	4088.296		0.1				24460.07			
4087.281	4088.434		0.4	0.3			24459.24	4f 3	2c 2	R 1
4087.372	4088.526		0.3	0.2			24458.69	3A 2	2B 2	P 1
4087.468	4088.622		0.2	0.2			24458.12			
4087.568	4088.722		0.5	0.4			24457.52			

λ_{air}	λ_{vac}	Ref.	I_1	I_2	I_3	I_4	ν	Upper	Lower	Br.
4087.657	4088.811		0.4	0.3			24456.99			
4087.754	4088.908	GK	1.5	1.5	0.8	2.29	24456.41	Z 2	2B 1	R 2
4087.856	4089.010		0.5	0.3			24455.80			
4087.961	4089.115						24455.17			
4088.026	4089.180		0.2	0.1d			24454.78			
4088.095	4089.249						24454.37			
4088.148	4089.302		0.0	0.1d			24454.05			
4088.292	4089.446			0.2d			24453.19			
4088.461	4089.615		0.1				24452.18	4e 3	2c 2	R 2
4088.583	4089.737			0.1d			24451.45			
4088.700	4089.854	GK	0.6	0.7	0.2	0.78	24450.75	4c 1	2a 0	Q 2
4088.799	4089.953			0.1			24450.16	4e 1	2c 0	P 6
4088.911	4090.065	GK	0.7	0.7	0.2b	0.73	24449.49			
4089.009	4090.164		0.1	0.1			24448.90			
4089.100	4090.254		0.2				24448.36			
4089.200	4090.354		0.1	0.1			24447.76			
4089.292	4090.446	GK	0.3	0.5		0.66	24447.21	4E 1	2B 4	Q 4
4089.489	4090.644		0.0	0.1d			24446.03			
4089.652	4090.806	G	0.2	0.4	0.2d		24445.06	5c 1	2a 1	Q 4
4089.782	4090.937		0.3	0.4			24444.28			
4090.000	4091.154			0.1			24442.98			
4090.190	4091.345			0.0			24441.84	3A 2	2B 2	P 3
4090.415	4091.569	G	0.2	0.2			24440.50	T 0	2B 2	P 3
4090.572	4091.727			0.1			24439.56			
4090.724	4091.879	GK	0.5	0.5		0.51	24438.65	3A 2	2B 2	P 4
4090.831	4091.986						24438.01			
4090.930	4092.085			0.1			24437.42			
4091.001	4092.155		0.2	0.2			24437.00	3A 2	2B 2	P 5
4091.079	4092.234	GK	0.3	0.4		0.53	24436.53	4E 1	2B 4	Q 1
4091.203	4092.358						24435.79			
4091.361	4092.515			0.2			24434.85			
4091.451	4092.606		0.1	0.2	0.1		24434.31			
4091.540	4092.695			0.2			24433.78			
4091.637	4092.792	GK	0.5	0.6	0.4	0.53	24433.20	5c 0	2a 0	P 5
4091.739	4092.894			0.1			24432.59			
4091.834	4092.989		0.2	0.2d			24432.02	4f 2	2c 1	P 5
4092.017	4093.172			0.2			24430.93			
4092.122	4093.278						24430.30			
4092.210	4093.365	G	0.4	0.4		0.34	24429.78	3A 2	2B 2	P 2
4092.337	4093.492		0.1	0.1			24429.02			
4092.551	4093.707		0.1	0.2d			24427.74			
4092.719	4093.874			0.0			24426.74			
4092.823	4093.978	GK	0.6	0.5	0.3		24426.12			
4092.953	4094.109	1	0.3	0.3			24425.34			
4093.191	4094.347			0.2			24423.92			
4093.356	4094.511			0.2	0.1		24422.94			
4093.500	4094.655		0.2	0.2			24422.08			
4093.632	4094.788			0.2			24421.29			
4093.711	4094.867		0.1				24420.82			
4093.802	4094.957	GK	0.6	0.5	0.2d	0.57	24420.28	4c 6	2a 4	Q 1
4093.909	4095.064		0.2				24419.64			
4094.021	4095.177		0.1				24418.97			
4094.105	4095.261		0.3	0.5	0.2		24418.47	U 1	2B 0	R 4
4094.105	4095.261		0.3	0.5	0.2		24418.47	4e 2	2c 1	Q 2
4094.212	4095.368	GK	0.6	0.7	0.3	0.97	24417.83	4c 1	2a 0	Q 3
4094.308	4095.464		0.1				24417.26	T 0	2B 2	P 4
4094.407	4095.563						24416.67			
4094.611	4095.767			0.1			24415.45			
4094.749	4095.905	G	0.3	0.4		0.52	24414.63	4E 1	2B 4	Q 3
4094.848	4096.004		0.0				24414.04			

λ_{air}	λ_{vac}	Ref.	I_1	I_2	I_3	I_4	ν	Upper	Lower	Br.
4094.915	4096.071						24413.64			
4095.135	4096.291		0.0	0.1d			24412.33			
4095.351	4096.507		0.2	0.2			24411.04	4d 1	2c 0	R 4
4095.351	4096.507		0.2	0.2			24411.04	4f 3	2c 2	P 5
4095.445	4096.601						24410.48			
4095.536	4096.692	GK	0.8	0.9	0.4	1.39	24409.94	Z 2	2B 1	R 3
4095.645	4096.801		0.2	0.2			24409.29			
4095.799	4096.955		0.2	0.3d			24408.37	3F 1	2B 0	P 5
4095.875	4097.031						24407.92			
4095.964	4097.120	GK	0.7	0.7	0.3		24407.39	4E 1	2B 4	Q 2
4096.083	4097.239	GK	0.7	0.7	0.3	1.29	24406.68	3D 2	2B 1	P 3
4096.083	4097.239	GK	0.7	0.7	0.3	1.29	24406.68	5c 1	2a 1	Q 5
4096.187	4097.343		0.1				24406.06			
4096.252	4097.408		0.2	0.2			24405.67			
4096.417	4097.573		0.4	0.2			24404.69			
4096.533	4097.689			0.1			24404.00			
4096.628	4097.785	G	0.6	0.5	0.3		24403.43	5c 1	2a 1	P 2
4096.758	4097.914		0.1	0.2			24402.66			
4096.835	4097.991						24402.20			
4097.035	4098.191		0.2	0.3d			24401.01			
4097.067	4098.223		0.2				24400.82			
4097.152	4098.309		0.1	0.1			24400.31			
4097.273	4098.430		0.3	0.3			24399.59	Z 2	2B 1	P 2
4097.355	4098.512			0.1			24399.10			
4097.441	4098.597	GK	0.9	0.9	0.3	1.51	24398.59	3F 1	2B 0	Q 2
4097.555	4098.712	GK	0.6	0.6	0.8		24397.91	3E 1	2B 0	Q10
4097.666	4098.823		0.2	0.2			24397.25			
4097.822	4098.979		0.1	0.1			24396.32			
4097.978	4099.135		0.4	0.5	0.3		24395.39	4e 2	2c 1	P 2
4098.081	4099.238		0.3	0.4			24394.78	4f 3	2c 2	Q 4
4098.081	4099.238		0.3	0.4			24394.78	4c 6	2a 4	Q 2
4098.430	4099.587			0.1d			24392.70	4e 1	2c 0	P 4
4098.551	4099.708						24391.98			
4098.632	4099.789	G	0.4	0.3			24391.50	4f 3	2c 2	Q 3
4098.990	4100.147			0.2d			24389.37			
4099.155	4100.312			0.1			24388.39			
4099.380	4100.537						24387.05			
4099.481	4100.638		0.2	0.2			24386.45	4e 1	2c 0	Q 6
4099.518	4100.675						24386.23	4e 2	2c 1	R 5
4099.585	4100.742			0.1			24385.83			
4099.693	4100.850		0.3b	0.1	0.3b		24385.19			
4099.983	4101.141			0.1			24383.46			
4100.369	4101.526		0.2				24381.17	4e 3	2c 2	R 2
4100.369	4101.526		0.2				24381.17	4e 2	2c 1	P 2
4100.644	4101.802			0.1d			24379.53			
4100.787	4101.945		0.1				24378.68	4d 1	2c 0	Q 3
4100.872	4102.029		0.3	0.3			24378.18	4D 0	2B 2	P 3
4100.998	4102.155		0.1b	0.1			24377.43			
4101.285	4102.443		0.3b				24375.72			
4101.380	4102.537						24375.16			
4101.529	4102.687		0.3	0.3			24374.27	4c 1	2a 0	Q 4
4101.615	4102.773			0.2			24373.76			
4101.736	4102.894	GK	1.5b	0.9b	1.5b		24373.04			
4102.081	4103.239			0.0d			24370.99			
4102.144	4103.301						24370.62			
4102.363	4103.520	K					24369.32			
4102.428	4103.586		0.3	0.4	0.2d		24368.93	W 2	2B 2	R 4
4102.489	4103.647						24368.57			
4102.595	4103.753						24367.94			
4102.846	4104.004			0.1			24366.45			

λ_{air}	λ_{vac}	Ref.	I_1	I_2	I_3	I_4	ν	Upper	Lower	Br.
4103.341	4104.499			0.1			24363.51	4f 3	2c 2	P 4
4103.501	4104.659						24362.56			
4103.614	4104.772		0.3	0.3			24361.89		2B 6	Q 2
4103.779	4104.937		0.3b	0.2d	0.4b		24360.91			
4103.925	4105.084			0.0d			24360.04			
4104.099	4105.257			0.1			24359.01			
4104.240	4105.399	K	0.3b	0.3d			24358.17	4c 6	2a 4	Q 3
4104.594	4105.753			0.0d			24356.07			
4104.867	4106.026		0.2	0.2	0.2d		24354.45			
4105.108	4106.267		0.0				24353.02			
4105.183	4106.341			0.1d			24352.58			
4105.486	4106.645		0.3	0.3			24350.78	U 1	2B 0	R 3
4105.486	4106.645		0.3	0.3			24350.78	4e 2	2c 1	P 3
4105.827	4106.985		0.3	0.2	0.2b		24348.76			
4106.044	4107.203		0.2	0.2			24347.47			
4106.145	4107.304			0.2			24346.87			
4106.233	4107.392	GK	0.9	0.9	0.5	1.37	24346.35	Z 2	2B 1	R 4
4106.338	4107.496		0.3	0.2			24345.73			
4106.446	4107.604		0.1				24345.09			
4106.530	4107.689		0.1				24344.59	4e 2	2c 1	Q 3
4106.807	4107.966		0.2				24342.95			
4107.114	4108.273			0.0			24341.13			
4107.208	4108.367	K					24340.57	7c 0	2a 1	P 3
4107.267	4108.426			0.2			24340.22			
4107.335	4108.494		0.1	0.1			24339.82			
4107.421	4108.580			0.2			24339.31			
4107.488	4108.647			0.1			24338.91			
4107.623	4108.782		0.2	0.2			24338.11			
4107.709	4108.869			0.0			24337.60			
4107.795	4108.955	GK	0.8	0.9	0.4	1.23	24337.09	3F 1	2B 0	P 4
4107.907	4109.066		0.2	0.1d			24336.43			
4108.086	4109.245		0.1				24335.37			
4108.155	4109.314			0.2			24334.96			
4108.282	4109.441		0.1	0.1			24334.21			
4108.545	4109.704			0.0d			24332.65			
4108.866	4110.025		0.2	0.1d			24330.75	4d 2	2c 1	P 1
4109.011	4110.171		0.1	0.2			24329.89			
4109.124	4110.284		0.2	0.0			24329.22			
4109.219	4110.378		0.0	0.1			24328.66			
4109.307	4110.466	GK	0.9	0.8	0.2	1.31	24328.14	3D 2	2B 1	P 4
4109.413	4110.573		0.2	0.1			24327.51			
4109.506	4110.666	GK	0.5	0.5	0.1	0.80	24326.96	W 2	2B 2	R 3
4109.641	4110.801		0.1	0.2			24326.16	5c 1	2a 1	P 3
4109.641	4110.801		0.1	0.2			24326.16	4f 3	2c 2	P 3
4109.827	4110.987		0.2	0.2			24325.06	U 1	2B 0	R 0
4109.955	4111.115		0.2	0.1			24324.30	4d 1	2c 0	P 3
4110.042	4111.201			0.0			24323.79	U 1	2B 0	R 1
4110.130	4111.289	GK	0.8	0.9	0.3	1.35	24323.27	W 2	2B 2	R 2
4110.233	4111.392		0.2	0.0			24322.66	4e 1	2c 0	Q 7
4110.319	4111.479		0.0				24322.15			
4110.445	4111.605			0.1			24321.40			
4110.598	4111.758	K	0.3	0.4	0.2		24320.50	4c 1	2a 0	Q 5
4110.743	4111.903		0.1	0.2			24319.64			
4110.870	4112.030			0.0			24318.89			
4110.934	4112.094	G	0.5	0.5			24318.51			
4111.049	4112.209	GK	0.6	0.6		0.96	24317.83	W 2	2B 2	R 1
4111.159	4112.319		0.1	0.1			24317.18			
4111.269	4112.429		0.3	0.3			24316.53	U 1	2B 0	R 2
4111.360	4112.520		0.2	0.2d			24315.99			
4111.505	4112.666		0.3	0.1			24315.13			

λ_{air}	λ_{vac}	Ref.	I_1	I_2	I_3	I_4	ν	Upper	Lower	Br.
4111.678	4112.838		0.1				24314.11			
4111.881	4113.041		0.1	0.0			24312.91			
4112.145	4113.305		0.4	0.2d			24311.35			
4112.346	4113.506		0.1				24310.16			
4112.407	4113.567			0.3d			24309.80			
4112.703	4113.863		0.2	0.2			24308.05			
4112.820	4113.980			0.2			24307.36			
4112.936	4114.097			0.1			24306.67			
4113.023	4114.183			0.1			24306.16	4e 3	2c 2	R 1
4113.101	4114.261		0.3	0.3			24305.70	Z 2	2B 1	P 3
4113.248	4114.409		0.3	0.3			24304.83			
4113.343	4114.503		0.2	0.2d			24304.27			
4113.439	4114.600						24303.70			
4113.525	4114.686	GK	0.8	0.8		1.35	24303.19	W 2	2B 2	R 0
4113.525	4114.686	GK	0.8	0.8		1.35	24303.19	4e 2	2c 1	P 4
4113.657	4114.818		0.3	0.2			24302.41	4D 0	2B 2	P 4
4113.729	4114.889		0.4	0.3	0.4		24301.99			
4113.876	4115.037			0.1			24301.12			
4113.964	4115.125		0.3	0.2	0.2		24300.60			
4114.098	4115.258			0.1			24299.81			
4114.159	4115.319			0.2			24299.45			
4114.245	4115.406		0.0				24298.94			
4114.541	4115.702			0.1d			24297.19			
4114.648	4115.809		0.2	0.1			24296.56	4d 2	2c 1	R 1
4114.943	4116.104		0.3	0.3			24294.82			
4115.195	4116.356			0.0			24293.33			
4115.393	4116.554		0.1	0.2			24292.16			
4115.552	4116.714		0.2	0.0d			24291.22			
4115.744	4116.905		0.4	0.3			24290.09	4d 2	2c 1	Q 1
4115.841	4117.002		0.1	0.2			24289.52			
4115.930	4117.092	GK	0.6	0.5	0.5		24288.99	3E 1	2B 0	Q 9
4116.051	4117.212		0.1				24288.28			
4116.110	4117.271		0.1				24287.93			
4116.217	4117.378			0.0			24287.30			
4116.313	4117.475	GK	0.7	0.6	0.2	0.87	24286.73	3F 1	2B 0	P 3
4116.410	4117.571		0.1				24286.16			
4117.007	4118.168			0.0d			24282.64			
4117.158	4118.319		0.4	0.4	0.2		24281.75	4F 0	2B 3	R 3
4117.305	4118.467			0.0			24280.88			
4117.444	4118.606			0.2			24280.06			
4117.505	4118.667		0.0	0.2			24279.70			
4117.626	4118.787			0.1			24278.99			
4117.726	4118.888						24278.40			
4117.807	4118.969			0.1d			24277.92			
4117.919	4119.081		0.1	0.2			24277.26			
4117.994	4119.156			0.1			24276.82			
4118.350	4119.512			0.1d			24274.72			
4118.691	4119.853			0.1d			24272.71			
4118.806	4119.969			0.1	0.3		24272.03			
4118.873	4120.035		0.2	0.2			24271.64			
4119.071	4120.233		0.1	0.2			24270.47			
4119.127	4120.289			0.2			24270.14			
4119.341	4120.503			0.0d			24268.88			
4119.467	4120.629		0.4	0.3			24268.14			
4119.545	4120.707		0.1	0.2	0.2		24267.68			
4119.642	4120.804			0.2			24267.11			
4119.737	4120.899		0.4	0.3	0.1d		24266.55			
4119.883	4121.045			0.1			24265.69	4e 2	2c 1	Q 4
4119.978	4121.140			0.0			24265.13			
4120.176	4121.339			0.2			24263.96			

λ_{air}	λ_{vac}	Ref.	I_1	I_2	I_3	I_4	ν	Upper	Lower	Br.
4120.258	4121.420			0.2			24263.48			
4120.350	4121.512			0.1			24262.94			
4120.465	4121.628		0.3	0.4	0.2d		24262.26			
4120.620	4121.782		0.1	0.2			24261.35			
4120.728	4121.891			0.1			24260.71			
4120.813	4121.976	1	0.3	0.3	0.6		24260.21			
4120.910	4122.073						24259.64			
4120.986	4122.149				0.2		24259.19	4e 1	2c 0	P 5
4121.326	4122.489		0.2	0.2d			24257.19			
4121.418	4122.581			0.1d			24256.65			
4121.651	4122.814			0.1d			24255.28			
4121.787	4122.950		0.2	0.1			24254.48			
4121.870	4123.033			0.0			24253.99			
4122.054	4123.217		0.2	0.2d			24252.91			
4122.108	4123.271			0.1			24252.59			
4122.159	4123.322						24252.29			
4122.390	4123.553			0.1d			24250.93	4e 2	2c 1	P 5
4122.689	4123.852			0.2			24249.17	4d 1	2c 0	Q 4
4122.745	4123.909			0.2			24248.84	4e 3	2c 2	Q 5
4122.885	4124.048			0.2d			24248.02			
4123.041	4124.205		0.1	0.3			24247.10	4e 2	2c 1	P 3
4123.140	4124.303						24246.52			
4123.264	4124.427	GK	0.5	0.5	0.3	0.30	24245.79	4D 0	2B 2	P 5
4123.397	4124.560						24245.01			
4123.487	4124.650						24244.48			
4123.563	4124.727	K		0.1d			24244.03			
4123.671	4124.834						24243.40			
4123.761	4124.924	GK	0.8	0.6	0.2	0.93	24242.87	W 2	2B 2	P 1
4123.761	4124.924	GK	0.8	0.6	0.2	0.93	24242.87	3E 1	2B 0	R 3
4123.868	4125.031		0.1	0.0d			24242.24	5c 1	2a 1	P 4
4124.070	4125.234	G	0.4	0.4			24241.05	3D 2	2B 1	P 5
4124.227	4125.390		0.1	0.0d			24240.13			
4124.305	4125.469			0.0d			24239.67			
4124.450	4125.613		0.2	0.3			24238.82			
4124.645	4125.809			0.0d			24237.67			
4124.795	4125.959			0.0d			24236.79			
4124.988	4126.151		0.3	0.3			24235.66	4e 3	2c 2	Q 4
4124.988	4126.151		0.3	0.3			24235.66	4d 2	2c 1	R 2
4125.108	4126.272						24234.95			
4125.178	4126.342			0.2d			24234.54			
4125.360	4126.524			0.2			24233.47			
4125.519	4126.682			0.1d			24232.54			
4125.686	4126.849		0.3	0.3d	0.2		24231.56			
4125.781	4126.945			0.1			24231.00			
4125.888	4127.052			0.1			24230.37			
4126.012	4127.176						24229.64			
4126.406	4127.570			0.1			24227.33			
4126.455	4127.619						24227.04			
4126.656	4127.820		0.1	0.0			24225.86	4D 0	2B 2	P 8
4126.782	4127.947			0.0			24225.12			
4127.041	4128.206			0.1b			24223.60			
4127.293	4128.458		0.3	0.2d			24222.12			
4127.471	4128.635			0.1			24221.08			
4127.585	4128.749			0.1			24220.41			
4127.725	4128.889			0.1d			24219.59			
4127.948	4129.112			0.1			24218.28			
4128.072	4129.237			0.0			24217.55	4e 3	2c 2	Q 3
4128.158	4129.322			0.3	0.1		24217.05			
4128.555	4129.719		0.2	0.0d			24214.72			
4128.602	4129.767						24214.44			

λ_{air}	λ_{vac}	Ref.	I_1	I_2	I_3	I_4	ν	Upper	Lower	Br.
4128.693	4129.858		0.3	0.3			24213.91	4D 0	2B 2	P 6
4128.906	4130.071		0.1	0.2			24212.66			
4128.976	4130.141		0.2	0.2			24212.25			
4129.078	4130.243			0.1			24211.65			
4129.251	4130.415						24210.64			
4129.390	4130.555		0.2	0.2			24209.82			
4129.453	4130.618						24209.45			
4129.546	4130.711	GK	0.7	0.7	0.2	0.97	24208.91	U 1	2B 0	P 2
4129.650	4130.815			0.1d			24208.30			
4129.773	4130.937			0.2			24207.58			
4129.837	4131.002		0.1	0.3			24207.20	4D 0	2B 2	P 7
4129.936	4131.101			0.1			24206.62			
4130.030	4131.195	GK	0.7	0.6		0.59	24206.07			
4130.143	4131.308		0.1	0.0			24205.41			
4130.221	4131.386			0.0d			24204.95			
4130.373	4131.538			0.1d			24204.06			
4130.435	4131.600			0.0			24203.70			
4130.547	4131.712			0.2			24203.04			
4130.617	4131.782		0.2	0.3	0.2		24202.63	Z 2	2B 1	P 4
4130.745	4131.910			0.2			24201.88			
4130.885	4132.050		0.1	0.2	0.2d		24201.06			
4131.010	4132.175			0.2			24200.33			
4131.122	4132.288			0.3			24199.67			
4131.276	4132.441		0.3	0.3			24198.77	5c 2	2a 2	Q 1
4131.365	4132.530			0.2			24198.25			
4131.459	4132.624	GK	0.9	0.9	0.3	1.43	24197.70	W 2	2B 2	P 2
4131.570	4132.735		0.2	0.1			24197.05	4e 2	2c 1	P 6
4131.684	4132.850			0.0			24196.38	4e 3	2c 2	Q 2
4131.793	4132.959		0.2	0.2			24195.74			
4131.875	4133.041			0.0			24195.26			
4131.966	4133.131	GK	0.8	0.9	1.5b	0.95	24194.73	3E 1	2B 0	Q 8
4132.067	4133.232		0.3	0.2			24194.14			
4132.193	4133.359			0.1			24193.40			
4132.318	4133.483			0.2d			24192.67			
4132.485	4133.651		0.1	0.2d			24191.69			
4132.550	4133.716						24191.31			
4132.665	4133.830			0.2			24190.64			
4132.699	4133.864	G	0.3	0.2		0.45	24190.44			
4132.777	4133.943		0.3	0.3	0.3		24189.98			
4132.895	4134.061		0.1				24189.29			
4132.945	4134.111			0.1			24189.00			
4133.150	4134.316			0.1			24187.80			
4133.208	4134.374		0.0				24187.46			
4133.266	4134.432			0.0			24187.12	4f 3	2c 2	P 3
4133.406	4134.572	GK	0.5	0.5	0.2		24186.30			
4133.517	4134.683		0.2	0.2			24185.65			
4133.637	4134.803						24184.95			
4133.724	4134.890		0.3	0.3			24184.44			
4133.820	4134.986		0.2	0.2			24183.88	4e 2	2c 1	Q 5
4133.910	4135.076		0.1	0.0			24183.35			
4133.997	4135.164	GK	0.9	0.9	0.5	1.40	24182.84	3E 1	2B 0	R 2
4134.104	4135.270		0.2	0.1			24182.22			
4134.163	4135.329			0.2			24181.87			
4134.273	4135.439		0.1	0.3			24181.23	4d 2	2c 1	Q 2
4134.531	4135.697			0.1			24179.72			
4134.639	4135.805		0.2	0.3	0.3		24179.09			
4134.753	4135.920			0.2			24178.42			
4134.835	4136.002		0.1	0.1d			24177.94			
4134.914	4136.080			0.3			24177.48			
4135.010	4136.176	G	0.5	0.6	0.5		24176.92			

λ_{air}	λ_{vac}	Ref.	I_1	I_2	I_3	I_4	ν	Upper	Lower	Br.
4135.143	4136.310		0.1				24176.14			
4135.270	4136.436			0.2d	0.1		24175.40	4d 2	2c 1	P 2
4135.530	4136.696		0.2	0.1			24173.88			
4135.682	4136.849		0.1	0.2			24172.99			
4136.079	4137.246			0.2			24170.67			
4136.225	4137.391		0.3	0.4	0.4		24169.82			
4136.346	4137.513			0.0			24169.11			
4136.503	4137.670		0.2	0.2d			24168.19			
4136.635	4137.802			0.2d			24167.42	4d 2	2c 1	R 3
4136.875	4138.042			0.2d			24166.02			
4137.072	4138.239			0.0			24164.87			
4137.176	4138.343			0.1			24164.26			
4137.312	4138.478			0.2			24163.47			
4137.438	4138.605	G	0.4	0.4	0.3		24162.73			
4137.548	4138.715			0.1			24162.09			
4137.645	4138.812			0.3	0.1		24161.52			
4137.883	4139.051			0.1			24160.13	5c 1	2a 1	P 5
4137.991	4139.159			0.2			24159.50			
4138.074	4139.241			0.0			24159.02			
4138.197	4139.364	G	0.5	0.5	0.3	0.32	24158.30	5c 2	2a 2	Q 3
4138.331	4139.498			0.2			24157.52			
4138.498	4139.666		0.1	0.3			24156.54			
4138.584	4139.751			0.0			24156.04			
4138.954	4140.122			0.0			24153.88			
4139.067	4140.235		0.3	0.4	0.2d		24153.22			
4139.285	4140.452			0.0			24151.95			
4139.386	4140.554			0.2			24151.36			
4139.475	4140.643	GK	0.5	0.5	0.3	0.43	24150.84			
4139.561	4140.728		0.4	0.5	0.4		24150.34			
4139.693	4140.860		0.2	0.4			24149.57	3D 2	2B 1	P 6
4139.827	4140.994	G	0.3	0.4			24148.79			
4139.969	4141.137			0.0			24147.96			
4140.158	4141.325			0.1			24146.86			
4140.243	4141.411			0.0			24146.36			
4140.367	4141.534		│			24145.64			
4140.574	4141.742			0.1d			24144.43			
4140.744	4141.912		0.2	0.3			24143.44			
4140.826	4141.994	GK	0.5	0.6	0.2b	0.88	24142.96	W 2	2B 2	P 3
4140.929	4142.097						24142.36			
4141.065	4142.233			0.1			24141.57			
4141.164	4142.332		0.5	0.0			24140.99	4d 1	2c 0	P 4
4141.543	4142.711			0.2			24138.78			
4141.737	4142.905			0.1			24137.65			
4141.835	4143.003			0.1			24137.08			
4141.953	4143.122		0.0	0.1			24136.39			
4142.041	4143.209	G	0.3	0.6	0.5	0.20	24135.88			
4142.309	4143.477			0.3	0.2d		24134.32			
4142.410	4143.578			0.3	0.2		24133.73			
4142.587	4143.755			0.2			24132.70			
4142.719	4143.887		0.0	0.2			24131.93			
4142.801	4143.970	GK	0.4	0.6	0.2	0.77	24131.45	U 1	2B 0	P 3
4142.928	4144.097		0.0	0.0			24130.71			
4143.059	4144.227			0.3	0.2		24129.95			
4143.179	4144.348		0.0	0.1			24129.25	4f 3	2c 2	P 5
4143.289	4144.458	GK	0.4	0.5		0.66	24128.61	Y 2	2B 2	R 0
4143.337	4144.506				0.1		24128.33			
4143.478	4144.647			0.1			24127.51			
4143.619	4144.787		0.0	0.4	0.2		24126.69			
4143.756	4144.925	1	0.4	0.3	0.3		24125.89			
4143.868	4145.036	GK	0.3	0.6	0.3	0.63	24125.24	5c 2	2a 2	Q 4

λ_air	λ_vac	Ref.	I₁	I₂	I₃	I₄	ν	Upper	Lower	Br.
4143.973	4145.141		0.0d				24124.63			
4144.091	4145.260		0.0				24123.94			
4144.165	4145.334		0.0	0.2			24123.51			
4144.265	4145.433			0.0			24122.93			
4144.340	4145.509		0.0	0.4	0.2		24122.49	4e 1	2c 0	P 6
4144.471	4145.640			0.1			24121.73			
4144.564	4145.732			0.2			24121.19			
4144.686	4145.854		0.0	0.3	0.3		24120.48			
4144.875	4146.044			0.1			24119.38			
4144.954	4146.123		0.0	0.2			24118.92			
4145.119	4146.288		0.0				24117.96	4e 2	2c 1	P 4
4145.193	4146.362		0.0d				24117.53			
4145.277	4146.446		0.0	0.1			24117.04	4d 1	2c 0	Q 5
4145.383	4146.552		0.2	0.3			24116.42			
4145.438	4146.607						24116.10			
4145.555	4146.724	GK	0.3	0.9	0.9	0.80	24115.42	3E 1	2B 0	Q 7
4145.657	4146.826		0.0	0.1			24114.83			
4145.885	4147.055			0.2			24113.50			
4145.947	4147.116			0.0			24113.14			
4146.062	4147.232		0.2	0.2	0.2d		24112.47			
4146.162	4147.331		0.2	0.0			24111.89			
4146.245	4147.414	GK	0.7	0.9	0.3	1.08	24111.41	3E 1	2B 0	R 1
4146.360	4147.529		0.2	0.1			24110.74			
4146.463	4147.633		0.0				24110.14			
4146.618	4147.787			0.2d			24109.24			
4146.826	4147.995		0.0	0.0			24108.03			
4146.929	4148.099	GK	0.4	0.5	0.2	0.54	24107.43	Y 2	2B 2	R 2
4147.041	4148.211		0.0				24106.78			
4147.110	4148.279		0.2	0.4			24106.38	4E 2	2B 6	Q 2
4147.335	4148.505			0.0			24105.07	3E 3	2B 3	Q 8
4147.435	4148.605		0.0				24104.49			
4147.810	4148.980		0.0				24102.31			
4147.864	4149.033		0.0	0.2			24102.00			
4148.222	4149.391			0.2			24099.92			
4148.320	4149.490						24099.35			
4148.377	4149.546			0.2			24099.02			
4148.490	4149.660						24098.36			
4148.929	4150.099		0.1	0.4			24095.81			
4149.124	4150.294			0.1			24094.68			
4149.213	4150.383			0.1			24094.16			
4149.392	4150.562			0.1			24093.12			
4149.921	4151.091			0.1			24090.05			
4150.192	4151.362				0.1		24088.48			
4150.281	4151.452			0.2			24087.96			
4150.440	4151.610		0.0	0.3	0.2b		24087.04	Z 2	2B 1	P 5
4150.617	4151.788			0.1			24086.01			
4150.834	4152.005			0.0			24084.75			
4151.050	4152.220			0.3	0.1d		24083.50	5c 2	2a 2	Q 5
4151.412	4152.582			0.1			24081.40			
4151.486	4152.657	K	0.0	0.0			24080.97			
4151.579	4152.750	GK	0.5	0.8	0.3d	1.09	24080.43	W 2	2B 2	P 4
4151.679	4152.850		0.2	0.1			24079.85			
4151.934	4153.105			0.0d			24078.37			
4152.053	4153.224						24077.68			
4152.114	4153.284		0.0				24077.33			
4152.193	4153.364	G	0.3	0.4		0.20	24076.87	Y 2	2B 2	P 1
41 2.291	4153.462			0.0			24076.30			
4152.459	4153.629		0.0d	0.2			24075.33			
4152.647	4153.818	GK	0.4	0.9	0.5	0.77	24074.24	3c 5	2a 1	Q 1
4152.769	4153.940		0.0	0.1			24073.53			

λ_{air}	λ_{vac}	Ref.	I_1	I_2	I_3	I_4	ν	Upper	Lower	Br.
4152.855	4154.026		0.0	0.0d			24073.03			
4152.892	4154.062		0.0				24072.82			
4153.005	4154.176			0.3d			24072.16			
4153.083	4154.254			0.0d			24071.71			
4153.168	4154.339		0.1d	0.4d	0.3		24071.22			
4153.269	4154.440			0.2			24070.63			
4153.452	4154.623			0.2			24069.57			
4153.563	4154.734			0.2			24068.93			
4153.652	4154.824			0.2			24068.41			
4153.753	4154.924			0.2			24067.83			
4154.153	4155.324			0.1			24065.51			
4154.357	4155.528		0.0d	0.3			24064.33	4d 2	2c 1	Q 3
4154.554	4155.725			0.2			24063.19			
4154.790	4155.962			0.2d			24061.82	4e 3	2c 2	P 2
4154.939	4156.110						24060.96			
4155.063	4156.234			0.3			24060.24			
4155.156	4156.328		0.0	0.2	0.2b		24059.70	4c 1	2a 0	P 5
4155.498	4156.670			0.1			24057.72			
4155.628	4156.799		0.1	0.2			24056.97	3D 2	2B 1	P 7
4155.714	4156.886		0.0				24056.47			
4155.811	4156.983	GK	0.6	0.7	0.2	1.02	24055.91	X 1	2B 1	R 0
4155.916	4157.088		0.1	0.2			24055.30			
4156.163	4157.335			0.2d			24053.87			
4156.281	4157.453						24053.19			
4156.452	4157.624		0.0	0.3	0.2		24052.20			
4156.635	4157.807	GK	0.4	1.5d	1.5b	1.46	24051.14	3E 1	2B 0	Q 6
4156.635	4157.807	GK	0.4	1.5d	1.5b	1.46	24051.14	4F 0	2B 3	R 2
4156.635	4157.807	GK	0.4	1.5d	1.5b	1.46	24051.14	3F 2	2B 2	R 2
4156.725	4157.897		0.0				24050.62			
4156.775	4157.947		0.2	0.1			24050.33			
4156.870	4158.042	GK	0.7	1.0d	0.7	1.49	24049.78	U 1	2B 0	P 4
4156.996	4158.168		0.2	0.1			24049.05			
4157.084	4158.257		0.1	0.0			24048.54			
4157.183	4158.355	GK	0.5	0.6	0.2	0.81	24047.97	X 1	2B 1	R 1
4157.352	4158.525		0.0	0.2			24046.99			
4157.480	4158.653			0.1			24046.25			
4157.558	4158.730			0.3d	0.2		24045.80			
4157.747	4158.919			0.1d			24044.71			
4157.894	4159.066			0.1d			24043.86			
4158.143	4159.315			0.1d	0.2d		24042.42			
4158.364	4159.536			0.2	0.2		24041.14			
4158.437	4159.609		0.0				24040.72			
4158.556	4159.729		0.0	0.2			24040.03	3F 2	2B 2	Q 4
4158.556	4159.729		0.0	0.2			24040.03	3c 8	2a 3	Q 1
4158.587	4159.760	1			1.5b		24039.85			
4158.710	4159.883			0.3			24039.14			
4158.791	4159.964	K					24038.67	3E 3	2B 3	Q 7
4158.878	4160.050			0.2			24038.17			
4158.951	4160.123		0.1	0.3			24037.75	3c 5	2a 1	Q 2
4159.030	4160.203		0.0				24037.29			
4159.139	4160.312		0.3	0.3d			24036.66			
4159.227	4160.400		0.3				24036.15			
4159.316	4160.488	GK	1.0b	1.0	0.3	1.41	24035.64	4c 2	2a 1	Q 1
4159.419	4160.592		0.4				24035.04			
4159.475	4160.648				0.2		24034.72			
4159.556	4160.729		0.0	0.3	0.3		24034.25			
4159.599	4160.772		0.1				24034.00			
4159.662	4160.835		0.0				24033.64	4e 3	2c 2	P 3
4159.731	4160.904		0.0	0.1			24033.24			
4159.883	4161.056		0.1	0.3d			24032.36			

λ_{air}	λ_{vac}	Ref.	I_1	I_2	I_3	I_4	ν	Upper	Lower	Br.
4159.998	4161.170		0.0	0.1			24031.70			
4160.072	4161.245		0.0	0.1			24031.27			
4160.136	4161.309		0.0				24030.90			
4160.600	4161.773			0.1			24028.22			
4160.664	4161.837				0.2		24027.85			
4160.765	4161.938		0.0	0.2			24027.27			
4160.888	4162.061		0.0				24026.56			
4161.036	4162.210		0.0	0.3			24025.70			
4161.307	4162.480			0.2d			24024.14			
4161.385	4162.558		0.1	0.0			24023.69			
4161.487	4162.660	GK	0.5	0.7	0.2d	0.79	24023.10	Y 2	2B 2	P 2
4161.580	4162.754		0.0				24022.56			
4161.639	4162.813		0.0	0.0			24022.22			
4161.735	4162.908	GK	0.6	0.8	0.3	1.24	24021.67	X 1	2B 1	R 2
4161.845	4163.019		0.2	0.2			24021.03			
4161.946	4163.119	GK	0.8	1.5	0.8	1.61	24020.45	3E 1	2B 0	R 0
4162.053	4163.227		0.2	0.2			24019.83			
4162.137	4163.310		0.0				24019.35			
4162.240	4163.414			0.1			24018.75			
4162.360	4163.534	G	0.1	0.4	0.1	0.43	24018.06	W 2	2B 2	P 5
4162.485	4163.658			0.0			24017.34			
4162.556	4163.729			0.0			24016.93			
4162.831	4164.005			0.0			24015.34			
4162.944	4164.118		0.0				24014.69			
4163.022	4164.196	GK	0.4	0.6	0.2b	0.79	24014.24	4c 2	2a 1	Q 2
4163.123	4164.296		0.0	0.0			24013.66			
4163.218	4164.392			0.2			24013.11			
4163.322	4164.496		0.0				24012.51			
4163.402	4164.576	GK	0.4	0.9	0.6	1.24	24012.05	U 1	2B 0	P 5
4163.513	4164.687		0.1	0.0			24011.41			
4163.605	4164.779	GK	0.5	1.0	1.0	1.55	24010.88	U 1	2B 0	P 6
4163.716	4164.890		0.1	0.2			24010.24	4d 2	2c 1	P 3
4164.021	4165.195		0.0	0.3			24008.48			
4164.174	4165.348	1			0.9		24007.60			
4164.305	4165.479			0.2			24006.84			
4164.448	4165.622	G	0.2	0.5		0.48	24006.02	3E 3	2B 3	R 3
4164.543	4165.717		0.0	0.0			24005.47			
4164.626	4165.800		0.5	0.4			24004.99			
4164.732	4165.906		0.1				24004.38			
4164.795	4165.969			0.4	0.3		24004.02			
4164.901	4166.075			0.2			24003.41			
4165.013	4166.187			0.3			24002.76			
4165.097	4166.271			0.2			24002.28			
4165.190	4166.365	GK	0.3	1.0	0.9	1.21	24001.74	4e 3	2c 2	P 4
4165.190	4166.365	GK	0.3	1.0	0.9	1.21	24001.74	3E 1	2B 0	Q 5
4165.322	4166.497			0.2			24000.98			
4165.471	4166.646			0.2			24000.12			
4165.676	4166.851			0.3			23998.94			
4165.753	4166.927						23998.50			
4165.827	4167.002	G					23998.07			
4165.978	4167.153			0.2			23997.20			
4166.136	4167.311		0.0	0.4			23996.29	W 2	2B 2	P 6
4166.227	4167.401			0.0			23995.77			
4166.388	4167.563			0.0			23994.84			
4166.534	4167.708		0.0d	0.2			23994.00			
4166.610	4167.785			0.1			23993.56			
4166.730	4167.905	GK	0.2	0.5	0.3	0.15	23992.87			
4166.855	4168.030		0.0	0.2			23992.15			
4166.998	4168.172			0.1			23991.33			
4167.128	4168.303			0.3			23990.58			

λ_{air}	λ_{vac}	Ref.	I_1	I_2	I_3	I_4	ν	Upper	Lower	Br.
4167.253	4168.428			0.2			23989.86			
4167.404	4168.579			0.2			23988.99			
4167.472	4168.647		0.0				23988.60			
4167.566	4168.741	GK	0.5	0.8b	0.4	1.00	23988.06	3E 3	2B 3	R 2
4167.651	4168.826		0.0				23987.57			
4167.719	4168.893			0.1			23987.18			
4167.781	4168.956			0.1d			23986.82			
4167.875	4169.050			0.1			23986.28	4e 1	2c 0	P 7
4168.007	4169.182		0.0	0.4	0.3		23985.52	4e 2	2c 1	P 5
4168.183	4169.358			0.1			23984.51	4F 0	2B 3	P 5
4168.341	4169.516			0.1			23983.60			
4168.473	4169.648		0.1	0.5			23982.84	3c 5	2a 1	Q 3
4168.579	4169.754	GK	0.4	0.8	0.3	0.77	23982.23	4c 2	2a 1	Q 3
4168.579	4169.754	GK	0.4	0.8	0.3	0.77	23982.23	3D 2	2B 1	P 8
4168.671	4169.846		0.1				23981.70			
4168.758	4169.933			0.0			23981.20			
4168.895	4170.070		0.0				23980.41			
4168.963	4170.138		0.2	0.1			23980.02			
4169.099	4170.274			0.1			23979.24			
4169.226	4170.401			0.4			23978.51			
4169.354	4170.530			0.2			23977.77			
4169.514	4170.690		0.0	0.2			23976.85			
4169.572	4170.747		0.0				23976.52			
4169.636	4170.811		0.0				23976.15			
4169.709	4170.884	G	0.2	0.4	0.2		23975.73			
4169.874	4171.050			0.1			23974.78	4d 3	2c 2	P 1
4169.959	4171.135						23974.29			
4170.039	4171.215						23973.83			
4170.132	4171.307			0.1			23973.30			
4170.267	4171.443		0.1	0.2			23972.52	4F 0	2B 3	Q 4
4170.417	4171.593			0.1			23971.66			
4170.581	4171.756			0.2			23970.72			
4170.652	4171.828			0.0			23970.31			
4170.725	4171.901			0.1			23969.89	4e 3	2c 2	P 5
4170.835	4172.010		0.0	0.4			23969.26			
4170.901	4172.076			0.3			23968.88			
4170.967	4172.143						23968.50			
4171.017	4172.193			0.2			23968.21			
4171.127	4172.303		0.1	0.5	0.2		23967.58			
4171.209	4172.385		0.1	0.4			23967.11			
4171.304	4172.480	GK	0.7	1.5b	1.5b	2.14	23966.56	3E 1	2B 0	Q 4
4171.409	4172.585		0.2	0.5			23965.96	3E 3	2B 3	Q 6
4171.597	4172.773			0.1			23964.88			
4171.675	4172.851		0.1	0.3			23964.43			
4171.895	4173.071			0.2			23963.17			
4171.969	4173.145						23962.74			
4172.050	4173.226		0.0				23962.28			
4172.121	4173.297		0.0		0.1		23961.87			
4172.279	4173.455						23960.96			
4172.346	4173.522			0.3			23960.58			
4172.422	4173.598		0.0				23960.14			
4172.520	4173.696		0.3	0.4			23959.58			
4172.661	4173.837		0.0	0.2			23958.77			
4172.816	4173.992		0.1	0.2			23957.88			
4172.938	4174.114		0.2	0.4			23957.18	Z 2	2B 1	P 6
4173.082	4174.259			0.3			23956.35			
4173.110	4174.286						23956.19			
4173.208	4174.384			0.2			23955.63	4d 1	2c 0	P 5
4173.385	4174.562		0.0	0.4			23954.61			
4173.500	4174.677			0.1			23953.95			

λ_{air}	λ_{vac}	Ref.	I_1	I_2	I_3	I_4	ν	Upper	Lower	Br.
4173.574	4174.750			0.2			23953.53			
4173.716	4174.893			0.1			23952.71			
4173.807	4174.984	GK	0.1d	0.5	0.5	0.28	23952.19			
4173.901	4175.078			0.1			23951.65			
4174.000	4175.177	GK	0.3	0.4		0.15	23951.08	Y 2	2B 2	P 3
4174.000	4175.177	GK	0.3	0.4		0.15	23951.08	4d 3	2c 2	R 1
4174.093	4175.269			0.1			23950.55			
4174.232	4175.409			0.2			23949.75	3F 2	2B 2	Q 3
4174.318	4175.494	G					23949.26	3F 2	2B 2	R 1
4174.431	4175.608			0.1			23948.61			
4174.574	4175.751			0.2			23947.79			
4174.710	4175.887			0.3d			23947.01			
4174.783	4175.960		0.0				23946.59			
4174.874	4176.051		0.1	0.1			23946.07			
4174.968	4176.145	GK	0.6r	0.6	0.2		23945.53	X 1	2B 1	P 2
4175.067	4176.244		0.2				23944.96			
4175.156	4176.333	GK	0.6	1.5	1.0	1.94	23944.45	3E 1	2B 0	Q 3
4175.266	4176.443		0.1	0.2			23943.82			
4175.338	4176.514		0.3	0.2			23943.41	3E 3	2B 3	R 1
4175.458	4176.635			0.2d			23942.72			
4175.512	4176.689		0.0d				23942.41			
4175.627	4176.804		0.0	0.2			23941.75	4d 3	2c 2	Q 1
4175.720	4176.897		0.0				23941.22			
4175.796	4176.973		0.1	0.1			23940.78			
4175.924	4177.101		0.2	0.4			23940.05	4c 2	2a 1	Q 4
4176.072	4177.249		0.1d	0.2			23939.20			
4176.196	4177.373		0.0	0.2			23938.49			
4176.325	4177.502			0.1			23937.75			
4176.393	4177.570		0.2	0.3			23937.36			
4176.501	4177.678		0.0	0.1			23936.74			
4176.642	4177.820		0.2	0.4			23935.93			
4176.702	4177.879		0.2				23935.59			
4176.726	4177.903		0.2	0.3			23935.45			
4176.817	4177.994		0.2	0.2			23934.93			
4176.880	4178.057						23934.57			
4176.923	4178.101		0.5	0.6			23934.32			
4176.955	4178.132		0.3				23934.14			
4177.025	4178.202		0.6				23933.74			
4177.107	4178.284	GK	1.5b	1.5b	1.5b	2.57	23933.27	3E 1	2B 0	Q 2
4177.218	4178.396		0.6	0.6			23932.63			
4177.274	4178.452		0.0				23932.31	4e 3	2c 2	P 6
4177.339	4178.516		0.1				23931.94			
4177.398	4178.576		0.2	0.1d			23931.60			
4177.513	4178.691		0.3b	0.4			23930.94			
4177.574	4178.752		0.0				23930.59			
4177.635	4178.813		0.2				23930.24			
4177.716	4178.893	GK	0.8	1.0d	0.9	1.76	23929.78	3E 1	2B 0	Q 1
4177.829	4179.007		0.2	0.4			23929.13			
4177.953	4179.131		0.0				23928.42			
4178.007	4179.185		0.3	0.4			23928.11			
4178.137	4179.314		0.1	0.3			23927.37			
4178.236	4179.414		0.1	0.2			23926.80			
4178.337	4179.515			0.2			23926.22			
4178.407	4179.585		0.0	0.0			23925.82			
4178.500	4179.678		0.2	0.2			23925.29			
4178.544	4179.721		0.2				23925.04			
4178.645	4179.823	G	0.4r	0.4b	0.2	0.32	23924.46	4c 7	2a 5	Q 1
4178.889	4180.067		0.0	0.2b			23923.06			
4178.956	4180.134						23922.68			
4179.110	4180.287		0.1				23921.80			

λ_{air}	λ_{vac}	Ref.	I_1	I_2	I_3	I_4	ν	Upper	Lower	Br.
4179.192	4180.370		0.4	0.5			23921.33			
4179.300	4180.478		0.2	0.2			23920.71			
4179.396	4180.574		0.4	0.3			23920.16			
4179.501	4180.679		0.4	0.2			23919.56			
4179.588	4180.766	GK	1.5b	1.5	0.8	1.73	23919.06	3E 1	2B 0	P 3
4179.697	4180.875		0.5	0.3			23918.44	4e 3	2c 2	R 1
4179.803	4180.981		0.1	0.3			23917.83			
4179.917	4181.095	G	0.4	0.5			23917.18			
4180.008	4181.186		0.5	0.2			23916.66			
4180.102	4181.280	GK	1.5b	1.5	0.9b	2.09	23916.12	3E 1	2B 0	P 4
4180.210	4181.389		0.5	0.5			23915.50			
4180.280	4181.459		0.0				23915.10			
4180.328	4181.506		0.1				23914.83			
4180.383	4181.562		0.0	0.2			23914.51			
4180.494	4181.672			0.1			23913.88			
4180.597	4181.775		0.0	0.0			23913.29			
4180.688	4181.866	GK	0.5	0.7	0.2	0.81	23912.77	3E 3	2B 3	Q 5
4180.688	4181.866	GK	0.5	0.7	0.2	0.81	23912.77	3A 1	2B 1	R 2
4180.800	4181.978		0.1	0.1			23912.13			
4180.967	4182.146			0.2			23911.17			
4181.120	4182.298			0.1			23910.30	3c 5	2a 1	Q 4
4181.120	4182.298			0.1			23910.30	4e 3	2c 2	R 2
4181.245	4182.424		0.0	0.2			23909.58			
4181.305	4182.483		0.2	0.3			23909.24			
4181.378	4182.557			0.1			23908.82			
4181.445	4182.623			0.2			23908.44			
4181.552	4182.730			0.2			23907.83			
4181.576	4182.755						23907.69			
4181.681	4182.860		0.2	0.4			23907.09			
4181.775	4182.954			0.3			23906.55			
4181.879	4183.057	1		0.2	1.5		23905.96			
4181.917	4183.096		0.0				23905.74			
4181.977	4183.155		0.3	0.4			23905.40			
4182.050	4183.229		0.1	0.2			23904.98			
4182.167	4183.346	GK	1.5b	1.5	1.5b	2.06	23904.31	3E 1	2B 0	P 2
4182.167	4183.346	GK	1.5b	1.5	1.5b	2.06	23904.31	3E 1	2B 0	P 5
4182.272	4183.451		0.3	0.4			23903.71			
4182.362	4183.540		0.0	0.1			23903.20			
4182.554	4183.733		0.1	0.4			23902.10			
4182.671	4183.850	G					23901.43			
4182.787	4183.966			0.2d			23900.77			
4182.932	4184.111			0.0d			23899.94			
4183.070	4184.249		0.0	0.3d			23899.15			
4183.200	4184.379		0.0	0.4	0.2		23898.41	4e 3	2c 2	R 3
4183.380	4184.559		0.0	0.3			23897.38			
4183.559	4184.738		0.2	0.4	0.2d		23896.36	4d 3	2c 2	R 2
4183.646	4184.825			0.2			23895.86			
4183.788	4184.967		0.0	0.2d			23895.05	5c 3	2a 3	Q 1
4183.998	4185.177		0.0	0.2d			23893.85			
4184.133	4185.312			0.1			23893.08			
4184.249	4185.428		0.2	0.4			23892.42			
4184.289	4185.468		0.1				23892.19			
4184.389	4185.568	GK	0.2	0.5	0.3	0.40	23891.62			
4184.499	4185.678		0.0				23890.99			
4184.562	4185.741		0.0	0.1			23890.63			
4184.641	4185.820	G	0.2	0.4	0.2b	0.34	23890.18	3E 3	2B 3	R 0
4184.746	4185.925		0.3	0.5			23889.58			
4184.884	4186.064		0.1	0.1			23888.79			
4184.956	4186.136						23888.38			
4185.058	4186.237	K	0.1	0.5			23887.80	4c 2	2a 1	Q 5

λ_{air}	λ_{vac}	Ref.	I_1	I_2	I_3	I_4	ν	Upper	Lower	Br.
4185.244	4186.423			0.1			23886.74			
4185.363	4186.542			0.2			23886.06			
4185.643	4186.823			0.3			23884.46	3f 3	2c 0	R 4
4185.813	4186.993			0.1			23883.49			
4185.946	4187.126			0.3			23882.73			
4186.148	4187.328			0.3			23881.58			
4186.365	4187.545			0.3			23880.34			
4186.509	4187.689			0.1			23879.52			
4186.721	4187.901						23878.31	5c 3	2a 3	Q 2
4186.721	4187.901						23878.31	3F 2	2B 2	Q 2
4187.272	4188.452			0.0			23875.17			
4187.396	4188.576			0.1			23874.46			
4187.588	4188.768			0.0			23873.37			
4187.714	4188.894	G	0.2d	0.5	0.2	0.20	23872.65			
4188.012	4189.192		0.0	0.2b			23870.95			
4188.217	4189.398	GK	0.3	0.7	0.3	0.94	23869.78	3E 3	2B 3	Q 4
4188.217	4189.398	GK	0.3	0.7	0.3	0.94	23869.78	4e 1	2c 0	P 8
4188.217	4189.398	GK	0.3	0.7	0.3	0.94	23869.78	4e 3	2c 2	Q 1
4188.416	4189.596			0.4	0.2		23868.65			
4188.574	4189.754			0.1			23867.75	3f 3	2c 0	R 4
4188.695	4189.875	G	0.1	0.4			23867.06			
4188.839	4190.019			0.2			23866.24			
4188.923	4190.103			0.1			23865.76			
4189.033	4190.214		0.3	0.4			23865.13	X 1	2B 1	P 3
4189.126	4190.307	G	0.4	0.5		0.49	23864.60	Y 2	2B 2	P 4
4189.239	4190.419		0.1	0.2			23863.96			
4189.334	4190.514	G	0.2	0.6	0.6		23863.42			
4189.458	4190.639	GK	0.3	0.8	0.9	0.79	23862.71	3D 1	2B 0	R 8
4189.572	4190.753		0.0	0.1			23862.06			
4189.737	4190.918			0.2			23861.12			
4189.834	4191.015			0.0			23860.57			
4189.983	4191.164			0.3			23859.72			
4190.227	4191.408			0.2			23858.33	3f 3	2c 0	R 3
4190.540	4191.721			0.2			23856.55			
4190.707	4191.888	1		0.1	1.5		23855.60			
4190.826	4192.007			0.1			23854.92			
4191.025	4192.206	1		0.2	1.5b		23853.79			
4191.129	4192.310			0.0			23853.20			
4191.220	4192.401		0.1d	0.4			23852.68	5c 3	2a 3	Q 3
4191.220	4192.401		0.1d	0.4			23852.68	4F 0	2B 3	R 1
4191.220	4192.401		0.1d	0.4			23852.68	4e 2	2c 1	P 6
4191.471	4192.652		0.1	0.3			23851.25	3f 3	2c 0	R 3
4191.549	4192.730			0.1			23850.81			
4191.779	4192.960			0.1			23849.50			
4191.897	4193.078						23848.83			
4191.984	4193.166						23848.33			
4192.106	4193.287	GK	0.3	0.7	0.3d	0.70	23847.64			
4192.213	4193.394			0.1			23847.03			
4192.594	4193.776						23844.86			
4192.791	4193.973		0.0				23843.74			
4192.865	4194.047	G		0.2			23843.32			
4192.974	4194.156		0.1d	0.3			23842.70			
4193.166	4194.348			0.2			23841.61	4d 3	2c 2	Q 2
4193.265	4194.446			0.1			23841.05			
4193.440	4194.622			0.2			23840.05			
4193.578	4194.759		0.2	0.4			23839.27			
4193.695	4194.877		0.0	0.1			23838.60			
4193.834	4195.016	G	0.0	0.2			23837.81	4d 2	2c 1	P 4
4194.019	4195.201	GK	0.2	0.8	0.2d	0.83	23836.76	3E 3	2B 3	Q 3
4194.186	4195.368		0.0	0.1			23835.81			

λ_{air}	λ_{vac}	Ref.	I_1	I_2	I_3	I_4	ν	Upper	Lower	Br.
4194.304	4195.486	GK	0.6	0.8	0.2	0.87	23835.14	3A 1	2B 1	R 1
4194.420	4195.602		0.0	0.1			23834.48			
4194.595	4195.777			0.2			23833.49			
4194.764	4195.946			0.2			23832.53			
4194.945	4196.127		0.0	0.2			23831.50	4d 3	2c 2	P 2
4195.012	4196.194						23831.12			
4195.058	4196.240		0.0d	0.1d			23830.86			
4195.186	4196.368		0.2	0.3			23830.13			
4195.262	4196.444		0.1	0.3			23829.70			
4195.292	4196.474		0.1				23829.53			
4195.383	4196.565		0.2d	0.2			23829.01			
4195.473	4196.655		0.4	0.4			23828.50			
4195.514	4196.696		0.2				23828.27			
4195.579	4196.761		0.5	0.3			23827.90			
4195.667	4196.849	GK	1.5b	1.5	1.0b	2.19	23827.40	3D 1	2B 0	R 0
4195.776	4196.958		0.5	0.4			23826.78			
4195.831	4197.013		0.0				23826.47			
4195.882	4197.064		0.0				23826.18			
4195.950	4197.133		0.1	0.2			23825.79			
4196.042	4197.224		0.0d				23825.27	3f 3	2c 0	R 2
4196.095	4197.277			0.1			23824.97			
4196.244	4197.427						23824.12			
4196.509	4197.691			0.0			23822.62	3f 3	2c 0	R 2
4196.694	4197.876		0.1	0.2b			23821.57			
4196.790	4197.973			0.1d			23821.02			
4196.912	4198.095			0.3	0.2		23820.33			
4197.007	4198.190			0.2			23819.79	5c 3	2a 3	Q 4
4197.007	4198.190			0.2			23819.79	4F 0	2B 3	Q 3
4197.145	4198.327		0.0	0.1			23819.01			
4197.215	4198.398			0.2			23818.61			
4197.275	4198.458		0.1	0.4			23818.27	4F 0	2B 3	P 4
4197.391	4198.574		0.0	0.1			23817.61			
4197.495	4198.678		0.1	0.4			23817.02			
4197.619	4198.801		0.0				23816.32			
4197.684	4198.867		0.2	0.2			23815.95			
4197.758	4198.941		0.2	0.4			23815.53			
4197.848	4199.031		0.1	0.2			23815.02			
4197.985	4199.168						23814.24			
4198.015	4199.198		0.1	0.3			23814.07			
4198.119	4199.302		0.0	0.0			23813.48			
4198.209	4199.392	GK	0.8	1.0	0.5	1.47	23812.97	3E 3	2B 3	Q 2
4198.310	4199.493	i	0.1	0.1	1.5		23812.40			
4198.400	4199.583	G	0.2	0.5			23811.89			
4198.507	4199.690			0.0			23811.28			
4198.620	4199.803		0.0d	0.2			23810.64			
4198.721	4199.904	GK	0.3	0.7	0.5	0.34	23810.07			
4198.795	4199.978		0.0				23809.65	4d 2	2c 1	Q 5
4198.876	4200.059		0.0	0.1			23809.19	4e 3	2c 2	Q 2
4198.996	4200.179		0.1	0.3			23808.51	3F 2	2B 2	P 4
4199.183	4200.366		0.0	0.2	0.1		23807.45			
4199.299	4200.482		0.2	0.2			23806.79			
4199.382	4200.565		0.1	0.2			23806.32			
4199.407	4200.590		0.1				23806.18			
4199.486	4200.669		0.2	0.1			23805.73	3E 2	2B 2	Q10
4199.599	4200.782		0.4	0.5	0.2b		23805.09			
4199.629	4200.812		0.2				23804.92			
4199.707	4200.890		0.5	0.0			23804.48			
4199.788	4200.971	GK	1.5b	1.5	0.8	1.95	23804.02	3D 1	2B 0	R 1
4199.895	4201.079		0.5	0.3			23803.41			
4199.966	4201.149		0.0	0.1			23803.01			

λ_{air}	λ_{vac}	Ref.	I_1	I_2	I_3	I_4	ν	Upper	Lower	Br.
4200.098	4201.282	GK	0.5	0.8	0.2	1.04	23802.26	3D 1	2B 0	P 1
4200.241	4201.425		0.3	0.2			23801.45			
4200.335	4201.518		0.0				23800.92			
4200.402	4201.585		0.0d	0.1d			23800.54			
4200.539	4201.723		0.1	0.3	0.2		23799.76			
4200.651	4201.834		0.0	0.3			23799.13			
4200.670	4201.854	1	0.0		1.5b		23799.02			
4200.788	4201.972		0.2	0.2			23798.35			
4200.833	4202.016		0.0				23798.10			
4200.894	4202.078	G	0.4	0.7			23797.75	3E 3	2B 3	Q 1
4200.972	4202.156	GK	0.7	0.9	0.3	1.49	23797.31	3D 3	2B 3	R 0
4201.080	4202.263		0.2				23796.70			
4201.194	4202.378			0.2			23796.05			
4201.269	4202.452		0.0d	0.1d			23795.63			
4201.352	4202.535			0.1d			23795.16			
4201.458	4202.641		0.0	0.1			23794.56			
4201.563	4202.747		0.1	0.4	0.2		23793.96			
4201.728	4202.911			0.0			23793.03			
4201.789	4202.973		0.0				23792.68			
4201.881	4203.065		0.3	0.3			23792.16			
4201.975	4203.159		0.1				23791.63			
4202.097	4203.281		0.1	0.2			23790.94			
4202.132	4203.316		0.0				23790.74			
4202.201	4203.385		0.2				23790.35			
4202.281	4203.464	GK	0.7	0.8	0.2	1.13	23789.90	3D 3	2B 3	R 1
4202.392	4203.576		0.0				23789.27			
4202.434	4203.618	GK	0.4	0.7	0.2		23789.03	3E 3	2B 3	P 2
4202.434	4203.618	GK	0.4	0.7	0.2		23789.03	3D 1	2B 0	R 7
4202.558	4203.742		0.0d	0.1			23788.33			
4202.646	4203.830						23787.83			
4202.752	4203.936				0.2		23787.23			
4202.818	4204.002		0.1	0.2d			23786.86			
4202.902	4204.086		0.0				23786.38			
4202.996	4204.180		0.3	0.5	0.1		23785.85			
4203.125	4204.309		0.1	0.2			23785.12			
4203.205	4204.389						23784.67			
4203.291	4204.475			0.1			23784.18	3f 3	2c 0	R 1
4203.404	4204.589			0.2			23783.54			
4203.486	4204.670		0.0				23783.08			
4203.569	4204.753			0.2			23782.61			
4203.675	4204.859			0.1			23782.01			
4203.726	4204.910		0.1	0.2			23781.72			
4203.828	4205.013		0.0	0.1d			23781.14			
4203.912	4205.096		0.0				23780.67			
4203.977	4205.161		0.2	0.3			23780.30			
4204.097	4205.282		0.0	0.2			23779.62			
4204.184	4205.368		0.1	0.3			23779.13			
4204.214	4205.398		0.0				23778.96			
4204.288	4205.473		0.0				23778.54			
4204.380	4205.565		0.2	0.3			23778.02			
4204.483	4205.667		0.2	0.3			23777.44			
4204.614	4205.798		0.3	0.4			23776.70			
4204.674	4205.858		0.1				23776.36			
4204.705	4205.890		0.3	0.4			23776.18			
4204.796	4205.980		0.5	0.4	0.2		23775.67	3E 3	2B 3	P 3
4204.896	4206.081	GK	0.9	0.9	0.3		23775.10	3A 1	2B 1	R 0
4205.004	4206.189		0.2	0.2			23774.49	3F 2	2B 2	P 3
4205.095	4206.279	GK	1.5b	1.5b	1.5b	2.53	23773.98	3D 1	2B 0	R 2
4205.204	4206.389	GK	0.6	0.4			23773.36			
4205.294	4206.479	GK	0.5	0.4			23772.85			

λ_{air}	λ_{vac}	Ref.	I_1	I_2	I_3	I_4	ν	Upper	Lower	Br.
4205.394	4206.578		0.3	0.2			23772.29			
4205.471	4206.656		0.2	0.3			23771.85			
4205.512	4206.697		0.1				23771.62			
4205.583	4206.768		0.1				23771.22			
4205.650	4206.835		0.1	0.2			23770.84			
4205.726	4206.911		0.1	0.2d			23770.41			
4205.818	4207.003		0.1	0.2			23769.89			
4205.917	4207.102		0.2	0.3			23769.33	4d 1	2c 0	P 6
4205.933	4207.118		0.2				23769.24			
4206.004	4207.189		0.3	0.1			23768.84			
4206.085	4207.270	GK	0.9	1.0	0.3	1.56	23768.38	3D 3	2B 3	R 2
4206.190	4207.375		0.4	0.4			23767.79	X 1	2B 1	P 4
4206.292	4207.477		0.0				23767.21			
4206.369	4207.554		0.0				23766.78			
4206.450	4207.635						23766.32			
4206.553	4207.738			0.2			23765.74			
4206.650	4207.835						23765.19			
4206.691	4207.876						23764.96			
4206.783	4207.968						23764.44			
4206.848	4208.033		0.0d				23764.07			
4206.976	4208.161		0.1d				23763.35			
4207.066	4208.251			0.3			23762.84			
4207.190	4208.375		0.4	0.5			23762.14			
4207.298	4208.483		0.1	0.2			23761.53			
4207.383	4208.568		0.1				23761.05			
4207.454	4208.639		0.0				23760.65			
4207.539	4208.724	GK	0.3	0.8	0.4	0.79	23760.17	W 1	2B 1	R 4
4207.539	4208.724	GK	0.3	0.8	0.4	0.79	23760.17	4e 3	2c 2	P 2
4207.826	4209.011			0.2d			23758.55			
4207.661	4208.846						23759.48			
4207.898	4209.084	G					23758.14			
4208.035	4209.220		0.0d	0.3			23757.37			
4208.098	4209.284			0.2	0.2		23757.01			
4208.233	4209.419	GK	0.3v	0.7	0.4	0.57	23756.25			
4208.345	4209.530		0.0				23755.62			
4208.439	4209.624	GK	0.5	0.6		0.72	23755.09	3E 3	2B 3	P 4
4208.665	4209.851						23753.81			
4208.727	4209.913			0.2b			23753.46			
4208.979	4210.165			0.3			23752.04			
4209.085	4210.271		0.0				23751.44			
4209.163	4210.349	GK	0.6	1.5d	1.5b	1.45	23751.00	3D 1	2B 0	R 6
4209.291	4210.477		0.1	0.3			23750.28			
4209.325	4210.510		0.0				23750.09			
4209.404	4210.590			0.0			23749.64			
4209.466	4210.652				0.1		23749.29			
4209.645	4210.831			0.3			23748.28			
4209.718	4210.904	G	0.0d	0.3			23747.87			
4209.844	4211.030	G	0.2	0.5	0.3		23747.16			
4209.950	4211.136		0.1	0.4			23746.56	4e 3	2c 2	Q 3
4210.048	4211.234		0.0				23746.01			
4210.126	4211.312	GK	0.7	1.5	1.0b	1.96	23745.57	3D 1	2B 0	R 3
4210.243	4211.429		0.1	0.3			23744.91			
4210.397	4211.583			0.4			23744.04			
4210.539	4211.725			0.2			23743.24			
4210.672	4211.858	G		0.2			23742.49			
4210.745	4211.931						23742.08			
4210.807	4211.993						23741.73			
4210.890	4212.076			0.2			23741.26			
4210.958	4212.144			0.1	0.1		23740.88			
4211.096	4212.282	G	0.1	0.4			23740.10			

λ_{air}	λ_{vac}	Ref.	I_1	I_2	I_3	I_4	ν	Upper	Lower	Br.
4211.273	4212.460			0.3			23739.10			
4211.412	4212.598		0.0	0.3			23738.32			
4211.591	4212.777		0.3	0.4			23737.31	4d 2	2c 1	P 4
4211.735	4212.921						23736.50			
4211.814	4213.001	GK	0.3	0.7	0.2	0.88	23736.05	3D 3	2B 3	R 3
4211.937	4213.123		0.0d				23735.36			
4212.036	4213.223	GK	0.4	1.0	1.0	1.14	23734.80	3D 1	2B 0	R 5
4212.146	4213.333		0.0				23734.18			
4212.208	4213.395			0.1			23733.83	4d 3	2c 2	Q 3
4212.315	4213.501		0.2d	0.5			23733.23			
4212.404	4213.590		0.0	0.2			23732.73			
4212.494	4213.681	GK	0.8b	1.5b	1.5b	2.20	23732.22	3D 1	2B 0	R 4
4212.494	4213.681	GK	0.8b	1.5b	1.5b	2.20	23732.22	4E 2	2B 2	R 4
4212.609	4213.796		0.2	0.5			23731.57			
4212.812	4213.999		0.0	0.1			23730.43			
4212.881	4214.068			0.2			23730.04	4E 0	2B 3	R 2
4213.059	4214.245			0.1			23729.04			
4213.250	4214.437			0.1			23727.96			
4213.350	4214.537			0.2			23727.40			
4213.527	4214.714		0.2b	0.4			23726.40			
4213.828	4215.015			0.1d			23724.71			
4213.893	4215.080						23724.34			
4214.128	4215.315			0.2b			23723.02			
4214.273	4215.461	GK	0.4	0.7	0.4b	0.83	23722.20	3D 3	2B 3	R 4
4214.371	4215.558		0.0				23721.65			
4214.503	4215.690		0.0d				23720.91			
4214.604	4215.791		0.3	0.5	0.2		23720.34			
4214.728	4215.916			0.1			23719.64			
4214.897	4216.084						23718.69			
4215.080	4216.268		0.2	0.4	0.1		23717.66			
4215.253	4216.440			0.0			23716.69			
4215.359	4216.547		0.0d				23716.09			
4215.512	4216.700			0.0			23715.23			
4215.615	4216.803			0.2			23714.65			
4215.766	4216.954			0.0			23713.80			
4215.937	4217.125		0.0	0.3			23712.84			
4216.074	4217.261	GK	0.3	0.6	0.4		23712.07			
4216.175	4217.363	G					23711.50			
4216.218	4217.406	GK	0.5	0.7		0.83	23711.26	3D 1	2B 0	P 2
4216.367	4217.555			0.0			23710.42			
4216.479	4217.667			0.1			23709.79			
4216.545	4217.733						23709.42	3E 3	2B 3	P 5
4216.678	4217.866	GK	0.3	0.8	0.3	0.89	23708.67	W 1	2B 1	R 3
4217.008	4218.195			0.0			23706.82			
4217.207	4218.395			0.1d	0.1		23705.70			
4217.369	4218.557		0.1	0.2d			23704.79			
4217.461	4218.649						23704.27			
4217.650	4218.838		0.2	0.3			23703.21	6c 0	2a 1	Q 1
4217.789	4218.977		0.0d				23702.43			
4217.901	4219.089		0.1	0.3			23701.80			
4218.129	4219.317			0.1			23700.52			
4218.307	4219.495		0.1	0.2			23699.52	4c 3	2a 2	R 1
4218.554	4219.742			0.0			23698.13			
4218.712	4219.901						23697.24			
4218.798	4219.986			0.2			23696.76			
4218.905	4220.093		0.2	0.3			23696.16	3D 3	2B 3	P 2
4219.008	4220.196			0.1d			23695.58	3E 2	2B 2	Q 9
4219.108	4220.296		0.0d				23695.02			
4219.193	4220.382						23694.54			
4219.328	4220.517		0.2	0.2			23693.78			

λ_{air}	λ_{vac}	Ref.	I_1	I_2	I_3	I_4	ν	Upper	Lower	Br.
4219.418	4220.606		0.2				23693.28	6c 0	2a 1	Q 2
4219.501	4220.690	GK	0.9b	0.7		0.98	23692.81	3A 1	2B 1	P 1
4219.610	4220.798		0.2	0.1			23692.20			
4219.715	4220.904			0.2			23691.61			
4219.818	4221.007			0.0			23691.03			
4220.066	4221.255		0.1	0.3			23689.64			
4220.260	4221.449		0.0				23688.55			
4220.413	4221.602		0.2	0.3			23687.69			
4220.547	4221.736		0.1				23686.94			
4220.650	4221.839		0.0				23686.36			
4220.752	4221.941		0.0				23685.79			
4220.820	4222.008						23685.41			
4220.937	4222.126		0.1	0.3			23684.75	4F 0	2B 3	P 3
4221.083	4222.272		0.0				23683.93			
4221.139	4222.327		0.3	0.4			23683.62	4e 3	2c 2	Q 4
4221.139	4222.327		0.3	0.4			23683.62	4F 0	2B 3	Q 2
4221.376	4222.565						23682.29			
4221.443	4222.632		0.1				23681.91			
4221.579	4222.768		0.1				23681.15			
4221.711	4222.900			0.2b			23680.41			
4221.809	4222.998		0.1	0.0d			23679.86			
4221.878	4223.067						23679.47			
4221.975	4223.164		0.3	0.3			23678.93			
4222.071	4223.260		0.2				23678.39	6c 0	2a 1	Q 3
4222.165	4223.355	GK	1.5b	1.5	0.7	1.92	23677.86	W 1	2B 1	R 2
4222.165	4223.355	GK	1.5b	1.5	0.7	1.92	23677.86	Z 1	2B 0	R 1
4222.165	4223.355	GK	1.5b	1.5	0.7	1.92	23677.86	4d 3	2c 2	P 3
4222.269	4223.458		0.2				23677.28			
4222.342	4223.531		0.3				23676.87			
4222.436	4223.626		0.4				23676.34			
4222.522	4223.711	GK	1.5b	1.5	0.6	1.90	23675.86	Z 1	2B 0	R 0
4222.633	4223.822		0.4	0.3			23675.24			
4222.729	4223.918		0.0				23674.70			
4222.825	4224.015		0.0				23674.16	4e 2	2c 1	P 7
4222.914	4224.104		0.1d				23673.66			
4223.212	4224.402		0.0d				23671.99			
4223.287	4224.477						23671.57			
4223.471	4224.661		0.0	0.2			23670.54			
4223.519	4224.709		0.0				23670.27			
4223.623	4224.812		0.2	0.2			23669.69			
4223.751	4224.941		0.2	0.4			23668.97			
4223.855	4225.044		0.3				23668.39			
4223.940	4225.130	GK	1.5b	1.5	0.7	1.75	23667.91	Z 1	2B 0	R 2
4224.049	4225.239		0.3	0.3			23667.30			
4224.162	4225.352	G		0.3	0.3		23666.67			
4224.269	4225.459		0.1	0.3b			23666.07			
4224.315	4225.505		0.2				23665.81			
4224.353	4225.543		0.1				23665.60			
4224.408	4225.598		0.2	0.1			23665.29			
4224.510	4225.700	GK	1.0	0.9	0.3	1.45	23664.72	3A 1	2B 1	P 2
4224.594	4225.784		0.8b	0.4			23664.25	X 2	2B 3	R 0
4224.729	4225.919						23663.49			
4224.849	4226.039				0.2		23662.82	3f 3	2c 0	Q 2
4225.008	4226.198			0.1			23661.93			
4225.113	4226.303			0.3	0.2		23661.34	3E 2	2B 2	R 3
4225.378	4226.568			0.2			23659.86	4d 2	2c 1	P 5
4225.472	4226.662		0.0				23659.33			
4225.547	4226.737	GK	0.4	0.6	0.1	0.30	23658.91	3A 1	2B 1	P 4
4225.656	4226.846		0.0				23658.30			
4225.710	4226.900			0.1d	0.1		23658.00			

λ_{air}	λ_{vac}	Ref.	I_1	I_2	I_3	I_4	ν	Upper	Lower	Br.
4225.756	4226.946						23657.74	6c 0	2a 1	Q 4
4226.040	4227.231		0.1	0.3			23656.15			
4226.226	4227.416			0.2			23655.11			
4226.385	4227.575			0.2d			23654.22			
4226.591	4227.781		0.1	0.2			23653.07			
4226.742	4227.933	GK	0.4	0.6		0.71	23652.22	3A 1	2B 1	P 3
4227.002	4228.192			0.1d			23650.77			
4227.180	4228.371		0.0	0.2			23649.77			
4227.293	4228.484		0.5	0.4			23649.14	X 2	2B 3	R 1
4227.386	4228.577	GK	0.6	0.8	0.3	1.23	23648.62	W 1	2B 1	R 1
4227.504	4228.695		0.1	0.2			23647.96			
4227.720	4228.911			0.2d			23646.75			
4228.167	4229.358			0.2			23644.25			
4228.527	4229.718			0.1			23642.24	3f 3	2c 0	Q 3
4228.709	4229.900			0.1			23641.22			
4228.820	4230.011	GK	0.2	0.5	0.2d	0.38	23640.60	4E 0	2B 3	R 0
4229.346	4230.537			0.1			23637.66			
4229.505	4230.696			0.2			23636.77			
4229.620	4230.811						23636.13			
4229.788	4230.979				0.2		23635.19			
4229.901	4231.092				0.2		23634.56			
4230.554	4231.746		0.1				23630.91	X 2	2B 3	P 1
4230.844	4232.036		0.0				23629.29			
4230.930	4232.122		0.1				23628.81			
4231.009	4232.201	GK	0.5	0.6	0.3	0.51	23628.37			
4231.127	4232.319		0.1	0.1			23627.71			
4231.303	4232.494		0.0	0.2			23626.73			
4231.571	4232.763			0.1			23625.23			
4231.731	4232.922			0.3			23624.34			
4231.765	4232.956		0.2	0.0			23624.15			
4231.969	4233.161		0.1				23623.01			
4232.039	4233.231		0.0				23622.62			
4232.118	4233.310		0.0				23622.18	3D 3	2B 3	P 3
4232.204	4233.396			0.0			23621.70			
4232.417	4233.609						23620.51			
4232.512	4233.704		0.1	0.2			23619.98			
4232.593	4233.784		0.1				23619.53	3f 3	2c 0	Q 4
4232.680	4233.872	GK	0.8	0.7	0.2	1.04	23619.04	Z 1	2B 0	P 1
4232.795	4233.987		0.2	0.2			23618.40			
4232.895	4234.087	GK	0.5	0.6	0.2d	0.88	23617.84	X 2	2B 3	R 2
4233.014	4234.206		0.1	0.1			23617.18			
4233.125	4234.317		0.0d	0.0			23616.56			
4233.222	4234.414		0.2	0.2			23616.02			
4233.336	4234.528		0.2				23615.38	3A 2	2B 3	R 4
4233.419	4234.611	GK	0.8b	0.9	0.3	1.48	23614.92	W 1	2B 1	R 0
4233.525	4234.717		0.3	0.2			23614.33	3f 3	2c 0	Q 4
4233.638	4234.830		0.1				23613.70	4f 1	2c 1	R 5
4233.699	4234.891		0.4r	0.3			23613.36			
4233.826	4235.018	GK	1.5b	0.8	0.2	1.36	23612.65	4c 3	2a 2	Q 1
4233.932	4235.124		0.3	0.0			23612.06			
4234.000	4235.192		0.2	0.3	0.2		23611.68	3D 1	2B 0	P 3
4234.131	423 .323		0.1				23610.95			
4234.202	4235.395		0.0	0.0			23610.55			
4234.498	4235.691			0.0			23608.90			
4234.636	4235.829		0.0d	0.2			23608.13			
4234.771	4235.963		0.0d	0.1			23607.38			
4234.952	4236.145			0.1			23606.37			
4235.437	4236.629			0.1			23603.67			
4235.526	4236.719		0.1	0.2			23603.17			
4235.787	4236.979			0.1			23601.72			

λ_{air}	λ_{vac}	Ref.	I_1	I_2	I_3	I_4	ν	Upper	Lower	Br.
4235.941	4237.134		0.2	0.2			23600.86			
4236.126	4237.319		0.0d	0.1	0.2		23599.83			
4236.248	4237.441				0.2		23599.15			
4236.336	4237.529		0.1	0.4			23598.66	3E 2	2B 2	Q 8
4236.758	4237.951			0.0			23596.31			
4236.927	4238.120			0.3			23595.37			
4237.156	4238.349		0.0				23594.09			
4237.226	4238.420				0.2		23593.70			
4237.298	4238.491			0.2			23593.30			
4237.386	4238.579		0.0				23592.81			
4237.559	4238.752	GK	0.4	0.6	0.2	0.70	23591.85	4c 3	2a 2	Q 2
4237.658	4238.851		0.0	0.0			23591.30			
4238.055	4239.248	K	0.0d	0.3	0.1d		23589.09			
4238.211	4239.404			0.1			23588.22			
4238.326	4239.519						23587.58			
4238.392	4239.586						23587.21			
4238.515	4239.708			0.2			23586.53	3E 2	2B 2	R 2
4238.829	4240.023		0.1	0.3			23584.78	X 2	2B 3	R 3
4239.036	4240.229						23583.63			
4239.221	4240.415						23582.60			
4239.383	4240.576			0.3			23581.70			
4239.624	4240.817			0.2			23580.36			
4239.737	4240.931			0.2			23579.73			
4239.807	4241.001			0.2			23579.34			
4239.994	4241.188		0.0				23578.30			
4240.086	4241.280		0.0				23577.79			
4240.248	4241.442		0.2	0.0			23576.89			
4240.399	4241.593		0.0				23576.05			
4241.244	4242.438		0.0	0.0			23571.35			
4241.377	4242.572		0.0				23570.61			
4241.843	4243.038		0.2	0.3b			23568.02			
4242.058	4243.252		0.0	0.2b			23566.83			
4242.122	4243.317				0.1b		23566.47			
4242.265	4243.459			0.0			23565.68			
4242.362	4243.556			0.0			23565.14			
4242.484	4243.679	GK	0.2	0.6	0.5b		23564.46			
4242.592	4243.787		0.0		0.3		23563.86			
4242.643	4243.837		0.0				23563.58			
4242.722	4243.917		0.2	0.2d			23563.14	X 2	2B 3	P 2
4242.936	4244.131		0.1d	0.1			23561.95			
4243.037	4244.232		0.1	0.0			23561.39			
4243.136	4244.331	GK	0.5	0.9		0.82	23560.84	4c 3	2a 2	Q 3
4243.136	4244.331	GK	0.5	0.9		0.82	23560.84	4f 0	2c 0	R 4
4243.255	4244.450		0.2				23560.18			
4243.338	4244.533	GK	0.9	1.0	0.4	1.37	23559.72	Z 1	2B 0	P 2
4243.439	4244.633		0.2	0.2	0.3		23559.16			
4243.660	4244.855			0.3d	0.2		23557.93			
4243.887	4245.082						23556.67			
4243.986	4245.181			0.0			23556.12			
4244.176	4245.371			0.3			23555.07			
4244.347	4245.542			0.1			23554.12			
4244.424	4245.619			0.1			23553.69			
4245.224	4246.420			0.0			23549.25	3F 1	2B 1	Q 6
4245.457	4246.652		0.1	0.1			23547.96			
4245.531	4246.726	G	0.3	0.3		0.23	23547.55			
4245.630	4246.825		0.0	0.5	0.3		23547.00			
4245.810	4247.006			0.1			23546.00			
4245.955	4247.150			0.1			23545.20			
4246.218	4247.414		0.0d				23543.74			
4246.510	4247.706			0.2d			23542.12			

λ_{air}	λ_{vac}	Ref.	I_1	I_2	I_3	I_4	ν	Upper	Lower	Br.
4246.611	4247.807		0.0	0.1			23541.56			
4246.698	4247.893	GK	0.7	0.8		0.82	23541.08	W 1	2B 1	P 1
4246.774	4247.969			0.2	0.2		23540.66			
4246.826	4248.022		0.1	0.2			23540.37			
4247.003	4248.198	GK	0.3	0.6	0.2	0.45	23539.39	4E 0	2B 3	P 2
4247.539	4248.734				0.2		23536.42			
4247.629	4248.825			0.1			23535.92			
4247.753	4248.949		0.0				23535.23	3D 3	2B 3	P 4
4247.889	4249.085			0.2			23534.48			
4248.241	4249.437		0.0	0.1			23532.53			
4248.727	4249.923				0.4		23529.84			
4248.932	4250.129				0.4		23528.70			
4249.033	4250.230			0.0d			23528.14			
4249.294	4250.490			0.0			23526.70			
4249.429	4250.625		0.0	0.2d			23525.95			
4249.492	4250.689				0.4		23525.60			
4249.637	4250.833				0.4		23524.80			
4249.778	4250.974						23524.02			
4249.908	4251.104						23523.30			
4250.276	4251.473		0.1	0.3	0.2		23521.26			
4250.396	4251.592	F					23520.60			
4250.520	4251.717			0.2			23519.91	4c 3	2a 2	Q 4
4250.788	4251.984						23518.43			
4250.911	4252.107			0.1			23517.75			
4251.025	4252.221			0.0			23517.12	4d 3	2c 2	P 4
4251.184	4252.380	1	0.0	0.2	0.6		23516.24			
4251.296	4252.493		0.0	0.2			23515.62			
4251.435	4252.632			0.3			23514.85	3E 2	2B 2	Q 7
4251.621	4252.818		0.0d				23513.82			
4252.015	4253.212			0.0			23511.64			
4252.126	4253.323		0.1	0.4	0.2		23511.03	3E 2	2B 2	R 1
4252.614	4253.811		0.2d	0.5	0.4		23508.33			
4252.793	4253.990						23507.34	3A 2	2B 3	R 3
4252.813	4254.010		0.3	0.3			23507.23	3D 1	2B 0	P 4
4252.907	4254.104			0.1			23506.71	6c 1	2a 2	Q 1
4253.003	4254.200		0.0d	0.0			23506.18			
4253.132	4254.329		0.3	0.4			23505.47	4E 0	2B 3	Q 1
4253.204	4254.401		0.1				23505.07			
4253.302	4254.499	GK	0.8	0.9	0.3	1.29	23504.53	W 1	2B 1	P 2
4253.407	4254.604		0.2	0.1			23503.95			
4253.553	4254.751			0.1			23503.14			
4253.676	4254.874			0.2d			23502.46			
4253.895	4255.093			0.2d	0.1		23501.25			
4254.044	4255.241	GK	0.3	0.7	0.8	0.45	23500.43			
4254.165	4255.363		0.0				23499.76			
4254.268	4255.466			0.1			23499.19	4D 0	2B 3	R 6
4254.326	4255.524		0.1	0.3			23498.87			
4254.424	4255.622			0.3			23498.33			
4254.529	4255.727			0.2			23497.75			
4254.614	4255.812			0.1			23497.28			
4254.743	4255.940		0.2	0.4			23496.57	4E 0	2B 3	P 4
4255.000	4256.198			0.0	0.4		23495.15			
4255.067	4256.265			0.0			23494.78			
4255.146	4256.344			0.0			23494.34			
4255.222	4256.420	GK	0.2	0.5	0.5	0.38	23493.92	4E 0	2B 3	Q 4
4255.382	4256.580			0.2d			23493.04			
4255.492	4256.690						23492.43			
4255.628	4256.826				0.4		23491.68			
4255.829	4257.027		0.0d	0.0d	0.4		23490.57			
4256.344	4257.542						23487.73			

λ_air	λ_vac	Ref.	I₁	I₂	I₃	I₄	ν	Upper	Lower	Br.
4256.522	4257.720			0.1			23486.75			
4256.730	4257.928	G	0.4	0.5		0.48	23485.60	Z 1	2B 0	P 3
4256.998	4258.197			0.1d			23484.12			
4257.232	4258.431			0.1			23482.83	3f 3	2c 0	P 3
4257.386	4258.585			0.1d			23481.98			
4257.491	4258.690			0.0			23481.40	X 2	2B 3	P 3
4257.582	4258.781			0.3			23480.90			
4257.829	4259.027			0.2d			23479.54			
4258.099	4259.298			0.1d			23478.05	4d 2	2c 1	P 6
4258.483	4259.682			0.1	0.2		23475.93			
4258.647	4259.845			0.2			23475.03			
4258.763	4259.962			0.4	0.4		23474.39			
4258.850	4260.049			0.2			23473.91			
4259.068	4260.266			0.0			23472.71			
4259.340	4260.539	i			0.7b		23471.21			
4259.547	4260.746			0.1			23470.07			
4259.668	4260.867		0.1	0.4			23469.40	4c 3	2a 2	Q 5
4259.933	4261.132						23467.94			
4260.340	4261.539	GK	0.5	0.6		0.56	23465.70	W 1	2B 1	P 3
4260.462	4261.661			0.2			23465.03			
4260.549	4261.748			0.1			23464.55			
4260.692	4261.892			0.0			23463.76			
4260.832	4262.031						23462.99			
4260.977	4262.177		0.3	0.3			23462.19			
4261.128	4262.328		0.1	0.2			23461.36			
4261.224	4262.424			0.1	0.3		23460.83			
4261.315	4262.515		0.0	0.0			23460.33			
4261.382	4262.582		0.1				23459.96			
4261.464	4262.664			0.1d			23459.51			
4261.526	4262.725				0.4		23459.17			
4261.626	4262.825		0.0	0.3			23458.62	4E 0	2B 3	Q 3
4261.688	4262.887	GK	0.4	0.5	0.4	0.41	23458.28			
4261.824	4263.023			0.0			23457.53			
4261.985	4263.185		0.0		0.4		23456.64			
4262.294	4263.494			0.2			23454.94			
4262.451	4263.650			0.1			23454.08			
4262.592	4263.792		0.0				23453.30			
4262.665	4263.865			0.1			23452.90			
4262.783	4263.983	GK	0.5r	0.7b	0.2	0.73	23452.25	4E 0	2B 3	Q 2
4262.951	4264.150			0.1d			23451.33			
4263.029	4264.229		0.0				23450.90			
4263.269	4264.469			0.1			23449.58			
4263.587	4264.787			0.0			23447.83			
4263.825	4265.025			0.2d			23446.52			
4263.963	4265.163			0.2d	0.2		23445.76			
4264.202	4265.402	G	0.1	0.5	0.2b		23444.45	3E 2	2B 2	Q 6
4264.471	4265.671				0.1		23442.97			
4264.547	4265.748			0.1d			23442.55			
4264.658	4265.859			0.0d			23441.94			
4264.760	4265.960						23441.38			
4264.898	4266.099		0.0d				23440.62			
4265.008	4266.208			0.1d			23440.02	3F 1	2B 1	R 3
4265.173	4266.374	GK	0.4	0.7	0.1	0.72	23439.11	3A 2	2B 3	R 2
4265.173	4266.374	GK	0.4	0.7	0.1	0.72	23439.11	3D 3	2B 3	P 5
4265.277	4266.477		0.0	0.1			23438.54			
4265.415	4266.616			0.1d			23437.78			
4265.553	4266.754			0.2			23437.02			
4265.679	4266.880			0.1			23436.33			
4265.796	4266.996			0.2			23435.69			
4266.161	4267.362			0.2			23433.68	3F 3	2B 4	Q 3

λ_{air}	λ_{vac}	Ref.	I_1	I_2	I_3	I_4	ν	Upper	Lower	Br.
4266.212	4267.413	1			0.2		23433.40			
4266.289	4267.490	1			1.0		23432.98			
4266.405	4267.606			0.1			23432.34			
4266.737	4267.938			0.2			23430.52	3E 2	2B 2	R 0
4266.848	4268.049			0.0d			23429.91			
4267.039	4268.240	G		0.6	0.3	0.23	23428.86			
4267.187	4268.388			0.3			23428.05			
4267.389	4268.590			0.1	0.5		23426.94			
4267.500	4268.701			0.2			23426.33			
4267.633	4268.834			0.2			23425.60			
4267.746	4268.947			0.2d	0.7b		23424.98			
4267.821	4269.022			0.2			23424.57			
4267.928	4269.129	GK	0.4	0.6		0.58	23423.98	W 1	2B 1	P 4
4268.110	4269.312				0.5		23422.98			
4268.170	4269.372			0.2d			23422.65			
4268.294	4269.496						23421.97			
4268.438	4269.640						23421.18			
4268.528	4269.729						23420.69			
4268.590	4269.791			0.1			23420.35			
4268.699	4269.900						23419.75			
4268.796	4269.997			0.2			23419.22			
4268.898	4270.099			0.1d			23418.66			
4269.049	4270.250		0.0	0.3			23417.83	4f 1	2c 1	R 4
4269.049	4270.250		0.0	0.3			23417.83	T 0	2B 3	R 2
4269.198	4270.400			0.1			23417.01			
4269.381	4270.582			0.2			23416.01	4e 3	2c 2	P 5
4269.477	4270.679			0.1	0.2b		23415.48			
4269.563	4270.765			0.5			23415.01	4D 0	2B 3	R 5
4269.685	4270.887			0.2d			23414.34			
4269.846	4271.047				0.3		23413.46			
4270.169	4271.370				0.2		23411.69			
4270.293	4271.494			0.1			23411.01			
4270.422	4271.624						23410.30			
4270.466	4271.668						23410.06			
4270.510	4271.712			0.3			23409.82			
4270.688	4271.890			0.2			23408.84			
4270.946	4272.148			0.1			23407.43	3f 3	2c 0	P 4
4271.048	4272.250			0.3			23406.87	3D 1	2B 0	P 5
4271.218	4272.420	G		0.4	0.1		23405.94	3F 1	2B 1	Q 5
4271.595	4272.798			0.1d			23403.87			
4271.761	4272.964						23402.96			
4271.944	4273.146			0.0			23401.96			
4272.085	4273.287		0.3	0.5	0.3		23401.19	Z 1	2B 0	P 4
4272.174	4273.376	1		0.1	1.5b	0.46	23400.70			
4272.305	4273.508				0.2d		23399.98			
4272.388	4273.590			0.3			23399.53			
4272.528	4273.731			0.2d			23398.76			
4272.660	4273.862			0.1			23398.04			
4272.753	4273.955			0.3			23397.53			
4273.272	4274.474		0.0				23394.69			
4273.394	4274.597		0.0	0.0			23394.02			
4273.480	4274.683		0.0		0.7		23393.55			
4273.670	4274.873						23392.51			
4273.790	4274.993				0.6		23391.85			
4273.920	4275.123				0.7		23391.14			
4273.966	4275.169				0.3		23390.89			
4274.085	4275.287			0.3			23390.24			
4274.191	4275.393				0.7		23389.66			
4274.439	4275.642						23388.30			
4274.611	4275.814			0.4			23387.36	3E 2	2B 2	Q 5

λ_{air}	λ_{vac}	Ref.	I_1	I_2	I_3	I_4	ν	Upper	Lower	Br.
4274.781	4275.984			0.2			23386.43	W 1	2B 1	P 5
4274.962	4276.165				0.3		23385.44			
4275.084	4276.287		0.1	0.2			23384.77	X 2	2B 3	P 4
4275.258	4276.461				0.3		23383.82			
4275.382	4276.586				0.3		23383.14			
4275.518	4276.721		0.0d	0.2d			23382.40			
4275.629	4276.833				0.3		23381.79			
4275.847	4277.050			0.1	0.2		23380.60			
4275.962	4277.165			0.3d			23379.97			
4276.355	4277.559	G	0.1d	0.5	0.4b		23377.82			
4276.482	4277.685			0.1			23377.13			
4276.600	4277.804			0.2			23376.48			
4276.716	4277.919			0.2d			23375.85			
4276.805	4278.009			0.0			23375.36			
4276.904	4278.108	G	0.3	0.5			23374.82	3F 3	2B 4	Q 2
4277.043	4278.247		0.0d	0.1			23374.06			
4277.192	4278.395			0.2			23373.25			
4277.274	4278.478		0.0	0.0			23372.80			
4277.316	4278.520		0.0				23372.57			
4277.395	4278.598		0.0				23372.14			
4277.475	4278.679	GK	0.5	0.7			23371.70	3A 2	2B 3	R 1
4277.561	4278.765		0.1	0.2			23371.23	W 1	2B 1	P 6
4277.616	4278.820		0.1				23370.93			
4277.706	4278.910		0.3	0.4			23370.44	T 0	2B 3	R 1
4277.830	4279.034			0.2			23369.76			
4277.931	4279.135			0.1			23369.21			
4278.123	4279.327		0.0	0.0			23368.16	4f 0	2c 0	R 3
4278.295	4279.499						23367.22			
4278.863	4280.067			0.3			23364.12	3D 3	2B 3	P 6
4279.024	4280.228			0.2d			23363.24			
4279.118	4280.322						23362.73			
4279.238	4280.443		0.1d	0.1			23362.07			
4279.339	4280.543		0.1				23361.52			
4279.418	4280.622	GK	0.8	0.9	0.3	1.10	23361.09	4D 0	2B 3	R 0
4279.486	4280.690				0.9		23360.72			
4279.537	4280.741		0.1	0.1d			23360.44			
4279.706	4280.910				0.9		23359.52	3E 2	2B 2	P 6
4279.960	4281.165		0.0		0.1		23358.13			
4280.030	4281.234			0.1	0.9		23357.75			
4280.222	4281.427			0.2	0.9		23356.70			
4280.255	4281.460			0.1	0.4		23356.52			
4280.402	4281.606				0.3		23355.72			
4280.613	4281.817			0.1	0.3		23354.57			
4280.752	4281.957	GK	0.2	0.8	0.4	0.83	23353.81	4D 0	2B 3	R 4
4280.869	4282.074		0.0	0.0			23353.17	4D 0	2B 3	P 1
4280.932	4282.136				0.3		23352.83	3E 2	2B 2	P 5
4281.109	4282.314				0.3		23351.86			
4281.406	4282.611			0.2			23350.24			
4281.560	4282.765			0.2d			23349.40			
4281.817	4283.022	GK	0.3	0.7	0.4b	0.34	23348.00			
4281.920	4283.125			0.1			23347.44			
4281.995	4283.200						23347.03			
4282.272	4283.477			0.1			23345.52			
4282.485	4283.690			0.2			23344.36			
4282.544	4283.749		0.0d				23344.04			
4282.650	4283.855		0.1	0.3			23343.46	3E 2	2B 2	P 4
4282.738	4283.943	G		0.6		0.48	23342.98	3E 2	2B 2	Q 4
4282.863	4284.068			0.0			23342.30			
4282.956	4284.162			0.1			23341.79			
4283.129	4284.334			0.2			23340.85			

λ_{air}	λ_{vac}	Ref.	I_1	I_2	I_3	I_4	ν	Upper	Lower	Br.
4283.226	4284.431			0.0			23340.32			
4283.388	4284.593		0.0	0.3	0.2		23339.44			
4283.505	4284.710		0.0				23338.80			
4283.586	4284.791			0.1			23338.36			
4283.692	4284.898			0.1	0.2		23337.78			
4283.876	4285.081						23336.78			
4284.043	4285.248		0.1	0.4b	0.3b		23335.87	3c 6	2a 2	Q 1
4284.076	4285.281	G	0.2				23335.69	3E 2	2B 2	P 3
4284.320	4285.526			0.2d			23334.36	T 0	2B 3	R 0
4284.429	4285.634		0.0	0.1			23333.77			
4284.539	4285.744			0.3d			23333.17			
4284.682	4285.888						23332.39			
4284.913	4286.119			0.0			23331.13			
4285.205	4286.411			0.0			23329.54			
4285.270	4286.475		0.0				23329.19			
4285.380	4286.586			0.1	1.0		23328.59			
4285.514	4286.720	G	0.2	0.5	1.0		23327.86			
4285.685	4286.891			0.1			23326.93			
4285.793	4286.999				0.3		23326.34	3D 2	2B 2	P 1
4285.843	4287.049		0.4	0.5	0.2		23326.07	3f 3	2c 0	P 5
4285.946	4287.152			0.2	0.3		23325.51			
4286.045	4287.251	GK	0.4	0.6	1.5	0.52	23324.97	3E 2	2B 2	P 2
4286.192	4287.398			0.2	1.5		23324.17	3D 1	2B 0	P 6
4286.286	4287.492			0.0			23323.66			
4286.468	4287.674			0.2	0.4		23322.67			
4286.576	4287.782			0.1	0.3		23322.08			
4286.725	4287.931						23321.27			
4286.826	4288.032		0.3	0.3			23320.72			
4286.968	4288.174		0.0	0.1			23319.95	3f 3	2c 0	P 5
4287.025	4288.231						23319.64			
4287.148	4288.354			0.1			23318.97			
4287.295	4288.501		0.0	0.1			23318.17			
4287.508	4288.715			0.2d			23317.01			
4287.615	4288.821	GK	0.3	0.6	0.3b	0.48	23316.43	4D 0	2B 3	R 3
4287.782	4288.989			0.1			23315.52			
4287.955	4289.161			0.1	0.2		23314.58			
4288.185	4289.391			0.2			23313.33			
4288.389	4289.596			0.1			23312.22			
4288.595	4289.802			0.2			23311.10			
4288.702	4289.909		0.1	0.5	0.2		23310.52	3E 2	2B 2	Q 3
4288.845	4290.052	G	0.5	0.5		0.52	23309.74	3D 2	2B 2	R 1
4288.980	4290.186		0.0	0.2			23309.01			
4289.208	4290.415	G	0.4	0.5	0.1	0.38	23307.77	4D 0	2B 3	R 1
4289.326	4290.533		0.0				23307.13			
4289.473	4290.680		0.0	0.1			23306.33			
4289.563	4290.770		0.1				23305.84			
4289.644	4290.851	GK	0.7	0.9	0.3b	1.10	23305.40	3D 2	2B 2	R 2
4289.753	4290.960		0.1	0.0			23304.81			
4289.821	4291.028		0.0				23304.44			
4289.917	4291.124		0.0d	0.1			23303.92			
4290.033	4291.240		0.1				23303.29			
4290.114	4291.321	GK	0.7	0.9	0.4b	1.13	23302.85	4D 0	2B 3	R 2
4290.226	4291.433		0.1	0.1			23302.24			
4290.311	4291.518						23301.78			
4290.406	4291.613			0.1	0.2b		23301.26			
4290.497	4291.704		0.0				23300.77			
4290.579	4291.787	G	0.4	0.6		0.56	23300.32	3D 2	2B 2	R 0
4290.579	4291.787	G	0.4	0.6		0.56	23300.32	3c 6	2a 2	Q 2
4290.695	4291.903						23299.69			
4290.938	4292.146				0.2		23298.37			

λ_{air}	λ_{vac}	Ref.	I_1	I_2	I_3	I_4	ν	Upper	Lower	Br.
4291.005	4292.212		0.0	0.2	0.2		23298.01			
4291.121	4292.328				0.9		23297.38			
4291.200	4292.407		0.0		0.9		23296.95			
4291.342	4292.549			0.1			23296.18			
4291.543	4292.750			0.1			23295.09			
4291.797	4293.004			0.1			23293.71			
4291.917	4293.124			0.2	0.3b		23293.06			
4291.981	4293.189				0.4		23292.71			
4292.040	4293.248			0.3	1.5		23292.39	3F 3	2B 4	P 4
4292.134	4293.342	G	0.2	0.6	1.5	0.65	23291.88	3D 2	2B 2	R 3
4292.245	4293.452			0.2	0.4	0.3	23291.28	3F 1	2B 1	R 2
4292.304	4293.511			0.0			23290.96			
4292.350	4293.557			0.0	0.1d		23290.71			
4292.422	4293.629				0.2d		23290.32			
4292.584	4293.792	G	0.1d	0.6	0.4b		23289.44			
4292.715	4293.922	G	0.5	0.7		0.73	23288.73	3E 2	2B 2	Q 2
4292.715	4293.922	G	0.5	0.7		0.73	23288.73	3A 2	2B 3	R 0
4292.842	4294.050			0.2	0.1		23288.04	X 2	2B 3	P 5
4292.940	4294.147			0.0d			23287.51			
4293.061	4294.269		0.1d				23286.85			
4293.113	4294.321			0.2	0.2b		23286.57			
4293.201	4294.409						23286.09			
4293.295	4294.503			0.1d			23285.58			
4293.413	4294.621				0.1		23284.94			
4293.506	4294.714		0.0	0.1			23284.44			
4293.592	4294.800		0.0				23283.97			
4293.780	4294.988			0.1			23282.95			
4294.136	4295.344			0.1			23281.02			
4294.230	4295.439	G	0.2	0.8	0.4b	0.52	23280.51	3F 1	2B 1	Q 4
4294.337	4295.546			0.2			23279.93			
4294.474	4295.682		0.1	0.4			23279.19	4E 1	2B 5	Q 4
4295.109	4296.317		0.1	0.2			23275.75	3E 2	2B 2	Q 1
4295.149	4296.358						23275.53			
4295.242	4296.450		0.0	0.2	0.1		23275.03			
4295.352	4296.561		0.0	0.1	0.3		23274.43			
4295.430	4296.638	GK	0.4	0.9	0.5	0.92	23274.01			
4295.555	4296.764		0.0d	0.2			23273.33			
4295.653	4296.862	GK	0.2	0.9	0.3	0.92	23272.80	3D 2	2B 2	R 4
4295.771	4296.980			0.2			23272.16			
4295.897	4297.105			0.1			23271.48			
4296.070	4297.279			0.1			23270.54			
4296.212	4297.421				0.4		23269.77			
4296.246	4297.454		0.0d	0.3			23269.59			
4296.384	4297.593			0.2			23268.84			
4296.637	4297.846			0.1	1.0b		23267.47			
4296.785	4297.994			0.0d			23266.67	4f 0	2c 0	Q 5
4296.971	4298.180			0.2	0.2		23265.66			
4297.062	4298.271			0.2	0.1		23265.17			
4297.141	4298.350			0.2			23264.74			
4297.197	4298.406			0.1	0.2		23264.44			
4297.296	4298.505				0.3		23263.90			
4297.513	4298.722			0.1	0.5		23262.73			
4297.583	4298.792			0.1			23262.35			
4297.701	4298.910			0.1			23261.71	3D 1	2B 0	P 7
4297.777	4298.986			0.1d	0.4		23261.30			
4297.853	4299.062		0.0				23260.89			
4297.901	4299.110			0.0			23260.63			
4298.026	4299.235	GK	0.4	0.6	1.5b	0.38	23259.95	4D 0	2B 3	P 2
4298.163	4299.372		0.0	0.2	0.2		23259.21	3e 3	2c 0	R 4
4298.315	4299.524			0.2			23258.39			

λ_{air}	λ_{vac}	Ref.	I_1	I_2	I_3	I_4	ν	Upper	Lower	Br.
4298.431	4299.640		0.2	0.3	0.1d		23257.76	4E 1	2B 5	Q 1
4298.472	4299.681	F					23257.54			
4298.520	4299.729			0.2			23257.28	3A 2	2B 3	P 6
4298.780	4299.990			0.3d	0.3		23255.87			
4299.021	4300.230		0.0				23254.57			
4299.130	4300.339		0.1	0.1d	0.2		23253.98			
4299.244	4300.454		0.0		0.3		23253.36			
4299.335	4300.544		0.1	0.3			23252.87			
4299.418	4300.628			0.1			23252.42			
4299.498	4300.707			0.1	0.3		23251.99			
4299.673	4300.883	G		0.5	0.1		23251.04	3D 2	2B 2	R 5
4299.814	4301.023		0.0				23250.28			
4299.895	4301.105			0.1			23249.84		2B 7	Q 2
4299.895	4301.105			0.1			23249.84	T 0	2B 3	P 1
4299.969	4301.179						23249.44			
4300.104	4301.314	1		0.2	1.5b		23248.71			
4300.308	4301.517		0.0d	0.2	1.0b		23247.61			
4300.446	4301.656			0.2			23246.86			
4300.580	4301.789			0.1	0.3		23246.14	3c 6	2a 2	Q 3
4300.724	4301.934						23245.36			
4300.855	4302.065	G	0.1	0.6			23244.65	3D 2	2B 2	R 6
4300.981	4302.191			0.1	0.2		23243.97			
4301.137	4302.347				0.5		23243.13	3f 3	2c 0	P 6
4301.286	4302.496			0.2			23242.32	3e 3	2c 0	R 3
4301.370	4302.580			0.3			23241.87	4E 1	2B 5	Q 3
4301.577	4302.787			0.1			23240.75			
4301.720	4302.930		0.0d	0.2			23239.98			
4301.816	4303.026			0.1d			23239.46			
4301.984	4303.194		0.0d	0.4	0.3b		23238.55			
4302.282	4303.493				0.4		23236.94			
4302.332	4303.543			0.2			23236.67			
4302.419	4303.630						23236.20			
4302.484	4303.695			0.2			23235.85			
4302.660	4303.870			0.2			23234.90			
4302.758	4303.969		0.0d		0.4		23234.37			
4302.836	4304.046			0.1d			23233.95			
4302.906	4304.117		0.0		0.4		23233.57			
4302.981	4304.191		0.1	0.3			23233.17	T 0	2B 3	P 2
4303.105	4304.315		0.1		0.5		23232.50			
4303.249	4304.460		0.1	0.3	0.1		23231.72			
4303.284	4304.495		0.1				23231.53			
4303.362	4304.573		0.5				23231.11	4E 1	2B 5	Q 2
4303.436	4304.647	GK	0.9	1.0d	0.9b	1.75	23230.71	X 0	2B 0	R 2
4303.544	4304.754		0.2	0.2			23230.13			
4303.620	4304.830				0.8		23229.72			
4303.692	4304.902		0.1d	0.2			23229.33			
4303.885	4305.095	GK	0.6	1.0	0.5	1.48	23228.29	4g 0	2a 0	P 1
4303.885	4305.095	GK	0.6	1.0	0.5	1.48	23228.29	X 0	2B 0	R 1
4303.885	4305.095	GK	0.6	1.0	0.5	1.48	23228.29	Z 2	2B 2	R 0
4303.985	4305.195	GK	0.4	0.9	1.5b		23227.75	4f 1	2c 1	R 3
4303.985	4305.195	GK	0.4	0.9	1.5b		23227.75	X 0	2B 0	R 3
4304.122	4305.332		0.1	0.2			23227.01			
4304.261	4305.471		0.0	0.1			23226.26	3c 9	2a 4	Q 1
4304.333	4305.544		0.0				23225.87	3f 3	2c 0	P 6
4304.389	4305.599				0.5		23225.57			
4304.580	4305.790			0.1	0.5		23224.54			
4304.630	4305.840						23224.27			
4304.852	4306.063				0.2		23223.07			
4304.911	4306.122			0.1			23222.75			
4305.087	4306.298			0.1	0.2b		23221.80			

λ_{air}	λ_{vac}	Ref.	I_1	I_2	I_3	I_4	ν	Upper	Lower	Br.
4305.319	4306.530			0.1	0.4		23220.55			
4305.443	4306.654	K			0.5		23219.88			
4305.588	4306.799		0.1d	0.3d			23219.10			
4305.638	4306.849				0.4		23218.83			
4305.825	4307.037	GK	0.2	0.9	0.9	0.65	23217.82	3D 1	2B 0	P10
4305.894	4307.105				0.7		23217.45			
4305.968	4307.179			0.2			23217.05			
4306.113	4307.324		0.2	0.2	0.6		23216.27	3e 3	2c 0	R 2
4306.113	4307.324		0.2	0.2	0.6		23216.27	3D 1	2B 0	P 8
4306.198	4307.409		0.2				23215.81			
4306.285	4307.497	GK	1.0b	1.0	0.4b	1.55	23215.34	X 0	2B 0	R 0
4306.391	4307.602		0.3	0.2			23214.77			
4306.497	4307.708		0.1				23214.20			
4306.612	4307.823		0.0		0.2		23213.58			
4306.677	4307.888		0.4	0.3	0.6		23213.23	3A 2	2B 3	P 1
4306.855	4308.066				0.6		23212.27			
4307.003	4308.215		0.0				23211.47			
4307.048	4308.259		0.0				23211.23	3A 2	2B 3	P 5
4307.104	4308.315		0.0	0.2d			23210.93			
4307.137	4308.348		0.0				23210.75	4f 2	2c 2	R 4
4307.308	4308.519				0.7		23209.83			
4307.375	4308.586			0.1			23209.47			
4307.432	4308.644		0.1				23209.16			
4307.510	4308.722	GK	0.5	0.6	0.3	0.59	23208.74	Z 2	2B 2	R 1
4307.579	4308.790				0.7		23208.37			
4307.636	4308.848		0.1	0.0	0.3		23208.06			
4307.746	4308.957				0.2		23207.47			
4307.829	4309.041			0.1			23207.02			
4307.902	4309.113			0.3			23206.63			
4308.007	4309.219	G	0.3	0.5	0.7		23206.06	3A 2	2B 3	P 4
4308.169	4309.381		0.0		0.7		23205.19			
4308.254	4309.466		0.0d	0.2	0.2		23204.73			
4308.399	4309.611		0.3	0.4	0.2		23203.95	3A 2	2B 3	P 3
4308.442	4309.654		0.2				23203.72			
4308.499	4309.711		0.1				23203.41			
4308.561	4309.773		0.3	0.2	0.7		23203.08			
4308.641	4309.853	GK	1.0b	0.9	0.5	1.21	23202.65	4c 4	2a 3	Q 1
4308.641	4309.853	GK	1.0b	0.9	0.5	1.21	23202.65	T 0	2B 3	P 3
4308.750	4309.962		0.2	0.2d			23202.06			
4308.826	4310.038		0.0				23201.65			
4308.902	4310.114		0.1		0.7		23201.24			
4308.990	4310.202			0.1	0.2		23200.77			
4309.071	4310.284		0.0	0.1			23200.33			
4309.103	4310.315		0.0		0.6		23200.16			
4309.153	4310.365			0.0			23199.89			
4309.254	4310.466		0.0	0.6			23199.35			
4309.281	4310.493		0.1				23199.20			
4309.372	4310.584			0.1	0.6		23198.71			
4309.551	4310.763			0.2d			23197.75			
4309.640	4310.852			0.1d	0.3		23197.27	3D 1	2B 0	P 9
4309.701	4310.913				0.7		23196.94			
4309.824	4311.036				0.2		23196.28			
4309.898	4311.110			0.1	0.2		23195.88			
4310.013	4311.226				0.2		23195.26			
4310.108	4311.320			0.1	1.5b		23194.75	3D 2	2B 2	P 2
4310.197	4311.410				0.4		23194.27			
4310.348	4311.560		0.0		0.2		23193.46			
4310.471	4311.683			0.3	1.0		23192.80			
4310.547	4311.759				0.2		23192.39			
4310.612	4311.824			0.1			23192.04			

λ_{air}	λ_{vac}	Ref.	I_1	I_2	I_3	I_4	ν	Upper	Lower	Br.
4310.688	4311.900				0.9		23191.63			
4310.733	4311.945				0.7		23191.39			
4310.820	4312.032			0.2d			23190.92			
4310.930	4312.142				0.4		23190.33			
4311.000	4312.213	G	0.1	0.6	0.9		23189.95			
4311.155	4312.367		0.0	0.2	1.0b		23189.12			
4311.337	4312.549			0.2	0.4		23188.14			
4311.433	4312.646				1.5		23187.62			
4311.486	4312.698		0.4	0.4	1.5		23187.34	3A 2	2B 3	P 2
4311.590	4312.802		0.0	0.0			23186.78			
4311.716	4312.929	GK	0.4	0.8	1.5b	0.85	23186.10	X 0	2B 0	R 4
4311.843	4313.055		0.0d				23185.42			
4311.883	4313.096				0.2		23185.20			
4312.092	4313.305				1.0		23184.08	T 0	2B 3	P 4
4312.151	4313.364				1.0		23183.76			
4312.218	4313.431		0.0				23183.40			
4312.298	4313.511		0.1	0.2	1.5b		23182.97			
4312.401	4313.613	GK	0.4	0.6		0.64	23182.42	4c 4	2a 3	Q 2
4312.455	4313.667				0.2		23182.13			
4312.540	4313.753				0.3		23181.67			
4312.598	4313.811			0.2	0.8		23181.36	4f 0	2c 0	R 2
4312.709	4313.922		0.0	0.2	0.7		23180.76			
4312.793	4314.006		0.0	0.1			23180.31			
4312.882	4314.095	GK	0.6	0.8	1.0	1.05	23179.83	Z 2	2B 2	R 2
4312.966	4314.179		0.1				23179.38			
4313.024	4314.237		0.0	0.2	1.0b		23179.07			
4313.214	4314.427				0.2		23178.05			
4313.322	4314.535			0.1d			23177.47			
4313.377	4314.591				0.2b		23177.17			
4313.433	4314.646			0.1			23176.87			
4313.597	4314.810			0.2	1.0		23175.99			
4313.658	4314.872				1.0		23175.66	3c 6	2a 2	Q 4
4313.822	4315.036		0.0				23174.78			
4313.882	4315.095			0.1	0.3		23174.46			
4314.012	4315.226			0.1			23173.76			
4314.183	4315.397	G	0.1	0.5			23172.84	3F 1	2B 1	Q 3
4314.219	4315.432				1.0b		23172.65			
4314.355	4315.568						23171.92			
4314.539	4315.753		0.0	0.2			23170.93			
4314.597	4315.810				0.2		23170.62			
4314.669	4315.883				0.4		23170.23	4g 0	2a 0	R 0
4314.725	4315.939		0.0	0.3			23169.93			
4314.859	4316.073			0.2d			23169.21	4f 2	2c 2	R 4
4315.033	4316.246			0.1d			23168.28			
4315.291	4316.505			0.2			23166.89			
4315.480	4316.693			0.1			23165.88	3E 1	2B 1	Q10
4315.603	4316.816			0.3			23165.22	3e 3	2c 0	R 1
4315.692	4316.906			0.0			23164.74			
4315.763	4316.977	G	0.2	0.6	0.3b		23164.36	4g 1	2a 1	P 1
4315.888	4317.101			0.2			23163.69			
4316.074	4317.288		0.2	0.3			23162.69	3F 1	2B 1	R 1
4316.227	4317.441		0.2	0.2			23161.87			
4316.314	4317.528			0.1			23161.40			
4316.452	4317.666			0.2			23160.66			
4316.726	4317.940			0.1			23159.19			
4316.816	4318.030			0.1			23158.71			
4316.862	4318.076						23158.46			
4316.970	4318.185	G	0.2d	0.3			23157.88			
4317.248	4318.462			0.1			23156.39			
4317.317	4318.531	K					23156.02			

λ_air	λ_vac	Ref.	I₁	I₂	I₃	I₄	ν	Upper	Lower	Br.
4317.616	4318.830			0.1			23154.42			
4317.804	4319.018			0.2d			23153.41			
4317.888	4319.102		0.0				23152.96			
4318.007	4319.222	GK	0.4	0.7		0.76	23152.32	4c 4	2a 3	Q 3
4318.108	4319.322		0.0	0.2			23151.78			
4318.227	4319.442	GK	0.8	0.7	0.2d	0.95	23151.14	X 0	2B 0	P 1
4318.330	4319.544		0.2				23150.59			
4318.578	4319.793			0.0d			23149.26			
4319.063	4320.278						23146.66			
4319.190	4320.405				0.4		23145.98			
4319.250	4320.464			0.0d			23145.66			
4319.653	4320.868			0.3	0.1d		23143.50			
4319.877	4321.092		0.0d	0.3			23142.30			
4319.989	4321.204			0.1			23141.70			
4320.086	4321.301						23141.18	4g 1	2a 1	R 0
4320.399	4321.615						23139.50	4D 0	2B 3	P 3
4320.547	4321.762			0.1			23138.71			
4320.652	4321.867	G	0.2	0.5		0.34	23138.15	Z 2	2B 2	R 3
4320.652	4321.867	G	0.2	0.5		0.34	23138.15	4f 1	2c 1	Q 5
4320.741	4321.956						23137.67			
4320.850	4322.065			0.1			23137.09			
4320.993	4322.209		0.0	0.4			23136.32	W 2	2B 3	R 4
4321.244	4322.459		0.1	0.1d			23134.98	3D 2	2B 2	P 3
4321.608	4322.823						23133.03			
4321.800	4323.016		0.0	0.1d			23132.00			
4322.075	4323.290			0.2			23130.53	4f 0	2c 0	Q 4
4322.219	4323.434			0.1			23129.76			
4322.353	4323.569						23129.04			
4322.645	4323.861			0.2			23127.48			
4322.787	4324.003			0.1			23126.72			
4322.892	4324.107			0.3			23126.16			
4323.224	4324.440		0.0				23124.38			
4323.290	4324.506		0.0				23124.03			
4323.367	4324.582			0.1			23123.62			
4323.497	4324.713	G	0.2	0.4			23122.92	Z 2	2B 2	P 2
4323.580	4324.796		0.2d	0.2			23122.48			
4323.705	4324.921				0.4		23121.81			
4323.823	4325.039		0.0d		0.2		23121.18			
4323.868	4325.084						23120.94			
4324.124	4325.340		0.0d	0.3			23119.57			
4324.270	4325.486			0.3			23118.79	U 1	2B 1	R 4
4324.481	4325.697			0.2			23117.66			
4324.597	4325.813			0.1			23117.04	3F 1	2B 1	P 5
4324.704	4325.920			0.2			23116.47			
4324.874	4326.090			0.2b			23115.56			
4324.973	4326.190			0.2			23115.03			
4325.108	4326.324			0.2			23114.31			
4325.166	4326.382			0.3			23114.00			
4325.265	4326.481			0.2			23113.47			
4325.336	4326.553			0.1			23113.09			
4325.437	4326.654			0.4			23112.55	4c 4	2a 3	Q 4
4325.503	4326.719		0.0d				23112.20			
4325.576	4326.792		0.0	0.2			23111.81			
4325.684	4326.901		0.0	0.2			23111.23			
4325.806	4327.023			0.4b	0.4b		23110.58	4f 0	2c 0	R 3
4326.038	4327.255			0.2			23109.34			
4326.173	4327.389			0.1	0.3b		23108.62			
4326.358	4327.575			0.1			23107.63			
4326.484	4327.700			0.2			23106.96			
4326.589	4327.805		0.0	0.2			23106.40			

λ_{air}	λ_{vac}	Ref.	I_1	I_2	I_3	I_4	ν	Upper	Lower	Br.
4326.727	4327.944		0.0d				23105.66			
4326.881	4328.097			0.2			23104.84			
4326.963	4328.180			0.1			23104.40			
4327.074	4328.290			0.4	0.2		23103.81			
4327.210	4328.427			0.2			23103.08			
4327.364	4328.581	GK	0.2	0.8	0.6b	0.51	23102.26	3e 3	2c 0	R 2
4327.495	4328.712			0.1			23101.56			
4327.638	4328.854			0.1			23100.80			
4327.755	4328.972		0.1d	0.2			23100.17			
4327.844	4329.061		0.1	0.1			23099.70	3e 3	2c 0	R 1
4327.932	4329.149	GK	1.0b	1.0	0.3b	1.45	23099.23	X 0	2B 0	P 2
4328.038	4329.255		0.2	0.2			23098.66			
4328.207	4329.424				0.3		23097.76			
4328.271	4329.488			0.1d			23097.42			
4328.385	4329.602						23096.81			
4328.529	4329.747			0.0d			23096.04			
4328.827	4330.045				0.2		23094.45			
4328.921	4330.138				0.2		23093.95			
4329.032	4330.249			0.1d			23093.36			
4329.167	4330.384				0.2		23092.64			
4329.240	4330.457		0.0	0.2d	0.2		23092.25			
4329.307	4330.525		0.0d				23091.89			
4329.373	4330.590			0.2			23091.54			
4329.474	4330.692			0.2			23091.00			
4329.547	4330.765		0.0				23090.61			
4329.654	4330.872	GK	0.2	0.8	0.9b	0.40	23090.04			
4329.778	4330.995		0.0	0.1			23089.38			
4329.885	4331.102		0.1d	0.3			23088.81	W 2	2B 3	R 3
4329.947	4331.164				0.2		23088.48			
4330.005	4331.222				0.2		23088.17	3e 3	2c 0	R 3
4330.050	4331.267		0.3	0.3			23087.93			
4330.121	4331.339		0.0				23087.55			
4330.177	4331.395		0.0	0.1			23087.25			
4330.277	4331.494	K	0.1	0.7	0.6b		23086.72			
4330.408	4331.626	GK	0.5	0.8		0.77	23086.02	3F 1	2B 1	Q 2
4330.408	4331.626	GK	0.5	0.8		0.77	23086.02	3c 3	2a 0	R 3
4330.558	4331.776		0.0	0.2			23085.22			
4331.025	4332.243			0.1	0.2b		23082.73			
4331.202	4332.420			0.2			23081.79			
4331.290	4332.508		0.0				23081.32			
4331.380	4332.598	GK	0.3	0.7	0.2b	0.45	23080.84	W 2	2B 3	R 2
4331.380	4332.598	GK	0.3	0.7	0.2b	0.45	23080.84	Z 2	2B 2	R 4
4331.586	4332.804			0.1	0.4		23079.74			
4331.673	4332.891				0.4		23079.28			
4331.787	4333.005				0.4		23078.67			
4331.812	4333.030			0.1			23078.54			
4331.913	4333.131			0.4			23078.00			
4332.050	4333.268			0.1	0.5		23077.27			
4332.116	4333.334			0.2			23076.92			
4332.185	4333.403				0.5		23076.55			
4332.238	4333.456			0.2			23076.27	4F 0	2B 4	R 3
4332.317	4333.535			0.1	0.5		23075.85			
4332.418	4333.636		0.2d	0.5	0.6		23075.31			
4332.521	4333.740		0.1	0.2	0.6		23074.76			
4332.619	4333.837	GK	0.9b	1.0	1.0b	1.29	23074.24	Y 1	2B 0	R 0
4332.619	4333.837	GK	0.9b	1.0	1.0b	1.29	23074.24	Y 1	2B 0	R 1
4332.640	4333.858	F					23074.13			
4332.734	4333.952		0.2	0.2			23073.63			
4332.782	4334.001				0.2		23073.37			
4332.824	4334.042			0.1			23073.15			

λ_{air}	λ_{vac}	Ref.	I_1	I_2	I_3	I_4	ν	Upper	Lower	Br.
4332.929	4334.147		0.0	0.3	0.2		23072.59	W 2	2B 3	R 1
4333.038	4334.256		0.0	0.1			23072.01			
4333.231	4334.450			0.0			23070.98			
4333.366	4334.585		0.0d	0.3	0.3		23070.26			
4333.483	4334.701			0.2			23069.64	4D 0	2B 3	P 4
4333.566	4334.784	1		0.2	1.5b		23069.20			
4333.650	4334.869			0.1			23068.75			
4333.768	4334.987			0.2			23068.12			
4333.840	4335.058				0.3		23067.74			
4334.018	4335.237			0.2	0.3		23066.79			
4334.214	4335.432				0.2		23065.75			
4334.285	4335.504			0.2			23065.37			
4334.405	4335.624		0.2	0.3			23064.73	3c 3	2a 0	R 1
4334.561	4335.780						23063.90			
4334.682	4335.900	GK		0.4	0.4		23063.26	4c 4	2a 3	Q 5
4334.806	4336.025		0.3	0.4	0.4		23062.60	3D 2	2B 2	P 4
4334.922	4336.141			0.1			23061.98			
4334.982	4336.201						23061.66			
4335.046	4336.265			0.2			23061.32			
4335.125	4336.344			0.1			23060.90			
4335.285	4336.504		0.0				23060.05			
4335.353	4336.572	1	0.3	0.4	1.0b		23059.69			
4335.435	4336.654		0.0	0.1			23059.25			
4335.522	4336.741	GK	0.8	1.0	0.5b	1.20	23058.79	Y 1	2B 0	R 2
4335.637	4336.856		0.2	0.2			23058.18			
4335.736	4336.955				0.2		23057.65			
4335.815	4337.034			0.2d			23057.23			
4335.954	4337.174		0.1	0.3			23056.49	W 2	2B 3	R 0
4336.056	4337.275			0.2			23055.95			
4336.094	4337.313		0.2				23055.75			
4336.144	4337.364		0.2	0.3	0.4b		23055.48			
4336.227	4337.446		0.0				23055.04			
4336.325	4337.544	GK	0.3	0.7	0.3b		23054.52			
4336.487	4337.706			0.0			23053.66			
4336.714	4337.934		0.0d				23052.45			
4336.761	4337.981			0.3	0.2		23052.20			
4336.792	4338.011		0.0				23052.04			
4336.982	4338.201		0.0		0.2b		23051.03			
4337.226	4338.446			0.2			23049.73			
4337.288	4338.508			0.1			23049.40			
4337.394	4338.613	G	0.0	0.5	0.2b		23048.84			
4337.508	4338.728		0.0				23048.23			
4337.552	4338.771			0.3	0.3b		23048.00			
4337.783	4339.003			0.2			23046.77	4g 0	2a 0	R 1
4337.971	4339.191		0.0				23045.77			
4338.045	4339.265						23045.38	4g 1	2a 1	R 1
4338.169	4339.389			0.2			23044.72			
4338.259	4339.479		0.3				23044.24	4f 1	2c 1	R 2
4338.259	4339.479		0.3				23044.24	3e 3	2c 0	Q 2
4338.278	4339.498			0.5d			23044.14			
4338.314	4339.534		0.3		0.9b		23043.95	U 1	2B 1	R 3
4338.369	4339.588			0.2d			23043.66	3E 1	2B 1	Q 9
4338.553	4339.773			0.2			23042.68			
4338.615	4339.835			0.1d	0.4		23042.35			
4338.698	4339.918			0.3d			23041.91	3e 3	2c 0	Q 1
4338.834	4340.054				0.4		23041.19			
4339.024	4340.244			0.0d			23040.18			
4339.079	4340.298				0.5		23039.89			
4339.143	4340.363				0.5		23039.55			
4339.252	4340.472				0.5		23038.97			

λ_{air}	λ_{vac}	Ref.	I_1	I_2	I_3	I_4	ν	Upper	Lower	Br.
4339.350	4340.570			0.2d			23038.45			
4339.444	4340.664				1.0		23037.95			
4339.546	4340.766	GK	0.4	0.7			23037.41	3F 1	2B 1	P 4
4339.677	4340.898		0.0	0.2			23036.71	3e 3	2c 0	Q 3
4339.830	4341.050	GK	0.5	0.7			23035.90	X 0	2B 0	P 3
4339.928	4341.148			0.0			23035.38			
4340.002	4341.222		0.4b	0.1			23034.99			
4340.064	4341.284						23034.66			
4340.113	4341.333				1.5b		23034.40			
4340.218	4341.438		0.4				23033.84	Z 2	2B 2	P 3
4340.463	4341.684	GK	1.5b	1.5b			23032.54			
4340.814	4342.034		0.4b	0.2			23030.68			
4341.202	4342.423		0.2	0.1			23028.62	3c 3	2a 0	R 0
4341.409	4342.630		0.0	0.3			23027.52			
4341.517	4342.737			0.2			23026.95			
4341.628	4342.849			0.2			23026.36	4D 0	2B 3	P 8
4341.804	4343.024	G	0.1	0.4			23025.43			
4342.250	4343.471			0.2d	0.2b		23023.06			
4342.592	4343.813			0.3d			23021.25			
4342.628	4343.849		0.4b		1.0b		23021.06			
4342.654	4343.875			0.4d			23020.92			
4342.731	4343.952						23020.51			
4342.782	4344.003		0.0				23020.24			
4342.850	4344.071			0.1			23019.88	4D 0	2B 3	P 5
4343.175	4344.396				0.4		23018.16			
4343.297	4344.519				0.2		23017.51	3e 3	2c 0	Q 1
4343.433	4344.654						23016.79	4f 0	2c 0	R 1
4343.488	4344.709			0.3b	0.5		23016.50			
4343.609	4344.830	GK	0.5	0.7	0.5	0.41	23015.86	Y 1	2B 0	P 1
4343.684	4344.906		0.0		0.5		23015.46			
4343.964	4345.185			0.2d	0.5		23013.98			
4344.000	4345.221		0.0d				23013.79			
4344.109	4345.330		0.0	0.1			23013.21			
4344.266	4345.487		0.2	0.3d			23012.38			
4344.370	4345.591			0.1			23011.83			
4344.506	4345.727						23011.11			
4344.589	4345.810		0.0				23010.67			
4344.759	4345.980		0.1				23009.77			
4344.796	4346.018			0.3b	0.4b		23009.57			
4344.823	4346.044		0.1				23009.43			
4344.868	4346.090			0.1			23009.19			
4345.155	4346.377	i			1.0b		23007.67			
4345.202	4346.424		0.0	0.2			23007.42	U 1	2B 1	R 1
4345.263	4346.484		0.0				23007.10			
4345.331	4346.552		0.2	0.3			23006.74	U 1	2B 1	R 0
4345.726	4346.947			0.5	0.4		23004.65	4f 0	2c 0	R 2
4345.726	4346.947			0.5	0.4		23004.65	4c 4	2a 3	Q 6
4345.750	4346.972				0.5		23004.52	3e 3	2c 0	Q 4
4345.890	4347.112			0.3d	0.4		23003.78	U 1	2B 1	R 2
4346.132	4347.353						23002.50			
4346.204	4347.425		0.0				23002.12			
4346.275	4347.497		0.0				23001.74			
4346.317	4347.539				0.4b		23001.52	4f 1	2c 1	Q 4
4346.351	4347.573			0.4			23001.34	4f 0	2c 0	Q 3
4346.436	4347.658						23000.89	4f 2	2c 2	R 3
4346.493	4347.715			0.1			23000.59			
4346.534	4347.756						23000.37	4f 2	2c 2	R 3
4346.674	4347.896	GK	0.2	0.6	0.3	0.26	22999.63			
4346.799	4348.021			0.1			22998.97			
4346.956	4348.178				0.2		22998.14	4D 0	2B 3	P 7

λ_{air}	λ_{vac}	Ref.	I_1	I_2	I_3	I_4	ν	Upper	Lower	Br.
4346.997	4348.219			0.1			22997.92	W 2	2B 3	P 1
4347.230	4348.452			0.3			22996.69			
4347.353	4348.575			0.1			22996.04			
4347.426	4348.648						22995.65	4D 0	2B 3	P 6
4347.472	4348.694	K		0.3			22995.41			
4347.576	4348.798		0.1	0.3	0.5		22994.86			
4347.674	4348.896		0.1	0.3			22994.34	4E 2	2B 7	Q 2
4347.850	4349.072		0.0d	0.4	0.5		22993.41			
4347.960	4349.182				0.7		22992.83			
4348.336	4349.558				0.5		22990.84			
4348.398	4349.621						22990.51			
4348.603	4349.825		0.0	0.0			22989.43			
4348.788	4350.010						22988.45			
4348.943	4350.166		0.0	0.2			22987.63			
4349.011	4350.234		0.0				22987.27			
4349.076	4350.298		0.1	0.3			22986.93			
4349.515	4350.737	F					22984.61	4e 0	2c 0	R 6
4349.675	4350.898			0.0d			22983.76	3A 0	2B 0	R 2
4349.833	4351.055			0.3			22982.93	3D 2	2B 2	P 5
4349.950	4351.173			0.1d			22982.31			
4350.219	4351.442		0.0		0.4		22980.89			
4350.349	4351.572		0.0	0.1	0.3b		22980.20			
4350.425	4351.648	G	0.3	0.5		0.20	22979.80	3F 1	2B 1	P 3
4350.571	4351.794			0.2d	0.2		22979.03			
4350.656	4351.879		0.0	0.1			22978.58	3e 3	2c 0	Q 2
4350.747	4351.970		0.5	0.3			22978.10			
4350.825	4352.048		0.1	0.6	0.5b		22977.69			
4351.048	4352.271			0.1	0.2		22976.51			
4351.292	4352.515		0.0d	0.3	0.2b		22975.22			
4351.455	4352.678			0.2			22974.36			
4351.628	4352.851			0.1			22973.45			
4351.709	4352.932		0.0d	0.2	0.4		22973.02			
4351.828	4353.052		0.0				22972.39			
4351.961	4353.184			0.2			22971.69			
4352.050	4353.273		0.0	0.2	0.5		22971.22			
4352.092	4353.315			0.2			22971.00	4g 0	2a 0	P 2
4352.156	4353.379	GK	0.4	0.5		0.23	22970.66			
4352.241	4353.465				0.4		22970.21			
4352.372	4353.596			0.1d			22969.52			
4352.550	4353.774				0.4		22968.58			
4352.740	4353.963			0.1			22967.58			
4353.003	4354.227		0.0d				22966.19			
4353.096	4354.320		0.1d	0.1			22965.70			
4353.217	4354.441			0.1			22965.06			
4353.295	4354.519			0.0			22964.65	3e 3	2c 0	Q 5
4353.415	4354.638	GK	0.4	0.7	0.3	0.76	22964.02	X 0	2B 0	P 4
4353.528	4354.752			0.0			22963.42			
4353.606	4354.830		0.1d	0.0			22963.01			
4353.652	4354.875						22962.77			
4353.801	4355.025			0.1d			22961.98			
4354.065	4355.289						22960.59			
4354.179	4355.403	G	0.2	0.4			22959.99	3c 3	2a 0	Q 1
4354.355	4355.579			0.1			22959.06			
4354.457	4355.681		0.0	0.2d	0.6b		22958.52			
4354.539	4355.763	GK	0.6	0.8	0.3	1.01	22958.09	Y 1	2B 0	P 2
4354.643	4355.867		0.1	0.0			22957.54			
4354.717	4355.941				0.3		22957.15	4e 0	2c 0	R 5
4354.882	4356.106			0.1d			22956.28			
4354.968	4356.192				0.2		22955.83			
4355.061	4356.285		0.1	0.2	0.1		22955.34	W 2	2B 3	P 2

λ_{air}	λ_{vac}	Ref.	I_1	I_2	I_3	I_4	ν	Upper	Lower	Br.
4355.328	4356.552				0.2		22953.93			
4355.512	4356.737			0.1			22952.96			
4355.634	4356.858			0.1			22952.32			
4355.715	4356.940			0.2	0.4		22951.89			
4355.808	4357.033	G	0.1	0.5	0.3	0.30	22951.40	4g 1	2a 1	P 2
4355.922	4357.147			0.1			22950.80			
4355.991	4357.215				0.8		22950.44			
4356.034	4357.259		0.0	0.4	0.9		22950.21			
4356.146	4357.371						22949.62			
4356.203	4357.428			0.0			22949.32			
4356.355	4357.579				0.4		22948.52			
4356.520	4357.745		0.0d				22947.65			
4356.600	4357.824				0.4		22947.23			
4356.928	4358.153			0.0			22945.50	4g 1	2a 1	R 2
4357.061	4358.286	G	0.1	0.4	0.2		22944.80	4f 2	2c 2	Q 5
4357.188	4358.413			0.1			22944.13			
4357.388	4358.613			0.1			22943.08			
4357.479	4358.704			0.1			22942.60			
4357.566	4358.791			0.0			22942.14			
4357.661	4358.886			0.3			22941.64			
4357.745	4358.970		0.0				22941.20			
4357.802	4359.027				0.2b		22940.90			
4357.859	4359.084		0.2	0.2d			22940.60			
4357.965	4359.190			0.0d			22940.04			
4358.049	4359.274						22939.60			
4358.163	4359.388	1					22939.00			
4358.224	4359.449	1					22938.68			
4358.328	4359.553	K		0.2d	0.2		22938.13			
4358.416	4359.641		0.0	0.1			22937.67			
4358.495	4359.721	1					22937.25			
4358.541	4359.766	G	0.2				22937.01	Z 2	2B 2	P 4
4358.541	4359.766	G	0.2				22937.01	4c 4	2a 3	Q 7
4358.573	4359.798		0.0	0.4	0.3		22936.84	3E 1	2B 1	Q 8
4358.678	4359.903		0.0		0.3		22936.29	3E 1	2B 1	R 4
4358.792	4360.017				0.1		22935.69	3E 3	2B 4	Q 8
4358.952	4360.177				0.2		22934.85			
4359.016	4360.241		0.0d				22934.51			
4359.140	4360.365		0.0				22933.86			
4359.334	4360.559				0.2		22932.84			
4359.444	4360.669		0.1	0.2			22932.26			
4359.495	4360.720				0.4		22931.99			
4359.602	4360.827		0.0	0.1			22931.43			
4359.739	4360.964				0.4		22930.71			
4360.029	4361.255			0.2b			22929.18			
4360.113	4361.339		0.0	0.1			22928.74			
4360.290	4361.516				0.4		22927.81			
4360.347	4361.573		0.0d	0.2			22927.51	3c 3	2a 0	Q 2
4360.459	4361.685		0.2	0.5	0.7b		22926.92	4f 1	2c 1	R 3
4360.697	4361.923			0.0d			22925.67	w 0	2c 1	P 1
4360.868	4362.094		0.0	0.4d	0.2b		22924.77			
4360.923	4362.149				.		22924.48			
4361.000	4362.225			0.1			22924.08			
4361.146	4362.372		0.0				22923.31	4g 0	2a 0	R 2
4361.243	4362.469		0.0				22922.80			
4361.331	4362.556						22922.34	3e 3	2c 0	Q 3
4361.443	4362.669			0.1			22921.75			
4361.544	4362.769		0.0	0.0			22921.22	3e 3	2c 0	P 2
4361.648	4362.874		0.0	0.4	0.2		22920.67			
4361.753	4362.979		0.0	0.0			22920.12			
4361.922	4363.148	GK	0.2	0.7	1.0b	0.45	22919.23	4e 0	2c 0	R 4

λ_{air}	λ_{vac}	Ref.	I_1	I_2	I_3	I_4	ν	Upper	Lower	Br.
4361.922	4363.148	GK	0.2	0.7	1.0b	0.45	22919.23	4f 0	2c 0	P 6
4362.261	4363.487		0.0				22917.45			
4362.353	4363.579		0.0				22916.97			
4362.453	4363.680		0.0d				22916.44			
4362.543	4363.769				0.2		22915.97			
4362.667	4363.893			0.1			22915.32			
4362.754	4363.980				0.2		22914.86			
4362.834	4364.060			0.4	0.7		22914.44			
4362.969	4364.196		0.2	0.2			22913.73			
4363.097	4364.323			0.2	0.3		22913.06			
4363.291	4364.517				0.3		22912.04			
4363.344	4364.571			0.1			22911.76			
4363.461	4364.687			0.1			22911.15			
4363.512	4364.738		0.0				22910.88			
4363.625	4364.851	K					22910.29			
4363.800	4365.026	1			0.2		22909.37			
4363.887	4365.114			0.2			22908.91			
4364.030	4365.257			0.3	0.5		22908.16			
4364.171	4365.398			0.2	0.4		22907.42			
4364.503	4365.729			0.1d			22905.68			
4364.667	4365.893		0.1d	0.3			22904.82	W 2	2B 3	P 3
4364.787	4366.013		0.0	0.1			22904.19			
4364.932	4366.158	G		0.6	0.6b		22903.43	4f 2	2c 2	Q 5
4365.105	4366.332		0.0				22902.52			
4365.210	4366.437						22901.97			
4365.313	4366.540						22901.43			
4365.549	4366.776			0.4			22900.19	3D 2	2B 2	P 6
4365.599	4366.826		0.0		0.4b		22899.93			
4365.787	4367.014			0.1			22898.94			
4365.839	4367.066		0.0				22898.67			
4365.923	4367.150		0.0		0.2b		22898.23			
4366.047	4367.274		0.0				22897.58			
4366.115	4367.342		0.0				22897.22			
4366.169	4367.396			0.1			22896.94			
4366.352	4367.579			0.3d			22895.98	4f 0	2c 0	Q 2
4366.352	4367.579			0.3d			22895.98	3e 3	2c 0	P 2
4366.489	4367.716			0.0d	0.3b		22895.26			
4366.579	4367.806		0.0				22894.79			
4366.667	4367.894	GK	0.2	0.7	0.6		22894.33			
4366.785	4368.012			0.1			22893.71			
4366.937	4368.165		0.0	0.4	0.6		22892.91			
4367.042	4368.270			0.1			22892.36			
4367.157	4368.384		0.2	0.4			22891.76	3A 0	2B 0	R 1
4367.304	4368.531			0.1	0.2		22890.99			
4367.359	4368.586						22890.70			
4367.468	4368.695			0.0	0.1		22890.13			
4367.573	4368.800			0.2			22889.58			
4367.615	4368.842		0.0		0.2		22889.36	X 0	2B 0	P 5
4367.733	4368.960	GK	0.4	0.8b	1.0b	0.83	22888.74	4f 0	2c 0	R 1
4367.760	4368.987		0.2				22888.60			
4368.000	4369.228		0.0	0.0d			22887.34			
4368.164	4369.392				0.2		22886.48			
4368.248	4369.476	K					22886.04			
4368.435	4369.663		0.0	0.3	0.5b		22885.06			
4368.500	4369.728	G	0.3	0.3		0.62	22884.72			
4368.626	4369.854		0.0	0.1			22884.06			
4368.781	4370.009		0.1	0.3			22883.25	4f 1	2c 1	R 1
4368.871	4370.098			0.0			22882.78	w 0	2c 1	Q 1
4368.947	4370.175		0.0				22882.38			
4369.029	4370.257	G	0.4	0.4	0.2	0.30	22881.95	Y 2	2B 3	R 0

λ_{air}	λ_{vac}	Ref.	I_1	I_2	I_3	I_4	ν	Upper	Lower	Br.
4369.029	4370.257	G	0.4	0.4	0.2	0.30	22881.95	Y 1	2B 0	P 3
4369.147	4370.375		0.0				22881.33			
4369.210	4370.438			0.2			22881.00			
4369.302	4370.530		0.0	0.3	0.2		22880.52	Y 2	2B 3	R 1
4369.438	4370.666			0.2			22879.81			
4369.552	4370.780		0.0	0.3	0.2		22879.21	3c 3	2a 0	Q 3
4369.703	4370.931		0.0		0.3		22878.42			
4369.850	4371.078				0.3		22877.65			
4369.957	4371.185			0.1d			22877.09			
4370.079	4371.308			0.2	0.2		22876.45			
4370.165	4371.394		0.0	0.2			22876.00			
4370.276	4371.504	GK	0.1	0.7	0.9b		22875.42	4e 0	2c 0	R 3
4370.341	4371.569		0.0				22875.08			
4370.429	4371.657			0.0			22874.62			
4370.595	4371.824			0.2			22873.75	w 0	2c 1	R 1
4370.767	4371.996	GK	0.2	0.8	1.5b	0.74	22872.85	4f 1	2c 1	Q 3
4370.884	4372.112			0.1			22872.24	4f 0	2c 0	Q 4
4371.029	4372.257				0.4		22871.48			
4371.260	4372.489		0.0d	0.1			22870.27	3E 1	2B 1	R 2
4371.312	4372.540				0.1		22870.00			
4371.404	4372.632				0.2		22869.52			
4371.547	4372.776						22868.77			
4371.622	4372.850		0.0	0.0			22868.38			
4371.679	4372.908		0.0				22868.08			
4371.820	4373.049			0.1			22867.34			
4371.908	4373.137			0.0			22866.88			
4371.998	4373.227	GK	0.2	0.6	0.3b	0.38	22866.41			
4372.098	4373.326			0.2	0.2		22865.89			
4372.180	4373.409			0.0			22865.46			
4372.254	4373.483			0.2			22865.07	Y 2	2B 3	R 2
4372.292	4373.521		0.0d		0.1		22864.87			
4372.450	4373.678		0.0d	0.4	0.4b		22864.05	3e 2	2c 0	P 3
4372.702	4373.931		0.0	0.2	0.2		22862.73			
4372.874	4374.103		0.1	0.3	0.2		22861.83			
4372.953	4374.181		0.0				22861.42			
4373.029	4374.258			0.0			22861.02			
4373.337	4374.566			0.1d			22859.41	3E 3	2B 4	Q 7
4373.379	4374.608				0.2		22859.19			
4374.034	4375.263			0.0			22855.77			
4374.227	4375.456				0.3		22854.76			
4374.286	4375.515			0.3			22854.45			
4374.422	4375.651		0.0	0.4	0.3b		22853.74	4g 1	2a 1	R 3
4374.583	4375.812			0.1			22852.90			
4374.792	4376.021			0.3			22851.81			
4374.832	4376.061		0.0d				22851.60			
4374.964	4376.193						22850.91			
4375.010	4376.239		0.0d	0.2			22850.67			
4375.058	4376.287				0.3		22850.42	3e 3	2c 0	Q 4
4375.230	4376.460			0.0			22849.52			
4375.303	4376.532						22849.14			
4375.389	4376.618			0.1			22848.69			
4375.458	4376.687		0.0				22848.33			
4375.548	4376.777	G	0.3	0.5			22847.86	W 2	2B 3	P 4
4375.573	4376.802				0.2	0.30	22847.73			
4375.661	4376.891		0.0		0.2		22847.27			
4375.755	4376.984		0.0	0.0			22846.78			
4375.987	4377.216			0.3	0.2b		22845.57	3E 1	2B 1	Q 7
4376.071	4377.301			0.2			22845.13			
4376.299	4377.529		0.0				22843.94			
4376.454	4377.684			0.0			22843.13			

λ_{air}	λ_{vac}	Ref.	I_1	I_2	I_3	I_4	ν	Upper	Lower	Br.
4376.561	4377.791			0.0	0.1b		22842.57	4F 0	2B 4	R 2
4376.877	4378.107		0.0				22840.92			
4377.219	4378.449		0.0	0.2	0.3		22839.14			
4377.370	4378.600			0.2			22838.35			
4377.493	4378.723			0.2			22837.71			
4377.665	4378.895	K	0.0	0.5	0.9b		22836.81	4f 0	2c 0	P 5
4377.780	4379.010		0.0d	0.4	0.4		22836.21			
4377.897	4379.127			0.1			22835.60			
4377.974	4379.204						22835.20			
4378.066	4379.296			0.3	0.2		22834.72			
4378.162	4379.392			0.2			22834.22			
4378.200	4379.430		0.0				22834.02			
4378.277	4379.507		0.0	0.3			22833.62			
4378.396	4379.626			0.1			22833.00			
4378.490	4379.720			0.2			22832.51			
4378.668	4379.898			0.1d			22831.58	Y 2	2B 3	P 1
4378.808	4380.038		0.0d				22830.85	4f 2	2c 2	R 2
4378.871	4380.102		0.1	0.4	0.1		22830.52			
4379.009	4380.240			0.2	0.2		22829.80			
4379.140	4380.370		0.0d	0.3			22829.12	3c 3	2a 0	P 2
4379.140	4380.370		0.0d	0.3			22829.12	3F 0	2B 0	R 4
4379.140	4380.370		0.0d	0.3			22829.12	Z 2	2B 2	P 5
4379.226	4380.457			0.2			22828.67			
4379.309	4380.539		0.0				22828.24	4f 1	2c 1	R 2
4379.399	4380.629	GK	0.6	1.0	1.5b	1.31	22827.77	4e 0	2c 0	R 2
4379.499	4380.729		0.0	0.1			22827.25			
4379.654	4380.885			0.1			22826.44			
4379.775	4381.006			0.2d			22825.81			
4379.817	4381.048		0.0				22825.59			
4379.957	4381.188	GK	0.2	0.8	0.9b	0.54	22824.86	4f 0	2c 0	Q 3
4380.068	4381.299			0.0			22824.28	U 1	2B 1	P 3
4380.163	4381.393		0.0				22823.79			
4380.366	4381.597			0.0			22822.73			
4380.475	4381.706						22822.16			
4380.596	4381.827		0.0d				22821.53			
4380.769	4382.000		0.0d				22820.63			
4381.032	4382.263						22819.26			
4381.109	4382.340			0.1			22818.86			
4381.372	4382.603			0.2			22817.49	3D 2	2B 2	P 7
4381.474	4382.705		0.0	0.1			22816.96			
4381.604	4382.835	G	0.2	0.7	0.4	0.18	22816.28			
4381.771	4383.003				0.2		22815.41			
4381.817	4383.049			0.2d			22815.17	3c 3	2a 0	Q 4
4382.102	4383.333		0.0				22813.69			
4382.182	4383.414	GK	0.6	0.6	0.2	0.72	22813.27	3A 0	2B 0	R 0
4382.303	4383.535			0.2			22812.64			
4382.576	4383.808			0.1d			22811.22			
4382.741	4383.973				0.2		22810.36			
4382.914	4384.146			0.0			22809.46			
4383.012	4384.244	G	0.2	0.5	0.3	0.23	22808.95	4f 3	2c 3	R 3
4383.012	4384.244	G	0.2	0.5	0.3	0.23	22808.95	3F 2	2B 3	R 2
4383.139	4384.371		0.0	0.2			22808.29			
4383.314	4384.546	G	0.2	0.7	0.3		22807.38	3F 2	2B 3	Q 4
4383.414	4384.646		0.1				22806.86			
4383.489	4384.721	GK	0.7	0.8		1.08	22806.47	4f 1	2c 1	P 6
4383.489	4384.721	GK	0.7	0.8		1.08	22806.47	4c 5	2a 4	Q 1
4383.610	4384.842		0.1	0.2			22805.84			
4383.729	4384.961		0.0	0.3	0.2		22805.22			
4383.893	4385.124			0.1			22804.37			
4384.018	4385.249			0.2			22803.72			

λ_{air}	λ_{vac}	Ref.	I_1	I_2	I_3	I_4	ν	Upper	Lower	Br.
4384.073	4385.305			0.5	0.8		22803.43			
4384.191	4385.422	GK	0.2	0.7		0.49	22802.82			
4384.296	4385.528			0.2			22802.27			
4384.360	4385.592		0.0d	0.0			22801.94			
4384.458	4385.690	GK	0.2	0.9	1.5	0.68	22801.43			
4384.591	4385.823		0.0	0.1			22800.74	3E 3	2B 4	R 3
4384.716	4385.948			0.2			22800.09	4g 0	2a 0	R 3
4384.819	4386.051			0.1	0.1		22799.55			
4384.912	4386.144			0.2	0.1		22799.07			
4385.000	4386.232			0.2d			22798.61	3e 3	2c 0	P 3
4385.000	4386.232			0.2d			22798.61	3e 3	2c 0	P 4
4385.100	4386.332		0.0	0.0			22798.09			
4385.145	4386.377			0.3	0.3		22797.86			
4385.337	4386.569			0.1			22796.86			
4385.452	4386.684			0.0			22796.26			
4385.525	4386.758			0.2			22795.88			
4385.625	4386.858						22795.36			
4385.693	4386.925	G	0.2	0.4			22795.01	3E 1	2B 1	R 1
4385.776	4387.008		0.0				22794.58			
4385.903	4387.135			0.2			22793.92			
4385.979	4387.212			0.1			22793.52			
4386.089	4387.321			0.1			22792.95			
4386.151	4387.383		0.0				22792.63			
4386.255	4387.487	GK	0.4	0.8	0.3	0.66	22792.09	W 2	2B 3	P 5
4386.255	4387.487	GK	0.4	0.8	0.3	0.66	22792.09	Y 1	2B 0	P 4
4386.370	4387.603			0.1			22791.49			
4386.513	4387.745			0.2			22790.75	4F 0	2B 4	P 5
4386.634	4387.866			0.1			22790.12			
4386.796	4388.028						22789.28			
4386.915	4388.147			0.2	0.2		22788.66			
4387.011	4388.244			0.0			22788.16			
4387.100	4388.332			0.1			22787.70			
4387.188	4388.421			0.1			22787.24			
4387.294	4388.527	GK	0.2	0.6		0.40	22786.69	4c 5	2a 4	Q 2
4387.346	4388.579				0.2		22786.42			
4387.421	4388.654			0.1			22786.03			
4387.525	4388.758		0.0	0.4			22785.49	4b 1	2a 0	R 2
4387.556	4388.789				0.2		22785.33	4f 2	2c 2	Q 4
4387.837	4389.070		0.0	0.2			22783.87			
4387.924	4389.157		0.5	0.4	1.5		22783.42			
4388.039	4389.272		0.1d	0.1			22782.82			
4388.134	4389.367						22782.33			
4388.209	4389.442				0.1		22781.94			
4388.267	4389.500			0.2			22781.64			
4388.409	4389.642		0.2d				22780.90			
4388.461	4389.694			0.4d	0.3		22780.63	Y 2	2B 3	P 2
4388.602	4389.835			0.1			22779.90			
4388.723	4389.956			0.3			22779.27	3E 3	2B 4	Q 6
4388.770	4390.003		0.1				22779.03	3E 3	2B 4	R 2
4388.881	4390.114	GK	0.2b	0.7b	0.5	0.63	22778.45	4e 1	2c 1	R 4
4388.881	4390.114	GK	0.2b	0.7b	0.5	0.63	22778.45	X 0	2B 0	P 6
4388.881	4390.114	GK	0.2b	0.7b	0.5	0.63	22778.45	W 2	2B 3	P 6
4388.997	4390.230		0.1				22777.85			
4389.078	4390.311	GK	0.6	0.9	1.5	1.13	22777.43	4e 0	2c 0	R 1
4389.159	4390.392		0.2	0.3			22777.01			
4389.276	4390.509		0.2	0.3			22776.40			
4389.373	4390.606		0.0	0.0			22775.90			
4389.452	4390.685	GK	0.2	0.6	0.3	0.34	22775.49			
4389.577	4390.810		0.1	0.2			22774.84	X 1	2B 2	R 0
4389.754	4390.988				0.3		22773.92			

λ_air	λ_vac	Ref.	I₁	I₂	I₃	I₄	ν	Upper	Lower	Br.
4389.878	4391.111		0.0d	0.1d			22773.28			
4390.097	4391.331		0.1d	0.4d	0.3		22772.14	4F 0	2B 4	Q 4
4390.219	4391.452			0.2d			22771.51			
4390.462	4391.695	G	0.1	0.7	0.4		22770.25	3E 1	2B 1	Q 6
4390.578	4391.811			0.2			22769.65			
4390.703	4391.936		0.1	0.4	0.3		22769.00			
4390.809	4392.042		0.0				22768.45	X 1	2B 2	R 1
4390.896	4392.129	GK	0.5b	1.0b	1.5b	1.35	22768.00	4f 0	2c 0	Q 2
4390.896	4392.129	GK	0.5b	1.0b	1.5b	1.35	22768.00	4f 1	2c 1	Q 2
4391.023	4392.257			0.0			22767.34	w 0	2c 1	P 2
4391.100	4392.334						22766.94			
4391.254	4392.488			0.2d			22766.14			
4391.588	4392.822			0.2d			22764.41			
4391.733	4392.967	GK	0.3	0.9	1.5b	0.86	22763.66	4f 0	2c 0	P 4
4391.876	4393.109			0.2			22762.92	4g 1	2a 1	R 4
4392.016	4393.250		0.0				22762.19			
4392.094	4393.328	GK	0.5	0.8	0.3	0.89	22761.79	4b 1	2a 0	R 1
4392.202	4393.436		0.1	0.1			22761.23			
4392.258	4393.492						22760.94			
4392.476	4393.710			0.0			22759.81			
4392.719	4393.953	G		0.4	0.2		22758.55	w 0	2c 1	Q 2
4392.895	4394.129	GK	0.3	0.6	0.2	0.49	22757.64	4c 5	2a 4	Q 3
4393.012	4394.247		0.0				22757.03			
4393.091	4394.326		0.0d	0.2	0.2d		22756.62			
4393.395	4394.629			0.0			22755.05			
4393.578	4394.812		0.0				22754.10			
4393.723	4394.957			0.3d			22753.35	3D 2	2B 2	P 8
4393.887	4395.121				0.1		22752.50			
4393.987	4395.222			0.0d			22751.98			
4394.181	4395.415			0.1d			22750.98			
4394.343	4395.577	GK	0.3	0.7	0.2	0.51	22750.14	U 1	2B 1	P 4
4394.412	4395.647		0.0				22749.78			
4394.778	4396.012	GK	0.2	0.7	0.4		22747.89			
4394.973	4396.207			0.1d			22746.88			
4395.170	4396.404		0.0	0.4	0.4		22745.86			
4395.272	4396.507		0.1	0.2			22745.33			
4395.326	4396.561			0.3			22745.05	X 1	2B 2	R 2
4395.597	4396.832			0.0			22743.65			
4396.068	4397.303			0.1d			22741.21			
4396.345	4397.580		0.0d	0.4	0.2		22739.78			
4396.536	4397.771			0.1			22738.79			
4396.751	4397.986			0.0			22737.68			
4396.902	4398.137	GK	0.1	0.6	0.8	0.26	22736.90	3c 3	2a 0	Q 5
4397.035	4398.270			0.1d			22736.21	4f 2	2c 2	R 3
4397.296	4398.532			0.2d			22734.86			
4397.356	4398.592		0.0				22734.55			
4397.478	4398.713		0.1	0.4	0.3		22733.92	3c 3	2a 0	P 3
4397.832	4399.068						22732.09			
4397.913	4399.149		0.0	0.2			22731.67	3E 3	2B 4	R 1
4398.089	4399.325	GK	0.2	0.8	1.5	0.61	22730.76	4g 1	2a 1	P 3
4398.089	4399.325	GK	0.2	0.8	1.5	0.61	22730.76	3F 0	2B 0	Q 6
4398.242	4399.478	GK	0.2	0.8	0.4	0.83	22729.97	U 1	2B 1	P 6
4398.391	4399.627			0.1			22729.20			
4398.463	4399.698						22728.83	3E 0	2B 0	Q12
4398.647	4399.882				0.2		22727.88			
4398.730	4399.966			0.1			22727.45			
4398.811	4400.047			0.0			22727.03			
4398.937	4400.173	G	0.1	0.6	0.6	0.28	22726.38			
4399.138	4400.374	F					22725.34			
4399.214	4400.450			0.1			22724.95	4e 1	2c 1	R 3

λ_{air}	λ_{vac}	Ref.	I$_1$	I$_2$	I$_3$	I$_4$	ν	Upper	Lower	Br.
4399.313	4400.548		0.0				22724.44			
4399.431	4400.667			0.1			22723.83			
4399.562	4400.798		0.0	0.3			22723.15			
4399.667	4400.903	GK	0.5	0.6	0.1	0.43	22722.61	4b 1	2a 0	R 0
4399.764	4401.000		0.1				22722.11			
4399.824	4401.060			0.2			22721.80			
4399.868	4401.104		0.0				22721.57			
4399.954	4401.190	G	0.1	0.6	0.2	0.34	22721.13	4f 1	2c 1	P 5
4399.954	4401.190	G	0.1	0.6	0.2	0.34	22721.13	U 1	2B 1	P 5
4399.990	4401.226		0.1				22720.94			
4400.083	4401.319			0.2			22720.46			
4400.172	4401.408						22720.00			
4400.225	4401.461			0.1d			22719.73			
4400.378	4401.614		0.0	0.3	0.1		22718.94	3E 3	2B 4	Q 5
4400.556	4401.792		0.1	0.5	0.2		22718.02			
4400.643	4401.879		0.1	0.2			22717.57			
4400.730	4401.966	G	0.6	0.9	0.4	1.27	22717.12	3A 0	2B 0	P 4
4400.810	4402.046	GK	0.5			0.52	22716.71	4f 1	2c 1	R 1
4400.827	4402.063			0.9			22716.62			
4400.843	4402.079		0.5				22716.54			
4400.934	4402.170	G	0.1	0.8	1.5	0.62	22716.07	4f 0	2c 0	P 3
4400.934	4402.170	G	0.1	0.8	1.5	0.62	22716.07	4g 0	2a 0	P 3
4401.011	4402.247		0.0				22715.67			
4401.093	4402.329	GK	0.6	0.6			22715.25	3A 0	2B 0	P 1
4401.093	4402.329	GK	0.6	0.6			22715.25	4c 5	2a 4	Q 4
4401.199	4402.435		0.0d	0.2			22714.70			
4401.284	4402.521			0.1			22714.26			
4401.412	4402.649	GK	0.2	0.7	1.0	0.45	22713.60			
4401.550	4402.786		0.1	0.2			22712.89	Y 2	2B 3	P 3
4401.661	4402.897	GK	0.2	0.7	1.5	0.41	22712.32			
4401.796	4403.032		0.0d	0.5			22711.62	3F 2	2B 3	Q 3
4401.953	4403.190			0.6	0.2		22710.81	3E 1	2B 1	Q 5
4402.127	4403.364			0.0			22709.91			
4402.269	4403.505			0.1			22709.18			
4402.393	4403.630		0.1				22708.54			
4402.478	4403.715	GK	0.1	0.6	0.3	0.30	22708.10	Z 2	2B 2	P 6
4402.678	4403.915		0.0	0.1d			22707.07	3E 0	2B 0	R 6
4402.748	4403.984		0.0				22706.71			
4402.901	4404.138						22705.92			
4403.002	4404.239			0.1d	0.2		22705.40			
4403.064	4404.301						22705.08			
4403.155	4404.392			0.1			22704.61			
4403.295	4404.532		0.2	0.4			22703.89	3E 0	2B 0	R 6
4403.295	4404.532		0.2	0.4			22703.89	3F 2	2B 3	R 1
4403.419	4404.656			0.2			22703.25			
4403.516	4404.753			0.0			22702.75			
4403.630	4404.867	GK	0.3	0.8	0.3	0.57	22702.16	3E 1	2B 1	R 0
4403.747	4404.984		0.0	0.0			22701.56			
4403.968	4405.205		0.0d				22700.42	4f 1	2c 1	Q 4
4404.084	4405.321		0.1				22699.82			
4404.162	4405.399	GK	0.5	0.8	0.3	0.83	22699.42	3A 0	2B 0	P 3
4404.294	4405.531		0.0d	0.2			22698.74			
4404.398	4405.636		0.1	0.1			22698.20			
4404.482	4405.719			0.8	0.5		22697.77			
4404.527	4405.764		0.1				22697.54			
4404.608	4405.845	GK	1.0	0.9	0.3	1.30	22697.12	3A 0	2B 0	P 2
4404.723	4405.960		0.2	0.2			22696.53			
4404.857	4406.094		0.0	0.1			22695.84			
4404.938	4406.175			0.2	0.2d		22695.42			
4405.055	4406.292		0.0d	0.1			22694.82			

λ_{air}	λ_{vac}	Ref.	I_1	I_2	I_3	I_4	ν	Upper	Lower	Br.
4405.132	4406.369			0.0			22694.42			
4405.214	4406.451			0.2	0.2d		22694.00			
4405.338	4406.575		0.0				22693.36			
4405.466	4406.703				0.1d		22692.70			
4405.635	4406.872		0.2	0.3			22691.83			
4405.728	4406.966			0.0			22691.35			
4405.837	4407.074		0.0	0.4	0.1		22690.79			
4405.973	4407.210			0.1			22690.09			
4406.060	4407.298			0.2			22689.64			
4406.196	4407.434		0.1	0.2			22688.94			
4406.480	4407.717						22687.48			
4406.620	4407.857						22686.76			
4406.779	4408.017				0.2		22685.94			
4406.876	4408.114	G	0.0d	0.5			22685.44	3e 3	2c 0	P 4
4407.037	4408.275			0.1			22684.61			
4407.154	4408.392	G	0.2	0.6	0.3		22684.01			
4407.292	4408.530			0.1			22683.30			
4407.636	4408.874		0.1	0.2			22681.53			
4407.832	4409.070			0.3			22680.52			
4408.067	4409.305			0.1d			22679.31	4f 2	2c 2	R 1
4408.291	4409.529			0.2			22678.16			
4408.374	4409.612			0.0			22677.73			
4408.481	4409.719		0.0d	0.3			22677.18	3E 3	2B 4	R 0
4408.584	4409.822			0.1			22676.65			
4408.722	4409.960						22675.94			
4409.090	4410.328			0.1d			22674.05			
4409.152	4410.390		0.0d				22673.73			
4409.265	4410.503			0.0			22673.15			
4409.319	4410.558			0.2d			22672.87			
4409.438	4410.676			0.1	0.2b		22672.26			
4409.644	4410.883			0.3			22671.20			
4409.786	4411.025			0.2			22670.47			
4409.846	4411.085			0.1			22670.16			
4409.944	4411.182	G		0.5			22669.66	3E 3	2B 4	Q 4
4410.047	4411.285		0.1d	0.3	0.2b		22669.13	X 1	2B 2	P 2
4410.268	4411.507		0.0d	0.3			22667.99	4f 2	2c 2	Q 3
4410.479	4411.717	GK	0.3	0.9	0.6	1.01	22666.91	3E 1	2B 1	Q 4
4410.613	4411.852	GK	0.6	0.9	0.6	1.06	22666.22	4e 1	2c 1	R 2
4410.743	4411.982		0.0				22665.55			
4410.846	4412.085		0.0d	0.1d			22665.02			
4411.107	4412.346			0.1d			22663.68			
4411.306	4412.545			0.3d	0.3		22662.66			
4411.491	4412.730			0.2d			22661.71			
4411.627	4412.866		0.0d	0.4			22661.01			
4411.783	4413.022		0.0d	0.5b	0.2b		22660.21			
4411.925	4413.164		0.0	0.1			22659.48			
4412.044	4413.283		0.2	0.7	0.4		22658.87			
4412.082	4413.322		0.2				22658.67			
4412.155	4413.394		0.2				22658.30			
4412.248	4413.487	GK	1.5b	1.5b	1.5b	1.93	22657.82	4e 0	2c 0	R 1
4412.380	4413.620		0.1	0.1			22657.14			
4412.482	4413.721	GK	0.6b	0.9b	1.5	1.03	22656.62	4f 1	2c 1	Q 3
4412.482	4413.721	GK	0.6b	0.9b	1.5	1.03	22656.62	4e 0	2c 0	Q 2
4412.591	4413.830		0.1				22656.06			
4412.682	4413.922			0.0			22655.59			
4412.809	4414.048			0.2			22654.94			
4412.918	4414.157			0.2	0.1		22654.38			
4413.047	4414.286			0.2			22653.72			
4413.152	4414.391			0.1			22653.18			
4413.296	4414.535		0.1	0.5	0.5		22652.44			

λ_{air}	λ_{vac}	Ref.	I_1	I_2	I_3	I_4	ν	Upper	Lower	Br.
4413.389	4414.629		0.0				22651.96			
4413.489	4414.728	GK	0.5	1.0b	1.5b	1.11	22651.45	4f 2	2c 2	R 2
4413.602	4414.841		0.0	0.2			22650.87			
4413.736	4414.976		0.1d	0.4	0.3		22650.18	4f 3	2c 3	R 2
4413.842	4415.081			0.0			22649.64			
4414.017	4415.257		0.1d	0.3			22648.74			
4414.109	4415.348		0.0	0.0			22648.27			
4414.212	4415.452	GK	0.7	1.5b	1.5b	1.52	22647.74	4e 0	2c 0	Q 3
4414.337	4415.576		0.0	0.2			22647.10			
4414.432	4415.672		0.1	0.2			22646.61			
4414.528	4415.767	GK	0.3	0.8	0.7	0.72	22646.12	4f 1	2c 1	P 4
4414.627	4415.867			0.1			22645.61			
4414.744	4415.984			0.3			22645.01			
4414.814	4416.054			0.2			22644.65			
4414.877	4416.117			0.0			22644.33			
4414.992	4416.232	GK	0.5	1.0b	1.5b	1.33	22643.74	3F 0	2B 0	R 3
4414.992	4416.232	GK	0.5	1.0b	1.5b	1.33	22643.74	4e 0	2c 0	R 2
4415.116	4416.356			0.2			22643.10			
4415.286	4416.526			0.1			22642.23	3c 3	2a 0	Q 6
4415.399	4416.639			0.1			22641.65			
4415.557	4416.797			0.2			22640.84	4F 0	2B 4	R 1
4415.703	4416.944			0.3	0.2		22640.09			
4415.807	4417.047			0.4	0.3		22639.56			
4416.014	4417.254			0.2			22638.50			
4416.131	4417.371		0.0	0.0			22637.90			
4416.216	4417.457	GK	0.3	0.9b	1.5b	1.10	22637.46	3E 1	2B 1	Q 3
4416.216	4417.457	GK	0.3	0.9b	1.5b	1.10	22637.46	4e 0	2c 0	Q 4
4416.372	4417.613		0.0	0.2			22636.66			
4416.476	4417.716	GK	0.5	0.7		0.85	22636.13	3F 2	2B 3	Q 2
4416.476	4417.716	GK	0.5	0.7		0.85	22636.13	3A 1	2B 2	R 2
4416.476	4417.716	GK	0.5	0.7		0.85	22636.13	4E 2	2C 0	R 1
4416.651	4417.892		0.3	0.3			22635.23			
4416.868	4418.109			0.2			22634.12			
4416.923	4418.163				0.2		22633.84			
4417.036	4418.276			0.1			22633.26			
4417.139	4418.380		0.1	0.4			22632.73			
4417.184	4418.425		0.1				22632.50			
4417.276	4418.517		0.3				22632.03	Y 2	2B 3	P 4
4417.331	4418.571	GK	0.5	1.5b	1.5b	1.67	22631.75	4e 0	2c 0	R 3
4417.331	4418.571	GK	0.5	1.5b	1.5b	1.67	22631.75	3E 3	2B 4	Q 3
4417.500	4418.741			0.3			22630.88			
4417.627	4418.868			0.2			22630.23			
4417.735	4418.975			0.2			22629.68	3d 3	2c 0	R 1
4417.895	4419.136			0.2			22628.86			
4418.014	4419.255			0.1d			22628.25	3E 2	2B 3	Q10
4418.098	4419.339			0.2			22627.82			
4418.217	4419.458			0.3	0.2		22627.21			
4418.401	4419.641			0.4	0.3		22626.27			
4418.662	4419.903	GK	0.3	0.9b	1.5	0.88	22624.93	4e 0	2c 0	Q 5
4418.828	4420.069			0.2			22624.08	3c 3	2a 0	P 4
4418.942	4420.182			0.1			22623.50			
4419.029	4420.271			0.1			22623.05			
4419.139	4420.380	G	0.3	0.5	0.2b		22622.49	4e 0	2c 0	Q 1
4419.281	4420.523		0.1	0.3			22621.76			
4419.393	4420.634		0.2	0.1			22621.19			
4419.485	4420.726	GK	0.9b	1.0	0.7	1.55	22620.72	3E 1	2B 1	Q 2
4419.604	4420.845		0.2	0.2			22620.11			
4419.684	4420.925			0.1			22619.70			
4419.823	4421.064	GK	0.2b	0.8	1.5	0.81	22618.99	4e 0	2c 0	R 4
4419.989	4421.230			0.3			22618.14	4F 0	2B 4	P 4

λ_{air}	λ_{vac}	Ref.	I_1	I_2	I_3	I_4	ν	Upper	Lower	Br.
4420.116	4421.357		0.2	0.3			22617.49	3c 7	2a 3	Q 1
4420.116	4421.357		0.2	0.3			22617.49	3e 3	2c 0	P 6
4420.245	4421.486		0.8	0.9			22616.83			
4420.319	4421.560	GK	0.8	1.0	0.6	1.52	22616.45	4e 2	2c 2	R 4
4420.319	4421.560	GK	0.8	1.0	0.6	1.52	22616.45	3E 1	2B 1	P 4
4420.436	4421.678		0.2	0.2			22615.85			
4420.565	4421.807			0.1			22615.19			
4420.671	4421.912			0.4			22614.65	4F 0	2B 4	Q 3
4420.720	4421.961				0.3		22614.40			
4420.755	4421.996		0.0	0.4			22614.22			
4420.821	4422.063		0.0				22613.88			
4420.915	4422.157	GK	0.3	0.8	0.3	0.87	22613.40	3E 1	2B 1	P 5
4420.915	4422.157	GK	0.3	0.8	0.3	0.87	22613.40	3E 1	2B 1	Q 1
4421.060	4422.301		0.1	0.2			22612.66			
4421.162	4422.403	GK	0.8	0.8	0.3	0.94	22612.14	3E 1	2B 1	P 3
4421.275	4422.517		0.0	0.1			22611.56			
4421.341	4422.583			0.0			22611.22			
4421.435	4422.677			0.1			22610.74			
4421.553	4422.794			0.5	0.8		22610.14			
4421.692	4422.933		0.0	0.2			22609.43			
4421.778	4423.019			0.2			22608.99			
4421.881	4423.123	F					22608.46	4e 0	2c 0	Q 6
4421.944	4423.186			0.1			22608.14			
4422.081	4423.323			0.1d			22607.44			
4422.169	4423.411						22606.99			
4422.222	4423.463			0.1d			22606.72			
4422.325	4423.567				0.2		22606.19			
4422.464	4423.706			0.2			22605.48			
4422.556	4423.798			0.0			22605.01	4e 1	2c 1	R 1
4422.644	4423.886	GK	0.1	0.8	1.5b		22604.56	3D 1	2B 1	R 8
4422.644	4423.886	GK	0.1	0.8	1.5b		22604.56	4e 0	2c 0	R 5
4422.644	4423.886	GK	0.1	0.8	1.5b		22604.56	4f 3	2c 3	Q 4
4422.685	4423.927		0.1				22604.35			
4422.777	4424.019	GK	0.5	0.7		1.08	22603.88	3E 3	2B 4	Q 2
4422.891	4424.133		0.1	0.1			22603.30			
4422.981	4424.223	GK	0.8	0.9	0.3	1.18	22602.84	4b 1	2a 0	P 1
4423.098	4424.340		0.1	0.2			22602.24			
4423.247	4424.489	GK	0.7b	1.0b	0.7	1.32	22601.48	4f 1	2c 1	Q 2
4423.331	4424.573				0.6		22601.05			
4423.370	4424.612		0.1	0.4			22600.85			
4423.587	4424.830			0.1d			22599.74			
4423.679	4424.922			0.4	0.3		22599.27			
4423.858	4425.100			0.2d			22598.36			
4423.979	4425.221						22597.74			
4424.100	4425.343			0.2			22597.12			
4424.214	4425.456			0.2			22596.54			
4424.298	4425.540	G	0.0	0.5	0.5		22596.11	4f 1	2c 1	P 3
4424.553	4425.795		0.0				22594.81			
4424.658	4425.901			0.2d			22594.27			
4424.780	4426.022			0.1			22593.65	X 1	2B 2	P 3
4424.917	4426.159	K	0.1	0.5	1.0		22592.95			
4425.046	4426.289		0.2	0.5			22592.29			
4425.146	4426.389	GK	0.6	0.9	0.4	1.12	22591.78	3E 1	2B 1	P 2
4425.258	4426.500		0.0d	0.1			22591.21			
4425.358	4426.600			0.3	0.3		22590.70			
4425.461	4426.704			0.2			22590.17			
4425.587	4426.830			0.1			22589.53			
4425.702	4426.945			0.3			22588.94			
4425.896	4427.139	GK	0.5b	1.0b	1.5b	1.18	22587.95	4f 0	2c 0	P 3
4426.010	4427.253		0.0	0.0			22587.37	w 0	2c 1	P 3

λ_{air}	λ_{vac}	Ref.	I_1	I_2	I_3	I_4	ν	Upper	Lower	Br.
4426.112	4427.355	GK	0.1	0.7	0.9		22586.85	4f 0	2c 0	P 4
4426.112	4427.355	GK	0.1	0.7	0.9		22586.85	4f 2	2c 2	P 6
4426.112	4427.355	GK	0.1	0.7	0.9		22586.85	3d 3	2c 0	R 2
4426.277	4427.519		0.2	0.5			22586.01	3E 3	2B 4	Q 1
4426.384	4427.627			0.1			22585.46	3d 3	2c 0	Q 1
4426.435	4427.678						22585.20			
4426.516	4427.759		0.0	0.0			22584.79			
4426.614	4427.857	G	0.4	0.7	0.1	0.52	22584.29	3D 3	2B 4	R 0
4426.614	4427.857	G	0.4	0.7	0.1	0.52	22584.29	4e 0	2c 0	Q 7
4426.804	4428.047			0.1			22583.32			
4426.929	4428.172			0.3	0.1		22582.68	3c 7	2a 3	Q 2
4427.010	4428.253			0.1			22582.27			
4427.104	4428.347			0.1d			22581.79			
4427.135	4428.378						22581.63			
4427.172	4428.415			0.0d			22581.44			
4427.351	4428.594			0.2			22580.53			
4427.468	4428.712	G	0.2	0.5			22579.93	3E 3	2B 4	P 2
4427.600	4428.843		0.0	0.1d			22579.26			
4427.733	4428.977	GK	0.2	0.9b	1.5b	0.80	22578.58	4f 0	2c 0	P 5
4427.812	4429.055		0.3				22578.18	3D 3	2B 4	R 1
4427.908	4429.151			0.2			22577.69			
4428.079	4429.322	GK	0.2	0.6	0.3	0.40	22576.82			
4428.188	4429.432		0.0				22576.26			
4428.257	4429.500		0.2	0.5	0.1		22575.91	3F 2	2B 3	P 4
4428.384	4429.628		0.0				22575.26			
4428.418	4429.661			0.1			22575.09			
4428.526	4429.769						22574.54			
4428.635	4429.879		0.1	0.2			22573.98			
4428.785	4430.028	GK	0.1d	0.8	1.0	0.45	22573.22	3F 0	2B 0	Q 5
4428.947	4430.191			0.1d			22572.39			
4429.012	4430.256		0.0d				22572.06			
4429.130	4430.374	G	0.1d	0.6	0.5		22571.46			
4429.246	4430.489		0.4	0.6	0.3		22570.87			
4429.291	4430.535	G				0.64	22570.64	3E 3	2B 4	P 3
4429.324	4430.568		0.4	0.6	0.2		22570.47			
4429.442	4430.686			0.1			22569.87			
4429.548	4430.792	K	0.0d	0.6	1.5		22569.33	4f 2	2c 2	Q 2
4429.666	4430.909			0.0d			22568.73			
4429.732	4430.976						22568.39			
4429.899	4431.143			0.4	0.1		22567.54			
4430.035	4431.279		0.1b	0.5			22566.85	3d 3	2c 0	P 1
4430.109	4431.353			0.1			22566.47			
4430.164	4431.408			0.1			22566.19			
4430.231	4431.475	G	0.2	0.5	0.7		22565.85			
4430.280	4431.524			0.4			22565.60			
4430.471	4431.715						22564.63			
4430.518	4431.762			0.1			22564.39			
4430.643	4431.887			0.2	0.2		22563.75			
4430.842	4432.086			0.0			22562.74			
4430.979	4432.223				0.4		22562.04			
4431.042	4432.286			0.1d			22561.72			
4431.130	4432.375						22561.27			
4431.176	4432.420					0.81	22561.04			
4431.229	4432.473			0.1			22560.77			
4431.364	4432.608	G	0.0	0.7	0.7		22560.08			
4431.464	4432.709	K	0.2	0.7			22559.57			
4431.511	4432.756	G	0.4				22559.33	3D 3	2B 4	R 2
4431.633	4432.878		0.0				22558.71			
4431.832	4433.076		0.3b	0.4	0.2		22557.70			
4432.026	4433.271		0.1	0.5	0.2		22556.71			

λ_{air}	λ_{vac}	Ref.	I_1	I_2	I_3	I_4	ν	Upper	Lower	Br.
4432.105	4433.349		0.1				22556.31	3e 3	2c 0	P 5
4432.225	4433.469		0.2	0.2			22555.70			
4432.295	4433.540		0.1	0.2			22555.34	3A 1	2B 2	R 1
4432.364	4433.609	GK	0.5	0.7	0.2	0.83	22554.99	4e 2	2c 2	R 3
4432.364	4433.609	GK	0.5	0.7	0.2	0.83	22554.99	3E 3	2B 4	P 4
4432.364	4433.609	GK	0.5	0.7	0.2	0.83	22554.99	3E 0	2B 0	Q11
4432.494	4433.738		0.0				22554.33			
4432.608	4433.852			0.1	0.1		22553.75			
4432.763	4434.008			0.1			22552.96			
4433.048	4434.293	K		0.5	0.7		22551.51	4f 2	2c 2	R 2
4433.190	4434.434			0.1d			22550.79			
4433.404	4434.649			0.5	0.9		22549.70	3E 0	2B 0	Q11
4433.709	4434.954			0.3	0.5		22548.15			
4433.884	4435.129	GK	0.3	0.6	0.3	0.43	22547.26			
4433.984	4435.229			0.1			22546.75			
4434.075	4435.319		0.1	0.5	0.8		22546.29			
4434.261	4435.506	GK	0.5	0.8	0.2	0.72	22545.34	4e 0	2c 0	P 2
4434.261	4435.506	GK	0.5	0.8	0.2	0.72	22545.34	u 2	2a 0	R 3
4434.385	4435.630			0.2			22544.71			
4434.460	4435.705			0.1			22544.33			
4434.564	4435.809	GK	0.3	0.8	0.7	0.76	22543.80			
4434.702	4435.947			0.2			22543.10			
4434.826	4436.071		0.2	0.5			22542.47			
4434.977	4436.223	GK	0.5	0.8b	0.9	0.94	22541.70	4f 2	2c 2	R 1
4435.217	4436.463			0.1d			22540.48			
4435.377	4436.622				0.2		22539.67			
4435.459	4436.705						22539.25			
4435.668	4436.913			0.1			22538.19			
4435.745	4436.990			0.1			22537.80			
4435.885	4437.130	GK	0.4	0.7	0.6	0.66	22537.09	4e 0	2c 0	Q 2
4436.064	4437.309		0.2	0.3			22536.18	3F 2	2B 3	P 3
4436.192	4437.437	GK	0.1d	0.6	0.5		22535.53			
4436.345	4437.591			0.0			22534.75			
4436.450	4437.695			0.2b			22534.22			
4436.625	4437.870						22533.33			
4436.668	4437.914			0.3	0.2		22533.11			
4436.810	4438.056			0.4	0.3		22532.39			
4436.904	4438.150			0.1			22531.91			
4437.097	4438.343		0.2b	0.5	0.2		22530.93	3D 3	2B 4	R 3
4437.157	4438.402	G					22530.63			
4437.210	4438.456			0.3d	0.4		22530.36	3c 7	2a 3	Q 3
4437.417	4438.662			0.2d			22529.31			
4437.543	4438.788		0.4	0.1			22528.67			
4437.645	4438.891	GK	0.2d	0.7	0.9	0.26	22528.15			
4437.789	4439.035			0.2			22527.42			
4437.984	4439.230		0.2	0.3			22526.43	4f 3	2c 3	R 3
4438.206	4439.453	GK	0.4	0.8	0.8	0.65	22525.30	4d 0	2c 0	P 1
4438.354	4439.600			0.3	0.7		22524.55			
4438.463	4439.709		0.0d				22524.00			
4438.543	4439.790		0.0				22523.59	3d 3	2c 0	R 3
4438.632	4439.878	GK	0.8	0.8	0.3	0.93	22523.14	4b 1	2a 0	P 2
4438.746	4439.993		0.1	0.1			22522.56	u 2	2a 0	R 2
4438.863	4440.109	F					22521.97	3D 3	2B 4	R 4
4438.936	4440.182			0.2b			22521.60			
4439.145	4440.391		0.0d	0.3b			22520.54	4E 0	2B 4	R 2
4439.259	4440.505			0.1			22519.96			
4439.422	4440.669	GK	0.2	0.7	0.5	0.46	22519.13	3D 1	2B 1	R 7
4439.422	4440.669	GK	0.2	0.7	0.5	0.46	22519.13	4f 2	2c 2	Q 4
4439.529	4440.775			0.1			22518.59			
4439.618	4440.864			0.1			22518.14	3E 0	2B 0	R 5

λ_{air}	λ_{vac}	Ref.	I_1	I_2	I_3	I_4	ν	Upper	Lower	Br.
4439.702	4440.949			0.3	0.2		22517.71	4f 2	2c 2	P 5
4439.886	4441.132						22516.78			
4439.939	4441.186			0.2b			22516.51			
4440.061	4441.308			0.1			22515.89			
4440.148	4441.395		0.0	0.4			22515.45	3E 3	2B 4	P 5
4440.237	4441.483						22515.00			
4440.448	4441.695			0.1d	0.1		22513.93			
4440.823	4442.069		0.0d	0.2			22512.03			
4441.097	4442.344		0.1	0.5	0.2		22510.64			
4441.223	4442.470			0.1			22510.00			
4441.312	4442.559		0.0				22509.55			
4441.405	4442.652	GK	0.6	0.9	0.5	0.93	22509.08	3d 3	2c 0	Q 2
4441.405	4442.652	GK	0.6	0.9	0.5	0.93	22509.08	4g 1	2a 1	P 4
4441.509	4442.756		0.0	0.1			22508.55			
4441.616	4442.863			0.3	0.6		22508.01			
4441.859	4443.106	GK	0.1d	0.6	0.2b	0.32	22506.78	3E 2	2B 3	Q 9
4441.937	4443.185			0.2			22506.38			
4442.068	4443.315		0.2	0.5	0.3		22505.72			
4442.198	4443.445			0.1			22505.06			
4442.439	4443.686			0.2			22503.84			
4442.565	4443.812			0.1			22503.20			
4442.603	4443.850			0.1			22503.01			
4442.662	4443.909			0.2			22502.71			
4442.723	4443.970		0.1	0.3	0.2		22502.40	X 1	2B 2	P 4
4442.853	4444.101	G	0.6b	0.9	0.4	0.97	22501.74	4e 0	2c 0	P 2
4442.958	4444.205		0.0				22501.21	3c 3	2a 0	P 5
4443.023	4444.271			0.1d			22500.88	4f 3	2c 3	R 1
4443.102	4444.350						22500.48	u 2	2a 0	R 1
4443.355	4444.602		0.2	0.4	0.2		22499.20	3E 2	2B 3	R 4
4443.460	4444.707	G	0.1d	0.6	0.3	0.26	22498.67	4e 1	2c 1	Q 5
4443.460	4444.707	G	0.1d	0.6	0.3	0.26	22498.67	4e 1	2c 1	Q 4
4443.592	4444.839			0.0			22498.00			
4443.732	4444.980			0.1d			22497.29			
4443.875	4445.122			0.2	0.2		22496.57			
4444.025	4445.272			0.3			22495.81			
4444.104	4445.351		0.0	0.0			22495.41	4f 3	2c 3	Q 3
4444.214	4445.462	GK	0.6b	1.0b	0.8	1.28	22494.85	2B 2	R 4	
4444.214	4445.462	GK	0.6b	1.0b	0.8	1.28	22494.85	4e 1	2c 1	Q 3
4444.343	4445.590		0.0	0.2d			22494.20	5c 0	2a 1	R 2
4444.449	4445.697		0.3	0.3			22493.66	3A 1	2B 2	R 0
4444.621	4445.869			0.2			22492.79			
4444.766	4446.013			0.4b			22492.06			
4445.056	4446.304		0.2	0.4			22490.59			
4445.080	4446.328		0.2				22490.47			
4445.149	4446.397		0.2				22490.12			
4445.246	4446.494	GK	1.0b	1.5b	0.9b	1.86	22489.63	4e 1	2c 1	R 1
4445.246	4446.494	GK	1.0b	1.5b	0.9b	1.86	22489.63	4e 2	2c 2	R 2
4445.246	4446.494	GK	1.0b	1.5b	0.9b	1.86	22489.63	4e 1	2c 1	Q 2
4445.359	4446.607		0.1	0.2			22489.06			
4445.469	4446.717			0.1			22488.50			
4445.538	4446.786		0.0	0.0			22488.15			
4445.637	4446.885	GK	0.5	0.7	0.2		22487.65	3D 1	2B 1	R 1
4445.637	4446.885	GK	0.5	0.7	0.2		22487.65	4f 2	2c 2	Q 3
4445.754	4447.002		0.3	0.4	0.2d		22487.06	3D 3	2B 4	P 2
4445.890	4447.138		0.0	0.2			22486.37			
4445.997	4447.245		0.1d	0.3			22485.83	3D 1	2B 1	P 1
4446.221	4447.469			0.1d			22484.70			
4446.404	4447.653		0.0	0.2d			22483.77			
4446.547	4447.795			0.3			22483.05			
4446.658	4447.906			0.3	0.4		22482.49			

λ_{air}	λ_{vac}	Ref.	I_1	I_2	I_3	I_4	ν	Upper	Lower	Br.
4446.713	4447.961			0.1			22482.21			
4446.814	4448.062	GK	0.2	0.7	0.4	0.48	22481.70	4e 1	2c 1	Q 1
4446.933	4448.181						22481.10			
4447.071	4448.319			0.2	0.2		22480.40			
4447.144	4448.393		0.0d	0.2			22480.03			
4447.221	4448.470			0.1			22479.64	4F 0	2B 4	P 3
4447.340	4448.589		0.1b	0.3			22479.04			
4447.435	4448.684		0.2	0.2			22478.56			
4447.540	4448.788	GK	1.0b	1.5d	1.5b	1.82	22478.03	4d 0	2c 0	R 1
4447.647	4448.895		0.3	1.0d			22477.49	4e 1	2c 1	R 2
4447.827	4449.075		0.2	0.5	1.0		22476.58	4c 0	2a 0	R 5
4447.827	4449.075		0.2	0.5	1.0		22476.58	4e 0	2c 0	P 3
4447.922	4449.170	GK	0.9	1.0d	1.5	1.45	22476.10	4d 0	2c 0	Q 1
4447.922	4449.170	GK	0.9	1.0d	1.5	1.45	22476.10	u 2	2a 0	R 0
4448.009	4449.258		0.5	0.9d			22475.66			
4448.094	4449.343		0.0	0.2d			22475.23			
4448.223	4449.471			0.2d			22474.58	4F 0	2B 4	Q 2
4448.341	4449.590				0.1		22473.98			
4448.585	4449.834						22472.75			
4448.658	4449.907				0.1		22472.38			
4448.737	4449.986			0.2d			22471.98			
4448.779	4450.028				0.2		22471.77			
4448.943	4450.192			0.0			22470.94			
4449.018	4450.267			0.1			22470.56			
4449.119	4450.368	G	0.1	0.6	0.3		22470.05	3D 1	2B 1	R 6
4449.236	4450.485			0.2			22469.46			
4449.361	4450.610	K	0.0	0.6	0.7		22468.83	4c 0	2a 0	R 4
4449.504	4450.753	GK	0.2b	0.7	0.5	0.65	22468.11			
4449.618	4450.867		0.0	0.2			22467.53			
4449.682	4450.931						22467.21			
4449.725	4450.974		0.1b	0.3			22466.99			
4449.773	4451.022						22466.75			
4449.815	4451.064		0.0				22466.54			
4449.916	4451.165	GK	0.5b	1.0b	1.0	1.46	22466.03	4e 1	2c 1	R 3
4450.044	4451.293		0.0	0.2			22465.38			
4450.179	4451.428		0.0d	0.2d			22464.70	3d 3	2c 0	P 2
4450.254	4451.504		0.0				22464.32			
4450.349	4451.599	GK	0.2	0.7	0.7	0.51	22463.84	4g 0	2a 0	P 4
4450.349	4451.599	GK	0.2	0.7	0.7	0.51	22463.84	4e 0	2c 0	Q 3
4450.680	4451.930		0.0d		0.2		22462.17			
4450.742	4451.991		0.3	0.3d			22461.86			
4450.831	4452.080	GK	0.7	0.9	0.4	1.20	22461.41	3D 1	2B 1	R 2
4450.940	4452.189		0.1	0.2			22460.86	4f 3	2c 3	R 2
4451.037	4452.286						22460.37			
4451.215	4452.465						22459.47			
4451.273	4452.522		0.0				22459.18			
4451.346	4452.596	GK	0.3	0.7	0.3	0.61	22458.81			
4451.433	4452.683		0.0	0.2			22458.37			
4451.725	4452.974						22456.90			
4451.951	4453.200			0.3			22455.76			
4452.004	4453.254			0.2			22455.49			
4452.109	4453.359	GK	0.2	0.7	0.6	0.76	22454.96	3F 0	2B 0	R 2
4452.254	4453.504		0.1d	0.5			22454.23			
4452.363	4453.613			0.2			22453.68			
4452.482	4453.732	GK	0.1	0.7	0.6		22453.08			
4452.593	4453.843		0.0	0.5			22452.52			
4452.671	4453.920		0.0				22452.13			
4452.762	4454.012	GK	0.6	0.9b	0.6	1.14	22451.67	4f 2	2c 2	P 4
4452.871	4454.121		0.2	0.2			22451.12	4e 1	2c 1	R 4
4452.871	4454.121		0.2	0.2			22451.12	X 2	2B 4	R 0

λ_{air}	λ_{vac}	Ref.	I_1	I_2	I_3	I_4	ν	Upper	Lower	Br.
4452.982	4454.232			0.2			22450.56			
4453.067	4454.317		0.0	0.1			22450.13			
4453.156	4454.406	GK	0.7	1.0b	1.5b	1.28	22449.68	·4c 0	2a 0	R 3
4453.270	4454.520		0.1	0.3			22449.11			
4453.462	4454.712			0.2			22448.14	4e 1	2c 1	Q 1
4453.539	4454.789						22447.75			
4453.642	4454.893			0.3	0.3		22447.23			
4453.841	4455.091			0.0			22446.23			
4453.972	4455.222			0.2			22445.57			
4454.031	4455.282				0.2b		22445.27			
4454.081	4455.331			0.3			22445.02			
4454.212	4455.462			0.1d			22444.36			
4454.311	4455.562			0.5	0.3		22443.86	3D 1	2B 1	R 5
4454.424	4455.675			0.2			22443.29	3d 3	2c 0	R 4
4454.635	4455.885			0.2			22442.23			
4454.780	4456.030		0.1d	0.3d	0.2		22441.50	5c 0	2a 1	R 1
4454.994	4456.245	G		0.2d	0.1d		22440.42	4e 3	2c 3	R 4
4455.068	4456.318	F					22440.05			
4455.159	4456.409			0.1d			22439.59			
4455.260	4456.511		0.0d				22439.08			
4455.349	4456.600	G	0.2	0.7			22438.63	3D 1	2B 1	R 3
4455.473	4456.723	GK	0.1	0.8	0.8		22438.01			
4455.588	4456.838		0.1	0.1			22437.43	X 2	2B 4	R 1
4455.683	4456.934	GK	0.7b	0.9	0.7	1.09	22436.95	W 1	2B 2	R 3
4455.796	4457.047		0.0				22436.38			
4455.850	4457.101						22436.11			
4455.959	4457.210	GK	0.5	0.9	0.4	0.94	22435.56			
4456.023	4457.273						22435.24			
4456.070	4457.321		0.1d	0.2	0.3		22435.00	4e 1	2c 1	R 5
4456.223	4457.474						22434.23			
4456.327	4457.577						22433.71			
4456.370	4457.621			0.1d			22433.49			
4456.480	4457.730		0.0				22432.94			
4456.559	4457.810	GK	0.3	0.9			22432.54	3D 1	2B 1	R 4
4456.664	4457.915	GK	0.5	1.0	1.5b	1.42	22432.01	4f 2	2c 2	Q 2
4456.664	4457.915	GK	0.5	1.0	1.5b	1.42	22432.01	3F 0	2B 0	Q 4
4456.768	4458.019		0.2				22431.49			
4456.857	4458.108	GK	0.8	1.0	0.5	1.49	22431.04	4b 1	2a 0	P 3
4456.960	4458.211		0.1	0.3	0.9		22430.52			
4457.046	4458.297	GK	0.5	0.9			22430.09	4f 1	2c 1	P 3
4457.046	4458.297	GK	0.5	0.9			22430.09	4f 1	2c 1	P 4
4457.167	4458.418		0.0	0.2			22429.48			
4457.255	4458.506			0.1			22429.04			
4457.388	4458.639			0.2			22428.37			
4457.477	4458.728			0.1			22427.92			
4457.576	4458.828	GK	0.3	0.7	0.2	0.66	22427.42	4E 0	2B 4	R 0
4457.753	4459.005			0.2			22426.53			
4457.942	4459.193			0.2			22425.58			
4458.010	4459.261						22425.24			
4458.073	4459.325						22424.92			
4458.117	4459.368			0.1			22424.70			
4458.173	4459.424						22424.42			
4458.219	4459.470		0.0				22424.19			
4458.308	4459.559	GK	0.6	0.8	0.2	0.77	22423.74	4c 6	2a 5	Q 1
4458.415	4459.667		0.1	0.2d			22423.20	3E 2	2B 3	R 3
4458.469	4459.720						22422.93			
4458.557	4459.808			0.2			22422.49			
4458.660	4459.911		0.1				22421.97			
4458.744	4459.995	GK	0.4	0.9	1.5b	1.30	22421.55	3F 0	2B 0	P 6
4458.869	4460.120	GK	0.5	0.9			22420.92	4e 2	2c 2	R 1

λ_{air}	λ_{vac}	Ref.	I_1	I_2	I_3	I_4	ν	Upper	Lower	Br.
4458.869	4460.120	GK	0.5	0.9			22420.92	4e 1	2c 1	R 6
4458.994	4460.246			0.1			22420.29	4f 1	2c 1	P 5
4459.131	4460.383	GK	0.5	0.9	1.0	1.11	22419.60	4c 0	2a 0	R 2
4459.271	4460.522		0.0	0.2			22418.90	X 2	2B 4	P 1
4459.342	4460.594			0.0			22418.54			
4459.456	4460.707	GK	0.1b	0.7	0.6	0.59	22417.97			
4459.523	4460.775			0.5d			22417.63			
4459.656	4460.908		0.2	0.2	0.1		22416.96	3D 3	2B 4	P 3
4459.847	4461.099						22416.00			
4459.947	4461.199						22415.50			
4460.007	4461.258			0.4			22415.20	3A 2	2B 4	R 4
4460.170	4461.422	GK	0.0	0.7	0.2	0.30	22414.38			
4460.311	4461.563			0.2			22413.67			
4460.462	4461.714		0.0	0.3			22412.91	3A 1	2B 2	P 1
4460.528	4461.780		0.1d	0.3			22412.58			
4460.602	4461.854						22412.21			
4460.653	4461.905		0.0	0.2			22411.95			
4460.771	4462.023		0.2b	0.6			22411.36			
4460.864	4462.116		0.2	0.2			22410.89			
4460.966	4462.218	GK	1.5b	1.5b	1.5b	1.95	22410.38	4d 0	2c 0	R 2
4461.083	4462.335		0.2	0.4			22409.79	4e 0	2c 0	P 4
4461.245	4462.497		0.0				22408.98			
4461.302	4462.554		0.1	0.3d			22408.69	X 2	2B 4	R 2
4461.380	4462.632			0.1			22408.30			
4461.493	4462.746	G	0.2	0.7	0.5		22407.73			
4461.722	4462.975	G	0.1	0.4b	0.1		22406.58	3d 3	2c 0	Q 3
4461.908	4463.160	GK	0.2	0.8	0.4	0.46	22405.65	4f 2	2c 2	P 3
4462.045	4463.297			0.2			22404.96			
4462.165	4463.417		0.0	0.1			22404.36	4c 6	2a 5	Q 2
4462.256	4463.509			0.1	0.2		22403.90			
4462.354	4463.606		0.0	0.2			22403.41			
4462.523	4463.776		0.3	0.4	0.2		22402.56			
4462.667	4463.919			0.2			22401.84			
4462.784	4464.037	G	0.3	0.5			22401.25	W 1	2B 2	R 2
4462.918	4464.170			0.4	0.5		22400.58			
4463.009	4464.262						22400.12			
4463.069	4464.322			0.1d			22399.82			
4463.212	4464.465	G	0.2d	0.7	0.5		22399.10	3E 2	2B 3	Q 8
4463.358	4464.611		0.0				22398.37	3D 1	2B 1	P 2
4463.489	4464.742			0.0d	0.1		22397.71			
4463.685	4464.937			0.3d			22396.73	4f 3	2c 3	Q 2
4463.872	4465.125	GK	0.2	0.9	1.5	0.54	22395.79	3E 0	2B 0	Q10
4464.024	4465.276			0.2			22395.03			
4464.227	4465.480	GK	0.3	0.9	1.5	0.93	22394.01	4e 0	2c 0	Q 4
4464.358	4465.611	GK	0.4	0.5			22393.35	3A 1	2B 2	P 4
4464.566	4465.819			0.1			22392.31			
4464.715	4465.968			0.1			22391.56			
4464.805	4466.058			0.2			22391.11			
4464.917	4466.170		0.1	0.4			22390.55			
4465.048	4466.302			0.2			22389.89			
4465.184	4466.437			0.3	0.2		22389.21			
4465.405	4466.659		0.2	0.4d	0.1b		22388.10	3A 1	2B 2	P 2
4465.405	4466.659		0.2	0.4d	0.1b		22388.10	4f 2	2c 2	Q 3
4465.755	4467.008						22386.35			
4466.008	4467.261						22385.08			
4466.064	4467.317			0.1			22384.80			
4466.235	4467.489			0.4	0.3		22383.94			
4466.381	4467.634			0.2			22383.21			
4466.449	4467.702			0.2			22382.87	5c 0	2a 1	R 0
4466.577	4467.830			0.3d	0.3		22382.23			

λ_{air}	λ_{vac}	Ref.	I_1	I_2	I_3	I_4	ν	Upper	Lower	Br.
4466.658	4467.912			0.2d			22381.82			
4466.708	4467.962			0.2d			22381.57			
4466.868	4468.122		0.1	0.2d			22380.77			
4466.938	4468.191	G	0.3	0.4			22380.42	3A 1	2B 2	P 3
4466.994	4468.247		0.1				22380.14	3E 0	2B 0	P 4
4467.050	4468.303		0.3	0.2			22379.86			
4467.141	4468.395	GK	1.5b	1.5b	1.5	1.85	22379.40	4c 0	2a 0	R 1
4467.141	4468.395	GK	1.5b	1.5b	1.5	1.85	22379.40	X 2	2B 4	R 3
4467.259	4468.513		0.3	0.4			22378.81			
4467.401	4468.655		0.0		0.2		22378.10			
4467.447	4468.701		0.0	0.1			22377.87			
4467.561	4468.814		0.0d	0.1d			22377.30			
4467.648	4468.902		0.0				22376.86	4c 6	2a 5	Q 3
4467.728	4468.982	GK	0.2	0.5	0.1	0.48	22376.46			
4467.914	4469.168			0.2b			22375.53			
4468.024	4469.278			0.0			22374.98			
4468.126	4469.380		0.3	0.4	0.1		22374.47	4e 1	2c 1	Q 2
4468.279	4469.533	GK	0.5	0.7	0.2	0.64	22373.70			
4468.423	4469.677		0.0d	0.1			22372.98			
4468.503	4469.757						22372.58			
4468.539	4469.793		0.2	0.3			22372.40			
4468.705	4469.959			0.2			22371.57			
4468.795	4470.049			0.2			22371.12			
4468.909	4470.163			0.3			22370.55			
4469.032	4470.287			0.2			22369.93			
4469.112	4470.367			0.2			22369.53			
4469.170	4470.425		0.0	0.1			22369.24			
4469.258	4470.513	G	0.3	0.5	0.4		22368.80	W 1	2B 2	R 1
4469.368	4470.622		0.2	0.4	0.2		22368.25			
4469.554	4470.808	GK	0.3	0.6	0.4	0.34	22367.32	4e 3	2c 3	R 3
4469.664	4470.918			0.2			22366.77	4e 1	2c 1	P 2
4469.756	4471.010				0.1		22366.31			
4469.850	4471.104			0.1d			22365.84			
4469.994	4471.248				0.2		22365.12			
4470.254	4471.508			0.1d	0.2b		22363.82			
4470.381	4471.636			0.1			22363.18			
4470.463	4471.718			0.2			22362.77			
4470.697	4471.952	GK	0.5	0.6	0.2	0.57	22361.60	4f 3	2c 3	R 1
4470.697	4471.952	GK	0.5	0.6	0.2	0.57	22361.60	Z 1	2B 1	R 1
4470.813	4472.068			0.1			22361.02			
4471.287	4472.542		0.2d				22358.65			
4471.385	4472.640		0.2	0.3			22358.16			
4471.485	4472.740	GK	1.0b	0.8		1.10	22357.66	Z 1	2B 1	R 0
4471.537	4472.792	1			1.5b		22357.40			
4471.607	4472.862	GK	0.3	0.7			22357.05	4e 0	2c 0	P 3
4471.607	4472.862	GK	0.3	0.7			22357.05	4d 0	2c 0	Q 2
4471.629	4472.884		0.1				22356.94			
4471.673	4472.928		0.4				22356.72			
4471.791	4473.046		0.0	0.1			22356.13			
4471.863	4473.118		0.1				22355.77			
4471.957	4473.212	GK	0.6b	0.9	0.9	1.10	22355.30	4d 0	2c 0	P 2
4471.957	4473.212	GK	0.6b	0.9	0.9	1.10	22355.30	Z 1	2B 1	R 2
4472.087	4473.342		0.0				22354.65			
4472.177	4473.432						22354.20			
4472.219	4473.474	G	0.2	0.4	0.2		22353.99	X 2	2B 4	P 2
4472.363	4473.618		0.1d	0.4	0.3		22353.27			
4472.481	4473.736			0.2	0.2		22352.68			
4472.547	4473.803						22352.35			
4472.611	4473.867		0.0d	0.2			22352.03			
4472.766	4474.021			0.1			22351.26			

λ_{air}	λ_{vac}	Ref.	I_1	I_2	I_3	I_4	ν	Upper	Lower	Br.
4472.886	4474.141						22350.66	4e 2	2c 2	Q 5
4472.886	4474.141						22350.66	4f 3	2c 3	P 5
4472.960	4474.215						22350.29			
4473.122	4474.377			0.0d			22349.48	3d 3	2c 0	R 5
4473.374	4474.629			0.3d	0.3		22348.22			
4473.612	4474.868		0.0				22347.03			
4473.642	4474.898		0.0				22346.88			
4473.694	4474.950		0.0				22346.62			
4473.736	4474.992		0.1	0.3d	0.4		22346.41	5c 1	2a 2	R 3
4473.784	4475.040						22346.17			
4473.840	4475.096		0.0	0.2			22345.89			
4473.917	4475.172		0.0	0.1			22345.51			
4473.965	4475.220						22345.27			
4474.059	4475.314		0.1d	0.4			22344.80			
4474.173	4475.428		0.0				22344.23			
4474.257	4475.512	GK	0.6	1.0d	1.5b	1.39	22343.81	4e 0	2c 0	P 5
4474.257	4475.512	GK	0.6	1.0d	1.5b	1.39	22343.81	4d 0	2c 0	R 3
4474.257	4475.512	GK	0.6	1.0d	1.5b	1.39	22343.81	3E 2	2B 3	R 2
4474.375	4475.631		0.0	0.2			22343.22			
4474.627	4475.883			0.1			22341.96			
4474.848	4476.103		0.1d	0.2			22340.86			
4474.992	4476.248			0.2	0.1		22340.14	4e 2	2c 2	Q 4
4475.096	4476.352			0.2			22339.62	3F 3	2B 5	R 2
4475.222	4476.478			0.2			22338.99			
4475.373	4476.628				0.1		22338.24			
4475.433	4476.689						22337.94			
4475.611	4476.867			0.2	0.2		22337.05			
4475.859	4477.115	G	0.0	0.5	0.3		22335.81	4d 1	2c 1	P 1
4475.934	4477.190						22335.44			
4475.990	4477.246		0.2	0.3			22335.16	3D 3	2B 4	P 4
4476.144	4477.400		0.0d	0.2			22334.39			
4476.248	4477.504		0.1				22333.87			
4476.326	4477.583	GK	0.6	0.6	0.2	0.83	22333.48	4E 2	2C 0	P 3
4476.326	4477.583	GK	0.6	0.6	0.2	0.83	22333.48	W 1	2B 2	R 0
4476.461	4477.717		0.0	0.1			22332.81	4e 1	2c 1	P 2
4476.573	4477.829		0.1	0.0			22332.25			
4476.631	4477.887		0.0				22331.96			
4476.717	4477.974		0.0	0.3			22331.53			
4476.767	4478.024		0.0				22331.28			
4476.876	4478.132		0.2	0.2d			22330.74			
4476.964	4478.220		0.2	0.1d			22330.30	4E 0	2B 4	P 2
4477.072	4478.328	GK	1.0b	0.9	1.5	1.33	22329.76	4c 0	2a 0	R 0
4477.190	4478.447		0.2	0.3			22329.17	3d 3	2c 0	P 3
4477.256	4478.512	F					22328.84			
4477.407	4478.663		0.0	0.2d			22328.09			
4477.612	4478.868			0.4b	0.2		22327.07	4e 2	2c 2	Q 3
4477.816	4479.073	GK	0.2	0.6	0.3	0.48	22326.05	4b 1	2a 0	P 4
4477.979	4479.235	K		0.5	0.7		22325.24	4e 0	2c 0	Q 5
4478.111	4479.368			0.1			22324.58			
4478.187	4479.444						22324.20			
4478.251	4479.508			0.1			22323.88			
4478.372	4479.628			0.3			22323.28			
4478.434	4479.691	F					22322.97			
4478.747	4480.004						22321.41			
4478.805	4480.062		0.0d	0.0			22321.12			
4478.871	4480.128						22320.79			
4478.986	4480.243	GK	0.9b	0.7b	0.3	1.04	22320.22	4f 3	2c 3	Q 4
4479.098	4480.355		0.0				22319.66			
4479.205	4480.461			0.1			22319.13			
4479.309	4480.566		0.1	0.1			22318.61			

λ_{air}	λ_{vac}	Ref.	I_1	I_2	I_3	I_4	ν	Upper	Lower	Br.
4479.389	4480.646	GK	0.4	0.4	0.2	0.43	22318.21	4c 1	2a 1	P 3
4479.389	4480.646	GK	0.4	0.4	0.2	0.43	22318.21	4e 1	2c 1	P 3
4479.389	4480.646	GK	0.4	0.4	0.2	0.43	22318.21	5c 0	2a 1	Q 1
4479.490	4480.746		0.0				22317.71			
4479.556	4480.813			0.0			22317.38			
4479.606	4480.863		0.1				22317.13			
4479.686	4480.943	GK	0.6	0.6	0.3	0.61	22316.73	4e 2	2c 2	R 1
4479.783	4481.040			0.2			22316.25			
4479.949	4481.206						22315.42			
4480.044	4481.301			0.1			22314.95			
4480.156	4481.413	G	0.2	0.4	0.1		22314.39	3c 4	2a 1	Q 1
4480.276	4481.534			0.2			22313.79			
4480.395	4481.652		0.0	0.4	0.2		22313.20			
4480.537	4481.795			0.2			22312.49	4D 0	2B 4	R 6
4480.720	4481.978	G	0.0	0.4	0.5		22311.58	4e 2	2c 2	Q 2
4480.837	4482.094		0.0	0.1			22311.00			
4480.927	4482.184						22310.55			
4481.032	4482.289						22310.03			
4481.090	4482.347			0.2			22309.74			
4481.259	4482.516	GK	0.2	0.5	0.9	0.93	22308.90	3F 0	2B 0	Q 3
4481.393	4482.651		0.1	0.2			22308.23	5c 0	2a 1	Q 2
4481.618	4482.876		0.0d				22307.11			
4481.713	4482.970				0.2		22306.64			
4481.907	4483.165	GK	0.4r	0.4b	0.5		22305.67	3E 2	2B 3	Q 7
4481.907	4483.165	GK	0.4r	0.4b	0.5		22305.67	4e 1	2c 1	Q 3
4482.008	4483.266		0.0				22305.17			
4482.074	4483.332	GK	0.2	0.5	1.0	0.85	22304.84	3D 1	2B 1	P 3
4482.074	4483.332	GK	0.2	0.5	1.0	0.85	22304.84	3F 0	2B 0	P 5
4482.074	4483.332	GK	0.2	0.5	1.0	0.85	22304.84	4f 3	2c 3	Q 3
4482.074	4483.332	GK	0.2	0.5	1.0	0.85	22304.84	4e 2	2c 2	R 2
4482.247	4483.505	GK	0.3	0.5	0.5		22303.98			
4482.390	4483.647		0.0	0.1			22303.27			
4482.510	4483.768	G	0.4	0.5		0.49	22302.67	Z 1	2B 1	P 1
4482.510	4483.768	G	0.4	0.5		0.49	22302.67	3A 2	2B 4	R 3
4482.685	4483.943		0.0d	0.4	0.2		22301.80			
4482.830	4484.088						22301.08	4b 2	2a 1	R 1
4482.896	4484.154			0.0			22300.75			
4483.031	4484.289	F					22300.08			
4483.071	4484.329	G	0.0	0.5b	0.1		22299.88	3F 1	2B 2	Q 6
4483.071	4484.329	G	0.0	0.5b	0.1		22299.88	5c 1	2a 2	R 2
4483.242	4484.500				0.1		22299.03			
4483.403	4484.661			0.2			22298.23			
4483.511	4484.769	G	0.0d	0.9	0.6		22297.69	4g 1	2a 1	P 5
4483.550	4484.808	F				0.36	22297.50			
4483.735	4484.993			0.1			22296.58	4E 0	2B 4	P 4
4483.841	4485.099		0.0d	0.3			22296.05			
4483.976	4485.234			0.1d			22295.38			
4484.069	4485.327		0.0	0.1	0.2		22294.92			
4484.163	4485.421	G	0.3	0.8	0.3	0.51	22294.45	3F 0	2B 0	R 1
4484.278	4485.536		0.0	0.2	0.3		22293.88	4E 0	2B 4	Q 1
4484.278	4485.536		0.0	0.2	0.3		22293.88	4E 0	2B 4	Q 4
4484.358	4485.616	G	0.1	0.5			22293.48	4e 2	2c 2	Q 1
4484.358	4485.616	G	0.1	0.5			22293.48	5c 0	2a 1	Q 3
4484.358	4485.616	G	0.1	0.5			22293.48	4E 0	2B 4	Q 1
4484.457	4485.715						22292.99	3c 4	2a 1	Q 2
4484.523	4485.781			0.2			22292.66			
4484.779	4486.037			0.4			22291.39			
4485.004	4486.262		0.0				22290.27			
4485.135	4486.393	G	0.2d	0.8	0.4		22289.62	4f 3	2c 3	P 4
4485.135	4486.393	G	0.2d	0.8	0.4		22289.62	4e 2	2c 2	R 3

λ_{air}	λ_{vac}	Ref.	I_1	I_2	I_3	I_4	ν	Upper	Lower	Br.
4485.302	4486.560			0.2			22288.79			
4485.511	4486.770	G	0.0d	0.8	0.6		22287.75	4e 3	2c 3	R 2
4485.618	4486.876		0.1				22287.22			
4485.668	4486.927			0.2			22286.97			
4485.757	4487.015		0.0				22286.53			
4485.831	4487.090	GK	0.8r	0.9	0.5		22286.16	4d 1	2c 1	R 1
4485.831	4487.090	GK	0.8r	0.9	0.5		22286.16	4d 0	2c 0	R 4
4485.954	4487.213		0.0				22285.55	3d 3	2c 0	Q 4
4486.077	4487.335	GK	0.9	1.5b	1.0	1.61	22284.94	4d 1	2c 1	Q 1
4486.107	4487.366		0.9				22284.79			
4486.206	4487.464		0.0	0.2			22284.30			
4486.357	4487.615			0.1			22283.55			
4486.481	4487.740			0.3			22282.93	3c 4	2a 1	Q 2
4486.618	4487.877		0.0				22282.25			
4486.715	4487.974	G	0.5	0.5		0.46	22281.77			
4486.828	4488.087			0.2	0.1		22281.21			
4486.953	4488.211		0.0				22280.59			
4487.005	4488.264		0.0d	0.0			22280.33			
4487.104	4488.363	GK	0.5	0.5	0.1		22279.84	4e 2	2c 2	Q 1
4487.220	4488.479		0.0				22279.26			
4487.285	4488.544			0.2			22278.94			
4487.372	4488.631		0.1	0.3			22278.51			
4487.504	4488.763			0.1			22277.85			
4487.615	4488.874		0.0	0.5	0.6		22277.30	4e 0	2c 0	P 6
4487.734	4488.993		0.0				22276.71			
4487.813	4489.072	GK	0.7	1.5	1.5b	1.56	22276.32	X 2	2B 4	P 3
4487.813	4489.072	GK	0.7	1.5	1.5b	1.56	22276.32	2u 2	2a 0	P 2
4487.936	4489.195		0.1	0.2			22275.71			
4488.044	4489.304		0.0	0.2			22275.17			
4488.123	4489.382			0.1			22274.78			
4488.210	4489.469		0.2	0.3			22274.35			
4488.332	4489.592		0.1	0.3b			22273.74	5c 0	2a 1	Q 4
4488.451	4489.711		0.0				22273.15			
4488.554	4489.813	GK	0.4	0.8	0.3	0.59	22272.64	4f 2	2c 2	P 4
4488.554	4489.813	GK	0.4	0.8	0.3	0.59	22272.64	4e 2	2c 2	R 4
4488.760	4490.019		0.1	0.3b			22271.62			
4488.945	4490.205		0.0	0.6	0.3		22270.70			
4489.032	4490.291		0.0				22270.27			
4489.149	4490.408		0.1	0.6	0.4		22269.69	7c 0	2a 2	R 1
4489.250	4490.509	GK	0.5	0.8		0.77	22269.19	4f 2	2c 2	P 3
4489.250	4490.509	GK	0.5	0.8		0.77	22269.19	3E 0	2B 0	R 3
4489.399	4490.658		0.0	0.2			22268.45			
4489.520	4490.779	G	0.1	0.6	0.6		22267.85	4e 1	2c 1	P 4
4489.689	4490.949		0.0	0.2			22267.01			
4489.818	4491.078		0.0	0.1			22266.37			
4489.905	4491.164						22265.94			
4489.947	4491.207		0.0	0.4			22265.73			
4490.048	4491.308		0.1b	0.5	0.3		22265.23	3E 2	2B 3	R 1
4490.155	4491.415		0.1	0.2			22264.70			
4490.256	4491.516		0.3	0.5			22264.20			
4490.298	4491.558		0.1				22263.99			
4490.365	4491.624		0.4				22263.66			
4490.451	4491.711	GK	1.5b	1.5	1.5b	2.05	22263.23	4c 0	2a 0	Q 1
4490.451	4491.711	GK	1.5b	1.5	1.5b	2.05	22263.23	3E 0	2B 0	Q 9
4490.570	4491.830		0.4	0.6			22262.64			
4490.651	4491.911		0.0				22262.24			
4490.708	4491.967		0.1	0.7	0.9		22261.96	3E 0	2B 0	Q 9
4490.766	4492.026		0.1				22261.67			
4490.857	4492.117	GK	0.6	0.8			22261.22	W 1	2B 2	P 1
4490.964	4492.224		0.0				22260.69	3F 3	2B 5	Q 3

λ_{air}	λ_{vac}	Ref.	I_1	I_2	I_3	I_4	ν	Upper	Lower	Br.
4491.040	4492.300		0.0				22260.31			
4491.107	4492.367		0.0	0.4	0.3		22259.98			
4491.258	4492.518		0.1d	0.5			22259.23			
4491.357	4492.617		0.0				22258.74			
4491.460	4492.720		0.0d	0.1			22258.23			
4491.569	4492.829		0.0	0.2			22257.69			
4491.603	4492.864						22257.52			
4491.724	4492.985	GK		0.7	1.0		22256.92	4f 3	2c 3	Q 2
4491.724	4492.985	GK		0.7	1.0		22256.92	4e 0	2c 0	Q 6
4491.833	4493.094		0.0	0.2			22256.38			
4491.971	4493.231	GK	0.3	0.6			22255.70	5c 1	2a 2	R 1
4492.080	4493.340		0.0	0.2			22255.16			
4492.175	4493.435		0.1	0.3	0.2		22254.69			
4492.312	4493.572		0.0	0.2			22254.01			
4492.441	4493.701	G	0.1	0.5	0.3		22253.37	4f 2	2c 2	P 5
4492.441	4493.701	G	0.1	0.5	0.3		22253.37	4E 0	2B 4	Q 3
4492.588	4493.849		0.0d	0.2			22252.64			
4492.691	4493.952		0.2	0.4	0.2		22252.13			
4492.845	4494.105	G	0.2	0.6	0.1		22251.37	4e 2	2c 2	R 5
4493.014	4494.275				0.1		22250.53			
4493.099	4494.360		0.0	0.3	0.1		22250.11			
4493.166	4494.426			0.3	0.2		22249.78	5c 0	2a 1	Q 5
4493.384	4494.645			0.2			22248.70			
4493.523	4494.784		0.0	0.4			22248.01			
4493.608	4494.869		0.0				22247.59			
4493.691	4494.952	GK	0.8b	1.5	1.5b	1.56	22247.18	Z 1	2B 1	P 2
4493.691	4494.952	GK	0.8b	1.5	1.5b	1.56	22247.18	4c 0	2a 0	Q 2
4493.818	4495.079		0.1	0.3			22246.55			
4494.034	4495.295		0.0d	0.3			22245.48	3D 3	2B 4	P 5
4494.186	4495.447				0.2		22244.73			
4494.426	4495.687			0.3			22243.54			
4494.507	4495.768		0.2	0.4	0.1		22243.14	4E 0	2B 4	Q 2
4494.616	4495.877						22242.60			
4494.780	4496.041			0.4	0.3		22241.79			
4494.934	4496.195	G	0.2b	0.6	0.3		22241.03	4f 3	2c 3	P 3
4495.091	4496.352						22240.25			
4495.140	4496.401			0.1d			22240.01			
4495.200	4496.461		0.0				22239.71			
4495.322	4496.583	GK	0.3	0.8	0.4	0.49	22239.11	4e 1	2c 1	Q 4
4495.433	4496.694		0.0	0.1			22238.56			
4495.485	4496.747		0.0				22238.30			
4495.643	4496.904	G	0.2d	0.6	0.6		22237.52	4d 0	2c 0	R 5
4495.853	4497.115		0.1d	0.2			22236.48			
4495.938	4497.200		0.0	0.4	0.2		22236.06	3c 4	2a 1	Q 3
4496.110	4497.371			0.2	0.1		22235.21			
4496.286	4497.547			0.4	0.3		22234.34			
4496.420	4497.681			0.1d			22233.68			
4496.492	4497.754				0.2		22233.32			
4496.587	4497.849		0.0				22232.85			
4496.674	4497.936	GK	0.5	0.9	0.4	1.15	22232.42			
4496.814	4498.076			0.1			22231.73			
4496.899	4498.160			0.1			22231.31			
4497.006	4498.268		0.0				22230.78			
4497.103	4498.365	GK	0.5r	0.9	1.0	0.95	22230.30	4d 0	2c 0	Q 3
4497.103	4498.365	GK	0.5r	0.9	1.0	0.95	22230.30	3A 2	2B 4	R 2
4497.263	4498.525		0.0	0.0			22229.51			
4497.374	4498.636		0.1	0.1			22228.96			
4497.439	4498.701		0.0				22228.64			
4497.494	4498.755		0.2				22228.37			
4497.579	4498.840	GK	0.9	1.0	0.4	1.52	22227.95	W 1	2B 2	P 2

λ_{air}	λ_{vac}	Ref.	I_1	I_2	I_3	I_4	ν	Upper	Lower	Br.
4497.696	4498.958		0.2	0.2			22227.37	4e 0	2c 0	P 7
4497.775	4499.037		0.0d	0.1			22226.98			
4497.886	4499.148	G	0.3	0.7	0.6		22226.43	3E 2	2B 3	Q 6
4498.000	4499.261		0.2	0.2			22225.87	4e 0	2c 0	P 4
4498.103	4499.365	GK	1.5b	1.5	1.5	1.79	22225.36	4d 1	2c 1	R 2
4498.224	4499.486		0.2	0.3			22224.76			
4498.326	4499.587		0.0	0.3			22224.26			
4498.433	4499.695		0.1	0.2			22223.73			
4498.524	4499.786	GK	0.9	1.5	1.5b	1.73	22223.28	4c 0	2a 0	Q 3
4498.643	4499.905		0.2	0.4			22222.69	3D 0	2B 0	R12
4498.678	4499.940		0.0				22222.52			
4498.708	4499.970		0.0				22222.37			
4498.769	4500.031		0.0				22222.07			
4498.842	4500.104		0.0				22221.71			
4498.904	4500.167		0.0	0.2			22221.40			
4498.973	4500.235			0.3	0.3		22221.06	4D 0	2B 4	R 5
4499.081	4500.343						22220.53			
4499.123	4500.385		0.0	0.2			22220.32			
4499.216	4500.478	GK	0.2	0.6	0.2		22219.86			
4499.352	4500.614		0.0	0.3			22219.19			
4499.471	4500.734		0.0	0.1			22218.60			
4499.554	4500.817	GK	0.4	0.9	0.7	0.93	22218.19	4e 1	2c 1	P 5
4499.690	4500.952			0.2			22217.52			
4499.850	4501.112		0.0d	0.3			22216.73			
4500.047	4501.309	GK	0.2b	0.8	1.5	0.54	22215.76	4g 0	2a 0	P 5
4500.152	4501.414		0.0	0.1			22215.24			
4500.346	4501.609		0.1	0.5			22214.28			
4500.387	4501.649		0.1				22214.08			
4500.462	4501.724			0.1			22213.71	4e 2	2c 2	R 4
4500.555	4501.818		0.0d	0.1			22213.25			
4500.677	4501.939			0.3	0.3d		22212.65			
4500.770	4502.032			0.4			22212.19			
4500.881	4502.144			0.2			22211.64			
4500.975	4502.237			0.2	0.3		22211.18			
4501.131	4502.393			0.3d			22210.41			
4501.175	4502.438						22210.19			
4501.281	4502.543			0.1d			22209.67			
4501.429	4502.691			0.2			22208.94			
4501.495	4502.758			0.1			22208.61	T 0	2B 4	R 2
4501.595	4502.857	GK	0.1	0.9	1.0	0.57	22208.12	4e 3	2c 3	R 1
4501.595	4502.857	GK	0.1	0.9	1.0	0.57	22208.12	4e 2	2c 2	R 5
4501.704	4502.967			0.1			22207.58	3D 1	2B 1	P 4
4501.769	4503.032		0.0d	0.2			22207.26			
4501.864	4503.127		0.2	0.4			22206.79	4e 2	2c 2	Q 2
4501.958	4503.220	GK	0.9	1.0	0.8	1.53	22206.33	3F 0	2B 0	Q 2
4502.063	4503.326		0.2	0.3			22205.81			
4502.193	4503.456			0.2			22205.17	5c 1	2a 2	R 0
4502.262	4503.525		0.1d	0.3	0.2d		22204.83			
4502.355	4503.618			0.1			22204.37			
4502.462	4503.725			0.1d			22203.84			
4502.602	4503.865	GK	0.4	0.8	0.3		22203.15	4e 1	2c 1	P 3
4502.746	4504.009	GK	0.5	0.8		0.95	22202.44			
4502.870	4504.133		0.1	0.1			22201.83			
4502.984	4504.247			0.3			22201.27			
4503.067	4504.330						22200.86			
4503.276	4504.539			0.1			22199.83			
4503.389	4504.653		0.0d	0.3			22199.27			
4503.546	4504.809	G	0.2	0.4			22198.50	4d 0	2c 0	R 6
4503.546	4504.809	G	0.2	0.4			22198.50	3F 3	2B 5	Q 2
4503.669	4504.933			0.2			22197.89			

λ_{air}	λ_{vac}	Ref.	I_1	I_2	I_3	I_4	ν	Upper	Lower	Br.
4503.748	4505.012			0.1			22197.50			
4503.860	4505.123	G	0.2	0.7	0.3		22196.95			
4503.966	4505.229			0.0			22196.43			
4504.217	4505.481			0.2			22195.19			
4504.298	4505.562		0.0	0.1			22194.79			
4504.469	4505.732	GK	0.5	0.9	0.4d	1.14	22193.95	W 1	2B 2	P 3
4504.605	4505.868			0.1			22193.28			
4504.704	4505.968			0.2b			22192.79	5c 0	2a 1	P 2
4504.818	4506.082			0.0			22192.23			
4504.922	4506.185	GK	0.2	0.9	1.0	0.81	22191.72	4c 0	2a 0	Q 4
4505.179	4506.443						22190.45			
4505.255	4506.518						22190.08			
4505.443	4506.707			0.4	0.4		22189.15			
4505.531	4506.794			0.1			22188.72			
4505.628	4506.892	GK	0.4	1.5	1.0	1.53	22188.24	3F 0	2B 0	P 4
4505.628	4506.892	GK	0.4	1.5	1.0	1.53	22188.24	4e 3	2c 3	Q 5
4505.754	4507.018			0.4			22187.62			
4506.053	4507.316						22186.15			
4506.142	4507.406						22185.71			
4506.296	4507.560	F					22184.95			
4506.380	4507.644			0.1			22184.54	X 2	2B 4	P 4
4506.532	4507.796	G		0.5			22183.79	4e 2	2c 2	P 2
4506.532	4507.796	G		0.5			22183.79	3E 2	2B 3	R 0
4506.640	4507.904			0.1			22183.26			
4506.721	4507.985						22182.86			
4506.879	4508.143			0.3			22182.08			
4506.997	4508.261			0.1			22181.50			
4507.184	4508.448			0.5	0.4		22180.58			
4507.292	4508.556			0.1			22180.05			
4507.398	4508.662			0.2			22179.53			
4507.572	4508.837		0.1	0.4			22178.67	Z 1	2B 1	P 3
4507.735	4508.999			0.1			22177.87			
4507.855	4509.119			0.4			22177.28	4d 0	2c 0	P 3
4507.855	4509.119			0.4			22177.28	3D 3	2B 4	P 6
4507.983	4509.247						22176.65			
4508.121	4509.386						22175.97			
4508.162	4509.426			0.1			22175.77			
4508.229	4509.493						22175.44			
4508.318	4509.583	G		0.4	0.3		22175.00			
4508.398	4509.662			0.2			22174.61			
4508.534	4509.798			0.1			22173.94	3c 4	2a 1	Q 4
4508.652	4509.916			0.2			22173.36			
4508.686	4509.951						22173.19			
4508.735	4510.000						22172.95	3E 0	2B 0	P 8
4508.790	4510.055	G		0.5	0.3d		22172.68	4f 2	2c 2	P 4
4508.906	4510.171		0.1	0.2			22172.11			
4509.008	4510.272		0.0				22171.61			
4509.101	4510.366	GK	0.5b	0.9	0.4	0.56	22171.15	4e 1	2c 1	Q 5
4509.101	4510.366	GK	0.5b	0.9	0.4	0.56	22171.15	4d 1	2c 1	Q 2
4509.235	4510.500		0.0d	0.1			22170.49	4e 2	2c 2	P 2
4509.235	4510.500		0.0d	0.1			22170.49	4e 1	2c 1	Q 5
4509.300	4510.565	F					22170.17			
4509.374	4510.639		0.1	0.5			22169.81	u 2	2a 0	P 3
4509.374	4510.639		0.1	0.5			22169.81	4d 1	2c 1	P 2
4509.449	4510.714			0.2			22169.44	3d 3	2c 0	P 4
4509.589	4510.854			0.2			22168.75			
4509.707	4510.972			0.1			22168.17	3F 1	2B 2	R 3
4509.799	4511.064			0.2			22167.72			
4509.939	4511.204	F		0.1			22167.03	4e 1	2c 1	P 6
4510.086	4511.351			0.1			22166.31			

λ_{air}	λ_{vac}	Ref.	I_1	I_2	I_3	I_4	ν	Upper	Lower	Br.
4510.169	4511.434			0.1			22165.90			
4510.261	4511.526			0.4	0.4		22165.45			
4510.434	4511.699			0.3			22164.60			
4510.592	4511.858			0.2			22163.82			
4510.725	4511.990	i		0.3	1.5b		22163.17			
4510.790	4512.055			0.1			22162.85	4d 1	2c 1	R 3
4510.900	4512.165	GK	0.5	1.5		1.17	22162.31			
4511.008	4512.273	F					22161.78			
4511.057	4512.322	G	0.1d	0.7	0.4		22161.54	3E 2	2B 3	Q 5
4511.057	4512.322	G	0.1d	0.7	0.4		22161.54	4e 3	2c 3	Q 4
4511.217	4512.483			0.0d	0.0		22160.75			
4511.354	4512.619	GK	0.4	0.6		0.63	22160.08	3A 2	2B 4	R 1
4511.496	4512.762			0.1d	0.3		22159.38			
4511.604	4512.870			0.1	0.2		22158.85	T 0	2B 4	R 1
4511.692	4512.957	GK	0.7	1.0	0.5	1.50	22158.42	W 1	2B 2	P 4
4511.808	4513.073			0.1	0.1		22157.85			
4511.895	4513.161	GK	0.0	0.7	0.7		22157.42			
4512.024	4513.289			0.0			22156.79			
4512.266	4513.532						22155.60			
4512.352	4513.617				0.0		22155.18			
4512.400	4513.666			0.2			22154.94			
4512.478	4513.743	G		0.2b	0.7	0.3	22154.56	7c 0	2a 2	Q 1
4512.549	4513.815	F				0.64	22154.21			
4512.655	4513.921			0.1b	0.6		22153.69	4D 0	2B 4	R 4
4512.753	4514.018				0.2		22153.21			
4512.867	4514.133	GK	0.3	1.0	1.0b	0.84	22152.65	4c 0	2a 0	Q 5
4512.995	4514.261				0.3		22152.02			
4513.134	4514.400				0.3d		22151.34	3E 0	2B 0	R 2
4513.293	4514.559				0.2		22150.56			
4513.364	4514.630						22150.21			
4513.452	4514.718	G	0.0d	0.4			22149.78			
4513.670	4514.936				0.2		22148.71	3d 3	2c 0	Q 5
4513.796	4515.062			0.4			22148.09	4D 0	2B 4	R 0
4513.827	4515.093	GK	0.4	0.8	0.4	0.97	22147.94	4e 2	2c 2	P 3
4513.827	4515.093	GK	0.4	0.8	0.4	0.97	22147.94	3F 1	2B 2	Q 5
4513.955	4515.221			0.0	0.2		22147.31			
4514.118	4515.384				0.2		22146.51			
4514.316	4515.582	GK	0.2	1.0	1.0b	0.66	22145.54	3E 0	2B 0	Q 8
4514.432	4515.698			0.0	0.2		22144.97	7c 0	2a 2	Q 3
4514.640	4515.906				0.3d		22143.95	V 1	2B 0	R 2
4514.832	4516.098			0.4			22143.01	4E 1	2B 6	Q 4
4514.974	4516.241			0.2			22142.31			
4515.082	4516.349			0.1			22141.78			
4515.184	4516.451	GK	0.3	0.9	0.4	0.88	22141.28	4D 0	2B 4	P 1
4515.184	4516.451	GK	0.3	0.9	0.4	0.88	22141.28	3E 2	2B 3	P 6
4515.184	4516.451	GK	0.3	0.9	0.4	0.88	22141.28	4e 2	2c 2	Q 3
4515.315	4516.581			0.1			22140.64			
4515.382	4516.649			0.0d	0.2		22140.31			
4515.460	4516.726				0.0		22139.93			
4515.568	4516.834	GK	0.2	1.0	1.0	0.64	22139.40	4c 1	2a 1	R 3
4515.790	4517.057			0.2			22138.31			
4516.010	4517.277			0.3d			22137.23			
4516.210	4517.477			0.4			22136.25			
4516.363	4517.630			0.3			22135.50			
4516.453	4517.720			0.1			22135.06			
4516.586	4517.852			0.4			22134.41			
4516.680	4517.946			0.4			22133.95			
4516.914	4518.181			0.2d			22132.80	4e 3	2c 3	Q 3
4517.006	4518.273			0.1d			22132.35			
4517.241	4518.508		0.1d	0.2			22131.20			

λ_{air}	λ_{vac}	Ref.	I_1	I_2	I_3	I_4	ν	Upper	Lower	Br.
4517.335	4518.602		0.1				22130.74			
4517.433	4518.700	GK	0.9b	0.9	0.6	0.98	22130.26	4c 0	2a 0	P 2
4517.559	4518.826		0.2	0.2			22129.64			
4517.655	4518.922		0.0	0.0			22129.17			
4517.737	4519.004	GK	0.4	0.6		0.58	22128.77	5c 0	2a 2	Q 1
4517.904	4519.171	GK	0.2	0.8	0.3	0.83	22127.95	W 1	2B 2	P 5
4518.027	4519.294		0.1	0.2			22127.35			
4518.125	4519.392	G	0.2	0.4		0.59	22126.87	3E 2	2B 3	P 5
4518.125	4519.392	G	0.2	0.4		0.59	22126.87	5c 0	2a 1	P 3
4518.321	4519.588			0.2d			22125.91			
4518.439	4519.707			0.1			22125.33			
4518.544	4519.811			0.3			22124.82			
4518.672	4519.939	G	0.2	0.7	0.5		22124.19	3D 0	2B 0	R11
4518.672	4519.939	G	0.2	0.7	0.5		22124.19	3F 3	2B 5	P 4
4518.956	4520.224		0.0d	0.2d			22122.80			
4519.038	4520.305		0.0	0.1			22122.40			
4519.128	4520.395	GK	0.3	0.9	0.6	1.03	22121.96	W 1	2B 2	P 6
4519.250	4520.518		0.2	0.2			22121.36	T 0	2B 4	R 0
4519.334	4520.602						22120.95			
4519.395	4520.663			0.1d			22120.65			
4519.526	4520.794		0.1	0.4			22120.01	5c 1	2a 2	Q 2
4519.571	4520.839	F					22119.79			
4519.665	4520.933		0.0	0.3			22119.33			
4519.773	4521.041			0.2			22118.80			
4519.851	4521.119		0.0d	0.1			22118.42			
4519.961	4521.229	GK	0.3	0.9	1.0	0.83	22117.88			
4520.096	4521.364		0.0d	0.1			22117.22			
4520.229	4521.497			0.1			22116.57			
4520.360	4521.628		0.0	0.0			22115.93	3D 1	2B 1	P 5
4520.462	4521.730	G	0.2	0.5	0.4		22115.43			
4520.558	4521.826		0.0d				22114.96			
4520.644	4521.912	G	0.3	0.6			22114.54	4d 2	2c 2	P 1
4520.761	4522.028		0.0d	0.0			22113.97			
4520.924	4522.192			0.2d			22113.17			
4521.006	4522.274		0.0				22112.77			
4521.102	4522.370		0.0	0.5	0.2d		22112.30			
4521.227	4522.495		0.1b	0.3			22111.69			
4521.315	4522.583		0.1	0.2			22111.26	4D 0	2B 4	R 3
4521.315	4522.583		0.1	0.2			22111.26	4E 1	2B 6	Q 1
4521.403	4522.671	G	0.6	0.9	0.5	1.47	22110.83	3E 2	2B 3	P 4
4521.499	4522.767	GK	0.3	1.0			22110.36	4e 2	2c 2	P 4
4521.499	4522.767	GK	0.3	1.0			22110.36	3E 2	2B 3	Q 4
4521.644	4522.912		0.0d	0.2			22109.65			
4521.715	4522.984		0.0d				22109.30			
4521.773	4523.041		0.2	0.4	0.2		22109.02			
4521.862	4523.130	F					22108.58			
4521.914	4523.182			0.3			22108.33			
4522.004	4523.272			0.0			22107.89			
4522.082	4523.350		0.1	0.4	0.3		22107.51			
4522.206	4523.475			0.2			22106.90			
4522.268	4523.536						22106.60			
4522.307	4523.575	GK1	0.2	0.7	1.0b		22106.41	5c 1	2a 2	Q 3
4522.429	4523.698	G	0.0d	0.6		0.49	22105.81	4d 0	2c 0	Q 4
4522.489	4523.757						22105.52	4c 1	2a 1	R 2
4522.568	4523.837			0.0			22105.13			
4522.673	4523.941			0.2			22104.62			
4522.753	4524.021		0.0	0.2			22104.23			
4522.869	4524.138	GK	0.5	0.6	0.2	1.14	22103.66	4e 3	2c 3	Q 2
4522.978	4524.246	GK	0.4b	0.7			22103.13	4e 1	2c 1	Q 6
4523.094	4524.363		0.1	0.1			22102.56			

λ_{air}	λ_{vac}	Ref.	I_1	I_2	I_3	I_4	ν	Upper	Lower	Br.
4523.191	4524.459	GK	0.6	0.9	0.6	1.05	22102.09	4f 3	2c 3	P 3
4523.191	4524.459	GK	0.6	0.9	0.6	1.05	22102.09	3F 0	2B 0	P 3
4523.279	4524.547		0.1	0.4			22101.66			
4523.330	4524.598		0.2	0.2			22101.41	Z 1	2B 1	P 4
4523.387	4524.656		0.0				22101.13	4E 1	2B 6	Q 3
4523.459	4524.727	G	0.2d	0.7	0.3		22100.78	4f 3	2c 3	P 4
4523.674	4524.942		0.1	0.5	0.4		22099.73	4e 0	2c 0	P 5
4523.833	4525.102		0.1	0.5			22098.95			
4523.956	4525.225		0.0d	0.4			22098.35			
4524.036	4525.305		0.0	0.2			22097.96			
4524.138	4525.407	GK	0.6b	1.5	0.9	1.43	22097.46	3E 2	2B 3	P 3
4524.138	4525.407	GK	0.6b	1.5	0.9	1.43	22097.46	4d 1	2c 1	R 4
4524.280	4525.548			0.5			22096.77	4c 0	2a 0	Q 6
4524.390	4525.659			0.2			22096.23	4D 0	2B 4	R 1
4524.482	4525.751		0.2	0.5	0.2d		22095.78			
4524.565	4525.834	F					22095.38			
4524.703	4525.972			0.4	0.3		22094.70			
4524.814	4526.083		0.0				22094.16	X 2	2B 4	P 5
4524.896	4526.165	GK	0.4	0.7		0.61	22093.76	4D 0	2B 4	R 2
4525.000	4526.269		0.1				22093.25			
4525.062	4526.331			0.1d			22092.95			
4525.129	4526.398			0.1			22092.62			
4525.228	4526.497	G		0.7	0.3		22092.14	4g 1	2a 1	P 6
4525.396	4526.665						22091.32			
4525.574	4526.843			0.2d			22090.45			
4525.685	4526.954		0.0d				22089.91			
4525.805	4527.075			0.1			22089.32			
4525.928	4527.198						22088.72			
4526.076	4527.345			0.3			22088.00	5c 1	2a 2	Q 4
4526.203	4527.472		0.0				22087.38			
4526.273	4527.542	G	0.3	0.4			22087.04	4E 1	2B 6	Q 2
4526.396	4527.665			0.3	0.2		22086.44			
4526.525	4527.794			0.1			22085.81			
4526.625	4527.895			0.3			22085.32			
4526.696	4527.965	F					22084.97			
4526.816	4528.085			0.2			22084.39			
4526.873	4528.143			0.1			22084.11			
4526.953	4528.223		0.0		0.2		22083.72			
4527.013	4528.282	G	0.3	0.3			22083.43			
4527.103	4528.372		0.0	0.2	0.4		22082.99			
4527.174	4528.444	GK	0.6	0.9		1.18	22082.64	3E 2	2B 3	P 2
4527.300	4528.569		0.0d	0.2			22082.03			
4527.371	4528.641		0.0				22081.68			
4527.466	4528.735		0.0				22081.22			
4527.539	4528.809		0.3	0.6	0.3		22080.86	3D 2	2B 3	P 1
4527.675	4528.944		0.0	0.1			22080.20			
4527.777	4529.047	G	0.4	0.8	0.3		22079.70	4d 2	2c 2	R 1
4527.855	4529.125			0.2			22079.32	4e 1	2c 1	P 4
4528.005	4529.275			0.1			22078.59			
4528.093	4529.363		0.0d				22078.16			
4528.159	4529.429			0.1			22077.84			
4528.337	4529.607			0.1d			22076.97			
4528.635	4529.905			0.1d			22075.52	3A 2	2B 4	R 0
4528.745	4530.015		0.0d	0.5	0.3d		22074.98			
4528.877	4530.147			0.3			22074.34			
4528.969	4530.239		0.0	0.1			22073.89			
4529.076	4530.346	GK	0.6b	1.0b	0.5	0.88	22073.37	4e 3	2c 3	Q 1
4529.076	4530.346	GK	0.6b	1.0b	0.5	0.88	22073.37	4d 2	2c 2	Q 1
4529.076	4530.346	GK	0.6b	1.0b	0.5	0.88	22073.37	4e 2	2c 2	Q 4
4529.187	4530.457		0.0				22072.83			

λ_{air}	λ_{vac}	Ref.	I_1	I_2	I_3	I_4	ν	Upper	Lower	Br.
4529.271	4530.541	GK	0.2	0.9	0.3	1.09	22072.42	4e 2	2c 2	P 5
4529.271	4530.541	GK	0.2	0.9	0.3	1.09	22072.42	3E 2	2B 3	Q 3
4529.419	4530.689						22071.70			
4529.484	4530.754			0.3d			22071.38			
4529.550	4530.820						22071.06			
4529.601	4530.871						22070.81			
4529.677	4530.947	G	0.1b	0.6	0.9		22070.44	3A 2	2B 4	P 6
4529.771	4531.042						22069.98			
4529.831	4531.101			0.2			22069.69			
4529.956	4531.227						22069.08			
4530.102	4531.372			0.2b			22068.37	4f 3	2c 3	P 5
4530.299	4531.569			0.1d			22067.41			
4530.398	4531.668			0.1			22066.93	3f 2	2c 0	R 5
4530.517	4531.787			0.5b	0.2d		22066.35			
4530.621	4531.892		0.0	0.1			22065.84			
4530.714	4531.984		0.0	0.0			22065.39			
4530.784	4532.054		0.0	0.3			22065.05	5c 1	2a 2	Q 5
4530.901	4532.171	GK	0.7	0.8	0.2d	0.82	22064.48	3D 2	2B 3	R 1
4531.007	4532.278		0.1	0.3b			22063.96			
4531.092	4532.362		0.1	0.1			22063.55			
4531.198	4532.469	GK	0.8b	1.0b	0.3	1.33	22063.03	3D 2	2B 3	R 2
4531.198	4532.469	GK	0.8b	1.0b	0.3	1.33	22063.03	u 2	2a 0	P 4
4531.198	4532.469	GK	0.8b	1.0b	0.3	1.33	22063.03	4c 1	2a 1	R 1
4531.322	4532.592		0.1	0.2			22062.43			
4531.509	4532.779			0.1			22061.52			
4531.576	4532.847			0.2	0.2		22061.19			
4531.763	4533.034		0.0d	0.2			22060.28			
4531.876	4533.147		0.1	0.1			22059.73			
4531.954	4533.225	GK	0.2	0.9b	0.9	0.68	22059.35			
4532.073	4533.344			0.2			22058.77			
4532.184	4533.455		0.0d	0.4	0.3		22058.23			
4532.373	4533.644		0.0d	0.3			22057.31	5c 0	2a 1	P 4
4532.667	4533.938			0.2d			22055.88			
4532.902	4534.173		0.4	0.6	0.3		22054.74	4c 7	2a 6	Q 1
4533.041	4534.312	GK	0.6	1.0			22054.06	4d 1	2c 1	Q 3
4533.126	4534.397	GK	0.8	1.0	0.9v	1.51	22053.65	3D 2	2B 3	R 0
4533.126	4534.397	GK	0.8	1.0	0.9v	1.51	22053.65	4c 0	2a 0	Q 7
4533.126	4534.397	GK	0.8	1.0	0.9v	1.51	22053.65	3D 2	2B 3	R 3
4533.196	4534.467			0.6d			22053.31			
4533.251	4534.522		0.1				22053.04			
4533.333	4534.604			0.1			22052.64			
4533.459	4534.730		0.1	0.2			22052.03			
4533.547	4534.818		0.0	0.2			22051.60			
4533.631	4534.903		0.0	0.2			22051.19			
4533.705	4534.977		0.2	0.4			22050.83	4D 0	2B 4	P 2
4533.839	4535.110		0.0d	0.2			22050.18			
4533.960	4535.232		0.2	0.4			22049.59			
4534.008	4535.279		0.1				22049.36			
4534.063	4535.335		0.2	0.3			22049.09			
4534.152	4535.423	GK	1.5b	1.5b	1.0	1.68	22048.66	4c 0	2a 0	P 3
4534.275	4535.546		0.3	0.4			22048.06			
4534.357	4535.629			0.0			22047.66			
4534.423	4535.695		0.1				22047.34			
4534.466	4535.738	GK	0.2	0.9	1.0		22047.13	3E 0	2B 0	Q 7
4534.542	4535.814		0.2				22046.76			
4534.625	4535.896	GK	1.0b	1.5b	0.4	1.79	22046.36	3E 2	2B 3	Q 2
4534.744	4536.016		0.2	0.3			22045.78			
4534.837	4536.108		0.0				22045.33			
4534.915	4536.186		0.0	0.2			22044.95	3E 0	2B 0	P 7
4535.018	4536.289			0.1			22044.45			

λ_{air}	λ_{vac}	Ref.	I_1	I_2	I_3	I_4	ν	Upper	Lower	Br.
4535.122	4536.394			0.4	0.3		22043.94	4e 2	2c 2	P 3
4535.285	4536.557		0.0	0.2			22043.15	3D 1	2B 1	P 6
4535.396	4536.668		0.0	0.2			22042.61			
4535.470	4536.742		0.0	0.3			22042.25			
4535.546	4536.818		0.0				22041.88			
4535.631	4536.902		0.2	0.2			22041.47			
4535.727	4536.999		0.0	0.1			22041.00			
4535.808	4537.079		0.0	0.0			22040.61			
4535.890	4537.162		0.2				22040.21			
4535.902	4537.174	GK		0.9b	0.5	1.06	22040.15	3E 0	2B 0	R 1
4535.902	4537.174	GK		0.9b	0.5	1.06	22040.15	3D 2	2B 3	R 4
4535.902	4537.174	GK		0.9b	0.5	1.06	22040.15	4e 1	2c 1	Q 7
4535.946	4537.217		0.4				22039.94	3E 0	2B 0	R 1
4536.071	4537.343		0.0				22039.33			
4536.236	4537.508		0.0	0.1			22038.53			
4536.333	4537.604		0.0	0.4			22038.06	T 0	2B 4	P 1
4536.419	4537.691		0.1d	0.3	0.2d		22037.64			
4536.586	4537.858			0.2			22036.83	V 1	2B 0	R 1
4536.783	4538.055		0.0d	0.2			22035.87			
4536.895	4538.167		0.2	0.5			22035.33			
4537.012	4538.284			0.2d			22034.76			
4537.170	4538.443			0.1d			22033.99	4e 2	2c 2	P 6
4537.286	4538.558			0.0			22033.43			
4537.376	4538.649	G	0.2d	0.6	0.3d		22032.99			
4537.481	4538.754		0.0d				22032.48			
4537.549	4538.822		0.0d	0.2			22032.15			
4537.652	4538.925		0.0	0.1			22031.65			
4537.724	4538.997		0.2		0.6d		22031.30			
4537.737	4539.009	GK		0.9b		0.51	22031.24			
4537.780	4539.052		0.2		0.3d		22031.03			
4537.891	4539.164	GK	0.4	0.9		1.08	22030.49	3E 2	2B 3	Q 1
4536.137	4537.409		0.2	0.1d			22039.01	3f 2	2c 0	Q 6
4538.011	4539.283			0.1			22029.91			
4538.122	4539.394			0.2			22029.37			
4538.211	4539.483			0.0			22028.94			
4538.314	4539.586	GK	0.3	1.0b	1.5b	0.86	22028.44	3D 0	2B 0	R10
4538.439	4539.712		0.0	0.1			22027.83			
4538.493	4539.765		0.0				22027.57			
4538.546	4539.819		0.0	0.2			22027.31			
4538.601	4539.873	F					22027.05			
4538.687	4539.959		0.0	0.6d			22026.63	3D 2	2B 3	R 6
4538.841	4540.114			0.1d			22025.88			
4538.979	4540.252		0.1d	0.7			22025.21	3D 2	2B 3	R 5
4539.080	4540.353		0.0				22024.72	5c 1	2a 2	P 2
4539.161	4540.433	GK	0.5b	1.0b	0.6	0.91	22024.33	4d 2	2c 2	R 2
4539.161	4540.433	GK	0.5b	1.0b	0.6	0.91	22024.33	T 0	2B 4	P 2
4539.292	4540.565		0.0	0.2			22023.69			
4539.418	4540.691		0.0d	0.1			22023.08			
4539.490	4540.763			0.2			22022.73			
4539.622	4540.895			0.4			22022.09			
4539.960	4541.233			0.2			22020.45			
4539.997	4541.270						22020.27			
4540.082	4541.355			0.1			22019.86			
4540.309	4541.582			0.1d			22018.76	3f 2	2c 0	R 5
4540.379	4541.652			0.0d			22018.42	3f 2	2c 0	R 4
4540.645	4541.918		0.0d	0.4d			22017.13	3A 2	2B 4	P 5
4540.816	4542.089						22016.30			
4540.930	4542.203		0.0d	0.2d	0.2d		22015.75			
4541.109	4542.382	GK	0.1	0.9b	0.4	0.95	22014.88	3F 1	2B 2	Q 4
4541.154	4542.428		0.3				22014.66	3F 1	2B 2	R 2

λ_{air}	λ_{vac}	Ref.	I_1	I_2	I_3	I_4	ν	Upper	Lower	Br.
4541.245	4542.518				0.2		22014.22			
4541.289	4542.562			0.1			22014.01			
4541.427	4542.700			0.3			22013.34			
4541.527	4542.800	F					22012.85			
4541.612	4542.886		0.1d	0.4	0.3		22012.44			
4541.685	4542.958	F		0.2			22012.09			
4541.765	4543.038			0.2			22011.70			
4541.864	4543.138			0.1			22011.22			
4541.928	4543.201			0.3	0.2d		22010.91			
4541.973	4543.247						22010.69			
4542.116	4543.389			0.2			22010.00			
4542.231	4543.505		0.0d	0.0			22009.44			
4542.341	4543.614		0.0	0.4			22008.91			
4542.409	4543.682		0.0				22008.58			
4542.452	4543.726		0.0	0.2			22008.37			
4542.562	4543.835			0.3	0.3d		22007.84			
4542.795	4544.069						22006.71			
4542.857	4544.130			0.1			22006.41			
4542.958	4544.232	GK	0.4	0.6	0.2d	0.66	22005.92	3A 2	2B 4	P 4
4543.069	4544.343		0.0	0.1			22005.38			
4543.183	4544.457		0.0	0.1			22004.83	4e 2	2c 2	Q 5
4543.274	4544.548						22004.39			
4543.390	4544.663		0.0d	0.2			22003.83			
4543.497	4544.771		0.0d	0.3			22003.31			
4543.594	4544.868		0.1	0.1			22002.84			
4543.693	4544.967	GK	0.7	1.5	1.0b	1.38	22002.36			
4543.813	4545.087		0.0	0.3			22001.78			
4543.881	4545.155		0.0d	0.2			22001.45	3A 2	2B 4	P 1
4543.955	4545.229			0.2			22001.09			
4544.040	4545.314			0.1			22000.68	3d 3	2c 0	Q 6
4544.150	4545.424			0.5	0.7		22000.15			
4544.257	4545.531			0.3			21999.63	4d 1	2c 1	P 3
4544.439	4545.713		0.2	0.3			21998.75	3A 2	2B 4	P 3
4544.519	4545.793	F					21998.36			
4544.596	4545.870			0.1d			21997.99			
4544.660	4545.934						21997.68	T 0	2B 4	P 3
4544.738	4546.012			0.2			21997.30			
4544.809	4546.083						21996.96			
4544.873	4546.147			0.4			21996.65			
4545.083	4546.358		0.0	0.2	0.8		21995.63			
4545.185	4546.459			0.1			21995.14			
4545.352	4546.626			0.2			21994.33			
4545.482	4546.757	K	0.1	0.4	0.5		21993.70			
4545.606	4546.881			0.1d			21993.10			
4545.718	4546.992		0.0	0.3			21992.56	4d 0	2c 0	P 4
4545.800	4547.075			0.0			21992.16	4d 0	2c 0	Q 5
4545.887	4547.162	G		0.5	0.3		21991.74	3d 3	2c 0	P 5
4545.887	4547.162	G		0.5	0.3		21991.74	3D 1	2B 1	P 7
4545.958	4547.232			0.2			21991.40			
4546.026	4547.300			0.1			21991.07			
4546.152	4547.426			0.2			21990.46			
4546.201	4547.476		0.0				21990.22			
4546.247	4547.522			0.2			21990.00			
4546.309	4547.584		0.0	0.1			21989.70			
4546.406	4547.681		0.0	0.1			21989.23			
4546.514	4547.788		0.0d	0.3b	0.4		21988.71			
4546.710	4547.985			0.1			21987.76			
4546.804	4548.079	F					21987.31			
4546.911	4548.186		0.0	0.1			21986.79	3f 2	2c 0	R 4
4547.045	4548.320		0.0	0.2	0.3		21986.14			

λ_{air}	λ_{vac}	Ref.	I_1	I_2	I_3	I_4	ν	Upper	Lower	Br.
4547.138	4548.413		0.0	0.1			21985.69			
4547.233	4548.508	GK	0.2	0.8	0.7	0.38	21985.23	3D 1	2B 1	P10
4547.339	4548.614			0.1			21984.72			
4547.455	4548.730		0.0	0.1			21984.16	T 0	2B 4	P 4
4547.535	4548.810		0.0				21983.77			
4547.606	4548.881						21983.43			
4547.674	4548.949		0.2	0.2			21983.10			
4547.703	4548.978	F					21982.96			
4547.773	4549.048		0.1	0.1			21982.62			
4547.879	4549.154		0.1	0.4	0.5		21982.11			
4547.964	4549.239	GK	0.8b	0.9		1.43	21981.70	Z 2	2B 3	R 0
4548.055	4549.330		0.1	0.2			21981.26			
4548.088	4549.363		0.1				21981.10			
4548.239	4549.514		0.0d	0.2			21980.37			
4548.320	4549.595			0.0			21979.98			
4548.425	4549.700		0.1	0.3			21979.47			
4548.549	4549.824		0.0	0.1			21978.87			
4548.678	4549.953		0.2	0.4			21978.25	3A 2	2B 4	P 2
4548.756	4550.031		0.0				21977.87			
4548.849	4550.125			0.2			21977.42			
4548.982	4550.257		0.0d	0.3d			21976.78	4e 0	2c 0	P 6
4549.131	4550.406		0.0	0.1			21976.06			
4549.249	4550.524			0.1			21975.49			
4549.369	4550.644			0.1			21974.91			
4549.472	4550.748		0.1	0.4			21974.41			
4549.605	4550.880		0.2	0.2			21973.77			
4549.698	4550.974		0.0d	0.4			21973.32			
4549.806	4551.081		0.1	0.1			21972.80			
4549.895	4551.170	GK	0.5	1.5b	1.0	1.36	21972.37	3E 0	2B 0	P 6
4550.017	4551.293		0.0	0.3			21971.78			
4550.148	4551.423		0.0d	0.1			21971.15	7c 0	2a 2	P 3
4550.230	4551.506		0.0	0.0			21970.75			
4550.321	4551.597	G	0.3	0.7			21970.31	4d 2	2c 2	Q 2
4550.471	4551.746		0.0d	0.2			21969.59			
4550.572	4551.848		0.0	0.2			21969.10	4e 3	2c 3	P 2
4550.661	4551.937		0.0	0.2			21968.67			
4550.790	4552.065		0.0	0.4			21968.05			
4550.872	4552.148		0.0	0.2			21967.65			
4550.978	4552.254	GK	0.6	1.5b	1.5	1.45	21967.14	3E 0	2B 0	Q 6
4551.102	4552.378		0.0	0.3			21966.54			
4551.254	4552.530		0.0	0.1			21965.81			
4551.378	4552.654	GK	0.2	0.7	0.4		21965.21			
4551.523	4552.799		0.1	0.2			21964.51			
4551.589	4552.865		0.1	0.2			21964.19			
4551.647	4552.923		0.1				21963.91			
4551.722	4552.998	GK	0.8b	1.0b	0.7	1.25	21963.55	Z 2	2B 3	R 1
4551.722	4552.998	GK	0.8b	1.0b	0.7	1.25	21963.55	4d 2	2c 2	R 3
4551.722	4552.998	GK	0.8b	1.0b	0.7	1.25	21963.55	4d 2	2c 2	P 2
4551.844	4553.120		0.0				21962.96			
4551.886	4553.162		0.2	0.3			21962.76			
4552.006	4553.282		0.1d				21962.18			
4552.037	4553.313		0.0	0.2			21962.03			
4552.130	4553.406		0.0	0.2			21961.58	3f 2	2c 0	R 3
4552.203	4553.479		0.0	0.3			21961.23	u 2	2a 0	P 5
4552.336	4553.612		0.1	0.1			21960.59			
4552.425	4553.701	GK	0.5	0.9	0.8		21960.16	4c 0	2a 0	P 4
4552.555	4553.831		0.0	0.2			21959.53	3E 1	2B 2	Q10
4552.626	4553.902	K	0.0d	0.6	0.4		21959.19	4e 1	2c 1	P 5
4552.713	4553.989		0.0	0.4d			21958.77			
4552.810	4554.087		0.0	0.1			21958.30	3D 1	2B 1	P 8

λ_{air}	λ_{vac}	Ref.	I_1	I_2	I_3	I_4	ν	Upper	Lower	Br.
4552.920	4554.197		0.2	0.3d			21957.77			
4553.047	4554.323		0.0	0.1d			21957.16			
4553.090	4554.367		0.0				21956.95	5c 1	2a 2	P 3
4553.182	4554.458	G	0.3	0.6			21956.51			
4553.246	4554.522			0.2			21956.20			
4553.393	4554.669		0.1	0.0			21955.49			
4553.513	4554.789	F					21954.91			
4553.607	4554.883		0.1d	0.3			21954.46			
4553.739	4555.016		0.1d	0.2			21953.82			
4553.820	4555.097		0.0				21953.43			
4553.918	4555.194		0.4	0.3			21952.96			
4553.957	4555.234		0.1				21952.77			
4554.017	4555.294	G	0.5	0.7	0.3		21952.48	3D 2	2B 3	P 2
4554.075	4555.352		0.1				21952.20			
4554.163	4555.439	GK	1.5b	1.5b	0.6	2.03	21951.78	4c 1	2a 1	Q 1
4554.163	4555.439	GK	1.5b	1.5b	0.6	2.03	21951.78	3D 1	2B 1	P 9
4554.281	4555.557		0.4	0.5			21951.21			
4554.351	4555.628		0.0				21950.87			
4554.457	4555.734		0.0	0.2			21950.36			
4554.497	4555.773		0.0				21950.17			
4554.586	4555.863		0.1	0.1			21949.74			
4554.643	4555.920	F					21949.46			
4554.727	4556.004		0.0	0.3			21949.06	4e 3	2c 3	P 3
4554.854	4556.130		0.0	0.2			21948.45			
4554.993	4556.269		0.1	0.2			21947.78			
4555.130	4556.406		0.0d	0.2			21947.12			
4555.196	4556.473	F					21946.80			
4555.291	4556.568		0.0	0.3			21946.34			
4555.377	4556.654		0.0	0.2			21945.93	5c 2	2a 3	Q 1
4555.482	4556.759		0.0				21945.42			
4555.549	4556.826			0.2			21945.10			
4555.622	4556.899		0.0				21944.75			
4555.700	4556.977	F					21944.37			
4555.773	4557.050						21944.02	3f 2	2c 0	R 3
4555.881	4557.158	G	0.1	0.4	0.2		21943.50			
4555.959	4557.236	F					21943.12			
4556.068	4557.345			0.2			21942.60			
4556.166	4557.443		0.0				21942.13			
4556.216	4557.493	F					21941.89			
4556.319	4557.596		0.0	0.2			21941.39			
4556.435	4557.713		0.2				21940.83			
4556.475	4557.752			0.6	0.3d		21940.64			
4556.662	4557.939			0.2	0.0d		21939.74			
4556.737	4558.014			0.1			21939.38			
4556.913	4558.191		0.2d	0.3	0.2d		21938.53			
4557.036	4558.313		0.1	0.1			21937.94			
4557.129	4558.407	GK	0.9b	1.0	0.4	1.55	21937.49	Z 2	2B 3	R 2
4557.237	4558.515		0.2	0.2			21936.97			
4557.314	4558.592		0.0				21936.60			
4557.395	4558.673	GK	0.8	1.0	0.5	1.51	21936.21	4d 1	2c 1	Q 4
4557.395	4558.673	GK	0.8	1.0	0.5	1.51	21936.21	W 2	2B 4	R 4
4557.395	4558.673	GK	0.8	1.0	0.5	1.51	21936.21	4c 1	2a 1	Q 2
4557.395	4558.673	GK	0.8	1.0	0.5	1.51	21936.21	5c 2	2a 3	Q 2
4557.503	4558.781		0.1	0.2d			21935.69			
4557.638	4558.916		0.0	0.2d			21935.04	4D 0	2B 4	P 3
4557.738	4559.016			0.1			21934.56			
4557.811	4559.088			0.2	0.2		21934.21			
4557.927	4559.205				0.2		21933.65			
4557.996	4559.273			0.1			21933.32			
4558.131	4559.408		0.0	0.3			21932.67			

λ_{air}	λ_{vac}	Ref.	I_1	I_2	I_3	I_4	ν	Upper	Lower	Br.
4558.226	4559.504		0.1	0.2	0.4		21932.21			
4558.322	4559.600	GK	0.8	1.0b	0.7	1.40	21931.75	3E 0	2B 0	R 0
4558.453	4559.731		0.1	0.3			21931.12			
4558.530	4559.808		0.0				21930.75			
4558.613	4559.891	GK	0.5	1.0b	0.9	1.50	21930.35	3E 0	2B 0	P 5
4558.736	4560.013		0.0	0.2			21929.76			
4558.794	4560.072		0.4				21929.48			
4558.854	4560.132	GK	0.1	0.8	0.8	0.91	21929.19	3D 0	2B 0	R 9
4558.912	4560.190		0.0				21928.91			
4558.997	4560.275		0.0	0.2			21928.50			
4559.122	4560.400	K	0.2	0.5	0.5		21927.90	4e 3	2c 3	P 4
4559.257	4560.535		0.0	0.1			21927.25			
4559.305	4560.583		0.0				21927.02			
4559.393	4560.671			0.2			21926.60			
4559.536	4560.814			0.2			21925.91	4e 2	2c 2	P 4
4559.625	4560.903	F					21925.48			
4559.675	4560.953		0.0	0.3			21925.24			
4559.748	4561.026			0.1			21924.89			
4559.912	4561.190	F					21924.10			
4559.960	4561.238		0.0d	0.3			21923.87			
4560.116	4561.395		0.0	0.2	0.3b		21923.12			
4560.210	4561.488	GK	0.8	0.8		1.12	21922.67	V 1	2B 0	R 0
4560.333	4561.611		0.1	0.1			21922.08			
4560.383	4561.661						21921.84			
4560.464	4561.742		0.0	0.0			21921.45			
4560.599	4561.877	GK	0.1	0.6	0.3	0.61	21920.80	X 0	2B 1	R 3
4560.599	4561.877	GK	0.1	0.6	0.3	0.61	21920.80	5c 2	2a 3	Q 3
4560.722	4562.000			0.2			21920.21			
4560.871	4562.150			0.3			21919.49			
4560.998	4562.277		0.0	0.6	0.3d		21918.88			
4561.059	4562.337		0.1				21918.59			
4561.144	4562.423	GK	0.6	0.9	0.4	1.32	21918.18	X 0	2B 1	R 2
4561.263	4562.541		0.1	0.3			21917.61			
4561.333	4562.611	F					21917.27			
4561.446	4562.724			0.2			21916.73			
4561.552	4562.831			0.2			21916.22	3c 8	2a 4	Q 1
4561.704	4562.983		0.0d	0.3			21915.49			
4561.773	4563.051			0.2			21915.16			
4561.846	4563.124			0.1			21914.81			
4561.939	4563.218		0.2d	0.2			21914.36			
4562.027	4563.305		0.0d	0.4			21913.94			
4562.135	4563.414		0.1	0.2			21913.42			
4562.226	4563.505	GK	0.8	1.5b	0.7	1.72	21912.98	4c 1	2a 1	Q 3
4562.226	4563.505	GK	0.8	1.5b	0.7	1.72	21912.98	4E 2	2B 8	Q 2
4562.351	4563.630		0.2	0.3			21912.38			
4562.458	4563.736	GK	0.4	0.7	0.2d	0.92	21911.87	X 0	2B 1	R 1
4562.576	4563.855		0.0	0.1			21911.30			
4562.678	4563.957						21910.81			
4562.789	4564.068						21910.28			
4562.874	4564.153			0.2b			21909.87			
4563.064	4564.343			0.1			21908.96	4e 3	2c 3	P 5
4563.178	4564.457			0.3d			21908.41			
4563.262	4564.541		0.0	0.1d			21908.01			
4563.299	4564.578						21907.83			
4563.439	4564.718		0.0	0.2			21907.16			
4563.582	4564.861		0.1	0.7	0.4		21906.47			
4563.636	4564.916		0.0				21906.21			
4563.736	4565.016	GK	0.3	1.5b	1.0b	1.28	21905.73	3E 0	2B 0	Q 5
4563.861	4565.141		0.0	0.3			21905.13			
4563.928	4565.207		0.0				21904.81			

λ_{air}	λ_{vac}	Ref.	I_1	I_2	I_3	I_4	ν	Upper	Lower	Br.
4563.997	4565.276		0.0				21904.48			
4564.039	4565.318		0.0	0.0d			21904.28			
4564.082	4565.362		0.0				21904.07			
4564.153	4565.432		0.0d	0.1			21903.73			
4564.257	4565.537	G	0.1	0.7	0.4		21903.23	4F 0	2B 5	R 3
4564.372	4565.651			0.1			21902.68			
4564.505	4565.785		0.0d	0.3			21902.04			
4564.610	4565.889		0.0	0.0			21901.54			
4564.703	4565.983	GK	0.1	0.7	0.2		21901.09	3F 1	2B 2	Q 3
4564.822	4566.102		0.0	0.1d			21900.52			
4564.874	4566.154		0.0				21900.27			
4564.926	4566.206	GK	0.2	0.7	0.2d	0.67	21900.02	Z 2	2B 3	R 3
4564.926	4566.206	GK	0.2	0.7	0.2d	0.67	21900.02	5c 2	2a 3	Q 4
4564.926	4566.206	GK	0.2	0.7	0.2d	0.67	21900.02	Z 0	2B 4	P 3
4565.072	4566.352			0.0d			21899.32	3f 2	2c 0	R 2
4565.210	4566.489			0.3			21898.66			
4565.309	4566.589	F					21898.18			
4565.358	4566.637		0.1	0.1			21897.95			
4565.452	4566.731		0.1	0.1			21897.50			
4565.546	4566.825	GK	0.8	0.9	0.3	1.32	21897.05	X 0	2B 1	R 0
4565.604	4566.884		0.4	0.2			21896.77	4d 2	2c 2	R 4
4565.604	4566.884		0.4	0.2			21896.77	3D 2	2B 3	P 3
4565.683	4566.963			0.1			21896.39			
4565.746	4567.025		0.0d	0.3	0.4		21896.09			
4565.856	4567.136		0.2	0.5			21895.56			
4565.994	4567.274		0.0	0.3			21894.90			
4566.123	4567.403			0.2			21894.28			
4566.246	4567.526		0.2	0.2			21893.69			
4566.394	4567.674		0.0				21892.98			
4566.478	4567.758		0.0	0.3	0.3d		21892.58	3f 2	2c 0	R 2
4566.611	4567.891			0.2			21891.94			
4566.701	4567.981	F					21891.51	4e 2	2c 2	Q 6
4566.789	4568.069			0.2			21891.09			
4566.878	4568.158			0.1			21890.66			
4566.972	4568.252			0.2	0.2d		21890.21			
4567.039	4568.319			0.1			21889.89			
4567.127	4568.407			0.3			21889.47	4d 0	2c 0	Q 6
4567.185	4568.465	F					21889.19			
4567.245	4568.526			0.2			21888.90			
4567.346	4568.626			0.3			21888.42			
4567.460	4568.741			0.1d			21887.87			
4567.617	4568.897		0.1	0.4b			21887.12			
4567.757	4569.037	GK	0.2	0.7			21886.45	X 0	2B 1	R 4
4567.821	4569.102		0.1				21886.14			
4567.926	4569.206		0.3	0.6			21885.64			
4567.965	4569.246		0.2				21885.45			
4568.022	4569.302		0.4	0.3			21885.18	5c 1	2a 2	P 4
4568.122	4569.402	GK	1.5b	1.5b	1.5	2.27	21884.70	3E 0	2B 0	P 4
4568.239	4569.519		0.4	0.7			21884.14			
4568.354	4569.634			0.1			21883.59	W 2	2B 4	R 3
4568.510	4569.791		0.3	0.4			21882.84	3F 1	2B 2	R 1
4568.627	4569.908	GK	0.2	0.8	0.4	0.83	21882.28	4c 1	2a 1	Q 4
4568.736	4570.016		0.1	0.1			21881.76			
4568.832	4570.112		0.0d				21881.30			
4568.951	4570.231			0.1			21880.73	Z 2	2B 3	P 2
4569.057	4570.338		0.0	0.2			21880.22			
4569.110	4570.391	F					21879.97			
4569.160	4570.440			0.3	0.2b		21879.73			
4569.247	4570.528			0.2			21879.31			
4569.343	4570.624			0.1			21878.85			

λ_{air}	λ_{vac}	Ref.	I_1	I_2	I_3	I_4	ν	Upper	Lower	Br.
4569.565	4570.846		0.0	0.1d			21877.79			
4569.649	4570.930	F					21877.39			
4569.759	4571.040	G		0.3			21876.86	V 1	2B 0	P 4
4569.832	4571.113	F					21876.51	4d 3	2c 3	P 1
4569.958	4571.238		0.0	0.2			21875.91			
4570.016	4571.297	F					21875.63			
4570.091	4571.372		0.0	0.2			21875.27			
4570.206	4571.487		0.0				21874.72			
4570.260	4571.541		0.0d	0.2			21874.46			
4570.409	4571.690		0.2	0.6			21873.75			
4570.467	4571.748	F					21873.47	5c 2	2a 3	Q 5
4570.570	4571.851			0.2			21872.98			
4570.708	4571.989		0.0	0.1			21872.32			
4570.749	4572.030		0.0				21872.12			
4570.808	4572.089	G	0.4	0.6	0.2b	0.75	21871.84	W 2	2B 4	R 2
4570.940	4572.221			0.0			21871.21			
4571.151	4572.432	G		0.1d			21870.20			
4571.270	4572.551	G	0.0	0.6	0.4		21869.63	4D 0	2B 4	P 4
4571.295	4572.576		0.1				21869.51			
4571.433	4572.714			0.2			21868.85			
4571.558	4572.840			0.1d			21868.25			
4571.730	4573.011			0.3b			21867.43			
4571.866	4573.147	F					21866.78			
4571.928	4573.210			0.2			21866.48			
4572.031	4573.312		0.0d	0.3			21865.99			
4572.131	4573.413		0.1	0.1			21865.51			
4572.225	4573.507	GK	0.7	1.0	1.0b	1.27	21865.06	4c 0	2a 0	P 5
4572.344	4573.626		0.0	0.2			21864.49			
4572.516	4573.797		0.1	0.6			21863.67			
4572.598	4573.879		0.0	0.2			21863.28			
4572.711	4573.992	GK	0.8	1.5b	1.0	2.13	21862.74	3E 0	2B 0	Q 4
4572.838	4574.120		0.2	0.6			21862.13			
4572.924	4574.205		0.0				21861.72			
4572.982	4574.264	G	0.3b	0.7	0.3		21861.44	4d 2	2c 2	Q 3
4573.110	4574.392		0.3	0.4			21860.83	W 2	2B 4	R 1
4573.213	4574.494		0.0				21860.34			
4573.315	4574.597	F					21859.85			
4573.372	4574.653			0.0d			21859.58	3F 1	2B 2	P 5
4573.526	4574.808	G	0.2	0.4			21858.84			
4573.606	4574.888		0.2	0.3			21858.46			
4573.658	4574.940				0.3		21858.21			
4573.729	4575.011			0.4			21857.87			
4573.809	4575.091			0.1			21857.49			
4573.863	4575.145	G		0.2			21857.23			
4573.993	4575.275			0.2			21856.61	4D 0	2B 4	P 8
4574.054	4575.336		0.0d	0.3	0.4		21856.32	4e 0	2c 0	P 7
4574.127	4575.409			0.2			21855.97			
4574.188	4575.470			0.1			21855.68			
4574.242	4575.524	G			0.2		21855.42			
4574.282	4575.564			0.1			21855.23			
4574.407	4575.689			0.1			21854.63			
4574.502	4575.784			0.3			21854.18			
4574.633	4575.916			0.2			21853.55	U 1	2B 2	R 4
4574.732	4576.014	G	0.2	0.5	0.2		21853.08	4d 3	2c 3	R 1
4574.860	4576.142			0.2			21852.47			
4574.985	4576.267			0.5			21851.87			
4575.100	4576.383		0.1	0.2			21851.32			
4575.173	4576.455	F					21850.97			
4575.255	4576.538			0.2d			21850.58			
4575.341	4576.623		0.0				21850.17			

λ_air	λ_vac	Ref.	I₁	I₂	I₃	I₄	ν	Upper	Lower	Br.
4575.410	4576.693		0.2	0.5			21849.84			
4575.488	4576.770			0.0			21849.47			
4575.561	4576.843		0.0	0.2			21849.12			
4575.664	4576.946		0.3	0.3			21848.63			
4575.722	4577.005		0.1				21848.35			
4575.760	4577.042	G	0.3	0.6			21848.17	Z 2	2B 3	R 4
4575.877	4577.160	GK	1.5b	1.5b	1.0	1.86	21847.61	3E 0	2B 0	P 3
4575.990	4577.273	G	0.3	0.4			21847.07	3b 6	2a 0	R 1
4576.039	4577.321		0.0				21846.84			
4576.129	4577.411			0.1			21846.41			
4576.261	4577.543		0.0	0.4	0.2		21845.78			
4576.346	4577.629		0.0	0.1d			21845.37			
4576.453	4577.736		0.0	0.0			21844.86			
4576.548	4577.830	GK	0.3	0.9	0.6	0.87	21844.41	V 1	2B 0	P 3
4576.548	4577.830	GK	0.3	0.9	0.6	0.87	21844.41	4c 1	2a 1	Q 5
4576.648	4577.931		0.0				21843.93			
4576.738	4578.021	GK	0.5	0.6	0.3	0.82	21843.50	W 2	2B 4	R 0
4576.738	4578.021	GK	0.5	0.6	0.3	0.82	21843.50	4d 3	2c 3	Q 1
4576.866	4578.149		0.0				21842.89			
4576.939	4578.222			0.2			21842.54			
4577.019	4578.302			0.1			21842.16			
4577.124	4578.407		0.0d	0.2			21841.66			
4577.224	4578.507	G	0.3b	0.7	0.3d	0.51	21841.18	3b 6	2a 0	R 2
4577.375	4578.658			0.0d			21840.46	3b 6	2a 0	R 0
4577.495	4578.778			0.2			21839.89			
4577.572	4578.855			0.2			21839.52	4e 1	2c 1	P 6
4577.627	4578.910		0.0	0.1	0.3		21839.26			
4577.706	4578.989		0.3	0.4			21838.88			
4577.828	4579.111		0.1	0.3			21838.30			
4577.922	4579.205		0.1	0.2			21837.85			
4578.013	4579.296	GK	0.6	1.5	1.0b	1.88	21837.42	3E 0	2B 0	Q 3
4578.143	4579.426		0.2	0.4			21836.80			
4578.317	4579.600			0.2			21835.97	3f 2	2c 0	R 1
4578.465	4579.749		0.2	0.2			21835.26			
4578.566	4579.849	GK	0.6	0.7		0.79	21834.78	X 0	2B 1	P 1
4578.683	4579.967		0.1	0.2			21834.22	3f 2	2c 0	R 1
4578.822	4580.105			0.2			21833.56			
4578.925	4580.208						21833.07			
4578.971	4580.254		0.0d	0.3d			21832.85			
4579.134	4580.418	G	0.1	0.6	0.5		21832.07			
4579.252	4580.535		0.0	0.3			21831.51			
4579.359	4580.642		0.1	0.2			21831.00			
4579.451	4580.735	GK	0.8	1.5	0.9b	1.78	21830.56	3E 0	2B 0	Q 1
4579.577	4580.860	G	0.6	0.6			21829.96	3D 2	2B 3	P 4
4579.674	4580.957		0.2	0.3			21829.50			
4579.789	4581.072		0.4	0.6			21828.95			
4579.825	4581.108		0.2				21828.78			
4579.890	4581.173		0.5	0.6			21828.47			
4579.988	4581.272	GK	1.5b	1.5b	1.0	2.46	21828.00	3E 0	2B 0	Q 2
4580.116	4581.400		0.4	0.6			21827.39	4d 1	2c 1	P 4
4580.255	4581.538		0.0				21826.73			
4580.303	4581.587		0.2	0.4			21826.50			
4580.397	4581.681		0.1				21826.05	4D 0	2B 4	P 5
4580.477	4581.761		0.0	0.4	0.3		21825.67			
4580.599	4581.883		0.0	0.2			21825.09	3E 1	2B 2	Q 9
4580.725	4582.008		0.0	0.1			21824.49			
4580.802	4582.086		0.1	0.1			21824.12			
4580.847	4582.130		0.0				21823.91			
4580.918	4582.202		0.0	0.0			21823.57			
4581.010	4582.294		0.0	0.0			21823.13			

λ_{air}	λ_{vac}	Ref.	I_1	I_2	I_3	I_4	ν	Upper	Lower	Br.
4581.094	4582.378		0.1	0.3	0.3b		21822.73			
4581.220	4582.504		0.0				21822.13			
4581.334	4582.617		0.0	0.3			21821.59			
4581.384	4582.668						21821.35			
4581.436	4582.720		0.1				21821.10			
4581.535	4582.819	GK	0.4	1.5b	1.0	1.13	21820.63	3D 0	2B 0	R 8
4581.535	4582.819	GK	0.4	1.5b	1.0	1.13	21820.63	4e 3	2c 3	R 1
4581.667	4582.951		0.0				21820.00			
4581.857	4583.140		0.0				21819.10	4D 0	2B 4	P 7
4581.968	4583.252		0.0	0.2			21818.57	5c 1	2a 2	P 5
4582.066	4583.351		0.1	0.2			21818.10			
4582.153	4583.437	G	0.0	0.5	0.3		21817.69	4e 3	2c 3	R 2
4582.153	4583.437	G	0.0	0.5	0.3		21817.69	4d 1	2c 1	Q 5
4582.195	4583.479		0.0				21817.49			
4582.272	4583.556		0.2	0.4			21817.12			
4582.390	4583.674		0.2	0.5			21816.56			
4582.430	4583.714		0.1				21816.37			
4582.493	4583.777		0.3	0.3			21816.07			
4582.589	4583.874	GK	1.5b	1.5b	1.0b	2.09	21815.61	3E 0	2B 0	P 2
4582.709	4583.993		0.3	0.5			21815.04			
4582.913	4584.197			0.2			21814.07			
4582.976	4584.260	G		0.3			21813.77	4e 3	2c 3	R 3
4583.102	4584.386		0.0d				21813.17			
4583.203	4584.487			0.2b			21812.69			
4583.398	4584.683			0.2b			21811.76	4d 0	2c 0	P 5
4583.524	4584.809			0.1			21811.16			
4583.594	4584.878						21810.83			
4583.697	4584.981		0.0d	0.3b			21810.34			
4583.793	4585.078		0.1	0.1			21809.88			
4583.884	4585.168	GK	0.6	0.9	0.3	0.97	21809.45	3F 1	2B 2	Q 2
4584.002	4585.286		0.1	0.1			21808.89	4D 0	2B 4	P 6
4584.170	4585.454			0.0			21808.09			
4584.300	4585.585		0.2				21807.47	4e 2	2c 2	P 5
4584.367	4585.652		0.1	0.4	0.3		21807.15			
4584.414	4585.698		0.2				21806.93			
4584.500	4585.784	GK	1.5b	1.0	0.4	1.73	21806.52	V 1	2B 0	P 2
4584.500	4585.784	GK	1.5b	1.0	0.4	1.73	21806.52	4d 2	2c 2	P 3
4584.617	4585.902		0.2	0.2			21805.96			
4584.688	4585.973	F					21805.62			
4584.742	4586.026		0.0	0.0			21805.37			
4584.876	4586.161		0.1b	0.4			21804.73			
4584.975	4586.260			0.1			21804.26			
4585.118	4586.403	G	0.3b	0.7b	0.4	0.30	21803.58	4d 3	2c 3	R 2
4585.225	4586.510	F					21803.07			
4585.339	4586.624		0.0d				21802.53			
4585.429	4586.714		0.0	0.1			21802.10			
4585.579	4586.863			0.0			21801.39	3d 3	2c 0	P 6
4585.661	4586.946	F					21801.00			
4585.740	4587.026		0.0d				21800.62			
4585.804	4587.089			0.0			21800.32			
4585.857	4587.142	F					21800.07			
4585.898	4587.183		0.0d	0.0			21799.87			
4585.995	4587.280		0.1d	0.4	0.4		21799.41			
4586.138	4587.423	G		0.1			21798.73			
4586.208	4587.493			0.3			21798.40			
4586.254	4587.539						21798.18			
4586.315	4587.600						21797.89			
4586.401	4587.686		0.0d	0.4			21797.48	4d 0	2c 0	Q 7
4586.487	4587.773			0.1			21797.07			
4586.599	4587.884	G	0.2	0.5	0.4		21796.54			

λ_{air}	λ_{vac}	Ref.	I_1	I_2	I_3	I_4	ν	Upper	Lower	Br.
4586.660	4587.945		0.0				21796.25			
4586.784	4588.069		0.1	0.2			21795.66	3E 3	2B 5	Q 8
4586.784	4588.069		0.1	0.2			21795.66	Z 2	2B 3	P 3
4586.870	4588.156		0.0				21795.25			
4586.971	4588.257		0.2	0.4			21794.77			
4587.167	4588.452		0.3b	0.2			21793.84			
4587.228	4588.514		0.0				21793.55			
4587.352	4588.638			0.1	0.2		21792.96			
4587.428	4588.714			0.1			21792.60			
4587.624	4588.909			0.1			21791.67			
4587.719	4589.004			0.2			21791.22			
4587.794	4589.080			0.0			21790.86			
4587.847	4589.133	F					21790.61			
4587.982	4589.267			0.0			21789.97			
4588.043	4589.329	F					21789.68			
4588.171	4589.457		0.1	0.2			21789.07			
4588.247	4589.533	F					21788.71			
4588.315	4589.600						21788.39			
4588.485	4589.771		0.1	0.2			21787.58			
4588.607	4589.893		0.0				21787.00			
4588.685	4589.971	GK	0.9b	0.9	0.3	1.49	21786.63	X 0	2B 1	P 2
4588.843	4590.129	G	0.5d	0.5			21785.88	W 2	2B 4	P 1
4588.883	4590.169	F					21785.69			
4588.989	4590.274			0.1d			21785.19			
4589.115	4590.401			0.2			21784.59			
4589.309	4590.595				0.3d		21783.67			
4589.399	4590.685		0.1d				21783.24			
4589.498	4590.784			0.1			21782.77			
4589.661	4590.947			0.3			21782.00			
4589.779	4591.065			0.0			21781.44			
4589.922	4591.208	GK	0.6	0.6		0.60	21780.76			
4589.998	4591.284						21780.40			
4590.059	4591.345		0.0d				21780.11			
4590.213	4591.499	K		0.4	0.4		21779.38			
4590.370	4591.656	F					21778.63			
4590.421	4591.708			0.1d			21778.39			
4590.565	4591.851			0.1d			21777.71			
4590.624	4591.910	F					21777.43			
4590.702	4591.988			0.2d			21777.06			
4590.885	4592.172		0.0d				21776.19			
4590.961	4592.247		0.0	0.3d			21775.83			
4591.029	4592.315	F					21775.51			
4591.250	4592.536			0.2d	0.2		21774.46			
4591.433	4592.720		0.0d	0.3d	0.3		21773.59			
4591.619	4592.905			0.2			21772.71			
4591.714	4593.000		0.0	0.1			21772.26	U 1	2B 2	R 3
4591.809	4593.095	GK	0.5	0.9	0.6	1.06	21771.81	3F 1	2B 2	P 4
4591.918	4593.205	G	0.1	0.5			21771.29	4e 3	2c 3	Q 1
4591.918	4593.205	G	0.1	0.5			21771.29	5c 3	2a 4	Q 1
4592.085	4593.372			0.3			21770.50			
4592.193	4593.480	F					21769.99			
4592.281	4593.568		0.0d	0.3			21769.57			
4592.440	4593.726			0.0d			21768.82			
4592.516	4593.803	F					21768.46			
4592.600	4593.887		0.0	0.2			21768.06			
4592.752	4594.039	F					21767.34			
4592.870	4594.157			0.2			21766.78			
4593.085	4594.372			0.1			21765.76			
4593.279	4594.566			0.2b			21764.84			
4593.433	4594.720			0.1			21764.11			

λ_{air}	λ_{vac}	Ref.	I_1	I_2	I_3	I_4	ν	Upper	Lower	Br.
4593.594	4594.881	GK	0.2	0.7	0.7		21763.35	4c 0	2a 0	P 6
4593.712	4594.999			0.1			21762.79			
4593.893	4595.181			0.2			21761.93			
4593.936	4595.223		0.0		0.2		21761.73			
4594.100	4595.388		0.0	0.2			21760.95			
4594.204	4595.491			0.2			21760.46	5c 3	2a 4	Q 2
4594.286	4595.573			0.1			21760.07			
4594.373	4595.660			0.1			21759.66			
4594.487	4595.774			0.4			21759.12			
4594.533	4595.820	F					21758.90			
4594.654	4595.941			0.1			21758.33			
4594.751	4596.038	GK	0.3	0.5	0.2	0.60	21757.87	Y 1	2B 1	R 1
4594.918	4596.205			0.4			21757.08	3D 2	2B 3	P 5
4595.068	4596.355		0.0	0.1			21756.37			
4595.173	4596.461	GK	0.6	0.8	0.3	0.88	21755.87	Y 1	2B 1	R 0
4595.302	4596.589		0.1	0.1			21755.26			
4595.391	4596.678				0.3		21754.84			
4595.574	4596.862			0.3	0.3		21753.97			
4595.756	4597.044			0.2			21753.11			
4595.898	4597.185			0.2d			21752.44			
4595.982	4597.270						21752.04			
4596.098	4597.386	G	0.1	0.4	0.8		21751.49	4c 2	2a 2	R 1
4596.208	4597.496			0.1			21750.97			
4596.312	4597.600	F					21750.48			
4596.411	4597.699		0.0	0.4d			21750.01			
4596.610	4597.898	G	0.2d	0.4			21749.07	4d 3	2c 3	Q 2
4596.636	4597.924	F					21748.95			
4596.703	4597.991	G	0.0	0.4			21748.63	4c 1	2a 1	P 3
4596.703	4597.991	G	0.0	0.4			21748.63	4d 2	2c 2	Q 4
4596.868	4598.156	K	0.2d	0.4	0.4		21747.85			
4597.022	4598.310		0.2d	0.2			21747.12			
4597.126	4598.414		0.2	0.2			21746.63			
4597.210	4598.498	GK	0.8b	1.0	0.4	1.36	21746.23	Y 1	2B 1	R 2
4597.210	4598.498	GK	0.8b	1.0	0.4	1.36	21746.23	W 2	2B 4	P 2
4597.335	4598.623		0.2	0.2			21745.64			
4597.493	4598.782			0.1			21744.89			
4597.629	4598.917			0.0			21744.25	5c 3	2a 4	Q 3
4597.868	4599.156			0.0			21743.12			
4597.933	4599.222						21742.81			
4598.088	4599.376		0.0d	0.3	0.3d		21742.08			
4598.303	4599.592			0.2			21741.06			
4598.354	4599.643		0.0				21740.82			
4598.498	4599.786	GK	0.3	1.0	0.9b	0.83	21740.14	3D 0	2B 0	R 7
4598.644	4599.932			0.2			21739.45			
4598.752	4600.040			0.2			21738.94	4d 3	2c 3	P 2
4598.834	4600.122	F					21738.55			
4599.120	4600.409	F					21737.20			
4599.281	4600.569			0.3			21736.44			
4599.363	4600.652	F					21736.05			
4599.461	4600.749			0.0			21735.59			
4599.617	4600.906			0.1			21734.85			
4599.710	4600.999						21734.41			
4599.791	4601.079		0.1	0.1			21734.03			
4599.878	4601.166				0.2d		21733.62			
4599.943	4601.232			0.2			21733.31			
4600.066	4601.355			0.1			21732.73			
4600.182	4601.471	G		0.4	0.4		21732.18			
4600.322	4601.611			0.1			21731.52			
4600.441	4601.729		0.1	0.2			21730.96			
4600.525	4601.814			0.1			21730.56			

λ_{air}	λ_{vac}	Ref.	I_1	I_2	I_3	I_4	ν	Upper	Lower	Br.
4600.602	4601.890			0.1	0.3		21730.20			
4600.769	4602.058		0.1	0.2			21729.41			
4600.868	4602.157	GK	0.2	0.7	0.3d	0.86	21728.94	X 0	2B 1	P 3
4601.004	4602.293		0.0	0.1			21728.30			
4601.082	4602.371	F					21727.93			
4601.163	4602.452			0.2			21727.55	U 1	2B 2	R 1
4601.239	4602.528			0.1			21727.19	U 1	2B 2	R 2
4601.336	4602.625			0.4			21726.73			
4601.402	4602.691			0.4	0.4		21726.42			
4601.650	4602.939		0.2	0.3			21725.25	U 1	2B 2	R 0
4601.932	4603.221			0.1d			21723.92			
4602.154	4603.443			0.0d			21722.87	4e 1	2c 1	P 7
4602.154	4603.443			0.0d			21722.87	5c 3	2a 4	Q 4
4602.336	4603.626		0.0				21722.01			
4602.429	4603.718	F					21721.57			
4602.533	4603.822	F					21721.08			
4602.707	4603.996		0.0	0.1			21720.26			
4602.826	4604.115			0.2			21719.70			
4602.860	4604.149	F					21719.54			
4603.004	4604.293			0.1			21718.86			
4603.135	4604.425			0.2			21718.24			
4603.350	4604.640	F					21717.23			
4603.385	4604.675		0.0d	0.1			21717.06			
4603.604	4604.893	F					21716.03	4e 3	2c 3	Q 2
4603.861	4605.151	F					21714.82			
4603.932	4605.222			0.2			21714.48			
4604.062	4605.352	F					21713.87			
4604.144	4605.434			0.2			21713.48			
4604.258	4605.548	F					21712.94			
4604.469	4605.759	F					21711.95			
4604.517	4605.807			0.2			21711.72	3E 3	2B 5	Q 7
4604.620	4605.910	F					21711.24			
4604.791	4606.081		0.0d	0.0			21710.43			
4604.833	4606.123		0.0d				21710.23			
4604.933	4606.223	G	0.3	0.4			21709.76			
4605.105	4606.395			0.2	0.2d		21708.95			
4605.234	4606.524		0.0				21708.34			
4605.311	4606.601		0.4	0.5			21707.98	3F 1	2B 2	P 3
4605.383	4606.673	GK	0.1	0.8	0.6	0.65	21707.64	3E 1	2B 2	Q 8
4605.466	4606.756		0.0				21707.25			
4605.608	4606.898		0.0				21706.58			
4605.767	4607.057	F					21705.83			
4605.905	4607.195						21705.18			
4606.148	4607.438	F					21704.03	Z 2	2B 3	P 4
4606.255	4607.545			0.2			21703.53			
4606.410	4607.700			0.1			21702.80			
4606.516	4607.807		0.1d	0.3			21702.30			
4606.629	4607.919			0.2			21701.77			
4606.750	4608.040			0.1			21701.20			
4606.898	4608.189		0.0	0.2			21700.50			
4606.998	4608.289		0.0d	0.2			21700.03			
4607.091	4608.382	GK	0.5b	0.7	0.4	0.84	21699.59	W 2	2B 4	P 3
4607.091	4608.382						21699.59	Y 1	2B 1	P 1
4607.227	4608.518		0.0d	0.2d			21698.95			
4607.410	4608.701	GK	0.5	1.5	1.0b	1.26	21698.09			
4607.535	4608.826		0.0	0.3			21697.50			
4607.605	4608.896		0.1				21697.17			
4608.042	4609.333	F					21695.11			
4608.325	4609.616		0.0d	0.2d			21693.78			
4608.410	4609.701						21693.38			

λ_{air}	λ_{vac}	Ref.	I_1	I_2	I_3	I_4	ν	Upper	Lower	Br.
4608.557	4609.848	G	0.2	0.5	0.2	0.23	21692.69			
4608.695	4609.986			0.2			21692.04			
4608.829	4610.120			0.2			21691.41			
4608.933	4610.224			0.3			21690.92	4e 2	2c 2	P 6
4609.133	4610.424	G	0.0	0.4	0.3		21689.98			
4609.258	4610.549	G	0.3	0.5	0.3d		21689.39	3c 5	2a 2	Q 1
4609.427	4610.718	F					21688.59			
4609.492	4610.783			0.1			21688.29			
4609.579	4610.870			0.1			21687.88			
4609.713	4611.004		0.0	0.3			21687.25			
4609.804	4611.095	F					21686.82			
4609.885	4611.176		0.0	0.2			21686.44			
4609.983	4611.274			0.1			21685.98			
4610.087	4611.378			0.4			21685.49			
4610.296	4611.587	F					21684.51			
4610.419	4611.710			0.0			21683.93			
4610.562	4611.854	F					21683.26			
4610.629	4611.921			0.2			21682.94			
4610.793	4612.084		0.0d	0.4			21682.17	3D 2	2B 3	P 6
4610.916	4612.208			0.1			21681.59			
4611.008	4612.299			0.1d			21681.16			
4611.095	4612.387	F					21680.75			
4611.330	4612.622	F					21679.64			
4611.418	4612.710			0.1d			21679.23			
4611.552	4612.844			0.1d			21678.60			
4611.612	4612.904						21678.32			
4611.669	4612.961			0.2			21678.05			
4611.769	4613.061	F					21677.58			
4611.973	4613.265			0.2			21676.62			
4612.077	4613.369	F					21676.13			
4612.156	4613.448			0.1			21675.76			
4612.325	4613.617			0.3			21674.97			
4612.417	4613.709	F					21674.54			
4612.484	4613.776		0.0				21674.22			
4612.648	4613.940		0.0	0.3	0.3d		21673.45			
4612.763	4614.055						21672.91			
4612.805	4614.098			0.1			21672.71			
4612.910	4614.202		0.1				21672.22			
4612.961	4614.253		0.3	0.3b			21671.98			
4613.033	4614.325		0.1				21671.64			
4613.114	4614.406	GK	0.6	0.9	0.4	1.38	21671.26	3A 0	2B 1	R 2
4613.244	4614.536		0.1	0.1			21670.65			
4613.448	4614.741		0.0	0.0			21669.69			
4613.642	4614.934	G	0.3	0.5b	0.4		21668.78	4c 1	2a 1	P 4
4613.642	4614.934	G	0.3	0.5b	0.4		21668.78	Y 2	2B 4	R 0
4613.642	4614.934	G	0.3	0.5b	0.4		21668.78	Y 2	2B 4	R 1
4613.746	4615.039		0.0				21668.29			
4613.974	4615.267			0.0			21667.22	4e 3	2c 3	P 2
4614.112	4615.404	F					21666.57			
4614.204	4615.497			0.0			21666.14	4F 0	2B 5	R 2
4614.402	4615.695		0.0	0.2d			21665.21			
4614.507	4615.799		0.1				21664.72			
4614.590	4615.882	GK	0.5	0.9	0.4	1.24	21664.33	X 0	2B 1	P 4
4614.590	4615.882	GK	0.5	0.9	0.4	1.24	21664.33	3E 1	2B 2	R 3
4614.722	4616.015		0.1	0.2			21663.71			
4614.779	4616.072	F		0.2			21663.44			
4614.867	4616.159			0.2			21663.03			
4614.984	4616.277			0.2			21662.48			
4615.103	4616.396			0.0			21661.92	4e 3	2c 3	Q 3
4615.227	4616.520		0.3	0.3			21661.34			

λ_{air}	λ_{vac}	Ref.	I_1	I_2	I_3	I_4	ν	Upper	Lower	Br.
4615.313	4616.606	F					21660.94			
4615.525	4616.818	F					21659.94			
4615.753	4617.046			0.3			21658.87	3c 5	2a 2	Q 2
4615.949	4617.242				0.3d		21657.95	3f 2	2c 0	Q 7
4616.011	4617.304		0.0	0.3			21657.66			
4616.169	4617.462			0.1	0.2d		21656.92			
4616.312	4617.605			0.3			21656.25			
4616.378	4617.671		0.2	0.2			21655.94	4d 1	2c 1	P 5
4616.378	4617.671		0.2	0.2			21655.94	Y 2	2B 4	R 2
4616.497	4617.790	GK		0.7	0.9		21655.38	4c 0	2a 0	P 7
4616.638	4617.931			0.1			21654.72			
4616.734	4618.027			0.1d			21654.27			
4616.800	4618.093			0.0			21653.96			
4616.866	4618.159		0.0d	0.1			21653.65			
4617.000	4618.293		0.1	0.2			21653.02			
4617.105	4618.398		0.2	0.1			21652.53			
4617.218	4618.511		0.2	0.0			21652.00			
4617.322	4618.616		0.4	0.4	0.2		21651.51			
4617.367	4618.660		0.2				21651.30			
4617.433	4618.726		0.4	0.2			21650.99			
4617.527	4618.820	GK	1.5b	1.5b	0.4	1.91	21650.55	4c 2	2a 2	Q 1
4617.644	4618.938		0.4	0.3			21650.00			
4617.727	4619.021		0.1d				21649.61			
4617.789	4619.083			0.2			21649.32			
4617.875	4619.168	G	0.2	0.5	0.2		21648.92	4d 3	2c 3	Q 3
4617.936	4619.230		0.1				21648.63			
4618.007	4619.300			0.2	0.2		21648.30			
4618.049	4619.343		0.0				21648.10			
4618.128	4619.422	GK	0.5	0.8		1.25	21647.73	W 2	2B 4	P 4
4618.222	4619.516		0.0				21647.29			
4618.303	4619.597	GK	0.4	1.5b	1.0b	1.55	21646.91	3D 0	2B 0	R 6
4618.401	4619.695		0.0				21646.45			
4618.438	4619.731		0.0	0.3			21646.28			
4618.506	4619.800		0.1				21645.96			
4618.602	4619.896	GK	0.6	0.8	0.3	1.18	21645.51	Y 1	2B 1	P 2
4618.602	4619.896	GK	0.6	0.8	0.3	1.18	21645.51	4d 2	2c 2	P 4
4618.719	4620.013		0.2	0.1			21644.96			
4618.788	4620.081		0.0				21644.64			
4618.820	4620.113		0.0	0.0			21644.49			
4618.935	4620.229		0.0d				21643.95	4d 0	2c 0	P 6
4619.043	4620.337	F					21643.44			
4619.236	4620.530		0.0d	0.0d			21642.54			
4619.332	4620.626	F					21642.09			
4619.422	4620.716		0.0d	0.2			21641.67			
4619.597	4620.891	G	0.1	0.4	0.3		21640.85	3b 6	2a 0	P 2
4619.714	4621.008			0.2			21640.30			
4619.825	4621.119		0.2	0.3			21639.78			
4620.064	4621.358			0.2			21638.66			
4620.177	4621.471		0.0d				21638.13			
4620.220	4621.514			0.1			21637.93			
4620.308	4621.602			0.3			21637.52			
4620.357	4621.651		0.0		0.2		21637.29			
4620.453	4621.747			0.1			21636.84			
4620.557	4621.852	G	0.2	0.6	0.4		21636.35			
4620.677	4621.971		0.1				21635.79			
4620.756	4622.050	GK	0.5	0.9	0.3d	1.28	21635.42	4c 2	2a 2	Q 2
4620.878	4622.172		0.1	0.1			21634.85			
4620.957	4622.251		0.0				21634.48			
4621.042	4622.337		0.0	0.2			21634.08			
4621.149	4622.443		0.0	0.4	0.4		21633.58			

λ_{air}	λ_{vac}	Ref.	I_1	I_2	I_3	I_4	ν	Upper	Lower	Br.
4621.203	4622.497		0.0				21633.33			
4621.354	4622.649		0.0d				21632.62			
4621.455	4622.749			0.0			21632.15			
4621.672	4622.967	F					21631.13	4d 2	2c 2	Q 5
4621.826	4623.121	G					21630.41			
4621.965	4623.260		0.0				21629.76			
4622.020	4623.315	F					21629.50			
4622.138	4623.433			0.2			21628.95			
4622.232	4623.527			0.0			21628.51	4F 0	2B 5	P 5
4622.318	4623.612	G		0.4			21628.11	3E 3	2B 5	R 3
4622.442	4623.737			0.2			21627.53			
4622.583	4623.878			0.1			21626.87			
4622.747	4624.042	K	0.1	0.3	0.3d		21626.10			
4622.901	4624.196			0.1			21625.38			
4623.017	4624.312			0.3			21624.84			
4623.143	4624.438		0.0	0.1			21624.25			
4623.248	4624.543			0.4			21623.76	3E 3	2B 5	Q 6
4623.357	4624.652		0.0d	0.3			21623.25			
4623.435	4624.730	F					21622.88			
4623.581	4624.876			0.2			21622.20			
4623.735	4625.030	F					21621.48			
4623.829	4625.124	F					21621.04			
4623.908	4625.204			0.1			21620.67			
4624.013	4625.308		0.0				21620.18			
4624.097	4625.392	GK	0.6	0.7	0.3d	0.88	21619.79	Y 2	2B 4	P 1
4624.097	4625.392	GK	0.6	0.7	0.3d	0.88	21619.79	U 1	2B 2	P 2
4624.216	4625.512		0.0				21619.23	3d 3	2c 0	P 7
4624.283	4625.578						21618.92			
4624.364	4625.659			0.1d			21618.54			
4624.529	4625.824			0.2d			21617.77			
4624.636	4625.931		0.0				21617.27			
4624.738	4626.034	G	0.4	0.5		0.68	21616.79			
4624.810	4626.105	F					21616.46			
4624.897	4626.192			0.2			21616.05			
4624.978	4626.273	F					21615.67			
4625.117	4626.413			0.4			21615.02			
4625.201	4626.496		0.0	0.2			21614.63			
4625.318	4626.614	GK	0.4	1.5b	1.0	1.59	21614.08	3D 0	2B 0	R 5
4625.460	4626.755	G	0.2b	0.7			21613.42			
4625.511	4626.806						21613.18	3c 5	2a 2	Q 3
4625.592	4626.888	GK	0.5	1.0	0.3	1.13	21612.80	4c 2	2a 2	Q 3
4625.689	4626.984		0.2	0.3			21612.35			
4625.759	4627.055		0.0				21612.02			
4625.815	4627.110		0.1d	0.2			21611.76			
4625.920	4627.215		0.0	0.1			21611.27			
4626.022	4627.318		0.0	0.3	0.4		21610.79			
4626.162	4627.457		0.0d				21610.14			
4626.254	4627.549			0.2			21609.71	4e 3	2c 3	Q 4
4626.332	4627.628	F					21609.34			
4626.429	4627.725		0.0	0.2			21608.89			
4626.607	4627.903		0.0d	0.3			21608.06	3D 2	2B 3	P 7
4626.774	4628.070	G		0.6	0.3d	0.61	21607.28	3F 2	2B 4	Q 4
4626.774	4628.070	G		0.6	0.3d	0.61	21607.28	4e 3	2c 3	Q 4
4626.830	4628.125		0.2				21607.02			
4626.930	4628.226			0.1			21606.55			
4626.982	4628.278						21606.31			
4627.061	4628.357	G	0.2b	0.7	0.5d		21605.94	3E 1	2B 2	Q 7
4627.207	4628.503		0.2	0.4			21605.26			
4627.324	4628.620			0.2			21604.71	4F 0	2B 5	Q 4
4627.408	4628.704	F					21604.32			

λ_{air}	λ_{vac}	Ref.	I_1	I_2	I_3	I_4	ν	Upper	Lower	Br.
4627.457	4628.753		0.1	0.4			21604.09			
4627.524	4628.820		0.0				21603.78			
4627.547	4628.843		0.1	0.3	0.4		21603.67			
4627.622	4628.918	F					21603.32			
4627.665	4628.961		0.2	0.2			21603.12	Z 2	2B 3	P 5
4627.766	4629.062		0.4	0.7			21602.65	3E 3	2B 5	R 2
4627.821	4629.117		0.2				21602.39			
4627.901	4629.197		0.4				21602.02			
4627.984	4629.280	GK	1.5b	1.5b	1.0b	2.36	21601.63	3D 0	2B 0	R 0
4628.108	4629.405		0.4	0.7			21601.05			
4628.185	4629.482		0.1				21600.69			
4628.250	4629.546		0.0	0.2			21600.39			
4628.314	4629.610		0.1				21600.09			
4628.363	4629.660	G	0.2	0.5			21599.86	3F 2	2B 4	R 2
4628.443	4629.739	1	0.0		0.8		21599.49			
4628.511	4629.807		0.1				21599.17			
4628.567	4629.863		0.0				21598.91			
4628.603	4629.900		0.1	0.1			21598.74			
4628.717	4630.013	GK	0.2b	0.7	0.3d	0.82	21598.21	W 2	2B 4	P 5
4628.717	4630.013	GK	0.2b	0.7	0.3d	0.82	21598.21	X 0	2B 1	P 5
4628.816	4630.112			0.1			21597.75			
4628.916	4630.213		0.1	0.2			21597.28			
4628.966	4630.262						21597.05			
4629.008	4630.305		0.0				21596.85			
4629.068	4630.365		0.0	0.1			21596.57			
4629.152	4630.449		0.2	0.4			21596.18			
4629.300	4630.596		0.1	0.0			21595.49			
4629.396	4630.693		0.2	0.5			21595.04			
4629.538	4630.834		0.0	0.5			21594.38			
4629.585	4630.882		0.1				21594.16			
4629.684	4630.980	GK	0.5	0.8	0.3	0.78	21593.70	3E 1	2B 2	R 2
4629.812	4631.109		0.1	0.2			21593.10	4d 3	2c 3	P 3
4629.878	4631.175	F					21592.79			
4629.941	4631.238		0.1	0.3	0.4		21592.50			
4630.147	4631.444	GK	0.2	0.7			21591.54	W 2	2B 4	P 6
4630.293	4631.589		0.3	0.5			21590.86			
4630.434	4631.731		0.0				21590.20			
4630.496	4631.793		0.1	0.1	0.3		21589.91			
4630.522	4631.819			0.2			21589.79			
4630.662	4631.958		0.1	0.3			21589.14			
4630.724	4632.021		0.0				21588.85			
4630.784	4632.081		0.1				21588.57			
4630.855	4632.152		0.1	0.2			21588.24			
4630.945	4632.242		0.2	0.2			21587.82			
4630.994	4632.291		0.0				21587.59			
4631.082	4632.379	GK	0.3	0.8	0.5	0.81	21587.18			
4631.155	4632.452		0.0				21586.84			
4631.202	4632.499		0.0				21586.62			
4631.256	4632.553		0.3	0.4			21586.37			
4631.299	4632.596		0.2				21586.17			
4631.361	4632.658		0.4	0.2			21585.88			
4631.462	4632.759	GK	1.5b	1.5d	1.0b	2.20	21585.41	3D 0	2B 0	R 1
4631.580	4632.877		0.4	0.1			21584.86			
4631.640	4632.937		0.1	0.3			21584.58	3a 2	2c 0	R 3
4631.726	4633.023		0.3	0.2			21584.18			
4631.758	4633.055		0.1				21584.03			
4631.846	4633.143	GK	0.9b	1.5b	1.0	2.37	21583.62	4c 1	2a 1	P 5
4631.846	4633.143	GK	0.9b	1.5b	1.0	2.37	21583.62	3D 0	2B 0	R 4
4631.973	4633.270		0.2	0.8			21583.03	4c 2	2a 2	Q 4
4632.091	4633.388		0.0	0.1			21582.48			

λ_{air}	λ_{vac}	Ref.	I_1	I_2	I_3	I_4	ν	Upper	Lower	Br.
4632.179	4633.476		0.0	0.2			21582.07			
4632.279	4633.577		0.1	0.4			21581.60			
4632.391	4633.688		0.1	0.1			21581.08			
4632.470	4633.768						21580.71			
4632.509	4633.807		0.0	0.2			21580.53			
4632.617	4633.914		0.1	0.3			21580.03			
4632.765	4634.062		0.0	0.2			21579.34			
4632.863	4634.161		0.1	0.1			21578.88			
4632.921	4634.219		0.0				21578.61			
4633.027	4634.324		0.1b	0.4			21578.12			
4633.087	4634.384			0.3			21577.84			
4633.179	4634.477		0.1		0.4		21577.41			
4633.248	4634.545		0.2	0.3			21577.09			
4633.364	4634.661		0.1d	0.2			21576.55			
4633.503	4634.801		0.3	0.4			21575.90			
4633.611	4634.908	G	0.5	0.5			21575.40	3A 0	2B 1	R 1
4633.707	4635.005	GK	0.4	0.3	0.3b		21574.95	Y 1	2B 1	P 3
4633.825	4635.123		0.5	0.7			21574.40			
4633.866	4635.164		0.4				21574.21			
4633.937	4635.235		0.6				21573.88			
4634.027	4635.325	GK	1.5b	1.5b	1.0	2.69	21573.46	3D 0	2B 0	R 2
4634.161	4635.458		0.6	0.8			21572.84			
4634.208	4635.506		0.0				21572.62			
4634.287	4635.585		0.2	0.2			21572.25			
4634.397	4635.695		0.4d	0.7			21571.74	Y 2	2B 4	P 2
4634.491	4635.789		0.2	0.2			21571.30			
4634.590	4635.888	GK	0.8	1.5	1.0		21570.84	3D 0	2B 0	R 3
4634.590	4635.888	GK	0.8	1.5	1.0		21570.84	3f 3	2c 1	R 4
4634.713	4636.011	GK	0.7	1.0		1.66	21570.27	3D 0	2B 0	P 1
4634.826	4636.124		0.2				21569.74			
4634.863	4636.161		0.1	0.2			21569.57			
4634.945	4636.243		0.0	0.2			21569.19			
4635.029	4636.327		0.0	0.1			21568.80			
4635.087	4636.385			0.0			21568.53			
4635.179	4636.477		0.1	0.3			21568.10			
4635.278	4636.576		0.1	0.2			21567.64			
4635.340	4636.638		0.1				21567.35			
4635.454	4636.752			0.2			21566.82			
4635.547	4636.845		0.0				21566.39			
4635.625	4636.923	F					21566.02			
4635.744	4637.042			0.1			21565.47			
4635.985	4637.283				0.3b		21564.35			
4636.022	4637.320			0.3b			21564.18			
4636.125	4637.423			0.1			21563.70			
4636.230	4637.528		0.1	0.0			21563.21			
4636.338	4637.636		0.4	0.6			21562.71			
4636.419	4637.717	F					21562.33			
4636.462	4637.761		0.1	0.3	0.2d		21562.13			
4636.555	4637.853		0.0	0.0			21561.70			
4636.624	4637.922			0.2			21561.38			
4636.701	4638.000		0.0	0.1			21561.02			
4636.791	4638.090			0.2			21560.60			
4636.920	4638.219	G	0.2	0.5	0.2		21560.00			
4637.024	4638.322			0.2			21559.52			
4637.118	4638.417	G					21559.08			
4637.217	4638.516			0.2			21558.62			
4637.269	4638.567			0.1			21558.38			
4637.398	4638.697	G		0.4d	0.4		21557.78			
4637.454	4638.753	F					21557.52			
4637.557	4638.856			0.1			21557.04			

λ_{air}	λ_{vac}	Ref.	I_1	I_2	I_3	I_4	ν	Upper	Lower	Br.
4637.624	4638.922		0.0				21556.73			
4637.665	4638.963			0.5			21556.54	3E 3	2B 5	Q 5
4637.787	4639.086			0.1			21555.97			
4637.891	4639.189			0.3	0.2		21555.49			
4638.033	4639.331	G		0.2			21554.83	3f 3	2c 1	R 4
4638.149	4639.448		0.0	0.1			21554.29			
4638.269	4639.568			0.2			21553.73	3D 2	2B 3	P 8
4638.327	4639.626		0.0				21553.46	3c 5	2a 2	Q 4
4638.429	4639.727		0.3	0.3			21552.99	3E 3	2B 5	R 1
4638.480	4639.779	GK	0.3	0.5		0.81	21552.75	U 1	2B 2	P 3
4638.622	4639.921		0.0				21552.09			
4638.655	4639.953		0.1	0.4			21551.94			
4638.788	4640.087			0.1d			21551.32			
4638.822	4640.121						21551.16			
4639.102	4640.401			0.1			21549.86			
4639.223	4640.522		0.0	0.3			21549.30			
4639.287	4640.587		0.0				21549.00	3a 2	2c 0	R 2
4639.380	4640.679			0.2			21548.57			
4639.490	4640.789		0.1	0.1			21548.06			
4639.617	4640.916		0.0				21547.47			
4639.675	4640.974			0.2			21547.20			
4639.793	4641.093			0.1			21546.65			
4639.884	4641.183	GK	0.2d	0.7	0.3		21546.23	4c 2	2a 2	Q 5
4640.022	4641.321	F		0.1			21545.59			
4640.204	4641.503	F					21544.74			
4640.246	4641.545	G		0.3	0.2b		21544.55			
4640.487	4641.786	GK	0.2	0.7	0.8		21543.43			
4640.659	4641.959						21542.63			
4640.709	4642.008			0.2			21542.40			
4640.838	4642.138			0.0			21541.80			
4640.989	4642.288		0.1d	0.2			21541.10			
4641.079	4642.379	G	0.1d	0.5	0.4		21540.68			
4641.209	4642.508						21540.08			
4641.295	4642.595	F					21539.68			
4641.558	4642.857			0.1d			21538.46			
4641.814	4643.114		0.0d				21537.27			
4642.118	4643.418			0.0d	0.2		21535.86			
4642.175	4643.475	F					21535.60			
4642.385	4643.685				0.3		21534.62			
4642.489	4643.789		0.0				21534.14			
4642.554	4643.854		0.0				21533.84	3f 2	2c 0	P 3
4642.664	4643.964	GK	0.2	0.7	0.7		21533.33	3e 2	2c 0	R 5
4642.782	4644.082			0.0			21532.78	3f 3	2c 1	R 3
4642.857	4644.157	F					21532.43			
4643.043	4644.343		0.0	0.2d			21531.57			
4643.170	4644.470		0.0				21530.98	4e 2	2c 2	P 7
4643.259	4644.559	G		0.5	0.5	0.60	21530.57			
4643.384	4644.684			0.1			21529.99			
4643.489	4644.790	G	0.0	0.5	0.5		21529.50	3F 0	2B 1	P 4
4643.662	4644.962						21528.70			
4643.791	4645.092		0.0				21528.10			
4643.869	4645.169	GK	0.5	0.5		0.72	21527.74	X 1	2B 3	R 0
4643.999	4645.299		0.0d	0.0			21527.14	3f 3	2c 1	R 3
4644.165	4645.465			0.1			21526.37			
4644.318	4645.618			0.2			21525.66			
4644.515	4645.815	F					21524.75			
4644.649	4645.950	F					21524.13			
4644.719	4646.020			0.0			21523.80			
4644.816	4646.117			0.0			21523.35			
4644.913	4646.214	G	0.3	0.4			21522.90	X 1	2B 3	R 1

λ_{air}	λ_{vac}	Ref.	I_1	I_2	I_3	I_4	ν	Upper	Lower	Br.
4645.013	4646.313			0.0			21522.44			
4645.140	4646.441		0.1	0.2			21521.85			
4645.190	4646.490		0.0d				21521.62			
4645.252	4646.553		0.1				21521.33			
4645.347	4646.648	GK	0.3	1.0b	0.9	1.01	21520.89	3E 1	2B 2	Q 6
4645.531	4646.832		0.1				21520.04			
4645.695	4646.996		0.1				21519.28	3e 2	2c 0	R 4
4645.915	4647.216			0.0			21518.26			
4646.040	4647.341		0.0d				21517.68			
4646.099	4647.399			0.3	0.3		21517.41			
4646.312	4647.613			0.2d			21516.42	3b 6	2a 0	P 3
4646.487	4647.788		0.0	0.1			21515.61			
4646.578	4647.879	G	0.5	0.7	0.3	0.45	21515.19	3E 1	2B 2	R 1
4646.690	4647.991						21514.67			
4646.785	4648.086		0.0				21514.23			
4646.841	4648.143		0.0	0.0d			21513.97			
4646.923	4648.224	F					21513.59			
4647.099	4648.400			0.0d			21512.78			
4647.284	4648.586			0.0			21511.92			
4647.364	4648.665			0.0			21511.55			
4647.466	4648.767			0.4	0.3		21511.08			
4647.543	4648.844	F					21510.72			
4647.660	4648.962			0.2d			21510.18			
4647.747	4649.048			0.1			21509.78			
4647.828	4649.129	F					21509.40			
4647.941	4649.243			0.2			21508.88			
4648.066	4649.368		0.0	0.2			21508.30	4d 0	2c 0	P 7
4648.200	4649.502	G	0.3	0.4	0.2		21507.68	Y 2	2B 4	P 3
4648.276	4649.578			0.1			21507.33			
4648.339	4649.640			0.3			21507.04			
4648.471	4649.772			0.4			21506.43	3F 2	2B 4	Q 3
4648.471	4649.772			0.4			21506.43	3a 2	2c 0	R 1
4648.570	4649.872	GK	0.1	0.6	0.8		21505.97			
4648.631	4649.932		0.1	0.5			21505.69			
4648.963	4650.265				0.2		21504.15			
4649.054	4650.356			0.0d			21503.73			
4649.223	4650.525				0.2d		21502.95			
4649.270	4650.572	GK	0.3	0.6		0.83	21502.73	X 1	2B 3	R 2
4649.372	4650.674		0.1	0.0	0.4		21502.26			
4649.446	4650.747		0.0				21501.92			
4649.510	4650.812	GK	0.2	0.9	0.3	1.13	21501.62	3E 3	2B 5	Q 4
4649.640	4650.942		0.1	0.1			21501.02			
4649.731	4651.033		0.0				21500.60			
4649.993	4651.295			0.1			21499.39			
4650.116	4651.418		0.0				21498.82			
4650.170	4651.472		0.0	0.1			21498.57			
4650.244	4651.546		0.1				21498.23			
4650.300	4651.602		0.1				21497.97			
4650.380	4651.682	GK	0.3	0.8	0.5	1.00	21497.60	X 0	2B 1	P 6
4650.488	4651.790		0.2	0.4			21497.10	3E 3	2B 5	R 0
4650.603	4651.905		0.1	0.0			21496.57			
4650.683	4651.985		0.1	0.3			21496.20			
4650.755	4652.057	F					21495.87			
4650.821	4652.124		0.1				21495.56			
4650.862	4652.165		0.2	0.1			21495.37			
4650.960	4652.262	GK	0.8	0.7	0.5	1.06	21494.92	3A 0	2B 1	R 0
4651.074	4652.377		0.1	0.1			21494.39			
4651.176	4652.478		0.0				21493.92			
4651.276	4652.578			0.1d			21493.46			
4651.390	4652.693		0.0	0.0d			21492.93			

λ_{air}	λ_{vac}	Ref.	I_1	I_2	I_3	I_4	ν	Upper	Lower	Br.
4651.432	4652.734		0.0				21492.74			
4651.512	4652.814	GK	0.5	0.7	0.3b	0.70	21492.37	Y 1	2B 1	P 4
4651.512	4652.814	GK	0.5	0.7	0.3b	0.70	21492.37	3F 2	2B 4	R 1
4651.637	4652.940		0.0	0.1			21491.79	3f 3	2c 1	R 2
4651.741	4653.044			0.1			21491.31			
4651.782	4653.085			0.2d			21491.12			
4651.868	4653.170	F					21490.72			
4651.912	4653.215	G		0.4	0.2b		21490.52	4c 3	2a 3	R 2
4652.007	4653.310			0.1			21490.08	Z 2	2B 3	P 6
4652.113	4653.416			0.0			21489.59	3f 3	2c 1	R 2
4652.213	4653.516			0.2b			21489.13	3e 2	2c 0	R 3
4652.356	4653.658			0.1			21488.47			
4652.473	4653.775			0.3			21487.93			
4652.568	4653.871		0.1	0.3			21487.49			
4652.691	4653.994		0.1d	0.0			21486.92			
4652.802	4654.105		0.2	0.4			21486.41			
4652.847	4654.150		0.2				21486.20			
4652.912	4654.215		0.3	0.2			21485.90	4d 1	2c 1	P 6
4653.003	4654.306	GK	1.0b	1.5b	0.9	1.94	21485.48	3D 0	2B 0	P 2
4653.120	4654.423		0.3	0.0			21484.94			
4653.209	4654.512	GK	0.7	1.0	0.5	1.29	21484.53	U 1	2B 2	P 4
4653.330	4654.633		0.1	0.2			21483.97			
4653.402	4654.705	F					21483.64			
4653.616	4654.919			0.1d			21482.65			
4653.774	4655.077		0.1	0.2	0.2b		21481.92	3f 2	2c 0	P 4
4653.861	4655.164			0.2			21481.52			
4653.963	4655.266		0.0	0.0			21481.05	4d 2	2c 2	P 5
4654.060	4655.363	GK	0.4	1.0	0.7	1.30	21480.60	U 1	2B 2	P 6
4654.212	4655.515		0.3	0.2d			21479.90			
4654.323	4655.626						21479.39			
4654.366	4655.669			0.0d			21479.19			
4654.539	4655.842	G	0.1	0.6	0.3		21478.39			
4654.622	4655.925						21478.01			
4654.741	4656.044		0.1	0.3			21477.46			
4654.888	4656.191			0.0			21476.78			
4655.038	4656.341			0.1			21476.09			
4655.144	4656.447	F					21475.60			
4655.205	4656.508			0.1			21475.32			
4655.326	4656.629		0.2	0.3			21474.76			
4655.443	4656.747			0.0			21474.22	3f 2	2c 0	P 4
4655.523	4656.827			0.2			21473.85			
4655.753	4657.057			0.2			21472.79			
4655.885	4657.189			0.2			21472.18			
4655.980	4657.284	F					21471.74			
4656.057	4657.360			0.1			21471.39			
4656.219	4657.523			0.1			21470.64			
4656.283	4657.587	F					21470.35			
4656.336	4657.640				0.5		21470.10			
4656.404	4657.707			0.2d			21469.79			
4656.546	4657.850	F					21469.13			
4656.647	4657.950			0.0d			21468.67			
4656.749	4658.052			0.2			21468.20			
4657.200	4658.504			0.0	0.2		21466.12			
4657.451	4658.755				0.2		21464.96			
4657.521	4658.825			0.0			21464.64			
4657.669	4658.973			0.2b			21463.96			
4657.864	4659.168	GK	0.3	0.9	0.6	1.02	21463.06	U 1	2B 2	P 5
4657.994	4659.298			0.1			21462.46			
4658.053	4659.357	F					21462.19			
4658.155	4659.459			0.2			21461.72	4F 0	2B 5	R 1

λ_{air}	λ_{vac}	Ref.	I_1	I_2	I_3	I_4	ν	Upper	Lower	Br.
4658.252	4659.556	F					21461.27			
4658.337	4659.641			0.1			21460.88			
4658.412	4659.716	F					21460.53			
4658.576	4659.880			0.1d			21459.78			
4658.687	4659.991			0.0			21459.27			
4658.791	4660.095	GK	0.2	0.8	0.3	0.93	21458.79	3E 3	2B 5	Q 3
4658.930	4660.234			0.1			21458.15			
4659.006	4660.310	F					21457.80	4e 3	2c 3	P 4
4659.079	4660.384		0.2	0.3			21457.46			
4659.182	4660.486			0.0			21456.99			
4659.286	4660.590	G	0.1	0.6	0.5		21456.51			
4659.431	4660.736			0.1			21455.84			
4659.507	4660.812			0.0			21455.49			
4659.564	4660.868			0.2			21455.23			
4659.618	4660.923			0.2			21454.98			
4659.677	4660.981			0.0			21454.71			
4659.787	4661.092			0.0			21454.20			
4659.887	4661.192		0.0	0.3d			21453.74			
4659.974	4661.279		0.0	0.1			21453.34			
4660.104	4661.409	GK	0.2	0.9	0.7	0.90	21452.74	4c 2	2a 2	P 3
4660.104	4661.409	GK	0.2	0.9	0.7	0.90	21452.74	3E 1	2B 2	Q 5
4660.213	4661.518		0.1	0.2			21452.24			
4660.311	4661.616		0.2	0.0			21451.79			
4660.402	4661.707	GK	0.9b	1.0b	0.8	1.76	21451.37	Z 0	2B 0	R 1
4660.520	4661.824		0.2	0.3			21450.83			
4660.622	4661.926			0.1			21450.36			
4660.739	4662.044	G	0.1	0.6	0.7		21449.82	3F 0	2B 1	Q 6
4660.739	4662.044	G	0.1	0.6	0.7		21449.8₂	4F 0	2B 5	P 4
4660.859	4662.163		0.1	0.1			21449.27			
4660.941	4662.246		0.1	0.3			21448.89			
4661.082	4662.387		0.1	0.1d			21448.24			
4661.129	4662.434	F					21448.03			
4661.193	4662.498		0.3	0.4			21447.73			
4661.243	4662.548		0.2				21447.50			
4661.306	4662.611		0.4	0.3			21447.21			
4661.402	4662.707	GK	1.5b	1.5b	0.9	2.10	21446.77	Z 0	2B 0	R 0
4661.402	4662.707	GK	1.5b	1.5b	0.9	2.10	21446.77	4c 3	2a 3	R 1
4661.526	4662.831		0.4	0.3			21446.20			
4661.724	4663.029		0.0	0.2			21445.29	3f 3	2c 1	R 1
4661.860	4663.166		0.0	0.2			21444.66	3f 3	2c 1	R 1
4662.004	4663.309	G	0.1	0.4	0.3b		21444.00	3c 4	2a 1	Q 5
4662.180	4663.485			0.2			21443.19	4d 3	2c 3	P 4
4662.337	4663.642		0.1	0.3d			21442.47			
4662.487	4663.792		0.0	0.2			21441.78	3e 2	2c 0	R 2
4662.487	4663.792		0.0	0.2			21441.78	4F 0	2B 5	Q 3
4662.608	4663.914		0.3	0.5			21441.22			
4662.654	4663.959		0.2				21441.01			
4662.721	4664.027		0.3	0.2			21440.70			
4662.815	4664.120	GK	1.5b	1.5	1.0b	2.14	21440.27	Z 0	2B 0	R 2
4662.937	4664.242		0.3	0.4			21439.71			
4663.056	4664.362		0.0				21439.16			
4663.143	4664.449		0.1	0.2			21438.76			
4663.192	4664.497	F					21438.54			
4663.270	4664.575			0.3d			21438.18			
4663.435	4664.740		0.0	0.1d			21437.42			
4663.507	4664.812			0.0d			21437.09			
4663.742	4665.047		0.2	0.4	0.2		21436.01			
4663.861	4665.167		0.0	0.1			21435.46			
4663.979	4665.284			0.1			21434.92			
4664.070	4665.376			0.1			21434.50			

λ_{air}	λ_{vac}	Ref.	I_1	I_2	I_3	I_4	ν	Upper	Lower	Br.
4664.231	4665.537			0.0			21433.76			
4664.342	4665.648			0.3			21433.25			
4664.451	4665.757			0.0	0.3		21432.75			
4664.556	4665.861		0.0				21432.27			
4664.645	4665.951	GK	0.5	0.7	0.2	0.72	21431.86	Y 2	2B 4	P 4
4664.780	4666.086		0.0	0.1			21431.24			
4664.887	4666.193	F					21430.75			
4664.982	4666.288						21430.31			
4665.041	4666.347			0.1			21430.04	3f 2	2c 0	P 5
4665.137	4666.443		0.1b	0.4			21429.60			
4665.240	4666.546	F					21429.13			
4665.317	4666.623		0.0	0.2	0.2d		21428.77			
4665.411	4666.717		0.2b	0.2			21428.34			
4665.500	4666.806		0.2	0.1			21427.93			
4665.592	4666.898	GK	0.8b	1.0	0.4	1.65	21427.51	3E 3	2B 5	Q 2
4665.711	4667.018		0.4	0.4			21426.96	3F 2	2B 4	Q 2
4665.801	4667.107	GK	0.5	0.4		0.86	21426.55	X 1	2B 3	P 2
4665.801	4667.107	GK	0.5	0.4		0.86	21426.55	3E 0	2B 1	R 6
4665.875	4667.181						21426.21			
4665.942	4667.249			0.0			21425.90			
4666.008	4667.314				0.2d		21425.60			
4666.071	4667.377		0.1d	0.2b			21425.31			
4666.212	4667.519	G	0.2d	0.6	0.5		21424.66			
4666.391	4667.697						21423.84			
4666.491	4667.797						21423.38			
4666.626	4667.933			0.2			21422.76			
4666.818	4668.124			0.1	0.3		21421.88			
4666.899	4668.205			0.2			21421.51			
4666.994	4668.301		0.0	0.1			21421.07			
4667.088	4668.395	GK	0.6	1.0	0.4	1.13	21420.64	3E 1	2B 2	R 0
4667.204	4668.510		0.0	0.1			21420.11			
4667.260	4668.567	F					21419.85	V 0	2B 0	R 2
4667.482	4668.789			0.2	0.2d		21418.83			
4667.598	4668.905		0.0	0.2			21418.30			
4667.663	4668.970		0.0				21418.00			
4667.709	4669.016		0.2	0.2			21417.79			
4667.798	4669.105	GK	0.8b	1.0	0.5	1.59	21417.38	3A 0	2B 1	P 4
4667.921	4669.227		0.2	0.2			21416.82			
4668.067	4669.373		0.0d				21416.15			
4668.134	4669.441			0.1			21415.84			
4668.258	4669.565	F					21415.27			
4668.341	4669.648			0.2			21414.89			
4668.424	4669.731			0.1			21414.51			
4668.528	4669.835	F					21414.03			
4668.618	4669.925			0.0			21413.62			
4668.725	4670.032			0.2			21413.13			
4668.780	4670.087	F					21412.88			
4668.823	4670.130			0.2			21412.68			
4668.923	4670.230	F					21412.22			
4668.982	4670.289		0.0	0.3			21411.95	3f 2	2c 0	P 5
4669.085	4670.392		0.0				21411.48			
4669.174	4670.481		0.3d	0.3d			21411.07			
4669.285	4670.592	GK	0.5	0.9	0.7	1.15	21410.56	Z 0	2B 0	R 3
4669.362	4670.669		0.1				21410.21			
4669.418	4670.726		0.0	0.2			21409.95			
4669.499	4670.806		0.0d		0.3		21409.58			
4669.630	4670.937			0.3			21408.98			
4669.787	4671.094		0.0	0.2			21408.26			
4669.933	4671.240		0.1d	0.4	0.5		21407.59			
4670.027	4671.334	GK	0.4	0.7		0.92	21407.16	3E 3	2B 5	Q 1

λ_{air}	λ_{vac}	Ref.	I_1	I_2	I_3	I_4	ν	Upper	Lower	Br.
4670.151	4671.459			0.1			21406.59			
4670.269	4671.576						21406.05			
4670.326	4671.633			0.1d			21405.79			
4670.468	4671.775		0.0d	0.2d			21405.14			
4670.520	4671.827		0.0				21404.90			
4670.577	4671.884		0.2				21404.64			
4670.662	4671.969	GK	0.9b	0.9	0.3d	1.49	21404.25	3D 3	2B 5	R 0
4670.729	4672.037				0.2d		21403.94			
4670.810	4671.118	GK	0.6	0.7			21403.57	3E 3	2B 5	P 2
4670.915	4672.223		0.0d				21403.09			
4670.963	4672.271		0.0	0.2			21402.87			
4671.103	4672.410		0.1	0.4			21402.23			
4671.214	4672.522		0.0	0.2			21401.72			
4671.310	4672.618	GK	0.6	1.5b	1.0b	1.78	21401.28	3E 1	2B 2	Q 4
4671.441	4672.749			0.3			21400.68			
4671.624	4672.932		0.2	0.0			21399.84			
4671.725	4673.033	GK	0.7b	0.7	0.2b	1.17	21399.38	3D 3	2B 5	R 1
4671.849	4673.157	G	0.6	0.4			21398.81	3A 0	2B 1	P 1
4671.978	4673.286		0.1d	0.2			21398.22			
4672.094	4673.402	GK	0.5	0.6	0.3		21397.69	3E 3	2B 5	P 3
4672.223	4673.531		0.1d	0.2			21397.10			
4672.298	4673.606	F					21396.75			
4672.371	4673.679			0.1	0.2		21396.42			
4672.461	4673.769			0.0			21396.01			
4672.603	4673.911			0.2			21395.36			
4672.648	4673.956		0.0d				21395.15			
4672.764	4674.072			0.1			21394.62			
4672.904	4674.212			0.3			21393.98	3A 1	2B 3	R 2
4673.020	4674.328		0.2	0.2			21393.45			
4673.109	4674.417	GK	0.8b	0.8	0.5	1.21	21393.04	3D 0	2B 0	P 3
4673.240	4674.549	GK	0.6	0.6	0.3	1.06	21392.44	3A 0	2B 1	P 3
4673.369	4674.677		0.1	0.1			21391.85			
4673.461	4674.769		0.0d				21391.43			
4673.524	4674.833			0.1			21391.14			
4673.636	4674.944			0.3	0.3d		21390.63			
4673.732	4675.040			0.0			21390.19			
4673.817	4675.126			0.1			21389.80			
4673.972	4675.280	F					21389.09			
4674.051	4675.359			0.1d			21388.73			
4674.191	4675.499	F					21388.09			
4674.283	4675.591		0.0d	0.2d			21387.67			
4674.342	4675.650		0.2				21387.40			
4674.442	4675.751	G	0.7	0.8			21386.94	3E 3	2B 5	P 4
4674.541	4675.849	GK	1.0b	0.9	0.4	1.55	21386.49	Z 0	2B 0	P 1
4674.659	4675.967		0.3	0.2			21385.95	3a 2	2c 0	Q 2
4674.706	4676.015	F					21385.73			
4674.753	4676.061		0.2b	0.4	0.2d		21385.52			
4674.871	4676.179		0.2	0.1			21384.98			
4674.967	4676.275	GK	0.9b	0.9	0.3	1.47	21384.54	3A 0	2B 1	P 2
4675.100	4676.409		0.3	0.2			21383.93			
4675.142	4676.450		0.1				21383.74			
4675.220	4676.529		0.2	0.0	0.3b		21383.38			
4675.317	4676.625	GK	0.9b	1.0		1.54	21382.94	3D 3	2B 5	R 2
4675.435	4676.743		0.2	0.2			21382.40			
4675.465	4676.774						21382.26			
4675.555	4676.864			0.1			21381.85	3b 6	2a 0	P 4
4675.673	4676.982				0.4		21381.31			
4675.767	4677.076		0.1d	0.2			21380.88			
4675.866	4677.174		0.1				21380.43			
4675.951	4677.260		0.1	0.2			21380.04			

λ_{air}	λ_{vac}	Ref.	I_1	I_2	I_3	I_4	ν	Upper	Lower	Br.
4676.008	4677.317		0.0				21379.78			
4676.090	4677.399	F					21379.40			
4676.348	4677.657	F					21378.22			
4676.463	4677.772		0.1d				21377.70	3e 2	2c 0	R 1
4676.563	4677.872		0.1				21377.24			
4676.611	4677.920						21377.02			
4676.730	4678.039		0.0d	0.4b	0.3		21376.48	3f 2	2c 0	P 6
4676.916	4678.225	G	0.2b	0.6d	0.5		21375.63	3F 2	2B 4	P 4
4676.916	4678.225	G	0.2b	0.6d	0.5		21375.63	3D 1	2B 2	R 8
4676.916	4678.225	G	0.2b	0.6d	0.5		21375.63	Z 0	2B 0	R 4
4676.990	4678.299		0.1				21375.29			
4677.115	4678.424		0.0d				21374.72			
4677.242	4678.551		0.0				21374.14			
4677.316	4678.625			0.2d			21373.80			
4677.366	4678.676		0.0				21373.57			
4677.458	4678.768			0.2			21373.15			
4677.520	4678.829		0.0				21372.87			
4677.636	4678.945		0.0	0.3			21372.34			
4677.780	4679.089			0.3	0.3		21371.68			
4677.970	4679.280		0.1	0.3			21370.81			
4678.146	4679.455	G	0.2b	0.5	0.3	0.43	21370.01			
4678.268	4679.578			0.2			21369.45			
4678.384	4679.694			0.1			21368.92	3a 2	2c 0	Q 3
4678.498	4679.808			0.1			21368.40			
4678.575	4679.884			0.2			21368.05			
4678.643	4679.952		0.1	0.2			21367.74			
4678.756	4680.066			0.2			21367.22			
4678.890	4680.200			0.3d			21366.61			
4679.006	4680.316		0.1	0.2			21366.08			
4679.098	4680.408	GK	0.4d	1.0	0.8	1.59	21365.66	3E 1	2B 2	Q 3
4679.188	4680.497		0.0		0.7		21365.25			
4679.225	4680.535		0.2	0.5			21365.08			
4679.335	4680.644			0.1			21364.58			
4679.442	4680.752		0.0	0.2d	0.2		21364.09			
4679.573	4680.883		0.0				21363.49			
4679.646	4680.955			0.0			21363.16			
4679.884	4681.194		0.1d	0.2			21362.07			
4679.998	4681.308		0.1d	0.3			21361.55			
4680.107	4681.417	F					21361.05			
4680.226	4681.536		0.3b	0.2			21360.51			
4680.270	4681.580		0.2				21360.31			
4680.338	4681.648		0.3				21360.00			
4680.428	4681.738	GK	1.5b	1.0	0.3d	1.59	21359.59	4c 3	2a 3	Q 1
4680.546	4681.856		0.3	0.2			21359.05			
4680.603	4681.913		0.0				21358.79			
4680.678	4681.988		0.3	0.3			21358.45	3D 3	2B 5	R 3
4680.756	4682.067	G	0.2	0.6	0.2d	0.89	21358.09			
4680.884	4682.194		0.1				21357.51			
4680.949	4682.259		0.0				21357.21			
4681.037	4682.347		0.0	0.0			21356.81			
4681.164	4682.474			0.2			21356.23			
4681.241	4682.551		0.1d	0.0			21355.88			
4681.289	4682.599		0.0				21355.66			
4681.364	4682.674	GK	0.4d	0.9	0.9b	1.03	21355.32	3E 1	2B 2	P 5
4681.364	4682.674	GK	0.4d	0.9	0.9b	1.03	21355.32	X 1	2B 3	P 3
4681.364	4682.674	GK	0.4d	0.9	0.9b	1.03	21355.32	4e 3	2c 3	P 5
4681.491	4682.801		0.1	0.3			21354.74			
4681.543	4682.854		0.0				21354.50			
4681.646	4682.957	G	0.2	0.5			21354.03	3D 3	2B 5	R 4
4681.722	4683.032	F					21353.69			

λ_{air}	λ_{vac}	Ref.	I_1	I_2	I_3	I_4	ν	Upper	Lower	Br.
4681.833	4683.143	G	0.2b	0.5			21353.18	3E 3	2B 5	P 5
4681.938	4683.248			0.0			21352.70			
4682.006	4683.316		0.0	0.1d			21352.39			
4682.135	4683.446		0.2	0.4			21351.80			
4682.184	4683.494		0.2				21351.58			
4682.249	4683.560		0.3	0.2			21351.28			
4682.346	4683.656	GK	1.0b	1.5b	0.8	1.85	21350.84	3E 1	2B 2	P 4
4682.469	4683.779		0.3	0.3d			21350.28			
4682.605	4683.915		0.0d	0.5	0.4		21349.66	V 0	2B 0	R 3
4682.769	4684.080		0.2	0.2d			21348.91			
4682.872	4684.183						21348.44			
4682.916	4684.227			0.0			21348.24			
4683.035	4684.345			0.2			21347.70			
4683.164	4684.475		0.0d	0.2			21347.11	3a 2	2c 0	Q 4
4683.329	4684.639		0.1d	0.4b	0.2		21346.36	3E 2	2B 4	Q 9
4683.436	4684.747		0.0				21345.87			
4683.487	4684.797		0.1	0.1			21345.64			
4683.640	4684.951	GK	0.5b	0.8			21344.94	4c 3	2a 3	Q 2
4683.737	4685.048	GK	0.5				21344.50	3f 2	2c 0	P 6
4683.822	4685.133	GK	1.5b	1.5b	1.0b	2.20	21344.11	3E 1	2B 2	Q 2
4683.822	4685.133	GK	1.5b	1.5b	1.0b	2.20	21344.11	4E 0	2B 5	R 2
4683.947	4685.258		0.4	0.5			21343.54			
4684.037	4685.348		0.1				21343.13			
4684.152	4685.463		0.0	0.1d			21342.61			
4684.244	4685.555		0.1	0.3d			21342.19			
4684.345	4685.656		0.0	0.1			21341.73			
4684.448	4685.759		0.3d	0.4			21341.26			
4684.571	4685.882		0.2				21340.70			
4684.656	4685.968	GK	0.9b	1.0	0.5	1.44	21340.31	3E 1	2B 2	P 3
4684.775	4686.086		0.3	0.2			21339.77			
4684.997	4686.308			0.1			21338.76			
4685.102	4686.413			0.1			21338.28			
4685.302	4686.613			0.1d	0.3		21337.37	3f 2	2c 0	P 7
4685.482	4686.793			0.4	0.2		21336.55	3E 0	2B 1	Q11
4685.482	4686.793			0.4	0.2		21336.55	3F 0	2B 1	R 3
4685.611	4686.923			0.2			21335.96			
4685.770	4687.081			0.3			21335.24			
4685.945	4687.257		0.1	0.4			21334.44			
4686.044	4687.356		0.0	0.1			21333.99			
4686.143	4687.455	GK	0.6	1.0	0.5	1.46	21333.54	3E 1	2B 2	Q 1
4686.273	4687.584		0.1	0.4	0.4		21332.95			
4686.328	4687.639		0.0				21332.70			
4686.448	4687.760			0.2			21332.15			
4686.580	4687.892		0.3	0.4			21331.55			
4686.628	4687.940		0.2				21331.33			
4686.686	4687.997		0.3	0.2			21331.07	3F 2	2B 4	P 3
4686.780	4688.092	GK	1.5b	1.5		1.94	21330.64	Z 0	2B 0	P 2
4686.897	4688.208		0.3	0.3			21330.11	3f 3	2c 1	Q 2
4686.949	4688.261		0.2				21329.87			
4686.998	4688.309		0.1	0.1d			21329.65	3f 3	2c 1	Q 2
4687.094	4688.406			0.0			21329.21			
4687.157	4688.469	F					21328.92			
4687.215	4688.527		0.1	0.2			21328.66			
4687.371	4688.683						21327.95			
4687.426	4688.738			0.2			21327.70			
4687.637	4688.949	G	0.1d	0.5	0.5		21326.74			
4687.727	4689.039		0.0				21326.33			
4687.831	4689.143			0.2			21325.86			
4687.881	4689.193		0.0				21325.63			
4687.980	4689.292		0.0	0.3	0.2		21325.18			

λ_{air}	λ_{vac}	Ref.	I_1	I_2	I_3	I_4	ν	Upper	Lower	Br.
4688.004	4689.316			0.2			21325.07			
4688.224	4689.536		0.1d	0.3			21324.07			
4688.343	4689.655		0.1d	0.1			21323.53			
4688.437	4689.750	GK	0.5	0.8	0.3	1.19	21323.10	4c 3	2a 3	Q 3
4688.501	4689.813						21322.81			
4688.556	4689.868		0.0	0.1d			21322.56			
4688.780	4690.093			0.3			21321.54			
4688.963	4690.275			0.0			21320.71	3f 3	2c 1	Q 3
4689.115	4690.427		0.2	0.3			21320.02			
4689.161	4690.473		0.1				21319.81			
4689.247	4690.559		0.0d				21319.42			
4689.372	4690.685			0.1			21318.85			
4689.475	4690.788			0.2			21318.38	3f 3	2c 1	Q 3
4689.693	4691.006			0.3b			21317.39			
4689.841	4691.153			0.2			21316.72			
4689.975	4691.287		0.1b	0.4			21316.11			
4690.085	4691.398		0.2	0.1			21315.61			
4690.180	4691.492	GK	0.9b	1.5	0.7	1.70	21315.18	3E 1	2B 2	P 2
4690.301	4691.613		0.2	0.3			21314.63	4d 2	2c 2	P 6
4690.446	4691.758		0.0				21313.97			
4690.499	4691.811			0.2b			21313.73			
4690.570	4691.883	F					21313.41			
4690.648	4691.961			0.1d			21313.05			
4690.758	4692.071			0.1d			21312.55			
4690.838	4692.150			0.2			21312.19			
4690.901	4692.214		0.0				21311.90			
4691.007	4692.320			0.1d			21311.42			
4691.157	4692.470	G	0.4	0.5			21310.74	3D 3	2B 5	P 2
4691.229	4692.542		0.1				21310.41			
4691.304	4692.617		0.0d	0.2			21310.07	3A 1	2B 3	R 1
4691.472	4692.784		0.0	0.1			21309.31			
4691.705	4693.018			0.1			21308.25			
4691.769	4693.082						21307.96			
4691.819	4693.132		0.1				21307.73			
4691.839	4693.152		0.0	0.3			21307.64			
4691.914	4693.227						21307.30			
4691.952	4693.265		0.2	0.1			21307.13			
4692.044	4693.357	GK	0.8	1.0	0.8	1.40	21306.71	3D 0	2B 0	P 4
4692.044	4693.357	GK	0.8	1.0	0.8	1.40	21306.71	4F 0	2B 5	P 3
4692.167	4693.480		0.1d	0.2			21306.15	3f 3	2c 1	Q 4
4692.244	4693.558				0.2		21305.80			
4692.288	4693.602			0.2			21305.60			
4692.377	4693.690			0.1			21305.20			
4692.434	4693.747			0.0			21304.94			
4692.518	4693.831			0.3			21304.56	V 0	2B 0	R 1
4692.918	4694.232						21302.74			
4692.973	4694.287			0.0d			21302.49			
4693.139	4694.452			0.3d			21301.74			
4693.275	4694.589			0.1d			21301.12			
4693.410	4694.723	G	0.0	0.4	0.3b		21300.51	3f 3	2c 1	Q 4
4693.690	4695.003			0.3			21299.24	3E 2	2B 4	R 4
4693.820	4695.133	G	0.2	0.5	0.2b	0.45	21298.65	4F 0	2B 5	Q 2
4694.040	4695.354			0.1			21297.65			
4694.089	4695.402						21297.43			
4694.247	4695.561	F					21296.71			
4694.380	4695.693			0.2			21296.11			
4694.488	4695.801						21295.62			
4694.569	4695.883			0.0			21295.25			
4694.635	4695.949			0.0			21294.95			
4694.783	4696.097			0.3			21294.28	4c 3	2a 3	Q 4

λ_{air}	λ_{vac}	Ref.	I_1	I_2	I_3	I_4	ν	Upper	Lower	Br.
4694.902	4696.216			0.0			21293.74			
4695.145	4696.458		0.0	0.3			21292.64			
4695.237	4696.551	F					21292.22			
4695.460	4696.774				0.2b		21291.21			
4695.517	4696.831			0.3			21290.95	3e 2	2c 0	R 2
4695.517	4696.831			0.3			21290.95	3f 3	2c 1	Q 5
4695.676	4696.990		0.0	0.4	0.4		21290.23			
4695.813	4697.127			0.1			21289.61			
4695.892	4697.206		0.0				21289.25			
4695.991	4697.306		0.0				21288.80			
4696.232	4697.546		0.0				21287.71			
4696.523	4697.837	F					21286.39			
4696.697	4698.012						21285.60	3e 2	2c 0	R 1
4696.898	4698.213						21284.69			
4696.987	4698.301			0.2d			21284.29			
4697.143	4698.458			0.3			21283.58			
4697.209	4698.524				0.3		21283.28			
4697.249	4698.564			0.1			21283.10			
4697.322	4698.637			0.0			21282.77			
4697.435	4698.749	G		0.5	0.5		21282.26	3F 0	2B 1	Q 5
4697.642	4698.957						21281.32			
4697.688	4699.003			0.1d			21281.11			
4697.761	4699.076	i					21280.78			
4697.836	4699.151			0.1d			21280.44			
4697.920	4699.235						21280.06			
4698.002	4699.317	F					21279.69			
4698.061	4699.376			0.3			21279.42	3D 1	2B 2	R 7
4698.155	4699.470	F					21279.00			
4698.203	4699.518			0.2	0.3b		21278.78			
4698.313	4699.628			0.3			21278.28			
4698.516	4699.831			0.0			21277.36			
4698.563	4699.878						21277.15			
4698.662	4699.977						21276.70			
4698.779	4700.094						21276.17			
4698.878	4700.193			0.1			21275.72			
4698.952	4700.267	F					21275.39			
4699.254	4700.569	F					21274.02	3f 3	2c 1	Q 5
4699.413	4700.728						21273.30			
4699.526	4700.841			0.1			21272.79			
4699.577	4700.892				0.2b		21272.56			
4699.630	4700.945		0.0	0.1			21272.32			
4699.771	4701.086		0.1	0.4	0.4		21271.68			
4699.853	4701.168	GK	0.5	0.5		0.76	21271.31	X 2	2B 5	R 0
4699.992	4701.307			0.1			21270.68	3f 2	2c 0	P 7
4700.158	4701.473		0.0				21269.93			
4700.213	4701.528	G	0.2	0.4			21269.68	X 1	2B 3	P 4
4700.345	4701.661			0.1			21269.08			
4700.518	4701.833		0.0				21268.30			
4700.587	4701.902	F					21267.99			
4700.717	4702.032						21267.40			
4700.923	4702.238	F					21266.47			
4700.975	4702.291			0.2d			21266.23			
4701.157	4702.472						21265.41			
4701.236	4702.552	G		0.4	0.3		21265.05			
4701.365	4702.680			0.1			21264.47			
4701.449	4702.764						21264.09			
4701.499	4702.815			0.1			21263.86			
4701.541	4702.857						21263.67			
4701.614	4702.930			0.0			21263.34			
4701.723	4703.038	GK	0.2	0.7	0.6	0.77	21262.85			

λ_{air}	λ_{vac}	Ref.	I$_1$	I$_2$	I$_3$	I$_4$	ν	Upper	Lower	Br.
4701.911	4703.226	G		0.4	0.2		21262.00	W 1	2B 3	R 4
4702.065	4703.381						21261.30	3e 2	2c 0	Q 3
4702.112	4703.428			0.0			21261.09			
4702.234	4703.549		0.0	0.1			21260.54			
4702.344	4703.660	G 1	0.2d	0.6	1.0b		21260.04			
4702.459	4703.775		0.1	0.1			21259.52			
4702.563	4703.879	GK	0.6	0.8	0.5	1.25	21259.05	Z 0	2B 0	P 3
4702.563	4703.879	GK	0.6	0.8	0.5	1.25	21259.05	4c 3	2a 3	Q 5
4702.656	4703.972		0.4	0.3d			21258.63	X 2	2B 5	R 1
4702.776	4704.091		0.0d	0.0d			21258.09			
4702.837	4704.153						21257.81	3e 2	2c 0	R 4
4702.917	4704.233			0.1d			21257.45			
4702.999	4704.315			0.0			21257.08			
4703.092	4704.408			0.2			21256.66	3e 2	2c 0	Q 2
4703.172	4704.488			0.1			21256.30			
4703.342	4704.658			0.1			21255.53			
4703.521	4704.837			0.3	0.2d		21254.72			
4703.672	4704.988			0.0			21254.04			
4703.884	4705.200			0.3b			21253.08			
4703.997	4705.313	F					21252.57			
4704.187	4705.504						21251.71			
4704.423	4705.739	F					21250.65	3e 2	2c 0	Q 4
4704.537	4705.854			0.2			21250.13			
4704.670	4705.986	F					21249.53			
4704.832	4706.148			0.2			21248.80			
4704.964	4706.281						21248.20			
4705.020	4706.336			0.0			21247.95			
4705.128	4706.445		0.0d				21247.46	4E 0	2B 5	R 0
4705.128	4706.445		0.0d				21247.46	4f 0	2c 1	R 4
4705.190	4706.507			0.4			21247.18	3A 2	2B 5	R 4
4705.190	4706.507			0.4			21247.18	4f 0	2c 1	R 4
4705.266	4706.582	GK	0.5	0.6	0.4d	0.91	21246.84	3A 1	2B 3	R 0
4705.332	4706.649		0.0				21246.54			
4705.387	4706.704		0.1				21246.29			
4705.458	4706.775		0.0				21245.97			
4705.505	4706.822		0.0	0.0d			21245.76			
4705.567	4706.883		0.1				21245.48			
4705.700	4707.016		0.0				21244.88			
4705.737	4707.054		0.0				21244.71			
4705.846	4707.163		0.2	0.3			21244.22	3D 3	2B 5	P 3
4705.910	4707.227	F					21243.93			
4706.147	4707.464			0.0			21242.86			
4706.353	4707.670			0.0			21241.93			
4706.424	4707.741						21241.61			
4706.630	4707.947						21240.68	3e 2	2c 0	Q 1
4706.761	4708.078			0.0			21240.09	X 2	2B 5	P 1
4706.810	4708.127				0.3		21239.87			
4707.040	4708.357						21238.83			
4707.100	4708.417			0.1			21238.56			
4707.191	4708.508		0.1	0.2			21238.15			
4707.315	4708.632			0.2			21237.59			
4707.397	4708.714						21237.22			
4707.506	4708.823				0.2		21236.73			
4707.550	4708.867						21236.53			
4707.745	4709.062			0.2d	0.2b		21235.65			
4707.882	4709.199	F					21235.03			
4707.973	4709.291			0.0d			21234.62			
4708.191	4709.508		0.2b	0.2	0.2		21233.64			
4708.341	4709.659			0.1			21232.96			
4708.461	4709.779		0.4	0.5			21232.42	X 2	2B 5	R 2

λ_{air}	λ_{vac}	Ref.	I_1	I_2	I_3	I_4	ν	Upper	Lower	Br.
4708.528	4709.845	GK		0.6	0.5	1.00	21232.12	3D 0	2B 0	P 5
4708.712	4710.029			0.1			21231.29			
4708.962	4710.280			0.1			21230.16			
4709.055	4710.373			0.2			21229.74	3c 9	2a ·5	Q 1
4709.138	4710.455	G	0.0	0.6	0.5		21229.37	3E 2	2B 4	Q 8
4709.238	4710.555		0.0	0.1			21228.92			
4709.331	4710.648		0.2	0.3			21228.50			
4709.366	4710.684		0.2				21228.34			
4709.442	4710.759		0.3	0.2			21228.00			
4709.537	4710.855	GK	1.5b	1.5	0.6	1.90	21227.57	3D 1	2B 2	R 0
4709.659	4710.977		0.3	0.3			21227.02	3E 0	2B 1	R 5
4709.786	4711.103			0.1			21226.45			
4709.885	4711.203						21226.00			
4709.961	4711.279			0.2			21225.66	3e 2	2c 0	Q 5
4710.141	4711.458			0.1			21224.85			
4710.298	4711.616			0.3			21224.14			
4710.513	4711.831			0.1			21223.17	3e 2	2c 0	R 5
4710.591	4711.909						21222.82			
4710.662	4711.980			0.1d			21222.50			
4710.782	4712.100						21221.96			
4710.869	4712.187			0.3	0.2		21221.57			
4710.962	4712.280						21221.15			
4711.064	4712.382	GK	0.3	1.0	0.8	1.05	21220.69	3D 1	2B 2	R 6
4711.190	4712.509			0.2			21220.12			
4711.286	4712.604		0.1d	0.3			21219.69			
4711.457	4712.775						21218.92			
4711.577	4712.895		0.1	0.1d			21218.38			
4711.659	4712.977	G		0.4	0.2b		21218.01	3E 2	2B 4	R 3
4711.781	4713.099		0.1d	0.1			21217.46			
4711.879	4713.197		0.1d	0.2			21217.02			
4711.959	4713.277		0.0	0.0			21216.66			
4712.028	4713.346		0.0				21216.35			
4712.111	4713.429	F					21215.97			
4712.196	4713.515						21215.59			
4712.345	4713.664		0.0				21214.92	4c 3	2a 3	Q 6
4712.441	4713.759		0.0	0.1d			21214.49			
4712.645	4713.964			0.1d			21213.57			
4712.825	4714.144	G	0.2	0.3			21212.76			
4712.950	4714.268		0.1d	0.0			21212.20			
4713.025	4714.344		0.2				21211.86			
4713.145	4714.464	1	0.8	0.5	1.5b		21211.32			
4713.270	4714.588		0.2	0.2			21210.76			
4713.378	4714.697		0.4	0.3			21210.27			
4713.438	4714.757	F					21210.00			
4713.587	4714.906		0.0d				21209.33			
4713.721	4715.039		0.2	0.3			21208.73			
4713.783	4715.102		0.2				21208.45			
4713.841	4715.159		0.3				21208.19			
4713.927	4715.246	GK	1.0b	0.9d	0.5	1.67	21207.80	3D 1	2B 2	R 1
4714.052	4715.371		0.5	0.3			21207.24			
4714.207	4715.526		0.0d				21206.54	X 2	2B 5	R 3
4714.328	4715.647	GK	0.4	0.5		0.86	21206.00	3D 1	2B 2	P 1
4714.416	4715.735		0.0				21205.60			
4714.494	4715.813		0.0	0.2			21205.25	3f 2	2c 0	P 8
4714.581	4715.900						21204.86			
4714.937	4716.256		0.0				21203.26			
4714.988	4716.307			0.1			21203.03			
4715.106	4716.425						21202.50			
4715.346	4716.665	G	0.1	0.3	0.4		21201.42			
4715.504	4716.823			0.1			21200.71			

λ_{air}	λ_{vac}	Ref.	I_1	I_2	I_3	I_4	ν	Upper	Lower	Br.
4715.599	4716.919		0.0	0.1			21200.28			
4715.797	4717.117						21199.39			
4715.949	4717.268	G	0.1b	0.4	0.2		21198.71	W 1	2B 3	R 3
4716.127	4717.446	G	0.0d	0.4	0.6		21197.91			
4716.271	4717.591		0.1	0.1d			21197.26			
4716.343	4717.662	F					21196.94			
4716.478	4717.798			0.1d			21196.33			
4716.554	4717.873		0.0d				21195.99	3F 3	2B 6	R 2
4716.690	4718.009		0.2	0.3			21195.38			
4716.753	4718.073	F					21195.10			
4716.834	4718.154						21194.73			
4716.943	4718.263			0.1			21194.24			
4717.146	4718.466	F					21193.33			
4717.602	4718.922		0.0				21191.28			
4717.725	4719.045			0.1			21190.73			
4717.843	4719.163		0.0		0.3b		21190.20			
4717.960	4719.280	F					21189.67			
4718.036	4719.356			0.1			21189.33			
4718.108	4719.428	F					21189.01			
4718.239	4719.559		0.1				21188.42			
4718.306	4719.626			0.3b			21188.12			
4718.355	4719.675		0.0				21187.90			
4718.486	4719.806		0.1	0.4			21187.31			
4718.640	4719.960		0.0	0.2	0.3		21186.62			
4718.704	4720.025		0.1				21186.33			
4718.756	4720.076		0.0				21186.10			
4718.827	4720.147	GK	0.3	0.9	0.5		21185.78	3D 1	2B 2	R 5
4718.885	4720.205		0.3				21185.52			
4718.958	4720.279		0.4				21185.19			
4719.039	4720.359	GK	1.5b	1.5b	0.9	2.27	21184.83	3D 1	2B 2	R 2
4719.039	4720.359	GK	1.5b	1.5b	0.9	2.27	21184.83	3e 2	2c 0	Q 6
4719.168	4720.488		0.4	0.4			21184.25			
4719.237	4720.557		0.1		0.2d		21183.94			
4719.301	4720.622	GK	0.7	0.8		1.09	21183.65	V 0	2B 0	R 0
4719.426	4720.747		0.1				21183.09			
4719.482	4720.802		0.0	0.0			21182.84			
4719.593	4720.914		0.0				21182.34			
4719.660	4720.981		0.0				21182.04			
4719.716	4721.036						21181.79			
4719.827	4721.148						21181.29			
4719.903	4721.224						21180.95			
4720.008	4721.328			0.0d			21180.48			
4720.055	4721.375		0.0				21180.27			
4720.163	4721.484	F					21179.78			
4720.318	4721.638						21179.09			
4720.382	4721.703		0.0				21178.80			
4720.484	4721.805	G	0.3	0.5	0.2b		21178.34			
4720.605	4721.926		0.2	0.2			21177.80	X 2	2B 5	P 2
4720.670	4721.990			0.3			21177.51	3e 2	2c 0	R 6
4720.761	4722.082	F					21177.10			
4720.906	4722.227						21176.45			
4721.006	4722.327	GK	0.2	0.8	0.7	0.64	21176.00	3D 0	2B 0	P 6
4721.006	4722.327	GK	0.2	0.8	0.7	0.64	21176.00	3D 0	2B 0	P10
4721.142	4722.463		0.0				21175.39			
4721.252	4722.572		0.1d		0.2		21174.90			
4721.401	4722.722		0.2b	0.4			21174.23			
4721.466	4722.787						21173.94			
4721.550	4722.871	GK	0.6	1.0	0.6		21173.56	Z 0	2B 0	P 4
4721.680	4723.001		0.1	0.2			21172.98			
4721.753	4723.074		0.0				21172.65			

λ_{air}	λ_{vac}	Ref.	I_1	I_2	I_3	I_4	ν	Upper	Lower	Br.
4722.025	4723.346		0.0d	0.1			21171.43			
4722.074	4723.396		0.1				21171.21			
4722.173	4723.494		0.0				21170.77			
4722.226	4723.547			0.2d			21170.53			
4722.275	4723.596		0.0				21170.31			
4722.507	4723.828			0.3			21169.27			
4722.596	4723.918			0.3	0.2d		21168.87			
4722.699	4724.020		0.0d	0.0			21168.41			
4722.853	4724.174		0.5	0.5			21167.72	3A 1	2B 3	P 1
4722.949	4724.270		0.0				21167.29	3D 3	2B 5	P 4
4723.036	4724.357	GK	0.6b	1.5b	1.0b	2.10	21166.90	3E 0	2B 1	Q10
4723.036	4724.357	GK	0.6b	1.5b	1.0b	2.10	21166.90	3D 1	2B 2	R 3
4723.036	4724.357	GK	0.6b	1.5b	1.0b	2.10	21166.90	3D 1	2B 2	R 4
4723.172	4724.494			0.3			21166.29			
4723.340	4724.661			0.1			21165.54	3e 2	2c 0	Q 2
4723.449	4724.770						21165.05			
4723.672	4724.994						21164.05			
4723.764	4725.085		0.1d				21163.64			
4723.855	4725.177	GK	0.3	0.7	0.9b	0.72	21163.23	4c 3	2a 3	P 3
4723.855	4725.177	GK	0.3	0.7	0.9b	0.72	21163.23	3E 0	2B 1	Q10
4724.025	4725.346		0.1b	0.4	0.4		21162.47			
4724.177	4725.498			0.0d			21161.79			
4724.415	4725.737		0.1d	0.3			21160.72	3A 1	2B 3	P 4
4724.641	4725.963			0.1d			21159.71			
4724.732	4726.054		0.1				21159.30			
4724.835	4726.157	GK	0.7	1.0d	0.4	1.44	21158.84	W 1	2B 3	R 2
4724.965	4726.287		0.1	0.2			21158.26	3f 3	2c 1	P 3
4725.246	4726.568			0.2			21157.00	4D 0	2B 5	R 6
4725.391	4726.713			0.3			21156.35			
4726.050	4727.372						21153.40			
4726.158	4727.480	GK	0.4	0.7	0.3	0.92	21152.92	V 0	2B 0	P 4
4726.372	4727.694						21151.96			
4726.473	4727.795				0.2		21151.51			
4726.575	4727.898						21151.05			
4726.864	4728.186				0.4		21149.76			
4726.945	4728.267	F					21149.40			
4727.112	4728.434		0.0				21148.65			
4727.197	4728.519		0.0	0.2	0.3		21148.27	3D 0	2B 0	P 9
4727.416	4728.738		0.2	0.3			21147.29			
4727.588	4728.911		0.0d				21146.52			
4727.662	4728.984		0.1				21146.19			
4727.758	4729.080	GK	0.5	0.6		0.78	21145.76	3A 1	2B 3	P 2
4727.870	4729.192		0.1				21145.26			
4728.008	4729.331		0.1d	0.2			21144.64			
4728.297	4729.619		0.0				21143.35	·		
4728.397	4729.720		0.1				21142.90			
4728.451	4729.774		0.0				21142.66			
4728.518	4729.841	G	0.2	0.7	0.5	0.53	21142.36	3A 1	2B 3	P 3
4728.518	4729.841	G	0.2	0.7	0.5	0.53	21142.36	3F 0	2B 1	R 2
4728.668	4729.991						21141.69			
4728.813	4730.136			0.4			21141.04	3D 0	2B 0	P 7
4728.916	4730.239	GK	0.2	0.7	0.9b	0.46	21140.58	3F 0	2B 1	P 6
4728.970	4730.293		0.1				21140.34			
4729.048	4730.371						21139.99			
4729.153	4730.476			0.2			21139.52			
4729.224	4730.547	F					21139.20			
4729.393	4730.716			0.2			21138.45			
4729.476	4730.799						21138.08			
4729.554	4730.877						21137.73			
4729.621	4730.944						21137.43			

λ_{air}	λ_{vac}	Ref.	I$_1$	I$_2$	I$_3$	I$_4$	ν	Upper	Lower	Br.
4729.670	4730.993			0.2			21137.21			
4729.755	4731.078						21136.83			
4729.975	4731.298			0.0			21135.85			
4730.060	4731.383						21135.47			
4730.125	4731.448						21135.18			
4730.223	4731.546	F	0.0	0.3			21134.74	3E 2	2B 4	R 2
4730.291	4731.614	F					21134.44			
4730.557	4731.880		0.0	0.1			21133.25			
4730.673	4731.996		0.0d				21132.73			
4730.767	4732.090	GK	0.3	0.8	0.8	0.77	21132.31	3F 0	2B 1	Q 4
4730.895	4732.218			0.1			21131.74			
4731.094	4732.417			0.1			21130.85			
4731.159	4732.482						21130.56			
4731.329	4732.652			0.2			21129.80	3A 2	2B 5	R 3
4731.452	4732.776						21129.25			
4731.562	4732.885		0.1d				21128.76			
4731.656	4732.979	G	0.5	0.5			21128.34	4E 0	2B 5	P 4
4731.808	4733.132						21127.66			
4731.943	4733.266	G		0.6	0.3		21127.06	3E 2	2B 4	Q 7
4732.106	4733.430						21126.33			
4732.232	4733.555			0.2			21125.77	4E 0	2B 5	Q 4
4732.348	4733.672						21125.25			
4732.534	4733.858			0.0d			21124.42			
4732.648	4733.972		0.0				21123.91			
4732.740	4734.064	GK	0.5	0.8	0.3b	1.13	21123.50	W 1	2B 3	R 1
4732.863	4734.187		0.1	0.2			21122.95			
4732.962	4734.286		0.1	0.0			21122.51			
4733.054	4734.378	G	0.5	0.6			21122.10	3D 1	2B 2	P 2
4733.054	4734.378	G	0.5	0.6			21122.10	3e 2	2c 0	R 7
4733.102	4734.426	F					21121.88			
4733.204	4734.528			0.2			21121.43			
4733.253	4734.577						21121.21			
4733.621	4734.945			0.0			21119.57	3e 2	2c 0	P 2
4733.903	4735.227						21118.31			
4734.004	4735.328						21117.86			
4734.121	4735.445						21117.34			
4734.284	4735.609			0.0			21116.61			
4734.421	4735.745						21116.00			
4734.470	4735.795			0.2	0.2	0.3	21115.78			
4734.526	4735.851						21115.53			
4734.627	4735.952			0.3	0.4		21115.08	3c 2	2a 0	R 3
4734.677	4736.001	G					21114.86	4E 0	2B 5	Q 1
4734.769	4736.093			0.3	0.5	0.4	21114.45			
4734.899	4736.223			0.1			21113.87			
4734.953	4736.277						21113.63			
4735.032	4736.356	F					21113.28			
4735.130	4736.454			0.2			21112.84	3c 2	2a 0	R 4
4735.240	4736.564	G		0.3	0.4	0.4b	21112.35	V 0	2B 0	P 3
4735.240	4736.564	G		0.3	0.4	0.4b	21112.35	3D 0	2B 0	P 8
4735.329	4736.654			0.3	0.4		21111.95			
4735.435	4736.759						21111.48			
4735.632	4736.957			0.1			21110.60	3e 2	2c 0	Q 3
4735.706	4737.031	G		0.2	0.4		21110.27			
4735.830	4737.154						21109.72			
4735.901	4737.226					0.7	21109.40			
4736.204	4737.529						21108.05			
4736.265	4737.590			0.2	0.2		21107.78			
4736.476	4737.801			0.2	0.3		21106.84			
4736.635	4737.960			0.1			21106.13			
4736.723	4738.047						21105.74			

λ_{air}	λ_{vac}	Ref.	I_1	I_2	I_3	I_4	ν	Upper	Lower	Br.
4736.830	4738.155						21105.26			
4737.169	4738.494	F					21103.75			
4737.223	4738.548			0.1	0.2		21103.51	X 2	2B 5	P 3
4737.387	4738.712		0.2	0.2			21102.78			
4737.445	4738.771	G					21102.52			
4737.499	4738.824		0.2	0.2			21102.28	3c 2	2a 0	R 2
4737.847	4739.172			0.1d			21100.73			
4738.000	4739.325						21100.05			
4738.056	4739.381			0.0d			21099.80			
4738.132	4739.458						21099.46			
4738.402	4739.727						21098.26			
4738.501	4739.826		0.2	0.3			21097.82			
4738.721	4740.046	GK	0.4	0.5		0.62	21096.84	4c 3	2a 3	P 4
4738.968	4740.293						21095.74			
4739.123	4740.449		0.1	0.2			21095.05	3c 2	2a 0	R 5
4739.123	4740.449		0.1	0.2			21095.05	4f 2	2c 3	R 4
4739.598	4740.923	F					21092.94			
4740.082	4741.408						21090.78			
4740.344	4741.669	F					21089.62			
4740.420	4741.746		0.2	0.2b			21089.28			
4740.559	4741.885		0.1				21088.66			
4740.788	4742.114		0.2	0.2			21087.64			
4740.905	4742.231		0.1				21087.12			
4740.993	4742.319	GK	0.9	1.0	0.3	1.57	21086.73	W 1	2B 3	R 0
4741.114	4742.440		0.2	0.2			21086.19			
4741.303	4742.629				0.2		21085.35			
4741.409	4742.735						21084.88			
4741.820	4743.147	G	0.3	0.4			21083.05	3c 6	2a 3	Q 1
4741.820	4743.147	G	0.3	0.4			21083.05	3e 2	2c 0	P 2
4741.820	4743.147	G	0.3	0.4			21083.05	3D 3	2B 5	P 5
4741.949	4743.275		0.1	0.1			21082.48			
4742.014	4743.340						21082.19			
4742.120	4743.446	GK	0.7	0.8	0.3	1.27	21081.72	3F 1	2B 3	Q 6
4742.120	4743.446	GK	0.7	0.8	0.3	1.27	21081.72	Z 1	2B 2	R 1
4742.241	4743.567		0.2	0.1			21081.18			
4742.309	4743.635						21080.88			
4742.365	4743.691		0.3	0.5	0.2		21080.63	4E 0	2B 5	Q 3
4742.365	4743.691		0.3	0.5	0.2		21080.63	3E 0	2B 1	R 4
4742.585	4743.912		0.3	0.5	0.2		21079.65			
4742.695	4744.022		0.2	0.2			21079.16			
4742.786	4744.112	GK	0.9	0.9b	0.5	1.77	21078.76	4c 4	2a 4	Q 1
4742.786	4744.112	GK	0.9	0.9b	0.5	1.77	21078.76	Z 1	2B 2	R 2
4742.921	4744.247		0.1	0.3			21078.16			
4743.105	4744.432						21077.34			
4743.215	4744.542		0.2	0.2b			21076.85	3e 2	2c 0	P 3
4743.274	4744.600						21076.59			
4743.314	4744.641		0.2				21076.41			
4743.393	4744.720	GK	0.9	1.0b	0.4	1.64	21076.06	Z 1	2B 2	R 0
4743.521	4744.848		0.2	0.2			21075.49			
4743.605	4744.931	G	0.4	0.5			21075.12	3c 2	2a 0	R 1
4743.728	4745.055		0.1				21074.57			
4743.792	4745.118						21074.29			
4743.963	4745.289						21073.53			
4744.194	4745.521						21072.50			
4744.330	4745.656	G	0.3	0.5	0.3		21071.90	Z 0	2B 0	P 5
4744.474	4745.801						21071.26			
4744.879	4746.206						21069.46			
4745.021	4746.348						21068.83			
4745.156	4746.483		0.2	0.1			21068.23			
4745.318	4746.645	GK	0.6	0.7	0.3	1.05	21067.51	V 0	2B 0	P 2

λ_{air}	λ_{vac}	Ref.	I_1	I_2	I_3	I_4	ν	Upper	Lower	Br.
4745.444	4746.772		0.2	0.1			21066.95	4E 0	2B 5	Q 2
4745.566	4746.893						21066.41			
4745.755	4747.083		0.1	0.1d			21065.57			
4745.978	4747.306	GK	0.3	0.6		0.54	21064.58	4c 4	2a 4	Q 2
4746.280	4747.608						21063.24			
4746.404	4747.732						21062.69			
4746.528	4747.855	F					21062.14			
4746.611	4747.939	G	0.2	0.4	0.3b		21061.77			
4746.756	4748.083						21061.13			
4746.882	4748.210						21060.57			
4746.996	4748.324	F					21060.06	3c 2	2a 0	R 6
4747.209	4748.537	F					21059.12			
4747.276	4748.604		0.2		0.2b		21058.82	4D 0	2B 5	R 5
4747.416	4748.744	F					21058.20			
4747.495	4748.823						21057.85			
4747.675	4749.003			0.1d			21057.05			
4747.767	4749.095	F					21056.64			
4748.036	4749.364		0.2	0.4	0.3b		21055.45			
4748.167	4749.495						21054.87			
4748.262	4749.590		0.1	0.1			21054.45	3F 3	2B 6	Q 2
4748.420	4749.748	F					21053.75	3A 2	2B 5	R 2
4748.505	4749.833		0.2b	0.2			21053.37	3c 6	2a 3	Q 2
4748.505	4749.833		0.2b	0.2			21053.37	3E 2	2B 4	R 1
4748.624	4749.952	F					21052.84			
4748.674	4750.003		0.2	0.3			21052.62			
4748.792	4750.120	F					21052.10			
4748.972	4750.300						21051.30			
4749.085	4750.413						21050.80			
4749.232	4750.560			0.0			21050.15			
4749.314	4750.642	F					21049.79			
4749.481	4750.809	F					21049.05			
4749.602	4750.930			0.1			21048.51			
4749.742	4751.070			0.0			21047.89			
4749.970	4751.298		0.1				21046.88			
4750.152	4751.481						21046.07			
4750.288	4751.616						21045.47			
4750.344	4751.673		0.2	0.3d			21045.22			
4750.400	4751.729	F					21044.97			
4750.529	4751.858			0.0d			21044.40			
4750.597	4751.926						21044.10			
4750.659	4751.988	F					21043.83			
4750.739	4752.068	GK	0.4	0.6		0.75	21043.47	4c 4	2a 4	Q 3
4750.739	4752.068	GK	0.4	0.6		0.75	21043.47	4f 0	2c 1	R 3
4750.739	4752.068	GK	0.4	0.6		0.75	21043.47	3e 2	2c 0	Q 4
4750.802	4752.131	F					21043.19			
4750.940	4752.269						21042.58			
4751.071	4752.400	F					21042.00			
4751.143	4752.472		0.0				21041.68			
4751.201	4752.530	F					21041.43			
4751.376	4752.705			0.1d			21040.65			
4751.503	4752.831		0.1				21040.09			
4751.581	4752.910	GK	0.6	0.8b	0.5	0.61	21039.74	3E 2	2B 4	Q 6
4751.807	4753.136		0.1	0.1b	0.2		21038.74	4E 1	2B 7	Q 4
4752.039	4753.368	F					21037.71			
4752.214	4753.543			0.1			21036.94			
4752.325	4753.654						21036.45			
4752.392	4753.722		0.2	0.1b			21036.15			
4752.569	4753.898	G	0.3	0.4	0.2		21035.37			
4752.636	4753.965	F					21035.07			
4752.731	4754.061			0.1			21034.65			

λ_{air}	λ_{vac}	Ref.	I_1	I_2	I_3	I_4	ν	Upper	Lower	Br.
4752.856	4754.185	G	0.4	0.4	0.3		21034.10	3c 2	2a 0	R 0
4753.115	4754.445		0.2	0.3			21032.95	3D 1	2B 2	P 3
4753.188	4754.517	F					21032.63			
4753.292	4754.621			0.2			21032.17	T 0	2B 5	R 2
4753.356	4754.685	F					21031.89			
4753.611	4754.940	F					21030.76			
4753.712	4755.042			0.0d			21030.31			
4753.792	4755.121	F					21029.96			
4753.977	4755.306		0.1	0.0d			21029.14			
4754.047	4755.376						21028.83			
4754.118	4755.448	F					21028.51			
4754.223	4755.553						21028.05			
4754.268	4755.598		0.2	0.2			21027.85			
4754.418	4755.747			0.1			21027.19			
4754.547	4755.877	F					21026.62			
4754.751	4756.081	F					21025.72			
4754.974	4756.304	F					21024.73			
4755.139	4756.469		0.2b	0.3b	0.2		21024.00	3e 2	2c 0	P 4
4755.404	4756.733	GK	0.6	0.6		0.74	21022.83	Z 1	2B 2	P 1
4755.530	4756.860		0.2	0.3			21022.27			
4755.583	4756.913	F					21022.04			
4755.652	4756.982			0.1			21021.73	3D 3	2B 5	P 6
4755.759	4757.089			0.1			21021.26			
4755.901	4757.231			0.0			21020.63	3f 3	2c 1	P 5
4756.107	4757.437	GK	0.3	0.5	0.3		21019.72			
4756.236	4757.566			0.1			21019.15			
4756.406	4757.736						21018.40			
4756.494	4757.824						21018.01			
4756.614	4757.944	F					21017.48	3E 0	2B 1	Q 9
4756.843	4758.173		0.3d	0.4	0.7		21016.47	X 2	2B 5	P 4
4756.956	4758.286	GK	0.8d	0.7		1.15	21015.97	W 1	2B 3	P 1
4756.956	4758.286	GK	0.8d	0.7		1.15	21015.97	4c 4	2a 4	Q 4
4757.015	4758.345	F					21015.71			
4757.078	4758.408		0.1	0.2			21015.43			
4757.173	4758.503	F					21015.01			
4757.259	4758.590			0.1			21014.63			
4757.320	4758.651			0.1			21014.36			
4757.436	4758.766	G	0.3	0.6	0.6d		21013.85	3F 0	2B 1	P 5
4757.558	4758.889			0.1			21013.31			
4757.664	4758.995						21012.84			
4757.764	4759.095			0.1			21012.40			
4757.863	4759.193	F					21011.96			
4758.049	4759.380			0.1			21011.14	V 0	2B 0	P 5
4758.149	4759.480			0.0			21010.70			
4758.255	4759.586			0.1			21010.23			
4758.446	4759.776	G	0.4	0.6	0.8		21009.39			
4758.588	4759.919	F					21008.76	3c 6	2a 3	Q 3
4758.728	4760.059	F					21008.14			
4758.851	4760.182			0.0d			21007.60			
4758.982	4760.313	F					21007.02			
4759.096	4760.427		0.1d	0.1			21006.52			
4759.152	4760.483	F					21006.27			
4759.325	4760.656		0.0d	0.0			21005.51			
4759.384	4760.715	F					21005.25			
4759.626	4760.957		0.1d				21004.18			
4759.687	4761.018						21003.91			
4759.848	4761.179	G	0.3	0.4			21003.20			
4759.966	4761.297		0.2	0.2			21002.68			
4760.136	4761.467	G	0.3	0.6	0.4	0.88	21001.93	3F 0	2B 1	Q 3
4760.258	4761.590			0.2			21001.39			

λ_{air}	λ_{vac}	Ref.	I_1	I_2	I_3	I_4	ν	Upper	Lower	Br.
4760.399	4761.730						21000.77			
4760.624	4761.955	F					20999.78			
4760.698	4762.029			0.1			20999.45			
4760.816	4762.147						20998.93			
4760.893	4762.224	F					20998.59			
4761.127	4762.458		0.2	0.4	0.3		20997.56			
4761.256	4762.587		0.1				20996.99	4E 1	2B 7	Q 1
4761.319	4762.651			0.1			20996.71			
4761.465	4762.796		0.2d	0.2			20996.07			
4761.526	4762.857						20995.80			
4761.582	4762.914	F					20995.55			
4761.641	4762.973		0.2	0.3b	0.2		20995.29			
4761.740	4763.072	F					20994.86			
4761.906	4763.238	F					20994.12			
4761.979	4763.311				0.2		20993.80			
4762.017	4763.349	F					20993.63			
4762.136	4763.468		0.2	0.3			20993.11			
4762.265	4763.597		0.1	0.1			20992.54	4E 1	2B 7	Q 3
4762.376	4763.708		0.3	0.5	0.2		20992.05			
4762.515	4763.847			0.1			20991.44			
4762.630	4763.962			0.0			20990.93			
4762.837	4764.169	F					20990.02			
4762.966	4764.298	F					20989.45			
4763.078	4764.409						20988.96			
4763.311	4764.643			0.2d			20987.93	3F 3	2B 6	P 4
4763.382	4764.714		0.2				20987.62			
4763.466	4764.798	F					20987.25			
4763.536	4764.868		0.1				20986.94			
4763.659	4764.991		0.2	0.2			20986.40			
4763.756	4765.088		0.2	0.3			20985.97			
4763.847	4765.179	G	1.0b	1.0	0.4b	1.89	20985.57	W 1	2B 3	P 2
4763.847	4765.179	G	1.0b	1.0	0.4b	1.89	20985.57	4D 0	2B 5	R 4
4763.847	4765.179	G	1.0b	1.0	0.4b	1.89	20985.57	3e 2	2c 0	P 3
4763.976	4765.309						20985.00			
4764.153	4765.486						20984.22			
4764.226	4765.558						20983.90			
4764.296	4765.629						20983.59			
4764.392	4765.724	F					20983.17			
4764.558	4765.890						20982.44			
4764.710	4766.042						20981.77			
4764.824	4766.156	F					20981.27			
4764.869	4766.201		0.1	0.3	0.7		20981.07	3A 2	2B 5	R 1
4764.869	4766.201		0.1	0.3	0.7		20981.07	4c 4	2a 4	Q 5
4764.960	4766.292			0.2			20980.67			
4765.037	4766.369			0.1			20980.33			
4765.130	4766.462			0.1			20979.92	T 0	2B 5	R 1
4765.207	4766.540		0.3	0.4d	0.4		20979.58			
4765.391	4766.724						20978.77			
4765.468	4766.801		0.2				20978.43			
4765.559	4766.892	G	0.3	0.5	0.3d		20978.03	3F 0	2B 1	R 1
4765.698	4767.030			0.0			20977.42			
4765.857	4767.190		0.1	0.2			20976.72			
4765.918	4767.251						20976.45			
4766.123	4767.455				0.2d		20975.55			
4766.223	4767.555		0.2	0.3b			20975.11	4E 1	2B 7	Q 2
4766.320	4767.653	F					20974.68			
4766.395	4767.728						20974.35			
4766.520	4767.853			0.1			20973.80			
4766.653	4767.986	F					20973.22			
4766.764	4768.096				0.2		20972.73			

λ_{air}	λ_{vac}	Ref.	I_1	I_2	I_3	I_4	ν	Upper	Lower	Br.
4766.811	4768.144			0.1			20972.52			
4766.950	4768.283						20971.91			
4767.004	4768.337				0.2		20971.67			
4767.059	4768.392			0.2d			20971.43			
4767.143	4768.476			0.0			20971.06			
4767.255	4768.588	G	0.7	0.8	0.3	1.12	20970.57	Z 1	2B 2	P 2
4767.255	4768.588	G	0.7	0.8	0.3	1.12	20970.57	3E 2	2B 4	R 0
4767.402	4768.735		0.1	0.1			20969.92			
4767.571	4768.904						20969.18			
4767.679	4769.012	F					20968.70			
4767.800	4769.133			0.0d			20968.17	4D 0	2B 5	R 0
4767.873	4769.206						20967.85	Z 0	2B 0	P 6
4767.946	4769.279	G	0.4	0.6	0.3	0.54	20967.53	3E 2	2B 4	Q 5
4767.946	4769.279	G	0.4	0.6	0.3	0.54	20967.53	4f 0	2c 1	Q 5
4768.080	4769.413		0.2	0.1			20966.94			
4768.187	4769.520	G	0.7	0.7	0.2	0.60	20966.47	3c 2	2a 0	Q 1
4768.234	4769.567	F					20966.26			
4768.319	4769.652			0.1			20965.89			
4768.444	4769.777						20965.34			
4768.584	4769.917	F					20964.72			
4768.644	4769.977			0.1	0.4		20964.46			
4768.767	4770.100						20963.92	3e 2	2c 0	Q 5
4768.842	4770.175			0.1			20963.59			
4768.908	4770.241	F					20963.30			
4769.058	4770.391	F					20962.64	4D 0	2B 5	P 1
4769.181	4770.514		0.1	0.2			20962.10	3E 0	2B 1	R 3
4769.370	4770.703		0.0				20961.27			
4769.534	4770.867	G	0.4	0.6	0.4	0.59	20960.55			
4769.617	4770.951	F					20960.18			
4769.665	4770.999			0.2d			20959.97			
4769.716	4771.049						20959.75			
4769.818	4771.152		0.2	0.3d			20959.30			
4769.966	4771.300						20958.65			
4770.126	4771.460	F					20957.95			
4770.280	4771.614	F					20957.27	3e 2	2c 0	P 5
4770.367	4771.700						20956.89			
4770.428	4771.762		0.1	0.2d			20956.62			
4770.510	4771.844			0.0			20956.26			
4770.621	4771.955	G	0.7	0.8	0.3	1.28	20955.77	W 1	2B 3	P 3
4770.756	4772.090		0.1	0.2			20955.18			
4770.901	4772.235	G	0.5	0.8	0.4	1.01	20954.54	3E 2	2B 4	P 6
4771.047	4772.381		0.1	0.1			20953.90			
4771.106	4772.440	F					20953.64			
4771.298	4772.632						20952.80			
4771.400	4772.734		0.1	0.2			20952.35			
4771.862	4773.196	F					20950.32			
4771.964	4773.298	F					20949.87			
4772.177	4773.511			0.1d			20948.94			
4772.311	4773.646			0.0			20948.35			
4772.376	4773.710	F					20948.07			
4772.452	4773.787		0.3	0.4	0.3		20947.73			
4772.533	4773.867	F					20947.38			
4772.692	4774.026			0.1			20946.68			
4772.765	4774.099	F					20946.36			
4773.031	4774.366			0.0d			20945.19	3e 3	2c 1	R 4
4773.284	4774.619			0.1			20944.08			
4773.376	4774.711	F					20943.68			
4773.469	4774.804						20943.27	3f 3	2c	P 6
4773.561	4774.896	F					20942.87			
4773.765	4775.100		0.3	0.3			20941.97	3D 1	2B 2	P 4

λ_{air}	$\lambda_{vac.}$	Ref.	I_1	I_2	I_3	I_4	ν	Upper	Lower	Br.
4773.765	4775.100		0.3	0.3			20941.97	4c 4	2a 4	Q 6
4773.906	4775.241		0.2	0.2			20941.35	T 0	2B 5	R 0
4774.093	4775.428		0.2	0.3	0.3b		20940.53			
4774.271	4775.606	G	0.5	0.7		1.11	20939.75	3c 2	2a 0	Q 2
4774.399	4775.734		0.2	0.1			20939.19			
4774.563	4775.898		0.2	0.2d			20938.47	4D 0	2B 5	R 3
4774.695	4776.030		0.3	0.5	0.6		20937.89			
4774.864	4776.199			0.0			20937.15			
4774.959	4776.294	F					20936.73			
4775.056	4776.391			0.2			20936.31			
4775.202	4776.537	F					20935.67			
4775.256	4776.592			0.2			20935.43			
4775.477	4776.812	F					20934.46			
4775.562	4776.897		0.2	0.2			20934.09			
4775.605	4776.941				0.3		20933.90			
4775.651	4776.986		0.3	0.5			20933.70			
4775.770	4777.105			0.0			20933.18			
4775.866	4777.201	G	0.4	0.7	0.3	0.94	20932.76	3E 2	2B 4	P 5
4776.002	4777.338			0.1			20932.16	X 2	2B 5	P 5
4776.119	4777.454			0.1d			20931.65			
4776.239	4777.574	F					20931.12			
4776.351	4777.686	F					20930.63			
4776.429	4777.765		0.2				20930.29	3F 1	2B 3	R 3
4776.619	4777.954						20929.46			
4776.708	4778.043						20929.07			
4777.018	4778.354		0.3	0.5	0.4		20927.71			
4777.203	4778.539						20926.90			
4777.246	4778.582		0.2	0.2			20926.71			
4777.310	4778.646						20926.43			
4777.367	4778.703		0.2				20926.18			
4777.450	4778.785	G	0.8	1.0	0.4	1.63	20925.82	W 1	2B 3	P 4
4777.577	4778.913		0.2	0.2			20925.26			
4777.643	4778.979	F					20924.97			
4777.854	4779.189						20924.05			
4777.981	4779.317	F					20923.49			
4778.182	4779.518	F					20922.61			
4778.310	4779.646		0.3	0.4			20922.05	3F 1	2B 3	Q 5
4778.525	4779.861		0.1	0.2			20921.11			
4778.715	4780.051		0.2	0.4			20920.28			
4778.824	4780.160		0.1	0.1			20919.80			
4778.975	4780.311		0.3	0.4	0.3		20919.14			
4779.203	4780.540		0.2	0.4			20918.14			
4779.363	4780.700		0.3	0.5	0.3b		20917.44	4D 0	2B 5	R 2
4779.363	4780.700		0.3	0.5	0.3b		20917.44	3e 3	2c 1	R 3
4779.363	4780.700		0.3	0.5	0.3b		20917.44	4D 0	2B 5	R 1
4779.558	4780.894		0.2	0.4			20916.59			
4779.704	4781.040		0.1				20915.95			
4779.839	4781.175		0.1				20915.36	4f 2	2c 3	R 3
4779.882	4781.219			0.2b			20915.17			
4779.937	4781.273		0.2d				20914.93	3E 0	2B 1	P 8
4779.937	4781.273		0.2d				20914.93	3A 2	2B 5	P 6
4780.111	4781.447			0.1			20914.17			
4780.192	4781.528	F					20913.81			
4780.298	4781.635		0.2	0.2			20913.35			
4780.472	4781.808						20912.59			
4780.543	4781.879		0.1	0.1d			20912.28			
4780.609	4781.945	F					20911.99			
4780.776	4782.113		0.3	0.6d	0.4		20911.26			
4780.904	4782.241	G	0.6	0.9		1.72	20910.70	3E 2	2B 4	P 4
4781.011	4782.348	G	0.6	1.0	0.6		20910.23	3E 2	2B 4	Q 4

λ_{air}	λ_{vac}	Ref.	I_1	I_2	I_3	I_4	ν	Upper	Lower	Br.
4781.133	4782.469			0.2			20909.70			
4781.215	4782.552		0.2	0.2			20909.34			
4781.348	4782.684			0.1			20908.76			
4781.414	4782.751		0.1				20908.47			
4781.554	4782.891	F					20907.86			
4781.622	4782.959		0.3	0.1			20907.56			
4781.717	4783.054	F					20907.14			
4781.782	4783.119	G	0.3	0.4			20906.86	Z 1	2B 2	P 3
4781.846	4783.183	F					20906.58			
4781.972	4783.309		0.1				20906.03			
4782.047	4783.384	F					20905.70			
4782.217	4783.554		0.2	0.1			20904.96			
4782.340	4783.677			0.0			20904.42			
4782.448	4783.785	G	0.5	0.8	0.5	0.90	20903.95	W 1	2B 3	P 6
4782.448	4783.785	G	0.5	0.8	0.5	0.90	20903.95	3D 0	2B 1	R11
4782.562	4783.899			0.1			20903.45			
4782.625	4783.962	F					20903.18			
4782.704	4784.041			0.1			20902.83			
4782.867	4784.204	G	0.4	0.7	0.3	0.74	20902.12	W 1	2B 3	P 5
4782.990	4784.327		0.1	0.2			20901.58			
4783.162	4784.499	F					20900.83			
4783.292	4784.629			0.1d			20900.26			
4783.434	4784.771	G	0.3	0.6	0.2		20899.64	3c 2	2a 0	Q 3
4783.567	4784.904		0.1	0.0			20899.06			
4783.633	4784.971				0.2		20898.77			
4783.681	4785.019		0.2	0.3			20898.56			
4783.828	4785.165			0.1			20897.92			
4783.884	4785.221	F					20897.67			
4783.933	4785.271						20897.46			
4784.057	4785.394			0.1			20896.92			
4784.133	4785.470	F					20896.59			
4784.212	4785.550			0.0			20896.24			
4784.373	4785.710		0.3	0.3			20895.54	3A 2	2B 5	R 0
4784.421	4785.758						20895.33			
4784.590	4785.928			0.2			20894.59	4c 4	2a 4	Q 7
4784.698	4786.035			0.0			20894.12			
4784.783	4786.120	G	0.8	0.9	0.5	1.02	20893.75	3F 0	2B 1	Q 2
4784.918	4786.255		0.1	0.2			20893.16			
4785.000	4786.338			0.1			20892.80			
4785.099	4786.436	G	0.5	0.7	0.2	0.83	20892.37	3E 2	2B 4	P 3
4785.291	4786.629		0.2	0.4	0.3		20891.53			
4785.472	4786.810		0.1d				20890.74			
4785.664	4787.002						20889.90			
4785.765	4787.103			0.2d			20889.46			
4785.978	4787.316	G	0.6	0.9	0.7	1.04	20888.53	3F 0	2B 1	P 4
4786.104	4787.442			0.1			20887.98	3d 2	2c 0	R 1
4786.201	4787.539	G	0.4	0.8	1.0b		20887.56	3E 0	2B 1	Q 8
4786.349	4787.688		0.2	0.2			20886.91			
4786.524	4787.862		0.3	0.4			20886.15			
4786.629	4787.967	F					20885.69			
4786.684	4788.022		0.2				20885.45			
4786.801	4788.139		0.2				20884.94			
4786.952	4788.291						20884.28			
4787.039	4788.377	F					20883.90			
4787.242	4788.580	F					20883.02	3e 3	2c 1	R 2
4787.357	4788.695	F					20882.51			
4787.496	4788.834	F					20881.91			
4787.730	4789.068	F					20880.89			
4788.028	4789.366						20879.59			
4788.319	4789.657						20878.32			

λ_{air}	λ_{vac}	Ref.	I_1	I_2	I_3	I_4	ν	Upper	Lower	Br.
4788.394	4789.733			0.2			20877.99			
4788.468	4789.806			0.1			20877.67	3e 3	2c 0	P 6
4788.596	4789.935		0.2	0.3			20877.11			
4788.759	4790.098	F					20876.40			
4788.933	4790.272			0.0d			20875.64	3e 2	2c 0	Q 6
4788.992	4790.331	F					20875.39			
4789.083	4790.421		0.1	0.3			20874.99			
4789.229	4790.568			0.1d			20874.35	4D 0	2B 5	P 2
4789.415	4790.754	G	0.8	0.9d	0.4	1.60	20873.54	3E 2	2B 4	P 2
4789.542	4790.881	F					20872.99			
4789.620	4790.958			0.0			20872.65	3e 2	2c 0	P 4
4789.692	4791.031	F					20872.33			
4790.053	4791.392			0.0d			20870.76			
4790.242	4791.581	F					20869.94			
4790.423	4791.762	G	0.4	0.6			20869.15	3D 2	2B 4	P 1
4790.547	4791.886			0.1			20868.61			
4790.611	4791.950		0.1d				20868.33			
4790.666	4792.005			0.1d			20868.09			
4790.866	4792.205	G	0.6	0.9	0.4	1.23	20867.22	3E 2	2B 4	Q 3
4790.999	4792.338		0.1	0.2			20866.64			
4791.078	4792.417	F					20866.30			
4791.224	4792.563	G	0.4	0.5		0.91	20865.66			
4791.425	4792.764	F					20864.78			
4791.509	4792.848			0.1			20864.42			
4791.613	4792.952	F					20863.97			
4791.676	4793.016			0.0			20863.69	4E 2	2B 9	Q 2
4791.819	4793.158			0.2			20863.07			
4791.884	4793.224	F					20862.79			
4792.016	4793.356		0.2d	0.2d			20862.21			
4792.214	4793.554	G					20861.35			
4792.287	4793.627			0.0			20861.03			
4792.348	4793.688	F					20860.77			
4792.414	4793.753		0.1d	0.3			20860.48			
4792.549	4793.889	G	0.4	0.7	0.9		20859.89			
4792.793	4794.133	F					20858.83			
4793.007	4794.346		0.1d	0.3d			20857.90	3D 1	2B 2	P 5
4793.100	4794.440	F					20857.49			
4793.225	4794.565			0.1			20856.95			
4793.285	4794.625						20856.69			
4793.363	4794.703			0.2			20856.35			
4793.450	4794.790			0.0			20855.97			
4793.646	4794.986						20855.12	3A 2	2B 5	P 5
4793.714	4795.055		0.2d	0.3			20854.82			
4793.827	4795.167		0.2d				20854.33			
4793.914	4795.255	G	0.9	1.0	0.4	1.69	20853.95	3D 2	2B 4	R 2
4794.041	4795.381		0.2	0.1			20853.40			
4794.184	4795.524	G	0.6	0.8	0.3	1.09	20852.78	3D 2	2B 4	R 1
4794.184	4795.524	G	0.6	0.8	0.3	1.09	20852.78	3d 2	2c 0	Q 1
4794.296	4795.636			0.2			20852.29			
4794.344	4795.685						20852.08			
4794.425	4795.765			0.0d			20851.73			
4794.531	4795.871			0.0d			20851.27			
4794.600	4795.940		0.2	0.3			20850.97			
4794.686	4796.026	F					20850.59			
4794.735	4796.076			0.1			20850.38			
4794.871	4796.211		0.2	0.3d			20849.79			
4794.924	4796.264	F					20849.56			
4795.156	4796.497	G	0.5	0.8	0.3b	1.06	20848.55	3D 2	2B 4	R 3
4795.156	4796.497	G	0.5	0.8	0.3b	1.06	20848.55	4f 0	2c 1	R 2
4795.329	4796.669		0.2b	0.2b			20847.80	T 0	2B 5	P 2

λ_{air}	λ_{vac}	Ref.	I_1	I_2	I_3	I_4	ν	Upper	Lower	Br.
4795.502	4796.842	F					20847.05			
4795.582	4796.922		0.2	0.3			20846.70	3c 2	2a 0	Q 4
4795.731	4797.072	F					20846.05			
4795.876	4797.217						20845.42			
4795.965	4797.306	F					20845.03			
4796.134	4797.475			0.1			20844.30			
4796.293	4797.633		0.1	0.2			20843.61	3d 2	2c 0	P 1
4796.431	4797.772	F					20843.01			
4796.541	4797.882			0.1			20842.53	4f 2	2c 3	Q 5
4796.606	4797.946		0.1				20842.25			
4796.672	4798.013	F					20841.96			
4796.781	4798.121		0.2	0.2			20841.49			
4796.900	4798.241		0.2	0.2			20840.97			
4796.988	4798.329	G	0.8	0.9d	0.3	1.65	20840.59	3D 2	2B 4	R 0
4797.117	4798.458	G	0.6	0.9d	0.5		20840.03	3D 2	2B 4	R 4
4797.117	4798.458	G	0.6	0.9d	0.5		20840.03	3D 2	2B 4	R 6
4797.273	4798.614		0.1				20839.35	3d 2	2c 0	R 2
4797.343	4798.684	F					20839.05			
4797.395	4798.736		0.2	0.3			20838.82	3E 0	2B 1	R 2
4797.483	4798.824	F					20838.44			
4797.556	4798.897		0.2d	0.3d			20838.12			
4797.662	4799.003		0.2	0.2			20837.66	3A 2	2B 5	P 4
4797.757	4799.098	G	0.9b	1.5b	0.6	1.86	20837.25	3E 2	2B 4	Q 2
4797.886	4799.227		0.2	0.3			20836.69	V 1	2B 1	R 2
4798.063	4799.404		0.2	0.4	0.2		20835.92	Z 1	2B 2	P 4
4798.282	4799.623		0.2		0.2		20834.97			
4798.362	4799.704		0.3	0.3			20834.62	3c 2	2a 0	P 2
4798.531	4799.872	F					20833.89			
4798.625	4799.966	F	0.1				20833.48			
4798.737	4800.078	F					20832.99			
4798.786	4800.128		0.3	0.4	0.3d		20832.78			
4798.913	4800.254	F					20832.23			
4799.136	4800.478	G	0.3	0.6	0.3d	0.49	20831.26	3D 2	2B 4	R 5
4799.281	4800.623			0.2			20830.63			
4799.383	4800.724		0.1	0.1			20830.19			
4799.498	4800.840	F					20829.69			
4799.676	4801.018	F					20828.92			
4799.791	4801.132		0.1	0.1d			20828.42			
4800.043	4801.385	F					20827.32			
4800.129	4801.471		0.2d	0.3			20826.95			
4800.235	4801.577			0.3			20826.49	3e 3	2c 1	R 1
4800.378	4801.720	F					20825.87	3A 2	2B 5	P 3
4800.673	4802.015		0.1	0.3			20824.59	T 0	2B 5	P 3
4800.962	4802.304			0.1			20823.34			
4801.093	4802.435		0.2	0.2d			20822.77	3A 2	2B 5	P 1
4801.215	4802.557			0.1			20822.24			
4801.630	4802.972		0.2	0.2	0.2b		20820.44			
4801.799	4803.141			0.1			20819.71			
4801.930	4803.272			0.0			20819.14			
4802.011	4803.353	G	0.6	0.8	0.3	1.09	20818.79	3E 2	2B 4	Q 1
4802.154	4803.496		0.1	0.1			20818.17			
4802.417	4803.759	G	0.4	0.6		1.13	20817.03	4f 0	2c 1	Q 4
4802.548	4803.891			0.1			20816.46	T 0	2B 5	P 4
4802.698	4804.041			0.0d			20815.81			
4802.777	4804.119		0.2	0.2			20815.47			
4802.920	4804.262		0.3	0.3			20814.85			
4803.100	4804.442				0.2		20814.07			
4803.261	4804.604			0.2			20813.37			
4803.481	4804.823						20812.42			
4803.619	4804.962						20811.82			

λ_{air}	λ_{vac}	Ref.	I_1	I_2	I_3	I_4	ν	Upper	Lower	Br.
4803.751	4805.093	F					20811.25			
4803.841	4805.183			0.2d			20810.86			
4803.898	4805.241		0.1	0.1			20810.61			
4804.062	4805.405	G	0.5	0.7		0.72	20809.90	4c 5	2a 5	Q 1
4804.201	4805.544		0.0	0.1			20809.30			
4804.249	4805.592	F					20809.09			
4804.365	4805.708						20808.59			
4804.418	4805.761	F					20808.36			
4804.596	4805.939	F					20807.59			
4804.764	4806.107	F					20806.86			
4804.898	4806.241			0.1			20806.28			
4804.988	4806.331						20805.89			
4805.041	4806.384		0.2	0.4			20805.66			
4805.309	4806.652		0.2	0.2			20804.50			
4805.545	4806.888	F					20803.48	3e 2	2c 0	P 7
4805.843	4807.186						20802.19			
4805.894	4807.237		0.2	0.3			20801.97	3A 2	2B 5	P 2
4806.021	4807.364				1.0		20801.42			
4806.189	4807.533			0.4			20800.69			
4806.453	4807.796			0.1d			20799.55			
4806.534	4807.877	F					20799.20			
4806.689	4808.032			0.0d			20798.53			
4806.900	4808.243	F					20797.62			
4807.054	4808.397						20796.95			
4807.116	4808.460			0.1			20796.68	3e 2	2c 0	Q 7
4807.227	4808.571		0.2				20796.20	4c 5	2a 5	Q 2
4807.306	4808.649	G	0.6	0.8d	1.5b	0.77	20795.86	3D 0	2B 1	R10
4807.472	4808.816		0.5	0.7			20795.14	3F 0	2B 1	P 3
4807.528	4808.871	G				0.64	20794.90			
4807.653	4808.996		0.2	0.3			20794.36	3D 1	2B 2	P 6
4807.784	4809.128			0.1			20793.79			
4807.983	4809.327						20792.93			
4808.166	4809.510		0.2	0.4	0.2		20792.14			
4808.270	4809.614						20791.69			
4808.469	4809.813			0.1			20790.83			
4808.594	4809.938			0.1			20790.29			
4808.679	4810.023			0.0			20789.92			
4808.779	4810.123	G	0.3	0.6	0.6		20789.49			
4808.918	4810.262			0.2			20788.89			
4809.070	4810.414	F					20788.23			
4809.211	4810.555						20787.62			
4809.348	4810.692	F					20787.03			
4809.523	4810.867	F					20786.27			
4809.711	4811.055		0.1				20785.46	4f 0	2c 1	R 3
4809.955	4811.299	F					20784.41			
4810.077	4811.421						20783.88			
4810.129	4811.473	F					20783.65			
4810.220	4811.565			0.1d			20783.26			
4810.371	4811.715	F					20782.61			
4810.433	4811.778	G	0.3	0.6	0.3		20782.34	3F 1	2B 3	Q 4
4810.433	4811.778	G	0.3	0.6	0.3		20782.34	3E 1	2B 3	Q10
4810.547	4811.891			0.1			20781.85			
4810.672	4812.016			0.2			20781.31	3c 2	2a 0	Q 5
4810.780	4812.125	G	0.3	0.4	0.4		20780.84			
4810.933	4812.278		0.1				20780.18			
4811.007	4812.352			0.1d			20779.86			
4811.179	4812.523	G	0.3	0.4	0.4		20779.12	3D 1	2B 2	P10
4811.179	4812.523	G	0.3	0.4	0.4		20779.12	4g 1	2a 2	P 1
4811.614	4812.959	G	0.4	0.7	0.9		20777.24	3E 0	2B 1	Q 7
4811.748	4813.093			0.0d			20776.66			

λ_{air}	λ_{vac}	Ref.	I_1	I_2	I_3	I_4	ν	Upper	Lower	Br.
4811.825	4813.170	F					20776.33			
4811.915	4813.260	G	0.3	0.5	0.2		20775.94	4c 5	2a 5	Q 3
4812.068	4813.413			0.1			20775.28	3E 0	2B 1	P 7
4812.068	4813.413			0.1			20775.28	3d 2	2c 0	R 3
4812.244	4813.589			0.2d			20774.52			
4812.492	4813.837						20773.45			
4812.568	4813.913			0.1d			20773.12			
4812.612	4813.957						20772.93			
4812.744	4814.089		0.3				20772.36	3F 1	2B 3	R 2
4812.784	4814.129	G		0.5b	0.3b		20772.19			
4812.809	4814.154		0.3				20772.08			
4812.918	4814.263						20771.61			
4813.028	4814.373	F					20771.14			
4813.083	4814.428			0.0d			20770.90			
4813.171	4814.516				0.2		20770.52			
4813.293	4814.638	F					20769.99			
4813.407	4814.752		0.1	0.2d			20769.50	3e 3	2c 1	R 2
4813.518	4814.864		0.2				20769.02			
4813.602	4814.947	G	0.8	1.0	0.3	1.49	20768.66	Z 2	2B 4	R 0
4813.736	4815.082		0.1	0.2			20768.08	W 2	2B 5	R 4
4813.835	4815.180	F					20767.65			
4813.878	4815.223			0.0			20767.47			
4813.970	4815.315	F					20767.07			
4814.126	4815.471						20766.40	3d 2	2c 0	Q 2
4814.270	4815.615	F					20765.78			
4814.501	4815.847						20764.78			
4814.779	4816.125						20763.58	3e 3	2c 1	R 3
4815.079	4816.424		0.2	0.2d			20762.29	4D 0	2B 5	P 3
4815.079	4816.424		0.2	0.2d			20762.29	4F 0	2B 6	R 3
4815.440	4816.786			0.1			20760.73	3e 3	2c 1	R 1
4815.587	4816.933	F					20760.10			
4815.659	4817.004			0.0			20759.79			
4815.715	4817.061	F					20759.55			
4815.890	4817.236	F					20758.79			
4816.007	4817.352		0.2	0.1			20758.29			
4816.065	4817.411	F					20758.04			
4816.231	4817.577	F					20757.32			
4816.524	4817.870	F					20756.06			
4816.650	4817.996	F					20755.52			
4816.836	4818.182	F					20754.72			
4817.058	4818.404	F					20753.76			
4817.211	4818.557			0.1			20753.10	4g 1	2a 2	R 0
4817.292	4818.638	F					20752.75			
4817.510	4818.857	G	0.8	0.8		1.25	20751.81	Z 2	2B 4	R 1
4817.510	4818.857	G	0.8	0.8		1.25	20751.81	3D 1	2B 2	P 7
4817.578	4818.924				0.3		20751.52			
4817.645	4818.991		0.1	0.2			20751.23			
4817.765	4819.111	F					20750.71			
4817.972	4819.319		0.2	0.4	0.4		20749.82			
4818.070	4819.416	F					20749.40			
4818.149	4819.495			0.2			20749.06	3e 2	2c 0	P 5
4818.255	4819.602	F					20748.60			
4818.372	4819.718			0.0			20748.10			
4818.485	4819.832	F					20747.61			
4818.660	4820.006		0.1	0.3			20746.86			
4818.733	4820.080	F					20746.54			
4818.836	4820.183		0.2	0.3			20746.10			
4819.062	4820.408		0.1	0.3			20745.13			
4819.238	4820.585	G	0.4	0.5	0.2d	0.89	20744.37	3c 2	2a 0	P 3
4819.238	4820.585	G	0.4	0.5	0.2d	0.89	20744.37	4c 5	2a 5	Q 4

λ_{air}	λ_{vac}	Ref.	I_1	I_2	I_3	I_4	ν	Upper	Lower	Br.
4819.467	4820.813	G	0.6	0.7	0.2 b		20743.39	3D 2	2B 4	P 2
4819.598	4820.945		0.1	0.2			20742.82			
4819.784	4821.131			0.1			20742.02			
4819.972	4821.319			0.1			20741.21			
4820.309	4821.656	F					20739.76			
4820.492	4821.839	F					20738.97			
4820.547	4821.894		0.2	0.1			20738.74			
4820.737	4822.084		0.2	0.2d	0.2b		20737.92	4f 2	2c 3	R 2
4820.888	4822.236				0.2d		20737.27			
4820.949	4822.296	F					20737.01			
4821.306	4822.653	F					20735.47			
4821.586	4822.933	F					20734.27	3D 1	2B 2	P 9
4821.789	4823.136	F					20733.40			
4822.040	4823.388	F					20732.32	3d 2	2c 0	P 2
4822.364	4823.712	F					20730.92			
4822.458	4823.806		0.2	0.2	0.3		20730.52			
4822.557	4823.905	F					20730.09			
4822.735	4824.083		0.2	0.3b			20729.33	3D 1	2B 2	P 8
4822.858	4824.206		0.2	0.0			20728.80			
4822.951	4824.299	G	0.9	1.0b	0.4	1.72	20728.40	Z 2	2B 4	R 2
4823.084	4824.432		0.1	0.2			20727.83			
4823.300	4824.648	F					20726.90			
4823.440	4824.788	F					20726.30			
4823.658	4825.006	F					20725.36			
4823.861	4825.209			0.2			20724.49	4f 1	2c 2	R 3
4823.908	4825.256	F					20724.29			
4824.075	4825.423	G	0.3	0.4	0.4		20723.57	3E 0	2B 1	R 1
4824.229	4825.577			0.1			20722.91			
4824.343	4825.691			0.0d			20722.42			
4824.571	4825.919	G	0.7	0.9	0.9	0.81	20721.44			
4824.704	4826.052		0.1	0.1			20720.87	V 1	2B 1	R 1
4824.915	4826.263	F					20719.96			
4825.128	4826.476	F					20719.05			
4825.260	4826.609			0.2			20718.48			
4825.363	4826.711			0.1d			20718.04			
4825.547	4826.895			0.0			20717.25	4D 0	2B 5	P 8
4825.645	4826.993	F					20716.83			
4825.726	4827.075			0.2			20716.48			
4825.844	4827.193	F					20715.97			
4825.910	4827.259		0.1d	0.2d			20715.69			
4826.175	4827.524	F					20714.55			
4826.297	4827.646	F					20714.03			
4826.917	4828.266		0.1				20711.37	4f 2	2c 3	Q 4
4827.024	4828.373	G	0.3	0.4			20710.91	W 2	2B 5	R 3
4827.107	4828.456	F					20710.55			
4827.523	4828.872	G	0.2	0.4			20708.77	w 0	2c 2	P 1
4827.620	4828.969	F					20708.35			
4827.760	4829.109	F					20707.75			
4827.870	4829.219	G	0.2	0.3			20707.28			
4828.017	4829.366			0.2			20706.65			
4828.080	4829.429		0.2d				20706.38			
4828.323	4829.672		0.2	0.1			20705.34			
4828.665	4830.014	F					20703.87	3c 2	2a 0	Q 6
4828.817	4830.166	F					20703.22			
4828.906	4830.255			0.1			20702.84	3e 3	2c 1	Q 1
4828.965	4830.314	F					20702.59			
4829.148	4830.498	F					20701.80	4D 0	2B 5	P 4
4829.337	4830.687		0.2	0.3			20700.99	4e 0	2c 1	R 6
4829.542	4830.892	F					20700.11			
4829.769	4831.119			0.0d			20699.14			

λ_{air}	λ_{vac}	Ref.	I_1	I_2	I_3	I_4	ν	Upper	Lower	Br.
4829.858	4831.208	F					20698.76			
4829.942	4831.291		0.2	0.2d			20698.40	3d 2	2c 0	R 4
4830.001	4831.351	F					20698.15			
4830.167	4831.517	F					20697.43			
4830.278	4831.627		0.2	0.3			20696.96			
4830.330	4831.680	F					20696.74			
4830.420	4831.770		0.1	0.0d			20696.35			
4830.532	4831.882		0.1				20695.87			
4830.616	4831.966	G	0.5	0.8	0.3b	0.96	20695.51	W 2	2B 5	R 2
4830.770	4832.120	G	0.4	0.7		0.90	20694.85	Z 2	2B 4	R 3
4830.968	4832.318	F					20694.00	4f 3	2c 4	R 3
4831.147	4832.497	F					20693.24			
4831.328	4832.678			0.0d			20692.46			
4831.562	4832.912	G	0.5	0.8	0.8	0.93	20691.46	3E 0	2B 1	P 6
4831.562	4832.912	G	0.5	0.8	0.8	0.93	20691.46	3D 2	2B 4	P 3
4831.688	4833.038			0.1			20690.92			
4831.755	4833.105	F					20690.63			
4831.954	4833.304	F					20689.78			
4832.121	4833.471	F					20689.06			
4832.330	4833.680	F					20688.17			
4832.592	4833.942		0.2	0.2d			20687.05	3E 3	2B 6	Q 8
4832.786	4834.136	G	0.8	1.0d	1.5b	1.02	20686.22	3E 0	2B 1	Q 6
4832.919	4834.269			0.1			20685.65			
4832.977	4834.328		0.1d				20685.40			
4833.187	4834.538		0.1d	0.2d			20684.50			
4833.374	4834.725	G	0.4	0.6	0.8		20683.70	3D 0	2B 1	R 9
4833.510	4834.860		0.1	0.0d			20683.12			
4833.746	4835.097	G	0.4	0.5			20682.11	W 2	2B 5	R 1
4833.940	4835.291	F					20681.28			
4834.059	4835.410			0.1			20680.77			
4834.168	4835.519	F					20680.30			
4834.508	4835.859			0.0d			20678.85	3e 3	2c 1	Q 1
4834.560	4835.911	F					20678.63			
4834.789	4836.140	F					20677.65	4f 0	2c 1	R 1
4834.971	4836.322			0.0			20676.87	4f 0	2c 1	Q 3
4835.118	4836.469	F					20676.24			
4835.193	4836.544		0.1	0.3			20675.92			
4835.271	4836.622	F					20675.59			
4835.371	4836.722	G					20675.16			
4835.467	4836.818			0.0			20674.75			
4835.704	4837.055	F					20673.74			
4835.836	4837.187			0.2			20673.17			
4835.974	4837.326			0.1			20672.58			
4836.024	4837.375	F					20672.37			
4836.335	4837.686	F					20671.04	4D 0	2B 5	P 7
4836.335	4837.686	F					20671.04	Z 2	2B 4	P 2
4836.491	4837.842	F					20670.37			
4836.679	4838.030			0.0			20669.57			
4836.793	4838.145		0.0	0.0			20669.08			
4836.899	4838.250	G	0.3	0.6	0.6		20668.63	4b 0	2a 0	R 4
4837.013	4838.365		0.2	0.2	0.4		20668.14			
4837.106	4838.458	F					20667.74			
4837.315	4838.667			0.1			20666.85			
4837.540	4838.892			0.0			20665.89	w 0	2c 2	Q 1
4837.811	4839.163	F					20664.73			
4838.006	4839.357		0.1	0.1			20663.90			
4838.104	4839.456	G	0.6	0.6			20663.48	4D 0	2B 5	P 5
4838.104	4839.456	G	0.6	0.6			20663.48	W 2	2B 5	R 0
4838.245	4839.596	G	0.8	1.0b	0.9b	1.54	20662.88	3F 1	2B 3	Q 3
4838.245	4839.596	G	0.8	1.0b	0.9b	1.54	20662.88	4b 0	2a 0	R 3

λ_{air}	λ_{vac}	Ref.	I_1	I_2	I_3	I_4	ν	Upper	Lower	Br.
4838.547	4839.899		0.2	0.3			20661.59			
4838.735	4840.087	F					20660.79			
4838.814	4840.166			0.1			20660.45	4g 1	2a 2	R 1
4839.001	4840.353		0.2	0.4			20659.65	3d 2	2c 0	Q 3
4839.113	4840.465	F					20659.17			
4839.184	4840.536		0.2	0.3			20658.87			
4839.289	4840.641		0.1	0.3			20658.42			
4839.448	4840.801			0.1d			20657.74			
4839.546	4840.898	F					20657.32			
4839.792	4841.144	F					20656.27			
4839.924	4841.276		0.2	0.4			20655.71			
4840.037	4841.389	F					20655.23			
4840.191	4841.544			0.1			20654.57			
4840.498	4841.851			0.0d			20653.26	4D 0	2B 5	P 6
4840.665	4842.017		0.2	0.2d			20652.55			
4840.824	4842.176			0.0			20651.87			
4840.880	4842.232	F					20651.63			
4841.007	4842.359			0.0			20651.09	4f 2	2c 3	R 3
4841.128	4842.481	F					20650.57			
4841.161	4842.514		0.2	0.3			20650.43			
4841.309	4842.662			0.1			20649.80			
4841.373	4842.725		0.1				20649.53			
4841.492	4842.845	G	0.4	0.7	0.3	0.70	20649.02	X 0	2B 2	R 3
4841.614	4842.967			0.0			20648.50			
4841.708	4843.061	G	0.5	0.7	0.3	0.79	20648.10	Z 2	2B 4	R 4
4841.877	4843.229				0.3		20647.38			
4841.902	4843.255	F					20647.27			
4842.068	4843.421	F					20646.56			
4842.179	4843.532			0.2d			20646.09	4g 0	2a 1	R 0
4842.379	4843.732	G	0.8	1.0	0.8	1.03	20645.24	4b 0	2a 0	R 2
4842.379	4843.732	G	0.8	1.0	0.8	1.03	20645.24	3e 3	2c 1	Q 2
4842.505	4843.858			0.2			20644.70			
4842.688	4844.041		0.1				20643.92			
4842.788	4844.141	F					20643.49			
4842.916	4844.269		0.2	0.3			20642.95	3c 2	2a 0	P 4
4843.042	4844.396			0.1			20642.41			
4843.237	4844.590	G	0.8	0.9	0.4	1.55	20641.58	X 0	2B 2	R 2
4843.439	4844.792	G	0.5	0.6		0.86	20640.72			
4843.584	4844.938		0.2	0.2			20640.10			
4843.748	4845.102	G	0.6	0.8	0.7	0.94	20639.40	3E 0	2B 1	P 5
4843.892	4845.245		0.1	0.2			20638.79			
4844.100	4845.453	F					20637.90			
4844.176	4845.529		0.2	0.3			20637.58	3F 1	2B 3	R 1
4844.270	4845.623	F					20637.18			
4844.457	4845.811		0.2	0.4	0.3		20636.38	3E 1	2B 3	Q 9
4844.535	4845.888	F					20636.05			
4844.658	4846.011	F					20635.53	4f 0	2c 1	P 6
4844.793	4846.147		0.2	0.1b			20634.95			
4844.977	4846.330			0.0d			20634.17			
4845.067	4846.421	F					20633.78			
4845.303	4846.657		0.1d	0.1d			20632.78	3F 1	2B 3	P 5
4845.481	4846.835	G	0.6	0.7	0.3b	1.00	20632.02	X 0	2B 2	R 1
4845.579	4846.933	F					20631.60			
4845.641	4846.995		0.1	0.2			20631.34			
4845.724	4847.078	F					20630.99			
4845.827	4847.180		0.1				20630.55			
4845.982	4847.336	G	0.4	0.5			20629.89	3D 2	2B 4	P 4
4846.111	4847.465			0.1			20629.34			
4846.164	4847.518	F					20629.11			
4846.247	4847.601		0.2	0.2			20628.76			

λ_{air}	λ_{vac}	Ref.	I_1	I_2	I_3	I_4	ν	Upper	Lower	Br.
4846.347	4847.701	F					20628.33			
4846.430	4847.784			0.2d			20627.98			
4846.592	4847.946	F					20627.29			
4846.682	4848.036			0.1			20626.91			
4846.755	4848.109					0.4b	20626.60			
4846.872	4848.226		0.2	0.3			20626.10			
4847.007	4848.361	F					20625.53			
4847.253	4848.607		0.1				20624.48			
4847.427	4848.781			0.2			20623.74			
4847.625	4848.979	F					20622.90			
4847.737	4849.091		0.1				20622.42			
4847.810	4849.164				0.5		20622.11			
4847.899	4849.254			0.2			20621.73			
4848.007	4849.362		0.2				20621.27			
4848.104	4849.458	G	0.4	0.7	0.5	0.77	20620.86	X 0	2B 2	R 4
4848.104	4849.458	G	0.4	0.7	0.5	0.77	20620.86	U 1	2B 3	R 4
4848.299	4849.653		0.2				20620.03			
4848.475	4849.830			0.0d			20619.28			
4848.743	4850.098				0.2		20618.14			
4848.816	4850.171		0.1	0.2b			20617.83			
4848.997	4850.352		0.2	0.4	0.6		20617.06	4f 1	2c 2	Q 3
4849.115	4850.470		0.2	0.3			20616.56	4E 1	2C 0	R 1
4849.211	4850.566		0.3				20616.15			
4849.305	4850.660	G	1.0b	1.5b	0.8	1.87	20615.75	4b 0	2a 0	R 1
4849.305	4850.660	G	1.0b	1.5b	0.8	1.87	20615.75	3e 2	2c 0	P 6
4849.423	4850.778		0.0				20615.25	X 0	2B 2	R 0
4849.529	4850.884	G	0.5	0.9	0.8	0.90	20614.80	3E 0	2B 1	Q 5
4849.529	4850.884	G	0.5	0.9	0.8	0.90	20614.80	3c 2	2a 0	Q 7
4849.670	4851.025		0.2				20614.20			
4849.849	4851.204	G	0.7	0.9	0.5	0.99	20613.44	3E 0	2B 1	R 0
4849.985	4851.340		0.2	0.1			20612.86			
4850.035	4851.390			0.0			20612.65			
4850.167	4851.522		0.3	0.3			20612.09			
4850.296	4851.651			0.1d			20611.54			
4850.524	4851.879			0.1			20610.57			
4850.759	4852.114	F					20609.57			
4851.024	4852.379	F					20608.45			
4851.150	4852.506			0.0			20607.91			
4851.332	4852.687	G		0.4			20607.14	W 2	2B 5	P 1
4851.562	4852.917	F					20606.16			
4851.621	4852.977				0.8b		20605.91			
4851.715	4853.071			0.3			20605.51	t 1	2a 0	R 1
4851.715	4853.071			0.3			20605.51	4e 0	2c 1	R 4
4851.833	4853.189			0.1			20605.01			
4851.991	4853.346		0.4	0.5			20604.34	V 1	2B 1	R 0
4852.311	4853.667		0.1	0.3			20602.98			
4852.460	4853.816	F					20602.35			
4852.599	4853.954		0.2d	0.2			20601.76			
4852.763	4854.119	F					20601.06			
4852.811	4854.166		0.2	0.3d			20600.86			
4852.983	4854.338	G					20600.13			
4853.334	4854.689		0.2d	0.2b			20598.64	4E 1	2C 0	R 2
4853.418	4854.774	F					20598.28			
4853.600	4854.956		0.2				20597.51	3e 3	2c 1	Q 3
4853.865	4855.221	F					20596.38			
4853.967	4855.323		0.2				20595.95			
4854.125	4855.481		0.3	0.3			20595.28			
4854.326	4855.682	F					20594.43	3E 3	2B 6	Q 7
4854.420	4855.776			0.1			20594.03	3E 3	2B 6	Q 7
4854.578	4855.934		0.0				20593.36			

λ_{air}	λ_{vac}	Ref.	I_1	I_2	I_3	I_4	ν	Upper	Lower	Br.
4854.873	4856.229	G					20592.11			
4855.240	4856.596	F					20590.55	Z 2	2B 4	P 3
4855.363	4856.719		0.1				20590.03			
4855.516	4856.873		0.2				20589.38			
4855.693	4857.050						20588.63			
4855.745	4857.102			0.1			20588.41			
4855.861	4857.217		0.2	0.2			20587.92	3e 3	2c 1	P 2
4856.005	4857.361		0.1	0.2			20587.31			
4856.092	4857.448		0.2	0.3			20586.94	3d 2	2c 0	P 3
4856.219	4857.576		0.2	0.1			20586.40	4e 1	2c 2	R 4
4856.340	4857.696		0.3	0.4d			20585.89			
4856.450	4857.807		0.2	0.2			20585.42			
4856.554	4857.911	G	0.9b	1.5b		1.78	20584.98	3E 0	2B 1	P 4
4856.689	4858.045		0.2	0.3			20584.41			
4856.894	4858.251		0.1	0.1d			20583.54	4f 2	2c 3	Q 3
4857.010	4858.366		0.0				20583.05			
4857.163	4858.520						20582.40			
4857.404	4858.761		0.0				20581.38			
4857.510	4858.867	F					20580.93	3c 3	2a 1	R 3
4857.510	4858.867	F					20580.93	4f 2	2c 3	R 1
4857.620	4858.977	F					20580.46			
4857.758	4859.115		0.2	0.3			20579.88	4E 1	2C 0	Q 1
4857.944	4859.301		0.1				20579.09			
4858.041	4859.398		0.2				20578.68			
4858.107	4859.464		0.3	0.2d			20578.40			
4858.216	4859.573		0.2				20577.94	4E 1	2C 0	R 3
4858.372	4859.729	G	0.4	0.3			20577.28	V 1	2B 1	P 4
4858.532	4859.889		0.2	0.1			20576.60			
4858.657	4860.015		0.2				20576.07			
4858.756	4860.114	G	0.7	0.7			20575.65	4b 0	2a 0	R 0
4858.867	4860.225			0.2			20575.18			
4858.938	4860.296		0.5b	0.4b			20574.88			
4859.125	4860.482	F					20574.09			
4859.295	4860.652		0.2d	0.2			20573.37			
4859.463	4860.820		0.1d	0.1			20572.66	4g 1	2a 2	P 2
4859.649	4861.007		0.2				20571.87			
4859.772	4861.130		0.2	0.1d			20571.35			
4859.926	4861.283		0.2	0.2d			20570.70			
4860.112	4861.470	G	0.6	0.7		1.34	20569.91	W 2	2B 5	P 2
4860.245	4861.602		0.2	0.2			20569.35			
4860.349	4861.706		0.3	0.4			20568.91	3c 3	2a 1	R 2
4860.450	4861.808		0.1				20568.48			
4860.552	4861.910		0.3				20568.05			
4860.661	4862.018		0.1				20567.59			
4860.791	4862.148	G	0.5d	0.5d			20567.04	4g 1	2a 2	R 2
4860.791	4862.148	G	0.5d	0.5d			20567.04	3F 1	2B 3	Q 2
4860.916	4862.274		0.1d				20566.51			
4861.315	4862.673	G	1.5b	1.5b			20564.82			
4861.637	4862.995		0.1				20563.46	3D 2	2B 4	P 5
4861.734	4863.092	G	1.0	1.0d			20563.05	3E 0	2B 1	Q 4
4861.734	4863.092	G	1.0	1.0d			20563.05	4f 0	2c 1	Q 2
4861.734	4863.092	G	1.0	1.0d			20563.05	3e 3	2c 1	P 2
4861.847	4863.205		0.6	0.9d			20562.57	3D 0	2B 1	R 8
4861.998	4863.357		0.1				20561.93			
4862.204	4863.562		0.2	0.2d			20561.06			
4862.367	4863.726		0.3				20560.37			
4862.587	4863.946		0.3	0.2d			20559.44			
4862.760	4864.118						20558.71	4f 0	2c 1	Q 4
4862.859	4864.218		0.2d				20558.29			
4863.153	4864.511						20557.05			

λ_{air}	λ_{vac}	Ref.	I_1	I_2	I_3	I_4	ν	Upper	Lower	Br.
4863.299	4864.658			0.1			20556.43	3b 5	2a 0	R 1
4863.460	4864.819		0.2	0.2			20555.75	w 0	2c 2	P 2
4863.657	4865.015	G	0.6	0.6			20554.92	X 0	2B 2	P 1
4863.725	4865.084		0.3				20554.63			
4863.789	4865.148		0.5	0.3d			20554.36			
4863.891	4865.250		0.3	0.4			20553.93	4c 6	2a 6	Q 1
4863.922	4865.280						20553.80			
4864.066	4865.425		0.0				20553.19			
4864.161	4865.519			0.2			20552.79			
4864.291	4865.650		0.2d				20552.24			
4864.414	4865.773		0.1d				20551.72			
4864.589	4865.948		0.2d	0.1			20550.98	4e 0	2c 1	R 3
4864.911	4866.270		0.2				20549.62	4f 0	2c 1	R 1
4865.134	4866.493		0.2				20548.68			
4865.295	4866.654		0.1	0.0d			20548.00	w 0	2c 2	Q 2
4865.460	4866.820		0.3	0.2			20547.30	3b 5	2a 0	R 0
4865.657	4867.016		0.0				20546.47			
4865.820	4867.180		0.2				20545.78			
4865.965	4867.324		0.1				20545.17			
4866.166	4867.525		0.4	0.2b			20544.32			
4866.318	4867.677	G	0.6	0.6		1.08	20543.68	3c 3	2a 1	R 1
4866.557	4867.916		0.1d	0.2			20542.67			
4866.649	4868.009			0.0			20542.28			
4866.720	4868.080			0.0			20541.98	4e 2	2c 3	R 4
4866.837	4868.196			0.2			20541.49			
4866.938	4868.298		0.2				20541.06			
4867.033	4868.393	G	0.9	1.0		1.22	20540.66	3E 0	2B 1	P 3
4867.033	4868.393	G	0.9	1.0		1.22	20540.66	4c 6	2a 6	Q 2
4867.173	4868.533		0.2	0.2			20540.07			
4867.263	4868.623			0.0			20539.69	3e 3	2c 1	P 3
4867.365	4868.725		0.4	0.6			20539.26	3F 1	2B 3	P 4
4867.505	4868.864			0.1d			20538.67			
4867.756	4869.116	G		0.2			20537.61	V 1	2B 1	P 3
4867.953	4869.312		0.2	0.2			20536.78	3e 3	2c 1	Q 4
4867.953	4869.312		0.2	0.2			20536.78	4f 0	2c 1	P 5
4868.159	4869.519		0.2	0.2			20535.91			
4868.285	4869.644		0.2	0.4			20535.38			
4868.458	4869.818			0.1d			20534.65			
4868.688	4870.048		0.3b				20533.68	U 1	2B 3	R 3
4868.688	4870.048		0.3b				20533.68	3b 5	2a 0	R 2
4868.742	4870.102			0.3b			20533.45			
4868.856	4870.216		0.0				20532.97			
4868.941	4870.301						20532.61			
4869.124	4870.484		0.1				20531.84			
4869.257	4870.617		0.2	0.2			20531.28			
4869.364	4870.724		0.1				20530.83			
4869.454	4870.814	G	0.8	1.0		1.34	20530.45	3c 2	2a 0	P 5
4869.454	4870.814	G	0.8	1.0		1.34	20530.45	3E 0	2B 1	Q 3
4869.603	4870.963		0.2	0.2			20529.82			
4869.779	4871.139		0.1				20529.08			
4869.968	4871.329		0.2				20528.28			
4870.108	4871.469			0.0			20527.69			
4870.210	4871.571		0.2				20527.26			
4870.293	4871.654	G	0.4	0.5		0.65	20526.91	W 2	2B 5	P 3
4870.441	4871.801		0.1	0.0			20526.29			
4870.595	4871.955		0.2	0.2			20525.64	4g 0	2a 1	R 1
4870.659	4872.019						20525.37			
4870.768	4872.129		0.1				20524.91			
4870.872	4872.233		0.2	0.2			20524.47			
4871.084	4872.444		0.0	0.1	0.5b		20523.58			

λ_{air}	λ_{vac}	Ref.	I_1	I_2	I_3	I_4	ν	Upper	Lower	Br.
4871.357	4872.717		0.2	0.0			20522.43			
4871.456	4872.817		0.3	0.4			20522.01	4F 0	2B 6	R 2
4871.456	4872.817		0.3	0.4			20522.01	4c 6	2a 6	Q 3
4871.592	4872.952		0.2	0.2			20521.44	4e 1	2c 2	R 3
4871.758	4873.119			0.0d			20520.74			
4871.862	4873.223		0.1	0.2			20520.30			
4871.950	4873.311		0.2				20519.93			
4872.005	4873.366			0.1			20519.70			
4872.192	4873.553		0.2	0.1			20518.91			
4872.304	4873.665		0.1	0.0			20518.44			
4872.518	4873.879		0.2	0.2d			20517.54			
4872.682	4874.043		0.2	0.1			20516.85			
4872.819	4874.180		0.3	0.3b			20516.27			
4872.926	4874.287						20515.82			
4873.021	4874.382	G	1.0b	1.5b		1.98	20515.42	3E 0	2B 1	Q 2
4873.157	4874.518		0.2	0.3			20514.85	3c 2	2a 0	Q 8
4873.320	4874.682	G	0.7	0.9	0.6	1.30	20514.16	3E 0	2B 1	Q 1
4873.475	4874.836		0.2	0.1			20513.51			
4873.715	4875.076		0.1				20512.50			
4873.848	4875.209		0.2	0.1			20511.94			
4874.000	4875.361		0.0				20511.30			
4874.100	4875.461		0.2	0.1			20510.88			
4874.290	4875.651	G	0.8	0.9	0.3	1.47	20510.08	X 0	2B 2	P 2
4874.432	4875.794		0.2	0.2			20509.48			
4874.547	4875.908			0.1			20509.00			
4874.637	4875.998		0.1	0.0			20508.62			
4874.768	4876.129	G	0.4	0.5	0.5		20508.07	3E 1	2B 3	Q 8
4874.768	4876.129	G	0.4	0.5	0.5		20508.07	4f 1	2c 2	Q 4
4874.915	4876.277		0.1	0.0			20507.45			
4875.115	4876.476		0.2	0.1			20506.61			
4875.269	4876.631		0.0				20505.96			
4875.460	4876.821		0.3	0.3d			20505.16			
4875.671	4877.033		0.4	0.4			20504.27	3c 3	2a 1	R 0
4875.671	4877.033		0.4	0.4			20504.27	Z 2	2B 4	P 4
4875.762	4877.123		0.1				20503.89			
4875.842	4877.204		0.3	0.3			20503.55			
4875.964	4877.326	G	0.9	1.0	0.9b	1.53	20503.04	3E 0	2B 1	P 2
4876.102	4877.463		0.2	0.2			20502.46			
4876.197	4877.559				0.3		20502.06			
4876.318	4877.680		0.1				20501.55			
4876.454	4877.816		0.1				20500.98			
4876.634	4877.996		0.0				20500.22	4f 0	2c 1	Q 3
4876.727	4878.089			0.1			20499.83	4f 1	2c 2	R 1
4876.829	4878.192			0.3			20499.40	3E 3	2B 6	Q 6
4877.070	4878.432		0.1				20498.39			
4877.229	4878.591						20497.72	4F 0	2B 6	P 5
4877.362	4878.725		0.2				20497.16			
4877.593	4878.956		0.1		0.3		20496.19			
4877.746	4879.108		0.2	0.2			20495.55	3D 2	2B 4	P 6
4877.934	4879.296		0.2	0.1			20494.76	4e 0	2c 1	R 2
4878.026	4879.389			0.0	0.3d		20494.37			
4878.124	4879.486	G	0.7	0.7		1.16	20493.96	V 1	2B 1	P 2
4878.124	4879.486	G	0.7	0.7		1.16	20493.96	3c 7	2a 4	Q 1
4878.248	4879.610			0.0			20493.44			
4878.350	4879.713		0.0	0.1d			20493.01			
4878.483	4879.846		0.3	0.4			20492.45			
4878.757	4880.120						20491.30			
4879.024	4880.387		0.2	0.3d			20490.18	Y 2	2B 5	R 1
4879.191	4880.553		0.2				20489.48			
4879.333	4880.696		0.3	0.4			20488.88	Y 2	2B 5	R 0

λ_air	λ_vac	Ref.	I₁	I₂	I₃	I₄	ν	Upper	Lower	Br.
4879.526	4880.889			0.1			20488.07			
4879.665	4881.027		0.2	0.3			20487.49	3E 3	2B 6	R 3
4879.857	4881.220				0.4		20486.68			
4879.915	4881.278						20486.44			
4880.015	4881.378		0.1				20486.02			
4880.065	4881.428						20485.81			
4880.260	4881.623		0.2	0.2			20484.99	U 1	2B 3	R 2
4880.260	4881.623		0.2	0.2			20484.99	3e 3	2c 1	P 4
4880.474	4881.838		0.1	0.2			20484.09	4g 1	2a 2	R 3
4880.651	4882.014			0.0			20483.35			
4880.746	4882.109			0.1	0.3b		20482.95			
4880.906	4882.269			0.2d			20482.28	U 1	2B 3	R 1
4881.132	4882.495						20481.33			
4881.206	4882.569				0.2b		20481.02			
4881.328	4882.691			0.1d			20480.51			
4881.428	4882.791		0.2				20480.09			
4881.518	4882.882	G	0.6	0.7	0.3	1.08	20479.71	W 2	2B 5	P 4
4881.518	4882.882	G	0.6	0.7	0.3	1.08	20479.71	Y 2	2B 5	R 2
4881.654	4883.018		0.1	0.1			20479.14			
4881.783	4883.146		0.2	0.2	0.2		20478.60	U 1	2B 3	R 0
4881.914	4883.277	G	0.4	0.4			20478.05	Y 1	2B 2	R 1
4882.062	4883.425			0.0d			20477.43			
4882.129	4883.492						20477.15			
4882.310	4883.673	G	0.3	0.3			20476.39	4f 2	2c 3	Q 2
4882.467	4883.831						20475.73			
4882.589	4883.952		0.1	0.0d			20475.22	3e 3	2c 0	P 7
4882.777	4884.141	G	0.6	0.6	0.3	0.85	20474.43	Y 1	2B 2	R 0
4882.777	4884.141	G	0.6	0.6	0.3	0.85	20474.43	3e 3	2c 1	P 3
4883.073	4884.437						20473.19			
4883.311	4884.675						20472.19			
4883.536	4884.899						20471.25			
4883.591	4884.954		0.2	0.0			20471.02			
4883.760	4885.124	G	0.4	0.6d	1.0b		20470.31	3D 0	2B 1	R 7
4883.760	4885.124	G	0.4	0.6d	1.0b		20470.31	4e 2	2c 3	R 3
4883.913	4885.277	G	0.5	0.6		0.95	20469.67	Y 1	2B 2	R 2
4883.913	4885.277	G	0.5	0.6		0.95	20469.67	3F 1	2B 3	P 3
4884.080	4885.444		0.2				20468.97			
4884.211	4885.575						20468.42	4F 0	2B 6	Q 4
4884.442	4885.806		0.2	0.1d			20467.45			
4884.555	4885.919				0.2		20466.98			
4884.662	4886.026						20466.53			
4884.755	4886.119		0.2				20466.14			
4884.960	4886.325						20465.28			
4885.101	4886.465		0.1	0.2d			20464.69	3c 7	2a 4	Q 2
4885.445	4886.809				0.2		20463.25			
4885.645	4887.010				0.3b		20462.41			
4885.801	4887.165						20461.76			
4886.187	4887.552		0.2	0.2			20460.14	4E 1	2C 0	P 2
4886.395	4887.760	G	0.8	0.8	0.4	0.70	20459.27	4b 0	2a 0	P 1
4886.529	4887.894		0.3	0.3			20458.71	3E 3	2B 6	R 2
4886.529	4887.894		0.3	0.3			20458.71	4f 2	2c 3	R 2
4886.689	4888.054		0.1	0.0d			20458.04			
4886.801	4888.166		0.1	0.0			20457.57	4f 2	2c 3	P 5
4886.897	4888.262	G	0.5	0.6	0.2b	0.88	20457.17	X 0	2B 2	P 3
4887.028	4888.393		0.1	0.1			20456.62			
4887.181	4888.546						20455.98			
4887.380	4888.744		0.2	0.1			20455.15	4e 1	2c 2	R 2
4887.881	4889.246				0.3		20453.05	4f 1	2c 2	Q 3
4888.149	4889.514						20451.93			
4888.472	4889.837	G	0.6	0.6		1.05	20450.58			

λ_{air}	λ_{vac}	Ref.	I_1	I_2	I_3	I_4	ν	Upper	Lower	Br.
4888.627	4889.992		0.2	0.1d			20449.93	4f 0	2c 1	P 4
4888.684	4890.050						20449.69			
4888.837	4890.203		0.2				20449.05			
4888.981	4890.346						20448.45			
4889.139	4890.504						20447.79			
4889.218	4890.583		0.1				20447.46			
4889.294	4890.659						20447.14			
4889.488	4890.853		0.2				20446.33			
4889.720	4891.085		0.1				20445.36			
4890.234	4891.600			0.1			20443.21			
4890.327	4891.693				0.2		20442.82			
4890.416	4891.782		0.2				20442.45			
4890.567	4891.932						20441.82			
4890.732	4892.097		0.3	0.3d			20441.13	Y 2	2B 5	P 1
4890.878	4892.243			0.1			20440.52			
4891.007	4892.373						20439.98			
4891.059	4892.425			0.1			20439.76			
4891.165	4892.531		0.2	0.2			20439.32	3F 2	2B 5	Q 4
4891.272	4892.638	G	0.8	0.8b	0.3	0.98	20438.87	3c 3	2a 1	Q 1
4891.272	4892.638	G	0.8	0.8b	0.3	0.98	20438.87	3b 5	2a 0	P 1
4891.430	4892.796		0.1				20438.21	4e 0	2c 1	R 1
4891.540	4892.906						20437.75			
4891.610	4892.976		0.1				20437.46			
4891.698	4893.065						20437.09			
4891.777	4893.143		0.2				20436.76			
4891.952	4893.318	G	0.4	0.6	0.2		20436.03	W 2	2B 5	P 5
4891.952	4893.318	G	0.4	0.6	0.2		20436.03	W 2	2B 5	P 6
4892.081	4893.448			0.1			20435.49			
4892.151	4893.517		0.2				20435.20			
4892.225	4893.591		0.1				20434.89	4f 0	2c 1	Q 2
4892.500	4893.867			0.0			20433.74			
4892.551	4893.917						20433.53			
4892.821	4894.188						20432.40			
4892.888	4894.255		0.2				20432.12			
4893.001	4894.367		0.3	0.3			20431.65			
4893.233	4894.600		0.2				20430.68			
4893.355	4894.722		0.3	0.3d			20430.17			
4893.561	4894.928		0.2				20429.31	3D 2	2B 4	P 7
4893.715	4895.081			0.0d			20428.67			
4893.959	4895.326						20427.65			
4894.021	4895.388		0.2	0.1d			20427.39			
4894.184	4895.551		0.3	0.3d			20426.71			
4894.410	4895.776		0.2d	0.4			20425.77	3E 3	2B 6	Q 5
4894.410	4895.776		0.2d	0.4			20425.77	3E 1	2B 3	R 3
4894.637	4896.004			0.1			20424.82			
4894.781	4896.148		0.1				20424.22			
4894.944	4896.311		0.3	0.4			20423.54	3F 2	2B 5	R 2
4895.085	4896.452		0.1	0.0			20422.95			
4895.232	4896.599		0.3	0.3			20422.34			
4895.414	4896.781		0.2	0.2			20421.58	3d 2	2c 0	P 4
4895.502	4896.869		0.2				20421.21	3c 7	2a 4	Q 3
4895.577	4896.944		0.3	0.4			20420.90			
4895.725	4897.092		0.2				20420.28			
4895.876	4897.244	G	0.4	0.5	0.2		20419.65	Y 1	2B 2	P 1
4896.020	4897.387		0.1				20419.05			
4896.099	4897.467						20418.72			
4896.167	4897.534		0.1				20418.44			
4896.229	4897.596						20418.18			
4896.286	4897.654		0.2	0.1			20417.94			
4896.468	4897.836	G	0.7	0.8	0.9	0.83	20417.18			

λ_{air}	λ_{vac}	Ref.	I_1	I_2	I_3	I_4	ν	Upper	Lower	Br.
4896.608	4897.975		0.1	0.1			20416.60			
4896.730	4898.098		0.1				20416.09			
4896.888	4898.256	G	0.4	0.3		0.58	20415.43			
4897.066	4898.433		0.1				20414.69			
4897.131	4898.498						20414.42			
4897.284	4898.652		0.2	0.1			20413.78			
4897.476	4898.844	G	0.4	0.4		0.40	20412.98	3c 3	2a 1	Q 2
4897.685	4899.053		0.1				20412.11			
4897.889	4899.257		0.0				20411.26	4E 1	2B 0	Q 3
4897.973	4899.341						20410.91	3e 3	2c 1	P 5
4898.098	4899.465		0.2				20410.39			
4898.474	4899.842						20408.82	4g 0	2a 1	R 2
4898.667	4900.034			0.0d			20408.02			
4898.842	4900.210	G	0.3	0.3d	0.9		20407.29	3c 2	2a 0	P 6
4899.010	4900.378		0.2	0.2			20406.59	3E 3	2B 6	R 1
4899.161	4900.529						20405.96	4g 1	2a 2	R 4
4899.221	4900.589						20405.71			
4899.413	4900.781						20404.91			
4899.560	4900.928						20404.30	3c 2	2a 0	Q 9
4899.644	4901.012		0.2				20403.95	3d 2	2c 0	Q 5
4899.706	4901.074						20403.69			
4899.953	4901.322		0.1				20402.66			
4900.227	4901.596				0.3b		20401.52			
4900.597	4901.966				0.2		20399.98			
4900.681	4902.050			0.0d			20399.63			
4900.890	4902.259	G	0.6	0.6	0.3	1.15	20398.76	X 0	2B 2	P 4
4901.032	4902.401		0.1	0.1			20398.17			
4901.130	4902.499						20397.76			
4901.378	4902.747	G	0.3	0.4	0.3		20396.73	3E 1	2B 3	Q 7
4901.378	4902.747	G	0.3	0.4	0.3		20396.73	4e 2	2c 3	R 2
4901.539	4902.908		0.2				20396.06			
4901.618	4902.987		0.2				20395.73			
4901.710	4903.078		0.3	0.4			20395.35	Y 2	2B 5	P 2
4901.871	4903.239	G	0.5	0.5	0.2b	1.08	20394.68	3A 0	2B 2	R 2
4902.013	4903.381		0.2	0.1d			20394.09			
4902.075	4903.444						20393.83	t 1	2a 0	R 3
4902.311	4903.679						20392.85			
4902.573	4903.942		0.1d	0.0d			20391.76			
4902.623	4903.992						20391.55	4f 0	2c 1	P 3
4902.957	4904.326		0.2				20390.16	4f 1	2c 2	Q 2
4903.342	4904.711		0.1				20388.56	t 1	2a 0	P 1
4903.429	4904.798		0.1				20388.20	4e 1	2c 2	R 1
4903.573	4904.942	G	0.4	0.5	0.3	0.45	20387.60			
4903.758	4905.127						20386.83			
4903.917	4905.286		0.2	0.2	0.2		20386.17			
4904.150	4905.520			0.1			20385.20	4f 3	2c 4	R 1
4904.292	4905.662		0.1				20384.61			
4904.401	4905.770						20384.16	w 0	2c 2	P 3
4904.401	4905.770						20384.16	3D 2	2B 4	P 8
4904.790	4906.160						20382.54			
4905.072	4906.441		0.1				20381.37			
4905.385	4906.754						20380.07			
4905.481	4906.851		0.2				20379.67			
4905.568	4906.937						20379.31			
4905.647	4907.017		0.1				20378.98			
4905.792	4907.161		0.2				20378.38			
4905.905	4907.275		0.2				20377.91			
4906.025	4907.395	G	0.4	0.4		0.81	20377.41	U 1	2B 3	P 2
4906.025	4907.395	G	0.4	0.4		0.81	20377.41	4f 2	2c 3	P 4
4906.186	4907.556		0.2	0.3d			20376.74			

λ_{air}	λ_{vac}	Ref.	I_1	I_2	I_3	I_4	ν	Upper	Lower	Br.
4906.237	4907.607		0.2		0.4b		20376.53			
4906.324	4907.694	G	0.7	0.7		0.78	20376.17	4b 0	2a 0	P 2
4906.466	4907.836		0.2	0.0d			20375.58			
4906.593	4907.963			0.1			20375.05			
4906.781	4908.151	G	0.3	0.5	0.2	0.63	20374.27	3c 3	2a 1	Q 3
4907.058	4908.428						20373.12			
4907.244	4908.614						20372.35			
4907.480	4908.850						20371.37	3e 3	2c 1	P 4
4907.694	4909.064			0.1			20370.48			
4907.928	4909.298		0.2	0.2			20369.51			
4908.065	4909.436	G	0.6	0.6	0.3d	1.08	20368.94	Y 1	2B 2	P 2
4908.203	4909.573		0.2	0.1			20368.37			
4908.318	4909.689						20367.89			
4908.415	4909.785		0.1				20367.49			
4908.605	4909.976		0.2	0.1			20366.70			
4908.776	4910.147	G	0.7	0.8	0.9b	1.13	20365.99	3D 0	2B 1	R 6
4908.906	4910.277		0.3	0.6			20365.45	3E 3	2B 6	Q 4
4909.049	4910.419		0.1				20364.86			
4909.157	4910.528		0.1				20364.41			
4909.208	4910.578		0.1				20364.20			
4909.429	4910.800						20363.28			
4909.627	4910.998						20362.46			
4909.699	4911.070		0.0				20362.16			
4910.245	4911.615						20359.90			
4910.348	4911.719						20359.47			
4910.558	4911.929		0.2				20358.60			
4910.756	4912.127						20357.78			
4910.975	4912.346		0.2	0.2			20356.87			
4911.195	4912.566		0.2	0.3			20355.96			
4911.441	4912.812						20354.94			
4911.574	4912.945						20354.39			
4911.636	4913.008						20354.13			
4911.902	4913.273						20353.03			
4911.991	4913.363						20352.66			
4912.083	4913.454						20352.28			
4912.312	4913.684	G	0.4	0.6		0.57	20351.33	3E 1	2B 3	R 2
4912.387	4913.759			0.2			20351.02			
4912.529	4913.901		0.1	0.0			20350.43			
4912.723	4914.094			0.1			20349.63	3E 3	2B 6	R 0
4912.892	4914.263		0.2	0.2			20348.93			
4913.164	4914.536		0.2	0.2			20347.80	3b 5	2a 0	P 2
4913.391	4914.763						20346.86			
4913.517	4914.889			0.0			20346.34			
4913.587	4914.959		0.1				20346.05			
4913.684	4915.055						20345.65	4f 3	2c 4	R 2
4913.756	4915.128						20345.35			
4913.845	4915.217		0.1				20344.98			
4914.019	4915.391						20344.26			
4914.070	4915.442		0.2				20344.05	3e 2	2c 0	P 8
4914.358	4915.730		0.2				20342.86			
4914.599	4915.971			0.0			20341.86			
4914.761	4916.133						20341.19			
4914.867	4916.240						20340.75			
4915.022	4916.394	G	0.3	0.4		0.49	20340.11	X 0	2B 2	P 5
4915.119	4916.491		0.1				20339.71			
4915.196	4916.568						20339.39	4f 2	2c 3	Q 2
4915.353	4916.725		0.1	0.1d			20338.74			
4915.539	4916.912		0.2				20337.97			
4915.626	4916.999						20337.61			
4915.718	4917.090						20337.23			

λ_{air}	λ_{vac}	Ref.	I_1	I_2	I_3	I_4	ν	Upper	Lower	Br.
4915.863	4917.236		0.1				20336.63			
4915.974	4917.347			0.1			20336.17			
4916.052	4917.424						20335.85			
4916.272	4917.644	G	0.3	0.3		0.26	20334.94	Y 2	2B 5	P 3
4916.383	4917.755		0.2	0.1			20334.48			
4916.569	4917.942			0.0			20333.71	3F 2	2B 5	Q 3
4916.743	4918.116						20332.99			
4916.975	4918.348						20332.03			
4917.103	4918.476						20331.50			
4917.319	4918.692						20330.61	4E 2	2C 1	R 1
4917.457	4918.829						20330.04			
4917.609	4918.982		0.2	0.3			20329.41			
4917.788	4919.161						20328.67			
4917.861	4919.234						20328.37			
4917.936	4919.309						20328.06			
4918.098	4919.471		0.3	0.3			20327.39			
4918.231	4919.604		0.2				20326.84			
4918.386	4919.759	G	0.3	0.3			20326.20	4e 0	2c 1	Q 6
4918.519	4919.892						20325.65	4e 0	2c 1	Q 5
4918.628	4920.001		0.2				20325.20			
4918.700	4920.073						20324.90			
4918.833	4920.207		0.3	0.3			20324.35	4e 0	2c 1	Q 4
4919.121	4920.495	G	0.6	0.8b	0.6	1.06	20323.16	3c 3	2a 1	Q 4
4919.121	4920.495	G	0.6	0.8b	0.6	1.06	20323.16	4e 0	2c 1	Q 3
4919.121	4920.495	G	0.6	0.8b	0.6	1.06	20323.16	4e 0	2c 1	Q 2
4919.121	4920.495	G	0.6	0.8b	0.6	1.06	20323.16	3D 0	2B 1	R 5
4919.259	4920.633			0.1			20322.59	4e 2	2c 3	R 1
4919.375	4920.749						20322.11			
4919.596	4920.969						20321.20	4e 0	2c 1	Q 7
4919.789	4921.163			0.2			20320.40	4e 1	2c 2	Q 5
4919.937	4921.311		0.2	0.1			20319.79			
4920.143	4921.517		0.2	0.1			20318.94			
4920.328	4921.701	G	0.4	0.5		0.80	20318.18	3E 3	2B 6	Q 3
4920.465	4921.839			0.2d			20317.61			
4920.807	4922.180		0.2	0.1			20316.20			
4920.947	4922.321						20315.62	4F 0	2B 6	R 1
4920.993	4922.367						20315.43			
4921.039	4922.413		0.2				20315.24			
4921.196	4922.570	G	0.4	0.5d		0.88	20314.59	3c 3	2a 1	P 2
4921.196	4922.570	G	0.4	0.5d		0.88	20314.59	X 1	2B 4	R 0
4921.196	4922.570	G	0.4	0.5d		0.88	20314.59	U 1	2B 3	P 3
4921.252	4922.626				0.3		20314.36	4E 1	2C 0	P 3
4921.475	4922.849						20313.44	3F 2	2B 5	R 1
4921.669	4923.043		0.2	0.3			20312.64	4c 7	2a 7	Q 1
4921.805	4923.179						20312.08			
4921.926	4923.300	1	0.2	0.3	1.5		20311.58			
4922.017	4923.392	G	0.2	0.2			20311.20	X 1	2B 4	R 1
4922.190	4923.564		0.1d				20310.49	4e 0	2c 1	R 2
4922.270	4923.644			0.0d			20310.16			
4922.454	4923.828	G	0.4	0.4		0.90	20309.40			
4922.597	4923.971		0.2	0.0			20308.81			
4922.760	4924.134		0.2	0.2	0.2		20308.14			
4922.992	4924.367		0.1	0.0			20307.18	4e 0	2c 1	R 3
4923.114	4924.488		0.1				20306.68	4e 1	2c 2	Q 4
4923.213	4924.587		0.1				20306.27			
4923.288	4924.662		0.2				20305.96	4e 0	2c 1	Q 1
4923.400	4924.774				0.1		20305.50			
4923.431	4924.806						20305.37	4e 0	2c 1	R 5
4923.509	4924.883						20305.05			
4923.581	4924.956						20304.75			

λ_{air}	λ_{vac}	Ref.	I_1	I_2	I_3	I_4	ν	Upper	Lower	Br.
4923.717	4925.092		0.2				20304.19			
4923.952	4925.327		0.3	0.2			20303.22	u 1	2a 0	R 0
4923.952	4925.327		0.3	0.2			20303.22	Z 2	2B 4	P 6
4923.952	4925.327		0.3	0.2			20303.22	Y 1	2B 2	P 3
4924.033	4925.407	G	0.5	0.7	0.6	0.90	20302.89	3E 1	2B 3	Q 6
4924.033	4925.407	G	0.5	0.7	0.6	0.90	20302.89	4f 2	2c 3	Q 3
4924.195	4925.570		0.1	0.0			20302.22			
4924.350	4925.725		0.2				20301.58			
4924.489	4925.863				0.1		20301.01	4F 0	2B 6	Q 3
4924.527	4925.902						20300.85			
4924.772	4926.147		0.2				20299.84			
4925.034	4926.409		0.2	0.1			20298.76			
4925.238	4926.613	G	0.6	0.8	0.4	1.08	20297.92			
4925.372	4926.747			0.1			20297.37			
4925.522	4926.897						20296.75			
4925.685	4927.060						20296.08			
4925.818	4927.193	G	0.3	0.3		0.43	20295.53	3A 0	2B 2	R 1
4925.818	4927.193	G	0.3	0.3		0.43	20295.53	4g 0	2a 1	Q 3
4925.891	4927.266		0.2				20295.23			
4925.942	4927.317			0.1			20295.02			
4926.061	4927.436		0.1	0.1			20294.53			
4926.168	4927.543		0.2	0.2			20294.09			
4926.236	4927.611				0.2		20293.81			
4926.277	4927.652	G	0.3d	0.5d		0.85	20293.64	X 1	2B 4	R 2
4926.476	4927.851			0.1			20292.82			
4926.588	4927.963		0.2d	0.2			20292.36			
4926.712	4928.087		0.1				20291.85	4e 1	2c 2	Q 3
4926.850	4928.225		0.1				20291.28	3d 3	2c 1	R 1
4927.049	4928.424						20290.46			
4927.209	4928.585			0.0d			20289.80	4e 2	2c 3	Q 5
4927.370	4928.745		0.1	0.1d			20289.14			
4927.574	4928.949		0.1	0.1d	0.2		20288.30			
4927.914	4929.289		0.1	0.0d			20286.90			
4928.062	4929.438	F					20286.29			
4928.183	4929.559		0.2	0.2b			20285.79			
4928.283	4929.659		0.2				20285.38			
4928.373	4929.749	G	0.9b	0.9b	0.9	1.35	20285.01	4b 0	2a 0	P 3
4928.550	4929.926		0.2				20284.28			
4928.628	4930.004	G	0.9	1.5b	1.5b		20283.96	3D 0	2B 1	R 4
4928.715	4930.091					1.94	20283.60	3E 3	2B 6	Q 2
4928.715	4930.091					1.94	20283.60	4e 0	2c 1	Q 1
4928.788	4930.164	G	1.0b	1.0b			20283.30	3D 0	2B 1	R 0
4928.939	4930.315		0.2	0.2			20282.68			
4929.107	4930.483		0.3	0.3	0.3		20281.99			
4929.182	4930.558		0.1				20281.68	4f 3	2c 4	Q 2
4929.318	4930.694		0.2	0.2			20281.12			
4929.471	4930.847		0.2				20280.49			
4929.607	4930.983		0.2				20279.93			
4929.799	4931.176		0.2	0.1			20279.14	4f 0	2c 1	P 5
4929.960	4931.336		0.1				20278.48	4e 1	2c 2	Q 2
4930.266	4931.642				0.1		20277.22			
4930.305	4931.681		0.2				20277.06			
4930.441	4931.818		0.2	0.1d			20276.50			
4930.575	4931.951		0.1				20275.95			
4930.665	4932.041		0.2				20275.58			
4930.864	4932.241		0.2	0.1d			20274.76	t 1	2a 0	P 3
4930.959	4932.336		0.1				20274.37			
4931.088	4932.465			0.3d			20273.84	3c 2	2a 0	P 7
4931.117	4932.494		0.3		0.1		20273.72	4f 0	2c 1	P 4
4931.268	4932.645		0.2	0.2b			20273.10			

λ_{air}	λ_{vac}	Ref.	I_1	I_2	I_3	I_4	ν	Upper	Lower	Br.
4931.322	4932.698			0.2			20272.88	4e 1	2c 2	R 1
4931.560	4932.937			0.2			20271.90			
4931.684	4933.061						20271.39			
4931.808	4933.185	G	0.5	0.6	0.3		20270.88	4b 1	2a 1	R 2
4931.930	4933.307		0.1	0.1			20270.38			
4932.049	4933.426		0.4	0.3			20269.89	3E 1	2B 3	R 1
4932.178	4933.555		0.2				20269.36			
4932.268	4933.645	G	1.0b	1.0	1.0	1.53	20268.99	3D 0	2B 1	R 1
4932.392	4933.769			0.1			20268.48			
4932.458	4933.835			0.2			20268.21			
4932.582	4933.959			0.1			20267.70			
4932.808	4934.185			0.2			20266.77			
4932.903	4934.280						20266.38	4e 1	2c 2	R 2
4932.903	4934.280						20266.38	4e 2	2c 3	Q 4
4933.086	4934.463		0.2	0.3	0.2		20265.63			
4933.203	4934.580		0.2				20265.15	4e 1	2c 2	Q 1
4933.339	4934.716		0.3	0.3			20264.59			
4933.424	4934.801		0.2				20264.24			
4933.522	4934.899	G	0.9	0.9	1.5	1.48	20263.84	3D 0	2B 1	R 3
4933.522	4934.899	G	0.9	0.9	1.5	1.48	20263.84	3F 0	2B 2	R 4
4933.522	4934.899	G	0.9	0.9	1.5	1.48	20263.84	Y 0	2B 5	P 4
4933.658	4935.035		0.2	0.2			20263.28	4f 0	2c 1	P 3
4933.699	4935.077						20263.11	4e 1	2c 2	R 3
4933.826	4935.203	G	0.4	0.6	0.4	1.00	20262.59	U 1	2B 3	P 6
4933.989	4935.366		0.3				20261.92			
4934.026	4935.403			0.3b			20261.77			
4934.072	4935.449		0.3				20261.58			
4934.145	4935.522		0.4				20261.28			
4934.247	4935.625	GK	1.0b	1.5d	1.5	2.04	20260.86	3D 0	2B 1	R 2
4934.247	4935.625	GK	1.0b	1.5d	1.5	2.04	20260.86	3E 3	2B 6	Q 1
4934.384	4935.761		0.3	0.3			20260.30			
4934.440	4935.817						20260.07			
4934.564	4935.941	G	0.5	0.5	0.2	0.70	20259.56	3E 3	2B 6	P 2
4934.564	4935.941	G	0.5	0.5	0.2	0.70	20259.56	3c 3	2a 1	Q 5
4934.717	4936.095		0.3	0.1			20258.93	4e 1	2c 2	R 6
4934.717	4936.095		0.3	0.1			20258.93	4e 1	2c 2	R 4
4934.808	4936.185		0.1				20258.56			
4934.883	4936.260						20258.25			
4934.976	4936.353		0.2				20257.87			
4935.054	4936.431		0.2				20257.55			
4935.158	4936.536		0.4	0.3			20257.12	3E 3	2B 6	P 3
4935.158	4936.536		0.4	0.3			20257.12	3e 3	2c 1	P 5
4935.239	4936.616	G	0.8	0.7	0.2	1.28	20256.79	3D 3	2B 6	R 0
4935.239	4936.616	G	0.8	0.7	0.2	1.28	20256.79	4e 1	2c 2	R 5
4935.380	4936.758			0.1			20256.21			
4935.424	4936.802		0.2				20256.03			
4935.495	4936.872						20255.74			
4935.714	4937.092						20254.84			
4935.940	4937.318	G	0.6	0.6	0.4	1.00	20253.91	3D 0	2B 1	P 1
4935.940	4937.318	G	0.6	0.6	0.4	1.00	20253.91	3d 3	2c 1	R 2
4936.055	4937.433		0.2				20253.44			
4936.145	4937.523	G	0.6	0.5		1.06	20253.07	3D 3	2B 6	R 1
4936.289	4937.667		0.2	0.1			20252.48			
4936.411	4937.789	G	0.6	0.6	0.3	1.05	20251.98	U 1	2B 3	P 4
4936.552	4937.930			0.1			20251.40			
4936.596	4937.974		0.2				20251.22			
4936.706	4938.084	G	0.6	0.6	0.2	0.90	20250.77	3E 3	2B 6	P 4
4936.706	4938.084	G	0.6	0.6	0.2	0.90	20250.77	3F 2	2B 5	Q 2
4936.852	4938.230			0.1			20250.17			
4937.006	4938.384		0.2	0.0d			20249.54			

λ_{air}	λ_{vac}	Ref.	I_1	I_2	I_3	I_4	ν	Upper	Lower	Br.
4937.172	4938.550		0.1	0.1			20248.86			
4937.323	4938.701	G	0.3	0.5	0.4	0.74	20248.24	X 0	2B 2	P 6
4937.484	4938.862		0.1	0.1d			20247.58			
4937.559	4938.937						20247.27			
4937.601	4938.979		0.2				20247.10	3d 3	2c 1	Q 1
4937.784	4939.162	G	0.4	0.4	0.4		20246.35			
4937.903	4939.281		0.1	0.1			20245.86			
4937.971	4939.350						20245.58			
4938.106	4939.484		0.1				20245.03			
4938.235	4939.613						20244.50			
4938.408	4939.786		0.2				20243.79			
4938.454	4939.833						20243.60	3d 2	2c 0	P 5
4938.606	4939.984		0.2d				20242.98			
4938.681	4940.060						20242.67	4f 1	2c 2	P 5
4938.884	4940.262				0.1		20241.84	4e 2	2c 3	Q 3
4938.945	4940.323		0.2d				20241.59			
4939.020	4940.399			0.0d			20241.28			
4939.162	4940.541	G	0.9	0.8	0.5	1.27	20240.70	4b 1	2a 1	R 1
4939.291	4940.670		0.1	0.1			20240.17			
4939.394	4940.772		0.2	0.1			20239.75			
4939.518	4940.897		0.2				20239.24			
4939.596	4940.975	G	0.8	0.7		1.46	20238.92	3D 3	2B 6	R 2
4939.672	4941.051				0.4		20238.61			
4939.709	4941.087		0.2	0.2d			20238.46	4f 1	2c 2	P 4
4939.806	4941.185		0.2				20238.06			
4939.884	4941.263		0.2				20237.74			
4940.004	4941.383	G	0.4	0.4	0.3	0.81	20237.25	U 1	2B 3	P 5
4940.131	4941.510		0.1	0.0			20236.73			
4940.282	4941.661		0.2	0.2			20236.11			
4940.353	4941.732				0.1		20235.82			
4940.504	4941.883			0.0			20235.20			
4940.837	4942.216			0.1			20233.84			
4941.103	4942.482			0.0d			20232.75			
4941.403	4942.782				0.2		20231.52			
4941.452	4942.831		0.2	0.3			20231.32	4e 1	2c 2	Q 1
4941.620	4943.000		0.2				20230.63			
4941.716	4943.095			0.1			20230.24			
4941.838	4943.217						20229.74			
4941.984	4943.364	G	0.4	0.9	0.1	0.63	20229.14	3c 3	2a 1	P 3
4942.131	4943.510		0.2				20228.54			
4942.226	4943.606			0.2d			20228.15	3d 3	2c 1	P 1
4942.302	4943.682						20227.84			
4942.410	4943.789		0.0				20227.40	4f 1	2c 2	P 3
4942.544	4943.924	G	0.5	1.0	0.6	1.03	20226.85	Y 1	2B 2	P 4
4942.544	4943.924	G	0.5	1.0	0.6	1.03	20226.85	3E 1	2B 3	Q 5
4942.695	4944.075		0.1				20226.23			
4942.815	4944.195			0.2			20225.74	3b 5	2a 0	P 3
4943.047	4944.427			0.1			20224.79			
4943.338	4944.718			0.1d			20223.60			
4943.399	4944.779		0.1				20223.35			
4943.614	4944.994	G	0.3	0.5	0.1	0.52	20222.47	3E 3	2B 6	P 5
4943.732	4945.112		0.2d				20221.99			
4943.876	4945.256	.		0.1			20221.40			
4943.993	4945.373				0.1		20220.92			
4944.094	4945.474		0.2	0.2			20220.51			
4944.280	4945.660		0.2	0.0			20219.75			
4944.502	4945.882		0.2	0.2			20218.84	4e 2	2c 3	Q 2
4944.588	4945.968			0.1			20218.49	4e 2	2c 3	R 1
4944.739	4946.119		0.3	0.7			20217.87	3D 3	2B 6	R 4
4944.835	4946.215	G	0.4	0.9	0.3	1.17	20217.48	X 1	2B 4	P 2

λ_{air}	λ_{vac}	Ref.	I_1	I_2	I_3	I_4	ν	Upper	Lower	Br.
4944.835	4946.215	G	0.4	0.9	0.3	1.17	20217.48	3D 3	2B 6	R 3
4944.986	4946.367		0.2	0.4	0.3		20216.86			
4945.202	4946.582		0.1	0.3			20215.98			
4945.366	4946.746						20215.31			
4945.488	4946.868						20214.81			
4945.593	4946.973			0.1			20214.38			
4945.745	4947.125				0.1		20213.76			
4945.821	4947.201	G	0.3	0.5		0.60	20213.45	3A 0	2B 2	R 0
4945.987	4947.367			0.0			20212.77			
4946.129	4947.509			0.0			20212.19	4e 2	2c 3	R 2
4946.295	4947.676		0.2	0.2			20211.51	4E 2	2C 1	Q 2
4946.501	4947.881		0.1d	0.1			20210.67			
4946.689	4948.070			0.1			20209.90			
4946.922	4948.303		0.1				20208.95			
4947.066	4948.447			0.1			20208.36			
4947.245	4948.626			0.2	0.1d		20207.63	3F 2	2B 5	P 4
4947.610	4948.991		0.1				20206.14			
4947.869	4949.250			0.1			20205.08	4e 2	2c 3	R 3
4948.242	4949.623						20203.56	4e 0	2c 1	Q 2
4948.340	4949.721		0.1	0.2			20203.16			
4948.516	4949.897						20202.44			
4948.687	4950.069						20201.74			
4948.756	4950.137			0.1			20201.46			
4948.991	4950.372		0.2	0.4	0.3		20200.50	3F 0	2B 2	Q 6
4949.097	4950.478			0.0			20200.07	4E 0	2B 6	R 2
4949.305	4950.686			0.2d			20199.22			
4949.366	4950.747						20198.97			
4949.457	4950.838		0.2				20198.60	3d 3	2c 1	R 3
4949.530	4950.912	G	0.5	0.9	0.2	0.74	20198.30	4b 1	2a 1	R 0
4949.530	4950.912	G	0.5	0.9	0.2	0.74	20198.30	4e 2	2c 3	R 4
4949.702	4951.083		0.1				20197.60			
4949.788	4951.169		0.1	0.0			20197.25			
4950.067	4951.449			0.1	0.1		20196.11			
4950.231	4951.613		0.1	0.1			20195.44	4e 2	2c 3	Q 1
4950.295	4951.677						20195.18			
4950.418	4951.799		0.1d				20194.68			
4950.555	4951.936			0.1			20194.12			
4950.753	4952.135		0.2	0.4	0.2		20193.31			
4950.908	4952.290		0.1				20192.68	4f 2	2c 3	P 5
4951.026	4952.407			0.2			20192.20			
4951.070	4952.451		0.2				20192.02			
4951.217	4952.599	G	0.3	0.5	0.4		20191.42			
4951.548	4952.930			0.1			20190.07	4f 3	2c 4	Q 3
4951.729	4953.111		0.2				20189.33			
4951.827	4953.209				0.1		20188.93			
4952.024	4953.406			0.1d			20188.13			
4952.355	4953.737			0.1d			20186.78	4d 0	2c 1	P 1
4952.595	4953.978	G	0.4	1.0	0.5	0.92	20185.80	4b 0	2a 0	P 4
4952.718	4954.100		0.1				20185.30			
4952.851	4954.233		0.2	0.3			20184.76	3A 1	2B 4	R 2
4952.851	4954.233		0.2	0.3			20184.76	3c 3	2a 1	Q 6
4952.949	4954.331		0.2				20184.36	4f 2	2c 3	P 3
4953.035	4954.417		0.1				20184.01			
4953.172	4954.554	G	0.3	0.6	0.5		20183.45			
4953.400	4954.783		0.2	0.2	0.2		20182.52			
4953.690	4955.072		0.2	0.1			20181.34			
4953.862	4955.244			0.1			20180.64			
4953.940	4955.323				0.2		20180.32			
4954.007	4955.389		0.1	0.2			20180.05			
4954.370	4955.753			0.1			20178.57	4E 1	2C 0	P 4

λ_{air}	λ_{vac}	Ref.	I_1	I_2	I_3	I_4	ν	Upper	Lower	Br.
4954.581	4955.964			0.3	0.3		20177.71			
4954.763	4956.146		0.1	0.1			20176.97	3E 0	2B 2	R 6
4954.994	4956.376	G	0.3	0.7	0.5	0.53	20176.03	3D 1	2B 3	R 8
4954.994	4956.376	G	0.3	0.7	0.5	0.53	20176.03	3d 3	2c 1	Q 2
4955.190	4956.573			0.1			20175.23			
4955.320	4956.703		0.1	0.2			20174.70			
4955.509	4956.892	G	0.4	1.0	0.3	0.54	20173.93	3E 1	2B 3	R 0
4955.610	4956.993						20173.52			
4955.669	4957.052		0.2				20173.28			
4955.760	4957.143	G	0.7	1.5	0.8	1.24	20172.91	3D 0	2B 1	P 2
4955.893	4957.276		0.2				20172.37			
4956.011	4957.394			0.1			20171.89			
4956.121	4957.504		0.1				20171.44			
4956.249	4957.632				0.1d		20170.92			
4956.340	4957.723		0.2	0.1			20170.55			
4956.590	4957.974			0.2			20169.53			
4956.635	4958.018		0.1				20169.35			
4956.792	4958.175	G	0.5	1.5	0.9	1.39	20168.71	3E 1	2B 3	Q 4
4956.792	4958.175	G	0.5	1.5	0.9	1.39	20168.71	4e 1	2c 1	P 2
4957.023	4958.406			0.1			20167.77			
4957.291	4958.674			0.2			20166.68	3D 3	2B 6	P 2
4957.446	4958.829		0.2	0.3			20166.05	4F 0	2B 6	P 3
4957.655	4959.038			0.2			20165.20			
4957.837	4959.220			0.2			20164.46			
4958.024	4959.407			0.2			20163.70			
4958.213	4959.597		0.2	0.3	0.1		20162.93	4e 1	2c 2	Q 2
4958.400	4959.784		0.2	0.0			20162.17			
4958.606	4959.990		0.1d	0.2			20161.33			
4958.840	4960.224	G	0.2d	0.5	0.3		20160.38			
4958.906	4960.290						20160.11			
4959.034	4960.418			0.1			20159.59			
4959.239	4960.623			0.2			20158.76			
4959.381	4960.765		0.1	0.2			20158.18	3F 2	2B 5	P 3
4959.536	4960.920			0.1			20157.55			
4959.802	4961.186		0.2	0.2d			20156.47			
4959.994	4961.378		0.1	0.0d			20155.69			
4960.117	4961.501	G	0.3	0.5		0.57	20155.19	4e 1	2c 2	P 2
4960.225	4961.610		0.1				20154.75			
4960.341	4961.725		0.2	0.1			20154.28	4F 0	2B 6	Q 2
4960.486	4961.870		0.1				20153.69			
4960.681	4962.065			0.2			20152.90			
4960.762	4962.146		0.2				20152.57			
4960.870	4962.255		0.2				20152.13	4g 1	2a 2	P 4
4960.946	4962.331	G	0.5	1.0	0.3	1.18	20151.82	3A 0	2B 2	P 4
4960.946	4962.331	G	0.5	1.0	0.3	1.18	20151.82	4e 0	2c 1	P 3
4961.146	4962.530		0.1	0.2			20151.01			
4961.404	4962.789		0.2	0.3			20149.96	X 1	2B 4	P 3
4961.693	4963.077			0.2			20148.79			
4961.968	4963.353			0.0			20147.67	4e 2	2c 3	R 5
4962.114	4963.498						20147.08			
4962.202	4963.587			0.1			20146.72			
4962.279	4963.663				0.1		20146.41			
4962.431	4963.816			0.1			20145.79			
4962.631	4964.016				0.1		20144.98			
4962.757	4964.141			0.0			20144.47			
4963.102	4964.487			0.1			20143.07			
4963.439	4964.824		0.2	0.0			20141.70			
4963.577	4964.962				0.1		20141.14			
4963.799	4965.184			0.0			20140.24			
4963.996	4965.381		0.2				20139.44			

λ_{air}	λ_{vac}	Ref.	I_1	I_2	I_3	I_4	ν	Upper	Lower	Br.
4964.043	4965.428			0.2			20139.25	4d 0	2c 1	R 1
4964.139	4965.524						20138.86	4e 0	2c 1	Q 3
4964.257	4965.643			0.2			20138.38			
4964.307	4965.692		0.1				20138.18			
4964.425	4965.810		0.2				20137.70			
4964.534	4965.919			0.1			20137.26	4d 0	2c 1	Q 1
4964.605	4965.990		0.2				20136.97			
4964.696	4966.082		0.0				20136.60			
4964.896	4966.281		0.1	0.1			20135.79			
4965.088	4966.474	G	0.4	1.0	0.3	0.95	20135.01	Z 0	2B 1	R 1
4965.258	4966.644			0.0			20134.32			
4965.404	4966.789			0.2			20133.73			
4965.643	4967.029		0.1	0.3	0.3		20132.76			
4965.823	4967.209		0.2	0.5			20132.03	3c 3	2a 1	P 4
4965.823	4967.209		0.2	0.5			20132.03	3d 3	2c 1	P 2
4966.025	4967.411		0.1	0.3			20131.21	3E 2	2B 5	R 4
4966.255	4967.641			0.1			20130.28			
4966.477	4967.863			0.2			20129.38	3E 1	2B 3	P 5
4966.558	4967.944						20129.05	3c 2	2a 0	P 8
4966.704	4968.090	G	0.6	1.5	0.3		20128.46	Z 0	2B 1	R 0
4966.894	4968.280	G	0.6b	1.5b	0.9	1.54	20127.69	3E 1	2B 3	Q 3
4966.894	4968.280	G	0.6b	1.5b	0.9	1.54	20127.69	Z 0	2B 1	R 2
4967.123	4968.510		0.1d	0.2			20126.76			
4967.346	4968.732			0.1			20125.86	4f 3	2c 4	P 3
4967.518	4968.905			0.1			20125.16			
4967.585	4968.971		0.1d				20124.89			
4967.674	4969.060			0.1			20124.53			
4967.864	4969.250	G	0.3d	0.6	0.3	0.81	20123.76	X 2	2B 6	R 0
4968.091	4969.477						20122.84			
4968.257	4969.643		0.1d	0.3d	0.2		20122.17	4e 1	2c 2	P 2
4968.481	4969.868			0.1			20121.26			
4968.629	4970.016	G	0.3	0.5	0.1d	0.61	20120.66	3A 0	2B 2	P 3
4968.825	4970.211			0.2			20119.87			
4968.891	4970.278						20119.60			
4969.034	4970.421		0.2	0.3			20119.02	3A 0	2B 2	P 1
4969.034	4970.421		0.2	0.3			20119.02	4d 1	2c 2	P 1
4969.220	4970.606	G	0.7	1.5	0.6	1.41	20118.27	3E 1	2B 3	P 4
4969.351	4970.737		0.0				20117.74			
4969.444	4970.831		0.0	0.0d			20117.36			
4969.541	4970.927						20116.97			
4969.620	4971.007		0.0				20116.65			
4969.684	4971.071						20116.39			
4969.766	4971.152			0.0d			20116.06	5c 0	2a 2	R 2
4970.060	4971.446			0.2	0.1		20114.87	4e 1	2c 2	P 3
4970.388	4971.775			0.1			20113.54			
4970.514	4971.901			0.2			20113.03			
4970.697	4972.084		0.2				20112.29	X 2	2B 6	R 1
4970.739	4972.126	G		0.6b	0.2	0.82	20112.12			
4971.006	4972.393		0.2	0.6	0.1		20111.04	3A 2	2B 6	R 4
4971.095	4972.482						20110.68			
4971.147	4972.534			0.1			20110.47			
4971.342	4972.729			0.0			20109.68			
4971.543	4972.930			0.1			20108.87	5c 1	2a 3	R 3
4971.765	4973.152	G	0.4	0.7	0.2	0.90	20107.97	3A 0	2B 2	P 2
4972.151	4973.538			0.1	0.1		20106.41			
4972.227	4973.615		0.2				20106.10			
4972.616	4974.003		0.1	0.2			20104.53			
4972.833	4974.221		0.2	0.6	0.3		20103.65	3D 3	2B 6	P 3
4972.833	4974.221		0.2	0.6	0.3		20103.65	Z 0	2B 1	R 3
4972.833	4974.221		0.2	0.6	0.3		20103.65	u 1	2a 0	P 2

λ_{air}	λ_{vac}	Ref.	I_1	I_2	I_3	I_4	ν	Upper	Lower	Br.
4972.908	4974.295		0.3				20103.35			
4973.007	4974.394			0.1			20102.95			
4973.106	4974.493		0.1				20102.55	4e 1	2c 2	Q 3
4973.190	4974.577		0.6				20102.21	3E 1	2B 3	P 3
4973.311	4974.699	GK	0.8	1.5b	1.0	1.75	20101.72	3E 1	2B 3	Q 2
4973.492	4974.879		0.1				20100.99			
4973.571	4974.959			0.2			20100.67			
4973.749	4975.137			0.1			20099.95	4E 0	2B 6	R 0
4973.870	4975.258		0.1				20099.46			
4973.925	4975.312			0.1			20099.24	4f 2	2c 3	P 4
4974.152	4975.540	G	0.2	0.6	0.1		20098.32	3A 1	2B 4	R 1
4974.410	4975.798			0.1			20097.28			
4974.581	4975.969			0.0			20096.59	4e 0	2c 1	P 4
4974.979	4976.367				0.1		20094.98			
4975.180	4976.568			0.1			20094.17	X 2	2B 6	P 1
4975.365	4976.754			0.4	0.1		20093.42			
4975.546	4976.934			0.1			20092.69	4e 3	2c.4	Q 1
4975.752	4977.140		0.1	0.0			20091.86	3f 1	2c 0	R 6
4975.895	4977.284	G	0.2	0.4	0.2		20091.28	4e 2	2c 3	P 2
4976.031	4977.420			0.1			20090.73			
4976.237	4977.626	G	0.2	0.6	0.3		20089.90	3E 2	2B 5	Q 8
4976.443	4977.831			0.2			20089.07			
4976.628	4978.017	G	0.5	1.5	0.4	1.26	20088.32	3E 1	2B 3	Q 1
4976.628	4978.017	G	0.5	1.5	0.4	1.26	20088.32	X 2	2B 6	R 2
4976.876	4978.265		0.1d	0.3			20087.32			
4977.015	4978.404			0.1			20086.76			
4977.171	4978.560	G	0.5	0.9	0.3	0.79	20086.13	3D 0	2B 1	P 3
4977.352	4978.741		0.1	0.0			20085.40			
4977.538	4978.927		0.0	0.1d	0.1		20084.65			
4977.805	4979.194			0.1d			20083.57	3F 3	2B 7	R 2
4978.058	4979.447		0.1	0.2			20082.55	3d 3	2c 1	Q 3
4978.259	4979.648	G	0.6	1.5	0.4	1.21	20081.74	4b 1	2a 1	P 1
4978.410	4979.799		0.1				20081.13			
4978.482	4979.871						20080.84			
4978.532	4979.921		0.1	0.0			20080.64			
4978.584	4979.973				0.7		20080.43	4e 0	2c 1	Q 4
4978.921	4980.310		0.1	0.1			20079.07			
4979.117	4980.506	G	0.4	1.5	0.8	0.98	20078.28	4b 0	2a 0	P 5
4979.315	4980.705						20077.48	4e 2	2c 3	P 2
4979.397	4980.787		0.2d	0.2	0.1		20077.15	4d 0	2c 1	R 2
4979.712	4981.102			0.1			20075.88	Z 0	2B 1	R 4
4979.876	4981.265			0.2			20075.22	4e 1	2c 2	P 4
4979.953	4981.342		0.1				20074.91			
4980.057	4981.447			0.1			20074.49	3b 5	2a 0	P 4
4980.285	4981.675			0.2			20073.57			
4980.382	4981.772		0.2				20073.18			
4980.481	4981.871	G	0.8	1.5	0.7	1.39	20072.78	3E 1	2B 3	P 2
4980.667	4982.057		0.1				20072.03			
4980.734	4982.124			0.1d			20071.76			
4980.938	4982.328			0.1d			20070.94			
4981.136	4982.526	G	0.3	0.5		0.67	20070.14	3D 1	2B 3	R 7
4981.136	4982.526	G	0.3	0.5		0.67	20070.14	Z 0	2B 1	P 1
4981.288	4982.678		0.2	0.4	0.2b		20069.53	X 1	2B 4	P 4
4981.288	4982.678		0.2	0.4	0.2b		20069.53	4d 1	2c 2	R 1
4981.615	4983.005		0.2d	0.3			20068.21	4d 1	2c 2	Q 1
4981.774	4983.164		0.0	0.0			20067.57			
4981.916	4983.306		0.1	0.2			20067.00			
4982.119	4983.510		0.1	0.3			20066.18	X 2	2B 6	R 3
4982.460	4983.850		0.1	0.2	0.2		20064.81	3F 0	2B 2	R 3
4982.611	4984.001		0.1	0.3			20064.20			

λ_{air}	λ_{vac}	Ref.	I_1	I_2	I_3	I_4	ν	Upper	Lower	Br.
4982.763	4984.153		0.0				20063.59			
4982.854	4984.245			0.1	0.9		20063.22	4e 2	2c 3	P 3
4982.956	4984.347		0.2	0.2			20062.81			
4983.197	4984.588	G	0.3	0.7	0.3	1.04	20061.84	W 1	2B 4	R 4
4983.376	4984.766		0.0	0.0			20061.12			
4983.538	4984.928	G	0.3	0.7	0.4	0.67	20060.47			
4983.731	4985.122			0.1			20059.69			
4983.875	4985.266			0.1			20059.11			
4984.072	4985.462			0.1			20058.32			
4984.288	4985.679			0.1			20057.45			
4984.487	4985.878			0.2			20056.65	5c 0	2a 2	R 1
4984.487	4985.878			0.2			20056.65	3d 2	2c 0	P 6
4984.720	4986.111	G	0.2	0.4	0.2b		20055.71			
4984.872	4986.263			0.1d			20055.10			
4985.163	4986.554			0.0d			20053.93			
4985.416	4986.807			0.0			20052.91	5c 1	2a 3	R 2
4985.416	4986.807			0.0			20052.91	3f 1	2c 0	R 4
4985.640	4987.031			0.2d			20052.01			
4986.006	4987.397			0.2d			20050.54			
4986.170	4987.561			0.1			20049.88			
4986.389	4987.780			0.2			20049.00	3d 3	2c 1	R 5
4986.545	4987.937			0.3	0.2		20048.37			
4987.028	4988.419		0.1				20046.43	4e 1	2c 2	Q 4
4987.742	4989.134		0.1	0.1			20043.56	4E 1	2C 1	P 3
4987.742	4989.134		0.1	0.1			20043.56	4E 1	2C 0	P 5
4988.081	4989.472			0.2			20042.20	3f 1	2c 0	R 5
4988.315	4989.706			0.0			20041.26	u 2	2a 1	R 3
4988.551	4989.943			0.1			20040.31			
4988.775	4990.167			0.1			20039.41	4e 1	2c 2	P 5
4988.942	4990.334			0.2			20038.74			
4989.066	4990.458			0.1			20038.24			
4989.243	4990.635	G	0.3	0.6	0.3		20037.53			
4989.407	4990.799		0.1	0.1			20036.87			
4989.664	4991.056			0.1			20035.84	4e 2	2c 3	P 4
4989.930	4991.323		0.1	0.2			20034.77			
4990.147	4991.539	G	0.5	1.0	0.2	1.23	20033.90	X 2	2B 6	P 2
4990.147	4991.539	G	0.5	1.0	0.2	1.23	20033.90	3A 1	2B 4	R 0
4990.336	4991.729						20033.14			
4990.446	4991.838		0.1	0.3			20032.70	4D 0	2B 6	R 6
4990.446	4991.838		0.1	0.3			20032.70	4e 0	2c 1	P 3
4990.635	4992.028			0.1			20031.94			
4990.820	4992.212			0.1			20031.20	3D 3	2B 6	P 4
4990.952	4992.344			0.3	0.2		20030.67			
4991.099	4992.491			0.0			20030.08			
4991.286	4992.678			0.2			20029.33			
4991.458	4992.850			0.1d			20028.64			
4991.542	4992.935						20028.30			
4991.627	4993.020			0.0d			20027.96			
4991.717	4993.109		0.1				20027.60			
4991.862	4993.254			0.0d			20027.02			
4992.038	4993.431			0.1			20026.31			
4992.163	4993.556		0.2				20025.81			
4992.238	4993.631				0.1		20025.51	4e 0	2c 1	Q 5
4992.320	4993.713			0.1			20025.18	3c 3	2a 1	P 5
4992.559	4993.952			0.3	0.2		20024.22	3F 0	2B 2	Q 5
4992.559	4993.952			0.3	0.2		20024.22	4d 0	2c 1	Q 2
4992.737	4994.129			0.0			20023.51			
4992.884	4994.277			0.1			20022.92			
4992.966	4994.359		0.1				20022.59			
4993.078	4994.471		0.1		0.1		20022.14	4d 0	2c 1	P 2

λ_{air}	λ_{vac}	Ref.	I_1	I_2	I_3	I_4	ν	Upper	Lower	Br.
4993.198	4994.591						20021.66			
4993.275	4994.668			0.1			20021.35			
4993.335	4994.728				0.1		20021.11			
4993.490	4994.883			0.1			20020.49			
4993.664	4995.057		0.2d	0.3	0.2		20019.79			
4993.861	4995.254			0.2			20019.00	4d 0	2c 1	R 3
4994.088	4995.482	G	0.4	0.9	0.2	0.89	20018.09	Z 0	2B 1	P 2
4994.258	4995.651		0.2				20017.41			
4994.315	4995.709			0.2			20017.18	7c 0	2a 3	R 1
4994.530	4995.923			0.2	0.2		20016.32	4d 2	2c 3	P 1
4994.819	4996.213			0.1d			20015.16			
4995.109	4996.502		0.2	0.2			20014.00	4d 1	2c 2	R 2
4995.229	4996.622	G	0.2	0.6	0.3	0.52	20013.52			
4995.453	4996.847			0.1			20012.62			
4995.583	4996.977			0.2			20012.10	4e 2	2c 3	P 5
4995.735	4997.129			0.1			20011.49	3F 3	2B 7	Q 3
4995.925	4997.319			0.2			20010.73			
4996.125	4997.519			0.2			20009.93	4E 0	2B 6	P 2
4996.280	4997.674			0.2			20009.31			
4996.377	4997.771		0.1				20008.92			
4996.467	4997.861	G	0.5	1.0	0.2	0.89	20008.56	4b 1	2a 1	P 2
4996.652	4998.046		0.2	0.1			20007.82	u 2	2a 1	R 2
4996.762	4998.156		0.1				20007.38			
4996.847	4998.241	G	0.5	1.5	0.5	1.03	20007.04	3D 0	2B 1	P 4
4997.096	4998.490		0.2	0.1			20006.04			
4997.264	4998.658			0.0			20005.37			
4997.506	4998.900			0.2			20004.40	3d 3	2c 1	P 3
4997.609	4999.003			0.1			20003.99	4e 1	2c 2	P 6
4997.719	4999.113		0.2d	0.2			20003.55			
4997.938	4999.333	G	0.5	1.5	1.0	1.07	20002.67	3D 1	2B 3	R 6
4997.938	4999.333	G	0.5	1.5	1.0	1.07	20002.67	5c 1	2a 3	R 1
4997.938	4999.333	G	0.5	1.5	1.0	1.07	20002.67	u 1	2a 0	P 3
4998.136	4999.530		0.1	0.1			20001.88			
4998.423	4999.818			0.1			20000.73			
4998.506	4999.900						20000.40	4c 0	2a 1	R 5
4998.583	4999.978			0.0			20000.09	4e 1	2c 2	P 3
4998.795	5000.190			0.2			19999.24			
4998.995	5000.390			0.1			19998.44			
4999.163	5000.558			0.2			19997.77			
4999.360	5000.755			0.2			19996.98			
4999.576	5000.970		0.0	0.3			19996.12			
4999.728	5001.123			0.0			19995.51			
4999.878	5001.273		0.1	0.2			19994.91			
5000.021	5001.415		0.1	0.2			19994.34	5c 0	2a 2	R 0
5000.138	5001.533		0.0				19993.87	4e 0	2c 1	P 6
5000.223	5001.618	G	0.3	0.9	0.2	0.85	19993.53	W 1	2B 4	R 3
5000.383	5001.778		0.2				19992.89			
5000.453	5001.848			0.2			19992.61	4E 0	2B 6	P 4
5000.656	5002.051		0.2	0.4	0.3		19991.80	4e 1	2c 2	Q 5
5000.806	5002.201			0.4			19991.20			
5001.111	5002.506			0.2			19989.98	4E 0	2B 6	Q 4
5001.314	5002.709		0.1	0.3			19989.17	3A 2	2B 6	R 3
5001.501	5002.897		0.1	0.1			19988.42			
5001.769	5003.164			0.2			19987.35			
5001.862	5003.257		0.2				19986.98			
5001.929	5003.325			0.4b	0.3		19986.71			
5002.185	5003.580			0.1d			19985.69			
5002.257	5003.653				0.1		19985.40			
5002.488	5003.883			0.0			19984.48			
5002.590	5003.986		0.1d				19984.07			

λ_{air}	λ_{vac}	Ref.	I_1	I_2	I_3	I_4	ν	Upper	Lower	Br.
5002.775	5004.171			0.2			19983.33	3f 1	2c 0	R 4
5002.996	5004.391		0.1d	0.1			19982.45			
5003.176	5004.572			0.2			19981.73	4d 2	2c 3	R 1
5003.309	5004.704		0.2				19981.20			
5003.394	5004.790	G	0.9	1.5	0.6	1.60	19980.86	3D 1	2B 3	R 0
5003.544	5004.940		0.2				19980.26			
5003.597	5004.992						19980.05			
5003.639	5005.035			0.1			19979.88			
5003.807	5005.203		0.2	0.4	0.2		19979.21	3E 2	2B 5	Q 7
5003.807	5005.203		0.2	0.4	0.2		19979.21	u 2	2a 1	R 1
5004.040	5005.436		0.1	0.1			19978.28			
5004.243	5005.639		0.0	0.1	0.1		19977.47			
5004.436	5005.832		0.0	0.0d			19976.70	4c 0	2a 1	R 4
5004.759	5006.155			0.2			19975.41	4d 2	2c 3	Q 1
5004.939	5006.336			0.2	0.1		19974.69			
5005.180	5006.576			0.1			19973.73	4e 0	2c 1	Q 6
5005.531	5006.927		0.1	0.1			19972.33	3d 3	2c 1	Q 4
5005.531	5006.927		0.1	0.1			19972.33	4d 0	2c 1	R 4
5005.739	5007.135			0.2			19971.50			
5005.934	5007.331		0.1	0.3			19970.72			
5006.280	5007.677		0.1	0.2			19969.34	3E 0	2B 2	R 5
5006.453	5007.850		0.0	0.3	0.2		19968.65	4E 0	2B 6	Q 1
5006.651	5008.048			0.1			19967.86			
5006.737	5008.133		0.2				19967.52			
5006.849	5008.246			0.1			19967.07			
5006.950	5008.346		0.1				19966.67			
5007.043	5008.439			0.2	0.1		19966.30			
5007.248	5008.645			0.2			19965.48			
5007.326	5008.723				0.1		19965.17			
5007.399	5008.795			0.1			19964.88			
5007.471	5008.868		0.2d				19964.59			
5007.567	5008.964			0.2d			19964.21			
5007.770	5009.167						19963.40	4E 1	2B 8	Q 4
5007.828	5009.224			0.3	0.4		19963.17	4b 0	2a 0	P 6
5007.898	5009.295		0.1				19962.89	X 2	2B 6	P 3
5007.986	5009.383	G	0.8	1.5	0.5	1.45	19962.54	3D 1	2B 3	R 1
5008.141	5009.538		0.1				19961.92			
5008.211	5009.608						19961.64			
5008.254	5009.651		0.2	0.2			19961.47			
5008.430	5009.827	G	0.4	0.8		0.93	19960.77	3D 1	2B 3	P 1
5008.430	5009.827	G	0.4	0.8		0.93	19960.77	3A 1	2B 4	P 4
5008.543	5009.940	G					19960.32			
5008.630	5010.028	G	0.4	1.5	0.7	0.92	19959.97	4d 1	2c 2	Q 2
5008.663	5010.060	G					19959.84	3D 1	2B 3	R 5
5008.831	5010.228		0.1	0.1			19959.17			
5008.947	5010.344			0.1			19958.71	3f 1	2c 0	R 3
5009.019	5010.417		0.1				19958.42	3E 2	2B 5	R 2
5009.019	5010.417		0.1				19958.42	4d 1	2c 2	P 2
5009.087	5010.484			0.1			19958.15			
5009.205	5010.602		0.1	0.0			19957.68			
5009.353	5010.751		0.2	0.5	0.6		19957.09	3E 0	2B 2	Q10
5009.409	5010.806						19956.87			
5009.509	5010.906		0.1				19956.47			
5009.617	5011.014	G	0.4	0.7		0.96	19956.04	4g 1	2a 2	P 5
5009.617	5011.014	G	0.4	0.7		0.96	19956.04	3A 1	2B 4	P 1
5009.752	5011.150		0.2				19955.50			
5009.823	5011.220				0.1		19955.22			
5009.868	5011.265			0.2			19955.04			
5010.071	5011.469			0.2			19954.23			
5010.234	5011.632		0.1	0.0			19953.58			

λ_{air}	λ_{vac}	Ref.	I_1	I_2	I_3	I_4	ν	Upper	Lower	Br.
5010.327	5011.725				0.1		19953.21			
5010.405	5011.803		0.1	0.0			19952.90	3D 3	2B 6	P 5
5010.596	5011.994		0.2b	0.4			19952.14	Z 0	2B 1	P 3
5010.669	5012.067				0.2		19951.85			
5010.739	5012.137			0.0			19951.57	u 2	2a 1	R 0
5010.970	5012.368		0.1	0.3			19950.65			
5011.096	5012.494		0.2				19950.15			
5011.186	5012.584	G	0.7	1.5	0.6	1.72	19949.79	W 1	2B 4	R 2
5011.352	5012.750		0.2				19949.13	5c 1	2a 3	R 0
5011.473	5012.871			0.1d			19948.65			
5011.651	5013.049				0.0		19947.94			
5011.895	5013.293			0.1d			19946.97			
5012.003	5013.401				0.1		19946.54			
5012.061	5013.459		0.1				19946.31			
5012.154	5013.552			0.2			19945.94			
5012.304	5013.702			0.1			19945.34	4e 0	2c 1	P 7
5012.433	5013.831		0.1	0.3	0.2		19944.83	4c 0	2a 1	R 3
5012.611	5014.009			0.2			19944.12			
5012.767	5014.165						19943.50			
5012.837	5014.235		0.2	0.3			19943.22	3D 0	2B 1	P10
5012.943	5014.341		0.2				19942.80			
5013.038	5014.437	GK	0.9b	1.5b	1.0	1.96	19942.42	3D 1	2B 3	R 2
5013.038	5014.437	GK	0.9b	1.5b	1.0	1.96	19942.42	3F 3	2B 7	Q 2
5013.202	5014.600		0.2				19941.77			
5013.332	5014.731		0.2	0.6	0.2		19941.25	3D 0	2B 1	P 5
5013.443	5014.841		0.2				19940.81			
5013.591	5014.990			0.1			19940.22	4E 0	2B 6	Q 3
5013.591	5014.990			0.1			19940.22	4e 1	2c 2	Q 6
5013.943	5015.342			0.1	0.1		19938.82			
5014.145	5015.543			0.0			19938.02	4d 0	2c 1	R 5
5014.270	5015.669		0.1				19937.52			
5014.379	5015.777		0.2				19937.09	3A 1	2B 4	P 3
5014.477	5015.875	G	0.4	1.0	0.2	1.41	19936.70	3A 1	2B 4	P 2
5014.650	5016.049		0.1				19936.01			
5014.708	5016.107			0.0			19935.78			
5014.806	5016.205						19935.39			
5014.869	5016.268		0.1	0.2			19935.14			
5015.068	5016.467	G	0.7	1.5b	1.5	1.74	19934.35	3D 1	2B 3	R 4
5015.214	5016.613		0.1				19933.77			
5015.327	5016.726			0.2			19933.32	5c 0	2a 2	Q 1
5015.523	5016.922		0.2	0.1			19932.54			
5015.679	5017.078	1		0.5	1.5b		19931.92			
5015.780	5017.179			0.2			19931.52	4d 2	2c 3	R 2
5016.039	5017.438			0.1			19930.49			
5016.170	5017.569			0.1			19929.97	5c 0	2a 2	Q 2
5016.341	5017.740	G	0.4	0.9			19929.29	3c 4	2a 2	Q 1
5016.492	5017.891	G	0.6	1.5	0.8	1.52	19928.69	3D 1	2B 3	R 3
5016.711	5018.110			0.0			19927.82	4D 0	2B 6	R 5
5016.918	5018.317			0.2			19927.00			
5017.038	5018.438		0.1				19926.52			
5017.124	5018.523	G	0.5	1.5	0.4	1.21	19926.18	4b 1	2a 1	P 3
5017.207	5018.606						19925.85			
5017.265	5018.664		0.1				19925.62			
5017.358	5018.758			0.2			19925.25			
5017.555	5018.954		0.2				19924.47			
5017.678	5019.077		0.2	0.1	0.1		19923.98	5c 0	2a 2	Q 3
5017.859	5019.259		0.2				19923.26			
5018.018	5019.418			0.2d			19922.63	4E 0	2B 6	Q 2
5018.144	5019.544				0.1		19922.13			
5018.408	5019.808			0.1			19921.08			

λ_{air}	λ_{vac}	Ref.	I_1	I_2	I_3	I_4	ν	Upper	Lower	Br.
5018.575	5019.974			0.1			19920.42			
5018.749	5020.148	G	0.2	0.4	0.0		19919.73	3c 8	2a 5	Q 1
5018.955	5020.355			0.1			19918.91			
5019.192	5020.592			0.1			19917.97			
5019.268	5020.668				0.0		19917.67			
5019.434	5020.834		0.1	0.1			19917.01	4E 1	2B 8	Q 1
5019.434	5020.834		0.1	0.1			19917.01	4E 1	2B 8	Q 3
5019.434	5020.834		0.1	0.1			19917.01	5c 0	2a 2	Q 4
5019.764	5021.164	G		0.2	0.1		19915.70	4b 2	2a 2	R 1
5019.764	5021.164	G		0.2	0.1		19915.70	3f 1	2c 0	R 3
5020.054	5021.454			0.2			19914.55	4d 0	2c 1	R 6
5020.394	5021.795			0.1			19913.20			
5020.566	5021.966		0.1	0.2			19912.52	4e 0	2c 1	P 4
5020.740	5022.140	G	0.4	1.5	0.3	1.31	19911.83	W 1	2B 4	R 1
5020.979	5022.380			0.2			19910.88			
5021.179	5022.579			0.2			19910.09	3A 2	2B 6	R 2
5021.494	5022.894			0.1d			19908.84			
5021.663	5023.063				0.2d		19908.17	5c 0	2a 2	Q 5
5021.847	5023.248						19907.44			
5022.064	5023.465			0.2			19906.58			
5022.172	5023.573			0.0			19906.15			
5022.286	5023.687		0.1	0.2			19905.70	4d 0	2c 1	Q 3
5022.357	5023.757				0.2b		19905.42			
5022.445	5023.846			0.3			19905.07	4c 0	2a 1	R 2
5022.445	5023.846			0.3			19905.07	4d 1	2c 2	R 4
5022.669	5024.070	G	0.2	0.6			19904.18	3c 4	2a 2	Q 2
5022.831	5024.232			0.1			19903.54			
5023.003	5024.404			0.2	0.2d		19902.86	3D 0	2B 1	P 9
5023.497	5024.898			0.1	0.0		19900.90			
5023.671	5025.073			0.1			19900.21			
5023.866	5025.267			0.0			19899.44			
5024.207	5025.608			0.1			19898.09			
5024.272	5025.674		0.1				19897.83			
5024.439	5025.840			0.3			19897.17	3D 3	2B 6	P 6
5024.689	5026.090			0.2d			19896.18			
5024.954	5026.356	G	0.2	0.7	0.5	0.52	19895.13	3D 0	2B 1	P 6
5024.954	5026.356	G	0.2	0.7	0.5	0.52	19895.13	3F 1	2B 4	Q 6
5025.159	5026.560			0.0			19894.32			
5025.333	5026.735			0.1			19893.63	4E 1	2B 8	Q 2
5025.467	5026.869				0.1		19893.10			
5025.505	5026.906			0.2			19892.95			
5025.737	5027.139			0.2			19892.03			
5025.917	5027.318	G	0.0	0.3	0.3d		19891.32	3F 0	2B 2	P 6
5026.154	5027.556			0.1			19890.38			
5026.387	5027.789			0.1			19889.46			
5026.566	5027.968			0.0			19888.75			
5026.786	5028.188			0.1			19887.88	4e 1	2c 2	P 4
5026.966	5028.367			0.2			19887.17			
5027.094	5028.496		0.0				19886.66			
5027.249	5028.651			0.1d			19886.05			
5027.517	5028.919			0.1			19884.99			
5027.600	5029.002		0.2				19884.66			
5027.701	5029.103	G	0.3	0.9	0.5	0.61	19884.26	3E 2	2B 5	Q 6
5027.769	5029.172						19883.99			
5027.873	5029.275			0.0			19883.58	3F 3	2B 7	P 4
5028.040	5029.442			0.1			19882.92			
5028.108	5029.511				0.1		19882.65			
5028.424	5029.827			0.2			19881.40			
5028.652	5030.055		0.2	0.6	0.3		19880.50	X 2	2B 6	P 4
5028.842	5030.244	G	0.4	0.9	0.2	1.02	19879.75	3D 1	2B 3	P 2

λ_{air}	λ_{vac}	Ref.	I_1	I_2	I_3	I_4	ν	Upper	Lower	Br.
5029.062	5030.464			0.1			19878.88	4d 2	2c 3	R 3
5029.272	5030.674			0.1			19878.05	4d 2	2c 3	Q 2
5029.487	5030.890			0.1			19877.20			
5029.662	5031.064			0.2			19876.51			
5029.892	5031.295			0.4	0.2		19875.60	5c 1	2a 3	Q 1
5030.125	5031.528			0.2			19874.68	3E 2	2B 5	R 1
5030.178	5031.581						19874.47			
5030.365	5031.768	GK	0.7	1.5	0.5	1.62	19873.73	Z 0	2B 1	P 4
5030.365	5031.768	GK	0.7	1.5	0.5	1.62	19873.73	W 1	2B 4	R 0
5030.492	5031.895		0.1				19873.23	5c 1	2a 3	Q 2
5030.611	5032.014			0.1			19872.76			
5030.727	5032.130				0.1		19872.30			
5030.844	5032.247			0.0			19871.84			
5030.998	5032.401			0.3	0.2		19871.23	3D 0	2B 1	P 7
5031.216	5032.619			0.0			19870.37			
5031.335	5032.738		0.2				19869.90			
5031.545	5032.948			0.1			19869.07			
5031.705	5033.108			0.1	0.0		19868.44	5c 1	2a 3	Q 3
5031.889	5033.293			0.2			19867.71			
5031.998	5033.402			0.0			19867.28			
5032.145	5033.549	G	0.3b	0.9	0.4	0.65	19866.70	3F 0	2B 2	Q 4
5032.145	5033.549	G	0.3b	0.9	0.4	0.65	19866.70	3c 4	2a 2	Q 3
5032.363	5033.766		0.2d	0.4	0.2		19865.84	3F 0	2B 2	R 2
5032.616	5034.020			0.1			19864.84			
5033.014	5034.418			0.1			19863.27			
5033.240	5034.643			0.2	0.1		19862.38	5c 1	2a 3	Q 4
5033.306	5034.709						19862.12	3f 1	2c 0	R 2
5033.397	5034.801			0.2			19861.76			
5033.706	5035.110			0.2	0.0		19860.54			
5034.046	5035.449			0.1			19859.20			
5034.261	5035.665	G	0.3	0.6	0.3	0.48	19858.35	4c 0	2a 1	R 1
5034.477	5035.881			0.2			19857.50			
5034.758	5036.162			0.1	0.1		19856.39			
5034.969	5036.373			0.2			19855.56	3d 3	2c 1	P 4
5035.103	5036.507		0.2				19855.03			
5035.270	5036.675			0.3	0.2		19854.37	3D 0	2B 1	P 8
5035.270	5036.675			0.3	0.2		19854.37	5c 1	2a 3	Q 5
5035.417	5036.822			0.2			19853.79			
5035.570	5036.974			0.2			19853.19	V 0	2B 1	P 4
5035.712	5037.116			0.4	0.2		19852.63	4d 0	2c 1	P 3
5035.795	5037.200		0.1d				19852.30			
5035.940	5037.344			0.5			19851.73			
5036.267	5037.672			0.2			19850.44			
5036.419	5037.824			0.2	0.0		19849.84	3d 3	2c 1	Q 5
5036.551	5037.956			0.3			19849.32	4D 0	2B 6	R 4
5036.711	5038.116		0.1	0.4	0.1		19848.69			
5036.861	5038.266		0.0				19848.10			
037.013	5038.418		0.2	0.1			19847.50			
5037.135	5038.540		0.2				19847.02			
5037.315	5038.720		0.2				19846.31			
5037.511	5038.916			0.0			19845.54			
5038.074	5039.479			0.1			19843.32			
5038.371	5039.776			0.2			19842.15	4E 2	2B10	Q 2
5038.371	5039.776			0.2			19842.15	3f 1	2c 0	R 2
5038.562	5039.967			0.4	0.5		19841.40			
5038.745	5040.150			0.2			19840.68			
5038.968	5040.373		0.2	0.4	0.2		19839.80			
5039.197	5040.602						19838.90			
5039.258	5040.663			0.2	0.1d		19838.66			
5039.593	5040.998		0.2	0.2			19837.34			

λ_{air}	λ_{vac}	Ref.	I_1	I_2	I_3	I_4	ν	Upper	Lower	Br.
5039.824	5041.230	GK	0.7	1.5	0.7	1.72	19836.43	Z 1	2B 3	R 1
5039.824	5041.230	GK	0.7	1.5	0.7	1.72	19836.43	Z 1	2B 3	R 2
5040.226	5041.631			0.1			19834.85	3A 2	2B 6	R 1
5040.434	5041.840		0.2d	0.5	0.2		19834.03	4b 1	2a 1	P 4
5040.653	5042.058			0.0			19833.17			
5040.813	5042.219			0.1d			19832.54			
5040.930	5042.335				0.1		19832.08			
5041.207	5042.613			0.0			19830.99			
5041.418	5042.824			0.2			19830.16			
5041.621	5043.027	G	0.7	1.5	0.4	1.56	19829.36	Z 1	2B 3	R 0
5041.865	5043.271			0.1			19828.40			
5042.020	5043.426			0.3	0.3		19827.79			
5042.234	5043.640						19826.95			
5042.529	5043.935			0.0			19825.79			
5042.839	5044.246			0.0			19824.57			
5043.048	5044.454			0.4	0.0		19823.75			
5043.335	5044.742			0.1	0.0		19822.62	4d 2	2c 3	R 4
5043.661	5045.068			0.2			19821.34			
5043.771	5045.177						19820.91	4D 0	2B 6	R 0
5043.928	5045.335	G	0.2	0.6	0.1	0.83	19820.29			
5044.076	5045.482			0.0			19819.71			
5044.137	5045.544						19819.47			
5044.264	5045.671			0.3	0.2		19818.97			
5044.427	5045.834		0.1				19818.33			
5044.531	5045.938			0.2			19817.92			
5044.735	5046.142		0.1	0.3	0.1		19817.12	3c 4	2a 2	Q 4
5044.880	5046.287			0.1			19816.55	3f 1	2c 0	Q 7
5044.992	5046.399		0.1	0.2			19816.11	4D 0	2B 6	P 1
5045.176	5046.582			0.3	0.2		19815.39			
5045.341	5046.748			0.3	0.2		19814.74	4E 0	2B 2	R 4
5045.670	5047.077			0.2			19813.45			
5046.131	5047.538			0.0d			19811.64			
5046.582	5047.989			0.1d			19809.87			
5047.027	5048.435			0.0			19808.12			
5047.542	5048.950		0.1				19806.10			
5047.733	5049.141	G	0.3d	0.8	0.9	0.62	19805.35	3E 2	2B 5	Q 5
5047.733	5049.141	G	0.3d	0.8	0.9	0.62	19805.35	V 0	2B 1	P 3
5047.733	5049.141	G	0.3d	0.8	0.9	0.62	19805.35	4c 0	2a 1	R 0
5047.996	5049.403	G	0.5	1.5		1.21	19804.32	W 1	2B 4	P 1
5048.230	5049.638			0.1			19803.40			
5048.705	5050.112			0.2	0.2		19801.54	X 2	2B 6	P 5
5048.896	5050.304			0.0			19800.79	4e 0	2c 1	P 5
5049.105	5050.513		0.2	0.5	0.3		19799.97	3E 0	2B 2	Q 9
5049.337	5050.745	G	0.3	0.9		1.00	19799.06	3E 2	2B 5	P 6
5049.418	5050.826				0.5		19798.74			
5049.515	5050.923		0.2	0.5			19798.36			
5049.768	5051.176			0.1			19797.37			
5049.811	5051.219		0.1		0.1		19797.20			
5049.929	5051.337			0.1			19796.74			
5050.056	5051.464		0.2	0.2			19796.24	4d 1	2c 2	P 3
5050.214	5051.622			0.2			19795.62			
5050.416	5051.824	G	0.3	0.5	0.1	0.64	19794.83	3D 1	2B 3	P 3
5050.648	5052.056		0.2	0.2			19793.92			
5050.794	5052.202			0.1	0.1		19793.35			
5051.151	5052.559			0.2	0.1		19791.95	3f 1	2c 0	Q 6
5051.286	5052.695			0.0			19791.42			
5051.462	5052.871		0.2	0.3			19790.73	3E 2	2B 5	R 0
5051.462	5052.871		0.2	0.3			19790.73	3A 2	2B 6	P 6
5051.628	5053.037				0.2b		19790.08			
5051.702	5053.111			0.0			19789.79			

λ$_{air}$	λ$_{vac}$	Ref.	I$_1$	I$_2$	I$_3$	I$_4$	ν	Upper	Lower	Br.
5051.945	5053.353			0.1d			19788.84			
5052.024	5053.432		0.2				19788.53			
5052.389	5053.798		0.1				19787.10	u 1	2a 0	P 5
5052.501	5053.910		0.1	0.0			19786.66			
5052.606	5054.015		0.4				19786.25			
5052.721	5054.130	G	0.2	0.4	0.2		19785.80			
5052.938	5054.347		0.1				19784.95			
5053.071	5054.480			0.2d			19784.43			
5053.160	5054.569						19784.08			
5053.216	5054.625						19783.86			
5053.357	5054.766				0.0		19783.31			
5053.607	5055.016				0.0		19782.33			
5053.686	5055.096			0.1			19782.02			
5053.957	5055.366		0.2	0.3			19780.96	Z 0	2B 1	P 5
5053.957	5055.366		0.2	0.3			19780.96	4e 1	2c 2	P 5
5054.118	5055.527				0.1		19780.33			
5054.269	5055.678		0.2				19779.74			
5054.366	5055.775			0.1			19779.36			
5054.650	5056.059		0.2	0.2			19778.25			
5054.824	5056.233	G	0.3	0.9	0.2		19777.57	Z 1	2B 3	P 1
5054.824	5056.233	G	0.3	0.9	0.2		19777.57	5c 1	2a 3	P 2
5055.094	5056.504	GK	0.7	1.5b	0.6	1.83	19776.51	W 1	2B 4	P 2
5055.258	5056.668		0.1				19775.87			
5055.358	5056.767			0.1d			19775.48			
5055.442	5056.852		0.2				19775.15			
5055.767	5057.176			0.2	0.0		19773.88			
5055.874	5057.284		0.1	0.2			19773.46	4D 0	2B 6	R 2
5056.132	5057.542		0.2	0.3			19772.45			
5056.291	5057.701	G	0.2	0.6	0.4		19771.83			
5056.483	5057.893			0.1			19771.08	4D 0	2B 6	R 1
5056.634	5058.044			0.3	0.2		19770.49	3E 2	2B 5	P 5
5056.634	5058.044			0.3	0.2		19770.49	3f 1	2c 0	R 1
5056.897	5058.307			0.3	0.2		19769.46	4g 1	2a 2	P 6
5057.094	5058.504			0.1			19768.69			
5057.332	5058.742	G	0.2	0.3	0.3		19767.76			
5057.509	5058.919			0.0			19767.07	3f 2	2c 1	R 5
5057.744	5059.154				0.0		19766.15			
5057.798	5059.208			0.1d			19765.94			
5058.013	5059.423	G		0.3	0.2		19765.10	3f 1	2c 0	R 1
5058.450	5059.861				0.0		19763.39			
5058.850	5060.260			0.0	0.0		19761.83	u 2	2a 1	P 2
5058.978	5060.388						19761.33			
5059.075	5060.485			0.0			19760.95	4d 3	2c 4	P 1
5059.303	5060.713			0.1d			19760.06			
5059.561	5060.972			0.0			19759.05	3f 1	2c 0	Q 5
5059.674	5061.085			0.0			19758.61			
5059.869	5061.279		0.2	0.1			19757.85			
5060.010	5061.420			0.1	0.1		19757.30	5c 0	2a 2	P 3
5060.109	5061.520						19756.91			
5060.194	5061.605			0.2			19756.58	3f 2	2c 1	R 6
5060.389	5061.799			0.4	0.3		19755.82	3F 0	2B 2	P 5
5060.627	5062.038		0.2				19754.89	V 0	2B 1	P 2
5060.752	5062.163						19754.40			
5060.804	5062.215			0.0d			19754.20			
5060.955	5062.366		0.2				19753.61			
5061.073	5062.484		0.2				19753.15			
5061.303	5062.714			0.1			19752.25			
5061.357	5062.768		0.2				19752.04			
5061.544	5062.955			0.1			19751.31			
5061.634	5063.045		0.2				19750.96			

λ_{air}	λ_{vac}	Ref.	I_1	I_2	I_3	I_4	ν	Upper	Lower	Br.
5061.731	5063.142	G	0.5	1.5	0.5	1.43	19750.58	W 1	2B 4	P 3
5061.964	5063.376			0.1			19749.67			
5062.105	5063.517			0.0			19749.12			
5062.387	5063.799		0.2	0.3			19748.02	3A 2	2B 6	R 0
5062.608	5064.019			0.1			19747.16			
5062.782	5064.194						19746.48			
5063.131	5064.543		0.1	0.1d			19745.12			
5063.439	5064.850			0.3	0.3		19743.92			
5063.490	5064.902		0.1				19743.72			
5063.575	5064.986		0.2	0.2			19743.39			
5063.685	5065.097		0.1				19742.96			
5063.752	5065.163	G	0.4				19742.70	3E 2	2B 5	P 4
5063.828	5065.240			1.5b		1.52	19742.40	3E 2	2B 5	Q 4
5063.882	5065.294	G	0.4		0.9		19742.19	4c 0	2a 1	Q 1
5064.111	5065.523			0.0			19741.30			
5064.416	5065.828				0.0		19740.11			
5065.196	5066.608			0.0			19737.07			
5065.312	5066.724						19736.62			
5065.396	5066.808			0.1			19736.29			
5065.555	5066.968						19735.67			
5065.635	5067.047		0.1				19735.36			
5065.768	5067.181		0.2				19734.84			
5065.853	5067.265			0.1			19734.51			
5065.979	5067.391				0.1		19734.02			
5066.035	5067.448			0.2			19733.80			
5066.120	5067.532						19733.47			
5066.169	5067.581			0.0			19733.28	7c 0	2a 3	P 3
5066.331	5067.743	G	0.2	0.5	0.3		19732.65	4c 0	2a 1	Q 2
5066.559	5067.972			0.0			19731.76			
5066.749	5068.162			0.2			19731.02	4D 0	2B 6	P 2
5066.978	5068.390			0.3	0.2		19730.13	3F 0	2B 2	Q 3
5067.196	5068.609						19729.28			
5067.219	5068.632		0.2	0.4			19729.19			
5067.299	5068.711						19728.88			
5067.473	5068.886	G	0.5	1.5	0.3	1.41	19728.20	Z 1	2B 3	P 2
5067.473	5068.886	G	0.5	1.5	0.3	1.41	19728.20	3F 1	2B 4	Q 5
5067.697	5069.110			0.0			19727.33			
5067.928	5069.341			0.2			19726.43			
5068.123	5069.536	G	0.6	1.5	0.7	1.68	19725.67	W 1	2B 4	P 4
5068.324	5069.737						19724.89	3F 1	2B 4	R 3
5068.457	5069.870			0.0d			19724.37	3A 2	2B 6	P 5
5068.779	5070.192				0.1		19723.12			
5068.833	5070.246			0.0			19722.91			
5069.025	5070.439		0.1				19722.16			
5069.185	5070.598			0.1			19721.54			
5069.280	5070.693				0.1		19721.17			
5069.434	5070.847			0.2	0.1		19720.57	3f 1	2c 0	Q 4
5069.493	5070.906		0.1				19720.34	V 0	2B 1	P 5
5069.668	5071.081	G	0.3	1.0	0.2	0.91	19719.66	3f 2	2c 1	R 5
5069.668	5071.081	G	0.3	1.0	0.2	0.91	19719.66	3E 2	2B 5	P 3
5069.835	5071.249			0.0			19719.01	5c 1	2a 3	P 3
5069.977	5071.390	G	0.2d	0.6	0.5		19718.46	4c 0	2a 1	Q 3
5070.290	5071.704	G	0.2d	0.9	0.4	0.95	19717.24	W 1	2B 4	P 6
5070.488	5071.902			0.1			19716.47			
5070.658	5072.072			0.1			19715.81			
5071.093	5072.506			0.2			19714.12			
5071.162	5072.576		0.2		0.2d		19713.85			
5071.391	5072.805		0.2	0.2d			19712.96	T 0	2B 6	P 1
5071.790	5073.204			0.2	0.1		19711.41	3f 1	2c 0	Q 7
5072.068	5073.482			0.2			19710.33			

λ_{air}	λ_{vac}	Ref.	I_1	I_2	I_3	I_4	ν	Upper	Lower	Br.
5072.153	5073.567		0.1				19710.00			
5072.307	5073.721	G	0.5	1.0	0.3	1.00	19709.40	3D 1	2B 3	P 4
5072.619	5074.033	G	0.2	0.9	0.3	1.01	19708.19	W 1	2B 4	P 5
5073.002	5074.416				0.2		19706.70			
5073.141	5074.555		0.1				19706.16			
5073.293	5074.707						19705.57			
5073.339	5074.754			0.1			19705.39			
5073.432	5074.846				0.2		19705.03	4e 3	2c 4	R 1
5073.520	5074.934						19704.69	3f 2	2c 1	R 4
5073.674	5075.088			0.2	0.1		19704.09	T 0	2B 6	P 2
5073.970	5075.385			0.0			19702.94	4e 3	2c 4	R 2
5074.215	5075.629			0.2	0.1		19701.99	3A 2	2B 6	P 4
5074.545	5075.959			0.1d			19700.71	5c 0	2a 2	P 4
5074.810	5076.225			0.2	0.2		19699.68	4c 0	2a 1	Q 4
5075.086	5076.500			0.4	0.2		19698.61	4e 3	2c 4	R 3
5075.204	5076.619			0.2			19698.15	3F 0	2B 2	R 1
5075.279	5076.694		0.1				19697.86			
5075.439	5076.853	G	0.8	1.5	0.4	1.37	19697.24	3E 2	2B 5	P 2
5075.670	5077.085		0.1				19696.34			
5075.722	5077.137			0.1	0.2		19696.14			
5075.928	5077.343			0.1			19695.34			
5076.140	5077.555	G	0.2	1.0	0.4	1.01	19694.52	3E 2	2B 5	Q 3
5076.140	5077.555	G	0.2	1.0	0.4	1.01	19694.52	3f 1	2c 0	Q 6
5076.307	5077.722						19693.87	4e 0	2c 1	P 6
5076.495	5077.910			0.0d			19693.14			
5076.884	5078.300			0.0d			19691.63	3d 3	2c 1	P 5
5077.204	5078.620	G	0.4	0.8	0.2	0.89	19690.39	3D 2	2B 5	P 1
5077.204	5078.620	G	0.4	0.8	0.2	0.89	19690.39	3E 0	2B 2	R 3
5077.452	5078.867		0.0	0.1	0.1		19689.43	3f 1	2c 0	Q 5
5077.694	5079.110			0.2			19688.49			
5077.911	5079.326			0.1			19687.65			
5078.125	5079.541			0.1			19686.82	Z 0	2B 1	P 6
5078.458	5079.873			0.2			19685.53	3A 2	2B 6	P 3
5078.458	5079.873			0.2			19685.53	3E 0	2B 2	P 8
5078.749	5080.165	G	0.2	0.7	0.4	0.66	19684.40	3D 2	2B 5	R 6
5078.876	5080.291				0.3		19683.91	T 0	2B 6	P 3
5078.925	5080.340		0.3	0.4			19683.72			
5079.180	5080.596		0.0				19682.73			
5079.301	5080.717			0.3	0.2		19682.26	3f 1	2c 0	Q 3
5079.562	5080.978		0.0				19681.25			
5079.704	5081.120			0.1			19680.70			
5079.750	5081.167			0.2			19680.52	T 0	2B 6	P 4
5080.047	5081.463			0.2			19679.37			
5080.091	5081.507		0.1				19679.20			
5080.213	5081.629			0.1			19678.73	4d 0	2c 1	P 4
5080.290	5081.706		0.1	0.0			19678.43	3f 1	2c 0	Q 4
5080.499	5081.915	G	1.0	1.5	0.5	1.65	19677.62	3D 2	2B 5	R 2
5080.499	5081.915	G	1.0	1.5	0.5	1.65	19677.62	3e 1	2c 0	R 5
5080.499	5081.915	G	1.0	1.5	0.5	1.65	19677.62	4e 1	2c 2	P 6
5080.682	5082.099		0.1				19676.91			
5080.791	5082.207		0.2	0.3	0.4		19676.49	3A 2	2B 6	P 1
5080.791	5082.207		0.2	0.3	0.4		19676.49	4c 0	2a 1	Q 5
5080.894	5082.311				0.4		19676.09			
5080.956	5082.373	G	0.2	1.0		1.22	19675.85	3D 2	2B 5	R 3
5081.225	5082.641		0.1	0.1			19674.81	4d 2	2c 3	Q 4
5081.434	5082.850	G	0.9	0.9	0.4	1.15	19674.00	3D 2	2B 5	R 1
5081.434	5082.850	G	0.9	0.9	0.4	1.15	19674.00	3f 2	2c 1	R 4
5081.700	5083.117		0.1	0.2			19672.97			
5081.927	5083.344	G	0.3	1.5	0.6	1.47	19672.09	3D 2	2B 5	R 4
5082.075	5083.491				0.4		19671.52			

λ_{air}	λ_{vac}	Ref.	I_1	I_2	I_3	I_4	ν	Upper	Lower	Br.
5082.119	5083.535			0.5			19671.35			
5082.258	5083.675						19670.81			
5082.330	5083.747			0.1			19670.53			
5082.514	5083.931			0.2			19669.82			
5082.770	5084.186	G	0.4	0.9	0.4	1.08	19668.83	Z 1	2B 3	P 3
5082.770	5084.186	G	0.4	0.9	0.4	1.08	19668.83	3D 2	2B 5	R 5
5082.979	5084.396			0.0			19668.02			
5083.186	5084.603		0.2	0.6	0.2b		19667.22			
5083.364	5084.781			0.1			19666.53			
5083.736	5085.153			0.0d			19665.09	u 2	2a 1	P 3
5083.910	5085.327				0.1		19664.42			
5084.220	5085.637				0.0		19663.22			
5084.468	5085.885			0.2	0.1		19662.26	3f 1	2c 0	Q 3
5084.613	5086.030		0.1	0.0			19661.70			
5084.843	5086.260	G	1.5	1.5b	0.6	1.66	19660.81	3D 2	2B 5	R 0
5084.843	5086.260	G	1.5	1.5b	0.6	1.66	19660.81	3E 2	2B 5	Q 2
5085.112	5086.529			0.0			19659.77			
5085.278	5086.695			0.1			19659.13	5c 1	2a 3	P 4
5085.443	5086.861	G		0.6	0.9		19658.49	3E 0	2B 2	Q 8
5085.575	5086.993		0.2	0.3			19657.98	3A 2	2B 6	P 2
5085.878	5087.295			0.1	0.1		19656.81			
5086.212	5087.629			0.1			19655.52			
5086.382	5087.800				0.1		19654.86			
5086.496	5087.914			0.1			19654.42	4F 0	2B 7	R 3
5086.820	5088.238			0.0			19653.17			
5087.084	5088.502			0.0			19652.15			
5087.376	5088.794			0.0			19651.02			
5087.539	5088.958			0.1			19650.39			
5087.726	5089.144			0.2			19649.67	3f 1	2c 0	Q 2
5087.884	5089.302			0.2	0.2		19649.06	3e 1	2c 0	R 4
5088.184	5089.602			0.2			19647.90			
5088.422	5089.841			0.1	0.0		19646.98			
5088.658	5090.077			0.1			19646.07			
5088.834	5090.253				0.1		19645.39			
5088.891	5090.310			0.1			19645.17			
5089.106	5090.525	G	0.1	0.4	0.2		19644.34	3f 1	2c 0	Q 2
5089.339	5090.758			0.1			19643.44			
5089.720	5091.139				0.0		19641.97			
5089.992	5091.411			0.2			19640.92			
5090.221	5091.639	G	0.2	0.9	0.3b	0.95	19640.04	3E 2	2B 5	Q 1
5090.436	5091.854			0.2			19639.21	4c 0	2a 1	Q 6
5090.858	5092.277			0.1	0.1		19637.58			
5091.071	5092.490			0.1			19636.76	3f 2	2c 1	R 3
5091.237	5092.656			0.1			19636.12			
5091.374	5092.793				0.0		19635.59			
5091.452	5092.871			0.0			19635.29	4d 1	2c 2	P 4
5091.958	5093.377				0.0		19633.34			
5092.105	5093.525			0.1			19632.77			
5092.287	5093.706	G		0.6	0.2	0.63	19632.07	3D 1	2B 3	P 5
5092.287	5093.706	G		0.6	0.2	0.63	19632.07	W 2	2B 6	R 4
5092.559	5093.979			0.4	0.5		19631.02			
5092.917	5094.337			0.1d			19629.64			
5093.322	5094.742	G		0.4	0.2		19628.08			
5093.709	5095.129			0.1d			19626.59			
5094.350	5095.770			0.2d			19624.12			
5094.643	5096.063		0.1	0.6			19622.99	3F 0	2B 2	P 4
5094.706	5096.126	G			0.5		19622.75			
5094.758	5096.178			0.5		0.51	19622.55			
5094.981	5096.401			0.1			19621.69	4D 0	2B 6	P 3
5095.235	5096.656		0.0	0.1d			19620.71			

λ_{air}	λ_{vac}	Ref.	I₁	I₂	I₃	I₄	ν	Upper	Lower	Br.
5095.355	5096.775						19620.25			
5095.576	5096.996			0.3	0.1		19619.40	3f 2	2c 1	R 3
5095.770	5097.191			0.2	0.2		19618.65			
5095.965	5097.385			0.1	0.2		19617.90	3F 3	2C 0	R 2
5095.965	5097.385			0.1	0.2		19617.90	4c 0	2a 1	Q 7
5096.147	5097.567	G	0.2	0.6	0.2	0.64	19617.20	4b 2	2a 2	P 3
5096.147	5097.567	G	0.2	0.6	0.2	0.64	19617.20	3F 0	2B 2	Q 2
5096.311	5097.731		0.1	0.6	0.2		19616.57			
5096.526	5097.947		0.2	0.3			19615.74	4c 0	2a 1	P 2
5096.677	5098.098			0.0			19615.16			
5096.890	5098.311			0.1			19614.34			
5097.002	5098.422						19613.91			
5097.184	5098.604						19613.21			
5097.426	5098.846			0.0			19612.28			
5097.636	5099.057				0.1		19611.47			
5097.849	5099.270			0.0			19610.65			
5098.192	5099.613		0.1	0.2			19609.33			
5098.541	5099.962			0.2	0.1		19607.99	3e 1	2c 0	R 3
5098.541	5099.962			0.2	0.1		19607.99	5c 1	2a 3	P 5
5098.861	5100.282			0.1			19606.76	4d 0	2c 1	Q 6
5099.269	5100.690				0.1		19605.19			
5099.539	5100.961		0.1	0.1			19604.15			
5099.745	5101.166	G	0.4	0.9	0.3	1.08	19603.36	Z 1	2B 3	P 4
5099.974	5101.395		0.0	0.1			19602.48			
5100.239	5101.661			0.1			19601.46	3D 1	2B 3	P10
5100.239	5101.661			0.1			19601.46	4e 3	2c 4	Q 2
5100.411	5101.833			0.3	0.3d		19600.80	3F 3	2C 0	R 2
5100.708	5102.129			0.1			19599.66			
5100.976	5102.397				0.1d		19598.63			
5101.468	5102.890			0.2	0.1		19596.74			
5101.637	5103.059			0.0			19596.09			
5102.231	5103.653						19593.81			
5102.418	5103.840			0.2	0.1		19593.09	4e 0	2c 1	P 7
5102.697	5104.119			0.1			19592.02			
5102.939	5104.361			0.1			19591.09			
5103.108	5104.531			0.1			19590.44			
5103.296	5104.718	G		0.7	0.9		19589.72	3D 0	2B 2	R10
5103.379	5104.802		0.1				19589.40			
5103.569	5104.992	G	1.0	1.5	0.3	1.40	19588.67	Z 2	2B 5	R 0
5103.747	5105.169		0.0				19587.99			
5103.846	5105.268			0.1			19587.61			
5104.158	5105.581			0.1			19586.41			
5104.356	5105.779				0.1		19585.65			
5104.518	5105.941			0.1			19585.03			
5104.755	5106.178			0.2d			19584.12	3F 3	2C 0	R 1
5105.128	5106.551			0.2			19582.69			
5105.201	5106.624				0.3b		19582.41			
5105.256	5106.678	G		0.3			19582.20	3F 1	2B 4	Q 4
5105.477	5106.900			0.0			19581.35			
5105.563	5106.986						19581.02			
5105.715	5107.137			0.1			19580.44			
5105.868	5107.291			0.1			19579.85			
5106.007	5107.430				0.0		19579.32	4e 1	2c 2	P 7
5106.359	5107.782			0.1			19577.97			
5106.575	5107.998			0.1			19577.14			
5106.787	5108.210	G	0.1	1.0	0.4	0.83	19576.33	3D 1	2B 3	P 6
5106.980	5108.403			0.0			19575.59			
5107.196	5108.620			0.1			19574.76			
5107.329	5108.753			0.1			19574.25			
5107.449	5108.873		0.1	0.2			19573.79			

λ_{air}	λ_{vac}	Ref.	I_1	I_2	I_3	I_4	ν	Upper	Lower	Br.
5107.648	5109.071	G	1.0	1.5	0.2	1.23	19573.03	Z 2	2B 5	R 1
5107.796	5109.220			0.1			19572.46			
5107.843	5109.267		0.0				19572.28			
5107.940	5109.363				0.1		19571.91			
5108.159	5109.583			0.0d			19571.07	u 2	2a 1	P 4
5108.368	5109.792	G		0.5	0.1	0.58	19570.27	W 2	2B 6	R 3
5108.368	5109.792	G		0.5	0.1	0.58	19570.27	4d 2	2c 3	Q 5
5108.647	5110.071			0.1			19569.20			
5108.893	5110.316			0.1			19568.26			
5108.934	5110.358				0.1		19568.10			
5109.141	5110.565		0.2	0.2			19567.31	3D 2	2B 5	P 2
5109.308	5110.732	G	0.9	1.0	0.3	1.16	19566.67	3f 2	2c 1	R 2
5109.308	5110.732	G	0.9	1.0	0.3	1.16	19566.67	4c 1	2a 2	Q 1
5109.504	5110.928		0.1				19565.92			
5109.566	5110.990			0.1			19565.68	4D 0	2B 6	P 4
5109.726	5111.150			0.1	0.1		19565.07			
5109.969	5111.393			0.0			19564.14			
5110.013	5111.437		0.0				19563.97			
5110.209	5111.633			0.4			19563.22	3F 1	2B 4	R 2
5110.454	5111.879		0.0	0.3	0.2		19562.28	3E 0	2B 2	R 2
5110.642	5112.067			0.1			19561.56			
5110.750	5112.174						19561.15			
5110.932	5112.357			0.1			19560.45			
5111.238	5112.663			0.2			19559.28	3f 2	2c 1	R 2
5111.460	5112.885			0.1			19558.43			
5111.735	5113.159	G	0.2	0.7	0.1	0.69	19557.38	4c 1	2a 2	Q 2
5112.001	5113.426			0.1			19556.36			
5112.613	5114.038			0.1			19554.02	V 1	2B 2	R 2
5112.613	5114.038			0.1			19554.02	4D 0	2B 6	P 7
5112.927	5114.352		0.1	0.1			19552.82			
5113.131	5114.556	GK	1.0	1.5	0.5	1.71	19552.04	Z 2	2B 5	R 2
5113.131	5114.556	GK	1.0	1.5	0.5	1.71	19552.04	4e 3	2c 4	P 2
5113.280	5114.705	G	0.5	1.0			19551.47	W 2	2B 6	R 2
5113.280	5114.705	G	0.5	1.0			19551.47	3e 1	2c 0	R 2
5113.447	5114.872			0.0			19550.83			
5113.701	5115.126			0.3	0.3		19549.86			
5113.950	5115.375			0.0			19548.91			
5114.339	5115.765			0.0			19547.42			
5114.557	5115.982			0.1			19546.59			
5115.090	5116.516		0.1d	0.1d			19544.55	3D 1	2B 3	P 9
5115.313	5116.738		0.3		0.5		19543.70	4c 0	2a 1	P 3
5115.339	5116.765	G		0.9		1.03	19543.60	4c 1	2a 2	Q 3
5115.478	5116.903			0.3			19543.07	3D 1	2B 3	P 7
5115.797	5117.223			0.0d			19541.85			
5116.245	5117.671			0.0	0.1		19540.14			
5116.496	5117.922				0.1		19539.18			
5116.551	5117.977		0.3	0.3			19538.97			
5116.748	5118.173			0.1			19538.22	3a 1	2c 0	R 5
5116.905	5118.331	G	0.1	0.5	0.6	0.68	19537.62	3E 0	2B 2	Q 7
5117.012	5118.438			0.2			19537.21			
5117.198	5118.624		0.1	0.1			19536.50			
5117.250	5118.677						19536.30			
5117.389	5118.815	G	0.7	0.8	0.2	0.90	19535.77	W 2	2B 6	R 1
5117.389	5118.815	G	0.7	0.8	0.2	0.90	19535.77	3E 0	2B 2	P 7
5117.617	5119.043			0.0			19534.90			
5117.947	5119.374				0.0		19533.64			
5118.201	5119.628				0.1		19532.67	4D 0	2B 6	P 5
5118.453	5119.879						19531.71			
5118.665	5120.092			0.1d			19530.90			
5118.783	5120.210			0.0			19530.45			

λ_{air}	λ_{vac}	Ref.	I_1	I_2	I_3	I_4	ν	Upper	Lower	Br.
5118.956	5120.383	G	0.3	0.6	0.3		19529.79	3D 1	2B 3	P 8
5119.258	5120.684		0.2	0.3			19528.64	4D 0	2B 6	P 6
5119.452	5120.878			0.0			19527.90			
5119.661	5121.088		0.2	0.3	0.1		19527.10			
5119.887	5121.314			0.1			19526.24			
5120.123	5121.550	G	0.2	0.6	0.3		19525.34	4c 1	2a 2	Q 4
5120.427	5121.854		0.1	0.2			19524.18			
5120.647	5122.074	G	0.4	0.4	0.2		19523.34	3F 0	2B 2	P 3
5120.797	5122.224		0.2	0.2			19522.77			
5120.973	5122.400	G	0.7	0.9	0.2	0.97	19522.10	Z 2	2B 5	R 3
5121.196	5122.623			0.1			19521.25			
5121.700	5123.127			0.1			19519.33			
5121.867	5123.295						19518.69	3D 2	2B 5	P 3
5122.098	5123.526			0.2			19517.81	3d 3	2c 1	P 6
5122.405	5123.833		0.2	0.2			19516.64			
5122.587	5124.014	G	1.5	1.0	0.2	1.25	19515.95	W 2	2B 6	R 0
5122.781	5124.208		0.2	0.1			19515.21			
5122.917	5124.345				0.0		19514.69			
5123.012	5124.439			0.1			19514.33			
5123.269	5124.697			0.0			19513.35			
5123.385	5124.812				0.0		19512.91			
5123.437	5124.865			0.1			19512.71			
5123.666	5125.093		0.2		0.2		19511.84			
5123.708	5125.135			0.3			19511.68			
5123.952	5125.380			0.1			19510.75			
5124.348	5125.776						19509.24			
5124.514	5125.942			0.1			19508.61			
5124.966	5126.394			0.1			19506.89			
5125.313	5126.741			0.1			19505.57			
5125.560	5126.988			0.2	0.2		19504.63	3a 1	2c 0	R 4
5125.833	5127.261			0.2			19503.59			
5126.033	5127.461		0.2	0.4	0.2		19502.83	4c 1	2a 2	Q 5
5126.230	5127.658			0.0			19502.08			
5126.677	5128.105						19500.38			
5126.795	5128.223			0.0	0.1		19499.93			
5126.840	5128.268		0.1				19499.76			
5127.153	5128.581			0.1			19498.57	4c 2	2a 3	R 1
5127.300	5128.729			0.0			19498.01			
5127.368	5128.797				0.2		19497.75			
5127.447	5128.876		0.3	0.4			19497.45	3f 2	2c 1	R 1
5127.644	5129.073			0.1			19496.70			
5127.731	5129.160		0.2				19496.37			
5127.897	5129.326	G	0.6	0.7	0.2		19495.74	3f 2	2c 1	R 1
5128.057	5129.486		0.6	0.4			19495.13	Z 2	2B 5	P 2
5128.326	5129.755			0.1			19494.11			
5128.636	5130.065			0.1			19492.93			
5128.841	5130.270	G	0.2	0.4	0.2		19492.15			
5129.170	5130.599			0.1			19490.90			
5129.383	5130.813						19490.09			
5129.665	5131.094			0.0			19489.02			
5130.410	5131.839			0.0			19486.19			
5130.668	5132.098			0.2			19485.21	u 2	2a 1	P 5
5131.263	5132.693			0.0			19482.95			
5131.595	5133.025			0.0			19481.69			
5131.851	5133.280		0.1	0.1			19480.72			
5132.024	5133.454	G	0.7	0.9	0.3	0.97	19480.06	Z 2	2B 5	R 4
5132.293	5133.723		0.2	0.2	0.2		19479.04			
5132.512	5133.942						19478.21	4d 3	2c 4	P 3
5132.620	5134.050		0.2	0.2	0.1		19477.80	3e 1	2c 0	R 1
5132.775	5134.206			0.0			19477.21	4d 1	2c 2	P 5

λ_{air}	λ_{vac}	Ref.	I_1	I_2	I_3	I_4	ν	Upper	Lower	Br.
5132.999	5134.430			0.2	0.2		19476.36	3E 1	2B 4	Q 9
5133.226	5134.656		0.1				19475.50			
5133.368	5134.799		0.3	0.2			19474.96			
5133.592	5135.023			0.1			19474.11			
5133.743	5135.173			0.1			19473.54			
5133.914	5135.345			0.1			19472.89			
5133.972	5135.403	G	0.1				19472.67			
5134.122	5135.553		0.6	0.9	0.7		19472.10			
5134.333	5135.764		0.1				19471.30			
5134.476	5135.906			0.1			19470.76			
5134.660	5136.091				0.1		19470.06			
5134.734	5136.165			0.2	0.2		19469.78	3f 1	2c 0	P 3
5134.850	5136.281				0.0		19469.34			
5134.935	5136.365			0.1			19469.02			
5135.167	5136.598		0.2	0.3	0.2		19468.14	4c 0	2a 1	P 4
5135.470	5136.901		0.1	0.1			19466.99			
5135.647	5137.078			0.1			19466.32	3D 0	2B 2	R 9
5135.774	5137.204	G	0.2	0.4	0.5		19465.84			
5135.937	5137.368			0.1			19465.22			
5136.085	5137.516			0.0			19464.66	3a 1	2c 0	R 3
5136.183	5137.614		0.2	0.3	0.2		19464.29	3f 1	2c 0	P 3
5136.478	5137.909		0.1				19463.17			
5136.526	5137.957			0.2			19462.99			
5136.824	5138.255		0.3	0.3			19461.86	3D 2	2B 5	P 4
5136.940	5138.371		0.1				19461.42			
5137.109	5138.540	G	1.5	0.8	0.1	0.91	19460.78	W 2	2B 6	P 1
5137.294	5138.725						19460.08			
5137.904	5139.335		0.1	0.3	0.1		19457.77	3F 1	2B 4	Q 3
5138.189	5139.620			0.1			19456.69			
5138.353	5139.784						19456.07			
5138.598	5140.030			0.2			19455.14			
5138.812	5140.244			0.1			19454.33			
5139.327	5140.759			0.1			19452.38			
5139.544	5140.976			0.1			19451.56			
5139.716	5141.148			0.1			19450.91			
5139.943	5141.375				0.1		19450.05			
5140.252	5141.684			0.1			19448.88			
5140.337	5141.769				0.1		19448.56			
5140.834	5142.266			0.0			19446.68			
5141.072	5142.504			0.2	0.1		19445.78	4e 3	2c 4	P 3
5141.238	5142.671			0.1			19445.15			
5141.407	5142.840		0.2	0.4	0.3		19444.51	3f 1	2c 0	P 4
5141.611	5143.043		0.4	0.3	0.2		19443.74	3E 0	2B 2	R 1
5141.857	5143.289		0.1	0.1			19442.81			
5142.050	5143.483		0.2	0.4	0.3		19442.08	3E 0	2B 2	P 6
5142.264	5143.697			0.1			19441.27			
5142.439	5143.872			0.0			19440.61	V 1	2B 2	R 1
5142.690	5144.123			0.2	0.2		19439.66	3F 1	2B 4	P 5
5142.883	5144.316		0.2	0.2			19438.93			
5143.153	5144.586						19437.91			
5143.203	5144.636			0.1			19437.72			
5143.256	5144.689		0.1				19437.52			
5143.428	5144.861	G	0.6	1.0	1.0	0.81	19436.87	3E 0	2B 2	Q 6
5143.563	5144.996	G	1.0	0.9		0.30	19436.36	3c 5	2a 3	Q 1
5143.775	5145.208		0.1				19435.56	3f 2	2c 1	Q 6
5143.913	5145.346			0.0			19435.04			
5144.180	5145.613				0.1		19434.03			
5144.333	5145.767			0.0d			19433.45			
5144.741	5146.174			0.0d			19431.91			
5145.075	5146.508			0.1			19430.65			

λ_air	λ_vac	Ref.	I₁	I₂	I₃	I₄	ν	Upper	Lower	Br.
5145.581	5147.014		0.1	0.2			19428.74			
5145.721	5147.155						19428.21			
5145.832	5147.266			0.0			19427.79			
5146.020	5147.454		0.1		0.2		19427.08	3f 1	2c 0	P 5
5146.100	5147.534			0.2			19426.78			
5146.155	5147.589		0.2				19426.57			
5146.344	5147.777	GK	1.5	1.5	0.3	1.51	19425.86	W 2	2B 6	P 2
5146.344	5147.777	GK	1.5	1.5	0.3	1.51	19425.86	3F 1	2B 4	R 1
5146.561	5147.995		0.2				19425.04	3f 2	2c 1	Q 5
5146.762	5148.196			0.1	0.1		19424.28	3f 1	2c 0	P 4
5147.083	5148.517			0.1	0.0		19423.07	3F 3	2C 0	Q 3
5147.276	5148.710			0.0			19422.34			
5147.504	5148.938			0.2	0.1		19421.48			
5147.613	5149.047			0.0			19421.07			
5147.759	5149.193	G	0.3	0.6	0.2		19420.52	U 1	2B 4	R 4
5147.960	5149.394			0.2			19419.76	3a 1	2c 0	R 2
5148.183	5149.617			0.1			19418.92			
5148.244	5149.678				0.0		19418.69			
5148.337	5149.771			0.1			19418.34			
5148.466	5149.901	G	0.6	0.3	0.0		19417.85	Z 2	2B 5	P 3
5148.697	5150.131			0.0			19416.98			
5148.846	5150.280				0.2		19416.42			
5148.925	5150.360			0.2			19416.12			
5149.246	5150.681			0.1			19414.91			
5149.485	5150.919			0.1			19414.01			
5149.721	5151.156		0.1	0.3	0.2		19413.12			
5149.872	5151.307			0.0			19412.55			
5150.029	5151.463	G	0.3	0.5			19411.96	3c 5	2a 3	Q 2
5150.175	5151.609			0.2	0.4		19411.41	3f 1	2c 0	P 6
5150.175	5151.609			0.2	0.4		19411.41	3f 1	2c 0	P 7
5150.302	5151.737		0.2	0.3			19410.93	X 0	2B 3	R 3
5150.474	5151.909	G		0.2			19410.28	3f 2	2c 1	Q 4
5150.615	5152.050			0.1			19409.75	4F 0	2B 7	R 2
5150.769	5152.204		0.2	0.0			19409.17			
5150.907	5152.342						19408.65			
5151.005	5152.440			0.1d			19408.28			
5151.225	5152.660			0.0			19407.45			
5151.361	5152.796				0.3		19406.94			
5151.409	5152.844		0.1				19406.76			
5151.573	5153.008			0.3			19406.14	3E 3	2B 7	Q 6
5151.762	5153.197			0.0			19405.43			
5151.871	5153.306		0.2		0.2		19405.02			
5152.099	5153.534			0.1			19404.16			
5152.255	5153.691			0.0			19403.57			
5152.348	5153.784				0.2		19403.22			
5152.415	5153.850		0.1	0.3			19402.97			
5152.595	5154.031			0.1			19402.29			
5152.784	5154.219			0.1			19401.58			
5152.965	5154.400		0.1	0.3			19400.90	3D 2	2B 5	P 5
5153.190	5154.626			0.2			19400.05			
5153.243	5154.679		0.1				19399.85			
5153.413	5154.849		0.7	0.6	0.4		19399.21	X 0	2B 3	R 2
5153.679	5155.115		0.2	0.2			19398.21	4F 0	2B 7	P 5
5153.870	5155.306	G	1.5	1.0	0.2	1.31	19397.49	4c 2	2a 3	Q 1
5154.080	5155.516		0.2				19396.70			
5154.205	5155.641			0.0			19396.23			
5154.304	5155.739		0.2				19395.86			
5154.490	5155.925	G	0.3	0.5	0.2		19395.16	3f 2	2c 1	Q 3
5154.731	5156.167			0.0			19394.25			
5155.013	5156.449			0.2			19393.19	3f 2	2c 1	Q 5

λ_{air}	λ_{vac}	Ref.	I_1	I_2	I_3	I_4	ν	Upper	Lower	Br.
5155.104	5156.540				0.3b		19392.85			
5155.202	5156.638	G	0.2	0.4			19392.48	3f 2	2c 1	Q 4
5155.391	5156.827			0.1			19391.77			
5155.516	5156.952				0.1		19391.30			
5155.630	5157.066			0.0			19390.87			
5155.784	5157.220			0.1	0.1		19390.29			
5155.917	5157.353			0.1			19389.79			
5156.141	5157.577		0.3	0.3	0.4		19388.95	4c 0	2a 1	P 5
5156.266	5157.702	G	0.5	0.6		0.94	19388.48	4c 2	2a 3	Q 2
5156.327	5157.763				0.3		19388.25	X 0	2B 3	R 4
5156.476	5157.912		0.1	0.3			19387.69	3f 2	2c 1	Q 3
5156.476	5157.912		0.1	0.3			19387.69	3f 2	2c 1	Q 6
5156.718	5158.154		0.3	0.2			19386.78	X 0	2B 3	R 1
5156.803	5158.239				0.2		19386.46			
5156.848	5158.285	G	1.0	0.8		1.03	19386.29	W 2	2B 6	P 3
5157.045	5158.481			0.1			19385.55			
5157.181	5158.617						19385.04			
5157.255	5158.692			0.1			19384.76			
5157.534	5158.971			0.3	0.2		19383.71	3f 1	2c 0	P 5
5157.729	5159.165			0.0			19382.98			
5157.878	5159.314		0.3	0.4			19382.42	3f 2	2c 1	Q 2
5158.027	5159.464			0.0			19381.86			
5158.162	5159.599	G	0.3	0.7	0.5		19381.35	3E 0	2B 2	P 5
5158.373	5159.810	G	0.6	0.7	0.2		19380.56	3f 2	2c 1	Q 2
5158.628	5160.065			0.1			19379.60			
5158.857	5160.294	G	0.3	0.5		0.60	19378.74	3E 3	2B 7	R 3
5159.174	5160.611			0.1			19377.55			
5159.262	5160.699						19377.22			
5159.408	5160.845						19376.67			
5159.464	5160.901			0.1			19376.46			
5159.677	5161.115		0.2	0.4			19375.66	3c 5	2a 3	Q 3
5159.677	5161.115		0.2	0.4			19375.66	3f 2	2c 1	Q 7
5159.837	5161.274	G	0.7	0.9	0.3	0.89	19375.06	4c 2	2a 3	Q 3
5160.090	5161.527			0.0d			19374.11			
5160.306	5161.743			0.1			19373.30			
5160.372	5161.810						19373.05			
5160.516	5161.954	G	1.0	0.6		0.96	19372.51			
5160.737	5162.175		0.1	0.2	0.1		19371.68	3e 1	2c 0	Q 3
5160.993	5162.431						19370.72			
5161.038	5162.476			0.1			19370.55	3a 1	2c 0	R 1
5161.212	5162.649		0.1	0.2			19369.90	3e 1	2c 0	Q 4
5161.377	5162.815			0.2	0.3		19369.28			
5161.486	5162.924	G	0.6	0.4			19368.87	X 0	2B 3	R 0
5161.702	5163.140		0.2	0.3	0.1		19368.06			
5162.008	5163.446		0.1	0.2			19366.91			
5162.262	5163.700				0.5		19365.96			
5162.520	5163.958			0.0			19364.99			
5162.643	5164.081						19364.53			
5162.904	5164.342			0.0	0.0		19363.55			
5163.051	5164.489			0.1	0.1		19363.00	3f 3	2c 2	R 4
5163.469	5164.908				0.1		19361.43			
5163.571	5165.009						19361.05			
5163.685	5165.124			0.1			19360.62	4d 0	2c 1	P 6
5163.864	5165.303	G	0.2	0.3	0.1		19359.95	3c 9	2a 6	Q 1
5164.067	5165.505						19359.19			
5164.139	5165.577			0.2			19358.92			
5164.384	5165.823	G	0.4	0.4			19358.00	3F 1	2B 4	Q 2
5164.542	5165.980		0.1	0.3			19357.41	4c 2	2a 3	Q 4
5164.721	5166.159	G	0.2	0.8	0.8		19356.74	3E 0	2B 2	Q 5
5164.721	5166.159	G	0.2	0.8	0.8		19356.74	3e 1	2c 0	Q 2

λ_{air}	λ_{vac}	Ref.	I$_1$	I$_2$	I$_3$	I$_4$	ν	Upper	Lower	Br.
5164.963	5166.402			0.1			19355.83			
5165.254	5166.693		0.1	0.3	0.3		19354.74	3e 1	2c 0	Q 5
5165.374	5166.813						19354.29			
5165.524	5166.963			0.0			19353.73			
5165.657	5167.096						19353.23			
5166.023	5167.462			0.1			19351.86			
5166.127	5167.566				0.1		19351.47			
5166.418	5167.857		0.2	0.2			19350.38			
5166.661	5168.100			0.0			19349.47			
5166.877	5168.317		0.1	0.3	0.1		19348.66			
5166.966	5168.405						19348.33			
5167.056	5168.495			0.1			19347.99			
5167.214	5168.653			0.1			19347.40			
5167.243	5168.683		0.2				19347.29			
5167.409	5168.848	G	0.9	0.7	0.2	0.77	19346.67	3E 3	2B 7	R 2
5167.433	5168.872						19346.58			
5167.636	5169.075		0.1	0.0			19345.82			
5167.719	5169.158						19345.51			
5167.767	5169.206			0.1			19345.33			
5167.876	5169.316				0.0		19344.92			
5168.010	5169.449			0.2			19344.42			
5168.063	5169.503		0.1				19344.22			
5168.248	5169.687	G	1.5	1.0	0.3	1.31	19343.53	W 2	2B 6	P 4
5168.248	5169.687	G	1.5	1.0	0.3	1.31	19343.53	Y 2	2B 6	R 1
5168.507	5169.946		0.1				19342.56			
5168.550	5169.989			0.2			19342.40			
5168.675	5170.115		0.1	0.0			19341.93			
5168.822	5170.262	G	0.7	0.5			19341.38	Y 2	2B 6	R 0
5169.012	5170.452		0.1	0.1			19340.67	3f 1	2c 0	P 6
5169.210	5170.649	G	0.2	0.4			19339.93	3D 2	2B 5	P 6
5169.426	5170.866		0.2	0.2			19339.12	3F 1	2B 4	P 4
5169.613	5171.053	G	0.3	0.6	0.5		19338.42	3E 1	2B 4	Q 8
5169.817	5171.256			0.1			19337.66			
5169.996	5171.436		0.1	0.1			19336.99	3d 3	2c 1	P 7
5170.124	5171.564				0.2		19336.51			
5170.177	5171.618	G	0.8	0.5		0.79	19336.31	Z 2	2B 5	P 4
5170.362	5171.802	G	0.5	0.5	0.2		19335.62	Y 2	2B 6	R 2
5170.362	5171.802	G	0.5	0.5	0.2		19335.62	4c 2	2a 3	Q 5
5170.536	5171.976			0.0			19334.97			
5170.608	5172.048		0.1				19334.70			
5170.699	5172.139			0.1			19334.36			
5170.921	5172.361	G	0.5	0.9	1.5		19333.53	3D 0	2B 2	R 8
5171.156	5172.597		0.1	0.0			19332.65			
5171.344	5172.784	G	0.8	0.8	0.4		19331.95	3E 0	2B 2	R 0
5171.622	5173.062			0.2d			19330.91			
5171.962	5173.402			0.0			19329.64	3f 3	2c 2	R 3
5172.061	5173.501		0.2	0.2	0.2b		19329.27	3e 1	2c 0	Q 6
5172.103	5173.544	G					19329.11			
5172.221	5173.662			0.3			19328.67	U 1	2B 4	R 3
5172.387	5173.828			0.0			19328.05			
5172.548	5173.988	G	0.3	0.4	0.1		19327.45	3e 1	2c 0	Q 1
5172.548	5173.988	G	0.3	0.4	0.1		19327.45	3c 5	2a 3	Q 4
5172.794	5174.235			0.3			19326.53	3E 3	2B 7	Q 5
5172.794	5174.235			0.3			19326.53	3b 6	2a 1	R 2
5172.914	5174.355	G	0.5	0.4			19326.08	3b 6	2a 1	R 1
5173.155	5174.596			0.1			19325.18			
5173.337	5174.778						19324.50			
5173.407	5174.848		0.1	0.3	0.1		19324.24	3f 3	2c 2	R 3
5173.602	5175.043		0.1	0.2			19323.51			
5173.779	5175.220	G	1.5	0.7	0.1	0.95	19322.85	4d 1	2c 2	P 6

λ_{air}	λ_{vac}	Ref.	I_1	I_2	I_3	I_4	ν	Upper	Lower	Br.
5173.779	5175.220	G	1.5	0.7	0.1	0.95	19322.85	4c 3	2a 4	R 1
5173.779	5175.220	G	1.5	0.7	0.1	0.95	19322.85	V 1	2B 2	R 0
5173.996	5175.437		0.2	0.4	0.2		19322.04	3e 1	2c 0	R 2
5174.186	5175.627						19321.33			
5174.481	5175.922			0.1			19320.23			
5174.526	5175.967		0.1				19320.06			
5174.700	5176.141	G	1.5	1.5	0.9	1.16	19319.41	3E 0	2B 2	P 4
5174.904	5176.345		0.1	0.0			19318.65			
5174.982	5176.423						19318.36			
5175.038	5176.479		0.2	0.2			19318.15			
5175.249	5176.691	G	0.3	0.5	0.3		19317.36	3e 1	2c 0	R 1
5175.477	5176.919						19316.51			
5175.544	5176.986		0.1	0.2d			19316.26	3b 6	2a 1	R 0
5175.823	5177.264	G	0.1	0.4	0.4		19315.22	3e 1	2c 0	R 3
5176.386	5177.827			0.2	0.3		19313.12			
5176.568	5178.010			0.1			19312.44			
5176.772	5178.213	G	0.2	0.4	0.2		19311.68	4c 1	2a 2	P 4
5176.772	5178.213	G	0.2	0.4	0.2		19311.68	W 2	2B 6	P 6
5177.013	5178.455			0.0			19310.78			
5177.241	5178.683			0.2d	0.1		19309.93			
5177.310	5178.752		0.2				19309.67	X 0	2B 3	P 1
5177.503	5178.945				0.1		19308.95			
5177.584	5179.026			0.1d			19308.65			
5177.847	5179.289						19307.67			
5178.102	5179.544						19306.72			
5178.279	5179.721			0.2	0.2		19306.06	4c 0	2a 1	P 6
5178.488	5179.930	G	0.4	0.6	0.2	0.70	19305.28	W 2	2B 6	P 5
5178.823	5180.266			0.1			19304.03			
5178.920	5180.362				0.0		19303.67			
5179.201	5180.644				0.1		19302.62			
5179.550	5180.993						19301.32			
5179.888	5181.331			0.1			19300.06			
5180.111	5181.554			0.2	0.2		19299.23	3e 1	2c 0	R 4
5180.245	5181.688		0.2	0.2			19298.73			
5180.377	5181.820		0.1	0.2			19298.24			
5180.581	5182.024	G	1.0	1.5	1.0	1.22	19297.48	3E 0	2B 2	Q 4
5180.820	5182.263		0.1	0.1			19296.59			
5181.072	5182.515				0.1		19295.65			
5181.131	5182.574		0.1	0.1			19295.43			
5181.303	5182.746	G	0.5	0.3			19294.79	Y 2	2B 6	P 1
5181.464	5182.907			0.1	0.2		19294.19	3f 1	2c 0	P 7
5181.685	5183.128		0.4	0.2			19293.37			
5181.975	5183.418		0.4	0.3			19292.29	3E 3	2B 7	R 1
5182.120	5183.563		0.0	0.1			19291.75			
5182.302	5183.746			0.2			19291.07			
5182.496	5183.939			0.0d			19290.35			
5182.751	5184.194		0.2	0.3d			19289.40			
5183.092	5184.536			0.1			19288.13			
5183.366	5184.810		0.2	0.3	0.2		19287.11			
5183.528	5184.971			0.1			19286.51			
5183.675	5185.119						19285.96			
5183.788	5185.232			0.1			19285.54			
5183.952	5185.396						19284.93			
5184.030	5185.474			0.0			19284.64			
5184.221	5185.665			0.1			19283.93			
5184.313	5185.756				0.1d		19283.59			
5184.541	5185.985			0.1			19282.74			
5184.794	5186.238			0.3			19281.80	3D 2	2B 5	P 7
5184.794	5186.238			0.3			19281.80	3F 3	2c 0	P 3
5185.111	5186.555	G	0.4	0.5	0.1		19280.62	3f 3	2c 2	R 2

λ_{air}	λ_{vac}	Ref.	I_1	I_2	I_3	I_4	ν	Upper	Lower	Br.
5185.283	5186.727		0.1				19279.98			
5185.415	5186.859		0.3	0.3			19279.49	3F 2	2B 6	R 2
5185.660	5187.104		0.1	0.3	0.1		19278.58	3f 3	2c 2	R 2
5185.907	5187.352			0.0			19277.66			
5186.036	5187.481				0.0		19277.18			
5186.131	5187.575			0.0			19276.83	3e 1	2c 0	R 5
5186.357	5187.801			0.2			19275.99	U 1	2B 4	R 2
5186.526	5187.971			0.3	0.4		19275.36			
5186.844	5188.288			0.1			19274.18			
5187.242	5188.687		0.1	0.0			19272.70			
5187.425	5188.870	G	0.4	0.5	0.2		19272.02			
5187.735	5189.179		0.1		0.6		19270.87			
5187.780	5189.225			0.1d			19270.70	U 1	2B 4	R 1
5188.079	5189.524		0.1	0.2d			19269.59			
5188.278	5189.723	G	1.5	0.9	0.5	0.88	19268.85	3E 0	2B 2	P 3
5188.478	5189.923		0.2	0.3			19268.11			
5188.626	5190.071		0.7	0.3			19267.56	X 0	2B 3	P 2
5188.874	5190.319		0.1	0.2			19266.64			
5189.156	5190.601		0.2	0.2	0.0		19265.59	U 1	2B 4	R 0
5189.399	5190.844		0.2				19264.69	3F 1	2B 4	P 3
5189.590	5191.035			0.0			19263.98			
5190.053	5191.499				0.0		19262.26			
5190.218	5191.663			0.1			19261.65			
5190.388	5191.833	G	0.4	0.8	0.2	0.91	19261.02	3E 3	2B 7	Q 4
5190.671	5192.116			0.0			19259.97	3a 2	2c 1	R 3
5190.832	5192.278		0.1	0.1			19259.37			
5191.024	5192.469	G	0.7	1.0	0.7	0.97	19258.66	3E 0	2B 2	Q 3
5191.218	5192.663		0.1	0.0			19257.94	3f 1	2c 0	P 8
5191.361	5192.806			0.0			19257.41			
5191.644	5193.089		0.2	0.2			19256.36	3a 1	2c 0	Q 1
5191.932	5193.378			0.1			19255.29			
5192.145	5193.591			0.0			19254.50			
5192.342	5193.788				0.2		19253.77			
5192.528	5193.974			0.1			19253.08			
5192.790	5194.236		0.1	0.2			19252.11			
5193.008	5194.454	G	1.0	0.7	0.1	0.88	19251.30	Y 2	2B 6	P 2
5193.227	5194.673		0.1				19250.49			
5193.297	5194.743						19250.23			
5193.432	5194.878		0.1				19249.73	3a 1	2c 0	Q 2
5193.634	5195.080			0.0			19248.98			
5193.815	5195.261			0.0			19248.31			
5193.963	5195.410		0.1				19247.76			
5194.114	5195.561		0.2	0.3	0.1		19247.20	Z 2	2B 5	P 5
5194.392	5195.839			0.2			19246.17			
5194.438	5195.885		0.1				19246.00			
5194.587	5196.033			0.2	0.1		19245.45			
5194.843	5196.290			0.2			19244.50	3D 2	2B 5	P 8
5195.156	5196.603		0.1				19243.34			
5195.235	5196.681			0.1			19243.05			
5195.353	5196.800		0.1				19242.61			
5195.559	5197.005		0.2	0.3	0.2		19241.85			
5195.721	5197.168			0.1			19241.25			
5195.910	5197.357		0.1	0.1			19240.55			
5195.953	5197.400						19240.39			
5196.064	5197.511		0.2				19239.98			
5196.115	5197.562			0.3			19239.79	3a 1	2c 0	Q 3
5196.191	5197.638		0.2				19239.51			
5196.377	5197.824	GK	1.5	1.5	1.0	1.46	19238.82	3E 0	2B 2	Q 2
5196.585	5198.032		0.2				19238.05			
5196.706	5198.154		0.2	0.2	0.2		19237.60			

λ_{air}	λ_{vac}	Ref.	I_1	I_2	I_3	I_4	ν	Upper	Lower	Br.
5196.836	5198.283		0.1				19237.12			
5197.039	5198.486		0.2	0.1			19236.37			
5197.220	5198.667	G	1.5	0.9		1.30	19235.70	4c 3	2a 4	Q 1
5197.290	5198.737				0.2		19235.44			
5197.325	5198.772			0.1			19235.31			
5197.436	5198.883		0.2	0.1			19234.90			
5197.517	5198.964	G				0.84	19234.60			
5197.593	5199.040		0.8	0.9	0.4		19234.32	3E 0	2B 2	Q 1
5197.593	5199.040		0.8	0.9	0.4		19234.32	3E 3	2B 7	R 0
5197.779	5199.227						19233.63			
5197.830	5199.278		0.1	0.1			19233.44	3e 2	2c 1	R 5
5198.014	5199.462		0.3	0.2			19232.76	Y 1	2B 3	R 1
5198.274	5199.721		0.1	0.1d			19231.80			
5198.422	5199.870		0.1				19231.25			
5198.512	5199.959						19230.92			
5198.587	5200.035	G	0.3	0.6	0.9		19230.64	3D 0	2B 2	R 7
5198.847	5200.295			0.1d			19229.68			
5198.990	5200.438		0.1				19229.15			
5199.196	5200.643		0.5	0.3			19228.39	3f 3	2c 2	R 1
5199.339	5200.787	G	0.9	0.8	0.2		19227.86	3f 3	2c 2	R 1
5199.339	5200.787	G	0.9	0.8	0.2		19227.86	Y 1	2B 3	R 0
5199.544	5200.992		0.5	0.5			19227.10	4c 3	2a 4	Q 2
5199.544	5200.992		0.5	0.5			19227.10	Y 1	2B 3	R 2
5199.717	5201.165	G	1.5	1.0	0.7	1.10	19226.46	3a 1	2c 0	Q 4
5199.717	5201.165	G	1.5	1.0	0.7	1.10	19226.46	3E 0	2B 2	P 2
5199.945	5201.393		0.1	0.0			19225.62	4d 0	2c 1	P 7
5200.142	5201.590			0.1			19224.89			
5200.469	5201.918			0.0			19223.68			
5200.735	5202.183			0.2			19222.70			
5200.900	5202.348			0.0			19222.09			
5201.086	5202.535		0.3	0.3			19221.40			
5201.276	5202.724			0.1			19220.70	3E 1	2B 4	R 3
5201.525	5202.973			0.2	0.2		19219.78	4c 3	2a 1	P 7
5201.776	5203.225			0.2			19218.85	X 0	2B 3	P 3
5202.012	5203.460	G		0.6	0.3	0.82	19217.98	3E 1	2B 4	Q 7
5202.066	5203.515		0.8				19217.78			
5202.312	5203.761		0.1	0.1	0.1		19216.87			
5202.556	5204.005			0.1			19215.97	3a 2	2c 1	R 2
5202.821	5204.270		0.2	0.2d			19214.99	4c 2	2a 3	P 3
5203.027	5204.476	G	0.7	0.8	0.1	0.99	19214.23	4c 3	2a 4	Q 3
5203.274	5204.723			0.1			19213.32			
5203.341	5204.790						19213.07			
5203.523	5204.972		0.1	0.2			19212.40			
5203.750	5205.199			0.0			19211.56			
5203.964	5205.413		0.2	0.2			19210.77	3f 2	2c 1	P 3
5204.146	5205.595		0.1	0.2			19210.10	3a 1	2c 0	Q 5
5204.230	5205.679				0.2b		19209.79	4F 0	2B 7	P 4
5204.316	5205.766	G	0.4	0.7		0.76	19209.47	3E 3	2B 7	Q 3
5204.390	5205.839		0.4				19209.20	3f 2	2c 1	P 3
5204.644	5206.094			0.2			19208.26			
5204.880	5206.329		0.1	0.1			19207.39			
5205.070	5206.519	G	0.9	0.8	0.4		19206.69	3e 1	2c 0	P 2
5205.297	5206.747		0.1	0.2			19205.85	3e 2	2c 1	R 4
5205.520	5206.969	G	0.2	0.4	0.1		19205.03			
5205.726	5207.175			0.0			19204.27			
5206.024	5207.474				0.1		19203.17			
5206.216	5207.666			0.0			19202.46			
5206.479	5207.929				0.0		19201.49	4F 0	2B 7	R 1
5206.713	5208.162						19200.63			
5206.989	5208.439				0.2		19199.61			

λ_{air}	λ_{vac}	Ref.	I_1	I_2	I_3	I_4	ν	Upper	Lower	Br.
5207.320	5208.770			0.0			19198.39			
5207.567	5209.017						19197.48	3F 3	2C 0	P 4
5207.567	5209.017						19197.48	4c 3	2a 4	Q 4
5207.619	5209.069			0.1			19197.29			
5207.800	5209.250	G	0.2	0.3	0.2		19196.62	3e 1	2c 0	Q 2
5207.952	5209.402		0.1				19196.06			
5208.034	5209.484			0.0			19195.76			
5208.237	5209.687		0.1	0.1			19195.01			
5208.365	5209.815				0.1		19194.54			
5208.422	5209.872	G	0.7	0.4		0.45	19194.33	Y 2	2B 6	P 3
5208.609	5210.059			0.0	0.2		19193.64			
5209.518	5210.969		0.1				19190.29			
5209.695	5211.145						19189.64	3a 1	2c 0	Q 6
5209.765	5211.216				0.1		19189.38			
5210.360	5211.811			0.1	0.2		19187.19			
5210.854	5212.305			0.0			19185.37			
5211.199	5212.650			0.1			19184.10			
5211.506	5212.957	G	0.8	0.4	0.0	0.53	19182.97			
5211.620	5213.071						19182.55			
5211.797	5213.248		0.1	0.1			19181.90			
5212.145	5213.596			0.0			19180.62			
5212.609	5214.061				0.2		19178.91			
5212.859	5214.311		0.1				19177.99			
5212.911	5214.362			0.1			19177.80			
5213.115	5214.566	G	0.4	0.4			19177.05	3e 1	2c 0	P 3
5213.167	5214.618				0.2		19176.86	4c 3	2a 4	Q 5
5213.207	5214.659		0.3	0.4			19176.71			
5213.476	5214.928			0.1			19175.72			
5213.846	5215.298		0.2	0.2			19174.36	Y 1	2B 3	P 1
5214.045	5215.497			0.0			19173.63	3F 3	2C 0	P 4
5214.107	5215.559						19173.40			
5214.197	5215.649				0.1		19173.07			
5214.366	5215.817			0.1d			19172.45			
5214.447	5215.899		0.2				19172.15			
5214.629	5216.081	G	1.5	1.0	0.3	1.22	19171.48	3E 3	2B 7	Q 2
5214.850	5216.302		0.2	0.2			19170.67			
5215.062	5216.514			0.2			19169.89			
5215.291	5216.743		0.1	0.3	0.3		19169.05			
5215.489	5216.941		0.5	0.5			19168.32	3f 2	2c 1	P 4
5215.489	5216.941		0.5	0.5			19168.32	U 1	2B 4	P 2
5215.614	5217.066	G	0.5	0.8	0.9	0.74	19167.86	3a 2	2c 1	R 1
5215.799	5217.252			0.1			19167.18			
5215.840	5217.292		0.1				19167.03	3F 2	2B 6	R 1
5215.963	5217.415			0.0			19166.58	X 0	2B 3	P 4
5216.338	5217.791				0.2b		19165.20			
5216.412	5217.864		0.1	0.1			19164.93	3a 1	2c 0	Q 7
5216.594	5218.047		0.3	0.4	0.2		19164.26	3e 2	2c 1	R 3
5216.787	5218.240			0.1			19163.55			
5217.038	5218.490			0.1d			19162.63			
5217.092	5218.545				0.1		19162.43			
5217.397	5218.850		0.1	0.2	0.1		19161.31	3f 2	2c 1	P 4
5217.781	5219.234			0.1			19159.90			
5218.538	5219.991			0.1d			19157.12			
5218.846	5220.299			0.0d			19155.99			
5219.091	5220.545		0.0	0.2			19155.09			
5219.279	5220.733		0.3	0.3	0.0		19154.40			
5219.451	5220.904			0.0			19153.77			
5219.680	5221.133		0.2	0.0			19152.93			
5219.854	5221.308	G	0.4	0.4	0.1		19152.29	3A 0	2B 3	R 2
5219.854	5221.308	G	0.4	0.4	0.1		19152.29	4c 3	2a 4	Q 6

λ_{air}	λ_{vac}	Ref.	I_1	I_2	I_3	I_4	ν	Upper	Lower	Br.
5220.102	5221.556		0.2	0.2			19151.38			
5220.397	5221.850		0.4	0.3			19150.30			
5220.721	5222.175		0.1	0.1			19149.11			
5220.912	5222.366		0.8	0.4			19148.41	3E 3	2B 7	P 3
5221.032	5222.486		0.1	0.1			19147.97	Z 2	2B 5	P 6
5221.149	5222.603	G	0.8	0.8		0.81	19147.54	3E 3	2B 7	P 2
5221.242	5222.696				0.3		19147.20			
5221.283	5222.737		0.1	0.0			19147.05			
5221.449	5222.903	G	0.9	0.8	0.2	0.92	19146.44	3E 3	2B 7	P 4
5221.449	5222.903	G	0.9	0.8	0.2	0.92	19146.44	3E 3	2B 7	Q 1
5221.657	5223.110						19145.68			
5221.738	5223.192		0.1	0.0			19145.38			
5221.908	5223.361		0.1				19144.76			
5222.109	5223.563		0.2	0.2d			19144.02			
5222.426	5223.880		0.2	0.2	0.1		19142.86			
5222.598	5224.052	G	1.5	1.0	0.2		19142.23	3E 1	2B 4	R 2
5222.712	5224.166			0.1			19141.81	3e 1	2c 0	Q 3
5222.854	5224.308	G	1.5	1.5	0.2	1.19	19141.29	3D 3	2B 7	R 0
5223.086	5224.540		0.1	0.0			19140.44			
5223.329	5224.783			0.1d			19139.55			
5223.375	5224.830		0.1				19139.38			
5223.542	5224.996	G	1.5	0.9		0.93	19138.77	3D 3	2B 7	R 1
5223.610	5225.064				0.1b		19138.52			
5223.747	5225.201			0.0			19138.02			
5223.979	5225.433						19137.17			
5224.167	5225.621			0.1			19136.48			
5224.224	5225.679		0.1				19136.27			
5224.399	5225.854	G	1.0	0.9	0.3	0.57	19135.63	3a 1	2c 0	P 2
5224.647	5226.102		0.7	0.3			19134.72	X 1	2B 5	R 0
5224.841	5226.296	G	0.5	0.6	0.3	0.60	19134.01	3e 1	2c 0	P 4
5224.981	5226.435			0.0			19133.50			
5225.112	5226.566		0.1	0.1			19133.02			
5225.270	5226.725		0.4	0.3			19132.44	X 1	2B 5	R 1
5225.579	5227.034			0.0			19131.31			
5225.969	5227.424		0.1	0.2	0.1		19129.88	3f 2	2c 1	P 5
5225.969	5227.424		0.1	0.2	0.1		19129.88	6c 1	2a 4	Q 1
5226.286	5227.741			0.1d			19128.72			
5226.371	5227.826		0.2				19128.41			
5226.565	5228.020		0.9	0.5			19127.70	Y 2	2B 6	P 4
5226.781	5228.236	G	1.5b	1.5	0.3	1.43	19126.91	Y 1	2B 3	P 2
5226.781	5228.236	G	1.5b	1.5	0.3	1.43	19126.91	3b 6	2a 1	P 2
5226.781	5228.236	G	1.5b	1.5	0.3	1.43	19126.91	3D 3	2B 7	R 2
5227.016	5228.471		0.2	0.1			19126.05			
5227.349	5228.805			0.1d			19124.83			
5227.809	5229.264		0.1	0.2			19123.15	3E 3	2B 7	P 5
5228.074	5229.529	G	0.4	0.4	0.2		19122.18			
5228.435	5229.890				0.1		19120.86			
5228.703	5230.158			0.0d			19119.88			
5228.957	5230.413		0.5	0.4			19118.95	3f 3	2c 2	Q 2
5229.165	5230.621		0.7	0.6			19118.19	3f 3	2c 2	Q 2
5229.274	5230.730	G	0.6	0.6			19117.79	3f 3	2c 2	Q 3
5229.274	5230.730	G	0.6	0.6			19117.79	3e 1	2c 0	P 2
5229.397	5230.853		0.5	0.4			19117.34	X 1	2B 5	R 2
5229.578	5231.034	G	0.6	0.9	1.0b		19116.68	3D 0	2B 2	R 6
5229.720	5231.176	G	0.3	0.9			19116.16	3E 1	2B 4	Q 6
5229.797	5231.253						19115.88			
5229.898	5231.354			0.0			19115.51			
5230.046	5231.502			0.3d	0.1		19114.97	3f 3	2c 2	Q 3
5230.158	5231.614			0.2d			19114.56	3f 3	2c 2	Q 4
5230.300	5231.756			0.0			19114.04	X 0	2B 3	P 5

λ_{air}	λ_{vac}	Ref.	I_1	I_2	I_3	I_4	ν	Upper	Lower	Br.
5230.653	5232.109		0.1	0.3			19112.75	3f 2	2c 1	P 5
5230.700	5232.156				0.3		19112.58	3f 3	2c 2	Q 5
5230.952	5232.408						19111.66			
5231.053	5232.509			0.0d			19111.29			
5231.209	5232.665				0.1		19110.72			
5231.247	5232.704			0.2d			19110.58			
5231.381	5232.838		0.0				19110.09			
5231.562	5233.018		0.7	0.5			19109.43	U 1	2B 4	P 3
5231.748	5233.205	G	0.5	0.7	0.2	0.85	19108.75	3e 2	2c 1	R 2
5231.748	5233.205	G	0.5	0.7	0.2	0.85	19108.75	3D 3	2B 7	R 3
5231.995	5233.451		0.1	0.4d	0.2		19107.85	3f 3	2c 2	Q 4
5232.353	5233.810		0.2	0.2d			19106.54			
5232.518	5233.974				0.1		19105.94			
5232.745	5234.202			0.0			19105.11			
5233.304	5234.761			0.1			19103.07			
5233.715	5235.172			0.0			19101.57			
5233.953	5235.410			0.2			19100.70			
5234.307	5235.764				0.2		19099.41			
5234.353	5235.810			0.1			19099.24			
5234.545	5236.002			0.0			19098.54			
5234.806	5236.263			0.2			19097.59			
5235.055	5236.512		0.1	0.2			19096.68			
5235.431	5236.888			0.1			19095.31	3f 3	2c 2	Q 5
5235.740	5237.198			0.0			19094.18			
5236.017	5237.475	G	0.2	0.3	0.2		19093.17	3f 2	2c 1	P 6
5236.355	5237.812			0.1			19091.94			
5236.684	5238.142			0.2			19090.74			
5236.769	5238.227						19090.43			
5237.027	5238.485		0.5	0.2			19089.49			
5237.356	5238.814			0.1			19088.29	4E 0	2B 7	R 2
5238.135	5239.594			0.2			19085.45			
5238.514	5239.972			0.0			19084.07			
5238.833	5240.291		0.2	0.1			19082.91			
5239.019	5240.478	G	1.5	1.0	0.0	1.27	19082.23	4c 4	2a 5	Q 1
5239.019	5240.478	G	1.5	1.0	0.0	1.27	19082.23	3b 4	2a 0	R 2
5239.247	5240.706		0.2	0.0			19081.40			
5239.459	5240.917			0.0			19080.63			
5239.670	5241.129		0.1	0.0			19079.86	6c 0	2a 3	Q 5
5239.832	5241.291	G	0.5	0.4		0.61	19079.27			
5240.063	5241.521			0.1			19078.43			
5240.321	5241.780			0.2	0.0		19077.49	3e 1	2c 0	Q 4
5240.760	5242.219	G	0.3	0.5	0.2		19075.89	U 1	2B 4	P 6
5240.760	5242.219	G	0.3	0.5	0.2		19075.89	3e 1	2c 0	P 5
5241.087	5242.546			0.1			19074.70	6c 0	2a 3	Q 4
5241.285	5242.744	G	0.6	0.5		0.80	19073.98	4c 4	2a 5	Q 2
5241.535	5242.994		0.2	0.1	0.0		19073.07	4F 0	2C 0	R 3
5241.920	5243.379		0.3	0.3	0.1		19071.67	3F 2	2B 6	P 4
5242.223	5243.682			0.1			19070.57	3b 4	2a 0	R 0
5242.431	5243.891	G	0.4	0.5	0.2		19069.81	3a 1	2c 0	P 3
5242.662	5244.122			0.1			19068.97			
5243.091	5244.551			0.0d			19067.41	6c 0	2a 3	Q 2
5243.476	5244.936			0.1			19066.01			
5243.724	5245.183	G	0.6	1.0	1.0	0.72	19065.11	3D 0	2B 2	R 5
5243.724	5245.183	G	0.6	1.0	1.0	0.72	19065.11	3b 4	2a 0	R 3
5243.724	5245.183	G	0.6	1.0	1.0	0.72	19065.11	6c 0	2a 3	Q 1
5243.724	5245.183	G	0.6	1.0	1.0	0.72	19065.11	Y 1	2B 3	P 3
5244.010	5245.470			0.1			19064.07			
5244.202	5245.662		0.2	0.2			19063.37			
5244.370	5245.830			0.1			19062.76			
5244.467	5245.926		0.1				19062.41			

λ_{air}	λ_{vac}	Ref.	I_1	I_2	I_3	I_4	ν	Upper	Lower	Br.
5244.640	5246.100	G	0.7	0.7	0.2	0.88	19061.78	4c 4	2a 5	Q 3
5244.640	5246.100	G	0.7	0.7	0.2	0.88	19061.78	3f 2	2c 1	P 6
5244.819	5246.279			0.1			19061.13			
5245.061	5246.521			0.1			19060.25			
5245.306	5246.766			0.2			19059.36			
5245.386	5246.846		0.2				19059.07			
5245.435	5246.895				0.1		19058.89			
5245.617	5247.077	G	1.0	0.6	0.1		19058.23	3E 1	2B 4	R 1
5245.856	5247.317		0.1	0.3	0.2		19057.36	4F 0	2B 7	P 3
5246.065	5247.526		0.2	0.2	0.2		19056.60			
5246.542	5248.002			0.2	0.1		19054.87	4f 0	2c 2	R 4
5246.542	5248.002			0.2	0.1		19054.87	3f 2	2c 1	P 7
5246.589	5248.049		0.2				19054.70	3D 3	2B 7	P 2
5246.856	5248.316		0.1				19053.73			
5247.049	5248.509		0.3	0.1	0.1		19053.03			
5247.214	5248.674		0.1				19052.43			
5247.373	5248.834	G	0.6	0.6			19051.85	U 1	2B 4	P 4
5247.605	5249.066	G	0.2	0.5	0.4		19051.01			
5247.798	5249.258		0.2	0.1			19050.31	3A 0	2B 3	R 1
5247.886	5249.347						19049.99			
5248.580	5250.041			0.0			19047.47			
5249.018	5250.479				0.1d		19045.88	4c 4	2a 5	Q 4
5249.073	5250.534		0.1	0.2			19045.68			
5249.547	5251.009			0.1			19043.96			
5249.727	5251.188	G	0.3	0.4	0.1		19043.31	U 1	2B 4	P 5
5250.002	5251.464		0.3	0.4	0.1		19042.31	4F 0	2B 7	Q 2
5250.182	5251.643	G	0.6	0.5	0.2		19041.66	3c 1	2a 0	R 5
5250.314	5251.775	G	0.7	0.3			19041.18	X 1	2B 5	P 2
5250.532	5251.993		0.1				19040.39			
5250.962	5252.424		0.3	0.3	0.0		19038.83	3e 2	2c 1	R 1
5251.276	5252.738		0.1				19037.69			
5251.334	5252.796			0.2			19037.48			
5251.635	5253.097		0.1	0.3	0.2		19036.39			
5251.864	5253.326		0.3	0.4			19035.56			
5252.046	5253.508		0.1	0.1			19034.90			
5252.203	5253.665			0.2			19034.33			
5252.372	5253.834		0.2	0.2			19033.72			
5252.576	5254.038	G	0.3	0.7	0.3	0.51	19032.98	3E 1	2B 4	Q 5
5252.708	5254.170				0.2		19032.50			
5252.802	5254.264			0.1			19032.16			
5253.037	5254.499			0.2			19031.31	3F 0	2B 3	R 4
5253.299	5254.761			0.2			19030.36			
5253.393	5254.855				0.1		19030.02	X 0	2B 3	P 6
5253.484	5254.946			0.1			19029.69			
5253.779	5255.242			0.0			19028.62			
5254.183	5255.645			0.0			19027.16			
5254.542	5256.004		0.2	0.3			19025.86			
5254.699	5256.162		0.2	0.3			19025.29			
5255.047	5256.510			0.1			19024.03			
5255.528	5256.991						19022.29			
5256.042	5257.505			0.1			19020.43			
5256.144	5257.607			0.1			19020.06			
5256.412	5257.875		1.5	0.2			19019.09			
5256.462	5257.925						19018.91			
5256.614	5258.077	GK	1.5	1.5	1.5b	1.38	19018.36	3D 0	2B 2	R 4
5256.879	5258.342		0.1	0.1			19017.40	3F 2	2B 6	P 3
5257.106	5258.569	G	0.4	0.6	0.3		19016.58	3e 1	2c 0	P 3
5257.335	5258.799		0.1	0.1			19015.75			
5257.637	5259.100			0.0			19014.66			
5257.800	5259.263			0.0			19014.07			

λ_{air}	λ_{vac}	Ref.	I_1	I_2	I_3	I_4	ν	Upper	Lower	Br.
5257.991	5259.454			0.1			19013.38			
5258.079	5259.543						19013.06			
5258.303	5259.767		0.2	0.3	0.2		19012.25			
5258.500	5259.963	G	0.8	0.6			19011.54	3b 6	2a 1	P 3
5258.666	5260.129		0.2	0.3	0.4		19010.94			
5258.865	5260.328			0.0			19010.22			
5259.205	5260.669		0.1				19008.99			
5259.393	5260.857	G	1.5	0.6		0.88	19008.31	3A 1	2B 5	R 2
5259.393	5260.857	G	1.5	0.6		0.88	19008.31	X 2	2B 7	R 0
5259.584	5261.048		0.1	0.2			19007.62	3f 2	2c 1	P 7
5259.637	5261.101						19007.43			
5259.703	5261.167		0.1	0.3	0.3		19007.19	3e 1	2c 0	P 6
5259.880	5261.344	G	0.3	0.5	0.3	0.65	19006.55	3A 2	2B 7	R 4
5259.880	5261.344	G	0.3	0.5	0.3	0.65	19006.55	3D 1	2B 4	R 8
5259.880	5261.344	G	0.3	0.5	0.3	0.65	19006.55	3c 1	2a 0	R 3
5260.154	5261.618		0.1				19005.56	3e 1	2c 0	Q 5
5260.777	5262.241		0.2	0.1d			19003.31			
5260.935	5262.399			0.0			19002.74			
5261.023	5262.488		0.1				19002.42	3F 3	2B 8	R 2
5261.187	5262.651	G	1.5	1.0	0.8	1.22	19001.83	3D 0	2B 2	R 0
5261.187	5262.651	G	1.5	1.0	0.8	1.22	19001.83	3a 1	2c 0	P 4
5261.464	5262.928		0.1	0.0			19000.83			
5261.605	5263.069			0.0			19000.32			
5261.849	5263.313			0.1d	0.1		18999.44			
5262.114	5263.579				0.1		18998.48			
5262.250	5263.715	G	1.0	0.4			18997.99	X 2	2B 7	R 1
5262.818	5264.283		0.1	0.1d	0.2		18995.94			
5263.076	5264.541	G	0.5	0.8	0.2		18995.01	3D 3	2B 7	P 3
5263.076	5264.541	G	0.5	0.8	0.2		18995.01	3E 2	2B 6	R 4
5263.306	5264.771		0.2	0.3			18994.18	Y 1	2B 3	P 4
5263.469	5264.934		0.2	0.0			18993.59			
5263.619	5265.084	G	0.7	1.0	0.8		18993.05			
5263.738	5265.203		0.1				18992.62			
5263.890	5265.355	G	0.9	1.5	0.8	1.07	18992.07	3D 0	2B 2	R 3
5264.096	5265.561		0.2	0.1			18991.33			
5264.290	5265.755	G	0.8	1.0	0.4	0.72	18990.63	3c 1	2a 0	R 2
5264.497	5265.963		0.2	0.1			18989.88			
5264.694	5266.160	G	1.5	1.0	0.5	1.05	18989.17	3D 0	2B 2	R 1
5264.936	5266.401		0.1	0.0			18988.30			
5265.185	5266.651			0.0			18987.40			
5265.571	5267.036		0.1	0.1			18986.01			
5265.673	5267.139			0.0			18985.64			
5265.801	5267.266			0.2			18985.18			
5265.865	5267.330		0.3				18984.95			
5266.050	5267.516	G	1.5b	1.5	1.0	1.53	18984.28	3D 0	2B 2	R 2
5266.050	5267.516	G	1.5b	1.5	1.0	1.53	18984.28	4E 0	2B 7	R 0
5266.261	5267.727		0.2				18983.52			
5266.311	5267.777			0.1			18983.34			
5266.361	5267.827		0.2				18983.16			
5266.586	5268.052				0.2		18982.35	3F 0	2B 3	Q 6
5266.694	5268.160		0.1				18981.96	3c 1	2a 0	R 4
5266.888	5268.354		0.2	0.1			18981.26			
5267.033	5268.498		0.3	0.4	0.3		18980.74	3E 2	2B 6	Q 8
5267.291	5268.757		0.1	0.0			18979.81	X 2	2B 7	P 1
5267.590	5269.056			0.0			18978.73			
5267.685	5269.151				0.0		18978.39			
5267.807	5269.273		0.2				18977.95			
5268.029	5269.495		0.3	0.2			18977.15	X 1	2B 5	P 3
5268.246	5269.712	G	1.5	0.7	0.2	1.00	18976.37	X 2	2B 7	R 2
5268.479	5269.945		0.1	0.0			18975.53			

λ_{air}	λ_{vac}	Ref.	I_1	I_2	I_3	I_4	ν	Upper	Lower	Br.
5268.676	5270.142		0.1	0.1			18974.82			
5268.879	5270.345	G	0.6	0.5	0.3		18974.09	3D 0	2B 2	P 1
5269.026	5270.492			0.0			18973.56			
5269.176	5270.642		0.5	0.2			18973.02			
5269.753	5271.220						18970.94			
5269.887	5271.353		0.1d				18970.46			
5270.192	5271.659		0.3	0.3			18969.36	3d 1	2c 0	R 1
5270.406	5271.873	G	0.9	1.5	0.8	1.20	18968.59	3E 1	2B 4	Q 4
5270.665	5272.132		0.1	0.1			18967.66			
5270.920	5272.387		0.3	0.1			18966.74	3A 0	2B 3	R 0
5271.207	5272.674			0.2	0.2b		18965.71			
5271.282	5272.749		0.2				18965.44			
5271.610	5273.077				0.1		18964.26			
5271.849	5273.316		0.1	0.1			18963.40			
5272.074	5273.541			0.2			18962.59			
5272.127	5273.594		0.3				18962.40			
5272.302	5273.769	G	1.5b	1.5	0.8	1.43	18961.77	3c 1	2a 0	R 1
5272.302	5273.769	G	1.5b	1.5	0.8	1.43	18961.77	3b 4	2a 0	P 1
5272.533	5274.000	G	1.5	0.9	0.3		18960.94	3E 1	2B 4	R 0
5272.811	5274.278		0.1	0.1			18959.94			
5273.025	5274.492	G	1.5	0.9	0.2b		18959.17	3c 6	2a 4	Q 1
5273.025	5274.492	G	1.5	0.9	0.2b		18959.17	3E 0	2B 3	R 6
5273.225	5274.693			0.1			18958.45			
5273.378	5274.846	G	0.2	0.5d	0.2		18957.90	3e 2	2c 1	R 2
5273.492	5274.960		0.2				18957.49	X 2	2B 7	R 3
5273.637	5275.105		0.1	0.1			18956.97			
5273.821	5275.288	G	0.3	0.7	0.4		18956.31	3e 2	2c 1	R 3
5273.951	5275.419		0.1				18955.84			
5274.110	5275.578	G	0.6	0.5			18955.27	3f 3	2c 2	P 3
5274.397	5275.864			0.1			18954.24			
5274.564	5276.031			0.0			18953.64			
5275.282	5276.750		0.2				18951.06			
5275.471	5276.939		0.1	0.2			18950.38			
5275.972	5277.440				0.2		18948.58			
5276.064	5277.532			0.1			18948.25	3e 1	2c 0	P 7
5276.301	5277.769	G	0.6	0.8	0.3		18947.40	3e 2	2c 1	R 1
5276.507	5277.975			0.0			18946.66			
5276.702	5278.170				0.1		18945.96			
5276.894	5278.362			0.0			18945.27			
5277.106	5278.574		0.1	0.3	0.2		18944.51	3e 2	2c 1	R 4
5277.106	5278.574		0.1	0.3	0.2		18944.51	3d 1	2c 0	P 1
5277.418	5278.886			0.2			18943.39			
5277.568	5279.037		0.2	0.2			18942.85			
5277.794	5279.262	G	0.6	0.7	0.4		18942.04	3d 1	2c 0	Q 1
5277.936	5279.405		0.3	0.2			18941.53			
5278.167	5279.636		0.2	0.0			18940.70			
5278.345	5279.814	G	1.5	0.9		1.10	18940.06	4c 5	2a 6	Q 1
5278.424	5279.892				0.1b		18939.78			
5278.552	5280.021		0.1	0.1			18939.32			
5278.750	5280.219			0.0			18938.61			
5278.942	5280.411			0.2	0.1b		18937.92	3e 2	2c 1	Q 4
5279.154	5280.623		0.2	0.3			18937.16	3e 2	2c 1	Q 3
5279.405	5280.874		0.1	0.2			18936.26	3F 3	2B 8	Q 3
5279.614	5281.083	G	0.3	0.5	0.0		18935.51	3E 1	2B 4	P 5
5279.614	5281.083	G	0.3	0.5	0.0		18935.51	3c 6	2a 4	Q 2
5279.817	5281.287		0.2	0.1			18934.78			
5280.082	5281.552		0.2	0.2d			18933.83			
5280.316	5281.786		0.5	0.3	0.1		18932.99	3a 1	2c 0	P 5
5280.420	5281.889		0.1	0.1			18932.62			
5280.537	5282.006	G	0.4	0.5		0.65	18932.20	4c 5	2a 6	Q 2

λ_{air}	λ_{vac}	Ref.	I_1	I_2	I_3	I_4	ν	Upper	Lower	Br.
5280.874	5282.344			0.0			18930.99			
5281.067	5282.536		0.1	0.1	0.2		18930.30			
5281.391	5282.860						18929.14			
5281.416	5282.885		0.1	0.1			18929.05	3e 1	2c 0	Q 6
5281.605	5283.075		0.4	0.3			18928.37			
5281.798	5283.268				0.2d		18927.68			
5281.932	5283.402		0.1				18927.20			
5282.096	5283.566		0.4	0.2			18926.61	4F 0	2C 0	R 2
5282.096	5283.566		0.4	0.2			18926.61	3D 3	2B 7	P 4
5282.300	5283.770		0.1	0.3	0.3		18925.88	3e 2	2c 1	Q 5
5282.504	5283.974			0.1			18925.15	4E 1	2B 9	Q 4
5282.716	5284.186			0.0			18924.39			
5282.906	5284.376			0.4	0.4		18923.71	3e 2	2c 1	Q 2
5282.906	5284.376			0.4	0.4		18923.71	3e 2	2c 1	R 5
5283.160	5284.630		0.7	0.3			18922.80			
5283.286	5284.756	G	0.7	1.0	0.4	0.97	18922.35	3E 1	2B 4	Q 3
5283.464	5284.935		0.9	0.3			18921.71	X 2	2B 7	P 2
5283.685	5285.155		0.6	0.7	0.2		18920.92	4c 5	2a 6	Q 3
5283.959	5285.429			0.1			18919.94			
5284.165	5285.636		0.3	0.2			18919.20	3A 0	2B 3	P 4
5284.165	5285.636		0.3	0.2			18919.20	3A 1	2B 5	R 1
5284.313	5285.784		0.2				18918.67			
5284.506	5285.977	G	1.5b	1.5	0.7	1.14	18917.98	3d 1	2c 0	R 2
5284.506	5285.977	G	1.5b	1.5	0.7	1.14	18917.98	3E 1	2B 4	P 4
5284.506	5285.977	G	1.5b	1.5	0.7	1.14	18917.98	3c 1	2a 0	R 0
5284.730	5286.200		0.2	0.0			18917.18			
5284.897	5286.368		0.2	0.2			18916.58			
5285.140	5286.611				0.1		18915.71			
5285.422	5286.893			0.1	0.0		18914.70			
5285.797	5287.268		0.2	0.2			18913.36			
5285.987	5287.458		0.2	0.3			18912.68			
5286.099	5287.570				0.1b		18912.28			
5286.146	5287.617			0.1			18912.11			
5286.328	5287.799		0.1				18911.46			
5286.658	5288.129		0.1				18910.28			
5286.820	5288.291			0.0			18909.70			
5286.982	5288.453						18909.12			
5287.030	5288.501		0.2	0.2	0.1		18908.95			
5287.343	5288.814			0.2	0.1		18907.83			
5287.558	5289.030		0.2	0.3			18907.06			
5287.726	5289.197			0.1			18906.46			
5287.891	5289.362			0.2			18905.87			
5287.969	5289.441				0.1		18905.59			
5288.087	5289.558						18905.17			
5288.227	5289.698			0.1			18904.67	3E 2	2B 6	R 3
5288.417	5289.888		0.1	0.3	0.2b		18903.99	3e 1	2c 0	R 3
5288.618	5290.090			0.1			18903.27	4F 0	2C 0	R 3
5288.806	5290.278			0.2			18902.60			
5288.929	5290.401		0.0	0.0			18902.16	3e 2	2c 1	Q 6
5289.108	5290.580	G	0.5	0.6	0.2b		18901.52	3f 3	2c 2	P 4
5289.108	5290.580	G	0.5	0.6	0.2b		18901.52	3e 2	2c 1	Q 1
5289.108	5290.580	G	0.5	0.6	0.2b		18901.52	X 1	2B 5	P 4
5289.318	5290.790		0.1	0.1			18900.77			
5289.472	5290.944	G	0.3	0.7	0.2		18900.22	3c 6	2a 4	Q 3
5289.472	5290.944	G	0.3	0.7	0.2		18900.22	3f 3	2c 2	P 4
5289.749	5291.221		0.1				18899.23			
5289.956	5291.428		0.0				18898.49			
5290.180	5291.652		0.1	0.0d			18897.69	4E 0	2B 7	P 2
5290.376	5291.848	G	1.5	0.7	0.2	0.62	18896.99	3E 1	2B 4	P 3
5290.558	5292.030	G	1.0	0.7	0.2	0.52	18896.34	3D 0	2B 2	P 2

λ_{air}	λ_{vac}	Ref.	I_1	I_2	I_3	I_4	ν	Upper	Lower	Br.
5290.779	5292.251		0.0				18895.55			
5290.942	5292.414			0.1d	0.1		18894.97	3e 2	2c 1	R 6
5291.135	5292.607		0.1				18894.28			
5291.297	5292.769						18893.70	W 1	2B 5	R 4
5291.353	5292.826			0.2			18893.50			
5291.398	5292.870		0.3				18893.34			
5291.594	5293.067	G	1.5b	1.5	0.8	1.42	18892.64	3E 1	2B 4	Q 2
5291.801	5293.274		0.2				18891.90			
5291.885	5293.358		0.2	0.0			18891.60	3D 1	2B 4	R 7
5292.084	5293.557			0.0	0.1		18890.89			
5292.146	5293.618						18890.67			
5292.378	5293.851						18889.84	3b 6	2a 1	P 4
5292.505	5293.977			0.0			18889.39			
5292.852	5294.325			0.0			18888.15	4E 0	2B 7	R 4
5293.000	5294.473		0.4				18887.62			
5293.099	5294.571	G		0.3b			18887.27			
5293.146	5294.619		0.1		0.2		18887.10			
5293.390	5294.863			0.1			18886.23			
5293.662	5295.135			0.1			18885.26	4E 0	2B 7	Q 4
5293.830	5295.303		0.0	0.1			18884.66			
5294.158	5295.631		0.1	0.2	0.2		18883.49			
5294.455	5295.929		0.2	0.1			18882.43	3A 0	2B 3	P 3
5294.769	5296.243			0.1			18881.31	3a 2	2c 1	P 3
5295.005	5296.478	G	0.3	0.3			18880.47	3A 2	2B 7	R 3
5295.182	5296.655				0.1		18879.84			
5295.468	5296.941		0.1	0.2			18878.82			
5295.917	5297.390		0.2	0.1			18877.22			
5296.096	5297.570	G	0.9	1.0	0.3	0.76	18876.58	3E 1	2B 4	Q 1
5296.307	5297.780		0.0	0.0			18875.83			
5296.570	5298.044			0.0	0.2		18874.89			
5296.711	5298.184		0.2				18874.39			
5296.879	5298.353	G	0.4	0.4	0.1		18873.79	3A 0	2B 3	P 1
5297.165	5298.639		0.1	0.1			18872.77			
5297.370	5298.844		0.1	0.2			18872.04			
5297.494	5298.968			0.0			18871.60			
5297.704	5299.178	G	0.5	0.3b			18870.85	3b 4	2a 0	P 2
5297.814	5299.288		0.2				18870.46			
5297.974	5299.448		0.1				18869.89			
5298.639	5300.114			0.1d			18867.52			
5298.856	5300.330		0.1	0.1	0.2		18866.75			
5298.996	5300.470		0.3	0.4			18866.25	4E 1	2B 9	Q 1
5299.173	5300.647		0.3	0.1			18865.62	3A 0	2B 3	P 2
5299.285	5300.760	G	0.3	0.5	0.4		18865.22	3e 2	2c 1	Q 1
5299.493	5300.968		0.2	0.2			18864.48	4E 1	2B 9	Q 3
5299.707	5301.181	G	1.5	0.9	0.3	0.83	18863.72	3E 1	2B 4	P 2
5299.901	5301.375			0.1			18863.03			
5300.063	5301.538		0.1	0.3	0.3b		18862.45	3a 1	2c 0	P 6
5300.131	5301.606						18862.21			
5300.184	5301.659		0.3	0.3			18862.02			
5300.243	5301.718						18861.81	3E 2	2B 6	Q 7
5300.375	5301.850		0.1				18861.34	3F 3	2B 8	Q 2
5300.569	5302.044		0.2	0.2d	0.2b		18860.65			
5300.822	5302.297		0.1	0.1			18859.75			
5301.056	5302.531		0.2	0.3	0.3		18858.92	3e 2	2c 1	R 7
5301.379	5302.854			0.1			18857.77			
5301.480	5302.955				0.1		18857.41			
5301.609	5303.085			0.0			18856.95			
5301.905	5303.380		0.2	0.2			18855.90	4E 2	2B11	Q 2
5302.110	5303.585	G	1.5	1.0	0.5	0.95	18855.17	Z 0	2B 2	R 1
5302.110	5303.585	G	1.5	1.0	0.5	0.95	18855.17	3d 1	2c 0	R 3

λ_{air}	λ_{vac}	Ref.	I_1	I_2	I_3	I_4	ν	Upper	Lower	Br.
5302.405	5303.880		0.5	0.2			18854.12	X 2	2B 7	P 3
5302.405	5303.880		0.5	0.2			18854.12	4E 0	2B 7	Q 1
5302.405	5303.880		0.5	0.2			18854.12	3A 1	2B 5	R 0
5302.630	5304.106		0.3	0.5	0.2		18853.32	3c 6	2a 4	Q 4
5302.630	5304.106		0.3	0.5	0.2		18853.32	3D 3	2B 7	P 5
5302.760	5304.235		0.2				18852.86			
5302.895	5304.370		0.4	0.2			18852.38			
5303.114	5304.590	G				1.64	18851.60	3c 1	2a 0	Q 1
5303.167	5304.643		1.5b	1.5b	1.5b		18851.41			
5303.235	5304.711	G					18851.17	Z 0	2B 2	R 2
5303.393	5304.868		0.2				18850.61			
5303.533	5305.009		0.2	0.1			18850.11			
5303.724	5305.200		0.1	0.2			18849.43			
5303.969	5305.445	G		0.4	0.3		18848.56	3d 1	2c 0	Q 2
5304.180	5305.656			0.2			18847.81			
5304.214	5305.690		0.2				18847.69			
5304.414	5305.890	G	1.5	1.5	0.5	1.30	18846.98	3f 3	2c 2	P 5
5304.414	5305.890	G	1.5	1.5	0.5	1.30	18846.98	Z 0	2B 2	R 0
5304.597	5306.073		0.5	0.1			18846.33			
5304.771	5306.247				0.1		18845.71			
5304.887	5306.363		0.2	0.2			18845.30			
5305.106	5306.582			0.1			18844.52			
5305.171	5306.647		0.1				18844.29	4E 1	2B 9	Q 2
5305.380	5306.856		0.2	0.2			18843.55			
5305.551	5307.027	G	1.0	0.4	0.0	0.36	18842.94			
5305.765	5307.242		0.2	0.3			18842.18	3f 3	2c 2	P 5
5305.999	5307.475		0.1	0.1			18841.35			
5306.058	5307.534						18841.14			
5306.154	5307.630				0.0		18840.80	4f 0	2c 2	R 3
5306.461	5307.937			0.2			18839.71	4F 0	2C 0	Q 4
5306.627	5308.104		0.2				18839.12			
5307.109	5308.585			0.1	0.0		18837.41			
5307.334	5308.811				0.1		18836.61			
5307.596	5309.073			0.1			18835.68			
5307.960	5309.437		0.1	0.0			18834.39			
5308.230	5309.707		0.2	0.3			18833.43			
5308.383	5309.859		0.2	0.2			18832.89			
5308.555	5310.031		0.5	0.5			18832.28	3e 2	2c 1	Q 2
5308.625	5310.102	G				0.65	18832.03			
5308.679	5310.156		0.7	0.7	0.4		18831.84	Z 0	2B 2	R 3
5308.839	5310.316		0.2	0.2			18831.27	4E 0	2B 7	Q 3
5309.034	5310.511	G	1.5	1.5	0.7	1.04	18830.58	3c 1	2a 0	Q 2
5309.034	5310.511	G	1.5	1.5	0.7	1.04	18830.58	V 0	2B 2	R 2
5309.262	5310.739		0.1	0.1			18829.77			
5309.474	5310.951			0.0			18829.02	4D 0	2B 7	R 5
5309.789	5311.267						18827.90			
5310.001	5311.478			0.0			18827.15			
5310.859	5312.336			0.0			18824.11			
5311.177	5312.655				0.1		18822.98			
5311.533	5313.011		0.4	0.1			18821.72			
5311.649	5313.126	G	0.4	0.4	0.2		18821.31	3d 1	2c 0	P 2
5311.951	5313.428				0.1		18820.24			
5312.258	5313.736				0.1		18819.15			
5312.312	5313.790						18818.96			
5312.597	5314.075			0.0			18817.95			
5312.775	5314.253						18817.32	3a 2	2c 1	P 4
5312.995	5314.473			0.1			18816.54			
5313.159	5314.637	G	0.4	0.7	0.4		18815.96	3D 1	2B 4	R 6
5313.159	5314.637	G	0.4	0.7	0.4		18815.96	3E 3	2C 0	R 2
5313.405	5314.883		0.1	0.1			18815.09			

λ_{air}	λ_{vac}	Ref.	I_1	I_2	I_3	I_4	ν	Upper	Lower	Br.
5313.597	5315.075	G	1.5	0.8	0.2	0.72	18814.41	3D 0	2B 2	P 3
5313.597	5315.075	G	1.5	0.8	0.2	0.72	18814.41	3E 2	2B 6	R 2
5313.831	5315.309		0.3	0.3	0.1		18813.58			
5313.989	5315.468	G	0.3	0.3			18813.02	3E 3	2C 0	R 1
5314.159	5315.637		0.2	0.1			18812.42			
5314.354	5315.832	G	1.5	0.8		0.79	18811.73	4c 6	2a 7	Q 1
5314.552	5316.030						18811.03			
5314.608	5316.087		0.1	0.2			18810.83	4E 0	2B 7	Q 2
5314.656	5316.135						18810.66	Z 0	2B 2	R 4
5314.975	5316.454		0.1				18809.53			
5315.173	5316.652	G	0.3	0.5	0.1b	0.80	18808.83	3a 1	2c 0	P 7
5315.371	5316.850			0.1			18808.13			
5315.594	5317.073			0.0			18807.34			
5315.798	5317.277	G	0.3	0.4			18806.62			
5316.035	5317.514			0.1			18805.78			
5316.236	5317.715		0.2	0.2	0.1		18805.07			
5316.439	5317.918		0.1	0.0			18804.35			
5316.668	5318.148		0.2		0.2		18803.54			
5316.702	5318.181			0.2			18803.42			
5316.900	5318.379			0.0			18802.72			
5317.084	5318.563						18802.07			
5317.426	5318.906		0.1	0.1			18800.86	3E 3	2C 0	R 3
5317.636	5319.115			0.2			18800.12			
5317.701	5319.180		0.2				18799.89			
5317.766	5319.245						18799.66			
5317.890	5319.370	G	1.5	1.5	1.5	1.20	18799.22	4F 0	2C 0	R 2
5317.890	5319.370	G	1.5	1.5	1.5	1.20	18799.22	3c 1	2a 0	Q 3
5318.133	5319.613		0.1	0.1			18798.36	3F 0	2B 3	Q 5
5318.329	5319.808		0.3	0.3			18797.67	3A 2	2B 7	R 2
5318.623	5320.103			0.1d			18796.63			
5318.880	5320.360			0.3b	0.1		18795.72	3f 3	2c 2	P 6
5318.965	5320.445		0.2				18795.42			
5319.161	5320.640	G	1.5b	1.0	0.3	1.34	18794.73			
5319.268	5320.748		0.4	0.5			18794.35			
5319.463	5320.943		0.1	0.1			18793.66			
5319.628	5321.108		0.4	0.2			18793.08			
5319.755	5321.235		0.4	0.3			18792.63	3A 1	2B 5	P 4
5319.973	5321.453		0.1	0.1			18791.86			
5320.194	5321.674						18791.08			
5320.228	5321.708		0.2	0.3	0.3		18790.96			
5320.412	5321.892						18790.31			
5320.440	5321.920	G	1.5	0.9	0.5	0.89	18790.21	Z 0	2B 2	P 1
5320.692	5322.172		0.2	0.0			18789.32	3E 3	2C 0	Q 1
5320.692	5322.172		0.2	0.0			18789.32	4f 0	2c 2	Q 5
5320.930	5322.410		0.1	0.0			18788.48			
5321.310	5322.790		0.1	0.0			18787.14			
5321.497	5322.977	G	0.9	0.9	0.4	0.30	18786.48	3e 2	2c 1	P 2
5321.497	5322.977	G	0.9	0.9	0.4	0.30	18786.48	3e 2	2c 1	Q 3
5321.726	5323.206		0.0	0.0			18785.67			
5321.885	5323.365			0.1d			18785.11			
5322.069	5323.549		0.3	0.3			18784.46	4F 0	2C 0	R 1
5322.287	5323.768	G	0.2	0.4			18783.69	3d 1	2c 0	R 4
5322.287	5323.768	G	0.2	0.4			18783.69	3E 0	2B 3	Q10
5322.358	5323.838				0.5b		18783.44			
5322.414	5323.895		0.2	0.4			18783.24	3e 1	2c 0	P 5
5322.633	5324.113		0.0	0.0			18782.47			
5322.806	5324.286			0.1			18781.86			
5323.046	5324.527						18781.01	3f 3	2c 2	P 6
5323.171	5324.652			0.2			18780.57			
5323.333	5324.814			0.1			18780.00	3E 0	2B 3	Q10

λ_{air}	λ_{vac}	Ref.	I_1	I_2	I_3	I_4	ν	Upper	Lower	Br.
5323.514	5324.995			0.1			18779.36			
5323.684	5325.165		0.2	0.4	0.4		18778.76			
5323.914	5325.395			0.1			18777.95			
5324.098	5325.579		0.2				18777.30	3A 1	2B 5	P 1
5324.223	5325.704			0.2	0.1		18776.86			
5324.271	5325.752		0.1				18776.69			
5324.447	5325.928		0.7	0.4	0.1		18776.07	T 0	2B 7	R 2
5324.447	5325.928		0.7	0.4	0.1		18776.07	X 2	2B 7	P 4
5324.645	5326.127			0.1			18775.37			
5324.864	5326.345						18774.60			
5324.946	5326.427			0.0	0.0		18774.31			
5325.017	5326.498						18774.06			
5325.250	5326.731			0.1			18773.24	W 1	2B 5	R 2
5325.349	5326.830				0.1		18772.89			
5325.545	5327.026		0.1				18772.20			
5325.695	5327.176			0.0	0.1		18771.67			
5325.922	5327.404			0.1			18770.87	V 0	2B 2	R 3
5326.214	5327.696			0.1			18769.84			
5326.450	5327.931				0.1		18769.01			
5326.566	5328.048		0.2	0.2			18768.60			
5326.776	5328.258	G	1.5	0.9	0.3	0.96	18767.86	3D 1	2B 4	R 0
5326.776	5328.258	G	1.5	0.9	0.3	0.96	18767.86	3E 3	2C 0	R 4
5326.958	5328.440			0.1	0.2		18767.22			
5327.040	5328.522		0.1				18766.93			
5327.290	5328.772	G	0.3	0.5	0.2		18766.05	3D 1	2B 4	R 5
5327.568	5329.050		0.1	0.1			18765.07			
5327.773	5329.255		0.2	0.2			18764.35	3A 1	2B 5	P 3
5327.926	5329.408			0.2			18763.81			
5328.304	5329.786		0.1				18762.48			
5328.346	5329.828			0.0			18762.33			
5328.764	5330.246						18760.86			
5328.891	5330.374		0.3	0.2			18760.41	3A 1	2B 5	P 2
5329.045	5330.527	G	0.4	0.7	0.3		18759.87	3E 2	2B 6	Q 6
5329.295	5330.777			0.2			18758.99			
5329.349	5330.831		0.1				18758.80			
5329.445	5330.928			0.0			18758.46			
5329.641	5331.124	G	0.3	0.9	0.6		18757.77	3c 1	2a 0	Q 4
5329.940	5331.422	G	0.4	0.5	0.3		18756.72	32 4	21 0	P 3
5330.181	5331.664		0.0	0.1			18755.87			
5330.440	5331.923			0.1			18754.96	3E 3	2C 0	Q 2
5330.781	5332.264			0.1			18753.76			
5330.940	5332.423						18753.20	3a 2	2c 1	P 5
5331.031	5332.514		0.2	0.2			18752.88	3e 3	2c 2	R 4
5331.256	5332.739	G	0.8	0.8	0.3		18752.09	3e 2	2c 1	P 3
5331.418	5332.901		0.2	0.1			18751.52			
5331.606	5333.089	G	1.5	0.9	0.2	0.78	18750.86	3D 1	2B 4	R 1
5331.796	5333.279		0.3	0.5	0.1		18750.19	3e 2	2c 1	P 2
5331.967	5333.450			0.1			18749.59			
5332.140	5333.623	G	0.6	0.6	0.2	0.53	18748.98	4F 0	2C 0	Q 3
5332.140	5333.623	G	0.6	0.6	0.2	0.53	18748.98	3D 1	2B 4	P 1
5332.459	5333.942		0.1	0.1d	0.1		18747.86			
5332.550	5334.033						18747.54			
5332.758	5334.241		0.2	0.1	0.0		18746.81			
5332.985	5334.468			0.2			18746.01			
5333.213	5334.696	G	0.3	0.3	0.1	0.54	18745.21	4D 0	2B 7	R 4
5333.361	5334.844		0.1	0.2			18744.69			
5333.398	5334.881						18744.56			
5333.546	5335.029		0.1	0.0			18744.04			
5333.805	5335.288		0.1	0.1			18743.13	3E 0	2B 3	R 5
5333.978	5335.462						18742.52			

λ air	λ vac	Ref.	I₁	I₂	I₃	I₄	ν	Upper	Lower	Br.
5334.058	5335.542		0.3	0.2			18742.24			
5334.268	5335.752	G	1.5b	1.5	0.5	1.32	18741.50	Z 0	2B 2	P 2
5334.268	5335.752	G	1.5b	1.5	0.5	1.32	18741.50	3D 0	2B 2	P 4
5334.499	5335.983		0.2				18740.69			
5334.573	5336.057			0.0			18740.43			
5334.764	5336.248		0.1				18739.76			
5334.812	5336.296			0.0	0.1b		18739.59	3F 1	2B 5	Q 6
5335.057	5336.541		0.1	0.2			18738.73			
5335.282	5336.766	G	0.8	0.8	0.6		18737.94	3d 1	2c 0	Q 3
5335.532	5337.017		0.1				18737.06	3D 0	2B 2	P10
5335.598	5337.082			0.1			18736.83			
5335.880	5337.364			0.1			18735.84			
5335.943	5337.427		0.1				18735.62			
5336.128	5337.612		0.2	0.2			18734.97			
5336.344	5337.828	G	1.0	1.0	0.6	1.10	18734.21	3D 1	2B 4	R 4
5336.586	5338.071	G	1.5b	1.5	0.6	1.41	18733.36	3D 1	2B 4	R 2
5336.800	5338.284		0.2				18732.61	W 1	2B 5	R 1
5336.880	5338.364		0.2	0.0			18732.33			
5337.014	5338.498		0.1	0.0			18731.86			
5337.196	5338.681			0.0			18731.22			
5337.307	5338.792				0.1		18730.83			
5337.495	5338.980			0.2			18730.17			
5337.692	5339.176		0.2	0.3	0.1b		18729.48	3e 2	2c 1	Q 4
5337.692	5339.176		0.2	0.3	0.1b		18729.48	3E 3	2C 0	Q 1
5337.965	5339.450	G	0.7	0.4			18728.52	3E 2	2B 6	R 1
5338.250	5339.735			0.1			18727.52			
5338.675	5340.160			0.1	0.1		18726.03			
5339.014	5340.500		0.1	0.1	0.1b		18724.84			
5339.206	5340.691			0.2			18724.17			
5339.397	5340.882	G	0.8	1.0	0.3	0.73	18723.50	3D 1	2B 4	R 3
5339.639	5341.124			0.3	0.3	0.2	18722.65			
5340.175	5341.661			0.0			18720.77			
5340.198	5341.683						18720.69	3A 2	2B 7	R 1
5340.364	5341.849		0.1				18720.11	3E 3	2C 0	R 5
5340.415	5341.900			0.0			18719.93			
5340.635	5342.120		0.2	0.1			18719.16	T 0	2B 7	R 1
5340.828	5342.314	G	1.5b	0.9	0.3	0.69	18718.48	3c 1	2a 0	P 2
5341.122	5342.608		0.2	0.2	0.3		18717.45			
5341.308	5342.794			0.0			18716.80			
5341.451	5342.936		0.1	0.1			18716.30			
5341.656	5343.142		0.4	0.4	0.1		18715.58			
5341.922	5343.407		0.1	0.2	0.1		18714.65	3e 3	2c 2	R 3
5342.156	5343.642			0.0			18713.83			
5342.313	5343.799				0.1		18713.28			
5342.398	5343.884			0.0			18712.98			
5342.601	5344.087						18712.27	4c 7	2a 8	Q 1
5342.664	5344.150		0.1	0.1d			18712.05			
5342.955	5344.441			0.0d			18711.03			
5343.158	5344.644	G	0.9	1.5	0.7	0.52	18710.32	3e 2	2c 1	P 4
5343.446	5344.933			0.0			18709.31			
5343.652	5345.138		0.1	0.1			18708.59	V 0	2B 2	R 1
5344.000	5345.487			0.1			18707.37			
5344.240	5345.727	G	0.5	1.0	0.9		18706.53	3c 1	2a 0	Q 5
5344.418	5345.904			0.1			18705.91			
5344.552	5346.038			0.2			18705.44			
5344.612	5346.098		0.2				18705.23	4D 0	2B 7	R 0
5344.612	5346.098		0.2				18705.23	3d 1	2c 0	R 5
5344.792	5346.278	G	1.5	1.0		1.39	18704.60			
5344.978	5346.464			0.0			18703.95			
5345.023	5346.510		0.1				18703.79			

λ_{air}	λ_{vac}	Ref.	I_1	I_2	I_3	I_4	ν	Upper	Lower	Br.
5345.175	5346.661						18703.26			
5345.375	5346.862						18702.56			
5345.646	5347.133		0.1	0.1			18701.61	X 2	2B 7	P 5
5345.695	5347.182						18701.44			
5345.844	5347.331	G	1.5	0.6		0.49	18700.92			
5346.012	5347.499			0.1	0.1		18700.33			
5346.141	5347.628		0.1				18699.88			
5346.418	5347.905		0.1				18698.91			
5346.661	5348.148			0.0			18698.06			
5346.844	5348.331		0.3	0.3			18697.42			
5346.901	5348.389	G			0.2		18697.22			
5347.122	5348.609		0.3	0.3			18696.45	3D 3	2C 0	Q 1
5347.422	5348.909				0.1b		18695.40			
5347.628	5349.115			0.0			18694.68			
5347.788	5349.276						18694.12			
5347.851	5349.339		0.1				18693.90	W 1	2B 5	R 0
5348.048	5349.536		0.2	0.2			18693.21			
5348.129	5349.616				0.1		18692.93			
5348.269	5349.756						18692.44			
5348.638	5350.126		0.2	0.2	0.1b		18691.15			
5349.113	5350.601						18689.49			
5349.233	5350.721				0.1		18689.07	4D 0	2B 7	R 3
5349.331	5350.818						18688.73			
5349.442	5350.930		0.2				18688.34	3E 3	2C 0	Q 2
5349.671	5351.159		0.1		0.0		18687.54			
5349.797	5351.285						18687.10			
5349.952	5351.440		0.1	0.0			18686.56			
5350.032	5351.520						18686.28			
5350.235	5351.723	G	0.3	0.4			18685.57	3E 3	2C 0	Q 3
5350.384	5351.872						18685.05	3D 0	2B 2	P 9
5350.459	5351.947			0.1d			18684.79			
5350.530	5352.018		0.2				18684.54			
5350.702	5352.190	G	1.0	0.7	0.3	0.67	18683.94	4F 0	2C 0	R 1
5350.974	5352.462		0.2	0.2	0.2		18682.99	3D 0	2B 2	P 5
5351.143	5352.631						18682.40			
5351.195	5352.683						18682.22			
5351.255	5352.743			0.0			18682.01			
5351.450	5352.938		0.2	0.2	0.1		18681.33			
5351.739	5353.227	G	1.5	0.8	0.3	0.71	18680.32	Z 0	2B 2	P 3
5351.982	5353.471		0.2				18679.47			
5352.054	5353.543			0.0			18679.22			
5352.275	5353.763	G	0.9	0.4		0.23	18678.45	T 0	2B 7	R 0
5352.513	5354.001						18677.62			
5352.561	5354.050		0.3	0.1			18677.45			
5352.664	5354.153		0.2				18677.09			
5352.748	5354.236						18676.80			
5352.842	5354.331				0.1		18676.47			
5352.965	5354.454		0.1	0.0			18676.04			
5353.146	5354.635			0.0			18675.41			
5353.378	5354.867	G	0.3	0.6	0.3		18674.60	3E 2	2B 6	Q 5
5353.378	5354.867	G	0.3	0.6	0.3		18674.60	3E 2	2B 6	P 6
5353.584	5355.073		0.1	0.1			18673.88			
5353.751	5355.240		0.2	0.3	0.2		18673.30	3F 0	2B 3	P 6
5353.977	5355.466		0.3	0.4	0.4		18672.51			
5354.169	5355.659	G	0.3	0.5			18671.84	3e 3	2c 2	R 2
5354.350	5355.839		0.1	0.0			18671.21			
5354.473	5355.963						18670.78			
5354.531	5356.020		0.2	0.2			18670.58	3D 1	2B 4	P 2
5354.565	5356.054						18670.46			
5354.749	5356.238	G	1.5	0.7	0.1	0.90	18669.82			

λ_{air}	λ_{vac}	Ref.	I_1	I_2	I_3	I_4	ν	Upper	Lower	Br.
5354.990	5356.479						18668.98			
5355.070	5356.559		0.2	0.2			18668.70	3d 1	2c 0	P 3
5355.156	5356.645						18668.40			
5355.377	5356.866		0.2	0.2			18667.63			
5355.509	5356.998		0.3	0.2			18667.17			
5355.624	5357.113						18666.77			
5355.713	5357.202		0.4	0.2			18666.46			
5355.911	5357.400	GK	1.5b	1.5	0.5	1.60	18665.77	3E 3	2C 0	R 6
5356.123	5357.613		0.2				18665.03			
5356.212	5357.702		0.2	0.1			18664.72			
5356.318	5357.808		0.1				18664.35	3e 2	2c 1	Q 5
5356.318	5357.808		0.1				18664.35	3E 3	2C 0	P 2
5356.421	5357.911		0.2	0.3	0.1		18663.99	4F 0	2C 0	Q 2
5356.571	5358.060						18663.47			
5356.628	5358.118		0.7	0.3			18663.27			
5356.812	5358.302		0.2	0.3	0.2		18662.63			
5356.993	5358.482		0.1				18662.00			
5357.059	5358.548			0.1			18661.77			
5357.153	5358.643		0.2				18661.44	4D 0	2B 7	R 2
5357.300	5358.790	G	0.9	0.9	0.4		18660.93	3e 2	2c 1	P 3
5357.475	5358.965						18660.32			
5357.662	5359.152		0.3	0.2			18659.67			
5357.854	5359.344			0.0			18659.00			
5358.006	5359.496		0.1	0.1			18658.47			
5358.170	5359.660		0.5				18657.90			
5358.239	5359.729	G		0.6	0.3	0.28	18657.66			
5358.282	5359.772		0.1				18657.51	3e 2	2c 1	P 5
5358.466	5359.956			0.2			18656.87	3e 1	2c 0	P 6
5358.514	5360.005		0.2				18656.70	4D 0	2B 7	R 1
5358.721	5360.212	G	1.0	1.0	0.3	1.03	18655.98			
5358.957	5360.447		0.1	0.0			18655.16			
5359.060	5360.551			0.1			18654.80			
5359.212	5360.703		0.3	0.5	0.3		18654.27	4F 0	2C 0	Q 4
5359.417	5360.907	G	0.3	0.6	0.2		18653.56			
5359.681	5361.171			0.1d			18652.64			
5359.813	5361.304			0.0d			18652.18			
5359.977	5361.467		0.3	0.3	0.1		18651.61			
5360.299	5361.789		0.1	0.1			18650.49			
5360.511	5362.002			0.1			18649.75			
5360.790	5362.281			0.2	0.1		18648.78			
5361.017	5362.508			0.0			18647.99			
5361.242	5362.732		0.2	0.1d			18647.21			
5361.325	5362.816						18646.92			
5361.411	5362.902			0.0			18646.62			
5361.552	5363.043		0.4				18646.13			
5361.581	5363.072	G		0.6	0.6	0.38	18646.03	3c 1	2a 0	Q 6
5361.716	5363.207		0.2				18645.56	3D 0	2B 2	P 6
5361.863	5363.354		0.1				18645.05			
5361.969	5363.460						18644.68			
5362.038	5363.529		0.1				18644.44			
5362.081	5363.572			0.1			18644.29			
5362.176	5363.667						18643.96			
5362.236	5363.728		0.2				18643.75	3D 3	2C 0	R 2
5362.369	5363.860	G	0.9	0.6	0.2	0.38	18643.29	3E 2	2B 6	R 0
5362.559	5364.050						18642.63			
5362.636	5364.128		0.1	0.2			18642.36			
5362.838	5364.329	G	1.0	1.5	0.6	1.20	18641.66			
5363.071	5364.562			0.0			18640.85			
5363.125	5364.617		0.1				18640.66			
5363.275	5364.767			0.1d			18640.14			

λ_{air}	λ_{vac}	Ref.	I_1	I_2	I_3	I_4	ν	Upper	Lower	Br.
5363.333	5364.824		0.1				18639.94	3E 2	2B 6	P 5
5363.333	5364.824		0.1				18639.94	4f 1	2c 3	R 3
5363.505	5364.997	G	0.4	0.7		0.92	18639.34			
5363.733	5365.224			0.2			18638.55			
5364.032	5365.523						18637.51			
5364.248	5365.739			0.0			18636.76	4f 0	2c 2	R 2
5364.547	5366.039		0.7	0.3d			18635.72			
5364.619	5366.111				0.2		18635.47			
5364.979	5366.471		0.3	0.3	0.2		18634.22	3F 0	2B 3	Q 4
5365.238	5366.730		0.1	0.2			18633.32			
5365.451	5366.943	G	0.7				18632.58	3A 2	2B 7	R 0
5365.543	5367.035		0.4	0.5d	0.3		18632.26			
5365.693	5367.185		0.2	0.2			18631.74	3D 0	2B 2	P 7
5365.895	5367.387	G	1.5b	1.5	0.8	1.15	18631.04	3c 1	2a 0	P 3
5366.111	5367.603		0.2				18630.29			
5366.183	5367.675		0.2	0.1			18630.04			
5366.488	5367.980						18628.98			
5366.684	5368.176			0.1d			18628.30	3E 3	2C 0	Q 3
5366.724	5368.217						18628.16			
5366.802	5368.295		0.1				18627.89			
5366.969	5368.462		0.3	0.3			18627.31	Z 1	2B 4	R 2
5367.027	5368.519				0.1		18627.11			
5367.182	5368.675			0.1			18626.57	Z 2	2B 6	R 1
5367.260	5368.753						18626.30			
5367.370	5368.862			0.1			18625.92			
5367.545	5369.038		0.1	0.2	0.2		18625.31	3D 0	2B 2	P 8
5367.545	5369.038		0.1	0.2	0.2		18625.31	4f 0	2c 2	Q 4
5367.545	5369.038		0.1	0.2	0.2		18625.31	W 1	2B 5	P 1
5367.675	5369.168		0.4	0.3			18624.86	Z 1	2B 4	R 1
5367.675	5369.168		0.4	0.3			18624.86	3A 2	2B 7	P 5
5367.765	5369.257						18624.55			
5367.886	5369.378		0.1	0.1			18624.13			
5368.084	5369.577		0.2	0.2			18623.44	3F 0	2B 3	R 2
5368.252	5369.745			0.4			18622.86			
5368.309	5369.802						18622.66	3b 4	2a 0	P 4
5368.355	5369.848	G	1.0	0.6	0.3	0.59	18622.50			
5368.615	5370.108		0.1	0.2			18621.60			
5368.670	5370.163				0.2		18621.41	3d 1	2c 0	R 6
5368.820	5370.313	G	0.3	0.6	0.3		18620.89	3c 2	2a 1	R 4
5369.108	5370.601			0.1			18619.89			
5369.382	5370.875	G	0.4	0.7	0.5		18618.94	3c 2	2a 1	R 5
5369.656	5371.149			0.1			18617.99			
5369.918	5371.412		0.3	0.4	0.3		18617.08	3d 1	2c 0	Q 4
5370.123	5371.616		0.6	0.3			18616.37	Z 1	2B 4	R 0
5370.371	5371.865			0.1			18615.51			
5370.492	5371.986						18615.09			
5370.683	5372.176				0.2		18614.43			
5370.761	5372.254			0.0			18614.16			
5371.121	5372.615			0.0	0.2		18612.91			
5371.245	5372.739						18612.48			
5371.364	5372.857						18612.07			
5371.413	5372.907		0.2	0.2			18611.90			
5371.563	5373.057		0.5	0.5	0.3		18611.38	4F 0	2C 0	Q 3
5371.696	5373.190		0.1	0.0			18610.92			
5371.748	5373.241						18610.74	3E 0	2B 3	Q 9
5371.748	5373.241						18610.74	w 0	2c 3	P 1
5371.889	5373.383	G	1.5	1.5	0.7	0.95	18610.25	3c 2	2a 1	R 3
5372.120	5373.614		0.5	0.3	0.2		18609.45	3e 3	2c 2	R 1
5372.273	5373.767		0.1				18608.92	3E 3	2C 0	P 2
5372.440	5373.934	G	1.5	1.0	0.5	0.94	18608.34	Z 0	2B 2	P 4

λ_air	λ_vac	Ref.	I₁	I₂	I₃	I₄	ν	Upper	Lower	Br.
5372.671	5374.166		0.2	0.3			18607.54			
5372.772	5374.267		0.1				18607.19			
5372.876	5374.371			0.1			18606.83			
5372.952	5374.446		0.2				18606.57	3E 2	2B 6	P 4
5373.087	5374.581	G	0.7	1.0	0.3	0.95	18606.10	3E 2	2B 6	Q 4
5373.255	5374.749		0.5	0.3			18605.52			
5373.411	5374.905	G	1.5	1.0	0.7	1.21	18604.98	4E 0	2C 0	R 2
5373.500	5374.995		0.6				18604.67			
5373.607	5375.101			0.0			18604.30			
5373.688	5375.182						18604.02			
5373.783	5375.278			0.1			18603.69			
5373.841	5375.335						18603.49			
5374.063	5375.558		0.2	0.3	0.2		18602.72	3c 2	2a 1	R 6
5374.315	5375.809	G	0.9	0.7		0.71	18601.85			
5374.451	5375.945						18601.38	3D 3	2C 0	Q 2
5374.578	5376.072		1.0	0.4			18600.94			
5374.771	5376.266		0.2	0.2			18600.27	W 1	2B 5	P 2
5374.930	5376.425		0.2	0.4	0.2		18599.72			
5375.306	5376.801		0.2	0.2			18598.42	T 0	2B 7	P 1
5375.436	5376.931		0.1				18597.97			
5375.615	5377.110	G	0.7	0.4		0.32	18597.35	3A 2	2B 7	P 4
5375.728	5377.223				0.1		18596.96			
5375.916	5377.411		0.1				18596.31			
5376.176	5377.671						18595.41			
5376.240	5377.735			0.2d			18595.19			
5376.431	5377.926		0.3	0.3			18594.53			
5376.517	5378.012	G	0.2	0.7	0.5	0.28	18594.23	3e 2	2c 1	P 6
5376.737	5378.232						18593.47			
5376.853	5378.348			0.1			18593.07			
5376.954	5378.449						18592.72			
5377.012	5378.507			0.0			18592.52			
5377.206	5378.701	G	1.0	0.5	0.2d	0.49	18591.85	T 0	2B 7	P 2
5377.206	5378.701	G	1.0	0.5	0.2d	0.49	18591.85	3e 2	2c 1	Q 6
5377.385	5378.880						18591.23			
5377.457	5378.953		0.2	0.1			18590.98			
5377.596	5379.091						18590.50			
5377.642	5379.138			0.0			18590.34			
5377.854	5379.349		0.3	0.2			18589.61	3D 1	2B 4	P 3
5377.990	5379.485			0.1			18589.14			
5378.160	5379.656		0.2	0.3			18588.55			
5378.238	5379.734						18588.28			
5378.383	5379.879	G	1.5	1.5	0.4	0.83	18587.78	3c 2	2a 1	R 2
5378.383	5379.879	G	1.5	1.5	0.4	0.83	18587.78	V 0	2B 2	P 4
5378.577	5380.072						18587.11			
5378.698	5380.194			0.1			18586.69			
5378.762	5380.258						18586.47			
5378.947	5380.443						18585.83			
5379.092	5380.588						18585.33			
5379.187	5380.683						18585.00			
5379.271	5380.767						18584.71			
5379.329	5380.825		0.1				18584.51			
5379.518	5381.013	G	1.5	0.7	0.2	0.69	18583.86	V 0	2B 2	R 0
5379.700	5381.196			0.1			18583.23			
5379.781	5381.277		0.2				18582.95	4f 0	2c 2	R 3
5379.949	5381.445	G	0.5	0.6		0.94	18582.37			
5380.016	5381.511		0.6		0.1		18582.14	3E 0	2B 3	R 4
5380.209	5381.706						18581.47			
5380.291	5381.787						18581.19			
5380.389	5381.885		0.1	0.2			18580.85	X 2	2C 0	R 1
5380.461	5381.957						18580.60	3E 3	2C 0	R 7

λ_{air}	λ_{vac}	Ref.	I_1	I_2	I_3	I_4	ν	Upper	Lower	Br.
5380.609	5382.105						18580.09			
5380.670	5382.166			0.1			18579.88			
5380.835	5382.331						18579.31			
5380.922	5382.418		0.2				18579.01	3E 2	2B 6	P 3
5380.988	5382.485	G		0.2	0.2b		18578.78	4F 0	2B 8	R 3
5381.197	5382.693			0.1			18578.06			
5381.249	5382.745		0.2				18577.88	W 1	2B 5	P 3
5381.382	5382.879			0.1			18577.42			
5381.530	5383.027		0.2	0.3			18576.91	3A 2	2B 7	P 3
5381.733	5383.229	G	0.3	0.5	0.8		18576.21	3c 1	2a 0	Q 7
5381.898	5383.395		0.2	0.0			18575.64	T 0	2B 7	P 4
5382.040	5383.537						18575.15	T 0	2B 7	P 3
5382.252	5383.748						18574.42			
5382.321	5383.818						18574.18			
5382.391	5383.887			0.1	0.0		18573.94			
5382.527	5384.023						18573.47			
5382.631	5384.128			0.0			18573.11			
5382.773	5384.270		0.2	0.2			18572.62			
5382.976	5384.473			0.1			18571.92			
5383.193	5384.690		0.1	0.2			18571.17	3D 3	2C 0	P 2
5383.463	5384.960			0.2			18570.24			
5383.614	5385.111		0.1	0.2			18569.72			
5383.753	5385.250	G	0.7	0.4			18569.24			
5383.912	5385.409						18568.69			
5384.014	5385.511		0.1				18568.34			
5384.107	5385.604			0.0			18568.02	w 0	2c 3	Q 1
5384.315	5385.813		0.2	0.1			18567.30	3E 3	2C 0	P 3
5384.443	5385.940						18566.86			
5384.547	5386.045		0.1				18566.50			
5384.611	5386.109						18566.28			
5384.727	5386.225				0.1		18565.88	Z 1	2B 4	P 1
5384.727	5386.225				0.1		18565.88	3F 1	2B 5	Q 5
5384.797	5386.294		0.1				18565.64			
5384.867	5386.364			0.1d			18565.40			
5384.988	5386.486						18564.98			
5385.052	5386.549		0.2				18564.76	3E 3	2C 0	Q 4
5385.165	5386.663		0.2				18564.37			
5385.337	5386.834		0.9	0.6			18563.78			
5385.409	5386.906				0.3		18563.53	X 2	2C 0	Q 1
5385.499	5386.996	G	1.5b	1.0			18563.22			
5385.606	5387.104			0.2	0.3		18562.85			
5385.667	5387.165						18562.64	X 2	2C 0	P 1
5385.795	5387.292		0.9	0.2			18562.20	3A 2	2B 7	P 1
5385.931	5387.429						18561.73			
5385.998	5387.496		0.2	0.1			18561.50	W 1	2B 5	P 6
5386.192	5387.690	G	1.5	1.5	0.6	1.03	18560.83	3e 3	2c 2	R 3
5386.425	5387.922		0.2				18560.03	3D 2	2B 6	R 6
5386.456	5387.954			0.5	0.2		18559.92	3e 2	2c 1	P 4
5386.520	5388.018		0.2				18559.70			
5386.735	5388.233	G	1.5	1.5	0.7	1.13	18558.96	4F 0	2C 0	Q 2
5386.735	5388.233	G	1.5	1.5	0.7	1.13	18558.96	w 0	2c 3	R 1
5386.956	5388.454		0.2	0.5			18558.20	3e 3	2c 2	R 2
5387.098	5388.596		0.3	0.2			18557.71	W 1	2B 5	P 4
5387.272	5388.770		0.2	0.3			18557.11			
5387.522	5389.020	G	1.5	1.0	0.2	1.04	18556.25			
5387.716	5389.214			0.1			18555.58			
5387.795	5389.293		0.2				18555.31			
5387.870	5389.368			0.2			18555.05			
5387.946	5389.444		0.4				18554.79			
5388.065	5389.563						18554.38			

λ_{air}	λ_{vac}	Ref.	I_1	I_2	I_3	I_4	ν	Upper	Lower	Br.
5388.155	5389.653	GK	1.5	1.5b	0.9	1.63	18554.07	3E 2	2B 6	Q 3
5388.155	5389.653	GK	1.5	1.5b	0.9	1.63	18554.07	3c 2	2a 1	R 1
5388.370	5389.868		0.2				18553.33	3E 2	2B 6	P 2
5388.492	5389.990		0.9	0.5	0.2		18552.91			
5388.695	5390.193						18552.21			
5388.742	5390.240		0.1				18552.05	3F 1	2B 5	R 3
5388.834	5390.333			0.1d			18551.73			
5388.983	5390.481		0.2				18551.22	3E 3	2C 0	Q 4
5389.070	5390.568						18550.92	3d 1	2c 0	R 7
5389.174	5390.673	G	1.0	0.3			18550.56			
5389.320	5390.818						18550.06			
5389.447	5390.946		0.2	0.1			18549.62			
5389.648	5391.146	G	0.3	0.3	0.1		18548.93	3d 2	2c 1	R 1
5389.938	5391.437						18547.93			
5390.011	5391.510		0.1				18547.68			
5390.133	5391.632			0.1d	0.3		18547.26			
5390.293	5391.792		0.2	0.0			18546.71			
5390.514	5392.013	G	1.5	0.7		0.72	18545.95	W 1	2B 5	P 5
5390.514	5392.013	G	1.5	0.7		0.72	18545.95	3A 2	2B 7	P 2
5390.726	5392.225		0.2				18545.22			
5390.895	5392.394		0.1				18544.64			
5391.101	5392.600		0.8				18543.93	3D 2	2B 6	P 1
5391.145	5392.644	G		1.0	0.4	0.97	18543.78	3e 3	2c 2	R 1
5391.200	5392.699		1.5				18543.59			
5391.369	5392.868		0.1	0.0			18543.01			
5391.546	5393.045	G	1.5	0.8	0.2b	0.95	18542.40			
5391.729	5393.228		0.1				18541.77			
5391.825	5393.324		0.1				18541.44			
5391.892	5393.391						18541.21			
5391.939	5393.438		0.1				18541.05			
5392.087	5393.586		0.3	0.1			18540.54			
5392.293	5393.793	G	1.5b	1.5	0.2	1.72	18539.83	4E 0	2C 0	Q 1
5392.509	5394.008		0.2				18539.09			
5392.587	5394.087		0.2				18538.82			
5392.730	5394.229		0.2	0.3	0.1d		18538.33	3D 2	2B 6	R 5
5392.936	5394.436		0.1				18537.62			
5392.986	5394.485						18537.45			
5393.184	5394.683						18536.77			
5393.271	5394.770			0.1			18536.47			
5393.425	5394.925	G	0.4	0.6	0.2	0.43	18535.94	3D 2	2B 6	R 4
5393.530	5395.029		0.2				18535.58			
5393.643	5395.143		0.3	0.5	0.3		18535.19	3D 2	2B 6	R 3
5393.716	5395.216		0.3				18534.94			
5393.899	5395.399		0.1	0.2			18534.31			
5394.103	5395.603	G	1.0	0.6		0.51	18533.61	V 0	2B 2	P 3
5394.103	5395.603	G	1.0	0.6		0.51	18533.61	3D 2	2B 6	R 2
5394.362	5395.862	G	8.	0.8		1.13	18532.72			
5394.534	5396.034		1.0	0.3			18532.13			
5394.674	5396.174		0.2	0.2			18531.65	4F 0	2C 0	P 5
5394.796	5396.296	G	0.9	0.9	0.4	0.56	18531.23	33 1	2a 0	P 4
5395.110	5396.610			0.2d			18530.15			
5395.195	5396.695						18529.86	3F 0	2B 3	P 5
5395.288	5396.788				0.3d		18529.54			
5395.358	5396.858	G	0.2	0.3			18529.30			
5395.611	5397.111						18528.43			
5395.661	5397.161		0.2	0.1			18528.26	3e 1	2c 0	P 7
5395.827	5397.327	G	0.7	0.3			18527.69	3D 2	2B 6	R 1
5395.827	5397.327	G	0.7	0.3			18527.69	W 2	2B 7	R 4
5395.827	5397.327	G	0.7	0.3			18527.69	3D 3	2C 0	R 3
5395.969	5397.470			0.2	0.4		18527.20			

λ_{air}	λ_{vac}	Ref.	I_1	I_2	I_3	I_4	ν	Upper	Lower	Br.
5396.191	5397.691		0.1	0.1			18526.44			
5396.299	5397.799						18526.07	4g 1	2a 3	P 1
5396.389	5397.889	G	0.6	0.3			18525.76			
5396.724	5398.224			0.0			18524.61	4f 1	2c 3	R 2
5396.922	5398.422						18523.93			
5397.030	5398.530			0.2			18523.56			
5397.207	5398.708		0.5	0.5	0.2b		18522.95	Z 0	2B 2	P 5
5397.409	5398.909						18522.26			
5397.467	5398.968		0.2	0.1			18522.06			
5397.682	5399.183		0.2	0.3	0.2		18521.32	3e 2	2c 1	P 7
5397.872	5399.373		0.1	0.2			18520.67			
5398.029	5399.530		1.0				18520.13			
5398.102	5399.603	G	1.0	0.9	0.1	1.33	18519.88			
5398.318	5399.819		0.6	0.3			18519.14	Z 1	2B 4	P 2
5398.525	5400.026						18518.43			
5398.569	5400.070		0.1				18518.28			
5398.691	5400.192			0.2d			18517.86			
5398.770	5400.271		0.2				18517.59			
5398.965	5400.466	G	1.5	1.0	0.4	1.28	18516.92	3E 2	2B 6	Q 2
5399.102	5400.603			0.5			18516.45			
5399.184	5400.685		0.1				18516.17			
5399.350	5400.851		0.2	0.3	0.1		18515.60			
5399.604	5401.105		0.1	0.2			18514.73			
5399.817	5401.318	G	0.9	0.7	0.5		18514.00	3d 2	2c 1	Q 1
5399.817	5401.318	G	0.9	0.7	0.5		18514.00	3e 2	2c 1	Q 7
5400.076	5401.578	G	1.0	0.4		0.59	18513.11	3D 2	2B 6	R 0
5400.161	5401.662		0.5				18512.82			
5400.310	5401.811						18512.31			
5400.380	5401.881		0.0				18512.07			
5400.505	5402.006			0.5			18511.64			
5400.701	5402.202		0.2				18510.97			
5400.835	5402.336		0.4	0.3	0.2		18510.51	3E 3	2C 0	P 3
5401.042	5402.544	G	1.5b	1.5	0.4	1.20	18509.80	3c 2	2a 1	R 0
5401.264	5402.765		0.2	0.0			18509.04	3D 1	2B 4	P 4
5401.264	5402.765		0.2	0.0			18509.04	3e 3	2c 2	Q 3
5401.360	5402.862		0.2	0.2			18508.71			
5401.544	5403.046		0.2	0.2			18508.08			
5401.599	5403.101			0.1			18507.89			
5401.754	5403.256	G	1.5	0.4	0.3	0.75	18507.36			
5402.055	5403.557	G	0.6	0.5	0.4		18506.33	3d 2	2c 1	R 2
5402.303	5403.805			0.1			18505.48			
5402.367	5403.869		0.0				18505.26			
5402.440	5403.942			0.0			18505.01			
5402.580	5404.082	G	0.4	0.3	0.2		18504.53	3d 2	2c 1	P 1
5402.682	5404.184		0.1				18504.18			
5402.860	5404.363		0.1	0.1			18503.57			
5403.053	5404.555		0.2	0.3	0.1		18502.91			
5403.205	5404.707	G	1.0	0.3		0.68	18502.39			
5403.237	5404.739						18502.28			
5403.421	5404.923		0.1				18501.65			
5403.535	5405.037						18501.26			
5403.663	5405.166						18500.82			
5403.827	5405.329		0.2	0.2			18500.26	3e 3	2c 2	Q 2
5403.827	5405.329		0.2	0.2			18500.26	3d 1	2c 0	P 4
5403.976	5405.478	G	0.6	0.5	0.3d		18499.75	4F 0	2C 0	P 4
5404.189	5405.692						18499.02	3e 3	2c 2	Q 4
5404.330	5405.832		0.2	0.1			18498.54			
5404.505	5406.007		0.2	0.2	0.4		18497.94	3c 1	2a 0	Q 8
5404.747	5406.250	G	1.0	0.8			18497.11	3c 7	2a 5	Q 1
5404.747	5406.250	G	1.0	0.8			18497.11	4g 1	2a 3	R 0

λ_{air}	λ_{vac}	Ref.	I_1	I_2	I_3	I_4	ν	Upper	Lower	Br.
5404.975	5406.478		0.1	0.1d			18496.33			
5405.136	5406.639		0.2				18495.78			
5405.332	5406.835	G	1.5	1.5	0.2	1.20	18495.11			
5405.548	5407.051		0.2				18494.37			
5405.732	5407.235	G	0.8	0.9	0.3	0.69	18493.74	3E 2	2B 6	Q 1
5405.960	5407.463						18492.96	X 2	2C 0	R 2
5406.182	5407.685		0.2	0.2d	0.1		18492.20	3F 0	2B 3	Q 3
5406.375	5407.878						18491.54			
5406.454	5407.957		0.3				18491.27			
5406.592	5408.095		0.1				18490.80			
5406.764	5408.267		0.5	0.5	0.2		18490.21			
5406.995	5408.498	G	0.3	0.5	0.6		18489.42	3d 1	2c 0	Q 5
5407.042	5408.545						18489.26			
5407.106	5408.610		0.2	0.3			18489.04			
5407.337	5408.841						18488.25			
5407.490	5408.993				0.0		18487.73			
5407.542	5409.046						18487.55			
5407.680	5409.183						18487.08	3e 3	2c 2	Q 5
5407.881	5409.385	G	0.5	0.5	0.2		18486.39	3e 3	2c 2	Q 1
5407.881	5409.385	G	0.5	0.5	0.2		18486.39	3E 0	2B 3	P 8
5408.078	5409.581		0.1				18485.72			
5408.241	5409.745						18485.16			
5408.329	5409.833		0.1		0.2		18484.86			
5408.385	5409.888						18484.67			
5408.543	5410.046			0.0d			18484.13			
5408.589	5410.093		0.2				18483.97			
5408.786	5410.289	G	1.5	1.5	0.2	1.22	18483.30			
5409.043	5410.547		0.2				18482.42			
5409.274	5410.778		0.1				18481.63			
5409.450	5410.954						18481.03			
5409.517	5411.021		0.2				18480.80			
5409.643	5411.147			0.0	0.7		18480.37	3b 4	2a 0	P 5
5409.702	5411.206	G	1.5	1.0		1.32	18480.17			
5409.942	5411.446		0.2	0.3			18479.35			
5410.073	5411.578		0.2	0.0			18478.90			
5410.229	5411.733	G	1.5	0.9	0.2	0.88	18478.37	V 0	2B 2	P 2
5410.439	5411.944		0.1				18477.65			
5410.489	5411.993			0.1d			18477.48			
5410.647	5412.152						18476.94			
5410.811	5412.316	G	0.5	0.5	0.2		18476.38	4F 0	2C 0	P 3
5411.055	5412.559		0.1	0.3			18475.55			
5411.286	5412.790			0.2	0.1		18474.76			
5411.470	5412.975	G	0.2	0.4			18474.13	3c 7	2a 5	Q 2
5411.608	5413.112						18473.66	4f 0	2c 2	Q 3
5411.722	5413.227						18473.27			
5411.878	5413.382			0.0			18472.74	4f 0	2c 2	P 6
5412.156	5413.661						18471.79			
5412.220	5413.725		0.1				18471.57			
5412.379	5413.883		0.2				18471.03			
5412.466	5413.971						18470.73			
5412.540	5414.044						18470.48			
5412.698	5414.203	G					18469.94			
5412.821	5414.326						18469.52			
5413.073	5414.578						18468.66			
5413.451	5414.956						18467.37			
5413.624	5415.129						18466.78			
5413.827	5415.332		0.2	0.0			18466.09	3D 3	2C 0	Q 3
5414.202	5415.707		0.2				18464.81			
5414.389	5415.895						18464.17			
5414.474	5415.980		0.2		0.0		18463.88			

λ_{air}	λ_{vac}	Ref.	I_1	I_2	I_3	I_4	ν	Upper	Lower	Br.
5414.562	5416.068	G	0.7	0.3			18463.58	Z 1	2B 4	P 3
5414.562	5416.068	G	0.7	0.3			18463.58	w 0	2c 3	P 2
5414.753	5416.258						18462.93			
5414.835	5416.341		0.1				18462.65	4f 2	2c 4	R 1
5414.947	5416.452			0.0			18462.27	V 0	2B 2	P 5
5415.152	5416.657	G	0.4	0.4	0.2b		18461.57	W 2	2B 7	R 3
5415.152	5416.657	G	0.4	0.4	0.2b		18461.57	3e 3	2c 2	Q 1
5415.331	5416.836						18460.96	4f 0	2c 2	R 1
5415.472	5416.977		0.2	0.0			18460.48	X 2	2C 0	Q 2
5415.472	5416.977		0.2	0.0			18460.48	4f 0	2c 2	R 2
5415.703	5417.209		0.1				18459.69	3E 3	2C 0	Q 5
5415.938	5417.444	G	0.3	0.4	0.6		18458.89	3E 0	2B 3	Q 8
5416.144	5417.649			0.0			18458.19			
5416.205	5417.711						18457.98			
5416.284	5417.790		0.1				18457.71			
5416.522	5418.028		0.2				18456.90			
5416.625	5418.131						18456.55			
5416.736	5418.242		0.1				18456.17			
5416.968	5418.474						18455.38	w 0	2c 3	Q 2
5417.139	5418.644		0.2		0.1d		18454.80			
5417.318	5418.823		0.3	0.1			18454.19			
5417.456	5418.961		0.2				18453.72			
5417.558	5419.064			0.2			18453.37			
5417.599	5419.105		0.4				18453.23	3F 0	2B 3	R 1
5417.799	5419.305	G	1.5b	1.5	1.0	1.66	18452.55	4E 0	2C 0	R 1
5418.025	5419.531		0.3				18451.78	3E 0	2B 3	R 3
5418.163	5419.669		0.4	0.3			18451.31			
5418.278	5419.784		0.2	0.2			18450.92			
5418.404	5419.910		0.4	0.4	0.2		18450.49	3d 2	2c 1	R 3
5418.642	5420.148	G	0.3	0.7	0.5		18449.68	3e 2	2c 1	P 5
5418.827	5420.333		0.2				18449.05			
5418.918	5420.424			0.0d			18448.74			
5418.980	5420.486						18448.53			
5419.109	5420.615		0.2				18448.09			
5419.279	5420.786		0.1				18447.51			
5419.429	5420.936		0.6	0.5	0.4		18447.00			
5419.541	5421.047		0.2				18446.62			
5419.629	5421.135			0.2			18446.32			
5419.682	5421.188		0.8	0.3			18446.14			
5419.899	5421.406	GK	1.5b	1.5b	1.0	1.78	18445.40	3c 2	2a 1	Q 1
5420.111	5421.617		0.4				18444.68			
5420.196	5421.703		0.4	0.3			18444.39			
5420.328	5421.835		0.2	0.0			18443.94			
5420.390	5421.897						18443.73			
5420.501	5422.008		0.2				18443.35			
5420.655	5422.161		0.3	0.1			18442.83			
5420.854	5422.361		0.2				18442.15			
5420.957	5422.464		0.2	0.2			18441.80			
5421.142	5422.649	G	1.5	0.9		0.91	18441.17	Z 2	2B 6	R 0
5421.331	5422.838	G	1.0	0.9	0.5	0.86	18440.53			
5421.528	5423.035		0.2	0.5			18439.86	3c 7	2a 5	Q 3
5421.651	5423.158		0.2	0.3	0.4		18439.44	W 2	2B 7	R 2
5421.766	5423.273		0.1				18439.05			
5421.857	5423.364			0.2d			18438.74			
5421.924	5423.432		0.3				18438.51	X 2	2C 0	P 2
5422.127	5423.635	G	1.0	1.0	0.6	1.16	18437.82	Z 0	2B 2	P 6
5422.127	5423.635	G	1.0	1.0	0.6	1.16	18437.82	4E 0	2C 0	R 2
5422.127	5423.635	G	1.0	1.0	0.6	1.16	18437.82	3D 1	2B 4	P 5
5422.304	5423.811		0.6	0.5			18437.22	4E 0	2C 0	Q 1
5422.513	5424.020			0.0			18436.51	3E 3	2C 0	P 4

λ_{air}	λ_{vac}	Ref.	I_1	I_2	I_3	I_4	ν	Upper	Lower	Br.
5422.592	5424.099		0.2				18436.24			
5422.704	5424.211			0.0			18435.86			
5422.821	5424.329		0.1				18435.46			
5422.957	5424.464			0.1d			18435.00			
5423.063	5424.570		0.1	0.1			18434.64			
5423.239	5424.747	G	0.9	1.0	0.4	1.06	18434.04	3e 3	2c 2	Q 2
5423.313	5424.820		0.6				18433.79	3d 2	2c 1	Q 2
5423.313	5424.820		0.6				18433.79	4f 1	2c 3	Q 4
5423.495	5425.003		0.1				18433.17			
5423.545	5425.053			0.1			18433.00			
5423.613	5425.121						18432.77			
5423.740	5425.247						18432.34			
5423.854	5425.362			0.1			18431.95			
5424.063	5425.571		0.3	0.4	0.5		18431.24			
5424.266	5425.774						18430.55			
5424.413	5425.921		0.1				18430.05			
5424.605	5426.113						18429.40			
5424.761	5426.269						18428.87			
5424.828	5426.336		0.2				18428.64			
5424.955	5426.463			0.2			18428.21			
5425.029	5426.537		0.2				18427.96			
5425.202	5426.710	G	1.5	1.5	0.9	1.34	18427.37	4E 0	2C 0	Q 3
5425.397	5426.905	G	1.5	0.6			18426.71	Z 2	2B 6	R 1
5425.482	5426.990		0.3				18426.42			
5425.559	5427.067						18426.16			
5425.659	5427.167		0.2	0.2			18425.82			
5425.732	5427.240						18425.57			
5425.874	5427.382		1.5				18425.09	3c 2	2a 1	Q 2
5425.933	5427.441	G		1.5b	1.5	1.59	18424.89			
5425.980	5427.488		1.5				18424.73	4E 0	2C 0	R 3
5426.254	5427.762	G	1.5	1.0		1.27	18423.80			
5426.477	5427.986	G	1.5	0.9	0.2	1.00	18423.04	3D 2	2B 6	P 2
5426.737	5428.245		0.2	0.0	0.0		18422.16	3c 1	2a 0	P 5
5426.940	5428.448		0.2	0.1			18421.47	W 2	2B 7	R 1
5427.084	5428.593	G	0.3	0.1			18420.98			
5427.279	5428.787		0.1	0.0			18420.32			
5427.405	5428.914						18419.89			
5427.529	5429.038		0.1	0.1			18419.47			
5427.600	5429.109						18419.23			
5427.903	5429.412		0.1	0.2			18418.20			
5428.113	5429.621	G	0.6	0.6	0.4		18417.49			
5428.239	5429.748			0.0			18417.06			
5428.378	5429.887	G	0.4	0.6	0.2b	0.62	18416.59			
5428.493	5430.002		0.1				18416.20			
5428.561	5430.070						18415.97			
5428.714	5430.223		0.2	0.1			18415.45			
5428.923	5430.432				0.5		18414.74			
5428.947	5430.456	G	1.5	0.9		1.08	18414.66	4E 0	2C 0	P 2
5428.988	5430.497						18414.52			
5429.109	5430.618						18414.11	3F 1	2B 5	Q 4
5429.165	5430.674		0.1	0.1			18413.92	3D 3	2C 0	P 3
5429.312	5430.822	G	0.6	0.7	0.4	0.63	18413.42			
5429.534	5431.043			0.1			18412.67			
5429.663	5431.173						18412.23	3D 0	2B 3	R10
5429.826	5431.335		0.3				18411.68			
5429.882	5431.391	G		0.6	1.0		18411.49	3c 1	2a 0	Q 9
5429.917	5431.426		0.3				18411.37			
5430.159	5431.668	G	0.5	0.7	0.4	0.54	18410.55			
5430.345	5431.854			0.0			18409.92			
5430.451	5431.960		0.1				18409.56			

λ_{air}	λ_{vac}	Ref.	I_1	I_2	I_3	I_4	ν	Upper	Lower	Br.
5430.590	5432.099			0.1			18409.09			
5430.775	5432.285	G	0.5	0.9	0.7	1.09	18408.46			
5430.899	5432.409	G	1.5	1.0			18408.04	Z 2	2B 6	R 2
5431.124	5432.633		0.7	0.4			18407.28	4g 1	2a 3	R 1
5431.259	5432.769						18406.82			
5431.321	5432.831			0.1			18406.61			
5431.371	5432.881		0.1				18406.44			
5431.560	5433.070		0.4	0.3			18405.80			
5431.764	5433.274		0.5	0.4	0.3		18405.11			
5431.973	5433.483						18404.40			
5432.068	5433.578			0.2d			18404.08			
5432.142	5433.652		0.1				18403.83	3D 3	2C 0	R 4
5432.316	5433.826	G	1.0	0.7	0.2b	0.46	18403.24	Z 1	2B 4	P 4
5432.511	5434.021						18402.58			
5432.596	5434.106		0.1	0.2			18402.29			
5432.759	5434.269						18401.74	4f 1	2c 3	R 1
5432.912	5434.422			0.2d			18401.22			
5432.971	5434.481		0.1				18401.02			
5433.128	5434.638	G	0.7	0.4			18400.49	W 2	2B 7	R 0
5433.284	5434.794			0.0			18399.96			
5433.379	5434.889		0.1	0.2			18399.64			
5433.612	5435.122	G	0.7	0.6	0.5		18398.85	3d 2	2c 1	P 2
5433.863	5435.373		0.1	0.1			18398.00			
5433.978	5435.489						18397.61			
5434.120	5435.630			0.0	0.1		18397.13			
5434.339	5435.849		0.1	0.1			18396.39			
5434.401	5435.911						18396.18			
5434.599	5436.109		0.3	0.2			18395.51	3E 3	2C 0	P 4
5434.817	5436.328	G	1.5b	1.5b	1.5	1.48	18394.77	3c 2	2a 1	Q 3
5434.817	5436.328	G	1.5b	1.5b	1.5	1.48	18394.77	3e 3	2c 2	Q 3
5435.036	5436.547		0.2				18394.03			
5435.136	5436.647		0.3	0.2			18393.69			
5435.379	5436.889						18392.87			
5435.470	5436.981				0.1		18392.56			
5435.547	5437.058						18392.30			
5435.645	5437.155			0.1			18391.97			
5435.801	5437.312						18391.44			
5435.925	5437.436			0.1			18391.02			
5436.112	5437.622	G	0.3	0.4	0.2	0.36	18390.39	3F 0	2B 3	P 4
5436.342	5437.853	G	0.3	0.5	0.2		18389.61	3D 1	2B 4	P 6
5436.591	5438.102			0.1d			18388.77			
5436.771	5438.282						18388.16			
5436.919	5438.430				0.2		18387.66			
5436.957	5438.468		0.2	0.0			18387.53			
5437.194	5438.705						18386.73	3F 1	2B 5	R 2
5437.274	5438.785				0.0		18386.46			
5437.345	5438.856						18386.22			
5437.555	5439.066			0.0			18385.51			
5437.806	5439.317	G	0.3	0.5	0.5b		18384.66	3d 2	2c 1	R 4
5437.806	5439.317	G	0.3	0.5	0.5b		18384.66	3D 1	2B 4	P 9
5437.995	5439.507						18384.02			
5438.102	5439.613						18383.66			
5438.282	5439.794		0.1	0.0			18383.05			
5438.566	5440.078		0.2	0.2d			18382.09			
5438.785	5440.297	G	1.5	1.0	0.5	1.03	18381.35	4F 0	2C 0	P 3
5438.785	5440.297	G	1.5	1.0	0.5	1.03	18381.35	Z 2	2B 6	R 3
5438.942	5440.454						18380.82			
5439.066	5440.578		0.2	0.2			18380.40			
5439.294	5440.806			0.1			18379.63			
5439.362	5440.874		0.1				18379.40			

λ_{air}	λ_{vac}	Ref.	I_1	I_2	I_3	I_4	ν	Upper	Lower	Br.
5439.409	5440.921			0.0			18379.24			
5439.566	5441.078		0.6	0.7		0.76	18378.71	4F 0	2C 0	P 4
5439.655	5441.167	G			0.3		18378.41			
5439.717	5441.229		1.5	0.7			18378.20	3D 2	2B 6	P 3
5439.924	5441.437		0.1	0.2	0.1		18377.50			
5440.034	5441.546						18377.13			
5440.191	5441.703	G	0.9	0.8	0.3	0.52	18376.60	3e 3	2c 2	P 2
5440.191	5441.703	G	0.9	0.8	0.3	0.52	18376.60	X 2	2C 0	R 3
5440.466	5441.978						18375.67			
5440.505	5442.017			0.1			18375.54			
5440.721	5442.233	G	0.4	0.3	0.1		18374.81	3F 0	2B 3	Q 2
5440.916	5442.429						18374.15			
5440.970	5442.482		0.2	0.0			18373.97			
5441.053	5442.565						18373.69			
5441.340	5442.852						18372.72			
5441.438	5442.950			0.0			18372.39			
5441.500	5443.012						18372.18			
5441.731	5443.243						18371.40			
5441.915	5443.427						18370.78			
5442.030	5443.543			0.0			18370.39			
5442.214	5443.726				0.1		18369.77			
5442.276	5443.789	G	0.3	0.3		0.54	18369.56			
5442.551	5444.064	G	0.3	0.5	0.3	0.32	18368.63	4F 0	2C 0	P 5
5442.551	5444.064	G	0.3	0.5	0.3	0.32	18368.63	4f 1	2c 3	Q 3
5442.741	5444.254			0.1			18367.99			
5442.812	5444.325						18367.75			
5442.990	5444.503						18367.15			
5443.159	5444.672			0.0			18366.58	4f 0	2c 2	Q 4
5443.387	5444.900	G	0.4	0.3	0.3		18365.81			
5443.565	5445.078		0.1	0.2			18365.21			
5443.790	5445.303		0.2	0.3			18364.45	3D 1	2B 4	P 7
5443.944	5445.457			0.0			18363.93			
5444.060	5445.573						18363.54			
5444.140	5445.653						18363.27			
5444.247	5445.760		0.1	0.1			18362.91			
5444.371	5445.884	G	0.6	0.4		0.36	18362.49	3E 0	2B 3	P 7
5444.644	5446.157		0.1	0.1d			18361.57	4f 2	2c 4	Q 2
5444.843	5446.356			0.0			18360.90			
5445.071	5446.585		0.2	0.3			18360.13	3D 1	2B 4	P 8
5445.380	5446.893		0.2	0.1			18359.09			
5445.599	5447.113		0.1	0.2			18358.35			
5445.759	5447.273						18357.81	4f 0	2c 2	P 5
5445.854	5447.368		0.2	0.1			18357.49			
5445.914	5447.427						18357.29			
5446.044	5447.558		0.1	0.3	0.2		18356.85	3d 1	2c 0	Q 6
5446.237	5447.751			0.2			18356.20			
5446.424	5447.938		0.2	0.3	0.2d		18355.57			
5446.676	5448.190	G	0.8	1.5	0.3	0.60	18354.72	3c 2	2a 1	Q 4
5446.860	5448.374		0.2	0.2			18354.10			
5446.917	5448.430						18353.91			
5447.098	5448.611						18353.30			
5447.172	5448.686			0.1			18353.05			
5447.258	5448.772						18352.76			
5447.409	5448.923		0.3	0.3d			18352.25	3e 3	2c 2	P 2
5447.507	5449.021	G	0.2	0.5	0.3		18351.92	4f 0	2c 2	Q 2
5447.667	5449.181						18351.38			
5447.733	5449.247		0.2	0.1			18351.16	Z 2	2B 6	P 2
5447.804	5449.318				0.1		18350.92			
5447.911	5449.425		0.2	0.2d			18350.56			
5447.982	5449.496						18350.32			

λ_{air}	λ_{vac}	Ref.	I_1	I_2	I_3	I_4	ν	Upper	Lower	Br.
5448.059	5449.573			0.2d			18350.06			
5448.258	5449.772		0.3	0.3	0.2		18349.39			
5448.502	5450.016		0.1				18348.57			
5448.591	5450.105				0.2d		18348.27			
5448.659	5450.173		0.2	0.3d			18348.04			
5448.849	5450.364		0.2				18347.40			
5448.965	5450.479		0.2	0.3			18347.01			
5449.110	5450.625		0.9	0.3			18346.52	W 2	2B 7	P 1
5449.238	5450.753			0.2			18346.09			
5449.431	5450.946		0.0	0.1			18345.44			
5449.577	5451.091			0.1			18344.95			
5449.722	5451.237		0.2	0.5			18344.46	3e 3	2c 2	Q 4
5449.886	5451.400	G	0.7	0.5		0.43	18343.91	Z 2	2B 6	R 4
5449.886	5451.400	G	0.7	0.5		0.43	18343.91	4f 2	2c 4	R 2
5450.121	5451.635			0.0			18343.12			
5450.230	5451.745						18342.75			
5450.415	5451.930			0.0			18342.13			
5450.495	5452.010						18341.86			
5450.611	5452.126						18341.47			
5450.658	5452.173		0.1	0.2			18341.31			
5450.849	5452.364		0.2	0.3	0.2d		18340.67			
5451.057	5452.572			0.1			18339.97			
5451.205	5452.720			0.2	0.1d		18339.47			
5451.500	5453.015			0.3d			18338.48			
5451.598	5453.113		0.2		0.5		18338.15			
5451.755	5453.270						18337.62			
5451.883	5453.398		0.1	0.2d			18337.19	4F 0	2B 8	P 5
5452.097	5453.612	G	0.9	0.8			18336.47	3e 3	2c 2	P 3
5452.234	5453.749		0.1				18336.01			
5452.314	5453.829						18335.74			
5452.383	5453.898		0.2	0.0			18335.51			
5452.525	5454.041	G	1.0	0.9	0.3	0.43	18335.03	3d 2	2c 1	Q 3
5452.728	5454.243						18334.35			
5452.796	5454.311		0.1				18334.12			
5452.849	5454.365			0.2			18333.94			
5453.037	5454.552		0.3				18333.31			
5453.114	5454.630			0.3	0.2		18333.05	3e 2	2c 1	P 6
5453.114	5454.630			0.3	0.2		18333.05	4f 0	2c 2	R 1
5453.203	5454.719		0.1				18332.75			
5453.284	5454.799						18332.48			
5453.453	5454.969			0.1			18331.91			
5453.656	5455.171						18331.23			
5453.870	5455.386		0.2				18330.51			
5454.054	5455.570			0.1	0.1		18329.89			
5454.144	5455.659		0.1				18329.59			
5454.254	5455.770						18329.22			
5454.328	5455.844		0.1	0.1			18328.97			
5454.498	5456.014		0.2	0.4	0.3		18328.40	4F 0	2B 8	R 2
5454.498	5456.014		0.2	0.4	0.3		18328.40	3E 0	2B 3	Q 7
5454.745	5456.261	G	1.0	0.5	0.1	0.54	18327.57	4E 0	2c 0	Q 2
5454.935	5456.451		0.1				18326.93			
5455.093	5456.609		0.2	0.2d			18326.40	3E 0	2B 3	P 7
5455.150	5456.666						18326.21			
5455.304	5456.820	G	1.5	1.0		1.06	18325.69	3D 2	2B 6	P 4
5455.304	5456.820	G	1.5	1.0		1.06	18325.69	4g 1	2a 3	P 2
5455.513	5457.029		0.2				18324.99			
5455.751	5457.267		0.1				18324.19			
5455.909	5457.425			0.0			18323.66			
5455.989	5457.505						18323.39	3d 1	2c 0	P 5
5456.156	5457.672						18322.83			

λ_{air}	λ_{vac}	Ref.	I_1	I_2	I_3	I_4	ν	Upper	Lower	Br.
5456.338	5457.854		0.1	0.1			18322.22	4g 0	2a 2	P 1
5456.477	5457.994		0.2	0.2			18321.75			
5456.582	5458.098						18321.40			
5456.722	5458.238	G	1.0	0.7	0.2	0.72	18320.93			
5456.984	5458.500	G	1.5b	1.5	0.2	1.00	18320.05	3c 2	2a 1	P 2
5456.984	5458.500	G	1.5b	1.5	0.2	1.00	18320.05	4g 1	2a 3	R 2
5456.984	5458.500	G	1.5b	1.5	0.2	1.00	18320.05	3E 0	2B 3	R 2
5457.201	5458.718		0.2	0.1			18319.32			
5457.300	5458.816		0.2				18318.99			
5457.383	5458.900				0.2		18318.71			
5457.446	5458.962		0.1	0.1			18318.50			
5457.523	5459.040						18318.24			
5457.642	5459.159		0.2	0.2	0.2		18317.84			
5457.773	5459.290		0.2	0.3			18317.40			
5457.922	5459.439	G	0.9	0.6	0.1	0.43	18316.90	4E 0	2C 0	P 2
5458.009	5459.526		0.2				18316.61			
5458.122	5459.639						18316.23			
5458.202	5459.719						18315.96			
5458.313	5459.830			0.2d	0.2		18315.59	X 2	2C 0	Q 3
5458.506	5460.023		0.1	0.1			18314.94			
5458.652	5460.169		0.9	0.6			18314.45	3A 2	2C 0	R 2
5458.819	5460.336	G	1.0	0.6	0.1	0.75	18313.89	W 2	2B 7	P 2
5459.094	5460.611		0.2	0.1			18312.97			
5459.237	5460.754						18312.49			
5459.374	5460.891		0.3	0.1			18312.03	V 1	2B 3	R 2
5459.600	5461.118	G	1.5b	1.5	0.2	1.63	18311.27	4E 1	2C 1	R 1
5459.684	5461.201						18310.99			
5459.809	5461.326		0.2				18310.57			
5459.860	5461.377			0.2	0.2		18310.40			
5459.905	5461.422		0.2				18310.25			
5460.012	5461.529		0.1				18309.89			
5460.054	5461.571			0.1			18309.75			
5460.206	5461.723			0.2			18309.24			
5460.352	5461.869			0.1			18308.75			
5460.522	5462.039		0.2	0.2			18308.18			
5460.731	5462.248		0.2	0.2			18307.48			
5460.993	5462.511			0.1			18306.60			
5461.109	5462.627			0.1			18306.21			
5461.157	5462.675		0.0				18306.05			
5461.220	5462.738			0.2			18305.84			
5461.408	5462.925	G	0.8	1.5	0.4	0.58	18305.21	3c 2	2a 1	Q 5
5461.620	5463.137		0.1	0.2			18304.50			
5461.793	5463.311	G	1.5	0.8		0.92	18303.92			
5461.927	5463.445		1.0	0.6			18303.47	3A 2	2C 0	R 1
5461.927	5463.445		1.0	0.6			18303.47	3E 3	2B 8	R 3
5462.133	5463.651		0.2	0.2			18302.78			
5462.300	5463.818	G	1.0	0.6		0.51	18302.22	T 0	2C 0	R 1
5462.479	5463.997						18301.62			
5462.581	5464.099		0.1	0.2			18301.28			
5462.721	5464.239						18300.81			
5462.787	5464.305		0.1	0.2			18300.59			
5462.996	5464.514	G	1.5	0.4		1.16	18299.89	4E 1	2C 1	R 2
5463.190	5464.708						18299.24	3A 2	2C 0	R 3
5463.190	5464.708						18299.24	w 0	2c 3	P 3
5463.252	5464.771						18299.03			
5463.396	5464.914			0.1			18298.55			
5463.605	5465.123		0.1	0.1			18297.85			
5463.891	5465.410		0.1	0.2			18296.89	3A 2	2C 0	R 4
5463.891	5465.410		0.1	0.2			18296.89	4f 0	2c 2	Q 3
5463.891	5465.410		0.1	0.2			18296.89	4f 1	2c 3	Q 2

λ_{air}	λ_{vac}	Ref.	I_1	I_2	I_3	I_4	ν	Upper	Lower	Br.
5464.103	5465.622			0.0			18296.18			
5464.342	5465.861		0.2	0.1			18295.38			
5464.536	5466.055						18294.73	3D 3	2C 0	Q 4
5464.859	5466.378						18293.65			
5464.964	5466.482			0.1			18293.30			
5465.029	5466.548		0.2				18293.08	T 0	2C 0	R 2
5465.185	5466.703	G	0.9	1.0	0.3	0.46	18292.56	3e 3	2c 2	P 4
5465.469	5466.987		0.1	0.1			18291.61			
5465.699	5467.217		0.3	0.3			18290.84			
5465.830	5467.349		0.2	0.3	0.2d		18290.40			
5466.036	5467.555			0.2			18289.71			
5466.216	5467.735		0.3	0.3			18289.11			
5466.320	5467.839			0.0			18288.76	4F 0	2B 8	Q 4
5466.458	5467.977	G	1.0	1.0	0.1	1.06	18288.30	4E 1	2C 1	R 3
5466.763	5468.282			0.1			18287.28			
5466.861	5468.380						18286.95			
5467.041	5468.560			0.0	0.2		18286.35			
5467.148	5468.668						18285.99			
5467.375	5468.895		0.3	0.3			18285.23	3F 0	2B 3	P 3
5467.375	5468.895		0.3	0.3			18285.23	3F 1	2B 5	Q 3
5467.630	5469.149			0.1			18284.38			
5467.872	5469.391		0.1	0.0			18283.57			
5468.156	5469.676			0.0			18282.62			
5468.596	5470.115		0.1	0.2	0.2		18281.15			
5468.662	5470.181						18280.93			
5468.928	5470.448		0.3	0.3			18280.04			
5469.021	5470.540						18279.73			
5469.119	5470.639						18279.40			
5469.242	5470.762			0.0			18278.99			
5469.398	5470.917		0.1				18278.47			
5469.448	5470.968			0.2			18278.30			
5469.652	5471.172		0.5	0.3			18277.62	W 2	2B 7	P 3
5469.766	5471.286		0.4	0.1			18277.24	Z 2	2B 6	P 3
5469.766	5471.286		0.4	0.1			18277.24	3F 1	2B 5	P 5
5469.832	5471.352						18277.02	3D 0	2B 3	R 9
5469.960	5471.480		0.3	0.3	0.3		18276.59	3c 1	2a 0	P 6
5470.116	5471.636	G	0.4	0.4		0.41	18276.07			
5470.185	5471.705				0.3		18275.84			
5470.271	5471.792			0.0			18275.55			
5470.406	5471.926		0.2	0.3		0.48	18275.10			
5470.592	5472.112	G	0.5	0.3			18274.48	4E 1	2C 1	Q 1
5470.885	5472.405		0.3	0.2			18273.50			
5471.029	5472.549			0.0			18273.02	X 2	2C 0	P 3
5471.200	5472.720		0.2	0.1			18272.45			
5471.298	5472.819						18272.12			
5471.361	5472.882		0.2	0.2			18271.91			
5471.592	5473.112	G	1.5	1.2	0.2	0.53	18271.14	3e 3	2c 2	P 3
5471.855	5473.376	G	0.4	0.5		0.45	18270.26	3D 2	2B 6	P 5
5472.080	5473.601		0.1	0.1			18269.51			
5472.302	5473.822			0.0			18268.77			
5472.541	5474.062		0.2				18267.97			
5472.820	5474.341			0.0			18267.04			
5472.931	5474.452						18266.67			
5473.015	5474.535			0.2			18266.39			
5473.131	5474.652		0.2	0.1			18266.00	3E 3	2C 0	P 5
5473.308	5474.829	G	1.5	0.9		0.71	18265.41	3E 3	2B 8	R 2
5473.383	5474.904				0.2		18265.16	3E 3	2B 8	Q 5
5473.524	5475.045			0.1			18264.69			
5473.584	5475.105		0.1				18264.49			
5473.734	5475.255			0.1			18263.99			

λ_{air}	λ_{vac}	Ref.	I_1	I_2	I_3	I_4	ν	Upper	Lower	Br.
5474.246	5475.768		0.3	0.3	0.2		18262.28	3d 2	2c 1	P 3
5474.369	5475.891		0.2				18261.87			
5474.501	5476.022		0.1				18261.43			
5474.579	5476.100			0.2			18261.17			
5474.660	5476.181		0.3				18260.90			
5474.855	5476.376	G	1.5b	1.5	0.3	1.48	18260.25	4D 0	2C 0	Q 1
5474.981	5476.502			0.3			18259.83	4e 1	2c 3	Q 5
5475.068	5476.589		0.2				18259.54			
5475.164	5476.685		0.2				18259.22			
5475.395	5476.916		0.2	0.3			18258.45			
5475.590	5477.111		0.3	0.2			18257.80	4g 0	2a 2	R 0
5475.590	5477.111		0.3	0.2			18257.80	4f 0	2c 2	P 4
5475.791	5477.312		0.1				18257.13			
5475.881	5477.402			0.1			18256.83			
5476.112	5477.633		0.2	0.1			18256.06	3E 3	2C 0	P 5
5476.274	5477.795	G	1.5	1.0		0.92	18255.52			
5476.547	5478.068			0.2			18254.61			
5476.697	5478.218	G	0.4	0.4	0.2		18254.11			
5476.934	5478.455			0.2			18253.32			
5476.979	5478.500		0.1				18253.17			
5477.189	5478.710	G	1.0	0.5		0.48	18252.47	U 1	2B 5	R 4
5477.483	5479.005		0.4	0.1			18251.49			
5477.696	5479.218		0.4	0.2			18250.78			
5477.858	5479.380	G	0.4	0.4	0.1	0.30	18250.24	4E 0	2C 0	Q 3
5478.074	5479.596						18249.52			
5478.107	5479.629			0.0			18249.41			
5478.302	5479.824		0.1				18248.76			
5478.416	5479.938						18248.38			
5478.653	5480.176			0.0			18247.59			
5478.764	5480.287		0.2				18247.22	3F 1	2B 5	R 1
5478.951	5480.473		0.2	0.5	0.2		18246.60	3c 2	2a 1	Q 6
5478.951	5480.473		0.2	0.5	0.2		18246.60	4g 1	2a 3	R 3
5479.206	5480.728		0.1	0.3			18245.75			
5479.428	5480.951		0.1	0.2			18245.01			
5479.759	5481.281		0.2	0.3	0.1		18243.91			
5479.942	5481.464	G	0.3	0.4		0.51	18243.30			
5480.008	5481.531				0.2		18243.08			
5480.071	5481.594			0.0			18242.87			
5480.167	5481.690		0.1	0.1			18242.55			
5480.347	5481.870		0.3	0.4			18241.95			
5480.549	5482.071						18241.28			
5480.876	5482.399		0.5	0.2d			18240.19			
5480.612	5482.135		0.3	0.2			18241.07			
5480.726	5482.249		0.1				18240.69			
5481.083	5482.606	GK	1.5b	1.9	0.3	1.51	18239.50	4D 0	2C 0	R 1
5481.083	5482.606	GK	1.5b	1.9	0.3	1.51	18239.50	3c 2	2a 1	P 3
5481.162	5482.685						18239.24	W 2	2B 7	P 4
5481.300	5482.823		0.3				18238.78			
5481.378	5482.901		0.3	0.2			18238.52			
5481.519	5483.042		0.1				18238.05			
5481.730	5483.253		0.1				18237.35			
5481.916	5483.439		0.3				18236.73			
5482.427	5483.950		0.3				18235.03			
5482.662	5484.185		0.1				18234.25			
5482.851	5484.374	G	1.5	0.4		0.51	18233.62	T 0	2C 0	Q 1
5483.101	5484.624		0.1	0.0			18232.79	4e 1	2c 3	Q 4
5483.221	5484.744						18232.39			
5483.293	5484.817			0.2			18232.15			
5483.341	5484.865		0.1				18231.99			
5483.519	5485.042	G	0.3	0.4		0.41	18231.40	3e 3	2c 2	P 5

λ_{air}	λ_{vac}	Ref.	I_1	I_2	I_3	I_4	ν	Upper	Lower	Br.
5483.741	5485.265		0.2	0.3			18230.66			
5484.000	5485.524	G	0.4	0.7	0.3		18229.80	t 0	2a 0	R 0
5484.000	5485.524	G	0.4	0.7	0.3		18229.80	Y 2	2B 7	R 1
5484.298	5485.822			0.2			18228.81			
5484.364	5485.888						18228.59			
5484.469	5485.993			0.1			18228.24			
5484.644	5486.168			0.1			18227.66			
5485.050	5486.574						18226.31			
5485.164	5486.688		0.2				18225.93	Y 2	2B 7	R 0
5485.246	5486.770			0.1	0.2b		18225.66			
5485.480	5487.005		0.1	0.2			18224.88			
5485.712	5487.236	G	0.6	1.0	0.3	0.38	18224.11	3d 2	2c 1	Q 4
5485.712	5487.236	G	0.6	1.0	0.3	0.38	18224.11	3E 0	2B 3	P 6
5485.860	5487.384		0.2				18223.62	Y 2	2B 7	R 2
5485.953	5487.477			0.1			18223.31	4f 0	2c 2	Q 2
5486.143	5487.667	G	0.4	0.3			18222.68			
5486.287	5487.812			0.1			18222.20			
5486.561	5488.086		0.2	0.1			18221.29			
5486.633	5488.158			0.3	0.3		18221.05	3d 1	2c 0	Q 7
5486.805	5488.330		0.1	0.0			18220.48			
5486.950	5488.474		0.2	0.2			18220.00			
5487.094	5488.619		0.2	0.1			18219.52			
5487.287	5488.812	G	0.6	0.9	0.4		18218.88	3E 0	2B 3	Q 6
5487.422	5488.947			0.0			18218.43			
5487.576	5489.101			0.0	0.2		18217.92			
5487.703	5489.227			0.2			18217.50			
5487.874	5489.399		0.3	0.3	0.3		18216.93	3D 3	2C 0	P 4
5487.971	5489.496	G					18216.61			
5488.100	5489.625			0.2			18216.18			
5488.275	5489.800	G	0.5	0.6		0.45	18215.60	3D 2	2B 6	P 6
5488.489	5490.014			0.2	0.4		18214.89			
5488.576	5490.101		0.2				18214.60	4e 2	2c 4	Q 1
5488.832	5490.358			0.2			18213.75			
5489.134	5490.659		0.1				18212.75			
5489.251	5490.777			0.2			18212.36	3E 3	2B 8	R 1
5489.435	5490.960		0.3				18211.75			
5489.483	5491.009	G		0.4	0.2		18211.59	3e 2	2c 1	P 7
5489.483	5491.009	G		0.4	0.2		18211.59	3c 3	2a 2	R 3
5489.676	5491.202			0.1			18210.95			
5489.839	5491.365			0.2	0.1		18210.41			
5489.981	5491.506		0.1	0.2			18209.94			
5490.174	5491.699	G	1.3	0.5	0.2	0.45	18209.30	t 0	2a 0	P 1
5490.409	5491.935			0.1			18208.52			
5490.482	5492.007		0.1				18208.28			
5490.693	5492.218			0.0			18207.58	4e 2	2c 4	R 1
5490.693	5492.218			0.0			18207.58	4e 1	2c 3	Q 3
5491.054	5492.580		0.1	0.2			18206.38			
5491.253	5492.779	G	1.3	0.7	0.2	0.51	18205.72	X 0	2B 4	R 3
5491.253	5492.779	G	1.3	0.7	0.2	0.51	18205.72	4f 2	2c 4	P 3
5491.253	5492.779	G	1.3	0.7	0.2	0.51	18205.72	W 2	2B 7	P 5
5491.471	5492.996		0.0	0.2			18205.00			
5491.655	5493.180			0.1			18204.39			
5491.796	5493.322		0.1	0.2			18203.92			
5491.959	5493.485			0.1			18203.38			
5492.068	5493.594		0.2				18203.02			
5492.149	5493.675			0.0			18202.75			
5492.530	5494.056			0.1			18201.49			
5492.792	5494.318		0.2	0.2	0.4		18200.62			
5492.976	5494.502	G	1.6	1.1		1.03	18200.01	Z 2	2B 6	P 4
5493.094	5494.620			0.2			18199.62			

λ_{air}	λ_{vac}	Ref.	I_1	I_2	I_3	I_4	ν	Upper	Lower	Br.
5493.169	5494.696		0.1				18199.37			
5493.284	5494.810		0.2	0.4	0.2		18198.99	3F 2	2B 7	Q 4
5493.284	5494.810		0.2	0.4	0.2		18198.99	3E 1	2B 5	Q 8
5493.444	5494.970		0.2	0.2			18198.46	3E 0	2B 3	R 1
5493.695	5495.221			0.1			18197.63			
5494.081	5495.608		0.2	0.2			18196.35			
5494.274	5495.801		0.1	0.1			18195.71	3C 2	2A 0	R 3
5494.501	5496.027		0.2	0.2			18194.96	V 1	2B 3	R 1
5494.655	5496.182			0.2			18194.45			
5494.827	5496.354		0.3	0.4			18193.88			
5495.039	5496.565		0.2				18193.18			
5495.180	5496.707		0.7	0.3			18192.71	4E 1	2C 1	Q 2
5495.334	5496.861			0.1			18192.20			
5495.479	5497.006		0.1	0.2			18191.72			
5495.628	5497.154		0.2	0.3			18191.23			
5495.757	5497.284		0.2	0.2			18190.80			
5495.821	5497.348				0.8		18190.59			
5495.960	5497.487	G	2.1	1.6		1.40	18190.13	X 0	2B 4	R 2
5495.960	5497.487	G	2.1	1.6		1.40	18190.13	3c 3	2a 2	R 2
5496.147	5497.674		0.4				18189.51			
5496.214	5497.741			0.2			18189.29			
5496.374	5497.901		0.1	0.2			18188.76			
5496.546	5498.073		0.7	0.8	0.3		18188.19	X 0	2B 4	R 4
5496.546	5498.073		0.7	0.8	0.3		18188.19	4f 0	2c 2	P 3
5496.628	5498.155	G	1.4	1.1		0.75	18187.92	3E 3	2B 8	Q 4
5496.628	5498.155	G	1.4	1.1		0.75	18187.92	4f 2	2c 4	Q 3
5496.691	5498.218		1.5	0.6			18187.71	3A 2	2C 0	Q 1
5496.879	5498.406			0.0			18187.09			
5497.066	5498.593			0.2			18186.47			
5497.272	5498.799		0.3	0.7			18185.79	3E 3	2B 8	Q 4
5497.272	5498.799		0.3	0.7			18185.79	4e 1	2c 3	Q 2
5497.531	5499.059			0.2			18184.93			
5497.731	5499.258	G	1.3	0.7		0.51	18184.27	t 0	2a 0	R 1
5498.061	5499.588		0.4	0.3			18183.18	3A 2	2C 0	Q 2
5498.321	5499.848		0.2	0.2			18182.32			
5498.517	5500.045		0.3	0.3			18181.67	3F 1	2B 5	Q 2
5498.517	5500.045		0.3	0.3			18181.67	T 0	2C 0	P 1
5498.517	5500.045		0.3	0.3			18181.67	T 0	2C 0	Q 2
5498.517	5500.045		0.3	0.3			18181.67	4f 1	2c 3	P 5
5498.898	5500.426			0.2			18180.41	Y 2	2B 7	P 1
5498.898	5500.426			0.2			18180.41	4g 1	2a 3	R 4
5499.037	5500.565			0.1			18179.95	4e 0	2c 2	Q 7
5499.219	5500.747	G	0.5	1.1	0.3		18179.35	3c 2	2a 1	Q 7
5499.219	5500.747	G	0.5	1.1	0.3		18179.35	3e 3	2c 2	P 4
5499.376	5500.904		0.3	0.2			18178.83			
5499.576	5501.104	G	2.3	1.7	0.4	1.66	18178.17	4D 0	2C 0	R 2
5499.576	5501.104	G	2.3	1.7	0.4	1.66	18178.17	4E 0	2C 0	P 4
5499.791	5501.319		0.2				18177.46			
5499.884	5501.412		0.2	0.2			18177.15			
5500.036	5501.564		0.3	0.3			18176.65	4b 0	2a 1	R 4
5500.375	5501.903	G	0.5	0.9b	0.3		18175.53	4E 0	2C 0	Q 4
5500.508	5502.036	G	1.5	1.3		1.00	18175.09	X 0	2B 4	R 1
5500.841	5502.369		0.2	0.2	0.1		18173.99			
5501.195	5502.723		0.1				18172.82			
5501.398	5502.926			0.2			18172.15	3C 2	2A 0	R 2
5501.704	5503.232		0.3	0.3			18171.14	3F 1	2B 5	P 4
5501.858	5503.387			0.1			18170.63	3e 3	2c 2	P 6
5501.946	5503.474						18170.34			
5502.167	5503.695		0.1				18169.61			
5502.321	5503.850		0.2	0.1			18169.10			

λ_{air}	λ_{vac}	Ref.	I_1	I_2	I_3	I_4	ν	Upper	Lower	Br.
5502.524	5504.053		0.2	0.2			18168.43			
5502.894	5504.422			0.1			18167.21			
5502.951	5504.480						18167.02	4e 1	2c 3	Q 1
5503.206	5504.735		0.1				18166.18			
5503.294	5504.823			0.1	0.1		18165.89			
5503.494	5505.022		0.1				18165.23			
5503.660	5505.189		0.2	0.3			18164.68	4f 1	2c 3	P 4
5503.851	5505.380			0.1			18164.05	3D 2	2B 6	P 7
5504.248	5505.777		0.1				18162.74			
5504.730	5506.259		0.2	0.2			18161.15			
5504.939	5506.468		0.6	0.4			18160.46	4E 1	2C 1	P 2
5505.021	5506.550		0.2				18160.19			
5505.157	5506.687		0.2	0.2			18159.74			
5505.345	5506.875		0.9	0.6			18159.12	4E 2	2C 2	R 1
5505.512	5507.042	GK	2.4	1.9	0.3	1.51	18158.57	3c 3	2a 2	R 1
5505.579	5507.108						18158.35			
5505.733	5507.263		0.3				18157.84	4b 0	2a 1	R 3
5505.843	5507.372		0.3	0.3			18157.48			
5505.979	5507.509		0.2				18157.03			
5506.064	5507.594			0.0	0.3		18156.75			
5506.134	5507.663		0.3	0.2			18156.52			
5506.343	5507.873	G	2.0	1.4	0.2	1.26	18155.83	U 1	2B 5	R 3
5506.343	5507.873	G	2.0	1.4	0.2	1.26	18155.83	X 0	2B 4	R 0
5506.561	5508.091		0.2				18155.11	3E 0	2B 3	P 5
5506.649	5508.179		0.2				18154.82			
5506.786	5508.316		0.1	0.2			18154.37			
5506.998	5508.528		0.2	0.4			18153.67			
5507.220	5508.750			0.0			18152.94			
5507.390	5508.919		0.1				18152.38			
5507.650	5509.181		0.1	0.2			18151.52			
5507.845	5509.375	G	1.6	1.3	0.2	0.86	18150.88	3c 2	2a 1	P 4
5507.845	5509.375	G	1.6	1.3	0.2	0.86	18150.88	3E 3	2B 8	R 0
5508.018	5509.548		0.7				18150.31			
5508.115	5509.645	G	1.3	0.9	0.1d	0.48	18149.99	4E 0	2C 0	P 3
5508.300	5509.830		0.0	0.2			18149.38			
5508.455	5509.985						18148.87			
5508.537	5510.067			0.1	0.3		18148.60			
5508.746	5510.276			0.2			18147.91			
5508.946	5510.477		0.2	0.1			18147.25	4e 0	2c 2	Q 5
5509.186	5510.717	G	0.7	0.3			18146.46			
5509.453	5510.984	G	0.3	0.6			18145.58	t 0	21 0	R 2
5509.633	5511.163		0.2				18144.99	3A 2	2C 0	P 1
5510.079	5511.610		0.1	0.1			18143.52	X 2	2C 0	Q 4
5510.258	5511.789	G	1.3	0.6		0.40	18142.93	4f 1	2c 3	P 3
5510.456	5511.986		0.1	0.1			18142.28			
5510.574	5512.105			0.2			18141.89	3d 1	2c 0	P 6
5510.750	5512.281			0.0			18141.31			
5510.866	5512.397	G	0.5	0.6	0.3		18140.93			
5510.993	5512.524		0.2	0.0			18140.51	4g 0	2a 2	R 1
5511.166	5512.697			0.0			18139.94			
5511.358	5512.889		0.2	0.1			18139.31	Y 2	2B 7	P 2
5511.823	5513.354		0.1	0.1			18137.78			
5512.096	5513.628		0.2	0.2			18136.88	3A 2	2C 0	Q 3
5512.482	5514.014			0.1			18135.61	3D 2	2B 6	P 8
5512.607	5514.138		0.2	0.2			18135.20	4D 0	2C 0	P 2
5512.607	5514.138		0.2	0.2			18135.20	3C 2	2A 0	R 1
5512.826	5514.357			0.1			18134.48	4F 0	2B 8	P 4
5512.975	5514.506	G	0.5	0.7	0.4		18133.99	3D 0	2B 3	R 8
5512.975	5514.506	G	0.5	0.7	0.4		18133.99	3E 3	2B 8	Q 3
5513.188	5514.719			0.1			18133.29	4e 1	2c 3	Q 1

λ_{air}	λ_{vac}	Ref.	I_1	I_2	I_3	I_4	ν	Upper	Lower	Br.
5513.321	5514.853		0.1	0.0			18132.85			
5513.531	5515.063	G	0.7	0.3			18132.16	4e 0	2c 2	Q 4
5513.720	5515.251		0.1	0.2			18131.54	3E 3	2B 8	Q 3
5513.914	5515.446	G	0.5	0.9	0.3	0.46	18130.90	3E 0	2B 3	Q 5
5513.914	5515.446	G	0.5	0.9	0.3	0.46	18130.90	4b 0	2a 1	R 2
5514.173	5515.705			0.0			18130.05			
5514.313	5515.845						18129.59			
5514.352	5515.884			0.1			18129.46			
5514.437	5515.969		0.2				18129.18			
5514.602	5516.134	G	1.5	0.7		0.90	18128.64			
5514.836	5516.368		0.1	0.1			18127.87			
5515.070	5516.602			0.1			18127.10			
5515.134	5516.666		0.2				18126.89	4e 0	2c 2	R 5
5515.216	5516.748	G					18126.62			
5515.444	5516.977						18125.87			
5515.566	5517.098						18125.47			
5515.715	5517.247			0.1	0.1		18124.98			
5516.102	5517.634		0.1				18123.71	4g 1	2a 3	P 3
5516.224	5517.756			0.0			18123.31			
5516.400	5517.932			0.0			18122.73	3E 3	2C 0	P 6
5516.558	5518.091	G	0.3	0.3			18122.21	4E 1	2C 1	Q 3
5516.851	5518.383			0.1			18121.25	4F 0	2B 8	R 1
5517.094	5518.627			0.3			18120.45	4e 0	2c 2	Q 3
5517.243	5518.776						18119.96			
5517.283	5518.816		0.1	0.2			18119.83			
5517.466	5518.998	G	1.3	0.6		0.45	18119.23	4D 0	2C 0	Q 2
5517.597	5519.129						18118.80			
5517.749	5519.282		0.1	0.2			18118.30	3D 3	2C 0	Q 5
5517.965	5519.498		0.5	0.6			18117.59			
5518.020	5519.553	G			0.3		18117.41			
5518.056	5519.589		0.4	0.6			18117.29	4F 0	2B 8	Q 3
5518.242	5519.775		0.4	0.2			18116.68	Z 2	2B 6	P 5
5518.446	5519.979				0.2		18116.01	t 0	2a 0	R 3
5518.489	5520.022	G	2.3	1.5		1.40	18115.87	3c 3	2a 2	R 0
5518.538	5520.071						18115.71			
5518.763	5520.296		0.3	0.1			18114.97	T 0	2C 0	Q 3
5518.955	5520.488		0.2	0.3			18114.34			
5519.394	5520.927			0.1			18112.90			
5519.610	5521.143						18112.19	4e 0	2c 2	Q 2
5519.903	5521.436			0.1			18111.23			
5520.110	5521.643		0.1				18110.55			
5520.424	5521.958		0.2	0.2			18109.52			
5520.601	5522.134						18108.94			
5520.689	5522.223		0.2	0.0			18108.65	·T 0	2C 0	P 2
5520.872	5522.406	G	1.5	1.2	0.3	1.06	18108.05	3d 2	2c 1	P 4
5520.872	5522.406	G	1.5	1.2	0.3	1.06	18108.05	4D 0	2C 0	R 3
5521.095	5522.628		0.3	0.2			18107.32			
5521.229	5522.763		0.2	0.2			18106.88			
5521.500	5523.034		0.2	0.3			18105.99			
5521.635	5523.168			0.0			18105.55			
5521.750	5523.284		0.2				18105.17			
5521.833	5523.367	G	1.3	1.1	0.3		18104.90	3d 2	2c 1	Q 5
5522.190	5523.723		0.3	0.3	0.2		18103.73	3c 2	2a 1	Q 8
5522.190	5523.723		0.3	0.3	0.2		18103.73	4e 2	2c 4	Q 2
5522.443	5523.977						18102.90			
5522.519	5524.053		0.2	0.2	0.3		18102.65	3f 0	2c 0	R 5
5522.696	5524.230	G	0.3	0.3			18102.07	4e 0	2c 2	R 1
5522.977	5524.511		0.2	0.2			18101.15	4f 0	2c 2	P 5
5523.264	5524.798		0.2	0.2			18100.21			
5523.444	5524.978		0.3	0.3			18099.62	U 1	2B 5	R 2

λ_{air}	λ_{vac}	Ref.	I$_1$	I$_2$	I$_3$	I$_4$	ν	Upper	Lower	Br.
5523.444	5524.978		0.3	0.3			18099.62	4e 0	2c 2	R 2
5523.743	5525.277		0.2	0.2			18098.64			
5523.947	5525.482	G	1.6	1.1		0.90	18097.97	3d 1	2c 0	Q 8
5523.947	5525.482	G	1.6	1.1		0.90	18097.97	X 0	2B 4	P 1
5524.155	5525.689	G	1.5	1.0		0.91	18097.29			
5524.399	5525.933		0.1	0.1			18096.49			
5524.515	5526.050				0.1		18096.11			
5524.561	5526.095			0.1			18095.96			
5524.924	5526.459		0.2		0.3		18094.77	4b 0	2a 1	R 1
5525.034	5526.569		0.3	0.1			18094.41			
5525.364	5526.899				0.1		18093.33			
5525.480	5527.015			0.1			18092.95			
5525.794	5527.329		0.4	0.3			18091.92	U 1	2B 5	R 1
5525.794	5527.329		0.4	0.3			18091.92	3F 1	2B 5	P 3
5526.085	5527.620		0.2	0.2			18090.97			
5526.167	5527.702						18090.70			
5526.311	5527.846	G	1.5	1.1		0.92	18090.23	3E 3	2B 8	Q 2
5526.494	5528.029						18089.63			
5526.567	5528.102		0.1	0.2			18089.39			
5526.839	5528.374			0.2	0.1		18088.50			
5527.123	5528.659		0.2	0.3	0.5		18087.57			
5527.209	5528.744						18087.29			
5527.343	5528.879	G	1.5	1.3	0.3	0.79	18086.85	3E 0	2B 3	P 4
5527.343	5528.879	G	1.5	1.3	0.3	0.79	18086.85	3C 2	2A 0	R 0
5527.576	5529.111		0.2	0.1			18086.09			
5527.744	5529.279		0.7	0.3			18085.54	U 1	2B 5	R 0
5527.744	5529.279		0.7	0.3			18085.54	Y 2	2B 7	P 3
5527.826	5529.362	G	1.3	0.5		0.34	18085.27	3E 0	2B 3	R 0
5528.025	5529.560		0.1	0.2			18084.62	t 1	2a 1	R 1
5528.211	5529.747	G	1.2	0.4			18084.01	3F 2	2B 7	Q 3
5528.416	5520.952			0.0			18083.34			
5528.817	5530.353			0.1			18082.03	4f 0	2c 2	P 4
5529.086	5530.622	G	0.2	0.3			18081.15			
5529.331	5530.866			0.2			18080.35			
5529.523	5531.059	G	1.3	0.9	0.2	0.62	18079.72			
5529.808	5531.344		0.3				18078.79	V 1	2B 3	P 4
5529.854	5531.390	G	1.3	1.2	0.3	0.45	18078.64	3e 3	2c 2	P 5
5529.899	5531.435		0.6				18078.49			
5530.132	5531.668			0.1			18077.73	4g 0	2a 2	P 2
5530.459	5531.995		0.1		0.2		18076.66			
5530.615	5532.151	G	0.8	0.3			18076.15	V 1	2B 3	R 0
5530.829	5532.366		0.1	0.1			18075.45			
5531.187	5532.724	G	0.3	0.3			18074.28	3d 3	2c 2	R 1
5531.365	5532.901			0.0			18073.70			
5531.708	5533.244			0.0			18072.58	3E 3	2B 8	P 3
5531.934	5533.471		0.3	0.2			18071.84			
5532.219	5533.756			0.1			18070.91	3E 3	2B 8	P 4
5532.421	5533.958	G	0.2	0.3			18070.25	3E 1	2B 5	Q 7
5532.421	5533.958	G	0.2	0.3			18070.25	4e 1	2c 3	Q 2
5532.559	5534.096		0.3	0.1			18069.80			
5533.021	5534.558		0.1				18068.29			
5533.242	5534.779		0.1	0.2			18067.57			
5533.324	5534.861		0.2				18067.30			
5533.600	5535.137		0.2	0.1			18066.40	X 2	2C 0	P 4
5533.600	5535.137		0.2	0.1			18066.40	3A 2	2C 0	Q 4
5533.600	5535.137		0.2	0.1			18066.40	4e 2	2c 4	Q 1
5533.600	5535.137		0.2	0.1			18066.40	3E 3	2B 8	P 2
5533.600	5535.137		0.2	0.1			18066.40	3E 3	2B 8	Q 1
5533.778	5535.315		0.2	0.2			18065.82			
5534.062	5535.600	G	1.3	1.2	0.4	0.72	18064.89	3E 0	2B 3	Q 4

λ_{air}	λ_{vac}	Ref.	I_1	I_2	I_3	I_4	ν	Upper	Lower	Br.
5534.225	5535.762		0.2	0.3			18064.36			
5534.344	5535.882			0.0			18063.97			
5534.479	5536.016	G	0.7	0.6		0.30	18063.53			
5534.694	5536.231		0.2				18062.83			
5534.749	5536.286			0.0			18062.65	3A 2	2C 0	P 2
5534.749	5536.286			0.0			18062.65	4e 1	2c 3	P 2
5534.954	5536.492			0.0			18061.98	3E 3	2B 8	P 5
5535.083	5536.620			0.0			18061.56			
5535.227	5536.764		0.3	0.3			18061.09			
5535.393	5536.930			0.1			18060.55	4f 0	2c 2	P 3
5535.595	5537.132		0.2				18059.89			
5535.766	5537.304		0.4	0.2			18059.33			
5535.984	5537.522	G	2.2	1.5	0.2	1.42	18058.62	X 0	2B 4	P 2
5535.984	5537.522	G	2.2	1.5	0.2	1.42	18058.62	3D 3	2B 8	R 1
5536.202	5537.739		0.3	0.1			18057.91	3D 3	2B 8	R 0
5536.358	5537.896	G	1.3	0.6		0.57	18057.40			
5536.567	5538.104			0.0			18056.72			
5536.683	5538.221		0.2	0.1			18056.34			
5536.870	5538.408		0.8	0.4			18055.73			
5536.974	5538.512		0.2				18055.39			
5537.112	5538.650		0.4	0.4			18054.94	4E 1	2C 1	Q 4
5537.287	5538.825	G	1.0		0.3		18054.37	3c 2	2a 1	P 5
5537.401	5538.939			1.5b			18054.00			
5537.468	5539.006	GK	2.8	2.4	0.3	1.87	18053.78	3c 3	2a 2	Q 1
5537.689	5539.227		0.4				18053.06			
5537.763	5539.301		0.4	0.3			18052.82			
5537.928	5539.466		0.2	0.1			18052.28			
5538.079	5539.617		0.1				18051.79			
5538.244	5539.783		0.2				18051.25	4b 0	2a 1	R 0
5538.465	5540.004		0.2	0.2			18050.53			
5538.665	5540.203	G	1.3	0.7			18049.88	3c 8	2a 6	Q 1
5538.926	5540.464		0.1	0.1			18049.03			
5539.131	5540.670		0.1	0.1			18048.36	4E 2	2C 2	Q 2
5539.131	5540.670		0.1	0.1			18048.36	3E 1	2B 5	R 3
5539.291	5540.829		0.2	0.1			18047.84			
5539.444	5540.983			0.0			18047.34			
5539.573	5541.112		0.2	0.3	0.1d		18046.92			
5539.763	5541.302		0.2	0.2			18046.30			
5539.963	5541.502	G	1.3	0.6		0.41	18045.65	3D 3	2B 8	R 2
5540.236	5541.775		0.2	0.2			18044.76			
5540.454	5541.993		0.2	0.0			18044.05			
5540.657	5542.196	G	0.7	0.4			18043.39			
5540.924	5542.463	G	1.3	0.7	0.2		18042.52	3d 3	2c 2	R 2
5541.302	5542.841			0.1	0.1		18041.29			
5541.498	5543.038		0.2				18040.65			
5541.584	5543.124			0.0			18040.37			
5541.796	5543.336			0.1			18039.68			
5542.039	5543.578		0.4	0.3			18038.89			
5542.279	5543.818	G	0.4	0.5			18038.11	3D 3	2B 8	R 4
5542.555	5544.095		0.4	0.5	0.2		18037.21			
5542.724	5544.264		0.1				18036.66			
5542.856	5544.396		0.8	0.3			18036.23			
5542.976	5544.516		0.2				18035.84			
5543.121	5544.660	G	1.6	1.5	0.5	1.30	18035.37	4D 0	2C 0	R 4
5543.121	5544.660	G	1.6	1.5	0.5	1.30	18035.37	3b 5	2a 1	R 1
5543.317	5544.857		0.2	0.0			18034.73			
5543.499	5545.038	G	2.2	1.6	0.2	1.30	18034.14	3c 3	2a 2	Q 2
5543.772	5545.312		0.1	0.2			18033.25	3D 3	2B 8	R 3
5543.975	5545.515		0.1				18032.59			
5544.347	5545.887		0.1				18031.38			

λ_{air}	λ_{vac}	Ref.	I_1	I_2	I_3	I_4	ν	Upper	Lower	Br.
5544.544	5546.084	G	1.5	0.7	0.2		18030.74	3E 0	2B 3	P 3
5544.679	5546.219	G				0.41	18030.30	t 0	2a 0	P 2
5544.679	5546.219	G				0.41	18030.30	4e 1	2c 3	P 3
5544.741	5546.281		0.9				18030.10			
5544.762	5546.302	G	1.3	1.1	0.2		18030.03	3d 3	2c 2	Q 1
5544.806	5546.346		0.9				18029.89	4g 0	2a 2	R 2
5545.008	5546.549			0.1			18029.23	4e 1	2c 3	P 2
5545.190	5546.730			0.1	0.1		18028.64			
5545.344	5546.884			0.1			18028.14			
5545.519	5547.059	G	0.2	0.4			18027.57	V 1	2B 3	P 3
5545.651	5547.192		0.2				18027.14			
5545.750	5547.290			0.0	0.1		18026.82			
5545.857	5547.398		0.2				18026.47			
5545.904	5547.444			0.1			18026.32			
5546.264	5547.804		0.1	0.0			18025.15			
5546.491	5548.032	G	1.3	0.6		0.46	18024.41	4E 1	2c 1	P 3
5546.645	5548.186			0.3			18023.91			
5546.771	5548.312						18023.50	Z 2	2B 6	P 6
5546.833	5548.374		0.3				18023.30	Y 2	2B 7	P 4
5546.935	5548.475		0.2	0.3	0.2d		18022.97	3b 5	2a 1	R 0
5547.156	5548.697		1.0	0.3			18022.25			
5547.350	5548.891						18021.62			
5547.393	5548.934		0.2	0.4	0.3		18021.48	3D 0	2B 3	R 7
5547.510	5549.051		1.0				18021.10	Y 1	2B 4	R 1
5547.584	5549.125	G	1.5	0.5			18020.86	4d 1	2c 3	P 1
5547.667	5549.208		0.4				18020.59	3E 0	2B 3	Q 3
5547.701	5549.242		1.6	1.2	0.3	1.08	18020.48	3c 2	2a 1	Q 9
5547.744	5549.285		0.4				18020.34			
5547.938	5549.479		0.4	0.2			18019.71	3C 2	2A 0	Q 1
5548.055	5549.596			0.2	0.2		18019.33	3b 5	2a 1	R 2
5548.163	5549.704	G	0.4	0.3			18018.98			
5548.403	5549.944	G	1.3	1.0	0.4	0.57	18018.20	Y 1	2B 4	R 2
5548.403	5549.944	G	1.3	1.0	0.4	0.57	18018.20	4e 1	2c 3	Q 3
5548.625	5550.166		0.2	0.2			18017.48	4E 0	2C 0	P 4
5548.874	5550.415			0.2			18016.67			
5549.108	5550.649		0.2	0.3			18015.91			
5549.296	5550.837		0.1	0.2			18015.30			
5549.484	5551.025	G	1.4	1.0		0.65	18014.69	Y 1	2B 4	R 0
5549.746	5551.287	G	1.5	1.1	0.1	0.91	18013.84	X 0	2B 4	P 3
5549.922	5551.463		0.1	0.1			18013.27			
5550.057	5551.599		0.7	0.5			18012.83			
5550.183	5551.725		0.2	0.2			18012.42			
5550.344	5551.885		1.5	0.5			18011.90			
5550.393	5551.934	G	1.3				18011.74			
5550.458	5551.999		0.4	0.4			18011.53	3d 3	2c 2	P 1
5550.584	5552.126		0.1				18011.12	3A 2	2C 0	Q 5
5550.750	5552.292			0.1			18010.58			
5550.926	5552.468		0.1	0.2			18010.01			
5551.108	5552.650		0.3	0.4			18009.42			
5551.459	5553.001		0.1	0.2			18008.28			
5551.654	5553.195			0.0			18007.65			
5551.845	5553.387		0.2	0.3	0.1		18007.03	4E 0	2B 8	R 2
5552.054	5553.596			0.2			18006.35			
5552.261	5553.803			0.4	0.2		18005.68			
5552.317	5553.859		0.2				18005.50			
5552.523	5554.065	G	2.2	1.8	0.3	1.42	18004.83	3c 3	2a 2	Q 3
5552.795	5554.337		0.2	0.2			18003.95			
5552.998	5554.540		0.1	0.2	0.1		18003.29			
5553.251	5554.793		0.1				18002.47			
5553.430	5554.972		0.3	0.2			18001.89			

λ_{air}	λ_{vac}	Ref.	I_1	I_2	I_3	I_4	ν	Upper	Lower	Br.
5553.621	5555.164						18001.27	4e 1	2c 3	P 4
5553.674	5555.216		0.2	0.2			18001.10			
5553.837	5555.380	G	1.3	0.9	0.2	0.69	18000.57	4D 0	2C 0	Q 3
5554.078	5555.620		0.1	0.1			17999.79	5c 1	2a 4	R 3
5554.411	5555.954		0.1	0.2			17998.71	3C 2	2A 0	Q 2
5554.461	5556.003						17998.55			
5554.670	5556.213		0.2		0.1d		17997.87			
5554.896	5556.438		0.2	0.2			17997.14			
5555.106	5556.648	G	1.6	1.3	0.3	0.96	17996.46	3E 0	2B 3	Q 2
5555.331	5556.874		0.2				17995.73			
5555.390	5556.932			0.4	0.2		17995.54	3d 3	2c 2	R 3
5555.390	5556.932			0.4	0.2		17995.54	3A 2	2C 0	P 3
5555.439	5556.982		0.2				17995.38			
5555.637	5557.179		0.1	0.2			17994.74	3F 2	2B 7	Q 2
5555.782	5557.325	G	0.2	0.6			17994.27	T 0	2C 0	P 3
5555.782	5557.325	G	0.2	0.6			17994.27	4d 2	2c 4	Q 1
5556.010	5557.553			0.1			17993.53			
5556.233	5557.776		0.2	0.2			17992.81			
5556.492	5558.035	G	1.3	0.6		0.49	17991.97	U 1	2B 5	P 2
5556.628	5558.171			0.2			17991.53			
5557.005	5558.548			0.0			17990.31			
5557.212	5558.755		0.2	0.3			17989.64			
5557.388	5558.931	G	1.2	0.6	0.2	0.26	17989.07	3E 0	2B 3	Q 1
5557.743	5559.286		0.1	0.2			17987.92			
5557.913	5559.456				0.1		17987.37			
5557.959	5559.503	G	1.3	0.4		0.43	17987.22	X 1	2B 6	R 0
5558.086	5559.630			0.2			17986.81			
5558.166	5559.710		0.1				17986.55	3D 3	2C 0	P 5
5558.305	5559.849	G	0.6	0.4			17986.10	X 1	2B 6	R 1
5558.506	5560.050						17985.45			
5558.673	5560.217				0.8		17984.91			
5558.729	5560.273		0.1	0.0			17984.73			
5558.927	5560.470	G	1.3	1.0	0.2	0.46	17984.09	3E 0	2B 3	P 2
5559.087	5560.631		0.1	0.2			17983.57			
5559.659	5561.203				0.2		17981.72	4F 0	2B 8	P 3
5559.733	5561.277	G	0.4	0.4			17981.48	u 0	2a 0	R 2
5559.879	5561.423			0.0			17981.01			
5560.046	5561.590		0.2	0.4	0.8		17980.47	3f 0	2c 0	R 4
5560.247	5561.791	G	0.3	0.5			17979.82			
5560.358	5561.902				0.2		17979.46			
5560.531	5562.076		0.1	0.2			17978.90	4e 1	2c 3	P 5
5560.782	5562.326		0.1	0.1	0.2		17978.09			
5560.943	5562.487		0.3	0.3			17977.57			
5561.159	5562.704				0.2		17976.87			
5561.230	5562.775		0.2	0.3			17976.64			
5561.320	5562.865						17976.35			
5561.444	5562.988	G	0.5	1.0	0.4		17975.95	3d 1	2c 0	P 7
5561.552	5563.097		0.2				17975.60			
5561.744	5563.288	G	1.9	1.2	0.1	1.07	17974.98	V 1	2B 3	P 2
5562.035	5563.580		0.2	0.2			17974.04			
5562.242	5563.787	G	1.4	1.0	0.1	0.83	17973.37	3D 3	2B 8	P 2
5562.242	5563.787	G	1.4	1.0	0.1	0.83	17973.37	X 1	2B 6	R 2
5562.443	5563.988			0.1			17972.72			
5562.592	5564.137			0.1			17972.24	4e 1	2c 3	Q 4
5562.759	5564.304			0.2			17971.70			
5563.010	5564.555			0.4	0.2		17970.89			
5563.381	5564.926			0.0			17969.69	4d 0	2c 2	P 1
5563.471	5565.016		0.5				17969.40			
5563.545	5565.090						17969.16			
5563.697	5565.242			0.1			17968.67	3E 3	2C 0	P 7

λ_{air}	λ_{vac}	Ref.	I_1	I_2	I_3	I_4	ν	Upper	Lower	Br.
5563.967	5565.512		0.0	0.2			17967.80	3C 2	2A 0	Q 3
5564.152	5565.697			0.1	0.1		17967.20	3F 2	2B 7	P 4
5564.307	5565.852		0.2	0.2			17966.70			
5564.515	5566.060	G	1.8	1.5	0.4	1.24	17966.03	X 0	2B 4	P 4
5564.555	5566.060	G	1.8	1.5	0.4	1.24	17966.03	3E 1	2B 5	R 2
5564.515	5566.060	G	1.8	1.5	0.4	1.24	17966.03	3c 3	2a 2	Q 4
5564.725	5566.271		0.2	0.2			17965.35			
5564.886	5566.432	G	0.7	0.6	0.1		17964.83	3d 3	2c 2	Q 2
5565.209	5566.754			0.2			17963.79			
5565.308	5566.853		0.1		0.1		17963.47			
5565.450	5566.996			0.2			17963.01			
5565.555	5567.101	G	0.8	0.4			17962.67	Y 1	2B 4	P 1
5565.555	5567.101	G	0.8	0.4			17962.67	4e 2	2c 4	P 2
5565.754	5567.299	G	0.5	0.9	0.2		17962.03	4D 0	2C 0	R 5
5566.008	5567.554			0.2			17961.21	4F 0	2B 8	Q 2
5566.181	5567.727	G	1.2	0.7	0.3	0.41	17960.65	3E 1	2B 5	Q 6
5566.377	5567.923			0.1			17960.02			
5566.615	5568.161		0.3	0.3	0.1		17959.25			
5566.885	5568.431	G	1.0	0.3			17958.38			
5567.087	5568.633		1.2	0.3	0.3		17957.73	4e 0	2c 2	P 2
5567.294	5568.840			0.1			17957.06			
5567.493	5569.039			0.1			17956.42			
5567.549	5569.095		0.1				17956.24	W 2	2C 0	R 2
5567.728	5569.275						17955.66			
5567.744	5569.290	G	1.2	0.3			17955.61	W 2	2C 0	Q 1
5567.998	5569.544		0.1	0.2			17954.79			
5568.153	5569.700			0.2			17954.29			
5568.426	5569.972		0.2	0.3			17953.41	3F 3	2B 9	R 2
5568.938	5570.484		0.1	0.1			17951.76			
5569.214	5570.761		0.1	0.2			17950.87			
5569.428	5570.975		0.3	0.8			17950.18	3c 2	2a 1	P 6
5569.537	5571.083	G	1.3	1.2	0.5	0.40	17949.83	u 0	2a 0	R 1
5569.807	5571.353		0.1	0.1			17948.96	4e 0	2c 2	P 3
5570.251	5571.797			0.0			17947.53			
5570.477	5572.024						17946.80			
5571.005	5572.552		0.1	0.1			17945.10			
5571.123	5572.670						17944.72			
5571.179	5572.726			0.0	0.3		17944.54			
5571.427	5572.974		0.1	0.0			17943.74	3d 2	2c 1	P 5
5571.601	5573.148	G	1.3	0.6		0.56	17943.18	3A 0	2B 4	R 2
5571.753	5573.300		0.2	0.3			17942.69			
5571.896	5573.443						17942.23			
5572.020	5573.568		0.1	0.1			17941.83			
5572.172	5573.720		0.1	0.0			17941.34	4e 2	2c 4	Q 3
5572.418	5573.965	G	0.3	0.6	1.0		17940.55	3f 0	2c 0	R 5
5572.629	5574.176			0.0			17939.87			
5572.778	5574.326			0.1			17939.39			
5572.946	5574.493			0.1			17938.85			
5573.145	5574.692		0.3	0.3			17938.21	4b 0	2a 1	P 1
5573.306	5574.854			0.2			17937.69			
5573.471	5575.018	G	0.2	0.4		0.28	17937.16	3d 3	2c 2	R 4
5573.636	5575.183		0.4	0.4			17936.63	U 1	2B 5	P 3
5573.726	5575.273		0.2				17936.34			
5573.819	5575.367				0.1		17936.04			
5573.924	5575.472	G	2.0	1.3		0.93	17935.70	3c 3	2a 2	P 2
5574.133	5575.681		0.2				17935.03	5c 1	2a 4	R 2
5574.232	5575.780		0.2	0.2			17934.71			
5574.509	5576.057	G	0.2	0.3	0.3		17933.82	3f 1	2c 1	R 6
5574.695	5576.243			0.0			17933.22			
5574.879	5576.427			0.0			17932.63			

λ_{air}	λ_{vac}	Ref.	I_1	I_2	I_3	I_4	ν	Upper	Lower	Br.
5575.034	5576.582		0.1	0.1d			17932.13			
5575.255	5576.803	G	0.4	0.4			17931.42	3A 2	2B 8	R 4
5575.255	5576.803	G	0.4	0.4			17931.42	4e 1	2c 3	Q 5
5575.255	5576.803	G	0.4	0.4			17931.42	4D 0	2C 3	P 3
5575.479	5577.027		0.1	0.1			17930.70			
5575.631	5577.180		0.1				17930.21			
5575.812	5577.360		0.2	0.2			17929.63	W 2	2C 0	P 1
5576.169	5577.718		0.3	0.1			17928.48			
5576.546	5578.094		0.3	0.2			17927.27	2K 6	2B 0	R 1
5576.885	5578.433		0.2				17926.18	4g 0	2a 2	R 3
5576.885	5578.433		0.2				17926.18	4g 1	2a 3	P 4
5576.963	5578.511			0.0			17925.93			
5577.435	5578.984		0.5	0.3			17924.41			
5577.709	5579.258			0.0			17923.53			
5577.968	5579.516		0.5	0.2			17922.70	4d 0	2c 2	R 1
5578.500	5580.049		0.3	0.2			17920.99	2K 6	2B 0	R 2
5578.711	5580.261	G	1.4	1.2	0.3	0.78	17920.31	3d 3	2c 2	P 2
5578.711	5580.261	G	1.4	1.2	0.3	0.78	17920.31	4d 0	2c 2	Q 1
5578.711	5580.261	G	1.4	1.2	0.3	0.78	17920.31	U 1	2B 5	P 6
5578.711	5580.261	G	1.4	1.2	0.3	0.78	17920.31	X 0	2B 4	P 5
5579.054	5580.603			0.1	0.2		17919.21	3D 3	2B 8	P 3
5579.231	5580.781		0.5	0.0			17918.64	2K 6	2B 0	R 0
5579.356	5580.905	G	1.5	1.2	0.3	0.40	17918.24	3c 3	2a 2	Q 5
5579.356	5580.905	G	1.5	1.2	0.3	0.40	17918.24	u 0	2a 0	R 0
5579.471	5581.021	G					17917.87	X 2	2B 8	R 1
5579.589	5581.139	G	1.5	1.1	0.3		17917.49	Y 1	2B 4	P 2
5579.589	5581.139	G	1.5	1.1	0.3		17917.49	3b 5	2a 1	P 1
5579.805	5581.354		0.1	0.2			17916.80			
5579.973	5581.522	G	1.4	0.7		0.61	17916.26			
5580.113	5581.662			0.2			17915.81			
5580.212	5581.762		0.1	0.1			17915.49	4e 1	2c 3	P 3
5580.384	5581.933		0.3	0.2			17914.94			
5580.586	5582.136			0.1			17914.29			
5580.751	5582.301	G	0.5	0.4	0.2		17913.76			
5580.876	5582.426			0.1			17913.36			
5581.063	5582.613	G	0.3	0.4	0.2		17912.76			
5581.184	5582.734						17912.37			
5581.278	5582.828		0.1				17912.07			
5581.571	5583.121			0.1			17911.13			
5581.739	5583.289		0.1				17910.59			
5581.833	5583.383				0.2		17910.29			
5581.904	5583.454	G	0.4	0.4			17910.06			
5582.063	5583.613			0.0			17909.55			
5582.229	5583.779		0.1				17909.02	3F 2	2B 7	P 3
5582.444	5583.994		0.2	0.0			17908.33			
5582.886	5584.436		0.1	0.2	0.1		17906.91			
5583.051	5584.602	G	1.5	1.0		0.83	17906.38			
5583.158	5584.708			0.1			17906.04	5c 3	2a 6	Q 2
5583.298	5584.848	G	0.3	0.5	0.2	0.32	17905.59			
5583.522	5585.073		0.1	0.0			17904.87	5c 3	2a 6	Q 1
5583.709	5585.260			0.0			17904.27	4e 0	2c 2	P 4
5583.897	5585.447	G	0.2	0.2	0.2		17903.67			
5584.037	5585.587			0.2			17903.22			
5584.296	5585.846	G	0.3	0.3	0.6		17902.39			
5584.458	5586.009		0.2	0.3			17901.87			
5584.642	5586.193		0.2	0.2	0.2		17901.28	4E 0	2B 8	R 0
5584.642	5586.193		0.2	0.2	0.2		17901.28	4E 1	2C 1	P 4
5584.761	5586.311			0.1			17900.90	4d 2	2c 4	P 1
5584.904	5586.455	G	0.3	0.3	0.2		17900.44			
5585.045	5586.595			0.0			17899.99			

λ_{air}	λ_{vac}	Ref.	I_1	I_2	I_3	I_4	ν	Upper	Lower	Br.
5585.219	5586.770		0.2	0.2	0.2		17899.43			
5585.322	5586.873			0.0			17899.10			
5585.466	5587.017	G	0.4	0.6	0.3		17898.64	3D 0	2B 3	R 6
5585.569	5587.120						17898.31			
5585.637	5587.188			0.1			17898.09			
5585.768	5587.319			0.1			17897.67			
5585.937	5587.488	G	1.3	0.6		0.41	17897.13	3C 2	2A 0	P 2
5585.937	5587.488	G	1.3	0.6		0.41	17897.13	X 1	2B 6	P 2
5586.074	5587.625			0.1			17896.69			
5586.237	5587.788				0.2		17896.17			
5586.296	5587.847		0.1	0.3			17895.98			
5586.424	5587.975		0.1				17895.57			
5586.568	5588.119		0.4	0.3	0.1		17895.11	X 2	2B 8	R 2
5586.998	5588.550			0.0			17893.73			
5587.517	5589.068		0.2	0.2			17892.07	4b 1	2a 2	R 2
5587.876	5589.427			0.2	0.2		17890.92	3E 2	2B 7	R 4
5588.013	5589.565		0.1				17890.48			
5588.088	5589.640						17890.24			
5588.210	5589.762		0.2	0.2			17889.85	4E 0	2C 0	P 5
5588.473	5590.024	G	0.8	0.3			17889.01	t 1	2a 1	R 3
5588.698	5590.249			0.1	0.2		17888.29			
5588.769	5590.321		0.1				17888.06			
5588.947	5590.499		0.3	0.3			17887.49	4E 2	2B12	Q 2
5588.947	5590.499		0.3	0.3			17887.49	4E 2	2C 2	P 3
5588.947	5590.499		0.3	0.3			17887.49	3A 2	2C 0	P 4
5589.129	5590.681		0.1	0.1			17886.91			
5589.366	5590.918		0.1				17886.15			
5589.482	5591.034			0.0			17885.78			
5589.604	5591.156		0.1				17885.39			
5589.923	5591.475		0.1	0.0d			17884.37			
5590.091	5591.643	G	1.3	1.0	0.2	0.54	17883.83	U 1	2B 5	P 4
5590.091	5591.643	G	1.3	1.0	0.2	0.54	17883.83	W 2	2C 0	Q 2
5590.288	5591.840			0.0			17883.20	3F 3	2B 9	Q 3
5590.338	5591.890		0.1				17883.04			
5590.516	5592.069	G	0.4	0.2			17882.47			
5590.751	5592.303			0.1			17881.72	X 2	2B 8	R 3
5590.957	5592.510	G	0.4	0.6	0.2		17881.06	U 1	2B 5	P 5
5591.217	5592.769		0.1	0.2			17880.23	W 2	2C 0	R 3
5591.420	5592.972	G	1.4	1.1	0.2	0.38	17879.58	3d 3	2c 2	Q 3
5591.420	5592.972	G	1.4	1.1	0.2	0.38	17879.58	3E 1	2B 5	R 1
5591.642	5593.195		0.1	0.1	0.1		17878.87	5c 1	2a 4	R 1
5592.077	5593.630		0.2	0.2	0.1		17877.48			
5592.286	5593.839		0.1	0.2			17876.81			
5592.621	5594.174		0.2	0.0			17875.74			
5592.765	5594.318		0.2	0.2			17875.28	4D 0	2C 0	Q 4
5593.069	5594.622		0.1	0.2	0.1		17874.31			
5593.391	5594.944			0.1			17873.28			
5593.494	5595.047		0.1				17872.95			
5593.648	5595.201			0.0			17872.46			
5593.857	5595.410		0.2	0.2	0.1		17871.79			
5594.058	5595.611			0.0			17871.15			
5594.186	5595.739	G	0.2	0.5			17870.74	3E 1	2B 5	Q 5
5594.277	5595.830	G					17870.45			
5594.399	5595.952	G	0.2	0.3			17870.06	3d 3	2c 2	R 5
5594.502	5596.055		0.2	0.2	0.2b		17869.73			
5594.831	5596.384		0.2	0.0			17868.68	5c 0	2a 3	R 2
5595.059	5596.613	G	0.2	0.5	0.4		17867.95	t 1	2a 1	P 1
5595.385	5596.939	G	0.3	0.4	0.2		17866.91	u 1	21 1	R 3
5595.385	5596.939	G	0.3	0.4	0.2		17866.91	3D 1	2B 5	R 8
5595.623	5597.177			0.3			17866.15	4d 2	2c 4	R 1

λ_{air}	λ_{vac}	Ref.	I_1	I_2	I_3	I_4	ν	Upper	Lower	Br.
5595.686	5597.239		0.1				17865.95	T 0	2C 0	P 4
5595.770	5597.324			0.2	0.3		17865.68	4d 0	2c 2	R 2
5595.770	5597.324			0.2	0.3		17865.68	4d 1	2c 3	P 2
5595.924	5597.478			0.2			17865.19			
5596.046	5597.600			0.2			17864.80	4e 0	2c 2	P 5
5596.134	5597.687				0.1		17864.52			
5596.190	5597.744	G	0.7	0.4			17864.34	3A 1	2B 6	R 2
5596.347	5597.901		0.1				17863.84			
5596.494	5598.048		0.3	0.3	0.2		17863.37			
5596.773	5598.327			0.1			17862.48			
5596.857	5598.411		0.1				17862.21			
5597.049	5598.603		0.2	0.4			17861.60	3c 3	2a 2	Q 6
5597.049	5598.603		0.2	0.4			17861.60	4b 0	2a 1	P 2
5597.158	5598.712		0.2				17861.25			
5597.287	5598.841		0.1				17860.84			
5597.393	5598.947			0.3			17860.50			
5597.434	5598.988		0.6		0.2		17860.37			
5597.647	5599.201	GR	2.5	1.8	0.3	1.48	17859.69	3c 3	2a 2	P 3
5597.647	5599.201	GR	2.5	1.8	0.3	1.48	17859.69	Y 1	2B 4	P 3
5597.719	5599.273						17859.46			
5597.867	5599.421		0.3				17858.99			
5597.945	5599.499		0.3	0.2			17858.74			
5598.080	5599.634		0.2	0.3			17858.31			
5598.309	5599.863	G	0.3	0.5	0.2d		17857.58			
5598.337	5599.891						17857.49			
5598.494	5600.048			0.0			17856.99			
5598.563	5600.117				0.1		17856.77			
5598.632	5600.186		0.1	0.0			17856.55			
5598.760	5600.314		0.7				17856.14	2K 6	2B 0	P 1
5598.826	5600.380	G		0.5	0.8		17855.93	3f 0	2c 0	R 4
5598.920	5600.474		0.7	0.4			17855.63	4b 1	2a 2	R 1
5599.086	5600.641			0.1			17855.10			
5599.212	5600.766		0.2	0.4	0.2		17854.70			
5599.325	5600.879			0.0			17854.34			
5599.472	5601.027	G	0.3	1.0	0.3	0.43	17853.87			
5599.591	5601.146		0.3	0.3			17853.49	t 0	2a 0	P 3
5599.839	5601.394				0.1		17852.70			
5599.989	5601.544			0.2d			17852.22			
5600.071	5601.626		0.2		0.1		17851.96			
5600.269	5601.824		1.5	0.4			17851.33	3D 3	2B 8	P 4
5600.404	5601.958	G	1.8	1.6		1.37	17850.90			
5600.444	5601.999			1.0b	0.8b		17850.77			
5600.535	5602.090		0.7				17850.48	3f 0	2c 0	R 3
5600.695	5602.250			0.0			17849.97			
5600.937	5602.492			0.1	0.2d		17849.20			
5601.345	5602.900		0.1				17847.90			
5601.433	5602.988			0.0			17847.62			
5601.618	5603.173		0.4	0.8	0.3		17847.03			
5601.734	5603.289	G	1.3	1.2		0.76	17846.66			
5602.095	5603.651						17845.51			
5602.268	5603.823			0.0d			17844.96			
5602.371	5603.927				0.1		17844.63			
5602.535	5604.090		0.6	0.4			17844.11			
5602.727	5604.282		1.2	0.9	0.2		17843.50	X 0	2B 4	P 6
5602.940	5604.495			0.2			17842.82	4E 1	2B10	Q 3
5603.232	5604.787		0.2	0.3	0.1		17841.89	4g 0	2a 2	P 3
5603.398	5604.954						17841.36			
5603.599	5605.155		0.1	0.3	0.2		17840.72	3f 1	2c 1	R 5
5603.832	5605.387		0.1	0.1			17839.98			
5604.058	5605.614	G	0.4	0.9	0.3	0.36	17839.26	3D 0	2B 3	R 5

λ_air	λ_vac	Ref.	I₁	I₂	I₃	I₄	ν	Upper	Lower	Br.
5604.262	5605.818		0.2	0.0			17838.61	3A 0	2B 4	R 1
5604.359	5605.915	G	0.5	1.0	0.2	0.43	17838.30	3c 2	2a 1	P 7
5604.510	5606.066		0.1	0.0			17837.82			
5604.689	5606.245	G	1.5	1.3	0.3	1.15	17837.25	u 1	2a 1	R 2
5604.689	5606.245	G	1.5	1.3	0.3	1.15	17837.25	4E 1	2B10	Q 1
5604.894	5606.450		0.4	0.2			17836.60	X 1	2B 6	P 3
5605.447	5607.003			0.0			17834.84	X 2	2C 0	P 5
5605.783	5607.339		0.2	0.2	0.2		17833.77			
5605.959	5607.515		0.3	0.3			17833.21	3b 5	2a 1	P 2
5606.129	5607.685			0.2			17832.67			
5606.292	5607.849		0.3	0.3			17832.15			
5606.456	5608.012		0.3				17831.63			
5606.585	5608.141		0.2	0.3			17831.22	3F 0	2B 4	R 4
5606.704	5608.261				0.8		17830.84	4e 0	2c 2	P 6
5606.808	5608.365			0.2			17830.51	W 2	2C 0	P 2
5607.019	5608.575		0.8	0.3			17829.84	4e 0	2c 2	P 3
5607.261	5608.818		0.2	0.3			17829.07			
5607.459	5609.016			0.1			17828.44			
5607.953	5609.510			0.1			17826.87			
5608.227	5609.783		0.1	0.2			17826.00			
5608.815	5610.372			0.0			17824.13			
5609.234	5610.791			0.1			17822.80	4E 1	2B10	Q 2
5609.501	5611.058	G	0.4	0.4			17821.95	5c 1	2a 4	R 0
5609.724	5611.282		0.1	0.0			17821.24			
5609.942	5611.499		0.3	0.2			17820.55	3C 2	2A 0	P 3
5610.134	5611.691		0.2	0.1			17819.94			
5610.575	5612.132			0.0			17818.54			
5610.893	5612.450		0.2	0.0			17817.53	W 2	2C 0	R 4
5611.378	5612.935		0.1	0.1			17815.99	4d 0	2c 2	R 3
5611.696	5613.254			0.1			17814.98			
5611.885	5613.443	G	0.2	0.4	0.8		17814.38			
5612.058	5613.616		0.2	0.0			17813.83	4e 1	2c 3	P 4
5612.168	5613.726		0.1	0.1	0.1		17813.48			
5612.301	5613.859			0.3			17813.06	4d 0	2c 2	Q 2
5612.345	5613.903		0.4				17812.92	4E 0	2B 8	P 4
5612.556	5614.114	GR	2.3	1.9	0.4	1.76	17812.25	u 1	2a 1	R 1
5612.556	5614.114	GR	2.3	1.9	0.4	1.76	17812.25	Y 2	2C 0	R 1
5612.556	5614.114	GR	2.3	1.9	0.4	1.76	17812.25	3F 3	2B 9	Q 2
5612.776	5614.335		0.3				17811.55			
5612.862	5614.420		0.2	0.1			17811.28			
5612.981	5614.539				0.1		17810.90			
5613.022	5614.580		0.1				17810.77	4d 0	2c 2	P 2
5613.167	5614.725			0.1			17810.31	4E 0	2B 8	Q 4
5613.255	5614.814		0.2				17810.03	4b 1	2a 2	R 0
5613.444	5615.003	G	0.2	0.4	0.3		17809.43	3f 1	2c 1	R 6
5613.719	5615.277			0.1			17808.56			
5614.043	5615.602			0.0			17807.53			
5614.772	5616.330						17805.22	3A 2	2B 8	R 3
5615.141	5616.700			0.0			17804.05			
5615.333	5616.892						17803.44	5c 0	2a 3	R 1
5615.425	5616.984		0.1	0.1			17803.15			
5615.617	5617.176	G	1.2	0.4			17802.54	2K 6	2B 0	P 2
5615.857	5617.416		0.1	0.1			17801.78			
5616.046	5617.605		0.3	0.3			17801.18	3d 3	2c 2	P 3
5616.229	5617.788	G	1.3	1.1	0.2	0.71	17800.60	3E 1	2B 5	Q 4
5616.500	5618.060		0.1	0.2	0.1		17799.74			
5616.649	5618.208						17799.27			
5616.750	5618.309	G	0.3	0.4	0.2		17798.95			
5616.904	5618.464		0.2	0.3			17798.46			
5617.132	5618.691				0.1		17797.74	7c 0	2a 4	Q 3

λ_{air}	λ_{vac}	Ref.	I_1	I_2	I_3	I_4	ν	Upper	Lower	Br.
5617.264	5618.824		0.1	0.1d			17797.32			
5617.507	5619.067	G	0.3	0.5	0.2		17796.55	3c 3	2a 2	Q 7
5617.804	5619.363			0.1			17795.61	3F 0	2B 4	Q 6
5617.962	5619.521			0.0	0.2		17795.11			
5618.123	5619.682		0.2	0.1			17794.60			
5618.284	5619.843	G	1.3	0.6		0.46	17794.09	Y 1	2B 4	P 4
5618.499	5620.058			0.1			17793.41			
5618.574	5620.134		0.1				17793.17			
5619.143	5620.703				0.1		17791.37			
5619.348	5620.908			0.1			17790.72			
5619.594	5621.154	G	0.3	0.4			17789.94			
5619.787	5621.347			0.2			17789.33			
5620.002	5621.562	G	0.3	0.4	0.2		17788.65			
5620.381	5621.941			0.1d			17787.45			
5620.716	5622.276		0.1				17786.39			
5620.906	5622.466	G	1.6	1.5	0.6	1.05	17785.79	3D 0	2B 3	R 4
5621.146	5622.706		0.1	0.1			17785.03			
5621.544	5623.105			0.0	0.2		17783.77			
5621.860	5623.421		0.1	0.1			17782.77			
5622.322	5623.883		0.2	0.3			17781.31	4E 1	2C 1	P 5
5622.439	5624.000	G	0.4	0.4			17780.94	3E 1	2B 5	R 0
5622.439	5624.000	G	0.4	0.4			17780.94	Y 2	2C 0	Q 1
5622.603	5624.164	G	0.3	0.5	0.2b		17780.42	3d 3	2c 2	Q 4
5622.603	5624.164	G	0.3	0.5	0.2b		17780.42	4b 0	2a 1	P 3
5622.603	5624.164	G	0.3	0.5	0.2b		17780.42	4d 0	2c 2	R 4
5622.863	5624.424		0.2	0.2			17779.60			
5623.084	5624.645	G	1.7	1.2	0.2	1.15	17778.90	u 1	2a 1	R 0
5623.084	5624.645	G	1.7	1.2	0.2	1.15	17778.90	W 2	2C 0	Q 3
5623.343	5624.904		0.1	0.2			17778.08			
5623.558	5625.120			0.1			17777.40	7c 0	2a 4	Q 1
5623.726	5625.287				0.2		17776.87			
5623.948	5625.509			0.1			17776.17			
5624.071	5625.632						17775.78			
5624.134	5625.695		0.2	0.2			17775.58			
5624.302	5625.863	G	1.7	1.3	0.2	0.89	17775.05	3c 3	2a 2	P 4
5624.625	5626.186			0.2d			17774.03	4E 0	2B 8	Q 1
5624.881	5626.442	G	0.8	0.5			17773.22	3d 2	2c 1	P 6
5624.881	5626.442	G	0.8	0.5			17773.22	3E 1	2B 5	P 5
5624.881	5626.442	G	0.8	0.5			17773.22	3A 1	2B 6	R 1
5625.191	5626.753	G	1.2	0.3			17772.24	3E 0	2B 4	R 6
5625.191	5626.753	G	1.2	0.3			17772.24	5c 1	2a 4	Q 5
5625.236	5626.797						17772.10			
5625.397	5626.959			0.0			17771.59			
5625.584	5627.145				0.2		17771.00			
5625.631	5627.193			0.2			17770.85			
5625.774	5627.335						17770.40	3F 3	2B 9	P 4
5625.907	5627.468	G	1.3	0.6	0.1	0.46	17769.98	t 1	2a 1	P 3
5626.163	5627.725		0.1	0.1	0.1		17769.17			
5626.442	5628.004			0.1			17768.29			
5626.613	5628.175			0.0			17767.75			
5626.717	5628.279	G	1.2	0.9	0.2	0.40	17767.42	4D 0	2B 8	R 5
5626.948	5628.510		0.1	0.0	0.1		17766.69	4d 1	2c 3	Q 3
5627.145	5628.707				0.1		17766.07			
5627.189	5628.751		0.1	0.2			17765.93	5c 1	2a 4	Q 4
5627.386	5628.948						17765.31	X 1	2B 6	P 4
5627.427	5628.989	G	1.3	1.4	1.5b	0.70	17765.18	3f 0	2c 0	R 3
5627.572	5629.135		0.2				17764.72			
5627.766	5629.328			0.0			17764.11			
5627.886	5629.448				0.1		17763.73	4d 2	2c 4	R 3
5628.219	5629.781			0.2	0.1		17762.68	4d 2	2c 4	Q 2

λ_{air}	λ_{vac}	Ref.	I_1	I_2	I_3	I_4	ν	Upper	Lower	Br.
5628.685	5630.247			0.1	0.2		17761.21			
5628.995	5630.558			0.1			17760.23	5c 1	2a 4	Q 3
5629.265	5630.827		0.2	0.2			17759.38			
5629.537	5631.100	B		0.2			17758.52	4d 0	2c 2	R 5
5629.797	5631.360	G	0.3	0.4	0.1		17757.70	W 1	2B 6	R 4
5629.797	5631.360	G	0.3	0.4	0.1		17757.70	3A 2	2C 0	P 5
5630.083	5631.645			0.1			17756.80			
5630.143	5631.706		0.1				17756.61			
5630.381	5631.943			0.2			17755.86	4E 0	2B 8	Q 3
5630.422	5631.985		0.2				17755.73			
5630.622	5632.185	G	1.8	1.5	0.3	1.11	17755.10	3D 0	2B 3	R 0
5630.622	5632.185	G	1.8	1.5	0.3	1.11	17755.10	5c 1	2a 4	Q 2
5630.812	5632.375			0.2			17754.50			
5630.859	5632.422		0.1				17754.35	3f 0	2c 0	Q 6
5630.989	5632.552	G	1.4	1.3	0.7	0.93	17753.94	3D 0	2B 3	R 3
5631.059	5632.622		0.5				17753.72	3A 0	2B 4	R 0
5631.202	5632.765		0.1	0.2			17753.27			
5631.313	5632.876				0.1		17752.92			
5631.529	5633.092			0.1			17752.24			
5631.649	5633.213						17751.86	5c 1	2a 4	Q 1
5631.649	5633.213						17751.86	4d 0	2c 2	P 6
5631.814	5633.378	G	0.5	0.4			17751.34	4D 0	2C 0	P 4
5631.871	5633.435				0.2		17751.16			
5631.986	5633.549		0.1	0.1			17750.80			
5632.192	5633.755	G	1.5	1.2		0.88	17750.15	3E 1	2B 5	P 4
5632.249	5633.812				0.2		17749.97			
5632.354	5633.917	G	0.4	0.8			17749.64	3E 1	2B 5	Q 3
5632.354	5633.917	G	0.4	0.8			17749.64	4D 0	2C 0	Q 5
5632.652	5634.215		1.5	0.1			17748.70			
5632.750	5634.314			0.0			17748.39			
5633.058	5634.622				0.1		17747.42			
5633.509	5635.073			0.2			17746.00			
5633.699	5635.263				0.1		17745.40	4g 1	2a 3	P 5
5633.757	5635.320						17745.22			
5633.957	5635.521		0.3	0.2			17744.59			
5634.169	5635.733	G	1.6	1.3	0.3	0.99	17743.92	3D 0	2B 3	R 1
5634.169	5635.733	G	1.6	1.3	0.3	0.99	17743.92	3D 1	2B 5	R 7
5634.442	5636.006	G	0.4	1.2	0.6		17743.06	3f 1	2c 1	R 5
5634.598	5636.162		0.3	0.0			17742.57			
5634.811	5636.375	GR	2.2	1.9	0.6	1.50	17741.90	3D 0	2B 3	R 2
5635.027	5636.591		0.2				17741.22			
5635.106	5636.670		0.2	0.1			17740.97			
5635.214	5636.778				0.2		17740.63			
5635.287	5636.852		0.1	0.2			17740.40	Y 2	2C 0	R 2
5635.564	5637.128	G	0.4	1.0	0.6		17739.53	3f 1	2c 1	R 4
5635.843	5637.408				0.1		17738.65			
5635.932	5637.497		0.1	0.1			17738.37	3F 2	2C 0	R 3
5635.932	5637.497		0.1	0.1			17738.37	5c 0	2a 3	R 0
5636.215	5637.779		0.1	0.0			17737.48			
5636.447	5638.011	G	0.2	0.2	0.2		17736.75			
5636.816	5638.380		0.1	0.1			17735.59	3C 2	2A 0	P 4
5637.044	5638.609		0.3	0.2	0.2		17734.87	2K 6	2B 0	P 3
5637.124	5638.689	G					17734.62			
5637.181	5638.746			0.2			17734.44			
5637.356	5638.921				0.2		17733.89			
5637.569	5639.134			0.2			17733.22			
5637.823	5639.388			0.0			17732.42			
5638.122	5639.687				0.2		17731.48			
5638.281	5639.846		0.4	0.2			17730.98			
5638.482	5640.047			0.1			17730.35			

λ_{air}	λ_{vac}	Ref.	I_1	I_2	I_3	I_4	ν	Upper	Lower	Br.
5638.711	5640.276		0.2	0.2d			17729.63	4E 0	2B 8	Q 2
5638.962	5640.527	G	0.9	0.6	0.2		17728.84	3D 0	2B 3	P 1
5639.165	5640.731			0.1			17728.20			
5639.713	5641.278			0.0			17726.48			
5639.789	5641.354		0.1				17726.24			
5640.212	5641.778		0.2	0.2			17724.91			
5640.419	5641.984	G	1.3	0.9	0.2	0.46	17724.26	3E 1	2B 5	P 3
5640.654	5642.220		0.2	0.2			17723.52			
5641.141	5642.707						17721.99			
5641.336	5642.901			0.0			17721.38			
5641.507	5643.073	G	0.3	0.4	0.2		17720.84	3b 5	2a 1	P 3
5641.507	5643.073	G	0.3	0.4	0.2		17720.84	4e 0	2c 2	P 4
5641.753	5643.319		0.0	0.1			17720.07	4e 1	2c 3	P 5
5641.928	5643.494			0.0			17719.52			
5642.052	5643.618	G	1.2	0.5	0.2	0.36	17719.13	3A 0	2B 4	P 4
5642.052	5643.618	G	1.2	0.5	0.2	0.36	17719.13	u 0	2a 0	P 2
5642.304	5643.870			0.2			17718.34			
5642.405	5643.972		0.1				17718.02			
5642.491	5644.057			0.4			17717.75	3c 2	2a 1	P 8
5642.717	5644.284	G	1.5	1.7	1.5b	0.77	17717.04	3f 0	2c 0	R 2
5642.944	5644.510	G	1.8	1.5		1.19	17716.33	3E 1	2B 5	Q 2
5642.944	5644.510	G	1.8	1.5		1.19	17716.33	3A 2	2B.8	R 2
5643.039	5644.606		0.3				17716.03			
5643.163	5644.730			0.1			17715.64			
5643.211	5644.778		0.2d				17715.49			
5643.938	5645.504						17713.21			
5644.448	5646.014		0.2	0.0			17711.61	4d 1	2c 3	P 3
5644.629	5646.196	G	0.3	0.3			17711.04			
5644.706	5646.272						17710.80			
5645.292	5646.859		0.2				17708.96			
5645.675	5647.242			0.0	0.1		17707.76			
5645.904	5647.471		0.2				17707.04			
5646.035	5647.602				0.4		17706.63			
5646.105	5647.672	G	1.5	1.0		0.94	17706.41	3A 1	2B 6	R 0
5646.207	5647.774		0.3	0.0	0.2		17706.09			
5646.430	5647.998		0.3	0.4	0.6		17705.39			
5646.612	5648.179			0.2			17704.82			
5646.740	5648.307		0.1				17704.42			
5646.871	5648.438			0.1			17704.01			
5647.132	5648.699		0.3	0.4	0.2		17703.19	4d 0	2c 2	Q 3
5647.406	5648.974	G	0.4	0.9	0.9		17702.33	3E 2	2B 7	R 2
5647.646	5649.213		0.2	0.2			17701.58			
5647.748	5649.315				0.1		17701.26			
5648.000	5649.568		0.2	0.2	0.2		17700.47			
5648.303	5649.871			0.2	0.2		17699.52			
5648.651	5650.219			0.2	0.2		17698.43			
5648.842	5650.410	G	0.7	1.1			17697.83	3E 1	2B 5	Q 1
5648.842	5650.410	G	0.7	1.1			17697.83	5c 0	2a 3	Q 5
5649.187	5650.755		0.2	0.3	0.1		17696.75	4b 1	2a 2	P 1
5649.187	5650.755		0.2	0.3	0.1		17696.75	W 2	2C 0	P 3
5649.522	5651.090		0.2	0.4	0.3		17695.70			
5649.797	5651.365			0.1			17694.84	T 0	2B 8	R 2
5649.982	5651.550			0 1			17694.26			
5650.116	5651.684		0.2				17693.84	4b 0	2a 1	P 4
5650.177	5651.745			0.2			17693.65			
5650.461	5652.029			0.2			17692.76			
5650.640	5652.208		0.3	0.3			17692.20	Y 2	2C 0	Q 2
5650.710	5652.279	G			0.8b		17691.98			
5650.777	5652.346		0.2	0.3			17691.77			
5651.011	5652.579				0.1		17691.04	5c 0	2a 3	Q 4

λ_{air}	λ_{vac}	Ref.	I_1	I_2	I_3	I_4	ν	Upper	Lower	Br.
5651.011	5652.579				0.1		17691.04	3F 2	2C 0	R 2
5651.349	5652.918				0.2d		17689.98			
5651.972	5653.541	G	0.2	0.2			17688.03			
5652.177	5653.745	G	1.4	1.1	1.2	0.78	17687.39	3E 1	2B 5	P 2
5652.266	5653.835		0.2				17687.11			
5652.368	5653.937			0.1			17686.79			
5652.563	5654.132	G	0.5	1.0	1.6		17686.18	3f 0	2c 0	Q 5
5652.563	5654.132	G	0.5	1.0	1.6		17686.18	t 0	2a 0	P 4
5652.563	5654.132	G	0.5	1.0	1.6		17686.18	5c 0	2a 3	Q 3
5652.848	5654.417		0.2	0.2			17685.29			
5653.155	5654.724		0.2	0.1			17684.33	3F 2	2C 0	R 2
5653.411	5654.980		0.2				17683.53	3c 3	2a 2	P 5
5653.717	5655.287		0.3	0.2			17682.57	5c 0	2a 3	Q 2
5653.941	5655.510			0.0			17681.87			
5654.181	5655.750			0.2			17681.12			
5654.485	5656.054	G	0.5	0.9			17680.17	W 1	2B 6	R 3
5654.485	5656.054	G	0.5	0.9			17680.17	5c 0	2a 3	Q 1
5654.734	5656.304		1.5	0.8			17679.39			
5654.779	5656.349	G	1.5	1.2	1.2	0.99	17679.25			
5654.850	5656.419		0.6	0.7	0.3		17679.03			
5655.035	5656.604			0.0			17678.45			
5655.240	5656.809		0.2	0.1			17677.81			
5655.384	5656.953		0.3		0.1		17677.36	3A 0	2B 4	P 3
5655.483	5657.053			0.3			17677.05			
5655.541	5657.110		0.5				17676.87			
5655.755	5657.324	GR	2.4	1.9	1.5	1.42	17676.20	3c 4	2a 3	Q 1
5655.985	5657.555		0.3				17675.48			
5656.100	5657.670		0.4	1.0	1.6		17675.12			
5656.363	5657.933			0.1			17674.30			
5656.555	5658.125			0.2			17673.70			
5656.782	5658.352			0.3	0.1		17672.99			
5657.067	5658.637		0.2	0.2			17672.10			
5657.285	5658.855	G	0.5	1.0	1.4		17671.42	3d 3	2c 2	Q 5
5657.285	5658.855	G	0.5	1.0	1.4		17671.42	u 2	2a 2	R 3
5657.528	5659.098			0.1			17670.66	4d 1	2c 3	Q 4
5657.794	5659.364	G	1.3	1.5	1.8	0.52	17669.83	3f 1	2c 1	R 4
5657.794	5659.364	G	1.3	1.5	1.8	0.52	17669.83	3f 0	2c 0	R 2
5657.794	5659.364	G	1.3	1.5	1.8	0.52	17669.83	4D 0	2B 8	R 4
5657.848	5659.418		0.8				17669.66			
5658.015	5659.585						17669.14			
5658.101	5659.672		0.1	0.2			17668.87			
5658.389	5659.960		0.2	0.3	0.6		17667.97			
5658.633	5660.203		0.1	0.2			17667.21	3c 4	2a 3	Q 2
5658.723	5660.293				0.1		17666.93			
5658.819	5660.389		0.3	0.5			17666.63			
5659.011	5660.581			0.0			17666.03			
5659.091	5660.661				0.2		17665.78			
5659.187	5660.758			0.2d			17665.48	4f 3	3b 0	R 3
5659.437	5661.008	G	0.5	1.2	1.7		17664.70			
5659.655	5661.226			0.1			17664.02			
5659.764	5661.334		0.2	0.2			17663.68			
5659.911	5661.482	G	0.3	0.4	0.1		17663.22	3d 3	2c 2	P 4
5660.062	5661.633			0.1			17662.75	4b 2	2a 3	R 1
5660.244	5661.815		0.2	0.2			17662.18	3A 0	2B 4	P 1
5660.366	5661.937	G	0.5	0.3			17661.80			
5660.488	5662.059		0.2	0.3	0.2		17661.42			
5660.610	5662.181			0.1			17661.04			
5660.805	5662.376	GR	1.2	1.1	1.4	0.45	17660.43	3D 1	2B 5	R 6
5660.978	5662.549		0.1				17659.89	5c 1	2a 4	P 2
5661.132	5662.703		0.2	0.3	0.3		17659.41			

λ_{air}	λ_{vac}	Ref.	I_1	I_2	I_3	I_4	ν	Upper	Lower	Br.
5661.276	5662.848		0.1	0.1			17658.96	4c 0	2a 2	R 5
5661.408	5662.979		0.2	0.2			17658.55			
5661.629	5663.200	GR	1.7	1.6	1.5	1.17	17657.86			
5661.828	5663.399	G	1.8	1.4	1.4	0.89	17657.24	3c 4	2a 3	Q 2
5661.915	5663.486		0.2		0.2		17656.97			
5662.062	5663.633		0.8	0.3			17656.51	3A 0	2B 4	P 2
5662.062	5663.633		0.8	0.3			17656.51	3A 1	2B 6	P 4
5662.219	5663.791			0.1	0.1		17656.02	Y 2	2C 0	P 2
5662.441	5664.012		0.2	0.4	0.8		17655.33			
5662.700	5664.272		0.2	0.2	0.4		17654.52	2K 6	2B 0	P 4
5662.877	5664.448	G	1.6	1.3	1.4	0.96	17653.97	3D 0	2B 3	P 2
5663.066	5664.638			0.1			17653.38			
5663.137	5664.708		0.1				17653.16			
5663.220	5664.792			0.2			17652.90			
5663.445	5665.016	G	0.3	0.5	0.9		17652.20			
5663.788	5665.360			0.0			17651.13			
5663.878	5665.450				0.1		17650.85			
5664.183	5665.754			0.0			17649.90			
5664.552	5666.124		0.2	0.2			17648.75			
5664.802	5666.374				0.1		17647.97	W 2	2C 0	Q 4
5665.036	5666.608			0.2	0.2		17647.24			
5665.274	5666.846			0.1			17646.50			
5665.560	5667.132				0.1		17645.61			
5665.656	5667.228			0.2			17645.31			
5665.797	5667.370			0.1			17644.87			
5666.128	5667.701		0.2	0.2			17643.84			
5666.424	5667.996		0.3	0.3			17642.92			
5666.510	5668.083				0.2		17642.65			
5666.873	5668.446			0.0			17641.52			
5667.060	5668.632		0.1	0.2	0.2		17640.94			
5667.352	5668.925	G	0.5	1.1	1.8		17640.03	3A 2	2B 8	R 1
5667.352	5668.925	G	0.5	1.1	1.8		17640.03	3C 2	2A 0	P 5
5667.574	5669.146		0.1				17639.34			
5667.625	5669.198			0.1			17639.18	T 0	2B 8	R 1
5667.795	5669.368	G	0.2	0.6	1.3		17638.65			
5668.136	5669.709		0.5	0.3	0.2		17637.59			
5668.370	5669.944		0.2	0.4	0.5		17636.86			
5668.505	5670.079		0.2				17636.44	3F 2	2C 0	R 1
5668.730	5670.304		0.2	0.0			17635.74	3F 2	2C 0	R 1
5668.952	5670.526			0.0			17635.05			
5669.123	5670.696			0.2			17634.52			
5669.325	5670.898	G	0.4	0.9	1.4		17633.89	3f 1	2c 1	R 3
5669.618	5671.191			0.1			17632.98			
5669.785	5671.358		0.1				17632.46			
5669.927	5671.500			0.1			17632.02			
5670.074	5671.648		0.2	0.0			17631.56			
5670.226	5671.799				0.3		17631.09			
5670.280	5671.854	G	1.4	0.9	1.2	0.48	17630.92	3A 1	2B 6	P 1
5670.361	5671.934		0.3				17630.67			
5670.496	5672.069			0.2b			17630.25			
5670.547	5672.121		0.2				17630.09	4b 1	2a 2	P 2
5670.763	5672.336	GR	1.5	1.0	0.2		17629.42	W 1	2B 6	R 2
5670.923	5672.497	GR	1.8	1.6	1.6	1.11	17628.92	3c 4	2a 3	Q 3
5670.923	5672.497	GR	1.8	1.6	1.6	1.11	17628.92	u 2	2a 2	R 2
5671.210	5672.784		0.1	0.1			17628.03			
5671.406	5672.980			0.0			17627.42			
5671.561	5673.134		0.2	0.2	0.2		17626.94			
5671.754	5673.328			0.1			17626.34			
5671.905	5673.479						17625.87			
5672.001	5673.575				0.2		17625.57			

λ_{air}	λ_{vac}	Ref.	I_1	I_2	I_3	I_4	ν	Upper	Lower	Br.
5672.046	5673.620			0.2			17625.43			
5672.217	5673.791		0.1	0.2			17624.90	4D 0	2C 0	Q 6
5672.368	5673.942	G	1.2	0.5	1.9	0.45	17624.43	7c 0	2a 4	P 3
5672.497	5674.071						17624.03			
5672.590	5674.164		0.6	0.3			17623.74	3A 1	2B 6	P 3
5672.671	5674.245			1.2			17623.49			
5672.758	5674.332		0.6	1.2	1.5		17623.22			
5673.076	5674.651		0.2	0.2	0.2		17622.23	4e 0	2c 2	P 5
5673.279	5674.854		0.1	0.2			17621.60	4D 0	2B 8	P 1
5673.279	5674.854		0.1	0.2			17621.60	3F 0	2B 4	R 3
5673.279	5674.854		0.1	0.2			17621.60	4D 0	2B 8	R 0
5673.450	5675.024			0.0			17621.07			
5673.627	5675.201			0.2			17620.52			
5673.688	5675.263		0.1				17620.33			
5673.859	5675.433			0.1			17619.80	4c 0	2a 2	R 4
5674.081	5675.656	G	0.3	1.0	1.6		17619.11	u 0	2a 0	P 3
5674.178	5675.752		0.2				17618.81			
5674.249	5675.823			0.1			17618.59			
5674.342	5675.917		0.2	0.2			17618.30			
5674.503	5676.078	G	0.8	0.6	0.2		17617.80	3c 9	2a 7	Q 1
5674.741	5676.316		0.2	0.2			17617.06			
5674.802	5676.377						17616.87			
5674.993	5676.567	GR	1.7	1.1		1.02	17616.28	3A 1	2B 6	P 2
5675.179	5676.754						17615.70			
5675.237	5676.812		0.2	0.1			17615.52	3F 1	2C 6	Q 6
5675.592	5677.167			0.2			17614.42	3E 2	2B 7	R 1
5675.953	5677.528			0.1			17613.30	4D 0	2B 8	R 3
5676.172	5677.747	G	0.4	1.0	1.5		17612.62	3f 0	2c 0	Q 4
5676.394	5677.969			0.2			17611.93			
5676.507	5678.082				0.1		17611.58			
5676.623	5678.198	G	0.3	0.4			17611.22			
5676.694	5678.269			0.2			17611.00			
5676.852	5678.427		0.1	0.1			17610.51	5c 1	2a 4	P 3
5677.032	5678.608		0.3	0.3			17609.95	Z 0	2B 3	R 1
5677.219	5678.795		0.1	0.2			17609.37			
5677.403	5678.979	G	0.6	0.4			17608.80	Z 0	2B 3	R 2
5677.716	5679.292		0.2	0.2			17607.83			
5677.819	5679.395						17607.51			
5678.071	5679.646		0.1				17606.73	4d 2	2c 4	P 3
5678.129	5679.704			0.1			17606.55			
5678.413	5679.988		0.2	0.2	0.2		17605.67			
5678.700	5680.276				0.1		17604.78			
5678.777	5680.353			0.2			17604.54	3F 0	2B 4	Q 5
5678.848	5680.424		0.4				17604.32			
5679.016	5680.592	G	1.2	0.9	1.2	0.36	17603.80	3D 1	2B 5	R 5
5679.371	5680.947		0.1	0.2			17602.70			
5679.800	5681.376		0.1	0.2			17601.37			
5679.987	5681.563			0.2	0.1		17600.79	4d 0	2c 2	Q 4
5680.142	5681.718	G	0.9	0.4			17600.31	Z 0	2B 3	R 0
5680.387	5681.964		0.2	0.1			17599.55			
5680.597	5682.173		0.1	0.2			17598.90			
5680.810	5682.386						17598.24			
5681.088	5682.664						17597.38			
5681.249	5682.826		0.1				17596.88			
5681.478	5683.055		0.2	0.3	0.1		17596.17			
5681.662	5683.239		0.1	0.0	0.1		17595.60			
5681.782	5683.358		0.2				17595.23	T 0	2B 8	R 0
5681.859	5683.436			0.1	0.2		17594.99			
5682.089	5683.665	G	0.3b	0.5			17594.28	u 2	2a 2	R 1
5682.089	5683.665	G	0.3b	0.5			17594.28	3f 2	2c 2	R 6

λ_{air}	λ_{vac}	Ref.	I_1	I_2	I_3	I_4	ν	Upper	Lower	Br.
5682.131	5683.707				0.2		17594.15			
5682.269	5683.846		0.2	0.3			17593.72	Z 0	2B 3	R 3
5682.502	5684.079	G	1.2	1.5	2.1	0.49	17593.00	3f 0	2c 0	Q 6
5682.773	5684.350		0.1	0.2			17592.16			
5682.848	5684.425				0.2		17591.93			
5683.077	5684.654	GR	1.4	1.6	2.0	0.85	17591.22	3f 1	2c 1	R 3
5683.077	5684.654	GR	1.4	1.6	2.0	0.85	17591.22	3c 4	2a 3	Q 4
5683.126	5684.703						17591.07			
5683.352	5684.929		0.3	0.3			17590.37			
5683.533	5685.110		0.3	0.2	0.1		17589.81			
5683.762	5685.339	GR	1.8	1.4	0.1	1.29	17589.10	u 1	2a 1	P 2
5683.956	5685.533		0.2				17588.50			
5684.072	5685.650	R					17588.14	V 0	2B 3	R 2
5684.072	5685.650	R					17588.14	3f 2	2c 2	R 5
5684.130	5685.708	GR	2.1	1.9	2.0	1.34	17587.96	3f 0	2c 0	R 1
5684.130	5685.708	GR	2.1	1.9	2.0	1.34	17587.96	3D 1	2B 5	R 0
5684.173	5685.750	R					17587.83			
5684.373	5685.950		0.2	0.1			17587.21			
5684.541	5686.118	GF	1.3	1.0		0.76	17586.69	W 1	2B 6	R 1
5684.758	5686.335		0.1	0.1			17586.02			
5685.246	5686.823		0.2				17584.51			
5685.482	5687.059	G	0.4	0.4	0.1		17583.78	3e 0	2c 0	R 5
5685.718	5687.295		0.1	0.1			17583.05			
5685.818	5687.396		0.1				17582.74			
5685.976	5687.554		0.2	0.3			17582.25	3b 5	2a 1	P 4
5686.093	5687.671		0.1				17581.89			
5686.235	5687.813	G	0.5	0.6			17581.45			
5686.436	5688.014		0.1	0.2			17580.83			
5686.581	5688.159			0.2			17580.38			
5686.733	5688.311		0.3	0.2			17579.91	4D 0	2B 8	R 2
5686.908	5688.486		0.2	0.4	0.2		17579.37	3e 0	2c 0	R 5
5687.092	5688.670		0.2	0.2			17578.80	4d 1	2c 3	Q 5
5687.251	5688.829		0.1	0.1			17578.31			
5687.461	5689.039	G	1.2	1.0	0.2	0.49	17577.66	Z 0	2B 3	R 4
5687.545	5689.124		0.2				17577.40			
5687.710	5689.289		0.2	0.1d			17576.89			
5687.762	5689.340						17576.73	4D 0	2B 8	R 1
5687.927	5689.505	G	1.3	1.0	0.2	0.53	17576.22	3D 0	2B 3	P 3
5687.927	5689.505	G	1.3	1.0	0.2	0.53	17576.22	4g 1	2a 3	P 6
5688.008	5689.586		0.2				17575.97			
5688.206	5689.784		0.2	0.3	0.3		17575.36	3E 2	2B 7	Q 5
5688.206	5689.784		0.2	0.3	0.3		17575.36	4c 0	2a 2	R 3
5688.490	5690.069		0.2	0.3	0.3		17574.48			
5688.749	5690.328		0.4	0.2			17573.68			
5688.833	5690.412				0.1		17573.42			
5688.950	5690.529		0.4	0.3			17573.06			
5689.177	5690.755	R					17572.36	3f 0	2c 0	R 1
5689.245	5690.823	GR	2.4	2.3	2.4	1.54	17572.15			
5689.264	5690.843	R					17572.09	3D 1	2B 5	R 1
5689.487	5691.066		0.3				17571.40			
5689.840	5691.419	G	0.7	0.5			17570.31	3D 1	2B 5	P 1
5690.074	5691.652		0.2	0.2			17569.59			
5690.174	5691.753			0.1			17569.28			
5690.320	5691.899		0.3	0.3			17568.83			
5690.469	5692.048		0.3	0.4			17568.37			
5690.715	5692.294			0.0	0.1		17567.61	5c 0	2a 3	P 2
5690.935	5692.514		0.2	0.2			17566.93	4D 0	2C 0	P 5
5691.159	5692.738	G	1.6	1.5	1.6	1.12	17566.24	3D 1	2B 5	R 4
5691.428	5693.007		0.1				17565.41			
5691.476	5693.056			0.1d	0.1		17565.26			

λ_{air}	λ_{vac}	Ref.	I_1	I_2	I_3	I_4	ν	Upper	Lower	Br.
5691.697	5693.276		0.1	0.0			17564.58			
5691.807	5693.386				0.1		17564.24			
5691.911	5693.490	G	0.2	0.4			17563.92	3A 2	2B 8	P 5
5692.111	5693.691	G	1.2	0.6		0.48	17563.30	u 2	2a 2	R 0
5692.277	5693.856			0.2			17562.79	Y 2	2C 0	Q 3
5692.277	5693.856			0.2			17562.79	5c 1	2a 4	P 4
5692.465	5694.044	G	0.6	1.2	1.8		17562.21	3f 0	2c 0	Q 5
5692.740	5694.320			0.1			17561.36			
5693.275	5694.855		0.2	0.2	0.1		17559.71	3D 0	2B 3	P10
5693.606	5695.186		0.2	0.2d			17558.69			
5693.797	5695.377		0.1	0.2			17558.10			
5693.959	5695.539		0.5	0.3			17557.60			
5694.144	5695.724	GR	2.2	1.7	1.5	1.40	17557.03	3D 1	2B 5	R 2
5694.144	5695.724	GR	2.2	1.7	1.5	1.40	17557.03	4b 1	2a 2	P 3
5694.378	5695.958		0.2	0.2			17556.31			
5694.452	5696.032		0.2				17556.08			
5694.602	5696.182	G	0.3	0.5	0.2		17555.62	3d 3	2c 2	Q 6
5694.686	5696.266		0.2				17555.36			
5694.877	5696.457			0.2			17554.77			
5694.936	5696.516		0.2		0.2		17554.59			
5695.101	5696.681		0.1	0.1			17554.08			
5695.306	5696.886	G	0.6	1.0	1.5		17553.45	3f 1	2c 1	Q 7
5695.565	5697.145		0.1	0.1			17552.65			
5695.672	5697.253				0.1		17552.32			
5695.870	5697.451						17551.71			
5695.945	5697.525			0.1d			17551.48			
5696.175	5697.756	G	1.5	1.2	1.3	1.00	17550.77	3D 1	2B 5	R 3
5696.380	5697.960		0.1	0.2			17550.14			
5696.571	5698.152		0.2	0.2			17549.55	3E 0	2B 4	R 5
5696.646	5698.227		0.2				17549.32	3A 2	2B 8	R 0
5696.867	5698.447	G	1.3	0.5	0.2	0.51	17548.64			
5697.065	5698.645			0.1			17548.03			
5697.182	5698.762		0.2	0.2			17547.67			
5697.428	5699.009	G	1.7	1.6	1.5b	0.78	17546.91	3e 0	2c 0	R 4
5697.532	5699.113	G			2.2	1.06	17546.59			
5697.649	5699.230		1.5	0.9			17546.23	W 1	2B 6	R 0
5697.844	5699.425		0.2	0.1			17545.63			
5698.013	5699.594	G	1.2	1.0	1.4		17545.11	Z 0	2B 3	P 1
5698.013	5699.594	G	1.2	1.0	1.4		17545.11	3c 4	2a 3	Q 5
5698.308	5699.889			0.1			17544.20			
5698.377	5699.958						17543.99			
5698.633	5700.214		0.4	0.2			17543.20			
5698.708	5700.289				0.1d		17542.97			
5698.952	5700.533		0.2	0.1			17542.22			
5699.052	5700.633		0.2				17541.91			
5699.195	5700.776	G	0.5	1.1	1.6		17541.47	3f 2	2c 2	R 5
5699.468	5701.049		0.1	0.2			17540.63	3E 2	2B 7	P 5
5699.592	5701.173				0.2		17540.25			
5699.640	5701.222			0.0			17540.10			
5699.819	5701.401		0.2				17539.55			
5699.868	5701.449			0.2			17539.40			
5700.040	5701.622		0.1		0.2b		17538.87			
5700.258	5701.839		0.3	0.2	0.1		17538.20			
5700.398	5701.979		0.2	0.2			17537.77			
5700.641	5702.223	GR	1.6	1.7	2.1	0.98	17537.02	3f 0	2c 0	Q 3
5700.641	5702.223	GR	1.6	1.7	2.1	0.98	17537.02	4D 0	2B 8	P 2
5700.713	5702.295	R					17536.80			
5700.879	5702.460						17536.29			
5700.937	5702.519		0.1	0.1			17536.11			
5701.136	5702.717			0.0			17535.50			

λ_{air}	λ_{vac}	Ref.	I_1	I_2	I_3	I_4	ν	Upper	Lower	Br.
5701.425	5703.007			0.1	0.1		17534.61			
5701.568	5703.150						17534.17			
5701.721	5703.303		0.2	0.3			17533.70			
5702.179	5703.761		0.1	0.2			17532.29	4e 0	2c 2	P 6
5702.179	5703.761		0.1	0.2			17532.29	W 2	2C 0	Q 5
5702.339	5703.921						17531.80			
5702.414	5703.996	G	1.2	0.5			17531.57	t 0	2a 0	P 5
5702.719	5704.302		0.1	0.2			17530.63			
5703.012	5704.594		0.2	0.3			17529.73	W 2	2C 0	P 4
5703.253	5704.835	GR	1.8	1.8	2.1	1.08	17528.99	3f 1	2c 1	R 2
5703.445	5705.027			0.0			17528.40			
5703.510	5705.092		0.2	0.2			17528.20			
5703.767	5705.349	GR	1.7	1.8	2.4	1.05	17527.41	3f 0	2c 0	Q 4
5704.031	5705.613		0.2	0.2	0.2		17526.60	4c 0	2a 2	R 2
5704.239	5705.822			0.1			17525.96	4F 0	2B 9	R 3
5704.239	5705.822			0.1			17525.96	5c 1	2a 4	P 5
5705.102	5706.684			0.1			17523.31			
5705.333	5706.916		0.1	0.0			17522.60			
5705.499	5707.082	G	1.2	0.4			17522.09	u 0	2a 0	P 4
5705.499	5707.082	G	1.2	0.4			17522.09	3A 2	2B 8	P 4
5705.772	5707.355			0.1			17521.25			
5705.981	5707.564			0.0			17520.61			
5706.202	5707.785		0.1	0.2	0.2		17519.93	5c 0	2a 3	P 3
5706.440	5708.023		0.2	0.2			17519.20			
5706.733	5708.316		0.2	0.3	0.2		17518.30	T 0	2B 8	P 1
5706.942	5708.525						17517.66			
5707.007	5708.590			0.0			17517.46			
5707.212	5708.796				0.1		17516.83			
5707.290	5708.874		0.1				17516.59			
5707.453	5709.037			0.2			17516.09			
5707.717	5709.301		0.2	0.2			17515.28	3F 2	2C 0	Q 2
5707.883	5709.467			0.0			17514.77			
5708.138	5709.721		0.2	0.2	0.1		17513.99	4d 0	2c 2	Q 5
5708.431	5710.015		0.1	0.2			17513.09	3d 3	2c 2	P 5
5708.646	5710.230	G	1.2	1.1	1.6		17512.43	3f 2	2c 2	R 4
5708.975	5710.559		0.2	0.2			17511.42	3F 2	2C 0	Q 2
5709.246	5710.830		0.2	0.1			17510.59	T 0	2B 8	P 2
5709.393	5710.977		0.1	0.2			17510.14			
5709.578	5711.163		0.3	0.4	0.4		17509.57	3f 1	2c 1	Q 6
5709.758	5711.342	G	1.8	1.7	1.8	1.08	17509.02	3f 1	2c 1	R 2
5709.758	5711.342	G	1.8	1.7	1.8	1.08	17509.02	4f 3	3b 0	R 2
5709.810	5711.394	R					17508.86	3D 0	2B 3	P 4
5710.048	5711.632		0.2	0.0			17508.13			
5710.420	5712.004			0.0			17506.99			
5710.498	5712.082		0.2				17506.75	3F 2	2C 0	Q 3
5710.599	5712.184			0.0			17506.44			
5710.782	5712.366		0.1	0.1			17505.88	4b 0	2a 1	P 6
5711.033	5712.618		0.3	0.3			17505.11			
5711.180	5712.764			0.0			17504.66			
5711.229	5712.813		0.1				17504.51	Y 2	2C 0	P 3
5711.408	5712.993			0.2			17503.96			
5711.467	5713.052		0.2				17503.78			
5711.673	5713.257		0.2	0.3	0.1		17503.15	3F 2	2C 0	Q 3
5711.833	5713.417		0.1	0.2			17502.66			
5712.019	5713.603	G	1.4	1.1	1.3		17502.09	3E 2	2B 7	P 4
5712.152	5713.737		0.5	1.0	1.2		17501.68	3E 2	2B 7	Q 4
5712.152	5713.737		0.5	1.0	1.2		17501.68	4D 0	2C 0	Q 7
5712.152	5713.737		0.5	1.0	1.2		17501.68	3A 2	2B 8	P 3
5712.410	5713.995		0.4	0.4	0.3		17500.89	T 0	2B 8	P 4
5712.600	5714.185		0.1				17500.31			

λ_{air}	λ_{vac}	Ref.	I_1	I_2	I_3	I_4	ν	Upper	Lower	Br.
5712.720	5714.305				0.2d	0.1	17499.94	T 0	2B 8	P 3
5712.773	5714.358		0.2				17499.78			
5712.982	5714.566	G	1.5	0.9			17499.14	Z 0	2B 3	P 2
5713.138	5714.723	G	1.7	1.6	1.9		17498.66	3e 0	2c 0	R 3
5713.311	5714.896		0.2				17498.13			
5713.452	5715.037	GR	2.2	1.8	1.5	1.53	17497.70	u 1	2a 1	P 3
5713.658	5715.243		0.2				17497.07			
5713.742	5715.328		0.3	0.2			17496.81			
5713.912	5715.497		0.1	0.1			17496.29			
5714.066	5715.651		0.2	0.0			17495.82	3D 0	2B 3	P 9
5714.242	5715.827		0.1				17495.28			
5714.353	5715.938		0.2	0.1			17494.94			
5714.552	5716.138	GR	1.4	0.8			17494.33	3D 1	2B 5	P 2
5714.634	5716.219		0.2				17494.08			
5714.745	5716.331			0.0			17493.74			
5714.807	5716.393		0.1				17493.55			
5714.928	5716.513			0.1			17493.18			
5715.196	5716.781		0.2	0.1			17492.36			
5715.493	5717.079		0.1				17491.45			
5715.644	5717.229				0.1		17490.99			
5715.801	5717.386		0.2	0.2			17490.51			
5716.003	5717.589	G	1.6	1.7	2.0	0.98	17489.89	3f 0	2c 0	Q 3
5716.281	5717.867		0.1	0.1	0.2		17489.04	3F 2	2c 0	Q 4
5716.480	5718.066		0.1				17488.43			
5716.686	5718.272		0.1				17487.80			
5716.837	5718.423		0.2	0.2			17487.34			
5717.082	5718.668		0.1	0.2			17486.59	3F 0	2B 4	P 6
5717.082	5718.668		0.1	0.2			17486.59	4d 0	2c 2	P 4
5717.739	5719.325		0.1				17484.58			
5717.863	5719.450						17484.20			
5718.060	5719.646			0.2			17483.60			
5718.266	5719.852		0.3	0.3	0.3d		17482.97			
5718.498	5720.084			0.1			17482.26	3A 2	2B 8	P 1
5718.701	5720.287	G	0.5	1.0	1.5		17481.64	3f 2	2c 2	R 4
5718.985	5720.572		0.2	0.1			17480.77			
5719.133	5720.719		0.2				17480.32			
5719.296	5720.883		0.2	0.2			17479.82	4c 1	2a 3	R 2
5719.515	5721.102	G	1.5	0.9			17479.15	W 1	2B 6	P 1
5719.699	5721.285			0.0			17478.59			
5719.741	5721.328		0.1				17478.46			
5719.895	5721.482			0.0			17477.99			
5720.091	5721.678			0.1			17477.39			
5720.235	5721.822		0.2	0.3	0.2d		17476.95	4b 1	2a 2	P 4
5720.487	5722.074				0.1		17476.18			
5720.527	5722.114		0.4	0.1			17476.06			
5720.949	5722.536			0.0			17474.77	5c 0	2a 3	P 4
5721.443	5723.031		0.1				17473.26	4c 0	2a 2	R 1
5721.519	5723.106						17473.03			
5721.669	5723.257			0.0	0.1d		17472.57			
5721.840	5723.427			0.0			17472.05			
5722.200	5723.787			0.1			17470.95			
5722.400	5723.987	G	1.2	0.5			17470.34	3E 2	2B 7	P 3
5722.616	5724.203		0.1	0.1			17469.68			
5722.845	5724.433		0.2	0.1	0.2		17468.98			
5723.015	5724.603		0.1	0.1			17468.46			
5723.189	5724.777		0.2	0.2			17467.93			
5723.422	5725.009	R					17467.22			
5723.474	5725.062	GR	1.8	1.7	1.9	1.00	17467.06	3f 0	2c 0	Q 2
5723.540	5725.128						17466.86			
5723.792	5725.380		0.1		0.1		17466.09			

λ_{air}	λ_{vac}	Ref.	I_1	I_2	I_3	I_4	ν	Upper	Lower	Br.
5724.051	5725.639		0.2				17465.30			
5724.103	5725.691			0.2			17465.14			
5724.261	5725.849	G	1.5	1.0	0.2		17464.66	3A 2	2B 8	P 2
5724.415	5726.003			0.1			17464.19			
5724.602	5726.190	G	1.5	1.1	1.5		17463.62			
5724.844	5726.432			0.1	0.2		17462.88	V 0	2B 3	R 1
5724.916	5726.504		0.2				17462.66			
5725.218	5726.806		0.1	0.1			17461.74			
5725.438	5727.026		0.3	0.3			17461.07			
5725.592	5727.180				0.2		17460.60			
5725.654	5727.242		0.1	0.2			17460.41			
5725.870	5727.459	G	1.2	1.3	1.9		17459.75	3f 1	2c 1	Q 5
5726.084	5727.672		0.1				17459.10			
5726.189	5727.777			0.2	0.2		17458.78			
5726.254	5727.843		0.3				17458.58			
5726.382	5727.971		0.1	0.2			17458.19			
5726.543	5728.131	G	0.5	0.3			17457.70			
5726.671	5728.259		0.3	0.4	1.2		17457.31	3D 0	2B 3	P 5
5726.848	5728.437		0.3	0.0			17456.77			
5727.058	5728.647	G	1.8	1.3	1.2	1.09	17456.13	W 1	2B 6	P 2
5727.235	5728.824		0.8	0.4			17455.59			
5727.524	5729.113		0.2	0.2			17454.71			
5727.714	5729.303		0.2				17454.13			
5727.881	5729.470		0.1	0.1			17453.62			
5728.055	5729.644		0.3	0.3	0.2d		17453.09			
5728.174	5729.763		0.2		0.2		17452.73	W 2	2B 8	R 4
5728.308	5729.897		0.6	0.4			17452.32			
5728.551	5730.140	GR	2.5	2.4	2.4	1.49	17451.58	3f 0	2c 0	Q 2
5728.600	5730.189	R					17451.43			
5728.751	5730.340		1.0	0.9			17450.97	Z 1	2B 5	R 2
5728.837	5730.426		0.5				17450.71			
5728.958	5730.547						17450.34	2E 0	2B 4	Q 9
5729.024	5730.613		0.2	0.2			17450.14			
5729.221	5730.810	G	1.2	1.1	1.3		17449.54	3b 3	2a 0	R 1
5729.221	5730.810	G	1.2	1.1	1.3		17449.54	4e 0	2c 2	P 7
5729.437	5731.027		0.1	0.0			17448.88			
5729.605	5731.194			0.1			17448.37			
5729.756	5731.345		0.1	0.2			17447.91			
5729.970	5731.559		0.2	0.4			17447.26	3b 3	2a 0	R 2
5730.111	5731.700		0.1	0.3	0.3		17446.83			
5730.249	5731.838		1.3	1.2	1.7		17446.41			
5730.318	5731.907	G					17446.20			
5730.380	5731.970		1.0	0.6			17446.01	Z 1	2B 5	R 1
5730.475	5732.065		0.2	0.0			17445.72			
5730.653	5732.242		0.4	0.9			17445.18	3E 2	2B 7	Q 3
5730.889	5732.479		0.1				17444.46	4d 0	2c 2	Q 6
5730.945	5732.535			0.2			17444.29			
5731.155	5732.745		0.2	0.2			17443.65			
5731.297	5732.886				0.2d		17443.22			
5731.435	5733.025		0.2	0.2			17442.80			
5731.645	5733.235		0.8	0.6	0.2		17442.16	Z 0	2B 3	P 3
5731.720	5733.310		0.1				17441.93			
5731.918	5733.508	GR	2.2	2.2	2.4	1.29	17441.33	3e 0	2c 0	R 2
5731.974	5733.564	R					17441.16	3E 2	2B 7	P 2
5732.158	5733.748		0.5	0.9			17440.60			
5732.444	5734.034		0.1	0.2			17439.73			
5732.670	5734.261		0.3	0.4	1.2		17439.04	3D 2	2B 7	R 5
5732.792	5734.382		0.2	0.3	0.3		17438.67			
5733.045	5734.635		0.2	0.2			17437.90			
5733.259	5734.849	G	1.3	0.8	1.4		17437.25	W 1	2B 6	P 6

λ_{air}	λ_{vac}	Ref.	I$_1$	I$_2$	I$_3$	I$_4$	ν	Upper	Lower	Br.
5733.259	5734.849	G	1.3	0.8	1.4		17437.25	W 1	2B 6	P 3
5733.259	5734.849	G	1.3	0.8	1.4		17437.25	4D 0	2B 8	P 3
5733.357	5734.948		0.2				17436.95			
5733.558	5735.148	G	1.5	0.7			17436.34	Z 1	2B 5	R 0
5733.650	5735.241		0.3				17436.06			
5733.703	5735.293			0.6	0.3		17435.90			
5733.811	5735.402		0.1				17435.57			
5734.002	5735.593			0.1			17434.99	3F 1	2B 6	Q 5
5734.055	5735.645		0.1				17434.83			
5734.268	5735.859		0.4	0.3	0.2d		17434.18	3F 0	2B 4	Q 4
5734.466	5736.056	GR	1.2	0.9	1.5		17433.58	3f 2	2c 2	R 3
5734.538	5736.129						17433.36			
5734.634	5736.224			0.2			17433.07			
5734.844	5736.435			0.1			17432.43			
5734.913	5736.504		0.4	0.2			17432.22			
5735.137	5736.728	GR	2.2	2.0	2.0	1.23	17431.54	3f 1	2c 1	R 1
5735.137	5736.728	GR	2.2	2.0	2.0	1.23	17431.54	3D 2	2B 7	R 4
5735.222	5736.813						17431.28			
5735.417	5737.007		0.4				17430.69			
5735.545	5737.136		0.2	0.1d			17430.30			
5735.729	5737.320		0.9	0.4			17429.74	3D 2	2B 7	P 1
5735.923	5737.514		0.3	0.2			17429.15			
5736.101	5737.692	G	0.6	0.8	1.6		17428.61	3f 1	2c 1	Q 7
5736.101	5737.692	G	0.6	0.8	1.6		17428.61	3b 3	2a 0	R 0
5736.243	5737.834		0.1				17428.18			
5736.397	5737.988	R	0.4b	0.5	0.1		17427.71	3D 0	2B 3	P 6
5736.638	5738.229		0.7	0.3			17426.98			
5736.875	5738.466	GR	2.7	2.5	2.4	1.61	17426.26	3f 1	2c 1	R 1
5736.875	5738.466	GR	2.7	2.5	2.4	1.61	17426.26	3D 2	2B 7	R 3
5736.937	5738.529	R					17426.07			
5737.174	5738.766		0.5	0.2			17425.35	3D 0	2B 3	P 8
5737.355	5738.947		0.1				17424.80			
5737.457	5739.049		0.2	0.2			17424.49			
5737.668	5739.260		0.2		0.1		17423.85			
5737.757	5739.349			0.3			17423.58			
5737.885	5739.477		0.1				17423.19			
5738.040	5739.632		0.4	0.5	1.3		17422.72			
5738.185	5739.777		0.2				17422.28	3D 0	2B 3	P 7
5738.294	5739.885			0.3	0.3		17421.95	3b 3	2a 0	R 3
5738.426	5740.017	GR	1.9	1.0	1.4	1.08	17421.55	W 1	2B 6	P 4
5738.426	5740.017	GR	1.9	1.0	1.4	1.08	17421.55	3D 2	2B 7	R 2
5738.633	5740.225		0.2				17420.92			
5738.742	5740.334		0.2	0.2			17420.59			
5738.920	5740.512		0.2	0.3			17420.05			
5739.183	5740.775			0.3			17419.25			
5739.220	5740.811		0.2				17419.14			
5739.378	5740.970			0.1			17418.66			
5739.480	5741.072				0.4		17418.35			
5739.753	5741.345		0.1				17417.52			
5739.826	5741.418			0.2			17417.30	4c 0	2a 2	R 0
5739.958	5741.550		0.9				17416.90	3D 1	2B 5	P 3
5740.096	5741.688	G	1.6	1.6	2.0		17416.48	3f 2	2c 2	R 3
5740.096	5741.688	G	1.6	1.6	2.0		17416.48	4d 1	2c 3	P 5
5740.337	5741.929		0.1	0.1			17415.75			
5740.515	5742.107	G	1.2	0.6	0.1		17415.21	W 1	2B 6	P 5
5740.693	5742.285			0.1			17414.67			
5740.838	5742.430		0.4	0.4	0.2		17414.23	3F 0	2B 4	R 2
5740.963	5742.555			0.0			17413.85			
5741.118	5742.711	G	1.2	0.7	0.7		17413.38	3D 2	2B 7	R 1
5741.197	5742.790		0.2				17413.14			

λ_{air}	λ_{vac}	Ref.	I_1	I_2	I_3	I_4	ν	Upper	Lower	Br.
5741.395	5742.988	G	1.2	0.9	1.9		17412.54			
5741.593	5743.185		0.1	0.2			17411.94			
5741.847	5743.439	GR	1.4	1.5	2.1		17411.17	3f 1	2c 1	Q 6
5741.847	5743.439	GR	1.4	1.5	2.1		17411.17	3F 1	2B 6	R 3
5742.087	5743.680		0.1				17410.44			
5742.150	5743.743			0.0			17410.25			
5742.328	5743.921		0.1	0.0			17409.71			
5742.516	5744.109		0.3	0.3	0.1		17409.14			
5742.632	5744.225		0.3				17408.79			
5742.853	5744.446		0.1	0.2			17408.12			
5743.067	5744.660	G	1.2	0.9	0.7		17407.47	3f 1	2c 1	Q 4
5743.417	5745.010		0.1				17406.41			
5743.654	5745.248			0.1			17405.69			
5743.714	5745.307		0.2				17405.51			
5743.928	5745.522	GR	1.5	0.9	1.7		17404.86	3E 2	2B 7	Q 2
5744.186	5745.779		0.2	0.1			17404.08			
5744.390	5745.984		0.1				17403.46			
5744.476	5746.070			0.0			17403.20			
5745.183	5746.776		0.1	0.0			17401.06			
5745.374	5746.968	G	1.3	0.9	1.2		17400.48	u 1	2a 1	P 4
5745.602	5747.196			0.1			17399.79			
5745.704	5747.298		0.1				17399.48			
5745.883	5747.477		0.3	0.3	0.1		17398.94			
5746.124	5747.718		0.1	0.1			17398.21	4g 0	2a 2	P 5
5746.322	5747.916	G	1.3	0.7	1.2		17397.61	3D 2	2B 7	R 0
5746.586	5748.180		0.2				17396.81			
5747.158	5748.752			0.0			17395.08			
5747.386	5748.980			0.0			17394.39			
5747.554	5749.149		0.3	0.3	0.0		17393.88			
5747.875	5749.469			0.0			17392.91			
5748.063	5749.658						17392.34	4F 0	2B 9	P 5
5748.364	5749.958						17391.43			
5748.440	5750.034			0.0			17391.20	4d 0	2c 2	Q 7
5748.678	5750.273				0.1		17390.48			
5748.777	5750.372		0.1	0.2b			17390.18			
5749.002	5750.597	G	1.3	1.0	1.8		17389.50	3f 1	2c 1	Q 5
5749.792	5751.387	G	1.0	0.3			17387.11	Z 1	2B 5	P 1
5749.640	5751.235		0.1				17387.57			
5749.276	5750.871			0.0			17388.67			
5749.554	5751.149			0.1d			17387.83			
5749.908	5751.503				0.1		17386.76			
5749.964	5751.559			0.0			17386.59			
5750.073	5751.668		0.1				17386.26	W 2	2B 8	R 3
5750.282	5751.877		0.3	0.1			17385.63	4D 0	2B 8	P 4
5750.530	5752.125			0.1			17384.88			
5750.652	5752.247		0.1				17384.51			
5750.804	5752.399		0.2	0.2			17384.05			
5751.139	5752.734	G	1.3	0.8	1.4		17383.04	u 2	2a 2	P 2
5751.367	5752.962		0.2	0.1			17382.35	3E 0	2B 4	R 4
5751.555	5753.151	G	1.4	0.6			17381.78			
5751.727	5753.323			0.0			17381.26			
5751.837	5753.432		0.1				17380.93			
5751.903	5753.498			0.0			17380.73			
5752.151	5753.746		0.1	0.1			17379.98			
5752.346	5753.942	G	0.5	0.5			17379.39	3E 2	2B 7	Q 1
5752.346	5753.942	G	0.5	0.5			17379.39	4b 2	2a 3	P 3
5752.641	5754.237			0.2			17378.50			
5752.866	5754.462	G	0.4	0.8	1.6		17377.82	3e 1	2c 1	R 5
5753.048	5754.644		0.1	0.0			17377.27			
5753.214	5754.809		0.1				17376.77			

λ_{air}	λ_{vac}	Ref.	I$_1$	I$_2$	I$_3$	I$_4$	ν	Upper	Lower	Br.
5753.323	5754.919		0.2	0.2			17376.44			
5753.558	5755.154	G	1.4	0.9	1.2		17375.73	Z 0	2B 3	P 4
5753.641	5755.237		0.2				17375.48			
5753.816	5755.412		0.2	0.3	0.2		17374.95			
5753.969	5755.565			0.0			17374.49	4c 2	2a 4	R 1
5754.184	5755.780		0.2	0.3			17373.84	3b 3	2a 0	R 4
5754.263	5755.859				0.1		17373.60			
5754.326	5755.922		0.1				17373.41			
5754.399	5755.995			0.2			17373.19	3f 0	2c 0	P 7
5754.638	5756.234		0.3	0.5	1.5		17372.47	4D 0	2B 8	P 5
5754.843	5756.439			0.0			17371.85			
5755.174	5756.771		0.1	0.1d			17370.85			
5755.410	5757.006			0.2			17370.14			
5755.473	5757.069		0.2				17369.95			
5755.691	5757.288	GR	1.8	1.7	1.9		17369.29	3e 0	2c 0	R 1
5755.807	5757.404		0.9				17368.94			
5756.000	5757.596		0.1				17368.36			
5756.192	5757.788			0.1d			17367.78			
5756.510	5758.107		0.5				17366.82			
5756.556	5758.153	G		1.0b	2.2		17366.68	3f 0	2c 0	P 8
5756.606	5758.203		0.5				17366.53			
5756.828	5758.425		0.1	0.2			17365.86			
5756.997	5758.594			0.1	0.1		17365.35			
5757.130	5758.727		0.2	0.2			17364.95			
5757.332	5758.929	GR	1.9	1.8	2.4	1.07	17364.34	3f 1	2c 1	Q 4
5757.378	5758.975	R					17364.20			
5757.611	5759.208		0.1	0.2	0.1		17363.50			
5757.810	5759.406		0.1	0.1			17362.90			
5757.989	5759.586	G	0.5	0.5	0.1		17362.36			
5758.314	5759.911		0.2	0.4	0.2		17361.38			
5758.490	5760.087		0.1	0.1			17360.85	4f 3	3b 0	R 1
5758.705	5760.302		0.1	0.1			17360.20			
5758.937	5760.535		0.2				17359.50			
5759.053	5760.651			0.2			17359.15			
5759.203	5760.800		0.2	0.1			17358.70			
5759.379	5760.976		1.0	0.6			17358.17	W 2	2B 8	R 2
5759.541	5761.138	GR	2.1	2.0	2.3	1.19	17357.68	3f 1	2c 1	Q 3
5759.607	5761.205	R					17357.48			
5759.787	5761.384						17356.94			
5759.840	5761.437		0.3	0.2	0.1		17356.78			
5760.115	5761.713			0.3	0.1		17355.95			
5760.165	5761.763		0.2				17355.80			
5760.384	5761.982	GR	2.1	2.0	2.3	1.21	17355.14	3f 2	2c 2	R 2
5760.384	5761.982	GR	2.1	2.0	2.3	1.21	17355.14	V 0	2B 3	P 4
5760.421	5762.018	R					17355.03	3d 3	2c 2	P 6
5760.676	5762.274		0.2	0.2			17354.26			
5760.819	5762.417			0.0			17353.83	4c 0	2a 2	Q 2
5760.879	5762.476		0.1				17353.65			
5760.968	5762.566			0.2			17353.38			
5761.048	5762.646				1.3		17353.14			
5761.181	5762.779		0.3	0.5	0.3		17352.74			
5761.400	5762.998		0.2	0.1	0.1		17352.08			
5761.699	5763.297		0.1	0.0			17351.18			
5761.978	5763.576			0.0			17350.34			
5762.343	5763.941				0.2d		17349.24			
5762.479	5764.077		0.2	0.2			17348.83	4c 0	2a 2	Q 3
5762.712	5764.310	GR	1.7	1.6	1.8	0.97	17348.13	3f 2	2c 2	R 2
5762.758	5764.356	R					17347.99			
5762.981	5764.579			0.1			17347.32			
5763.054	5764.652		0.1				17347.10			

λ_{air}	λ_{vac}	Ref.	I_1	I_2	I_3	I_4	ν	Upper	Lower	Br.
5763.193	5764.792			0.2d			17346.68			
5763.522	5765.121		0.2	0.3			17345.69			
5763.655	5765.254			0.1			17345.29			
5763.815	5765.413			0.2			17344.81			
5764.021	5765.619	G	1.2	0.8	1.3		17344.19			
5764.267	5765.865		0.3	0.2			17343.45			
5764.483	5766.082	GR	1.3	0.7	0.2		17342.80	Z 1	2B 5	P 2
5764.483	5766.082	GR	1.3	0.7	0.2		17342.80	4c 0	2a 2	Q 4
5764.696	5766.294		0.1	0.1			17342.16			
5764.845	5766.444		0.3				17341.71			
5764.991	5766.590	G	1.3	0.8	0.1		17341.27	3D 1	2B 5	P 4
5764.991	5766.590	G	1.3	0.8	0.1		17341.27	W 2	2B 8	P 1
5765.068	5766.557		0.2				17341.04			
5765.271	5766.870		0.1	0.1			17340.43			
5765.543	5767.142				0.2		17339.61			
5765.623	5767.222			0.2			17339.37			
5765.683	5767.282		0.1				17339.19	W 2	2C 0	P 5
5765.902	5767.502	G	1.4	1.5	2.3	0.94	17338.53	3f 0	2c 0	P 6
5766.112	5767.711		0.1				17337.90			
5766.295	5767.894	G	1.8	1.7	2.0	1.06	17337.35	3f 1	2c 1	Q 3
5766.361	5767.961						17337.15	V 0	2B 3	R 0
5766.455	5768.054		0.3				17336.87			
5766.578	5768.177		0.1	0.1b			17336.50			
5766.667	5768.267				0.2		17336.23			
5766.721	5768.320			0.2			17336.07	3F 0	2B 4	P 5
5766.784	5768.383		0.2				17335.88			
5766.994	5768.593	G	1.5	1.6	2.2	0.95	17335.25	3e 1	2c 1	R 4
5766.994	5768.593	G	1.5	1.6	2.2	0.95	17335.25	4c 0	2a 2	Q 5
5767.223	5768.822			0.1			17334.56			
5767.369	5768.969			0.3	0.6		17334.12			
5767.416	5769.015		0.2b				17333.98	3F 2	2C 0	P 3
5767.726	5769.325			0.1			17333.05	t 1	2a 1	P 5
5767.726	5769.325			0.1			17333.05	4d 0	2c 2	P 5
5767.805	5769.405				0.1		17332.81			
5768.065	5769.665			0.2			17332.03			
5768.251	5769.851	G	0.3	0.6	1.4		17331.47			
5768.494	5770.094			0.2			17330.74			
5768.667	5770.267			0.1			17330.22			
5768.830	5770.430		0.2				17329.73			
5768.937	5770.537			0.2			17329.41			
5769.123	5770.723			0.1			17328.85			
5769.287	5770.887		0.3	0.2			17328.36			
5769.346	5770.946						17328.18			
5769.496	5771.096		0.1	0.0			17327.73	3F 2	2C 0	P 3
5769.643	5771.243		0.1				17327.29			
5769.739	5771.340						17327.00			
5769.803	5771.403		0.2				17326.81			
5769.959	5771.559		0.2	0.2			17326.34	4c 3	2a 5	R 1
5770.012	5771.613						17326.18			
5770.179	5771.779	G	1.5	0.7	1.3		17325.68	Z 2	2B 7	R 0
5770.269	5771.869		0.3				17325.41			
5770.332	5771.932			0.1			17325.22			
5770.475	5772.076		1.2	0.3			17324.79			
5770.652	5772.252		0.1	0.3	1.3		17324.26			
5770.878	5772.479		0.1	0.2			17323.58			
5771.028	5772.629		0.0		0.1		17323.13			
5771.102	5772.702			0.0			17322.91			
5771.188	5772.789		0.1				17322.65			
5771.288	5772.889						17322.35			
5771.428	5773.029		0.1	0.2			17321.93			

λ_{air}	λ_{vac}	Ref.	I_1	I_2	I_3	I_4	ν	Upper	Lower	Br.
5771.611	5773.212		1.2	0.4			17321.38			
5771.811	5773.412		0.2	0.3	0.2		17320.78			
5772.065	5773.665		0.3d	0.4	0.3		17320.02	3E 3	2B 9	Q 5
5772.231	5773.832		0.1	0.0			17319.52			
5772.451	5774.052	G	1.2	1.0	1.9		17318.86	3f 0	2c 0	P 5
5772.488	5774.089						17318.75	3F 3	2C 1	R 2
5772.721	5774.322	G	1.2	0.7	0.2		17318.05	3b 3	2a 0	P 1
5772.988	5774.589		0.3	0.2			17317.25	W 2	2B 8	R 0
5773.218	5774.819	GR	2.1	1.9	2.0	1.19	17316.56	3f 1	2c 1	Q 2
5773.218	5774.819	GR	2.1	1.9	2.0	1.19	17316.56	3E 0	2B 4	P 8
5773.271	5774.872	R					17316.40			
5773.451	5775.052						17315.86	4c 0	2a 2	Q 6
5773.518	5775.119		0.2	0.1			17315.66			
5773.641	5775.243				0.1		17315.29			
5773.721	5775.323		0.2	0.1			17315.05			
5773.915	5775.516		0.1				17314.47			
5773.978	5775.579			0.2	0.1		17314.28			
5774.058	5775.659		0.2				17314.04			
5774.205	5775.806		0.4	0.2	0.1		17313.60	Y 2	2C 0	P 4
5774.205	5775.806		0.4	0.2	0.1		17313.60	4c 1	2a 3	Q 1
5774.362	5775.963		0.4	0.2			17313.13			
5774.602	5776.203	GR	2.2	1.8	1.4	1.21	17312.41	Z 2	2B 7	R 1
5774.602	5776.203	GR	2.2	1.8	1.4	1.21	17312.41	3c 5	2a 4	Q 1
5774.805	5776.407		0.6	0.2			17311.80			
5774.855	5776.457						17311.65			
5775.052	5776.654	GR	2.6	2.5	2.4	1.57	17311.06	u 1	2a 1	P 5
5775.052	5776.654	GR	2.6	2.5	2.4	1.57	17311.06	3f 1	2c 1	Q 2
5775.052	5776.654	GR	2.6	2.5	2.4	1.57	17311.06	3D 2	2B 7	P 2
5775.096	5776.697	R					17310.93			
5775.339	5776.941		0.5	0.3			17310.20	4c 1	2a 3	Q 2
5775.469	5777.071			0.1			17309.81			
5775.529	5777.131		0.2	0.0			17309.63			
5775.723	5777.325		0.1				17309.05			
5775.890	5777.491		0.2	0.0			17308.55			
5776.080	5777.682		0.1	0.0			17307.98			
5776.210	5777.812		0.2				17307.59			
5776.370	5777.972		0.6	0.1			17307.11			
5776.524	5778.126		0.5	0.2			17306.65			
5776.624	5778.226		0.3				17306.35			
5776.764	5778.366		0.3	0.5	1.3		17305.93	4c 1	2a 3	Q 3
5776.988	5778.590		0.1				17305.26			
5777.195	5778.797		0.1				17304.64			
5777.405	5779.007		0.2	0.3	1.2		17304.01	3f 3	2c 3	R 4
5777.615	5779.218		0.1	0.3	0.2		17303.38	3b 3	2a 0	R 5
5777.842	5779.445						17302.70			
5777.929	5779.532		0.1	0.2			17302.44			
5778.213	5779.816		0.2	0.2			17301.59	3F 3	2C 1	R 2
5778.477	5780.080		0.2	0.2	0.2		17300.80			
5778.708	5780.310			0.3	0.2		17300.11			
5778.771	5780.373		0.2				17299.92	4c 1	2a 3	Q 4
5778.995	5780.597	GR	1.9	2.0	2.5	1.16	17299.25	3f 0	2c 0	P 4
5779.195	5780.798		0.2				17298.65			
5779.302	5780.905			0.1	0.3		17298.33			
5779.442	5781.045		0.2	0.1			17297.91			
5779.716	5781.319		0.5	0.4	0.2		17297.09	Z 0	2B 3	P 5
5779.914	5781.516			0.5	0.4		17296.50			
5779.997	5781.600	GR	1.8	1.5	1.8	1.08	17296.25			
5780.064	5781.667	R					17296.05	Z 2	2B 7	R 2
5780.278	5781.881	G	0.9	1.0	0.3		17295.41	u 2	2a 2	P 3
5780.278	5781.881	G	0.9	1.0	0.3		17295.41	V 0	2B 3	P 3

λ_{air}	λ_{vac}	Ref.	I_1	I_2	I_3	I_4	ν	Upper	Lower	Br.
5780.512	5782.115		0.2	0.1			17294.71			
5780.726	5782.329	GR	1.4	1.0	1.2		17294.07	3c 5	2a 4	Q 2
5780.806	5782.409		0.2				17293.83			
5780.923	5782.526		0.2	0.2			17293.48	4e 3	3b 0	R 4
5781.087	5782.690		0.5	0.3			17292.99			
5781.391	5782.994		0.3	0.2			17292.08	4c 1	2a 3	Q 5
5781.618	5783.222		0.2	0.2	0.2		17291.40			
5781.812	5783.416		1.2	0.3			17290.82	Z 1	2B 5	P 3
5782.100	5783.703		0.2	0.2			17289.96			
5782.307	5783.911		0.4	0.3			17289.34	3E 0	2B 4	Q 8
5782.434	5784.038		1.1	0.5	1.3		17288.96	3f 3	2c 3	R 4
5782.622	5784.225		0.1				17288.40			
5782.719	5784.322			0.2	0.2		17288.11			
5782.789	5784.393		0.2				17287.90			
5783.010	5784.613	GR	1.7	1.7	1.8	0.97	17287.24	3f 0	2c 0	P 3
5783.137	5784.741		0.9				17286.86	3F 0	2B 4	Q 3
5783.271	5784.874			0.2			17286.46			
5783.334	5784.938		0.2				17286.27			
5783.505	5785.109			0.0			17285.76			
5783.568	5785.172		0.1				17285.57			
5783.876	5785.480		0.2	0.1			17284.65			
5784.020	5785.624			0.1			17284.22			
5784.144	5785.748		0.1	0.2			17283.85			
5784.408	5786.012	G	1.5	1.5	1.8		17283.06	3e 1	2c 1	R 3
5784.586	5786.190		0.2				17282.53			
5784.746	5786.351		0.3	0.5	1.3		17282.05	3a 1	2c 1	R 6
5784.977	5786.582		0.4	0.2			17281.36			
5785.215	5786.819	GR	2.2	2.0	1.9	1.24	17280.65	3f 2	2c 2	R 1
5785.249	5786.853	R					17280.55			
5785.312	5786.916						17280.36			
5785.446	5787.050		0.2				17279.96			
5785.526	5787.131		0.6	0.4	0.2		17279.72			
5785.768	5787.372	GR	2.8	2.5	2.4	1.67	17279.00	3f 2	2c 2	R 1
5785.768	5787.372	GR	2.8	2.5	2.4	1.67	17279.00	4F 0	2B 9	R 2
5785.828	5787.432	R					17278.82	3F 3	2C 1	R 1
5786.065	5787.670		0.5	0.3			17278.11	3F 1	2B 6	Q 4
5786.263	5787.868		0.2	0.2			17277.52			
5786.464	5788.069		0.2	0.2			17276.92			
5786.678	5788.283	G	1.7	0.8			17276.28			
5786.806	5788.410		0.2	0.3			17275.90	3D 1	2B 5	P 5
5786.940	5788.544		0.2				17275.50			
5787.144	5788.749		0.2	0.1			17274.89			
5787.278	5788.883			0.2			17274.49			
5787.348	5788.953		0.1				17274.28			
5787.479	5789.084		0.2				17273.89			
5787.516	5789.121	G	1.2	0.7	1.5		17273.78	3f 2	2c 2	Q 6
5787.560	5789.164		0.6				17273.65	4c 2	2a 4	Q 1
5787.697	5789.302			0.1			17273.24			
5787.854	5789.459		0.6	0.5			17272.77	Z 2	2B 7	R 3
5788.035	5789.640		0.2	0.3			17272.23			
5788.267	5789.872	GR					17271.54	3f 0	2c 0	P 3
5788.317	5789.922	R	2.1	2.0	2.1	1.14	17271.39			
5788.548	5790.153		0.2	0.2			17270.70	4c 2	2a 4	Q 2
5788.655	5790.260			0.3	0.2		17270.38	3F 2	2C 0	P 4
5788.773	5790.378		0.2	0.2	0.4		17270.03			
5788.940	5790.546		0.2	0.2			17269.53	3D 2	2B 7	P 3
5789.094	5790.700			0.1			17269.07			
5789.316	5790.921			0.1			17268.41			
5789.624	5791.230		0.1	0.2			17267.49			
5789.862	5791.468	G	1.6	1.5	1.5	0.97	17266.78			

λ_{air}	λ_{vac}	Ref.	I_1	I_2	I_3	I_4	ν	Upper	Lower	Br.
5789.939	5791.545		0.2				17266.55	3c 5	2a 4	Q 3
5790.050	5791.656		0.3	0.2			17266.22	W 2	2B 8	P 1
5790.050	5791.656		0.3	0.2			17266.22	4c 2	2a 4	Q 3
5790.318	5791.924			0.2			17265.42			
5790.543	5792.149		0.1				17264.75			
5790.650	5792.256			0.1			17264.43			
5790.774	5792.380		0.1				17264.06			
5790.979	5792.585		1.2	0.4	1.3		17263.45			
5791.060	5792.665		0.2				17263.21			
5791.184	5792.790		0.1	0.2			17262.84			
5791.341	5792.947		0.1				17262.37			
5791.439	5793.045			0.3			17262.08			
5791.502	5793.108		0.3				17261.89			
5791.697	5793.303	GR	2.1	2.1	2.4	1.23	17261.31	3e 0	2c 0	Q 3
5791.751	5793.357	R					17261.15			
5791.804	5793.410			1.5b	1.5b		17260.99			
5791.922	5793.528	G	2.1		0.8		17260.64	3e 0	2c 0	Q 4
5791.922	5793.528	G	2.1		0.8		17260.64	4c 2	2a 4	Q 4
5792.170	5793.776		0.1	0.1			17259.90			
5792.351	5793.958		0.2	0.3			17259.36			
5792.643	5794.250		0.2	0.5			17258.49			
5792.929	5794.535		0.3	0.4			17257.64	3F 2	2C 0	P 4
5793.080	5794.686						17257.19			
5793.194	5794.800			0.2			17256.85			
5793.486	5795.092			0.1			17255.98			
5793.540	5795.146			0.0			17255.82			
5793.627	5795.234		0.1				17255.56			
5793.869	5795.475		0.2	0.3			17254.84			
5794.084	5795.690		0.2	0.2			17254.20	2A 4	2B 0	R 1
5794.339	5795.946			0.2			17253.44	4c 2	2a 4	Q 5
5794.389	5795.996		0.2				17253.29			
5794.611	5796.218	GR	1.7	1.8	2.5	1.04	17252.63	3e 0	2c 0	Q 5
5794.658	5796.265	R					17252.49			
5794.863	5796.470	G	1.4	1.5			17251.88	3f 0	2c 0	P 4
5795.138	5796.745			0.2			17251.06			
5795.300	5796.907			0.1			17250.58	3E 3	2B 9	R 3
5795.300	5796.907			0.1			17250.58	4F 0	2B 9	Q 4
5795.471	5797.078		0.3	0.6	1.4		17250.07			
5795.756	5797.364		0.1	0.2			17249.22			
5795.992	5797.599	GR	1.3	0.9	1.5		17248.52	3e 0	2c 0	Q 2
5796.210	5797.817			0.0			17247.87			
5796.358	5797.965		0.3	0.3			17247.43			
5796.607	5798.214	G	1.4	1.5	2.0		17246.69	3f 2	2c 2	Q 5
5796.607	5798.214	G	1.4	1.5	2.0		17246.69	3E 0	2B 4	R 3
5796.792	5798.399		0.3	0.0			17246.14			
5796.939	5798.547		0.1	0.2			17245.70	2A 4	2B 0	R 0
5797.225	5798.833	G	1.2	0.8	1.4		17244.85	3f 3	2c 3	R 3
5797.417	5799.024			0.1			17244.28			
5797.477	5799.085		0.1		0.1		17244.10			
5797.612	5799.219			0.0			17243.70			
5797.733	5799.341		0.1				17243.34	3F 1	2B 6	R 2
5797.988	5799.596		0.2	0.2			17242.58			
5798.194	5799.801		0.1	0.1			17241.97			
5798.419	5800.027		0.2		0.1		17241.30			
5798.486	5800.094			0.2			17241.10	3F 0	2B 4	R 1
5798.637	5800.245			0.2	0.2		17240.65			
5798.728	5800.336		0.2				17240.38			
5798.964	5800.572	G	1.6	1.6	2.2	1.00	17239.68	3f 3	2c 3	R 3
5798.964	5800.572	G	1.6	1.6	2.2	1.00	17239.68	Z 2	2B 7	R 4
5799.145	5800.753		0.7				17239.14	Z 2	2B 7	P 2

λ_air	λ_vac	Ref.	I₁	I₂	I₃	I₄	ν	Upper	Lower	Br.
5799.145	5800.753		0.7				17239.14	4c 3	2a 5	Q 1
5799.435	5801.043		0.2	0.3	0.2		17238.28	3a 1	2c 1	R 5
5799.839	5801.447			0.1	0.1		17237.08	4c 0	2a 2	P 2
5800.017	5801.625		0.2	0.1			17236.55	4c 3	2a 5	Q 2
5800.205	5801.813	G	1.4	0.7			17235.99	V 0	2B 3	P 2
5800.205	5801.813	G	1.4	0.7			17235.99	V 0	2B 3	P 5
5800.293	5801.901		0.2				17235.73			
5800.390	5801.999				0.2		17235.44			
5800.454	5802.063		0.4	0.3			17235.25	Z 1	2B 5	P 4
5800.599	5802.207		0.4		0.2		17234.82			
5800.710	5802.318		0.1	0.2			17234.49			
5800.848	5802.457		0.3	0.5	0.2		17234.08	3D 1	2B 5	P 6
5801.158	5802.766	G	1.7	1.7	2.3	1.01	17233.16	3e 0	2c 0	Q 6
5801.158	5802.766	G	1.7	1.7	2.3	1.01	17233.16	4c 3	2a 5	Q 3
5801.158	5802.766	G	1.7	1.7	2.3	1.01	17233.16	3f 0	2c 0	P 5
5801.350	5802.958		1.5	0.8			17232.59	W 2	2B 8	P 2
5801.552	5803.160			0.1			17231.99			
5801.659	5803.268		0.2	0.2			17231.67			
5801.872	5803.480	G	1.2	0.9	1.8		17231.04	3b 3	2c 0	R 6
5801.872	5803.480	G	1.2	0.9	1.8		17231.04	3c 5	2a 4	Q 4
5802.020	5803.628		0.3	0.8			17230.60			
5802.205	5803.814			0.1			17230.05			
5802.303	5803.911		0.2				17229.76			
5802.353	5803.962			0.2			17229.61			
5802.498	5804.107	G	1.2	0.7	0.2		17229.18	3b 3	2a 0	P 2
5802.686	5804.295			0.1			17228.62			
5802.754	5804.363		0.1				17228.42			
5802.828	5804.437			0.1d			17228.20			
5803.104	5804.713		0.1	0.0			17227.38	4c 3	2a 5	Q 4
5803.330	5804.939		0.1	0.2	0.2		17226.71			
5803.411	5805.020						17226.47			
5803.606	5805.215		0.2	0.4	0.4		17225.89			
5803.714	5805.323		0.2				17225.57			
5803.818	5805.428		0.4				17225.26			
5803.930	5805.539	G	1.3	1.0d	2.0		17224.93	3f 2	2c 2	Q 6
5804.246	5805.856		0.2	0.4	0.2		17223.99			
5804.408	5806.018		0.3				17223.51			
5804.448	5806.058			0.2			17223.39			
5804.678	5806.287	G	1.3	1.5b	2.2		17222.71			
5804.971	5806.581		0.1				17221.84			
5805.048	5806.658			0.2			17221.61	3D 2	2B 7	P 4
5805.176	5806.786		0.1	0.1	0.1		17221.23			
5805.365	5806.975		0.3	0.3			17220.67	3D 1	2B 5	P 8
5805.557	5807.167		0.2	0.3			17220.10			
5805.692	5807.302		0.1				17219.70	2A 4	2B 0	R 3
5805.760	5807.369				0.2		17219.50	Z 0	2B 3	P 6
5805.868	5807.477		0.4	0.4			17219.18	3E 2	2C 0	R 2
5805.952	5807.562						17218.93			
5806.097	5807.707	GR	2.3	2.2	2.4	1.36	17218.50	3f 2	2c 2	Q 4
5806.097	5807.707	GR	2.3	2.2	2.4	1.36	17218.50	3e 1	2c 1	R 2
5806.137	5807.747	R	1.5b	1.5b			17218.38			
5806.384	5807.994		0.3	0.3			17217.65			
5806.461	5808.071				0.2		17217.42			
5806.572	5808.182		0.1				17217.09			
5806.765	5808.375		0.2	0.3	0.3		17216.52	3D 1	2B 5	P 7
5806.950	5808.560		0.3	0.4	0.3		17215.97	3E 3	2B 9	R 2
5807.149	5808.759		0.2	0.2			17215.38	3a 0	2c 0	R 3
5807.355	5808.965		0.8	0.5			17214.77	3E 2	2C 0	R 3
5807.541	5809.151	G	1.3	0.9	1.9		17214.22	3f 2	2c 2	Q 5
5807.729	5809.340	G	1.2	0.9	1.5d		17213.66	u 2	2a 2	P 4

λ_{air}	λ_{vac}	Ref.	I_1	I_2	I_3	I_4	ν	Upper	Lower	Br.
5807.729	5809.340	G	1.2	0.9	1.5d		17213.66	3f 0	2c 0	P 6
5807.922	5809.532		0.2				17213.09			
5808.003	5809.613			0.2	0.1		17212.85			
5808.141	5809.752		0.5				17212.44	4c 4	2a 6	Q 1
5808.283	5809.893	GR	1.4	1.0	1.5		17212.02	3e 0	2c 0	Q 1
5808.347	5809.957	R					17211.83			
5808.523	5810.133			0.0			17211.31	4c 6	2a 8	Q 1
5808.755	5810.366			0.0			17210.62			
5808.917	5810.528		0.2	0.2			17210.14	4c 4	2a 6	Q 2
5809.231	5810.842		0.3	0.5	1.4		17209.21			
5809.525	5811.136		0.2	0.4	1.4		17208.34			
5809.913	5811.524				0.1		17207.19	4c 4	2a 6	Q 3
5810.042	5811.652		0.2	0.2			17206.81			
5810.133	5811.744				0.2		17206.54			
5810.251	5811.862			0.1			17206.19			
5810.328	5811.940						17205.96			
5810.420	5812.031		0.1	0.1	0.2		17205.69			
5810.524	5812.135						17205.38			
5810.636	5812.247		0.2	0.2			17205.05			
5810.936	5812.548			0.2	0.2b		17204.16			
5811.041	5812.652		0.1	0.0			17203.85			
5811.207	5812.818		0.4	0.4	0.2		17203.36	3F 2	2C 0	P 5
5811.504	5813.115	G	1.9	2.0	2.4	1.16	17202.48	3e 0	2c 0	R 2
5811.700	5813.311		0.5				17201.90	W 2	2B 8	P 3
5811.906	5813.518		0.1				17201.29			
5812.014	5813.626			0.3	0.2		17200.97			
5812.075	5813.686		0.2				17200.79			
5812.305	5813.916	G	1.7				17200.11	3f 2	2c 2	Q 4
5812.467	5814.079			1.5b			17199.63			
5812.548	5814.160	R	2.6	2.6	2.9	1.64	17199.39			
5812.602	5814.214	GR					17199.23	3e 0	2c 0	R 1
5812.602	5814.214	GR					17199.23	3e 0	2c 0	R 3
5812.673	5814.285	R	1.5b				17199.02	4f 3	3b 0	P 5
5812.896	5814.508		0.3				17198.36			
5813.042	5814.653		0.5	0.3			17197.93	4c 5	2a 7	Q 1
5813.042	5814.653		0.5	0.3			17197.93	4d 0	2c 2	P 6
5813.271	5814.883			0.2			17197.25	3E 2	2C 0	R 1
5813.464	5815.076		0.1	0.2	0.3		17196.68			
5813.650	5815.262		0.2	0.3			17196.13			
5813.887	5815.499			0.2			17195.43			
5813.944	5815.556				0.3		17195.26			
5814.059	5815.671			0.2			17194.92			
5814.204	5815.816		0.2	0.1			17194.49			
5814.323	5815.935				0.2		17194.14			
5814.431	5816.043		0.2	0.3			17193.82			
5814.563	5816.175				0.2		17193.43			
5814.661	5816.273			0.3			17193.14			
5814.715	5816.327		0.3				17192.98			
5814.871	5816.483				1.5b		17192.52			
5814.949	5816.561	GR	2.2	2.2	2.5	1.36	17192.29	3f 0	2c 0	P 7
5814.949	5816.561	GR	2.2	2.2	2.5	1.36	17192.29	3f 2	2c 2	Q 3
5814.999	5816.612	R	1.5b				17192.14			
5815.243	5816.855	G	1.4	1.5	1.5b		17191.42	3e 0	2c 0	R 4
5815.507	5817.119		0.4	0.8	0.4		17190.64	3a 1	2c 1	R 4
5815.767	5817.380			0.2			17189.87	3F 0	2B 4	P 4
5815.913	5817.525		0.2	0.2			17189.44			
5816.082	5817.694		0.2				17188.94			
5816.126	5817.738			0.4	0.3		17188.81			
5816.234	5817.847		0.2				17188.49			
5816.430	5818.043						17187.91			

λ_{air}	λ_{vac}	Ref.	I_1	I_2	I_3	I_4	ν	Upper	Lower	Br.
5816.447	5818.060	G	2.1	1.9	1.9	1.14	17187.86	3f 3	2c 3	R 2
5816.698	5818.310		0.3	0.2			17187.12			
5816.762	5818.375						17186.93			
5816.847	5818.459			0.3	0.3		17186.68			
5816.921	5818.534		0.2				17186.46			
5817.070	5818.683	GR	1.7	1.6	1.7	0.95	17186.02	3f 3	2c 3	R 2
5817.114	5818.727	R					17185.89			
5817.348	5818.961		0.2	0.2	0.2		17185.20			
5817.608	5819.221	GR	1.8	1.7	1.9	1.03	17184.43	3f 2	2c 2	Q 3
5817.642	5819.255	R					17184.33			
5817.774	5819.387		0.1				17183.94			
5817.913	5819.526		0.5	0.2			17183.53	2A 4	2B 0	P 1
5818.235	5819.848		0.8	0.4			17182.58			
5818.411	5820.024		0.5	0.3	0.3		17182.06			
5818.529	5820.142		0.4				17181.71			
5818.594	5820.207			0.1			17181.52			
5818.766	5820.380		0.2	0.3	0.3		17181.01	3E 2	2C 0	R 4
5818.925	5820.539				0.2		17180.54			
5818.990	5820.603			0.3			17180.35			
5819.054	5820.667		0.1				17180.16			
5819.284	5820.898	G	1.5	1.7	2.6	0.96	17179.48	3e 0	2c 0	R 5
5819.410	5821.023				1.5b		17179.11			
5819.593	5821.206		0.7	0.9	2.0		17178.57			
5819.938	5821.552			0.1			17177.55			
5819.989	5821.603		0.1				17177.40			
5820.098	5821.711				0.2		17177.08			
5820.264	5821.877		0.2				17176.59			
5820.409	5822.023			0.2d	0.2		17176.16			
5820.731	5822.345		0.2	0.2	0.2		17175.21			
5820.840	5822.454						17174.89			
5821.002	5822.616		0.1	0.2	0.2		17174.41	4c 0	2a 2	P 3
5821.097	5822.711						17174.13			
5821.345	5822.959				0.1		17173.40	4f 3	3b 0	Q 4
5821.416	5823.030		0.2	0.2			17173.19			
5821.633	5823.247		0.2	0.1	0.2		17172.55			
5821.830	5823.444		0.3	0.2	0.1		17171.97			
5822.057	5823.671	R					17171.30	3f 2	2c 2	Q 2
5822.118	5823.732	GR	2.1	1.9	1.8	1.14	17171.12	3D 2	2B 7	P 5
5822.321	5823.935	R	0.2		0.1		17170.52			
5822.514	5824.129		0.7				17169.95			
5822.755	5824.370	GR	2.6	2.4	2.3	1.58	17169.24	3f 2	2c 2	Q 2
5822.806	5824.420	R					17169.09			
5823.050	5824.665		0.6	0.3			17168.37	Z 2	2B 7	P 3
5823.227	5824.841		0.3				17167.85			
5823.291	5824.906			0.2			17167.66			
5823.417	5825.031		0.1				17167.29			
5823.576	5825.191		0.2	0.0			17166.82			
5823.715	5825.330				0.1		17166.41			
5823.780	5825.394		0.1				17166.22			
5823.936	5825.550		0.3	0.2			17165.76	3F 0	2B 4	Q 2
5824.122	5825.737		0.1				17165.21			
5824.221	5825.835			0.2	0.2		17164.92			
5824.295	5825.910		0.2				17164.70			
5824.543	5826.158	G	1.6	1.5b	2.2	0.97	17163.97	W 2	2B 8	P 4
5824.584	5826.199		1.5d				17163.85			
5824.852	5826.467		0.1	0.1			17163.06			
5825.120	5826.735	G	0.3	0.7	1.3		17162.27	3e 0	2c 0	R 6
5825.337	5826.952		0.1	0.2			17161.63	3E 3	2B 9	R 1
5825.337	5826.952		0.1	0.2			17161.63	4f 3	3b 0	Q 3
5825.575	5827.190	G	1.3	1.0	1.4		17160.93	3a 0	2c 0	R 2

λ_{air}	λ_{vac}	Ref.	$I_1^{\cdot}$	I_2	I_3	I_4	ν	Upper	Lower	Br.
5825.853	5827.468		0.3	0.2			17160.11	3F 3	2C 1	Q 2
5825.942	5827.557				0.1		17159.85			
5826.047	5827.662			0.2			17159.54			
5826.268	5827.883			0.1			17158.89			
5826.594	5828.209						17157.93			
5826.692	5828.307		0.2				17157.64			
5826.818	5828.433				0.1		17157.27			
5826.937	5828.552		0.2				17156.92			
5827.259	5828.875		0.1	0.0			17155.97			
5827.409	5829.024						17155.53			
5827.860	5829.476		0.2	0.2	0.1d		17154.20			
5828.054	5829.670						17153.63			
5828.312	5829.928		0.1				17152.87			
5828.367	5829.982			0.1	0.2		17152.71			
5828.560	5830.176		0.3	0.2			17152.14			
5828.761	5830.377		0.3	0.5	1.4		17151.55			
5829.121	5830.737			0.2			17150.49			
5829.342	5830.958		0.2	0.4	0.2		17149.84	3E 0	2B 4	Q 7
5829.522	5831.138						17149.31	Y 2	2B 8	R 1
5829.662	5831.278			0.2	0.2		17148.90			
5829.750	5831.366		0.1				17148.64			
5829.995	5831.611		0.2	0.3			17147.92	3E 0	2B 4	P 7
5829.995	5831.611		0.2	0.3			17147.92	3E 3	2B 9	Q 4
5830.148	5831.764		0.2	0.2			17147.47			
5830.297	5831.914		0.3	0.3			17147.03	4e 3	3b 0	R 2
5830.495	5832.111		0.3	0.3			17146.45	3F 1	2B 6	P 5
5830.644	5832.261				0.2		17146.01			
5830.777	5832.393		0.2	0.2			17145.62			
5830.838	5832.455						17145.44			
5830.988	5832.604	GR	1.8	1.8	2.4	1.01	17145.00	3e 0	2c 0	R 7
5831.035	5832.652	R					17144.86	3f 1	2c 1	P 3
5831.069	5832.686	R					17144.76	W 2	2B 8	P 5
5831.284	5832.900		0.3	0.5			17144.13	3F 1	2C 6	Q 3
5831.498	5833.115	G	0.5	0.9	1.9		17143.50	u 2	2a 2	P 5
5831.736	5833.353		0.1	0.2			17142.80	Y 2	2B 8	R 0
5831.736	5833.353		0.1	0.2			17142.80	4f 3	2b 0	P 4
5831.882	5833.499		0.4	0.2			17142.37	Y 2	2B 8	R 2
5832.005	5833.622		0.5	0.3			17142.01	Y 2	2B 8	R 0
5832.046	5833.662				0.2		17141.89			
5832.226	5833.843		0.2				17141.36			
5832.287	5833.904			0.2			17141.18	3F 3	2C 1	Q 3
5832.318	5833.935			0.2			17141.09			
5832.376	5833.992		0.1				17140.92			
5832.593	5834.210		0.8	0.6			17140.28			
5832.641	5834.258						17140.14			
5832.764	5834.381	GR	2.2	2.0	2.0	1.25	17139.78	3f 1	2c 1	P 3
5832.764	5834.381	GR	2.2	2.0	2.0	1.25	17139.78	3a 1	2c 1	R 3
5832.825	5834.442	R	1.5				17139.60			
5833.070	5834.687	GR	2.0	1.8	1.7		17138.88	3e 1	2c 1	R 1
5833.179	5834.796		1.0				17138.56			
5833.288	5834.905		0.2				17138.24			
5833.339	5834.956			0.3			17138.09			
5833.383	5835.000		0.2				17137.96			
5833.608	5835.225		0.2				17137.30			
5833.747	5835.365		0.1		0.1		17136.89			
5833.832	5835.450			0.2b			17136.64			
5833.900	5835.518		0.2				17136.44			
5834.122	5835.739		0.2	0.4	1.2		17135.79			
5834.425	5836.042		0.2	0.2	0.2		17134.90			
5834.690	5836.308		0.3	0.4			17134.12			

λ_{air}	λ_{vac}	Ref.	I_1	I_2	I_3	I_4	ν	Upper	Lower	Br.
5834.837	5836.454			0.2			17133.69			
5834.942	5836.560		0.2				17133.38	6c 1	2a 5	Q 1
5835.068	5836.686		0.2	0.3			17133.01	3F 3	2C 1	Q 3
5835.259	5836.877		0.2	0.3			17132.45			
5835.416	5837.034				0.1		17131.99			
5835.470	5837.088		0.3				17131.83			
5835.545	5837.163			0.4			17131.61			
5835.610	5837.228		0.4				17131.42			
5835.825	5837.442	GR	2.7		2.4		17130.79	3f 1	2c 1	P 4
5835.865	5837.483	R			1.5		17130.67			
5835.998	5837.616	R		1.5b			17130.28	3f 3	2c 3	R 1
5836.056	5837.674	R					17130.11	2A 4	2B 0	P 2
5836.121	5837.739	GR	1.5b				17129.92	3f 3	2c 3	R 1
5836.182	5837.800	R	2.7	2.5	2.3	1.66	17129.74			
5836.332	5837.950						17129.30			
5836.421	5838.039		0.4				17129.04			
5836.557	5838.175		0.3	0.7	0.7		17128.64	3f 1	2c 1	P 7
5836.799	5838.417	G	1.4	1.0b	1.5		17127.93	3f 1	2c 1	P 6
5837.082	5838.700	G	1.3	1.0	2.0		17127.10	3f 1	2c 1	P 5
5837.245	5838.864		0.1				17126.62			
5837.409	5839.027	G	1.2	1.0	1.8		17126.14			
5837.784	5839.402	GR	1.7	1.6	1.5		17125.04	3e 0	2c 0	Q 1
5837.876	5839.495		0.7				17124.77			
5838.077	5839.696		0.1				17124.18			
5838.180	5839.798			0.2	0.2		17123.88			
5838.237	5839.856		0.2				17123.71	3F 2	2B 8	Q 4
5838.558	5840.177	G	0.5	0.8	1.8		17122.77	3e 0	2c 0	R 8
5838.725	5840.344		0.1				17122.28			
5838.865	5840.484		0.2	0.2			17121.87			
5839.063	5840.681	G	0.6	0.8	1.7		17121.29			
5839.336	5840.954		0.1	0.2	0.2		17120.49			
5839.598	5841.217		0.4	0.8	1.8		17119.72			
5839.912	5841.531	G	1.3	0.9	1.4		17118.80	3b 3	2a 0	P 3
5840.205	5841.824		0.1	0.2	0.2		17117.94			
5840.431	5842.050		0.3	0.6	1.5		17117.28			
5840.611	5842.231				0.1		17116.75	3E 3	2B 9	P 5
5840.611	5842.231				0.1		17116.75	3D 0	2B 4	R 9
5840.714	5842.333		0.2	0.3			17116.45	U 1	2B 6	R 4
5840.936	5842.555		0.3	0.3	0.2		17115.80	4f 3	3b 0	Q 2
5841.178	5842.797		0.3	0.5	1.4		17115.09			
5841.448	5843.067	G	1.4	1.0	1.4		17114.30			
5841.796	5843.415		0.2				17113.28			
5841.949	5843.569		0.3	0.2	0.1		17112.83			
5842.154	5843.774		0.2				17112.23			
5842.298	5843.917			0.2			17111.81			
5842.359	5843.979		0.2				17111.63			
5842.578	5844.197	G	1.5	1.5	1.7		17110.99	3E 0	2B 4	R 2
5842.578	5844.197	G	1.5	1.5	1.7		17110.99	3f 1	2c 1	P 4
5842.578	5844.197	G	1.5	1.5	1.7		17110.99	4c 0	2a 2	P 4
5842.861	5844.481		0.1	0.1			17110.16			
5843.062	5844.682						17109.57			
5843.189	5844.809		0.1	0.2			17109.20			
5843.452	5845.072	G	1.5	1.0	1.3		17108.43	3a 0	2c 0	R 1
5843.571	5845.191		0.4	0.3			17108.08			
5843.722	5845.342		0.1	0.2	0.2		17107.64			
5844.022	5845.642				0.1		17106.76			
5844.207	5845.827		0.3		0.1		17106.22	4c 2	2a 4	P 3
5844.381	5846.001			0.1			17105.71			
5844.429	5846.049				0.1		17105.57			
5844.736	5846.357		0.3	0.6	1.4		17104.67			

λ_{air}	λ_{vac}	Ref.	I_1	I_2	I_3	I_4	ν	Upper	Lower	Br.
5845.133	5846.753		0.2	0.3	1.3		17103.51			
5845.232	5846.852		0.3		0.2		17103.22			
5845.461	5847.081			0.1	0.2		17102.55	V 1	2B 4	R 2
5845.645	5847.266		0.3				17102.01			
5845.796	5847.416		0.4	0.5	0.3		17101.57			
5846.097	5847.717		0.1	0.2			17100.69	3F 1	2B 6	R 1
5846.233	5847.854				0.5		17100.29	Y 2	2B 8	P 1
5846.849	5848.470		0.2				17098.49			
5846.955	5848.576			0.2			17098.18	3E 3	2B 9	R 0
5847.023	5848.644		0.2				17097.98			
5847.280	5848.901	G	0.6	0.9	2.0		17097.23	3e 0	2c 0	R 9
5847.280	5848.901	G	0.6	0.9	2.0		17097.23	4f 3	3b 0	P 3
5847.533	5849.154		0.1				17096.49			
5847.622	5849.243				0.2		17096.23	4F 0	2B 9	P 4
5847.786	5849.407		0.4	0.3			17095.75	Z 2	2B 7	P 4
5848.063	5849.684				0.2		17094.94			
5848.107	5849.729		0.1				17094.81			
5848.309	5849.930			0.1			17094.22			
5848.515	5850.136			0.2			17093.62			
5848.792	5850.413		0.2		0.2		17092.81			
5848.939	5850.560		0.2				17092.38			
5848.977	5850.598			0.5	0.5		17092.27			
5849.079	5850.701		0.3				17091.97			
5849.315	5850.937	GR	2.2	2.2	2.3	1.23	17091.28	3e 0	2c 0	P 2
5849.374	5850.995	R					17091.11			
5849.617	5851.238		0.3	0.2	0.2		17090.40			
5849.822	5851.444		0.2	0.4	0.2		17089.80	3E 1	2B 6	Q 8
5850.065	5851.687			0.2			17089.09			
5850.106	5851.728		0.1				17088.97			
5850.181	5851.803		1.1				17088.75			
5850.318	5851.940		0.9	0.6			17088.35			
5850.479	5852.101		0.2	0.5	1.3		17087.88			
5850.674	5852.296		0.2				17087.31			
5850.880	5852.502	GR	1.8	1.6	1.8	0.96	17086.71	3a 1	2c 1	R 2
5850.959	5852.581	R					17086.48	3F 2	2B 8	R 2
5851.133	5852.755						17085.97	4c 1	2a 3	P 4
5851.181	5852.803		0.2				17085.83			
5851.373	5852.995		0.2				17085.27			
5851.613	5853.235	GR	1.7	1.7	2.3	0.99	17084.57	3f 1	2c 1	P 5
5851.640	5853.262	R					17084.49			
5851.763	5853.386		0.3				17084.13			
5851.911	5853.533		0.5	0.7	1.5		17083.70			
5852.233	5853.855		0.1	0.2			17082.76	3E 2	2C 0	Q 1
5852.473	5854.095		0.3		0.7		17082.06			
5852.750	5854.372		0.1	0.3			17081.25	3E 3	2B 9	Q 3
5852.911	5854.534		0.2	0.2			17080.78			
5853.168	5854.791		0.4	0.6	1.2		17080.03	3F 0	2B 4	P 3
5853.387	5855.010			0.2	0.2		17079.39			
5853.470	5855.092		0.1				17079.15			
5853.699	5855.322		0.8	0.7	1.5		17078.48	3e 0	2c 0	Q 2
5853.857	5855.480						17078.02			
5853.922	5855.545			0.2			17077.83			
5854.042	5855.665		0.2	0.1	0.1		17077.48			
5854.340	5855.963		0.2	0.1			17076.61	3E 2	2C 0	Q 2
5854.560	5856.183			0.2			17075.97			
5854.615	5856.238		0.1				17075.81			
5854.834	5856.457			0.2			17075.17			
5854.910	5856.532				0.2		17074.95			
5855.126	5856.749		0.2	0.3	0.2		17074.32			
5855.475	5857.099						17073.30			

λ_{air}	λ_{vac}	Ref.	I_1	I_2	I_3	I_4	ν	Upper	Lower	Br.
5855.801	5857.424		0.2	0.4	1.3		17072.35			
5856.017	5857.641			0.1			17071.72			
5856.165	5857.788		0.2		0.1		17071.29			
5856.323	5857.946		0.1	0.1	0.2		17070.83	4F 0	2B 9	R 1
5856.546	5858.169		0.1				17070.18			
5856.707	5858.330		0.1		0.2d		17069.71			
5856.789	5858.413			0.2			17069.47			
5856.878	5858.502		0.2				17069.21			
5857.105	5858.729	GR	1.6	1.6	1.9		17068.55	3e 0	2c 0	P 3
5857.314	5858.938		0.2	0.4			17067.94			
5857.506	5859.130		0.2	0.3	0.1		17067.38			
5857.668	5859.291		0.1	0.3	0.2		17066.91			
5857.819	5859.443			0.2			17066.47			
5857.970	5859.594			0.2	0.4		17066.03			
5858.128	5859.752			0.1			17065.57			
5858.323	5859.947			0.1	0.2		17065.00			
5858.564	5860.188		0.3	0.6	1.4		17064.30	4F 0	2B 9	Q 3
5858.825	5860.449		0.1				17063.54			
5858.883	5860.507				0.2		17063.37			
5859.051	5860.675			0.1			17062.88			
5859.141	5860.765				0.2		17062.62			
5859.268	5860.892		0.1	0.1			17062.25			
5859.398	5861.022		0.2				17061.87	2A 4	2B 0	P 3
5859.515	5861.139			0.2			17061.53			
5859.570	5861.194		0.3				17061.37			
5859.814	5861.438	GR	2.2	1.9	1.9	1.31	17060.66	3F 4	2c 4	R 3
5860.092	5861.716		0.2	0.1			17059.85			
5860.288	5861.912				0.5		17059.28			
5860.387	5862.012			0.1			17058.99			
5860.466	5862.091		0.1				17058.76			
5860.638	5862.263	G	1.2	0.9	1.7		17058.26	3f 1	2c 1	P 6
5860.721	5862.345		0.7				17058.02	Y 2	2B 8	P 2
5860.896	5862.520			0.2			17057.51			
5860.961	5862.586		0.1				17057.32			
5861.181	5862.806	G	1.5	1.7	1.8	0.97	17056.68	3e 1	2c 1	Q 4
5861.181	5862.806	G	1.5	1.7	1.8	0.97	17056.68	3a 2	2c 2	R 3
5861.391	5863.015		0.1				17056.07			
5861.593	5863.218	G	1.7	1.5b	2.3		17055.48	3e 1	2c 1	Q 5
5861.865	5863.490		0.3				17054.69			
5861.927	5863.552			0.5	1.5		17054.51			
5862.064	5863.689		0.2				17054.11	3e 2	2c 2	R 5
5862.130	5863.755			0.3	0.3		17053.92			
5862.174	5863.799		0.1				17053.79			
5862.446	5864.071		0.2	0.2	0.2		17053.00			
5862.628	5864.253		0.2	0.3			17052.47			
5862.800	5864.425		0.4	0.6	1.4		17051.97	3f 3	2c 3	Q 5
5862.800	5864.425		0.4	0.6	1.4		17051.97	3E 2	2C 0	R 2
5863.127	5864.752		0.1	0.2	0.3		17051.02			
5863.405	5865.030		0.2	0.2			17050.21			
5863.584	5865.209		0.1	0.1			17049.69			
5863.742	5865.367		0.2	0.4			17049.23			
5863.808	5865.433				1.0		17049.04			
5863.907	5865.533		0.3	0.4			17048.75			
5864.179	5865.804		0.4	0.7	2.4		17047.96			
5864.447	5866.073	GR	2.1	2.0	2.3	1.18	17047.18	3e 1	2c 1	Q 3
5864.447	5866.073	GR	2.1	2.0	2.3	1.18	17047.18	4c 0	2a 2	P 5
5864.506	5866.131	R					17047.01			
5864.664	5866.290		0.4	1.0	0.9		17046.55	3E 2	2C 0	R 1
5864.664	5866.290		0.4	1.0	0.9		17046.55	3e 1	2c 1	Q 6
5864.905	5866.531				0.3		17045.85			

λ_{air}	λ_{vac}	Ref.	I_1	I_2	I_3	I_4	ν	Upper	Lower	Br.
5864.957	5866.582		0.1	0.2			17045.70			
5865.111	5866.737		0.1	0.1			17045.25			
5865.266	5866.892		0.2	0.3	0.2		17044.80			
5865.597	5867.222		0.2d	0.0			17043.84			
5865.727	5867.353				0.3		17043.46			
5865.772	5867.398		0.2	0.1			17043.33			
5865.955	5867.581		0.1				17042.80			
5866.085	5867.711		0.2	0.2	0.3		17042.42			
5866.251	5867.877			0.0	0.2		17041.94			
5866.388	5868.014		0.2	0.2	0.2		17041.54	3E 2	2C 0	Q 3
5866.388	5868.014		0.2	0.2	0.2		17041.54	3E 2	2C 0	R 3
5866.640	5868.266		1.5				17040.81	3E 3	2B 9	Q 2
5866.660	5868.286	G	1.5	1.5	1.6		17040.75	3f 3	2c 3	Q 4
5866.726	5868.352		0.8				17040.56			
5866.956	5868.583		0.2	0.1	0.2		17039.89			
5867.263	5868.889			0.1	0.3		17039.00			
5867.456	5869.082		0.2	0.2			17038.44			
5867.542	5869.168				0.4		17038.19			
5867.759	5869.385		0.2				17037.56	3F 1	2B 6	Q 2
5867.828	5869.454			0.3	0.4		17037.36	3E 0	2B 4	P 6
5868.003	5869.630			0.2			17036.85			
5868.227	5869.854		0.5	0.9	1.8		17036.20			
5868.461	5870.088		0.5	0.4	0.3		17035.52			
5868.703	5870.329		0.1	0.2			17034.82	3f 3	2c 3	Q 5
5868.703	5870.329		0.1	0.2			17034.82	3F 1	2B 6	P 4
5868.823	5870.450				0.2		17034.47			
5868.995	5870.622				0.4		17033.97			
5869.061	5870.688	R	0.8d				17033.78			
5869.119	5870.746	R		1.5b			17033.61	3f 3	2c 3	Q 4
5869.257	5870.884	GR	2.3	2.1	2.2	1.31	17033.21	3f 3	2c 3	Q 3
5869.312	5870.939	R					17033.05	3E 3	2B 9	P 4
5869.312	5870.939	R					17033.05	X 0	2B 5	R 3
5869.609	5871.236		0.8	1.7	2.2		17032.19	3E 0	2B 4	Q 6
5869.685	5871.311				1.5		17031.97			
5869.802	5871.429	G	1.8	1.6			17031.63	3a 1	2c 1	R 1
5869.915	5871.542		0.4				17031.30			
5870.077	5871.704	G	1.7	1.0	2.2		17030.83	3f 1	2c 1	P 7
5870.326	5871.953	G	1.5	1.0	0.2		17030.11	3f 3	2c 3	Q 3
5870.560	5872.187	G	1.5b	0.9	0.3		17029.43			
5870.657	5872.284		0.4				17029.15			
5870.870	5872.498		0.2	0.4	0.3		17028.53			
5871.056	5872.684		0.2	0.3			17027.99			
5871.146	5872.773				0.4		17027.73			
5871.377	5873.005		0.3	0.3	0.2		17027.06			
5871.536	5873.163				0.2		17026.60			
5871.619	5873.246	GR	2.1	1.9	1.7		17026.36	3f 3	2c 3	Q 2
5871.681	5873.308	R					17026.18			
5871.767	5873.394						17025.93			
5871.946	5873.574	GR	2.5	2.2	2.0	1.41	17025.41	3f 3	2c 3	Q 2
5872.008	5873.636	R					17025.23			
5872.177	5873.805						17024.74			
5872.274	5873.901		0.4d	0.3	0.3		17024.46			
5872.439	5874.067		0.5				17023.98			
5872.560	5874.188	GR	1.7	1.5	1.8		17023.63	3e 1	2c 1	Q 2
5872.591	5874.219	R					17023.54	3e 0	2c 0	P 4
5872.895	5874.523		0.2	1.6	0.6		17022.66	3e 0	2c 0	Q 3
5873.116	5874.743		0.1	0.2			17022.02			
5873.164	5874.792						17021.88			
5873.288	5874.916		0.1	0.2			17021.52			
5873.426	5875.054		0.2	0.2			17021.12			

λ_{air}	λ_{vac}	Ref.	I_1	I_2	I_3	I_4	ν	Upper	Lower	Br.
5873.585	5875.213			0.2			17020.66			
5873.754	5875.382		0.2	0.3			17020.17	3E 3	2B 9	P 3
5873.754	5875.382		0.2	0.3			17020.17	X 0	2B 5	R 4
5873.871	5875.499						17019.83			
5873.989	5875.617			0.1			17019.49			
5874.213	5875.841			0.3			17018.84			
5874.261	5875.889		0.2				17018.70			
5874.458	5876.086		0.2				17018.13			
5874.600	5876.228		0.2	0.2			17017.72			
5874.765	5876.394		0.1				17017.24	Z 2	2B 7	P 5
5874.890	5876.518	G	1.2	0.8	1.3		17016.88	3E 2	2C 0	R 4
5874.890	5876.518	G	1.2	0.8	1.3		17016.88	3E 3	2B 9	P 2
5874.969	5876.597		0.5				17016.65			
5875.097	5876.725		0.3	0.2			17016.28			
5875.207	5876.836		0.2				17015.96	3E 3	2B 9	Q 1
5875.297	5876.925		1.1	0.6			17015.70			
5875.397	5877.026		0.6				17015.41	U 1	2B 6	R 3
5875.622	5877.250	i	1.5b	1.5	1.5b		17014.76			
5875.829	5877.457		0.4	0.1			17014.16			
5875.953	5877.582		1.0b				17013.80	X 0	2B 5	R 2
5876.050	5877.678	G	1.4	1.5b	2.1		17013.52	3e 2	2c 2	R 4
5876.271	5877.900		0.2				17012.88	3D 2	2C 0	P 1
5876.381	5878.010			0.2			17012.56			
5876.444	5878.072		0.2				17012.38			
5876.613	5878.242		0.4	0.5			17011.89			
5876.813	5878.442		0.3	0.7	1.4		17011.31			
5876.983	5878.611		0.2				17010.82			
5877.135	5878.763		0.2				17010.38			
5877.186	5878.815			0.2			17010.23	Y 2	2B 8	P 3
5877.366	5878.995		0.2				17009.71			
5877.484	5879.113			0.2			17009.37			
5877.570	5879.199		0.2				17009.12			
5877.632	5879.261						17008.94	3F 2	2B 8	Q 3
5877.784	5879.413		0.2	0.2			17008.50			
5877.902	5879.531		0.2				17008.16	3D 3	2B 9	R 1
5878.102	5879.731	G	1.6	1.0	1.5		17007.58	3f 2	2c 2	P 3
5878.285	5879.915		0.3				17007.05			
5878.455	5880.084	GR	2.6	2.4	2.0	1.49	17006.56			
5878.527	5880.157	R					17006.35	3f 2	2c 2	P 3
5878.586	5880.215	R					17006.18			
5878.769	5880.399		0.3		0.6		17005.65			
5878.883	5880.513		0.2	0.2			17005.32	3D 3	2B 9	R 0
5878.956	5880.585						17005.11			
5879.139	5880.769	R					17004.58			
5879.219	5880.848	G	2.1	1.9	1.8	1.17	17004.35	3e 0	2c 0	P 2
5879.219	5880.848	G	2.1	1.9	1.8	1.17	17004.35	3a 2	2c 2	R 2
5879.271	5880.900	R					17004.20			
5879.599	5881.229		0.2	0.1	0.5		17003.25	3e 0	2c 0	R12
5879.779	5881.408				0.2		17002.73			
5879.903	5881.533		0.1				17002.37			
5879.986	5881.616			0.1	0.3		17002.13			
5880.225	5881.855	G	0.5	0.9	1.6		17001.44			
5880.488	5882.118		0.2	0.5	1.4		17000.68			
5880.637	5882.266		0.1		0.6		17000.25	3D 3	2B 9	R 4
5880.699	5882.329			0.2			17000.07			
5880.827	5882.457		0.2		0.3		16999.70			
5880.913	5882.543			0.2			16999.45			
5881.090	5882.720		0.2	0.2	0.3		16998.94			
5881.180	5882.810						16998.68			
5881.259	5882.889			0.2			16998.45			

λ_{air}	λ_{vac}	Ref.	I_1	I_2	I_3	I_4	ν	Upper	Lower	Br.
5881.460	5883.090		0.4	0.3	1.0		16997.87			
5881.588	5883.218		0.3	0.2			16997.50	6c 0	2a 4	Q 5
5881.796	5883.426		0.4	0.1	0.8		16996.90			
5881.969	5883.599	R					16996.40	3D 2	2C 0	R 1
5881.969	5883.599	R					16996.40	3D 3	2B 9	R 2
5882.031	5883.661	GR	1.9	1.0	1.3	1.04	16996.22	X 0	2B 5	R 1
5882.263	5883.893		0.3	0.3	0.3		16995.55			
5882.346	5883.976		0.3				16995.31			
5882.592	5884.222			0.2	0.7		16994.60			
5882.824	5884.454			0.2	0.3		16993.93			
5882.876	5884.506		0.1				16993.78			
5883.069	5884.700				0.4		16993.22			
5883.132	5884.762		0.2	0.2			16993.04			
5883.371	5885.001	G	0.7	1.0	1.8		16992.35			
5883.620	5885.251		0.3	0.3			16991.63	3a 0	2c 0	Q 1
5883.620	5885.251		0.3	0.3			16991.63	3F 3	2C 1	P 3
5883.855	5885.486	R					16990.95			
5883.938	5885.569	GR	2.2	2.2	1.5b	1.33	16990.71	3e 1	2c 1	R 3
5884.011	5885.642	R					16990.50			
5884.223	5885.853		0.1	0.3			16989.89			
5884.333	5885.964		0.2		0.2		16989.57			
5884.417	5886.047		0.3				16989.33			
5884.579	5886.210	GR	2.4	2.3	2.3	1.36	16988.86	3e 1	2c 1	R 2
5884.659	5886.290	R					16988.63	3e 1	2c 1	Q 1
5884.735	5886.366	R					16988.41			
5884.964	5886.595		0.6	0.7	0.3		16987.75	3b 3	2a 0	P 4
5885.064	5886.695		0.4				16987.46	3a 0	2c 0	Q 2
5885.161	5886.792		0.2				16987.18			
5885.283	5886.914		0.3	0.3	0.2		16986.83	3E 0	2B 4	R 1
5885.283	5886.914		0.3	0.3	0.2		16986.83	4e 3	3b 0	R 0
5885.553	5887.184	G	1.3	1.5	2.2		16986.05	3e 1	2c 1	R 4
5885.820	5887.451		0.1	0.2	0.2		16985.28			
5886.024	5887.655		0.1	0.1	0.2		16984.69			
5886.232	5887.863		0.6	0.7			16984.09			
5886.360	5887.992				0.4		16983.72			
5886.409	5888.040		1.1	0.4			16983.58	V 1	2B 4	R 1
5886.672	5888.304			0.1			16982.82	4c 0	2a 2	P 6
5886.759	5888.390				0.2		16982.57			
5886.884	5888.515		0.3	0.3			16982.21	3E 2	2C 0	Q 4
5886.977	5888.609				0.2		16981.94			
5887.081	5888.713		0.2	0.3			16981.64			
5887.300	5888.931	R					16981.01			
5887.341	5888.973	GR	2.0	1.8	1.7	1.04	16980.89	3a 0	2c 0	Q 3
5887.341	5888.973	GR	2.0	1.8	1.7	1.04	16980.89	3f 4	2c 4	R 1
5887.404	5889.035	R					16980.71	3E 2	2C 0	R 5
5887.404	5889.035	R					16980.71	4f 0	2c 3	R 4
5887.404	5889.035	R					16980.71	3D 3	2B 9	R 3
5887.633	5889.264		0.2	0.3			16980.05			
5887.733	5889.365		0.2		0.2		16979.76			
5887.900	5889.531		0.4	1.5			16979.28	2A 4	2B 0	P 4
5888.156	5889.788	GR	2.5	2.4	2.5	1.45	16978.54	3e 1	2c 1	R 1
5888.225	5889.857	R					16978.34	6c 0	2a 4	Q 4
5888.368	5890.000		0.4		0.9		16977.93			
5888.399	5890.031			0.9			16977.84			
5888.447	5890.079		0.4				16977.70			
5888.534	5890.166				0.9		16977.45	3a 0	2c 0	Q 4
5888.624	5890.256	G	1.4	0.8b			16977.19	3e 1	2c 1	R 5
5889.002	5890.635	R					16976.10			
5889.061	5890.693	GR	2.1	2.1	2.7	1.24	16975.93	3f 2	2c 2	P 4
5889.061	5890.693	GR	2.1	2.1	2.7	1.24	16975.93	X 0	2B 5	R 0

λ_{air}	λ_{vac}	Ref.	I_1	I_2	I_3	I_4	ν	Upper	Lower	Br.
5889.360	5890.992		0.2				16975.07			
5889.509	5891.141		0.3	0.7	1.5		16974.64			
5889.603	5891.235						16974.37			
5889.787	5891.419			0.3			16973.84			
5889.981	5891.613	i			1.5b		16973.28			
5890.078	5891.710		0.7	0.9	1.7		16973.00	3F 2	2B 8	R 1
5890.439	5892.071			0.2	0.2		16971.96			
5890.515	5892.148		0.2				16971.74			
5890.803	5892.436		0.3	0.5	1.4		16970.91			
5890.998	5892.630		0.3	0.5	0.3		16970.35			
5891.338	5892.971	G	1.6	1.6	2.0		16969.37	3f 2	2c 2	P 4
5891.588	5893.221						16968.65			
5891.647	5893.280		0.2	0.3			16968.48			
5891.824	5893.457		0.6	0.7	1.5		16967.97			
5892.005	5893.638		0.1	0.1			16967.45			
5892.182	5893.815		0.2	0.5	1.5		16966.94	3e 0	2c 0	P 5
5892.401	5894.034			0.1			16966.31			
5892.741	5894.374				0.2d		16965.33			
5892.807	5894.440			0.2d			16965.14			
5893.088	5894.721		0.3	0.4	0.1		16964.33	3D 0	2B 4	R 8
5893.262	5894.895		0.3	0.4	0.8		16963.83			
5893.450	5895.083		0.3	0.2			16963.29			
5893.540	5895.173						16963.03			
5893.689	5895.323	GR	1.9	1.6	0.2		16962.60	3c 6	2a 5	Q 1
5893.689	5895.323	GR	1.9	1.6	0.2		16962.60	4e 3	3b 0	Q 2
5893.689	5895.323	GR	1.9	1.6	0.2		16962.60	3E 2	2C 0	Q 1
5894.005	5895.639	G	1.3	1.6	2.2		16961.69	3E 0	2B 4	P 5
5894.005	5895.639	G	1.3	1.6	2.2		16961.69	3e 0	2c 0	Q 4
5894.005	5895.639	G	1.3	1.6	2.2		16961.69	6c 0	2a 4	Q 3
5894.193	5895.827	G	0.5	1.0			16961.15	3e 2	2c 2	R 3
5894.357	5895.990				0.2		16960.68			
5894.496	5896.129		0.1	0.2			16960.28			
5894.735	5896.369				0.2		16959.59			
5894.888	5896.522			0.1	0.2		16959.15			
5894.996	5896.630			0.1			16958.84			
5895.156	5896.790			0.2			16958.38			
5895.337	5896.971		0.5	0.7	1.4		16957.86	3E 2	2C 0	P 2
5895.337	5896.971		0.5	0.7	1.4		16957.86	4E 2	2B13	Q 2
5895.337	5896.971		0.5	0.7	1.4		16957.86	3E 2	2C 0	P 2
5895.591	5897.224			0.1d			16957.13			
5895.733	5897.367						16956.72			
5895.910	5897.544	i			1.5		16956.21			
5896.109	5897.743		0.1	0.2			16955.64	U 1	2B 6	R 2
5896.203	5897.836				0.3		16955.37			
5896.335	5897.969						16954.99			
5896.439	5898.073		0.1	0.3	0.1		16954.69			
5896.815	5898.449		0.1				16953.61			
5896.995	5898.630		0.2	1.6			16953.09			
5897.107	5898.741		0.2				16952.77	3D 2	2C 0	Q 1
5897.107	5898.741		0.2				16952.77	3E 1	2B 6	Q 7
5897.343	5898.978			0.1d			16952.09			
5897.736	5899.371	G	1.2	1.0	1.6		16950.96	3f 2	2c 2	P 5
5897.736	5899.371	G	1.2	1.0	1.6		16950.96	3a 2	2c 2	R 1
5897.736	5899.371	G	1.2	1.0	1.6		16950.96	3F 1	2B 6	P 3
5897.980	5899.615		0.1	0.1			16950.26			
5898.140	5899.775		0.5	0.5	0.4		16949.80	6c 0	2a 4	Q 2
5898.432	5900.067		0.2	0.4	0.6		16948.96			
5898.749	5900.384		0.5	0.3			16948.05	Y 2	2B 8	P 4
5898.812	5900.446				0.1		16947.87	3b 6	2a 2	R 2
5898.954	5900.589		0.1				16947.46			

λ_{air}	λ_{vac}	Ref.	I_1	I_2	I_3	I_4	ν	Upper	Lower	Br.
5899.101	5900.736			0.2	0.3		16947.04			
5899.216	5900.850		0.1	0.1			16946.71			
5899.414	5901.049			0.0	0.1		16946.14			
5899.630	5901.265		0.1	0.2			16945.52	U 1	2B 6	R 1
5899.835	5901.470	G	1.3	0.9			16944.93	3c 6	2a 5	Q 2
5899.926	5901.561				0.2		16944.67			
5900.197	5901.832			0.2			16943.89			
5900.483	5902.118	G	1.2	1.5	2.3		16943.07			
5900.758	5902.393		0.1	0.1	0.2		16942.28			
5900.911	5902.547			0.2			16941.84			
5900.981	5902.616		0.2				16941.64			
5901.075	5902.710		0.2	0.0			16941.37			
5901.204	5902.839	R					16941.00	6c 0	2a 4	Q 1
5901.228	5902.864	GR	1.8	1.5	1.3	0.89	16940.93	3b 6	2a 2	R 1
5901.531	5903.167		0.2	0.2			16940.06			
5901.817	5903.453			0.3	0.2		16939.24			
5902.172	5903.808				0.2		16938.22	3D 2	2C 0	R 2
5902.228	5903.864		0.4	0.3			16938.06	U 1	2B 6	R 0
5902.601	5904.237		0.1	0.3	0.2b		16936.99	3E 0	2B 4	Q 5
5902.810	5904.446			0.2			16936.39			
5903.030	5904.666		0.3	0.3			16935.76			
5903.211	5904.847			0.3			16935.24			
5903.277	5904.913		0.2				16935.05			
5903.493	5905.129	GR	1.4	1.7	2.2	0.99	16934.43	3f 2	2c 2	P 5
5903.528	5905.164	R					16934.33			
5903.797	5905.433		0.1	0.2b			16933.56			
5903.884	5905.520						16933.31			
5904.250	5905.886		0.2	0.2b			16932.26			
5904.487	5906.123		0.2	0.2b			16931.58	3F 3	2B10	R 2
5904.717	5906.354		0.1	0.1b	0.3		16930.92			
5904.867	5906.504						16930.49			
5905.003	5906.640	GR	1.3	1.0b	2.0		16930.10	3a 0	2c 0	P 1
5905.003	5906.640	GR	1.3	1.0b	2.0		16930.10	3f 2	2c 2	P 6
5905.097	5906.734		0.4				16929.83			
5905.359	5906.996		0.3	0.2	0.2		16929.08	3F 0	2B 9	P 3
5905.547	5907.184		0.1	0.1			16928.54			
5905.750	5907.386		1.2	0.7			16927.96	3b 6	2a 2	R 0
5905.865	5907.502		0.2	0.2			16927.63			
5906.039	5907.676			0.1			16927.13			
5906.130	5907.767						16926.87			
5906.238	5907.875			0.2			16926.56			
5906.350	5907.987			0.1			16926.24			
5906.507	5908.144				0.3		16925.79			
5906.702	5908.339			0.2			16925.23			
5906.811	5908.447				0.2		16924.92			
5906.859	5908.496			0.1			16924.78			
5906.936	5908.573		0.1				16924.56			
5907.107	5908.744		0.3	0.2			16924.07	3D 3	2B 9	P 2
5907.345	5908.982		0.3	1.7	1.2		16923.39			
5907.400	5909.037						16923.23			
5907.648	5909.285		0.1	0.2			16922.52			
5907.938	5909.575		0.3	0.2			16921.69	3E 2	2C 0	Q 2
5908.127	5909.764		0.3	0.2			16921.15	4E 1	2B11	Q 4
5908.242	5909.879						16920.82			
5908.294	5909.931	G	0.5	0.8	1.3		16920.67			
5908.472	5910.110		0.2	0.1			16920.16	3F 3	2C 1	P 4
5908.574	5910.211			0.2			16919.87			
5908.633	5910.270		0.2				16919.70			
5908.801	5910.438		0.6				16919.22			
5908.829	5910.466			0.4	1.4		16919.14	X 0	2B 5	P 1

λ_{air}	λ_{vac}	Ref.	I_1	I_2	I_3	I_4	ν	Upper	Lower	Br.
5908.905	5910.543		0.1				16918.92			
5909.045	5910.683	G	1.3	1.0			16918.52	3c 6	2a 5	Q 3
5909.150	5910.787		0.2				16918.22			
5909.195	5910.833						16918.09			
5909.391	5911.028	G	2.1	1.8	1.6	1.04	16917.53	3a 1	2c 1	Q 1
5909.457	5911.095	R					16917.34			
5909.709	5911.346	GR	1.4	1.6	0.2		16916.62	3a 1	2c 1	Q 2
5909.765	5911.402	R					16916.46			
5909.925	5911.563		0.2	0.2			16916.00			
5910.152	5911.790	GR	2.0	1.9	2.0	1.11	16915.35	3a 1	2c 1	Q 3
5910.219	5911.857	R					16915.16			
5910.460	5912.098		0.2	0.2			16914.47			
5910.628	5912.266		0.2	0.2			16913.99			
5910.806	5912.444	G	1.3	1.0	1.7		16913.48	3a 1	2c 1	Q 4
5910.806	5912.444	G	1.3	1.0	1.7		16913.48	3F 2	2B 8	Q 2
5911.071	5912.709						16912.72			
5911.134	5912.772		0.1	0.2			16912.54			
5911.264	5912.902						16912.17			
5911.351	5912.989				0.1		16911.92			
5911.404	5913.042		0.2				16911.77			
5911.491	5913.129		0.2	0.2			16911.52	4F 0	2B 9	Q 2
5911.736	5913.374	G	1.5	1.6	2.2		16910.82	3a 1	2c 1	Q 5
5911.875	5913.514		0.3				16910.42			
5911.963	5913.601			0.2			16910.17			
5912.061	5913.699				1.0		16909.89			
5912.145	5913.783		0.2	0.2			16909.65			
5912.337	5913.975		0.2	1.7	0.5		16909.10			
5912.673	5914.311				0.1		16908.14			
5912.732	5914.371		0.2	0.2			16907.97			
5913.029	5914.668		0.5				16907.12			
5913.057	5914.696			0.9	1.6		16907.04	3a 1	2c 1	Q 6
5913.148	5914.787		0.3				16906.78			
5913.397	5915.035			0.1			16906.07			
5913.764	5915.403		0.2	0.3	0.4		16905.02	3e 0	2c 0	P 6
5913.862	5915.501		0.1				16904.74			
5914.019	5915.658		0.2	0.3			16904.29			
5914.170	5915.809			0.1			16903.86			
5914.373	5916.012		0.1	0.2			16903.28			
5914.607	5916.246		0.2	0.2	0.1		16902.61			
5914.772	5916.411		0.1				16902.14			
5914.943	5916.582	G	1.2	1.0	2.0		16901.65	3a 1	2c 1	Q 7
5915.258	5916.897		0.2	0.2	0.1		16900.75			
5915.556	5917.195		0.6				16899.90			
5915.594	5917.233	G	1.1	0.9	1.6		16899.79	3e 1	2c 1	Q 1
5915.594	5917.233	G	1.1	0.9	1.6		16899.79	3f 2	2c 2	P 6
5915.594	5917.233	G	1.1	0.9	1.6		16899.79	3F 3	2C 1	P 4
5915.685	5917.324	R	0.4				16899.53			
5915.783	5917.422			0.1			16899.25			
5915.944	5917.583						16898.79			
5916.070	5917.709	GR	2.2	2.2	2.3	1.21	16898.43	3e 0	2c 0	P 3
5916.133	5917.773	R			1.5b		16898.25			
5916.312	5917.951		0.2				16897.74	3e 0	2c 0	Q 5
5916.494	5918.133	GR	2.0	2.0	2.2	1.04	16897.22	3e 2	2c 2	R 2
5916.781	5918.420		0.2				16896.40			
5916.841	5918.480			0.2			16896.23			
5916.963	5918.603		0.2	0.1			16895.88			
5917.187	5918.827	G	1.4	0.9	0.8		16895.24	3E 2	2C 0	Q 5
5917.187	5918.827	G	1.4	0.9	0.8		16895.24	Z 2	2C 0	R 1
5917.401	5919.041	G	0.6	0.9	2.0		16894.63			
5917.548	5919.188		0.1				16894.21			

λ_{air}	λ_{vac}	Ref.	I_1	I_2	I_3	I_4	ν	Upper	Lower	Br.
5917.720	5919.359		0.5				16893.72			
5917.737	5919.377			0.7	1.5		16893.67			
5917.818	5919.457		0.4				16893.44	3A 2	2B 9	R 4
5918.066	5919.706	GR	2.4	1.9	1.4	1.18	16892.73			
5918.137	5919.776	R					16892.53			
5918.368	5920.008		0.5	0.2			16891.87	3F 2	2B 8	P 4
5918.553	5920.193		0.1				16891.34			
5918.771	5920.411		0.1	0.2			16890.72			
5918.897	5920.537		0.3	0.3			16890.36			
5919.048	5920.688		0.5				16889.93			
5919.128	5920.768		0.4	0.4			16889.70			
5919.300	5920.940		0.3	0.3			16889.21	3E 2	2C 0	P 3
5919.538	5921.179	G	1.1	0.8b	0.7		16888.53			
5919.745	5921.385			0.1			16887.94			
5919.955	5921.596		0.3	0.2			16887.34			
5920.169	5921.810		0.6	0.6	0.2		16886.73	3E 0	2B 4	P 4
5920.260	5921.901		0.4				16886.47			
5920.495	5922.136	GR	2.5	2.2	1.7	1.35	16885.80			
5920.681	5922.322	R					16885.27			
5920.751	5922.392	R					16885.07			
5920.814	5922.455	GR	2.0	2.0	1.5		16884.89	3d 0	2c 0	R 1
5920.863	5922.504	R					16884.75			
5921.004	5922.644		0.2				16884.35			
5921.120	5922.760		0.2	0.1			16884.02			
5921.253	5922.893		0.2	0.5			16883.64	3c 6	2a 5	Q 4
5921.512	5923.153		0.2	0.2			16882.90			
5921.712	5923.353		1.0	0.4	1.8		16882.33	X 0	2B 5	P 2
5921.828	5923.469		0.1				16882.00			
5921.996	5923.637	G	1.2	0.9b	2.1		16881.52			
5922.235	5923.876		0.2				16880.84	Z 2	2C 0	Q 1
5922.235	5923.876		0.2				16880.84	2A 4	2B 0	P 5
5922.382	5924.023		0.2	0.5d	0.2		16880.42			
5922.607	5924.248		0.1		0.4		16879.78			
5922.698	5924.339			0.2d			16879.52			
5922.817	5924.458		0.2				16879.18			
5922.933	5924.574				0.1		16878.85	V 1	2B 4	P 4
5923.024	5924.666		0.3	0.2d	1.9		16878.59			
5923.214	5924.855				0.2		16878.05			
5923.351	5924.992		0.3	0.3d	1.3		16877.66			
5923.449	5925.090		0.3				16877.38			
5923.589	5925.231				0.1		16876.98			
5923.635	5925.276		0.2				16876.85			
5923.768	5925.410		0.2	0.2			16876.47			
5923.968	5925.610		0.3				16875.90	3D 2	2C 0	Q 2
5924.144	5925.785		0.2				16875.40			
5924.295	5925.936		0.3	0.2			16874.97			
5924.400	5926.042		0.2				16874.67	3E 2	2C 0	R 7
5924.492	5926.133		0.2				16874.41			
5924.572	5926.214		0.9	0.4			16874.18			
5924.632	5926.274				0.2		16874.01			
5924.829	5926.470	GR	2.8	2.5	1.0	1.60	16873.45	3e 1	2c 1	P 2
5924.892	5926.534	R					16873.27			
5925.050	5926.692				0.9		16872.82			
5925.110	5926.751		0.4				16872.65			
5925.257	5926.899		0.3	0.1			16872.23	3E 0	2B 4	R 0
5925.257	5926.899		0.3	0.1			16872.23	X 2	2B 9	R 0
5925.355	5926.997				0.3		16871.95	X 1	2B 7	R 1
5925.450	5927.092		1.2	0.3			16871.68	X 1	2B 7	R 0
5925.629	5927.271						16871.17			
5925.780	5927.422	GR	2.0	1.8	2.2	1.02	16870.74	3f 3	2c 3	P 3

λ_air	λ_vac	Ref.	I₁	I₂	I₃	I₄	ν	Upper	Lower	Br.
5925.833	5927.475	R					16870.59	3a 0	2c 0	P 2
5926.072	5927.714		0.2		0.2		16869.91			
5926.209	5927.851		0.1				16869.52			
5926.371	5928.013		0.3	0.1d			16869.06			
5926.550	5928.192		0.3		0.4		16868.55			
5926.627	5928.269			0.1			16868.33			
5926.813	5928.455		0.2	0.2	1.6		16867.80			
5927.091	5928.733				0.2		16867.01	X 2	2B 9	R 1
5927.210	5928.853		0.2				16866.67	3D 3	2B 9	P 3
5927.344	5928.986			0.1			16866.29			
5927.467	5929.109		0.2		0.2		16865.94			
5927.720	5929.362		0.3	0.5	1.2		16865.22			
5927.875	5929.517		0.6	0.6	0.2		16864.78	3E 0	2B 4	Q 4
5928.138	5929.781	G	1.3	1.5	1.5		16864.03	3f 2	2c 2	P 7
5928.138	5929.781	G	1.3	1.5	1.5		16864.03	3E 2	2C 0	Q 3
5928.374	5930.016	G	1.4	0.9	2.0		16863.36	3e 1	2c 1	Q 2
5928.374	5930.016	G	1.4	0.9	2.0		16863.36	V 1	2B 4	R 0
5928.732	5930.375		0.3	0.6	1.5		16862.34	4E 1	2B11	Q 3
5928.905	5930.547		0.2	0.1			16861.85	3F 3	2B10	Q 3
5929.102	5930.744		0.6	0.2			16861.29	X 1	2B 7	R 2
5929.214	5930.857		0.2	0.5	0.7		16860.97			
5929.344	5930.987		0.3				16860.60			
5929.516	5931.159	G	2.1	1.5	1.8		16860.11	3a 1	2c 1	P 1
5929.516	5931.159	G	2.1	1.5	1.8		16860.11	3c 0	2a 0	R 6
5929.798	5931.441		0.1	0.1			16859.31			
5929.967	5931.610		0.2				16858.83			
5930.030	5931.673			0.2d			16858.65			
5930.093	5931.736		0.3				16858.47			
5930.326	5931.969			0.2d			16857.81			
5930.653	5932.296	G	0.9	1.5	1.9		16856.88	3d 0	2c 0	P 1
5930.653	5932.296	G	0.9	1.5	1.9		16856.88	3c 0	2a 0	R 7
5930.811	5932.454		0.2				16856.43			
5930.927	5932.570		0.2	0.2			16856.10			
5930.990	5932.634				0.3		16855.92			
5931.092	5932.736		0.7	0.4			16855.63			
5931.360	5933.003	GR	2.7	2.7	2.6	1.57	16854.87	3d 0	2c 0	Q 1
5931.430	5933.074	R					16854.67			
5931.670	5933.313		0.5				16853.99			
5931.744	5933.387			0.3			16853.78			
5931.849	5933.493		0.2				16853.48			
5932.036	5933.679	G	1.5	1.8	2.3	1.03	16852.95	3c 0	2a 0	R 5
5932.310	5933.954	GR	2.0	1.8	0.8	1.01	16852.17	3e 1	2c 1	P 3
5932.370	5934.014	R					16852.00			
5932.578	5934.222		0.2				16851.41			
5932.708	5934.352		0.3	0.2	0.3		16851.04			
5933.021	5934.665		0.2	0.2			16850.15			
5933.127	5934.771				1.3		16849.85			
5933.236	5934.880			0.1			16849.54			
5933.472	5935.116	·	0.3	0.2	1.3		16848.87	X 2	2B 9	P 1
5933.786	5935.430		0.8	0.3			16847.98	U 1	2B 6	P 2
5934.000	5935.645		1.1	0.2	0.2		16847.37	4E 1	2B11	Q 1
5934.254	5935.898		0.2	0.2			16846.65	3D 3	2B 9	P 5
5934.349	5935.993				0.1		16846.38			
5934.444	5936.089						16846.11			
5934.571	5936.215	·	0.8	0.6	0.2		16845.75	X 2	2B 9	R 2
5934.659	5936.303		0.5				16845.50	3D 2	2C 0	R 3
5934.832	5936.476		0.2				16845.01			
5934.994	5936.638				1.2		16844.55			
5935.089	5936.733		0.2	0.1d			16844.28			
5935.244	5936.889			0.1			16843.84			

λ_{air}	λ_{vac}	Ref.	I_1	I_2	I_3	I_4	ν	Upper	Lower	Br.
5935.325	5936.970		0.1				16843.61			
5935.484	5937.128		0.2				16843.16			
5935.614	5937.259		0.5		0.2		16842.79	3D 0	2B 4	R 7
5935.695	5937.340		0.6	0.3			16842.56			
5935.787	5937.431		0.4				16842.30	Y 1	2B 5	R 1
5935.787	5937.431		0.4				16842.30	3E 2	2C 0	P 2
5935.967	5937.611	R					16841.79	Y 1	2B 5	R 2
5936.034	5937.678	GR	2.0	2.0	1.8	1.01	16841.60	3a 2	2c 2	Q·2
5936.034	5937.678	GR	2.0	2.0	1.8	1.01	16841.60	3a 2	2c 2	Q 3
5936.034	5937.678	GR	2.0	2.0	1.8	1.01	16841.60	3a 2	2c 2	Q 4
5936.153	5937.798	R					16841.26			
5936.238	5937.883		0.3				16841.02	X 0	2B 5	P 3
5936.340	5937.985		0.3				16840.73			
5936.389	5938.034			0.5			16840.59	3c 6	2a 5	Q 5
5936.492	5938.137		0.2				16840.30			
5936.650	5938.295		0.6	0.7	1.4		16839.85			
5936.865	5938.510		0.2		0.1		16839.24			
5937.003	5938.648		0.3	0.1			16838.85			
5937.183	5938.828		0.5				16838.34			
5937.281	5938.927		0.3	0.2	0.2		16838.06			
5937.405	5939.050		0.1				16837.71			
5937.549	5939.195		0.5	0.6	1.6		16837.30	3b 3	2a 0	P 5
5937.733	5939.378		0.3	0.3	0.2		16836.78	4E 1	2B11	Q 2
5937.916	5939.561						16836.26	3E 1	2B 6	Q 6
5938.022	5939.667	G	1.8	1.7	2.0	1.00	16835.96	3c 0	2a 0	R 4
5938.216	5939.861		0.1				16835.41			
5938.347	5939.992		0.6	0.4			16835.04			
5938.417	5940.062						16834.84	Y 1	2B 5	R 0
5938.618	5940.264	GR	2.7	2.7	2.8	1.58	16834.27	3d 0	2c 0	R 2
5938.661	5940.306	R					16834.15			
5938.696	5940.341	R					16834.05			
5938.907	5940.553		0.6	0.4			16833.45	3F 2	2B 8	P 3
5939.094	5940.740		0.2				16832.92			
5939.207	5940.853		0.4	0.8	0.9		16832.60			
5939.440	5941.086	G	0.8	1.5	2.1		16831.94	3e 0	2c 0	Q 6
5939.638	5941.284		0.1				16831.38			
5939.747	5941.393				0.2		16831.07			
5939.807	5941.453		0.2				16830.90			
5939.991	5941.636		0.1		0.2		16830.38			
5940.393	5942.039		0.2				16829.24	X 2	2B 9	R 3
5940.541	5942.187		0.5	0.2			16828.82			
5940.714	5942.360		0.2				16828.33			
5940.792	5942.438				0.2		16828.11			
5940.891	5942.537		0.3				16827.83	3D 2	2C 0	P 2
5940.891	5942.537		0.3				16827.83	4e 3	3b 0	P 2
5941.145	5942.791	GR	1.7	1.6	1.6	0.90	16827.11	3f 3	2c 3	P 4
5941.343	5942.989		0.2		0.1		16826.55			
5941.530	5943.176	GR	1.4	1.0	1.6		16826.02			
5941.569	5943.215	R					16825.91	3f 3	2c 3	P 4
5941.696	5943.342		0.4				16825.55	3E 0	2B 4	P 3
5941.756	5943.402						16825.38			
5941.978	5943.625	GR	2.0	1.7	1.3	0.94	16824.75			
5942.042	5943.688						16824.57			
5942.254	5943.900	G	0.8	0.7	0.2		16823.97			
5942.551	5944.197		0.3	0.2			16823.13			
5942.639	5944.285				0.5		16822.88			
5942.741	5944.388		0.8	0.8			16822.59	V 1	2B 4	P 3
5942.943	5944.589	G	1.8	1.6	1.6	0.83	16822.02	3e 2	2c 2	R 1
5942.943	5944.589	G	1.8	1.6	1.6	0.83	16822.02	4D 0	2B 9	R 5
5942.967	5944.614	R					16821.95	3E 1	2B 6	R 2

λ_{air}	λ_{vac}	Ref.	I_1	I_2	I_3	I_4	ν	Upper	Lower	Br.
5943.028	5944.674	R					16821.78			
5943.059	5944.706		0.8				16821.69			
5943.314	5944.960		0.2	0.2			16820.97			
5943.561	5945.208	GR	2.0	1.8	2.0	1.02	16820.27	3e 1	2c 1	P 4
5943.805	5945.452		0.1		0.1		16819.58			
5943.946	5945.593		0.3	0.2			16819.18			
5944.197	5945.844						16818.47			
5944.268	5945.915			0.1			16818.27			
5944.399	5946.046		0.1		0.3		16817.90			
5944.522	5946.169		0.3	0.2	0.3		16817.55			
5944.575	5946.222						16817.40	3c 0	2a 0	R 9
5944.575	5946.222						16817.40	3e 1	2c 1	Q 3
5944.830	5946.477		0.3	0.3	0.3		16816.68			
5945.127	5946.774		0.1				16815.84	3E 2	2B 8	R 4
5945.311	5946.958		0.4	0.4	0.2		16815.32	3E 0	2B 4	Q 3
5945.551	5947.198		0.2	0.2			16814.64			
5945.770	5947.418		0.2	0.3	0.3		16814.02			
5945.926	5947.573		0.2				16813.58	3D 3	2B 9	P 4
5946.082	5947.729		0.5	0.5	0.7		16813.14			
5946.248	5947.895		0.2				16812.67	Z 2	2C 0	R 2
5946.442	5948.090		0.2				16812.12			
5946.594	5948.242		0.2		0.5		16811.69			
5946.768	5948.415		0.3				16811.20			
5946.874	5948.521		0.2				16810.90			
5947.001	5948.649		0.2	0.5			16810.54			
5947.072	5948.720		0.6				16810.34			
5947.313	5948.960	GR	2.5	2.5	2.6	1.58	16809.66	3c 0	2a 0	R 3
5947.536	5949.183		0.4				16809.03			
5947.610	5949.258		0.4	0.3			16808.82			
5947.783	5949.431		0.2	0.2	0.3		16808.33			
5948.010	5949.658	GR	2.0	1.8	1.5	1.00	16807.69	3a 0	2c 0	P 3
5948.070	5949.718	R					16807.52			
5948.236	5949.884						16807.05			
5948.307	5949.955		0.2	0.2			16806.85			
5948.495	5950.143		0.3	0.2			16806.32			
5948.693	5950.341		0.2	0.1	0.2d		16805.76			
5948.841	5950.490		0.2				16805.34			
5949.022	5950.670		0.2				16804.83			
5949.185	5950.833		0.6	0.3	0.3		16804.37			
5949.362	5951.010		0.4	0.2			16803.87			
5949.479	5951.127		0.2				16803.54			
5949.553	5951.201			0.9	0.6		16803.33			
5949.645	5951.293		0.9				16803.07			
5949.897	5951.545	GR	3.0	2.7	2.4	1.72	16802.36	3a 1	2c 1	P 2
5949.950	5951.598	R					16802.21			
5950.024	5951.672	R					16802.00			
5950.201	5951.849		0.7	0.3			16801.50			
5950.431	5952.080		0.3	0.2	0.2		16800.85			
5950.669	5952.317			0.1			16800.18			
5950.732	5952.381		0.2				16800.00			
5950.927	5952.576		0.2				16799.45			
5951.083	5952.732		0.2	0.1			16799.01			
5951.228	5952.877		0.4				16798.60			
5951.420	5953.068		0.6		1.3		16798.06	X 0	2B 5	P 4
5951.586	5953.235			0.2d			16797.59			
5951.806	5953.455		0.3	0.4	0.5		16796.97			
5951.919	5953.568						16796.65			
5952.139	5953.788		0.5	0.3			16796.03	U 1	2B 6	P 3
5952.369	5954.018		0.1				16795.38	U 1	2B 6	P 6
5952.497	5954.146			0.1			16795.02			

λ_{air}	λ_{vac}	Ref.	I_1	I_2	I_3	I_4	ν	Upper	Lower	Br.
5952.731	5954.380		0.2	0.2			16794.36			
5952.858	5954.508		0.2		0.1		16794.00			
5952.965	5954.614						16793.70			
5953.068	5954.717		0.1				16793.41			
5953.206	5954.855		0.3	0.2	0.2		16793.02			
5953.422	5955.071		0.3				16792.41	3E 2	2C 0	P 4
5953.511	5955.160						16792.16	3E 2	2C 0	Q 4
5953.635	5955.284		0.2		0.2		16791.81			
5953.897	5955.547		0.8	0.3	0.2		16791.07	X 2	2B 9	P 2
5954.163	5955.813				0.2		16790.32	3F 3	2B10	Q 2
5954.436	5956.086		0.2	0.2			16789.55			
5954.812	5956.462		0.2				16788.49			
5954.982	5956.632		0.2				16788.01			
5955.096	5956.746						16787.69			
5955.199	5956.849	G	1.2	0.9	0.3		16787.40	3E 0	2B 4	Q 2
5955.209	5956.859				1.4		16787.37			
5955.280	5956.930		0.2				16787.17			
5955.458	5957.108			0.2			16786.67	3f 3	2c 3	P 5
5955.639	5957.289						16786.16			
5955.681	5957.331		0.2				16786.04			
5955.791	5957.441		0.4	0.7d	0.5		16785.73			
5955.901	5957.551						16785.42			
5955.990	5957.640		0.7	0.2			16785.17	X 1	2B 7	P 2
5956.174	5957.825		0.7				16784.65	3e 0	2c 0	P 4
5956.228	5957.878	G	1.2	1.6	1.9		16784.50	3e 1	2c 1	P 2
5956.274	5957.924		0.7				16784.37			
5956.519	5958.169	G	1.9	1.7	1.8		16783.68	4F 0	2C 1	R 3
5956.519	5958.169	G	1.9	1.7	1.8		16783.68	Y 1	2B 5	P 1
5956.781	5958.432		0.9	0.3			16782.94			
5957.001	5958.652		0.2	0.2			16782.32			
5957.232	5958.883	G	1.3	1.5	1.7		16781.67	3f 3	2c 3	P 5
5957.495	5959.145		0.2		0.2		16780.93	3e 0	2c 0	Q 7
5957.552	5959.202			0.2			16780.77	4e 3	3b 0	P 4
5957.829	5959.479			0.2	0.1		16779.99			
5958.102	5959.753		0.1	0.2	0.3		16779.22			
5958.351	5960.001			0.1			16778.52			
5958.432	5960.083		0.2				16778.29			
5958.564	5960.214			0.2			16777.92			
5958.635	5960.285		0.2				16777.72			
5958.759	5960.410		0.4	0.3			16777.37	3E 0	2B 4	Q 1
5958.844	5960.495						16777.13			
5958.951	5960.602		0.2	0.1			16776.83			
5959.061	5960.712						16776.52			
5959.139	5960.790		0.4	0.6	0.3		16776.30	3e 1	2c 1	P 5
5959.345	5960.996		0.6	0.8	1.0		16775.72			
5959.569	5961.220	R					16775.09	Z 2	2C 0	Q 2
5959.569	5961.220	R					16775.09	3E 0	2B 4	P 2
5959.647	5961.298	GR	2.5		2.5		16774.87	3d 0	2c 0	R 3
5959.647	5961.298	GR	2.5		2.5		16774.87	3E 0	2B 4	P 2
5959.690	5961.341	R		1.5b	1.5b		16774.75			
5959.825	5961.476	GR	1.5	2.5	2.5	1.54	16774.37	3c 0	2a 0	R 2
5960.031	5961.682						16773.79			
5960.123	5961.774		0.3	0.3	0.3		16773.53			
5960.411	5962.062	G	1.7	0.8	1.4		16772.72			
5960.521	5962.172						16772.41			
5960.624	5962.276		0.8	0.9	1.0		16772.12	4E 0	2B 9	Q 4
5961.310	5962.962						16770.19			
5961.527	5963.179		0.2				16769.58			
5961.683	5963.335		0.1				16769.14			
5961.858	5963.509		0.3	0.4	0.4		16768.65			

λ_{air}	λ_{vac}	Ref.	I_1	I_2	I_3	I_4	ν	Upper	Lower	Br.
5962.171	5963.822		0.2	0.1			16767.77			
5962.476	5964.128		0.3	0.1			16766.91	3A 0	2B 5	R 2
5962.818	5964.470		0.8	0.2			16765.95	V 1	2B 4	P 2
5963.013	5964.665		0.2				16765.40			
5963.156	5964.808		0.2	0.6	1.5		16765.00			
5963.216	5964.868		0.3				16764.83			
5963.462	5965.114	GR	2.1	2.1	2.1	1.08	16764.14	3d 0	2c 0	Q 2
5963.462	5965.114	GR	2.1	2.1	2.1	1.08	16764.14	3d 0	2c 0	P 2
5963.529	5965.181	R					16763.95	3e 1	2c 1	Q 4
5963.814	5965.466		0.3	0.3	0.3		16763.15			
5964.098	5965.751		0.1				16762.35			
5964.173	5965.825			0.1	0.4		16762.14			
5964.294	5965.946						16761.80			
5964.387	5966.039		0.2				16761.54			
5964.711	5966.363		0.2	0.2			16760.63	3D 2	2C 0	Q 3
5964.885	5966.537		0.2				16760.14			
5965.095	5966.747		0.3	0.3	0.2		16759.55			
5965.227	5966.879		0.3				16759.18			
5965.373	5967.025		0.3	0.7	1.3		16758.77			
5965.586	5967.239		0.2	0.2			16758.17	X 0	2B 5	P 5
5965.743	5967.395		0.3	0.3			16757.73	3D 1	2B 6	R 8
5965.914	5967.566		0.1				16757.25			
5966.074	5967.727		0.1				16756.80	4d 3	3b 0	R 0
5966.227	5967.880			0.1			16756.37			
5966.409	5968.061		0.2		0.2		16755.86	Z 2	2C 0	P 2
5966.555	5968.207		0.6	0.2			16755.45	4f 0	2c 3	R 3
5966.797	5968.450		0.1				16754.77			
5967.000	5968.653		0.2	0.3			16754.20			
5967.267	5968.920	GR	2.0	2.0	2.5	1.18	16753.45	3e 2	2c 2	R 3
5967.324	5968.977	R					16753.29			
5967.445	5969.098		0.3				16752.95			
5967.545	5969.198	G	1.3	1.5			16752.67	3e 2	2c 2	R 4
5967.648	5969.301		0.5				16752.38	3A 1	2B 7	R 2
5967.648	5969.301		0.5				16752.38	3A 2	2B 9	R 3
5967.940	5969.593		0.1				16751.56			
5968.189	5969.843		0.1				16750.86	3F 3	2B10	P 4
5968.368	5970.021		0.3	0.3			16750.36	U 1	2B 6	P 5
5968.571	5970.224		0.1				16749.79			
5968.731	5970.385		0.2	0.2			16749.34			
5968.813	5970.466						16749.11			
5969.034	5970.687		0.2	0.3			16748.49			
5969.319	5970.973	G	1.4	1.5	2.1		16747.69	U 1	2B 6	P 4
5969.319	5970.973	G	1.4	1.5	2.1		16747.69	3e 2	2c 2	Q 5
5969.319	5970.973	G	1.4	1.5	2.1		16747.69	3b 6	2a 2	P 2
5969.479	5971.133		0.5				16747.24			
5969.683	5971.336	G	1.8	1.8	2.0		16746.67	3e 2	2c 2	R 2
5969.911	5971.565	G	0.5	0.8			16746.03	3e 2	2c 2	Q 4
5970.107	5971.761	G	0.9		0.3		16745.48	3e 2	2c 2	R 5
5970.250	5971.903			1.5b			16745.08			
5970.310	5971.964	G	2.3	2.1	2.5	1.22	16744.91	3a 1	2c 1	P 3
5970.610	5972.264		0.2	0.3			16744.07	3E 2	2C 0	P 3
5970.731	5972.385		0.2				16743.73			
5970.938	5972.592	G	2.1	2.0	2.0	1.18	16743.15	3a 0	2c 0	P 4
5970.938	5972.592	G	2.1	2.0	2.0	1.18	16743.15	3b 6	2a 2	P 2
5971.145	5972.799						16742.57			
5971.237	5972.891		0.2	0.2			16742.31			
5971.441	5973.095		0.2	0.2			16741.74			
5971.640	5973.295		0.4	0.2			16741.18	Y 1	2B 5	P 2
5971.794	5973.448		0.2	0.1			16740.75			
5972.033	5973.687		0.5	1.0	1.7		16740.08	3e 2	2c 2	Q 6

λ_{air}	λ_{vac}	Ref.	I_1	I_2	I_3	I_4	ν	Upper	Lower	Br.
5972.033	5973.687		0.5	1.0	1.7		16740.08	3E 1	2B 6	Q 5
5972.233	5973.887		0.1				16739.52			
5972.347	5974.001			0.1			16739.20			
5972.404	5974.058		0.2				16739.04			
5972.522	5974.176		0.2				16738.71			
5972.682	5974.337		0.3	0.4	1.7		16738.26			
5972.843	5974.497		0.2	0.2			16737.81			
5972.932	5974.587		0.3				16737.56			
5973.064	5974.719		0.2				16737.19			
5973.128	5974.783			0.2			16737.01			
5973.196	5974.851		0.2				16736.82	4d 3	3b 0	P 1
5973.417	5975.072		0.3	0.6	1.4		16736.20			
5973.646	5975.301		0.2	0.2			16735.56			
5973.839	5975.493			0.2			16735.02			
5973.931	5975.586		0.2				16734.76			
5974.153	5975.808	G	2.1	2.0	2.0	1.04	16734.14	3d 0	2c 0	P 2
5974.153	5975.808	G	2.1	2.0	2.0	1.04	16734.14	3e 2	2c 2	Q 3
5974.367	5976.022		0.2				16733.54			
5974.495	5976.150		0.3	0.1			16733.18	3E 1	2B 6	R 1
5974.617	5976.272		0.3	0.9			16732.84	3e 2	2c 2	R 6
5974.645	5976.300				1.5		16732.76			
5974.738	5976.393		0.2				16732.50			
5974.910	5976.565		0.6	0.4	1.8		16732.02			
5975.017	5976.672		0.3				16731.72			
5975.120	5976.775		0.8	0.9			16731.43			
5975.213	5976.868		1.0				16731.17			
5975.381	5977.036	R					16730.70			
5975.445	5977.101	GR	3.3	3.1	2.8	2.08	16730.52	3e 2	2c 2	R 1
5975.445	5977.101	GR	3.3	3.1	2.8	2.08	16730.52	3c 0	2a 0	R 1
5975.495	5977.151	R		1.5b			16730.38	4e 3	3b 0	P 6
5975.753	5977.408		0.9				16729.66			
5975.920	5977.576		0.4	0.2			16729.19			
5976.128	5977.783		0.3	0.4d	0.2d		16728.61	4f 0	2c 3	Q 5
5976.199	5977.854				1.4		16728.41			
5976.292	5977.947		0.3	0.2			16728.15			
5976.370	5978.026		0.3				16727.93	X 1	2B 7	P 3
5976.474	5978.130		0.2				16727.64			
5976.571	5978.226			0.3	0.3		16727.37			
5976.628	5978.283		0.3				16727.21			
5976.810	5978.466		0.7	0.4	0.2		16726.70			
5976.892	5978.548		0.6				16726.47			
5977.096	5978.752		0.2				16725.90	X 2	2B 9	P 3
5977.214	5978.870		0.2	0.2			16725.57			
5977.389	5979.045		0.4	0.5	0.8		16725.08			
5977.546	5979.202		0.3				16724.64			
5977.678	5979.334			0.2			16724.27			
5977.786	5979.441		0.2				16723.97			
5977.879	5979.534		0.3	0.4			16723.71	4E 0	2B 9	Q 1
5977.879	5979.534		0.3	0.4			16723.71	3e 1	2c 1	P 6
5978.036	5979.692	G	1.2	1.5	2.0		16723.27			
5978.254	5979.910		0.2				16722.66			
5978.400	5980.056		0.2	0.2			16722.25			
5978.636	5980.293		0.3		0.1d		16721.59	3D 2	2C 0	R 4
5978.783	5980.439			0.2d			16721.18			
5978.847	5980.504		0.2				16721.00	3E 2	2B 8	R 3
5979.112	5980.768	G	1.9	1.5	0.2d		16720.26			
5979.380	5981.037		0.2	0.2			16719.51			
5979.609	5981.266		0.3	0.6	1.4		16718.87			
5979.920	5981.577		0.2	0.2			16718.00			
5980.149	5981.806		0.7	0.3			16717.36			

λ_{air}	λ_{vac}	Ref.	I_1	I_2	I_3	I_4	ν	Upper	Lower	Br.
5980.418	5982.074			0.2			16716.61			
5980.475	5982.131						16716.45			
5980.571	5982.228		0.1	0.2			16716.18			
5980.865	5982.521	G	0.7	1.5	2.0		16715.36	3e 2	2c 2	R 7
5981.090	5982.747		0.1				16714.73			
5981.158	5982.815						16714.54			
5981.266	5982.922		0.2				16714.24			
5981.369	5983.026		0.7	0.3	0.3		16713.95			
5981.477	5983.134		0.2				16713.65			
5981.720	5983.377		0.2		0.2d		16712.97			
5981.849	5983.506		0.4	0.2			16712.61	3e 2	2c 2	Q 2
5982.089	5983.746		0.4	0.4			16711.94	3D 0	2B 4	R 6
5982.117	5983.774				0.9		16711.86			
5982.278	5983.936		1.5	0.8			16711.41			
5982.415	5984.072						16711.03			
5982.572	5984.229	GR	2.2	2.4	2.9	1.34	16710.59	3d 0	2c 0	R 4
5982.636	5984.294			0.4b			16710.41			
5982.869	5984.526		0.3	0.2			16709.76			
5982.307	5983.964						16711.33			
5981.606	5983.262		0.2				16713.29	4d 3	3b 0	R 1
5983.066	5984.723		0.1	0.1			16709.21			
5983.245	5984.902		0.2				16708.71			
5983.381	5985.039	G	0.8	0.3	0.2		16708.33	3E 2	2C 0	Q 5
5983.503	5985.160		0.3				16707.99			
5983.661	5985.318			0.1d			16707.55			
5983.818	5985.476						16707.11			
5983.901	5985.558			0.0	0.2		16706.88			
5984.101	5985.759			0.1			16706.32			
5984.338	5985.995		0.3	0.5	1.3		16705.66	3e 1	2c 1	Q 5
5984.520	5986.178		0.3	0.2			16705.15			
5985.043	5986.701		0.2	0.2d			16703.69			
5985.154	5986.812						16703.38			
5985.283	5986.941		0.1				16703.02	4E 0	2B 9	Q 3
5985.423	5987.081			0.3b			16702.63			
5985.563	5987.221		0.1				16702.24			
5985.739	5987.396						16701.75			
5985.807	5987.465		0.2	0.2			16701.56			
5985.943	5987.601				0.4		16701.18			
5986.104	5987.762		0.1				16700.73			
5986.161	5987.820			0.2			16700.57			
5986.294	5987.952						16700.20			
5986.463	5988.121		0.1	0.1			16699.73			
5986.595	5988.253						16699.36			
5986.703	5988.361		0.1				16699.06			
5986.921	5988.580		0.2				16698.45			
5987.058	5988.716		0.2	0.2			16698.07			
5987.291	5988.949		0.6	0.7	1.7		16697.42			
5987.438	5989.096		0.3	0.2			16697.01			
5987.624	5989.283	G	0.7	0.7	1.5		16696.49			
5987.854	5989.512		0.2	0.1			16695.85			
5988.098	5989.756		0.5	0.2	0.1d		16695.17	X 2	2B 9	P 5
5988.187	5989.846		0.2				16694.92			
5988.342	5990.000		0.1				16694.49			
5988.485	5990.144		0.2	0.1			16694.09			
5988.650	5990.309		0.5	0.7	1.5		16693.63			
5988.794	5990.452		0.1				16693.23			
5988.962	5990.621		0.5	0.3			16692.76			
5989.034	5990.693						16692.56			
5989.224	5990.883	GR	2.3	2.2	2.3	1.15	16692.03	3e 1	2c 1	P 3
5989.278	5990.937	R					16691.88	Z 2	2C 0	R 3

λ_{air}	λ_{vac}	Ref.	I_1	I_2	I_3	I_4	ν	Upper	Lower	Br.
5989.544	5991.203		0.3	0.2			16691.14			
5989.691	5991.350		0.1				16690.73			
5989.780	5991.439			0.1			16690.48			
5989.838	5991.497		0.2				16690.32			
5989.974	5991.633						16689.94			
5990.060	5991.719		0.3	0.3			16689.70			
5990.225	5991.885						16689.24			
5990.268	5991.928		0.4	0.3			16689.12			
5990.538	5992.197	G	2.2	2.1	2.2	1.28	16688.37	3a 1	2c 1	P 4
5990.538	5992.197	G	2.2	2.1	2.2	1.28	16688.37	3D 2	2C 0	P 3
5990.825	5992.484		0.3	0.1			16687.57	X 0	2B 5	P 6
5990.997	5992.657		0.3	0.2	0.1		16687.09	Y 1	2B 5	P 3
5991.116	5992.775						16686.76			
5991.274	5992.933		0.4	0.4	1.3		16686.32			
5991.446	5993.106						16685.84			
5991.529	5993.188		0.2				16685.61			
5991.590	5993.249						16685.44			
5991.780	5993.440	GR	1.9	1.7	1.8		16684.91	3e 2	2c 2	Q 1
5991.845	5993.504	R					16684.73			
5991.934	5993.594		0.8				16684.48			
5992.057	5993.716	G	1.3		1.8		16684.14			
5992.125	5993.784			1.0b			16683.95			
5992.193	5993.853		0.7				16683.76	3a 0	2c 0	P 5
5992.505	5994.165		0.2	0.2			16682.89			
5992.664	5994.323		0.3	0.1			16682.45			
5992.753	5994.413						16682.20			
5992.858	5994.517		0.2				16681.91			
5992.998	5994.658		0.2	0.1			16681.52			
5993.192	5994.852		0.3	0.2			16680.98			
5993.353	5995.013		0.3	0.2			16680.53	4e 3	3b 0	R 1
5993.522	5995.182		0.6	0.3			16680.06	4E 0	2B 9	Q 2
5993.630	5995.290		0.4				16679.76			
5993.741	5995.401		0.6				16679.45			
5993.820	5995.481		1.0	0.4			16679.23			
5994.072	5995.732	GR	2.8	2.6	2.1	1.60	16678.53	3e 3	2c 3	R 4
5994.072	5995.732	GR	2.8	2.6	2.1	1.60	16678.53	3c 0	2a 0	R 0
5994.291	5995.952		0.6				16677.92	3a 2	2c 2	P 3
5994.374	5996.034		0.8	0.4	0.3		16677.69			
5994.435	5996.095						16677.52			
5994.557	5996.218		0.4	0.2			16677.18			
5994.762	5996.423		0.2	0.1			16676.61			
5994.906	5996.566		0.3	0.1	0.0		16676.21			
5995.168	5996.829		0.2	0.2			16675.48			
5995.247	5996.908		0.3				16675.26			
5995.438	5997.099		0.3				16674.73			
5995.546	5997.206		0.4	0.2			16674.43			
5995.704	5997.365		0.2				16673.99			
5995.852	5997.512		0.2				16673.58	3E 2	2C 0	P 5
5996.042	5997.703		0.2	0.2d			16673.05			
5996.211	5997.872		0.2				16672.58			
5996.359	5998.019		0.3	0.3	0.2		16672.17			
5996.470	5998.131		0.3				16671.86			
5996.618	5998.279		0.2	0.2			16671.45			
5996.833	5998.494		0.7	0.6	1.2		16670.85			
5997.053	5998.714		0.4	0.2			16670.24	4e 3	3b 0	R 3
5997.276	5998.937		0.2	0.3	0.0		16669.62			
5997.441	5999.103			0.1			16669.16			
5997.517	5999.178						16668.95			
5997.585	5999.247			0.2			16668.76			
5997.812	5999.473	G	1.2	1.0	1.8		16668.13	3b 3	2a 0	P 6

λ_{air}	λ_{vac}	Ref.	I_1	I_2	I_3	I_4	ν	Upper	Lower	Br.
5998.096	5999.758	G	1.2	1.5	2.0		16667.34	3e 0	2c 0	P 5
5998.096	5999.758	G	1.2	1.5	2.0		16667.34	3A 2	2B 9	R 2
5998.377	6000.038		0.2	0.2			16666.56			
5998.525	6000.186			0.1			16666.15			
5998.571	6000.233		0.2				16666.02			
5998.658	6000.319			0.1			16665.78	3e 1	2c 1	P 7
5998.762	6000.424		0.2	0.3	0.3		16665.49			
5998.942	6000.604		0.2	0.2			16664.99			
5999.137	6000.798		0.5	0.7	1.2		16664.45	3E 1	2B 6	Q 4
5999.385	6001.047		0.1	0.1d			16663.76			
5999.623	6001.284		0.1	0.2			16663.10	3F 0	2B 5	R 4
5999.806	6001.468				0.2		16662.59	X 2	2B 9	P 4
5999.806	6001.468				0.2		16662.59	4d 3	3b 0	R 2
5999.867	6001.529		0.1				16662.42			
6000.019	6001.680		0.5	0.3			16662.00			
6000.116	6001.778		0.2				16661.73			
6000.213	6001.875		0.2	0.2			16661.46			
6000.390	6002.051		0.5	0.3	0.1d		16660.97	X 1	2B 7	P 4
6000.599	6002.260			0.2			16660.39			
6000.779	6002.441		0.2	0.3			16659.89			
6000.883	6002.545		0.2		0.2		16659.60	3A 0	2B 5	R 1
6001.042	6002.704		0.4				16659.16			
6001.124	6002.786		0.5	0.3			16658.93	3A 1	2B 7	R 1
6001.276	6002.938			0.1			16658.51			
6001.380	6003.042						16658.22			
6001.485	6003.147		1.1	0.8	1.6		16657.93			
6001.820	6003.482			0.2			16657.00			
6002.011	6003.673						16656.47			
6002.126	6003.788			0.2			16656.15			
6002.198	6003.860						16655.95			
6002.267	6003.929		0.2	0.3			16655.76			
6002.335	6003.997				0.3		16655.57			
6002.418	6004.080		1.5				16655.34			
6002.454	6004.116			0.4			16655.24			
6002.569	6004.232		0.2	0.4			16654.92			
6002.811	6004.473	GR	2.3	2.4	2.5	1.26	16654.25	3d 0	2c 0	Q 3
6002.872	6004.535	R					16654.08			
6003.045	6004.708						16653.60			
6003.142	6004.805		0.5	0.4			16653.33	W 1	2B 7	R 4
6003.492	6005.155		0.1				16652.36			
6003.813	6005.476						16651.47			
6003.892	6005.555						16651.25			
6003.975	6005.638		0.3				16651.02			
6004.163	6005.826			0.1d	0.1d		16650.50			
6004.430	6006.093		0.3	0.2			16649.76			
6004.513	6006.176						16649.53			
6004.653	6006.316		0.2	0.2			16649.14			
6004.916	6006.580	GR	1.7	1.6	1.6		16648.41	3e 2	2c 2	Q 1
6005.021	6006.684		0.8				16648.12			
6005.277	6006.940		0.2	0.2	0.2		16647.41			
6005.645	6007.308		0.3	0.1			16646.39			
6005.840	6007.503		0.1	0.2			16645.85	T 0	2B 9	R 2
6006.038	6007.702		0.5	0.4	0.3		16645.30	3e 1	2c 1	Q 6
6006.038	6007.702		0.5	0.4	0.3		16645.30	3D 0	2B 4	R 5
6006.118	6007.781						16645.08			
6006.223	6007.886		0.4	0.2			16644.79			
6006.410	6008.074		0.3	0.7	1.5		16644.27			
6006.609	6008.272		0.1				16643.72			
6006.832	6008.496	G	1.3	1.6	1.7		16643.10	3d 0	2c 0	R 5
6006.955	6008.619		0.3				16642.76	3E 1	2B 6	P 5

λ_air	λ_vac	Ref.	I₁	I₂	I₃	I₄	ν	Upper	Lower	Br.
6007.197	6008.861	GR	1.	2.0	2.4	0.93	16642.09	3b 6	2a 2	P 3
6007.396	6009.059		0.2				16641.54			
6007.442	6009.106			0.2			16641.41			
6007.507	6009.171		0.3				16641.23			
6007.670	6009.334		0.2				16640.78			
6007.720	6009.384			0.1	0.1		16640.64			
6007.854	6009.518		0.2				16640.27	3F 0	2B 5	Q 6
6007.948	6009.612			0.2			16640.01			
6008.204	6009.868		0.2	0.4	0.6		16639.30			
6008.399	6010.063			0.1			16638.76			
6008.446	6010.110		0.3		0.1		16638.63			
6008.544	6010.208			0.2			16638.36			
6008.789	6010.453	G	1.1	0.9	1.3		16637.68			
6009.056	6010.721			0.1			16636.94			
6009.215	6010.880		0.2	0.2			16636.50			
6009.425	6011.089			0.2	1.4		16635.92			
6009.667	6011.331			0.4	0.2d		16635.25	Z 2	2C 0	Q 3
6009.764	6011.429		1.0				16634.98			
6010.003	6011.667		0.2				16634.32			
6010.101	6011.765			0.2			16634.05			
6010.310	6011.975	G	0.4	0.1			16633.47	3E 1	2B 6	R 0
6010.451	6012.116	G	1.2	1.0	1.6		16633.08	3a 1	2c 1	P 5
6010.451	6012.116	G	1.2	1.0	1.6		16633.08	3e 3	2c 3	R 3
6010.671	6012.336		0.1				16632.47			
6010.849	6012.513		0.2	0.2			16631.98	4D 0	2B 9	R 4
6010.997	6012.661		0.2				16631.57	4e 3	3b 0	Q 1
6011.087	6012.752			0.3			16631.32	3E 2	2C 0	P 4
6011.148	6012.813		0.3				16631.15			
6011.250	6012.915		0.3				16630.87			
6011.398	6013.063	GR	2.0	1.9	1.9	0.92	16630.46	3d 1	2c 1	R 1
6011.474	6013.139						16630.25			
6011.727	6013.392		0.3	0.4	1.5		16629.55	3e 3	2c 3	R 3
6011.857	6013.522		0.2				16629.19			
6011.900	6013.565			0.2			16629.07			
6012.099	6013.764	G	1.2	1.0	1.7		16628.52	3a 0	2c 0	P 6
6012.294	6013.960		0.8	0.4			16627.98			
6012.490	6014.155				0.2		16627.44	4F 0	2C 1	R 2
6012.562	6014.227	GR	1.6	1.0	1.2		16627.24	3c 7	2a 6	Q 1
6012.645	6014.310		0.5				16627.01			
6012.757	6014.423			0.0	0.1d		16626.70			
6012.837	6014.502		0.2				16626.48	3D 1	2B 6	R 7
6012.982	6014.647		0.2	0.2			16626.08	Y 1	2B 5	P 4
6013.155	6014.821		0.2				16625.60	3a 2	2c 2	P 4
6013.437	6015.103		0.1				16624.82			
6013.506	6015.171			0.1			16624.63			
6013.665	6015.331		0.1		0.1		16624.19			
6013.864	6015.530		0.8	0.1d			16623.64			
6014.023	6015.689						16623.20			
6014.096	6015.761		0.2				16623.00			
6014.291	6015.957		0.4	0.2	0.5		16622.46			
6014.494	6016.159		0.2				16621.90			
6014.638	6016.304		0.2	0.1			16621.50			
6014.845	6016.510	G	1.3	1.0	1.4		16620.93	3e 2	2c 2	Q 2
6014.845	6016.510	G	1.3	1.0	1.4		16620.93	3E 2	2B 8	R 2
6015.069	6016.735		0.1	0.1			16620.31			
6015.217	6016.883		0.3	0.3	0.3		16619.90			
6015.485	6017.151		0.2	0.2			16619.16			
6015.666	6017.332		0.2	0.2			16618.66			
6015.771	6017.437		0.2				16618.37	3A 2	2B 9	P 5
6015.905	6017.571		0.2	0.3			16618.00			

λ_{air}	λ_{vac}	Ref.	I_1	I_2	I_3	I_4	ν	Upper	Lower	Br.
6016.017	6017.684		0.2				16617.69			
6016.159	6017.825			0.2			16617.30			
6016.231	6017.897		0.2		1.5		16617.10	3E 0	2B 5	R 6
6016.459	6018.125		0.5b	0.6			16616.47			
6016.684	6018.350	G	1.1	0.2			16615.85			
6016.886	6018.553		0.3	0.3			16615.29			
6016.959	6018.625				0.0		16615.09			
6017.031	6018.698			0.2			16614.89			
6017.089	6018.756		0.3				16614.73			
6017.198	6018.864		0.2	0.3			16614.43			
6017.357	6019.024	G	1.1	0.5			16613.99	3E 1	2B 6	P 4
6017.441	6019.107		0.4		0.2		16613.76	4F 0	2C 1	R 3
6017.585	6019.252		0.3	0.2			16613.36			
6017.745	6019.411		0.6	0.6	0.4		16612.92			
6017.850	6019.516		0.3				16612.63	3D 2	2C 0	Q 4
6017.959	6019.625		0.3	0.9			16612.33			
6018.049	6019.716		0.9				16612.08			
6018.299	6019.966	GR	3.5	3.3	2.8	2.24	16611.39	3c 0	2a 0	Q 1
6018.506	6020.172		0.7				16610.82	2K 6	2B 1	R 1
6018.600	6020.267		0.8	0.7			16610.56			
6018.716	6020.383			0.8	1.0		16610.24	3c 7	2a 6	Q 2
6018.788	6020.455		0.5				16610.04			
6018.973	6020.640		0.3	0.2			16609.53			
6019.140	6020.807		0.8	0.6	0.6		16609.07	3E 1	2B 6	Q 3
6019.357	6021.024	G	1.4	0.8			16608.47	2K 6	2B 1	R 2
6019.604	6021.271			0.5	1.5		16607.79			
6019.669	6021.336		0.3				16607.61			
6019.825	6021.492		0.2	0.3			16607.18			
6019.981	6021.648		0.3	0.2			16606.75			
6020.042	6021.709						16606.58			
6020.097	6021.764		0.3	0.3			16606.43			
6020.296	6021.963	G	1.6	0.8	1.3		16605.88	3d 1	2c 1	P 1
6020.383	6022.050		0.5		1.5		16605.64			
6020.546	6022.214		0.2				16605.19			
6020.662	6022.330		0.2	0.2			16604.87			
6020.807	6022.475		0.3	0.4	0.4		16604.47			
6020.909	6022.576			0.2			16604.19			
6020.996	6022.663		0.4	0.4			16603.95			
6021.217	6022.885	R					16603.34			
6021.253	6022.921	GR	2.6	2.5	2.5	1.43	16603.24	3d 1	2c 1	Q 1
6021.322	6022.990	R					16603.05			
6021.453	6023.120	G	1.0				16602.69			
6021.591	6023.258		0.9				16602.31			
6021.772	6023.440		0.2	0.2			16601.81			
6021.964	6023.632		0.3	0.4	0.2		16601.28			
6022.106	6023.773		0.3	0.2			16600.89			
6022.233	6023.900				1.0		16600.54			
6022.298	6023.966	G	1.7	0.9			16600.36	2K 6	2B 1	R 0
6022.487	6024.154			0.2			16599.84			
6022.588	6024.256		0.2				16599.56			
6022.672	6024.340			0.2			16599.33			
6022.806	6024.474		0.2	0.3			16598.96			
6023.042	6024.710		0.1	0.2			16598.31			
6023.209	6024.877		0.3	0.4	0.2		16597.85	4d 3	3b 0	P 2
6023.318	6024.986		0.2	0.2			16597.55			
6023.426	6025.094		0.3	0.5			16597.25			
6023.514	6025.182		0.6				16597.01			
6023.757	6025.425	GR	2.8	2.7	2.4	1.67	16596.34	3c 0	2a 0	Q 2
6023.989	6025.657		0.3				16595.70			
6024.069	6025.737		0.4	0.4			16595.48			

λ_{air}	λ_{vac}	Ref.	I_1	I_2	I_3	I_4	ν	Upper	Lower	Br.
6024.254	6025.922		0.9	0.5	0.2		16594.97			
6024.370	6026.039		0.3				16594.65			
6024.465	6026.133			0.2			16594.39			
6024.563	6026.231				0.8		16594.12			
6024.624	6026.293		0.3	0.3			16593.95			
6024.777	6026.445			0.2			16593.53			
6024.915	6026.583		0.3	0.4			16593.15			
6025.133	6026.801		0.3		0.3		16592.55			
6025.180	6026.848			0.2			16592.42			
6025.361	6027.030		0.3	0.1			16591.92			
6025.438	6027.106			0.3			16591.71			
6025.485	6027.154		0.2				16591.58			
6025.543	6027.212			0.2			16591.42			
6025.721	6027.390		1.5b				16590.93	3A 1	2B 7	R 0
6025.779	6027.448	G	1.7	1.6	1.8	0.90	16590.77	3e 1	2c 1	P 4
6025.899	6027.568		0.3				16590.44			
6026.037	6027.706		0.2				16590.06	3A 2	2B 9	R 1
6026.153	6027.822			0.2	0.2		16589.74			
6026.266	6027.935	G	0.2	0.2			16589.43			
6026.418	6028.087			0.1			16589.01			
6026.618	6028.287		0.2	0.3d			16588.46	T 0	2B 9	R 1
6026.774	6028.443		0.2				16588.03			
6026.909	6028.578			0.1			16587.66			
6026.974	6028.643		0.2	0.2	0.6		16587.48	Z 2	2C 0	P 3
6027.101	6028.771		0.3				16587.13			
6027.232	6028.901		0.3	0.3			16586.77			
6027.428	6029.098		0.3	0.3			16586.23			
6027.519	6029.188		0.2				16585.98			
6027.632	6029.301		0.7	0.9	0.6		16585.67	3D 0	2B 4	R 4
6027.734	6029.403		0.5				16585.39			
6027.977	6029.647	GR	2.7	2.6	2.7	1.54	16584.72	3d 1	2c 1	R 2
6027.977	6029.647	GR	2.7	2.6	2.7	1.54	16584.72	3c 7	2a 6	Q 3
6028.017	6029.687	R					16584.61			
6028.341	6030.010		1.5	0.5			16583.72	3E 1	2B 6	P 3
6028.461	6030.130		0.3				16583.39			
6028.650	6030.319		0.2	0.1			16582.87	3e 2	2c 2	Q 3
6028.832	6030.501		0.2	0.2			16582.37			
6029.061	6030.730				0.3		16581.74			
6029.144	6030.814		0.2b				16581.51			
6029.199	6030.868			0.2			16581.36			
6029.417	6031.087		0.3	0.6	1.4		16580.76			
6029.592	6031.261		0.1	0.1			16580.28			
6029.748	6031.418		0.3	0.3			16579.85			
6029.875	6031.545						16579.50			
6030.068	6031.738	GR	1.9	1.8	2.1	1.03	16578.97	3e 3	2c 3	R 2
6030.068	6031.738	GR	1.9	1.8	2.1	1.03	16578.97	3a 1	2c 1	P 6
6030.133	6031.803	R					16578.79			
6030.341	6032.011		0.4	0.3			16578.22			
6030.501	6032.171		0.2	0.3	0.2		16577.78			
6030.697	6032.367		0.2		0.2		16577.24			
6030.795	6032.466			0.2			16576.97			
6030.919	6032.589		0.3				16576.63			
6031.036	6032.706		0.2	0.2d	0.3		16576.31			
6031.199	6032.870		0.4	0.1			16575.86			
6031.465	6033.135	GR	2.3	2.2	2.4	1.39	16575.13	3e 2	2c 2	P 2
6031.538	6033.208						16574.93			
6031.680	6033.350		0.4				16574.54			
6031.909	6033.579	GR	2.9	2.9	1.5b	1.91	16573.91	3c 0	2a 0	Q 3
6031.909	6033.579	GR	2.9	2.9	1.5b	1.91	16573.91	3A 0	2B 5	R 0
6031.909	6033.579	GR	2.9	2.9	1.5b	1.91	16573.91	3a 2	2c 2	P 5

λ_{air}	λ_{vac}	Ref.	I_1	I_2	I_3	I_4	ν	Upper	Lower	Br.
6032.171	6033.841	G	1.3				16573.19	3d 0	2c 0	R 6
6032.342	6034.013		0.2				16572.72			
6032.480	6034.151	G	1.5	0.8			16572.34	3E 1	2B 6	Q 2
6032.586	6034.257		0.4	0.1			16572.05	3D 2	2C 0	R 5
6032.768	6034.439		0.5	0.5	0.3		16571.55	W 1	2B 7	R 3
6032.990	6034.661		0.3	0.3	0.3		16570.94	4D 0	2B 9	P 1
6033.110	6034.781		0.2				16570.61			
6033.292	6034.963		0.2	0.2	0.2		16570.11			
6033.522	6035.192		0.2	0.2	0.4		16569.48	3E 3	2C 1	R 1
6033.686	6035.356		0.3				16569.03	4D 0	2B 9	R 0
6033.747	6035.418			0.2			16568.86			
6033.966	6035.637		0.4d	0.3	0.2b		16568.26			
6034.257	6035.928		0.3d	0.4			16567.46	3b 4	2a 1	R 2
6034.454	6036.125			0.2			16566.92			
6034.643	6036.314		0.6	0.4	0.2		16566.40	3b 4	2a 1	R 1
6034.786	6036.457			0.1			16566.01			
6035.026	6036.697	G	1.2	1.0	1.6		16565.35	3c 1	2a 1	R 5
6035.132	6036.803		0.3				16565.06			
6035.372	6037.043		0.4	0.4	0.2		16564.40	3E 3	2C 1	R 2
6035.543	6037.215		0.2				16563.93			
6035.594	6037.266			0.2			16563.79			
6035.831	6037.503		0.8	0.7	1.3		16563.14			
6036.024	6037.696		0.2				16562.61	4F 0	2C 1	Q 4
6036.090	6037.761			0.0			16562.43			
6036.152	6037.823		0.1				16562.26			
6036.254	6037.925			0.2			16561.98			
6036.312	6037.984		0.3				16561.82			
6036.462	6038.133		0.4	0.5	1.5		16561.41			
6036.677	6038.348		0.2	0.2			16560.82	4D 0	2B 9	R 3
6036.859	6038.531		0.6	0.8	1.6		16560.32	3b 4	2a 1	R 3
6037.114	6038.786		0.1	0.1			16559.62			
6037.311	6038.983		0.2				16559.08			
6037.493	6039.165		0.2	0.2d	0.2		16558.58			
6037.760	6039.431		0.2				16557.85			
6037.887	6039.559		0.4	0.4	0.3		16557.50			
6038.015	6039.687		0.1				16557.15			
6038.193	6039.866		0.2	0.2	0.3		16556.66			
6038.328	6040.001		0.1				16556.29			
6038.452	6040.125		0.3	0.1			16555.95			
6038.514	6040.187				0.1b		16555.78			
6038.624	6040.296			0.1			16555.48			
6038.715	6040.387		0.1				16555.23			
6039.000	6040.672		0.1	0.1	0.1		16554.45			
6039.098	6040.770		0.2	0.2			16554.18			
6039.313	6040.986		0.1	0.2			16553.59			
6039.390	6041.062				0.2		16553.38			
6039.613	6041.285		0.1	0.2			16552.77			
6039.879	6041.551		0.7	0.3			16552.04	3A 1	2B 7	P 4
6040.083	6041.756		0.5	0.5	0.2		16551.48	3E 1	2B 6	Q 1
6040.083	6041.756		0.5	0.5	0.2		16551.48	4f 0	2c 3	Q 4
6040.211	6041.884		0.2				16551.13	3A 0	2B 5	P 4
6040.459	6042.132		0.2	0.1	0.2d		16550.45			
6040.755	6042.427		0.3	0.2			16549.64			
6040.828	6042.500						16549.44	3e 0	2c 0	P 6
6040.963	6042.636				1.5b		16549.07			
6041.029	6042.701	GR	2.0	1.8	2.0	1.05	16548.89	3D 0	2B 4	R 3
6041.029	6042.701	GR	2.0	1.8	2.0	1.05	16548.89	3e 2	2c 2	P 3
6041.185	6042.858						16548.46			
6041.310	6042.983		0.4	0.3	0.2		16548.12			
6041.496	6043.169		0.1				16547.61			

λ_{air}	λ_{vac}	Ref.	I_1	I_2	I_3	I_4	ν	Upper	Lower	Br.
6041.605	6043.278			0.2			16547.31			
6041.810	6043.483		0.2	0.2			16546.75			
6041.978	6043.651		0.7	0.3	0.2b		16546.29	3b 4	2a 1	R 0
6042.197	6043.870		0.2	0.2			16545.69			
6042.412	6044.086		0.2	0.3			16545.10			
6042.507	6044.181						16544.84			
6042.716	6044.389	GR	1.9	2.0	2.4	1.27	16544.27	3c 0	2a 0	Q 4
6042.800	6044.473		0.4				16544.04	4f 0	2c 3	R 2
6043.037	6044.710		0.9	0.6			16543.39	3E 1	2B 6	P 2
6043.183	6044.856			0.2	1.0		16542.99			
6043.355	6045.028		0.7	0.4			16542.52	T 0	2B 9	R 0
6043.490	6045.163		0.7	0.3	0.5		16542.15	3D 0	2B 4	R 0
6043.611	6045.284		1.2	0.9	1.5		16541.82			
6043.822	6045.496		0.1	0.2			16541.24			
6043.965	6045.639		0.1	0.2			16540.85			
6044.137	6045.810		0.3	0.3	0.2b		16540.38			
6044.356	6046.030	G	1.4	0.6			16539.78	2K 6	2B 1	P 1
6044.520	6046.194						16539.33			
6044.641	6046.315	G	1.8	1.6	1.6	0.87	16539.00	3e 2	2c 2	P 2
6044.747	6046.421				0.7		16538.71			
6044.948	6046.622		0.2				16538.16			
6044.985	6046.658			0.2			16538.06			
6045.208	6046.881		0.5	0.4			16537.45			
6045.368	6047.042	G	1.5	1.6	1.9	0.94	16537.01	3d 0	2c 0	Q 4
6045.368	6047.042	G	1.5	1.6	1.9	0.94	16537.01	3e 2	2c 2	Q 4
6045.463	6047.137		1.0				16536.75			
6045.690	6047.364		1.1	0.8	0.4		16536.13	3D 1	2B 6	R 6
6045.928	6047.602		0.1				16535.48			
6046.100	6047.774				0.2		16535.01			
6046.169	6047.843		0.8				16534.82			
6046.319	6047.993		0.3	0.1			16534.41			
6046.429	6048.103		0.2	0.2			16534.11	2K 5	2B 0	R 1
6046.429	6048.103		0.2	0.2			16534.11	3E 2	2B 8	R 1
6046.648	6048.323		0.2	0.2			16533.51			
6046.875	6048.549	G	1.3	0.8	1.3		16532.89	3D 0	2B 4	R 2
6046.875	6048.549	G	1.3	0.8	1.3		16532.89	3b 6	2a 2	P 4
6046.974	6048.648		0.3				16532.62			
6047.105	6048.780		0.9	0.4			16532.26	3D 0	2B 4	R 1
6047.193	6048.868		0.2				16532.02			
6047.252	6048.926			0.4	0.5d		16531.86			
6047.394	6049.069			0.2			16531.47			
6047.489	6049.164		0.1				16531.21			
6047.559	6049.234			0.3			16531.02			
6047.654	6049.329		0.3				16530.76	4D 0	2B 9	R 2
6047.654	6049.329		0.3				16530.76	2K 5	2B 0	R 0
6047.848	6049.523	GR	2.1	2.0	2.3	1.15	16530.23			
6047.892	6049.567	R					16530.11	3d 1	2c 1	R 3
6048.152	6049.826		0.2	0.2			16529.40	Z 2	2C 0	R 4
6048.353	6050.028		0.2	0.2	0.2		16528.85			
6048.477	6050.152			0.1			16528.51			
6048.664	6050.339		0.2				16528.00			
6048.715	6050.390			0.2d	0.2d		16527.86			
6048.935	6050.610		0.1	0.2			16527.26			
6049.081	6050.756						16526.86			
6049.184	6050.859		0.2	0.2			16526.58			
6049.337	6051.012		0.3	0.4	1.3		16526.16	3a 1	2c 1	P 7
6049.466	6051.141						16525.81	4D 0	2B 9	R 1
6049.608	6051.283		0.6	0.5	1.4		16525.42			
6049.806	6051.481			0.1			16524.88			
6050.018	6051.694		0.2	0.2d	0.3		16524.30			

λ_{air}	λ_{vac}	Ref.	I_1	I_2	I_3	I_4	ν	Upper	Lower	Br.
6050.205	6051.880		0.1				16523.79			
6050.370	6052.045		0.2	0.1d			16523.34			
6050.425	6052.100						16523.19			
6050.571	6052.247			0.1d			16522.79			
6050.872	6052.547		0.2	0.2			16521.97			
6051.000	6052.675				0.3		16521.62			
6051.172	6052.847		0.2	0.2			16521.15			
6051.381	6053.056		0.2	0.2	0.6		16520.58			
6051.571	6053.247		0.4				16520.06	2K 5	2B 0	R 2
6051.773	6053.448			0.2			16519.51			
6051.886	6053.562		0.1	0.1			16519.20			
6052.066	6053.741		0.2	0.3			16518.71			
6052.271	6053.947	R					16518.15	4e 3	3b 0	Q 3
6052.319	6053.994	R	1.5b				16518.02	3e 2	2c 2	P 4
6052.362	6054.038	GR	2.1	2.0	2.2	1.23	16517.90			
6052.403	6054.079	R					16517.79			
6052.535	6054.211		1.2				16517.43	W 1	2B 7	R 2
6052.788	6054.464			0.2	0.4		16516.74	3E 3	2C 1	R 2
6052.788	6054.464			0.2	0.4		16516.74	3D 0	2B 4	P 1
6052.824	6054.500		1.5	0.4			16516.64	3A 1	2B 7	P 1
6052.919	6054.596		0.3				16516.38			
6053.004	6054.680		0.2	0.2			16516.15			
6053.092	6054.768		0.2				16515.91			
6053.249	6054.925	GR	2.2	2.0	2.1	1.18	16515.48	3d 1	2c 1	Q 2
6053.315	6054.991	R					16515.30	3A 1	2B 7	P 3
6053.565	6055.241		0.3	0.2	0.3		16514.62			
6053.700	6055.376		0.2				16514.25			
6053.829	6055.505		0.2	0.2			16513.90	3E 2	2B 8	Q 5
6054.085	6055.761		0.2				16513.20			
6054.144	6055.820			0.2			16513.04			
6054.294	6055.971		0.1				16512.63			
6054.371	6056.047			0.3	0.2		16512.42			
6054.481	6056.158		0.3				16512.12			
6054.522	6056.198			0.1			16512.01			
6054.719	6056.396	GR	1.4	1.5	1.4		16511.47	3e 3	2c 3	R 1
6054.719	6056.396	GR	1.4	1.5	1.4		16511.47	3D 2	2C 0	P 4
6054.775	6056.451	R					16511.32			
6054.841	6056.517		0.9	0.8	0.6		16511.14	3E 3	2C 1	R 3
6055.028	6056.704		0.2	0.2			16510.63			
6055.127	6056.803			0.1			16510.36			
6055.266	6056.943		0.2	0.4	0.5		16509.98			
6055.559	6057.236		0.2	0.3	0.4		16509.18			
6055.831	6057.508		0.8	0.4	1.4		16508.44	3E 3	2C 1	R 3
6056.091	6057.768	G	1.9	2.1	2.8	1.33	16507.73	3c 0	2a 0	Q 5
6056.091	6057.768	G	1.9	2.1	2.8	1.33	16507.73	3E 3	2C 1	R 1
6056.106	6057.783	R					16507.69			
6056.388	6058.065		0.2	0.3			16506.92			
6056.590	6058.267		0.2	0.3			16506.37	3E 2	2C 0	P 5
6056.678	6058.355						16506.13			
6056.807	6058.484		0.2	0.1	0.2		16505.78			
6056.953	6058.631		0.2	0.2			16505.38			
6057.122	6058.799		0.2	0.2	0.3		16504.92	3A 0	2B 5	P 3
6057.343	6059.020	GR	1.5	0.8			16504.32	3A 1	2B 7	P 2
6057.343	6059.020	GR	1.5	0.8			16504.32	4F 0	2B10	R 3
6057.493	6059.170			0.2			16503.91			
6057.559	6059.236		0.2				16503.73			
6057.677	6059.354		0.3	0.3			16503.41			
6057.761	6059.438				0.2d		16503.18			
6057.842	6059.519		0.2	0.1			16502.96			
6057.996	6059.673		0.2	0.2			16502.54			

λ_{air}	λ_{vac}	Ref.	I_1	I_2	I_3	I_4	ν	Upper	Lower	Br.
6058.113	6059.791			0.1			16502.22			
6058.315	6059.993	G	0.7	1.0	2.1		16501.67	3d 0	2c 0	R 7
6058.315	6059.993	G	0.7	1.0	2.1		16501.67	3c 1	2a 1	R 3
6058.532	6060.209		0.2				16501.08			
6058.697	6060.375		0.1				16500.63			
6058.859	6060.536		0.3				16500.19			
6058.969	6060.646		0.3	0.2			16499.89	4F 0	2C 1	R 2
6058.969	6060.646		0.3	0.2			16499.89	3B 4	2A 0	R 1
6059.274	6060.951		0.2	0.1			16499.06			
6059.358	6061.036			0.1	0.5		16498.83			
6059.578	6061.256		0.1	0.1			16498.23	4f 0	2c 3	R 3
6059.769	6061.447			0.2			16497.71			
6059.843	6061.521				0.1d		16497.51			
6059.920	6061.598		0.2	0.1			16497.30			
6060.078	6061.756			0.2			16496.87			
6060.240	6061.918		0.3	0.3			16496.43	3A 2	2B 9	R 0
6060.453	6062.131		0.9	0.3			16495.85			
6060.556	6062.234		0.7	0.4			16495.57			
6060.706	6062.384			0.1			16495.16			
6060.791	6062.469		0.2				16494.93	3B 4	2A 0	R 2
6060.853	6062.531			0.2			16494.76			
6061.063	6062.741		0.2	0.3			16494.19			
6061.243	6062.921			0.0			16493.70			
6061.415	6063.094		0.2	0.3	1.2		16493.23			
6061.665	6063.344		0.2	0.3			16492.55			
6061.879	6063.557		0.1	0.2			16491.97			
6062.070	6063.748		0.2				16491.45			
6062.140	6063.818			0.2	0.3		16491.26			
6062.268	6063.947			0.2			16490.91	3E 3	2C 1	R 4
6062.375	6064.053		0.4				16490.62			
6062.459	6064.138						16490.39			
6062.625	6064.303	GR	2.0	1.5	1.2	1.08	16489.94	3c 1	2a 1	R 4
6062.625	6064.303	GR	2.0	1.5	1.2	1.08	16489.94	2K 6	2B 1	P 2
6062.625	6064.303	GR	2.0	1.5	1.2	1.08	16489.94	2K 5	2B 0	R 3
6062.834	6064.513		0.2	0.0			16489.37			
6062.930	6064.609		0.2				16489.11			
6062.982	6064.660			0.4			16488.97			
6063.037	6064.715		0.3				16488.82			
6063.257	6064.936	R					16488.22			
6063.283	6064.962	GR	2.4	2.3	2.5	1.49	16488.15	3d 1	23 1	P 2
6063.283	6064.962	GR	2.4	2.3	2.5	1.49	16488.15	3c 1	2a 1	R 3
6063.320	6064.999	R					16488.05			
6063.357	6065.035	R					16487.95	4D 0	2B 9	P 2
6063.507	6065.186						16487.54			
6063.574	6065.252		0.3	0.3			16487.36			
6063.776	6065.455		0.1	0.2			16486.81			
6063.952	6065.631		0.2	0.3	0.2		16486.33			
6064.103	6065.782		0.1				16485.92			
6064.357	6066.036		0.2	0.4	0.3b		16485.23	3e 2	2c 2	Q 5
6064.515	6066.194			0.2			16484.80			
6064.574	6066.253						16484.64			
6064.802	6066.481	G	1.5	1.6	2.1		16484.02	3e 1	2c 1	P 5
6064.802	6066.481	G	1.5	1.6	2.1		16484.02	3A 2	2B 9	P 4
6064.802	6066.481	G	1.5	1.6	2.1		16484.02	3E 3	2C 1	Q 1
6065.089	6066.768		0.2	0.2			16483.24	3A 0	2B 5	P 1
6065.244	6066.923		0.2	0.1			16482.82			
6065.453	6067.133		0.3	0.4	0.4		16482.25	3B 4	2A 0	R 0
6065.453	6067.133		0.3	0.4	0.4		16482.25	3b 3	2a 0	P 7
6065.707	6067.387		0.3	0.2	0.1		16481.56			
6065.917	6067.597		0.2	0.2	0.2		16480.99			

λ_{air}	λ_{vac}	Ref.	I_1	I_2	I_3	I_4	ν	Upper	Lower	Br.
6066.083	6067.762		0.4				16480.54			
6066.193	6067.873		0.2	0.2	0.2		16480.24	3A 0	2B 5	P 2
6066.300	6067.980		0.3				16479.95			
6066.385	6068.064		0.8	0.3			16479.72			
6066.631	6068.311	GR	2.6	2.3	3.1	1.39	16479.05	3c 0	2a 0	P 2
6066.631	6068.311	GR	2.6	2.3	3.1	1.39	16479.05	3E 2	2B 8	P 5
6066.782	6068.462						16478.64	4F 0	2C 1	R 1
6066.863	6068.543		0.4				16478.42	3e 2	2c 2	P 5
6066.948	6068.628		0.7				16478.19			
6067.158	6068.838		0.3	0.3	0.6		16477.62			
6067.272	6068.952		0.2		0.6		16477.31			
6067.401	6069.081		0.2	0.4			16476.96			
6067.489	6069.169		0.4				16476.72			
6067.673	6069.353	GR	2.5	2.4	2.5	1.51	16476.22	3e 3	2c 3	R 3
6067.736	6069.416	R					16476.05	3c 1	2a 1	R 2
6067.968	6069.648		0.3				16475.42			
6068.053	6069.733		0.4	0.3			16475.19			
6068.307	6069.987	G	0.7b	1.5	2.1		16474.50			
6068.487	6070.168						16474.01			
6068.579	6070.260		0.9	0.3			16473.76			
6068.826	6070.507		0.4	0.6	0.6		16473.09	3D 1	2B 6	R 5
6069.062	6070.742	G	1.2	0.6			16472.45	W 1	2B 7	R 1
6069.154	6070.834		0.2				16472.20			
6069.287	6070.967		0.2	0.3			16471.84			
6069.434	6071.115		0.2	0.3			16471.44			
6069.707	6071.387		0.4	0.5			16470.70	2K 5	2B 0	P 1
6069.969	6071.649	GR	2.3	2.4	2.8	1.35	16469.99	3d 1	2c 1	R 4
6070.005	6071.686	R					16469.89			
6070.212	6071.892		0.2				16469.33			
6070.300	6071.981			0.4			16469.09			
6070.352	6072.033		0.4				16468.95	3B 4	2A 0	R 3
6070.496	6072.176	G	0.9	0.7	1.3		16468.56			
6070.698	6072.379			0.2			16468.01	T 0	2B 9	P 1
6070.846	6072.527		0.1	0.2	0.1		16467.61			
6071.026	6072.707		0.2	0.2			16467.12			
6071.159	6072.840		0.2	0.3			16466.76			
6071.355	6073.036		0.2	0.2			16466.23			
6071.568	6073.249	GR	1.8	1.6	1.7		16465.65	3e 3	2c 3	R 2
6071.631	6073.312	R					16465.48			
6071.727	6073.408		0.2				16465.22			
6071.849	6073.530		0.2				16464.89			
6071.996	6073.677	G	0.9	1.5	2.1		16464.49	3c 0	2a 0	Q 6
6072.111	6073.792		0.8				16464.18			
6072.159	6073.840			0.6			16464.05			
6072.435	6074.116		0.1	0.1d			16463.30			
6072.649	6074.330		0.3	0.2			16462.72	T 0	2B 9	P 4
6072.811	6074.493		0.2	0.2	0.2		16462.28			
6073.011	6074.692		0.2	0.2			16461.74			
6073.236	6074.917		0.3	0.3			16461.13	T 0	2B 9	P 2
6073.590	6075.271		0.3	0.2			16460.17			
6073.811	6075.493		0.3	0.3			16459.57			
6073.955	6075.637		0.2				16459.18	4F 0	2C 1	Q 3
6074.118	6075.799		0.4	0.3			16458.74	Z 2	2C 0	Q 4
6074.206	6075.888						16458.50			
6074.376	6076.058	GR	2.2	2.1	2.1	1.32	16458.04	3e 2	2c 2	P 3
6074.376	6076.058	GR	2.2	2.1	2.1	1.32	16458.04	3E 3	2C 1	R 5
6074.435	6076.117	R					16457.88			
6074.501	6076.183	R					16457.70			
6074.697	6076.379		0.8	0.4			16457.17			
6074.863	6076.545		0.2	0.4	0.8		16456.72			

λ_{air}	λ_{vac}	Ref.	I_1	I_2	I_3	I_4	ν	Upper	Lower	Br.
6075.081	6076.763		0.2	0.2			16456.13			
6075.254	6076.936		0.2	0.2	0.1		16455.66	3E 3	2C 1	Q 2
6075.450	6077.132		0.2	0.2			16455.13			
6075.598	6077.280		0.2	0.1			16454.73	X 1	2C 0	R 1
6075.771	6077.454		0.1	0.1			16454.26			
6076.015	6077.697			0.3	0.2b		16453.60			
6076.122	6077.804		0.2b				16453.31			
6076.189	6077.871			0.2			16453.13			
6076.399	6078.081		0.2				16452.56			
6076.595	6078.277		0.3	0.2			16452.03			
6076.735	6078.418		0.3				16451.65			
6076.868	6078.551			0.2			16451.29			
6076.990	6078.673		0.2b				16450.96			
6077.171	6078.854		0.1				16450.47			
6077.411	6079.094		0.3b	0.4	1.2		16449.82	4d 3	3b 0	P 3
6077.596	6079.279			0.2			16449.32			
6077.651	6079.334		0.3				16449.17			
6077.773	6079.456			0.2	0.2		16448.84	3F 0	2B 5	R 3
6078.014	6079.696		0.4	0.7	1.6		16448.19	3A 2	2B 9	P 3
6078.228	6079.911		0.3	0.0			16447.61			
6078.357	6080.040		0.2	0.2			16447.26	T 0	2B 9	P 3
6078.620	6080.303		0.3	0.3			16446.55	2K 5	2B 0	R 4
6078.712	6080.395						16446.30			
6078.890	6080.572	GR	2.0	1.9	1.9	1.16	16445.82	3e 3	2c 3	R 1
6078.938	6080.620	R					16445.69			
6079.207	6080.890		0.4	0.2			16444.96	3D 0	2B 4	P 2
6079.366	6081.049		0.4	0.2			16444.53	3E 2	2B 8	R 0
6079.599	6081.282		0.4	0.3	0.2b		16443.90			
6079.695	6081.379		0.2				16443.64			
6079.899	6081.582		0.4	0.3			16443.09			
6080.065	6081.748		0.3	0.2			16442.64			
6080.220	6081.904		0.6	0.5			16442.22	3F 0	2B 5	Q 5
6080.328	6082.011		0.4	0.2			16441.93			
6080.439	6082.122		0.4	0.6			16441.63			
6080.542	6082.226		1.0				16441.35			
6080.783	6082.466	GR	3.3	3.1	2.6		16440.70	3c 1	2a 1	R 1
6080.783	6082.466	GR	3.3	3.1	2.6		16440.70	3b 4	2a 1	P 1
6080.897	6082.581	R					16440.39	3D 1	2B 6	R 0
6081.008	6082.692		0.4				16440.09			
6081.071	6082.755			0.7			16439.92			
6081.104	6082.788		0.8				16439.83			
6081.197	6082.880			0.2			16439.58			
6081.289	6082.973		0.4	0.2			16439.33			
6081.478	6083.162		1.0	0.2	0.8		16438.82			
6081.626	6083.310		0.3	0.2			16438.42			
6081.833	6083.517		0.2	0.3			16437.86			
6081.907	6083.591				0.1		16437.66			
6081.996	6083.680		0.3	0.1			16437.42			
6082.107	6083.791			0.2			16437.12			
6082.170	6083.854		0.3				16436.95			
6082.296	6083.980		0.2	0.3	0.1b		16436.61			
6082.425	6084.109		0.2				16436.26			
6082.477	6084.161			0.2			16436.12	3D 2	2C 0	Q 5
6082.614	6084.298		0.3	0.2			16435.75			
6082.773	6084.457		0.3	0.2			16435.32			
6082.936	6084.620		0.3	0.2			16434.88			
6083.054	6084.739		0.2				16434.56			
6083.121	6084.805				1.0		16434.38			
6083.195	6084.879		0.2	0.2			16434.18			
6083.432	6085.116		0.4	0.6	1.3		16433.54			

λ_{air}	λ_{vac}	Ref.	I_1	I_2	I_3	I_4	ν	Upper	Lower	Br.
6083.639	6085.323		0.2	0.2			16432.98			
6083.780	6085.464		0.4	0.3	0.2		16432.60			
6083.921	6085.605		0.2	0.2			16432.22			
6084.002	6085.686				1.0		16432.00			
6084.058	6085.742	G	1.3				16431.85			
6084.169	6085.853			1.5b			16431.55	3A 2	2B 9	P 1
6084.258	6085.942	G	1.6		2.0		16431.31	3e 2	2c 2	P 6
6084.258	6085.942	G	1.6		2.0		16431.31	3e 0	2c 0	P 7
6084.458	6086.142	GR	1.4	0.9			16430.77	W 1	2B 7	R 0
6084.554	6086.238		0.4				16430.51			
6084.702	6086.386	G	1.2	1.5	1.5		16430.11	3D 1	2B 6	R 4
6084.883	6086.568		0.4	0.2d			16429.62			
6085.202	6086.887	G	1.4	1.0	2.4		16428.76	3e 3	2c 3	Q 4
6085.202	6086.887	G	1.4	1.0	2.4		16428.76	3d 0	2c 0	R 8
6085.202	6086.887	G	1.4	1.0	2.4		16428.76	3e 2	2c 2	Q 6
6085.535	6087.220	GR	1.0	0.7			16427.86	2K 6	2B 1	P 3
6085.680	6087.365		0.7	0.2			16427.47	4D 0	2B 9	P 5
6085.906	6087.591	G	1.2	0.5	0.2b		16426.86	3e 3	2c 3	Q 5
6085.906	6087.591	G	1.2	0.5	0.2b		16426.86	X 1	2C 0	Q 1
6085.906	6087.591	G	1.2	0.5	0.2b		16426.86	3E 2	2B 8	P 4
6085.906	6087.591	G	1.2	0.5	0.2b		16426.86	3E 2	2B 8	Q 4
6086.058	6087.743		0.5	0.5			16426.45			
6086.173	6087.857			0.0			16426.14			
6086.343	6088.028	GR	1.4	0.7	0.2		16425.68	3D 1	2B 6	R 1
6086.588	6088.273		0.3	0.4	1.5b		16425.02	3e 3	2c 3	Q 4
6086.814	6088.499	G	1.1	0.7	1.3		16424.41	3e 3	2c 3	Q 3
6086.980	6088.666		0.5	0.3			16423.96	3D 1	2B 6	P 1
6086.980	6088.666		0.5	0.3			16423.96	3E 3	2C 1	Q 1
6087.155	6088.840		0.2	0.2			16423.49			
6087.284	6088.970		0.3				16423.14			
6087.392	6089.077		0.3	0.3	0.3b		16422.85			
6087.573	6089.259		0.1	0.1			16422.36			
6087.737	6089.422		0.1	0.2	0.2b		16421.92	3E 3	2C 1	R 6
6087.959	6089.644		0.6	0.3			16421.32			
6088.219	6089.904		0.2				16420.62	4f 1	2c 4	R 1
6088.452	6090.138		0.2				16419.99			
6088.671	6090.357			0.1d			16419.40			
6088.764	6090.449		0.1				16419.15			
6088.979	6090.664		0.7	0.3d			16418.57			
6089.309	6090.995		0.1	0.2			16417.68	4e 0	2c 3	R 5
6089.554	6091.239		0.3	0.5			16417.02	4e 3	3b 0	P 3
6089.813	6091.499	G	1.6	1.7	2.2		16416.32	3d 0	2c 0	P 4
6089.813	6091.499	G	1.6	1.7	2.2		16416.32	3d 0	2c 0	Q 5
6089.813	6091.499	G	1.6	1.7	2.2		16416.32	3D 3	2C 1	R 1
6089.988	6091.674						16415.85			
6090.106	6091.792						16415.53			
6090.188	6091.874		0.9	0.2			16415.31	3A 2	2B 9	P 2
6090.374	6092.060	G	1.3	1.7	2.4		16414.81	3c 0	2a 0	Q 7
6090.374	6092.060	G	1.3	1.7	2.4		16414.81	2K 5	2B 0	P 2
6090.626	6092.312		0.4	0.3			16414.13	W 2	2B 9	R 4
6090.886	6092.572	GR	2.5	2.4	2.4	1.42	16413.43	3d 1	2c 1	Q 3
6090.952	6092.638	R	1.5b	1.5b			16413.25			
6091.027	6092.713						16413.05	3D 1	2B 6	R 2
6091.171	6092.858		0.2		0.3		16412.66			
6091.257	6092.943		0.2	0.1			16412.43			
6091.390	6093.077		0.3	0.2			16412.07			
6091.576	6093.262		0.2	0.2			16411.57			
6091.732	6093.418		0.2	0.2			16411.15			
6091.906	6093.593		0.3	0.2			16410.68			
6092.125	6093.812	GR	1.3	0.9	0.3		16410.09	3D 1	2B 6	R 3

λ_{air}	λ_{vac}	Ref.	I$_1$	I$_2$	I$_3$	I$_4$	ν	Upper	Lower	Br.
6092.248	6093.934						16409.76			
6092.385	6094.072		0.2				16409.39			
6092.433	6094.120			0.2	0.1		16409.26	4f 1	2c 4	R 2
6092.556	6094.243		0.2	0.1			16408.93			
6092.697	6094.384		0.2				16408.55			
6092.753	6094.439			0.1			16408.40			
6092.924	6094.610		0.3	0.3			16407.94			
6093.080	6094.766		0.3	0.1			16407.52	3e 3	2c 3	Q 2
6093.284	6094.971			0.2			16406.97			
6093.380	6095.067		0.2				16406.71			
6093.544	6095.231		0.2	0.3			16406.27			
6093.637	6095.323		0.2				16406.02			
6093.841	6095.528	GR	1.4	1.5	2.3		16405.47	3d 1	2c 1	R 5
6094.019	6095.706						16404.99			
6094.127	6095.814		0.3	0.3			16404.70			
6094.346	6096.033		0.2	0.2d			16404.11			
6094.513	6096.200		0.4	0.4	0.3		16403.66			
6094.606	6096.293		0.4		0.4		16403.41			
6094.699	6096.386		0.2	0.1			16403.16			
6094.851	6096.539		0.3	0.2			16402.75			
6095.056	6096.743		0.3	0.2			16402.20	4g 1	2a 4	P 1
6095.223	6096.910		0.2	0.1			16401.75			
6095.390	6097.078		0.5	0.4	0.2		16401.30			
6095.498	6097.185		0.3	0.3			16401.01			
6095.613	6097.301		0.4				16400.70			
6095.714	6097.401		0.8	0.6			16400.43			
6095.955	6097.643	GR	3.1	3.0	2.6	1.83	16399.78	Z 0	2B 4	R 2
6095.955	6097.643	GR	3.1	3.0	2.6	1.83	16399.78	3c 0	2a 0	P 3
6096.253	6097.940		0.8	0.5			16398.98			
6096.390	6098.078		0.4		0.3		16398.61			
6096.494	6098.182			0.3			16398.33	Z 0	2B 4	R 1
6096.647	6098.334		0.2				16397.92			
6096.810	6098.498		0.5	0.2			16397.48			
6097.000	6098.688		0.4	0.2			16396.97			
6097.149	6098.837		0.3	0.2			16396.57			
6097.350	6099.037		0.4	0.2			16396.03	3E 3	2C 1	Q 3
6097.524	6099.212		0.3	0.1			16395.56			
6097.669	6099.357		0.6	0.3			16395.17			
6097.785	6099.473		0.4	0.2			16394.86	3E 2	2B 8	P 3
6097.889	6099.577		0.4				16394.58			
6097.985	6099.674		0.9	0.4			16394.32			
6098.235	6099.923	GR	3.0	2.6	2.0	1.64	16393.65	3c 1	2a 1	R 0
6098.462	6100.150		0.7				16393.04			
6098.547	6100.235		0.8	0.3			16392.81			
6098.741	6100.429		0.5	0.2	0.4		16392.29			
6098.941	6100.630		0.7	0.4			16391.75			
6099.075	6100.764		0.3	0.1			16391.39			
6099.224	6100.913		0.4				16390.99	3D 3	2C 1	Q 1
6099.332	6101.021		0.2	0.2			16390.70			
6099.477	6101.166		1.0	0.1			16390.31			
6099.626	6101.315		0.2	0.2			16389.91			
6099.727	6101.415		0.2				16389.64			
6099.816	6101.505		0.3				16389.40	4f 0	2c 3	Q 3
6099.902	6101.590		0.5	0.8	1.5		16389.17	3E 3	2C 1	Q 2
6100.036	6101.724		0.2				16388.81			
6100.199	6101.888	GR	1.6	1.5	1.8		16388.37	Z 0	2B 4	R 3
6100.199	6101.888	GR	1.6	1.5	1.8		16388.37	3e 3	2c 3	Q 1
6100.278	6101.966	R					16388.16			
6100.415	6102.104		0.2				16387.79			
6100.594	6102.283		0.9	0.4	0.1		16387.31	3E 0	2B 5	R 5

λ_{air}	λ_{vac}	Ref.	I_1	I_2	I_3	I_4	ν	Upper	Lower	Br.
6100.594	6102.283		0.9	0.4	0.1		16387.31	Z 0	2B 4	R 0
6100.784	6102.473		0.2	0.5	1.5		16386.80			
6100.940	6102.629		0.2				16386.38			
6101.149	6102.838		0.3	0.3d	0.2		16385.82	Z 2	2C 0	P 4
6101.275	6102.964		0.2				16385.48			
6101.495	6103.184		0.4	0.2d			16384.89			
6101.651	6103.341		0.2				16384.47	4D 0	2B 9	P 3
6101.830	6103.519		0.2	0.2d			16383.99			
6102.002	6103.691		0.7				16383.53			
6102.091	6103.780		0.5	0.6	1.3		16383.29			
6102.348	6104.037		0.2	0.1			16382.60			
6102.534	6104.223		0.1	0.2			16382.10			
6102.780	6104.469		0.2	0.3			16381.44			
6102.966	6104.656		0.4	0.4	0.1		16380.94			
6103.134	6104.823		0.2		1.4		16380.49			
6103.208	6104.898			0.1			16380.29	4F 0	2C 1	Q 4
6103.410	6105.099		0.3	0.1d			16379.75			
6103.577	6105.267		0.9				16379.30			
6103.678	6105.368		0.2	0.1d			16379.03	V 0	2B 4	R 2
6103.939	6105.629		0.7	0.4	0.3		16378.33	4F 0	2C 1	R 1
6103.939	6105.629		0.7	0.4	0.3		16378.33	3B 4	2A 0	P 1
6104.118	6105.807			0.2			16377.85	X 1	2C 0	R 2
6104.174	6105.863		0.4				16377.70	3D 2	2B 8	R 5
6104.174	6105.863		0.4				16377.70	Z 0	2B 4	R 4
6104.282	6105.972		0.7	0.3	0.1		16377.41			
6104.379	6106.068		0.2				16377.15			
6104.509	6106.199		0.2	0.4	0.3		16376.80			
6104.617	6106.307		0.2				16376.51			
6104.770	6106.460			0.1			16376.10			
6104.848	6106.538		0.1				16375.89			
6105.061	6106.751		0.2	0.3			16375.32			
6105.184	6106.874		0.2	0.2			16374.99			
6105.382	6107.072		1.2	0.9	1.5		16374.46	3e 1	2c 1	P 6
6105.564	6107.254				0.4		16373.97			
6105.751	6107.441		0.2	0.2d			16373.47			
6105.978	6107.668		0.1				16372.86			
6106.280	6107.971		0.2b	0.1d			16372.05	3E 2	2C 0	P 6
6106.676	6108.366		0.4	0.3d			16370.99	3D 0	2B 4	P 3
6106.776	6108.467		0.2				16370.72			
6106.829	6108.519			0.1			16370.58			
6106.978	6108.668		0.2	0.3	1.2		16370.18			
6107.131	6108.821		0.7	0.4			16369.77	3E 2	2B 8	Q 3
6107.131	6108.821		0.7	0.4			16369.77	4g 1	2a 4	R 0
6107.220	6108.911				0.2		16369.53			
6107.328	6109.019		0.1	0.1			16369.24			
6107.586	6109.277		0.1	0.2			16368.55			
6107.828	6109.519	G	1.5	1.5	1.7	0.92	16367.90	4f 0	2c 3	R 2
6107.828	6109.519	G	1.5	1.5	1.7	0.92	16367.90	3e 2	2c 2	P 4
6108.108	6109.799		0.3	0.4			16367.15			
6108.224	6109.915				0.2b		16366.84			
6108.310	6110.001		0.3	0.4			16366.61			
6108.523	6110.214		0.1	0.1	0.1		16366.04			
6108.691	6110.382		0.2	0.2			16365.59			
6108.814	6110.505		0.3	0.2	0.1		16365.26	3E 3	2C 1	P 2
6108.937	6110.628	G	1.3	0.5			16364.93	W 1	2B 7	P 1
6109.030	6110.721		0.3				16364.68	4F 0	2C 1	Q 2
6109.030	6110.721		0.3				16364.68	3c 0	2a 0	Q 8
6109.135	6110.826						16364.40			
6109.217	6110.908		0.2	0.2			16364.18			
6109.467	6111.158	G	1.4	1.0	1.3		16363.51	3e 3	2c 3	Q 1

λ_{air}	λ_{vac}	Ref.	I_1	I_2	I_3	I_4	ν	Upper	Lower	Br.
6109.564	6111.255		0.8				16363.25	4f 0	2c 3	R 1
6109.699	6111.390		0.5				16362.89			
6109.766	6111.457			0.1			16362.71			
6109.994	6111.685		0.1	0.1			16362.10			
6110.225	6111.917			0.1			16361.48			
6110.266	6111.958		0.1				16361.37			
6110.416	6112.107		0.2	0.2			16360.97			
6110.602	6112.294		0.2	0.2			16360.47			
6110.651	6112.342						16360.34			
6110.797	6112.488	G	1.3	0.7			16359.95	3E 2	2B 8	P 2
6110.897	6112.589		0.2				16359.68			
6111.088	6112.780		1.0	0.8	1.7		16359.17	3c 0	2a 0	Q 8
6111.323	6113.015			0.1			16358.54			
6111.379	6113.071		0.1				16358.39			
6111.521	6113.213		0.1	0.1			16358.01	3E 3	2C 1	R 7
6111.727	6113.419		0.2	0.2			16357.46			
6111.876	6113.568		0.2	0.1			16357.06			
6112.059	6113.751		1.5	0.9	0.9		16356.57	3b 4	2a 1	P 2
6112.123	6113.815	G	1.4	1.3	1.4		16356.40	3D 2	2B 8	R 4
6112.175	6113.867		0.9	0.9			16356.26			
6112.403	6114.095		0.1				16355.65			
6112.549	6114.241		0.2	0.2			16355.26			
6112.669	6114.361						16354.94			
6112.784	6114.476	G	1.6	1.4	1.8	0.95	16354.63	2K 6	2B 1	P 4
6112.784	6114.476	G	1.6	1.4	1.8	0.95	16354.63	3d 0	2c 0	R 9
6112.878	6114.570		0.3				16354.38			
6113.087	6114.779		0.2	0.1			16353.82			
6113.263	6114.955			0.1			16353.35			
6113.461	6115.153			0.1			16352.82			
6113.629	6115.321	G	1.2	1.0			16352.37			
6113.704	6115.396		0.2	0.1			16352.17			
6113.984	6115.677		0.3	0.2			16351.42			
6114.198	6115.890		0.2	0.1			16350.85	3D 2	2B 8	R 3
6114.392	6116.084		1.0	0.5			16350.33	3D 1	2B 6	P 2
6114.485	6116.178		0.4				16350.08	3D 2	2B 8	P 1
6114.714	6116.406		0.1				16349.47			
6114.904	6116.597		0.3				16348.96			
6114.979	6116.672			0.2			16348.76			
6115.129	6116.821			0.1			16348.36			
6115.286	6116.979		1.2	0.8			16347.94	3d 1	2c 1	R 6
6115.286	6116.979		1.2	0.8			16347.94	4D 0	2B 9	P 4
6115.465	6117.158		0.2	0.1			16347.46			
6115.589	6117.282		0.2	0.2			16347.13			
6115.709	6117.401		1.0	0.5	0.5		16346.81			
6115.858	6117.551			0.1			16346.41			
6115.937	6117.630		0.2				16346.20			
6116.049	6117.742		0.3				16345.90			
6116.251	6117.944		0.2				16345.36			
6116.453	6118.146		0.3	0.2			16344.82	3D 3	2C 1	R 2
6116.719	6118.412	G	1.9	1.3	1.3	0.99	16344.11	W 1	2B 7	P 2
6116.828	6118.521		0.8	0.6	0.3d		16343.82	3d 1	2c 1	P 3
6117.026	6118.719		0.2				16343.29			
6117.172	6118.865		0.2				16342.90			
6117.359	6119.052		0.2				16342.40			
6117.531	6119.224		0.3	0.2			16341.94	2K.5	2B 0	P 3
6117.748	6119.442	G	1.3	1.2	1.3		16341.36	3e 3	2c 3	Q 2
6117.868	6119.562		0.3				16341.04			
6117.947	6119.640			0.2			16340.83			
6118.122	6119.816	G	1.7	1.2	1.3		16340.36	3D 2	2B 8	R 2
6118.250	6119.944		0.7				16340.02			

λ_{air}	λ_{vac}	Ref.	I_1	I_2	I_3	I_4	ν	Upper	Lower	Br.
6118.332	6120.026			0.4	0.3		16339.80			
6118.445	6120.138		0.3	0.1			16339.50	3E 3	2C 1	Q 3
6118.677	6120.370		0.3	0.3			16338.88	4e 0	2c 3	R 4
6118.800	6120.494		0.2				16338.55			
6118.995	6120.689	G	1.4	1.8	2.6		16338.03	3D 1	2C 1	R 6
6119.220	6120.914		0.3				16337.43			
6119.272	6120.966			0.2			16337.29			
6119.437	6121.131		0.5	0.4			16336.85			
6119.617	6121.311		0.2	0.2			16336.37			
6119.729	6121.423		0.2				16336.07			
6119.804	6121.498			0.1			16335.87	3F 1	2B 7	Q 5
6119.935	6121.629		0.3	0.2			16335.52	3D 0	2B 4	P 9
6120.119	6121.813		0.2	0.3			16335.03			
6120.359	6122.053		0.5	0.3	0.4		16334.39	X 1	2C 0	Q 2
6120.546	6122.240		0.4	0.1			16333.89			
6120.704	6122.398		0.4	0.3			16333.47	3D 2	2B 8	R 1
6120.704	6122.398		0.4	0.3			16333.47	W 2	2B 9	R 3
6120.895	6122.589		0.4	0.3			16332.96	Z 0	2B 4	P 1
6121.041	6122.735		0.4				16332.57			
6121.168	6122.863		1.0				16332.23			
6121.198	6122.893	G	2.0	1.6	1.6		16332.15	3d 2	2c 2	R 1
6121.318	6123.013		0.5				16331.83			
6121.449	6123.144						16331.48			
6121.494	6123.189		0.8	1.0			16331.36			
6121.547	6123.241						16331.22	3F 0	2B 5	P 6
6121.787	6123.481	GF	3.6	3.2	2.7	2.18	16330.58	3c 1	2a 1	Q 1
6122.045	6123.740						16329.89			
6122.090	6123.785		1.0	0.9			16329.77			
6122.289	6123.984		0.7	0.3			16329.24			
6122.529	6124.224	G	1.3	1.2	0.1		16328.60	W 1	2B 7	P 3
6122.660	6124.355		0.5	0.2			16328.25	3e 0	2c 0	P 8
6122.837	6124.531		0.3	0.3			16327.78			
6122.998	6124.693		0.4	0.9			16327.35	V 0	2B 4	R 3
6123.122	6124.816			0.3			16327.02			
6123.174	6124.869		0.4				16326.88			
6123.358	6125.053		0.3	0.2			16326.39			
6123.478	6125.173		0.2				16326.07			
6123.609	6125.304		0.3	0.2			16325.72			
6123.793	6125.488		0.4	0.2			16325.23			
6123.966	6125.661		0.4	0.1			16324.77			
6124.074	6125.769		0.2				16324.48			
6124.183	6125.878		0.2	0.2			16324.19			
6124.371	6126.066	G	1.3	1.0	0.2		16323.69	3E 2	2B 8	Q 2
6124.468	6126.163		0.3				16323.43			
6124.656	6126.351		0.3	0.2	1.5		16322.93			
6124.828	6126.524		0.3				16322.47			
6124.904	6126.599			0.2			16322.27	4F 0	2C 1	Q 3
6125.069	6126.764		0.4	0.3	0.1		16321.83	4e 1	2c 4	R 3
6125.215	6126.910		0.2				16321.44			
6125.301	6126.997			0.1			16321.21			
6125.399	6127.094		0.2				16320.95			
6125.508	6127.203			0.2			16320.66			
6125.557	6127.252		0.3				16320.53			
6125.819	6127.515		0.3	0.3			16319.83			
6125.951	6127.646		0.3				16319.48			
6126.120	6127.815		0.2	0.2			16319.03			
6126.356	6128.052		0.3	0.9	0.1		16318.40			
6126.514	6128.209		0.1	0.2			16317.98			
6126.675	6128.371		0.3	0.3			16317.55			
6126.833	6128.529	G	1.5	1.3	0.2		16317.13	W 1	2B 7	P 4

λ_{air}	λ_{vac}	Ref.	I_1	I_2	I_3	I_4	ν	Upper	Lower	Br.
6126.994	6128.690		0.3	0.4			16316.70			
6127.246	6128.942	GR	2.9	2.6	2.4	1.71	16316.03	W 1	2B 7	P 5
6127.246	6128.942	GR	2.9	2.6	2.4	1.71	16316.03	3c 1	2a 1	Q 2
6127.475	6129.171		0.3				16315.42			
6127.648	6129.344	GR	2.5	2.4	2.2	1.47	16314.96	3c 0	2a 0	P 4
6127.858	6129.554		0.8	0.2			16314.40	3D 2	2B 8	R 0
6127.960	6129.656		0.3	0.2			16314.13			
6128.091	6129.787		1.0	0.7			16313.78			
6128.192	6129.889		1.4	0.8			16313.51			
6128.343	6130.039		0.3	0.2			16313.11			
6128.493	6130.189		0.2				16312.71			
6128.636	6130.332			0.2			16312.33			
6128.700	6130.396		0.3d				16312.16			
6128.824	6130.520			0.1			16311.83			
6129.030	6130.727		0.3	0.2			16311.28			
6129.256	6130.952		0.5	0.4			16310.68			
6129.414	6131.110		0.2	0.1			16310.26			
6129.515	6131.212		0.3		0.3		16309.99	3E 3	2C 1	P 2
6129.692	6131.388		0.2				16309.52	3e 3	2c 3	Q 3
6129.804	6131.501			0.0			16309.22			
6129.951	6131.648	G	1.3	0.9			16308.83	3D 0	2B 4	P 4
6129.951	6131.648	G	1.3	0.9			16308.83	W 2	2B 9	R 2
6130.143	6131.839		0.2	0.1			16308.32			
6130.357	6132.054	G	1.4	1.2	1.4		16307.75	3c 8	2a 7	Q 1
6130.485	6132.182						16307.41			
6130.639	6132.336	G	1.4	1.1	0.2b		16307.00	Z 1	2B 6	R 2
6130.744	6132.441		0.3				16306.72			
6130.883	6132.580		0.1	0.0			16306.35			
6131.034	6132.731		0.2				16305.95	4E 0	2C 1	R 2
6131.286	6132.983		0.5	0.3			16305.28			
6131.440	6133.137			0.2	0.2		16304.87			
6131.564	6133.261		0.2	0.3			16304.54			
6131.816	6133.513	G	1.7	1.8	1.9		16303.87	3d 1	2c 1	Q 4
6131.816	6133.513	G	1.7	1.8	1.9		16303.87	3D 2	2C 0	P 5
6132.106	6133.803		0.1	0.2			16303.10			
6132.256	6133.953		0.2				16302.70	3F 1	2B 7	R 3
6132.448	6134.145		0.3	0.2			16302.19	3D 3	2C 1	Q 2
6132.542	6134.239		0.2				16301.94	X 1	2C 0	P 2
6132.726	6134.424		0.2	0.2	0.1		16301.45			
6132.862	6134.559		0.2	0.1			16301.09			
6133.012	6134.710		0.5	0.4			16300.69			
6133.133	6134.830		0.4	0.4	0.2b		16300.37			
6133.208	6134.905						16300.17	3b 3	2a 0	P 8
6133.377	6135.075		1.5	0.6			16299.72	Z 1	2B 6	R 1
6133.554	6135.252			0.1			16299.25	3E 2	2B 8	Q 1
6133.735	6135.432		0.3	0.3			16298.77			
6133.866	6135.564		0.2				16298.42			
6134.077	6135.775		1.0	1.0	1.5b		16297.86	3c 0	2a 0	Q 9
6134.284	6135.982	GR	2.4	2.2	2.3	1.35	16297.31	4f 0	2c 3	P 5
6134.284	6135.982	GR	2.4	2.2	2.3	1.35	16297.31	3d 2	2c 2	Q 1
6134.344	6136.042	R					16297.15			
6134.619	6136.317		0.6	0.4			16296.42	4d 3	3b 0	P 4
6134.841	6136.539		0.5	0.6			16295.83			
6134.939	6136.637		0.4	0.3	0.6		16295.57			
6135.135	6136.833	GR	1.5b				16295.05			
6135.195	6136.893						16294.89	3d 2	2c 2	R 2
6135.195	6136.893						16294.89	3d 0	2c 0	Q 6
6135.293	6136.991			1.5b	1.5b		16294.63			
6135.395	6137.093	G	3.0	2.9	2.9	1.89	16294.36	3c 1	2a 1	Q 3
6135.481	6137.179						16294.13			

λ_{air}	λ_{vac}	Ref.	I_1	I_2	I_3	I_4	ν	Upper	Lower	Br.
6135.624	6137.323		0.5				16293.75			
6135.696	6137.394		0.7	0.9			16293.56			
6135.896	6137.594		0.3	0.3			16293.03			
6136.080	6137.778		0.5	0.3			16292.54	3B 4	2A 0	P 2
6136.253	6137.952		0.3	0.2			16292.08	4f 0	2c 3	Q 4
6136.487	6138.185		0.5	0.6			16291.46			
6136.638	6138.336		0.2				16291.06			
6136.792	6138.490		0.2	0.2			16290.65	W 2	2B 9	R 1
6137.019	6138.717		0.8	0.3			16290.05	Z 0	2B 4	P 2
6137.127	6138.826			0.1			16289.76			
6137.229	6138.928		0.2	0.2			16289.49			
6137.316	6139.014				0.2		16289.26			
6137.455	6139.154	G	1.4	1.0			16288.89	Z 1	2B 6	R 0
6137.553	6139.252		0.4				16288.63			
6137.704	6139.402			0.3	1.5b		16288.23			
6137.892	6139.591	G	1.5	1.4	1.5		16287.73	3d 2	2c 2	P 1
6138.017	6139.715		0.8				16287.40	3E 3	2C 1	Q 4
6138.118	6139.817			0.1			16287.13			
6138.231	6139.930		0.8	0.4			16286.83			
6138.480	6140.179		0.2				16286.17			
6138.638	6140.337		0.1	0.2			16285.75			
6138.770	6140.469		0.2				16285.40			
6138.816	6140.515			0.1			16285.28			
6138.963	6140.662		0.2	0.1			16284.89			
6139.091	6140.790		0.3	0.2			16284.55			
6139.351	6141.050	G	1.9	1.7	1.8	0.93	16283.86	3e 3	2c 3	P 2
6139.686	6141.386		0.2b	0.2			16282.97	4g 1	2a 4	R 1
6139.894	6141.593		0.1b	0.1			16282.42			
6140.071	6141.770			0.2			16281.95			
6140.131	6141.831		0.3				16281.79			
6140.226	6141.925				0.2		16281.54			
6140.395	6142.095		0.3	0.2			16281.09			
6140.576	6142.276		0.1	0.1			16280.61			
6140.724	6142.423		0.1	0.2			16280.22			
6140.954	6142.653	G	1.1	1.4	2.1		16279.61	3c 2	2a 2	R 6
6140.954	6142.653	G	1.1	1.4	2.1		16279.61	3d 0	2c 0	R10
6141.206	6142.906		0.2	0.2			16278.94			
6141.327	6143.027			0.2			16278.62			
6141.504	6143.204		0.5	0.4			16278.15	3E 3	2C 1	P 3
6141.787	6143.487	G	1.5	1.5	2.0		16277.40	3c 2	2a 2	R 5
6141.787	6143.487	G	1.5	1.5	2.0		16277.40	3E 3	2C 1	Q 4
6142.044	6143.744		0.2	0.2			16276.72			
6142.199	6143.899		1.0	0.3			16276.31	3D 1	2B 6	P 3
6142.293	6143.993		0.2				16276.06			
6142.391	6144.091			0.1			16275.80			
6142.455	6144.155		0.1				16275.63			
6142.648	6144.348		0.2	0.1			16275.12	X 2	2C 1	R 1
6142.731	6144.431			0.2			16274.90			
6142.874	6144.574		0.1		0.2		16274.52			
6143.059	6144.759		0.7	0.3	0.3		16274.03	4F 0	2C 1	P 5
6143.206	6144.907			0.2			16273.64			
6143.365	6145.065		0.2	0.3	0.2		16273.22			
6143.520	6145.220			0.2			16272.81			
6143.686	6145.386		0.4	0.3			16272.37	3D 3	2C 1	P 2
6143.743	6145.443			0.7			16272.22			
6143.811	6145.511			0.2			16272.04			
6143.878	6145.579		0.1				16271.86			
6144.060	6145.760	G	1.6	1.7	2.1		16271.38	3e 2	2c 2	P 5
6144.267	6145.968		0.1				16270.83			
6144.381	6146.081		0.4	0.4			16270.53			

λ_air	λ_vac	Ref.	I₁	I₂	I₃	I₄	ν	Upper	Lower	Br.
6144.490	6146.191		0.5	0.2			16270.24	3e 3	2c 3	Q 4
6144.717	6146.417		0.1	0.1			16269.64			
6144.891	6146.591		0.2	0.2			16269.18	3A 1	2C 0	R 2
6145.117	6146.818		0.4	1.1	2.0		16268.58	3d 1	2c 1	R 7
6145.299	6146.999		0.1	0.1			16268.10	4F 0	2B10	P 5
6145.378	6147.079				0.6		16267.89			
6145.431	6147.132		0.2	0.2			16267.75			
6145.650	6147.351		0.5	0.9	1.3		16267.17			
6145.861	6147.562						16266.61			
6145.941	6147.642		0.2	0.3	1.5		16266.40			
6146.186	6147.887	GR	2.1	2.0	2.3	1.25	16265.75	3F 0	2B 5	Q 4
6146.186	6147.887	GR	2.1	2.0	2.3	1.25	16265.75	3c 1	2a 1	Q 4
6146.353	6148.054		0.5				16265.31			
6146.515	6148.216		0.1	0.2			16264.88	3e 1	2c 1	P 7
6146.670	6148.371		0.2	0.2			16264.47	W 2	2B 9	R 0
6146.901	6148.602	G	1.7	1.5	2.0	1.05	16263.86	3c 2	2a 2	R 4
6146.931	6148.632						16263.78	3D 0	2B 4	P 5
6147.222	6148.923		0.2	0.2			16263.01			
6147.430	6149.131		0.3	0.2			16262.46	4e 0	2c 3	R 3
6147.664	6149.366		0.2	0.3			16261.84			
6147.891	6149.592		0.2	0.2			16261.24			
6148.091	6149.793	G	0.8	1.0	0.9		16260.71			
6148.231	6149.933		0.3				16260.34			
6148.288	6149.990			0.2			16260.19			
6148.526	6150.228	G	1.4	1.2	1.6		16259.56	3e 3	2c 3	P 2
6148.526	6150.228	G	1.4	1.2	1.6		16259.56	4F 0	2C 1	Q 2
6148.655	6150.357				0.2		16259.22	4f 0	2c 3	Q 2
6148.863	6150.565		0.3	0.1			16258.67			
6149.086	6150.788		0.2				16258.08			
6149.218	6150.920			0.2	0.1		16257.73	X 2	2C 1	Q 1
6149.275	6150.977		0.2				16257.58	4F 0	2B10	R 2
6149.487	6151.189		0.2	0.2			16257.02	X 2	2C 1	P 1
6149.763	6151.465		0.1				16256.29	3D 0	2B 4	P 8
6150.119	6151.821		0.3	0.2			16255.35			
6150.368	6152.071		0.1				16254.69			
6150.542	6152.245		0.1	0.1			16254.23			
6150.716	6152.419		0.2	0.1			16253.77	4f 1	2c 4	Q 3
6150.887	6152.589		0.5	0.3			16253.32	2K 5	2B 0	P 4
6151.186	6152.888		0.4d	0.2			16252.53			
6151.424	6153.127	R					16251.90			
6151.447	6153.149	R					16251.84	3b 4	2a 1	P 3
6151.504	6153.206	GR	2.1	1.9	2.0	1.08	16251.69	3e 3	2c 3	P 3
6151.625	6153.327						16251.37	V 0	2B 4	R 1
6151.757	6153.460		0.4d	0.2			16251.02			
6152.038	6153.740		0.2d	0.2d			16250.28			
6152.265	6153.967						16249.68			
6152.306	6154.009		0.4	0.8	1.3		16249.57			
6152.518	6154.221		0.1	0.1d			16249.01			
6152.658	6154.361			0.1			16248.64			
6152.855	6154.558			0.2			16248.12			
6152.908	6154.611		0.2d				16247.98			
6153.170	6154.873	G	1.8	1.7	2.0		16247.29	3d 2	2c 2	R 3
6153.450	6155.153		0.2	0.2d			16246.55			
6153.791	6155.494		0.3	0.2			16245.65			
6153.973	6155.676		0.1	0.1			16245.17			
6154.177	6155.880		0.3	0.3			16244.63			
6154.227	6155.930				1.0		16244.50			
6154.363	6156.066		0.1				16244.14	Z 2	2C 0	Q 5
6154.533	6156.237		0.1				16243.69	3D 0	2B 4	P 7
6154.730	6156.434		0.2	0.1			16243.17	3d 0	2c 0	P 5

λ_{air}	λ_{vac}	Ref.	I_1	I_2	I_3	I_4	ν	Upper	Lower	Br.
6154.920	6156.623		0.1				16242.67			
6155.060	6156.763		0.3	0.2			16242.30	Z 2	2B 8	R 0
6155.249	6156.953		1.5	0.7	0.3		16241.80	3A 1	2C 0	R 1
6155.386	6157.089		0.4	0.2			16241.44			
6155.466	6157.169		0.3				16241.23			
6155.629	6157.332	GR	2.4	2.2	2.3	1.45	16240.80	3c 2	2a 2	R 3
6155.629	6157.332	GR	2.4	2.2	2.3	1.45	16240.80	Z 1	2B 6	P 1
6155.629	6157.332	GR	2.4	2.2	2.3	1.45	16240.80	3D 0	2B 4	P 6
6155.860	6157.563		0.3				16240.19			
6155.936	6157.639		0.4	0.3			16239.99			
6156.118	6157.821		0.2				16239.51			
6156.174	6157.878			0.1			16239.36			
6156.326	6158.030		0.2	0.2			16238.96			
6156.500	6158.204		0.2				16238.50	3D 3	2C 1	R 3
6156.660	6158.364		0.2	0.2			16238.08	3F 0	2B 5	R 2
6156.861	6158.565		1.0				16237.55			
6157.058	6158.762		0.2				16237.03	Z 0	2B 4	P 3
6157.111	6158.815			0.2			16236.89			
6157.282	6158.986		0.3				16236.44			
6157.430	6159.133		0.6	0.3			16236.05			
6157.638	6159.342		1.0	0.2			16235.50			
6157.794	6159.498			0.4			16235.09	3D 2	2C 0	Q 6
6157.794	6159.498			0.4			16235.09	4f 0	2c 3	R 1
6157.843	6159.547		0.2				16234.96			
6158.048	6159.752			0.1			16234.42	4E 0	2C 1	Q 1
6158.712	6160.416		0.1	0.2			16232.67			
6159.026	6160.731		0.2	0.2			16231.84	Z 2	2B 8	R 1
6159.269	6160.974		1.9	1.3			16231.20	3c 0	2a 0	Q10
6159.269	6160.974		1.9	1.3			16231.20	4F 0	2B10	Q 4
6159.436	6161.141				1.5b		16230.76			
6159.565	6161.270	G	1.2	2.0	2.7	1.29	16230.42	3c 1	2a 1	Q 5
6159.766	6161.471		0.2				16229.89	3D 2	2B 8	P 2
6159.945	6161.650		1.5	0.5	0.2		16229.42			
6160.051	6161.756		0.3				16229.14	3E 2	2C 0	P 7
6160.051	6161.756		0.3				16229.14	3E 3	2B10	R 3
6160.169	6161.874		0.2	0.2			16228.83			
6160.305	6162.010		0.1	0.2			16228.47			
6160.480	6162.185		0.2	0.2			16228.01			
6160.697	6162.401		0.2	0.2			16227.44			
6160.757	6162.462				1.5b		16227.28			
6160.867	6162.572		0.2	0.2			16226.99			
6161.046	6162.751		0.5	0.3			16226.52			
6161.148	6162.853		0.2				16226.25			
6161.270	6162.975		0.2	0.5			16225.93			
6161.365	6163.070		0.6				16225.68	4F 0	2C 1	P 4
6161.604	6163.309	GR	2.5	2.6	2.6	1.56	16225.05	3c 0	2a 0	P 5
6161.820	6163.526		0.3				16224.48			
6161.915	6163.621		0.5	0.4			16224.23			
6162.121	6163.826		0.3	0.3			16223.69			
6162.227	6163.932		0.3				16223.41			
6162.310	6164.016		0.2				16223.19			
6162.474	6164.179	GR	1.9	1.7	1.8	1.00	16222.76	3d 2	2c 2	Q 2
6162.538	6164.244	R					16222.59			
6162.679	6164.384						16222.22			
6162.797	6164.502		0.3	0.2			16221.91			
6162.975	6164.681		0.4	0.3	0.3d		16221.44			
6163.192	6164.897		0.4	0.2			16220.87	3E 3	2C 1	P 3
6163.462	6165.167		0.5				16220.16			
6163.568	6165.274			0.3d			16219.88			
6163.796	6165.502			1.0	0.6		16219.28			

λ_{air}	λ_{vac}	Ref.	I_1	I_2	I_3	I_4	ν	Upper	Lower	Br.
6163.891	6165.597		0.3b	0.2			16219.03			
6164.161	6165.867	G	1.8	1.6	1.8		16218.32	3e 3	2c 3	P 4
6164.446	6166.152		0.3b				16217.57			
6164.647	6166.353		0.2	0.2			16217.04			
6164.788	6166.494		0.1				16216.67			
6165.066	6166.772		0.2	0.1			16215.94	W 2	2B 9	P 1
6165.286	6166.992		0.2	0.2			16215.36			
6165.503	6167.209	G	1.5	1.0			16214.79	Z 2	2B 8	R 2
6165.606	6167.312		0.5				16214.52			
6165.670	6167.376			0.1			16214.35			
6165.799	6167.506		0.3	0.2			16214.01	3E 0	2B 5	R 4
6166.031	6167.738		0.2				16213.40			
6166.092	6167.799			0.1			16213.24			
6166.176	6167.882		0.1				16213.02			
6166.294	6168.000		0.2	0.1			16212.71			
6166.446	6168.152		0.2	0.2			16212.31	4f 0	2c 3	Q 3
6166.629	6168.335		0.6	0.4			16211.83			
6166.838	6168.544			0.2			16211.28			
6166.998	6168.704			0.0			16210.86			
6167.173	6168.879			0.2			16210.40			
6167.431	6169.138			0.3			16209.72			
6167.568	6169.275						16209.36			
6167.732	6169.439	GR	2.4	2.1	1.9	1.34	16208.93	3c 2	2a 2	R 2
6167.797	6169.503		0.3				16208.76			
6167.903	6169.610						16208.48			
6168.032	6169.739		0.2	0.2			16208.14			
6168.192	6169.899		0.2	0.1			16207.72	4g 1	2a 4	P 2
6168.364	6170.071		0.1	0.0			16207.27			
6168.497	6170.204		0.1				16206.92			
6168.569	6170.276			0.2			16206.73			
6168.725	6170.432		0.2	0.1d			16206.32			
6168.885	6170.592		0.2				16205.90			
6169.034	6170.741		0.3	0.1			16205.51			
6169.171	6170.878		0.9	0.7	0.2		16205.15	3D 1	2B 6	P 4
6169.273	6170.981		0.2				16204.88			
6169.365	6171.072		0.8	0.3			16204.64			
6169.605	6171.312		1.5b				16204.01			
6169.639	6171.346	GR	2.7	2.4	1.9	1.44	16203.92	3c 1	2a 1	P 2
6169.707	6171.415				1.5b		16203.74	4e 3	3b 0	P 5
6169.852	6171.559		0.5				16203.36			
6169.932	6171.639		0.6	0.2			16203.15			
6170.111	6171.818		0.5	1.2	2.2		16202.68	3d 0	2c 0	R11
6170.320	6172.028		0.2	0.1			16202.13	4g 1	2a 4	R 2
6170.480	6172.188		0.2				16201.71	3E 3	2C 1	Q 5
6170.545	6172.253			0.3d	1.5		16201.54			
6170.614	6172.321		0.2				16201.36			
6170.770	6172.477		0.2	0.8	1.4		16200.95			
6170.861	6172.569		0.1				16200.71			
6171.025	6172.733		0.2	0.1			16200.28	X 1	2C 0	Q 3
6171.189	6172.897		0.9	0.2			16199.85			
6171.406	6173.114		0.1	0.1			16199.28	W 2	2B 9	P 5
6171.585	6173.293		1.0	0.8			16198.81	Z 1	2B 6	P 2
6171.840	6173.548		0.1				16198.14			
6172.218	6173.926			0.2			16197.15	Z 2	2B 8	R 3
6172.625	6174.333			0.2			16196.08	3E 3	2B10	Q 5
6172.820	6174.528		0.4	0.4			16195.57			
6172.976	6174.684		0.1	0.0			16195.16			
6173.056	6174.764				0.4		16194.95			
6173.128	6174.837		0.2	0.1			16194.76			
6173.304	6175.012						16194.30	3E 3	2B10	R 2

λ_{air}	λ_{vac}	Ref.	I_1	I_2	I_3	I_4	ν	Upper	Lower	Br.
6173.350	6175.058		0.3	0.3			16194.18	X 2	2C 1	R 2
6173.350	6175.058		0.3	0.3			16194.18	3D 2	2B 8	P 3
6173.514	6175.222			0.2			16193.75	4e 1	2c 4	R 1
6173.739	6175.447			0.3d			16193.16			
6173.872	6175.580		0.2				16192.81			
6174.055	6175.763	GR	2.1	2.0	2.5	1.20	16192.33	3d 2	2c 2	R 4
6174.101	6175.809						16192.21			
6174.394	6176.103		0.1	0.2			16191.44			
6174.600	6176.309		0.2	0.3			16190.90			
6174.833	6176.542						16190.29	4e 0	2c 3	R 2
6174.879	6176.587	G	1.8	1.8	2.3		16190.17	3d 1	2c 1	Q 5
6175.188	6176.896			0.2			16189.36			
6175.298	6177.007		0.1				16189.07			
6175.462	6177.171		0.5	1.4	2.1		16188.64	3c 1	2a 1	Q 6
6175.607	6177.316		0.1				16188.26			
6175.653	6177.362			0.2			16188.14			
6175.748	6177.457		0.2				16187.89			
6175.870	6177.579	GR	1.6	1.5	1.6		16187.57			
6175.909	6177.618						16187.47	3d 2	2c 2	P 2
6175.935	6177.644	R					16187.40			
6176.065	6177.774						16187.06	4F 0	2C 1	P 3
6176.214	6177.923	GR	2.1	1.9	1.9	1.19	16186.67	3e 3	2c 3	P 3
6176.263	6177.973	R					16186.54	3d 1	2c 1	P 4
6176.515	6178.224		0.2	0.2			16185.88	4e 1	2c 4	Q 1
6176.714	6178.423		0.3	0.4			16185.36	3B 4	2A 0	P 3
6176.817	6178.526		0.3	0.3			16185.09			
6177.153	6178.862			0.1			16184.21			
6177.336	6179.045		0.1d	0.2			16183.73			
6177.527	6179.236	G	1.5	1.0	1.3		16183.23	W 2	2B 9	P 2
6177.527	6179.236	G	1.5	1.0	1.3		16183.23	4f 0	2c 3	P 4
6177.626	6179.335		0.3	0.8	0.4		16182.97			
6177.935	6179.645			0.1			16182.16			
6178.092	6179.801			0.1			16181.75			
6178.474	6180.183				1.3		16180.75			
6178.531	6180.240			0.3d	0.3		16180.60			
6178.783	6180.493			0.2			16179.94			
6178.859	6180.569		0.1				16179.74			
6178.955	6180.664			0.2			16179.49	3e 0	2c 0	R 5
6179.157	6180.867				0.9		16178.96			
6179.207	6180.917		1.5	1.0	1.6		16178.83			
6179.424	6181.134						16178.26			
6179.665	6181.375		0.4	0.3			16177.63			
6179.864	6181.574			0.1			16177.11	3E 0	2B 5	P 8
6180.146	6181.857		0.2				16176.37			
6180.204	6181.914			0.2d			16176.22	3D 3	2C 1	Q 3
6180.318	6182.029		0.3				16175.92			
6180.456	6182.166		0.2	0.5	0.2		16175.56	Z 0	2B 4	P 4
6180.800	6182.510		0.1	0.2			16174.66	4e 1	2c 4	R 1
6181.033	6182.743			0.2	0.1d		16174.05	3F 1	2B 7	Q 4
6181.224	6182.935		0.3	0.3			16173.55	3F 0	2B 5	P 5
6181.526	6183.236		0.3	1.1	1.8		16172.76	3d 0	2c 0	Q 7
6181.767	6183.477		0.2				16172.13			
6181.847	6183.558			0.1			16171.92			
6182.027	6183.737			0.1			16171.45			
6182.080	6183.791		0.2				16171.31	3e 3	2c 3	P 5
6182.268	6183.978		0.2	1.1	1.5		16170.82	3e 2	2c 2	P 6
6182.424	6184.135		0.4	0.3			16170.41			
6182.535	6184.246		0.3	0.2			16170.12			
6182.642	6184.353		0.4				16169.84			
6182.730	6184.441		1.0	0.7			16169.61			

λ_{air}	λ_{vac}	Ref.	I_1	I_2	I_3	I_4	ν	Upper	Lower	Br.
6182.990	6184.701	GF	2.9	2.8	2.4	1.84	16168.93	3c 2	2a 2	R 1
6183.220	6184.931		0.6				16168.33			
6183.296	6185.007		0.8	0.6			16168.13			
6183.487	6185.198		0.3	0.3			16167.63			
6183.721	6185.432		0.2	0.3	0.1		16167.02			
6183.878	6185.589		0.2	0.3			16166.61			
6184.073	6185.784		0.2	0.2			16166.10			
6184.222	6185.933		0.2	0.1			16165.71			
6184.394	6186.105		0.2	0.1			16165.26			
6184.589	6186.300		0.1	0.2			16164.75			
6184.788	6186.499		0.3	0.4			16164.23	Z 2	2B 8	R 4
6184.983	6186.695		0.2	0.3			16163.72			
6185.174	6186.886		0.2	0.2			16163.22			
6185.374	6187.085		0.3	0.4			16162.70	3E 3	2C 1	P 4
6185.374	6187.085		0.3	0.4			16162.70	5c 3	2c 7	Q 1
6185.618	6187.330		0.2	0.2			16162.06			
6185.794	6187.506			0.2			16161.60			
6186.020	6187.732			0.2			16161.01	X 2	2C 1	Q 2
6186.212	6187.923			0.2			16160.51			
6186.315	6188.027		0.1	0.1			16160.24			
6186.530	6188.241		0.9	1.4	1.7		16159.68	3c 0	2a 0	Q11
6186.755	6188.467		0.1	0.1			16159.09			
6186.970	6188.682		0.2	0.3			16158.53			
6187.085	6188.797			0.2			16158.23			
6187.226	6188.938		0.3	0.3			16157.86	Z 2	2B 8	P 2
6187.556	6189.268			0.2d			16157.00			
6187.632	6189.344		0.1				16156.80			
6187.766	6189.478			0.2			16156.45			
6188.004	6189.716			0.3	1.0b		16155.83			
6188.272	6189.984			0.1			16155.13	V 0	2B 4	P 4
6188.575	6190.287			0.2d			16154.34			
6188.870	6190.582		0.9	0.4	1.3		16153.57			
6189.130	6190.843		0.6	0.4			16152.89			
6189.279	6190.992			0.1			16152.50			
6189.402	6191.115		0.1	0.2			16152.18			
6189.628	6191.341			0.2			16151.59			
6189.805	6191.517			0.3			16151.13	2K 5	2B 0	P 5
6190.088	6191.801		0.1	0.2			16150.39	Z 2	2C 0	P 5
6190.157	6191.870		0.6	0.4			16150.21	Z 1	2B 6	P 3
6190.326	6192.039			0.8	1.3		16149.77	3E 0	2B 5	Q 8
6190.402	6192.115		0.1				16149.57	W 2	2B 9	P 3
6190.659	6192.372		0.2	0.2			16148.90			
6190.836	6192.549		0.2				16148.44			
6190.908	6192.621			0.2d			16148.25			
6191.119	6192.832		0.2				16147.70			
6191.173	6192.886			0.1			16147.56			
6191.365	6193.078		0.2	0.2			16147.06	4E 0	2C 1	R 1
6191.545	6193.258			0.2			16146.59	X 1	2C 0	P 3
6191.764	6193.477		0.5	0.4			16146.02	3A 1	2C 0	Q 1
6191.764	6193.477		0.5	0.4			16146.02	3D 2	2B 8	P 4
6191.913	6193.626			0.1			16145.63			
6192.093	6193.807		0.2	0.4			16145.16	3D 1	2B 6	P 5
6192.366	6194.079			0.2			16144.45			
6192.561	6194.275			0.1			16143.94			
6192.734	6194.447		0.1	0.2			16143.49			
6192.922	6194.635			0.2	0.8		16143.00			
6192.979	6194.693		0.5				16142.85			
6193.206	6194.919		0.3	0.3			16142.26			
6193.367	6195.081			0.1			16141.84			
6193.574	6195.288		0.2d	0.2			16141.30			

λ_{air}	λ_{vac}	Ref.	I_1	I_2	I_3	I_4	ν	Upper	Lower	Br.
6193.812	6195.526	G	1.0	1.6	2.3		16140.68	3c 1	2a 1	Q 7
6194.134	6195.848		0.3	0.2			16139.84	4E 1	2C 2	R 1
6194.276	6195.990			0.2			16139.47			
6194.399	6196.113			0.2			16139.15	X 2	2C 1	P 2
6194.583	6196.297		0.1				16138.67	4E 0	2C 1	R 2
6194.787	6196.501		0.1				16138.14			
6194.844	6196.559			0.2			16137.99			
6195.002	6196.716		0.3	0.3			16137.58	4g 1	2a 4	R 3
6195.002	6196.716		0.3	0.3			16137.58	4E 0	2C 1	Q 3
6195.109	6196.823		0.1	0.2			16137.30			
6195.248	6196.962		0.2				16136.94			
6195.378	6197.092		0.2				16136.60			
6195.612	6197.327		0.2	0.2d			16135.99			
6195.866	6197.580		0.4	0.4	0.1		16135.33			
6195.973	6197.688		0.2	0.1			16135.05	4E 0	2C 1	R 3
6196.227	6197.941			0.2			16134.39	4E 1	2C 2	R 2
6196.507	6198.222						16133.66			
6196.572	6198.287			0.3d			16133.49			
6196.853	6198.567		0.3	0.4d			16132.76			
6196.914	6198.629						16132.60	3E 3	2B10	R 1
6197.095	6198.809	GR	2.3	2.1	2.5	1.26	16132.13	3d 2	2c 2	Q 3
6197.160	6198.875	R					16131.96	4E 1	2C 2	R 3
6197.371	6199.086						16131.41	4E 0	2C 1	Q 1
6197.441	6199.155		0.2	0.3d			16131.23			
6197.552	6199.267						16130.94	3F 1	2B 7	R 2
6197.721	6199.436	G	1.9	1.9	2.3	1.11	16130.50	3b 4	2a 1	P 4
6197.721	6199.436	G	1.9	1.9	2.3	1.11	16130.50	3c 0	2a 0	P 6
6197.721	6199.436	G	1.9	1.9	2.3	1.11	16130.50	4f 0	2c 3	Q 2
6197.763	6199.478			0.80			16130.39			
6197.825	6199.540		0.3				16130.23			
6197.952	6199.666		0.2				16129.90	3D 3	2C 1	R 4
6198.036	6199.751			0.3d			16129.68			
6198.148	6199.862		0.2				16129.39			
6198.278	6199.993		0.2	0.2			16129.05			
6198.505	6200.220		0.2	0.1			16128.46	3E 3	2B10	Q 4
6198.655	6200.370		0.2	0.2	0.2		16128.07			
6198.839	6200.554		0.5	0.4			16127.59			
6198.943	6200.658		0.3	0.2			16127.32			
6199.055	6200.770		0.5				16127.03			
6199.147	6200.862		1.0	1.0			16126.79			
6199.239	6200.954				1.5b		16126.55			
6199.378	6201.093						16126.19			
6199.397	6201.112	GR	3.2	2.9	2.5	1.84	16126.14	3c 1	2a 1	P 3
6199.639	6201.354		0.7				16125.51	W 2	2B 9	P 4
6199.712	6201.427		0.8	0.7			16125.32			
6199.923	6201.639		0.3	0.5			16124.77	3d 0	2c 0	R12
6199.923	6201.639		0.3	0.5			16124.77	3D 3	2C 1	P 3
6200.135	6201.850		0.2	0.2			16124.22	V 0	2B 4	R 0
6200.285	6202.000		0.3	0.9			16123.83	4e 0	2c 3	R 1
6200.469	6202.185		0.2	0.2			16123.35			
6200.627	6202.343		0.4	0.3			16122.94			
6200.750	6202.466		0.2				16122.62			
6200.839	6202.554		0.4				16122.39			
6200.927	6202.643		1.0	0.7			16122.16			
6200.954	6202.670						16122.09			
6201.185	6202.901	GR	2.9	2.5	1.9	1.57	16121.49	3c 2	2a 2	R 0
6201.185	6202.901	GR	2.9	2.5	1.9	1.57	16121.49	3A 1	2C 0	Q 2
6201.423	6203.139		0.6				16120.87			
6201.504	6203.220		0.7	0.4			16120.66			
6201.700	6203.416		0.3				16120.15			

λ_{air}	λ_{vac}	Ref.	I_1	I_2	I_3	I_4	ν	Upper	Lower	Br.
6201.758	6203.474			1.0	1.7		16120.00			
6201.889	6203.605		0.1				16119.66			
6201.973	6203.689			0.2			16119.44			
6202.077	6203.793		0.2	0.1			16119.17			
6202.166	6203.882				0.3		16118.94			
6202.262	6203.978		0.2	0.2			16118.69	4E 2	2C 3	R 1
6202.439	6204.155		0.2	0.2			16118.23	3E 3	2C 1	P 4
6202.635	6204.351		0.2	0.3			16117.72			
6202.735	6204.452		0.2				16117.46			
6202.812	6204.529			0.2			16117.26			
6202.986	6204.702		0.1	0.4	0.4		16116.81			
6203.243	6204.960		0.2	0.2			16116.14			
6203.636	6205.352						16115.12			
6203.786	6205.503		0.2	0.5			16114.73	3C 1	2A 0	R 3
6204.010	6205.726			0.2d			16114.15	3F 0	2B 5	Q 3
6204.256	6205.973			0.1			16113.51			
6204.452	6206.169			0.0			16113.00			
6204.533	6206.250		0.1				16112.79			
6204.718	6206.435		0.1	0.2			16112.31			
6205.038	6206.754		0.2	0.4			16111.48	3D 1	2B 6	P 8
6205.246	6206.962		0.1	0.1			16110.94			
6205.369	6207.086			0.2			16110.62			
6205.554	6207.271		0.4	0.3			16110.14			
6205.716	6207.432		0.2	0.7	0.2		16109.72	3D 2	2B 8	P 5
6205.773	6207.490			0.8	1.3		16109.57	3D 1	2B 6	P 6
6205.889	6207.606		0.1				16109.27			
6205.951	6207.668			0.2			16109.11			
6206.120	6207.837		0.1	0.2			16108.67			
6206.413	6208.130			0.0	0.2		16107.91			
6206.771	6208.488			0.2			16106.98			
6207.056	6208.774			0.2			16106.24	4F 0	2C 1	P 5
6207.318	6209.036	G	1.4	1.4	1.6		16105.56	3e 3	2c 3	P 4
6207.318	6209.036	G	1.4	1.4	1.6		16105.56	3d 4	2c 4	R 3
6207.627	6209.344		0.1				16104.76			
6208.039	6209.757				0.1		16103.69			
6208.155	6209.873			0.0			16103.39	4E 1	2C 2	Q 1
6208.155	6209.873			0.0			16103.39	4f 0	2c 3	P 3
6208.220	6209.938						16103.22	Z 0	2B 4	P 5
6208.834	6210.551			0.1			16101.63			
6209.061	6210.779			0.0	0.1		16101.04	4F 0	2C 1	P 4
6209.524	6211.242		0.1	0.3			16099.84			
6209.663	6211.381						16099.48	3A 1	2C 0	P 1
6209.759	6211.477		0.3				16099.23			
6209.821	6211.539	G	1.3	1.0	1.4		16099.07	Z 1	2B 6	P 4
6209.821	6211.539	G	1.3	1.0	1.4		16099.07	3D 1	2B 6	P 7
6209.994	6211.713		0.2	0.2			16098.62	Y 2	2B 9	R 1
6210.318	6212.037				0.1		16097.78			
6211.287	6213.005						16095.27			
6211.445	6213.164			0.1			16094.86			
6211.781	6213.500			0.2			16093.99			
6211.850	6213.569		0.1				16093.81			
6212.179	6213.897		0.2	0.2			16092.96	Y 2	2B 9	R 2
6212.179	6213.897		0.2	0.2			16092.96	Z 2	2B 8	P 3
6212.356	6214.075		0.3	0.3			16092.50			
6212.464	6214.183						16092.22			
6212.526	6214.245			0.1			16092.06			
6212.738	6214.457		0.2	0.2			16091.51	3A 1	2C 0	Q 3
6212.738	6214.457		0.2	0.2			16091.51	4F 0	2C 1	P 3
6212.908	6214.627			0.1			16091.07			
6213.097	6214.816			0.2			16090.58	V 0	2B 4	P 3

λ_{air}	λ_{vac}	Ref.	I_1	I_2	I_3	I_4	ν	Upper	Lower	Br.
6213.360	6215.079		0.3	0.2			16089.90	Y 2	2B 9	R 0
6213.642	6215.361		0.3	0.3			16089.17			
6213.885	6215.604			0.3			16088.54			
6214.117	6215.836			0.0			16087.94			
6214.287	6216.006		0.1	0.2			16087.50			
6214.507	6216.226		0.6	1.1	1.7		16086.93	3c 1	2a 1	Q 8
6214.507	6216.226		0.6	1.1	1.7		16086.93	X 2	2C 1	R 3
6214.507	6216.226		0.6	1.1	1.7		16086.93	4e 0	2c 3	Q 5
6214.774	6216.493		0.1				16086.24			
6215.268	6216.988			0.2			16084.96			
6215.485	6217.204			0.2			16084.40			
6215.790	6217.510		0.3	1.0	1.3		16083.61	3c 0	2a 0	Q12
6215.790	6217.510		0.3	1.0	1.3		16083.61	4g 1	2a 4	R 4
6215.902	6217.622		0.1				16083.32	3C 1	2A 0	R 2
6216.068	6217.788		0.2	0.3			16082.89			
6216.134	6217.854						16082.72			
6216.281	6218.001			0.2			16082.34			
6216.486	6218.206	G	0.2	0.2			16081.81			
6217.371	6219.091			0.2b			16079.52			
6217.758	6219.478			0.2			16078.52			
6218.001	6219.722			0.2			16077.89			
6218.164	6219.884			0.2			16077.47			
6218.280	6220.000						16077.17			
6218.609	6220.329		0.6	0.4			16076.32			
6218.992	6220.712		0.2	0.2			16075.33			
6219.142	6220.863			0.2			16074.94			
6219.390	6221.111	G	1.1	1.1	1.7		16074.30	3d 1	2c 1	Q 6
6219.390	6221.111	G	1.1	1.1	1.7		16074.30	3E 0	2B 5	R 3
6219.483	6221.204		0.2				16074.06	3E 3	2B10	R 0
6219.731	6221.451		0.1d	0.2d			16073.42			
6220.280	6222.001			0.2			16072.00			
6220.481	6222.202			0.1			16071.48			
6220.663	6222.384		0.2	0.3d			16071.01	4e 1	2c 4	Q 2
6220.807	6222.528			0.3d			16070.64			
6220.927	6222.648		0.1				16070.33			
6221.027	6222.748						16070.07			
6221.236	6222.957		0.1d				16069.53			
6221.751	6223.472	G	1.7	1.7	2.5		16068.20	3e 2	2c 2	P 7
6222.088	6223.809		0.2				16067.33			
6222.224	6223.945		0.2				16066.98			
6222.293	6224.015			0.1			16066.80			
6222.363	6224.084		0.3				16066.62			
6222.475	6224.197		0.2	0.2			16066.33	4e 0	2c 3	R 5
6222.599	6224.321			0.2			16066.01			
6222.735	6224.456		0.2	0.3			16065.66			
6222.948	6224.669		0.2	0.2			16065.11			
6223.103	6224.824			0.2			16064.71			
6223.169	6224.890		0.1				16064.54			
6223.355	6225.076		0.5	0.8	1.4		16064.06			
6223.455	6225.177						16063.80			
6223.544	6225.266		0.3	0.2			16063.57			
6223.703	6225.425		0.2	0.3			16063.16	4E 2	2B14	Q 2
6223.905	6225.627		0.3	0.2			16062.64	3F 0	2B 5	R 1
6224.091	6225.813		0.7	0.3			16062.16			
6224.257	6225.979		0.7	0.5			16061.73			
6224.366	6226.088		0.5	0.3			16061.45			
6224.486	6226.208			1.1	1.1		16061.14			
6224.513	6226.235		0.8	1.0			16061.07			
6224.579	6226.301						16060.90			
6224.815	6226.538	GR	3.6	3.2	2.5	2.14	16060.29	3c 2	2a 2	Q 1

λ_{air}	λ_{vac}	Ref.	I$_1$	I$_2$	I$_3$	I$_4$	ν	Upper	Lower	Br.
6225.040	6226.763		1.0				16059.71	3E 3	2B10	Q 3
6225.141	6226.863		1.0	0.8			16059.45			
6225.312	6227.034		0.7	1.0			16059.01	3d 2	2c 2	P 3
6225.533	6227.255		0.3	0.3			16058.44	4e 1	2c 4	R 2
6225.533	6227.255		0.3	0.3			16058.44	4e 0	2c 3	Q 4
6225.684	6227.406		0.4	0.3			16058.05	3B 4	2A 0	P 4
6225.901	6227.623		0.3	0.8			16057.49			
6226.068	6227.790		0.3	0.2			16057.06			
6226.246	6227.969		0.4	0.2			16056.60			
6226.386	6228.108			0.2			16056.24			
6226.432	6228.155		0.2				16056.12			
6226.545	6228.267			0.2			16055.83			
6226.684	6228.407		0.2	0.2			16055.47	W 1	2C 0	R 1
6226.839	6228.562		0.2	0.3			16055.07			
6227.026	6228.748		0.2	0.2			16054.59			
6227.239	6228.962			0.2			16054.04			
6227.344	6229.066		0.3b				16053.77			
6227.429	6229.152			0.2			16053.55			
6227.600	6229.323		0.1				16053.11			
6227.704	6229.427			0.8	1.3		16052.84			
6227.813	6229.536		0.2				16052.56			
6227.902	6229.625			0.2			16052.33			
6228.023	6229.746		0.3				16052.02			
6228.112	6229.835			0.9	1.4		16051.79			
6228.162	6229.885		0.2				16051.66			
6228.337	6230.060		0.1				16051.21			
6228.515	6230.238		0.2	0.2d			16050.75			
6228.609	6230.332						16050.51			
6228.799	6230.522			0.2			16050.02			
6228.915	6230.638		0.2				16049.72	Y 2	2B 9	P 1
6229.043	6230.766		0.2	0.2			16049.39			
6229.264	6230.988		0.2	0.9	1.4		16048.82	3d 0	2c 0	Q 8
6229.524	6231.248		0.2	0.2			16048.15			
6229.707	6231.430		0.2	0.3			16047.68			
6229.827	6231.551		0.1				16047.37			
6229.932	6231.656		0.5	1.3	1.9		16047.10	3F 1	2B 7	P 5
6229.975	6231.698			1.0	1.5b		16046.99			
6230.029	6231.753		0.6				16046.85			
6230.258	6231.982	GF	2.8	2.6	2.2	1.65	16046.26	3c 2	2a 2	Q 2
6230.487	6232.211		0.2				16045.67			
6230.584	6232.308		0.3	0.4			16045.42			
6230.755	6232.479			0.2d			16044.98			
6230.969	6232.693		0.7	0.9	1.5		16044.43			
6231.105	6232.828		0.2				16044.08			
6231.202	6232.926			0.1			16043.83			
6231.334	6233.058		0.1				16043.49			
6231.392	6233.116			0.2			16043.34			
6231.590	6233.314			0.1	0.1d		16042.83	4F 0	2B10	Q 3
6231.742	6233.466			0.1			16042.44			
6231.920	6233.644		0.1	0.2			16041.98	V 0	2B 4	P 5
6232.134	6233.858		0.9	0.9	0.1d		16041.43	3C 1	2A 0	R 1
6232.134	6233.858		0.9	0.9	0.1d		16041.43	4F 0	2B10	R 1
6232.426	6234.150		0.2	0.2			16040.68			
6232.538	6234.262		0.1	0.2	0.3		16040.39			
6232.670	6234.394			0.8			16040.05	4f 0	2c 3	P 5
6232.740	6234.464		0.2	0.3			16039.87	4e 0	2c 3	R 4
6232.810	6234.534						16039.69			
6233.004	6234.729	GR	2.4	2.2	2.1	1.31	16039.19	2K 5	2B 0	P 6
6233.004	6234.729	GR	2.4	2.2	2.1	1.31	16039.19	3c 1	2a 1	P 4
6233.195	6234.919						16038.70			

λ_{air}	λ_{vac}	Ref.	I_1	I_2	I_3	I_4	ν	Upper	Lower	Br.
6233.315	6235.040		0.2	0.3			16038.39			
6233.552	6235.277		0.2	0.3			16037.78			
6233.875	6235.600			0.2			16036.95	3F 2	2B 9	R 2
6234.066	6235.790			0.3			16036.46			
6234.256	6235.981			0.2			16035.97	4e 0	3c 3	Q 3
6234.470	6236.195			0.2			16035.42	3F 1	2B 7	Q 3
6234.587	6236.311						16035.12			
6234.723	6236.447			0.2			16034.77			
6234.898	6236.622			0.1			16034.32	W 1	2C 0	R 2
6235.041	6236.766		0.2	0.2			16033.95			
6235.193	6236.918			0.1			16033.56			
6235.318	6237.042		0.3				16033.24			
6235.380	6237.105			0.2			16033.08			
6235.563	6237.288		0.6	0.8			16032.61	Z 0	2B 4	P 6
6235.652	6237.377				0.6		16032.38			
6235.734	6237.459	GR	1.7	0.8			16032.17	3d 3	2c 2	Q 4
6235.920	6237.646	GR	1.8	1.9	2.4	1.07	16031.69	3c 0	2a 0	P 7
6236.103	6237.828						16031.22			
6236.166	6237.891		0.2				16031.06			
6236.232	6237.957			0.2	0.4		16030.89			
6236.481	6238.206		0.1	0.3			16030.25			
6236.640	6238.365			0.2			16029.84			
6236.948	6238.673		0.1	0.3			16029.05	4E 1	2C 2	Q 2
6236.948	6238.673		0.1	0.3			16029.05	3D 2	2C 0	Q 7
6237.126	6238.852		0.2	0.2	0.2		16028.59	4E 0	2C 1	Q 2
6237.263	6238.988		0.2	0.2			16028.24			
6237.434	6239.159						16027.80			
6237.457	6239.183	G	1.3	1.6	2.0		16027.74	3c 1	2a 1	Q 9
6237.687	6239.412			0.2			16027.15			
6237.808	6239.533		0.2	0.4	0.2		16026.84	V 0	2B 4	P 2
6237.928	6239.654			0.2			16026.53			
6238.127	6239.852		0.6	0.8	0.2		16026.02	X 2	2C 1	Q 3
6238.391	6240.117	G	2.9	2.8	2.7	1.83	16025.34	3c 2	2a 2	Q 3
6238.621	6240.347		0.3				16024.75			
6238.711	6240.437		0.3	0.7	0.2		16024.52			
6238.898	6240.623		0.2	0.2			16024.04			
6239.193	6240.919			0.8	0.1		16023.28	3d 1	2c 1	P 5
6239.365	6241.091			0.2			16022.84	3A 2	2C 1	R 4
6239.501	6241.227			0.2			16022.49	3F 0	2B 5	P 4
6239.797	6241.523			0.2			16021.73			
6240.000	6241.726			0.2			16021.21	3A 1	2C 0	P 2
6240.132	6241.858			0.1			16020.87			
6240.210	6241.936						16020.67	Z 2	2B 8	P 4
6240.331	6242.057		1.2	0.8	0.2		16020.36			
6240.607	6242.334			0.2			16019.65	4e 0	2c 3	Q 2
6240.607	6242.334			0.2			16019.65	4e 0	2c 3	R 3
6240.806	6242.532		0.2		0.2		16019.14	3E 3	2B10	Q 2
6240.864	6242.591			0.5			16018.99			
6240.927	6242.653		0.2				16018.83	5c 1	2a 5	R 3
6241.180	6242.906	G	1.5	1.5	1.9		16018.18	3e 3	2c 3	P 5
6241.180	6242.906	G	1.5	1.5	1.9		16018.18	4e 1	2c 4	Q 1
6241.496	6243.222		0.1	0.2			16017.37	3D 3	2C 1	Q 4
6241.496	6243.222		0.1	0.2			16017.37	3c 9	2a 8	Q 1
6241.761	6243.487			0.1			16016.69			
6241.811	6243.538				0.1		16016.56			
6242.131	6243.858				0.2		16015.74			
6242.248	6243.975			0.3			16015.44	3A 2	2C 1	R 2
6242.462	6244.189				0.1		16014.89			
6242.563	6244.290			0.0			16014.63	4g 1	2a 4	P 3
6242.712	6244.438				1.2		16014.25			

λ_{air}	λ_{vac}	Ref.	I_1	I_2	I_3	I_4	ν	Upper	Lower	Br.
6242.797	6244.524			0.2	0.1		16014.03			
6243.059	6244.786		0.2	0.4	0.3		16013.36	3E 3	2B10	P 4
6243.292	6245.020		0.2	0.3	0.1		16012.76			
6243.464	6245.191		0.1	0.2			16012.32			
6243.620	6245.347		1.2	0.4	0.2		16011.92	U 1	2B 7	R 4
6243.788	6245.515		1.1	0.4			16011.49			
6244.041	6245.768		0.5	0.4	0.2		16010.84			
6244.193	6245.921			0.1			16010.45			
6244.345	6246.073		0.2	0.3			16010.06	3A 2	2C 1	R 3
6244.591	6246.319		0.2	0.4	0.1		16009.43			
6244.895	6246.623		0.2	0.3			16008.65	Y 2	2B 9	P 2
6245.192	6246.919			0.2	0.4		16007.89	4f 0	2c 3	P 4
6245.629	6247.357		0.4				16006.77	4e 0	2c 3	R 2
6245.879	6247.606		1.2	0.9	0.2		16006.13	3c 9	2a 8	Q 1
6246.078	6247.805			0.0	0.1		16005.62			
6246.148	6247.876						16005.44			
6246.296	6248.024			0.1	0.9		16005.06			
6246.550	6248.278	G	1.3	1.4	2.3		16004.41	3b 4	2a 1	P 5
6246.694	6248.422		0.4	0.1			16004.04	4e 0	2c 3	R 1
6246.901	6248.629		1.1	1.4	1.8		16003.51	3E 3	2C 1	P 5
6246.952	6248.680			1.0b	1.5b		16003.38	3c 0	2a 0	Q13
6247.225	6248.953				0.3		16002.68			
6247.292	6249.020			0.2			16002.51			
6247.498	6249.227			0.2	0.1		16001.98	3E 0	2B 5	Q 7
6247.670	6249.398			0.2			16001.54	4g 0	2a 3	R 0
6247.850	6249.578			0.2			16001.08			
6248.111	6249.840			0.2d			16000.41			
6248.193	6249.922				0.3		16000.20			
6248.330	6250.059			0.3	0.2		15999.85	3E 0	2B 5	P 7
6248.607	6250.336			0.2	0.1		15999.14	3E 3	2B10	P 3
6248.877	6250.606			0.3	1.3		15998.45	3E 3	2C 1	P 5
6248.939	6250.668		0.2				15998.29			
6249.150	6250.879	GR	2.0	1.9	2.2	1.20	15997.75	3c 2	2a 2	Q 4
6249.150	6250.879	GR	2.0	1.9	2.2	1.20	15997.75	3A 2	2C 1	R 1
6249.338	6251.067		0.1				15997.27			
6249.494	6251.223		0.2	0.2	0.2		15996.87	T 0	2C 1	R 1
6249.686	6251.414		0.3	0.3			15996.38			
6249.846	6251.575			0.1			15995.97			
6250.037	6251.766			0.2			15995.48	3E 3	2B10	P 2
6250.604	6252.333		0.2	0.2			15994.03	T 0	2C 1	R 2
6250.874	6252.603		0.1	0.3	0.3		15993.34	3E 3	2B10	P 5
6251.163	6252.892			0.2			15992.60			
6251.346	6253.076			0.1			15992.13			
6251.534	6253.263			0.2			15991.65			
6251.765	6253.494		1.1	0.4	0.1b		15991.06	3C 1	2A 0	R 0
6251.913	6253.643		0.2	0.7			15990.68			
6252.081	6253.811		0.2	0.3			15990.25	W 1	2C 0	R 3
6252.559	6254.288			0.3	0.3		15989.03	3F 0	2B 5	Q 2
6252.559	6254.288			0.3	0.3		15989.03	4d 1	2c 4	Q 1
6253.173	6254.902				0.1		15987.46			
6253.630	6255.360				0.1		15986.29	3F 1	2B 7	R 1
6253.779	6255.509			0.3	0.3		15985.91	W 1	2C 0	Q 1
6253.869	6255.599				0.2		15985.68			
6254.111	6255.841			0.1			15985.06			
6254.186	6255.916				0.2		15984.87			
6254.252	6255.982		0.2				15984.70			
6254.546	6256.276			0.2			15983.95			
6254.698	6256.428			0.1			15983.56	X 2	2C 1	P 3
6254.953	6256.683		0.1	0.0			15982.91			
6255.039	6256.769				0.3		15982.69			

λ_{air}	λ_{vac}	Ref.	I_1	I_2	I_3	I_4	ν	Upper	Lower	Br.
6255.148	6256.879	G	1.0	0.7			15982.41		2c	Q 2
6255.387	6257.117			0.2	0.1d		15981.80			
6255.481	6257.211		0.1				15981.56	3D 3	2B10	R 0
6255.904	6257.634		0.2				15980.48	3D 3	2B10	R 4
6255.959	6257.689		1.3	0.8	1.2		15980.34			
6256.233	6257.963			0.2	0.1		15979.64			
6256.389	6258.120						15979.24	4D 0	2C 1	P 1
6256.456	6258.186			0.1			15979.07	3D 3	2B10	R 1
6256.659	6258.390		0.1	0.4	1.3		15978.55			
6256.906	6258.637			0.2			15977.92			
6257.063	6258.794			0.2	0.1d		15977.52			
6257.212	6258.942		0.2	0.3			15977.14			
6257.388	6259.119		0.1	0.1			15976.69			
6257.584	6259.315	G	1.5	1.2	0.4		15976.19	3d 3	2c 3	R 1
6257.690	6259.420	G					15975.92			
6257.748	6259.479		1.0	0.9	1.4		15975.77	4f 0	2c 3	P 3
6257.956	6259.687			0.1			15975.24	4E 1	2C 2	Q 3
6258.113	6259.844			0.2			15974.84			
6258.167	6259.898						15974.70	3D 3	2B10	R 2
6258.324	6260.055			0.1			15974.30			
6258.453	6260.185			0.2			15973.97	3c 3	2a 3	R 3
6258.653	6260.384		0.2b	0.8	1.4		15973.46			
6258.892	6260.623			0.0			15972.85			
6259.073	6260.804			0.2			15972.39			
6259.245	6260.976			0.1			15971.95			
6259.425	6261.157			0.3	1.2		15971.49			
6259.590	6261.321			0.1			15971.07			
6259.782	6261.513		0.1	0.2d			15970.58			
6259.884	6261.615				0.2		15970.32			
6259.954	6261.686			0.1			15970.14			
6260.131	6261.862		0.2	0.9	1.5		15969.69			
6260.221	6261.952		0.7				15969.46			
6260.366	6262.098			0.1			15969.09			
6260.452	6262.184				0.2		15968.87	4e 0	2c 3	Q 1
6260.895	6262.627		0.2	0.6	0.2		15967.74			
6260.962	6262.694						15967.57			
6261.056	6262.788			0.1			15967.33			
6261.723	6263.455			0.2			15965.63			
6261.993	6263.725		0.2	0.3			15964.94			
6262.201	6263.933		0.1	0.3	0.5		15964.41			
6262.358	6264.090						15964.01			
6262.495	6264.228	GR	1.9	2.0	2.5	1.21	15963.66	3c 1	2a 1	Q10
6262.495	6264.228	GR	1.9	2.0	2.5	1.21	15963.66	3c 2	2a 2	Q 5
6262.578	6264.310		0.4				15963.45			
6262.833	6264.565		0.1	0.3	0.2		15962.80			
6263.001	6264.734			0.2			15962.37			
6263.166	6264.899			0.2			15961.95			
6263.355	6265.087			0.2			15961.47			
6263.570	6265.303		0.1	0.2			15960.92	4E 0	2C 1	Q 3
6264.238	6265.970		0.0	0.1	0.8		15959.22	3D 3	2B10	R 3
6264.473	6266.206			0.2			15958.62			
6264.552	6266.284				0.2		15958.42			
6264.615	6266.347			0.2			15958.26			
6264.850	6266.583		0.2	1.0	1.9		15957.66	3d 1	2c 1	Q 7
6265.046	6266.779		0.2	0.1			15957.16	Y 2	2B 9	P 3
6265.160	6266.893		1.1				15956.87			
6265.254	6266.987			0.3	0.2		15956.63			
6265.408	6267.141			0.1			15956.24	3F 2	2B 9	Q 3
6265.561	6267.294			0.2			15955.85	Z 2	2B 8	P 5
6265.722	6267.455			0.1			15955.44	5c 2	2a 6	Q 1

λ_{air}	λ_{vac}	Ref.	I_1	I_2	I_3	I_4	ν	Upper	Lower	Br.
6265.792	6267.526						15955.26			
6265.902	6267.635		0.8	0.4			15954.98	4D 0	2C 1	Q 1
6266.209	6267.942		0.1	0.2			15954.20			
6266.452	6268.186				0.6		15953.58			
6266.609	6268.343			0.2b			15953.18			
6266.704	6268.437				0.1		15952.94			
6267.089	6268.822		0.1	0.1			15951.96			
6267.395	6269.129		0.1	0.2			15951.18	X 1	2C 0	P 4
6267.643	6269.376			0.2	0.1		15950.55			
6267.792	6269.526		0.2				15950.17			
6267.949	6269.683	GR	1.9	1.7	1.9		15949.77	3d 3	2c 3	R 2
6268.000	6269.734	R					15949.64			
6268.201	6269.934						15949.13			
6268.252	6269.986		0.1	0.2			15949.00			
6268.409	6270.143		0.2	0.2	0.1		15948.60			
6268.558	6270.292			0.2			15948.22	4f 2	3b 0	R 4
6268.731	6270.465		0.3	0.3	0.1		15947.78	W 1	2C 0	P 1
6268.841	6270.575		0.1				15947.50			
6268.987	6270.721		0.3	0.2			15947.13			
6269.117	6270.851			0.2			15946.80			
6269.419	6271.153		0.1	0.1			15946.03	3c 1	2a 1	P 5
6269.844	6271.578		0.2		0.1		15944.95	3F 3	2B11	R 2
6269.934	6271.669			0.1			15944.72	5c 1	2a 5	R 2
6270.080	6271.814		0.1				15944.35			
6270.229	6271.964			0.3			15943.97			
6270.300	6272.034		0.2				15943.79			
6270.375	6272.109						15943.60	W 1	2C 0	R 4
6270.540	6272.274	GR	2.2	1.9	1.6	1.18	15943.18	3c 3	2a 3	R 2
6270.627	6272.361		0.5				15942.96	3D 3	2C 1	P 4
6270.745	6272.479		0.2				15942.66			
6270.843	6272.577		0.2	0.2			15942.41			
6270.969	6272.703		0.2				15942.09			
6271.067	6272.802		0.8	0.3			15941.84			
6271.118	6272.853				0.2		15941.71			
6271.311	6273.046	GR	2.7	2.3	1.7	1.42	15941.22	3c 2	2a 2	P 2
6271.390	6273.124						15941.02			
6271.547	6273.282		0.3				15940.62			
6271.626	6273.360		0.4	0.3			15940.42	4E 1	2B12	Q 4
6271.815	6273.549		0.2				15939.94			
6271.897	6273.632				0.1b		15939.73			
6271.960	6273.695		0.1				15939.57			
6272.074	6273.809			0.2			15939.28			
6272.184	6273.919		0.1				15939.00			
6272.287	6274.022			0.0			15938.74			
6272.369	6274.104		0.1				15938.53			
6272.425	6274.159			0.2			15938.39			
6272.503	6274.238				0.2		15938.19			
6272.649	6274.384		1.2	0.9	1.2		15937.82	2A 4	2B 1	R 1
6272.893	6274.628			0.3d	0.5		15937.20			
6273.157	6274.892			0.2	0.2		15936.53			
6273.511	6275.246			0.2			15935.63			
6273.574	6275.309				0.1		15935.47			
6273.633	6275.368			0.1			15935.32			
6273.763	6275.498		0.2	0.2			15934.99	W 1	2C 0	Q 2
6273.826	6275.561				0.3		15934.83			
6273.964	6275.699		0.2	0.3			15934.48	3E 0	2B 5	R 2
6274.090	6275.825			0.1			15934.16	4D 0	2C 1	R 1
6274.247	6275.982			0.2			15933.76	3A 1	2C 0	P 3
6274.377	6276.113		0.3				15933.43			
6274.480	6276.215	G	1.6	1.0	1.5		15933.17	2A 4	2B 1	R 2

λ_air	λ_vac	Ref.	I_1	I_2	I_3	I_4	ν	Upper	Lower	Br.
6274.661	6276.396		0.2				15932.71			
6274.802	6276.538		1.5				15932.35			
6274.838	6276.573	G	2.2	1.3	2.4	1.30	15932.26			
6274.877	6276.613			1.5b	1.5b		15932.16			
6274.917	6276.652						15932.06			
6274.944	6276.680	G	1.5b	2.1			15931.99	3d 3	2c 3	Q 1
6274.988	6276.723						15931.88			
6275.251	6276.987		0.1	0.2			15931.21			
6275.488	6277.223		0.2	0.2			15930.61	3F 1	2B 7	P 4
6275.744	6277.480			0.1			15929.96			
6275.898	6277.633				0.3		15929.57			
6275.949	6277.685			0.2			15929.44			
6276.161	6277.897		0.5	1.2	1.8		15928.90	3c 0	2a 0	P 8
6276.370	6278.106		0.2	0.2			15928.37			
6276.591	6278.327	GR	2.0	1.3	1.2		15927.81	2A 4	2B 1	R 0
6276.808	6278.544			0.2	0.1		15927.26			
6276.863	6278.599		0.2				15927.12			
6277.056	6278.792	G	1.8	1.8	2.1		15926.63	3d 2	2c 2	Q 5
6277.111	6278.847			1.5b			15926.49	V 1	2B 5	R 2
6277.218	6278.954		1.0		0.3		15926.22	3C 1	2A 0	Q 1
6277.533	6279.269		0.2	0.2d			15925.42	3F 1	2B 7	Q 2
6277.699	6279.435						15925.00			
6277.773	6279.510		0.6	0.1			15924.81			
6278.093	6279.829			0.2d			15924.00			
6278.262	6279.999						15923.57			
6278.369	6280.105	R	0.4	1.2	1.9		15923.30	3c 2	2a 2	Q 6
6278.369	6280.105	R	0.4	1.2	1.9		15923.30	3d 0	2c 0	Q 9
6278.609	6280.346		0.1	0.0			15922.69			
6278.700	6280.437			0.2	0.1		15922.46			
6278.897	6280.634		0.1	0.1			15921.96	3F 2	2B 9	R 1
6279.106	6280.843		0.2	0.3			15921.43			
6279.292	6281.028			0.3	0.2		15920.96			
6279.481	6281.218		0.2	0.5	0.2		15920.48			
6279.674	6281.411			0.2			15919.99			
6279.792	6281.529		0.3				15919.69			
6279.891	6281.628			1.1	1.9		15919.44	3c 1	2a 1	P 6
6279.891	6281.628			1.1	1.9		15919.44	3c 0	2a 0	Q14
6279.927	6281.664		0.2				15919.35			
6280.175	6281.912			0.1	0.2d		15918.72			
6280.467	6282.204			0.3	0.1d		15917.98			
6280.637	6282.374						15917.55			
6280.743	6282.480			0.1			15917.28			
6280.802	6282.540		0.1				15917.13			
6280.972	6282.709				0.1		15916.70			
6281.051	6282.788			0.2			15916.50			
6281.122	6282.859		0.1				15916.32			
6281.268	6283.005						15915.95			
6281.347	6283.084		0.3d	0.4	0.2d		15915.75	3d 2	2c 2	P 4
6281.639	6283.376			0.1			15915.01	4e 1	2c 4	P 3
6281.864	6283.602		0.4	0.2			15914.44			
6282.034	6283.771		0.1	0.1			15914.01	3d 0	2c 0	P 7
6282.034	6283.771		0.1	0.1			15914.01	4e 1	2c 4	P 2
6282.267	6284.004	G	1.3	1.0	1.3		15913.42	3d 3	2c 3	P 1
6282.373	6284.111		0.2				15913.15	3B 4	2A 0	P 5
6282.543	6284.281		0.2	0.4	0.2		15912.72	2A 4	2B 1	R 3
6282.879	6284.616		0.1	0.2	0.1		15911.87			
6283.084	6284.822		0.2	0.6	0.2		15911.35	3C 1	2A 0	Q 2
6283.321	6285.059	G	1.3	1.2	1.6		15910.75	3d 3	2c 3	R 3
6283.582	6285.320			0.2			15910.09	Y 2	2B 9	P 4
6283.921	6285.659		0.2	0.3			15909.23			

λ_{air}	λ_{vac}	Ref.	I_1	I_2	I_3	I_4	ν	Upper	Lower	Br.
6284.091	6285.829			0.0			15908.80			
6284.154	6285.892		0.1				15908.64			
6284.257	6285.995			0.2			15908.38			
6284.470	6286.208		0.2				15907.84	4F 0	2B10	P 3
6284.526	6286.264			0.2			15907.70			
6284.656	6286.394		0.2	0.2			15907.37	3F 0	2B 5	P 3
6284.814	6286.552		0.4	0.3			15906.97			
6284.925	6286.663		0.3	0.2	0.2		15906.69			
6285.043	6286.782		0.3	0.3			15906.39	U 1	2B 7	R 3
6285.162	6286.900		1.0	0.4			15906.09			
6285.205	6286.944						15905.98			
6285.384	6287.122	GR	2.9	2.6	2.1	1.67	15905.53	3c 3	2a 3	R 1
6285.384	6287.122	GR	2.9	2.6	2.1	1.67	15905.53	4d 1	2c 4	P 1
6285.458	6287.197						15905.34			
6285.616	6287.355		0.4				15904.94			
6285.711	6287.450		0.5	0.4	0.1		15904.70			
6285.905	6287.643		0.2	0.2	0.1		15904.21	4E 0	2C 1	P 4
6286.102	6287.841		0.2	1.0	1.2		15903.71			
6286.284	6288.023		0.2	0.2			15903.25			
6286.462	6288.201		0.1	0.2	0.1		15902.80	4e 1	2c 4	Q 3
6286.648	6288.387		0.1	0.2			15902.33	3D 3	2B10	P 2
6286.727	6288.466				0.1		15902.13			
6286.810	6288.549		0.2	0.3			15901.92			
6287.012	6288.750		0.3	1.0	1.7		15901.41	4E 0	2C 1	Q 4
6287.241	6288.980		0.1	0.1			15900.83			
6287.360	6289.099			0.2			15900.53			
6287.455	6289.193		0.1				15900.29			
6287.557	6289.296			0.3			15900.03			
6287.664	6289.403		0.1		0.4		15899.76	4e 0	2c 3	Q 2
6287.791	6289.530			0.1			15899.44			
6287.949	6289.688		0.2				15899.04			
6288.091	6289.830			0.1			15898.68			
6288.277	6290.016						15898.21			
6288.677	6290.416				0.6		15897.20			
6288.744	6290.483		0.1	0.2			15897.03			
6289.025	6290.764			0.2			15896.32	7c 0	2a 5	R 1
6289.171	6290.911			0.2			15895.95			
6289.440	6291.180			0.2			15895.27			
6289.709	6291.449	G	1.5	1.7	2.1		15894.59	3c 1	2a 1	Q11
6289.737	6291.477						15894.52			
6289.963	6291.702		0.1				15893.95			
6290.034	6291.773			0.2	0.2		15893.77			
6290.291	6292.031			0.2	0.1		15893.12			
6290.350	6292.090						15892.97			
6290.489	6292.229			0.3	1.2		15892.62	X 0	2B 6	R 3
6290.707	6292.446			0.1			15892.07			
6290.841	6292.581		0.1	0.2			15891.73			
6291.257	6292.997			0.1			15890.68			
6291.396	6293.135			0.2			15890.33			
6291.558	6293.298						15889.92	4E 1	2C 2	Q 4
6291.633	6293.373		0.3	0.3	0.1		15889.73	4F 0	2B10	Q 2
6291.843	6293.583		0.3	0.9	1.4		15889.20	3C 2	2A 1	R 3
6291.843	6293.583		0.3	0.9	1.4		15889.20	3C 1	2A 0	Q 3
6292.088	6293.829		0.3	0.3			15888.58			
6292.180	6293.920				0.1		15888.35			
6292.635	6294.375		0.2	0.2	0.0		15887.20	4E 1	2B12	Q 3
6292.635	6294.375		0.2	0.2	0.0		15887.20	4g 0	2a 3	R 1
6292.952	6294.692			0.1			15886.40			
6293.130	6294.871		0.3	0.3			15885.95			
6293.320	6295.061			0.1			15885.47			

λ_{air}	λ_{vac}	Ref.	I_1	I_2	I_3	I_4	ν	Upper	Lower	Br.
6293.911	6295.651		0.1	0.2	0.1		15883.98	4E 1	2B12	Q 1
6293.911	6295.651		0.1	0.2	0.1		15883.98	3A 2	2C 1	Q 2
6293.911	6295.651		0.1	0.2	0.1		15883.98	X 0	2B 6	R 4
6294.236	6295.976		0.2	0.2			15883.16			
6294.588	6296.329		0.2	0.1			15882.27	3A 2	2C 1	Q 1
6294.588	6296.329		0.2	0.1			15882.27	5c 1	2a 5	R 1
6294.802	6296.543			0.0			15881.73	3F 3	2B11	Q 3
6294.802	6296.543			0.0			15881.73	3E 0	2B 5	P 6
6295.012	6296.753			0.1			15881.20			
6295.211	6296.952		0.2	0.3	1.3		15880.70			
6295.441	6297.182		0.1	0.1			15880.12			
6295.845	6297.586		0.1				15879.10	3E 3	2C 1	P 6
6295.936	6297.677			0.1			15878.87	4D 0	2C 1	R 2
6296.043	6297.784		0.1				15878.60			
6296.134	6297.876				0.1d		15878.37			
6296.241	6297.983		0.1	0.2			15878.10			
6296.420	6298.161				0.2		15877.65			
6296.475	6298.217		0.2	0.7			15877.51			
6296.622	6298.364	G	1.2	1.5	2.2		15877.14	3c 2	2a 2	Q 7
6296.836	6298.578		0.1	0.1	0.4		15876.60	3E 0	2B 5	Q 6
6296.995	6298.736			0.2			15876.20	T 0	2C 1	P 1
6297.193	6298.935		0.3	0.4	0.3		15875.70			
6297.451	6299.193			0.1			15875.05			
6297.546	6299.288		0.2				15874.81			
6297.689	6299.431			0.2d	0.5		15874.45			
6297.959	6299.701		0.3	0.3			15873.77	3A 2	2B10	R 4
6298.086	6299.828			0.1			15873.45			
6298.149	6299.891		0.2		0.2		15873.29			
6298.296	6300.038		0.2	0.3			15872.92			
6298.383	6300.125						15872.70			
6298.614	6300.356	GR	1.7	1.4	1.6		15872.12	3d 3	2c 3	Q 2
6298.733	6300.475		0.1				15871.82			
6298.844	6300.586		0.4	0.3	0.1		15871.54	4d 0	2c 3	P 1
6298.959	6300.701		0.3	0.2			15871.25			
6299.078	6300.820		0.4				15870.95			
6299.126	6300.868			0.6			15870.83			
6299.173	6300.915		0.9		0.2		15870.71			
6299.423	6301.165	GR	3.1	2.8	2.3	1.81	15870.08	X 0	2B 6	R 2
6299.423	6301.165	GR	3.1	2.8	2.3	1.81	15870.08	3c 2	2a 2	P 3
6299.654	6301.396		0.6				15869.50			
6299.745	6301.487		0.7	0.5	0.1		15869.27			
6299.935	6301.678		0.3	0.2			15868.79			
6300.150	6301.892		0.2	0.1			15868.25	4E 1	2B12	Q 2
6300.150	6301.892		0.2	0.1			15868.25	4E 0	2C 1	P 3
6300.336	6302.079		0.3	0.2			15867.78			
6300.428	6302.170						15867.55			
6300.587	6302.329	G	1.8	1.1	1.3		15867.15	2A 4	2B 1	P 1
6300.682	6302.424		0.4				15866.91			
6300.861	6302.603		0.2	0.1			15866.46			
6301.071	6302.814		0.1				15865.93	X 2	2C 1	Q 4
6301.282	6303.024		0.2	0.2			15865.40			
6301.397	6303.139				0.3		15865.11			
6301.480	6303.223		0.1				15864.90	4e 0	2c 3	P 2
6301.647	6303.390		0.1				15864.48			
6301.822	6303.565		0.2	0.2	0.1d		15864.04	4e 0	2c 3	P 3
6301.822	6303.565		0.2	0.2	0.1d		15864.04	3F 2	2B 9	Q 2
6302.012	6303.755		0.2	0.2	0.2d		15863.56			
6302.271	6304.014	G	1.5	1.5	2.0		15862.91	3d 3	2c 3	R 4
6302.549	6304.292		0.2		0.1d		15862.21	4E 2	2C 3	P 3
6302.755	6304.498		0.2	0.2			15861.69			

λ_{air}	λ_{vac}	Ref.	I$_1$	I$_2$	I$_3$	I$_4$	ν	Upper	Lower	Br.
6302.914	6304.657		0.3				15861.29			
6302.974	6304.717				0.7		15861.14			
6303.030	6304.773		0.7	0.9	1.4		15861.00			
6303.065	6304.808						15860.91	W 1	2C 0	P 2
6303.149	6304.892		0.2				15860.70			
6303.220	6304.964		0.7	0.3			15860.52			
6303.284	6305.027						15860.36			
6303.479	6305.222	GR	2.6	2.2	1.6	1.33	15859.87	3c 3	2a 3	R 0
6303.479	6305.222	GR	2.6	2.2	1.6	1.33	15859.87	3C 1	2A 0	Q 4
6303.550	6305.293						15859.69			
6303.709	6305.453		0.3		0.1		15859.29			
6303.805	6305.548		0.4	0.3			15859.05			
6303.995	6305.739		0.2				15858.57			
6304.115	6305.858			0.5	1.5		15858.27	3d 1	2c 1	P 6
6304.182	6305.926		0.1				15858.10			
6304.349	6306.093		0.1	0.1			15857.68			
6304.453	6306.196			0.1			15857.42			
6304.560	6306.303		0.1	0.2			15857.15			
6304.751	6306.494		0.2				15856.67	W 1	2C 0	Q 3
6305.017	6306.761		0.3	0.3			15856.00	3D 3	2C 1	Q 5
6305.017	6306.761		0.3	0.3			15856.00	4d 1	2c 4	R 1
6305.311	6307.055		0.2	0.3			15855.26	3C 2	2A 1	R 2
6305.451	6307.194			0.1			15854.91			
6305.681	6307.425		0.3	0.3	0.2		15854.33			
6305.912	6307.656			0.2d			15853.75	3F 2	2B 9	P 4
6306.103	6307.847	G	1.2	0.9	1.4		15853.27			
6306.362	6308.105			0.2			15852.62			
6306.684	6308.428			0.2			15851.81			
6306.950	6308.694		0.1				15851.14	4e 0	2c 3	Q 3
6307.046	6308.790			0.1			15850.90			
6307.273	6309.017			0.2			15850.33			
6307.408	6309.152		0.2	0.2			15849.99	X 0	2B 6	R 1
6307.627	6309.371		0.3	0.2	0.4		15849.44			
6307.806	6309.550			0.1			15848.99			
6307.965	6309.710			0.2			15848.59	X 2	2B10	R 0
6308.578	6310.323		0.2	0.2			15847.05	3A 2	2C 1	Q 3
6308.841	6310.586			0.1			15846.39			
6308.996	6310.741						15846.00			
6309.064	6310.809		1.2	0.8	1.2		15845.83			
6309.315	6311.059			0.2			15845.20	3D 3	2B10	P 3
6309.785	6311.529		0.3	0.3	0.1		15844.02			
6309.964	6311.709		0.1	0.2			15843.57	U 1	2B 7	R 2
6310.466	6312.211		0.1	0.2			15842.31	3A 1	2C 0	P 4
6310.466	6312.211		0.1	0.2			15842.31	3F 1	2B 7	P 3
6310.573	6312.318				0.1		15842.04			
6310.705	6312.450		0.2				15841.71			
6310.816	6312.561			0.4	1.3		15841.43			
6311.091	6312.836						15840.74			
6311.215	6312.960		0.2b	0.4b	1.2		15840.43			
6311.318	6313.064						15840.17			
6311.665	6313.410		0.2	0.2			15839.30	3A 2	2C 1	P 1
6311.956	6313.701		0.2	0.3	1.2		15838.57			
6312.107	6313.853		0.1				15838.19	X 2	2B10	R 1
6312.271	6314.016		0.5				15837.78			
6312.318	6314.064			0.2d	0.2		15837.66			
6312.498	6314.243		0.1				15837.21			
6312.705	6314.451			0.1d	0.1		15836.69			
6312.924	6314.670			0.2			15836.14	4D 0	2C 1	P 2
6313.183	6314.929			0.3	0.2		15835.49	3F 1	2C 0	R 5
6313.415	6315.161			0.2			15834.91			

λ_{air}	λ_{vac}	Ref.	I_1	I_2	I_3	I_4	ν	Upper	Lower	Br.
6313.634	6315.380		0.2	0.4	1.2		15834.36			
6313.885	6315.631			0.1	0.1		15833.73			
6314.057	6315.803			0.1			15833.30			
6314.156	6315.902		0.1		0.1		15833.05			
6314.240	6315.986			0.2			15832.84			
6314.440	6316.186		0.6	1.2	1.4		15832.34	3c 0	2a 0	Q15
6314.647	6316.393			0.2			15831.82			
6314.938	6316.684		0.2	0.8	1.5		15831.09	U 1	2B 7	R 1
6315.185	6316.932			0.2			15830.47	4g 0	2a 3	P 2
6315.325	6317.071				0.2		15830.12			
6315.429	6317.175			0.1			15829.86	4g 1	2a 4	P 4
6315.429	6317.175			0.1			15829.86	4e 0	2c 3	P 4
6315.716	6317.463			0.2			15829.14			
6315.940	6317.686						15828.58			
6316.031	6317.778		0.3	0.3			15828.35	X 0	2B 6	R 0
6316.367	6318.113	GR	1.4	1.2	1.4		15827.51	3d 3	2c 3	P 2
6316.750	6318.496			0.2			15826.55			
6316.905	6318.652		0.1				15826.16			
6316.989	6318.736			0.1			15825.95			
6317.233	6318.979		0.5	1.0	1.5		15825.34	3c 2	2a 2	Q 8
6317.440	6319.187			0.2	0.2		15824.82	3D 0	2B 5	R 8
6317.688	6319.435			0.3	1.2		15824.20	3C 1	2A 0	Q 5
6317.688	6319.435			0.3	1.2		15824.20	X 2	2B10	R 2
6317.688	6319.435			0.3	1.2		15824.20	4d 0	2c 3	R 1
6317.879	6319.626			0.1			15823.72			
6318.151	6319.898			0.2	0.2		15823.04			
6318.343	6320.090		0.1		0.3		15822.56	U 1	2B 7	R 0
6318.398	6320.146			0.2			15822.42	4d 0	2c 3	Q 1
6318.398	6320.146			0.2			15822.42	5c 1	2a 5	R 0
6318.634	6320.381	G	1.3	1.5	2.1		15821.83	3c 0	2a 0	P 9
6318.890	6320.637		0.3	0.6	0.7		15821.19	3e 1	2a 1	Q12
6319.241	6320.989			0.2	0.1		15820.31	4D 0	2C 1	Q 2
6319.816	6321.564		0.1	0.2			15818.87	4D 0	2C 1	R 3
6320.020	6321.768			0.2			15818.36			
6320.136	6321.884		0.3				15818.07			
6320.208	6321.956						15817.89			
6320.244	6321.992		0.5		0.8b		15817.80			
6320.396	6322.144	GR	2.2	1.8	1.7	1.14	15817.42	2A 4	2B 1	P 2
6320.552	6322.299						15817.03			
6320.612	6322.359		0.2	0.4	0.3		15816.88			
6320.739	6322.487		0.5	0.3			15816.56	7c 0	2a 5	Q 3
6320.891	6322.639		0.1				15816.18			
6320.951	6322.699			0.0			15816.03			
6321.047	6322.795				0.1		15815.79			
6321.127	6322.875		0.1	0.1			15815.59			
6321.491	6323.239			0.2			15814.68			
6321.611	6323.359				1.2		15814.38			
6321.691	6323.439		0.1	0.3			15814.18			
6321.839	6323.587			0.0			15813.81	4e 0	2c 3	Q 4
6322.043	6323.791			0.1	0.2		15813.30			
6322.226	6323.975			0.1			15812.84			
6322.550	6324.299		0.1				15812.03			
6322.674	6324.423			0.1	0.1		15811.72			
6322.854	6324.603		0.2				15811.27			
6323.054	6324.803		0.1	0.2			15810.77			
6323.354	6325.103			0.2	0.1		15810.02			
6323.414	6325.163		0.1				15809.87			
6323.482	6325.231			0.0			15809.70	3d 3	2c 3	P 5
6323.626	6325.375		0.9	0.9			15809.34	3C 2	2A 1	R 1
6323.626	6325.375		0.9	0.9			15809.34	T 0	2C 1	P 2

λ_{air}	λ_{vac}	Ref.	I_1	I_2	I_3	I_4	ν	Upper	Lower	Br.
6323.838	6325.587	G	1.3	1.2	1.6		15808.81			
6324.130	6325.879		0.2	0.2	0.2		15808.08	3E 0	2B 5	R 1
6324.270	6326.019						15807.73	X 2	2B10	R 3
6324.386	6326.135		0.3	0.3	0.2		15807.44			
6324.674	6326.423		0.3	0.8	1.3		15806.72			
6324.914	6326.663		0.2				15806.12			
6325.050	6326.799				0.1		15805.78			
6325.170	6326.919		0.2	0.2			15805.48			
6325.335	6327.084			0.1			15805.07	V 1	2B 5	R 1
6325.587	6327.336		0.3	0.3			15804.44	4e 0	2c 3	P 5
6325.783	6327.532		0.2				15803.95	3F 3	2B11	Q 2
6325.935	6327.684		0.2	0.2			15803.57			
6326.267	6328.017	GR	1.6	1.2	1.3		15802.74	2K 4	2B 0	R 2
6326.267	6328.017	GR	1.6	1.2	1.3		15802.74	3f 0	2c 1	R 5
6326.343	6328.093		0.2				15802.55			
6326.487	6328.237		0.5	0.3	0.1		15802.19			
6326.604	6328.353	R	0.4	0.2			15801.90			
6326.724	6328.473		0.8	0.6			15801.60			
6326.896	6328.645	R	1.0	0.7	0.2		15801.17	3C 1	2A 0	P 2
6327.060	6328.809	GR	3.3	3.0	2.2	1.96	15800.76	3c 3	2a 3	Q 1
6327.104	6328.854			1.5b			15800.65			
6327.280	6329.030		0.9				15800.21	2K 4	2B 0	R 1
6327.388	6329.138		1.0	0.7	0.2		15799.94			
6327.577	6329.326		0.4	0.2			15799.47	3E 0	2B 5	P 5
6327.777	6329.526		0.2		0.1		15798.97			
6327.845	6329.595			0.2			15798.80	3E 1	2B 7	R 3
6327.941	6329.691		0.3				15798.56			
6328.049	6329.799		1.0	0.8	0.2		15798.29	2K 4	2B 0	R 3
6328.165	6329.915			0.2			15798.00			
6328.326	6330.075		0.2				15797.60			
6328.506	6330.256		0.3	0.2			15797.15			
6328.690	6330.440		0.2				15796.69			
6328.750	6330.500			0.0			15796.54			
6328.939	6330.689		0.2				15796.07			
6329.079	6330.829		0.2	0.2			15795.72	5c 1	2a 5	Q 4
6329.327	6331.077	GR	1.9	1.8	0.9		15795.10	3d 3	2c 3	Q 3
6329.383	6331.133	R					15794.96			
6329.511	6331.262						15794.64			
6329.648	6331.398	G	2.4		2.0		15794.30			
6329.696	6331.446				1.5b		15794.18			
6329.752	6331.502			1.5b			15794.04			
6329.816	6331.566	GR	1.5b	2.3	2.4	1.40	15793.88	3c 2	2a 2	P 4
6329.816	6331.566	GR	1.5b	2.3	2.4	1.40	15793.88	3D 3	2B10	P 4
6329.980	6331.731						15793.47			
6330.105	6331.855		0.2				15793.16			
6330.165	6331.915			0.2			15793.01			
6330.285	6332.035		0.1				15792.71			
6330.337	6332.088			0.1			15792.58			
6330.502	6332.252		0.3b	1.1	1.7		15792.17	X 2	2C 1	P 4
6330.742	6332.493		0.1				15791.57	X 1	2B 8	R 1
6330.922	6332.673			0.2			15791.12			
6331.015	6332.765		0.2	0.0			15790.89			
6331.219	6332.970		0.3	0.1			15790.38			
6331.375	6333.126		0.6	0.7	1.2		15789.99			
6331.536	6333.287			0.0			15789.59			
6331.672	6333.423		0.2	0.1			15789.25	3A 2	2C 1	Q 4
6331.793	6333.543			0.1			15788.95			
6331.913	6333.664		0.2	0.3	0.1		15788.65	4b 0	2a 2	R 3
6332.017	6333.768		0.2				15788.39	X 1	2B 8	R 0
6332.145	6333.896		0.2	0.2			15788.07			

λ_{air}	λ_{vac}	Ref.	I_1	I_2	I_3	I_4	ν	Upper	Lower	Br.
6332.270	6334.021		1.5	1.0	0.4		15787.76	2K 4	2B 0	R 0
6332.270	6334.021		1.5	1.0	0.4		15787.76	2K 4	2B 0	R 4
6332.482	6334.233	GR	2.6	2.4	1.9	1.51	15787.23	3c 3	2a 3	Q 2
6332.587	6334.338						15786.97			
6332.795	6334.546		0.2	0.3			15786.45			
6332.980	6334.731		0.2	0.2	0.3		15785.99	4e 0	2c 3	Q 5
6333.188	6334.940		0.5	0.2	0.1		15785.47			
6333.341	6335.092		0.5	0.3			15785.09			
6333.485	6335.237		0.6	0.7	1.2		15784.73	3c 1	2a 1	P 7
6333.714	6335.465		0.1				15784.16			
6333.782	6335.534				0.1		15783.99			
6333.887	6335.638			0.2			15783.73			
6334.007	6335.758			0.1			15783.43			
6334.412	6336.164				0.3		15782.42			
6335.312	6337.063			0.8	1.2		15780.18	X 1	2B 8	R 2
6335.536	6337.288	G	0.1	0.5	1.2		15779.62			
6335.127	6336.879			0.0			15780.64	3F 2	2B 9	P 3
6334.252	6336.003			0.2			15782.82	4g 0	2a 3	P2
6335.765	6337.517			0.2			15779.05			
6335.954	6337.706			0.3	0.3		15778.58	5c 1	2a 5	Q 3
6336.773	6338.525			0.1d	0.2d		15776.54			
6337.038	6338.791	G	1.3	1.0	1.3		15775.88			
6337.123	6338.875		0.2				15775.67			
6337.223	6338.975			0.0			15775.42			
6337.372	6339.124		0.1				15775.05			
6337.488	6339.241			0.2	0.2		15774.76	3E 0	2B 5	Q 5
6337.689	6339.442			0.0			15774.26			
6337.910	6339.663		0.1	0.2	0.1		15773.71			
6338.236	6339.988		0.4	0.2			15772.90	X 0	2B 6	P 1
6338.236	6339.988		0.4	0.2			15772.90	4d 0	2c 3	R 2
6338.400	6340.153				0.1d		15772.49			
6338.513	6340.266		0.1	0.1			15772.21	4f 2	3b 0	R 3
6338.758	6340.511		0.2	0.3	0.7		15771.60			
6339.092	6340.844			0.2			15770.77	3f 1	2c 2	R 6
6339.164	6340.917		0.1		0.2		15770.59			
6339.285	6341.037			0.0			15770.29			
6339.417	6341.170		0.3	0.4	0.2		15769.96			
6339.711	6341.464		0.2b	0.2			15769.23	X 2	2B10	P 2
6339.835	6341.588			0.1	0.2		15768.92			
6340.085	6341.838	G	1.3	1.4	1.7		15768.30	3c 2	2a 2	Q 9
6340.310	6342.063		0.3	0.4			15767.74			
6340.575	6342.328	GR	2.6	2.5	2.3	1.65	15767.08	3c 3	2a 3	Q 3
6340.881	6342.634		0.2	0.4			15766.32	3F 3	2B11	P 4
6341.086	6342.839		0.1	0.2			15765.81			
6341.693	6343.447			0.1			15764.30	5c 1	2a 5	Q 2
6341.693	6343.447			0.1			15764.30	3d 2	2c 2	P 5
6342.293	6344.047			0.1			15762.81			
6342.502	6344.256			0.2			15762.29			
6343.134	6344.888			0.2	0.1		15760.72			
6343.323	6345.077			0.2			15760.25			
6343.557	6345.310			0.4	1.2		15759.67	4d 1	2c 4	R 3
6343.854	6345.608			0.3	1.2		15758.93			
6344.096	6345.850			0.1			15758.33			
6344.176	6345.931				0.1		15758.13			
6344.309	6346.064			0.2d			15757.80			
6344.378	6346.132		0.1				15757.63			
6344.458	6346.213			0.1			15757.43	W 1	2C 0	Q 4
6344.639	6346.394		0.8	0.6	0.2		15756.98	3C 2	2A 1	R 0
6344.639	6346.394		0.8	0.6	0.2		15756.98	4E 1	2C 2	P 4
6344.829	6346.583			0.1			15756.51			

λ_{air}	λ_{vac}	Ref.	I_1	I_2	I_3	I_4	ν	Upper	Lower	Br.
6344.885	6346.639		0.2				15756.37			
6344.962	6346.716			0.2			15756.18			
6345.231	6346.986		0.1	0.2			15755.51	5c 1	2a 5	Q 1
6345.465	6347.220	G	1.6	1.1	1.2		15754.93	2A 4	2B 1	P 3
6345.779	6347.534		0.1				15754.15			
6345.840	6347.594				0.2		15754.00			
6346.114	6347.868		0.4	0.7	0.3		15753.32			
6346.637	6348.392			0.2	0.2d		15752.02	4E 0	2B10	Q 4
6346.637	6348.392			0.2	0.2d		15752.02	4b 0	2a 2	R 2
6346.637	6348.392			0.2	0.2d		15752.02	4d 0	2c 4	Q 2
6346.891	6348.646				0.9d		15751.39			
6346.931	6348.686	G	1.3	1.3	1.3		15751.29			
6347.097	6348.852				0.9		15750.88	5c 0	2a 4	R 2
6347.258	6349.013		0.1				15750.48			
6347.387	6349.142			0.1d	0.1		15750.16			
6347.552	6349.307			0.2			15749.75			
6347.673	6349.428						15749.45			
6347.762	6349.517		0.2	0.7	1.3		15749.23	3A 2	2C 1	Q 5
6348.286	6350.041		0.1	0.2			15747.93			
6348.495	6350.251		0.3	0.3			15747.41	W 1	2C 0	P 3
6348.588	6350.343				0.2		15747.18			
6349.036	6350.791			0.2			15746.07	3E 3	2C 1	P 7
6349.475	6351.231		0.1				15744.98	4e 0	2c 3	P 3
6349.685	6351.440			0.2			15744.46			
6349.886	6351.642		0.7	1.2	1.3		15743.96	3e 1	2a 1	Q13
6350.068	6351.824			0.1			15743.51			
6350.241	6351.997			0.3			15743.08			
6350.370	6352.126				0.2b		15742.76			
6350.463	6352.219			0.2			15742.53			
6350.661	6352.417		0.2	0.8	0.6		15742.04	3c 0	2a 0	Q16
6350.802	6352.558			0.1			15741.69			
6351.008	6352.764			0.2	0.2		15741.18			
6351.097	6352.853		0.1				15740.96			
6351.282	6353.038	G	1.5	1.4	1.8		15740.50	3c 3	2a 3	Q 4
6351.585	6353.341			0.1	0.2		15739.75			
6351.964	6353.720			0.2			15738.81			
6352.013	6353.769		0.1				15738.69			
6352.170	6353.926		1.2	0.7	0.1		15738.30	X 0	2B 6	P 2
6352.368	6354.124			0.1			15737.81			
6352.675	6354.431			0.2			15737.05			
6352.868	6354.625			0.0			15736.57			
6353.276	6355.033						15735.56	U 1	2B 7	P 2
6353.502	6355.259			0.2	0.0		15735.00			
6353.741	6355.497			0.6	1.2		15734.41			
6353.906	6355.663			0.1			15734.00			
6354.108	6355.865		0.3	0.4	0.2		15733.50			
6354.310	6356.067			0.1			15733.00			
6354.669	6356.426			0.2			15732.11			
6354.948	6356.705		1.1	0.7	1.3		15731.42			
6355.170	6356.927			0.6	0.5		15730.87	3A 2	2B10	R 3
6355.170	6356.927			0.6	0.5		15730.87	4d 0	2c 3	R 3
6355.409	6357.166			0.2			15730.28			
6355.619	6357.376		0.1	0.3	0.3		15729.76			
6355.776	6357.534			0.1			15729.37			
6355.885	6357.643		0.2				15729.10			
6356.144	6357.901	G	1.2	0.9	1.3		15728.46	3D 3	2C 1	P 5
6356.326	6358.083			0.0			15728.01			
6356.391	6358.148				0.2		15727.85			
6356.496	6358.253			0.3			15727.59			
6356.702	6358.459				0.3b		15727.08			

λ_{air}	λ_{vac}	Ref.	I_1	I_2	I_3	I_4	ν	Upper	Lower	Br.
6356.754	6358.512	GR	1.2	1.0	1.2		15726.95	3C 1	2A 0	P 3
6356.896	6358.653						15726.60			
6357.045	6358.803	GR	1.2	1.0	1.2		15726.23			
6357.288	6359.046			0.1	0.1		15725.63			
6357.474	6359.232				0.2		15725.17			
6357.733	6359.491				0.2		15724.53			
6357.789	6359.547		0.1	0.2			15724.39			
6358.016	6359.774	GR	1.5	0.9	0.2		15723.83	2K 4	2B 0	P'1
6358.097	6359.854		0.2				15723.63			
6358.178	6359.935						15723.43			
6358.226	6359.984			0.1			15723.31	3D 3	2B10	P 5
6358.436	6360.194			0.1	0.1		15722.79			
6358.647	6360.405			0.6	0.8		15722.27			
6358.901	6360.660			0.1			15721.64	3F 1	2C 0	R 3
6359.039	6360.797			0.2	0.1		15721.30			
6359.278	6361.036			0.0			15720.71			
6359.456	6361.214			0.1	0.2		15720.27	4d 0	2c 3	Q 2
6359.743	6361.501			0.1d			15719.56			
6360.071	6361.829		0.2	0.3	1.3		15718.75	3E 0	2B 5	P 4
6360.734	6362.493			0.1			15717.11			
6361.022	6362.780		0.3	0.7	0.2		15716.40	3d 3	2c 3	P 3
6361.269	6363.027			0.2			15715.79			
6361.406	6363.165			0.1			15715.45			
6361.580	6363.339			0.1			15715.02			
6361.722	6363.481			0.2	0.2		15714.67			
6361.888	6363.647		0.2d	0.3			15714.26			
6361.949	6363.708				0.1		15714.11			
6361.993	6363.752			0.2			15714.00			
6362.123	6363.882			0.3			15713.68			
6362.224	6363.983		0.3	0.5	0.2		15713.43			
6362.337	6364.096		0.3				15713.15	3F 1	2C 0	R 3
6362.479	6364.238	GR	2.4	2.4	2.4	1.51	15712.80	3c 2	2a 2	P 5
6362.706	6364.465		0.2				15712.24			
6362.807	6364.566		0.2	0.4	0.2		15711.99			
6363.042	6364.801			0.2			15711.41			
6363.119	6364.878						15711.22			
6363.277	6365.036			0.2			15710.83	V 1	2B 5	P 4
6363.342	6365.101		0.1				15710.67	4D 0	2C 1	Q 3
6363.398	6365.158			0.2			15710.53			
6363.504	6365.263		0.1				15710.27			
6363.641	6365.401		0.3	0.4			15709.93	3E 1	2B 7	R 2
6363.702	6365.461			0.7	1.2		15709.78	3c 0	2a 0	P10
6363.759	6365.518		0.5	0.5			15709.64	4b 0	2a 2	R 1
6363.986	6365.745			0.1	0.1		15709.08			
6364.184	6365.944			0.1			15708.59			
6364.289	6366.049			0.2	0.2		15708.33			
6364.362	6366.122		0.2				15708.15			
6364.557	6366.317	G	1.5	1.7	2.2	0.98	15707.67			
6364.618	6366.377			1.5b	1.5b		15707.52	3c 3	2a 3	Q 5
6364.666	6366.426		0.3	1.3	1.5		15707.40			
6364.828	6366.588		0.2				15707.00			
6364.954	6366.714		0.5				15706.69			
6364.986	6366.746	G	1.3	1.2			15706.61	3d 3	2c 3	Q 4
6365.051	6366.811		0.5		0.7b		15706.45	3c 2	2a 2	Q10
6365.051	6366.811		0.5		0.7b		15706.45	3A 2	2C 1	P 3
6365.371	6367.131			0.1			15705.66	4d 0	2c 3	R 4
6365.448	6367.208		0.1				15705.47	t 0	2a 1	R 0
6365.598	6367.358				0.1		15705.10	T 0	2C 1	P 3
6365.809	6367.569				0.3		15704.58	X 2	2B10	P 3
6365.943	6367.703		0.5				15704.25	4D 0	2C 1	R 5

λ_air	λ_vac	Ref.	I₁	I₂	I₃	I₄	ν	Upper	Lower	Br.
6366.162	6367.922			0.2	0.1		15703.71	X 1	2B 8	P 2
6366.871	6368.632				0.1		15701.96			
6367.253	6369.013			0.2			15701.02			
6367.463	6369.224		0.3	0.3			15700.50	X 0	2B 6	P 3
6367.577	6369.337			0.3	0.3		15700.22			
6367.614	6369.374			0.2			15700.13			
6367.792	6369.552			0.2	0.1		15699.69			
6367.991	6369.751		0.2	0.7	0.3		15699.20	t 1	2a 2	R 1
6368.234	6369.995			0.6	1.0		15698.60	4D 0	2B10	R 5
6368.234	6369.995			0.6	1.0		15698.60	W 2	2C 1	R 1
6368.400	6370.161			0.2			15698.19	4d 0	2c 3	R 5
6368.615	6370.376			0.5	1.2		15697.66	Y 1	2B 6	R 2
6368.798	6370.559			0.2			15697.21			
6368.964	6370.725		0.2	0.6	1.2		15696.80	3E 0	2B 5	Q 4
6369.139	6370.900			0.1			15696.37			
6369.309	6371.070		0.1	0.2	0.2		15695.95	Y 1	2B 6	R 1
6369.545	6371.306			0.2	0.3		15695.37			
6369.666	6371.427			0.2			15695.07	3D 0	2B 5	R 7
6369.788	6371.549			0.2			15694.77			
6369.955	6371.716		0.4				15694.36	4E 0	2B10	Q 1
6370.044	6371.805		0.5		1.5		15694.14			
6370.089	6371.850	G	1.7	1.4	1.8	1.01	15694.03			
6370.154	6371.915		1.5				15693.87	3C 2	2A 1	Q 1
6370.255	6372.016		0.3				15693.62			
6370.393	6372.154		0.5	0.4	0.1		15693.28	3d 1	2c 1	P 7
6370.568	6372.329			0.1			15692.85			
6370.681	6372.443				0.1		15692.57			
6370.799	6372.560			0.2			15692.28	3E 0	2B 5	R 0
6370.957	6372.719				0.2b		15691.89			
6371.010	6372.771			0.2			15691.76			
6371.347	6373.109				0.8		15690.93			
6371.416	6373.178		0.2	0.2			15690.76			
6371.530	6373.291			0.6			15690.48			
6371.635	6373.397		0.2	0.1			15690.22			
6371.753	6373.515		0.2				15689.93			
6371.822	6373.584			0.2			15689.76			
6371.871	6373.632		0.2				15689.64			
6371.948	6373.710		0.6	0.3			15689.45			
6372.058	6373.819		0.4				15689.18			
6372.212	6373.974	GR	2.4	1.8	1.4	1.20	15688.80	3c 3	2a 3	P 2
6372.297	6374.059						15688.59			
6372.452	6374.214		0.3				15688.21	4g 0	2a 3	R 3
6372.452	6374.214		0.3				15688.21	t 0	2a 1	P 1
6372.529	6374.291		0.4	0.2			15688.02			
6372.732	6374.494		0.2	0.1			15687.52	U 1	2B 7	P 3
6372.732	6374.494		0.2	0.1			15687.52	Y 1	2B 6	R 0
6372.988	6374.750			0.2			15686.89			
6373.155	6374.916			0.3	0.1		15686.48			
6373.569	6375.331			0.1			15685.46			
6373.727	6375.490			0.2			15685.07			
6373.927	6375.689			0.3	0.5		15684.58			
6374.199	6375.961			0.2d			15683.91			
6374.508	6376.270		0.9	0.7	0.1		15683.15	V 1	2B 5	R 0
6374.776	6376.538			0.2			15682.49			
6374.861	6376.624		0.1				15682.28			
6375.069	6376.831		1.2	0.7	0.2		15681.77	3C 2	2A 1	Q 2
6375.069	6376.831		1.2	0.7	0.2		15681.77	4E 0	2B10	Q 3
6375.264	6377.026			0.2			15681.29			
6375.455	6377.218		0.2	0.7	1.2		15680.82			
6375.670	6377.433			0.0			15680.29			

λ_{air}	λ_{vac}	Ref.	I_1	I_2	I_3	I_4	ν	Upper	Lower	Br.
6375.829	6377.592		0.2	0.2	0.1		15679.90			
6376.073	6377.836	G	1.7	1.3	1.3	0.98	15679.30	2A 4	2B 1	P 4
6376.073	6377.836	G	1.7	1.3	1.3	0.98	15679.30	5c 0	2a 4	R 1
6376.162	6377.925		0.2				15679.08			
6376.284	6378.047		0.1				15678.78			
6376.402	6378.165		0.5	0.4	0.2		15678.49			
6376.549	6378.312			0.1			15678.13			
6376.720	6378.483			0.1			15677.71			
6376.858	6378.621			0.2			15677.37			
6377.053	6378.816			0.4			15676.89			
6377.114	6378.877				0.2		15676.74			
6377.208	6378.971						15676.51			
6377.415	6379.178	GR	1.5	1.4	1.7		15676.00	3b 2	2a 0	R 2
6377.700	6379.463			0.1			15675.30			
6377.908	6379.671			0.2			15674.79			
6378.229	6379.992			0.4d	0.3b		15674.00			
6378.319	6380.082				1.2		15673.78			
6378.542	6380.306		0.2d	0.2d			15673.23			
6378.713	6380.477			0.1			15672.81			
6378.856	6380.619		0.2				15672.46			
6378.953	6380.717		0.3	0.2			15672.22			
6379.072	6380.835		0.2				15671.93			
6379.206	6380.969	G	2.1	1.5	1.2	1.12	15671.60	2K 4	2B 0	P 2
6379.422	6381.185		0.2				15671.07	3A 1	2B 8	R 2
6379.544	6381.307		0.2	0.3	0.3		15670.77			
6379.731	6381.495			0.2			15670.31			
6379.849	6381.613		0.2	0.3	0.2		15670.02			
6379.983	6381.747		0.2				15669.69			
6380.114	6381.878	GR	2.0	2.0	2.1	1.08	15669.37	3b 2	2a 0	R 1
6380.114	6381.878	GR	2.0	2.0	2.1	1.08	15669.37	5c 1	2a 5	P 2
6380.317	6382.081	G	0.2		0.6		15668.87	3c 3	2a 3	Q 6
6380.448	6382.211		0.2	0.2			15668.55			
6380.659	6382.423			0.2	0.1		15668.03	3E 2	2B 9	R 3
6380.761	6382.525						15667.78			
6380.843	6382.607			0.2			15667.58			
6380.957	6382.721			0.1			15667.30			
6381.099	6382.863			0.2			15666.95			
6381.295	6383.059		0.5	0.7	1.2		15666.47	3f 0	2c 1	R 4
6381.535	6383.299		0.2	0.2			15665.88			
6381.804	6383.568	GR	1.2	0.9	1.2		15665.22			
6382.052	6383.817			0.2	0.1		15664.61			
6382.268	6384.033			0.1			15664.08			
6382.395	6384.159			0.2	0.2		15663.77			
6382.668	6384.432		0.6	0.7	0.5		15663.10	3c 0	2a 0	Q14
6382.668	6384.432		0.6	0.7	0.5		15663.10	4b 0	2a 2	R 0
6382.668	6384.432		0.6	0.7	0.5		15663.10	t 0	2a 1	R 1
6382.668	6384.432		0.6	0.7	0.5		15663.10	4g 1	2a 4	P 5
6382.835	6384.599						15662.69			
6382.900	6384.665	GR	1.7	1.7	2.2	0.95	15662.53	3b 2	2a 0	R 3
6382.977	6384.742		0.2				15662.34			
6383.055	6384.819						15662.15			
6383.149	6384.913		0.3	0.4			15661.92	X 0	2B 6	P 4
6383.202	6384.966				0.2		15661.79	3f 1	2c 2	R 5
6383.275	6385.040			0.1			15661.61			
6383.446	6385.211		0.3	0.8	1.2		15661.19	3C 2	2A 1	Q 3
6383.589	6385.354			0.1			15660.84			
6383.723	6385.488			0.2			15660.51			
6383.947	6385.712			0.1			15659.96			
6384.131	6385.896			0.2	0.1		15659.51			
6384.298	6386.063			0.2			15659.10			

λ_{air}	λ_{vac}	Ref.	I_1	I_2	I_3	I_4	ν	Upper	Lower	Br.
6384.498	6386.263			0.4	1.2		15658.61	4E 0	2B10	Q 2
6384.681	6386.446			0.2	0.5		15658.16			
6384.893	6386.659			0.1			15657.64	W 1	2C 0	Q 5
6385.069	6386.834				0.2d		15657.21			
6385.142	6386.907			0.1			15657.03	W 2	2C 1	R 2
6385.407	6387.172			0.2	0.1		15656.38			
6385.574	6387.340			0.1			15655.97			
6385.750	6387.515			0.2			15655.54			
6386.146	6387.911			0.3	0.3		15654.57			
6386.370	6388.135			0.2			15654.02			
6386.643	6388.409			0.2	0.2		15653.35			
6386.966	6388.731			0.2			15652.56	X 1	2B 8	P 3
6386.966	6388.731			0.2			15652.56	3E 0	2B 5	P 3
6387.227	6388.992			0.2			15651.92	4E 1	2C 2	P 5
6387.488	6389.254		0.1	0.3	0.2		15651.28			
6387.545	6389.311						15651.14	U 1	2B 7	P 5
6387.741	6389.507		0.4	0.4			15650.66			
6387.900	6389.666	GR	1.5	1.3	1.6		15650.27	3b 5	2a 2	R 1
6387.974	6389.740		0.2				15650.09	W 2	2C 1	Q 1
6388.161	6389.927			0.2	0.2		15649.63	V 1	2B 5	P 3
6388.402	6390.168		0.3	1.0	1.2		15649.04	3c 0	2a 0	Q17
6388.574	6390.340			0.2			15648.62			
6388.647	6390.413				0.2		15648.44			
6388.753	6390.520		0.1	0.2			15648.18			
6388.872	6390.638				0.4		15647.89			
6388.962	6390.728		0.2	0.4			15647.67	3F 1	2C 0	R 2
6389.141	6390.908			0.2			15647.23	3f 1	2c 2	R 6
6389.370	6391.136		0.2	0.4	0.2		15646.67	3C 1	2A 0	P 4
6389.370	6391.136		0.2	0.4	0.2		15646.67	4e 0	2c 3	P 4
6389.648	6391.414		0.4	0.4			15645.99			
6389.938	6391.704			0.3	0.1		15645.28	3A 2	2B10	R 2
6389.938	6391.704			0.3	0.1		15645.28	4b 1	2a 3	R 2
6390.183	6391.949			0.1			15644.68			
6390.354	6392.121				0.5		15644.26			
6390.465	6392.231			0.1			15643.99			
6390.726	6392.493			0.2	0.1		15643.35	U 1	2B 7	P 4
6390.726	6392.493			0.2	0.1		15643.35	X 2	2B10	P 4
6390.824	6392.591		0.2				15643.11	7c 0	2a 5	P 3
6391.073	6392.840	GR	1.7	1.4	1.4		15642.50	3b 2	2a 0	R 0
6391.073	6392.840	GR	1.7	1.4	1.4		15642.50	4D 0	2C 1	P 3
6391.073	6392.840	GR	1.7	1.4	1.4		15642.50	3E 0	2B 5	Q 3
6391.155	6392.922		0.2				15642.30	3F 1	2C 0	R 2
6391.343	6393.110		0.1	0.1	0.3		15641.84			
6391.596	6393.363		0.1	0.2	0.1		15641.22	3f 0	2c 1	R 5
6391.723	6393.490			0.1			15640.91	U 1	2B 7	Q 5
6391.882	6393.649		1.1	1.0	0.5		15640.52	3c 2	2a 2	Q11
6391.882	6393.649		1.1	1.0	0.5		15640.52	3b 5	2a 2	R 2
6391.993	6393.760						15640.25			
6392.095	6393.862		0.6	1.2	1.4		15640.00			
6392.532	6394.299				0.4		15638.93			
6392.663	6394.430			0.2			15638.61			
6392.745	6394.512				0.2		15638.41			
6392.827	6394.594			0.1			15638.21			
6393.023	6394.790		0.3	0.9	1.6		15637.73	Y 1	2B 6	P 1
6393.239	6395.007			0.5	0.3d		15637.20	3C 2	2A 1	Q 4
6393.427	6395.195			0.2			15636.74			
6393.534	6395.301				0.2		15636.48			
6393.595	6395.363			0.2d			15636.33			
6393.783	6395.551		0.1				15635.87			
6394.045	6395.812			0.4	0.8		15635.23			

λ_{air}	λ_{vac}	Ref.	I_1	I_2	I_3	I_4	ν	Upper	Lower	Br.
6394.249	6396.017		0.5	0.7	0.3		15634.73	3b 5	2a 2	R 0
6394.483	6396.250			0.1			15634.16			
6394.634	6396.402				0.1		15633.79			
6394.744	6396.512			0.1			15633.52			
6395.166	6396.934			0.2			15632.49			
6395.444	6397.212		0.7	0.3	0.5		15631.81			
6395.669	6397.437			0.0			15631.26			
6395.837	6397.605			0.2			15630.85	t 0	2a 1	R 2
6396.107	6397.875			0.2	0.5		15630.19			
6396.270	6398.039			1.7			15629.79			
6396.553	6398.321	GR	1.1	1.0	1.6		15629.10	3b 2	2a 0	R 4
6396.553	6398.321	GR	1.1	1.0	1.6		15629.10	5c 1	2a 5	P 3
6396.852	6398.620			0.3	0.4b		15628.37			
6396.995	6398.763			0.1			15628.02			
6397.159	6398.927			0.3			15627.62			
6397.216	6398.984				0.2		15627.48	X 0	2B 6	P 5
6397.445	6399.214	GR	1.7	1.5	1.8	0.91	15626.92	3c 2	2a 2	P 6
6397.662	6399.431			0.1			15626.39			
6397.797	6399.566		0.2	0.7	0.5		15626.06			
6397.982	6399.750		0.2	0.2			15625.61			
6398.154	6399.922		0.2	0.2	0.2		15625.19			
6398.260	6400.029			0.1			15624.93			
6398.452	6400.221		0.5	1.0	1.7		15624.46	3c 3	2a 3	Q 7
6398.702	6400.471		0.3	0.2			15623.85	T 0	2B10	R 2
6398.702	6400.471		0.3	0.2			15623.85	W 2	2C 1	P 1
6398.882	6400.651		0.4	0.3			15623.41			
6399.018	6400.786		0.3				15623.08			
6399.120	6400.889		0.4	0.4			15622.83	3A 0	2B 6	R 2
6399.218	6400.987		0.9	0.5	0.1		15622.59			
6399.313	6401.081		0.8				15622.36			
6399.558	6401.327		0.5				15621.76			
6399.714	6401.483		3.0				15621.38			
6399.468	6401.237	GR	3.0	2.7	2.0	1.69	15621.98	3c 3	2a 3	P 3
6399.796	6401.565		0.7	0.4	0.2		15621.18			
6400.021	6401.790		0.3	0.8	0.5		15620.63	4f 3	3b 1	R 3
6400.210	6401.979		0.1	0.2			15620.17			
6400.337	6402.106			0.7	1.6		15619.86			
6400.378	6402.147		0.2	0.2	0.2		15619.76			
6400.566	6402.336		0.1	0.2			15619.30			
6400.742	6402.512		0.3	0.3	0.2		15618.87	3E 1	2B 7	R 1
6400.951	6402.721			0.2			15618.36	4d 0	2c 3	Q 3
6401.173	6402.942			0.3			15617.82			
6401.570	6403.340				0.2		15616.85			
6401.857	6403.627			0.2			15616.15			
6402.235	6404.004				0.8		15615.23			
6402.403	6404.172			0.2			15614.82			
6402.567	6404.336			0.1	0.1		15614.42			
6402.886	6404.656		0.1	0.2	0.2		15613.64	3A 2	2C 1	P 4
6403.707	6405.477			0.2	0.2		15611.64	4D 0	2B10	R 4
6403.707	6405.477			0.2	0.2		15611.64	5c 0	2a 4	R 0
6403.941	6405.711	R					15611.07	3E 0	2B 5	Q 2
6403.941	6405.711	R					15611.07	t 0	2a 1	R 3
6404.010	6405.781	GR	1.4	1.3	1.7		15610.90	3d 3	2c 3	Q 5
6404.055	6405.826	R					15610.79			
6404.277	6406.047		0.5	0.3	0.1		15610.25	3d 2	2c 2	P 6
6404.548	6406.318				0.2b		15609.59			
6404.638	6406.408			0.1			15609.37			
6404.839	6406.609			0.2			15608.88			
6404.892	6406.663				0.4b		15608.75			
6404.983	6406.753			0.2			15608.53			

λ_{air}	λ_{vac}	Ref.	I_1	I_2	I_3	I_4	ν	Upper	Lower	Br.
6405.085	6406.856			0.0			15608.28			
6405.250	6407.020	GR	1.0	1.0	1.2		15607.88	2K 4	2B 0	P 3
6405.438	6407.209		0.6	0.7	1.2		15607.42	3C 2	2A 1	Q 5
6405.438	6407.209		0.6	0.7	1.2		15607.42	W 1	2C 0	P 4
6405.598	6407.369			0.2			15607.03			
6405.783	6407.554	G	1.6	1.0	1.7		15606.58			
6405.984	6407.755			0.1			15606.09			
6406.115	6407.886				0.2		15605.77			
6406.177	6407.948			0.1			15605.62			
6406.386	6408.157			0.2	0.2b		15605.11			
6406.637	6408.408			0.1			15604.50			
6406.793	6408.564				0.5		15604.12			
6406.859	6408.630			0.1			15603.96	4g 0	2a 3	P 3
6406.928	6408.699		0.1				15603.79			
6407.187	6408.958			0.2	0.2		15603.16			
6407.413	6409.184		0.2	0.4	0.3		15602.61	4b 1	2a 3	R 1
6407.663	6409.435				0.1		15602.00			
6408.025	6409.796			0.2	0.2		15601.12			
6408.354	6410.125			0.2d			15600.32			
6408.653	6410.425						15599.59			
6408.797	6410.569			0.2	0.2		15599.24			
6408.879	6410.651						15599.04			
6409.027	6410.799		0.2	0.3	0.2		15598.68	3E 0	2B 5	Q 1
6409.027	6410.799		0.2	0.3	0.2		15598.68	3E 0	2B 5	P 2
6409.216	6410.988			0.6	1.2		15598.22	4D 0	2C 1	Q 4
6409.520	6411.292		0.2	0.2			15597.48			
6409.668	6411.440			0.2	0.2		15597.12	4f 2	3b 0	R 2
6409.668	6411.440			0.2	0.2		15597.12	Y 1	2B 6	P 2
6409.972	6411.744			0.1			15596.38	4d 1	2c 4	P 3
6410.314	6412.086			0.2			15595.55			
6410.531	6412.303				0.2b		15595.02	5c 0	2a 4	Q 4
6411.156	6412.928			0.1	0.2		15593.50			
6411.522	6413.294			0.2	0.2		15592.61	5c 1	2a 5	P 4
6411.650	6413.422						15592.30			
6411.785	6413.558	GR	1.5	1.5	2.0		15591.97	T 0	2C 1	P 4
6411.991	6413.763		0.1				15591.47			
6412.040	6413.813			0.2	0.2		15591.35	W 2	2C 1	R 3
6412.263	6414.035		0.1	0.3	0.2		15590.81			
6412.579	6414.352		1.3	0.8	0.3		15590.04	2A 4	2B 1	P 5
6412.752	6414.525		0.9	0.5			15589.62	V 1	2B 5	P 2
6412.995	6414.767		0.2	0.3	0.2		15589.03	3d 3	2c 3	P 4
6413.797	6415.570			0.2			15587.08			
6414.044	6415.817			0.2			15586.48			
6414.357	6416.130				0.2		15585.72	X 1	2B 8	P 4
6414.702	6416.475		0.2	0.2	0.6		15584.88	W 2	2C 1	Q 2
6415.258	6417.031		0.2		0.2		15583.53			
6415.616	6417.390				0.2		15582.66			
6416.028	6417.801			0.1			15581.66			
6416.123	6417.896				0.2		15581.43			
6416.279	6418.053				0.9		15581.05			
6416.456	6418.230		0.2	0.2			15580.62			
6416.679	6418.452		1.1	0.7	0.2		15580.08	3C 2	2A 1	P 2
6416.790	6418.564		0.1				15579.81			
6416.872	6418.646			0.2			15579.61			
6416.942	6418.716		0.2				15579.44			
6417.103	6418.877		0.5	1.1	0.5		15579.05	3c 1	2a 1	Q15
6417.280	6419.054		0.1				15578.62	3A 1	2B 8	R 1
6417.482	6419.256			0.2	0.1		15578.13	W 1	2B 8	R 4
6417.552	6419.326		0.1				15577.96			
6417.696	6419.470	G	1.2	0.7	0.2		15577.61	5c 0	2a 4	Q 3

λ_{air}	λ_{vac}	Ref.	I_1	I_2	I_3	I_4	ν	Upper	Lower	Br.
6418.013	6419.787		0.1	0.3	0.3		15576.84			
6418.289	6420.063	G	1.0	1.0	2.0		15576.17	3b 2	2a 0	R 5
6418.405	6420.179		0.2				15575.89			
6418.664	6420.438			0.2	0.2		15575.26			
6418.908	6420.682		0.2	0.7	1.2		15574.67	3c 3	2a 3	Q 8
6419.138	6420.913		0.2	0.1			15574.11			
6419.208	6420.983				0.1		15573.94			
6419.365	6421.139			0.1			15573.56			
6419.518	6421.292			0.2	0.2b		15573.19			
6419.736	6421.510			0.2			15572.66			
6419.851	6421.626				0.1		15572.38			
6419.950	6421.725						15572.14	3E 2	2B 9	R 2
6420.260	6422.034			0.1			15571.39	X 2	2B10	P 5
6420.507	6422.282			0.2			15570.79			
6420.767	6422.542		0.2	0.7	0.4		15570.16	3c 2	2a 2	Q12
6421.076	6422.851		1.0	0.8	0.3		15569.41	3F 1	2C 0	R 1
6421.348	6423.123		0.1				15568.75	3E 2	2B 9	Q 5
6421.427	6423.202			0.2d			15568.56			
6421.666	6423.441			0.3	0.1		15567.98			
6421.728	6423.503		0.2				15567.83			
6421.905	6423.680	G	1.2	0.8	0.2		15567.40	3F 1	2C 0	R 1
6422.087	6423.862			0.2			15566.96			
6422.215	6423.990				0.1		15566.65			
6422.669	6424.444			0.1	0.1		15565.55			
6422.904	6424.679			0.2			15564.98			
6422.994	6424.770				0.1		15564.76	5c 0	2a 4	Q 2
6422.994	6424.770				0.1		15564.76	3f 1	2c 2	R 5
6422.994	6424.770				0.1		15564.76	4d 0	2c 3	P 3
6423.234	6425.009			0.2			15564.18			
6423.428	6425.203			0.2	0.2b		15563.71			
6423.572	6425.348			0.2			15563.36	X 0	2B 6	P 6
6423.783	6425.558			0.2	0.1		15562.85			
6423.935	6425.711		0.2	0.3	0.2		15562.48			
6424.113	6425.889		0.2	0.3			15562.05			
6424.373	6426.149		1.1	0.6	0.2		15561.42	4e 0	2c 3	P 5
6424.646	6426.421		0.1	0.2	0.3		15560.76	3A 2	2B10	R 1
6424.786	6426.562			0.1			15560.42			
6424.947	6426.723		0.2	0.7	0.3		15560.03	3E 1	2B 7	Q 4
6425.091	6426.867			0.2			15559.68			
6425.261	6427.037		0.2	0.5	0.3		15559.27	T 0	2B10	R 1
6425.261	6427.037		0.2	0.5	0.3		15559.27	3C 1	2A 0	P 5
6425.480	6427.256			0.1			15558.74			
6425.752	6427.528			0.2			15558.08			
6426.128	6427.904			0.1	0.2		15557.17			
6426.434	6428.210		0.2	0.5	0.5		15556.43	3D 0	2B 5	R 6
6426.611	6428.388			0.2			15556.00	5c 0	2a 4	Q 1
6427.115	6428.892			0.2			15554.78			
6427.368	6429.144			0.2			15554.17			
6427.562	6429.338		0.3	0.4			15553.70	4b 1	2a 3	R 0
6427.665	6429.442		0.2				15553.45			
6427.760	6429.537				1.4		15553.22	3c 0	2a 0	Q18
6427.810	6429.586		1.0	1.0			15553.10	4b 0	2a 2	P 1
6427.847	6429.624						15553.01			
6428.119	6429.896	GR	2.7	2.4	1.8	1.47	15552.35	3c 4	2a 4	Q 1
6428.206	6429.983						15552.14			
6428.364	6430.140		0.3				15551.76			
6428.442	6430.219		0.4	0.4	0.2		15551.57			
6428.570	6430.347			0.1			15551.26			
6428.682	6430.459		0.2	0.3	0.2		15550.99			
6429.046	6430.823		0.2	0.7	0.1		15550.11			

λ_air	λ_vac	Ref.	I₁	I₂	I₃	I₄	ν	Upper	Lower	Br.
6429.120	6430.897						15549.93			
6429.174	6430.951		0.2				15549.80			
6429.310	6431.087	G	2.2	2.0	1.7	1.21	15549.47	3c 3	2a 3	P 4
6429.331	6431.108			1.5b			15549.42	3c 4	2a 4	Q 2
6429.492	6431.269				0.5		15549.03			
6429.633	6431.410		0.2	0.2			15548.69			
6429.881	6431.658			0.2			15548.09			
6430.051	6431.828			0.1	0.1b		15547.68			
6430.204	6431.981			0.2			15547.31	3f 1	2c 2	R 4
6430.481	6432.258			0.2			15546.64			
6430.547	6432.324				0.1		15546.48	Y 1	2B 6	P 3
6431.047	6432.825		0.1	0.6			15545.27			
6431.093	6432.870			0.3	0.4		15545.16	4D 0	2B10	R 0
6431.292	6433.069			0.2			15544.68			
6431.362	6433.139		0.2		0.2		15544.51			
6431.416	6433.193			0.3			15544.38			
6431.556	6433.334				0.3		15544.04			
6431.610	6433.388			0.2			15543.91	W 2	2C 1	R 4
6432.156	6433.934				0.1		15542.59	3f 0	2c 1	R 4
6432.421	6434.199				0.1		15541.95	4D 0	2B10	P 1
6432.670	6434.447			0.2	0.2		15541.35			
6432.885	6434.663		0.1	0.4	0.3		15540.83			
6433.125	6434.903						15540.25			
6433.187	6434.965			0.3	0.2		15540.10			
6433.253	6435.031		0.1				15539.94			
6433.510	6435.288	GR	2.0	1.7	1.5	1.11	15539.32	3c 4	2a 4	Q 2
6433.510	6435.288	GR	2.0	1.7	1.5	1.11	15539.32	4D 0	2B10	R 3
6433.580	6435.358		0.2				15539.15	4b 2	2a 4	R 1
6433.729	6435.508		0.1				15538.79			
6433.895	6435.673		0.3	0.4	0.2		15538.39			
6434.036	6435.814			0.1			15538.05			
6434.193	6435.971			0.2	0.2		15537.67			
6434.396	6436.174		0.4	0.6	0.2		15537.18			
6434.520	6436.299		0.3	0.4			15536.88			
6434.628	6436.406		0.2				15536.62			
6434.835	6436.613	GR	1.9	1.8	2.2	1.21	15536.12	3c 2	2a 2	P 7
6434.897	6436.676	R					15535.97	2K 4	2B 0	P 4
6435.154	6436.932			0.2	0.2		15535.35			
6435.311	6437.090		0.3	0.2			15534.97			
6435.390	6437.169				0.1		15534.78			
6435.514	6437.293			0.1			15534.48			
6435.684	6437.463			0.2			15534.07	3E 2	2B 9	P 5
6435.821	6437.600			0.1	0.3		15533.74			
6435.974	6437.753			0.2			15533.37			
6436.111	6437.890		0.1	0.2	0.2		15533.04			
6436.293	6438.072		1.0	0.8	0.4		15532.60	3b 5	2a 2	P 1
6436.505	6438.284			0.2			15532.09			
6436.617	6438.396				0.2		15531.82			
6436.728	6438.507		1.0	0.3			15531.55	W 2	2C 1	P 2
6436.907	6438.686			0.2			15531.12			
6437.077	6438.856			0.2	0.3		15530.71			
6437.246	6439.026		0.1	0.4	0.3		15530.30			
6437.549	6439.328		0.2	0.3	0.2		15529.57			
6437.831	6439.610	GR	1.9	1.8	1.7		15528.89	3b 2	2a 0	P 1
6437.918	6439.697		0.4				15528.68			
6438.142	6439.921		0.2	0.2			15528.14			
6438.357	6440.137			0.1			15527.62	3f 2	2c 3	R 5
6438.598	6440.378			0.1			15527.04	3F 0	2B 6	R 4
6438.702	6440.481				0.2		15526.79	4d 0	2c 3	Q 4
6438.892	6440.672			0.2			15526.33			

λ_{air}	λ_{vac}	Ref.	I_1	I_2	I_3	I_4	ν	Upper	Lower	Br.
6439.079	6440.859		0.2	0.3	0.2		15525.88			
6439.150	6440.929						15525.71	3f 0	2c 1	R 4
6439.199	6440.979			0.1			15525.59			
6439.490	6441.269			0.1	0.1		15524.89			
6439.809	6441.589			0.1			15524.12	4F 0	2B11	R 3
6440.000	6441.780			0.1	0.1b		15523.66			
6440.232	6442.012						15523.10			
6440.631	6442.411	G	1.3	1.1	1.4		15522.14	3d 4	2c 4	R 2
6440.826	6442.606						15521.67			
6440.909	6442.689		0.1	0.2			15521.47			
6441.021	6442.801			0.1	0.1		15521.20			
6441.133	6442.913			0.2			15520.93			
6441.245	6443.025		0.2	0.3			15520.66			
6441.299	6443.079				0.2		15520.53			
6441.369	6443.149						15520.36			
6441.498	6443.278	G	2.0	1.9	2.0	1.26	15520.05	3d 4	2c 4	Q 1
6441.498	6443.278	G	2.0	1.9	2.0	1.26	15520.05	3c 3	2a 3	Q 9
6441.498	6443.278	G	2.0	1.9	2.0	1.26	15520.05	3c 4	2a 4	Q 3
6441.514	6443.295	R		1.5b	0.9		15520.01			
6441.593	6443.374		0.7				15519.82			
6441.722	6443.502						15519.51	t 1	2a 2	R 3
6441.838	6443.619		0.1	0.2			15519.23			
6442.054	6443.835		0.1				15518.71			
6442.212	6443.992		0.2				15518.33	T 0	2B10	R 0
6442.353	6444.134		0.2	0.3			15517.99	3E 1	2B 7	R 0
6442.457	6444.237				0.2b		15517.74			
6442.648	6444.428			0.2			15517.28			
6442.880	6444.661			0.1			15516.72			
6443.146	6444.927				0.2b		15516.08			
6443.217	6444.997			0.1			15515.91	3F 0	2B 6	Q 6
6443.217	6444.997			0.1			15515.91	t 0	2a 1	P 2
6443.557	6445.338			0.3b	0.2b		15515.09			
6443.790	6445.571			0.2			15514.53			
6444.027	6445.807	G	1.2	0.8	0.3b		15513.96	3C 2	2A 1	P 3
6444.297	6446.078			0.2			15513.31	3A 0	2B 6	R 1
6444.629	6446.410				0.2b		15512.51			
6444.832	6446.614			0.1			15512.02			
6445.086	6446.867		0.1	0.2			15511.41			
6445.306	6447.087	G	0.9	0.8	1.2		15510.88			
6445.572	6447.353		0.1	0.2	0.1		15510.24			
6445.851	6447.632		0.2	0.2	0.1		15509.57	3E 1	2B 7	P 4
6446.146	6447.927			0.1	0.2		15508.86	4D 0	2B10	R 2
6446.349	6448.131			0.2			15508.37			
6446.466	6448.247		0.1				15508.09			
6446.524	6448.305			0.1			15507.95			
6446.736	6448.518		0.9	0.9	1.2		15507.44	3A 1	2B 8	R 0
6446.894	6448.676		0.4	0.2			15507.06			
6447.077	6448.859				0.2		15506.62	Y 2	2C 1	R 1
6447.148	6448.929			0.1			15506.45			
6447.314	6449.096			0.0			15506.05			
6447.518	6449.299			0.2			15505.56			
6447.680	6449.462			0.1			15505.17			
6447.854	6449.636		0.0	0.3	0.5		15504.75			
6448.108	6449.890		0.1	0.7	1.4		15504.14	3b 2	2a 0	R 6
6448.337	6450.119		0.2	0.6	0.4		15503.59			
6448.512	6450.294			0.2			15503.17			
6448.657	6450.439			0.2	0.1		15502.82			
6449.036	6450.818			0.1	0.2b		15501.91			
6449.227	6451.009			0.2			15501.45			
6449.477	6451.259			0.2	0.1		15500.85			

λ_air	λ_vac	Ref.	I₁	I₂	I₃	I₄	ν	Upper	Lower	Br.
6449.681	6451.463		0.3	0.7	0.3d		15500.36	3E 1	2B 7	Q 3
6449.681	6451.463		0.3	0.7	0.3d		15500.36	3A 2	2C 1	P 5
6449.893	6451.675			0.2			15499.85			
6450.105	6451.888		0.2	0.7	0.5		15499.34			
6450.380	6452.162		0.4	0.3			15498.68			
6450.683	6452.466			0.2	0.2		15497.95			
6450.900	6452.683			0.1			15497.43	u 1	2a 2	R 3
6450.900	6452.683			0.1			15497.43	4c 5	2a 8	Q 1
6451.166	6452.949			0.1			15496.79	4D 0	2B10	R 1
6451.337	6453.120			0.3	0.2		15496.38			
6451.520	6453.303			0.3	0.7		15495.94	3c 2	2a 2	Q13
6451.520	6453.303			0.3	0.7		15495.94	W 1	2B 8	R 3
6451.628	6453.411			0.1			15495.68			
6451.857	6453.641		0.4	1.1	1.2		15495.13			
6452.107	6453.891	G	1.2	1.1	1.5		15494.53	3c 4	2a 4	Q 4
6452.107	6453.891	G	1.2	1.1	1.5		15494.53	3A 2	2B10	P 5
6452.461	6454.245		0.1	0.5d	0.2		15493.68			
6452.611	6454.395			0.1			15493.32			
6452.794	6454.578		0.1	0.3	0.1		15492.88			
6452.944	6454.728			0.1			15492.52	3E 0	2B 6	R 6
6453.057	6454.840		0.3				15492.25			
6453.124	6454.907		0.2	0.8	0.2		15492.09	3c 1	2a 1	Q16
6453.286	6455.069			0.1			15491.70			
6453.432	6455.215			0.2			15491.35			
6453.669	6455.453		0.9	0.7	0.4		15490.78			
6454.044	6455.828			0.1			15489.88	Y 1	2B 6	P 4
6454.211	6455.995			0.1	0.1		15489.48			
6454.473	6456.257		0.2	0.2	0.2		15488.85	4F 1	2C 0	Q 4
6454.515	6456.299						15488.75	W 2	2C 1	Q 3
6454.794	6456.578			0.2			15488.08			
6455.011	6456.795	G	1.1	1.1	1.4		15487.56	4D 0	2C 1	Q 5
6455.307	6457.091		0.2	0.7	1.2		15486.85			
6455.465	6457.249			0.1			15486.47			
6455.661	6457.445			0.2	0.2		15486.00			
6455.861	6457.646		0.1	0.2			15485.52			
6455.928	6457.712						15485.36			
6456.112	6457.896	GR	1.4	1.1	1.3		15484.92			
6456.166	6457.950		0.8				15484.79			
6456.483	6458.267			0.2			15484.03			
6456.554	6458.338				0.3		15483.86			
6456.662	6458.446			0.1			15483.60	3E 2	2B 9	R 1
6456.825	6458.609		0.1	0.2			15483.21	3D 0	2B 5	R 5
6456.904	6458.688				0.2		15483.02			
6456.983	6458.768		0.5	0.4			15482.83	4b 0	2a 2	P 2
6456.983	6458.768		0.5	0.4			15482.83	t 1	2a 2	P 1
6457.137	6458.922			0.1			15482.46			
6457.317	6459.101			0.0			15482.03			
6457.630	6459.414			0.1			15481.28			
6457.780	6459.564				0.4b		15480.92	3f 2	2c 3	R 5
6458.068	6459.852				0.3d		15480.23			
6458.118	6459.902			0.2			15480.11			
6458.376	6460.161			0.1			15479.49			
6458.560	6460.345			0.2			15479.05			
6458.760	6460.545		1.1	0.6	0.3b		15478.57			
6459.002	6460.787			0.1			15477.99	3f 1	2c 2	R 4
6459.232	6461.017		0.1				15477.44	4D 0	2C 1	P 4
6459.516	6461.301		0.2				15476.76	3A 1	2B 8	P 4
6459.561	6461.347			0.3	0.3d		15476.65			
6459.862	6461.647			0.2			15475.93			
6459.991	6461.777		0.2		0.2b		15475.62			

λ_{air}	λ_{vac}	Ref.	I_1	I_2	I_3	I_4	ν	Upper	Lower	Br.
6460.033	6461.818			0.3			15475.52	Y 2	2C 1	Q 1
6460.279	6462.065		0.1				15474.93	3E 1	2B 7	P 3
6460.413	6462.198			0.1	0.2		15474.61			
6460.676	6462.461		0.3	0.4			15473.98			
6460.814	6462.599		0.5	0.3			15473.65			
6460.910	6462.695		1.0	0.6	0.8		15473.42			
6460.981	6462.766			0.1			15473.25	3c 3	2a 3	P 5
6461.223	6463.009		0.1	0.4	1.2		15472.67	3A 2	2B10	R 0
6461.457	6463.242		0.2	0.2			15472.11			
6461.774	6463.560			0.1	0.2		15471.35	3f 0	2c 1	Q 6
6462.046	6463.832		0.1	0.3	0.3		15470.70			
6462.313	6464.099		0.3	0.7	1.2		15470.06	3F 1	2C 0	Q 3
6462.313	6464.099		0.3	0.7	1.2		15470.06	4c 4	2a 7	Q 1
6462.505	6464.291		0.1	0.2			15469.60			
6462.693	6464.479		0.2	0.2			15469.15	3F 1	2C 0	Q 5
6462.844	6464.630		0.2	0.7	0.2		15468.79			
6462.965	6464.751			0.1			15468.50			
6463.103	6464.889		0.9	0.7	1.6		15468.17			
6463.316	6465.102						15467.66			
6463.412	6465.198			0.2	0.1		15467.43			
6463.746	6465.533				0.1		15466.63			
6463.838	6465.625		0.9	0.2			15466.41	4D 0	2B10	P 2
6464.068	6465.854		1.0	0.9	1.3		15465.86			
6464.252	6466.038		0.1	0.1			15465.42			
6464.411	6466.197		0.1	0.1	0.3b		15465.04			
6464.561	6466.348		0.5	0.3			15464.68	3A 2	2B10	P 4
6464.866	6466.653		1.1	0.7	0.3b		15463.95	3F 1	2C 0	Q 4
6465.034	6466.820			0.2			15463.55			
6465.251	6467.038		1.1	1.1	1.7		15463.03	3c 4	2a 4	Q 5
6465.490	6467.276		0.1	0.1			15462.46			
6465.665	6467.452		0.1	0.2			15462.04			
6465.824	6467.611			0.2	0.2b		15461.66			
6466.075	6467.862		0.2				15461.06			
6466.121	6467.908			0.3			15460.95			
6466.217	6468.004						15460.72			
6466.384	6468.171		1.2	0.9	0.3		15460.32	3E 1	2B 7	Q 2
6466.518	6468.305		0.3	0.3	0.3		15460.00			
6466.677	6468.464		1.3	0.9	1.2		15459.62	2K 4	2B 0	P 5
6467.008	6468.795		0.1	0.2	0.2		15458.83			
6467.208	6468.996			0.1			15458.35	u 1	2a 2	R 2
6467.334	6469.121				0.3b		15458.05			
6467.455	6469.243		0.1	0.1			15457.76	Y 2	2C 1	P 1
6467.681	6469.469			0.2			15457.22			
6467.811	6469.598				0.1		15456.91			
6467.882	6469.669			0.2			15456.74			
6468.129	6469.917		0.2	0.9	1.3		15456.15	4c 3	2a 6	R 1
6468.355	6470.143		0.2	0.6d	1.2		15455.61	3c 0	2a 0	Q19
6468.355	6470.143		0.2	0.6d	1.2		15455.61	4f 3	3b 1	R 2
6468.623	6470.410		0.9	0.5			15454.97			
6468.836	6470.624		1.3	0.8	0.4		15454.46	3F 1	2C 0	Q 3
6468.836	6470.624		1.3	0.8	0.4		15454.46	3b 5	2a 2	P 2
6469.054	6470.842			0.1			15453.94			
6469.217	6471.005			0.2	0.2b		15453.55	4d 0	2c 3	Q 5
6469.519	6471.307			0.2	0.3b		15452.83			
6469.602	6471.390		0.1				15452.63	3d 3	2c 3	P 5
6469.791	6471.579		0.2	0.2			15452.18			
6469.837	6471.625				0.2b		15452.07			
6469.946	6471.734		0.3	0.6			15451.81			
6470.113	6471.901			0.3	0.4		15451.41			
6470.385	6472.174			0.1			15450.76			

λ_{air}	λ_{vac}	Ref.	I_1	I_2	I_3	I_4	ν	Upper	Lower	Br.
6470.884	6472.672			0.2	0.2		15449.57	5c 0	2a 4	P 2
6471.102	6472.890		1.0	0.7	0.3		15449.05	3F 1	2C 0	Q 2
6471.102	6472.890		1.0	0.7	0.3		15449.05	W 1	2C 0	P 5
6471.382	6473.171		0.2	0.4	0.2		15448.38	3F 2	2C 1	R 3
6471.382	6473.171		0.2	0.4	0.2		15448.38	4c 0	2a 3	R 5
6471.529	6473.317			0.2			15448.03			
6471.659	6473.447		0.2	0.2			15447.72			
6472.011	6473.799			0.2			15446.88			
6472.191	6473.979	G	1.3	0.8	0.3d		15446.45			
6472.417	6474.206			0.4	0.2d		15445.91	3C 2	2A 1	P 4
6472.576	6474.365			0.2			15445.53			
6472.694	6474.482			0.2			15445.25			
6472.845	6474.633		0.2	0.5	0.4		15444.89			
6473.016	6474.805			0.1			15444.48			
6473.188	6474.977		0.1	0.3	0.2b		15444.07			
6473.390	6475.178		0.2	0.2			15443.59	4b 1	2a 3	P 1
6473.641	6475.430	GF	1.7	1.5	1.5		15442.99	3b 2	2a 0	P 2
6473.641	6475.430	GF	1.7	1.5	1.5		15442.99	T 0	2B10	P 4
6473.880	6475.669		1.4	0.8	0.2		15442.42	3F 1	2C 0	Q 2
6473.976	6475.765		0.2				15442.19			
6474.211	6476.000			0.3			15441.63			
6474.366	6476.155		0.1	0.2			15441.26	Y 2	2C 1	R 2
6474.551	6476.340		0.2	0.5	0.4		15440.82	3c 0	2a 0	P 8
6474.551	6476.340		0.2	0.5	0.4		15440.82	4f 2	3b 0	R 1
6474.551	6476.340		0.2	0.5	0.4		15440.82	3f 0	2c 1	R 3
6474.773	6476.562			0.2	0.2		15440.29			
6474.832	6476.621		0.1				15440.15			
6474.987	6476.776			0.2	0.4		15439.78	3A 1	2B 8	P 3
6475.176	6476.965		1.1	1.1	0.7		15439.33	T 0	2B10	P 2
6475.356	6477.145		1.3	1.0	0.3		15438.90	T 0	2B10	P 1
6475.469	6477.259	G	0.7				15438.63			
6475.490	6477.280		1.1	0.9	0.3		15438.58			
6475.541	6477.330		0.5				15438.46			
6475.746	6477.536			0.1			15437.97	3f 2	2c 3	R 4
6475.947	6477.737		0.2	0.2			15437.49			
6476.040	6477.829			0.3	0.2b		15437.27	3E 1	2B 7	Q 1
6476.291	6478.081			0.1			15436.67	3A 1	2B 8	P 1
6476.501	6478.291		0.7	0.7	0.3b		15436.17	W 1	2B 8	R 2
6476.870	6478.660		0.1	0.2d			15435.29			
6477.022	6478.811				0.2		15434.93			
6477.147	6478.937		0.3	0.2			15434.63			
6477.403	6479.193			0.2			15434.02			
6477.471	6479.260				0.5b		15433.86			
6477.592	6479.382		0.5	0.2			15433.57	u 2	2a 3	R 3
6477.815	6479.605		1.0	0.6	0.3		15433.04			
6478.163	6479.953		0.1				15432.21	3D 2	2B 9	R 5
6478.293	6480.083			0.1	0.3b		15431.90			
6478.528	6480.319		0.1	0.2			15431.34	3E 1	2B 7	P 2
6478.742	6480.533			0.2	0.2b		15430.83	3f 1	2c 2	R 3
6478.965	6480.755			0.2			15430.30			
6479.141	6480.932			0.2			15429.88			
6479.309	6481.100		0.2	0.3	0.5		15429.48			
6479.755	6481.545			0.2			15428.42			
6479.998	6481.789				0.3b		15427.84			
6480.107	6481.898		0.2	0.1			15427.58			
6480.355	6482.146				0.1		15426.99	3A 2	2B10	P 3
6480.355	6482.146				0.1		15426.99	u 1	2a 2	R 1
6480.632	6482.423			0.1			15426.33	3A 0	2B 6	R 0
6480.868	6482.659			0.7	1.3		15425.77	3c 4	2a 4	Q 6
6480.910	6482.701		0.2				15425.67	T 0	2B10	P 3

λ_{air}	λ_{vac}	Ref.	I_1	I_2	I_3	I_4	ν	Upper	Lower	Br.
6481.137	6482.928				0.1		15425.13			
6481.221	6483.012			0.1			15424.93			
6481.452	6483.243			0.1			15424.38			
6481.649	6483.440			0.1	0.2d		15423.91			
6481.822	6483.613		0.1	0.2			15423.50			
6481.889	6483.680		1.0				15423.34			
6482.006	6483.798		0.7	0.4	0.2d		15423.06	3A 1	2B 8	P 2
6482.851	6484.643				0.3b		15421.05			
6483.326	6485.118		0.0		0.3b		15419.92			
6483.738	6485.530				0.2		15418.94			
6484.054	6485.846			0.2			15418.19	3c 2	2a 2	Q14
6484.159	6485.951				0.5b		15417.94			
6484.273	6486.064		0.2	0.7			15417.67	3D 0	2B 5	R 4
6484.386	6486.178		0.2	0.8			15417.40			
6484.794	6486.586			0.1d	0.2b		15416.43			
6485.097	6486.889		0.3	0.3	0.3b		15415.71			
6485.417	6487.209		0.2	0.2	0.2		15414.95	3A 0	2B 6	P 4
6485.779	6487.571			0.2			15414.09	U 1	2C 0	R 1
6485.976	6487.769		0.3	0.7	1.7		15413.62	3b 2	2a 0	R 7
6486.140	6487.933		0.2	0.3			15413.23			
6486.284	6488.076			0.2			15412.89			
6486.439	6488.232		1.0	0.6	1.4		15412.52			
6486.633	6488.425			0.2			15412.06	4d 0	2c 3	P 4
6486.869	6488.661		0.2	0.3			15411.50			
6487.033	6488.825			0.2			15411.11	5c 0	2a 4	P 3
6487.188	6488.981		0.3	0.3			15410.74	4b 0	2a 2	P 3
6487.336	6489.128			0.1			15410.39			
6487.466	6489.259		0.3	0.3	0.2d		15410.08	3f 1	2c 2	Q 7
6487.597	6489.390			0.2			15409.77			
6487.769	6489.562		0.9	1.0	1.4		15409.36			
6488.035	6489.827			0.1			15408.73			
6488.435	6490.228			0.1	0.2d		15407.78	3f 2	2c 3	R 4
6488.666	6490.459		0.2	0.2			15407.23	W 2	2C 1	P 3
6488.970	6490.763			0.2			15406.51			
6489.050	6490.843		0.1		0.3b		15406.32			
6489.248	6491.041		0.5	0.3			15405.85			
6489.505	6491.298			0.1			15405.24			
6489.656	6491.449				0.2d		15404.88			
6489.879	6491.673		0.1	0.2			15404.35			
6490.200	6491.993			0.1			15403.59			
6490.444	6492.238			0.2			15403.01			
6490.634	6492.427		0.3	1.0	0.4b		15402.56	3c 1	2a 1	Q17
6490.840	6492.634			0.1			15402.07	3A 2	2B10	P 1
6490.920	6492.714						15401.88			
6490.996	6492.790			0.2			15401.70			
6491.161	6492.954						15401.31			
6491.257	6493.051			0.2			15401.08			
6491.375	6493.169		0.2	0.3			15400.80			
6491.439	6493.232				0.2b		15400.65	t 1	2a 2	P 3
6492.122	6493.916				0.3b		15399.03			
6492.214	6494.008			0.1			15398.81			
6492.522	6494.316			0.1			15398.08			
6492.590	6494.384		0.1				15397.92			
6492.670	6494.464			0.2	0.2		15397.73			
6493.037	6494.831			0.1	0.3		15396.86			
6493.150	6494.945		0.3	0.9			15396.59			
6493.399	6495.194		0.2	0.3	0.3		15396.00			
6493.644	6495.438		0.1				15395.42	4e 2	3b 0	R 4
6493.707	6495.502			0.2			15395.27			
6493.922	6495.717		0.8	0.6	0.2		15394.76	W 2	2B10	R 4

λ_{air}	λ_{vac}	Ref.	I_1	I_2	I_3	I_4	ν	Upper	Lower	Br.
6494.163	6495.957			0.1			15394.19	4c 0	2a 3	R 4
6494.256	6496.050						15393.97			
6494.357	6496.151			0.2	0.2b		15393.73	3A 2	2B10	P 2
6494.956	6496.751				0.2d		15392.31	W 1	2B 8	R 1
6495.023	6496.818			0.2			15392.15			
6495.251	6497.046		1.2	0.4			15391.61	3F 2	2C 1	R 2
6495.251	6497.046		1.2	0.4			15391.61	3E 2	2B 9	R 0
6495.500	6497.295		0.2	0.3	0.2d		15391.02			
6495.783	6497.578	G	0.9	1.3	1.5		15390.35	u 1	2a 2	R 0
6496.020	6497.814		0.1	0.1			15389.79			
6496.129	6497.924				0.2d		15389.53			
6496.184	6497.979			0.3	0.3		15389.40	4g 0	2a 3	P 4
6496.454	6498.249				0.2		15388.76	3E 2	2B 9	P 4
6496.454	6498.249				0.2		15388.76	3E 2	2B 9	Q 4
6496.708	6498.503		1.1	0.2	0.4		15388.16	3f 1	2c 2	R 3
6496.910	6498.705				0.1		15387.68			
6497.096	6498.891				0.2	0.3	15387.24			
6497.307	6499.102				0.2	0.4	15386.74	3f 0	2c 1	Q 5
6497.582	6499.377		0.0	0.1			15386.09			
6497.776	6499.571		0.4d	0.6	0.2d		15385.63			
6497.983	6499.778		0.7	0.8	0.3d		15385.14	3F 2	2C 1	R 2
6498.156	6499.952			0.3	0.3		15384.73	4f 2	3b 0	P 6
6498.359	6500.154		0.1				15384.25			
6498.600	6500.395			0.2			15383.68	3f 0	2c 1	R 2
6498.794	6500.590		1.3	0.9	1.4		15383.22	3c 4	2a 4	Q 7
6498.933	6500.729		0.2	0.3	0.4		15382.89	4b 1	2a 3	P 2
6499.132	6500.928			0.1			15382.42			
6499.335	6501.131		0.1	0.2			15381.94	u 2	2a 3	R 2
6499.440	6501.236						15381.69			
6499.580	6501.376		0.1	0.3	0.2		15381.36	4D 0	2C 1	Q 6
6499.838	6501.633	G	1.2	1.4	1.7		15380.75			
6500.087	6501.883		0.4	0.9	0.4		15380.16	2K 4	2B 0	P 6
6500.353	6502.149			0.2			15379.53			
6500.408	6502.204		0.1		0.2		15379.40			
6500.598	6502.394			0.2			15378.95			
6500.704	6502.500		0.2	0.3	0.2b		15378.70	5c 0	2a 4	P 4
6500.979	6502.775		0.2	0.4			15378.05	4c 2	2a 5	R 1
6501.122	6502.919		0.2				15377.71	U 1	2C 0	Q 1
6501.122	6502.919		0.2				15377.71	4c 3	2a 6	Q 3
6501.224	6503.020			0.1	0.3b		15377.47			
6501.499	6503.295			0.2			15376.82			
6501.681	6503.477		0.2	0.7	0.3d		15376.39			
6501.862	6503.659			0.2			15375.96	3D 0	2B 5	R 3
6502.044	6503.841		0.9				15375.53	3d 4	2c 4	Q 3
6502.087	6503.883	G	1.4	1.1	1.3		15375.43			
6502.125	6503.921		0.7				15375.34			
6502.374	6504.171			0.1	0.2		15374.75			
6502.603	6504.399			0.2			15374.21			
6502.784	6504.581		0.1	0.4	0.2		15373.78	3D 1	2B 7	R 5
6502.966	6504.763		0.1				15373.35			
6503.034	6504.831			0.2			15373.19			
6503.317	6505.114			0.3	0.3		15372.52	4c 3	2a 6	Q 2
6503.571	6505.368						15371.92			
6503.652	6505.449			0.2			15371.73			
6503.749	6505.546				0.6b		15371.50			
6503.838	6505.635		0.2	0.5			15371.29	3C 2	2A 1	P 5
6504.011	6505.808			0.2			15370.88			
6504.193	6505.990		0.2	0.3	0.2		15370.45	W 2	2C 1	Q 4
6504.371	6506.168			0.1			15370.03			
6504.536	6506.333			0.2			15369.64			

λ air	λ vac	Ref.	I₁	I₂	I₃	I₄	ν	Upper	Lower	Br.
6504.680	6506.477			0.1			15369.30	4c 3	2a 6	Q 1
6504.845	6506.642		0.2	0.2	0.4		15368.91			
6505.129	6506.926		0.1	0.2			15368.24			
6505.404	6507.201				0.3		15367.59			
6505.505	6507.303			0.2			15367.35			
6505.776	6507.574			0.4			15366.71			
6506.026	6507.824				0.3b		15366.12			
6506.221	6508.018			0.2			15365.66			
6506.538	6508.336				0.4		15364.91			
6506.665	6508.463			0.2			15364.61			
6506.865	6508.662			0.4			15364.14			
6506.945	6508.743		0.1		0.6b		15363.95	3A 0	2B 6	P 3
6507.013	6508.811			0.4			15363.79			
6507.144	6508.942		1.0		1.2		15363.48			
6507.263	6509.061		0.2				15363.20	4D 0	2B10	P 3
6507.555	6509.353				0.2		15362.51			
6507.678	6509.476			0.1			15362.22	3D 0	2B 5	R 0
6507.898	6509.696			0.1	0.2b		15361.70	4c 1	2a 4	R 0
6508.165	6509.963			0.2			15361.07			
6508.330	6510.128				0.4b		15360.68			
6508.462	6510.260		0.2	0.3			15360.37	U 1	2C 0	R 2
6508.834	6510.633			0.2			15359.49			
6509.182	6510.980			0.1	0.2b		15358.67			
6509.250	6511.048		0.2				15358.51			
6509.474	6511.273	G	1.5	0.9	0.2		15357.98			
6509.576	6511.375		0.2				15357.74			
6509.733	6511.532		0.1	0.1			15357.37			
6509.915	6511.714		0.2	0.3	0.3b		15356.94	Y 2	2C 1	P 2
6510.114	6511.913		0.1				15356.47	3D 0	2B 5	R 2
6510.322	6512.121		0.2	0.3	0.8b		15355.98			
6510.577	6512.375		0.1				15355.38			
6510.920	6512.719		0.1		0.2		15354.57			
6511.170	6512.969		0.2				15353.98			
6511.336	6513.135		1.1	0.7	0.2		15353.59	3D 0	2B 5	R 1
6511.722	6513.521			0.1	0.2		15352.68			
6512.002	6513.801		0.1	0.0	0.2		15352.02			
6512.252	6514.051	G	1.3	1.0	1.4		15351.43	3b 5	2a 2	P 3
6512.434	6514.234		0.1	0.1			15351.00			
6512.574	6514.374				0.2		15350.67			
6513.330	6515.129				0.3b		15348.89	3f 2	2c 3	R 3
6513.330	6515.129				0.3b		15348.89	t 0	2a 1	P 3
6513.571	6515.371						15348.32			
6513.720	6515.520		0.1				15347.97			
6513.839	6515.638				0.2b		15347.69	W 1	2B 8	R 0
6513.839	6515.638				0.2b		15347.69	3f 1	2c 2	Q 6
6514.195	6515.995		0.6	0.6	0.1b		15346.85			
6514.318	6516.118		0.2				15346.56			
6514.522	6516.322		1.0	0.2	0.3b		15346.08			
6514.726	6516.526		0.3	0.3			15345.60			
6515.180	6516.980				0.3b		15344.53			
6515.426	6517.227		0.2	0.1			15343.95			
6515.758	6517.558				0.2b		15343.17			
6516.199	6518.000		0.1		0.3b		15342.13	3E 2	2B 9	P 3
6516.510	6518.310		0.1				15341.40	u 2	2a 3	R 1
6516.798	6518.599		0.2	0.2			15340.72			
6516.871	6518.671				0.6b		15340.55			
6517.125	6518.926		0.1	0.2			15339.95			
6517.385	6519.185			0.3b			15339.34	3c 2	2a 2	P 9
6517.453	6519.253		0.3		0.1		15339.18			
6517.555	6519.355		0.2				15338.94			

λ_{air}	λ_{vac}	Ref.	I_1	I_2	I_3	I_4	ν	Upper	Lower	Br.
6517.695	6519.495	G	2.0	2.0	2.0		15338.61	3b 2	2a 0	P 3
6517.865	6519.666						15338.21	3D 0	2B 5	P 1
6518.022	6519.823		0.2	0.2	0.2		15337.84	3E 3	2B11	Q 6
6518.022	6519.823		0.2	0.2	0.2		15337.84	4c 0	2a 3	R 3
6518.264	6520.065		0.4	0.5			15337.27			
6518.396	6520.197		0.2	0.1	0.3b		15336.96	3A 0	2B 6	P 1
6518.396	6520.197		0.2	0.1	0.3b		15336.96	4b 0	2a 2	P 4
6518.540	6520.341		1.5	1.2			15336.62	3f 0	2c 1	R 2
6518.753	6520.554		0.2	0.1			15336.12	3A 0	2B 6	P 2
6519.008	6520.809		0.1	0.8	0.2b		15335.52	3c 4	2a 4	Q 8
6519.297	6521.098			0.1			15334.84			
6519.565	6521.366				0.2b		15334.21			
6519.659	6521.460			0.2			15333.99			
6519.901	6521.702				0.2b		15333.42			
6519.973	6521.775			0.0			15333.25			
6520.258	6522.060			0.1d	0.3b		15332.58			
6520.569	6522.370		1.3	0.9	0.3b		15331.85	2K 6	2B 2	R 2
6520.569	6522.370		1.3	0.9	0.3b		15331.85	3f 2	2c 3	R 3
6520.666	6522.468			0.2			15331.62			
6520.777	6522.578		0.1	0.2			15331.36			
6520.977	6522.779	G	1.6	1.1	0.3b		15330.89	2K 6	2B 2	R 1
6520.977	6522.779	G	1.6	1.1	0.3b		15330.89	3F 2	2C 1	R 1
6521.168	6522.970		0.1	0.0			15330.44			
6521.300	6523.102		0.9	0.7	0.2		15330.13	3F 2	2C 1	R 1
6521.534	6523.336			0.1			15329.58			
6521.811	6523.612		0.1	0.2			15328.93			
6522.121	6523.923		0.1	0.1	0.3b		15328.20	4D 0	2B10	P 4
6522.360	6524.162		0.2				15327.64			
6522.500	6524.302		0.5	1.0	1.6		15327.31			
6522.815	6524.617		0.1	0.2	0.2		15326.57	4e 2	3b 0	R 3
6523.049	6524.851		0.1				15326.02			
6523.211	6525.013		0.3	0.7	0.3		15325.64	3D 1	2B 7	R 4
6523.211	6525.013		0.3	0.7	0.3		15325.64	U 1	2C 0	R 3
6523.534	6525.337		0.3	0.2	0.3		15324.88	3D 1	2B 7	R 0
6523.751	6525.554			0.2	0.3		15324.37			
6523.960	6525.762			0.1			15323.88			
6524.164	6525.967		0.1	0.3	0.2		15323.40			
6524.398	6526.201			0.1			15322.85			
6524.875	6526.678			0.1	0.3b		15321.73			
6525.041	6526.844		0.2	0.3			15321.34			
6525.216	6527.019			0.2			15320.93			
6525.408	6527.211		0.1	0.3			15320.48			
6525.625	6527.428			0.1			15319.97			
6525.774	6527.577		0.2	0.3	0.3d		15319.62			
6526.089	6527.892		1.3	0.8	0.2d		15318.88	2K 6	2B 2	R 0
6526.089	6527.892		1.3	0.8	0.2d		15318.88	4b 1	2a 3	P 3
6526.396	6528.199			0.2			15318.16	3D 2	2B 9	R 4
6526.592	6528.395		0.2	0.3	0.3b		15317.70	3f 1	2c 2	R 2
6526.771	6528.574		0.2	0.2			15317.28			
6526.869	6528.672		0.1	0.1			15317.05	3E 2	2B 9	Q 3
6527.035	6528.839		0.4	0.3	0.2b		15316.66			
6527.163	6528.966		0.3				15316.36			
6527.355	6529.158	G	2.3	1.5b	1.5		15315.91	3c 5	2a 5	Q 1
6527.598	6529.401		0.2		0.2d		15315.34			
6527.704	6529.508		0.3	0.2			15315.09			
6527.824	6529.627	G					15314.81			
6527.930	6529.734		0.4	1.4	1.4		15314.56			
6528.114	6529.917			0.0			15314.13			
6528.386	6530.190				0.3b		15313.49			
6528.791	6530.595						15312.54			

λ_{air}	λ_{vac}	Ref.	I_1	I_2	I_3	I_4	ν	Upper	Lower	Br.
6529.120	6530.924				0.3b		15311.77	W 2	2B10	R 3
6529.260	6531.064		0.3				15311.44	3F 0	2B 6	Q 5
6529.260	6531.064		0.3				15311.44	3D 1	2B 7	R 1
6529.508	6531.312		0.2	0.2			15310.86			
6529.580	6531.384				0.3b		15310.69			
6529.649	6531.453		1.0	0.8			15310.53	3E 2	2B 9	P 2
6529.768	6531.572		0.2				15310.25			
6529.904	6531.708		0.2	0.1			15309.93			
6529.998	6531.802				0.8b		15309.71	3D 1	2B 7	P 1
6530.054	6531.858		0.2	0.2			15309.58	3f 0	2c 1	Q 6
6530.399	6532.203			0.1			15308.77	4D 0	2C 1	P 5
6530.510	6532.314				0.2		15308.51			
6530.924	6532.728			0.1d			15307.54	3F 0	2B 6	R 3
6531.031	6532.835				0.3b		15307.29	u 2	2a 3	R 0
6531.410	6533.215		0.1	0.0			15306.40			
6531.611	6533.415		0.1	0.2	0.3b		15305.93			
6531.876	6533.680		0.2	0.3	1.2		15305.31	3b 2	2a 0	R 8
6531.876	6533.680		0.2	0.3	1.2		15305.31	4f 2	3b 0	P 5
6531.987	6533.791		0.4				15305.05			
6532.298	6534.103		0.2	0.2	0.2		15304.32			
6532.490	6534.295		0.2				15303.87			
6532.644	6534.449	G	1.7	1.4	1.3		15303.51	3c 5	2a 5	Q 2
6532.644	6534.449	G	1.7	1.4	1.3		15303.51	4D 0	2B10	P 5
6533.015	6534.820		0.1	0.2			15302.64	4F 0	2B11	P 5
6533.203	6535.008		0.1	0.3			15302.20	U 1	2C 0	R 4
6533.370	6535.175		0.3	0.5	1.2		15301.81			
6533.536	6535.341		0.2	0.3			15301.42	3D 1	2B 7	R 3
6533.536	6535.341		0.2	0.3			15301.42	4f 3	3b 1	R 1
6533.711	6535.516		1.0	0.7	1.2		15301.01	3D 1	2B 7	R 2
6533.839	6535.644		0.2	0.1			15300.71			
6534.027	6535.832		1.2	0.7	0.4b		15300.27			
6534.283	6536.089			0.1			15299.67			
6534.437	6536.242			0.2			15299.31	3f 0	2c 1	Q 4
6534.437	6536.242			0.2			15299.31	3D 2	2B 9	P 1
6534.655	6536.460		0.2	0.7	0.4b		15298.80	3D 2	2B 9	R 3
6535.351	6537.157				0.3b		15297.17			
6535.612	6537.418				0.2		15296.56			
6535.898	6537.704		0.1		0.3		15295.89			
6536.261	6538.067		0.9	0.4	0.3		15295.04			
6536.467	6538.272		0.1	0.2			15294.56			
6536.565	6538.371				0.6b		15294.33			
6536.633	6538.439		0.9	1.0	1.3		15294.17			
6536.809	6538.614			0.2			15293.76	U 1	2C 0	Q 2
6537.116	6538.922				0.3		15293.04			
6537.646	6539.453				0.3		15291.80			
6537.758	6539.564		0.2	0.1			15291.54			
6537.817	6539.624						15291.40			
6538.023	6539.829	G	1.5	1.0			15290.92	3D 2	2B 9	R 2
6538.121	6539.927		0.2		0.4		15290.69	4c 2	2a 5	Q 4
6538.249	6540.056		0.1	0.2			15290.39			
6538.326	6540.133				0.4		15290.21			
6538.741	6540.547				0.2b		15289.24			
6538.908	6540.714		0.1	0.0			15288.85			
6539.066	6540.873				0.3b		15288.48			
6539.152	6540.958		1.1	0.7			15288.28			
6539.511	6541.318			0.2	0.3b		15287.44	W 2	2B10	R 2
6539.939	6541.745			0.1			15286.44			
6540.242	6542.049			0.3			15285.73			
6540.525	6542.332	G	1.8	1.6	1.6		15285.07	3c 5	2a 5	Q 3
6540.628	6542.435		0.2				15284.83	W 1	2B 8	P 1

λ_{air}	λ_{vac}	Ref.	I_1	I_2	I_3	I_4	ν	Upper	Lower	Br.
6540.628	6542.435		0.2				15284.83	4c 2	2a 5	Q 3
6540.777	6542.584		0.3	0.7	0.2		15284.48			
6541.086	6542.893			0.2d	0.2		15283.76			
6541.394	6543.201		0.3	0.8	1.2		15283.04	3c 4	2a 4	Q 9
6541.561	6543.368		0.1				15282.65	3D 2	2B 9	R 1
6541.749	6543.556		0.1	0.1			15282.21			
6541.839	6543.646				0.5		15282.00			
6541.899	6543.706		0.3	0.3	1.0		15281.86			
6542.237	6544.044			0.2			15281.07	3f 1	2c 2	Q 5
6542.357	6544.164				0.3b		15280.79			
6542.464	6544.271			0.1			15280.54			
6542.751	6544.558		0.3	0.9	1.3		15279.87	4c 2	2a 5	Q 2
6543.033	6544.841		0.2	0.2			15279.21	4D 0	2C 1	Q 7
6543.033	6544.841		0.2	0.2			15279.21	4c 0	2a 3	R 2
6543.136	6544.944		1.1	0.7	0.6b		15278.97			
6543.252	6545.059		0.8	0.5			15278.70			
6543.372	6545.179		0.2				15278.42			
6543.526	6545.334			0.1	0.3b		15278.06			
6543.787	6545.595			0.2			15277.45	4c 2	2a 5	Q 1
6544.361	6546.169				0.3b		15276.11			
6544.794	6546.602				0.2b		15275.10			
6544.905	6546.713		0.1	0.1			15274.84			
6545.111	6546.919		1.3	1.1	1.3		15274.36	3F 1	2B 8	Q 5
6545.167	6546.975		0.5				15274.23	3E 2	2B 9	Q 2
6545.411	6547.219		0.1	0.1d	0.3		15273.66			
6545.921	6547.729				0.3		15272.47	4d 0	2c 3	P 5
6546.114	6547.922		0.3	0.2			15272.02			
6546.260	6548.068		0.1				15271.68			
6546.397	6548.205		0.4	1.0	1.3		15271.36	4F 0	2B11	R 2
6546.607	6548.415			0.1			15270.87			
6546.924	6548.733		0.1	0.3b	1.1		15270.13	W 2	2C 1	Q 5
6547.203	6549.011			0.1			15269.48			
6547.391	6549.200		0.1	0.0			15269.04			
6547.512	6549.320				0.2b		15268.76			
6547.602	6549.410		0.2	0.3			15268.55	3D 0	2B 5	P 2
6547.953	6549.762			0.3			15267.73			
6548.133	6549.942			0.2	0.3b		15267.31			
6548.344	6550.153		0.2	0.4			15266.82			
6548.464	6550.273				0.2b		15266.54			
6548.575	6550.384		0.1	0.2			15266.28			
6548.725	6550.534			0.2			15265.93			
6548.910	6550.719		0.2	0.2	0.2b		15265.50			
6549.064	6550.873		0.5	0.4			15265.14			
6549.223	6551.032		1.0	0.5			15264.77	3F 1	2C 0	P 3
6549.390	6551.200		0.2	0.2			15264.38			
6549.583	6551.393		0.6				15263.93			
6549.661	6551.470			0.7b	1.2		15263.75			
6549.729	6551.539		0.6				15263.59	6c 1	2a 6	Q 1
6549.909	6551.719		0.1				15263.17			
6550.034	6551.843		0.7	0.5			15262.88	Z 1	2B 8	P 2
6550.223	6552.032		0.2	0.4			15262.44	3f 2	2c 3	R 2
6550.223	6552.032		0.2	0.4			15262.44	3f 0	2c 1	Q 5
6550.403	6552.213		0.2	0.2	0.2b		15262.02	3f 0	2c 1	Q 5
6550.536	6552.346		0.1				15261.71	3D 2	2B 9	R 0
6550.639	6552.449		0.4	0.4			15261.47	3F 1	2C 0	P 3
6550.639	6552.449		0.4	0.4			15261.47	W 2	2B10	R 1
6550.880	6552.689		0.8	0.3	0.3b		15260.91	4e 3	3b 1	R 4
6550.880	6552.689		0.8	0.3	0.3b		15260.91	3c 5	2a 5	Q 4
6550.987	6552.797		1.0	0.9	1.4		15260.66	4b 0	2a 2	P 5
6551.159	6552.968		0.2				15260.26			

λ_{air}	λ_{vac}	Ref.	I_1	I_2	I_3	I_4	ν	Upper	Lower	Br.
6551.287	6553.097		0.5	0.2			15259.96	2K 6	2B 2	P 1
6551.429	6553.239		0.2				15259.63			
6551.648	6553.458		0.1	0.1			15259.12			
6551.858	6553.668		0.1				15258.63			
6552.026	6553.836		1.2	0.5			15258.24			
6552.253	6554.063		0.3	0.5			15257.71			
6552.477	6554.287		0.2	0.2			15257.19			
6552.704	6554.515		0.1	0.1			15256.66	3E 0	2B 6	R 5
6552.945	6554.755		0.2	0.4			15256.10	4e 2	3b 0	R 2
6553.224	6555.034		0.4	0.5			15255.45	3f 2	2c 3	R 2
6553.224	6555.034		0.4	0.5			15255.45	W 2	2C 1	P 4
6553.417	6555.228		0.1	0.1			15255.00			
6553.546	6555.357			0.1d			15254.70	W 1	2B 8	P 5
6553.705	6555.516		0.2				15254.33			
6553.842	6555.653		0.5	0.4			15254.01			
6554.019	6555.829	G	1.1	0.2			15253.60			
6554.195	6556.006		0.2	0.4			15253.19	W 1	2B 8	P 3
6554.397	6556.208		0.3	0.2			15252.72	U 1	2C 0	P 2
6554.565	6556.375		0.2	0.2			15252.33			
6554.706	6556.517				0.2b		15252.00			
6554.810	6556.620		0.2b	0.2			15251.76			
6555.042	6556.852			0.1			15251.22	4b 1	2a 3	P 4
6555.257	6557.067				0.2b		15250.72			
6555.355	6557.166		0.2b	0.2			15250.49			
6555.605	6557.416		0.4	0.4			15249.91			
6555.820	6557.631		0.2	0.2	0.2b		15249.41			
6555.949	6557.760		0.2	0.2			15249.11	3E 3	2B11	R 3
6555.949	6557.760		0.2	0.2			15249.11	3f 0	2c 1	R 1
6556.151	6557.962		0.9				15248.64	3E 2	2B 9	Q 1
6556.271	6558.082			1.0b	1.0b		15248.36			
6556.327	6558.138		0.9				15248.23	3f 1	2c 2	Q 6
6556.585	6558.396		0.2				15247.63			
6556.723	6558.534		0.3	0.4	0.2b		15247.31			
6556.955	6558.766		0.2				15246.77	4F 0	2B11	Q 4
6557.114	6558.925		0.3	0.4b			15246.40			
6557.239	6559.050		0.2				15246.11			
6557.467	6559.278		0.2	0.2b			15245.58			
6557.647	6559.459		0.2				15245.16			
6557.832	6559.644		0.3b	0.5			15244.73			
6557.987	6559.799		0.2	0.2			15244.37			
6558.146	6559.958		0.3	0.5			15244.00			
6558.336	6560.147		0.3	0.4			15243.56			
6558.473	6560.285		0.3	0.6			15243.24			
6558.727	6560.539		0.2				15242.65			
6558.861	6560.672		0.2	0.3			15242.34			
6559.071	6560.883		0.6	0.7			15241.85	W 1	2B 8	P 4
6559.132	6560.944	G	1.3				15241.71			
6559.205	6561.017		0.8	0.7			15241.54			
6559.519	6561.331		0.2	0.2			15240.81			
6559.674	6561.486		0.1				15240.45	W 2	2B10	R 0
6559.829	6561.641			0.7			15240.09			
6559.980	6561.792		0.4b	0.7			15239.74			
6560.178	6561.990			0.5			15239.28			
6560.389	6562.201			0.5			15238.79			
6560.531	6562.343		0.4b	0.7			15238.46			
6560.673	6562.485						15238.13	3D 1	2B 7	P 2
6560.772	6562.584			0.4			15237.90			
6560.910	6562.722		0.6				15237.58			
6561.065	6562.877	G	1.8	0.9			15237.22	2A 3	2B 0	R 2
6561.340	6563.153		0.3	0.8			15236.58			

$$\text{THE SPECTRUM OF } H_2$$

λ_{air}	λ_{vac}	Ref.	I_1	I_2	I_3	I_4	ν	Upper	Lower	Br.
6561.568	6563.381		0.3				15236.05			
6561.672	6563.484	G		0.7			15235.81			
6562.154	6563.967						15234.69			
6562.387	6564.200						15234.15			
6562.796	6564.609	G	1.5b	1.5b	1.5b		15233.20	3c 5	2a 5	Q 5
6562.796	6564.609	G	1.5b	1.5b	1.5b		15233.20	3e 0	2c 1	R 4
6562.796	6564.609	G	1.5b	1.5b	1.5b		15233.20	3f 0	2c 1	R 1
6563.240	6565.053						15232.17			
6563.464	6565.277						15231.65	2A 3	2B 0	R 1
6563.934	6565.747			1.0			15230.56	3E 3	2B11	Q 5
6563.934	6565.747			1.0			15230.56	4f 2	3b 0	P 4
6564.020	6565.833		0.3				15230.36			
6564.205	6566.018		0.3				15229.93			
6564.408	6566.221		0.3	0.9			15229.46			
6564.649	6566.462		0.2				15228.90			
6564.834	6566.648		0.2	0.5			15228.47			
6565.037	6566.851		0.5	0.5			15228.00			
6565.222	6567.036		0.4	0.5			15227.57	3F 1	2B 8	R 3
6565.438	6567.252		0.2	0.4			15227.07			
6565.610	6567.424		0.4	0.5			15226.67	3E 1	2C 0	R 2
6565.800	6567.614		0.2	0.4			15226.23			
6565.934	6567.748			0.4			15225.92			
6566.007	6567.821		0.2				15225.75			
6566.154	6567.968			0.7			15225.41	3b 5	2a 2	P 4
6566.231	6568.045		0.3				15225.23			
6566.391	6568.205		0.3	0.3			15224.86			
6566.559	6568.373		0.2	0.2			15224.47			
6566.771	6568.585		0.2	0.3			15223.98	2A 3	2B 0	R 3
6566.909	6568.723						15223.66			
6567.034	6568.848	G	1.4	0.9			15223.37	Z 0	2B 5	R 2
6567.189	6569.003		0.3	0.7			15223.01			
6567.340	6569.154		0.2				15222.66			
6567.457	6569.271			0.3			15222.39			
6567.694	6569.508		0.2	0.4			15221.84	3F 1	2C 0	P 4
6567.866	6569.681		0.2	0.2			15221.44			
6568.138	6569.953		0.5	0.7			15220.81	3F 1	2C 0	P 4
6568.337	6570.151		0.1				15220.35			
6568.475	6570.289		0.2	0.2			15220.03			
6568.725	6570.540		0.7	0.7			15219.45	Z 0	2B 5	R 1
6568.945	6570.760		0.1	0.1			15218.94			
6569.243	6571.058		0.9				15218.25			
6569.377	6571.192		0.8	1.0b	0.9b		15217.94	4e 3	2a 6	P 3
6569.476	6571.291		0.8				15217.71	3E 1	2C 0	R 3
6569.476	6571.291		0.8				15217.71	2K 5	2B 1	R 1
6569.597	6571.412		0.2				15217.43			
6569.731	6571.546		0.4	0.6			15217.12			
6569.843	6571.658		0.7				15216.86	3F 2	2C 1	Q 3
6569.960	6571.775	G	1.5	1.0			15216.59	3b 2	2a 0	P 4
6570.158	6571.973		0.8	0.5			15216.13	3F 2	2C 1	Q 2
6570.305	6572.120		0.3	0.5			15215.79	Z 0	2B 5	R 3
6570.305	6572.120		0.3	0.5			15215.79	3f 1	2c 2	Q 4
6570.500	6572.315		0.2	0.2			15215.34	Y 2	2C 1	P 3
6570.651	6572.466		0.3	0.8			15214.99	3F 2	2C 1	Q 4
6570.754	6572.569		0.3		0.3b		15214.75	3f 1	2c 2	R 1
6570.893	6572.708		0.2	0.2			15214.43			
6571.083	6572.898		0.3	0.6			15213.99	3F 2	2C 1	Q 3
6571.212	6573.027			0.2			15213.69	3f 0	2c 1	Q 4
6571.346	6573.161		1.0	0.6			15213.38	2K 6	2B 2	P 2
6571.458	6573.274		0.2				15213.12			
6571.519	6573.334	G	1.4	0.2			15212.98			

λ_{air}	λ_{vac}	Ref.	I_1	I_2	I_3	I_4	ν	Upper	Lower	Br.
6571.575	6573.390		0.1				15212.85			
6571.666	6573.481	G					15212.64			
6571.769	6573.585						15212.40	2K 5	2B 1	R 0
6571.769	6573.585						15212.40	3f 0	2c 1	Q 3
6571.834	6573.649		1.5b	1.0			15212.25	3F 2	2C 1	Q 2
6572.042	6573.857	G	2.3	1.5	0.2b		15211.77	2A 3	2B 0	R 0
6572.318	6574.134		0.3	0.3			15211.13			
6572.417	6574.233		0.3	0.3			15210.90			
6572.608	6574.423		0.1		0.2b		15210.46	u 1	2a 2	P 2
6572.608	6574.423		0.1		0.2b		15210.46	3f 1	2c 2	Q 5
6572.997	6574.812		0.4				15209.56	Z 0	2B 5	R 4
6573.044	6574.860			0.6	0.3b		15209.45	3f 1	2c 2	R 1
6573.096	6574.912		0.3				15209.33			
6573.260	6575.076			0.1			15208.95			
6573.373	6575.188		0.1	0.2			15208.69			
6573.525	6575.340				0.3b		15208.34	3E 3	2B11	R 2
6573.727	6575.543		0.2	0.2d			15207.87			
6573.883	6575.698		0.2				15207.51	2K 5	2B 1	R 2
6573.986	6575.802	G	1.5	1.3	0.3b		15207.27	Z 0	2B 5	R 0
6574.177	6575.992		0.4	0.2			15206.83			
6574.371	6576.187		0.2				15206.38	3F 0	2B 6	P 6
6574.635	6576.451		0.1				15205.77			
6574.778	6576.594		0.2	0.3	0.4b		15205.44			
6575.089	6576.905			0.2			15204.72			
6575.171	6576.987		0.1				15204.53	u 0	2a 1	P 2
6575.314	6577.130			0.2			15204.20			
6575.444	6577.260		0.3	0.2			15203.90			
6575.595	6577.411			0.3			15203.55			
6575.729	6577.545		0.2				15203.24			
6575.859	6577.675		0.5				15202.94	4c 1	2a 4	Q 4
6575.932	6577.749			0.7b	0.8b		15202.77	V 0	2B 5	R 2
6575.993	6577.809		0.5				15202.63			
6576.235	6578.052		0.1	0.1			15202.07			
6576.408	6578.225		0.3	0.3			15201.67	3E 1	2C 0	R 1
6576.650	6578.467		0.1				15201.11			
6576.771	6578.588				0.2b		15200.83			
6577.057	6578.874			0.2			15200.17			
6577.170	6578.986				0.3b		15199.91			
6577.317	6579.133			0.1			15199.57			
6577.607	6579.424			0.2	0.0d		15198.90	3e 1	2c 2	R 5
6577.884	6579.701		0.3	0.3			15198.26	3D 0	2B 5	P 3
6578.100	6579.917			0.2			15197.76			
6578.256	6580.073				0.3b		15197.40			
6578.329	6580.146		0.1				15197.23			
6578.654	6580.471		0.2d	0.3			15196.48			
6579.005	6580.822			0.2d			15195.67			
6579.104	6580.922		0.1				15195.44			
6579.321	6581.138	G	1.3	1.0	1.2		15194.94	Z 1	2B 7	R 2
6579.321	6581.138	G	1.3	1.0	1.2		15194.94	3c 5	2a 5	Q 6
6579.321	6581.138	G	1.3	1.0	1.2		15194.94	Z 1	2B 7	R 2
6579.654	6581.472			0.1			15194.17	t 0	2a 1	P 4
6579.888	6581.706			0.2	0.3b		15193.63			
6580.139	6581.957		0.1	0.9			15193.05			
6580.330	6582.147		1.3	1.0	1.3		15192.61	4c 1	2a 4	Q 2
6580.421	6582.238		0.3				15192.40			
6580.594	6582.412		0.1	0.3			15192.00			
6580.681	6582.498				0.2b		15191.80			
6580.767	6582.585			0.2			15191.60			
6580.962	6582.780			0.2			15191.15			
6581.170	6582.988			0.2			15190.67			

λ_{air}	λ_{vac}	Ref.	I_1	I_2	I_3	I_4	ν	Upper	Lower	Br.
6581.383	6583.200		1.1	0.5	0.3b		15190.18			
6581.569	6583.387		0.2	0.2			15189.75	Z 2	2B 9	R 0
6581.569	6583.387		0.2	0.2			15189.75	4c 1	2a 4	Q 1
6581.729	6583.547		1.0	0.8	1.1		15189.38			
6581.907	6583.725		0.4	0.3			15188.97			
6582.059	6583.877				0.2b		15188.62			
6582.158	6583.976		0.1				15188.39			
6582.414	6584.232		0.3				15187.80			
6582.527	6584.345		0.2	0.4b	0.8b		15187.54	4g 0	2a 3	P 5
6582.613	6584.431		0.3				15187.34			
6582.860	6584.679			0.2			15186.77	W 2	2B10	P 1
6583.129	6584.947		0.1	0.3b			15186.15	3f 2	2c 3	Q 5
6583.268	6585.086				0.3b		15185.83			
6583.367	6585.186			0.8	1.0		15185.60			
6583.463	6585.281		0.5	0.4			15185.38	Z 1	2B 7	R 1
6583.684	6585.503			0.1			15184.87			
6583.771	6585.589				0.2b		15184.67			
6583.853	6585.672			0.2			15184.48			
6584.165	6585.984		1.0	0.8	1.2		15183.76			
6584.352	6586.170		0.2	0.2			15183.33			
6584.508	6586.327			0.1			15182.97	2K 5	2B 1	R 3
6584.508	6586.327			0.1			15182.97	4e 2	3b 0	R 1
6584.664	6586.483		0.4	0.8	0.2b		15182.61	3f 2	2c 3	R 1
6584.664	6586.483		0.4	0.8	0.2b		15182.61	4b 0	2a 2	P 6
6584.664	6586.483		0.4	0.8	0.2b		15182.61	U 1	2C 0	Q 3
6584.816	6586.635			0.2			15182.26			
6584.963	6586.782						15181.92			
6585.011	6586.830		0.2	0.2			15181.81	Z 2	2B 9	R 1
6585.185	6587.003		0.2	0.1			15181.41			
6585.328	6587.147						15181.08	4f 3	3b 1	P 5
6585.358	6587.177		0.6	1.0	1.2		15181.01	3f 2	2c 3	R 1
6585.449	6587.268		0.5				15180.80			
6585.644	6587.463		0.3	0.3d			15180.35	3F 1	2C 0	P 5
6585.644	6587.463		0.3	0.3d			15180.35	3D 2	2B 9	P 2
6585.814	6587.633		1.2	0.9	1.3		15179.96	3b 2	2a 0	R 9
6585.978	6587.798		0.1	0.1			15179.58			
6586.148	6587.967		0.1	0.2	0.1b		15179.19			
6586.339	6588.158			0.0			15178.75			
6586.430	6588.249		0.3				15178.54			
6586.551	6588.370		0.2	0.8	0.2b		15178.26	3F 1	2C 0	P 5
6586.785	6588.605		0.5	0.3			15177.72			
6586.985	6588.805				0.2b		15177.26	4f 2	3b 0	P 3
6587.293	6589.113						15176.55			
6587.471	6589.291				0.2b		15176.14			
6587.801	6589.621						15175.38			
6587.923	6589.742				0.3b		15175.10			
6588.122	6589.942		0.2				15174.64			
6588.470	6590.290		0.2	0.1	0.3b		15173.84	3e 0	2c 1	R 3
6588.674	6590.494		0.8	0.9			15173.37	Z 1	2B 7	R 0
6588.782	6590.602		0.2		0.3b		15173.12			
6588.930	6590.750		1.2	1.1	1.2		15172.78			
6589.130	6590.950			0.2	0.6b		15172.32			
6589.277	6591.097		0.1	0.4			15171.98	3f 1	2c 2	Q 4
6589.490	6591.310		0.1	0.1	0.3b		15171.49			
6589.638	6591.458			0.1			15171.15			
6589.764	6591.584		0.2				15170.86			
6589.920	6591.740		0.1	0.1			15170.50			
6590.112	6591.932		1.1	0.7	0.3b		15170.06			
6590.324	6592.145			0.2			15169.57			
6590.403	6592.223				0.3b		15169.39			

λ_{air}	λ_{vac}	Ref.	I_1	I_2	I_3	I_4	ν	Upper	Lower	Br.
6590.563	6592.384			0.1			15169.02			
6590.802	6592.623			0.1			15168.47			
6591.072	6592.892		0.1	0.2	0.3b		15167.85			
6591.276	6593.096		0.3	0.9			15167.38	3D 1	2B 7	P 3
6591.472	6593.292			0.2			15166.93			
6591.650	6593.470			0.0			15166.52			
6591.893	6593.714		0.1	0.1			15165.96			
6592.158	6593.979		1.0	0.9			15165.35	Z 2	2B 9	R 2
6592.254	6594.075				0.4b		15165.13	3f 0	2c 1	Q 3
6592.323	6594.144			0.2			15164.97			
6592.628	6594.449			0.2			15164.27	3D 2	2B 9	P 5
6592.906	6594.727		0.2	0.3			15163.63			
6593.089	6594.910			0.2			15163.21			
6593.306	6595.127	G	1.4	0.9	0.3b		15162.71			
6593.402	6595.223		0.2				15162.49			
6593.763	6595.584						15161.66	Z 2	2B10	P 2
6594.032	6595.854		0.1	0.2d	0.4b		15161.04	4c 0	2a 3	R 0
6594.298	6596.119			0.2			15160.43			
6594.585	6596.406		0.3	0.8	0.3		15159.77			
6594.859	6596.680		0.1				15159.14			
6594.963	6596.785			0.2d			15158.90			
6595.229	6597.050		1.1	1.2	1.3		15158.29			
6595.537	6597.359		0.2	0.2	0.5		15157.58			
6595.786	6597.607	G	1.1	1.2	1.4		15157.01			
6596.147	6597.969		0.4	0.4	0.2		15156.18	2K 6	2B 2	P 3
6596.377	6598.199			0.1	0.1		15155.65			
6596.591	6598.413			0.2			15155.16			
6596.834	6598.656	G	1.6	1.1	1.3		15154.60	Z 0	2B 5	P 1
6596.834	6598.656	G	1.6	1.1	1.3		15154.60	V 0	2B 5	R 3
6596.834	6598.656	G	1.6	1.1	1.3		15154.60	3c 5	2a 5	Q 7
6596.834	6598.656	G	1.6	1.1	1.3		15154.60	3f 1	2c 2	Q 3
6596.948	6598.770		1.5	0.8			15154.34	2K 5	2B 1	P 1
6597.091	6598.913		0.2				15154.01			
6597.170	6598.992			0.1	0.2		15153.83	3f 2	2c 3	Q 5
6597.357	6599.179			0.3			15153.40	3F 3	2C 2	R 2
6597.527	6599.349			0.2	0.3		15153.01			
6597.805	6599.628	G	1.3	1.5	1.4		15152.37			
6597.979	6599.802			0.1			15151.97			
6598.175	6599.998			0.2	0.2b		15151.52			
6598.380	6600.203		0.7	0.7	0.1b		15151.05			
6599.173	6600.995			0.4			15149.23			
6599.360	6601.183			0.2			15148.80	3f 3	2c 4	R 1
6599.578	6601.401		0.4	0.5	0.2b		15148.30			
6599.713	6601.536		0.8				15147.99			
6599.739	6601.562	G	1.3	1.0	1.1		15147.93			
6599.805	6601.627		0.7				15147.78			
6599.970	6601.793			0.2			15147.40			
6600.075	6601.898				0.1b		15147.16			
6600.166	6601.989			0.2			15146.95	2K 5	2B 1	R 4
6600.419	6602.242				0.2b		15146.37			
6600.506	6602.329			0.2d			15146.17			
6600.619	6602.443		0.1				15145.91			
6600.898	6602.721		0.2	0.2	0.3b		15145.27			
6601.103	6602.926		0.4	0.3			15144.80	3f 2	2c 3	Q 4
6601.295	6603.118			0.3	0.4b		15144.36	Z 2	2B 9	R 3
6601.496	6603.319		0.1	0.1			15143.90	3E 3	2B11	Q 4
6601.722	6603.546		0.2				15143.38			
6601.823	6603.646		0.3	0.3			15143.15	3e 1	2c 2	R 4
6602.067	6603.890	G	2.0	1.4	1.2		15142.59	2A 3	2B 0	P 1
6602.067	6603.890	G	2.0	1.4	1.2		15142.59	3E 3	2B11	R 1

λ_{air}	λ_{vac}	Ref.	I_1	I_2	I_3	I_4	ν	Upper	Lower	Br.
6602.367	6604.191		0.2	0.2			15141.90			
6602.629	6604.453		0.1	0.1			15141.30	3D 2	2B 9	P 3
6602.629	6604.453		0.1	0.1			15141.30	4f 3	3b 1	Q 4
6602.873	6604.697		0.4	0.4	1.1		15140.74	3D 0	2B 5	P 4
6603.152	6604.976			0.1			15140.10			
6603.527	6605.351				0.3b		15139.24			
6603.685	6605.508		0.2				15138.88			
6603.833	6605.657		0.5	0.3			15138.54			
6603.981	6605.805		0.2	0.3			15138.20	3F 3	2C 2	R 2
6604.217	6606.041		0.1		0.2b		15137.66			
6604.404	6606.228		0.1	0.2			15137.23			
6604.583	6606.407		0.5	0.3	0.2b		15136.82			
6604.849	6606.674				0.4		15136.21	u 2	2a 3	P 2
6605.107	6606.931			0.1			15135.62			
6605.386	6607.211						15134.98			
6605.478	6607.302			0.2			15134.77			
6605.727	6607.551		0.1	0.8	0.3b		15134.20	3f 1	2c 2	Q 3
6605.727	6607.551		0.1	0.8	0.3b		15134.20	3f 0	2c 1	Q 2
6605.923	6607.747		0.5	0.4			15133.75			
6606.041	6607.865		0.2				15133.48			
6606.168	6607.992						15133.19			
6606.242	6608.066			0.1			15133.02			
6606.443	6608.267				0.2b		15132.56			
6606.630	6608.455			0.8			15132.13	3F 1	2C 0	P 6
6607.036	6608.861		0.3	0.2	0.3b		15131.20			
6607.185	6609.010			0.1			15130.86			
6607.351	6609.176				0.2b		15130.48			
6607.630	6609.455						15129.84	3F 0	2B 6	Q 4
6607.630	6609.455						15129.84	3f 3	2c 4	R 3
6607.822	6609.647			0.1	0.3b		15129.40			
6608.023	6609.848		0.1	0.3	1.0		15128.94			
6608.626	6610.451		0.1				15127.56	W 2	2B10	P 3
6608.714	6610.539				0.5b		15127.36			
6608.792	6610.617			0.1			15127.18			
6609.098	6610.923		0.4	0.3			15126.48	Z 1	2B 7	P 1
6609.295	6611.120		0.2				15126.03	Z 2	2B 9	R 4
6609.356	6611.181			0.8	1.1		15125.89	3f 2	2c 3	Q 4
6609.552	6611.378			0.2			15125.44			
6609.710	6611.535				0.2b		15125.08	4c 2	2a 5	P 3
6609.985	6611.811			0.1			15124.45	4c 0	2a 3	Q 5
6610.291	6612.117			0.2	1.0		15123.75			
6610.566	6612.392			0.1			15123.12	4c 0	2a 3	Q 6
6610.724	6612.549				0.1		15122.76			
6610.820	6612.646		0.2	0.3			15122.54			
6611.039	6612.864			0.2	0.3d		15122.04			
6611.240	6613.066		0.5	0.4	0.2d		15121.58			
6611.358	6613.184		0.1				15121.31			
6611.541	6613.367			0.1			15120.89			
6611.747	6613.573			0.2	0.2b		15120.42			
6612.092	6613.918			0.3	0.4	1.0	15119.63			
6612.372	6614.198			0.2d	0.2		15118.99	3a 1	2c 2	R 6
6612.652	6614.478		0.3	0.4	1.0		15118.35	3f 0	2c 1	Q 2
6612.980	6614.807			0.1			15117.60			
6613.059	6614.885				0.1		15117.42	4c 0	2a 3	Q 4
6613.365	6615.192			0.2	0.2		15116.72	3D 0	2B 5	P 8
6613.365	6615.192			0.2	0.2		15116.72	4f 3	3b 1	Q 3
6613.658	6615.485			0.3	1.0		15116.05			
6613.986	6615.813		0.2	0.1	0.2b		15115.30			
6614.271	6616.098		0.1	0.2	0.2		15114.65			
6614.433	6616.260		0.2				15114.28	u 0	2a 1	P 3

λ_{air}	λ_{vac}	Ref.	I_1	I_2	I_3	I_4	ν	Upper	Lower	Br.
6614.669	6616.496	G	1.7	1.2	1.0		15113.74	Z 0	2B 5	P 2
6614.779	6616.605		0.3				15113.49			
6614.936	6616.763		0.1	0.2	0.2		15113.13			
6615.155	6616.982		0.3	0.7			15112.63			
6615.291	6617.118				0.9b		15112.32			
6615.343	6617.170			0.2			15112.20			
6615.549	6617.376			0.1			15111.73			
6615.912	6617.740				0.3d		15110.90	4c 0	2a 3	Q 3
6616.000	6617.827		0.2	0.2			15110.70			
6616.223	6618.050			0.1	0.2b		15110.19	4f 3	3b 1	P 4
6616.433	6618.261		0.1				15109.71			
6616.679	6618.506			0.3	0.2d		15109.15	3c 5	2a 5	Q 8
6616.854	6618.681			0.1			15108.75	4e 2	3b 0	R 0
6616.998	6618.826						15108.42	Z 2	2B 9	P 2
6617.117	6618.944		0.1	0.2			15108.15	3e 0	2c 1	R 2
6617.117	6618.944		0.1	0.2			15108.15	3D 2	2B 9	P 4
6617.283	6619.111		0.4	0.7	0.4b		15107.77			
6617.362	6619.189		0.2				15107.59	3f 2	2c 3	Q 3
6617.528	6619.356	G	1.6	1.1	1.0		15107.21	3F 3	2C 2	R 1
6617.620	6619.448		0.2				15107.00	4c 0	2a 3	Q 2
6617.734	6619.562		0.2	0.4	0.3b		15106.74	3E 1	2C 0	R 2
6617.852	6619.680			0.2			15106.47	W 2	2B10	P 4
6617.984	6619.812		0.3	0.8	0.2		15106.17	U 1	2C 0	P 3
6618.164	6619.991			0.2			15105.76			
6618.321	6620.149		0.2	0.3	0.2b		15105.40	3f 1	2c 2	Q 2
6618.404	6620.232		0.2				15105.21			
6618.571	6620.399			0.2			15104.83			
6618.777	6620.605		1.2	0.8	0.2b		15104.36			
6618.996	6620.824			0.1			15103.86			
6619.128	6620.956				0.3		15103.56			
6619.198	6621.026			0.2			15103.40			
6619.443	6621.271		0.7	0.8	0.2		15102.84			
6619.548	6621.376		0.2				15102.60			
6619.605	6621.433			0.1			15102.47			
6619.789	6621.618	G	2.0	1.4	1.2		15102.05	2K 5	2B 1	P 2
6619.789	6621.618	G	2.0	1.4	1.2		15102.05	3E 1	2C 0	R 1
6619.895	6621.723		0.4				15101.81			
6619.991	6621.819				0.3		15101.59			
6620.118	6621.946		0.2d	0.2			15101.30	3D 0	2B 5	P 5
6620.346	6622.174		0.2d	0.2			15100.78	3D 1	2B 7	P 4
6620.574	6622.402		0.1d	0.2			15100.26			
6620.785	6622.613		0.5	0.7			15099.78			
6620.842	6622.670		1.2	1.0	1.2		15099.65	3f 2	2c 3	Q 3
6620.842	6622.670		1.2	1.0	1.2		15099.65	3f 1	2c 2	Q 2
6620.894	6622.723		0.3	0.8			15099.53	3E 1	2C 0	R 3
6621.078	6622.907		0.2	0.1			15099.11			
6621.245	6623.074			0.2			15098.73	3F 1	2B 8	Q 4
6621.403	6623.231				0.2b		15098.37			
6621.491	6623.319		0.1	0.3			15098.17			
6621.784	6623.613			0.2d			15097.50			
6621.850	6623.679				0.6b		15097.35			
6622.034	6623.863		0.2	0.2			15096.93			
6622.135	6623.964		0.2				15096.70			
6622.254	6624.083		0.2				15096.43			
6622.359	6624.188		0.5	0.3			15096.19			
6622.460	6624.289		0.4				15095.96	3D 0	2B 5	P 7
6622.600	6624.429	G	2.6	1.5b	1.6		15095.64	2A 3	2B 0	P 2
6622.697	6624.526						15095.42			
6622.859	6624.688		0.2				15095.05			
6622.960	6624.789		0.3	0.2	0.3b		15094.82			

λ_{air}	λ_{vac}	Ref.	I_1	I_2	I_3	I_4	ν	Upper	Lower	Br.
6623.166	6624.995						15094.35			
6623.289	6625.118		0.1	0.3	0.2b		15094.07	3F 0	2B 6	R 2
6623.596	6625.426		0.2	0.2			15093.37	4e 3	3b 1	R 2
6623.842	6625.671	G	1.9	1.6	1.3		15092.81	3c 6	2a 6	Q 1
6623.947	6625.777		0.4				15092.57	U 1	2C 0	Q 4
6624.189	6626.018		0.1	0.2			15092.02	4F 0	2B11	P 4
6624.426	6626.255		0.2	0.2			15091.48			
6624.562	6626.391		0.3	0.3	1.0		15091.17			
6624.795	6626.624			0.2	0.4		15090.64	3f 0	2c 1	P 7
6625.023	6626.852			0.1			15090.12			
6625.247	6627.076			0.2	0.3b		15089.61			
6625.348	6627.177		0.1				15089.38			
6625.506	6627.336		1.2	0.8			15089.02	2K 6	2B 2	P 4
6625.712	6627.542			0.1	0.1		15088.55			
6626.217	6628.047			0.2			15087.40			
6626.314	6628.144				0.3b		15087.18			
6626.490	6628.320		1.1	0.7			15086.78	Z 1	2B 7	P 2
6626.705	6628.535			0.2	0.2b		15086.29			
6626.973	6628.803		0.2	0.2			15085.68	3D 0	2B 5	P 6
6627.294	6629.124			0.2	0.2b		15084.95			
6627.500	6629.330						15084.48	3F 1	2C 0	P 7
6627.645	6629.475		0.0	0.2b			15084.15			
6628.045	6629.875		0.2	0.2			15083.24			
6628.225	6630.055			0.2			15082.83			
6628.405	6630.236		0.2d	0.5	1.0		15082.42	3E 1	2C 0	R 4
6628.607	6630.438			0.1			15081.96	4E 2	2B15	Q 2
6628.761	6630.592		0.1	0.2			15081.61	3E 1	2C 0	Q 2
6628.761	6630.592		0.1	0.2			15081.61	W 2	2C 1	P 5
6629.007	6630.838	G	1.4	1.1	1.1		15081.05	3c 6	2a 6	Q 2
6629.205	6631.036		0.2				15080.60			
6629.324	6631.155			0.1			15080.33			
6629.544	6631.374			0.2	0.2b		15079.83	3E 3	2B11	R 0
6629.544	6631.374			0.2	0.2b		15079.83	3E 3	2B11	Q 3
6629.544	6631.374			0.2	0.2b		15079.83	3e 1	2c 2	R 3
6629.702	6631.533		0.1	0.2			15079.47			
6629.922	6631.753		0.2	0.2			15078.97			
6630.071	6631.902		1.2	1.0	1.4		15078.63	3f 2	2c 3	Q 2
6630.383	6632.215	G	1.5	1.3	1.8		15077.92	3E 0	2B 6	R 4
6630.383	6632.215	G	1.5	1.3	1.8		15077.92	3b 2	2a 0	P 5
6630.599	6632.430		0.1		0.1		15077.43			
6630.876	6632.707		0.3	0.3			15076.80			
6631.017	6632.848	G	1.3	1.0	1.1		15076.48	3f 2	2c 3	Q 2
6631.263	6633.094		0.1	0.2			15075.92	Z 2	2B10	P 5
6631.426	6633.257			0.2			15075.55			
6631.615	6633.446			0.4b	0.3d		15075.12			
6631.773	6633.605			0.2			15074.76			
6632.095	6633.926			0.2b	0.2		15074.03			
6632.486	6634.318		0.3	0.4	1.1		15073.14	3E 1	2C 0	Q 1
6632.781	6634.613		0.1	0.1			15072.47	V 0	2B 5	R 1
6632.988	6634.819		0.2	0.3	0.3		15072.00			
6633.269	6635.101			0.2	0.2		15071.36	Y 2	2B10	R 2
6633.626	6635.458		0.2	0.2	0.1		15070.55	3f 3	2c 4	R 2
6633.868	6635.700	G	1.7	1.1	0.3d		15070.00			
6633.969	6635.801		0.3				15069.77			
6634.198	6636.030		0.1	0.2			15069.25	Y 2	2B10	R 1
6634.383	6636.215			0.2	0.2		15068.83			
6634.572	6636.405			0.1			15068.40			
6634.819	6636.651			0.1			15067.84	3E 0	2B 6	P 8
6634.925	6636.757				0.3b		15067.60			
6635.110	6636.942			0.1			15067.18			

λ_{air}	λ_{vac}	Ref.	I_1	I_2	I_3	I_4	ν	Upper	Lower	Br.
6635.347	6637.180						15066.64			
6635.515	6637.347			0.2	0.2		15066.26	3F 2	2B10	Q 4
6635.946	6637.779			0.2	0.2b		15065.28			
6636.123	6637.955			0.1			15064.88			
6636.277	6638.109		0.5				15064.53			
6636.339	6638.171	G	1.3	1.0	1.1		15064.39			
6636.383	6638.215		0.7				15064.29	Z 0	2B 5	P 3
6636.537	6638.370		0.2				15063.94			
6636.713	6638.546	G	1.5	1.3	1.4		15063.54	3c 6	2a 6	Q 3
6636.889	6638.722		0.7	0.4			15063.14			
6636.991	6638.823		0.2		0.1		15062.91	4F 0	2B11	Q 3
6637.171	6639.004		0.1	0.1			15062.50	4f 3	3b 1	Q 2
6637.458	6639.291		0.1	0.2			15061.85			
6637.810	6639.643		0.3d	0.8	1.1		15061.05			
6637.982	6639.815			0.1			15060.66			
6638.097	6639.930		0.3	0.3	0.2		15060.40			
6638.304	6640.137				0.2		15059.93			
6638.370	6640.203		0.1	0.2			15059.78			
6638.577	6640.410		0.2	0.3	1.1		15059.31	3F 2	2C 1	P 3
6638.577	6640.410		0.2	0.3	1.1		15059.31	3c 5	2a 5	Q 9
6638.577	6640.410		0.2	0.3	1.1		15059.31	3a 1	2c 2	R 5
6639.040	6640.873				0.2b		15058.26			
6639.358	6641.191			0.2	0.1		15057.54	u 2	2a 3	P 3
6639.666	6641.500		0.2	0.4	1.1		15056.84	3E 1	2C 0	R 5
6639.904	6641.738		0.1	0.1			15056.30			
6640.173	6642.007		0.2	0.2			15055.69	t 0	2a 1	P 5
6640.407	6642.241		0.1	0.4	0.7d		15055.16	3f 0	2c 1	P 6
6640.676	6642.510			0.2			15054.55			
6641.206	6643.040			0.2			15053.35			
6641.492	6643.326			0.2	0.4b		15052.70	4f 3	3b 1	P 3
6641.925	6643.759				0.1d		15051.72	4F 0	2B11	R 1
6642.265	6644.099			0.8	0.3b		15050.95			
6642.587	6644.421			0.2			15050.22			
6642.816	6644.651		1.0	0.8	0.4b		15049.70	3F 1	2B 8	R 2
6642.896	6644.730		0.4				15049.52			
6643.099	6644.933		0.2	0.2			15049.06	3E 1	2C 0	Q 3
6643.253	6645.088		0.6				15048.71			
6643.297	6645.132	G	1.4	1.0	1.0		15048.61			
6643.359	6645.194		0.8				15048.47			
6643.554	6645.388		0.1	0.1			15048.03			
6643.682	6645.516				0.1d		15047.74			
6643.814	6645.649			0.2			15047.44			
6643.986	6645.821				0.3d		15047.05			
6644.128	6645.962		0.1				15046.73			
6644.357	6646.192		0.2	0.4	0.1d		15046.21			
6644.596	6646.430			0.2	0.3		15045.67	3D 1	2B 7	P 5
6644.839	6646.673				0.4		15045.12			
6645.170	6647.005			0.2	0.5		15044.37			
6645.338	6647.173		1.2	0.8	1.1		15043.99	3F 2	2C 1	P 3
6645.550	6647.385			0.1			15043.51	u 1	2a 2	P 4
6645.837	6647.672			0.2d	0.2		15042.86	3F 0	2B 6	P 5
6646.089	6647.924		0.2	0.3	0.2		15042.29	4e 2	3b 0	Q 1
6646.398	6648.233		0.4	0.8	1.0		15041.59	Z 1	2B 7	P 3
6646.645	6648.481			0.2			15041.03			
6646.897	6648.733			0.7	1.1		15040.46	3c 6	2a 6	Q 4
6646.897	6648.733			0.7	1.1		15040.46	3E 0	2B 6	Q 8
6646.959	6648.795		0.1				15040.32	Z 2	2B 9	P 3
6647.140	6648.976		0.2	0.2			15039.91			
6647.269	6649.104		0.4				15039.62	Y 2	2C 1	P 4
6647.410	6649.246	G	1.9	1.4	1.3		15039.30	2A 3	2B 0	P 3

λ_{air}	λ_{vac}	Ref.	I_1	I_2	I_3	I_4	ν	Upper	Lower	Br.
6647.618	6649.453		0.2				15038.83			
6647.742	6649.577		0.5	0.3			15038.55	3F 2	2C 1	P 3
6647.883	6649.719		0.2	0.3	0.5		15038.23			
6648.139	6649.975			0.2	0.4		15037.65	4E 1	2B13	Q 4
6648.489	6650.325				0.1		15036.86			
6648.754	6650.590		0.1	0.8	0.4		15036.26			
6648.940	6650.776			0.1			15035.84			
6649.170	6651.006		0.1	0.2	0.1		15035.32			
6649.382	6651.218		1.3	0.9	0.3		15034.84	2K 5	2B 1	P 3
6649.506	6651.342		0.2				15034.56			
6649.594	6651.430			0.1			15034.36			
6649.922	6651.758			0.1	0.1d		15033.62			
6650.205	6652.041		1.0	0.3			15032.98	3E 3	2B11	Q 2
6650.306	6652.143		0.5	0.4	0.2d		15032.75			
6650.528	6652.364			0.1			15032.25			
6650.682	6652.519		1.0	0.3	0.1d		15031.90			
6650.873	6652.709				0.1d		15031.47			
6651.041	6652.877			0.2			15031.09			
6651.346	6653.183		0.1	0.2	0.2d		15030.40	3e 0	2c 1	R 1
6651.944	6653.781			0.2	0.3		15029.05	3E 3	2B11	P 4
6652.355	6654.192			0.1			15028.12			
6652.444	6654.281				0.2b		15027.92			
6652.559	6654.396						15027.66	3E 3	2B11	P 5
6652.904	6654.741			0.2			15026.88			
6653.077	6654.914		0.2	0.3			15026.49			
6653.294	6655.131		0.2	0.5	0.2b		15026.00			
6653.489	6655.326			0.2			15025.56			
6653.653	6655.490		0.2	0.3	0.3b		15025.19			
6653.905	6655.742			0.2			15024.62			
6653.998	6655.835		0.1				15024.41			
6654.118	6655.955			0.2	0.2		15024.14			
6654.304	6656.141			0.?	0.2		15023.72	3E 1	2B 0	R 6
6654.521	6656.358		0.1				15023.23			
6654.636	6656.473				0.3b		15022.97			
6654.902	6656.739			0.1			15022.37			
6655.451	6657.289				0.2b		15021.13			
6655.708	6657.546			0.2			15020.55	Y 2	2B10	P 1
6655.930	6657.767		0.5	0.5	0.2		15020.05	3E 1	2C 0	Q 1
6656.160	6657.998			0.2	0.1		15019.53			
6656.368	6658.206		0.1	0.2			15019.06	3f 0	2c 1	P 5
6656.408	6658.246				0.5		15018.97			
6656.523	6658.361			0.3			15018.71	3E 3	2B11	P 3
6656.683	6658.521			0.2			15018.35			
6656.900	6658.738		0.?	0.3	0.3		15017.86			
6657.082	6658.920						15017.45			
6657.162	6659.000			0.2d	0.1		15017.27			
6657.490	6659.328		0.2d	0.4	0.3		15016.53	U 1	2C 0	Q 5
6657.592	6659.430			0.4			15016.30			
6657.960	6659.798			0.1			15015.47	3F 2	2B10	R 2
6658.150	6659.989		0.3		0.3b		15015.04	3f 3	2c 4	R 1
6658.306	6660.144			0.1d			15014.69			
6658.807	6660.645			0.1			15013.56			
6659.055	6660.894			0.2	0.1		15013.00			
6659.295	6661.133			0.2			15012.46			
6659.379	6661.218		0.2b				15012.27			
6659.485	6661.324			0.7	1.2		15012.03	3c 6	2a 6	Q 5
6659.720	6661.559			0.1d			15011.50			
6660.262	6662.101		0.1				15010.28	Z 2	2B 9	P 5
6660.532	6662.372			0.0			15009.67			
6660.657	6662.496				0.3		15009.39			

λ_air	λ_vac	Ref.	I₁	I₂	I₃	I₄	ν	Upper	Lower	Br.
6660.843	6662.682			0.2			15008.97	3E 3	2B11	P 2
6661.114	6662.953		0.2	0.3	0.4b		15008.36	6c 0	2a 5	Q 4
6661.300	6663.140		0.4	0.3			15007.94			
6661.442	6663.282	G	1.4	1.1	1.1		15007.62	Z 0	2B 5	P 4
6661.646	6663.486		0.2	0.3	0.1		15007.16	3e 1	2c 2	R 2
6661.962	6663.801			0.1			15006.45			
6662.184	6664.023			0.1	0.1b		15005.95			
6662.375	6664.214		0.1				15005.52			
6662.468	6664.307			0.2			15005.31			
6662.774	6664.614		0.2				15004.62			
6662.956	6664.796		0.2	0.2			15004.21			
6663.232	6665.071		0.2	0.2			15003.59			
6663.445	6665.285		0.3	0.2			15003.11			
6663.769	6665.609		0.1	0.3d	1.0		15002.38			
6663.956	6665.796			0.1			15001.96			
6664.031	6665.871	1			0.4		15001.79			
6664.169	6666.009			0.2d			15001.48			
6664.458	6666.298			0.2			15000.83			
6664.640	6666.480		0.1	0.3d			15000.42			
6664.786	6666.627				0.5		15000.09			
6664.889	6666.729		0.2	0.3d			14999.86	3f 2	2c 3	Q 3
6665.102	6666.942			0.1	0.2		14999.38			
6665.288	6667.129			0.2	0.3		14998.96			
6665.582	6667.422				0.2		14998.30	3a 1	2c 2	R 4
6665.640	6667.480			0.2			14998.17			
6665.982	6667.822				0.2		14997.40			
6666.093	6667.934		0.2				14997.15			
6666.217	6668.058						14996.87	3E 3	2B11	Q 1
6666.337	6668.178	G	1.4	1.0	1.0		14996.60	3F 3	2C 2	Q 2
6666.444	6668.285		0.2				14996.36	3D 3	2B11	R 4
6666.613	6668.454		0.1	0.1			14995.98			
6666.826	6668.667		0.1	0.2	0.2d		14995.50			
6667.049	6668.890			0.2	0.3		14995.00			
6667.222	6669.063		1.1	0.5			14994.61	Z 1	2B 7	P 4
6667.333	6669.174		0.2	0.4	0.1		14994.36	3F 3	2C 2	Q 3
6667.471	6669.312			0.2			14994.05			
6667.582	6669.423		0.2	0.3			14993.80	3e 2	2c 3	R 5
6667.733	6669.575			0.2	0.4b		14993.46			
6667.880	6669.721		0.2	0.4			14993.13	3F 2	2C 1	P 4
6668.103	6669.944			0.1			14992.63			
6668.236	6670.077		0.1				14992.33			
6668.334	6670.175			0.2			14992.11			
6668.454	6670.295		0.1		0.3		14991.84			
6668.570	6670.411		0.8	0.5			14991.58	t 1	2a 2	P 5
6668.748	6670.589			0.2			14991.18			
6668.903	6670.745			0.2			14990.83			
6669.148	6670.989		0.8	0.9			14990.28			
6669.317	6671.159		0.2	0.3	0.2d		14989.90	4c 0	2a 3	P 2
6669.709	6671.550				0.1		14989.02	3D 3	2B11	R 1
6669.860	6671.702			0.1d			14988.68	3D 3	2B11	R 2
6670.060	6671.902		0.1	0.2	0.2d		14988.23	u 2	2a 3	P 4
6670.269	6672.111			0.2			14987.76			
6670.461	6672.302		1.5	0.9	1.1		14987.33			
6670.563	6672.405		0.2				14987.10	3D 3	2B11	R 0
6670.652	6672.494			0.1	0.2		14986.90	V 0	2B 5	P 4
6670.652	6672.494			0.1	0.2		14986.90	Y 2	2B10	P 2
6670.826	6672.668		0.1	0.2			14986.51			
6671.035	6672.877		0.2	0.2	0.2		14986.04	3F 1	2B 8	P 5
6671.240	6673.082		1.4	0.5	1.1		14985.58	3f 0	2c 1	P 4
6671.485	6673.327		0.2	0.2	0.7		14985.03	3E 1	2C 0	Q 4

λ_{air}	λ_{vac}	Ref.	I_1	I_2	I_3	I_4	ν	Upper	Lower	Br.
6671.680	6673.523			0.2			14984.59			
6671.810	6673.652			0.2			14984.30			
6671.890	6673.732				0.3		14984.12			
6671.965	6673.808			0.2			14983.95			
6672.166	6674.008		0.5	0.6	0.3		14983.50	3F 2	2C 1	P 4
6672.166	6674.008		0.5	0.6	0.3		14983.50	3E 1	2C 0	R 7
6672.295	6674.137		0.1				14983.21			
6672.509	6674.351		0.1	0.2	0.2		14982.73			
6672.718	6674.560				0.1		14982.26	Z 2	2B 9	P 4
6672.718	6674.560				0.1		14982.26	4e 3	3b 1	Q 4
6672.945	6674.788		0.2	0.3			14981.75			
6673.048	6674.890				0.3b		14981.52			
6673.132	6674.975		0.2	0.6			14981.33			
6673.355	6675.198			0.2			14980.83	6c 0	2a 5	Q 3
6673.538	6675.380		0.2	0.4			14980.42	3F 1	2C 0	P 2
6673.711	6675.554			0.1	0.3		14980.03			
6673.965	6675.808				0.2b		14979.46			
6674.135	6675.977		0.1				14979.08	3D 3	2B11	R 3
6674.242	6676.084		0.2	0.2			14978.84			
6674.420	6676.263		0.2	0.6	1.1		14978.44	3B 3	2A 0	R 2
6674.420	6676.263		0.2	0.6	1.1		14978.44	3c 6	2a 6	Q 6
6674.629	6676.472			0.2			14977.97			
6674.803	6676.646		0.2	0.2	0.2		14977.58			
6675.030	6676.873		1.1	0.7	0.3		14977.07	3E 1	2C 0	Q 2
6675.030	6676.873		1.1	0.7	0.3		14977.07	3F 3	2C 2	Q 3
6675.030	6676.873		1.1	0.7	0.3		14977.07	3F 3	2B12	R 2
6675.191	6677.034		0.2	0.2			14976.71			
6675.409	6677.252	G	1.4	1.3	1.1		14976.22	3B 3	2A 0	R 1
6675.521	6677.364		0.2				14975.97			
6675.623	6677.466			0.1	0.2		14975.74			
6675.819	6677.663				0.3		14975.30			
6676.078	6677.921			0.2	0.3		14974.72			
6676.332	6678.175						14974.15			
6676.599	6678.443		0.2		0.5		14973.55	3F 0	2B 6	Q 3
6676.711	6678.555						14973.30			
6676.836	6678.679		0.2		0.4		14973.02	4e 2	3b 0	Q 2
6676.947	6678.791		0.2				14972.77			
6677.184	6679.027		0.1				14972.24	4E 1	2B13	Q 3
6677.259	6679.103				1.0		14972.07			
6677.349	6679.192		0.2	0.1			14971.87	3a 2	2c 3	R 3
6677.540	6679.384		0.3		0.5		14971.44			
6677.688	6679.531		0.3	0.2			14971.11			
6677.732	6679.576				0.7		14971.01			
6677.808	6679.652		0.3				14970.84			
6677.960	6679.804	G	1.9	1.6	1.6		14970.50	2A 3	2B 0	P 4
6678.129	6679.973	1		1.5	1.5b		14970.12			
6678.397	6680.241		0.3	0.2			14969.52	u 1	2a 2	P 5
6678.522	6680.366		0.5	0.3			14969.24			
6678.723	6680.567		0.2		0.5		14968.79			
6678.807	6680.651			0.2			14968.60			
6678.923	6680.768				0.3		14968.34			
6679.030	6680.875			0.2			14968.10			
6679.106	6680.951		0.2		0.4		14967.93			
6679.187	6681.031			0.2			14967.75			
6679.321	6681.165		0.2	0.2			14967.45			
6679.477	6681.321				0.4		14967.10			
6679.539	6681.384		0.3	0.8	1.1		14966.96			
6679.695	6681.540				0.4		14966.61			
6679.771	6681.616		0.2	0.3			14966.44			
6679.950	6681.794			0.1	0.2		14966.04			

λ_{air}	λ_{vac}	Ref.	I_1	I_2	I_3	I_4	ν	Upper	Lower	Br.
6680.155	6682.000			0.2	0.2		14965.58			
6680.454	6682.299		0.2	0.4d	1.1		14964.91	3f 1	2c 2	P 6
6680.664	6682.509						14964.44			
6680.780	6682.625			0.2d	0.3b		14964.18			
6681.267	6683.112			0.2	0.2b		14963.09			
6681.526	6683.371				0.3		14962.51	3f 0	2c 1	P 3
6681.606	6683.451		0.2	0.3			14962.33			
6681.727	6683.572				0.3		14962.06			
6681.937	6683.782			0.2	0.2		14961.59			
6682.097	6683.942				0.1		14961.23			
6682.200	6684.045			0.2			14961.00			
6682.410	6684.255		0.2	0.8	1.2		14960.53	3B 3	2A 0	R 3
6682.410	6684.255		0.2	0.8	1.2		14960.53	3F 1	2B 8	Q 3
6682.678	6684.523			0.2	0.3		14959.93			
6682.933	6684.778		0.1	0.3	1.0		14959.36	6c 0	2a 5	Q 2
6683.295	6685.140			0.2	0.1		14958.55			
6683.353	6685.198						14958.42			
6683.621	6685.466			0.2	0.5		14957.82			
6683.943	6685.788			0.2	0.4		14957.10			
6684.144	6685.989			0.2			14956.65			
6684.430	6686.275		0.1		0.2		14956.01			
6684.725	6686.570			0.2			14955.35			
6684.796	6686.642		0.2		0.8		14955.19			
6684.881	6686.727		0.2				14955.00			
6685.100	6686.946			0.1	0.3		14954.51			
6685.297	6687.143			0.2			14954.07			
6685.511	6687.357	G	1.3	0.9	1.2		14953.59	2K 5	2B 1	P 4
6685.511	6687.357	G	1.3	0.9	1.2		14953.59	3B 3	2A 0	R 0
6685.583	6687.429				0.8		14953.43	3e 0	2c 1	Q 5
6685.735	6687.581			0.1			14953.09			
6685.963	6687.809		0.2	0.2	0.2		14952.58			
6686.267	6688.113			0.2			14951.90			
6686.325	6688.171				0.2		14951.77			
6686.522	6688.368			0.2			14951.33			
6686.768	6688.614			0.3	0.5		14950.78			
6687.117	6688.963		0.1	0.2	0.2		14950.00	4E 1	2B13	Q 1
6687.117	6688.963		0.1	0.2	0.2		14950.00	4e 2	3b 0	P 2
6687.752	6689.599		0.2	0.2			14948.58			
6687.958	6689.804		1.3	1.0	1.2		14948.12	3E 1	2C 0	P 2
6687.958	6689.804		1.3	1.0	1.2		14948.12	3f 1	2c 2	P 5
6688.150	6689.997						14947.69			
6688.253	6690.100			0.2b	0.5		14947.46	3e 0	2c 1	Q 4
6688.455	6690.301			0.3	0.4		14947.01	3f 0	2c 1	P 3
6688.544	6690.391		0.2				14946.81			
6688.929	6690.776		0.2	0.2			14945.95			
6689.184	6691.031		0.1	0.2			14945.38			
6689.354	6691.201			0.1			14945.00			
6689.426	6691.273				0.1b		14944.84			
6689.506	6691.353		0.2	0.8			14944.66	6c 0	2a 5	Q 1
6689.744	6691.591		0.3	0.2			14944.13	V 0	2B 5	R 0
6689.918	6691.765			0.3	0.2b		14943.74			
6690.093	6691.940		0.1	0.2			14943.35			
6690.361	6692.209			0.2			14942.75			
6690.523	6692.370				0.1b		14942.39			
6690.594	6692.441			0.1d			14942.23			
6690.814	6692.661		0.3	0.3	0.2b		14941.74	3f 1	2c 2	P 3
6690.979	6692.827		0.3	0.8			14941.37	3F 2	2C 1	P 5
6691.181	6693.028		0.3	0.4	0.3		14940.92	Z 0	2B 5	P 5
6691.333	6693.181		0.1	0.2			14940.58			
6691.499	6693.346			0.3	0.4		14940.21	3c 6	2a 6	Q 7

λ_{air}	λ_{vac}	Ref.	I_1	I_2	I_3	I_4	ν	Upper	Lower	Br.
6691.978	6693.826			0.1			14939.14	3e 2	2c 3	R 4
6692.310	6694.157		0.4	0.9	1.2		14938.40	3f 0	2c 1	P 4
6692.310	6694.157		0.4	0.9	1.2		14938.40	3f 1	2c 2	P 4
6692.310	6694.157		0.4	0.9	1.2		14938.40	4E 1	2B13	Q 2
6692.569	6694.417		0.3	0.1			14937.82			
6692.780	6694.628		0.1	0.2			14937.35			
6693.009	6694.856		0.6				14936.84	3f 1	2c 2	P 3
6693.009	6694.856		0.6				14936.84	U 1	2B 8	R 4
6693.044	6694.892			0.9	1.2		14936.76	3e 0	2c 1	Q 3
6693.085	6694.933		0.5				14936.67	4c 0	2a 3	P 3
6693.085	6694.933		0.5				14936.67	3a 1	2c 2	R 3
6693.367	6695.215		0.2	0.2	0.1		14936.04	Y 2	2B10	P 3
6693.609	6695.457			0.3	0.3		14935.50			
6693.793	6695.641			0.2			14935.09			
6693.990	6695.838				0.2		14934.65	3F 2	2B10	Q 3
6694.290	6696.139		0.1	0.3	0.5		14933.98	3f 0	2c 1	P 5
6694.501	6696.349		0.4	0.5			14933.51	U 1	2C 0	P 4
6694.501	6696.349		0.4	0.5			14933.51	3E 0	2B 6	R 3
6694.640	6696.488		0.1	0.2			14933.20			
6694.869	6696.717	G	1.7	1.5	1.8		14932.69	3b 3	2a 1	R 2
6694.869	6696.717	G	1.7	1.5	1.8		14932.69	u 2	2a 3	P 5
6695.160	6697.009		0.2	0.2	0.1		14932.04			
6695.348	6697.197		1.3	0.9	0.2		14931.62			
6695.573	6697.421				0.3		14931.12	3f 0	2c 1	P 6
6695.967	6697.816		0.2	0.4	0.2		14930.24			
6696.160	6698.009		0.2	0.2	0.1		14929.81			
6696.407	6698.256						14929.26			
6696.456	6698.305		0.3	0.4	0.5		14929.15	3f 0	2c 1	P 7
6696.761	6698.610	G	2.6	1.5b	2.1		14928.47	3b 3	2a 1	R 1
6697.098	6698.947		0.3	0.2			14927.72	4F 0	2B11	P 3
6697.192	6699.041				0.3		14927.51			
6697.304	6699.153		0.2	0.1			14927.26			
6697.497	6699.346		0.1		0.1		14926.83			
6697.789	6699.638		0.2	0.4	0.2		14926.18			
6698.004	6699.853			0.2	0.4		14925.70	3E 2	2B 1	R 3
6698.516	6700.365			0.1			14924.56			
6698.722	6700.572			0.2			14924.10	4e 3	3b 1	R 0
6698.929	6700.778		0.2	0.8	1.3		14923.64	3B 3	2A 0	R 4
6698.929	6700.778		0.2	0.8	1.3		14923.64	3b 2	2a 0	P 6
6699.265	6701.115		0.3b	0.4b	0.2		14922.89	3f 2	2c 3	P 3
6699.642	6701.492	G	1.7	1.0			14922.05	3e 1	2c 2	R 1
6699.791	6701.640		0.8	1.3	1.2		14921.72	3f 2	2c 3	P 3
6700.100	6701.950			0.2	0.1		14921.03			
6700.302	6702.152			0.2			14920.58			
6700.451	6702.300				0.4		14920.25			
6700.531	6702.381		0.3	0.3			14920.07	3E 2	2C 1	R 2
6700.738	6702.588			0.1	0.1		14919.61	4e 3	3b 0	P 3
6700.927	6702.777		0.3	0.6	0.3		14919.19	3E 1	2C 0	Q 3
6700.927	6702.777		0.3	0.6	0.3		14919.19	3f 1	2c 2	P 4
6701.124	6702.974		0.1	0.1			14918.75			
6701.273	6703.123			0.2			14918.42			
6701.461	6703.311		0.2	0.4	0.2d		14918.00			
6701.641	6703.491		0.2	0.2			14917.60	V 0	2B 5	P 3
6701.884	6703.734	G	1.6	1.6	2.3		14917.06	3b 3	2a 1	R 3
6701.991	6703.842		0.2				14916.82			
6702.221	6704.071		0.1	0.2			14916.31	3D 3	2B11	P 2
6702.427	6704.278			0.2	0.1		14915.85	3F 0	2B 6	R 1
6702.593	6704.444		0.2	0.3	1.2		14915.48	3e 0	2c 1	Q 2
6702.769	6704.619		0.3	0.3			14915.09	3f 3	2c 4	Q 3
6703.029	6704.880		0.2	0.3	1.1		14914.51			

λ_{air}	λ_{vac}	Ref.	I_1	I_2	I_3	I_4	ν	Upper	Lower	Br.
6703.389	6705.240		0.2	0.2	0.1		14913.71			
6703.609	6705.460		0.2	0.2	0.2		14913.22	4e 2	3b 0	Q 3
6703.888	6705.739			0.2	0.1		14912.60			
6704.135	6705.986			0.3	0.3		14912.05			
6704.495	6706.346		0.2	0.2			14911.25	3f 3	2c 4	Q 2
6704.495	6706.346		0.2	0.2			14911.25	3a 2	2c 3	R 2
6704.783	6706.634		0.2	0.1			14910.61			
6705.052	6706.904		0.3	0.8	0.5		14910.01			
6705.322	6707.174			0.1			14909.41			
6705.579	6707.430			0.2	0.1		14908.84	4e 3	3b 1	Q 2
6705.952	6707.803		0.1		0.2d		14908.01			
6706.195	6708.046			0.1			14907.47			
6706.438	6708.289		0.2	0.3	0.2		14906.93	3E 2	2C 1	R 4
6706.586	6708.438			0.1			14906.60	3F 3	2B12	Q 3
6706.748	6708.600		0.2	0.4	1.1		14906.24	3f 1	2c 2	P 5
6706.748	6708.600		0.2	0.4	1.1		14906.24	3F 1	2B 8	R 1
6707.005	6708.856		0.1				14905.67			
6707.356	6709.207		0.2	0.2			14904.89			
6707.594	6709.446	G	1.6	1.3	1.5		14904.36	3b 3	2a 1	R 0
6707.761	6709.613		0.3				14903.99			
6707.941	6709.793		0.2	0.3	0.4		14903.59	4F 0	2B11	Q 2
6708.215	6710.067		0.1	0.1			14902.98			
6708.310	6710.162				0.1		14902.77			
6708.422	6710.274		0.1	0.1			14902.52			
6708.652	6710.504		1.2	1.0	1.1		14902.01			
6708.828	6710.680		1.3	1.0	1.2		14901.62	3f 2	2c 3	P 4
6709.179	6711.031			0.1			14900.84			
6709.408	6711.261			0.1			14900.33			
6709.670	6711.522		0.3	0.4	0.1		14899.75	3E 1	2C 0	P 2
6709.832	6711.684			0.1			14899.39			
6710.003	6711.855			0.2			14899.01			
6710.219	6712.072		0.1	0.1			14898.53			
6710.399	6712.252			0.2	0.1		14898.13			
6710.570	6712.423			0.1			14897.75			
6710.724	6712.576			0.2	0.1		14897.41			
6710.895	6712.747			0.1			14897.03			
6711.070	6712.923			0.2	0.3		14896.64			
6711.170	6713.022		0.2				14896.42			
6711.300	6713.153		0.2	0.3	0.3		14896.13	3f 1	2c 2	P 6
6711.566	6713.419		1.2	0.5	0.3		14895.54	3f 2	2c 3	P 4
6711.764	6713.617		0.1	0.1			14895.10			
6712.044	6713.897				0.2		14894.48			
6712.102	6713.955		0.4	0.3			14894.35	3D 1	2C 0	R 1
6712.355	6714.208		1.0	0.9	1.2		14893.79	3E 1	2C 0	P 3
6712.652	6714.505		0.2	0.5	0.7		14893.13	3F 2	2B10	R 1
6712.941	6714.794		0.2	0.2			14892.49	3D 1	2C 0	P 1
6713.080	6714.934				0.1		14892.18			
6713.148	6715.001		0.1	0.1			14892.03			
6713.364	6715.218		0.2	0.2			14891.55	3E 2	2C 1	R 1
6713.576	6715.430				0.5		14891.08			
6713.653	6715.506		0.2	0.5	1.1		14890.91	3a 0	2c 1	R 3
6713.653	6715.506		0.2	0.5	1.1		14890.91	3f 2	2c 3	P 5
6714.072	6715.926		0.2	0.4	0.2		14889.98			
6714.271	6716.124		0.1	0.3			14889.54	3A 2	2B11	R 4
6714.271	6716.124		0.1	0.3			14889.54	4e 2	3b 0	P 4
6714.492	6716.345		0.2	0.3			14889.05			
6714.672	6716.526		0.3	0.8	1.2		14888.65	4d 2	3b 0	R 0
6714.848	6716.702			0.2			14888.26			
6715.001	6716.855		0.1	0.2	0.1		14887.92			
6715.276	6717.130		0.2	0.4	0.5		14887.31	3f 1	2c 2	P 7

λ_{air}	λ_{vac}	Ref.	I_1	I_2	I_3	I_4	ν	Upper	Lower	Br.
6715.376	6717.229		0.2				14887.09			
6715.538	6717.392						14886.73			
6715.606	6717.460			0.1d	0.1		14886.58			
6715.967	6717.821			0.2d	0.1b		14885.78	3F 0	2B 6	P 4
6716.021	6717.875		0.2d				14885.66	4c 0	2a 3	P 4
6716.269	6718.123	G	1.6	1.3	1.3		14885.11	2A 3	2B 0	P 5
6716.269	6718.123	G	1.6	1.3	1.3		14885.11	3c.7	2a 7	Q 1
6716.386	6718.240		0.4				14884.85			
6716.594	6718.448		0.4	1.0	1.3		14884.39	4E 0	2B 6	Q 7
6716.594	6718.448		0.4	1.0	1.3		14884.39	3e 1	2c 2	Q 6
6716.842	6718.696		0.1	0.1			14883.84			
6717.248	6719.103			0.1	0.2		14882.94			
6717.501	6719.355		0.1	0.2	0.2		14882.38	3E 0	2B 6	P 7
6717.772	6719.626		1.3	1.0	1.9		14881.78	3b 3	2a 1	R 4
6717.943	6719.798				0.6		14881.40			
6717.993	6719.848		0.2	0.2			14881.29			
6718.173	6720.028		0.3	0.9	1.1		14880.89			
6718.399	6720.254		0.1	0.2			14880.39			
6718.453	6720.308				0.8d		14880.27	3e 0	2c 1	R 5
6718.557	6720.412		1.3	0.9			14880.04	3D 1	2C 0	Q 1
6718.557	6720.412		1.3	0.9			14880.04	V 0	2B 5	P 5
6718.851	6720.706				0.4		14879.39	3e 0	2c 1	R 6
6719.063	6720.918		0.1	0.2			14878.92			
6719.185	6721.040				0.2		14878.65			
6719.375	6721.230		0.1	0.2	0.4		14878.23	3e 0	2c 1	R 4
6719.609	6721.465		1.0	0.9	1.4		14877.71			
6719.840	6721.695		0.3	1.0	0.7		14877.20	3e 1	2c 2	Q 5
6719.840	6721.695		0.3	1.0	0.7		14877.20	Z 0	2B 5	P 6
6720.088	6721.943				0.2		14876.65	4d 2	3b 0	P 1
6720.197	6722.052			0.2			14876.41	3e 2	2c 3	R 3
6720.409	6722.264		0.1	0.2	0.1		14875.94			
6720.680	6722.535		0.3	0.5	0.2		14875.34	3a 1	2c 2	R 2
6720.919	6722.775		0.2	0.6	1.2		14874.81	3e 0	2c 1	R 3
6721.082	6722.938			0.2			14874.45			
6721.267	6723.123		1.1	1.0	1.2		14874.04			
6721.322	6723.177				0.4		14873.92	3f 2	2c 3	P 5
6721.855	6723.711			0.1d	0.4		14872.74	3e 0	2c 1	Q 1
6722.420	6724.276		0.2	0.4	0.2d		14871.49	4D 0	2B11	R 6
6722.723	6724.579		0.2	0.2			14870.82			
6723.003	6724.859		0.2	0.3	0.5		14870.20			
6723.193	6725.049			0.1			14869.78			
6723.356	6725.212		1.5	0.4	0.5		14869.42	3e 0	2c 1	R 2
6723.541	6725.397			0.2			14869.01			
6723.763	6725.619		0.3b	0.5b	0.2		14868.52			
6724.066	6725.922		1.5	0.9	0.2		14867.85			
6724.287	6726.144				0.2		14867.36			
6724.396	6726.252			0.2			14867.12			
6724.645	6726.501		0.4				14866.57			
6724.713	6726.569			0.3	0.2		14866.42	3B 3	2A 0	R 5
6724.943	6726.800			0.1			14865.91			
6725.169	6727.026			0.2b			14865.41	3D 3	2B11	P 3
6725.450	6727.307		0.2	0.6	1.4		14864.79	3e 1	2c 2	Q 4
6725.812	6727.669			0.1	0.1		14863.99			
6726.192	6728.049		0.2	0.2	0.2		14863.15			
6726.436	6728.293		0.3	0.6	0.6		14862.61			
6726.699	6728.556		0.2	0.3	0.1		14862.03			
6726.889	6728.746			0.2	0.3		14861.61			
6727.138	6728.995			0.2	0.1		14861.06			
6727.441	6729.298		0.4	0.9	1.2		14860.39	3e 0	2c 1	R 1
6727.527	6729.385		0.3				14860.20	4e 2	3b 0	P 5

λ_{air}	λ_{vac}	Ref.	I_1	I_2	I_3	I_4	ν	Upper	Lower	Br.
6727.527	6729.385		0.3				14860.20	2K 5	2B 1	P 5
6727.677	6729.534		0.1	0.1			14859.87			
6727.894	6729.751			0.1			14859.39			
6728.193	6730.050		0.3	0.6	0.2		14858.73			
6728.419	6730.277			0.2			14858.23			
6728.650	6730.508		0.5	1.0	0.3		14857.72			
6728.859	6730.716		0.3	0.5	0.3		14857.26			
6729.121	6730.979			0.1			14856.68			
6729.339	6731.196			0.1			14856.20			
6729.434	6731.292				0.1		14855.99			
6729.515	6731.373			0.1			14855.81			
6729.683	6731.541		0.1	0.2			14855.44	3F 1	2B 8	P 4
6729.855	6731.713		0.1	0.3	0.3		14855.06			
6730.091	6731.949			0.2			14854.54			
6730.240	6732.098		0.1	0.2			14854.21			
6730.408	6732.266		1.0	0.9	0.2		14853.84	X 2	2B11	R 0
6730.485	6732.343		0.2				14853.67			
6730.630	6732.488			0.2			14853.35			
6730.979	6732.837			0.2			14852.58	3a 2	2c 3	R 1
6730.979	6732.837			0.2			14852.58	4e 2	3b 0	Q 4
6731.405	6733.263				0.2		14851.64			
6731.881	6733.739		0.3	0.3			14850.59	V 0	2B 5	P 2
6732.030	6733.889			0.3			14850.26	3E 1	2C 0	Q 4
6732.144	6734.002			0.3	0.2b		14850.01			
6732.357	6734.215			0.2	0.1		14849.54			
6732.588	6734.447			0.2	0.4		14849.03			
6732.937	6734.796		0.1	0.0			14848.26	X 2	2B11	R 1
6733.164	6735.023		0.2	0.2			14847.76			
6733.350	6735.209		0.3	0.4			14847.35			
6733.495	6735.354	G	1.4	1.0	0.3		14847.03	3B 3	2A 0	P 1
6733.871	6735.730			0.2			14846.20			
6734.171	6736.030			0.2			14845.54	3F 0	2B 6	Q 2
6734.370	6736.229			0.3			14845.10			
6734.561	6736.420		0.1	0.2			14844.68			
6734.738	6736.597	G	0.8	1.0	1.3		14844.29	3e 1	2c 2	Q 3
6734.738	6736.597	G	0.8	1.0	1.3		14844.29	3F 1	2B 8	Q 2
6734.810	6736.670		0.6				14844.13			
6735.151	6737.010			0.2			14843.38			
6735.423	6737.282		1.5b	1.0	0.2		14842.78			
6735.523	6737.382	G	1.4				14842.56			
6735.686	6737.546			0.1			14842.20	3F 2	2B10	Q 2
6735.813	6737.673		0.2				14841.92	4d 2	3b 0	R 1
6735.968	6737.827			0.1	0.2		14841.58			
6736.385	6738.245				0.3		14840.66			
6736.471	6738.331			0.2			14840.47			
6736.540	6738.399						14840.32			
6736.612	6738.472		0.1	0.2			14840.16			
6737.470	6739.330						14838.27			
6737.729	6739.589		0.3	0.9	1.3		14837.70	X 2	2B11	R 2
6737.820	6739.680		0.3		0.5		14837.50			
6738.020	6739.880		0.2	0.2			14837.06			
6738.133	6739.993		0.2				14836.81	4c 0	2a 3	P 5
6738.274	6740.134		0.2	0.6	1.1		14836.50			
6738.419	6740.279			0.4			14836.18			
6738.533	6740.393			0.1			14835.93	3F 3	2B12	Q 2
6738.674	6740.534		0.7	0.5			14835.62	3F 3	2C 2	P 3
6738.837	6740.697			0.1	0.2b		14835.26			
6739.410	6741.270				0.1		14834.00			
6739.687	6741.547				0.2		14833.39			
6739.732	6741.593			0.1			14833.29			

λ_air	λ_vac	Ref.	I₁	I₂	I₃	I₄	ν	Upper	Lower	Br.
6739.873	6741.734						14832.98			
6739.955	6741.815			0.1			14832.80			
6741.168	6743.029			0.1			14830.13	4E 0	2B11	R 0
6741.168	6743.029			0.1			14830.13	X 2	2B11	P 1
6741.741	6743.602		0.1	0.2	0.3		14828.87			
6741.955	6743.816			0.1			14828.40			
6742.137	6743.998			0.2	0.1		14828.00			
6742.278	6744.139		0.2	0.2			14827.69	X 2	2B11	R 3
6742.278	6744.139		0.2	0.2			14827.69	3a 0	2c 1	R 2
6742.455	6744.316		0.4	1.2	2.2		14827.30	3b 3	2a 1	R 5
6742.792	6744.653			0.0	0.1		14826.56			
6743.087	6744.949			0.2			14825.91			
6743.333	6745.194			0.6	0.2		14825.37			
6743.565	6745.426	G	1.8	1.3	0.3		14824.86			
6744.365	6746.227			0.2			14823.10			
6744.506	6746.368			0.2			14822.79			
6744.739	6746.600			0.1	0.2		14822.28			
6744.948	6746.810			0.2			14821.82			
6745.276	6747.138			0.1			14821.10			
6745.631	6747.493			0.2			14820.32			
6745.844	6747.707		0.2	0.3	0.3		14819.85			
6746.404	6748.267			0.0	0.1		14818.62			
6746.778	6748.640		0.7	0.5	0.3		14817.80	3D 1	2C 0	R 2
6747.033	6748.895			0.2			14817.24			
6747.320	6749.182		0.5	0.3			14816.61			
6747.616	6749.478				0.3		14815.96			
6747.884	6749.747			0.2			14815.37			
6748.130	6749.993		0.3	0.5	0.2d		14814.83	3a 1	2c 2	R 1
6748.595	6750.458		0.2	0.1			14813.81			
6748.841	6750.704	G	1.5	1.2	1.2		14813.27			
6748.955	6750.818		0.9				14813.02			
6749.096	6750.959		0.3	0.3	0.2		14812.71	3e 1	2c 2	Q 2
6749.187	6751.050		0.2				14812.51			
6749.406	6751.269				0.1		14812.03			
6749.497	6751.360			0.2			14811.83			
6749.793	6751.656			0.2	0.2b		14811.18			
6749.966	6751.830			0.1			14810.80			
6750.131	6751.994			0.1			14810.44			
6750.308	6752.172		0.2	0.4	0.4		14810.05			
6750.450	6752.313				1.3		14809.74			
6750.495	6752.359		0.1	0.3	0.1		14809.64			
6750.664	6752.527			0.1			14809.27	3D 3	2B11	P 4
6750.851	6752.714			0.2			14808.86			
6751.088	6752.951			0.1			14808.34			
6751.174	6753.038						14808.15			
6751.498	6753.362		0.1				14807.44			
6751.625	6753.490			0.0			14807.16			
6751.822	6753.686		0.2				14806.73			
6751.895	6753.759			0.3b	0.1b		14806.57			
6752.178	6754.041						14805.95			
6752.419	6754.283			0.1			14805.42			
6752.711	6754.575		0.2				14804.78			
6752.857	6754.721		0.4	0.6	1.2		14804.46	3e 2	2c 3	R 2
6753.135	6755.000		0.1	0.1	0.2		14803.85			
6753.478	6755.342		0.1				14803.10			
6753.701	6755.565		0.7	0.9	0.2b		14802.61			
6753.788	6755.652		0.5				14802.42			
6753.934	6755.798			0.2			14802.10			
6754.130	6755.994			0.2	0.2		14801.67			
6754.386	6756.250			0.3	1.2		14801.11			

λ_air	λ_vac	Ref.	I₁	I₂	I₃	I₄	ν	Upper	Lower	Br.
6754.564	6756.428			0.1			14800.72			
6754.751	6756.615			0.2			14800.31			
6754.965	6756.830		1.4	0.9	1.4		14799.84	3E 1	2C 0	P 4
6755.130	6756.994		0.9	1.2			14799.48	3E 1	2C 0	P 3
6755.403	6757.268			0.1d			14798.88	3e 1	2c 2	R 5
6755.664	6757.528		0.1	0.2d			14798.31			
6755.887	6757.752			0.6d	1.4		14797.82			
6756.020	6757.885		0.3	0.2d			14797.53			
6756.262	6758.127	G	2.0	1.8	2.0		14797.00	3b 3	2a 1	P 1
6756.490	6758.355		0.2				14796.50			
6756.613	6758.478		0.2	0.2			14796.23			
6756.960	6758.825		0.1	0.2			14795.47			
6757.321	6759.186	G	1.5	1.0			14794.68			
6757.540	6759.406			0.3			14794.20	3e 1	2c 2	R 4
6757.705	6759.570		0.2	0.2	1.3		14793.84			
6758.299	6760.164						14792.54	4e 2	3b 0	Q 5
6758.888	6760.754		0.2	0.2d			14791.25			
6759.130	6760.996	G	1.2	1.0			14790.72	4d 2	3b 0	R 2
6759.267	6761.133		0.4	0.2			14790.42	3E 0	2B 6	R 2
6759.414	6761.279			0.1			14790.10	4c 0	2a 3	P 6
6759.807	6761.673			0.1			14789.24			
6760.104	6761.970			0.2d			14788.59			
6760.465	6762.331		0.3	1.0	1.4		14787.80	3e 1	2c 2	R 3
6760.671	6762.537			0.1			14787.35			
6760.895	6762.761			0.1			14786.86			
6761.183	6763.049		0.2	0.3	1.4		14786.23	3e 0	2c 1	Q 1
6761.407	6763.273		0.2	0.2			14785.74			
6761.759	6763.625		0.1	0.1			14784.97			
6762.084	6763.950		0.3	0.4			14784.26	U 1	2C 0	P 5
6762.290	6764.156			0.1			14783.81	X 0	2B 7	R 3
6762.509	6764.376			0.1			14783.33	X 2	2B11	P 2
6762.729	6764.595			0.0			14782.85	V 1	2B 6	R 2
6762.994	6764.861		0.2	0.3			14782.27			
6763.250	6765.117			0.0			14781.71			
6763.612	6765.479			0.0			14780.92			
6764.037	6765.904		0.1	0.1			14779.99			
6764.257	6766.124			0.0			14779.51	X 0	2B 7	R 4
6764.463	6766.330			0.1			14779.06			
6764.761	6766.628						14778.41			
6764.866	6766.733			0.2			14778.18			
6765.104	6766.971		0.3d	0.7	0.1		14777.66	3e 1	2c 2	R 2
6765.104	6766.971		0.3d	0.7	0.1		14777.66	3E 2	2C 1	Q 2
6765.369	6767.237			0.1			14777.08	3E 2	2C 1	Q 1
6765.585	6767.452			0.2			14776.61			
6765.786	6767.654		0.4	1.0	0.2		14776.17	3F 3	2C 2	P 4
6766.024	6767.892			0.1			14775.65			
6766.780	6768.648		0.6	0.4			14774.00	3b 3	2a 1	R 6
6766.780	6768.648		0.6	0.4			14774.00	3D 1	2C 0	Q 2
6766.780	6768.648		0.6	0.4			14774.00	4e 3	3b 1	P 2
6766.977	6768.845		0.6	0.6			14773.57	3E 1	2C 0	Q 5
6767.105	6768.973		1.3	1.0			14773.29			
6767.123	6768.991			0.1			14773.25			
6767.953	6769.821		1.2	1.0			14771.44	3e 1	2c 2	Q 1
6768.168	6770.036		0.1	0.1			14770.97			
6768.429	6770.297	G	1.4	1.1	0.1		14770.40	4E 0	2B11	P 4
6768.814	6770.682		0.2	0.2			14769.56	3a 0	2c 1	R 1
6769.117	6770.985		0.1	0.1			14768.90			
6769.346	6771.214			0.3			14768.40			
6769.428	6771.297		0.2				14768.22	Z 1	2C 0	R 1
6769.580	6771.448			0.0			14767.89	4E 0	2B11	Q 4

λ_{air}	λ_{vac}	Ref.	I_1	I_2	I_3	I_4	ν	Upper	Lower	Br.
6769.580	6771.448			0.0			14767.89	3a 2	2c 3	Q 4
6769.800	6771.668						14767.41	3F 1	2B 8	P 3
6770.079	6771.948	G	1.7				14766.80	3F 0	2B 6	P 3
6770.414	6772.283		0.2	0.1			14766.07			
6770.662	6772.530		0.3	0.8			14765.53			
6770.799	6772.668		0.2	0.2			14765.23			
6771.001	6772.870		1.5	1.1			14764.79			
6771.239	6773.108		0.1	0.2			14764.27			
6771.450	6773.319		0.8	0.6			14763.81	3B 3	2A 0	P 2
6771.657	6773.526			0.2			14763.36			
6771.831	6773.700		0.1				14762.98			
6771.914	6773.783			0.1			14762.80			
6772.143	6774.012		0.2	0.3			14762.30	U 1	2B 8	R 2
6772.372	6774.242	G	0.7	1.1	1.2		14761.80	3e 1	2c 2	R 1
6772.464	6774.333		0.5				14761.60			
6772.634	6774.503			0.1			14761.23			
6773.221	6775.091			0.0			14759.95			
6773.363	6775.233						14759.64			
6773.497	6775.366		0.1				14759.35	4E 0	2B11	P 2
6773.855	6775.724		0.1				14758.57			
6774.093	6775.963		0.5	1.0	1.2		14758.05	2K 5	2B 1	P 6
6774.093	6775.963		0.5	1.0	1.2		14758.05	3e 0	2c 1	R 2
6774.093	6775.963		0.5	1.0	1.2		14758.05	3E 2	2B10	R 4
6774.093	6775.963		0.5	1.0	1.2		14758.05	3a 1	2c 2	Q 7
6774.231	6776.101		0.2	0.2			14757.75	3D 3	2B11	P 5
6774.231	6776.101		0.2	0.2			14757.75	X 0	2B 7	R 2
6774.938	6776.808		0.2	0.4			14756.21			
6775.076	6776.946			0.1			14755.91			
6775.292	6777.161			0.2			14755.44	3f 3	2c 4	P 3
6775.526	6777.396		0.2	0.7	1.6		14754.93	3D 1	2C 0	P 2
6775.526	6777.396		0.2	0.7	1.6		14754.93	3b 2	2a 0	P 7
6775.673	6777.543			0.2			14754.61			
6775.842	6777.713		0.2	0.6	1.5		14754.24			
6776.201	6778.071		0.1	0.2			14753.46			
6776.453	6778.324			0.2d			14752.91			
6776.678	6778.549		0.2	0.5			14752.42	3E 2	2C 1	R 2
6776.678	6778.549		0.2	0.5			14752.42	3E 0	2B 6	Q 6
6776.977	6778.848		0.2	0.2			14751.77	3E 2	2C 1	Q 3
6777.175	6779.045		0.3	0.7			14751.34	3E 2	2C 1	R 3
6777.175	6779.045		0.3	0.7			14751.34	U 1	2B 8	R 1
6777.979	6779.849			0.2			14749.59			
6778.236	6780.107		0.1	0.3			14749.03	3a 2	2c 3	Q 2
6778.411	6780.282			0.1			14748.65			
6778.636	6780.507			0.1			14748.16	4e 3	3b 1	P 4
6778.875	6780.746			0.2			14747.64			
6779.119	6780.990		0.2	0.3			14747.11			
6779.555	6781.426		0.1	0.2			14746.16			
6779.781	6781.652			0.1			14745.67			
6779.987	6781.859		0.2	0.2			14745.22	3e 0	2c 1	Q 2
6779.987	6781.859		0.2	0.2			14745.22	3a 1	2c 2	Q 6
6780.277	6782.149			0.1			14744.59			
6780.369	6782.240		0.2				14744.39			
6780.466	6782.337			0.2			14744.18			
6780.622	6782.494		0.3				14743.84	3e 0	2c 1	P 3
6780.774	6782.645	G	1.4	0.9b			14743.51			
6780.935	6782.806		0.3				14743.16			
6781.082	6782.954		0.1	0.1			14742.84			
6781.639	6783.510			0.2			14741.63			
6781.846	6783.717	G	1.3	1.2			14741.18	3E 2	2C 1	R 1
6781.846	6783.717	G	1.3	1.2			14741.18	X 1	2B 9	R 1

λ_{air}	λ_{vac}	Ref.	I_1	I_2	I_3	I_4	ν	Upper	Lower	Br.
6782.053	6783.924			0.0			14740.73			
6782.310	6784.182			0.1			14740.17			
6782.563	6784.435		0.2	1.0			14739.62	3E 2	2C 1	R 4
6782.752	6784.624			0.1			14739.21	U 1	2B 8	R 0
6783.001	6784.873		0.2	0.2			14738.67			
6783.231	6785.103	G	1.0	1.2			14738.17			
6783.392	6785.264		0.3	0.2			14737.82			
6783.608	6785.480			0.1			14737.35			
6783.755	6785.628		0.1				14737.03			
6783.820	6785.692			0.1			14736.89			
6784.050	6785.922		0.2	0.3			14736.39	3D 3	2B11	P 6
6784.368	6786.240		0.2	0.2			14735.70	X 1	2B 9	R 0
6784.368	6786.240		0.2	0.2			14735.70	X 0	2B 7	R 1
6784.662	6786.535						14735.06	4d 2	3b 0	R 3
6784.791	6786.664			0.1			14734.78	3F 3	2C 2	P 4
6785.473	6787.346		0.2	0.2			14733.30	4D 0	2B11	R 5
6785.643	6787.516			0.1			14732.93			
6785.768	6787.641			0.2			14732.66	3a 1	2c 2	Q 5
6785.910	6787.783		0.3	0.6			14732.35			
6786.113	6787.986			0.2			14731.91			
6786.440	6788.313			0.0			14731.20			
6786.735	6788.608		0.1	0.0			14730.56	X 1	2B 9	R 2
6787.094	6788.968		0.2	0.2			14729.78	4d 2	3b 0	P 2
6787.094	6788.968		0.2	0.2			14729.78	4e 3	3b 1	P 6
6787.352	6789.226			0.2			14729.22			
6787.670	6789.544		0.7	0.4			14728.53	Z 1	2C 0	Q 1
6787.859	6789.733			0.1			14728.12			
6788.154	6790.028			0.1	0.1		14727.48			
6788.399	6790.272			0.2			14726.95			
6788.569	6790.443			0.2			14726.58			
6788.873	6790.747		0.1				14725.92			
6788.970	6790.844			0.0			14725.71			
6789.210	6791.084		0.1	0.0			14725.19			
6789.570	6791.444		0.1	0.1			14724.41	X 2	2B11	P 3
6789.667	6791.540	G					14724.20			
6789.791	6791.665		0.3	0.3			14723.93	3e 2	2c 3	R 1
6790.907	6792.781		0.3	0.4	0.1		14721.51	3a 1	2c 2	Q 4
6791.059	6792.934			0.1			14721.18			
6791.184	6793.058			0.2			14720.91			
6791.322	6793.197			0.1			14720.61			
6791.484	6793.358		0.2	0.3			14720.26	3D 1	2C 0	R 3
6791.668	6793.543			0.1			14719.86			
6791.839	6793.714			0.2			14719.49			
6792.074	6793.949			0.2			14718.98			
6792.296	6794.171		0.3	0.4			14718.50	3E 2	2C 1	R 5
6792.831	6794.706						14717.34			
6793.081	6794.955			0.2			14716.80			
6792.527	6794.401			0.2			14718.00			
6792.725	6794.600			0.1d			14717.57			
6793.413	6795.288		0.2	0.2			14716.08			
6793.607	6795.482						14715.66	3D 0	2B 6	R 8
6793.828	6795.703		0.2	0.2			14715.18			
6794.092	6795.967	G	1.5b	1.5	1.6		14714.61	3b 3	2a 1	P 2
6794.429	6796.304		0.1				14713.88			
6794.683	6796.558			0.2			14713.33			
6794.891	6796.766		0.3	0.3			14712.88	X 0	2B 7	R 0
6795.112	6796.988			0.2			14712.40	3a 1	2c 2	Q 3
6795.417	6797.292		0.2	0.2			14711.74	Z 1	2C 0	R 2
6795.653	6797.528			0.1			14711.23			
6795.990	6797.865		0.2	0.2			14710.50			

λ_{air}	λ_{vac}	Ref.	I_1	I_2	I_3	I_4	ν	Upper	Lower	Br.
6796.336	6798.212	G	1.5	1.0	0.2		14709.75	3e 0	2c 1	P 4
6796.530	6798.406		0.2	0.1			14709.33	Z 1	2C 0	P 1
6796.863	6798.739		0.1	0.2			14708.61			
6797.103	6798.979		0.1	0.2			14708.09			
6797.367	6799.243	G	0.9	1.0			14707.52	3c 8	2a 8	Q 1
6797.579	6799.455		0.3	0.2			14707.06	3D 2	2C 1	P 1
6797.880	6799.756			0.0			14706.41			
6798.268	6800.144		0.1	0.2			14705.57	3a 1	2c 2	Q 2
6798.527	6800.403		0.2	0.5			14705.01	3E 2	2C 1	Q 4
6798.689	6800.565			0.1			14704.66	4E 0	2B11	Q 1
6799.091	6800.968		0.2				14703.79			
6799.373	6801.250		0.2	0.3			14703.18			
6799.632	6801.509		0.1	0.0			14702.62			
6800.261	6802.138		0.1	0.0			14701.26	4E 0	2B11	Q 3
6800.479	6802.356	G	0.9	1.1			14700.79	3b 6	2a 3	R 2
6800.479	6802.356	G	0.9	1.1			14700.79	3a 1	2c 2	Q 1
6800.553	6802.430		0.6				14700.63			
6800.682	6802.559			0.3			14700.35			
6800.844	6802.721		0.1	0.2			14700.00			
6801.015	6802.892			0.1			14699.63			
6801.242	6803.119		0.2b	0.3d			14699.14			
6801.612	6803.489		0.1				14698.34			
6801.760	6803.637			0.0			14698.02			
6801.973	6803.850		0.2	0.3			14697.56	3e 0	2c 1	Q 3
6802.154	6804.031			0.1			14697.17			
6802.371	6804.249		0.2	0.4	0.2		14696.70			
6802.533	6804.411			0.1			14696.35			
6802.672	6804.549	G	1.5	1.0			14696.05	3c 8	2a 8	Q 1
6802.987	6804.864		0.2	0.2			14695.37			
6803.283	6805.161			0.2d			14694.73			
6803.593	6805.471		0.1				14694.06	4d 3	3b 1	R 0
6804.005	6805.883			0.0d			14693.17			
6804.302	6806.180		0.2	0.2			14692.53			
6804.872	6806.750		0.1	0.2			14691.30			
6805.131	6807.009		0.5	0.3			14690.74	3D 2	2C 1	R 1
6805.381	6807.259	G	1.9	1.3			14690.20			
6805.742	6807.621		0.1	0.1			14689.42	4e 2	3b 0	Q 6
6806.007	6807.885		0.2	0.4b	1.1		14688.85	3E 2	2C 1	R 6
6806.192	6808.070		0.4	0.2d			14688.45			
6806.433	6808.311	G	2.5	1.2	1.6		14687.93	3b 6	2a 3	R 1
6806.771	6808.650		0.3	0.7	1.1		14687.20	3e 2	2c 3	Q 5
6807.119	6808.997		0.1	0.3			14686.45	3E 1	2C 0	P 4
6807.351	6809.229		0.2	0.5			14685.95			
6807.582	6809.461		0.1	0.2			14685.45			
6807.828	6809.707		0.2	0.7	1.2		14684.92	3e 2	2c 3	R 5
6808.074	6809.953			0.2			14684.39			
6808.273	6810.152		0.2	0.4	0.1		14683.96			
6808.454	6810.333			0.1			14683.57			
6808.602	6810.481			0.2			14683.25	3e 1	2c 2	Q 1
6808.927	6810.806		0.1	0.2			14682.55			
6809.173	6811.052	G	1.5	0.8			14682.02			
6809.354	6811.233		0.3	0.2			14681.63			
6809.609	6811.488		0.1				14681.08			
6809.817	6811.697			0.0			14680.63			
6810.082	6811.961			0.2			14680.06			
6810.295	6812.175		0.5	0.5			14679.60			
6810.472	6812.351			0.2			14679.22			
6810.657	6812.537		0.1	0.4			14678.82	3e 2	2c 3	R 4
6810.894	6812.773			0.1			14678.31			
6811.752	6813.632			0.1			14676.46	3E 1	2C 0	P 5

λ_{air}	λ_{vac}	Ref.	I_1	I_2	I_3	I_4	ν	Upper	Lower	Br.
6812.198	6814.078			0.2			14675.50	4d 2	3b 0	R 4
6812.374	6814.254			0.3			14675.12			
6812.592	6814.473		0.2	0.3	0.2		14674.65			
6812.773	6814.654			0.1			14674.26			
6812.880	6814.760		0.1				14674.03			
6813.252	6815.132		0.2	0.2			14673.23			
6813.735	6815.615		0.2b	0.4d			14672.19	3e 2	2c 3	Q 4
6813.735	6815.615		0.2b	0.4d			14672.19	4E 0	2B11	Q 2
6813.939	6815.820			0.2			14671.75	3b 6	2a 3	R 0
6814.078	6815.959		0.2	0.5	0.1b		14671.45	3e 0	2c 1	P 2
6814.315	6816.196		0.3	0.7			14670.94			
6814.696	6816.577		0.1	0.1			14670.12			
6815.012	6816.893			0.2			14669.44			
6815.249	6817.130	G	0.5	1.1	1.3		14668.93	3e 2	2c 3	R 3
6815.463	6817.344			0.1			14668.47	3E 0	2B 6	P 5
6815.602	6817.483		0.1	0.2			14668.17			
6815.811	6817.692	G		1.1	0.2b		14667.72			
6815.774	6817.655		1.0				14667.80			
6815.890	6817.771		0.9				14667.55			
6816.155	6818.036		0.1				14666.98	3e 0	2c 1	P 5
6816.778	6818.659			0.0			14665.64			
6817.252	6819.133		0.1	0.2			14664.62			
6817.498	6819.380	G	1.5	1.4			14664.09	3a 0	2c 1	Q 4
6817.698	6819.580	G	2.1	1.4	1.6		14663.66			
6818.070	6819.952		0.1	0.2			14662.86			
6818.414	6820.296	G	2.0	1.6	1.1		14662.12	3e 1	2c 2	P 2
6818.619	6820.501	G	1.0	1.0			14661.68	3E 0	2B 6	R 1
6818.619	6820.501	G	1.0	1.0			14661.68	3B 3	2A 0	P 3
6818.838	6820.719			0.2			14661.21			
6819.349	6821.231						14660.11			
6819.740	6821.622			0.2			14659.27	3A 2	2B11	R 2
6819.940	6821.822		0.9	0.8			14658.84	3E 2	2c 1	P 2
6819.940	6821.822		0.9	0.8			14658.84	X 2	2B11	P 4
6820.061	6821.943		0.4	0.1			14658.58	X 0	2B 7	P 1
6820.061	6821.943		0.4	0.1			14658.58	V 1	2B 6	R 1
6820.336	6822.218	G	1.8	1.2			14657.99	2A 4	2B 2	R 1
6820.573	6822.455			0.1			14657.48			
6820.722	6822.604		0.2	0.2			14657.16			
6820.959	6822.841		1.0				14656.65	3E 2	2c 1	Q 1
6820.996	6822.879	G	1.9	1.4	1.4		14656.57	2A 4	2B 2	R 2
6821.038	6822.921		1.0				14656.48	3a 0	2c 1	Q 3
6821.201	6823.084		0.2	0.8	0.3		14656.13			
6821.378	6823.260			0.1			14655.75			
6821.485	6823.368						14655.52			
6821.569	6823.451			0.2b	0.1		14655.34			
6821.685	6823.568		0.1		0.1		14655.09			
6821.890	6823.773		0.3	0.3			14654.65	U 1	2B 8	P 2
6821.974	6823.856		0.2				14654.47	X 1	2B 9	P 2
6821.974	6823.856		0.2				14654.47	3a 0	2c 1	Q 2
6822.165	6824.047		0.2	0.7			14654.06	3e 2	23 3	R 2
6822.449	6824.331			0.0			14653.45	4d 3	3b 1	R 1
6822.858	6824.741			0.2			14652.57	3a 0	2c 1	Q 1
6823.077	6824.960		0.9	0.7	0.1b		14652.10	3e 1	2c 2	Q 2
6823.077	6824.960		0.9	0.7	0.1b		14652.10	3E 2	2C 1	R 7
6823.231	6825.114		0.2	0.1			14651.77			
6823.413	6825.296			0.2			14651.38			
6823.576	6825.459		0.3	0.4			14651.03			
6823.776	6825.659			0.1			14650.60			
6823.930	6825.813			0.2			14650.27			
6824.163	6826.046		0.4	0.7			14649.77	3e 2	2c 3	Q 3

λ_{air}	λ_{vac}	Ref.	I_1	I_2	I_3	I_4	ν	Upper	Lower	Br.
6824.354	6826.237			0.2			14649.36			
6824.521	6826.404		0.7	0.8	0.1b		14649.00	3e 1	2c 2	P 3
6824.782	6826.665		0.1	0.1			14648.44			
6825.024	6826.908		0.2	0.2			14647.92	Z 1	2C 0	Q 2
6825.024	6826.908		0.2	0.2			14647.92	3e 0	2c 1	Q 4
6825.318	6827.201		0.8	0.6			14647.29	3D 2	2C 1	Q 1
6825.565	6827.448		0.6	0.6			14646.76			
6825.761	6827.644	G	2.3	1.5	0.1b		14646.34	2A 4	2B 2	R 0
6825.761	6827.644	G	2.3	1.5	0.1b		14646.34	3E 2	2B10	R 3
6826.124	6828.008		0.2				14645.56			
6826.931	6828.815			0.2			14643.83	3E 0	2B 6	Q 5
6827.192	6829.076						14643.27	3a 1	2c 2	P 1
6827.350	6829.234			0.2			14642.93			
6827.719	6829.603			0.1			14642.14			
6828.073	6829.957			0.2			14641.38			
6828.311	6830.195		0.3	0.6	1.6		14640.87	2A 4	2B 2	R 3
6828.469	6830.354			0.2			14640.53	4f 0	2c 4	R 3
6828.633	6830.517		0.8	0.4			14640.18	3D 1	2C 0	Q 3
6828.857	6830.741		0.1	0.2			14639.70			
6829.067	6830.951		0.6	0.4			14639.25	3D 2	2C 1	R 2
6829.295	6831.180			0.2			14638.76			
6829.561	6831.446			0.2			14638.19	T 0	2B11	R 2
6830.527	6832.412			0.2			14636.12			
6830.770	6832.655	G	0.3	0.9			14635.60			
6831.236	6833.121			0.2			14634.60			
6831.334	6833.219						14634.39			
6831.586	6833.472			0.1			14633.85			
6831.857	6833.742		0.1	0.2			14633.27	3E 2	2C 1	Q 5
6832.114	6833.999		0.9				14632.72			
6832.142	6834.027	G		1.0	0.2		14632.66	3e 2	2c 3	R 1
6832.212	6834.097		0.7				14632.51			
6832.371	6834.256		0.1	0.2			14632.17			
6832.614	6834.499			0.1			14631.65			
6832.815	6834.700						14631.22			
6833.216	6835.102			0.2			14630.36			
6833.384	6835.270		0.3	0.8	0.3		14630.00			
6833.641	6835.527			0.1			14629.45	3A 2	2B11	P 6
6833.973	6835.859						14628.74			
6834.038	6835.924		0.2b	0.2b			14628.60	3E 1	2B 8	R 2
6834.361	6836.247		0.7	0.8d	0.2		14627.91	3e 1	2c 2	P 4
6834.361	6836.247		0.7	0.8d	0.2		14627.91	4D 0	2B11	R 4
6834.562	6836.448		0.2	0.2			14627.48			
6834.744	6836.630			0.2			14627.09			
6834.949	6836.836			0.2			14626.65			
6835.118	6837.004		0.7	0.5			14626.29	X 0	2B 7	P 2
6835.295	6837.181			0.2			14625.91	4e 3	3b 1	R 3
6835.898	6837.784			0.1			14624.62			
6836.225	6838.112		0.1	0.1			14623.92			
6836.450	6838.336			0.1			14623.44			
6836.679	6838.565		0.2	0.2			14622.95			
6836.852	6838.738		0.8	0.7			14622.58	3E 2	2C 1	Q 2
6837.362	6839.248			0.2			14621.49	3A 1	2B 9	R 2
6837.362	6839.248			0.2			14621.49	3e 0	2c 1	P 6
6837.801	6839.688			0.1			14620.55			
6838.133	6840.020			0.2			14619.84	3e 2	2c 3	Q 2
6838.428	6840.315						14619.21			
6838.573	6840.460			0.1			14618.90			
6838.905	6840.792			0.2			14618.19			
6839.106	6840.993			0.2			14617.76			
6839.251	6841.138		0.8	0.8	0.2b		14617.45			

λ_{air}	λ_{vac}	Ref.	I_1	I_2	I_3	I_4	ν	Upper	Lower	Br.
6839.509	6841.396		0.1	0.3			14616.90			
6839.691	6841.578		0.2				14616.51			
6839.963	6841.850		0.2	0.2			14615.93	3D 1	2C 0	R 4
6840.290	6842.178		0.1	0.1			14615.23			
6840.594	6842.482		0.3	0.3			14614.58			
6840.889	6842.777	G	2.1	1.5b	0.9		14613.95	3b 3	2a 1	P 3
6840.889	6842.777	G	2.1	1.5b	0.9		14613.95	3e 1	2c 2	Q 3
6841.156	6843.044						14613.38			
6841.231	6843.119		0.2	0.2			14613.22			
6841.470	6843.358		0.1				14612.71			
6841.671	6843.559						14612.28	U 1	2B 8	P 3
6842.107	6843.995		0.1	0.2			14611.35			
6842.270	6844.158			0.1			14611.00			
6842.444	6844.332			0.2			14610.63			
6842.697	6844.585		0.3	0.8	0.2		14610.09			
6843.085	6844.974			0.2			14609.26			
6843.338	6845.227		0.1	0.1			14608.72	4d 3	3b 1	R 2
6843.563	6845.452			0.1			14608.24			
6844.022	6845.911						14607.26			
6844.060	6845.948	G	0.9b	0.9b			14607.18			
6844.430	6846.319		0.1	0.2			14606.39	X 2	2B11	P 5
6844.683	6846.572		0.6	0.6			14605.85			
6844.974	6846.862		0.1	0.2			14605.23			
6845.138	6847.026			0.2			14604.88			
6845.391	6847.280			0.2			14604.34			
6845.559	6847.448			0.1			14603.98			
6845.761	6847.650		0.3	0.3			14603.55	Z 1	2c 0	P 2
6846.024	6847.913		0.1	0.2			14602.99			
6846.356	6848.246		0.3	0.4d			14602.28			
6846.563	6848.452			0.2			14601.84			
6846.971	6848.860			0.1			14600.97			
6847.200	6849.090		0.1	0.2			14600.48			
6847.440	6849.329		0.5	0.4			14599.97	X 1	2B 9	P 3
6847.440	6849.329		0.5	0.4			14599.97	3E 2	2C 1	P 3
6847.716	6849.606			0.1			14599.38			
6847.956	6849.845		0.1d	0.1			14598.87			
6848.246	6850.136		0.2	0.2			14598.25	3e 0	2c 1	Q 5
6848.692	6850.582			0.2			14597.30	3e 1	2c 2	P 5
6849.021	6850.910		0.1	0.1			14596.60			
6849.546	6851.436			0.1			14595.48			
6849.795	6851.685		0.2	0.2			14594.95			
6850.006	6851.896			0.2			14594.50			
6850.297	6852.187						14593.88			
6850.386	6852.277			0.1			14593.69			
6850.480	6852.370		0.1				14593.49			
6850.767	6852.657			0.1d			14592.88	3a 2	2c 3	P 3
6850.983	6852.873			0.1			14592.42			
6851.283	6853.173		0.2	0.3			14591.78	X 0	2B 7	P 3
6851.424	6853.314			0.2			14591.48	3a 0	2c 1	P 1
6851.626	6853.516	G	0.9	0.8	0.1d		14591.05	3a 1	2c 2	P 2
6851.893	6853.784		0.2	0.2			14590.48			
6852.072	6853.963		0.1				14590.10			
6852.222	6854.113		0.3	0.1			14589.78	U 1	2B 8	P 5
6852.349	6854.240	G	0.8	0.8	0.2		14589.51	Z 2	2C 1	R 1
6852.640	6854.531			0.1			14588.89			
6852.819	6854.710		0.1				14588.51			
6853.138	6855.029		0.2	0.2d			14587.83			
6853.382	6855.273	G	1.9	0.8			14587.31	2A 4	2B 2	P 1
6853.584	6855.476		0.2d	0.3			14586.88	3e 2	2c 3	Q 1
6853.852	6855.743		0.1	0.1			14586.31	3D 1	2C 0	P 3

λ_{air}	λ_{vac}	Ref.	I_1	I_2	I_3	I_4	ν	Upper	Lower	Br.
6854.073	6855.964			0.2			14585.84	Y 1	2B 7	R 2
6854.120	6856.011		0.2				14585.74			
6854.388	6856.279		0.2	0.4			14585.17			
6854.623	6856.514			0.2			14584.67			
6854.938	6856.829		0.2d	0.1			14584.00			
6855.192	6857.083						14583.46			
6855.662	6857.553						14582.46	3E 0	2B 6	P 4
6856.033	6857.925			0.2			14581.67	Y 1	2B 7	R 1
6856.315	6858.207		0.6	0.4			14581.07			
6856.503	6858.395			0.1			14580.67			
6856.706	6858.598			0.1			14580.24			
6856.927	6858.819		0.2d	0.2			14579.77			
6856.983	6858.875						14579.65	3E 1	2C 8	Q 5
6857.261	6859.153			0.1			14579.06			
6857.416	6859.308						14578.73			
6857.660	6859.553			0.1			14578.21	4d 2	3b 0	P 3
6858.164	6860.056						14577.14			
6858.253	6860.146		0.2				14576.95			
6858.347	6860.240		0.3	0.2			14576.75	3D 2	2C 1	Q 2
6858.733	6860.626			0.1			14575.93			
6859.001	6860.894		0.9	0.6			14575.36	Z 2	2C 1	Q 1
6859.246	6861.139			0.2			14574.84	3E 2	2C 1	Q 3
6859.458	6861.351			0.2			14574.39	V 1	2B 6	P 4
6859.712	6861.605		0.2	0.3	0.2		14573.85	3e 0	2c 1	P 3
6860.009	6861.901		0.2	0.2			14573.22	3e 1	2c 2	P 2
6860.150	6862.043			0.1	0.1		14572.92	3b 2	2a 0	P 8
6860.348	6862.240		0.2	0.3			14572.50			
6860.734	6862.627		0.1				14571.68	Y 1	2B 7	R 0
6860.734	6862.627		0.1				14571.68	3e 1	2c 2	Q 4
6860.734	6862.627		0.1				14571.68	4e 3	3b 1	Q 1
6861.261	6863.154			0.3b			14570.56	3A 2	2B11	R 1
6861.779	6863.672			0.2d			14569.46	T 0	2B11	R 1
6862.099	6863.993						14568.78			
6862.467	6864.360		0.1	0.2			14568.00	U 1	2B 8	P 4
6862.782	6864.676			0.1			14567.33			
6863.107	6865.001						14566.64			
6863.244	6865.138		0.0				14566.35			
6863.428	6865.322			0.1			14565.96			
6863.772	6865.666			0.1			14565.23			
6863.946	6865.840		0.3				14564.86			
6864.059	6865.953		0.2	0.4d			14564.62	3E 1	2C 0	P 5
6864.191	6866.085		0.2	0.2			14564.34			
6864.385	6866.279			0.1			14563.93			
6864.573	6866.467			0.3	0.1		14563.53			
6865.945	6867.839		0.2d	0.3	0.1		14560.62	3E 0	2B 6	Q 4
6865.945	6867.839		0.2d	0.3	0.1		14560.62	3e 1	2c 2	P 6
6866.152	6868.047			0.2			14560.18			
6866.332	6868.226		0.3	0.3	0.4		14559.80	4D 0	2B11	R 3
6866.733	6868.627		0.2	0.2			14558.95			
6867.016	6868.910						14558.35			
6867.105	6869.000		0.1				14558.16			
6867.445	6869.340		0.3	0.3			14557.44	X 0	2B 7	P 4
6867.709	6869.604			0.1			14556.88			
6867.973	6869.868		0.5	0.4	0.2b		14556.32	3D 2	2C 1	R 3
6868.228	6870.123			0.2			14555.78			
6868.398	6870.293			0.3			14555.42			
6868.780	6870.675		0.2d	0.2			14554.61			
6869.030	6870.926		0.4	0.3			14554.08			
6869.266	6871.162			0.2			14553.58			
6869.752	6871.648		0.2	0.1			14552.55	4D 0	2B11	P 1

λ_{air}	λ_{vac}	Ref.	I_1	I_2	I_3	I_4	ν	Upper	Lower	Br.
6870.310	6872.205						14551.37			
6870.583	6872.479			0.2			14550.79	4D 0	2B11	R 0
6870.583	6872.479			0.2			14550.79	3a 2	2c 3	P 4
6870.810	6872.706		0.5	0.6			14550.31	3e 2	2c 3	Q 1
6870.810	6872.706		0.5	0.6			14550.31	3E 2	2B10	R 2
6871.112	6873.008		0.2				14549.67			
6871.216	6873.112			0.1			14549.45			
6871.273	6873.169	1			1.0		14549.33			
6871.944	6873.840			0.0d			14547.91	X 1	2B 9	P 4
6872.274	6874.170		0.3	0.3			14547.21			
6872.458	6874.355			0.2			14546.82			
6872.742	6874.638			0.3d			14546.22	3d 0	2c 1	R 1
6872.813	6874.709		0.2		0.2		14546.07	3d 0	2c 1	Q 1
6872.936	6874.832			0.2d			14545.81			
6873.186	6875.083			0.1			14545.28			
6873.422	6875.319		0.1	0.1			14544.78	3E 0	2B 6	R 0
6873.852	6875.749		0.6	0.4			14543.87	2K 3	2B 0	R 0
6874.094	6875.990		0.1	0.1			14543.36			
6874.377	6876.274		0.2	0.3			14542.76	3E 2	2C 1	P 2
6874.566	6876.463			0.2			14542.36			
6874.689	6876.586		0.1	0.2			14542.10			
6874.892	6876.789		0.2	0.4			14541.67	3B 3	2A 0	P 4
6874.892	6876.789		0.2	0.4			14541.67	3a 1	2c 2	P 3
6875.063	6876.960		0.2	0.2			14541.31			
6875.280	6877.177	G	2.5	1.8	0.2		14540.85	2A 4	2B 2	P 2
6875.346	6877.243		1.5				14540.71			
6875.630	6877.527		0.2	0.1			14540.11	W 1	2B 9	R 4
6875.904	6877.801			0.2			14539.53			
6876.338	6878.235						14538.61	3E 1	2B 8	R 1
6876.533	6878.431						14538.20			
6876.652	6878.549		0.2	0.2			14537.95	2K 3	2B 0	R 1
6876.954	6878.852	G	1.8	1.4	0.3		14537.31	3a 0	2c 1	P 2
6877.186	6879.084			0.2			14536.82			
6877.399	6879.297				0.1b		14536.37			
6877.456	6879.353			0.2b			14536.25			
6877.744	6879.642		0.6	0.3			14535.64	V 1	2B 6	R 0
6877.910	6879.808		0.2	0.5			14535.29			
6878.222	6880.120			0.2			14534.63			
6878.450	6880.347		0.2	0.2			14534.15	Z 1	2C 0	Q 3
6878.748	6880.646			0.2d			14533.52			
6879.022	6880.920			0.2			14532.94			
6879.396	6881.294						14532.15			
6880.173	6882.071		0.4	0.4			14530.51			
6880.428	6882.327		0.1	0.2			14529.97			
6880.712	6882.611		0.2	0.4	0.2		14529.37	3A 2	2B11	P 5
6880.869	6882.767		0.1	0.2			14529.04			
6881.044	6882.942		0.7	0.4			14528.67	3D 2	2C 1	P 2
6881.286	6883.184		0.2	0.3			14528.16	X 0	2B 7	P 5
6881.286	6883.184		0.2	0.3			14528.16	3A 1	2B 9	R 1
6881.286	6883.184		0.2	0.3			14528.16	3e 2	2c 3	Q 2
6881.584	6883.483		0.1				14527.53			
6882.081	6883.980						14526.48	3e 1	2c 2	Q 5
6882.688	6884.587		0.2				14525.20			
6883.129	6885.028			0.1			14524.27	T 0	2B11	R 0
6883.749	6885.649			0.2			14522.96	4D 0	2B11	R 2
6883.749	6885.649			0.2			14522.96	Y 1	2B 7	P 1
6884.148	6886.047						14522.12			
6884.404	6886.303			0.2			14521.58			
6884.816	6886.716			0.3			14520.71			
6885.489	6887.389			0.1			14519.29			

λ_{air}	λ_{vac}	Ref.	I_1	I_2	I_3	I_4	ν	Upper	Lower	Br.
6885.684	6887.584			0.2			14518.88			
6885.774	6887.674		0.2				14518.69			
6885.902	6887.802		0.7	0.6			14518.42	3E 2	2C 1	P 4
6886.101	6888.001		0.1	0.2			14518.00	3E 2	2C 1	Q 4
6886.101	6888.001		0.1	0.2			14518.00	3d 0	2c 1	P 1
6886.400	6888.300		0.2	0.2			14517.37			
6886.756	6888.656			0.2			14516.62			
6887.031	6888.931		0.3	0.5	0.3		14516.04	3d 0	2c 1	Q 1
6887.126	6889.026		0.2				14515.84			
6887.264	6889.164			0.1			14515.55			
6887.434	6889.335		0.1	0.1			14515.19			
6887.971	6889.871			0.1			14514.06			
6888.146	6890.047		0.3	0.3			14513.69	Z 2	2C 1	R 2
6888.383	6890.284			0.2			14513.19			
6888.459	6890.360						14513.03			
6888.782	6890.683			0.1			14512.35	3E 0	2B 6	P 3
6889.480	6891.381			0.2			14510.88	3A 0	2B 7	R 2
6889.722	6891.623			0.1			14510.37			
6890.373	6892.274			0.2d			14509.00	V 1	2B 6	P 3
6890.477	6892.378		0.1				14508.78	2K 3	2B 0	R 2
6890.596	6892.497		0.4	0.4			14508.53			
6890.838	6892.739		0.1	0.2			14508.02			
6891.066	6892.967		0.3	0.7	0.2		14507.54			
6891.375	6893.276		0.1	0.2			14506.89	4D 0	2B11	R 1
6891.375	6893.276		0.1	0.2			14506.89	3D 1	2C 0	R 5
6891.698	6893.599		0.2	0.2			14506.21			
6892.026	6893.927			0.2			14505.52			
6892.591	6894.493			0.1			14504.33			
6892.720	6894.621						14504.06			
6892.815	6894.716			0.1			14503.86			
6893.333	6895.234		0.2	0.2			14502.77			
6893.580	6895.482		0.2	0.3	0.2		14502.25			
6893.803	6895.705			0.2			14501.78	3E 0	2B 6	Q 3
6893.989	6895.891		0.1	0.1			14501.39	3a 1	2c 2	P 4
6894.160	6896.062		0.3	0.5	0.4		14501.03	3d 0	2c 1	R 2
6894.388	6896.290		0.1	0.1			14500.55	3b 6	2a 3	P 2
6895.201	6897.103						14498.84	3e 0	2c 1	Q 7
6895.515	6897.417			0.1			14498.18	3e 2	2c 3	Q 3
6895.976	6897.879			0.2			14497.21			
6896.367	6898.269		0.1	0.2			14496.39			
6896.642	6898.545	G	0.4	1.3	0.4		14495.81	3b 3	2a 1	P 4
6896.880	6898.783			0.1			14495.31			
6897.294	6899.197						14494.44			
6897.356	6899.259						14494.31			
6897.751	6899.654						14493.48			
6898.218	6900.121		0.1				14492.50			
6898.517	6900.421			0.1			14491.87			
6898.779	6900.683	G	0.3	1.3	0.2		14491.32	2K 4	2B 1	R 3
6898.936	6900.840			0.1			14490.99			
6899.089	6900.992		0.1	0.2			14490.67			
6899.317	6901.221	G	2.0	1.6	0.3		14490.19	2K 4	2B 1	R 2
6899.608	6901.511		0.2	0.2			14489.58			
6899.870	6901.773	G	1.7	0.6	0.2b		14489.03	3e 1	2c 2	P 3
6900.084	6901.988			0.1			14488.58			
6900.308	6902.211	G	0.5	1.3	0.4		14488.11	2K 4	2B 1	R 4
6900.308	6902.211	G	0.5	1.3	0.4		14488.11	2K 3	2B 0	P 1
6900.532	6902.435		0.1	0.1			14487.64			
6900.636	6902.540			0.2			14487.42	3D 1	2C 0	Q 4
6900.727	6902.631						14487.23			
6900.813	6902.717			0.1			14487.05			

λ_{air}	λ_{vac}	Ref.	I_1	I_2	I_3	I_4	ν	Upper	Lower	Br.
6901.037	6902.940		0.2	0.4	0.1		14486.58			
6901.308	6903.212		0.1	0.1			14486.01			
6901.437	6903.341			0.1			14485.74			
6901.580	6903.484		0.4	0.2			14485.44	Y 1	2B 7	P 2
6901.918	6903.822			0.2			14484.73	3E 1	2B 8	Q 4
6902.118	6904.022			0.2			14484.31			
6902.337	6904.242	G	1.8	1.3			14483.85	2K 4	2B 1	R 1
6902.509	6904.413		0.1	0.1			14483.49			
6902.681	6904.585	G	1.9	1.3	0.1		14483.13	2A 4	2B 2	P 3
6902.823	6904.728			0.2			14482.83	3a 0	2c 1	P 3
6903.033	6904.938	G	0.5	1.3	0.2		14482.39	3e 1	2c 2	Q 6
6903.033	6904.938	G	0.5	1.3	0.2		14482.39	3e 2	2c 3	P 2
6903.205	6905.109			0.1			14482.03	3E 1	2B 8	P 5
6903.696	6905.600						14481.00			
6904.225	6906.130			0.1			14479.89	4D 0	2B11	P 2
6904.225	6906.130			0.1			14479.89	3A 2	2B11	P 4
6904.502	6906.406						14479.31			
6904.821	6906.726		0.2	0.2			14478.64			
6905.241	6907.146			0.2			14477.76			
6905.451	6907.356		0.2				14477.32			
6905.637	6907.542			0.2			14476.93			
6905.937	6907.842		0.2	0.4	0.8		14476.30			
6906.195	6908.100		0.3	0.2			14475.76	Z 2	2C 1	Q 2
6906.410	6908.315			0.2			14475.31			
6906.600	6908.506			0.1			14474.91			
6906.849	6908.754		0.2	0.2			14474.39			
6907.292	6909.198			0.2			14473.46	4e 3	3b 1	Q 3
6907.622	6909.527			0.1			14472.77			
6907.875	6909.780			0.3	0.3		14472.24	3e 0	2c 1	P 4
6908.285	6910.191						14471.38			
6908.486	6910.392		0.2	0.2			14470.96	3D 2	2C 1	Q 3
6908.758	6910.664			0.2			14470.39			
6908.987	6910.893			0.2			14469.91			
6909.226	6911.132	G	2.1	1.5	0.1		14469.41	2K 4	2B 1	R 0
6909.321	6911.227	G					14469.21			
6909.474	6911.380		0.1	0.2			14468.89			
6909.670	6911.576	G	2.0	1.4			14468.48			
6909.942	6911.848			0.1			14467.91			
6910.372	6912.278			0.2			14467.01	3E 0	2B 6	Q 2
6910.678	6912.584			0.2			14466.37			
6911.363	6913.270						14464.93	3e 3	2c 4	R 1
6911.700	6913.607		0.2	0.2			14464.23	3e 2	2c 3	P 3
6911.872	6913.779	G	0.4	1.3	0.2d		14463.87			
6912.221	6914.128			0.2			14463.14			
6912.431	6914.338						14462.70	3e 2	2c 3	Q 4
6912.646	6914.553			0.1d			14462.25			
6913.588	6915.495			0.2			14460.28	Z 1	2C 0	P 3
6913.918	6915.825		0.2	0.3			14459.59			
6914.296	6916.203				0.2		14458.80			
6914.540	6916.447			0.1			14458.29	T 0	2B11	P 4
6914.793	6916.701		0.3	0.3			14457.76			
6915.286	6917.194		0.2				14456.73	Z 2	2C 1	P 2
6915.286	6917.194		0.2				14456.73	2K 3	2B 0	R 3
6915.520	6917.428			0.1			14456.24			
6915.870	6917.777	G	0.3b	0.6	0.1		14455.51			
6916.056	6917.964			0.2			14455.12			
6916.180	6918.088			0.2			14454.86	3E 0	2B 6	P 2
6916.180	6918.088			0.2			14454.86	3A 1	2B 9	R 0
6916.463	6918.371	G	0.5	0.5			14454.27	3E 2	2C 1	P 3
6916.463	6918.371	G	0.5	0.5			14454.27	3E 2	2B10	R 1

λ_{air}	λ_{vac}	Ref.	I$_1$	I$_2$	I$_3$	I$_4$	ν	Upper	Lower	Br.
6916.463	6918.371	G	0.5	0.5			14454.27	3a 1	2c 2	P 5
6916.913	6918.821		0.1	0.1			14453.33	T 0	2B11	P 2
6917.008	6918.917						14453.13			
6917.396	6919.304						14452.32	3E 0	2B 6	Q 1
6917.650	6919.558			0.1			14451.79			
6918.291	6920.200						14450.45	3E 2	2C 1	Q 5
6918.483	6920.391				0.2		14450.05	3d 0	2c 1	R 3
6918.646	6920.554		0.2	0.2	0.4		14449.71			
6918.962	6920.870			0.2			14449.05	T 0	2B11	P 1
6919.527	6921.435		0.1	0.2			14447.87	3D 2	2C 1	R 4
6920.073	6921.982			0.2			14446.73	3A 2	2B11	P 3
6920.317	6922.226	G	0.3b	0.4	0.3		14446.22	3e 2	2c 3	P 2
6920.628	6922.537		0.6	0.4			14445.57	T 0	2B11	P 3
6920.628	6922.537		0.6	0.4			14445.57	V 1	2B 6	P 2
6920.844	6922.753			0.1			14445.12	3E 2	2B10	Q 5
6921.199	6923.108		0.1				14444.38			
6921.529	6923.439	G	0.3b	1.3	0.2b		14443.69	3e 2	2c 3	P 4
6921.860	6923.769			0.1			14443.00	W 1	2B 9	R 3
6922.219	6924.129			0.2d			14442.25			
6922.684	6924.594			0.1			14441.28			
6922.957	6924.867			0.1			14440.71			
6923.816	6925.726						14438.92	3A 1	2B 9	P 4
6924.334	6926.244						14437.84	Y 1	2B 7	P 3
6924.737	6926.647			0.2			14437.00	3E 1	2C 0	P 6
6924.818	6926.728				0.1		14436.83			
6925.614	6927.525						14435.17	3E 1	2B 8	R 0
6925.984	6927.894			0.1			14434.40	3E 1	2B 8	P 4
6926.113	6928.024						14434.13	3c 0	2a 1	R 9
6926.445	6928.355		0.1				14433.44			
6926.565	6928.475						14433.19			
6926.996	6928.907						14432.29	3c 0	2a 1	R 8
6927.121	6929.032			0.2d	0.1		14432.03	3D 0	2B 6	R 6
6927.260	6929.171				0.1		14431.74			
6927.601	6929.512		0.2b	0.3d	0.1b		14431.03	3d 0	2c 1	Q 2
6927.601	6929.512		0.2b	0.3d	0.1b		14431.03	3d 0	2c 1	P 2
6927.942	6929.853			0.2d			14430.32			
6928.326	6930.237			0.2d			14429.52	3a 0	2c 1	P 4
6928.586	6930.497			0.2d			14428.98			
6928.667	6930.578						14428.81			
6928.946	6930.857			0.2d			14428.23			
6929.191	6931.102	G	0.5	0.4			14427.72	2K 3	2B 0	P 2
6929.364	6931.275			0.2			14427.36			
6929.757	6931.669			0.1d			14426.54			
6930.017	6931.928			0.1d			14426.00			
6931.035	6932.947			0.1d			14423.88	4d 2	3b 0	P 4
6931.497	6933.409			0.1			14422.92	3F 0	2B 7	R 4
6932.232	6934.144		0.1	0.2	0.5		14421.39	3c 0	2a 1	R 7
6933.097	6935.010			0.1			14419.59			
6933.540	6935.452			0.2	0.2		14418.67	4F 0	2C 2	Q 4
6934.093	6936.006			0.2			14417.52			
6934.468	6936.381						14416.74			
6934.583	6936.496			0.1			14416.50			
6934.819	6936.732		0.2				14416.01	3E 2	2C 1	P 5
6934.819	6936.732		0.2				14416.01	3a 1	2c 2	P 6
6935.098	6937.011						14415.43			
6935.575	6937.488						14414.44			
6935.642	6937.555		0.1	0.2			14414.30			
6935.926	6937.839	G	1.9	1.5	0.3		14413.71	3d 1	2c 2	R 1
6935.926	6937.839	G	1.9	1.5	0.3		14413.71	3a 2	2c 3	P 5
6935.926	6937.839	G	1.9	1.5	0.3		14413.71	2A 4	2B 2	P 4

λ_{air}	λ_{vac}	Ref.	I_1	I_2	I_3	I_4	ν	Upper	Lower	Br.
6936.138	6938.051		0.1				14413.27			
6936.787	6938.701						14411.92	3A 2	2B11	P 1
6937.013	6938.927						14411.45			
6937.582	6939.495						14410.27	3E 2	2B10	P 5
6937.582	6939.495						14410.27	W 2	2B11	R 4
6937.654	6939.567	1			0.5		14410.12			
6938.140	6940.054						14409.11			
6938.424	6940.338		0.1	0.2			14408.52			
6938.829	6940.743	G					14407.68			
6938.944	6940.858	G	1.7	0.3			14407.44	2K 4	2B 1	P 1
6938.944	6940.858	G	1.7	0.3			14407.44	3A 2	2B11	P 2
6939.450	6941.364		0.2	0.2			14406.39			
6939.821	6941.735			0.2			14405.62	4g 1	2a 5	P 1
6940.110	6942.024		0.2b	0.3	0.1		14405.02	3B 3	2A 0	P 5
6940.428	6942.342	G	2.3	2.0	0.3		14404.36	3b 6	2a 3	P 3
6940.770	6942.685		0.1	0.1			14403.65			
6941.127	6943.041				0.3		14402.91	3c 0	2a 1	R 6
6941.368	6943.282						14402.41	Z 2	2C 1	R 3
6941.479	6943.393		0.1				14402.18			
6941.681	6943.596						14401.76			
6942.139	6944.054		0.2	0.2			14400.81	3E 3	2C 2	R 2
6942.467	6944.382		0.1				14400.13			
6942.930	6944.845		0.2	0.2			14399.17	3A 0	2B 7	R 1
6942.930	6944.845		0.2	0.2			14399.17	3D 2	2C 1	P 3
6942.930	6944.845		0.2	0.2			14399.17	3E 1	2B 8	P 3
6943.079	6944.994						14398.86	3e 1	2c 2	P 4
6943.175	6945.091			0.2			14398.66	3E 3	2C 2	R 1
6943.619	6945.534		0.2	0.2			14397.74			
6944.015	6945.930			0.3	0.4		14396.92	3d 0	2c 1	R 4
6944.126	6946.041		0.1				14396.69			
6944.265	6946.181			0.2			14396.40			
6944.536	6946.451		0.2	0.2			14395.84			
6944.830	6946.745			0.1			14395.23			
6944.926	6946.842						14395.03			
6945.727	6947.643		0.1				14393.37			
6946.084	6948.000						14392.63			
6946.480	6948.396				0.2		14391.81			
6947.195	6949.111			0.1			14390.33			
6947.446	6949.362						14389.81			
6947.523	6949.439						14389.65			
6948.184	6950.101		0.1	0.1			14388.28			
6948.769	6950.686			0.2			14387.07	3A 1	2B 9	P 3
6948.769	6950.686			0.2			14387.07	3A 1	2B 9	R 2
6949.063	6950.980	G	0.3	1.3	0.3		14386.46	3d 1	2c 2	Q 1
6949.150	6951.067		0.2				14386.28			
6949.353	6951.270		0.2	0.2			14385.86	3A 1	2B 9	P 1
6949.513	6951.430			0.2			14385.53	Y 1	2B 7	P 4
6949.513	6951.430			0.2			14385.53	2K 3	2B 0	R 4
6949.749	6951.666		0.2	0.3			14385.04			
6949.890	6951.807			0.1			14384.75			
6950.068	6951.985						14384.38			
6950.184	6952.101			0.1			14384.14	3a 0	2c 1	P 5
6950.629	6952.546		0.2	0.2			14383.22	4D 0	2B11	P 3
6951.257	6953.174			0.2			14381.92			
6951.479	6953.397		0.1	0.2	0.2		14381.46			
6952.151	6954.069			0.1			14380.07			
6952.548	6954.466			0.2			14379.25	3E 1	2B 8	Q 2
6952.799	6954.717			0.3			14378.73			
6952.886	6954.804				0.2		14378.55	3b 2	2a 0	P 9
6953.085	6955.003			0.1			14378.14			

λ_air	λ_vac	Ref.	I₁	I₂	I₃	I₄	ν	Upper	Lower	Br.
6953.307	6955.225						14377.68			
6953.520	6955.438						14377.24			
6953.728	6955.646			0.3	0.9		14376.81	3c 0	2a 1	R 5
6954.357	6956.275						14375.51			
6954.710	6956.628			0.2	0.1		14374.78			
6954.952	6956.870			0.3	0.1		14374.28			
6955.126	6957.044			0.2			14373.92	3A 1	2B 9	P 2
6955.368	6957.286	G	0.3	1.4	0.4		14373.42	3e 2	2c 3	P 3
6955.368	6957.286	G	0.3	1.4	0.4		14373.42	3d 1	2c 2	R 2
6955.697	6957.616			0.2			14372.74	4e 3	3b 1	P 3
6955.978	6957.896		0.3	0.3			14372.16			
6956.742	6958.661			0.1			14370.58	4g 1	2a 5	R 0
6957.372	6959.291			0.2			14369.28	3E 2	2B10	P 4
6957.372	6959.291			0.2			14369.28	3E 2	2B10	Q 4
6957.919	6959.838			0.1	0.2		14368.15	3e 0	2c 1	P 5
6957.991	6959.910		0.1				14368.00	4D 0	2B11	P 6
6957.991	6959.910		0.1				14368.00	3E 2	2B10	R 0
6959.057	6960.977						14365.80			
6959.275	6961.195						14365.35			
6959.421	6961.340			0.2			14365.05			
6959.672	6961.592		0.1	0.2			14364.53			
6959.983	6961.902			0.1			14363.89			
6960.080	6961.999						14363.69			
6960.239	6962.159				0.1		14363.36			
6960.361	6962.280			0.2			14363.11			
6960.777	6962.697			0.1			14362.25			
6961.092	6963.012			0.1			14361.60	3E 3	2C 2	R 3
6961.296	6963.216	G	0.2	1.4	0.8		14361.18	3b 3	2a 1	P 5
6961.325	6963.245						14361.12			
6961.616	6963.536			0.1d			14360.52	3E 3	2C 2	R 6
6961.800	6963.720		0.1				14360.14			
6962.052	6963.973		0.2	0.2d			14359.62			
6962.343	6964.264	G	2.4	1.7	0.2b		14359.02	2K 4	2B 1	P 2
6962.731	6964.652			0.1			14358.22			
6963.221	6965.141			0.1	0.1		14357.21	3E 1	2B 8	Q 1
6963.541	6965.462			0.1			14356.55			
6964.016	6965.937			0.2			14355.57			
6964.521	6966.442			0.2			14354.53	3E 3	2C 2	R 3
6964.817	6966.738			0.2	0.2		14353.92	3E 2	2C 1	P 4
6964.943	6966.864				0.1		14353.66			
6965.176	6967.097				0.5		14353.18			
6965.433	6967.354	i			1.5b		14352.65			
6965.555	6967.476			0.1			14352.40	3D 0	2B 6	R 5
6965.715	6967.636				0.2		14352.07			
6965.783	6967.704				0.3		14351.93			
6966.069	6967.990			0.2			14351.34	3E 3	2C 2	R 2
6966.598	6968.520			0.1			14350.25	3E 1	2B 8	P 2
6966.598	6968.520			0.1			14350.25	3e 3	2c 4	R 2
6966.807	6968.728						14349.82			
6967.001	6968.923			0.1			14349.42			
6967.953	6969.875				0.1b		14347.46			
6968.264	6970.186			0.2			14346.82	3E 3	2C 2	R 4
6968.643	6970.565			0.1			14346.04			
6968.939	6970.861		0.2	0.2			14345.43	2K 3	2B 0	P 3
6968.939	6970.861		0.2	0.2			14345.43	Z 2	2C 1	Q 3
6969.294	6971.216			0.1			14344.70			
6969.658	6971.580			0.3	0.5		14343.95	3c 0	2a 1	R 4
6969.658	6971.580			0.3	0.5		14343.95	4D 0	2B11	P 4
6970.013	6971.935			0.2	0.2		14343.22	3d 0	2c 1	R 5
6970.202	6972.125			0.1			14342.83			

λ_{air}	λ_{vac}	Ref.	I_1	I_2	I_3	I_4	ν	Upper	Lower	Br.
6970.421	6972.344		0.2	0.3			14342.38			
6971.301	6973.223		0.1				14340.57			
6971.714	6973.637			0.1			14339.72			
6971.889	6973.812						14339.36			
6972.142	6974.065			0.2			14338.84	4E 2	2B16	Q 2
6972.229	6974.152						14338.66			
6972.560	6974.483			0.2			14337.98	4D 0	2B11	P 5
6973.124	6975.047			0.2			14336.82			
6973.304	6975.227	G					14336.45			
6973.401	6975.325		0.2	0.3			14336.25	3E 3	2C 2	R 1
6973.893	6975.816			0.2			14335.24	3D 2	2C 1	Q 4
6974.749	6976.673						14333.48			
6975.109	6977.033			0.2			14332.74			
6975.489	6977.413			0.3			14331.96	2A 4	2B 2	P 5
6975.489	6977.413			0.3			14331.96	W 2	2B11	R 3
6975.679	6977.603			0.2			14331.57			
6975.917	6977.841			0.2			14331.08	4F 0	2B12	P 5
6976.307	6978.231			0.2			14330.28	3D 1	2C 0	Q 5
6976.599	6978.523		0.2	0.3	0.3		14329.68	3d 0	2c 1	Q 3
6976.599	6978.523		0.2	0.3	0.3		14329.68	3F 1	2B 9	Q 5
6976.939	6978.864			0.2			14328.98	3E 3	2C 2	R 5
6977.475	6979.400			0.2			14327.88			
6977.680	6979.604		0.2	0.2			14327.46			
6977.913	6979.838	G	1.9	1.5	0.2		14326.98	3d 1	2c 2	R 3
6978.152	6980.077						14326.49			
6979.146	6981.071						14324.45			
6979.545	6981.471						14323.63			
6980.462	6982.387			0.1			14321.75			
6980.632	6982.558				0.1		14321.40			
6981.368	6983.294		0.2	0.1			14319.89			
6981.836	6983.762						14318.93			
6982.314	6984.240			0.1			14317.95			
6984.051	6985.977			0.2			14314.39	3D 2	2C 1	R 5
6984.334	6986.260			0.2			14313.81			
6985.090	6987.017			0.1			14312.26	3D 1	2B 8	R 5
6985.090	6987.017			0.1			14312.26	3E 3	2C 2	Q 1
6985.778	6987.705			0.1			14310.85	3A 0	2B 7	R 0
6985.778	6987.705			0.1			14310.85	3A 0	2B 7	P 4
6986.887	6988.814				0.1		14308.58	3D 2	2B10	R 5
6987.629	6989.556	G	0.2	0.3			14307.06	3b 6	2a 3	P 4
6987.873	6989.800			0.2			14306.56			
6988.293	6990.221			0.2	0.2b		14305.70	3e 1	2c 2	P 5
6988.293	6990.221			0.2	0.2b		14305.70	3E 1	2C 0	P 7
6988.723	6990.651	G	0.3	1.6	1.0		14304.82	3c 0	2a 1	R 3
6988.982	6990.910		0.2	0.2			14304.29	3d 1	2c 2	Q 2
6989.456	6991.384						14303.32	4F 0	2B12	R 2
6989.862	6991.789		0.1	0.2			14302.49	4F 0	2C 2	Q 3
6990.614	6992.542	G	1.9	1.4			14300.95	2K 4	2B 1	P 3
6990.614	6992.542	G	1.9	1.4			14300.95	W 2	2B11	R 2
6990.825	6992.753						14300.52			
6991.015	6992.943		0.2	0.2			14300.13			
6991.392	6993.320			0.2			14299.36			
6991.993	6993.922		0.2	0.2	0.3		14298.13	Z 2	2C 1	P 3
6992.292	6994.220			0.2			14297.52			
6992.614	6994.543			0.2			14296.86			
6993.006	6994.934			0.2			14296.06			
6993.421	6995.350						14295.21	3E 2	2B10	Q 3
6993.524	6995.453			0.2			14295.00	W 1	2B 9	R 0
6993.984	6995.913			0.1			14294.06	3e 2	2c 3	P 4
6995.139	6997.068			0.2			14291.70			

λ_{air}	λ_{vac}	Ref.	I_1	I_2	I_3	I_4	ν	Upper	Lower	Br.
6995.624	6997.553			0.2			14290.71			
6995.839	6997.768						14290.27	3E 3	2C 2	Q 2
6996.094	6998.023		0.2	0.2	0.5		14289.75	3d 0	2c 1	R 6
6996.686	6998.616						14288.54	3E 2	2B10	P 2
6996.867	6998.797			0.1			14288.17			
6997.029	6998.958						14287.84			
6997.210	6999.140						14287.47			
6997.680	6999.610		0.1				14286.51	4g 1	2a 5	R 1
6998.229	7000.159			0.1			14285.39			
6998.542	7000.473			0.2	0.2		14284.75	Z 1	2C 0	P 4
6998.812	7000.742			0.2			14284.20			
6999.169	7001.100			0.2			14283.47			
6999.498	7001.428		0.2b	0.2b			14282.80			
6999.880	7001.811			0.2			14282.02			
7000.135	7002.066			0.2			14281.50	3D 0	2B 6	R 4
7000.463	7002.394			0.1			14280.83			
7000.772	7002.703			0.2			14280.20			
7001.287	7003.218			0.2			14279.15			
7001.542	7003.473	G	1.7	0.3			14278.63			
7001.768	7003.699			0.1			14278.17			
7002.067	7003.998		0.2	0.3	0.4		14277.56	3d 1	2c 2	R 4
7002.474	7004.405						14276.73	3d 1	2c 2	P 2
7002.474	7004.405						14276.73	3D 2	2B10	R 3
7002.650	7004.582			0.2	0.2		14276.37			
7003.852	7005.784						14273.92	3E 3	2B12	R 3
7004.147	7006.078	G	0.2	1.2	0.5		14273.32			
7004.294	7006.226			0.2			14273.02			
7004.682	7006.614			0.1			14272.23			
7005.565	7007.497			0.2			14270.43			
7005.880	7007.812			0.1			14269.79	3D 2	2B10	P 1
7005.880	7007.812			0.1			14269.79	3D 2	2B10	R 2
7006.754	7008.686			0.2			14268.01	X 0	2C 0	Q 1
7006.754	7008.686			0.2			14268.01	3E 3	2C 2	R 7
7008.006	7009.939			0.2	0.2		14265.46	4F 0	2B12	Q 4
7009.112	7011.044			0.1			14263.21			
7009.215	7011.148			0.1			14263.00			
7009.741	7011.674			0.2			14261.93			
7010.227	7012.160			0.2d			14260.94			
7010.493	7012.426			0.3			14260.40			
7010.763	7012.696	G	1.7	1.5	0.5		14259.85	3c 0	2a 1	R 2
7010.965	7012.898			0.2d			14259.44	3d 0	2c 1	P 3
7011.909	7013.843			0.1			14257.52			
7012.248	7014.182		0.2	0.4			14256.83			
7012.701	7014.635			0.4	0.3		14255.91	Z 2	2C 1	R 4
7012.991	7014.925						14255.32	3A 0	2B 7	P 3
7013.498	7015.432			0.1			14254.29	4f 0	2c 4	R 1
7013.783	7015.717			0.3			14253.71	3B 3	2A 0	P 6
7014.049	7015.983			0.2			14253.17	3D 2	2B10	R 1
7014.221	7016.155			0.3			14252.82	3E 3	2C 2	Q 1
7014.221	7016.155			0.3			14252.82	3E 2	2B10	Q 2
7014.221	7016.155			0.3			14252.82	4f 0	2c 4	R 2
7014.546	7016.480			0.2			14252.16			
7014.802	7016.736	G	1.8	1.2			14251.64			
7015.053	7016.987			0.1			14251.13			
7015.397	7017.332			0.0			14250.43	3D 1	2B 8	R 4
7016.756	7018.691			0.1			14247.67	4f 0	2c 4	R 1
7017.062	7018.997			0.3			14247.05			
7017.500	7019.435			0.1			14246.16	W 2	2B11	R 0
7017.835	7019.770			0.3	0.2		14245.48			
7018.052	7019.987			0.2			14245.04	3D 3	2C 2	R 1

λ_{air}	λ_{vac}	Ref.	I_1	I_2	I_3	I_4	ν	Upper	Lower	Br.
7018.372	7020.308			0.4			14244.39	3E 2	2C 1	P 5
7019.166	7021.101			0.0			14242.78			
7019.520	7021.456			0.3			14242.06	2K 3	2B 0	P 4
7019.846	7021.782			0.1			14241.40	X 0	2C 0	P 1
7019.846	7021.782			0.1			14241.40	3D 1	2B 8	R 0
7020.077	7022.013			0.3			14240.93			
7020.585	7022.521			0.2			14239.90	3E 3	2B12	R 2
7020.585	7022.521			0.2			14239.90	3E 3	2C 2	Q 3
7020.832	7022.768			0.1			14239.40			
7021.187	7023.123			0.2d			14238.68			
7021.759	7023.695		0.2	0.6			14237.52	3D 2	2B10	R 0
7021.759	7023.695		0.2	0.6			14237.52	3D 2	2C 1	P 4
7021.877	7023.814				0.2b		14237.28			
7022.065	7024.001			0.2d			14236.90			
7022.341	7024.277	G	2.1	1.7	0.3		14236.34	2K 4	2B 1	P 4
7022.528	7024.465						14235.96			
7022.775	7024.711			0.2d			14235.46	3D 0	2B 6	R 3
7023.170	7025.106			0.3d			14234.66			
7023.456	7025.393		0.2	0.5			14234.08	3d 2	2c 3	R 1
7023.456	7025.393		0.2	0.5			14234.08	W 1	2B 9	P 1
7023.826	7025.763	G		1.3	0.9		14233.33			
7023.999	7025.935			0.0			14232.98			
7024.161	7026.098			0.1			14232.65			
7024.522	7026.459			0.1			14231.92			
7024.793	7026.730			0.4			14231.37	3D 1	2B 8	R 1
7025.109	7027.046			0.2d			14230.73			
7025.371	7027.308			0.3			14230.20			
7026.126	7028.064			0.2			14228.67			
7026.437	7028.375		0.1	0.7			14228.04			
7026.650	7028.587			0.2			14227.61			
7026.911	7028.849			0.8	0.3		14227.08			
7027.168	7029.106			0.3			14226.56	3D 1	2B 8	R 3
7027.168	7029.106			0.3			14226.56	3d 1	2c 2	R 5
7027.613	7029.551			0.7			14225.66	3E 3	2C 2	Q 2
7027.899	7029.837			0.1			14225.08			
7028.092	7030.030			0.1			14224.69			
7028.334	7030.272			0.2			14224.20	3A 0	2B 7	P 2
7028.591	7030.529			0.3	0.1		14223.68	3c 1	2a 2	R 5
7028.591	7030.529			0.3	0.1		14223.68	3d 0	2c 1	Q 4
7028.873	7030.811			0.1			14223.11			
7029.298	7031.236			0.2			14222.25	3A 0	2B 7	P 1
7029.481	7031.419			0.0			14221.88			
7029.876	7031.815			0.2			14221.08			
7030.232	7032.171	i			0.7		14220.36			
7030.321	7032.260			0.1d			14220.18			
7030.460	7032.398						14219.90	3D 1	2B 8	R 2
7030.682	7032.621	G	0.2	1.2			14219.45	3D 3	2C 2	Q 1
7030.682	7032.621	G	0.2	1.2			14219.45	3E 2	2B10	Q 1
7030.939	7032.878			0.1			14218.93	3d 0	2c 1	R 7
7031.132	7033.071			0.2			14218.54			
7031.444	7033.383	G	0.2	1.1			14217.91			
7031.760	7033.699			0.1			14217.27	4g 1	2a 5	P 2
7031.923	7033.862			0.2			14216.94			
7032.210	7034.149			0.2			14216.36			
7032.433	7034.372						14215.91			
7032.814	7034.753			0.2			14215.14	4F 0	2C 2	Q 4
7032.814	7034.753			0.2			14215.14	3D 0	2B 6	R 0
7033.388	7035.327			0.2			14213.98			
7033.620	7035.560			0.2			14213.51	W 1	2B 9	P 2
7033.818	7035.758			0.4			14213.11	5c 2	2a 7	Q 1

λ_{air}	λ_{vac}	Ref.	I_1	I_2	I_3	I_4	ν	Upper	Lower	Br.
7034.105	7036.045			0.4			14212.53	3D 0	2B 6	R 2
7034.105	7036.045			0.4			14212.53	3e 1	2c 2	P 6
7034.318	7036.258			0.0			14212.10	3F 0	2B 7	Q 5
7034.575	7036.515			0.2			14211.58	4g 1	2a 5	R 2
7034.823	7036.763			0.7	0.3		14211.08	3b 3	2a 1	P 6
7035.115	7037.055	G	0.2	1.3	0.3		14210.49	3d 1	2c 2	Q 3
7035.115	7037.055	G	0.2	1.3	0.3		14210.49	3e 2	2c 3	P 5
7035.363	7037.303			0.2			14209.99			
7035.615	7037.555	G	2.2	1.9	0.9		14209.48	3c 0	2a 1	R 1
7035.922	7037.862			0.2			14208.86			
7036.180	7038.120			1.1	0.4		14208.34			
7036.452	7038.392			0.2			14207.79			
7036.710	7038.650			0.7			14207.27	3D 0	2B 6	R 1
7037.007	7038.947		0.2	0.1			14206.67	4F 0	2C 2	R 1
7037.462	7039.403			0.1			14205.75			
7037.680	7039.621			0.0			14205.31	4E 2	2C 4	R 1
7037.854	7039.795			0.2			14204.96			
7038.171	7040.112			0.2			14204.32			
7038.439	7040.379			0.2			14203.78	W 1	2B 9	P 4
7038.671	7040.612			0.3			14203.31			
7038.924	7040.865			0.2			14202.80			
7039.192	7041.133	G		1.3	0.2		14202.26	3d 2	2c 3	R 2
7039.474	7041.415			0.7			14201.69	3E 3	2C 2	P 2
7039.727	7041.668			0.2			14201.18			
7039.995	7041.936			0.1			14200.64			
7040.198	7042.139			0.2			14200.23	W 1	2B 9	P 3
7040.382	7042.323			0.4b			14199.86			
7040.704	7042.645	G	1.6	1.3	0.2		14199.21	3d 2	2c 3	Q 1
7040.704	7042.645	G	1.6	1.3	0.2		14199.21	4F 0	2C 2	Q 2
7040.704	7042.645	G	1.6	1.3	0.2		14199.21	3F 0	2B 7	R 3
7040.922	7042.864			0.2			14198.77			
7041.125	7043.067		0.2	0.7			14198.36	X 0	2C 0	Q 2
7041.413	7043.355			0.1			14197.78			
7041.661	7043.603			0.2			14197.28			
7041.745	7043.687		0.1				14197.11	W 2	2B11	P 1
7042.167	7044.109			0.1			14196.26			
7042.569	7044.511			0.3			14195.45			
7042.772	7044.714			0.3			14195.04			
7042.916	7044.858			0.1			14194.75	3e 3	2c 4	Q 3
7043.035	7044.977			0.3			14194.51			
7043.209	7045.151			0.2			14194.16			
7043.397	7045.340			0.4			14193.78			
7043.675	7045.618			0.2			14193.22			
7043.834	7045.776			0.1			14192.90			
7043.958	7045.901			0.2			14192.65			
7044.127	7046.069			0.3			14192.31	3E 3	2C 2	Q 3
7044.519	7046.462			0.2d			14191.52	3D 0	2B 6	P 1
7044.847	7046.789	G		1.2	0.9		14190.86	3b 4	2a 2	R 3
7045.244	7047.187		0.2	0.3			14190.06			
7045.457	7047.400			0.5			14189.63	3d 2	2c 3	P 1
7045.681	7047.624			0.3			14189.18			
7046.009	7047.951	G	1.6	1.6	0.5		14188.52	3b 4	2a 2	R 2
7046.386	7048.329			0.2			14187.76			
7046.624	7048.567			0.3			14187.28			
7046.803	7048.746			0.1			14186.92			
7046.942	7048.885			0.1			14186.64			
7047.111	7049.054			0.3			14186.30			
7047.315	7049.258			0.1			14185.89	4e 3	3b 1	P 5
7047.504	7049.447			0.2	0.2		14185.51			
7047.787	7049.730			0.2			14184.94			

λ_{air}	λ_{vac}	Ref.	I_1	I_2	I_3	I_4	ν	Upper	Lower	Br.
7048.005	7049.949			0.2			14184.50			
7048.378	7050.322			0.5			14183.75			
7048.617	7050.560			0.1			14183.27			
7048.791	7050.734			0.1			14182.92			
7049.034	7050.978			0.2			14182.43			
7049.288	7051.231			0.3			14181.92			
7049.611	7051.555	G	2.0	1.9	0.9		14181.27	3b 4	2a 2	R 1
7049.611	7051.555	G	2.0	1.9	0.9		14181.27	3D 3	2C 2	R 2
7049.611	7051.555	G	2.0	1.9	0.9		14181.27	Z 2	2C 1	Q 4
7049.974	7051.918			0.2			14180.54			
7050.098	7052.042			0.1			14180.29			
7050.227	7052.171			0.4	0.1b		14180.03			
7050.491	7052.435			0.1			14179.50	3E 3	2B12	R 1
7050.700	7052.644			0.1			14179.08	3D 1	2C 0	P 5
7050.918	7052.863			0.2			14178.64			
7051.073	7053.017			0.1	0.2		14178.33			
7051.311	7053.256			0.4			14177.85	3B 4	2A 1	R 2
7051.913	7053.858			0.2			14176.64			
7052.177	7054.121			0.1			14176.11			
7052.346	7054.291			0.1			14175.77			
7052.500	7054.445			0.2			14175.46	4F 0	2C 2	Q 3
7052.500	7054.445			0.2			14175.46	W 2	2B11	P 2
7052.734	7054.679			0.4	0.2b		14174.99	3d 1	2c 2	R 6
7052.734	7054.679			0.4	0.2b		14174.99	3F 1	2B 9	R 3
7053.192	7055.137	G		1.2			14174.07	3B 4	2A 1	R 1
7053.192	7055.137	G		1.2			14174.07	3D 2	2C 1	Q 5
7053.451	7055.395			0.1			14173.55			
7053.674	7055.619			0.2d			14173.10	3b 2	2a 0	P10
7053.958	7055.903			0.1			14172.53			
7054.162	7056.107			0.3			14172.12			
7054.336	7056.282			0.2			14171.77			
7054.804	7056.750			0.2			14170.83			
7055.038	7056.984			0.3			14170.36			
7055.213	7057.158			0.3			14170.01	3D 1	2C 0	Q 6
7055.367	7057.312			0.2			14169.70			
7055.586	7057.532			0.7			14169.26			
7055.895	7057.840			1.2	0.1		14168.64	2K 4	2B 1	P 5
7056.144	7058.089			0.2			14168.14	3e 0	2c 1	P 7
7056.343	7058.289			0.3			14167.74			
7056.617	7058.563			0.2			14167.19			
7056.851	7058.797			0.1			14166.72			
7057.060	7059.006			0.1			14166.30			
7057.289	7059.235		0.1	0.2			14165.84	Z 2	2B10	R 0
7057.838	7059.784			0.4			14164.74	Y 1	2C 0	R 1
7058.151	7060.098			0.2	0.3		14164.11			
7058.386	7060.332			0.1			14163.64	5c 1	2a 6	R 3
7058.625	7060.571			0.2			14163.16	3E 3	2B12	Q 4
7058.954	7060.900			0.5	0.1		14162.50	3d 2	2c 3	R 3
7059.133	7061.080			0.4			14162.14	3B 4	2A 1	R 3
7059.358	7061.304			0.1d			14161.69			
7059.572	7061.519			0.3			14161.26			
7059.861	7061.808			0.1			14160.68			
7060.051	7061.997			0.2			14160.30			
7060.275	7062.222			0.3			14159.85			
7060.998	7062.945			0.2			14158.40			
7061.222	7063.169	G	0.2	1.3	0.4		14157.95	3b 4	2a 2	R 0
7061.367	7063.314						14157.66			
7061.517	7063.464			0.2			14157.36	3E 0	2B 7	R 5
7061.691	7063.638			0.4	0.1		14157.01	3D 1	2B 8	P 2
7062.155	7064.102			0.2	0.2		14156.08	4g 1	2a 5	R 3

λ_{air}	λ_{vac}	Ref.	I_1	I_2	I_3	I_4	ν	Upper	Lower	Br.
7062.435	7064.382			0.2			14155.52			
7062.669	7064.617			0.2			14155.05			
7062.804	7064.751				0.2		14154.78			
7062.953	7064.901						14154.48			
7063.073	7065.021	G	2.3	1.7	0.6		14154.24	3c 0	2a 1	R 0
7063.393	7065.340			0.1			14153.60			
7063.562	7065.510				0.3		14153.26			
7063.722	7065.670			0.2	0.2		14152.94			
7064.031	7065.979			0.6			14152.32	3B 4	2A 1	R 0
7064.031	7065.979			0.6			14152.32	Z 2	2B10	R 1
7064.251	7066.199	G		0.2	0.2		14151.88			
7064.456	7066.404		0.2b	0.6	0.3		14151.47			
7064.690	7066.638			0.2	0.6		14151.00			
7064.910	7066.858		0.2b	0.2			14150.56			
7065.195	7067.143	i		1.0	1.5b		14149.99			
7065.489	7067.437			0.2			14149.40			
7065.709	7067.657			0.6			14148.96	X 1	2C 1	R 1
7065.919	7067.867			0.2			14148.54			
7066.019	7067.967				0.4		14148.34			
7066.114	7068.062			0.1			14148.15			
7066.223	7068.172			0.2	0.3		14147.93	W 2	2B11	P 3
7066.508	7068.457			0.6			14147.36	4e 0	2c 4	R 3
7066.593	7068.542				0.4		14147.19			
7066.803	7068.751			0.2			14146.77			
7066.963	7068.911			0.1	0.4		14146.45			
7067.227	7069.176	i		0.3d	1.5b		14145.92			
7067.422	7069.371			0.1			14145.53			
7067.552	7069.501				0.4		14145.27			
7067.662	7069.611			0.2			14145.05			
7067.752	7069.701				0.2		14144.87			
7067.932	7069.881		0.2	0.5	0.1		14144.51	4E 1	2B14	Q 4
7067.932	7069.881		0.2	0.5	0.1		14144.51	3E 3	2C 2	P 2
7068.207	7070.156			0.1			14143.96	4f 0	2c 4	Q 2
7068.432	7070.381			0.2			14143.51	Z 2	2B10	R 2
7068.572	7070.521		0.2				14143.23	3E 3	2C 2	Q 4
7068.672	7070.621		0.3	0.4	0.7		14143.03	X 0	2C 0	P 2
7068.922	7070.871		0.2				14142.53			
7069.031	7070.981		0.3	0.5			14142.31	4E 0	2C 2	R 2
7069.681	7071.631			0.2			14141.01			
7069.906	7071.856			0.3			14140.56	3d 1	2c 2	P 3
7070.076	7072.026			0.2			14140.22			
7070.256	7072.206			0.2			14139.86			
7070.516	7072.466			0.1			14139.34			
7070.756	7072.706			0.2			14138.86			
7071.352	7073.301			0.3			14137.67			
7071.502	7073.451			0.2			14137.37			
7071.647	7073.596		0.2	0.2			14137.08			
7071.812	7073.762		0.5	0.5			14136.75	3D 3	2C 2	Q 2
7071.812	7073.762		0.5	0.5			14136.75	3e 3	2c 4	P 3
7072.242	7074.192				0.4		14135.89			
7072.612	7074.562			0.3			14135.15			
7073.488	7075.438			0.2			14133.40			
7073.763	7075.714			0.2			14132.85	3c 1	2a 2	R 4
7073.923	7075.874			0.2			14132.53			
7074.154	7076.104			0.5	0.2		14132.07	3c 1	2a 2	R 3
7074.359	7076.310			0.2			14131.66			
7074.464	7076.415		0.2				14131.45			
7074.574	7076.525		0.5	0.5			14131.23	3E 3	2C 2	P 3
7074.789	7076.740			0.3			14130.80			
7075.015	7076.966			0.2			14130.35			

λ_{air}	λ_{vac}	Ref.	I_1	I_2	I_3	I_4	ν	Upper	Lower	Br.
7075.130	7077.081		0.5	1.3	0.1		14130.12	3d 2	2c 3	Q 2
7075.410	7077.361			0.1			14129.56			
7075.521	7077.471				0.2		14129.34			
7075.621	7077.572			1.3			14129.14			
7075.876	7077.827			0.2			14128.63			
7076.041	7077.992			0.1			14128.30			
7076.227	7078.178			0.4			14127.93	3E 2	2C 1	P 6
7076.492	7078.443			0.2			14127.40			
7076.768	7078.719			0.3			14126.85	Y 1	2C 0	Q 1
7076.963	7078.914			0.2			14126.46			
7077.179	7079.130			0.4			14126.03			
7077.399	7079.350			0.1			14125.59			
7077.604	7079.556			0.2			14125.18			
7077.980	7079.932			0.2			14124.43	3D 0	2B 6	P 2
7078.236	7080.187			0.3	0.1		14123.92			
7078.536	7080.488			0.2			14123.32			
7078.792	7080.744			0.4			14122.81	Z 2	2B10	R 3
7079.023	7080.974			0.1			14122.35			
7079.183	7081.135			0.2			14122.03			
7079.338	7081.290			0.2	0.1		14121.72	W 2	2B11	P 4
7079.338	7081.290			0.2	0.1		14121.72	3e 1	2c 2	P 7
7079.514	7081.466		0.2	0.4			14121.37	X 1	2C 1	Q 1
7079.689	7081.641			0.2			14121.02			
7079.870	7081.822		0.2	0.4			14120.66			
7080.040	7081.992			0.3			14120.32	3E 3	2B12	R 0
7080.331	7082.283			0.3			14119.74	3D 2	2B10	P 3
7080.547	7082.499			0.2			14119.31			
7080.657	7082.609				0.1		14119.09			
7080.823	7082.775			0.2			14118.76			
7080.988	7082.940			0.2			14118.43	2K 3	2B 0	P 5
7081.194	7083.146		0.3b	0.7d	0.2b		14118.02	3d 2	2c 3	R 4
7081.379	7083.332			0.2d			14117.65			
7081.495	7083.447			0.1			14117.42			
7081.630	7083.583			0.4	0.2b		14117.15	3d 0	2c 1	Q 5
7081.976	7083.929			0.1			14116.46			
7082.603	7084.556			0.1			14115.21			
7082.844	7084.797			0.0			14114.73			
7083.050	7085.003			0.2			14114.32			
7083.311	7085.264			0.4			14113.80	4g 1	2a 5	R 4
7083.482	7085.435		0.2d	0.4	0.2		14113.46			
7083.778	7085.731			0.2			14112.87			
7084.044	7085.997			0.4			14112.34	3E 3	2C 2	Q 4
7084.044	7085.997			0.4			14112.34	4d 2	3b 0	P 6
7084.114	7086.067						14112.20			
7084.255	7086.208		0.2	0.5			14111.92	3d 1	2c 2	Q 4
7084.255	7086.208		0.2	0.5			14111.92	Z 2	2C 1	P 4
7084.571	7086.524			0.2			14111.29	4F 0	2B12	P 4
7084.897	7086.851			0.2			14110.64			
7085.158	7087.112		0.1	0.3			14110.12	W 2	2B11	P 5
7085.460	7087.413		0.1d	0.2d			14109.52			
7085.801	7087.755		0.5	0.6			14108.84	3D 3	2C 2	P 2
7086.088	7088.041			0.5	0.3		14108.27			
7086.630	7088.584			0.1			14107.19			
7086.906	7088.860			0.2			14106.64	Z 2	2B10	R 4
7087.163	7089.117		0.2	0.3			14106.13	Y 1	2C 0	P 1
7087.379	7089.333			0.2			14105.70			
7087.886	7089.840			0.2			14104.69	3E 3	2B12	Q 3
7088.132	7090.087			0.3			14104.20	X 2	2C 2	R 1
7088.379	7090.333			0.1			14103.71			
7088.610	7090.564			0.3			14103.25			

λ_{air}	λ_{vac}	Ref.	I_1	I_2	I_3	I_4	ν	Upper	Lower	Br.
7088.967	7090.921			0.2d			14102.54	Y 1	2C 0	R 2
7088.967	7090.921			0.2d			14102.54	3d 0	2c 1	P 4
7089.751	7091.706			0.2			14100.98			
7090.369	7092.324			0.3			14099.75			
7090.636	7092.591	G	0.8	1.4	0.3		14099.22	2K 4	2B 1	P 6
7090.998	7092.953			0.1			14098.50			
7091.285	7093.240			0.2			14097.93			
7091.390	7093.346						14097.72			
7091.622	7093.577	G	1.8	1.2	0.2		14097.26	3c 1	2a 2	R 2
7091.622	7093.577	G	1.8	1.2	0.2		14097.26	4f 0	2c 4	Q 3
7091.622	7093.577	G	1.8	1.2	0.2		14097.26	X 0	2C 0	Q 3
7092.482	7094.438				0.1		14095.55			
7092.900	7094.855		0.3	0.6			14094.72	3d 2	2c 3	P 2
7093.121	7095.077			0.1			14094.28			
7093.338	7095.293		0.2	0.4	0.2		14093.85			
7093.514	7095.470		0.3	0.4			14093.50			
7093.821	7095.777			0.1			14092.89			
7094.078	7096.033			0.1			14092.38			
7094.284	7096.240			0.2			14091.97	3D 1	2B 8	P 3
7094.475	7096.431			0.3			14091.59	3D 3	2C 2	R 3
7094.777	7096.733			0.3			14090.99			
7094.878	7096.834		0.2				14090.79			
7095.095	7097.051	G	2.3	2.0	1.0		14090.36	3c 0	2a 1	Q 1
7095.351	7097.308		0.2	0.1			14089.85			
7095.538	7097.494	G	1.9	1.3	0.2		14089.48	2K 6	2B 3	R 2
7095.845	7097.801		0.2	0.4	0.1		14088.87			
7096.011	7097.968			0.4	0.2b		14088.54	3D 2	2B10	P 4
7096.011	7097.968			0.4	0.2b		14088.54	4F 0	2B12	R 1
7096.318	7098.275		0.1	0.2			14087.93	4F 0	2B12	Q 3
7096.656	7098.613		0.1	0.4			14087.26	3B 3	2A 0	P 7
7096.908	7098.864		0.1	0.5	0.1		14086.76	Z 2	2B10	P 2
7096.908	7098.864		0.1	0.5	0.1		14086.76	3c 2	2a 3	R 6
7097.074	7099.031		0.3	0.3			14086.43	X 2	2C 2	Q 1
7097.240	7099.197		0.1				14086.10			
7097.407	7099.363	G	1.7	1.2			14085.77	2K 6	2B 3	R 1
7097.407	7099.363	G	1.7	1.2			14085.77	X 2	2C 2	P 1
7097.684	7099.641		0.1	0.4			14085.22			
7097.800	7099.756			0.1			14084.99	Y 2	2B11	R 2
7097.986	7099.943		0.2	1.1			14084.62			
7098.374	7100.331			0.2			14083.85			
7098.707	7100.664		0.1	0.5	0.1		14083.19			
7099.045	7101.002		0.1	0.3			14082.52			
7099.125	7101.082				0.1		14082.36			
7099.408	7101.365	G	1.7	1.5	0.6		14081.80	3c 0	2a 1	Q 2
7099.675	7101.632			0.2			14081.27	3F 2	2B11	Q 4
7099.992	7101.950			0.2			14080.64	5c 1	2a 6	R 2
7100.325	7102.283			0.3			14079.98	Y 2	2B11	R 1
7100.648	7102.606		0.3	0.5			14079.34	Z 0	2B 6	R 2
7100.930	7102.888			0.2			14078.78	X 1	2C 1	R 2
7101.127	7103.085			0.2			14078.39			
7101.253	7103.211			0.3			14078.14			
7101.576	7103.534			0.2			14077.50			
7101.823	7103.781			0.2			14077.01	4E 1	2B14	Q 3
7102.131	7104.089		0.0	0.3			14076.40			
7102.484	7104.442			0.2			14075.70			
7102.807	7104.765			0.3b			14075.06	Z 0	2B 6	R 3
7103.160	7105.119			0.2			14074.36			
7103.574	7105.533			0.1			14073.54	Z 0	2B 6	R 4
7103.857	7105.815		0.3	0.7	0.2b		14072.98	Z 0	2B 6	R 1
7103.953	7105.911		0.2				14072.79			

λ_{air}	λ_{vac}	Ref.	I$_1$	I$_2$	I$_3$	I$_4$	ν	Upper	Lower	Br.
7104.276	7106.235	G	2.2	1.5	0.1		14072.15	2K 6	2B 3	R 0
7104.660	7106.618		0.1	0.2			14071.39	4E 1	2C 3	Q 1
7104.660	7106.618		0.1	0.2			14071.39	Y 2	2B11	R 0
7104.887	7106.846		0.1	0.4			14070.94			
7105.054	7107.012		0.2b	0.3			14070.61			
7105.256	7107.214			0.1			14070.21			
7105.468	7107.427		0.1	0.4			14069.79			
7105.685	7107.644		0.1				14069.36			
7105.836	7107.795	G	1.8	1.7	1.0		14069.06	3c 0	2a 1	Q 3
7106.266	7108.225		0.2	0.2			14068.21			
7106.508	7108.467		0.1	0.4	0.1		14067.73			
7106.715	7108.675		0.3	0.2			14067.32			
7106.993	7108.953	G	0.4	1.3	0.4		14066.77	3c 2	2a 3	R 5
7107.286	7109.246		0.1	0.5			14066.19			
7107.458	7109.417	i		0.0	0.2		14065.85			
7107.716	7109.675			0.2			14065.34			
7108.064	7110.024	G	0.9	1.1			14064.65	3E 3	2B12	Q 2
7108.064	7110.024	G	0.9	1.1			14064.65	3E 3	2C 2	P 3
7108.433	7110.393		0.2	0.1			14063.92			
7108.676	7110.636		0.1	0.1			14063.44			
7109.045	7111.005			0.2			14062.71	4E 0	2C 2	Q 1
7109.318	7111.278			0.2			14062.17			
7109.515	7111.475			0.1			14061.78			
7110.319	7112.279			0.2			14060.19	3F 1	2B 9	Q 4
7110.537	7112.497		0.3	0.4			14059.76	Z 0	2B 6	R 0
7110.906	7112.866		0.1				14059.03	V 0	2B 6	R 2
7111.123	7113.084			0.2			14058.60			
7111.310	7113.271		0.3	0.6			14058.23			
7111.518	7113.478			0.2			14057.82	3D 0	2B 6	P 3
7111.776	7113.737			0.2			14057.31			
7112.090	7114.050		0.1	0.3			14056.69			
7112.413	7114.374	G	1.5b	1.5	0.5		14056.05			
7112.651	7114.612	G	2.3	2.1	0.6		14055.58	3b 4	2a 2	P 1
7112.651	7114.612	G	2.3	2.1	0.6		14055.58	3E 3	2B12	P 5
7112.651	7114.612	G	2.3	2.1	0.6		14055.58	3c 1	2a 2	R 1
7113.041	7115.002		0.2	0.3			14054.81			
7113.284	7115.245			0.1			14054.33			
7113.562	7115.523			0.2			14053.78			
7113.881	7115.842			0.3			14053.15			
7114.241	7116.202	G	1.7	1.4			14052.44	3B 4	2A 1	P 1
7114.322	7116.283				0.6		14052.28	3c 0	2a 1	Q 4
7114.565	7116.526			0.2			14051.80			
7114.737	7116.698			0.2			14051.46			
7115.076	7117.038		0.2	0.3			14050.79			
7115.466	7117.428			0.1			14050.02			
7115.679	7117.640			0.2			14049.60			
7116.423	7118.385			0.2	0.2		14048.13	3E 3	2B12	P 4
7116.707	7118.669	G	0.7	1.3	0.6		14047.57	3d 2	2c 3	Q 3
7116.803	7118.765		0.5				14047.38	3F 3	2B13	R 2
7116.970	7118.932			0.1			14047.05			
7117.168	7119.130		0.2	0.7	0.4		14046.66	3b 3	2a 1	P 7
7117.457	7119.419			0.2			14046.09	4E 1	2B14	Q 1
7117.457	7119.419			0.2			14046.09	3D 2	2C 1	P 5
7117.736	7119.698			0.0			14045.54			
7118.491	7120.453			0.2			14044.05	Y 1	2C 0	Q 2
7118.840	7120.803		0.0				14043.36	3E 3	2B12	P 3
7118.840	7120.803		0.0				14043.36	4E 1	2B14	Q 2
7119.074	7121.036		0.0	0.3			14042.90			
7119.494	7121.457			0.2			14042.07			
7119.799	7121.761			0.4			14041.47			

λ_{air}	λ_{vac}	Ref.	I_1	I_2	I_3	I_4	ν	Upper	Lower	Br.
7120.098	7122.061			0.0			14040.88	3E 3	2B12	P 2
7120.098	7122.061			0.0			14040.88	3D 2	2B10	P 5
7120.204	7122.167						14040.67	3F 1	2B 9	P 5
7120.433	7122.396			0.1			14040.22	4F 0	2C 2	P 3
7121.168	7123.131			0.2			14038.77			
7121.305	7123.268		0.1				14038.50			
7121.427	7123.390	G	0.3	1.2	0.3		14038.26	3c 2	2a 3	R 4
7121.716	7123.679			0.2			14037.69			
7121.944	7123.907			0.2			14037.24			
7122.183	7124.146			0.3			14036.77			
7122.502	7124.466			0.1			14036.14			
7122.771	7124.735			0.1			14035.61			
7123.030	7124.994			0.2			14035.10	X 1	2C 1	Q 2
7123.279	7125.242			0.2			14034.61			
7123.396	7125.359		0.0				14034.38			
7123.670	7125.633			0.2d			14033.84			
7123.939	7125.903		0.3	0.7	0.4		14033.31	3E 3	2B12	Q 1
7123.939	7125.903		0.3	0.7	0.4		14033.31	4g 1	2a 5	P 3
7124.233	7126.197		0.6	1.2			14032.73			
7124.533	7126.497			0.2			14032.14			
7124.629	7126.593		0.1				14031.95			
7124.807	7126.771		0.3	1.2	1.0		14031.60	3c 0	2a 1	Q 5
7125.106	7127.071		0.1	0.2			14031.01	Y 2	2B11	P 1
7125.350	7127.314			0.2			14030.53	X 2	2C 2	R 2
7125.574	7127.538		0.1	0.4			14030.09			
7125.782	7127.746			0.2	0.2		14029.68			
7125.985	7127.950		0.2	0.3			14029.28			
7126.173	7128.138			0.3d			14028.91	3F 2	2B11	R 2
7126.402	7128.366			0.2d			14028.46			
7126.645	7128.610		0.1	0.3			14027.98			
7126.833	7128.798	G	1.7	1.2	0.1		14027.61	3A 0	2C 0	R 2
7127.057	7129.022						14027.17	3D 3	2B12	R 0
7127.250	7129.215			0.2			14026.79	7c 0	2a 6	R 1
7127.550	7129.515		0.0	0.3			14026.20			
7127.733	7129.698			0.2			14025.84	3D 3	2B12	R 1
7127.956	7129.921		0.3	0.4			14025.40	3F 0	2B 7	Q 4
7127.956	7129.921		0.3	0.4			14025.40	3D 1	2B 8	P 4
7128.236	7130.201		0.1	0.2			14024.85			
7128.338	7130.303		0.3				14024.65			
7128.516	7130.481		0.3	0.6	0.2		14024.30			
7128.800	7130.765			0.1			14023.74			
7128.907	7130.872						14023.53			
7129.004	7130.969			0.2			14023.34	4e 0	2c 4	R 1
7129.161	7131.126		0.3	0.3			14023.03			
7129.349	7131.315			0.4	0.2b		14022.66			
7129.471	7131.437		0.2				14022.42			
7129.644	7131.610			0.2			14022.08			
7129.903	7131.869		0.2	0.5	0.1		14021.57			
7130.107	7132.072			0.2			14021.17			
7130.285	7132.250			0.2			14020.82			
7130.453	7132.418			0.2			14020.49	3D 3	2B12	R 2
7130.661	7132.627		0.6	0.6			14020.08	3D 3	2C 2	Q 3
7131.027	7132.993			0.2			14019.36			
7131.297	7133.263		0.1	0.1			14018.83	Z 2	2B10	P 3
7131.597	7133.563		0.2	0.1			14018.24			
7131.851	7133.818			0.2			14017.74			
7132.126	7134.092		0.2	0.5			14017.20			
7132.411	7134.377			0.1			14016.64			
7133.114	7135.080		0.2	0.2			14015.26	3D 3	2B12	R 4
7133.419	7135.385	G	1.9	1.3	0.1		14014.66	2K 6	2B 3	P 1

λ_{air}	λ_{vac}	Ref.	I_1	I_2	I_3	I_4	ν	Upper	Lower	Br.
7133.729	7135.696		0.0	0.2			14014.05	V 0	2B 6	R 3
7133.969	7135.935			0.1			14013.58			
7134.137	7136.103			0.2			14013.25			
7134.366	7136.332			0.1			14012.80			
7134.569	7136.536		0.1	0.2			14012.40	5c 1	2a 6	R 1
7134.569	7136.536		0.1	0.2			14012.40	3d Q	2c 1	Q 6
7134.844	7136.811		0.3	1.3	0.3		14011.86	3d 1	2c 2	Q 5
7135.282	7137.249			0.2			14011.00			
7135.486	7137.453		0.2	0.3			14010.60	X 0	2C 0	P 3
7135.792	7137.759			0.2			14010.00			
7136.046	7138.013		0.1	0.3			14009.50			
7136.393	7138.360			0.2			14008.82			
7136.469	7138.436		0.2				14008.67			
7136.698	7138.666		0.3	0.3			14008.22	Z 0	2B 6	P 1
7136.698	7138.666		0.3	0.3			14008.22	4e 0	2c 4	R 1
7136.999	7138.966		0.1				14007.63	3D 0	2B 6	P 8
7137.213	7139.180			0.4	0.5		14007.21	3c 0	2a 1	Q 6
7137.524	7139.491		0.2	0.4			14006.60	3E 2	2C 1	P 7
7137.830	7139.797			0.2			14006.00			
7137.988	7139.955		0.2				14005.69			
7138.146	7140.113	G	2.2	1.7	0.3		14005.38	3c 1	2a 2	R 0
7138.146	7140.113	G	2.2	1.7	0.3		14005.38	3D 1	2C 0	Q 7
7138.528	7140.496		0.2	0.2			14004.63	3D 0	2B 6	P 4
7138.732	7140.700			0.2			14004.23	3D 3	2B12	R 3
7139.002	7140.970			0.3			14003.70			
7139.109	7141.077		0.2				14003.49			
7139.338	7141.306	G	2.0	1.8	1.8		14003.04	3c 2	2a 3	R 3
7139.685	7141.653		0.2	0.6			14002.36	X 1	2C 1	P 2
7139.935	7141.903		0.2	0.3			14001.87	Y 1	2C 0	P 2
7140.144	7142.112			0.3			14001.46			
7140.363	7142.332			0.2			14001.03			
7140.567	7142.536			0.2			14000.63	Y 2	2B11	P 2
7140.787	7142.755			0.2			14000.20	3F 1	2B 9	R 2
7141.006	7142.974			0.2			13999.77			
7141.200	7143.168			0.1			13999.39			
7141.383	7143.352			0.2			13999.03			
7141.664	7143.633		0.2	0.4			13998.48			
7141.878	7143.847			0.2			13998.06			
7142.098	7144.067		0.8	1.1			13997.63	3E 3	2C 2	P 4
7142.332	7144.301			0.1			13997.17			
7142.491	7144.459		0.1	0.2			13996.86			
7142.705	7144.674			0.3			13996.44			
7142.955	7144.924		0.2	0.2			13995.95	X 2	2C 2	Q 2
7143.149	7145.118			0.3			13995.57			
7143.333	7145.302			0.2			13995.21			
7143.593	7145.562			0.2			13994.70			
7143.884	7145.853		0.3	0.5			13994.13	3d 1	2c 2	P 4
7144.272	7146.241			0.2			13993.37			
7144.558	7146.527			0.2			13992.81			
7144.869	7146.839			0.2			13992.20			
7145.212	7147.181		0.2	0.6	1.2		13991.53	3F 3	2B13	Q 3
7145.212	7147.181		0.2	0.6	1.2		13991.53	3D 2	2C 1	Q 6
7145.508	7147.477			0.2			13990.95			
7145.911	7147.881			0.1			13990.16			
7146.233	7148.203			0.2			13989.53			
7146.427	7148.397			0.2			13989.15			
7146.672	7148.642			0.2			13988.67			
7146.815	7148.786			0.2			13988.39	4f 0	2c 4	P 3
7146.964	7148.934			0.3			13988.10			
7147.050	7149.021	1			1.5		13987.93			

λ_{air}	λ_{vac}	Ref.	I₁	I₂	I₃	I₄	ν	Upper	Lower	Br.
7147.163	7149.133			0.2			13987.71			
7147.342	7149.312		0.1	0.3	0.2b		13987.36			
7147.659	7149.629			0.2			13986.74			
7148.047	7150.017		0.1	0.1			13985.98			
7148.809	7150.779			0.1			13984.49	3D 1	2B 8	P 5
7149.069	7151.040		0.2	0.4	0.2		13983.98	4E 1	2C 3	Q 2
7149.499	7151.470			0.2			13983.14			
7149.754	7151.725			0.2			13982.64			
7150.066	7152.037		0.4	0.5			13982.03	3A 0	2C 0	R 1
7150.066	7152.037		0.4	0.5			13982.03	Z 2	2C 1	Q 5
7150.066	7152.037		0.4	0.5			13982.03	3F 0	2B 7	R 2
7150.378	7152.349			0.2d			13981.42	4E 0	2C 2	Q 3
7150.629	7152.600		0.2	0.3			13980.93			
7151.181	7153.153			0.2			13979.85			
7151.463	7153.434		0.2	0.5	0.8		13979.30	3c 0	2a 1	Q 7
7151.954	7153.925			0.2d			13978.34	3D 0	2B 6	P 7
7151.954	7153.925			0.2d			13978.34	4E 0	2C 2	R 3
7152.066	7154.038		0.1				13978.12	2K 3	2B 0	P 6
7152.281	7154.253	G	1.5	1.3	1.3		13977.70	3b 4	2a 2	P 2
7152.281	7154.253	G	1.5	1.3	1.3		13977.70	3D 3	2C 2	P 3
7152.281	7154.253	G	1.5	1.3	1.3		13977.70	4F 0	2C 2	P 5
7152.711	7154.683		0.2	0.2			13976.86			
7153.177	7155.149			0.2			13975.95	X 2	2C 2	P 2
7153.453	7155.425	G	1.0	0.8	1.4		13975.41	3B 4	2A 1	P 2
7153.453	7155.425	G	1.0	0.8	1.4		13975.41	4E 0	2C 2	R 1
7153.796	7155.768			0.2			13974.74			
7154.073	7156.045		0.2	0.5			13974.20	3E 3	2C 2	P 4
7154.073	7156.045		0.2	0.5			13974.20	3d 2	2c 3	P 3
7154.436	7156.408			0.2			13973.49	3E 0	2B 7	R 4
7154.708	7156.680		0.4	0.6			13972.96	4E 0	2C 2	R 2
7155.035	7157.008		0.1	0.3	0.2		13972.32			
7155.235	7157.207		0.1	0.2			13971.93			
7155.476	7157.448		1.0	0.7	1.1		13971.46			
7155.711	7157.684	G	1.8	1.9	1.2		13971.00	2K 6	2B 3	P 2
7155.711	7157.684	G	1.8	1.9	1.2		13971.00	3C 0	2A 0	R 4
7155.998	7157.971			0.1			13970.44	3D 0	2B 6	P 5
7156.229	7158.201		0.5	0.6	1.3		13969.99	3A 1	2C 1	R 2
7156.377	7158.350		0.4	0.4			13969.70	Z 0	2B 6	P 2
7156.377	7158.350		0.4	0.4			13969.70	X 0	2C 0	Q 4
7156.669	7158.642		0.2	0.3			13969.13			
7156.915	7158.888		0.2				13968.65			
7157.381	7159.354			0.2			13967.74			
7157.602	7159.575		0.2b	0.3			13967.31			
7157.781	7159.754			0.2			13966.96			
7158.017	7159.990			0.2			13966.50	4E 2	2C 4	P 3
7159.037	7161.010	G	0.9	0.7	1.4		13964.51	3c 0	2a 1	P 2
7159.037	7161.010	G	0.9	0.7	1.4		13964.51	3D 3	2C 2	R 4
7159.360	7161.333			0.2			13963.88			
7159.565	7161.539		0.1				13963.48			
7159.673	7161.646			0.2			13963.27			
7159.862	7161.836			0.1			13962.90	Z 2	2B10	P 4
7160.011	7161.985		0.1	0.2	0.1		13962.61			
7160.314	7162.287	G	2.0	1.6	0.3		13962.02	3c 2	2a 3	R 2
7160.647	7162.621			0.2			13961.37	7c 0	2a 6	Q 3
7160.806	7162.780			0.1			13961.06	3D 0	2B 6	P 6
7160.980	7162.954		0.1	0.2			13960.72	4E 1	2C 3	P 2
7161.211	7163.185			0.2			13960.27	4E 0	2C 2	Q 1
7161.401	7163.375			0.2			13959.90			
7161.622	7163.596		0.3	0.4			13959.47			
7161.909	7163.883			0.2			13958.91			

λ_{air}	λ_{vac}	Ref.	I_1	I_2	I_3	I_4	ν	Upper	Lower	Br.
7162.191	7164.165		0.2	0.6	0.1d		13958.36	3d 2	2c 3	Q 4
7162.504	7164.478			0.1			13957.75	3b 2	2a 0	P11
7162.638	7164.612			0.2			13957.49			
7162.792	7164.766		0.2	0.5			13957.19	4F 0	2C 2	P 4
7163.084	7165.059			0.2			13956.62			
7163.361	7165.336			0.2			13956.08	Y 2	2B11	P 3
7163.638	7165.613		0.1	0.4			13955.54			
7163.864	7165.839			0.1			13955.10			
7164.121	7166.096			0.2			13954.60	3F 2	2B11	Q 3
7164.301	7166.275			0.2			13954.25			
7164.542	7166.517			0.3			13953.78	3c 0	2a 1	Q 8
7164.717	7166.692		0.1	0.3			13953.44			
7164.963	7166.938			0.1			13952.96			
7165.163	7167.138			0.2			13952.57	4F 0	2B12	P 3
7165.466	7167.441			0.2			13951.98	4E 0	2C 2	P 2
7165.723	7167.698			0.1			13951.48			
7166.129	7168.104		0.2	0.2			13950.69	3d 3	2c 4	Q 1
7166.324	7168.299		0.1				13950.31			
7166.468	7168.443	G	0.8	0.9	0.2		13950.03			
7166.550	7168.526		0.7				13949.87			
7166.704	7168.680			0.2			13949.57	5c 1	2a 6	R 0
7166.874	7168.849			0.1			13949.24			
7167.007	7168.983			0.2			13948.98			
7167.275	7169.250		0.4	0.9	1.3		13948.46	3C 2	2K 2	R 3
7167.465	7169.440			0.2	0.3		13948.09	3D 3	2B12	P 2
7167.732	7169.708			0.2			13947.57			
7167.963	7169.939			0.2			13947.12			
7168.153	7170.129		0.2	0.2			13946.75			
7168.339	7170.314			0.3			13946.39			
7168.518	7170.494			0.3			13946.04			
7168.631	7170.607		0.3				13945.82			
7168.816	7170.792	G	2.5	2.4	1.8		13945.46	3c 1	2a 2	Q 1
7168.816	7170.792	G	2.5	2.4	1.8		13945.46	4g 0	2a 4	P 1
7169.181	7171.158		0.2	0.3			13944.75			
7169.392	7171.368			0.2			13944.34			
7169.541	7171.518			0.2			13944.05			
7169.680	7171.656		0.1	0.3			13943.78	3F 0	2B 7	P 5
7169.953	7171.929	G	0.5	1.4	1.8		13943.25	3C 0	2A 0	R 3
7169.953	7171.929	G	0.5	1.4	1.8		13943.25	3d 0	2c 1	P 5
7170.292	7172.269			0.2			13942.59			
7170.462	7172.438			0.1			13942.26			
7170.688	7172.665			0.4b			13941.82			
7170.806	7172.783		0.2				13941.59			
7171.023	7172.999			0.2			13941.17			
7171.275	7173.251			0.2			13940.68			
7171.444	7173.421			0.2			13940.35	X 2	2C 2	R 3
7171.732	7173.709			0.2			13939.79			
7172.257	7174.234			0.2			13938.77			
7172.406	7174.383		0.1	0.3			13938.48			
7172.695	7174.672	G	2.0	1.4			13937.92	2K 5	2B 2	R 1
7173.055	7175.032	G	2.2	1.8	1.6		13937.22	3c 1	2a 2	Q 2
7173.240	7175.217		0.2				13936.86			
7173.477	7175.454		0.4	0.4b			13936.40			
7173.549	7175.526		0.2				13936.26	3A 1	2C 1	R 1
7173.909	7175.887		0.1				13935.56	4F 0	2B12	Q 2
7174.079	7176.057			0.2			13935.23	4F 0	2C 2	P 3
7174.398	7176.376		0.4	0.7	0.3		13934.61			
7174.610	7176.587			0.2			13934.20			
7174.918	7176.896			0.2			13933.60			
7175.109	7177.087			0.1			13933.23			

λ_{air}	λ_{vac}	Ref.	I_1	I_2	I_3	I_4	ν	Upper	Lower	Br.
7175.258	7177.236			0.3	0.1		13932.94			
7175.536	7177.514			0.1			13932.40			
7175.706	7177.684			0.2			13932.07			
7176.021	7177.999		0.4	0.4			13931.46			
7176.299	7178.277	G	2.1	2.1	1.7		13930.93	2K 5	2B 2	R 0
7176.299	7178.277	G	2.1	2.1	1.7		13930.93	2K 5	2B 2	R 2
7176.649	7178.627		0.2	0.2			13930.24			
7176.922	7178.900		0.1	0.2			13929.71			
7177.206	7179.184			0.2			13929.16			
7177.463	7179.442			0.1			13928.66			
7177.741	7179.720			0.2			13928.12			
7178.035	7180.014		0.1	0.2			13927.55			
7178.473	7180.452			0.1			13926.70	V 0	2B 6	R 1
7178.844	7180.823		0.1	0.2			13925.98			
7179.087	7181.066			0.2			13925.51			
7179.174	7181.153		0.2				13925.34			
7179.386	7181.365	G	1.5		2.3		13924.94	3c 1	2a 2	Q 3
7179.386	7181.365	G	1.5		2.3		13924.94	Y 1	2C 0	Q 3
7179.535	7181.514	G	2.4	2.4	2.3		13924.65	2A 3	2B 1	R 2
7179.881	7181.860		0.2	0.2			13923.97			
7180.051	7182.030		0.2	0.3			13923.64	Z 0	2B 6	P 3
7180.051	7182.030		0.2	0.3			13923.64	5c 1	2a 6	Q 3
7180.448	7182.427			0.2			13922.87			
7180.603	7182.582			0.2			13922.57			
7180.897	7182.876			0.2			13922.00			
7181.201	7183.180			0.2			13921.41			
7181.382	7183.361		0.1	0.2			13921.06			
7181.542	7183.521			0.2			13920.75			
7181.686	7183.665			0.2			13920.47			
7181.861	7183.841			0.2			13920.13			
7182.032	7184.011			0.2			13919.80			
7182.290	7184.269		0.2	0.4			13919.30			
7182.651	7184.631		0.2	0.2			13918.60			
7182.945	7184.925	G	1.8	1.3			13918.03	2K 6	2B 3	P 3
7183.162	7185.142			0.2			13917.61			
7183.430	7185.410		0.2	0.3			13917.09	2A 3	2B 1	R 3
7183.699	7185.679			0.3			13916.57			
7183.828	7185.808		0.3				13916.32			
7184.050	7186.030	G	2.4	2.4	2.0		13915.90	3c 2	2a 3	R 1
7184.215	7186.195						13915.57	4f 2	3b 1	R 4
7184.380	7186.360	G	1.5b	2.1	1.8		13915.25	2A 3	2B 1	R 1
7184.674	7186.655		0.4	0.7	1.8		13914.68			
7184.917	7186.897			0.1			13914.21			
7185.088	7187.068		0.2	0.3	0.6		13913.88			
7185.356	7187.336			0.1			13913.36			
7185.707	7187.688			0.2			13912.68			
7185.966	7187.946			0.3	0.1		13912.18	3d 1	2c 2	Q 6
7186.260	7188.241		0.2d	0.3			13911.61			
7186.518	7188.499		0.3	0.7	1.2		13911.11	2K 5	2B 2	R 3
7186.518	7188.499		0.3	0.7	1.2		13911.11	7c 0	2a 6	Q 1
7186.694	7188.675			0.2			13910.77	X 1	2C 1	Q 3
7187.428	7189.409			0.2	0.2		13909.35	3d 0	2c 1	Q 7
7187.624	7189.605		0.1				13908.97			
7187.732	7189.714		0.3	0.9	1.5		13908.76	3c 1	2a 2	Q 4
7188.022	7190.003		0.1	0.1			13908.20	3A 2	2B12	R 4
7188.254	7190.236			0.2d			13907.75	3F 1	2B 9	Q 3
7188.570	7190.551	G	1.5	1.3	1.3		13907.14			
7188.890	7190.872		0.1	0.2			13906.52			
7189.128	7191.109			0.2			13906.06	Y 2	2B11	P 4
7189.361	7191.342			0.2			13905.61			

λ_{air}	λ_{vac}	Ref.	I_1	I_2	I_3	I_4	ν	Upper	Lower	Br.
7189.536	7191.518			0.2			13905.27			
7189.681	7191.663			0.2			13904.99			
7190.064	7192.046		0.2	0.5b			13904.25			
7190.348	7192.330			0.1			13903.70			
7190.581	7192.563			0.2			13903.25	3F 2	2B11	R 1
7190.865	7192.847		0.1	0.2			13902.70			
7191.207	7193.189		0.2	0.2			13902.04			
7191.517	7193.499			0.2			13901.44			
7191.750	7193.732			0.0			13900.99	5c 1	2a 6	Q 2
7192.066	7194.048		0.3				13900.38			
7193.039	7195.021			0.1d			13898.50	U 1	2B 9	R 4
7193.334	7195.316			0.2d			13897.93			
7193.717	7195.699		0.1	0.2			13897.19			
7193.986	7195.969		0.1	0.2			13896.67			
7194.234	7196.217		0.1				13896.19			
7194.307	7196.290			0.2			13896.05			
7194.566	7196.549		0.5	0.7	1.1		13895.55			
7194.628	7196.611		0.4				13895.43			
7194.887	7196.870	G	1.9	1.6	1.3		13894.94	3c 0	2a 1	P 3
7195.140	7197.123		0.2	0.1			13894.44	X 2	2B12	R 0
7195.348	7197.331		0.2	0.4			13894.04			
7195.441	7197.424		0.4				13893.86			
7195.663	7197.647	G	3.5	2.5	2.0		13893.43	2A 3	2B 1	R 0
7195.974	7197.958		0.3				13892.83			
7196.036	7198.020		0.3	0.3			13892.71	Z 2	2C 1	P 5
7196.249	7198.232		0.2	0.1d			13892.30			
7196.544	7198.527			0.1			13891.73	4e 0	2c 4	R 2
7196.746	7198.730			0.1			13891.34			
7197.031	7199.015			0.1			13890.79			
7197.394	7199.377			0.2			13890.09	3D 3	2B12	P 3
7197.699	7199.683		0.6	1.3	1.5		13889.50			
7197.850	7199.833		0.1				13889.21	3b 3	2a 1	P 8
7198.041	7200.025		0.4	1.3	1.8		13888.84	3c 1	2a 2	Q 5
7198.404	7200.388			0.2			13888.14	3C 0	2A 0	R 2
7198.643	7200.627				0.1		13887.68	4E 1	2C 3	Q 4
7199.706	7201.690			0.1			13885.63	5c 1	2a 6	Q 1
7199.970	7201.954		0.1	0.3			13885.12	X 2	2B12	R 1
7200.214	7202.198			0.2			13884.65			
7200.514	7202.499	G	1.4	1.3	1.8		13884.07			
7200.779	7202.764		0.1	0.1			13883.56			
7201.080	7203.064		0.2	0.2			13882.98			
7201.370	7203.355	G	2.0	1.6	1.9		13882.42	3b 4	2a 2	P 3
7201.635	7203.620			0.2			13881.91			
7201.827	7203.812		0.1				13881.54			
7201.915	7203.900		0.3	0.7	1.3		13881.37	2K 5	2B 2	R 4
7202.211	7204.196			0.1			13880.80			
7202.481	7204.466		0.1	0.2			13880.28			
7202.657	7204.642			0.1			13879.94			
7202.787	7204.772		0.2	0.2			13879.69			
7203.005	7204.990		0.1	0.2			13879.27			
7203.311	7205.296	G	1.9	1.5	1.4		13878.68	3B 4	2A 1	P 3
7203.311	7205.296	G	1.9	1.5	1.4		13878.68	3F 0	2C 0	R 7
7203.591	7205.577			0.1			13878.14			
7203.737	7205.722			0.2			13877.86			
7204.058	7206.044		0.3	0.7	1.7		13877.24			
7204.287	7206.272		0.1	0.3	0.5		13876.80			
7204.505	7206.490			0.2			13876.38			
7204.707	7206.693			0.1			13875.99			
7205.014	7206.999			0.3d			13875.40			
7205.247	7207.233		0.2				13874.95			

λ$_{air}$	λ$_{vac}$	Ref.	I$_1$	I$_2$	I$_3$	I$_4$	ν	Upper	Lower	Br.
7205.341	7207.327			0.3			13874.77	3E 3	2C 2	P 5
7205.341	7207.327			0.3			13874.77	4g 0	2a 4	R 0
7205.486	7207.472	G	2.2	1.6	1.2		13874.50	2K 5	2B 2	P 1
7205.860	7207.846		0.3	0.3			13873.77			
7206.026	7208.012			0.2			13873.45	3D 3	2C 2	Q 4
7206.239	7208.225			0.2			13873.04			
7206.463	7208.449		0.2	0.2			13872.61			
7206.733	7208.719		0.4	1.3	1.2		13872.09			
7206.961	7208.948	1		0.1	0.4		13871.65			
7207.076	7209.062		0.3	1.3			13871.43	Z 0	2B 6	P 4
7207.304	7209.291			0.2d			13870.99			
7207.481	7209.467		0.1				13870.65	4E 0	2B12	R 0
7207.668	7209.655			0.2			13870.29			
7207.845	7209.831			0.2			13869.95	4c 4	2a 8	Q 1
7208.037	7210.024		0.1	0.2			13869.58	X 2	2B12	R 2
7208.037	7210.024		0.1	0.2			13869.58	X 2	2C 2	Q 3
7208.255	7210.242		0.1	0.3	0.1		13869.16			
7208.624	7210.611		0.2	0.5	0.1		13868.45			
7208.895	7210.882			0.2			13867.93			
7209.103	7211.090			0.1			13867.53			
7209.331	7211.318			0.2			13867.09			
7209.602	7211.589			0.3			13866.57	X 2	2B12	P 1
7209.742	7211.729		0.2				13866.30			
7209.955	7211.942	G	2.7	2.0	1.6		13865.90	3A 0	2C 0	Q 1
7209.955	7211.942	G	2.7	2.0	1.6		13865.90	3d 2	2c 3	Q 5
7210.215	7212.202	G	2.6	2.0	1.6		13865.39	3c 1	2a 2	Q 6
7210.215	7212.202	G	2.6	2.0	1.6		13865.39	3c 2	2a 3	R 0
7210.559	7212.546		0.2	0.2			13864.73	3c 3	2a 4	R 3
7210.559	7212.546		0.2	0.2			13864.73	3F 0	2B 7	Q 3
7210.559	7212.546		0.2	0.2			13864.73	4E 0	2C 2	Q 2
7210.808	7212.795			0.1			13864.25			
7211.313	7213.300		0.2	0.4			13863.28	3A 2	2C 2	R 3
7211.458	7213.446			0.2			13863.00			
7211.620	7213.607			0.2	0.1		13862.69			
7211.869	7213.857		0.2	0.2			13862.21			
7212.129	7214.117	G	1.9	1.5	0.2		13861.72	3A 0	2C 0	Q 2
7212.275	7214.263		0.2				13861.43			
7212.463	7214.450			0.1d			13861.07	3d 3	2c 4	R 1
7212.806	7214.794			0.1			13860.41			
7212.952	7214.940			0.2			13860.13			
7213.103	7215.091			0.1			13859.84	4g 1	2a 5	P 4
7213.254	7215.241			0.1			13859.55			
7213.363	7215.351			0.2			13859.34			
7213.597	7215.585			0.0			13858.89			
7213.826	7215.814			0.2			13858.45			
7214.118	7216.106		0.2	0.4			13857.89	3A 2	2C 2	R 4
7214.326	7216.314			0.2			13857.49	X 1	2C 1	P 3
7214.565	7216.554			0.2d			13857.03			
7214.675	7216.663		0.2				13856.82	Y 1	2C 0	P 3
7214.878	7216.866	G	2.1	1.6	1.8		13856.43	2K 6	2B 3	P 4
7215.034	7217.022		0.3	0.6			13856.13	3F 2	2B11	Q 2
7215.221	7217.210		0.1	0.3			13855.77	3F 1	2B 9	R 1
7215.617	7217.606		0.1	0.3			13855.01			
7215.893	7217.882			0.2			13854.48			
7216.117	7218.106		0.2	0.4			13854.05			
7216.424	7218.413			0.2			13853.46	4e 0	2c 4	Q 1
7216.628	7218.617		0.2	0.4			13853.07			
7216.836	7218.825			0.1			13852.67	X 2	2B12	R 3
7217.003	7218.992		0.1	0.2			13852.35	4E 0	2C 2	P 2
7217.294	7219.284	G	1.7	0.8			13851.79	3A 2	2C 2	R 2

λ_{air}	λ_{vac}	Ref.	I_1	I_2	I_3	I_4	ν	Upper	Lower	Br.
7217.586	7219.575			0.2			13851.23	3C 2	2K 2	Q 4
7217.857	7219.846		0.3b	0.6			13850.71	V 0	2B 6	P 4
7218.222	7220.211	G	2.1	1.7	1.4		13850.02	3A 0	2C 0	Q 3
7218.222	7220.211	G	2.1	1.7	1.4		13850.02	3F 2	2B11	P 4
7218.597	7220.587		0.1	0.2			13849.29			
7218.785	7220.774				0.2		13848.93			
7218.863	7220.853		0.1	0.3d			13848.78			
7219.124	7221.113			0.2d			13848.28			
7219.421	7221.411		0.2b	0.4			13847.71	X 0	2C 0	P 4
7219.713	7221.703			0.2			13847.15			
7219.890	7221.880			0.2			13846.81			
7220.078	7222.068			0.3			13846.45			
7220.500	7222.490			0.1			13845.64	4D 0	2B12	R 6
7220.699	7222.689			0.2			13845.26			
7221.121	7223.111			0.2			13844.45	3d 1	2c 2	P 5
7221.392	7223.382			0.2			13843.93			
7221.653	7223.643		0.3	0.5	0.2		13843.43			
7221.867	7223.857			0.2			13843.02	4E 1	2C 3	P 3
7222.086	7224.076		0.1	0.3			13842.60			
7222.279	7224.269			0.2			13842.23			
7222.514	7224.504		0.3	0.8	0.3		13841.78			
7222.597	7224.588						13841.62	4d 0	2c 4	Q 1
7222.644	7224.635	G	1.7	1.3	1.5		13841.53	3d 2	2c 3	P 4
7222.890	7224.880		0.1	0.2			13841.06			
7223.140	7225.131	G	1.8	1.3			13840.58	3A 1	2C 1	Q 1
7223.464	7225.454		0.2	0.4			13839.96			
7223.672	7225.663			0.2			13839.56			
7223.855	7225.846		0.1	0.3d			13839.21			
7224.038	7226.029			0.1			13838.86			
7224.210	7226.201		0.3	1.2	1.5		13838.53	3c 1	2a 2	Q 7
7224.456	7226.446		0.2	0.5			13838.06			
7224.670	7226.661			0.1			13837.65			
7224.878	7226.869		0.1	0.3	0.4		13837.25			
7225.145	7227.136		1.6	1.3	1.2		13836.74	3C 2	2A 2	R 2
7225.145	7227.136		1.6	1.3	1.2		13836.74	X 2	2C 2	P 3
7225.474	7227.465		0.1	0.2			13836.11			
7225.625	7227.616		0.1				13835.82			
7225.771	7227.763			0.1			13835.54			
7226.080	7228.071		0.3	0.4			13834.95			
7226.393	7228.384			0.1			13834.35			
7226.602	7228.593			0.2			13833.95			
7226.858	7228.849			0.2			13833.46			
7227.030	7229.022			0.2			13833.13			
7227.265	7229.257		0.1	0.3			13832.68			
7227.568	7229.560			0.4			13832.10			
7227.882	7229.874	G	2.5	2.2	1.9		13831.50	3C 0	2A 0	R 1
7228.232	7230.224		0.2	0.3			13830.83			
7228.462	7230.454		0.2	0.6			13830.39	T 0	2C 2	R 2
7228.692	7230.684			0.1			13829.95			
7228.844	7230.836			0.1			13829.66			
7228.995	7230.987		0.1	0.2			13829.37			
7229.272	7231.265		0.4	0.4			13828.84			
7229.622	7231.615		0.2	0.3d			13828.17	3D 3	2B12	P 4
7229.811	7231.803			0.2			13827.81			
7230.009	7232.002		0.2	0.2			13827.43			
7230.145	7232.138			0.2			13827.17			
7230.323	7232.316			0.4			13826.83			
7230.412	7232.405		0.4d				13826.66	3A 2	2C 2	R 1
7230.653	7232.645	G	3.1	2.3	1.8		13826.20	2A 3	2B 1	P 1
7231.050	7233.043	G	3.2	2.4	1.8		13825.44	2K 5	2B 2	P 2

λ_{air}	λ_{vac}	Ref.	I_1	I_2	I_3	I_4	ν	Upper	Lower	Br.
7231.050	7233.043	G	3.2	2.4	1.8		13825.44	T 0	2C 2	R 1
7231.050	7233.043	G	3.2	2.4	1.8		13825.44	3c 3	2a 4	R 2
7231.217	7233.210	G			0.2		13825.12	3c 1	2a 2	P 2
7231.448	7233.440		0.2	0.2			13824.68	3E 0	2B 7	R 3
7231.610	7233.603		0.2				13824.37			
7231.704	7233.697			0.1			13824.19			
7231.856	7233.849		0.1	0.0			13823.90			
7232.034	7234.027		0.2	0.2d			13823.56			
7232.353	7234.346		1.7	0.8	0.3		13822.95	3c 0	2a 1	P 4
7232.541	7234.534		0.2				13822.59			
7232.719	7234.712		0.8	0.8	1.3		13822.25	3A 1	2C 1	Q 2
7233.064	7235.058		0.1	0.1			13821.59			
7233.363	7235.356		0.2	0.3			13821.02			
7233.504	7235.497		0.2	0.5	0.2b		13820.75			
7233.734	7235.728		0.1	0.2			13820.31			
7234.043	7236.037			0.2			13819.72			
7234.315	7236.309			0.2			13819.20			
7234.561	7236.555			0.2			13818.73			
7234.776	7236.770		0.1	0.3			13818.32			
7235.080	7237.073			0.3			13817.74	3F 1	2B 9	P 4
7235.080	7237.073			0.3			13817.74	3E 3	2C 2	P 6
7235.509	7237.503		0.1				13816.92			
7235.708	7237.702			0.1d			13816.54			
7235.991	7237.985			0.0			13816.00			
7236.190	7238.184		0.1				13815.62			
7236.399	7238.394			0.2d			13815.22	X 2	2B12	P 2
7236.656	7238.650			0.2d			13814.73			
7236.913	7238.907			0.3	0.2		13814.24	4E 0	2C 0	Q 3
7236.913	7238.907			0.3	0.2		13814.24	3d 1	2c 2	Q 7
7237.075	7239.070			0.4			13813.93			
7237.321	7239.316			0.1			13813.46			
7237.541	7239.536		0.3	0.2			13813.04			
7237.819	7239.814		0.2	0.3			13812.51			
7238.008	7240.003			0.2			13812.15			
7238.244	7240.238		0.2	0.4	0.2		13811.70	3F 0	2C 0	R 6
7238.479	7240.474			0.2			13811.25			
7238.794	7240.789		0.1	0.2			13810.65			
7239.067	7241.061			0.3			13810.13	Z 0	2B 6	P 5
7239.245	7241.240			0.2			13809.79			
7239.512	7241.507		0.2	0.3			13809.28			
7239.811	7241.806		0.2				13808.71			
7239.890	7241.885		0.3	0.3	0.2		13808.56			
7240.136	7242.131	G	1.8	1.7	1.9		13808.09	3C 1	2A 1	R 3
7240.136	7242.131	G	1.8	1.7	1.9		13808.09	4D 0	2C 2	P 1
7240.314	7242.310		0.4				13807.75			
7240.571	7242.567	G	3.3	2.5	1.9		13807.27	3c 2	2a 3	Q 1
7240.875	7242.871		0.3				13806.68	3D 2	2C 1	Q 7
7240.944	7242.939		0.4	0.3			13806.55			
7241.206	7243.201		0.2	0.2			13806.05			
7241.489	7243.485	G	1.9	1.5			13805.51	3A 0	2C 0	P 1
7241.489	7243.485	G	1.9	1.5			13805.51	5c 1	2a 6	P 2
7241.904	7243.899		0.1	0.2			13804.72			
7242.056	7244.051			0.2			13804.43			
7242.349	7244.345			0.2			13803.87			
7242.517	7244.513			0.2			13803.55			
7242.780	7244.776		0.1	0.1			13803.05			
7243.053	7245.049			0.1			13802.53			
7243.199	7245.195		0.2	0.2			13802.25			
7243.488	7245.484	G	2.0	1.5	1.3		13801.70	3A 1	2C 1	Q 3
7243.488	7245.484	G	2.0	1.5	1.3		13801.70	3F 0	2B 7	R 1

λ_{air}	λ_{vac}	Ref.	I_1	I_2	I_3	I_4	ν	Upper	Lower	Br.
7243.835	7245.831		0.2	0.3			13801.04			
7244.086	7246.083	G	2.7	2.1	2.3		13800.56	2A 2	2B 0	R 4
7244.123	7246.120				0.9		13800.49			
7244.459	7246.456		0.2	0.2			13799.85			
7244.606	7246.603						13799.57			
7244.722	7246.718	G	2.5	2.0	1.5		13799.35	3c 2	2a 3	Q 2
7245.094	7247.091			0.2			13798.64			
7245.168	7247.164	1					13798.50			
7245.373	7247.369			0.2			13798.11	3d 3	2c 4	P 1
7245.609	7247.606			0.2			13797.66			
7245.835	7247.832			0.2			13797.23			
7246.055	7248.052		0.2	0.3			13796.81	V 0	2B 6	R 0
7246.208	7248.205		0.2	0.3			13796.52			
7246.423	7248.420		0.2	0.3			13796.11			
7246.617	7248.614			0.2			13795.74	3d 3	2c 4	R 3
7246.828	7248.825		0.1	0.3	0.2		13795.34			
7247.069	7249.066		0.1	0.3			13794.88	3F 1	2B 9	Q 2
7247.216	7249.213			0.2			13794.60			
7247.342	7249.340			0.2			13794.36			
7247.553	7249.550		0.9	0.4			13793.96	3A 1	2C 1	P 1
7247.789	7249.786		0.2	0.3			13793.51			
7247.973	7249.970			0.1			13793.16			
7248.141	7250.138			0.2			13792.84			
7248.441	7250.438			0.2			13792.27			
7248.619	7250.617			0.2			13791.93			
7248.987	7250.985			0.2			13791.23	4E 0	2B12	P 2
7249.203	7251.200			0.1			13790.82			
7249.413	7251.411			0.2			13790.42			
7249.697	7251.695		0.3	0.6	1.2		13789.88			
7249.928	7251.926			0.2			13789.44	4E 0	2B12	P 4
7250.096	7252.094			0.1			13789.12			
7250.270	7252.268		0.2	0.3			13788.79			
7250.580	7252.578		0.2	0.3			13788.20	7c 0	2a 6	P 3
7250.896	7252.894	G	2.5	2.2	2.0		13787.60	3c 2	2a 3	Q 3
7251.253	7253.252		0.1	0.1			13786.92	4E 0	2B12	Q 4
7251.432	7253.430		1.6	0.5			13786.58			
7251.622	7253.620		0.2	0.1			13786.22			
7251.790	7253.788			0.2			13785.90			
7251.885	7253.883		0.2				13785.72	3D 3	2B12	P 5
7252.042	7254.041		0.3	0.5			13785.42	3d 0	2c 1	P 6
7252.258	7254.257		0.2	0.2			13785.01			
7252.469	7254.467		0.7	1.2	1.3		13784.61			
7252.690	7254.688		0.4	0.5			13784.19	X 0	2C 0	Q 5
7252.774	7254.772		0.2				13784.03			
7252.979	7254.978		0.9	0.6			13783.64			
7253.042	7255.041						13783.52			
7253.289	7255.288	G	3.6	2.8	2.3		13783.06	2A 3	2B 1	P 2
7253.289	7255.288	G	3.6	2.8	2.3		13783.06	4D 0	2C 2	Q 1
7253.605	7255.604		0.4				13782.45			
7253.679	7255.678		0.4	0.5			13782.31			
7254.026	7256.025	G	3.0	2.4	1.8		13781.65	3c 3	2a 4	R 1
7254.026	7256.025	G	3.0	2.4	1.8		13781.65	3F 0	2B 7	P 4
7254.347	7256.346		0.3				13781.04			
7254.395	7256.394		0.3	0.3			13780.95			
7254.642	7256.641		0.2	0.2			13780.48			
7254.869	7256.868			0.2			13780.05			
7255.079	7257.078		0.2b	0.3			13779.65			
7255.237	7257.236			0.0			13779.35	3F 2	2B11	P 3
7255.548	7257.547		0.2	0.4			13778.76			
7255.806	7257.805		0.1	0.2			13778.27	U 1	2B 9	R 3

λ_{air}	λ_{vac}	Ref.	I_1	I_2	I_3	I_4	ν	Upper	Lower	Br.
7256.006	7258.005		0.5	0.4			13777.89	3D 3	2C 2	P 4
7256.227	7258.227			0.2			13777.47			
7256.485	7258.485		0.1	0.3			13776.98	V 0	2B 6	P 3
7256.775	7258.775			0.2			13776.43			
7256.891	7258.891			0.1			13776.21			
7257.159	7259.159		0.4	0.6	1.2		13775.70	3A 2	2B12	R 3
7257.375	7259.375			0.1			13775.29			
7257.523	7259.523			0.2			13775.01			
7257.718	7259.718			0.1			13774.64			
7257.897	7259.897			0.1			13774.30			
7258.071	7260.071			0.2			13773.97	5c 1	2a 6	P 3
7258.329	7260.329		0.4	0.9	1.4		13773.48	3b 4	2a 2	P 4
7258.329	7260.329		0.4	0.9	1.4		13773.48	3E 2	2B11	R 4
7258.640	7260.640		0.2	0.4			13772.89			
7259.046	7261.046	G	1.7	1.4	1.5		13772.12	3c 2	2a 3	Q 4
7259.420	7261.421			0.1d			13771.41			
7259.589	7261.589		0.2	0.2			13771.09			
7259.889	7261.890	G	2.1	1.9	1.5		13770.52	3C 0	2A 0	R 0
7260.237	7262.238		0.1	0.1			13769.86			
7260.485	7262.486		0.3	0.3	0.1		13769.39			
7260.707	7262.707			0.2			13768.97			
7260.991	7262.992		0.4	0.8	0.2		13768.43	3B 4	2A 1	P 4
7261.255	7263.256		0.2	0.3	1.2		13767.93			
7261.603	7263.604	G	2.2	2.0	2.2		13767.27	3b 1	2a 0	R 2
7261.883	7263.884			0.2			13766.74			
7262.099	7264.100	G	0.8	1.0			13766.33	3C 1	2A 1	R 2
7262.310	7264.311	G	2.2	2.1	2.8		13765.94	3b 1	2a 0	R 3
7262.595	7264.596		0.1	0.2	0.1		13765.39			
7262.727	7264.728		0.2	0.4			13765.14			
7263.017	7265.018		0.2	0.2			13764.59			
7263.344	7265.346		0.1	0.3			13763.97			
7263.587	7265.589	G	0.2	0.2			13763.51	4g 0	2a 4	R 1
7263.840	7265.842		2.1	1.6	1.3		13763.04	2K 5	2B 2	P 3
7263.840	7265.842		2.1	1.6	1.3		13763.04	4D 0	2C 2	R 1
7264.215	7266.217		0.1				13762.32			
7264.379	7266.380			0.2	0.2		13762.01			
7265.149	7267.151			0.2	0.1		13760.55			
7265.376	7267.378			0.1			13760.12	5c 0	2a 5	R 2
7265.530	7267.532			0.2			13759.83			
7265.709	7267.711			0.2	0.1b		13759.49			
7265.968	7267.970			0.2			13759.00			
7266.137	7268.139		0.1	0.2			13758.68			
7266.338	7268.340		0.2	0.3			13758.30			
7266.633	7268.636	G	2.4	1.9	1.6		13757.74	3C 2	2A 2	R 1
7266.913	7268.916		0.2	0.2			13757.21	3d 3	2c 4	Q 2
7267.161	7269.164	G	2.7	2.1	1.7		13756.74	3c 1	2a 2	P 3
7267.161	7269.164	G	2.7	2.1	1.7		13756.74	T 0	2C 2	Q 1
7267.552	7269.555		0.2	0.3			13756.00	4d 0	2c 4	P 1
7267.732	7269.735		0.1	0.1			13755.66			
7267.928	7269.930		0.3	0.6	0.2b		13755.29			
7268.313	7270.316		0.2	0.3			13754.56			
7268.683	7270.686		0.3	0.4			13753.86			
7268.863	7270.866			0.2			13753.52			
7268.963	7270.966		0.1				13753.33			
7269.101	7271.104	G	0.6	1.5	1.9		13753.07	3c 2	2a 3	Q 5
7269.101	7271.104	G	0.6	1.5	1.9		13753.07	Z 0	2B 6	P 6
7269.354	7271.358		0.3	0.2			13752.59			
7269.608	7271.611		0.4	0.6			13752.11			
7269.735	7271.738		0.5				13751.87			
7269.989	7271.992	G	3.1	2.9	2.7		13751.40	3b 1	2a 0	R 1

λ_{air}	λ_{vac}	Ref.	I_1	I_2	I_3	I_4	ν	Upper	Lower	Br.
7270.280	7272.283		0.3				13750.84			
7270.364	7272.368		0.3	0.4			13750.68			
7270.607	7272.611		0.2	0.2			13750.22			
7270.856	7272.859		0.4	0.5			13749.75	W 1	2C 1	R 1
7270.856	7272.859		0.4	0.5			13749.75	4e 0	2c 4	P 2
7271.036	7273.039		0.2	0.2			13749.41	V 0	2B 6	P 5
7271.036	7273.039		0.2	0.2			13749.41	X 2	2B12	P 3
7271.284	7273.288		0.6	1.3	1.3		13748.94	3c 0	2a 1	P 5
7271.284	7273.288		0.6	1.3	1.3		13748.94	4e 0	2c 4	P 3
7271.591	7273.595		0.2	0.5b			13748.36			
7271.665	7273.669		0.2				13748.22			
7272.025	7274.029		0.4	1.4	2.1		13747.54	3b 1	2a 0	R 4
7272.131	7274.134		0.2				13747.34			
7272.490	7274.494		0.2	0.2d			13746.66			
7272.633	7274.637			0.2			13746.39			
7272.861	7274.865			0.3d			13745.96			
7272.982	7274.986	1		0.1d	1.5		13745.73			
7273.199	7275.203			0.2			13745.32			
7273.326	7275.331		0.3	0.5	0.2		13745.08			
7273.681	7275.685			0.2			13744.41			
7273.898	7275.902		0.2	0.2			13744.00			
7274.067	7276.072			0.2			13743.68			
7274.316	7276.320		0.7	1.3	1.6		13743.21	3C 2	2A 2	R 3
7274.692	7276.696		0.2	0.2			13742.50			
7274.935	7276.940		0.8	0.7	0.1		13742.04	4E 1	2C 3	P 4
7275.174	7277.178			0.3			13741.59			
7275.306	7277.311		0.2				13741.34	4E 0	2B12	Q 1
7275.428	7277.432				0.2		13741.11			
7275.528	7277.533	G	2.7	2.0	1.4		13740.93	3A 0	2C 0	P 2
7275.894	7277.898		0.2	0.3	0.3		13740.23			
7275.989	7277.994		0.3				13740.05			
7276.164	7278.169			0.1			13739.72			
7276.339	7278.344		0.1	0.2			13739.39			
7276.603	7278.608		0.5	0.8	1.2		13738.89	4E 0	2C 2	P 4
7276.863	7278.868			0.2			13738.40			
7276.942	7278.947		0.1				13738.25			
7277.022	7279.027			0.1			13738.10			
7277.234	7279.239		0.2	0.4			13737.70			
7277.578	7279.583		0.1	0.2			13737.05			
7277.695	7279.700			0.1			13736.83			
7277.901	7279.907		0.3	0.6	1.3		13736.44	3F 0	2C 0	R 5
7277.901	7279.907		0.3	0.6	1.3		13736.44	3E 0	2C 2	Q 4
7278.044	7280.050		0.1				13736.17	4e 0	2c 4	Q 3
7278.214	7280.219			0.1			13735.85			
7278.389	7280.394			0.2			13735.52			
7278.484	7280.490		0.1				13735.34			
7278.638	7280.643		0.7	0.9	0.1		13735.05	W 1	2C 1	R 2
7279.088	7281.094			0.2			13734.20			
7279.422	7281.428			0.3			13733.57			
7279.512	7281.518		0.3				13733.40	3F 0	2B 7	Q 2
7279.740	7281.746	G	2.8	1.9	1.9		13732.98	3c 3	2a 4	R 0
7280.069	7282.075	G	2.9	2.3	1.9		13732.35	2A 3	2B 1	P 3
7280.445	7282.451			0.2			13731.64			
7280.652	7282.658			0.2			13731.25			
7280.774	7282.780			0.2	0.2		13731.02			
7280.986	7282.992		0.1	0.8	1.3		13730.62	3c 2	2a 3	Q 6
7281.166	7283.173		0.1	0.2			13730.28			
7281.352	7283.358		1.5b	0.9	1.5b		13729.93			
7281.702	7283.708		0.2	0.2			13729.27			
7281.983	7283.990			0.2	0.2		13728.74			

λ_{air}	λ_{vac}	Ref.	I_1	I_2	I_3	I_4	ν	Upper	Lower	Br.
7282.185	7284.191		0.1	0.2			13728.36			
7282.482	7284.488			0.1			13727.80			
7282.710	7284.717		0.1				13727.37	4f 2	3b 1	R 3
7282.832	7284.839			0.3			13727.14	3D 3	2C 2	Q 5
7283.214	7285.221		0.1	0.2			13726.42	4E 0	2B12	Q 3
7283.426	7285.433			0.2			13726.02			
7283.644	7285.651		0.2	0.3			13725.61			
7283.898	7285.906	G	1.9	1.6	1.8		13725.13	2A 2	2B 0	R 3
7284.291	7286.298			0.2			13724.39			
7284.525	7286.532			0.2			13723.95			
7284.726	7286.734			0.2			13723.57			
7284.933	7286.941			0.2			13723.18			
7285.141	7287.148		0.2	0.6			13722.79			
7285.332	7287.339			0.2			13722.43			
7285.576	7287.583	G	1.8	1.2			13721.97	3A 1	2C 1	P 2
7285.576	7287.583	G	1.8	1.2			13721.97	X 2	2C 2	Q 4
7285.852	7287.859			0.1			13721.45			
7286.043	7288.051		0.2	0.3			13721.09			
7286.314	7288.322	G	1.8	1.7	1.7		13720.58	3C 2	2K 2	Q 3
7286.691	7288.699			0.2			13719.87			
7286.813	7288.821		0.1	0.1			13719.64			
7286.988	7288.996		0.2	0.3	0.1		13719.31			
7287.180	7289.188			0.2			13718.95			
7287.498	7289.506	G	2.7	2.2	1.9		13718.35	3b 1	2a 0	R 0
7287.498	7289.506	G	2.7	2.2	1.9		13718.35	3A 2	2C 2	Q 2
7287.892	7289.900		0.2	0.3			13717.61			
7288.162	7290.171		0.2	0.6	1.2		13717.10	T 0	2C 2	Q 2
7288.418	7290.426		0.1	0.3			13716.62			
7288.646	7290.654		0.3	0.3	0.1		13716.19			
7288.944	7290.952	G	2.8	2.3	1.9		13715.63	3C 1	2A 1	R 1
7288.944	7290.952	G	2.8	2.3	1.9		13715.63	4D 0	2C 2	R 2
7289.220	7291.228			0.2			13715.11			
7289.342	7291.351		0.2	0.3			13714.88			
7289.587	7291.595		0.1	0.2	0.2		13714.42	3F 1	2B 9	P 3
7289.789	7291.797			0.1			13714.04	4c 3	2a 7	R 1
7290.028	7292.037			0.2			13713.59			
7290.198	7292.207			0.2			13713.27			
7290.427	7292.435		0.3	0.6	0.4		13712.84	3F 0	2C 0	R 4
7290.427	7292.435		0.3	0.6	0.4		13712.84	4g 0	2a 4	P 2
7290.427	7292.435		0.3	0.6	0.4		13712.84	U 1	2B 9	R 2
7290.666	7292.675		0.6	1.5	2.6		13712.39	3b 1	2a 0	R 5
7290.666	7292.675		0.6	1.5	2.6		13712.39	4f 3	3b 2	R 3
7291.006	7293.015			0.1	0.2		13711.75	X 1	2B10	R 0
7291.006	7293.015			0.1	0.2		13711.75	X 1	2B10	R 1
7291.331	7293.340		0.2				13711.14			
7291.432	7293.441			0.3	0.3		13710.95	3D 3	2B12	P 6
7291.575	7293.584	G	2.0	1.4	1.4		13710.68	3A 2	2C 2	Q 1
7291.804	7293.813			0.1			13710.25			
7291.979	7293.989			0.2			13709.92			
7292.346	7294.356			0.2			13709.23	X 1	2B10	R 2
7292.538	7294.547			0.1			13708.87	4d 0	2c 4	R 1
7292.926	7294.936		0.3	0.7	1.2		13708.14	X 0	2B 8	R 3
7293.272	7295.282			0.2			13707.49			
7293.410	7295.420			0.1			13707.23			
7293.538	7295.548			0.2			13706.99			
7293.767	7295.777		0.3	0.3			13706.56	V 0	2B 6	P 2
7294.161	7296.171			0.2			13705.82			
7294.363	7296.373			0.2	0.2		13705.44			
7294.602	7296.612		0.4	1.0	1.6		13704.99	3c 2	2a 3	Q 7
7294.602	7296.612		0.4	1.0	1.6		13704.99	T 0	2C 2	P 1

λ_{air}	λ_{vac}	Ref.	I_1	I_2	I_3	I_4	ν	Upper	Lower	Br.
7294.773	7296.783		0.2	0.3			13704.67			
7294.922	7296.932		0.1	0.2	0.1		13704.39	X 0	2B 8	R 4
7295.055	7297.065			0.3			13704.14	3d 2	2c 3	P 5
7295.177	7297.187		0.3	0.4			13703.91	4E 0	2B12	Q 2
7295.177	7297.187		0.3	0.4			13703.91	4E 0	2C 2	P 3
7295.433	7297.443	G	2.9	2.4	2.2		13703.44	3C 0	2A 0	Q 1
7295.816	7297.826		0.3	0.3			13702.71			
7296.045	7298.055		0.2	0.8	0.4		13702.28			
7296.189	7298.199		0.2d	0.1			13702.01			
7296.391	7298.402			0.2			13701.63			
7296.668	7298.679		0.1	0.6			13701.11	W 1	2C 1	R 3
7296.908	7298.918			0.2			13700.66	U 1	2B 9	R 1
7297.100	7299.110			0.2			13700.30			
7297.478	7299.489			0.2d			13699.59			
7297.739	7299.750			0.2			13699.10			
7297.957	7299.968			0.2	0.1		13698.69			
7298.218	7300.229			0.2			13698.20			
7298.421	7300.432		0.1	0.2			13697.82			
7298.591	7300.602			0.2			13697.50			
7298.842	7300.853			0.2			13697.03			
7299.023	7301.034			0.2d			13696.69			
7299.241	7301.253			0.2			13696.28			
7299.412	7301.423			0.3			13695.96			
7299.551	7301.562			0.1			13695.70			
7299.790	7301.802		0.2	0.4b			13695.25	3d 1	2c 2	P 6
7300.062	7302.074	G	2.2	1.9	1.8		13694.74	3C 0	2A 0	Q 2
7300.291	7302.303	G	2.2	1.0			13694.32	3c 2	2a 3	P 2
7300.681	7302.692		0.2	0.2			13693.58			
7300.825	7302.836			0.2			13693.31			
7301.011	7303.023		0.2	0.7	0.2		13692.96			
7301.224	7303.236		0.2	0.2			13692.56			
7301.475	7303.487	G	2.1	1.6	1.3		13692.09	3C 2	2A 2	R 0
7301.774	7303.786			0.2			13691.53			
7301.896	7303.908		0.2	0.3			13691.30			
7302.163	7304.175	G	2.0	1.4	1.3		13690.80	3A 2	2C 2	Q 3
7302.163	7304.175	G	2.0	1.4	1.3		13690.80	3A 2	2B12	R 2
7302.483	7304.495			0.1d			13690.20			
7302.702	7304.714		0.2	0.3			13689.79			
7303.097	7305.109		0.7	0.7			13689.05			
7303.347	7305.360		0.1	0.2			13688.58			
7303.635	7305.648	G	2.3	1.8	1.6		13688.04	2K 5	2B 2	P 4
7304.041	7306.053		0.1	0.2			13687.28			
7304.265	7306.278			0.1			13686.86			
7304.505	7306.518			0.2			13686.41	U 1	2B 9	R 0
7304.708	7306.721		0.2d	0.2			13686.03			
7304.858	7306.870			0.2			13685.75			
7305.146	7307.159		0.1				13685.21			
7305.285	7307.297			0.2			13684.95			
7305.578	7307.591			0.3			13684.40			
7305.765	7307.778			0.2			13684.05			
7305.952	7307.965			0.6			13683.70			
7306.139	7308.152			0.2			13683.35			
7306.352	7308.365		0.1	0.4			13682.95	5c 0	2a 5	R 1
7306.529	7308.542			0.2			13682.62			
7306.753	7308.766	G	1.5		0.6		13682.20	3c 1	2a 2	P 4
7306.945	7308.958	G	2.2	2.2	2.3		13681.84	3C 0	2A 0	Q 3
7307.335	7309.348		0.2	0.2	0.2		13681.11			
7307.490	7309.503		0.2	0.3			13680.82			
7307.602	7309.616				0.4		13680.61			
7307.720	7309.733	G	2.5	1.9	1.6		13680.40	W 1	2C 1	Q 1

λ_{air}	λ_{vac}	Ref.	I$_1$	I$_2$	I$_3$	I$_4$	ν	Upper	Lower	Br.
7308.126	7310.139		0.2	0.2			13679.63			
7308.398	7310.412			0.1			13679.12			
7308.612	7310.625		0.1	0.2			13678.72			
7308.783	7310.797		0.2	0.2			13678.40	3E 0	2B 7	R 2
7308.975	7310.989		0.3	0.3			13678.04			
7309.066	7311.080		0.1				13677.87	X 2	2B12	P 4
7309.280	7311.294		0.6	0.4			13677.47	X 1	2C 1	P 4
7309.606	7311.620	G	3.2	2.4	1.8		13676.87	3c 3	2a 4	Q 1
7309.788	7311.802				0.3		13676.52	X 0	2B 8	R 2
7309.985	7311.999		0.3	0.4			13676.15			
7310.023	7312.037		0.4				13676.08			
7310.301	7312.315	G	2.8	2.2	2.0		13675.57	2A 2	2B 0	R 2
7310.301	7312.315	G	2.8	2.2	2.0		13675.57	Y 1	2C 0	P 4
7310.686	7312.700		0.2	0.3			13674.84			
7310.835	7312.850			0.2			13674.56			
7311.071	7313.085	G	1.9	1.3	1.3		13674.12	3A 0	2C 0	P 3
7311.333	7313.347			0.2			13673.63			
7311.530	7313.545		0.2	0.7	0.3		13673.26	3c 0	2a 1	P 6
7311.712	7313.727	1		0.2	0.5		13672.92			
7311.878	7313.892		0.4	0.6			13672.61	4D 0	2C 2	P 2
7312.049	7314.064			0.2			13672.29			
7312.268	7314.283		0.2	0.4	0.1		13671.88	4D 0	2C 2	R 3
7312.546	7314.561			0.4			13671.36			
7312.648	7314.663		0.3				13671.17			
7312.857	7314.872	G	3.0	2.4	2.4		13670.78	2A 3	2B 1	P 4
7312.857	7314.872	G	3.0	2.4	2.4		13670.78	3E 3	2B 9	R 3
7312.857	7314.872	G	3.0	2.4	2.4		13670.78	V 1	2B 7	R 2
7313.247	7315.262		0.2	0.3			13670.05			
7313.616	7315.631	G	2.4	1.9	1.5		13669.36	3c 3	2a 4	Q 2
7313.616	7315.631	G	2.4	1.9	1.5		13669.36	T 0	2B12	R 2
7313.616	7315.631	G	2.4	1.9	1.5		13669.36	W 1	2C 1	R 4
7313.616	7315.631	G	2.4	1.9	1.5		13669.36	T 0	2C 2	Q 3
7313.895	7315.910		0.1	0.2			13668.84			
7314.135	7316.151		1.7	0.9			13668.39	3A 2	2C 2	P 1
7314.424	7316.440			0.1			13667.85			
7314.590	7316.606		0.2	0.2			13667.54			
7314.665	7316.681				0.1		13667.40			
7315.008	7317.023		0.1	0.2			13666.76			
7315.259	7317.275		0.1	0.2			13666.29	3E 2	2B11	R 3
7315.597	7317.612		0.3	0.8	1.7		13665.66			
7315.789	7317.805			0.2			13665.30			
7315.907	7317.923		0.2				13665.08	4g 0	2a 4	R 2
7316.025	7318.041		0.4	1.3	1.3		13664.86	3C 0	2A 0	Q 4
7316.400	7318.415		0.3	0.7	0.3		13664.16			
7316.625	7318.640		0.2	0.2			13663.74			
7316.908	7318.924	G	2.0	1.6	1.5		13663.21	3C 2	2A 2	Q 2
7316.908	7318.924	G	2.0	1.6	1.5		13663.21	3b 4	2a 2	P 5
7317.112	7319.128				0.2		13662.83			
7317.198	7319.214		0.1	0.2			13662.67			
7317.439	7319.455		0.1	0.3	0.2		13662.22			
7317.610	7319.626			0.2			13661.90			
7317.755	7319.771		0.2	0.3			13661.63			
7318.049	7320.065	G	2.3	1.9	2.0		13661.08	3C 1	2A 1	R 0
7318.049	7320.065	G	2.3	1.9	2.0		13661.08	X 0	2C 0	P 5
7318.280	7320.296		0.9	0.9			13660.65	3b 1	2a 0	R 6
7318.280	7320.296		0.9	0.9			13660.65	2K 2	2B 0	R 1
7318.698	7320.714		0.1	0.2			13659.87	3C 2	2K 2	P 4
7318.901	7320.917		0.2				13659.49			
7318.955	7320.971		0.3	0.7	0.1b		13659.39			
7319.362	7321.378		0.3	0.7			13658.63			

λ_{air}	λ_{vac}	Ref.	I_1	I_2	I_3	I_4	ν	Upper	Lower	Br.
7319.592	7321.609	G	2.5	2.1	1.9		13658.20	3c 3	2a 4	Q 3
7319.968	7321.984		0.1	0.2			13657.50	3F 0	2B 7	P 3
7320.177	7322.193		0.1	0.3			13657.11			
7320.471	7322.488			0.2			13656.56			
7320.648	7322.665			0.2			13656.23			
7320.922	7322.939		0.2d	0.6			13655.72	3E 3	2C 2	P 7
7320.922	7322.939		0.2d	0.6			13655.72	X 0	2B 8	R 1
7321.270	7323.287			0.1			13655.07			
7321.506	7323.523		0.3	0.3			13654.63	4D 0	2C 2	Q 2
7321.656	7323.673		1.8	0.9	0.1		13654.35	2K 2	2B 0	R 0
7321.930	7323.947		0.2	1.2	0.3		13653.84	3F 0	2C 0	R 4
7322.262	7324.279		0.4	1.1	1.5		13653.22			
7322.616	7324.634			0.2d			13652.56			
7322.922	7324.939			0.1			13651.99	4E 1	2C 3	P 5
7323.115	7325.133			0.0			13651.63			
7323.281	7325.299			0.1			13651.32			
7323.469	7325.487			0.2			13650.97			
7323.727	7325.744		0.3	0.7	0.1		13650.49			
7324.183	7326.201			0.2			13649.64			
7324.515	7326.533			0.2			13649.02			
7324.794	7326.812		1.8	1.1	0.1		13648.50			
7325.063	7327.081			0.2			13648.00			
7325.369	7327.387			0.2			13647.43			
7325.589	7327.607			0.1			13647.02	4D 0	2B12	R 4
7325.798	7327.816			0.3			13646.63			
7326.088	7328.106		0.2				13646.09			
7326.152	7328.171		0.3	0.4			13645.97	T 0	2C 2	P 2
7326.346	7328.364		0.2				13645.61			
7326.603	7328.622		0.6	1.1	0.5		13645.13	3c 2	2a 3	Q 9
7326.603	7328.622		0.6	1.1	0.5		13645.13	3A 2	2C 2	Q 4
7326.743	7328.762						13644.87	3A 1	2C 1	P 3
7326.872	7328.891		0.7	0.7			13644.63	U 1	2B 9	P 5
7327.022	7329.041		0.5		1.4		13644.35	3B 4	2A 1	P 5
7327.167	7329.186		0.5	1.6	2.2		13644.08	3C 0	2A 0	Q 5
7327.328	7329.347						13643.78	3c 3	2a 4	Q 4
7327.484	7329.503		1.8	1.1	0.3		13643.49			
7327.651	7329.669		0.2				13643.18			
7327.769	7329.788						13642.96			
7327.892	7329.911		0.4	0.4			13642.73			
7328.155	7330.174	G	3.1	2.4	1.9		13642.24	3C 2	2A 2	Q 1
7328.155	7330.174	G	3.1	2.4	1.9		13642.24	W 1	2C 1	P 1
7328.478	7330.497		0.3				13641.64			
7328.564	7330.583		0.3	0.4	0.2		13641.48			
7328.849	7330.868		0.1	0.2			13640.95			
7329.026	7331.045			0.2			13640.62			
7329.241	7331.260			0.2			13640.22			
7329.509	7331.529			0.3			13639.72			
7329.891	7331.910			0.2			13639.01			
7330.074	7332.093			0.3			13638.67			
7330.224	7332.244			0.2			13638.39			
7330.568	7332.588			0.1			13637.75			
7330.756	7332.776			0.2			13637.40			
7331.095	7333.115			0.3			13636.77			
7331.375	7333.394		1.7	0.4			13636.25			
7331.643	7333.663		1.8	1.2	0.1		13635.76	W 1	2C 1	Q 2
7332.004	7334.024			0.1			13635.08			
7332.251	7334.271			0.2			13634.62	X 2	2B12	P 5
7332.638	7334.658		0.1	0.2			13633.90	4F 0	2B13	R 3
7332.638	7334.658		0.1	0.2			13633.90	3E 1	2B 9	Q 5
7332.810	7334.831			0.1			13633.58			

λ_{air}	λ_{vac}	Ref.	I_1	I_2	I_3	I_4	ν	Upper	Lower	Br.
7332.977	7334.997			0.2			13633.27			
7333.079	7335.100				0.1		13633.08			
7333.208	7335.229		0.2	0.3			13632.84	X 1	2B10	P 2
7333.483	7335.503	G	2.6	2.0	1.5		13632.33	3c 2	2a 3	P 3
7333.865	7335.885		0.2	0.2			13631.62			
7334.075	7336.095			0.1			13631.23	3d 0	2c 1	P 7
7334.295	7336.316			0.2			13630.82			
7334.435	7336.456		0.2	0.3d			13630.56			
7334.580	7336.601		0.1				13630.29			
7334.876	7336.897			0.3			13629.74	X 0	2B 8	R 0
7335.151	7337.172			0.1			13629.23			
7335.474	7337.495		0.2	0.6	1.2		13628.63			
7335.694	7337.715			0.1			13628.22			
7335.867	7337.888			0.3			13627.90			
7336.012	7338.033			0.2			13627.63			
7336.244	7338.265		0.1	0.5	1.2		13627.20	X 2	2C 2	P 4
7336.244	7338.265		0.1	0.5	1.2		13627.20	4c 3	2a 7	Q 1
7336.669	7338.690			0.2			13626.41			
7336.949	7338.970		0.1	0.3	0.2		13625.89			
7337.180	7339.202		0.4	1.1	1.6		13625.46	3c 3	2a 4	Q 5
7337.550	7339.572						13624.77	5c 0	2a 5	Q 4
7338.031	7340.053		0.3	0.7	0.1		13623.88	3C 2	2K 2	R 2
7338.031	7340.053		0.3	0.7	0.1		13623.88	3f 0	2c 2	R 5
7338.317	7340.338			0.2			13623.35			
7338.882	7340.904			0.1			13622.30			
7339.249	7341.271		0.1				13621.62			
7339.378	7341.400			0.2			13621.38			
7339.707	7341.729			0.2			13620.77			
7339.976	7341.998		0.7	1.1	1.3		13620.27	3A 2	2C 2	Q 5
7340.267	7342.289		0.2	0.8	1.5		13619.73	3C 0	2A 0	Q 6
7340.332	7342.354						13619.61			
7340.585	7342.607			0.2			13619.14			
7340.833	7342.855		0.1	0.3	0.1		13618.68			
7341.097	7343.120		0.2	0.7	0.3		13618.19	3F 0	2C 0	R 3
7341.286	7343.308		0.2				13617.84			
7341.399	7343.422		0.1	0.4			13617.63			
7341.717	7343.740		0.8b	0.9	0.2		13617.04			
7342.046	7344.069			0.1			13616.43			
7342.240	7344.263			0.1			13616.07	4d 0	2c 4	R 3
7342.472	7344.495		0.0	0.2			13615.64			
7342.742	7344.765		0.1	0.4			13615.14			
7343.162	7345.185			0.2			13614.36			
7343.405	7345.428		0.1	0.3			13613.91			
7343.589	7345.612			0.1			13613.57			
7343.783	7345.806			0.3			13613.21			
7344.058	7346.081			0.2			13612.70			
7344.355	7346.378		0.1	0.3			13612.15			
7344.662	7346.686	G	1.5b	1.3	0.2		13611.58	3C 0	2A 0	Q 6
7344.662	7346.686	G	1.5b	1.3	0.2		13611.58	5c 0	2a 5	R 0
7345.051	7347.074		0.2	0.3			13610.86			
7345.391	7347.414			0.2			13610.23			
7345.617	7347.641			0.2			13609.81			
7345.877	7347.900			0.5	0.2		13609.33			
7346.098	7348.122			0.2			13608.92			
7346.217	7348.240						13608.70			
7346.346	7348.370			0.2			13608.46			
7346.729	7348.753			0.2			13607.75			
7346.989	7349.013			0.2			13607.27	3A 2	2B12	R 1
7347.145	7349.169			0.2			13606.98			
7347.383	7349.407		0.3	0.5			13606.54			

λ_{air}	λ_{vac}	Ref.	I_1	I_2	I_3	I_4	ν	Upper	Lower	Br.
7347.637	7349.661			0.2			13606.07	T 0	2B12	R 1
7347.842	7349.866		0.1	0.2			13605.69			
7348.058	7350.082		0.3	0.3			13605.29	U 1	2B 9	P 2
7348.058	7350.082		0.3	0.3			13605.29	4d 0	2c 4	Q 2
7348.474	7350.498		0.6	0.8	1.3		13604.52			
7348.636	7350.660		0.2	0.6	0.2		13604.22	3c 3	2a 4	Q 6
7348.636	7350.660		0.2	0.6	0.2		13604.22	3c 1	2a 2	P 5
7348.879	7350.903				0.1		13603.77	3A 2	2B12	P 6
7349.036	7351.060		0.1	0.3			13603.48			
7349.235	7351.260			0.2			13603.11			
7349.414	7351.438		0.3	0.4	0.1		13602.78			
7349.581	7351.606		0.2				13602.47			
7349.760	7351.784		0.9	0.9	0.3		13602.14	2K 5	2B 2	P 5
7350.041	7352.065	G	2.9	2.6	2.3		13601.63	3b 1	2a 0	P 1
7350.306	7352.331						13601.13	3d 3	2c 4	P 3
7350.306	7352.331						13601.13	3f 1	2c 3	R 5
7350.489	7352.514				0.5		13600.79	3A 0	2C 0	P 4
7350.716	7352.741	G	3.1	2.6	2.1		13600.38	3C 1	2A 1	Q 1
7350.878	7352.903						13600.07	3A 1	2B10	R 2
7350.878	7352.903						13600.07	3A 2	2C 2	P 2
7351.116	7353.141		0.3	0.4			13599.63			
7351.332	7353.357		0.2d	0.2			13599.23			
7351.624	7353.649		0.6	0.7			13598.69	3E 2	2B11	Q 6
7352.095	7354.120			0.1			13597.82			
7352.268	7354.293		0.0	0.2			13597.50			
7352.392	7354.417			0.2			13597.27			
7352.543	7354.569		0.1	0.2			13596.99			
7352.716	7354.742		0.2	0.3			13596.67			
7352.971	7354.996		1.0	1.2	1.3		13596.20	2K 2	2B 0	P 1
7352.971	7354.996		1.0	1.2	1.3		13596.20	4D 0	2C 2	R 4
7352.971	7354.996		1.0	1.2	1.3		13596.20	4E 0	2C 2	P 4
7352.971	7354.996		1.0	1.2	1.3		13596.20	5c 0	2a 5	Q 3
7352.971	7354.996		1.0	1.2	1.3		13596.20	3c 0	2a 1	P 7
7353.290	7355.315	i		0.2	0.9		13595.61			
7353.490	7355.516			0.2			13595.24			
7353.717	7355.743	G	2.0	0.3	0.1		13594.82			
7354.036	7356.062	G	1.8	1.9	1.8		13594.23	2A 3	2B 1	P 5
7354.036	7356.062	G	1.8	1.9	1.8		13594.23	4b 0	2a 3	R 4
7354.036	7356.062	G	1.8	1.9	1.8		13594.23	3C 1	2A 1	Q 2
7354.361	7356.387		0.5		1.5b		13593.63			
7354.480	7356.506		1.0	1.6	2.2		13593.41			
7354.891	7356.917			0.2			13592.65			
7355.194	7357.220		0.5	1.1	1.8		13592.09	3b 1	2a 0	R 7
7355.194	7357.220		0.5	1.1	1.8		13592.09	3C 0	2A 0	Q 7
7355.443	7357.469			0.2			13591.63			
7355.622	7357.648		0.1	0.3			13591.30	4D 0	2B12	R 0
7355.914	7357.940	G	1.9	1.8	2.0		13590.76	3c 1	2a 2	P 5
7356.157	7358.184		0.3	0.5	0.4		13590.31			
7356.341	7358.368		0.1				13589.97			
7356.531	7358.557			0.2			13589.62			
7356.796	7358.823		0.8	0.9	1.3		13589.13			
7357.105	7359.131		0.1	0.1			13588.56	4D 0	2B12	P 1
7357.208	7359.234		0.2				13588.37			
7357.354	7359.381		0.2	0.3			13588.10			
7357.587	7359.613		0.7	1.3	1.7		13587.67	3C 2	2A 2	Q 5
7357.782	7359.808		0.2				13587.31			
7358.031	7360.058		0.2b	0.3			13586.85			
7358.345	7360.372			0.2			13586.27			
7358.637	7360.664			0.2			13585.73			
7358.838	7360.865			0.1			13585.36			

λ_{air}	λ_{vac}	Ref.	I_1	I_2	I_3	I_4	ν	Upper	Lower	Br.
7359.038	7361.065			0.2			13584.99			
7359.569	7361.596		0.1b	0.3			13584.01	4D 0	2B12	R 3
7359.715	7361.743			0.3			13583.74			
7359.851	7361.878			0.2			13583.49			
7360.051	7362.079			0.4			13583.12			
7360.344	7362.371	G	2.1	2.1	2.2		13582.58	3C 1	2A 1	Q 3
7360.734	7362.762		0.1	0.4			13581.86			
7360.962	7362.989			0.2			13581.44			
7361.113	7363.141		0.2	0.4			13581.16			
7361.341	7363.369	G	1.9	1.6	1.4		13580.75	3C 0	2A 0	P 2
7361.721	7363.749		0.3	0.8	1.3		13580.04	3c 3	2a 4	Q 7
7361.997	7364.025			0.1			13579.53	4g 0	2a 4	R 3
7362.170	7364.199		0.0	0.2			13579.21	3E 1	2B 9	R 2
7362.333	7364.361			0.1			13578.91			
7362.528	7364.557			0.2			13578.55	X 0	2B 8	P 1
7362.734	7364.763		0.3	0.3			13578.17	X 1	2B10	P 3
7362.903	7364.931			0.3			13577.86			
7363.065	7365.094			0.2			13577.56			
7363.288	7365.316		0.9	0.7			13577.15			
7363.580	7365.609		0.1	0.3			13576.61			
7363.814	7365.842				0.1		13576.18			
7363.911	7365.940		0.5	0.7			13576.00			
7363.982	7366.010		0.3				13575.87			
7364.210	7366.238		0.2	0.6	0.2		13575.45			
7364.492	7366.520			0.2			13574.93			
7364.752	7366.781			0.2d			13574.45	5c 0	2a 5	Q 2
7365.034	7367.063			0.2d			13573.93	3F 1	2C 1	R 5
7365.132	7367.161		0.1				13573.75			
7365.344	7367.372	G	2.0	1.6	1.4		13573.36	3C 2	2K 2	P 3
7365.604	7367.633			0.0			13572.88			
7365.745	7367.774		0.1	0.2			13572.62			
7365.989	7368.019			0.2			13572.17			
7366.277	7368.306		0.3	0.8	1.3		13571.64			
7366.445	7368.475		0.2	0.1			13571.33			
7366.662	7368.692	G	2.0	1.3	0.1		13570.94	3c 3	2a 4	P 2
7366.825	7368.855						13570.63			
7366.966	7368.996		0.5	1.3	1.7		13570.37	3C 1	2A 1	Q 4
7367.298	7369.327			0.1			13569.76			
7367.531	7369.561		0.1	0.1			13569.33	3E 0	2B 7	P 5
7367.672	7369.702			0.2			13569.07			
7367.846	7369.876		0.1	0.2			13568.75			
7368.085	7370.115	G	2.1	1.6	1.4		13568.31	3c 2	2a 3	P 4
7368.085	7370.115	G	2.1	1.6	1.4		13568.31	3A 1	2C 1	P 4
7368.443	7370.473		0.1	0.1			13567.65			
7368.666	7370.696		0.2	0.2			13567.24			
7368.916	7370.946	G	1.9	1.4	1.3		13566.78	W 1	2C 1	Q 3
7369.242	7371.272		0.1	0.2	0.1		13566.18			
7369.388	7371.419			0.1			13565.91			
7369.589	7371.620			0.2			13565.54			
7369.921	7371.951	G	1.8	1.5	1.6		13564.93	3F 0	2C 0	R 3
7370.258	7372.288			0.1			13564.31	T 0	2B12	R 0
7370.502	7372.533			0.2			13563.86	3E 2	2B11	R 2
7370.730	7372.761			0.1			13563.44			
7371.095	7373.125		0.2	0.2			13562.77			
7371.328	7373.359		0.3	0.3			13562.34			
7371.638	7373.669	G	2.9	2.1	1.6		13561.77	W 1	2C 1	P 2
7371.752	7373.783						13561.56	3C 2	2A 2	P 2
7372.024	7374.055		0.2				13561.06			
7372.122	7374.153	1		0.1	1.0		13560.88			
7372.345	7374.376		0.0	0.1			13560.47			

λ_{air}	λ_{vac}	Ref.	I_1	I_2	I_3	I_4	ν	Upper	Lower	Br.
7372.584	7374.615		0.1	0.2			13560.03			
7372.850	7374.881		1.5	1.0	0.1b		13559.54	3A 2	2C 2	P 3
7372.850	7374.881		1.5	1.0	0.1b		13559.54	5c 0	2a 5	Q 1
7373.046	7375.077			0.2			13559.18	U 1	2B 9	P 3
7373.253	7375.284		0.1	0.2			13558.80	3c 4	2a 5	Q 2
7373.503	7375.534		0.4	1.1	1.4		13558.34	3C 2	2A 2	Q 4
7373.503	7375.534		0.4	1.1	1.4		13558.34	T 0	2C 2	P 3
7373.911	7375.942			0.1d			13557.59	.		
7374.080	7376.111		0.1				13557.28	3A 2	2B12	P 5
7374.237	7376.269			0.2			13556.99			
7374.417	7376.448		0.3	0.2			13556.66			
7374.634	7376.666		0.4	0.3			13556.26			
7374.896	7376.927	G	3.0	2.2	2.1		13555.79	3c 4	2a 5	Q 1
7375.157	7377.188		0.4	0.5	1.0		13555.30	3C 1	2A 1	Q 5
7375.265	7377.297		0.3				13555.10			
7375.548	7377.580		0.2	0.3			13554.58	4D 0	2C 2	Q 3
7375.548	7377.580		0.2	0.3			13554.58	4D 0	2B12	R 2
7375.946	7377.977		0.0				13553.85			
7376.359	7378.391		0.2	0.6	0.1		13553.09	3c 3	2a 4	Q 8
7376.620	7378.652				0.2		13552.61			
7376.751	7378.783			0.2			13552.37			
7377.089	7379.121		1.7	1.0	0.1		13551.75			
7377.317	7379.349			0.1			13551.33			
7377.519	7379.551		0.0	0.3			13550.96	4b 0	2a 3	P 3
7377.704	7379.736			0.1			13550.62			
7377.900	7379.932		0.0	0.3d			13550.26			
7378.063	7380.096			0.1			13549.96			
7378.188	7380.221			0.2			13549.73			
7378.363	7380.395		0.1	0.4			13549.41			
7378.728	7380.760	G	1.9	1.5	1.3		13548.74	3c 4	2a 5	Q 2
7379.060	7381.092			0.2			13548.13			
7379.332	7381.365		0.6	0.9	0.3		13547.63			
7379.822	7381.855			0.3			13546.73	3E 0	2B 7	R 1
7380.149	7382.182		0.1	0.5	0.2		13546.13			
7380.432	7382.466			0.2			13545.61			
7380.700	7382.733			0.2			13545.12	X 0	2B 8	P 2
7380.847	7382.880			0.1			13544.85	3E 0	2B 7	Q 5
7381.059	7383.092		0.8	0.7			13544.46			
7381.261	7383.294			0.1			13544.09	V 1	2B 7	R 1
7381.501	7383.534			0.2	0.2		13543.65	4D 0	2B12	R 1
7381.642	7383.676			0.1			13543.39	4f 2	3b 1	R 2
7381.871	7383.905		0.4	0.7	0.2		13542.97			
7382.215	7384.248		0.1	0.3	0.2		13542.34	4b 2	2a 5	R 1
7382.465	7384.499			0.1	0.1		13541.88			
7382.722	7384.755						13541.41			
7382.880	7384.914			0.1	0.1		13541.12			
7383.114	7385.148			0.1			13540.69			
7383.354	7385.388		0.0		0.4		13540.25			
7383.414	7385.448			0.3	0.3		13540.14			
7383.774	7385.808		0.9	0.9			13539.48			
7383.943	7385.977	1			1.5b		13539.17			
7383.998	7386.032			0.2			13539.07			
7384.123	7386.157			0.2			13538.84			
7384.232	7386.266		0.2				13538.64	3c 4	2a 5	Q 3
7384.287	7386.321	1			1.0		13538.54			
7384.374	7386.408						13538.38			
7384.456	7386.490	G	2.3	1.9	1.8		13538.23	2K 2	2B 0	P 2
7384.538	7386.572				0.4		13538.08			
7384.810	7386.845		0.1	0.3	0.2		13537.58	4f 3	3b 2	R 2
7385.089	7387.123		0.1	0.2	0.2		13537.07			

λ_{air}	λ_{vac}	Ref.	I_1	I_2	I_3	I_4	ν	Upper	Lower	Br.
7385.269	7387.303			0.1			13536.74	3E 1	2B 9	P 5
7385.443	7387.478		0.0	0.2	0.1		13536.42			
7385.711	7387.745				0.1		13535.93			
7385.803	7387.838			0.2			13535.76			
7386.082	7388.116			0.2			13535.25			
7386.398	7388.433		0.3	0.8	1.5		13534.67			
7386.677	7388.711			0.1			13534.16			
7386.928	7388.962			0.3	0.2		13533.70			
7387.157	7389.192			0.2			13533.28			
7387.375	7389.410			0.3			13532.88			
7387.719	7389.754		0.1	0.3	0.1		13532.25			
7387.970	7390.005			0.6	0.2		13531.79			
7388.085	7390.120			0.3			13531.58			
7388.254	7390.289			0.1			13531.27			
7388.478	7390.513		0.2	0.5			13530.86			
7388.680	7390.715			0.3			13530.49			
7388.915	7390.950			0.1			13530.06	U 1	2B 9	P 4
7389.521	7391.557			0.2d			13528.95			
7389.794	7391.830		0.3	1.0	1.5		13528.45	3C 0	2A 0	Q 9
7389.996	7392.032		0.1	0.6			13528.08	X 1	2B10	P 4
7390.204	7392.240		0.4	0.4			13527.70	W 2	2C 2	R 1
7390.401	7392.436		0.0	0.3			13527.34	4b 1	2a 4	R 2
7390.739	7392.775		0.7	0.7	0.2		13526.72			
7390.947	7392.983			0.1			13526.34			
7391.275	7393.311		0.0	0.2			13525.74			
7391.362	7393.398				0.3		13525.58			
7391.532	7393.568			0.4	0.3		13525.27			
7391.685	7393.721			0.2			13524.99			
7391.925	7393.961		0.5	1.1	1.4		13524.55	3c 4	2a 5	Q 4
7392.226	7394.262		0.0	0.3			13524.00			
7392.395	7394.432		0.2	0.8	0.2b		13523.69	3c 3	2a 4	Q 9
7392.587	7394.623			0.1			13523.34			
7392.751	7394.787				0.1		13523.04			
7392.811	7394.847		0.4	0.7			13522.93			
7392.969	7395.006		0.3		0.4		13522.64			
7393.117	7395.153			0.0d			13522.37			
7393.330	7395.367			0.2			13521.98			
7393.549	7395.585			0.1			13521.58			
7393.773	7395.810			0.2			13521.17	6c 1	2a 7	Q 1
7394.036	7396.072		0.9	1.2	1.2		13520.69			
7394.107	7396.143		0.5				13520.56	W 1	2B10	R 4
7394.451	7396.488		0.2	0.4	0.2		13519.93			
7394.506	7396.543		0.2				13519.83			
7394.725	7396.761		0.5	0.6			13519.43			
7395.042	7397.079	G	3.1	3.1	2.0		13518.85	3b 1	2a 0	P 2
7395.324	7397.361						13518.33	3A 2	2B12	R 0
7395.452	7397.489		0.5	0.4	0.4		13518.10	4E 2	2B17	Q 2
7395.452	7397.489		0.5	0.4	0.4		13518.10	3c 0	2a 1	P 8
7395.808	7397.845		0.2	0.3			13517.45			
7396.081	7398.119	G	2.3	2.2	1.9		13516.95	3C 0	2A 0	P 3
7396.475	7398.513		0.2	0.2			13516.23	X 0	2B 8	P 3
7396.722	7398.759		0.2	0.2			13515.78			
7396.968	7399.005	G	1.9	1.7	1.8		13515.33	3C 2	2A 2	Q 3
7397.357	7399.394		0.1	0.2			13514.62			
7397.559	7399.597		0.1	0.2			13514.25			
7397.866	7399.903		0.2	0.3			13513.69			
7398.194	7400.232	G	2.5	1.9	1.5		13513.09	3c 3	2a 4	P 3
7398.194	7400.232	G	2.5	1.9	1.5		13513.09	3E 2	2B11	P 6
7398.451	7400.489			0.1			13512.62			
7398.572	7400.610		0.2	0.2			13512.40			

λ_{air}	λ_{vac}	Ref.	I_1	I_2	I_3	I_4	ν	Upper	Lower	Br.
7398.742	7400.780			0.2			13512.09			
7398.955	7400.993		0.1	0.4	0.2		13511.70	4D 0	2B12	P 2
7399.212	7401.251			0.6b			13511.23			
7399.262	7401.300		0.3	1.0	1.5		13511.14			
7399.569	7401.607			0.1	0.1		13510.58			
7400.297	7402.335		0.1	0.4			13509.25			
7400.571	7402.609		0.5	1.0	1.4		13508.75	2K 5	2B 2	P 6
7400.724	7402.763		0.2				13508.47	4c 2	2a 6	R 1
7400.894	7402.933			0.2			13508.16			
7401.135	7403.174		0.3	1.0	1.5		13507.72	3c 4	2a 5	Q 5
7401.535	7403.574			0.2			13506.99			
7401.738	7403.777			0.1			13506.62			
7401.990	7404.029		0.1	0.2			13506.16			
7402.357	7404.396		1.5	1.2	0.2		13505.49	3C 2	2K 2	R 1
7402.566	7404.605		0.2				13505.11			
7402.703	7404.742			0.3			13504.86	4b 0	2a 3	R 2
7402.873	7404.912			0.2			13504.55	Y 1	2B 8	R 2
7403.042	7405.082			0.0			13504.24	3f 1	2c 3	R 5
7403.212	7405.252			0.2			13503.93			
7403.514	7405.553		0.2b	0.4	0.0		13503.38			
7403.854	7405.893		0.1	0.2			13502.76			
7404.161	7406.200	G	1.8	1.5	1.6		13502.20	3c 2	2a 3	P 5
7404.539	7406.579			0.2			13501.51	Y 1	2B 8	R 1
7404.775	7406.815		0.3	0.4			13501.08			
7404.995	7407.034		0.1	0.5			13500.68			
7405.269	7407.309		0.5	0.9	0.3		13500.18	3F 1	2C 1	R 4
7405.549	7407.588		0.1				13499.67			
7405.806	7407.846		0.1	0.2	0.2		13499.20	3A 2	2B12	P 4
7406.053	7408.093	G	1.8	1.7	1.7		13498.75	3F 0	2C 0	R 2
7406.053	7408.093	G	1.8	1.7	1.7		13498.75	4E 0	2C 2	P 5
7406.053	7408.093	G	1.8	1.7	1.7		13498.75	3A 1	2B10	R 1
7406.191	7408.231						13498.50			
7406.306	7408.346			0.2	0.3		13498.29			
7406.465	7408.505		1.5	1.1	1.3		13498.00			
7406.728	7408.768		0.2	0.6	0.3		13497.52			
7407.052	7409.092			0.2			13496.93			
7407.250	7409.290			0.2			13496.57			
7407.420	7409.460			0.1			13496.26	3c 1	2a 2	P 6
7407.722	7409.762		0.3	0.2			13495.71	4D 0	2C 2	P 3
7407.985	7410.026		1.7	1.5	1.9		13495.23	4g 0	2a 4	P 3
7408.364	7410.404			0.1			13494.54			
7408.589	7410.630			0.2			13494.13			
7408.885	7410.926		0.7	0.7			13493.59	W 2	2C 2	R 2
7409.089	7411.129		0.1	0.4	0.4		13493.22			
7409.303	7411.344			0.1			13492.83			
7409.473	7411.514		0.1	0.2			13492.52			
7409.709	7411.750		0.3	0.7	0.3		13492.09			
7409.973	7412.014		0.4	0.1			13491.61			
7410.176	7412.217		0.1	0.1			13491.24			
7410.423	7412.464			0.2			13490.79			
7410.648	7412.690			0.2			13490.38			
7410.940	7412.981		0.2	0.6	1.3		13489.85			
7411.220	7413.261			0.2			13489.34			
7411.456	7413.497		0.2	0.4	0.2		13488.91			
7411.841	7413.882		0.5	1.1	1.5		13488.21	3E 1	2B 9	R 1
7411.841	7413.882		0.5	1.1	1.5		13488.21	Y 1	2B 8	R 0
7412.000	7414.042			0.3			13487.92			
7412.330	7414.371			0.1	0.3		13487.32			
7412.627	7414.668			0.1			13486.78			
7413.110	7415.152		0.2	0.2			13485.90	T 0	2B12	P 1

λ_{air}	λ_{vac}	Ref.	I_1	I_2	I_3	I_4	ν	Upper	Lower	Br.
7413.424	7415.465		0.1	0.3			13485.33	T 0	2B12	P 2
7413.748	7415.790		0.2	0.2			13484.74			
7414.056	7416.098	G	2.0	1.6	1.4		13484.18	3C 1	2A 1	P 2
7414.380	7416.422	G	0.3	0.6	1.4		13483.59			
7414.578	7416.620			0.1			13483.23			
7414.738	7416.780		0.3	0.6			13482.94			
7414.925	7416.967				0.1		13482.60	3c 1	2a 2	P 7
7415.106	7417.149		0.1	0.3			13482.27	X 0	2B 8	P 4
7415.387	7417.429			0.2			13481.76			
7415.689	7417.732			0.2			13481.21	w 0	3b 0	R 0
7415.981	7418.023			0.2	0.1		13480.68			
7416.234	7418.276		0.2	0.4			13480.22	W 1	2C 1	Q 4
7416.393	7418.436		0.2	0.4			13479.93			
7416.718	7418.761			0.3	0.1		13479.34	3E 2	2B11	Q 5
7416.856	7418.898		0.5	0.4			13479.09			
7417.120	7419.163		2.0	1.4	0.2		13478.61	W 2	2C 2	Q 1
7417.120	7419.163		2.0	1.4	0.2		13478.61	4b 1	2a 4	R 1
7417.472	7419.515		0.1	0.2			13477.97	3E 0	2B 7	P 4
7417.758	7419.801		0.2	0.2			13477.45	T 0	2B12	P 4
7418.039	7420.082			0.3	0.2		13476.94			
7418.198	7420.242		0.1				13476.65			
7418.452	7420.495		0.2	0.3			13476.19			
7418.964	7421.007			0.1			13475.26			
7419.250	7421.293		0.2	0.5	0.2		13474.74			
7419.492	7421.536			0.2			13474.30	3f 0	2c 2	R 4
7419.657	7421.701			0.1			13474.00			
7419.894	7421.938			0.3			13473.57			
7420.236	7422.279			0.2			13472.95	3f 1	2c 3	R 4
7420.555	7422.599		0.3	0.5	0.1		13472.37			
7420.935	7422.979			0.2			13471.68	3A 2	2B12	P 3
7421.216	7423.260			0.9	1.4	1.5	13471.17	3F 0	2C 0	R 2
7421.420	7423.464		0.2	0.2			13470.80	w 0	3b 0	P 1
7421.679	7423.723	G	2.6	2.1	2.0		13470.33	2K 2	2B 0	R 2
7421.679	7423.723	G	2.6	2.1	2.0		13470.33	V 1	2B 7	P 4
7421.679	7423.723	G	2.6	2.1	2.0		13470.33	T 0	2B12	P 3
7421.954	7423.999			0.2			13469.83			
7422.147	7424.192		0.2b	0.3	0.1		13469.48			
7422.505	7424.550		0.2	0.4			13468.83			
7422.798	7424.842	G	1.9	1.4	1.5		13468.30	2K 2	2B 0	P 3
7423.101	7425.145		0.2	0.2d			13467.75			
7423.294	7425.338			0.1d			13467.40			
7423.481	7425.526		0.1	0.3	0.2		13467.06			
7423.580	7425.625		0.1				13466.88	X 0	2B 8	P 5
7423.751	7425.796			0.1			13466.57			
7424.005	7426.049			0.2d			13466.11			
7424.236	7426.281		0.3				13465.69			
7424.369	7426.413		0.2	0.6	1.3		13465.45			
7424.760	7426.805			0.2			13464.74			
7425.074	7427.120			0.5	0.2		13464.17	3E 2	2B11	R 1
7425.284	7427.329			0.1	0.3		13463.79			
7425.620	7427.666			0.1			13463.18	3f 0	2c 2	R 5
7425.852	7427.897			0.4	1.3		13462.76			
7426.222	7428.267			0.2			13462.09			
7426.492	7428.537		0.1	0.4	0.1b		13461.60			
7426.702	7428.747			0.2			13461.22			
7426.989	7429.034			0.1			13460.70			
7427.342	7429.387			0.1			13460.06			
7427.645	7429.691			0.2			13459.51			
7427.915	7429.961			0.2			13459.02			
7428.109	7430.155		0.1	0.2			13458.67	5c 0	2a 5	P 2

λ_{air}	λ_{vac}	Ref.	I_1	I_2	I_3	I_4	ν	Upper	Lower	Br.
7428.379	7430.425	G	1.9	1.4	1.3		13458.18	W 1	2C 1	P 3
7428.628	7430.674			0.1			13457.73			
7428.793	7430.839		0.3	0.6	0.1		13457.43			
7429.047	7431.093		0.1	0.2			13456.97			
7429.262	7431.309		0.6	0.9	1.3		13456.58	3E 0	2B 7	Q 4
7429.262	7431.309		0.6	0.9	1.3		13456.58	4b 0	2a 3	R 1
7429.665	7431.712		0.1b	0.5	0.1		13455.85			
7429.969	7432.016		0.1				13455.30			
7430.063	7432.109		0.2	0.2			13455.13			
7430.383	7432.430	G	2.0	1.4	0.2		13454.55	3F 1	2C 1	R 3
7430.770	7432.817		0.1	0.2			13453.85	4D 0	2C 2	Q 4
7431.107	7433.154		0.7	0.5			13453.24			
7431.383	7433.430	G	2.1	1.7	1.6		13452.74	3c 3	2a 4	P 4
7431.383	7433.430	G	2.1	1.7	1.6		13452.74	W 2	2C 2	P 1
7431.532	7433.579		1.5	1.5b			13452.47			
7431.820	7433.866		0.3	0.3			13451.95			
7431.947	7433.993		0.2				13451.72	3C 0	2A 0	P 4
7432.355	7434.402			0.2d			13450.98			
7432.682	7434.729		0.9	1.2	1.3		13450.39	3C 2	2K 2	Q 2
7433.068	7435.116				0.1		13449.69	4d 0	2c 4	P 3
7433.389	7435.436		0.2	0.2			13449.11	3A 2	2B12	P 1
7433.654	7435.702		1.9	1.4	1.4		13448.63	3A 2	2C 2	P 4
7434.107	7436.155	G	2.8	2.2	1.8		13447.81	2A 2	2B 0	R 1
7434.212	7436.260						13447.62			
7434.395	7436.443		0.2	0.3			13447.29			
7434.500	7436.548		0.2				13447.10			
7434.666	7436.714			0.1			13446.80	3E 1	2B 9	Q 4
7434.843	7436.890		0.3	0.3	0.4		13446.48	t 1	2a 3	R 1
7435.053	7437.101	G	2.5	2.0	1.5		13446.10	3c 5	2a 6	Q 1
7435.103	7437.150						13446.01			
7435.352	7437.399		0.2		0.5		13445.56			
7435.440	7437.488		0.2	0.3			13445.40			
7435.667	7437.715			0.2			13444.99	3E 2	2B11	P 5
7435.932	7437.980		0.7	0.7	0.2		13444.51	W 2	2C 2	R 3
7436.264	7438.312			0.2	0.3		13443.91			
7436.480	7438.528		0.3b	0.4			13443.52			
7436.745	7438.794			0.3			13443.04	Y 1	2B 8	P 1
7437.050	7439.098		0.3	0.5	0.3		13442.49			
7437.221	7439.269		0.2	0.7			13442.18			
7437.481	7439.530		0.6	0.9	1.3		13441.71			
7437.835	7439.884		0.4	0.9	0.1		13441.07			
7438.162	7440.210		0.1	0.6			13440.48			
7438.367	7440.415		0.2	0.3			13440.11			
7438.632	7440.681	G	1.8	1.4	1.3		13439.63	3c 5	2a 6	Q 2
7438.914	7440.963			0.4	0.2		13439.12	3A 2	2B12	P 2
7439.429	7441.478			0.4			13438.19			
7439.590	7441.639		0.2		0.5		13437.90			
7439.651	7441.700			0.4			13437.79			
7440.016	7442.065			0.2b			13437.13			
7440.315	7442.364		0.7	1.0	1.3		13436.59			
7440.515	7442.564			0.1			13436.23			
7440.736	7442.785		0.7	1.1	0.2		13435.83			
7441.024	7443.073		0.2	0.3			13435.31			
7441.246	7443.295		0.2	0.3	0.1		13434.91			
7441.428	7443.478			0.1			13434.58			
7441.633	7443.683		0.3	0.9	0.3		13434.21	3c 2	2a 3	P 6
7441.711	7443.760		0.2				13434.07			
7441.955	7444.004			0.2			13433.63			
7442.221	7444.270		0.4	0.8	0.2		13433.15			
7442.426	7444.475		0.2	0.4			13432.78			

λ_{air}	λ_{vac}	Ref.	I_1	I_2	I_3	I_4	ν	Upper	Lower	Br.
7442.620	7444.669			1.6			13432.43	3F 1	2C 1	R 3
7442.791	7444.841			0.2			13432.12	2K 2	2B 0	R 3
7443.030	7445.079			0.1	0.1		13431.69			
7443.312	7445.362			0.2			13431.18	4F 0	2B13	P 5
7443.606	7445.656			0.3			13430.65	3A 1	2B10	R 0
7443.767	7445.817		0.2				13430.36			
7443.905	7445.955	G	1.9	1.6	1.6		13430.11	3c 5	2a 6	Q 3
7443.905	7445.955	G	1.9	1.6	1.6		13430.11	4c 2	2a 6	Q 3
7444.243	7446.294		0.4	0.6			13429.50	W 2	2B12	R 4
7444.243	7446.294		0.4	0.6			13429.50	5c 0	2a 5	P 3
7444.243	7446.294		0.4	0.6			13429.50	3A 0	2B 8	R 2
7444.532	7446.582		0.1	0.2			13428.98	3E 0	2B 7	R 0
7444.759	7446.809			0.1			13428.57			
7444.870	7446.920			0.2			13428.37			
7445.058	7447.109			0.2			13428.03			
7445.252	7447.303		0.1	0.3			13427.68			
7445.507	7447.558		1.0	1.4	1.7		13427.22	3F 0	2C 0	Q 5
7445.507	7447.558		1.0	1.4	1.7		13427.22	T 0	2C 2	P 4
7445.507	7447.558		1.0	1.4	1.7		13427.22	4b 1	2a 4	R 0
7445.785	7447.835			0.3			13426.72			
7445.918	7447.968			0.0			13426.48			
7446.073	7448.124		0.1	0.2			13426.20			
7446.428	7448.479		0.3b	0.8	0.3		13425.56			
7446.661	7448.712		0.1				13425.14			
7446.833	7448.884		0.1	0.2			13424.83			
7447.110	7449.161		0.9	1.1	1.3		13424.33			
7447.427	7449.478		0.2	0.2			13423.76			
7447.676	7449.727	G	1.8	1.7	1.8		13423.31	3F 1	2C 1	R 3
7448.020	7450.071		0.4	0.5			13422.69			
7448.292	7450.343		0.2	0.3			13422.20			
7448.508	7450.560		0.3	0.5			13421.81	W 1	2B10	R 3
7448.636	7450.688		0.2	0.3			13421.58			
7448.836	7450.887		1.0	0.9			13421.22			
7449.152	7451.204	G	3.2	2.9	2.8		13420.65	3b 1	2a 0	P 3
7449.374	7451.426	G	1.5b	1.5b			13420.25	3C 1	2A 1	P 3
7449.813	7451.865		1.0	1.2	1.3		13419.46	W 2	2C 2	Q 2
7449.813	7451.865		1.0	1.2	1.3		13419.46	w 0	3b 0	R 1
7449.813	7451.865		1.0	1.2	1.3		13419.46	V 1	2B 7	R 0
7450.085	7452.137		0.7	1.1	0.3		13418.97	3A 1	2B10	P 4
7450.202	7452.253		0.3				13418.76			
7450.468	7452.520			0.2			13418.28			
7450.635	7452.687		0.1	0.2			13417.98			
7450.912	7452.964		0.3	1.0	1.3		13417.48	3c 5	2a 6	Q 4
7451.345	7453.398			0.3			13416.70			
7451.590	7453.642			0.2			13416.26	4c 2	2a 6	Q 2
7451.895	7453.948		0.3	0.5	0.2		13415.71			
7452.029	7454.081		0.3	0.3			13415.47			
7452.184	7454.237		0.2	0.2			13415.19			
7452.429	7454.481		0.2	0.4	0.1		13414.75			
7452.751	7454.803		1.0	1.5	1.7		13414.17	2A 4	2B 3	R 2
7453.045	7455.098		0.1	0.1			13413.64			
7453.156	7455.209			0.2	0.1		13413.44			
7453.323	7455.376		1.0	1.1	0.1		13413.14	3C 2	2K 2	R 0
7453.551	7455.604		0.7	1.1			13412.73	2A 4	2B 3	R 1
7453.768	7455.820		0.1	0.2			13412.34			
7454.040	7456.093		0.7	1.1	1.4		13411.85			
7454.418	7456.471		0.3	0.3			13411.17			
7454.685	7456.738		0.2	1.9	0.2		13410.69			
7455.085	7457.138		0.2	0.3			13409.97			
7455.413	7457.466		0.4	0.4			13409.38			

λ_{air}	λ_{vac}	Ref.	I_1	I_2	I_3	I_4	ν	Upper	Lower	Br.
7455.713	7457.767	G	2.9	2.3	2.2		13408.84	2A 2	2B 0	P 4
7456.052	7458.106		0.7	0.8	0.6		13408.23	5c 0	2a 5	P 4
7456.336	7458.390		0.1	0.2			13407.72	4D 0	2B12	P 3
7456.636	7458.690		0.8	1.1	0.2		13407.18	4c 2	2a 6	Q 1
7456.787	7458.840		0.8	0.4			13406.91	4b 0	2a 3	R 0
7457.170	7459.224			0.2			13406.22			
7457.426	7459.480		0.2	0.2			13405.76			
7457.582	7459.636			0.1			13405.48			
7457.732	7459.786		0.2	0.3			13405.21			
7457.999	7460.053		0.2	0.2			13404.73			
7458.277	7460.331		0.7	1.2	1.7		13404.23	3f 1	2c 3	R 4
7458.444	7460.498		0.2				13403.93			
7458.533	7460.588			0.2			13403.77	Y 1	2B 8	P 2
7458.801	7460.855		0.4	0.4			13403.29	3E 0	2B 7	P 3
7458.956	7461.011		0.3	0.2			13403.01			
7459.129	7461.183			0.7			13402.70	2A 4	2B 3	R 3
7459.452	7461.506	G	3.5	2.8	2.3		13402.12	2A 2	2B 0	R 0
7459.452	7461.506	G	3.5	2.8	2.3		13402.12	3c 5	2a 6	Q 5
7459.758	7461.812		0.8				13401.57			
7459.830	7461.885		0.8	0.5			13401.44			
7459.925	7461.979				0.1		13401.27			
7460.058	7462.113		0.3	0.2			13401.03			
7460.270	7462.325		0.9	1.0	0.1		13400.65			
7460.532	7462.586		0.2	0.2			13400.18	V 1	2B 7	P 3
7460.843	7462.898		1.9	1.3	0.2		13399.62	2A 4	2B 3	R 0
7461.133	7463.188		0.2	0.2			13399.10			
7461.356	7463.411		0.4	1.0	0.5		13398.70	3D 2	2B11	R 6
7461.545	7463.600				0.1		13398.36			
7461.612	7463.667			0.2			13398.24			
7461.712	7463.767		0.1				13398.06			
7461.901	7463.956		1.8	1.1	1.6		13397.72			
7462.174	7464.229		1.9	1.2	1.7		13397.23	3b 5	2a 3	R 1
7462.419	7464.475		0.4	0.6	0.3		13396.79			
7462.587	7464.642			0.2			13396.49			
7462.720	7464.776			0.1			13396.25	3E 1	2B 9	P 4
7462.865	7464.920		0.1	0.3			13395.99	W 1	2C 1	Q 5
7463.021	7465.076				0.1		13395.71			
7463.099	7465.154			0.3b			13395.57			
7463.211	7465.266		0.2				13395.37			
7463.506	7465.561	G	2.0	1.5	1.4		13394.84			
7463.902	7465.957		0.2	0.2			13394.13	3F 0	2C 0	Q 6
7464.225	7466.280		0.4	0.9	0.4		13393.55	3b 5	2a 3	R 2
7464.587	7466.643	G	2.0	1.9	2.5		13392.90	3E 0	2B 7	Q 3
7464.587	7466.643	G	2.0	1.9	2.5		13392.90	2A 2	2B 0	P 6
7464.972	7467.027			0.3			13392.21			
7465.083	7467.139				0.2		13392.01			
7465.206	7467.262		0.3	0.8	1.7		13391.79			
7465.523	7467.579		0.1	0.2			13391.22			
7465.813	7467.869		0.1	0.4			13390.70	3c 3	2a 4	P 5
7465.947	7468.003		0.2				13390.46			
7466.193	7468.249	G	2.0	1.5	1.3		13390.02	3C 2	2K 2	Q 1
7466.494	7468.550			0.2			13389.48			
7466.706	7468.762			0.5			13389.10			
7466.856	7468.912		0.8	1.3	1.7		13388.83			
7467.062	7469.119			0.1			13388.46			
7467.453	7469.509		1.5	1.7	2.0		13387.76	3C 0	2A 0	P 5
7467.765	7469.822		0.1		0.2		13387.20			
7467.877	7469.933			0.3			13387.00			
7468.178	7470.235	G	2.1	1.8	2.0		13386.46	2A 2	2B 0	P 5
7468.418	7470.475		0.3	0.7	0.5		13386.03			

λ_air	λ_vac	Ref.	I₁	I₂	I₃	I₄	ν	Upper	Lower	Br.
7468.719	7470.776		0.7	1.2	1.5		13385.49	3F 0	2C 0	Q 7
7468.719	7470.776		0.7	1.2	1.5		13385.49	3F 0	2C 0	Q 4
7468.909	7470.966		0.2	0.2			13385.15			
7469.110	7471.167						13384.79	3E 2	2B11	P 4
7469.160	7471.217	G	2.2	2.0	1.7		13384.70	3F 0	2C 0	R 1
7469.238	7471.295						13384.56	3E 2	2B11	Q 4
7469.238	7471.295						13384.56	4f 2	3b 1	P 6
7469.489	7471.546		1.0	0.7	0.2		13384.11			
7469.718	7471.775		0.1				13383.70			
7469.813	7471.870		0.1	0.2			13383.53			
7470.081	7472.138		0.6	1.0	1.3		13383.05	3c 5	2a 6	Q 6
7470.388	7472.445		0.1	0.2			13382.50			
7470.740	7472.797		0.4	0.9	1.3		13381.87	3E 1	2B 9	R 0
7470.941	7472.998		0.1	0.2			13381.51			
7471.022	7473.080		0.1				13381.36	4f 2	3b 1	R 1
7471.231	7473.289		0.2	0.8	0.4		13380.99			
7471.454	7473.512			0.1			13380.59			
7471.577	7473.635		0.2				13380.37			
7471.706	7473.763		0.9	1.3	1.9		13380.14			
7472.041	7474.098			0.2	0.2		13379.54			
7472.387	7474.445		0.5	1.1	0.3		13378.92	W 2	2C 2	R 4
7472.583	7474.640		0.8	0.8	0.2		13378.57	3b 5	2a 3	R 0
7472.845	7474.903		0.3	0.5	0.2		13378.10			
7473.102	7475.160		0.1				13377.64			
7473.191	7475.249			0.3			13377.48			
7473.264	7475.322		0.4				13377.35	4f 3	3b 2	R 1
7473.476	7475.534			0.7	0.1		13376.97			
7473.856	7475.914		0.2	0.3			13376.29			
7473.990	7476.049			0.2			13376.05			
7474.214	7476.272		0.7	0.6			13375.65			
7474.516	7476.574	G	2.9	2.6	2.2		13375.11	3F 0	2C 0	R 1
7474.823	7476.881		0.4d				13374.56			
7474.901	7476.960		0.4	0.5	0.2		13374.42			
7475.175	7477.234		0.5	0.9	1.3		13373.93			
7475.455	7477.513		0.2	0.8	0.2		13373.43	3E 2	2B11	R 0
7475.455	7477.513		0.2	0.8	0.2		13373.43	4F 0	2B13	R 2
7475.583	7477.642		0.2				13373.20			
7475.656	7477.714			0.2			13373.07			
7475.907	7477.966		0.7	0.8	0.1		13372.62			
7476.204	7478.263			0.2			13372.09	3E 1	2B 9	Q 3
7476.438	7478.497		0.1	0.2			13371.67			
7476.601	7478.660			0.2	0.1		13371.38			
7476.712	7478.771		0.5	0.3			13371.18	4c 1	2a 5	R 2
7477.059	7479.118			0.2			13370.56			
7477.255	7479.314		0.0	0.2			13370.21			
7477.512	7479.571			0.1			13369.75			
7477.747	7479.806			0.2			13369.33			
7478.044	7480.103		0.5	1.1	1.7		13368.80			
7478.452	7480.511	G	2.5	2.0	1.6		13368.07	3C 2	2A 2	P 3
7478.452	7480.511	G	2.5	2.0	1.6		13368.07	W 2	2C 2	P 2
7478.760	7480.819		0.5	0.5	1.3		13367.52			
7478.995	7481.054		0.9	1.3	0.3		13367.10	3C 2	2A 2	P 4
7478.995	7481.054		0.9	1.3	0.3		13367.10	3D 1	2B 9	R 5
7479.235	7481.295		0.4	0.7	0.3		13366.67			
7479.465	7481.524		0.2	0.2			13366.26			
7479.688	7481.748		0.1	0.2	0.1		13365.86	4D 0	2B12	P 5
7479.951	7482.011		0.7	0.8			13365.39	3A 1	2B10	P 3
7480.047	7482.106			0.9	1.3		13365.22	4e 1	3b 0	R 4
7480.153	7482.213		0.4	0.8	0.3		13365.03	W 1	2B10	R 2
7480.394	7482.454			0.2			13364.60	4e 3	3b 2	R 4

λ_{air}	λ_{vac}	Ref.	I_1	I_2	I_3	I_4	ν	Upper	Lower	Br.
7480.690	7482.750		0.5	1.0	1.3		13364.07			
7480.914	7482.974		0.1	0.3			13363.67			
7481.021	7483.081			0.0			13363.48			
7481.155	7483.215			0.1			13363.24	4F 0	2B13	Q 4
7481.418	7483.478			0.3	0.1		13362.77	4D 0	2B12	P 4
7481.418	7483.478			0.3	0.1		13362.77	4e 2	3b 1	R 4
7481.737	7483.797		0.9	1.0	0.4		13362.20	Y 1	2B 8	P 3
7481.961	7484.022		0.2	0.3			13361.80			
7482.297	7484.358			0.1			13361.20			
7482.577	7484.638		0.2	0.8	0.2		13360.70			
7482.768	7484.828		0.1	0.1			13360.36			
7482.936	7484.996		0.3	0.7	0.4		13360.06			
7483.160	7485.220		0.1	0.1			13359.66			
7483.412	7485.472		0.1	0.2			13359.21			
7483.686	7485.747			0.2			13358.72	4D 0	2C 2	Q 5
7483.686	7485.747			0.2			13358.72	3E 3	2B13	Q 5
7483.882	7485.943			0.1			13358.37	3E 3	2B13	R 3
7484.017	7486.078			0.2			13358.13			
7484.347	7486.408			0.3	0.4		13357.54			
7484.684	7486.745		1.5	1.5	1.5		13356.94	3C 1	2A 1	P 4
7484.684	7486.745		1.5	1.5	1.5		13356.94	3A 1	2B10	P 1
7484.684	7486.745		1.5	1.5	1.5		13356.94	W 2	2B12	R 3
7485.076	7487.137		0.1	0.3			13356.24			
7485.339	7487.401		0.4	0.8			13355.77			
7485.614	7487.675		0.3	0.2			13355.28	3E 0	2B 7	Q 2
7485.917	7487.978		0.8	1.0	1.3		13354.74			
7486.079	7488.141			0.3			13354.45			
7486.286	7488.348			0.1			13354.08			
7486.432	7488.494			0.2			13353.82			
7486.634	7488.696			0.4			13353.46			
7486.746	7488.808				0.2		13353.26			
7486.802	7488.864			0.2			13353.16			
7486.914	7488.976		0.3				13352.96			
7487.066	7489.128			0.2			13352.69			
7487.357	7489.419			0.3			13352.17	3A 1	2B10	P 2
7487.699	7489.761	G	1.8	1.4	1.6		13351.56	3C 2	2A 2	P 5
7487.963	7490.025		0.2	0.2			13351.09			
7488.227	7490.289	G	2.2	1.8	1.4		13350.62	3c 6	2a 7	Q 1
7488.227	7490.289	G	2.2	1.8	1.4		13350.62	3f 0	2c 2	R 4
7488.535	7490.597			0.2			13350.07			
7488.765	7490.827			0.2			13349.66			
7488.894	7490.956				0.3		13349.43			
7489.029	7491.091		0.4	0.6			13349.19			
7489.309	7491.372						13348.69	3C 2	2K 2	P 2
7489.416	7491.478	G	2.3	2.0	1.9		13348.50	3F 1	2C 1	R 2
7489.753	7491.815		0.1	0.3	0.1		13347.90			
7489.983	7492.045		0.1	0.2			13347.49	3F 0	2B 8	R 4
7490.263	7492.326		0.2	0.6	0.2		13346.99	3E 1	2B 9	P 3
7490.572	7492.635			0.1			13346.44			
7490.836	7492.898		0.3	0.2			13345.97	3f 1	2c 3	R 3
7491.167	7493.230		0.9	1.3	1.6		13345.38	3F 0	2C 0	Q 5
7491.268	7493.331		0.2				13345.20			
7491.442	7493.505		0.9	1.3	0.1		13344.89	3c 6	2a 7	Q 2
7491.734	7493.797			0.2	0.1		13344.37			
7491.975	7494.038		0.2	0.2			13343.94			
7492.222	7494.286			0.3			13343.50	3D 2	2B11	R 5
7492.542	7494.606		1.9	1.6	1.5		13342.93	3F 1	2C 1	R 2
7492.823	7494.887		0.2	0.2			13342.43	4D 0	2B12	P 6
7492.823	7494.887		0.2	0.2			13342.43	3E 0	2B 7	P 2
7493.054	7495.117		1.5b	1.2			13342.02	2A 4	2B 3	P 1

λ_{air}	λ_{vac}	Ref.	I_1	I_2	I_3	I_4	ν	Upper	Lower	Br.
7493.267	7495.330		0.8	1.2	0.2		13341.64			
7493.542	7495.606		0.1	0.2			13341.15			
7493.733	7495.797			0.1			13340.81			
7493.924	7495.988		0.2	0.1			13340.47	3E 2	2B11	P 3
7494.205	7496.269		0.3	0.8	0.3		13339.97			
7494.413	7496.477		0.3	1.2	0.2		13339.60			
7494.907	7496.971			0.2			13338.72			
7495.177	7497.241			0.1			13338.24			
7495.295	7497.359		0.1				13338.03	3E 0	2B 7	Q 1
7495.430	7497.494			0.2			13337.79			
7495.750	7497.814		1.0	1.2	1.6		13337.22			
7496.003	7498.067		0.6	1.4	0.7		13336.77			
7496.217	7498.281		1.0	1.5			13336.39	3c 6	2a 7	Q 3
7496.430	7498.495		1.0	1.0	0.6		13336.01			
7496.723	7498.787		0.2	0.4			13335.49	Y 2	2C 2	R 1
7496.852	7498.916				0.2		13335.26			
7497.009	7499.074			0.3			13334.98			
7497.257	7499.321			0.2			13334.54			
7497.453	7499.518		0.1				13334.19			
7497.549	7499.614			0.3	0.1		13334.02			
7497.645	7499.709		0.2				13333.85			
7497.887	7499.951	G	2.2	1.8	1.5		13333.42	W 1	2C 1	P 4
7497.887	7499.951	G	2.2	1.8	1.5		13333.42	V 1	2B 7	P 2
7498.145	7500.210		0.2	0.1			13332.96			
7498.331	7500.396		1.5	1.2	0.2		13332.63	W 2	2C 2	Q 3
7498.331	7500.396		1.5	1.2	0.2		13332.63	W 2	2B12	R 2
7498.640	7500.705			0.2			13332.08			
7498.938	7501.003		0.3	0.5	0.3		13331.55			
7499.270	7501.335		0.1	0.2			13330.96			
7499.490	7501.555		0.3	0.3			13330.57			
7499.703	7501.768		0.3	0.1	0.3		13330.19			
7500.007	7502.072		0.2	0.3	0.2		13329.65	3E 1	2B 9	Q 2
7500.266	7502.331		0.1				13329.19			
7500.367	7502.433			0.3			13329.01			
7500.457	7502.523				0.2		13328.85			
7500.581	7502.647			0.1			13328.63			
7500.727	7502.793			0.2	0.3		13328.37			
7500.998	7503.063		0.2	0.4			13327.89			
7501.076	7503.142				0.2		13327.75			
7501.279	7503.345			0.0			13327.39			
7501.391	7503.457				0.3		13327.19			
7501.493	7503.559			0.0			13327.01			
7501.667	7503.733			0.2	0.4b		13326.70			
7501.836	7503.902			0.0			13326.40			
7502.067	7504.133			0.2			13325.99			
7502.151	7504.217		0.1		0.5		13325.84			
7502.337	7504.403		0.2	0.6			13325.51			
7502.484	7504.550		0.1	0.9	1.4		13325.25	3c 6	2a 7	Q 4
7502.805	7504.871			0.2	0.4b		13324.68	u 2	2a 4	R 3
7502.957	7505.023			0.1			13324.41			
7503.114	7505.180		0.1	0.2	0.3		13324.13			
7503.283	7505.349			0.2	0.9		13323.83			
7503.401	7505.468		0.7				13323.62			
7503.525	7505.592		0.2	0.7			13323.40			
7503.711	7505.778			0.2			13323.07			
7503.863	7505.930	i	1.9	1.6	1.5b		13322.80			
7504.015	7506.082				0.2		13322.53	3f 0	2c 2	R 3
7504.201	7506.268		0.9	1.4			13322.20	w 0	3b 0	P 2
7504.500	7506.566		2.0	0.9	0.7		13321.67			
7504.787	7506.854	G	2.1	2.0	2.0		13321.16	3F 0	2C 0	Q 3

λ_{air}	λ_{vac}	Ref.	I_1	I_2	I_3	I_4	ν	Upper	Lower	Br.
7505.114	7507.181		0.1				13320.58			
7505.215	7507.282		0.3	0.5	0.3		13320.40			
7505.373	7507.440			0.2	0.4		13320.12	4D 0	2C 2	Q 6
7505.536	7507.603			0.2			13319.83			
7505.632	7507.699				0.4		13319.66	4b 1	2a 4	P 1
7505.700	7507.767			0.4	0.3		13319.54			
7505.891	7507.958		0.2	0.2			13319.20			
7506.044	7508.111		0.1	0.5			13318.93	3A 0	2B 8	R 1
7506.173	7508.240		0.3	0.6	0.4		13318.70			
7506.325	7508.392		0.4	0.1	0.3		13318.43			
7506.444	7508.511			0.2			13318.22			
7506.517	7508.584		0.2		0.3		13318.09			
7506.658	7508.725			0.5			13317.84			
7506.985	7509.052	G	3.3	2.5	1.7		13317.26	2A 2	2B 0	P 1
7506.985	7509.052	G	3.3	2.5	1.7		13317.26	t 0	2a 2	R 0
7507.160	7509.227				0.4		13316.95			
7507.272	7509.340		0.7				13316.75			
7507.374	7509.441		0.7	0.2			13316.57			
7507.560	7509.627		0.4	0.3			13316.24			
7507.853	7509.921	G	2.1	2.0	2.2		13315.72	3F 0	2C 0	Q 4
7507.853	7509.921	G	2.1	2.0	2.2		13315.72	3E 2	2B11	Q 3
7508.141	7510.208		0.3	0.3			13315.21			
7508.220	7510.287				0.2		13315.07			
7508.378	7510.445		1.0	0.9	0.2		13314.79			
7508.710	7510.778		0.2				13314.20	3D 2	2B11	R 4
7508.851	7510.919			0.1d	0.2b		13313.95			
7509.066	7511.133			0.1	0.2b		13313.57			
7509.207	7511.274		0.3	0.2			13313.32			
7509.387	7511.455			0.1			13313.00			
7509.596	7511.664		0.3	0.4	0.1b		13312.63			
7509.720	7511.788		0.3	0.4			13312.41	4D 0	2C 2	P 4
7509.720	7511.788		0.3	0.4			13312.41	W 1	2B10	R 1
7509.872	7511.940		0.1	0.2			13312.14			
7510.104	7512.172		0.3	0.8	1.3		13311.73			
7510.341	7512.409			0.1			13311.31			
7510.419	7512.488				0.3		13311.17			
7510.521	7512.589		0.2	0.3			13310.99			
7510.792	7512.860		0.7	0.9	0.9		13310.51			
7510.882	7512.950			1.1	1.4		13310.35	Y 1	2B 8	P 4
7510.927	7512.996		0.2				13310.27	3E 3	2B13	R 2
7511.023	7513.091				0.2		13310.10			
7511.147	7513.216		0.2	0.3			13309.88			
7511.367	7513.436				0.8		13309.49	3f 0	2c 2	Q 6
7511.452	7513.521			0.2	0.4		13309.34			
7511.672	7513.741		0.2	0.4			13308.95			
7511.768	7513.837				0.2		13308.78			
7511.977	7514.046		0.2	0.4			13308.41	W 2	2B12	R 1
7512.310	7514.379	G	2.1	2.0	2.2		13307.82	3b 1	2a 0	P 4
7512.688	7514.757		0.3	0.5	0.3		13307.15			
7512.959	7515.028		0.8	1.0	0.8		13306.67			
7513.112	7515.181			1.2	1.7		13306.40	3E 1	2B 9	Q 1
7513.360	7515.429		0.1	0.1			13305.96			
7513.597	7515.666				0.2		13305.54			
7513.682	7515.751			0.2			13305.39			
7513.818	7515.887			0.2			13305.15			
7514.032	7516.101				0.4		13304.77			
7514.140	7516.209		0.8	1.2	1.4		13304.58			
7514.439	7516.508		1.0	0.8			13304.05	Y 2	2C 2	Q 1
7514.631	7516.700	1		0.3	1.5b		13303.71			
7514.829	7516.898		0.3	0.5			13303.36			

λ_{air}	λ_{vac}	Ref.	I_1	I_2	I_3	I_4	ν	Upper	Lower	Br.
7514.959	7517.028				1.0		13303.13	t 0	2a 2	P 1
7515.060	7517.130			0.1			13302.95			
7515.275	7517.344		0.2	0.3	0.4		13302.57	3E 2	2B11	P 2
7515.591	7517.661		1.9	1.2	0.3		13302.01			
7515.693	7517.763				0.3		13301.83			
7515.896	7517.966		0.1	0.2	0.2		13301.47			
7516.134	7518.203			0.2	0.2		13301.05	3C 0	2A 0	P 6
7516.399	7518.469			0.1	0.3		13300.58	3E 1	2B 9	P 2
7516.699	7518.769		1.5b	1.2	0.4		13300.05	4b 0	2a 3	P 1
7516.936	7519.006		0.1	0.1			13299.63	4f 3	3b 2	P 5
7517.179	7519.249		0.6	0.6	0.3		13299.20			
7517.258	7519.328		0.3				13299.06			
7517.598	7519.668		2.2	1.6	0.3		13298.46	2A 4	2B 3	P 2
7517.965	7520.035		0.2	0.2	0.1		13297.81	4c 0	2a 4	R 4
7518.282	7520.352		1.0	1.2	0.3		13297.25			
7518.598	7520.669		0.2	0.3			13296.69	3D 2	2B11	R 3
7518.689	7520.759				0.2		13296.53			
7518.825	7520.895		0.2	0.3			13296.29			
7518.915	7520.985				0.3		13296.13			
7519.011	7521.082			0.3			13295.96			
7519.232	7521.302		0.1	0.0			13295.57			
7519.390	7521.461			0.2	0.1		13295.29			
7519.645	7521.715		0.2	0.2			13294.84			
7519.775	7521.845		0.1	0.3			13294.61			
7520.097	7522.168		0.6	0.5			13294.04			
7520.301	7522.371	G	1.5b	1.6	1.7		13293.68	4e 1	3b 0	R 3
7520.612	7522.683		0.1	0.1			13293.13			
7520.861	7522.932		0.2	0.3			13292.69	4g 0	2a 4	P 4
7521.166	7523.237	G	2.0	1.8	1.8		13292.15	3F 2	2C 2	R 3
7521.529	7523.600		0.1	0.3			13291.51			
7521.862	7523.934		0.2b	0.4			13290.92			
7522.179	7524.251	G	1.9	1.8	1.9		13290.36	3E 0	2C 0	R 5
7522.179	7524.251	G	1.9	1.8	1.9		13290.36	3C 1	2A 1	P 5
7522.457	7524.528		0.7	0.9	0.5		13289.87			
7522.751	7524.822			0.2			13289.35			
7522.927	7524.998		0.5	0.4			13289.04	4b 2	2a 5	P 3
7523.193	7525.264		0.3	0.1			13288.57			
7523.357	7525.428		0.3	0.4			13288.28			
7523.634	7525.706		0.3	0.3			13287.79	4f 2	3b 1	P 5
7523.764	7525.836		0.2				13287.56			
7524.014	7526.085		0.8	0.5			13287.12	4f 1	3b 0	Q 4
7524.150	7526.221			0.3			13286.88			
7524.240	7526.312		0.9				13286.72	Y 2	2C 2	P 1
7524.410	7526.482						13286.42	W 2	2B12	R 0
7524.608	7526.680	G	3.9	3.1	2.6		13286.07			
7524.665	7526.737						13285.97	2A 2	2B 0	P 2
7524.733	7526.805						13285.85			
7524.965	7527.037		1.0				13285.44			
7525.056	7527.128		1.0	0.9	0.1		13285.28			
7525.231	7527.303		0.6	0.2			13284.97	3c 7	2a 8	Q 1
7525.339	7527.411			0.2			13284.78			
7525.503	7527.575		0.4	0.2	0.1		13284.49			
7525.730	7527.802		0.4	0.3			13284.09			
7526.007	7528.080		1.9	1.9	1.8		13283.60	3F 0	2C 0	Q 3
7526.268	7528.340		0.6	0.7	0.2		13283.14	3D 2	2B11	R 2
7526.347	7528.420		0.4				13283.00			
7526.461	7528.533			0.1			13282.80			
7526.625	7528.698		0.2	0.2			13282.51	4e 2	3b 1	R 3
7526.903	7528.975		0.3	0.3			13282.02			
7527.084	7529.157			0.1			13281.70	t 1	2a 3	R 3

λ_{air}	λ_{vac}	Ref.	I_1	I_2	I_3	I_4	ν	Upper	Lower	Br.
7527.282	7529.355		0.9	1.0	0.3		13281.35			
7527.577	7529.650		0.7	1.0	0.2		13280.83			
7527.792	7529.865		0.3	0.2			13280.45			
7527.878	7529.950		0.1				13280.30			
7528.025	7530.098			0.1			13280.04	3D 2	2B11	P 1
7528.325	7530.398		1.0	1.3	0.3		13279.51	3b 5	2a 3	P 1
7528.530	7530.602		0.9	1.5	1.5		13279.15	3B 2	2A 0	R 2
7528.700	7530.773		0.1				13278.85			
7528.802	7530.875			0.2			13278.67			
7529.062	7531.135		0.3	0.9	0.2		13278.21	t 0	2a 2	R 1
7529.329	7531.402		1.5	1.1	0.1		13277.74	Y 2	2C 2,	R 2
7529.482	7531.555			0.3			13277.47			
7529.686	7531.759		0.2				13277.11			
7529.851	7531.924			0.2			13276.82			
7529.936	7532.009		0.2				13276.67			
7530.140	7532.213		0.4	0.9	0.3		13276.31			
7530.486	7532.559			0.1			13275.70			
7530.616	7532.690		0.1				13275.47			
7530.826	7532.900		0.3	0.2			13275.10			
7531.087	7533.161		0.2	0.1	0.1		13274.64			
7531.292	7533.365			0.2			13274.28			
7531.405	7533.479		0.2				13274.08			
7531.609	7533.683		1.0	1.8	2.0		13273.72	3B 2	2A 0	R 3
7531.762	7533.836	G	2.1				13273.45	3c 7	2a 8	Q 1
7532.041	7534.114		0.5	1.1	0.1		13272.96			
7532.137	7534.211		0.7				13272.79			
7532.432	7534.506		0.7	1.0	0.3		13272.27	4E 1	2B15	Q 4
7532.812	7534.886		1.5	1.0			13271.60			
7533.136	7535.210		0.1				13271.03			
7533.323	7535.398		0.5	0.2	0.1		13270.70	W 1	2B10	R 0
7533.891	7535.965			0.1			13269.70	4c 2	2a 6	P 3
7534.237	7536.312		0.1	0.2			13269.09			
7534.504	7536.579		0.7	1.2			13268.62			
7534.589	7536.664				0.1		13268.47			
7534.862	7536.937	G	2.2	1.9	1.8		13267.99	3B 2	2A 0	R 1
7535.288	7537.363		1.0	0.3			13267.24			
7535.504	7537.579			0.1			13266.86			
7535.782	7537.857		0.1	0.2			13266.37	3E 2	2B11	Q 2
7535.998	7538.073		0.2	0.1			13265.99			
7536.185	7538.260		0.2	0.4			13265.66	4f 2	3b 1	Q 4
7536.362	7538.437		0.3	0.4	0.2		13265.35			
7536.583	7538.658		0.4	0.3			13264.96	4b 1	2a 4	P 2
7536.896	7538.971	G	2.4	1.0	1.7		13264.41	3F 0	2C 0	Q 2
7537.060	7539.136						13264.12	u 2	2a 4	R 2
7537.219	7539.295	G	2.6	2.0	1.9		13263.84	3F 1	2C 1	R 1
7537.219	7539.295		2.6	2.0	1.9		13263.84	3E 0	2C 0	R 4
7537.219	7539.295		2.6	2.0	1.9		13263.84	3D 2	2B11	R 1
7537.492	7539.568		0.2				13263.36			
7537.657	7539.732		0.6	0.4			13263.07			
7537.822	7539.897		0.2				13262.78			
7538.038	7540.113		1.0	0.5			13262.40			
7538.299	7540.375						13261.94			
7538.333	7540.409	G	3.3	2.3	2.2		13261.88	3F 1	2C 1	R 1
7538.362	7540.437						13261.83			
7538.629	7540.705		0.7				13261.36			
7538.703	7540.779		0.7	0.3			13261.23			
7538.919	7540.995		0.3	0.2			13260.85			
7539.158	7541.234		1.5	0.7	0.1		13260.43	W 2	2C 2	P 3
7539.249	7541.325			1.1			13260.27	3E 3	2B13	Q 4
7539.402	7541.478		0.2				13260.00			

λ_air	λ_vac	Ref.	I₁	I₂	I₃	I₄	ν	Upper	Lower	Br.
7539.561	7541.637		0.6				13259.72	u 1	2a 3	R 3
7539.681	7541.757		0.4	0.3d			13259.51			
7540.073	7542.149		0.2	0.2			13258.82			
7540.301	7542.377		0.6	1.0	1.4		13258.42			
7540.778	7542.855		0.5	1.0	1.6		13257.58			
7541.074	7543.151		0.1	0.2			13257.06			
7541.302	7543.378		0.2	0.3			13256.66			
7541.569	7543.646		0.3				13256.19			
7541.728	7543.805		0.4	0.4	0.1		13255.91			
7541.973	7544.050	G	3.0	2.4	2.1		13255.48	2A 2	2B 0	P 3
7542.297	7544.374		0.4				13254.91			
7542.337	7544.414		0.4	0.4	0.2		13254.84			
7542.684	7544.761		0.7	1.2	1.7		13254.23			
7542.861	7544.938			0.2			13253.92			
7543.054	7545.131		0.7	0.4			13253.58			
7543.419	7545.495		0.3d	0.4	0.2		13252.94	3D 0	2B 7	R 5
7543.783	7545.860		0.3	1.1	1.5		13252.30	3B 2	2A 0	R 4
7543.783	7545.860		0.3	1.1	1.5		13252.30	t 0	2a 2	R 2
7543.931	7546.008		0.2				13252.04			
7544.130	7546.207			0.1			13251.69			
7544.307	7546.384			0.2			13251.38			
7544.369	7546.446		0.3				13251.27			
7544.472	7546.549		0.3	0.3			13251.09			
7544.705	7546.782		0.8	0.4	0.2		13250.68			
7545.013	7547.090	G	3.0	2.6	2.2		13250.14	3F 0	2C 0	Q 2
7545.320	7547.398		0.5				13249.60			
7545.383	7547.460		0.5	0.3			13249.49			
7545.605	7547.682		0.2	0.2			13249.10			
7545.827	7547.905		0.4	0.3			13248.71			
7546.089	7548.167		0.1	0.2			13248.25			
7546.459	7548.537		0.0	0.2			13247.60			
7546.801	7548.879			0.2			13247.00			
7546.989	7549.067		0.1	0.2			13246.67			
7547.240	7549.318		0.8	1.1	0.4		13246.23			
7547.627	7549.705			0.2	0.1		13245.55			
7547.946	7550.025		1.5	1.2	0.1		13244.99	2A 4	2B 3	P 3
7547.946	7550.025		1.5	1.2	0.1		13244.99	3E 3	2B13	R 1
7548.317	7550.395		0.1d	0.3			13244.34	4f 3	3b 2	Q 4
7548.653	7550.731		1.5	1.6	1.6		13243.75	3E 0	2C 0	R 3
7548.756	7550.834		0.5				13243.57	3D 2	2B11	R 0
7549.058	7551.136		0.1	0.2	0.1		13243.04			
7549.434	7551.513		0.2	0.2			13242.38			
7549.605	7551.684		0.7	0.4			13242.08	t 0	2a 2	R 3
7549.890	7551.969			0.1			13241.58			
7550.135	7552.214		0.1	0.7	0.2		13241.15			
7550.420	7552.499			0.1			13240.65			
7550.648	7552.727		0.2b	0.3	0.1		13240.25			
7550.979	7553.058		1.5b	1.4	0.3		13239.67	3B 2	2A 0	R 0
7551.932	7554.011		0.2	0.4d			13238.00	3f 0	2c 2	R 3
7552.297	7554.376			0.2			13237.36			
7552.519	7554.599		0.3	0.3			13236.97			
7552.759	7554.839		0.6	0.9	0.2		13236.55			
7553.130	7555.210		1.0	1.0	0.1		13235.90	4b 0	2a 3	P 2
7553.393	7555.472		0.8	1.2	1.5		13235.44	3A 0	2B 8	P 4
7553.729	7555.809		0.1	0.1			13234.85			
7553.998	7556.077		0.2	0.2			13234.38			
7554.174	7556.254			0.1			13234.07			
7554.403	7556.483		0.8	0.9	0.3		13233.67	3f 2	2c 4	R 3
7554.403	7556.483		0.8	0.9	0.3		13233.67	W 2	2B12	P 1
7554.403	7556.483		0.8	0.9	0.3		13233.67	3F 1	2C 1	Q 4

λ_{air}	λ_{vac}	Ref.	I_1	I_2	I_3	I_4	ν	Upper	Lower	Br.
7554.700	7556.780		0.2	0.3			13233.15	4c 1	2a 5	Q 4
7554.974	7557.054		0.3	0.1			13232.67			
7555.259	7557.339		0.7	1.0	1.4		13232.17			
7555.579	7557.659		0.6	1.0	1.5		13231.61			
7555.853	7557.933		0.1	0.2			13231.13			
7556.081	7558.162				0.5		13230.73			
7556.150	7558.230		1.0	1.3	1.4		13230.61			
7556.458	7558.539		0.1	0.1			13230.07	t 1	2a 3	P 1
7556.847	7558.927			0.2d			13229.39	3E 2	2B11	Q 1
7557.155	7559.236			0.4			13228.85	4c 0	2a 4	R 3
7557.430	7559.510			0.2			13228.37			
7557.595	7559.676		0.2	0.3			13228.08			
7557.767	7559.848			0.1			13227.78			
7557.938	7560.019		0.4	0.4			13227.48	Y 2	2C 2	Q 2
7557.938	7560.019		0.4	0.4			13227.48	3A 0	2B 8	R 0
7558.275	7560.356		0.3	0.7	0.1		13226.89			
7558.401	7560.482		0.2				13226.67			
7558.641	7560.722	G	2.1	1.8	1.7		13226.25	3F 2	2C 2	R 2
7558.641	7560.722	G	2.1	1.8	1.7		13226.25	W 2	2C 2	Q 4
7558.870	7560.951		0.9	0.5	0.5		13225.85			
7559.052	7561.134	G	2.3	1.9	1.6		13225.53	2K 3	2B 1	R 0
7559.230	7561.311			0.8			13225.22	4f 1	3b 0	Q 3
7559.384	7561.465						13224.95	3f 1	2c 3	R 2
7559.584	7561.665	G	1.8	1.5	0.3		13224.60	3c 0	2a 0	P 6
7560.036	7562.117		0.1				13223.81			
7560.201	7562.283			0.2			13223.52			
7560.424	7562.506			0.1			13223.13			
7560.636	7562.718		0.2	0.3			13222.76			
7560.790	7562.872		0.2				13222.49	2K 4	2B 2	R 4
7560.962	7563.044		0.3	0.4	0.1		13222.19			
7561.282	7563.364	G	2.5	2.2	1.9		13221.63	3F 2	2C 2	R 2
7561.282	7563.364	G	2.5	2.2	1.9		13221.63	2K 3	2B 1	R 1
7561.677	7563.759		0.3	0.3			13220.94	4e 1	3b 0	R 2
7561.677	7563.759		0.3	0.3			13220.94	3f 1	2c 3	Q 5
7561.980	7564.062		1.5b	1.4	0.2		13220.41			
7562.323	7564.405		0.1				13219.81			
7562.472	7564.554			0.3	0.1		13219.55	2K 4	2B 2	R 3
7562.718	7564.800			0.1			13219.12			
7562.964	7565.046			0.2			13218.69			
7563.193	7565.275		0.4	1.0	1.4		13218.29			
7563.462	7565.544		0.1				13217.82			
7563.696	7565.779		0.8	1.1	0.5		13217.41	u 2	2a 4	R 1
7563.799	7565.882		0.2				13217.23			
7564.034	7566.116		0.7	1.0	1.7		13216.82			
7564.303	7566.386		0.1	0.1			13216.35			
7564.435	7566.517		0.1				13216.12			
7564.612	7566.695		0.2	0.2			13215.81	4c 1	2a 5	Q 3
7564.864	7566.947		0.5	1.2	2.0		13215.37	3B 2	2A 0	R 5
7565.018	7567.101		0.2				13215.10			
7565.127	7567.210	G		1.6			13214.91			
7565.305	7567.388		1.5	1.6	1.0		13214.60			
7565.602	7567.685		0.2	0.2	0.1		13214.08			
7565.877	7567.960		1.0	1.1	0.2		13213.60	2K 4	2B 2	R 2
7565.877	7567.960		1.0	1.1	0.2		13213.60	4f 3	3b 2	P 4
7566.559	7568.642		0.2	0.2			13212.41	3D 1	2B 9	R 4
7566.839	7568.923		0.2				13211.92			
7567.028	7569.112		0.5				13211.59	u 1	2a 3	R 2
7567.137	7569.221		1.0	1.5	1.9		13211.40	3F 1	2C 1	Q 4
7567.326	7569.410		0.7				13211.07	3F 1	2C 1	Q 5
7567.424	7569.507			0.3			13210.90			

λ_{air}	λ_{vac}	Ref.	I_1	I_2	I_3	I_4	ν	Upper	Lower	Br.
7567.658	7569.742		0.9	0.8	0.1b		13210.49			
7567.876	7569.960		0.1	0.1			13210.11	4b 1	2a 4	P 3
7568.065	7570.149			0.2			13209.78			
7568.386	7570.470		0.2	0.7	0.2		13209.22			
7568.672	7570.756			0.1			13208.72	4F 0	2B13	P 4
7568.672	7570.756			0.1			13208.72	3f 0	2c 2	Q 5
7568.844	7570.928		0.2	0.3			13208.42			
7569.131	7571.215			0.1			13207.92	4f 3	3b 2	Q 3
7569.360	7571.444		1.5	1.1	0.4		13207.52	3b 5	2a 3	P 2
7569.360	7571.444		1.5	1.1	0.4		13207.52	W 2	2B12	P 2
7569.847	7571.931		0.1	0.2			13206.67			
7570.254	7572.339			0.2			13205.96	3F 1	2B10	Q 5
7570.598	7572.683			0.2			13205.36	3f 1	2c 3	R 2
7570.713	7572.797		0.2				13205.16			
7570.902	7572.987		0.8	1.0	0.5		13204.83	W 1	2B10	P 1
7571.246	7573.331		0.2				13204.23			
7571.309	7573.394			0.8	0.4		13204.12			
7571.367	7573.451		1.5				13204.02	2K 4	2B 2	R 1
7571.602	7573.686	G	2.7	2.4	2.3		13203.61	2K 2	2B 0	P 4
7571.986	7574.071		0.2b	0.3			13202.94			
7572.244	7574.329		0.1	0.2			13202.49	4e 2	3b 1	R 2
7572.571	7574.656		0.9b	1.2	1.4		13201.92	4c 1	2a 5	Q 2
7572.955	7575.040			0.1			13201.25			
7573.173	7575.258			0.3			13200.87			
7573.420	7575.505			0.2			13200.44			
7573.632	7575.717		1.5	1.1	0.2		13200.07	3f 2	2c 4	R 1
7573.902	7575.987			0.1			13199.60			
7574.068	7576.154			0.2			13199.31			
7574.223	7576.309		0.2	0.2			13199.04			
7574.458	7576.544		0.3	0.2			13198.63	4E 1	2B15	Q 3
7574.768	7576.854		0.9	1.4	1.5		13198.09	4f 2	3b 1	P 4
7575.142	7577.227			0.1			13197.44			
7575.400	7577.485			0.5	0.2		13196.99			
7575.848	7577.933			1.6	1.6		13196.21	2K 3	2B 1	R 2
7576.169	7578.255			0.7			13195.65			
7576.416	7578.502	G	2.5	2.4	2.2		13195.22	3E 0	2C 0	R 2
7576.910	7578.996			0.7	0.1		13194.36			
7577.059	7579.145			1.0	0.3		13194.10			
7577.432	7579.519		0.5	1.1	0.1		13193.45	Y 2	2C 2	P 2
7577.725	7579.812			0.2			13192.94	4c 1	2a 5	Q 1
7578.007	7580.093		0.1	0.5	0.2		13192.45			
7578.386	7580.472			0.3			13191.79	W 1	2B10	P 2
7578.524	7580.610				0.1		13191.55			
7578.777	7580.863			0.3			13191.11	W 1	2C 1	P 5
7579.144	7581.231		0.1	0.4			13190.47			
7579.443	7581.530	G	2.1	2.0	2.1		13189.95	3F 1	2C 1	Q 4
7579.937	7582.024		0.2	0.4	0.1		13189.09	3D 1	2B 9	R 0
7579.937	7582.024		0.2	0.4	0.1		13189.09	3E 3	2B13	Q 3
7580.242	7582.329		0.1	0.4	0.2		13188.56	4D 0	2C 2	Q 7
7580.604	7582.691		2.1	1.6	0.2		13187.93	2K 4	2B 2	R 0
7580.817	7582.904			0.4			13187.56			
7581.012	7583.099		0.1	0.1			13187.22	V 1	2C 0	R 2
7581.242	7583.329		0.3	0.9	0.2		13186.82			
7581.553	7583.640		0.5	0.9	0.1		13186.28			
7581.840	7583.927		0.1	0.3			13185.78	W 1	2B10	P 5
7582.013	7584.100			0.2			13185.48			
7582.231	7584.319		0.3b	0.8	0.5		13185.10			
7582.628	7584.716			0.2			13184.41	W 1	2B10	P 4
7582.927	7585.015		0.3b	0.8	0.3		13183.89			
7583.209	7585.297		0.1	0.2			13183.40			

λ_{air}	λ_{vac}	Ref.	I_1	I_2	I_3	I_4	ν	Upper	Lower	Br.
7583.410	7585.498			0.1			13183.05			
7583.566	7585.653		0.2				13182.78			
7583.692	7585.780			0.2			13182.56			
7583.893	7585.981			0.1	0.1		13182.21			
7583.980	7586.068		0.1				13182.06			
7584.112	7586.200			0.4			13181.83			
7584.411	7586.499	G	2.1	2.0	2.6		13181.31	3b 1	2a 0	P 5
7584.503	7586.591						13181.15	2A 4	2B 3	P 4
7584.733	7586.822			0.3			13180.75	3E 3	2B13	R 0
7584.733	7586.822			0.3			13180.75	3D 1	2B 9	R 1
7585.039	7587.127	G	2.5	2.3	2.1		13180.22	3F 1	2C 1	Q 3
7585.039	7587.127	G	2.5	2.3	2.1		13180.22	u 2	2a 4	R 0
7585.039	7587.127	G	2.5	2.3	2.1		13180.22	3A 0	2B 8	P 3
7585.482	7587.570		0.3	1.3	1.5		13179.45			
7585.862	7587.950		0.1	0.2			13178.79	3D 1	2B 9	P 1
7585.862	7587.950		0.1	0.2			13178.79	W 1	2B10	P 3
7586.380	7588.468		0.1	0.2			13177.89			
7586.495	7588.583						13177.69			
7586.714	7588.802		0.3b	0.4			13177.31			
7586.858	7588.946			0.4	0.1b		13177.06	3D 0	2B 7	R 4
7587.094	7589.182			0.2			13176.65			
7587.416	7589.505		0.7	1.2	1.4		13176.09			
7587.923	7590.012		0.1	0.2			13175.21	4e 3	3b 2	R 2
7588.245	7590.334		0.2	0.3			13174.65			
7588.539	7590.628			0.2			13174.14	u 1	2a 3	R 1
7588.908	7590.997		0.4	0.3			13173.50	3D 1	2B 9	R 3
7589.184	7591.274		0.7	0.6	0.1		13173.02	4b 0	2a 3	P 3
7589.351	7591.441			0.2			13172.73	W 2	2B12	P 3
7589.507	7591.596		0.2	0.3			13172.46	3D 2	2B11	P 2
7589.507	7591.596		0.2	0.3			13172.46	4F 0	2B13	Q 3
7589.507	7591.596		0.2	0.3			13172.46	3f 0	2c 2	R 2
7589.841	7591.931	G	2.0	1.5	0.2		13171.88	2K 3	2B 1	P 1
7590.124	7592.213		0.1	0.3	0.3		13171.39	Z 2	2B11	R 0
7590.458	7592.547			0.3	0.1		13170.81			
7590.729	7592.818		0.3	0.2			13170.34	3D 1	2B 9	R 2
7591.149	7593.239		0.1	0.2			13169.61	4E 1	2B15	Q 1
7591.576	7593.666		0.1	0.2			13168.87			
7591.870	7593.960			0.2			13168.36			
7592.147	7594.237		0.3	1.1	1.4		13167.88			
7592.389	7594.479			0.1			13167.46			
7592.579	7594.669		0.3	0.6	0.2		13167.13			
7593.092	7595.183		0.4	0.5	0.1		13166.24			
7593.317	7595.408			0.2	0.2		13165.85			
7593.600	7595.690	G	2.1	1.8	1.7		13165.36	3F 1	2C 1	Q 3
7593.946	7596.037	G	2.0	1.1	1.5		13164.76			
7594.223	7596.314			0.2			13164.28			
7594.477	7596.568		0.4	1.0	0.5		13163.84			
7594.673	7596.764			0.4	0.4		13163.50			
7594.933	7597.023		0.2b				13163.05			
7595.042	7597.133	G					13162.86	t 1	2a 3	P 3
7595.273	7597.364		0.2	0.2			13162.46	Z 2	2B11	R 1
7595.521	7597.612			0.4			13162.03			
7595.590	7597.682		0.2		0.2		13161.91			
7595.844	7597.935	G	2.1	2.0	2.2		13161.47	3b 2	2a 1	R 2
7595.844	7597.935	G	2.1	2.0	2.2		13161.47	4c 0	2a 4	R 2
7596.214	7598.305			0.1			13160.83			
7596.370	7598.461		0.2	0.3			13160.56			
7596.531	7598.622			0.2			13160.28			
7596.745	7598.836			0.6			13159.91			
7597.056	7599.148	G	3.2	2.7	2.1		13159.37	3F 2	2C 2	R 1

λ_{air}	λ_{vac}	Ref.	I_1	I_2	I_3	I_4	ν	Upper	Lower	Br.
7597.253	7599.344	G	3.3		0.6		13159.03	3F 2	2C 2	R 1
7597.513	7599.604			0.6	0.3		13158.58			
7597.720	7599.812		0.2	0.4			13158.22			
7597.969	7600.060				1.5		13157.79			
7598.038	7600.130	G	2.2	2.2	2.7		13157.67	3b 2	2a 1	R 3
7598.107	7600.199				1.5		13157.55	Z 2	2B11	R 2
7598.408	7600.500		0.2	0.3	0.1		13157.03			
7598.656	7600.748		0.2	0.5	0.2		13156.60			
7598.933	7601.025		0.4	0.9	0.2		13156.12	4f 1	3b 0	Q 2
7599.176	7601.268			0.2			13155.70	w 0	3b 0	P 3
7599.361	7601.453		0.2b	0.3	0.1		13155.38	3E 3	2B13	P 5
7599.673	7601.765		0.7	1.1	0.3		13154.84	4b 1	2a 4	P 4
7599.892	7601.984		0.2	0.4			13154.46			
7600.083	7602.175			0.1			13154.13	4F 0	2B13	R 1
7600.273	7602.366		0.2	0.3			13153.80			
7600.539	7602.632		0.2	0.2			13153.34			
7600.834	7602.926		0.1				13152.83	3F 1	2B10	R 3
7600.927	7603.019			0.3			13152.67	3F 3	2B14	R 2
7601.117	7603.210		0.1	0.2			13152.34			
7601.285	7603.377			0.4			13152.05			
7601.556	7603.649		0.4	1.1	1.4		13151.58			
7601.845	7603.938		0.1	0.2			13151.08	3F 0	2B 8	Q 5
7602.244	7604.337		0.3	0.3			13150.39	4e 1	3b 0	R 1
7602.557	7604.649	G	2.4	2.1	1.7		13149.85	3F 1	2C 1	Q 2
7602.557	7604.649	G	2.4	2.1	1.7		13149.85	3f 1	2c 3	Q 5
7602.557	7604.649	G	2.4	2.1	1.7		13149.85	2K 3	2B 1	R 3
7602.898	7604.991		0.4	0.4			13149.26			
7603.135	7605.228		0.4	0.6			13148.85			
7603.383	7605.477				1.5		13148.42			
7603.441	7605.534	G	2.8	2.6	2.6		13148.32	3b 2	2a 1	R 1
7603.511	7605.604				1.5		13148.20			
7603.817	7605.910		0.3	0.4	0.2		13147.67			
7604.054	7606.148		0.3	1.0	0.4		13147.26			
7604.320	7606.414			0.1			13146.80			
7604.453	7606.547				0.1		13146.57	3f 0	2c 2	Q 6
7604.604	7606.697		0.3	0.3			13146.31			
7604.881	7606.975		0.8	1.3	0.4		13145.83			
7605.252	7607.345		0.2	0.4			13145.19	3E 3	2B13	P 4
7605.419	7607.513			0.2			13144.90			
7605.686	7607.779		0.3	0.4			13144.44	4f 2	3b 1	Q 2
7605.836	7607.930			0.2			13144.18	4f 3	3b 2	Q 2
7605.836	7607.930			0.2			13144.18	4f 3	3b 2	P 3
7606.044	7608.138			0.6			13143.82			
7606.368	7608.462	G	3.2	2.7	2.0		13143.26	3F 1	2C 1	Q 2
7606.368	7608.462	G	3.2	2.7	2.0		13143.26	Z 2	2B11	R 3
7606.739	7608.833		1.0	1.2	0.2		13142.62	3A 0	2B 8	P 1
7606.739	7608.833		1.0	1.2	0.2		13142.62	3A 0	2B 8	P 2
7607.040	7609.134			0.1			13142.10			
7607.277	7609.371			0.2			13141.69	3f 1	2c 3	Q 4
7607.486	7609.580		0.6	1.2	0.4		13141.33	W 2	2C 2	Q 5
7607.752	7609.846			0.2			13140.87	W 2	2B12	P 4
7607.960	7610.055		0.2	0.4			13140.51			
7608.122	7610.217			0.2			13140.23	3f 2	2c 4	R 2
7608.371	7610.466			0.1			13139.80	3D 2	2B11	P 3
7608.458	7610.553		0.2b				13139.65			
7608.748	7610.842		0.2	0.4			13139.15			
7609.292	7611.387			0.2	0.1		13138.21	W 2	2B12	P 5
7609.292	7611.387			0.2	0.1		13138.21	3e 1	2c 3	R 5
7609.628	7611.723			0.3	0.1		13137.63			
7609.941	7612.036		0.5	1.5	2.0		13137.09	3b 2	2a 1	R 4

λ_{air}	λ_{vac}	Ref.	I_1	I_2	I_3	I_4	ν	Upper	Lower	Br.
7609.941	7612.036		0.5	1.5	2.0		13137.09	t 0	2a 2	P 2
7610.346	7612.441		0.3	0.9	0.3		13136.39			
7611.018	7613.114		0.2	0.4			13135.23	3E 3	2B13	Q 2
7611.192	7613.288			0.2			13134.93			
7611.383	7613.479		0.6	0.7			13134.60	u 1	2a 3	R 0
7611.685	7613.780			0.2			13134.08			
7611.899	7613.995		0.2	0.3			13133.71			
7612.120	7614.215			0.2			13133.33			
7612.299	7614.395			0.2			13133.02	4f 2	3b 1	P 3
7612.572	7614.667		0.3	0.7	0.2		13132.55			
7612.792	7614.888		0.2	0.3	0.3		13132.17			
7613.221	7615.317			0.3			13131.43			
7613.314	7615.410		0.2				13131.27			
7613.540	7615.636			0.3			13130.88			
7613.859	7615.955						13130.33			
7613.911	7616.007	G	2.2	2.0	1.6		13130.24	3E 0	2C 0	R 1
7613.911	7616.007	G	2.2	2.0	1.6		13130.24	3B 2	2A 0	P 1
7614.311	7616.407		0.4	0.4			13129.55			
7614.520	7616.616			0.1			13129.19			
7614.740	7616.837		0.1	0.2			13128.81			
7614.914	7617.011			0.2			13128.51			
7615.059	7617.156		0.2	0.3			13128.26	3E 3	2B13	P 3
7615.436	7617.533		1.5b	1.4	0.2		13127.61	2K 4	2B 2	P 1
7615.715	7617.811			0.1			13127.13	V 1	2C 0	R 1
7615.964	7618.061		0.1	0.4	0.1		13126.70	3D 0	2B 7	R 3
7616.434	7618.531			0.2b			13125.89	6c 0	2a 6	Q 3
7616.434	7618.531			0.2b			13125.89	3f 0	2c 2	R 2
7616.788	7618.885		0.1	0.3			13125.28	W 2	2B12	P 6
7617.142	7619.239		0.2	0.4	0.2		13124.67			
7617.485	7619.582		0.2	0.3			13124.08	3F 0	2B 8	R 3
7617.833	7619.930		0.2	0.3			13123.48	4e 2	3b 1	R 1
7618.112	7620.209			0.2			13123.00			
7618.228	7620.325		0.1				13122.80			
7618.326	7620.424			0.3	0.2		13122.63			
7618.681	7620.778			0.3			13122.02	Z 2	2B11	R 4
7618.924	7621.022			0.1			13121.60			
7619.174	7621.272		0.1	0.2			13121.17			
7619.691	7621.789			0.3			13120.28			
7619.854	7621.951		0.1		0.3		13120.00			
7620.005	7622.102			0.3			13119.74			
7620.318	7622.416		0.8	1.5	0.3		13119.20	3F 3	2C 3	R 2
7620.609	7622.706		0.2	0.3			13118.70			
7620.905	7623.003	G	2.3	2.0	1.8		13118.19	3b 2	2a 1	R 0
7621.213	7623.311			0.2			13117.66			
7621.410	7623.508		0.2	0.4	0.2		13117.32			
7621.742	7623.840		0.3	0.4	0.1		13116.75	Y 2	2C 2	Q 3
7621.742	7623.840		0.3	0.4	0.1		13116.75	Y 2	2B12	R 2
7621.742	7623.840		0.3	0.4	0.1		13116.75	3f 1	2c 3	R 1
7622.090	7624.189		1.0	1.3	0.4		13116.15	Y 2	2B12	R 1
7622.381	7624.479		0.2	0.3			13115.65			
7622.671	7624.770	G	2.7	2.1	1.7		13115.15	2K 3	2B 1	P 2
7622.671	7624.770	G	2.7	2.1	1.7		13115.15	4f 2	3b 1	Q 3
7622.985	7625.084			0.2			13114.61			
7623.195	7625.293		0.5	1.2	1.6		13114.25			
7623.311	7625.409			0.7	1.0		13114.05			
7623.532	7625.630	G	2.1	1.5	1.7		13113.67	3b 5	2a 3	P 3
7623.828	7625.927			0.3	0.3		13113.16			
7624.241	7626.340			0.2	0.2		13112.45	3D 3	2B13	R 4
7624.520	7626.619		0.2	0.3			13111.97	Y 2	2B12	R 0
7624.741	7626.840			0.1			13111.59	3f 1	2c 3	R 1

λ_{air}	λ_{vac}	Ref.	I_1	I_2	I_3	I_4	ν	Upper	Lower	Br.
7624.974	7627.073		0.1	0.3			13111.19	3E 3	2B13	P 2
7624.974	7627.073		0.1	0.3			13111.19	4b 0	2a 3	P 4
7625.305	7627.404		0.2	0.3			13110.62			
7625.538	7627.637						13110.22			
7626.195	7628.295		0.6	1.2	1.5		13109.09			
7626.515	7628.615		0.2	0.4			13108.54	U 1	2C 1	R 1
7626.998	7629.098		0.2	0.4			13107.71	3D 1	2B 9	P 2
7626.998	7629.098		0.2	0.4			13107.71	3f 0	2c 2	Q 4
7627.336	7629.435		0.1				13107.13			
7627.504	7629.604				0.6		13106.84			
7627.644	7629.744			0.1			13106.60			
7627.918	7630.017		0.2	0.5	0.2		13106.13	2A 4	2B 3	P 5
7628.133	7630.233			0.1			13105.76			
7628.302	7630.402			0.1			13105.47			
7628.500	7630.600		0.3	0.4	0.1		13105.13	4g 0	2a 4	P 5
7628.750	7630.850			0.1			13104.70			
7628.896	7630.996			0.2	0.4		13104.45			
7629.187	7631.287		0.3	0.5			13103.95	3D 2	2B11	P 4
7629.256	7631.357				0.2		13103.83			
7629.472	7631.572		0.5	1.1	1.4		13103.46			
7629.722	7631.823		0.5	1.2	0.5		13103.03			
7630.048	7632.149		0.3	0.8	0.5		13102.47			
7630.345	7632.446			0.1			13101.96			
7630.549	7632.650			0.2			13101.61			
7630.805	7632.906			0.4			13101.17			
7630.875	7632.976		0.1		0.3b		13101.05			
7631.202	7633.302		0.4	0.6			13100.49	3F 2	2B12	Q 4
7631.202	7633.302		0.4	0.6			13100.49	3D 0	2B 7	R 2
7631.202	7633.302		0.4	0.6			13100.49	Z 2	2B11	P 2
7631.475	7633.576		0.7	1.6	2.5		13100.02	3b 2	2a 1	R 5
7631.807	7633.908			0.2			13099.45	3E 3	2B13	Q 1
7631.994	7634.095		0.2	0.2	0.2b		13099.13	3D 0	2B 7	R 0
7632.279	7634.380			0.2	0.2b		13098.64			
7632.559	7634.660				0.3b		13098.16			
7632.862	7634.963		0.2	0.2	0.4b		13097.64	3f 1	2c 3	Q 4
7633.154	7635.255		0.2	0.6	1.4		13097.14			
7633.352	7635.453				0.4d		13096.80			
7633.550	7635.651		0.3	0.8	0.8d		13096.46	4c 0	2a 4	R 1
7633.754	7635.855			0.3			13096.11	3E 0	2B 8	R 5
7634.039	7636.141			0.2	0.4d		13095.62	3F 3	2B14	Q 3
7634.139	7636.240		0.1				13095.45			
7634.261	7636.362				0.3		13095.24	6c 0	2a 6	Q 2
7634.453	7636.555				1.0		13094.91			
7634.675	7636.777		0.1	0.3			13094.53			
7635.106	7637.208	1		0.4	1.5b		13093.79			
7635.223	7637.325		0.4	1.3			13093.59	2K 2	2B 0	P 5
7635.456	7637.558			0.2			13093.19			
7635.666	7637.768		0.2d	0.3	0.9		13092.83	3F 3	2C 3	R 2
7635.666	7637.768		0.2d	0.3	0.9		13092.83	3D 0	2B 7	R 1
7635.999	7638.101			0.2	0.4		13092.26			
7636.209	7638.311				0.5		13091.90			
7636.430	7638.532			0.2	0.3		13091.52	3D 3	2B13	R 1
7636.576	7638.678				0.3		13091.27			
7636.664	7638.766				0.4		13091.12			
7636.774	7638.877			0.2			13090.93			
7636.891	7638.993				0.4		13090.73	3D 3	2B13	R 2
7637.078	7639.180		1.0	1.3	0.6		13090.41	W 2	2C 2	P 4
7637.224	7639.326		0.6				13090.16			
7637.346	7639.449				0.5b		13089.95			
7637.626	7639.729			0.1	0.4		13089.47			

λ_{air}	λ_{vac}	Ref.	I_1	I_2	I_3	I_4	ν	Upper	Lower	Br.
7637.854	7639.956				0.3		13089.08			
7638.023	7640.126		0.3				13088.79	3D 3	2B13	R 3
7638.198	7640.301			0.1	0.3b		13088.49			
7638.414	7640.517			0.2			13088.12	3D 3	2B13	R 0
7638.414	7640.517			0.2			13088.12	u 0	2a 2	R 2
7638.507	7640.610		0.1				13087.96			
7638.834	7640.937		0.2	0.3			13087.40			
7638.957	7641.060				0.2		13087.19			
7639.138	7641.241		0.2	0.7	0.3		13086.88			
7639.681	7641.784		0.4	0.9	0.4		13085.95	2K 3	2B 1	R 4
7639.961	7642.064			0.2			13085.47	4e 3	3b 2	Q 4
7640.247	7642.350			0.3	0.2b		13084.98			
7640.574	7642.677			0.1			13084.42			
7640.884	7642.987				0.2		13083.89			
7641.094	7643.197		0.2	0.2			13083.53	3f 0	2c 2	Q 5
7641.433	7643.536			0.3			13082.95			
7641.736	7643.840	G	2.5	1.9	0.5		13082.43	2K 4	2B 2	P 2
7642.122	7644.226		0.2	0.2			13081.77			
7642.455	7644.559			0.1			13081.20			
7642.741	7644.845			0.2	0.8		13080.71			
7642.817	7644.921		0.1		0.3		13080.58			
7643.080	7645.184		0.4	0.6	0.5		13080.13	4e 1	3b 0	R 0
7643.401	7645.505		0.9	1.4	1.7		13079.58			
7643.699	7645.804		1.0	1.5	1.6		13079.07	3F 0	2C 0	P 5
7643.951	7646.055			0.1			13078.64			
7644.126	7646.230		0.3	0.9	0.2		13078.34			
7644.360	7646.464			0.1			13077.94			
7644.617	7646.721			0.3			13077.50	3D 0	2B 7	P 1
7645.032	7647.137	G	2.1	1.8	1.8		13076.79	3F 0	2C 0	P 3
7645.032	7647.137	G	2.1	1.8	1.8		13076.79	3F 0	2C 0	P 5
7645.336	7647.441				0.1		13076.27			
7645.406	7647.511			0.3d			13076.15			
7645.746	7647.850		0.2	0.7	0.3		13075.57	3D 2	2B11	P 5
7646.131	7648.236		0.1	0.3			13074.91	6c 0	2a 6	Q 1
7646.465	7648.570		0.1	0.1	0.1b		13074.34			
7646.617	7648.722			0.2			13074.08			
7646.728	7648.833			0.7			13073.89			
7646.839	7648.944		0.4	0.6	0.4		13073.70			
7647.243	7649.348		0.2	0.5	0.1		13073.01			
7647.535	7649.640	G	2.4	2.2	2.0		13072.51	3F 0	2C 0	P 3
7647.646	7649.751						13072.32	U 1	2C 1	Q 1
7647.880	7649.986	G	2.2	2.1	2.1		13071.92	3F 0	2C 0	P 4
7648.255	7650.360			0.3			13071.28			
7648.389	7650.495		0.2b	0.3	0.2		13071.05			
7648.893	7650.998			0.2			13070.19	3f 1	2c 3	Q 3
7648.986	7651.092				0.1		13070.03			
7649.191	7651.297			0.3			13069.68			
7649.361	7651.466		0.2	0.2			13069.39			
7649.654	7651.759		0.1	0.2			13068.89			
7649.917	7652.023		2.1	0.7	0.1		13068.44	Y 2	2C 2	P 3
7649.917	7652.023		2.1	0.7	0.1		13068.44	3e 1	2c 3	R 4
7650.110	7652.216		0.1	0.3			13068.11			
7650.385	7652.491			0.4	0.3		13067.64	Y 2	2B12	P 1
7650.678	7652.784				0.6		13067.14	3F 2	2C 2	Q 3
7650.678	7652.784				0.6		13067.14	3f 2	2c 4	R 1
7650.783	7652.889	G	3.2	2.6	1.9		13066.96	3F 3	2C 3	R 1
7651.088	7653.194						13066.44			
7651.199	7653.305		0.5	0.4			13066.25			
7651.433	7653.540		0.2	0.2			13065.85			
7651.744	7653.850		0.9	1.3	1.5		13065.32			

λ_{air}	λ_{vac}	Ref.	I_1	I_2	I_3	I_4	ν	Upper	Lower	Br.
7651.826	7653.932	G					13065.18			
7651.931	7654.038		0.3	0.6	0.6		13065.00			
7652.306	7654.412		0.1	0.1			13064.36	Z 1	2B 9	R 2
7652.669	7654.776		0.2	0.3d			13063.74			
7652.898	7655.004			0.2			13063.35			
7652.980	7655.086		0.2				13063.21			
7653.103	7655.209			0.3			13063.00			
7653.437	7655.543		0.1	0.2			13062.43	4E 1	2B15	Q 2
7653.437	7655.543		0.1	0.2			13062.43	4D 0	2B13	R 6
7653.841	7655.948		0.5	0.5	0.1		13061.74			
7654.163	7656.270		2.2	1.0	0.4		13061.19	U 1	2C 1	R 2
7654.216	7656.323						13061.10	3F 2	2B12	R 2
7654.492	7656.599	G	2.7	2.3	2.0		13060.63	3F 0	2C 0	P 6
7654.492	7656.599	G	2.7	2.3	2.0		13060.63	3F 2	2C 2	Q 3
7654.779	7656.886		0.3	0.5	0.4		13060.14			
7654.902	7657.009			0.3			13059.93			
7655.130	7657.238		0.1	0.2			13059.54			
7655.541	7657.648		0.2	0.3d			13058.84			
7655.975	7658.082		0.2	0.8	0.2		13058.10			
7656.332	7658.440		0.1	0.1			13057.49			
7656.667	7658.774		0.1	0.2			13056.92			
7657.018	7659.126		0.2	0.2			13056.32			
7657.265	7659.372			0.3			13055.90			
7657.570	7659.677			0.2			13055.38			
7657.793	7659.900			0.2			13055.00			
7658.080	7660.188		0.4	0.8	0.2		13054.51	Z 1	2B 9	R 1
7658.186	7660.293						13054.33			
7658.367	7660.475			0.2			13054.02	3D 2	2B11	P 6
7658.543	7660.651		0.3b	0.5	0.3		13053.72			
7658.784	7660.892			0.2			13053.31			
7659.007	7661.115		0.4	0.4			13052.93			
7659.312	7661.420		0.5	1.0	0.8		13052.41			
7659.676	7661.784		0.2	0.2			13051.79	4E 0	2B13	R 2
7659.928	7662.037		0.4	0.3			13051.36			
7660.292	7662.401	G	2.5	2.2	1.7		13050.74	3F 2	2C 2	Q 2
7660.292	7662.401	G	2.5	2.2	1.7		13050.74	3F 0	2C 0	P 4
7660.768	7662.876	G	2.2	2.1	2.0		13049.93	3F 2	2C 2	Q 4
7660.768	7662.876	G	2.2	2.1	2.0		13049.93	4b 0	2a 3	P 5
7660.768	7662.876	G	2.2	2.1	2.0		13049.93	3B 2	2A 0	P 2
7660.885	7662.994						13049.73			
7661.041	7663.149						13049.47	3f 1	2c 3	Q 3
7661.190	7663.299			0.4			13049.21			
7661.466	7663.575	G	3.1	2.6	1.9		13048.74	3F 2	2C 2	Q 2
7661.772	7663.880		0.6	0.5	0.3		13048.22			
7662.089	7664.198		0.3	0.4			13047.68			
7662.324	7664.433			0.2			13047.28			
7662.564	7664.674		0.3	1.0	1.8		13046.87	3b 2	2a 1	R 6
7662.864	7664.973		0.1	0.1			13046.36			
7662.987	7665.097			0.2			13046.15	4e 2	3b 1	R 0
7663.181	7665.290				0.1		13045.82			
7663.240	7665.349			0.2			13045.72			
7663.557	7665.667		0.9	1.4	1.7		13045.18			
7663.810	7665.919			0.1			13044.75			
7663.962	7666.072			0.2			13044.49			
7664.139	7666.248			0.3	0.0		13044.19			
7664.456	7666.566			0.2			13043.65	u 0	2a 2	R 1
7664.750	7666.860		0.3	0.9	0.5		13043.15			
7664.909	7667.018			0.6	0.4		13042.88			
7665.008	7667.118		0.3	0.5			13042.71			
7665.396	7667.506	G	0.6	1.5	2.0		13042.05	3b 1	2a 0	P 6

λ_{air}	λ_{vac}	Ref.	I_1	I_2	I_3	I_4	ν	Upper	Lower	Br.
7665.626	7667.735		0.2				13041.66			
7665.837	7667.947			0.2			13041.30			
7666.184	7668.294		0.4	0.4			13040.71	3F 1	2B10	Q 4
7666.184	7668.294		0.4	0.4			13040.71	3e 0	2c 2	R 4
7666.537	7668.647		0.5	1.2	1.5		13040.11			
7666.660	7668.770			0.4			13039.90			
7666.901	7669.012		0.2	0.1			13039.49			
7667.013	7669.123				0.1		13039.30	3D 1	2B 9	P 3
7667.013	7669.123				0.1		13039.30	3D 1	2B 9	P 5
7667.125	7669.235			0.2			13039.11			
7667.436	7669.547		1.5	1.6	1.5		13038.58	2K 3	2₂ 1	P 3
7667.436	7669.547		1.5	1.6	1.5		13038.58	Z 2	2B11	P 3
7667.813	7669.923			0.2			13037.94			
7668.083	7670.194				0.4		13037.48	4F 0	2B13	P 3
7668.171	7670.282		0.3	0.9			13037.33	Z 1	2B 9	R 0
7668.518	7670.629			0.1			13036.74			
7668.813	7670.923		0.9	1.4	1.4		13036.24	U 1	2C 1	R 3
7669.148	7671.259		1.0	1.5	1.6		13035.67			
7669.466	7671.577			0.2	0.2		13035.13			
7669.760	7671.871		1.0	1.5	1.7		13034.63			
7670.031	7672.142			0.2	0.3		13034.17	4c 0	2a 4	R 0
7670.172	7672.283		0.2	0.3			13033.93			
7670.496	7672.607		0.2	0.4	0.2		13033.38			
7670.660	7672.772		0.2	0.4			13033.10			
7670.931	7673.042		0.1	0.1			13032.64	3f 0	2c 2	R 1
7670.931	7673.042		0.1	0.1			13032.64	Y 2	2B12	P 2
7671.749	7673.861		0.1	0.3			13031.25			
7671.920	7674.032			0.2			13030.96			
7672.138	7674.250		0.1	0.3			13030.59			
7672.438	7674.550	G	1.5	1.9	2.0		13030.08	3F 0	2C 0	P 5
7673.004	7675.115		1.0	1.3	0.2		13029.12	2K 4	2B 2	P 3
7673.528	7675.640		1.0	1.7	1.8		13028.23	U 1	2C 1	R 4
7673.952	7676.064		0.8	1.3	1.6		13027.51			
7674.287	7676.400			0.2	0.2		13026.94			
7674.576	7676.688		0.6	1.1	0.9		13026.45			
7674.883	7676.995			0.2	0.2		13025.93			
7675.154	7677.266		0.4	0.8	0.3		13025.47			
7675.578	7677.691		0.2	0.4	0.2b		13024.75			
7675.796	7677.909			0.2			13024.38			
7676.067	7678.180			0.2			13023.92			
7676.391	7678.504			0.2			13023.37			
7676.780	7678.893		0.2	0.3d			13022.71			
7677.146	7679.259			0.1			13022.09			
7677.376	7679.489			0.3			13021.70	3f 0	2c 2	Q 4
7677.959	7680.073		0.2	0.4b			13020.71			
7678.207	7680.320			0.1			13020.29	4c 0	2a 4	Q 4
7678.455	7680.568		0.3	1.0	0.4		13019.87			
7678.714	7680.828		0.4	0.7	0.2		13019.43	4c 1	2a 5	P 4
7679.121	7681.235		0.1	0.3			13018.74			
7679.430	7681.544			0.3			13018.22	u 2	2a 4	P 2
7679.511	7681.624			0.3			13018.08	3D 3	2B13	P 2
7679.688	7681.801						13017.78			
7679.794	7681.908	G	2.4	2.2	1.9		13017.60	3E 0	2C 0	Q 3
7680.053	7682.167			0.4			13017.16			
7680.148	7682.262		0.3				13017.00			
7680.219	7682.332			0.2			13016.88			
7680.390	7682.504			0.4	0.3		13016.59	3f 0	2c 2	R 1
7680.803	7682.917			0.3	0.1		13015.89			
7681.133	7683.247			0.2			13015.33			
7681.369	7683.483		0.1	0.3			13014.93			

λ_{air}	λ_{vac}	Ref.	I_1	I_2	I_3	I_4	ν	Upper	Lower	Br.
7681.599	7683.714			0.2			13014.54			
7681.771	7683.885		0.3				13014.25			
7681.906	7684.021			0.2			13014.02			
7682.219	7684.334		0.4	0.8	0.3		13013.49			
7682.538	7684.653			0.1			13012.95	3f 1	2c 3	Q 2
7682.792	7684.907		0.2	0.3			13012.52	3D 0	2B 7	P 2
7683.093	7685.208			0.2			13012.01			
7683.335	7685.450		0.2	0.3	0.1		13011.60			
7683.695	7685.810		1.0	1.7	1.7		13010.99	3E 0	2C 0	Q 4
7683.695	7685.810		1.0	1.7	1.7		13010.99	3F 3	2B14	Q 2
7683.961	7686.076			0.1			13010.54			
7684.150	7686.265		0.3	0.6	0.3		13010.22			
7684.339	7686.454		1.0	1.2	0.3		13009.90	3E 0	2C 0	Q 2
7684.339	7686.454		1.0	1.2	0.3		13009.90	V 1	2C 0	Q 3
7684.676	7686.791			0.1			13009.33	3f 0	2c 2	Q 3
7684.900	7687.015		0.4	0.4	0.1		13008.95			
7685.213	7687.329		0.9	1.0	1.5		13008.42	3F 0	2C 0	P 6
7685.432	7687.547				1.5		13008.05			
7685.562	7687.677	G	3.0	2.5	2.4		13007.83	3b 2	2a 1	P 1
7685.869	7687.985			0.2			13007.31	V 1	2C 0	Q 2
7685.993	7688.109			0.5	0.2b		13007.10	3f 1	2c 3	Q 2
7686.401	7688.517			0.1			13006.41			
7686.667	7688.783		0.1	0.3			13005.96	u 0	2a 2	R 0
7686.874	7688.990			0.2			13005.61	3A 2	2B13	R 4
7686.874	7688.990			0.2			13005.61	4F 0	2B13	Q 2
7687.098	7689.214		0.9	0.9	0.2		13005.23			
7687.424	7689.539		0.1	0.2			13004.68			
7687.802	7689.918		0.1	0.3d	0.2b		13004.04			
7688.375	7690.492				0.1		13003.07			
7688.677	7690.793		0.2	0.4	0.2b		13002.56	3C 2	2K 3	Q 5
7688.677	7690.793		0.2	0.4	0.2b		13002.56	4c 0	2a 4	Q 3
7689.197	7691.314			0.2			13001.68			
7689.440	7691.556			0.2			13001.27			
7689.694	7691.811		0.1	0.3			13000.84			
7690.085	7692.201			0.4	0.2		13000.18			
7690.369	7692.485	G	2.2	2.0	1.9		12999.70	3E 0	2C 0	R 2
7690.369	7692.485	G	2.2	2.0	1.9		12999.70	3b 5	2a 3	P 4
7690.753	7692.870		0.1	0.3	0.1		12999.05	3a 1	2c 3	R 5
7691.025	7693.142		0.3	0.2	0.1		12998.59			
7691.144	7693.260						12998.39			
7691.428	7693.545		0.1	0.3			12997.91			
7691.824	7693.941		0.5	0.5			12997.24			
7691.948	7694.065						12997.03			
7692.108	7694.225	G	2.7	2.5	2.1		12996.76	3E 0	2C 0	R 1
7692.108	7694.225	G	2.7	2.5	2.1		12996.76	4e 3	3b 2	R 0
7692.232	7694.350						12996.55			
7692.481	7694.598		0.2	0.4			12996.13			
7692.682	7694.800				1.5		12995.79	Z 1	2B 9	P 1
7692.765	7694.883	G	2.4	2.4	2.4		12995.65	3E 0	2C 0	R 3
7692.842	7694.959				1.5		12995.52			
7692.901	7695.019						12995.42			
7693.174	7695.291		0.3	0.4	0.2		12994.96	3e 1	2c 3	R 3
7693.422	7695.540		0.3	0.5	0.1		12994.54	U 1	2C 1	Q 2
7693.825	7695.943			0.2			12993.86			
7694.133	7696.250			0.2			12993.34	4e 1	3b 0	Q 1
7694.476	7696.594			0.2			12992.76			
7694.802	7696.920	G	1.5	1.3	1.5		12992.21			
7695.140	7697.258		0.1	0.2			12991.64			
7695.519	7697.637			0.1			12991.00	4e 3	3b 2	Q 2
7695.945	7698.063		0.1	0.2			12990.28			

λ_{air}	λ_{vac}	Ref.	I_1	I_2	I_3	I_4	ν	Upper	Lower	Br.
7696.242	7698.360		0.1	0.2			12989.78	4b 0	2a 3	P 6
7696.449	7698.567			0.2			12989.43	3F 3	2B14	P 4
7696.686	7698.804			0.2			12989.03	4c 0	2a 4	Q 2
7696.881	7699.000			0.3			12988.70			
7697.101	7699.219			0.2			12988.33			
7697.367	7699.486		0.7	1.0	0.5		12987.88			
7697.693	7699.812			0.2			12987.33	3D 1	2B 9	P 4
7697.788	7699.907		0.1				12987.17			
7698.031	7700.150			0.9			12986.76			
7698.096	7700.215		0.7				12986.65			
7698.245	7700.363		0.9	1.8	1.9		12986.40	3E 0	2C 0	R 4
7698.529	7700.648			0.1			12985.92			
7698.725	7700.844		0.3	0.6	0.2		12985.59			
7699.004	7701.122		0.8	1.4	1.9		12985.12	3F 0	2C 0	P 7
7699.229	7701.348			0.1			12984.74	3f 2	2c 4	Q 3
7699.442	7701.561	G	1.5	1.6	0.6		12984.38	3E 0	2C 0	Q 1
7699.733	7701.852		0.3	0.5	0.3		12983.89			
7700.195	7702.315			0.2d			12983.11			
7700.539	7702.659		0.5	1.0	0.5		12982.53	4e 2	3b 1	Q 1
7700.771	7702.890		0.2	0.3			12982.14			
7700.990	7703.110			0.2			12981.77			
7701.210	7703.329			0.1			12981.40			
7701.465	7703.585			0.1			12980.97			
7701.690	7703.810			0.2			12980.59	Y 2	2B12	P 3
7702.236	7704.356			0.2			12979.67	3F 2	2B12	Q 3
7702.521	7704.641			0.2			12979.19	t 0	2a 2	P 3
7702.586	7704.706	G					12979.08			
7702.907	7705.027			0.3			12978.54	3F 1	2B10	R 2
7702.907	7705.027			0.3			12978.54	Z 2	2B11	P 4
7703.198	7705.318		1.5	1.7	2.2		12978.05	3E 0	2C 0	Q 5
7703.198	7705.318		1.5	1.7	2.2		12978.05	3b 2	2a 1	R 7
7703.613	7705.733		0.1	0.2	0.1		12977.35			
7704.367	7706.488		0.1	0.3			12976.08			
7704.670	7706.790			0.2			12975.57			
7704.830	7706.951			0.6	0.2		12975.30	V 1	2C 0	Q 1
7705.032	7707.153	G	2.5	2.1	1.7		12974.96	3F 1	2C 1	P 3
7705.032	7707.153	G	2.5	2.1	1.7		12974.96	3D 3	2B13	P 3
7705.347	7707.468		0.2		0.1		12974.43			
7705.424	7707.545			0.3			12974.30			
7705.644	7707.765			0.2			12973.93			
7705.982	7708.103			0.4			12973.36			
7706.315	7708.436	G	1.5b	2.0	2.3		12972.80	3E 0	2C 0	R 5
7706.648	7708.769	G	1.5b	1.7	1.5		12972.24	3F 1	2C 1	P 3
7706.761	7708.882						12972.05			
7707.004	7709.125		0.2	0.3			12971.64			
7707.260	7709.381			0.2			12971.21			
7707.551	7709.672		1.0	1.4	0.4		12970.72	2K 4	2B 2	P 4
7707.551	7709.672		1.0	1.4	0.4		12970.72	3e 0	2c 2	R 3
7707.789	7709.910		0.8	0.9	0.5		12970.32			
7708.151	7710.273			0.2	0.2b		12969.71			
7708.395	7710.516			0.1			12969.30	Z 0	2B 7	R 4
7708.692	7710.814			0.2			12968.80			
7708.817	7710.938				0.1		12968.59			
7709.055	7711.176			0.2			12968.19			
7709.393	7711.515		0.8	1.3	1.6		12967.62	Z 0	2B 7	R 2
7709.744	7711.866			0.1d			12967.03			
7710.083	7712.205			0.2			12966.46	Z 0	2B 7	R 3
7710.422	7712.544			0.3	0.1		12965.89			
7710.535	7712.657						12965.70			
7710.850	7712.973		0.2				12965.17			

λ_{air}	λ_{vac}	Ref.	I_1	I_2	I_3	I_4	ν	Upper	Lower	Br.
7711.219	7713.341			0.3			12964.55			
7711.463	7713.585		0.2	0.2			12964.14			
7711.677	7713.799		0.1				12963.78			
7712.082	7714.204			0.2	0.1		12963.10	u 1	2a 3	P 2
7712.421	7714.543		0.4	0.3			12962.53			
7712.706	7714.829			0.3			12962.05	3f 0	2c 2	Q 3
7712.932	7715.055			0.1			12961.67	3E 0	2C 0	Q 7
7713.182	7715.305			0.2			12961.25			
7713.504	7715.627			0.2			12960.71			
7713.855	7715.978		0.2	0.6	0.6		12960.12			
7714.075	7716.198		0.3	0.9	0.7		12959.75			
7714.278	7716.401			0.1			12959.41			
7714.498	7716.621		0.5	1.0	0.6		12959.04	Z 0	2B 7	R 1
7714.914	7717.038			0.1			12958.34			
7715.123	7717.246		0.1	0.2			12957.99			
7715.498	7717.621		0.1	0.3			12957.36			
7715.760	7717.884			0.2			12956.92			
7716.028	7718.152			0.3	0.2		12956.47			
7716.362	7718.485			0.3			12955.91	Z 1	2B 9	P 2
7716.683	7718.807		0.9	1.4	1.7		12955.37	3E 0	2C 0	R 6
7717.082	7719.206		0.3	0.5	0.2		12954.70	X 2	2B13	R 0
7717.297	7719.421			0.1			12954.34			
7717.547	7719.671			0.2	0.1		12953.92			
7717.845	7719.969	G	1.5b	1.7	1.6		12953.42	3B 2	2A 0	P 3
7717.845	7719.969	G	1.5b	1.7	1.6		12953.42	U 1	2C 1	P 2
7717.845	7719.969	G	1.5b	1.7	1.6		12953.42	3E 0	2C 0	Q 6
7718.208	7720.332			0.2			12952.81			
7718.411	7720.535				0.1		12952.47			
7718.500	7720.625		0.2	0.3			12952.32			
7718.828	7720.952		1.0	1.5	0.3		12951.77	3F 3	2C 3	Q 3
7719.114	7721.239	G	2.8	2.3	1.6		12951.29	3F 3	2C 3	Q 2
7719.543	7721.668		0.3	0.3			12950.57	X 2	2B13	R 1
7719.770	7721.894			0.3			12950.19	3F 0	2B 8	Q 4
7720.074	7722.198		0.1	0.2			12949.68			
7720.300	7722.425		0.1	0.2			12949.30			
7720.563	7722.687		0.2	0.3			12948.86	3D 0	2B 7	P 3
7720.563	7722.687		0.2	0.3			12948.86	u 2	2a 4	P 3
7720.843	7722.968		0.4	1.0	0.5		12948.39			
7721.123	7723.248						12947.92			
7721.225	7723.350			0.2			12947.75			
7721.445	7723.570		0.4	0.6	0.5		12947.38			
7721.767	7723.892	G	2.2	2.0	1.8		12946.84	3F 1	2C 1	P 4
7721.767	7723.892	G	2.2	2.0	1.8		12946.84	V 0	2B 7	R 2
7722.155	7724.280		0.2	0.3			12946.19			
7722.489	7724.614			0.2	0.2		12945.63			
7722.692	7724.817			0.2	0.3		12945.29			
7722.901	7725.026		0.1	0.3	0.1		12944.94			
7723.014	7725.140	G					12944.75			
7723.205	7725.331		1.0	1.5	1.5		12944.43	3F 1	2C 1	P 4
7723.432	7725.557		0.3	0.7			12944.05	Z 0	2B 7	R 0
7723.605	7725.730			0.1			12943.76			
7723.802	7725.927			0.4			12943.43			
7723.998	7726.124	1	0.2	0.2	1.5b		12943.10			
7724.201	7726.327			0.3			12942.76			
7724.452	7726.578	G	1.5b	1.8	1.7		12942.34	3E 0	2C 0	Q 7
7724.452	7726.578	G	1.5b	1.8	1.7		12942.34	2K 3	2B 1	P 4
7724.607	7726.733				1.0		12942.08			
7724.834	7726.960			0.2	0.4		12941.70			
7725.079	7727.205		0.2	0.8	0.3		12941.29			
7725.318	7727.444			0.1	0.3		12940.89			

λ_{air}	λ_{vac}	Ref.	I_1	I_2	I_3	I_4	ν	Upper	Lower	Br.
7725.521	7727.647		0.1	0.3	0.2		12940.55			
7725.807	7727.933		0.5	1.1	0.4		12940.07			
7725.921	7728.047				0.2		12939.88	X 2	2B13	R 2
7726.016	7728.142				0.1		12939.72	3F 2	2B12	R 1
7726.142	7728.268			0.2			12939.51			
7726.434	7728.561	G	1.5	1.0	0.3b		12939.02			
7726.792	7728.919		0.2	0.3	0.2b		12938.42			
7727.103	7729.230			0.2			12937.90			
7727.354	7729.480		0.2	0.5	0.2b		12937.48			
7727.521	7729.648			0.2			12937.20	X 2	2B13	R 3
7727.730	7729.857		0.7	1.0	0.3		12936.85			
7728.113	7730.239			0.2			12936.21			
7728.274	7730.401			0.1			12935.94			
7728.531	7730.658			0.2			12935.51			
7728.650	7730.777		0.1				12935.31			
7728.907	7731.034		0.4	0.7	0.4		12934.88			
7729.170	7731.297		0.9	1.6	2.0		12934.44	3E 0	2C 0	R 7
7729.439	7731.566			0.1			12933.99			
7729.606	7731.734			0.1			12933.71			
7729.857	7731.985		0.0	0.2			12933.29			
7730.049	7732.176			0.2	0.1		12932.97			
7730.270	7732.397		0.1	0.2			12932.60	X 2	2B13	P 1
7730.593	7732.720		0.9	1.2	1.6		12932.06			
7730.957	7733.085		0.1	0.2	0.1		12931.45	4E 0	2B13	R 0
7731.214	7733.342			0.2			12931.02			
7731.543	7733.671			0.2	0.2		12930.47			
7731.926	7734.054				0.3		12929.83			
7732.081	7734.209		0.2d	0.3			12929.57			
7732.368	7734.496				0.1		12929.09	4e 1	3b 0	Q 2
7732.458	7734.586		0.4	0.6			12928.94			
7732.751	7734.879	G	2.5	2.2	2.1		12928.45	3E 1	2C 1	R 3
7732.751	7734.879	G	2.5	2.2	2.0		12928.45	3b 2	2a 1	P 2
7733.056	7735.184			0.2			12927.94			
7733.319	7735.448	G	2.4	2.0	1.8		12927.50	3E 1	2C 1	R 2
7733.319	7735.448	G	2.4	2.0	1.8		12927.50	3D 3	2B13	P 6
7733.732	7735.861		0.2	0.3			12926.81			
7733.948	7736.076		0.2	0.3	0.1		12926.45			
7734.259	7736.387			0.2			12925.93	3D 3	2B13	P 4
7734.552	7736.680			0.2			12925.44			
7734.875	7737.004		0.2	0.3	0.1		12924.90	Y 2	2B12	P 4
7735.055	7737.183			0.3			12924.60			
7735.300	7737.429			0.2			12924.19			
7735.497	7737.626			0.1			12923.86	3a 1	2c 3	R 4
7735.677	7737.806			0.1			12923.56			
7735.881	7738.010			0.2			12923.22			
7736.132	7738.261			0.2			12922.80	3f 0	2c 2	Q 2
7736.360	7738.489		0.2	0.6	0.5		12922.42	3F 1	2C 1	P 5
7736.641	7738.770			0.1			12921.95			
7736.839	7738.968			0.2			12921.62	Z 2	2B11	P 5
7737.048	7739.177			0.2			12921.27	4e 1	3b 0	P 2
7737.329	7739.459		1.0	1.6	0.7		12920.80	3E 0	2C 0	Q 1
7737.737	7739.866		0.1	0.3			12920.12			
7738.024	7740.154			0.2			12919.64			
7738.276	7740.405			0.2			12919.22	4e 2	3b 1	Q 2
7738.456	7740.585		0.1	0.3			12918.92			
7738.749	7740.879			0.2			12918.43			
7738.947	7741.076		0.1	0.3			12918.10			
7739.108	7741.238			0.2			12917.83			
7739.372	7741.502		0.9	1.1	1.5		12917.39			
7739.732	7741.862		0.2	0.3	0.2		12916.79	3F 1	2B10	P 5

λ_{air}	λ_{vac}	Ref.	I_1	I_2	I_3	I_4	ν	Upper	Lower	Br.
7740.031	7742.161	G	1.5	1.8	1.9		12916.29	3F 1	2C 1	P 5
7740.439	7742.569		0.2	0.3	0.1		12915.61			
7740.684	7742.815			0.1			12915.20			
7740.960	7743.090			0.2			12914.74			
7741.230	7743.360		0.8	0.9	0.2		12914.29	3e 1	2c 3	R 2
7741.643	7743.774			0.2			12913.60			
7741.997	7744.128		0.2	0.8	0.2		12913.01	Z 1	2B 9	P 3
7742.117	7744.248			0.4	0.3		12912.81			
7742.363	7744.494		0.4	0.9	0.1		12912.40			
7742.741	7744.872			0.2			12911.77			
7743.077	7745.207		0.1	0.2b			12911.21			
7743.275	7745.405	G		1.0			12910.88			
7743.466	7745.597		0.2	0.5			12910.56	2K 4	2B 2	P 5
7743.712	7745.843		0.3	0.9	1.5		12910.15	3E 0	2C 0	R 8
7743.712	7745.843		0.3	0.9	1.5		12910.15	3C 2	2K 3	R 3
7744.114	7746.245		0.1	0.2			12909.48			
7744.780	7746.911	G	2.1	0.1			12908.37			
7744.918	7747.049		0.2				12908.14			
7745.062	7747.194			0.2			12907.90			
7745.344	7747.476		0.6	1.1	1.5		12907.43			
7745.536	7747.668			0.1			12907.11			
7745.740	7747.872		0.9	1.3	1.7		12906.77	3f 0	2c 2	Q 2
7746.088	7748.220			0.1			12906.19			
7746.388	7748.520			0.2			12905.69			
7746.683	7748.814			0.2			12905.20	V 0	2B 7	R 3
7746.917	7749.049			0.3			12904.81			
7747.151	7749.283			0.1			12904.42			
7747.391	7749.523			0.2			12904.02			
7747.487	7749.619		0.1				12903.86			
7747.739	7749.871			0.2			12903.44			
7747.937	7750.069			0.2			12903.11			
7748.117	7750.250			0.2			12902.81			
7748.280	7750.412			0.2			12902.54			
7748.640	7750.772		0.6	1.1	1.4		12901.94			
7748.982	7751.115			0.2			12901.37			
7749.415	7751.547		0.1	0.2			12900.65	3F 0	2B 8	R 2
7749.727	7751.860		0.2b	0.3			12900.13	3D 0	2B 7	P 4
7750.100	7752.232			0.2			12899.51			
7750.442	7752.575		0.1	0.4			12898.94			
7751.109	7753.242		0.4	1.2	0.3		12897.83			
7751.524	7753.657		0.3	0.5	0.3		12897.14			
7751.842	7753.976		0.2	0.2			12896.61	3e 0	2c 2	R 2
7752.131	7754.264	G	1.5b	1.6	0.4		12896.13	3E 1	2C 1	R 1
7752.131	7754.264	G	1.5b	1.6	0.4		12896.13	4e 2	3b 1	P 2
7752.419	7754.553		0.2	0.3			12895.65			
7752.726	7754.860		0.3	1.1	0.9		12895.14			
7752.949	7755.082			0.1			12894.77			
7753.165	7755.299		0.5	1.2	1.0		12894.41	3b 2	2a 1	R 8
7753.484	7755.617			0.1			12893.88	Z 0	2B 7	P 1
7753.754	7755.888			0.2			12893.43			
7754.067	7756.201		1.0	1.3	1.5		12892.91	U 1	2C 1	Q 3
7754.211	7756.345						12892.67			
7754.326	7756.460			0.1			12892.48			
7754.560	7756.694		0.9	0.6			12892.09	3f 0	2c 2	P 6
7754.789	7756.923		1.5b	1.5			12891.71	3F 2	2C 2	P 3
7754.789	7756.923		1.5b	1.5			12891.71	u 2	2a 4	P 4
7755.114	7757.248		1.5	1.5	2.2		12891.17	3b 1	2a 0	P 7
7755.505	7757.639			0.2			12890.52	u 1	2a 3	P 3
7755.769	7757.904			0.2			12890.08			
7756.022	7758.156			0.2			12889.66			

λ_{air}	λ_{vac}	Ref.	I$_1$	I$_2$	I$_3$	I$_4$	ν	Upper	Lower	Br.
7756.203	7758.337			0.2			12889.36			
7756.455	7758.590			0.3			12888.94			
7756.714	7758.849		0.8	1.5	1.5		12888.51	3F 1	2C 1	P 6
7756.714	7758.849		0.8	1.5	1.5		12888.51	3F 2	2B12	Q 2
7757.166	7759.300	G	2.4	1.9	1.5		12887.76	3F 2	2C 2	P 3
7757.166	7759.300	G	2.4	1.9	1.5		12887.76	3f 1	2c 3	P 5
7757.551	7759.686		0.2	0.2			12887.12	4E 0	2B13	P 4
7757.846	7759.981			0.2			12886.63	4e 1	3b 0	P 3
7758.267	7760.402		0.1	0.2			12885.93	3F 1	2B10	Q 3
7758.574	7760.709			0.1			12885.42	3D 3	2B13	P 5
7758.773	7760.908			0.1			12885.09	X 2	2B13	P 2
7758.972	7761.107			0.2			12884.76			
7759.140	7761.276			0.1			12884.48	4E 0	2B13	Q 4
7759.351	7761.486		0.2	0.6	0.2		12884.13			
7759.592	7761.727			0.2			12883.73			
7759.881	7762.017			0.3b			12883.25	4d 1	3b 0	R 0
7760.218	7762.354		0.9	1.3	1.7		12882.69	3E 0	2C 0	R 9
7760.218	7762.354		0.9	1.3	1.7		12882.69	3F 0	2B 8	P 5
7760.628	7762.764			0.2			12882.01			
7760.887	7763.023			0.2			12881.58			
7761.164	7763.300			0.2			12881.12	Z 1	2B 9	P 4
7761.164	7763.300			0.2			12881.12	4d 1	3b 0	P 1
7761.448	7763.583			0.2			12880.65			
7761.586	7763.722		0.2	0.4			12880.42	2K 6	2B 4	R 2
7761.851	7763.987			0.3			12879.98			
7762.159	7764.295		1.0	1.6			12879.47			
7762.327	7764.463			0.4			12879.19	U 1	2B10	R 4
7762.605	7764.741		0.2	0.3			12878.73			
7762.840	7764.976		0.4	0.2			12878.34			
7763.039	7765.175			0.3			12878.01			
7763.298	7765.434			0.1			12877.58			
7763.660	7765.796			0.1			12876.98			
7763.967	7766.104			0.2			12876.47			
7764.299	7766.435		0.6	0.9			12875.92			
7764.474	7766.610			0.5			12875.63			
7764.854	7766.990		0.2	0.3			12875.00	4e 2	3b 1	P 3
7765.107	7767.244		1.5	1.2			12874.58	Y 2	2C 2	P 4
7765.300	7767.437			0.1			12874.26	4e 1	3b 0	Q 3
7765.493	7767.630		0.5	1.0			12873.94	2K 6	2B 4	R 1
7765.831	7767.968			0.1			12873.38			
7766.072	7768.209			0.2			12872.98			
7766.265	7768.402			0.2			12872.66			
7766.524	7768.662		0.2	0.4			12872.23	4c 0	2a 4	P 2
7766.784	7768.921		1.0	1.3			12871.80	3E 0	2C 0	Q 2
7767.182	7769.320			0.2			12871.14	3D 0	2B 7	P 5
7767.466	7769.603			0.1			12870.67			
7767.755	7769.893			0.2			12870.19			
7767.924	7770.062			0.3			12869.91			
7768.124	7770.261	G					12869.58			
7768.353	7770.491		0.5	1.0	1.4		12869.20	3F 2	2B12	P 4
7768.770	7770.907			0.3			12868.51	4e 2	3b 1	Q 3
7769.096	7771.233		0.5	1.1	1.7		12867.97			
7769.724	7771.862		0.1	0.4			12866.93			
7770.013	7772.152		0.1	0.2			12866.45			
7770.249	7772.387		0.1	0.5			12866.06	4e 3	3b 2	P 6
7770.533	7772.671		0.3	0.5			12865.59			
7770.847	7772.985			0.2			12865.07			
7771.070	7773.209		0.2	0.4			12864.70			
7771.391	7773.529			0.2d			12864.17	3f 1	2c 3	P 4
7771.656	7773.795			0.1d			12863.73			

λ_{air}	λ_{vac}	Ref.	I_1	I_2	I_3	I_4	ν	Upper	Lower	Br.
7772.001	7774.139		0.2	0.3			12863.16			
7772.152	7774.291			0.2			12862.91			
7772.424	7774.562			0.5			12862.46			
7772.533	7774.671		0.6				12862.28	Z 2	2B11	P 6
7772.877	7775.016			0.0			12861.71	4E 0	2B13	P 2
7772.877	7775.016			0.0			12861.71	3F 1	2C 1	P 7
7773.270	7775.409			0.1			12861.06	4D 0	2B13	R 5
7773.590	7775.729			0.3			12860.53			
7773.729	7775.868			0.2			12860.30	3A 2	2B13	R 3
7774.044	7776.183						12859.78			
7774.267	7776.406	G	2.3	2.1	1.9		12859.41	3E 0	2C 0	P 2
7774.267	7776.406	G	2.3	2.1	1.9		12859.41	2K 6	2B 4	R 0
7774.636	7776.775		0.2	0.3			12858.80			
7774.975	7777.114		0.4	0.7			12858.24			
7775.307	7777.447		0.3	0.4			12857.69	Z 0	2B 7	P 2
7775.567	7777.707			0.2			12857.26			
7775.839	7777.979		0.1	0.3			12856.81	3a 2	2c 4	R 3
7775.839	7777.979		0.1	0.3			12856.81	3f 1	2c 3	P 3
7775.839	7777.979		0.1	0.3			12856.81	4e 2	3b 1	P 4
7776.081	7778.221			0.2			12856.41	4e 3	3b 2	P 2
7776.378	7778.518			0.3			12855.92			
7776.795	7778.935		0.1	0.2			12855.23			
7777.031	7779.171			0.3			12854.84	4e 1	3b 0	P 4
7777.364	7779.504		0.9	1.3	1.7		12854.29			
7777.697	7779.837			0.1			12853.74			
7778.302	7780.442			0.2d			12852.74			
7778.653	7780.793		0.3	0.9			12852.16	3E 0	2C 0	R10
7778.653	7780.793		0.3	0.9			12852.16	3f 1	2c 3	P 3
7778.883	7781.023			0.3			12851.78	3a 1	2c 3	P 3
7779.179	7781.320			0.2			12851.29	4e 3	3b 2	P 4
7779.506	7781.647		0.9	1.2			12850.75	u 2	2a 4	P 5
7779.767	7781.907		1.5b	1.5			12850.32	V 1	2C 0	P 2
7779.888	7782.028						12850.12			
7780.106	7782.247		0.2	0.3			12849.76	2K 4	2B 2	P 6
7780.523	7782.664		1.0	1.5			12849.07	3F 2	2C 2	P 4
7780.959	7783.100		0.1	0.2			12848.35			
7781.789	7783.930			0.2			12846.98			
7782.038	7784.179		0.2	0.4			12846.57			
7782.492	7784.633			0.2			12845.82	3f 1	2c 3	P 5
7783.273	7785.415			0.3			12844.53			
7783.813	7785.955			0.2			12843.64			
7784.104	7786.246		0.6	1.1	0.7		12843.16			
7784.376	7786.518			0.1			12842.71	4e 2	3b 1	P 5
7784.583	7786.725		0.3	1.0			12842.37	3B 2	2A 0	P 4
7784.868	7787.010			0.1			12841.90			
7785.189	7787.331			0.2d			12841.37			
7785.577	7787.719		0.3	0.5			12840.73			
7786.062	7788.204			0.1			12839.93	3f 0	2c 2	P 5
7786.365	7788.508			0.2			12839.43			
7786.662	7788.805		0.2	0.5			12838.94			
7786.771	7788.914						12838.76			
7786.953	7789.096		0.1	0.5			12838.46			
7787.202	7789.345						12838.05			
7787.396	7789.539		0.5	0.8	0.2b		12837.73	t 0	2a 2	P 4
7787.639	7789.782			0.1			12837.33			
7787.845	7789.988			0.1			12836.99			
7788.009	7790.152		0.2	0.3			12836.72			
7788.215	7790.358			0.2			12836.38			
7788.373	7790.516		0.1	0.3			12836.12			
7788.525	7790.668			0.2			12835.87			

λ_{air}	λ_{vac}	Ref.	I_1	I_2	I_3	I_4	ν	Upper	Lower	Br.
7788.749	7790.892		0.1	0.3			12835.50			
7788.901	7791.044			0.3			12835.25			
7789.107	7791.251		0.2	0.4			12834.91			
7789.253	7791.396			0.3			12834.67			
7789.532	7791.675			0.7			12834.21			
7789.805	7791.949	G	3.0	2.7	2.7		12833.76	3b 2	2a 1	P 3
7789.805	7791.949	G	3.0	2.7	2.7		12833.76	X 2	2B13	P 3
7790.200	7792.343		0.5	0.6			12833.11			
7790.455	7792.598		0.2	0.3			12832.69			
7790.734	7792.878			0.2			12832.23			
7790.874	7793.017		0.1	0.3			12832.00	4e 2	3b 1	P 6
7791.244	7793.388			0.3d			12831.39	4d 1	3b 0	R 1
7791.602	7793.746			0.2			12830.80			
7791.961	7794.105		0.1b	0.2			12830.21			
7792.137	7794.281			0.2			12829.92			
7792.331	7794.475			0.2			12829.60			
7792.538	7794.682		0.3	0.8	1.3		12829.26			
7792.938	7795.083			0.2			12828.60			
7793.321	7795.466			0.2			12827.97	4c 0	2a 4	P 3
7793.674	7795.818		0.5	1.0	1.3		12827.39	2K 3	2B 1	P 5
7793.984	7796.128			0.2			12826.88	4e 1	3b 0	P 5
7794.220	7796.365			0.3			12826.49	3F 1	2B10	R 1
7794.220	7796.365			0.3			12826.49	3C 2	2K 3	Q 4
7794.573	7796.718		0.2	0.2			12825.91	4d 2	3b 1	R 0
7794.573	7796.718		0.2	0.2			12825.91	4e 1	3b 0	Q 4
7794.895	7797.040		0.3	0.6	1.1		12825.38	u 0	2a 2	P 2
7795.272	7797.417		0.1				12824.76			
7795.491	7797.636			0.3			12824.40			
7795.685	7797.830			0.2			12824.08	3e 1	2c 3	R 1
7796.044	7798.189			0.3			12823.49			
7796.226	7798.371			0.2			12823.19			
7796.488	7798.633		0.6	0.8	1.3		12822.76			
7796.755	7798.901	G	1.5b	1.7	1.5		12822.32	3E 0	2C 0	P 3
7797.157	7799.302		0.2	0.2			12821.66			
7797.339	7799.485			0.1			12821.36			
7797.576	7799.722		0.2	0.4			12820.97	3A 2	2B13	P 6
7797.978	7800.123			0.2			12820.31			
7798.440	7800.586			0.2d			12819.55	4e 2	3b 1	Q 4
7798.568	7800.714				0.2		12819.34			
7798.787	7800.933			0.3d			12818.98	V 1	2C 0	P 3
7798.860	7801.006	G					12818.86			
7799.103	7801.249		1.5b	1.8	1.7		12818.46	3E 0	2C 0	R11
7799.103	7801.249		1.5b	1.8	1.7		12818.46	3F 2	2C 2	P 4
7799.407	7801.553			0.2			12817.96	u 1	2a 3	P 4
7799.578	7801.724			0.3			12817.68			
7799.803	7801.949			0.1			12817.31			
7800.028	7802.174		0.3	0.5			12816.94	U 1	2C 1	P 3
7800.028	7802.174		0.3	0.5			12816.94	4d 2	3b 1	P 1
7800.387	7802.533	G	0.1	0.2			12816.35	3e 1	2c 3	Q 5
7800.667	7802.814			0.2			12815.89			
7800.996	7803.142		0.7	1.2	1.3		12815.35	U 1	2C 1	Q 4
7801.227	7803.374		0.4	1.0	1.2		12814.97	Z 0	2B 7	P 3
7801.684	7803.830			0.1d			12814.22			
7802.019	7804.165						12813.67	3e 0	2c 2	R 1
7802.128	7804.275			0.2			12813.49			
7802.487	7804.634			0.3			12812.90			
7802.554	7804.701		0.1				12812.79			
7802.628	7804.774	G					12812.67			
7802.810	7804.957		1.5	1.7	1.8		12812.37	3E 0	2C 0	Q 3
7802.810	7804.957		1.5	1.7	1.8		12812.37	3F 2	2C 2	P 5

λ_{air}	λ_{vac}	Ref.	I_1	I_2	I_3	I_4	ν	Upper	Lower	Br.
7802.810	7804.957		1.5	1.7	1.8		12812.37	V 0	2B 7	R 1
7803.230	7805.378			0.2			12811.68			
7803.474	7805.621		0.9	0.7			12811.28	4E 0	2B13	Q 3
7803.724	7805.871			0.2			12810.87			
7803.986	7806.133	G	2.2	1.7	1.4		12810.44	3F 3	2C 3	P 3
7804.339	7806.487	G	2.2				12809.86			
7804.449	7806.596		1.5b	2.2	2.2		12809.68	3E 1	2C 1	R 3
7804.857	7807.005		0.2	0.3			12809.01			
7805.003	7807.151			0.2	0.2b		12808.77			
7805.174	7807.321			0.1			12808.49			
7805.424	7807.571		0.1	0.4			12808.08			
7805.753	7807.901	G	2.1	1.9	1.8		12807.54	3E 1	2C 1	R 2
7805.753	7807.901	G	2.1	1.9	1.8		12807.54	3f 2	2c 4	P 3
7806.033	7808.181			0.2			12807.08	4E 0	2B13	Q 1
7806.259	7808.407		0.6	0.9	0.2		12806.71			
7806.600	7808.748		0.6	0.9	1.2		12806.15			
7806.960	7809.108			0.2			12805.56			
7807.277	7809.425	G	1.0	1.5	1.8		12805.04	3E 1	2C 1	R 4
7807.667	7809.815		0.1	0.2			12804.40			
7807.984	7810.133			0.2			12803.88	3F 2	2B12	P 3
7808.283	7810.431		0.2	0.4	0.1		12803.39			
7808.539	7810.688		0.2	0.2			12802.97	2K 6	2B 4	P 1
7808.740	7810.889		0.1	0.3			12802.64			
7808.936	7811.084			0.2			12802.32			
7809.247	7811.395		1.0	1.4	1.5		12801.81	4e 1	3b 0	P 6
7809.692	7811.841		0.2	0.2			12801.08			
7810.046	7812.195		1.5	1.6	1.4		12800.50	3E 1	2C 0	P 2
7810.394	7812.543			0.2			12799.93			
7810.619	7812.768			0.2			12799.56			
7810.857	7813.007		0.2	0.3	0.1		12799.17			
7811.089	7813.239			0.2			12798.79			
7811.309	7813.458		0.1	0.2			12798.43			
7811.510	7813.660			0.2			12798.10			
7811.767	7813.916			0.3			12797.68	3F 1	2B10	P 4
7811.907	7814.057		0.2	0.2			12797.45			
7812.145	7814.295			0.5			12797.06			
7812.243	7814.393		0.4				12796.90			
7812.463	7814.612	G	2.7	2.4	2.1		12796.54	3E 1	2C 1	R 1
7812.463	7814.612	G	2.7	2.4	2.1		12796.54	3b 2	2a 1	R 9
7812.780	7814.930		0.4	0.9	1.5		12796.02			
7813.092	7815.241			0.2			12795.51			
7813.238	7815.388			0.2			12795.27			
7813.470	7815.620		1.5	1.8	2.2		12794.89	3E 1	2C 1	R 5
7813.867	7816.017		0.5	1.8	0.1		12794.24			
7814.228	7816.378			0.1			12793.65			
7814.472	7816.622			0.1			12793.25	3f 0	2c 2	P 4
7814.759	7816.909			0.1			12792.78			
7815.028	7817.178			0.2			12792.34	3E 2	2B12	R 4
7815.339	7817.490		0.9	0.2			12791.83			
7815.596	7817.746			0.3	0.3		12791.41			
7815.816	7817.966			1.8			12791.05	3e 1	2c 3	Q 4
7816.097	7818.248		0.5	0.8	1.1		12790.59			
7816.653	7818.804		0.8	1.1	1.2		12789.68			
7816.971	7819.122			0.2			12789.16	3F 0	2B 8	Q 3
7816.971	7819.122			0.2			12789.16	4c 0	2a 4	P 4
7817.711	7819.862			0.2d			12787.95			
7818.151	7820.302		0.1				12787.23			
7818.438	7820.589			0.2			12786.76			
7818.762	7820.914		0.0	0.2			12786.23			
7819.086	7821.238		0.1	0.4	0.1		12785.70	3f 0	2c 2	P 7

λ_{air}	λ_{vac}	Ref.	I_1	I_2	I_3	I_4	ν	Upper	Lower	Br.
7819.264	7821.415			0.2			12785.41			
7819.472	7821.623		0.6	0.5			12785.07			
7820.248	7822.400		0.1	0.3			12783.80			
7820.395	7822.547			0.2			12783.56			
7820.573	7822.725			0.5	0.2		12783.27			
7820.946	7823.098			0.2			12782.66			
7821.087	7823.239						12782.43	3a 1	2c 3	R 2
7821.276	7823.428		0.3	0.5	1.0		12782.12	3E 1	2C 1	Q 2
7821.276	7823.428		0.3	0.5	1.0		12782.12	4d 2	3b 1	R 1
7821.497	7823.649			0.1			12781.76			
7821.711	7823.863			0.4	0.1		12781.41			
7822.121	7824.273			0.2			12780.74	t 1	2a 3	P 5
7822.268	7824.420			0.1			12780.50	4d 1	3b 0	R 2
7822.537	7824.689		0.9	1.3	1.6		12780.06	3E 1	2C 1	R 6
7822.935	7825.087			0.1			12779.41	4e 1	3b 0	Q 5
7823.596	7825.749			0.2d			12778.33			
7824.019	7826.171		0.1	0.3			12777.64			
7824.355	7826.508			0.1			12777.09			
7824.674	7826.827			0.2			12776.57			
7825.005	7827.158		0.9	1.1	1.2		12776.03			
7825.360	7827.513			0.2			12775.45	X 2	2B13	P 4
7825.660	7827.813		0.3	0.7	0.2		12774.96	3e 0	2c 2	Q 5
7825.660	7827.813		0.3	0.7	0.2		12774.96	4e 2	3b 1	Q 5
7825.966	7828.120		0.9	0.3			12774.46	4E 0	2B13	Q 2
7826.144	7828.297			0.3			12774.17			
7826.383	7828.536		0.1	0.4	0.1		12773.78			
7826.720	7828.873			0.1			12773.23	3F 1	2B10	Q 2
7827.008	7829.161			0.2			12772.76			
7827.265	7829.419			0.2			12772.34			
7827.602	7829.756		0.2	0.5	0.1b		12771.79			
7828.062	7830.216			0.2			12771.04			
7828.252	7830.406		0.2	0.3			12770.73			
7828.473	7830.627		0.1	0.4			12770.37			
7828.718	7830.872		0.2	0.2			12769.97			
7829.043	7831.197		0.3	0.6	0.2		12769.44			
7829.368	7831.522			0.2			12768.91	3f 0	2c 2	P 6
7829.687	7831.841	G	1.5b	1.8	1.6		12768.39	3E 0	2C 0	P 4
7830.171	7832.326		1.5	1.5	1.3		12767.60	3E 1	2C 1	Q 1
7830.300	7832.454						12767.39			
7830.533	7832.687		0.2	0.3			12767.01	4d 3	3b 2	R 0
7830.533	7832.687		0.2	0.3			12767.01	Z 0	2B 7	P 4
7830.889	7833.043		0.2	0.6			12766.43	3C 2	2K 3	P 5
7831.128	7833.283			0.1			12766.04			
7831.361	7833.516			0.2			12765.66			
7831.668	7833.823			0.1			12765.16			
7831.852	7834.007			0.1			12764.86			
7832.036	7834.191			0.2			12764.56			
7832.318	7834.473		0.0	0.2			12764.10			
7832.705	7834.860			0.2			12763.47			
7832.993	7835.148		0.4	1.0	0.9		12763.00			
7833.159	7835.314			0.3			12762.73			
7833.454	7835.609			0.1			12762.25			
7833.650	7835.805		0.3	0.3			12761.93	2K 6	2B 4	P 2
7833.859	7836.014			0.2			12761.59			
7834.153	7836.309	G	1.0	1.5	1.9		12761.11	3E 1	2C 1	R 7
7834.153	7836.309	G	1.0	1.5	1.9		12761.11	3F 3	2C 3	P 4
7834.153	7836.309	G	1.0	1.5	1.9		12761.11	3A 2	2B13	R 2
7834.153	7836.309	G	1.0	1.5	1.9		12761.11	3e 2	2c 4	R 3
7834.436	7836.591			0.1			12760.65			
7834.657	7836.812		0.3	0.6	0.3		12760.29	V 1	2C 0	P 4

λ_{air}	λ_{vac}	Ref.	I_1	I_2	I_3	I_4	ν	Upper	Lower	Br.
7835.025	7837.181		0.1	0.2			12759.69	3e 1	2c 3	Q 3
7835.308	7837.464		1.5	1.2	1.0		12759.23	3E 1	2C 1	Q 3
7835.308	7837.464		1.5	1.2	1.0		12759.23	3f 0	2c 2	P 3
7835.308	7837.464		1.5	1.2	1.0		12759.23	u 1	2a 3	P 5
7835.848	7838.004			0.1			12758.35			
7836.045	7838.201			0.1			12758.03			
7836.315	7838.471			0.2			12757.59			
7836.647	7838.803			0.3			12757.05	U 1	2B10	R 3
7836.954	7839.110	G	1.5	1.5	1.3		12756.55	3E 2	2C 2	R 2
7837.273	7839.430						12756.03			
7837.415	7839.571			0.2			12755.80	3f 0	2c 2	P 5
7837.415	7839.571			0.2			12755.80	3e 0	2c 2	Q 4
7837.869	7840.026			0.2			12755.06			
7838.152	7840.309	G	1.0	1.5	1.6		12754.60	U 1	2C 1	Q 5
7838.416	7840.573			0.1			12754.17			
7838.595	7840.751			0.1			12753.88			
7838.797	7840.954			0.1			12753.55			
7839.031	7841.188		0.2	0.5	0.3		12753.17	4d 3	3b 2	P 1
7839.283	7841.440			0.1			12752.76			
7839.510	7841.667			0.2			12752.39			
7839.744	7841.901			0.1			12752.01			
7840.187	7842.344			0.2			12751.29			
7840.316	7842.473		0.2	0.2			12751.08			
7840.482	7842.639		0.2	0.3			12750.81			
7840.752	7842.910		0.4	0.8	0.2		12750.37			
7841.091	7843.248			0.1			12749.82			
7841.275	7843.432			0.2			12749.52	3E 0	2B 8	R 3
7841.521	7843.679			0.2			12749.12			
7841.767	7843.925			0.2			12748.72			
7841.939	7844.097			0.2			12748.44			
7842.382	7844.540		0.7	0.2			12747.72			
7842.776	7844.934			0.2b			12747.08			
7842.868	7845.026		0.2				12746.93	3f 0	2c 2	P 4
7843.170	7845.328		0.8	1.2	1.1		12746.44	3E 0	2C 0	Q 4
7843.170	7845.328		0.8	1.2	1.1		12746.44	V 0	2B 7	P 4
7843.170	7845.328		0.8	1.2	1.1		12746.44	3C 2	2K 3	R 2
7843.527	7845.685		0.1	0.2			12745.86			
7843.896	7846.054		0.1	0.3			12745.26			
7844.118	7846.276			0.2			12744.90	u 0	2a 2	P 3
7844.395	7846.553			0.3			12744.45	4D 0	2B13	R 4
7844.789	7846.947			0.2			12743.81	3f 0	2c 2	P 3
7845.226	7847.384			0.2			12743.10			
7845.521	7847.680		0.4	0.9	0.9		12742.62			
7845.952	7848.111		1.5b	1.4	1.4		12741.92	3E 2	2C 2	R 4
7846.223	7848.382			0.2			12741.48			
7846.420	7848.579		0.2	0.4	0.1		12741.16			
7846.901	7849.059			0.1			12740.38			
7847.356	7849.515			0.2			12739.64	T 0	2B13	R 2
7847.726	7849.885			0.2			12739.04			
7848.046	7850.206		1.0	1.3	1.6		12738.52	3E 1	2C 1	R 8
7848.046	7850.206		1.0	1.3	1.6		12738.52	4F 0	2B14	R 3
7848.385	7850.544			0.2			12737.97	4e 1	3b 0	Q 6
7848.385	7850.544			0.2			12737.97	3e 1	2c 3	R 5
7848.632	7850.791		0.2	0.4			12737.57	3a 2	2c 4	R 1
7848.829	7850.988			0.2			12737.25			
7849.069	7851.229			0.2			12736.86	4d 2	3b 1	R 2
7849.316	7851.475			0.1			12736.46			
7849.537	7851.697		0.1	0.4	0.3		12736.10			
7849.704	7851.864			0.2			12735.83			
7849.889	7852.049			0.3			12735.53			

λ_{air}	λ_{vac}	Ref.	I_1	I_2	I_3	I_4	ν	Upper	Lower	Br.
7850.222	7852.381		0.9	0.9	0.7		12734.99			
7850.487	7852.647			0.1			12734.56			
7850.678	7852.838		0.1	0.2			12734.25	X 2	2B13	P 5
7850.912	7853.072			0.3			12733.87	3e 0	2c 2	Q 3
7851.270	7853.430			0.2			12733.29			
7851.442	7853.603			0.1			12733.01			
7851.621	7853.781		0.2	0.3			12732.72			
7851.899	7854.059						12732.27	3F 3	2C 3	P 4
7852.250	7854.411			0.1d			12731.70			
7852.688	7854.849			0.2			12730.99	4d 1	3b 0	R 3
7853.027	7855.188			0.2			12730.44			
7853.219	7855.379			0.2			12730.13			
7853.509	7855.669		0.2	0.7	1.5		12729.66	3b 1	2a 0	P 8
7853.509	7855.669		0.2	0.7	1.5		12729.66	4d 3	3b 2	R 1
7853.836	7855.996			0.1			12729.13			
7855.008	7857.169			0.2			12727.23			
7855.255	7857.416			0.2			12726.83			
7855.594	7857.756			0.4			12726.28			
7855.699	7857.861		0.2b				12726.11			
7855.959	7858.120		0.2	0.3			12725.69	4E 2	2B18	Q 2
7856.107	7858.268			0.2			12725.45			
7856.323	7858.484		0.2	0.4			12725.10			
7856.650	7858.812	G	2.2	2.1	2.2		12724.57	3b 2	2a 1	P 4
7856.650	7858.812	G	2.2	2.1	2.2		12724.57	4d 1	3b 0	P 2
7857.058	7859.219		0.1	0.3			12723.91			
7857.422	7859.584			0.3			12723.32			
7858.447	7860.609			0.3			12721.66	3F 0	2B 8	R 1
7858.824	7860.986			0.2			12721.05			
7859.096	7861.258		0.6				12720.61			
7859.182	7861.345		0.9	1.2	0.8		12720.47	3E 2	2C 2	R 1
7859.182	7861.345		0.9	1.2	0.8		12720.47	3e 1	2c 3	R 4
7859.541	7861.703			0.2			12719.89	3e 1	2c 3	Q 2
7859.875	7862.037			0.6	1.0		12719.35			
7860.010	7862.173		0.2	0.2			12719.13			
7860.245	7862.408			0.4			12718.75			
7860.573	7862.735		0.4	1.0	0.4		12718.22	3B 2	2A 0	P 5
7861.030	7863.193			0.1	0.1		12717.48	3e 0	2c 2	R 6
7861.525	7863.688			0.1			12716.68	3a 1	2c 3	R 1
7861.525	7863.688			0.1			12716.68	4e 3	3b 2	R 3
7862.217	7864.380			0.1			12715.56			
7862.440	7864.603			0.2			12715.20			
7862.644	7864.807		0.2	0.8	0.2d		12714.87			
7862.854	7865.017		1.5	1.5	1.3		12714.53	3E 1	2C 1	Q 1
7863.244	7865.407			0.2			12713.90	t 0	2a 2	P 5
7863.491	7865.655	G					12713.50			
7863.844	7866.007			0.2			12712.93	2K 6	2B 4	P 3
7864.141	7866.304		0.6	1.1	1.4		12712.45	3E 1	2C 1	R 9
7864.456	7866.620		0.1	0.2			12711.94			
7865.038	7867.202			0.2			12711.00	Z 0	2B 7	P 5
7865.298	7867.462			0.2			12710.58			
7865.644	7867.808			0.2d			12710.02			
7865.861	7868.025			0.1			12709.67			
7866.189	7868.353		0.1				12709.14			
7866.325	7868.489			0.2			12708.92			
7866.573	7868.737			0.1			12708.52			
7867.241	7869.406			0.2			12707.44	3E 1	2C 1	Q 4
7867.489	7869.653		0.2	0.6	0.3		12707.04			
7867.675	7869.839			0.2			12706.74	3F 0	2B 8	P 4
7867.885	7870.050			0.3			12706.40			
7868.120	7870.285			0.2	0.2		12706.02			

λ_{air}	λ_{vac}	Ref.	I$_1$	I$_2$	I$_3$	I$_4$	ν	Upper	Lower	Br.
7868.350	7870.514			0.2			12705.65			
7868.560	7870.725			0.4			12705.31			
7868.777	7870.942			0.2			12704.96			
7868.907	7871.072						12704.75			
7869.000	7871.165		0.3	0.6			12704.60			
7869.266	7871.431			0.2			12704.17	4e 1	3b 0	Q 7
7869.266	7871.431			0.2			12704.17	3e 0	2c 2	Q 2
7869.440	7871.605		0.1	0.3			12703.89			
7869.768	7871.933			0.2			12703.36			
7869.960	7872.125			0.3			12703.05	3e 1	2c 3	R 4
7870.288	7872.454		0.3	0.7	0.3		12702.52	3E 0	2C 0	P 5
7870.673	7872.838			0.2			12701.90	3e 0	2c 2	R 5
7871.007	7873.173		0.1	0.4			12701.36			
7871.484	7873.650			0.2			12700.59			
7871.720	7873.885		0.2	0.4	0.3		12700.21			
7872.135	7874.301		0.1	0.2			12699.54			
7872.433	7874.599		0.4	0.7	0.3		12699.06			
7872.606	7874.772		0.2	0.1			12698.78			
7872.780	7874.946		0.2	0.5	0.2		12698.50			
7873.276	7875.442		0.1	0.2			12697.70			
7873.648	7875.814		0.7	1.0	1.2		12697.10	2K 3	2B 1	P 6
7873.648	7875.814		0.7	1.0	1.2		12697.10	4e 3	3b 2	R 1
7874.076	7876.242		0.1	0.2			12696.41			
7874.262	7876.428			0.1			12696.11			
7874.473	7876.639		0.5	1.0	0.3		12695.77			
7874.646	7876.813			0.2			12695.49			
7874.833	7876.999		0.4	0.6	0.2		12695.19			
7875.143	7877.309		0.3	0.5	0.3		12694.69			
7875.453	7877.620	G	2.3	2.1	1.7		12694.19	3E 0	2C 0	P 3
7875.869	7878.035		0.2	0.3			12693.52			
7876.098	7878.265			0.2			12693.15	3F 1	2B10	P 3
7876.409	7878.575			0.3			12692.65	2K 5	2B 3	R 1
7876.576	7878.743		0.1	0.3			12692.38			
7876.830	7878.997			0.2			12691.97			
7877.042	7879.209		0.2	0.3	0.1		12691.63	U 1	2B10	R 2
7877.457	7879.625			0.2			12690.96	3E 2	2B12	R 3
7877.457	7879.625			0.2			12690.96	4d 3	3b 2	R 2
7877.457	7879.625			0.2			12690.96	4d 2	3b 1	R 3
7877.774	7879.941			0.2			12690.45			
7878.264	7880.432		0.1	0.2			12689.66			
7878.581	7880.748		0.2	0.4	0.1		12689.15	4e 2	3b 1	Q 6
7878.792	7880.960			0.2			12688.81			
7878.985	7881.152		0.4	0.5			12688.50	2K 5	2B 3	R 2
7879.419	7881.587		0.1				12687.80			
7879.655	7881.823			0.2			12687.42	3a 0	2c 2	R 3
7880.034	7882.202			0.3			12686.81			
7880.699	7882.867			0.2			12685.74			
7881.202	7883.370						12684.93	3e 1	2c 3	R 2
7881.618	7883.787		0.1	0.2			12684.26	2K 5	2B 3	R 0
7881.618	7883.787		0.1	0.2			12684.26	4d 1	3b 0	R 4
7881.854	7884.023		0.2	0.7	0.5		12683.88			
7882.159	7884.327		0.3	0.6	0.3		12683.39			
7882.532	7884.700			0.1			12682.79			
7882.718	7884.887			0.2			12682.49			
7882.930	7885.098		0.1	0.4			12682.15	3C 2	2K 3	Q 3
7883.284	7885.453			0.3			12681.58			
7883.607	7885.776		0.2	0.3			12681.06	V 0	2B 7	R 0
7884.484	7886.653			0.2			12679.65			
7884.801	7886.970			0.2			12679.14			
7885.081	7887.250			0.1			12678.69			

λ_air	λ_vac	Ref.	I_1	I_2	I_3	I_4	ν	Upper	Lower	Br.
7885.255	7887.424	G		0.2			12678.41			
7885.560	7887.729		1.5	1.3	0.9		12677.92	3E 0	2C 0	Q 5
7885.560	7887.729		1.5	1.3	0.9		12677.92	3E 1	2C 1	Q 2
7885.964	7888.134			0.2			12677.27			
7886.319	7888.488		0.2	0.4			12676.70			
7886.561	7888.731			0.4b			12676.31			
7886.692	7888.862		0.3				12676.10	4d 2	3b 1	P 2
7886.953	7889.123	G	2.1	1.7	1.2		12675.68	3D 0	2C 0	R 1
7887.370	7889.540		0.2	0.2			12675.01			
7887.582	7889.752		0.1				12674.67			
7887.775	7889.945		0.1	0.2			12674.36			
7888.074	7890.244		0.6	0.7	0.2		12673.88	3e 1	2c 3	Q 1
7888.391	7890.561			0.2			12673.37	3A 2	2B13	R 1
7888.627	7890.798		0.1	0.4			12672.99	2K 5	2B 3	R 3
7888.627	7890.798		0.1	0.4			12672.99	u 0	2a 2	P 4
7888.970	7891.140			0.1			12672.44			
7889.250	7891.420			0.2			12671.99	T 0	2B13	R 1
7889.250	7891.420			0.2			12671.99	3e 0	2c 2	R 3
7889.250	7891.420			0.2			12671.99	3a 1	2c 3	Q 5
7889.511	7891.682			0.2			12671.57			
7889.804	7891.975			0.3			12671.10	U 1	2B10	R 1
7890.115	7892.286			0.2			12670.60			
7890.389	7892.560			0.2			12670.16			
7890.645	7892.815		0.1	0.4	0.2		12669.75			
7891.068	7893.239	i			0.8		12669.07			
7891.187	7893.358			0.1d			12668.88			
7891.791	7893.962			0.2			12667.91	V 0	2B 7	P 3
7892.003	7894.174			0.2			12667.57			
7892.277	7894.448			0.4			12667.13			
7892.433	7894.604		0.2b	0.3			12666.88			
7892.669	7894.841			0.2			12666.50	X 0	2B 9	R 4
7892.931	7895.102		0.4	0.3			12666.08			
7893.417	7895.589			0.2			12665.30			
7894.178	7896.349			0.2			12664.08			
7894.383	7896.555		0.2	0.4	0.2b		12663.75	3e 1	2c 3	R 1
7894.539	7896.711			0.2			12663.50			
7894.714	7896.886			0.2			12663.22			
7894.832	7897.004		0.2	0.3			12663.03			
7895.100	7897.272		0.2	0.4			12662.60	U 1	2B10	R 0
7895.587	7897.759			0.2			12661.82			
7895.898	7898.071	G	1.0	1.4	1.3		12661.32	3B 3	2A 1	R 2
7896.173	7898.345			0.2			12660.88			
7896.366	7898.539		1.5	1.3	0.4		12660.57	3D 0	2C 0	P 1
7896.697	7898.869		0.2	0.2			12660.04			
7897.003	7899.175		1.5	1.1	0.3		12659.55	U 1	2C 1	P 4
7897.389	7899.562			0.2			12658.93			
7897.614	7899.786		0.2	0.3			12658.57			
7897.882	7900.055		0.8	0.8	0.4		12658.14	3e 0	2c 2	R 2
7898.232	7900.404			0.2			12657.58			
7898.481	7900.654			0.2			12657.18	3A 2	2B13	P 5
7898.712	7900.885			0.2			12656.81			
7898.955	7901.128			0.1			12656.42	2K 6	2B 4	P 4
7898.955	7901.128			0.1			12656.42	3e 0	2c 2	Q 1
7899.199	7901.372		0.1	0.2			12656.03			
7899.442	7901.615			0.2			12655.64			
7899.598	7901.771		0.2	0.2			12655.39	X 0	2B 9	R 3
7899.786	7901.959		0.2	0.2			12655.09			
7900.010	7902.184		0.8	0.7	0.3		12654.73	4D 0	2B13	P 1
7900.304	7902.477	G	2.7	2.4	1.9		12654.26	3D 0	2C 0	Q 1
7900.541	7902.714		1.5		1.0		12653.88	3B 3	2A 1	R 3

λ_{air}	λ_{vac}	Ref.	I_1	I_2	I_3	I_4	ν	Upper	Lower	Br.
7900.947	7903.120		0.2	0.3			12653.23			
7901.184	7903.358		0.1	0.2			12652.85			
7901.396	7903.570		0.6	0.7	0.3		12652.51			
7901.684	7903.857		0.1	0.2			12652.05	3F 0	2B 8	Q 2
7901.946	7904.120		0.6	0.9	0.7		12651.63	4D 0	2B13	R 0
7901.946	7904.120		0.6	0.9	0.7		12651.63	3e 2	2c 4	R 1
7902.165	7904.339			0.2			12651.28			
7902.302	7904.476		0.2				12651.06			
7902.402	7904.576		0.4	0.5			12650.90			
7902.721	7904.895	G	2.4	2.1	1.8		12650.39	3B 3	2A 1	R 1
7903.033	7905.207		0.2	0.2			12649.89	V 0	2B 7	P 5
7903.158	7905.332		0.2	0.3			12649.69			
7903.314	7905.488		0.3	0.3			12649.44			
7903.595	7905.769	G	2.4	2.1	1.7		12648.99	3E 1	2C 1	P 2
7903.595	7905.769	G	2.4	2.1	1.7		12648.99	3E 1	2B10	R 3
7903.595	7905.769	G	2.4	2.1	1.7		12648.99	2K 5	2B 3	R 4
7903.908	7906.082			0.2			12648.49			
7904.176	7906.351		2.0	1.7	1.5		12648.06	2A 3	2B 2	R 2
7904.483	7906.658		0.2				12647.57	3a 1	2c 3	Q 4
7904.483	7906.658		0.2				12647.57	4e 3	3b 2	Q 1
7904.595	7906.770			0.3			12647.39			
7904.776	7906.951		0.1				12647.10			
7904.939	7907.113			0.2			12646.84			
7905.208	7907.382	G	0.2	0.5			12646.41	X 1	2B11	P 2
7905.545	7907.720		0.1	0.2			12645.87			
7905.820	7907.995		0.2	0.4			12645.43			
7906.033	7908.208			0.1			12645.09	2A 3	2B 2	R 3
7906.283	7908.458			0.2			12644.69			
7906.539	7908.714		0.5	0.8	1.4		12644.28			
7906.915	7909.090			0.1			12643.68	3e 0	2c 2	R 1
7907.133	7909.309		0.2	0.1			12643.33	4d 2	3b 1	R 4
7907.290	7909.465			0.2			12643.08			
7907.640	7909.815			0.2			12642.52			
7907.853	7910.028		0.2				12642.18			
7908.172	7910.347		0.2	0.3			12641.67			
7908.416	7910.591			0.2			12641.28			
7908.585	7910.760		0.1	0.1			12641.01			
7908.835	7911.011		1.0	1.0	0.8		12640.61			
7909.179	7911.355			0.1			12640.06			
7909.467	7911.643			0.3			12639.60			
7909.924	7912.100		0.1				12638.87			
7910.187	7912.363			0.2			12638.45			
7910.594	7912.770		0.2	0.4			12637.80			
7911.101	7913.277			0.2			12636.99			
7911.464	7913.640		0.8	0.8	0.4		12636.41			
7911.796	7913.972			0.2			12635.88			
7911.877	7914.053		0.3				12635.75			
7912.090	7914.266		1.7	1.3	0.3		12635.41	2A 3	2B 2	R 1
7912.090	7914.266		1.7	1.3	0.3		12635.41	3C 2	2K 3	P 4
7912.472	7914.648		0.1	0.2			12634.80			
7912.898	7915.074				1.8		12634.12			
7913.129	7915.306		1.0	1.4	1.8		12633.75	3B 3	2A 1	R 4
7913.129	7915.306		1.0	1.4	1.8		12633.75	3a 2	2c 4	Q 2
7913.436	7915.613		0.3	0.9	0.7		12633.26			
7913.743	7915.920		0.2	0.3			12632.77			
7914.232	7916.409			0.2			12631.99			
7914.545	7916.722			0.2			12631.49			
7914.909	7917.086			0.3			12630.91			
7915.047	7917.224			0.2			12630.69			
7915.216	7917.393		0.2	0.4			12630.42			

λ_{air}	λ_{vac}	Ref.	I_1	I_2	I_3	I_4	ν	Upper	Lower	Br.
7915.592	7917.769		0.2	0.3			12629.82	3E 1	2C 1	Q 3
7916.112	7918.290			0.2			12628.99	2K 5	2B 3	P 1
7916.488	7918.666		0.1	0.2			12628.39			
7916.751	7918.929			0.2			12627.97	3E 2	2C 2	R 6
7916.978	7919.156		0.1				12627.61	3a 1	2c 3	Q 3
7917.059	7919.236			0.2			12627.48			
7917.303	7919.481		0.4	0.6			12627.09	X 0	2B 9	R 2
7917.567	7919.745			0.1			12626.67			
7917.830	7920.008		0.2	0.4			12626.25	4d 3	3b 2	P 2
7918.200	7920.378		0.1	0.2			12625.66			
7918.482	7920.660		0.6	0.9	0.8		12625.21	T 0	2B13	R 0
7918.783	7920.961			0.2			12624.73	4D 0	2B13	R 2
7918.902	7921.081			0.2			12624.54			
7919.053	7921.231		0.2	0.6			12624.30			
7919.222	7921.401		0.2	0.2			12624.03			
7919.486	7921.664		1.7	1.3	0.3		12623.61	3B 3	2A 1	R 0
7919.906	7922.085		0.1	0.2b			12622.94			
7920.176	7922.355			0.1			12622.51			
7920.427	7922.606		0.1	0.3			12622.11			
7920.647	7922.825			0.1			12621.76			
7920.810	7922.989			0.2			12621.50			
7920.985	7923.164			0.2			12621.22			
7921.180	7923.359		0.1	0.3			12620.91			
7921.375	7923.554		0.1	0.4			12620.60			
7921.676	7923.855			0.2			12620.12			
7922.078	7924.257			0.2			12619.48			
7922.366	7924.546			0.3			12619.02			
7922.699	7924.878		0.9	1.1	1.4		12618.49			
7922.925	7925.105		0.2	0.2			12618.13			
7923.057	7925.237			0.2			12617.92			
7923.183	7925.362		0.9	0.6			12617.72			
7923.478	7925.657	G	2.8	2.5	2.0		12617.25	3D 0	2C 0	R 2
7923.823	7926.003		0.4				12616.70			
7923.892	7926.072		0.4	0.4			12616.59			
7924.112	7926.292		0.2	0.2			12616.24	3a 0	2c 2	R 2
7924.257	7926.436			0.2			12616.01	3C 2	2K 3	R 1
7924.501	7926.681		0.2	0.4			12615.62			
7924.784	7926.964		0.2	0.1			12615.17			
7924.929	7927.109			0.2			12614.94			
7925.042	7927.222			0.2			12614.76			
7925.299	7927.479		0.1	0.3			12614.35			
7925.576	7927.756			0.2			12613.91	3A 1	2B11	R 2
7925.771	7927.951		0.1	0.2			12613.60			
7925.940	7928.120			0.1			12613.33			
7926.122	7928.303		0.2	0.3			12613.04	5c 2	2a 8	Q 1
7926.122	7928.303		0.2	0.3			12613.04	3a 1	2c 3	Q 2
7926.550	7928.730	G	0.2	0.3			12612.36			
7926.801	7928.982	G	2.3	1.7	1.1		12611.96	2A 3	2B 2	R 0
7926.801	7928.982	G	2.3	1.7	1.1		12611.96	3E 2	2C 2	Q 2
7927.084	7929.265			0.2			12611.51			
7927.260	7929.441		0.6	0.8			12611.23			
7927.354	7929.535		0.5		0.3		12611.08			
7927.556	7929.736		0.2	0.4	0.3		12610.76			
7927.776	7929.956			0.1			12610.41			
7928.014	7930.195			0.2			12610.03			
7928.197	7930.378			0.1			12609.74			
7928.467	7930.648			0.2			12609.31	4D 0	2B13	R 1
7928.775	7930.956			0.1			12608.82			
7929.077	7931.258		0.2	0.4	0.1		12608.34	3e 2	2c 4	R 1
7929.392	7931.573		0.1				12607.84			

λ_{air}	λ_{vac}	Ref.	I_1	I_2	I_3	I_4	ν	Upper	Lower	Br.
7929.574	7931.755		0.2	0.4			12607.55			
7930.417	7932.598		0.1	0.3			12606.21			
7930.599	7932.781		0.2	0.4			12605.92	3E 2	2C 2	Q 1
7930.599	7932.781		0.2	0.4			12605.92	3e 2	2c 4	Q 1
7931.178	7933.360		0.2	0.3			12605.00	X 0	2B 9	R 1
7931.442	7933.624		2.1	1.6	1.2		12604.58	3E 1	2C 1	P 3
7931.442	7933.624		2.1	1.6	1.2		12604.58	3E 0	2C 0	Q 6
7931.549	7933.731						12604.41			
7931.801	7933.983		0.2	0.2			12604.01			
7932.154	7934.335			0.2			12603.45			
7932.317	7934.499			0.2	0.3		12603.19			
7932.506	7934.688			0.3			12602.89			
7932.600	7934.782		0.1	0.2			12602.74	3a 1	2c 3	Q 1
7932.777	7934.959			0.3			12602.46			
7932.877	7935.060		0.2	0.5			12602.30			
7933.198	7935.381	G	2.2	2.2	2.6		12601.79	3b 2	2a 1	P 5
7933.595	7935.777		1.5	0.7b	0.2d		12601.16	3E 3	2B14	Q 6
7933.702	7935.884		0.4				12600.99			
7933.985	7936.168		1.8	1.5	1.1		12600.54	3E 1	2C 1	P 2
7934.370	7936.552		0.1	0.2			12599.93			
7934.678	7936.861		0.2	0.2			12599.44			
7935.119	7937.302		0.2	0.5	0.2		12598.74			
7935.365	7937.547			0.2			12598.35			
7935.440	7937.623		0.1				12598.23	X 1	2B11	P 3
7935.648	7937.831		0.2	0.5			12597.90			
7935.913	7938.095		0.6	1.0	1.6		12597.48	3B 3	2A 1	R 5
7936.240	7938.423		0.2	0.3			12596.96	3E 0	2B 8	R 2
7936.486	7938.669			0.1			12596.57			
7936.650	7938.833			0.2			12596.31	3A 2	2B13	P 4
7936.807	7938.990		0.2	0.3			12596.06	V 0	2B 7	P 2
7937.122	7939.306		1.6	1.5	1.5		12595.56	3E 2	2C 2	R 4
7937.122	7939.306		1.6	1.5	1.5		12595.56	3E 2	2C 2	Q 3
7937.122	7939.306		1.6	1.5	1.5		12595.56	3E 2	2B12	R 2
7937.387	7939.570	G	2.1	1.9	1.9		12595.14	3E 2	2C 2	R 3
7937.765	7939.949		0.9	0.5			12594.54			
7937.979	7940.163		0.2	0.3			12594.20			
7938.314	7940.497		0.3	0.6	0.2		12593.67			
7939.209	7941.393			0.2			12592.25			
7939.341	7941.525			0.2			12592.04			
7939.404	7941.588		0.2				12591.94			
7939.663	7941.847		0.3	0.3			12591.53			
7939.965	7942.149			0.2			12591.05			
7940.243	7942.427			0.2	0.4		12590.61			
7940.394	7942.578			0.1			12590.37			
7940.621	7942.805		0.6	1.2	0.4		12590.01			
7940.949	7943.133		1.6	1.5	1.7		12589.49	3E 2	2C 2	R 5
7941.075	7943.260		0.4				12589.29	V 1	2B 8	R 2
7941.397	7943.581	G	2.1	1.7	1.0		12588.78	3D 1	2C 1	R 1
7941.820	7944.004		0.2	0.2			12588.11			
7942.154	7944.339		0.1	0.3			12587.58			
7942.508	7944.692		2.0	1.7	1.4		12587.02	3D 1	2C 1	P 1
7942.508	7944.692		2.0	1.7	1.4		12587.02	3E 2	2C 2	R 2
7942.747	7944.932	G	0.2	0.2			12586.64			
7942.949	7945.134		0.7	1.2	1.6		12586.32			
7943.202	7945.387		0.3	0.8	0.4		12585.92			
7943.391	7945.576		0.1				12585.62			
7943.587	7945.772			0.2			12585.31			
7943.991	7946.176		0.2	0.3b	0.2b		12584.67	3e 1	2c 3	Q 1
7944.332	7946.517			0.2			12584.13			
7944.691	7946.877		0.2	0.4			12583.56	U 1	2B10	P 2

λ_{air}	λ_{vac}	Ref.	I_1	I_2	I_3	I_4	ν	Upper	Lower	Br.
7944.988	7947.174			0.3			12583.09	2K 5	2B 3	P 2
7944.988	7947.174			0.3			12583.09	3b 6	2a 4	R 2
7945.241	7947.426			0.2			12582.69	3F 0	2B 8	P 3
7945.418	7947.603			0.5	0.3b		12582.41	3B 2	2A 0	P 6
7945.639	7947.824		0.2	0.7			12582.06	4D 0	2B13	P 2
7946.106	7948.292		0.3	0.3			12581.32			
7946.346	7948.532				0.4		12580.94			
7946.630	7948.816			0.2	0.3		12580.49			
7946.997	7949.182		0.1	0.3			12579.91			
7947.066	7949.252				0.3		12579.80			
7947.325	7949.511		1.6	1.4	1.1		12579.39	3E 0	2C 0	P 4
7947.470	7949.656				0.5		12579.16	3A 2	2B13	R 0
7947.660	7949.846				0.3		12578.86			
7947.988	7950.175		0.2	0.4			12578.34			
7948.102	7950.288	1			1.5b		12578.16			
7948.247	7950.434		2.0	1.2	1.1		12577.93	3E 2	2C 2	R 6
7948.507	7950.693				1.0		12577.52			
7948.576	7950.763		0.1	0.2	1.0		12577.41			
7948.804	7950.990				0.4		12577.05			
7948.898	7951.085			0.1			12576.90	X 0	2B 9	R 0
7949.101	7951.287		0.1		0.2		12576.58			
7949.278	7951.464		0.4	0.8	0.4		12576.30	3E 1	2C 1	Q 4
7949.391	7951.578		0.6				12576.12			
7949.562	7951.749				0.2d		12575.85			
7949.701	7951.888		0.2	0.2			12575.63			
7949.758	7951.945				0.2d		12575.54			
7949.853	7952.040		0.2	0.2			12575.39			
7950.081	7952.267			0.3	0.3		12575.03			
7950.137	7952.324		0.4		0.2		12574.94	T 0	2B13	P 4
7950.397	7952.583	G	2.6	2.1	1.5		12574.53	3D 1	2C 1	Q 1
7950.713	7952.900		0.2				12574.03			
7950.820	7953.007		0.3	0.3	0.2b		12573.86			
7951.061	7953.248		0.1	0.2			12573.48	3E 2	2B12	Q 6
7951.295	7953.482		0.5	0.7	0.5		12573.11	3C 2	2K 3	Q 2
7951.756	7953.944		0.2	0.3	0.2		12572.38			
7952.016	7954.203			0.2			12571.97			
7952.307	7954.494		0.2	0.3			12571.51			
7952.579	7954.766			0.2			12571.08			
7952.794	7954.981		0.2	0.3			12570.74			
7952.971	7955.158		0.1	0.1			12570.46			
7953.167	7955.355		0.4	0.4			12570.15			
7953.483	7955.671	G	2.5	2.1	1.6		12569.65	3E 2	2C 2	R 1
7953.578	7955.766		2.5				12569.50	3e 1	2c 3	P 2
7953.578	7955.766		2.5				12569.50	3e 0	2c 2	Q 1
7953.888	7956.076		0.3	0.3	0.1		12569.01			
7954.186	7956.374		0.2	0.3			12568.54			
7954.407	7956.595		0.2	0.2			12568.19	4d 1	3b 0	P 3
7954.654	7956.842		0.3	0.6	0.7		12567.80			
7955.002	7957.190			0.2			12567.25			
7955.198	7957.387			0.1			12566.94			
7955.439	7957.627		0.3	0.7	0.5		12566.56			
7955.730	7957.919		0.3	1.0	0.2		12566.10			
7956.078	7958.267			0.2	0.4		12565.55			
7956.344	7958.533		0.2	0.3			12565.13	4e 3	3b 2	Q 3
7956.509	7958.698		0.1		0.1		12564.87			
7956.718	7958.907		0.4	0.4			12564.54			
7956.783	7958.972						12564.44	3e 1	2c 3	P 3
7957.041	7959.230	G	2.5	2.0	1.4		12564.03	3b 6	2a 4	R 1
7957.383	7959.572		0.3				12563.49			
7957.484	7959.673		0.3	0.3	0.1		12563.33			

λ_{air}	λ_{vac}	Ref.	I_1	I_2	I_3	I_4	ν	Upper	Lower	Br.
7957.820	7960.009		0.2				12562.80			
7958.238	7960.427		0.2	0.2			12562.14			
7958.555	7960.744		1.0	1.3	1.4		12561.64	3E 2	2C 2	R 7
7958.726	7960.915		0.3	0.4			12561.37			
7958.941	7961.131			0.1			12561.03			
7959.208	7961.397		0.1	0.1			12560.61	3E 2	2C 2	Q 4
7959.600	7961.790			0.3d			12559.99			
7959.797	7961.986		0.1	0.2			12559.68			
7960.146	7962.335		0.3	0.2			12559.13	3e 1	2c 3	Q 2
7960.266	7962.455		0.2				12558.94	4D 0	2B13	P 6
7960.551	7962.741		0.6d	1.4	1.6		12558.49			
7960.862	7963.051			0.1			12558.00			
7961.115	7963.305		0.1				12557.60	3E 1	2B10	R 2
7961.534	7963.724		0.2	0.5	0.2		12556.94	3b 1	2a 0	P 9
7961.851	7964.041		0.2	0.3	0.1		12556.44	3A 2	2B13	P 3
7962.219	7964.409		0.3	0.3			12555.86			
7962.523	7964.713	G	2.1	1.7	1.3		12555.38	3D 0	2C 0	Q 2
7962.523	7964.713	G	2.1	1.7	1.3		12555.38	T 0	2B13	P 2
7962.523	7964.713	G	2.1	1.7	1.3		12555.38	T 0	2B13	P 3
7962.878	7965.068		0.2	0.7	1.4		12554.82			
7963.011	7965.202		0.5				12554.61			
7963.176	7965.367		0.3	0.3	0.2		12554.35			
7963.481	7965.671	G	1.7	1.7	1.9		12553.87	3b 3	2a 2	R 2
7963.690	7965.881		0.3				12553.54	3e 1	2c 3	P 4
7963.938	7966.128			0.2			12553.15			
7964.229	7966.420		0.7	0.7	0.2b		12552.69	3a 0	2c 2	R 1
7964.458	7966.648			0.2			12552.33			
7964.674	7966.864		0.1	0.2			12551.99			
7964.953	7967.143		0.1	0.3			12551.55	T 0	2B13	P 1
7965.264	7967.455		0.1	0.5			12551.06			
7965.594	7967.785			0.2			12550.54			
7966.082	7968.274			0.1			12549.77			
7966.349	7968.540			0.2			12549.35			
7966.520	7968.712			0.2			12549.08			
7966.736	7968.928		0.1	0.3			12548.74			
7966.889	7969.080			0.2			12548.50			
7967.130	7969.321		0.3	0.7	0.3		12548.12			
7967.441	7969.633	G	1.7	1.9	2.4		12547.63	3b 3	2a 2	R 3
7967.847	7970.039			0.3	0.1		12546.99			
7968.063	7970.255		0.2				12546.65	3e 0	2c 2	P 2
7968.247	7970.439		1.8	1.2	0.1		12546.36	2A 3	2B 2	P 1
7968.495	7970.687			0.2			12545.97			
7968.781	7970.973	G	2.0	1.9	1.6		12545.52	3D 0	2C 0	R 3
7968.781	7970.973	G	2.0	1.9	1.6		12545.52	3a 1	2c 3	P 1
7969.194	7971.386		0.1	0.2			12544.87	3b 6	2a 4	R 0
7969.461	7971.653		0.2	0.3			12544.45			
7969.594	7971.786			0.2			12544.24			
7969.823	7972.015		0.5	0.5			12543.88	X 1	2B11	P 4
7970.160	7972.352	G	2.5	2.2	2.3		12543.35	3b 3	2a 2	R 1
7970.560	7972.752		0.7	0.5			12542.72			
7970.859	7973.051			0.2			12542.25			
7971.227	7973.420		1.5	1.1	0.2		12541.67	Z 0	2C 0	R 1
7971.469	7973.661		0.2	0.6	0.7		12541.29	3E 2	2C 2	R 8
7971.774	7973.967		0.2	0.3			12540.81	3e 0	2c 2	P 3
7972.073	7974.265		0.2	0.2			12540.34			
7972.384	7974.577			0.2			12539.85			
7972.607	7974.800		0.1	0.3			12539.50			
7972.899	7975.092		0.1	0.4			12539.04	3e 2	2c 4	R 2
7973.115	7975.308			0.1			12538.70			
7973.389	7975.582		0.3	0.4	0.2		12538.27			

λ_{air}	λ_{vac}	Ref.	I_1	I_2	I_3	I_4	ν	Upper	Lower	Br.
·7973.707	7975.900		0.2	0.4			12537.77	U 1	2B10	P 3
7974.127	7976.320			0.2			12537.11			
7974.305	7976.498		0.1				12536.83	3e 1	2c 3	P 5
7974.496	7976.689			0.2			12536.53			
7974.845	7977.039		1.5	0.9			12535.98	3D 2	2C 2	P 1
7974.845	7977.039		1.5	0.9			12535.98	4g 1	2a 6	P 1
7974.845	7977.039		1.5	0.9			12535.98	W 1	2B11	R 4
7975.202	7977.395		0.1	0.2			12535.42	4F 0	2B14	P 5
7975.507	7977.701		0.4	0.5			12534.94	3C 2	2K 3	P 3
7975.793	7977.987			0.2			12534.49			
7975.972	7978.165			0.1			12534.21			
7976.207	7978.401		0.2	0.5			12533.84	4d 2	3b 1	P 3
7976.207	7978.401		0.2	0.5			12533.84	3e 0	2c 2	Q 2
7976.271	7978.464						12533.74			
7976.411	7978.605			0.1			12533.52			
7976.678	7978.872		0.1	0.3			12533.10			
7977.054	7979.248			0.2			12532.51			
7977.264	7979.458			0.2			12532.18			
7977.505	7979.700		0.9	0.7	0.2		12531.80			
7977.926	7980.120		0.2	0.2			12531.14			
7978.098	7980.292		0.2	0.3			12530.87			
7978.581	7980.776		0.1	0.2			12530.11			
7978.823	7981.018		0.2	0.4			12529.73			
7979.104	7981.298		1.5	1.0	0.1		12529.29	3D 0	2C 0	P 2
7979.104	7981.298		1.5	1.0	0.1		12529.29	3e 1	2c 3	Q 3
7979.416	7981.610			0.2			12528.80			
7979.664	7981.859		0.2	0.5			12528.41			
7979.804	7981.999			0.1			12528.19			
7980.046	7982.241	G		0.2			12527.81	X 0	2B 9	P 1
7980.276	7982.471		0.2	0.4	0.3		12527.45			
7980.607	7982.802		0.2	0.2			12526.93	W 2	2B13	R 4
7980.970	7983.165		0.2	0.3			12526.36	U 1	2C 1	P 5
7981.282	7983.477	G	2.1	1.7	1.4		12525.87	3E 1	2C 1	P 4
7981.505	7983.701			0.5	0.3		12525.52			
7981.690	7983.885		0.2	0.2			12525.23			
7981.964	7984.159		0.4	1.1	1.8		12524.80	2K 5	2B 3	P 3
7981.964	7984.159		0.4	1.1	1.8		12524.80	3b 3	2a 2	R 4
7982.193	7984.389			0.2			12524.44			
7982.397	7984.593			0.2			12524.12			
7982.557	7984.752		0.4	0.3			12523.87			
7982.735	7984.931			0.2			12523.59			
7982.856	7985.052			0.2			12523.40			
7983.041	7985.237		0.1	0.3			12523.11			
7983.366	7985.562		0.3	0.2			12522.60			
7983.551	7985.747		0.2				12522.31			
7983.653	7985.849		0.2	0.3			12522.15			
7983.940	7986.136		0.3	0.4			12521.70	X 0	2B 9	P 5
7984.259	7986.455	G	2.3	1.9	1.6		12521.20	3B 3	2A 1	P 1
7984.259	7986.455	G	2.3	1.9	1.6		12521.20	3C 2	2K 3	R 0
7984.565	7986.761		0.2	0.1			12520.72	U 1	2B10	P 5
7984.737	7986.933		0.3	0.5			12520.45			
7984.967	7987.163		0.2	0.2			12520.09			
7985.254	7987.450		1.5b	1.3	0.7		12519.64	3D 2	2C 2	R 1
7985.573	7987.769		0.9	0.4			12519.14			
7985.892	7988.088	G	2.8	2.3	1.8		12518.64	3D 1	2C 1	R 2
7986.179	7988.375		0.4	0.2			12518.19			
7986.313	7988.509		0.4	0.3			12517.98			
7986.587	7988.784		0.4	0.9	0.8		12517.55	3E 2	2C 2	R 9
7986.587	7988.784		0.4	0.9	0.8		12517.55	3e 0	2c 2	P 4
7986.925	7989.122		1.5	0.8			12517.02			

λ_{air}	λ_{vac}	Ref.	I_1	I_2	I_3	I_4	ν	Upper	Lower	Br.
7987.225	7989.422		0.2	0.2			12516.55			
7987.544	7989.741	G	1.8	1.6	1.6		12516.05	3E 1	2C 1	Q 5
7987.544	7989.741	G	1.8	1.6	1.6		12516.05	3b 3	2a 2	R 0
7988.049	7990.246		0.3	0.3d	0.1		12515.26			
7988.285	7990.482		0.2	0.2			12514.89	3A 2	2B13	P 1
7988.559	7990.756		0.6	0.5			12514.46			
7988.974	7991.171			0.2			12513.81			
7989.236	7991.433		0.2	0.3			12513.40			
7989.836	7992.034		0.1	0.2			12512.46			
7990.206	7992.404		0.2	0.2			12511.88			
7990.455	7992.653			0.2			12511.49			
7990.717	7992.915		0.1	0.2			12511.08			
7990.947	7993.145		0.3	0.4			12510.72	U 1	2B10	P 4
7991.305	7993.503		0.4	0.4			12510.16	3E 1	2B10	Q 5
7991.637	7993.835	G	2.5	2.1	1.8		12509.64	3E 1	2C 1	P 3
7991.637	7993.835	G	2.5	2.1	1.8		12509.64	3A 2	2B13	P 2
7991.874	7994.072		1.0	1.0	0.4		12509.27			
7992.072	7994.270			0.1			12508.96	3A 1	2B11	R 1
7992.251	7994.449		0.2	0.2			12508.68			
7992.551	7994.749		0.1	0.1			12508.21			
7992.774	7994.973			0.2			12507.86	3E 0	2B 8	P 5
7992.979	7995.177		0.1	0.2			12507.54	3E 2	2B12	Q 5
7993.190	7995.388			0.2			12507.21			
7993.362	7995.561		0.6	0.4			12506.94			
7993.663	7995.861	G	2.5	1.8	1.3		12506.47	2A 3	2B 2	P 2
7994.002	7996.200		0.2				12505.94			
7994.085	7996.283		0.2	0.2			12505.81			
7994.283	7996.482		0.1	0.2			12505.50			
7994.967	7997.166		0.9	0.2			12504.43	3e 2	2c 4	Q 2
7995.044	7997.242	G	0.8	0.9	0.7		12504.31	3E 2	2C 2	Q 5
7995.440	7997.639			0.2			12503.69			
7995.510	7997.709		0.2				12503.58			
7995.939	7998.138		0.1				12502.91			
7996.105	7998.304			0.1			12502.65			
7996.335	7998.535			0.2			12502.29			
7996.553	7998.752		0.1	0.3			12501.95			
7996.681	7998.880			0.2			12501.75			
7996.854	7999.053			0.2			12501.48			
7996.949	7999.149		0.2	0.3			12501.33	3E 2	2B12	R 1
7997.276	7999.475	G	1.8	1.9	2.3		12500.82	2A 2	2B 1	R 4
7997.436	7999.635		0.4				12500.57	3C 2	2K 3	Q 1
7997.711	7999.910		0.4	0.7	0.4		12500.14			
7997.903	8000.102		0.2	0.1			12499.84			
7998.165	8000.365	G	2.2	1.6	1.1		12499.43	Z 0	2C 0	Q 1
7998.440	8000.640		0.7	0.4			12499.00			
7998.767	8000.966				0.2		12498.49			
7998.831	8001.031			0.2			12498.39			
7998.984	8001.184		0.2b	0.2			12498.15	3a 1	2c 3	P 2
7999.323	8001.523		0.1	0.1			12497.62	4g 1	2a 6	R 0
7999.605	8001.805		0.2	0.2			12497.18	3e 1	2c 3	Q 4
7999.957	8002.157		0.3				12496.63			
8000.047	8002.247			0.2			12496.49	4d 3	3b 2	P 3
8000.213	8002.414		0.2	0.3			12496.23			
8000.617	8002.817		0.9	0.2			12495.60	X 0	2B 9	P 2
8000.789	8002.990	G	2.1	1.7	1.3		12495.33	3E 2	2C 2	P 2
8001.148	8003.349		0.2b	0.2			12494.77	3e 0	2c 2	Q 3
8001.449	8003.650			0.2			12494.30			
8001.744	8003.944		0.5	0.7	0.4		12493.84			
8001.833	8004.034			0.7	1.1		12493.70			
8001.910	8004.111		0.3	0.4			12493.58			

λ_{air}	λ_{vac}	Ref.	I$_1$	I$_2$	I$_3$	I$_4$	ν	Upper	Lower	Br.
8002.192	8004.393		0.1	0.2			12493.14			
8002.403	8004.604		0.1	0.4			12492.81			
8002.660	8004.861			0.2			12492.41	4D 0	2B13	P 3
8003.063	8005.264			0.2			12491.78			
8003.191	8005.392		0.1				12491.58			
8003.416	8005.617		0.5	0.7	0.3		12491.23			
8003.698	8005.899			0.2			12490.79			
8004.313	8006.514				0.2		12489.83			
8004.447	8006.649		0.2				12489.62			
8004.582	8006.783		0.2		0.2		12489.41			
8004.825	8007.027			0.1			12489.03			
8005.011	8007.213			0.2			12488.74			
8005.191	8007.392			0.2			12488.46			
8005.370	8007.572		0.2	0.3			12488.18			
8005.447	8007.649				0.3		12488.06	3E 2	2B12	P 6
8005.447	8007.649				0.3		12488.06	3e 0	2c 2	P 5
8005.633	8007.835			0.1	0.2		12487.77			
8005.883	8008.085		0.2	0.3	1.0		12487.38			
8006.152	8008.354	1	0.2	0.3	1.5b		12486.96			
8006.486	8008.688				0.5		12486.44			
8006.569	8008.771				0.5		12486.31			
8006.646	8008.848			0.3			12486.19			
8006.986	8009.188	G	1.8	1.6	2.2		12485.66	3E 2	2C 2	Q 1
8006.986	8009.188	G	1.8	1.6	2.2		12485.66	3b 3	2a 2	R 5
8007.345	8009.547		0.2	0.4	0.2		12485.10			
8007.659	8009.862		0.2	0.3			12484.61			
8007.980	8010.182	G	1.5b	1.7	1.3		12484.11	Z 0	2C 0	R 2
8008.378	8010.580		0.2				12483.49	3E 0	2B 8	Q 5
8008.654	8010.856		0.5				12483.06			
8008.859	8011.062				0.2		12482.74			
8009.129	8011.331			0.2			12482.32			
8009.353	8011.556			0.2			12481.97			
8009.449	8011.652				0.1		12481.82			
8009.610	8011.813		0.9	0.6			12481.57			
8010.033	8012.236		0.2	0.5			12480.91	3e 1	2c 3	P 2
8010.457	8012.660				0.1		12480.25			
8010.778	8012.981		0.2b	0.3	0.1		12479.75			
8011.092	8013.296			0.2			12479.26			
8011.298	8013.501			0.2			12478.94			
8011.516	8013.719		0.1	0.3			12478.60	4F 0	2B14	R 2
8011.876	8014.079		0.1	0.2			12478.04	3a 2	2c 4	P 3
8012.255	8014.458		0.2	0.2			12477.45			
8012.409	8014.612		0.2				12477.21			
8012.678	8014.882		1.5	1.0	0.3		12476.79	Z 0	2C 0	P 1
8012.903	8015.107			0.2	0.2		12476.44			
8013.089	8015.293		0.9	0.6	0.2		12476.15			
8013.346	8015.550	G	2.8	2.2	1.6		12475.75	3D 2	2C 2	R 2
8013.346	8015.550	G	2.8	2.2	1.6		12475.75	3D 2	2C 2	Q 1
8013.667	8015.871		0.3	0.2	0.3		12475.25			
8013.963	8016.167		2.0	1.5	1.0		12474.79	3D 1	2C 1	Q 2
8014.085	8016.289				0.5		12474.60			
8014.278	8016.482		0.2	0.2	0.3		12474.30			
8014.573	8016.778			0.3			12473.84			
8014.702	8016.906	1			1.5b		12473.64			
8014.785	8016.990			0.3			12473.51			
8014.972	8017.176			0.2			12473.22			
8015.107	8017.311				1.0		12473.01			
8015.197	8017.401				1.0		12472.87	3E 2	2B12	P 5
8015.274	8017.478			0.2			12472.75			
8015.409	8017.613			0.4			12472.54			

λ_{air}	λ_{vac}	Ref.	I_1	I_2	I_3	I_4	ν	Upper	Lower	Br.
8015.499	8017.703		0.1				12472.40	3a 0	2c 2	Q 4
8015.608	8017.812			0.3			12472.23			
8015.724	8017.928		0.1		0.2		12472.05			
8015.820	8018.024			0.2			12471.90			
8015.929	8018.134				0.3		12471.73			
8016.103	8018.307		0.3	0.3			12471.46			
8016.173	8018.378				0.1		12471.35			
8016.456	8018.661			0.2	0.2		12470.91			
8016.668	8018.873				0.3		12470.58			
8016.733	8018.937		0.6	0.6			12470.48			
8017.131	8019.336		0.1				12469.86	4F 0	2B14	Q 4
8017.318	8019.523		0.2	0.2			12469.57			
8017.691	8019.896			0.3			12468.99			
8018.051	8020.256		0.1	0.3			12468.43			
8018.205	8020.410			0.2			12468.19			
8018.430	8020.635		0.2	0.6	0.2		12467.84			
8018.758	8020.964	G	2.2	2.3	2.2		12467.33	3D 0	2C 0	R 4
8018.758	8020.964	G	2.2	2.3	2.2		12467.33	3E 0	2B 8	R 1
8019.093	8021.298			0.2			12466.81			
8019.189	8021.395		0.2				12466.66			
8019.363	8021.568	G	0.7	1.7	2.1		12466.39	3b 2	2a 1	P 6
8019.556	8021.761		1.5		1.3		12466.09	4D 0	2B13	P 5
8019.556	8021.761		1.5		1.3		12466.09	3e 1	2c 3	Q 5
8019.871	8022.077		1.5	1.0	0.4		12465.60			
8020.302	8022.508		0.1	0.2			12464.93			
8020.566	8022.772			0.2			12464.52			
8020.669	8022.875		0.2				12464.36			
8020.791	8022.997			0.2			12464.17	V 1	2B 8	R 1
8020.997	8023.203			0.2			12463.85	4e 3	3b 2	P 3
8021.139	8023.345			0.2			12463.63	V 0	2C 0	R 2
8021.139	8023.345			0.2			12463.63	X 0	2B 9	P 3
8021.345	8023.551		0.2	0.3			12463.31	3E 3	2B14	Q 5
8021.518	8023.725			0.3			12463.04	3E 3	2B14	R 3
8021.737	8023.943		1.5	1.2	1.7		12462.70	Z 1	2C 1	R 1
8021.918	8024.124		0.6	0.6			12462.42			
8022.181	8024.388			0.1			12462.01			
8022.304	8024.510		0.2	0.2			12461.82			
8022.632	8024.838		0.2	0.3			12461.31			
8022.799	8025.006				0.4		12461.05			
8022.967	8025.173		1.5	1.7	1.7		12460.79	3E 0	2C 0	P 5
8023.121	8025.328	G	1.8				12460.55	2A 3	2B 2	P 3
8023.398	8025.605		0.2	0.1			12460.12	4D 0	2B13	P 4
8023.398	8025.605		0.2	0.1			12460.12	3e 0	2c 2	P 2
8023.553	8025.760		0.7	0.2			12459.88			
8023.856	8026.062		1.0	0.7			12459.41			
8024.100	8026.307		1.5	1.2	0.8		12459.03	3E 2	2C 2	Q 2
8024.293	8026.500			0.1			12458.73	3E 1	2B10	R 1
8024.461	8026.668		0.1	0.2			12458.47	3e 0	2c 2	P 6
8024.719	8026.925		0.2	0.4			12458.07			
8025.047	8027.254			0.2			12457.56			
8025.427	8027.634			0.2			12456.97	3a 1	2c 3	P 3
8025.833	8028.040		0.1	0.2			12456.34			
8026.117	8028.324		0.3	0.4			12455.90	3D 1	2C 1	P 2
8026.271	8028.479		0.1				12455.66	3e 0	2c 2	Q 4
8026.484	8028.691		0.3	0.5	0.9		12455.33	Y 1	2B 9	R 2
8026.484	8028.691		0.3	0.5	0.9		12455.33	2K 5	2B 3	P 4
8026.742	8028.949			0.3			12454.93			
8027.161	8029.368			0.2			12454.28			
8027.277	8029.484		0.2	0.2			12454.10			
8027.419	8029.626		0.1		0.1		12453.88	3F 1	2B11	Q 6

λ_{air}	λ_{vac}	Ref.	I_1	I_2	I_3	I_4	ν	Upper	Lower	Br.
8027.663	8029.871		0.3	0.4			12453.50	3a 0	2c 2	Q 3
8027.896	8030.103		1.5	1.3	1.3		12453.14	3E 2	2C 2	P 3
8028.231	8030.439		0.1	0.2			12452.62			
8028.424	8030.632			0.1			12452.32			
8028.624	8030.832		0.1	0.2			12452.01			
8029.017	8031.225		0.1	0.2			12451.40			
8029.391	8031.599	G		0.2			12450.82	Y 1	2B 9	R 1
8029.391	8031.599	G		0.2			12450.82	3b 3	2a 2	R 6
8029.707	8031.916			0.2			12450.33			
8030.062	8032.270			0.2			12449.78			
8030.430	8032.638		1.0	0.8			12449.21			
8030.701	8032.909		0.1	0.2			12448.79			
8031.391	8033.600		0.1	0.2			12447.72			
8031.701	8033.909		0.2	0.2			12447.24			
8032.049	8034.258		1.8	1.6	1.2		12446.70	3B 3	2A 1	P 2
8032.475	8034.684		0.2	0.2			12446.04			
8032.701	8034.910		0.1	0.1			12445.69			
8032.959	8035.168		0.2	0.3			12445.29			
8033.495	8035.704		0.2	0.2			12444.46			
8033.773	8035.982		1.5	0.8			12444.03	X 0	2B 9	P 4
8034.315	8036.524						12443.19	3a 0	2c 2	Q 2
8034.741	8036.951		0.1				12442.53			
8034.954	8037.164		0.2				12442.20			
8035.497	8037.706		0.1				12441.36	W 2	2B13	R 3
8035.910	8038.120		0.2	0.2			12440.72			
8036.065	8038.275		0.4	0.3			12440.48			
8036.291	8038.501		0.3	0.2			12440.13			
8036.608	8038.818	G	2.4	2.1	1.6		12439.64	3D 0	2C 0	Q 3
8036.730	8038.941		1.5				12439.45			
8037.034	8039.244		0.3	0.2			12438.98			
8037.176	8039.387				0.1		12438.76			
8037.357	8039.568			0.1			12438.48			
8037.661	8039.871			0.2			12438.01			
8037.842	8040.052		0.3	0.5	0.2b		12437.73			
8038.062	8040.272			0.1			12437.39	4E 1	2B16	Q 4
8038.482	8040.692		0.1	0.2d			12436.74	3A 1	2B11	R 0
8038.811	8041.022		0.2	0.3	0.2		12436.23	3B 2	2A 0	P 7
8038.857	8041.067						12436.16			
8039.167	8041.378		0.3				12435.68	Y 1	2B 9	R 0
8039.167	8041.378		0.3				12435.68	3a 0	2c 2	Q 1
8039.380	8041.591		0.1				12435.35			
8039.592	8041.803						12435.02	3e 2	2c 4	Q 1
8039.872	8042.083		0.4	0.6	1.0		12434.59	3A 1	2B11	P 4
8040.305	8042.516			0.0			12433.92			
8040.725	8042.936		0.2				12433.27			
8041.074	8043.286		0.2	0.1			12432.73			
8041.385	8043.596		0.2	0.3			12432.25	4F 0	2C 3	R 3
8041.831	8044.043		0.2	0.2			12431.56			
8042.161	8044.373	G	2.0	1.8	1.7		12431.05	3D 1	2C 1	R 3
8042.575	8044.787			0.2b			12430.41			
8042.659	8044.871		0.1				12430.28			
8042.866	8045.078		0.1	0.2			12429.96			
8043.287	8045.499		0.1				12429.31			
8043.585	8045.797			0.0d			12428.85			
8043.727	8045.939		0.1				12428.63			
8043.960	8046.172		0.1				12428.27			
8044.174	8046.386			0.1d			12427.94	3E 2	2C 2	Q 3
8044.730	8046.943		0.1	0.2b			12427.08	3E 1	2B10	Q 4
8045.203	8047.415			0.1d			12426.35			
8045.339	8047.551		0.1				12426.14			

λ_{air}	λ_{vac}	Ref.	I_1	I_2	I_3	I_4	ν	Upper	Lower	Br.
8045.753	8047.966			0.2			12425.50			
8046.122	8048.335			0.1	0.4		12424.93			
8046.634	8048.847		0.2	0.2			12424.14			
8047.049	8049.262		0.3	0.2			12423.50			
8047.372	8049.585	G	2.3	1.7	1.1		12423.00	Z 1	2C 1	Q 1
8047.729	8049.942		0.2	0.2			12422.45			
8048.027	8050.240		1.5	0.9			12421.99			
8048.234	8050.447			0.0			12421.67	3a 1	2c 3	P 4
8048.441	8050.655		1.5	1.2	0.8		12421.35	Z 0	2C 0	Q 2
8048.914	8051.128		0.6	0.9	1.1		12420.62			
8049.167	8051.381		0.3	0.4	0.4		12420.23			
8049.472	8051.685			0.2			12419.76			
8049.783	8051.997		0.1	0.2b			12419.28			
8049.951	8052.165		0.2				12419.02	3e 0	2c 2	Q 5
8050.211	8052.424		2.0	1.6	1.4		12418.62	3E 1	2C 1	P 5
8050.211	8052.424		2.0	1.6	1.4		12418.62	4F 0	2C 3	R 2
8050.211	8052.424		2.0	1.6	1.4		12418.62	Z 2	2C 2	R 1
8050.502	8052.716		1.9	1.7	1.7		12418.17	2A 2	2B 1	R 3
8050.885	8053.099		0.1	0.2			12417.58			
8051.028	8053.242			0.1			12417.36			
8051.196	8053.410			0.1d			12417.10	4g 1	2a 6	R 1
8051.481	8053.696		0.2	0.3	0.2		12416.66			
8051.812	8054.026		0.2				12416.15			
8052.305	8054.519			0.2			12415.39	3E 3	2B14	R 2
8052.545	8054.759		0.1	0.1			12415.02			
8052.727	8054.941			0.2			12414.74			
8053.038	8055.253		0.2	0.2			12414.26			
8053.317	8055.531	i	0.2		0.6		12413.83			
8053.434	8055.648			0.1			12413.65	3E 2	2B12	R 0
8053.518	8055.733		0.1				12413.52			
8053.654	8055.869			0.2			12413.31			
8053.920	8056.135		0.6	0.2	0.1		12412.90	3E 1	2B10	P 5
8054.134	8056.349		1.9	1.6	1.3		12412.57	Z 1	2C 1	R 2
8054.310	8056.525		1.0				12412.30			
8054.608	8056.823	G	2.5	2.1	2.1		12411.84	3b 3	2a 2	P 1
8054.933	8057.148		1.5	0.5			12411.34	3D 2	2C 2	Q 2
8055.309	8057.524		0.2	0.3	0.1d		12410.76			
8055.380	8057.596						12410.65			
8055.861	8058.076		0.9	0.4			12409.91			
8056.244	8058.459		1.5	1.6	1.4		12409.32	3D 2	2C 2	R 3
8056.432	8058.648		1.9	1.5	1.4		12409.03	3E 1	2C 1	P 4
8056.926	8059.141			0.0d			12408.27			
8057.361	8059.576		0.1	0.2b			12407.60			
8057.867	8060.083		0.2	0.2			12406.82			
8058.237	8060.453		0.1	0.2			12406.25			
8058.601	8060.817		0.2	0.2			12405.69			
8058.926	8061.142		1.9	1.6	1.4		12405.19	2A 3	2B 2	P 4
8059.303	8061.519		0.2	0.1			12404.61	3e 1	2c 3	P 3
8059.550	8061.766		0.4	0.2	0.2		12404.23			
8059.829	8062.045	G	2.5	1.8	1.2		12403.80	Z 2	2C 2	Q 1
8059.829	8062.045	G	2.5	1.8	1.2		12403.80	Z 1	2C 1	P 1
8059.829	8062.045	G	2.5	1.8	1.2		12403.80	3E 2	2B12	P 4
8060.160	8062.377		0.3	0.3			12403.29	4F 0	2C 3	Q 4
8060.160	8062.377		0.3	0.3			12403.29	3E 2	2B12	Q 4
8060.251	8062.468		0.2				12403.15	3E 0	2B 8	P 4
8060.251	8062.468		0.2				12403.15	W 2	2B13	R 2
8060.466	8062.682		0.2	0.2			12402.82			
8060.771	8062.988		0.1	0.2			12402.35			
8061.057	8063.274			0.1			12401.91			
8061.285	8063.502		0.2	0.3			12401.56			

λ_{air}	λ_{vac}	Ref.	I_1	I_2	I_3	I_4	ν	Upper	Lower	Br.
8061.779	8063.996			0.2			12400.80			
8062.136	8064.354		0.1				12400.25			
8062.579	8064.796		0.1	0.3			12399.57			
8062.930	8065.147		0.0	0.2			12399.03			
8063.242	8065.459			0.2			12398.55			
8063.593	8065.811			0.2			12398.01			
8063.769	8065.986			0.1			12397.74			
8064.003	8066.220		0.4	0.5			12397.38			
8064.309	8066.526			0.3			12396.91			
8064.400	8066.617		0.2				12396.77			
8064.940	8067.157			0.1			12395.94			
8065.688	8067.906		0.1	0.2d			12394.79	V 1	2B 8	P 4
8065.688	8067.906		0.1	0.2d			12394.79	V 0	2C 0	R 1
8066.007	8068.225		0.7	0.9	0.3		12394.30			
8066.345	8068.563			0.0			12393.78	3a 1	2c 3	P 5
8066.560	8068.778		0.1	0.1			12393.45			
8066.898	8069.117		0.1	0.0			12392.93			
8067.152	8069.371		0.1	0.0			12392.54	Y 1	2B 9	P 1
8067.380	8069.599		0.3	0.2			12392.19			
8067.862	8070.081		0.1	0.1d			12391.45	4d 2	3b 1	P 4
8068.201	8070.419		0.1				12390.93			
8068.389	8070.608		1.0	0.7	0.9		12390.64			
8068.813	8071.031		0.1	0.3	0.3		12389.99			
8069.047	8071.266			0.0			12389.63			
8069.138	8071.357		0.1				12389.49			
8069.659	8071.878				0.1		12388.69			
8070.076	8072.295						12388.05			
8070.239	8072.458			0.1d			12387.80			
8070.747	8072.967			0.2	0.2		12387.02			
8070.949	8073.169		0.1	0.2			12386.71			
8071.256	8073.475	G	1.5	1.5	1.6		12386.24	3D 0	2C 0	R 5
8071.607	8073.827			0.0			12385.70			
8071.881	8074.101		0.7	1.0	0.7		12385.28	Z 0	2C 0	R 3
8071.881	8074.101		0.7	1.0	0.7		12385.28	3A 1	2B11	P 3
8071.881	8074.101		0.7	1.0	0.7		12385.28	3e 0	2c 2	Q 6
8072.103	8074.322		0.9	0.6			12384.94			
8072.377	8074.596		0.1	0.2			12384.52			
8073.172	8075.392		0.4	0.3			12383.30	3b 6	2a 4	P 2
8073.328	8075.548		0.4	0.6			12383.06	3e 2	2c 4	Q 3
8073.622	8075.842			0.0			12382.61			
8073.993	8076.214		0.1	0.0			12382.04			
8074.202	8076.422				0.1		12381.72			
8074.606	8076.827			0.0			12381.10	3E 0	2B 8	Q 4
8074.737	8076.957		0.1				12380.90			
8074.867	8077.088			0.3	0.3		12380.70			
8075.109	8077.329		0.2	0.5	0.3		12380.33	3A 0	2B 9	R 2
8075.409	8077.629			0.1			12379.87			
8075.591	8077.812			0.0			12379.59			
8075.761	8077.982		0.4				12379.33			
8075.859	8078.079			0.2			12379.18			
8076.048	8078.269		0.2	0.3	0.8		12378.89	W 1	2B11	R 2
8076.244	8078.464		0.3	0.8	1.0		12378.59			
8076.492	8078.712		0.2	0.3			12378.21			
8076.759	8078.980			0.2			12377.80			
8077.059	8079.280		1.6	1.4	0.8		12377.34	3E 2	2C 2	P 2
8077.425	8079.646		0.2	0.1			12376.78	3E 1	2B10	P 4
8077.705	8079.927			0.2	0.2b		12376.35	2K 5	2B 3	P 5
8078.032	8080.253			0.0d			12375.85			
8078.326	8080.547			0.6	0.6		12375.40			
8078.567	8080.788			0.2	0.3		12375.03			

λ_{air}	λ_{vac}	Ref.	I_1	I_2	I_3	I_4	ν	Upper	Lower	Br.
8078.965	8081.187		0.7	1.1	0.3		12374.42	Z 0	2C 0	P 2
8078.965	8081.187		0.7	1.1	0.3		12374.42	W 2	2B13	R 1
8078.965	8081.187		0.7	1.1	0.3		12374.42	3a 0	2c 2	P 1
8079.233	8081.455			0.2			12374.01			
8079.514	8081.735		0.2	0.6			12373.58	3D 2	2B12	R 6
8079.749	8081.971		0.4	0.4	0.2		12373.22			
8079.938	8082.160						12372.93			
8080.101	8082.323		0.3	0.3			12372.68			
8080.356	8082.578			0.1	0.1		12372.29			
8080.728	8082.950		0.9	0.8			12371.72			
8081.055	8083.277			0.1			12371.22	3D 2	2B12	R 5
8081.055	8083.277			0.1			12371.22	3e 0	2c 2	P 3
8081.447	8083.669			0.1			12370.62			
8081.852	8084.074			0.1			12370.00			
8082.113	8084.336		0.2	0.4	0.2		12369.60			
8082.460	8084.682		0.1				12369.07			
8082.708	8084.931			0.2			12368.69			
8083.054	8085.277			0.1			12368.16			
8083.505	8085.728		0.0	0.3			12367.47	3D 0	2C 0	P 3
8083.852	8086.075			0.1			12366.94	3E 3	2B14	Q 4
8083.852	8086.075			0.1			12366.94	4d 3	3b 2	P 4
8084.172	8086.395		1.0	0.8	0.9		12366.45	3E 3	2C 3	R 1
8084.172	8086.395		1.0	0.8	0.9		12366.45	3A 2	2B11	P 1
8084.590	8086.814		0.1				12365.81	3A 1	2B11	P 2
8084.767	8086.990		0.2	0.2d			12365.54	3E 2	2B12	P 3
8085.035	8087.258	G	2.0	1.5	0.3		12365.13	3D 2	2C 2	P 2
8085.284	8087.507			0.7			12364.75			
8085.552	8087.775		0.3				12364.34			
8085.663	8087.886		0.3	0.5	0.2		12364.17			
8086.101	8088.324		0.2	0.2			12363.50			
8086.441	8088.665	G	2.3	1.8	1.6		12362.98	2A 2	2B 1	R 2
8086.840	8089.064		0.1	0.2d			12362.37			
8087.108	8089.332			0.1			12361.96			
8087.377	8089.600		0.4	0.8	1.6		12361.55	3D 0	2C 0	R 6
8087.750	8089.973				0.2		12360.98			
8087.946	8090.170		0.2	0.1			12360.68			
8088.797	8091.021			0.1			12359.38			
8089.418	8091.643		0.1	0.3			12358.43	4E 1	2B16	Q 3
8089.798	8092.023			0.2			12357.85	3E 1	2B10	R 0
8090.387	8092.612		0.6	0.5	0.2		12356.95			
8090.453	8092.677			0.7	0.8		12356.85			
8090.590	8092.815		0.1	0.0			12356.64			
8090.918	8093.142		1.5	1.2	1.4		12356.14			
8091.022	8093.247						12355.98			
8091.343	8093.568		1.5b	1.4	0.3		12355.49	3E 3	2C 3	R 2
8091.677	8093.902	G	2.4	1.9	1.7		12354.98	3B 3	2A 1	P 3
8091.972	8094.197			0.1			12354.53	Y 1	2B 9	P 2
8092.129	8094.354		0.2b	0.2			12354.29			
8092.404	8094.629			0.2			12353.87			
8092.470	8094.695		0.2		0.2		12353.77			
8092.745	8094.970	G	1.7	1.5	1.2		12353.35	3E 2	2C 2	P 4
8092.745	8094.970	G	1.7	1.5	1.2		12353.35	4g 1	2a 6	P 2
8093.033	8095.258			0.2			12352.91	3E 2	2C 2	Q 4
8093.249	8095.475			0.0			12352.58			
8093.741	8095.966			0.1			12351.83			
8094.134	8096.360		0.2	0.1			12351.23			
8094.429	8096.655		0.3	0.2			12350.78	3E 1	2B10	Q 3
8094.730	8096.956		1.5				12350.32	3D 1	2C 1	Q 3
8094.848	8097.074	G	2.7	2.1	1.5		12350.14	Z 2	2C 2	R 2
8095.097	8097.323			0.3			12349.76			

λ_{air}	λ_{vac}	Ref.	I_1	I_2	I_3	I_4	ν	Upper	Lower	Br.
8095.176	8097.402		0.2		0.1		12349.64			
8095.301	8097.527			0.1			12349.45			
8095.491	8097.717		0.1	0.2			12349.16	3e 2	2c 4	P 3
8095.648	8097.874				0.5		12348.92			
8095.799	8098.025		1.6	1.2	0.2		12348.69	Z 1	2C 1	Q 2
8096.022	8098.248						12348.35			
8096.205	8098.432		0.2	0.1			12348.07	4g 1	2a 6	R 2
8096.514	8098.740			0.1			12347.60			
8096.933	8099.160		0.1	0.2			12346.96	W 2	2B13	R 0
8097.130	8099.356			0.0			12346.66			
8098.002	8100.229			0.1d			12345.33			
8098.448	8100.675		0.1	0.2	0.2b		12344.65			
8098.724	8100.951		0.3	0.3			12344.23	2K 2	2B 1	R 1
8099.013	8101.240			0.2			12343.79			
8099.124	8101.351				0.7		12343.62			
8099.301	8101.528			0.1			12343.35			
8099.505	8101.732			0.2	0.2		12343.04			
8099.918	8102.145		0.2	0.3			12342.41			
8100.233	8102.460	G	2.1	2.0	2.0		12341.93	3D 1	2C 1	R 4
8100.233	8102.460	G	2.1	2.0	2.0		12341.93	W 2	2B13	P 6
8100.574	8102.802		0.2	0.1			12341.41	3E 3	2B14	R 1
8100.660	8102.887			0.2			12341.28			
8100.863	8103.091		0.3	0.8	0.8		12340.97	3E 0	2C 0	P 6
8101.310	8103.537				0.3b		12340.29	3E 2	2B12	Q 3
8101.546	8103.774				0.3b		12339.93			
8101.789	8104.017			0.2b			12339.56			
8101.855	8104.082				0.5		12339.46			
8102.137	8104.365			0.1d	0.4b		12339.03			
8102.393	8104.621		0.1	0.2d			12338.64			
8102.570	8104.798			0.1	0.6		12338.37			
8102.899	8105.127		0.7	0.4	0.3		12337.87			
8103.017	8105.245				0.8		12337.69			
8103.102	8105.330			0.2			12337.56			
8103.181	8105.409				0.5		12337.44			
8103.358	8105.587			0.1			12337.17			
8103.562	8105.790		0.1	0.2			12336.86	V 1	2B 8	R 0
8103.707	8105.935		1.9	0.9			12336.64			
8103.844	8106.073				1.5b		12336.43			
8103.996	8106.224		1.0	0.4			12336.20	2K 2	2B 1	R 0
8103.996	8106.224		1.0	0.4			12336.20	2A 3	2B 2	P 5
8104.298	8106.526	G	2.1	1.9	1.8		12335.74	3b 3	2a 2	P 2
8104.462	8106.691						12335.49			
8104.567	8106.796		0.2	0.5	0.4		12335.33			
8104.745	8106.973			0.1			12335.06			
8104.869	8107.098				0.4		12334.87			
8104.981	8107.210			0.1			12334.70			
8105.099	8107.328				0.3		12334.52	3E 2	2B12	P 2
8105.237	8107.466			0.1			12334.31			
8105.428	8107.657			0.2	0.4		12334.02			
8105.592	8107.821				0.4		12333.77			
8105.770	8107.999		0.1	0.6	0.5b		12333.50	3D 2	2B12	R 4
8106.131	8108.360		0.2	0.5	0.6b		12332.95			
8106.427	8108.656			0.1d	0.2		12332.50			
8106.637	8108.866				0.3		12332.18			
8106.821	8109.050				0.2		12331.90			
8106.979	8109.208				0.2		12331.66	3E 3	2C 3	R 4
8107.249	8109.478			0.2	1.0		12331.25	3e 2	2c 4	P 2
8107.374	8109.603		0.3	0.2			12331.06			
8107.683	8109.912		0.3	0.6	0.9		12330.59			
8108.057	8110.287		0.1		0.2		12330.02			

λ_{air}	λ_{vac}	Ref.	I_1	I_2	I_3	I_4	ν	Upper	Lower	Br.
8108.189	8110.418			0.2d			12329.82			
8108.255	8110.484				0.2		12329.72			
8108.531	8110.760		0.9	1.4	1.3		12329.30	3d 0	2c 2	Q 1
8108.531	8110.760		0.9	1.4	1.3		12329.30	3E 3	2C 3	R 3
8108.531	8110.760		0.9	1.4	1.3		12329.30	3d 0	2c 2	R 1
8108.820	8111.050		0.4	0.8	0.2d		12328.86	3E 3	2C 3	R 5
8109.024	8111.254			0.0			12328.55			
8109.261	8111.491			0.0	0.2d		12328.19	3E 0	2B 8	P 3
8109.636	8111.866			0.1			12327.62			
8109.735	8111.965				0.2d		12327.47			
8109.952	8112.182				0.4		12327.14			
8110.077	8112.307		0.8	0.8			12326.95			
8110.208	8112.438		0.2	0.4			12326.75			
8110.274	8112.504				0.2		12326.65			
8110.406	8112.636			0.1			12326.45	4E 1	2B16	Q 1
8110.643	8112.873		0.2	0.3	0.3		12326.09	3a 0	2c 2	P 2
8110.965	8113.195		0.1	0.2			12325.60			
8111.090	8113.320						12325.41			
8111.241	8113.472			0.1	0.4		12325.18	3E 1	2B10	P 3
8111.241	8113.472			0.1	0.4		12325.18	3e 1	2c 3	P 4
8111.557	8113.788			0.2	0.2		12324.70	V 1	2B 8	P 3
8111.821	8114.051		0.2	0.3	0.7		12324.30	V 0	2C 0	R 3
8112.354	8114.584			0.2			12323.49			
8112.663	8114.894			0.1			12323.02			
8112.913	8115.144			0.1			12322.64	W 1	2B11	R 1
8113.243	8115.473		0.7	0.2			12322.14			
8113.651	8115.882		0.1	0.2			12321.52	3D 2	2B12	R 3
8114.000	8116.231			0.2			12320.99			
8114.309	8116.541			0.3			12320.52			
8114.593	8116.824		0.1	0.2			12320.09			
8114.751	8116.982			0.2			12319.85	4E 1	2B16	Q 2
8115.054	8117.285	G	2.7	1.5	2.3		12319.39	3b 3	2a 1	P 7
8115.179	8117.410		0.9				12319.20			
8115.311	8117.542		1.5	1.8	1.0		12319.00	3E 3	2C 3	R 3
8116.207	8118.438		0.5	0.8	0.8		12317.64	3E 0	2B 8	Q 3
8116.562	8118.794	G	0.8	1.4	1.0		12317.10	3E 3	2C 3	R 2
8116.648	8118.880		0.2				12316.97	3D 2	2B12	P 1
8116.991	8119.223			0.1			12316.45			
8117.360	8119.592	G					12315.89	3d 1	2c 3	R 1
8117.703	8119.935			0.2			12315.37	4F 0	2B14	P 4
8117.821	8120.053		0.1				12315.19	3D 2	2B12	R 2
8118.131	8120.363		1.6	1.1			12314.72	3D 2	2C 2	Q 3
8118.408	8120.640			0.1			12314.30			
8118.533	8120.766		0.1				12314.11			
8119.047	8121.280			0.1			12313.33			
8119.206	8121.438	1			0.4		12313.09			
8119.417	8121.649	G	1.0	1.3	1.0		12312.77	3D 0	2C 0	Q 4
8119.700	8121.933			1.3	0.9		12312.34			
8119.885	8122.118			0.2	0.3		12312.06			
8120.360	8122.593				0.3b		12311.34			
8120.709	8122.942		0.2		0.3		12310.81			
8120.881	8123.114				0.3		12310.55			
8121.020	8123.252		1.5b	1.0	0.3		12310.34	3D 0	2C 0	R 6
8121.020	8123.252		1.5b	1.0	0.3		12310.34	Z 2	2C 2	Q 2
8121.402	8123.635		0.1	0.1			12309.76			
8121.468	8123.701				0.3		12309.66	Y 1	2B 9	P 3
8121.607	8123.840		1.6	1.2	0.9		12309.45	3F 0	2B 9	R 4
8121.824	8124.058		0.2	0.2d	0.2		12309.12			
8122.161	8124.394			0.0			12308.61			
8122.267	8124.500		0.1		0.2		12308.45			

λ_{air}	λ_{vac}	Ref.	I_1	I_2	I_3	I_4	ν	Upper	Lower	Br.
8122.405	8124.638			0.1			12308.24			
8122.755	8124.988		0.3	0.3	0.3b		12307.71	3E 1	2B10	Q 2
8123.032	8125.266			0.2			12307.29			
8123.217	8125.450		0.1	0.2	0.2		12307.01			
8123.435	8125.668				1.0		12306.68			
8123.573	8125.807	G	1.6	1.4	1.4		12306.47	Z 0	2C 0	Q 3
8123.804	8126.038			0.2	0.3		12306.12			
8124.009	8126.243			0.1			12305.81			
8124.253	8126.487			0.2	0.2		12305.44			
8124.577	8126.811		1.0	1.5	1.0		12304.95			
8124.748	8126.982	G		1.3			12304.69			
8124.966	8127.200		1.0	0.5	1.1		12304.36	Z 1	2C 1	P 2
8124.966	8127.200		1.0	0.5	1.1		12304.36	4e 3	3b 2	P 5
8125.171	8127.405		0.6	1.0	0.9		12304.05			
8125.369	8127.603		0.5	0.3			12303.75			
8125.515	8127.749		0.4	0.9	0.4		12303.53			
8125.581	8127.815		0.2				12303.43			
8125.765	8128.000			0.2			12303.15			
8125.865	8128.099		0.2		0.2		12303.00			
8126.089	8128.323	G	1.5	1.6	1.5		12302.66	3E 1	2C 1	P 5
8126.241	8128.475		0.2				12302.43			
8126.532	8128.766		0.1	0.2			12301.99			
8126.836	8129.070		0.1	0.2	0.2b		12301.53			
8127.087	8129.321		0.1				12301.15	4g 1	2a 6	R 3
8127.087	8129.321		0.1				12301.15	3d 0	2c 2	P 1
8127.291	8129.526			0.1			12300.84			
8127.780	8130.015		0.1				12300.10	3D 2	2B12	R 1
8127.780	8130.015		0.1				12300.10	4F 0	2C 3	R 2
8127.998	8130.233			0.1d			12299.77			
8128.408	8130.643		0.6	0.6	0.1		12299.15	W 2	2B13	P 1
8128.560	8130.795		0.2				12298.92			
8128.706	8130.941			0.2			12298.70			
8128.858	8131.093		0.3	0.2			12298.47			
8129.148	8131.384	G	2.4	2.0	1.5		12298.03	3E 2	2C 2	P 3
8129.148	8131.384	G	2.4	2.0	1.5		12298.03	3E 2	2B12	Q 2
8129.552	8131.787		0.2	0.1			12297.42			
8129.717	8131.952		0.8	0.3			12297.17	3D 1	2C 1	P 3
8129.836	8132.071				0.2d		12296.99			
8129.948	8132.184			0.1			12296.82			
8130.140	8132.375		0.3	0.3			12296.53			
8130.497	8132.733	G	2.3	1.7	1.3		12295.99	3E 3	2C 3	R 1
8130.854	8133.090	G	2.7	2.3	1.6		12295.45	3b 6	2a 4	P 3
8131.205	8133.440		0.3				12294.92			
8131.264	8133.500		0.3	0.3			12294.83			
8131.463	8133.698		0.2		0.2		12294.53			
8131.615	8133.851			0.2	0.4		12294.30			
8132.038	8134.274		0.2	0.2			12293.66	3E 3	2B14	Q 3
8132.349	8134.585		1.0	0.8	0.1		12293.19	3Z 2	2C 2	P 2
8132.746	8134.982		0.2	0.3d	0.2b		12292.59			
8133.196	8135.432			0.2d			12291.91			
8133.699	8135.935	G					12291.15	3d 1	2c 3	P 1
8133.983	8136.220		0.1	0.3			12290.72	2K 5	2B 3	P 6
8134.215	8136.452		0.1	0.3			12290.37			
8134.652	8136.888		0.2				12289.71	3d 0	2c 2	R 2
8134.936	8137.173			0.2			12289.28			
8135.267	8137.504			0.1			12288.78			
8135.519	8137.756			0.2			12288.40	3d 1	2c 3	Q 1
8135.731	8137.968		0.6	0.2			12288.08			
8135.883	8138.120		0.5	0.3b			12287.85			
8135.956	8138.193			0.7	0.3d		12287.74			

λ_{air}	λ_{vac}	Ref.	I_1	I_2	I_3	I_4	ν	Upper	Lower	Br.
8136.420	8138.657		0.2b		0.2		12287.04			
8136.658	8138.895		0.1				12286.68			
8136.744	8138.981			0.1			12286.55			
8136.910	8139.147		0.2	0.2			12286.30			
8137.208	8139.445		0.1	0.2			12285.85	V 0	2C 0	Q 3
8137.386	8139.624			0.2			12285.58			
8137.605	8139.842			0.1			12285.25			
8137.771	8140.008		0.1	0.1			12285.00			
8137.910	8140.147		0.1				12284.79			
8138.128	8140.366		0.1	0.2d			12284.46	7c 0	2a 7	R 1
8138.387	8140.624		0.6	0.9	0.9		12284.07			
8138.672	8140.909		0.1	0.0			12283.64	3D 2	2B12	R 0
8138.983	8141.221		0.7	0.3			12283.17			
8139.255	8141.493	G	1.8	1.7	1.5		12282.76	3D 2	2C 2	R 4
8139.520	8141.758		0.3	0.2			12282.36			
8139.686	8141.924		0.1	0.2	0.3		12282.11			
8139.864	8142.102			0.1			12281.84			
8140.163	8142.401		0.3	0.8	0.9		12281.39			
8140.521	8142.759			0.2			12280.85			
8140.779	8143.017		0.1				12280.46	3d 1	2c 3	R 2
8140.985	8143.223		0.2	0.2			12280.15			
8141.203	8143.442		0.8	0.3			12279.82	2K 2	2B 1	P 1
8141.203	8143.442		0.8	0.3			12279.82	3a 0	2c 2	P 3
8141.396	8143.634			0.1			12279.53	3e 0	2c 2	P 4
8141.767	8144.005		0.8	1.2	1.4		12278.97	3E 1	2B10	P 2
8142.045	8144.284			0.0			12278.55			
8142.291	8144.529		1.0	0.8			12278.18			
8142.589	8144.828		0.2	0.1			12277.73	4F 0	2C 3	Q 3
8142.589	8144.828		0.2	0.1			12277.73	W 2	2B13	P 2
8143.087	8145.326		0.1				12276.98	3E 1	2B10	Q 1
8143.087	8145.326		0.1				12276.98	4F 0	2B14	Q 3
8143.471	8145.710		0.2	0.1	0.2		12276.40	W 1	2B11	R 0
8143.810	8146.049			0.1			12275.89	W 1	2B11	P 6
8144.141	8146.381		0.2	0.2			12275.39	4F 0	2C 3	R 1
8144.327	8146.567			0.1			12275.11			
8144.646	8146.885			0.1			12274.63	V 0	2C 0	Q 2
8145.070	8147.310			0.2			12273.99			
8145.362	8147.602		0.2	0.3			12273.55	3E 3	2B14	R 0
8145.362	8147.602		0.2	0.3			12273.55	3E 0	2B 8	Q 2
8145.781	8148.020		0.2	0.5	0.2		12272.92			
8146.086	8148.326			0.2			12272.46	Y 1	2B 9	P 4
8146.378	8148.618		0.9	0.8			12272.02	3E 3	2C 3	Q 1
8146.703	8148.943			0.0			12271.53			
8146.929	8149.169			0.0			12271.19			
8147.135	8149.375			0.0			12270.88			
8147.334	8149.574			0.3	0.1		12270.58	5c 1	2a 7	R 1
8147.619	8149.860						12270.15			
8147.825	8150.065			0.3	0.3		12269.84			
8148.098	8150.338			0.1			12269.43			
8148.370	8150.610			0.3			12269.02			
8148.642	8150.883			0.3			12268.61	3A 0	2B 9	R 1
8149.373	8151.613			0.2			12267.51			
8149.625	8151.866			0.1			12267.13			
8149.884	8152.125			0.1			12266.74			
8150.256	8152.497		0.6	0.4			12266.18	3E 2	2B12	Q 1
8151.625	8153.867						12264.12			
8151.825	8154.066			0.2			12263.82			
8152.403	8154.645		0.3	0.4			12262.95			
8153.022	8155.263		0.6	0.5			12262.02			
8153.135	8155.376		0.1				12261.85			

λ_{air}	λ_{vac}	Ref.	I_1	I_2	I_3	I_4	ν	Upper	Lower	Br.
8153.288	8155.529		0.1				12261.62			
8153.507	8155.749		0.2	0.3			12261.29	3E 0	2B 8	P 2
8153.946	8156.188				0.1		12260.63			
8154.417	8156.659						12259.92	3E 3	2B14	P 5
8154.764	8157.006			0.1			12259.40	Z 0	2C 0	R 4
8156.299	8158.543						12257.09	W 2	2B13	P 3
8156.574	8158.816		0.2	0.3	0.2		12256.68			
8157.000	8159.242	G	0.3	0.2			12256.04	3E 3	2C 3	Q 2
8157.319	8159.562		1.0	1.1	0.1		12255.56	Z 2	2C 2	R 3
8157.552	8159.795			0.2			12255.21			
8157.878	8160.121			0.0			12254.72			
8158.091	8160.334		0.6				12254.40			
8158.278	8160.521			0.1			12254.12			
8158.504	8160.747		0.2	0.3			12253.78			
8158.897	8161.140		1.0	0.6			12253.19			
8159.117	8161.360		0.9	1.4	1.5		12252.86	3D 1	2C 1	R 5
8159.256	8161.500	G					12252.65			
8159.503	8161.746	G	1.6	0.7	0.2b		12252.28	3D 2	2C 2	P 3
8159.503	8161.746	G	1.6	0.7	0.2b		12252.28	3E 3	2B14	P 4
8159.503	8161.746	G	1.6	0.7	0.2b		12252.28	V 1	2B 8	P 2
8159.749	8161.993		0.8	1.3	1.1		12251.91	3B 3	2A 1	P 4
8160.089	8162.332		0.1				12251.40			
8160.176	8162.419			0.2	0.1		12251.27	4d 2	3b 1	P 5
8160.529	8162.772		0.1	0.3d	0.1		12250.74	4F 0	2B14	R 1
8161.035	8163.279		0.3	0.4	0.2		12249.98			
8161.348	8163.592			0.1			12249.51			
8161.828	8164.072			0.1d			12248.79	3D 1	2C 1	R 5
8162.381	8164.625			0.2d			12247.96			
8162.534	8164.778						12247.73			
8162.821	8165.065			0.2			12247.30			
8163.141	8165.385		0.3	0.3			12246.82	3d 0	2c 2	R 3
8163.421	8165.665		0.7	0.3			12246.40			
8163.774	8166.018		0.7	0.9	1.1		12245.87			
8163.974	8166.218			0.1			12245.57			
8164.101	8166.345		0.2b	0.2			12245.38			
8164.287	8166.532			0.3			12245.10	3e 1	2c 3	P 5
8164.381	8166.625		0.6	0.4			12244.96			
8164.707	8166.952	G	2.6	2.4	2.4		12244.47	3b 3	2a 2	P 3
8164.807	8167.052		1.5				12244.32	Z 1	2C 1	Q 3
8165.027	8167.272		0.3	0.3	0.2		12243.99			
8165.367	8167.612		0.1	0.1			12243.48	3D 1	2B10	R 5
8165.674	8167.919		0.1	0.1			12243.02			
8165.995	8168.240		0.3	0.7	0.3		12242.54			
8166.295	8168.540		0.2	0.3			12242.09	3d 1	2c 3	R 3
8166.762	8169.007			0.2b			12241.39			
8167.375	8169.621			0.1d			12240.47			
8167.496	8169.741		0.1				12240.29	3E 3	2B14	Q 2
8167.696	8169.941		0.3	0.3			12239.99	3F 1	2B11	Q 5
8168.103	8170.349			0.2			12239.38			
8168.323	8170.569		0.4	0.6	0.3d		12239.05			
8168.764	8171.009			0.1			12238.39	W 2	2B13	P 5
8168.970	8171.216						12238.08	W 2	2B13	P 4
8169.137	8171.383			0.1			12237.83			
8169.651	8171.897			0.1d			12237.06	3a 0	2c 2	P 4
8169.999	8172.245		0.1	0.1			12236.54			
8170.232	8172.479		0.8	0.9	0.3		12236.19	V 0	2C 0	Q 1
8170.292	8172.539		0.2				12236.10			
8170.419	8172.666		0.1				12235.91			
8171.127	8173.374		0.2				12234.85			
8171.849	8174.095		0.2	0.3			12233.77	Z 0	2C 0	P 3

λ_{air}	λ_{vac}	Ref.	I_1	I_2	I_3	I_4	ν	Upper	Lower	Br.
8172.303	8174.549		0.2	0.3	0.1d		12233.09			
8172.496	8174.743			0.2			12232.80	3E 3	2B14	P 3
8172.831	8175.077		0.2	0.3			12232.30			
8174.361	8176.608			0.3			12230.01			
8175.216	8177.464				0.2		12228.73			
8175.310	8177.558		0.1	0.3	0.2		12228.59			
8175.464	8177.711		0.1	0.2			12228.36			
8175.778	8178.026		0.4	0.5			12227.89			
8175.892	8178.139				0.1		12227.72			
8176.280	8178.527			0.2d			12227.14			
8176.995	8179.243		0.1	0.2			12226.07			
8177.289	8179.538		0.9	0.7			12225.63	2K 2	2B 1	P 2
8177.597	8179.845		0.1	0.0			12225.17			
8177.784	8180.033			0.2			12224.89			
8178.045	8180.294		0.3	0.3			12224.50			
8178.199	8180.448			0.2	0.5		12224.27			
8178.474	8180.722			0.1	0.1		12223.86			
8179.143	8181.391		0.7	0.6			12222.86			
8179.377	8181.626				0.1		12222.51			
8179.725	8181.974			0.2			12221.99			
8180.046	8182.295		0.9	1.2	1.4		12221.51	3E 0	2C 0	P 7
8180.046	8182.295		0.9	1.2	1.4		12221.51	3B 4	2K 2	R 3
8180.561	8182.810		0.9	1.0	0.2		12220.74	3E 3	2C 3	Q 1
8181.023	8183.273			0.1			12220.05	W 1	2B11	P 5
8181.023	8183.273			0.1			12220.05	3d 0	2c 2	Q 2
8181.398	8183.648				0.2b		12219.49	3d 0	2c 2	P 2
8181.398	8183.648				0.2b		12219.49	3D 3	2B14	R 4
8181.720	8183.969				0.2d		12219.01			
8182.021	8184.270		0.7	0.9	1.4		12218.56			
8182.322	8184.572			0.1			12218.11	3d 2	2c 4	Q 1
8182.557	8184.806				0.3		12217.76			
8182.858	8185.108		0.2	0.5			12217.31	3F 3	2B15	Q 3
8183.052	8185.302			0.2d			12217.02			
8183.213	8185.463	1			1.5b		12216.78			
8183.602	8185.852				0.6		12216.20	3E 3	2B14	P 2
8183.709	8185.959				0.6		12216.04			
8183.896	8186.146				0.4		12215.76			
8183.950	8186.200		0.2				12215.68			
8184.251	8186.502				0.2		12215.23			
8184.365	8186.616		0.1				12215.06	W 1	2B11	P 1
8184.660	8186.910		0.2	0.2			12214.62	3E 3	2C 3	Q 3
8184.841	8187.091		0.2	0.1			12214.35			
8185.116	8187.366		0.8	1.0	1.2		12213.94			
8185.505	8187.755			0.2	0.2b		12213.36			
8185.779	8188.030		1.7	1.3			12212.95	3D 3	2C 3	R 1
8186.054	8188.305		0.1	0.2			12212.54	4F 0	2C 3	Q 4
8186.269	8188.519				0.1	0.1d	12212.22			
8186.517	8188.767		0.1	0.2			12211.85			
8186.664	8188.915				0.3		12211.63	Z 2	2B12	R 0
8186.664	8188.915				0.3		12211.63	3d 1	2c 3	Q 2
8186.926	8189.177			0.1	0.1		12211.24	4E 1	2C 4	R 3
8187.261	8189.512		0.1	0.3	0.2		12210.74			
8187.482	8189.733		0.2	0.2			12210.41	3b 6	2a 4	P 4
8187.657	8189.908		1.0	0.6	1.1		12210.15	3D 1	2C 1	Q 4
8187.864	8190.115	G	1.7	1.5	1.2		12209.84	3E 2	2C 2	P 4
8188.274	8190.525			0.1			12209.23			
8188.508	8190.760			0.0			12208.88			
8188.777	8191.028		0.8	0.6			12208.48			
8189.146	8191.397		0.1				12207.93			
8189.293	8191.544		0.1	0.2			12207.71			

λ_{air}	λ_{vac}	Ref.	I_1	I_2	I_3	I_4	ν	Upper	Lower	Br.
8189.414	8191.665				0.1		12207.53			
8189.608	8191.860			0.0			12207.24			
8189.971	8192.222			0.0			12206.70			
8190.246	8192.497				0.2d		12206.29			
8190.373	8192.625			0.2			12206.10			
8190.722	8192.974		0.4	0.5	0.2b		12205.58	3F 0	2B 9	Q 5
8190.722	8192.974		0.4	0.5	0.2b		12205.58	W 1	2B11	P 2
8191.058	8193.310	G	1.8	1.5	1.2		12205.08	2A 4	2B 4	R 2
8191.058	8193.310	G	1.8	1.5	1.2		12205.08	3a 0	2c 2	P 5
8191.440	8193.692		0.2	0.2	0.3		12204.51	3D 2	2B12	P 2
8191.440	8193.692		0.2		0.3		12204.51	3d 0	2c 2	R 4
8191.716	8193.968			0.1	0.2		12204.10			
8191.937	8194.189		0.4	0.4			12203.77			
8192.138	8194.391		0.1	0.2	0.3		12203.47			
8192.373	8194.626			0.2	0.3		12203.12	3d 1	2c 3	R 4
8192.561	8194.814			0.1	0.3		12202.84			
8192.958	8195.210		1.5	0.8	0.4		12202.25			
8193.246	8195.499				0.4		12201.82			
8193.387	8195.640		0.2	0.2			12201.61			
8193.474	8195.727				0.2		12201.48			
8193.689	8195.942				0.4		12201.16			
8193.783	8196.036	G	1.6	1.2			12201.02	2A 4	2B 4	R 1
8193.898	8196.150				0.4		12200.85			
8194.126	8196.379				0.4		12200.51			
8194.233	8196.486			0.2	0.4		12200.35			
8194.603	8196.856			0.0			12199.80	W 1	2B11	P 4
8194.771	8197.024	i			1.5b		12199.55			
8194.851	8197.104			0.2d			12199.43	Z 2	2B12	R 1
8195.033	8197.286			0.1			12199.16			
8195.167	8197.420				1.0		12198.96			
8195.234	8197.487			0.3			12198.86	W 1	2B11	P 3
8195.275	8197.528		0.1		1.0		12198.80			
8195.436	8197.689			0.2	0.5		12198.56			
8195.590	8197.843		0.1				12198.33			
8195.819	8198.072		0.2	0.8	0.7		12197.99	3F 2	2B13	Q 4
8196.121	8198.374		0.2	0.8	0.2d		12197.54	2A 4	2B 4	R 3
8196.215	8198.469				0.2d		12197.40			
8196.316	8198.569		0.1	0.2			12197.25	3A 0	2B 9	P 4
8196.484	8198.737			0.0	0.3		12197.00			
8196.692	8198.946		0.2	0.3	0.3		12196.69			
8196.948	8199.201		0.0	0.2	0.3		12196.31	4E 1	2B 4	R 2
8197.250	8199.504		0.0	0.1	0.3		12195.86	3E 3	2B14	Q 1
8197.250	8199.504		0.0	0.1	0.3		12195.86	3D 3	2B14	R 2
8197.714	8199.968				0.2		12195.17			
8198.622	8200.875			0.1			12193.82			
8198.931	8201.185		0.3	0.7	0.9		12193.36	3E 1	2C 1	P 6
8199.240	8201.494				0.2		12192.90	3D 1	2B10	R 4
8199.240	8201.494				0.2		12192.90	3D 3	2B14	R 3
8199.738	8201.992		0.6	0.8			12192.16			
8199.899	8202.154			0.1			12191.92	3D 0	2B 8	R 5
8200.088	8202.342		0.3	0.9	1.0		12191.64			
8200.384	8202.638		0.3	0.4			12191.20	3D 2	2C 2	Q 4
8201.009	8203.264			0.1			12190.27	3D 0	2C 0	P 4
8201.292	8203.546			0.1			12189.85	3e 0	2c 2	P 5
8201.433	8203.688		0.2				12189.64			
8201.594	8203.849	G					12189.40	Z 2	2B12	R 2
8201.742	8203.997		1.9	1.6	0.9		12189.18	Z 2	2C 2	Q 3
8202.113	8204.367		0.1				12188.63			
8202.624	8204.879	G					12187.87	3D 3	2B14	R 1
8202.779	8205.034		0.1	0.1			12187.64			

λ_{air}	λ_{vac}	Ref.	I_1	I_2	I_3	I_4	ν	Upper	Lower	Br.
8203.008	8205.263		0.1		0.4		12187.30	Y 2	2B13	R 2
8203.237	8205.492		0.3	0.2			12186.96			
8203.492	8205.748	G	2.2	1.6	1.0		12186.58	2A 4	2B 4	R 0
8203.883	8206.138		0.2	0.2			12186.00	4E 1	2C 4	R 1
8204.273	8206.529		0.2	0.3			12185.42			
8204.691	8206.946			0.2			12184.80			
8204.974	8207.229		1.5	0.9			12184.38			
8205.310	8207.566			0.1			12183.88	3d 1	2c 3	P 2
8205.566	8207.822			0.1			12183.50			
8205.863	8208.118		0.2	0.3			12183.06			
8206.119	8208.374		0.2	0.5	0.6		12182.68			
8206.489	8208.745		0.1	0.2			12182.13	Y 2	2B13	R 1
8206.583	8208.839		0.1				12181.99			
8206.785	8209.042	G	1.5	1.6	1.4		12181.69	3D 0	2C 0	Q 5
8206.785	8209.042	G	1.5	1.6	1.4		12181.69	3C 2	2A 3	R 3
8207.143	8209.399		0.2	0.2			12181.16			
8207.446	8209.702		0.3b	1.0	1.2		12180.71	3D 3	2B14	R 0
8207.695	8209.951		1.5	1.0	1.3		12180.34	3E 3	2C 3	Q 2
8208.045	8210.302		0.2				12179.82			
8208.180	8210.437		0.3	0.2			12179.62			
8208.463	8210.720	G	1.5b	1.9	1.1		12179.20	3D 3	2C 3	Q 1
8208.888	8211.145			0.1			12178.57			
8208.942	8211.198		0.3				12178.49			
8209.110	8211.367		0.1				12178.24	4g 1	2a 6	P 3
8209.306	8211.563			0.1d			12177.95			
8209.703	8211.960		0.4	0.8	1.1		12177.36			
8210.317	8212.574			0.0			12176.45			
8210.971	8213.228			0.1			12175.48			
8211.187	8213.444		0.2	0.3			12175.16			
8211.689	8213.937						12174.43	3A 0	2B 9	R 0
8212.833	8215.091			0.1			12172.72	3F 1	2B11	R 3
8212.961	8215.219						12172.53	Y 2	2B13	R 0
8213.299	8215.557		1.0b	0.9			12172.03			
8213.433	8215.691			0.3			12171.83			
8213.946	8216.204			0.0			12171.07	Z 1	2C 1	P 3
8214.284	8216.542			0.1			12170.57			
8214.493	8216.751			0.1			12170.26			
8214.763	8217.021		0.6	0.9	1.0		12169.86			
8214.999	8217.258		0.1				12169.51			
8215.209	8217.467		0.7	0.8	1.5		12169.20	7c 0	2a 7	Q 1
8216.269	8218.527			0.2			12167.63	Z 2	2B12	R 3
8216.654	8218.912		0.1				12167.06			
8216.762	8219.020			0.1d			12166.90	4F 0	2C 3	R 1
8217.417	8219.676		0.2	0.2			12165.93	3d 1	2c 3	R 5
8217.883	8220.142		0.1				12165.24	4F 0	2C 3	Q 2
8218.187	8220.446		0.6	0.9	0.8		12164.79	3D 1	2B10	R 0
8218.187	8220.446		0.6	0.9	0.8		12164.79	3D 2	2B12	P 3
8218.383	8220.642			0.3			12164.50	3d 0	2c 2	R 5
8218.802	8221.061			0.1			12163.88			
8219.079	8221.338		0.5	0.3			12163.47			
8219.200	8221.460		0.1				12163.29			
8219.288	8221.548			0.2			12163.16			
8219.450	8221.710			0.1			12162.92			
8219.619	8221.879		0.4	0.5	0.6		12162.67			
8219.876	8222.136			0.1			12162.29			
8220.187	8222.447		0.2	0.6	1.6		12161.83	3b 2	2a 1	P 8
8220.511	8222.771		0.2	0.3			12161.35	2K 2	2B 1	P 3
8220.653	8222.913			0.2	0.2		12161.14			
8220.856	8223.116			0.1			12160.84			
8221.032	8223.292			0.2			12160.58			

λ_{air}	λ_{vac}	Ref.	I_1	I_2	I_3	I_4	ν	Upper	Lower	Br.
8221.262	8223.522		1.5	0.9			12160.24			
8221.478	8223.738		0.1				12159.92			
8221.580	8223.840			0.2			12159.77			
8221.884	8224.144		0.7	1.0	1.0		12159.32	3B 4	2A 2	R 2
8222.222	8224.482		0.3	0.4	0.4		12158.82			
8222.371	8224.631		0.2				12158.60			
8222.499	8224.760			0.3			12158.41			
8222.607	8224.868		0.4	0.4	0.2		12158.25			
8222.932	8225.193	G	2.9	2.5	2.3		12157.77	2K 2	2B 1	R 2
8223.250	8225.510		0.3	0.2			12157.30	3f 3	3b 0	R 4
8223.372	8225.632		0.3	0.3			12157.12			
8223.595	8225.856		0.2	0.2			12156.79	4E 0	2B14	R 2
8223.879	8226.140		1.6	1.4	0.7		12156.37	3E 3	2C 3	P 2
8224.265	8226.526		0.1	0.2			12155.80			
8224.704	8226.965		0.1				12155.15			
8224.840	8227.101			0.2d			12154.95			
8225.178	8227.439			0.1d			12154.45			
8225.496	8227.757			0.2d			12153.98			
8225.896	8228.157		0.1	0.2b			12153.39			
8226.187	8228.448			0.2			12152.96			
8226.444	8228.705		0.3	0.5	0.6		12152.58	Z 0	2C 0	Q 4
8226.857	8229.118		0.2	0.2			12151.97			
8227.067	8229.328		0.1				12151.66	3D 1	2B10	R 3
8227.276	8229.538		1.0	0.4			12151.35	Z 2	2C 2	P 3
8227.276	8229.538		1.0	0.4			12151.35	3D 1	2B10	R 1
8227.365	8229.626			0.6	0.6		12151.22			
8227.453	8229.714		0.2	0.4	0.2		12151.09			
8227.723	8229.985		0.2	0.1			12150.69	3E 0	2B 9	R 5
8227.913	8230.175			0.2	0.2b		12150.41			
8228.326	8230.588		0.1	0.3			12149.80	3D 1	2B10	P 1
8228.651	8230.913		0.2				12149.32			
8228.773	8231.035			0.2			12149.14			
8229.132	8231.394			0.2			12148.61	3D 1	2B10	R 2
8229.342	8231.604		0.1	0.3			12148.30			
8229.708	8231.970			0.2			12147.76			
8230.142	8232.404			0.2	0.1		12147.12			
8230.602	8232.865		0.2d	0.4	0.6		12146.44			
8230.867	8233.129			0.1			12146.05			
8231.104	8233.366		0.1	0.2	0.2		12145.70			
8231.856	8234.119		0.1	0.3			12144.59			
8232.595	8234.858		0.1	0.2			12143.50	5c 1	2a 7	Q 1
8232.900	8235.163			0.1			12143.05			
8233.042	8235.306				0.2		12142.84			
8233.158	8235.421		0.3	0.2			12142.67			
8233.354	8235.618			0.2	0.2		12142.38			
8233.653	8235.916		1.5	1.7	1.2		12141.94	4F 0	2B14	P 3
8233.978	8236.242			0.1			12141.46			
8234.216	8236.479		0.1	0.2			12141.11	Z 2	2B12	R 4
8234.629	8236.893			0.1d			12140.50			
8235.138	8237.402		0.3	0.6	0.9		12139.75			
8235.457	8237.721		0.2	0.3			12139.28			
8235.776	8238.040	G	1.8	1.8	1.9		12138.81	3b 3	2a 2	P 4
8236.196	8238.460		0.1	0.2			12138.19			
8237.092	8239.357				0.1		12136.87			
8237.255	8239.520		0.1	0.2b			12136.63			
8237.370	8239.635				0.2		12136.46			
8237.601	8239.866		1.0	2.0	1.5		12136.12	3B 3	2A 1	P 5
8237.846	8240.110	G	2.4	2.0	1.4		12135.76	3D 3	2C 3	R 2
8238.185	8240.450			0.2			12135.26			
8238.321	8240.585		0.2	0.2			12135.06			

λ_{air}	λ_{vac}	Ref.	I_1	I_2	I_3	I_4	ν	Upper	Lower	Br.
8238.525	8240.789		1.5	1.7	1.7		12134.76			
8238.769	8241.034	G	1.7	1.7	1.9		12134.40	4D 0	2B14	R 6
8239.197	8241.462		0.1	0.1			12133.77			
8239.686	8241.951			0.1			12133.05	Y 2	2B13	P 1
8239.896	8242.161			0.0			12132.74			
8240.073	8242.338		0.1	0.2			12132.48			
8240.209	8242.474		0.2				12132.28	Z 2	2B12	P 2
8240.324	8242.589		0.3	0.4	0.2b		12132.11			
8240.480	8242.745		0.4	0.4			12131.88			
8240.800	8243.065	G	2.9	2.4	2.0		12131.41	2A 2	2B 1	R 1
8240.901	8243.167						12131.26	3F 2	2B13	R 2
8241.112	8243.377			0.2			12130.95			
8241.221	8243.486		0.3	0.3			12130.79			
8241.268	8243.534		0.3				12130.72			
8241.533	8243.799		1.7	1.2			12130.33	2A 4	2B 4	P 1
8241.751	8244.016		0.4	0.2			12130.01			
8242.009	8244.275		0.1	0.3			12129.63			
8242.219	8244.485			0.2			12129.32			
8242.451	8244.716		0.7	0.9			12128.98			
8242.587	8244.852		0.2	0.4	0.2b		12128.78			
8242.702	8244.968		0.1				12128.61			
8242.933	8245.199	G	1.5	1.4	1.7		12128.27	3E 3	2C 3	Q 4
8243.273	8245.539			0.2	0.2		12127.77			
8243.463	8245.729			0.1			12127.49			
8243.660	8245.927		0.6	0.9	1.4		12127.20	3A 0	2B 9	P 3
8244.000	8246.266		0.1	0.1d			12126.70	3d 0	2c 2	Q 3
8244.000	8246.266		0.1	0.1d			12126.70	3d 0	2c 2	R 6
8244.279	8246.545		0.0	0.1			12126.29			
8244.497	8246.763		0.1	0.2	0.1		12125.97	3d 1	2c 3	Q 3
8244.918	8247.185			0.2			12125.35	2K 2	2B 1	R 3
8245.442	8247.708			0.1			12124.58			
8245.789	8248.055			0.2			12124.07			
8245.897	8248.164		0.2				12123.91			
8246.163	8248.429			0.2			12123.52	3D 3	2B14	P 2
8246.496	8248.763		0.3	0.3			12123.03	3D 2	2B12	P 4
8246.564	8248.831		0.2				12122.93			
8246.891	8249.157	G	2.2	1.9	2.0		12122.45	3B 4	2A 2	R 1
8247.095	8249.362		1.0		0.8		12122.15			
8247.442	8249.709		0.2	0.4			12121.64			
8247.734	8250.001			0.1			12121.21			
8248.047	8250.315		0.2	0.3	0.6		12120.75	4d 1	3b 0	P 6
8248.490	8250.757		0.1	0.1d			12120.10			
8249.055	8251.322		0.1	0.2b	0.2		12119.27			
8249.381	8251.649		0.0	0.2			12118.79			
8249.436	8251.703				0.1		12118.71	3d 2	2c 4	R 1
8249.715	8251.983			0.0			12118.30			
8249.960	8252.228		0.3	0.6	0.6		12117.94			
8250.355	8252.623		0.2	0.2	0.2b		12117.36	3F 3	2B15	P 4
8250.607	8252.875		1.0	0.8			12116.99	3D 1	2C 1	P 4
8250.981	8253.249		0.1	0.1			12116.44			
8251.158	8253.426			0.2			12116.18			
8251.458	8253.726		0.2	0.4			12115.74			
8251.724	8253.992		1.5	1.5	1.4		12115.35	3E 2	2C 2	P 5
8252.037	8254.305			0.1			12114.89			
8252.357	8254.625		0.1				12114.42			
8253.297	8255.566			0.2			12113.04			
8253.706	8255.975			0.3	0.2		12112.44	3A 2	2B14	R 4
8253.706	8255.975			0.3	0.2		12112.44	4d 2	3b 1	P 6
8254.033	8256.302	G	1.9	1.9	2.2		12111.96	2A 2	2B 1	P 6
8254.163	8256.432		0.6				12111.77			

λ_{air}	λ_{vac}	Ref.	I_1	I_2	I_3	I_4	ν	Upper	Lower	Br.
8254.306	8256.575				0.2		12111.56			
8254.469	8256.738		0.4	0.3			12111.32	V 0	2C 0	P 2
8254.715	8256.984		0.7	0.4			12110.96	Z 1	2C 1	P 4
8254.715	8256.984		0.7	0.4			12110.96	4F 0	2B14	Q 2
8254.926	8257.195			0.1			12110.65			
8255.144	8257.413	G	1.5	1.3	0.8		12110.33	3E 3	2C 3	P 2
8255.471	8257.740			0.1			12109.85	3E 3	2C 3	Q 4
8255.628	8257.897		0.2b	0.3			12109.62			
8255.948	8258.218	G	2.4	2.0	1.7		12109.15	2A 2	2B 1	P 4
8256.289	8258.559		0.1b		0.3		12108.65			
8256.371	8258.641			0.2			12108.53			
8256.617	8258.886			0.2			12108.17			
8256.733	8259.002		0.2	0.2			12108.00			
8256.965	8259.234		0.2	0.6	0.2		12107.66			
8257.306	8259.575			0.2			12107.16			
8258.636	8260.906		0.2	0.3	0.7		12105.21			
8259.168	8261.438			0.2			12104.43	3e 0	2c 2	P 6
8259.373	8261.643				0.1		12104.13			
8259.598	8261.868			0.2			12103.80			
8259.871	8262.141		0.2	0.5	0.3		12103.40	3D 2	2B12	P 5
8260.144	8262.414		0.1	0.1			12103.00	Y 2	2B13	P 2
8260.451	8262.721		1.6	1.2	1.0		12102.55	3D 3	2C 3	Q 2
8260.451	8262.721		1.6	1.2	1.0		12102.55	4c 3	2a 8	R 1
8260.963	8263.234		0.1	0.2	0.2		12101.80	3D 0	2B 8	R 4
8261.202	8263.472		0.3	0.6	0.3b		12101.45	3f 3	3b 0	R 3
8261.468	8263.739		0.2	0.3			12101.06			
8261.816	8264.087			0.1			12100.55			
8262.000	8264.272				0.2		12100.28			
8262.219	8264.490			0.1			12099.96			
8262.342	8264.613				0.2		12099.78			
8262.608	8264.879				0.3		12099.39			
8262.902	8265.173				0.3		12098.96			
8263.243	8265.515		0.1		0.3		12098.46			
8263.339	8265.610				0.3		12098.32			
8263.599	8265.870				0.2		12097.94			
8263.803	8266.075				0.4		12097.64			
8263.974	8266.246			0.1	0.4		12097.39			
8264.227	8266.499			0.2			12097.02	4E 0	2C 3	R 2
8264.459	8266.731	i			1.5b		12096.68			
8264.541	8266.813		1.7	0.6	0.9		12096.56			
8264.740	8267.011		0.1	0.5			12096.27			
8264.849	8267.120				0.8		12096.11			
8264.937	8267.209				0.8		12095.98			
8265.033	8267.305		0.4	0.2			12095.84			
8265.259	8267.531	G	1.8	1.6	1.7		12095.51	2A 2	2B 1	P 5
8265.539	8267.811		0.1				12095.10			
8265.696	8267.968				0.2		12094.87			
8265.785	8268.057		0.0				12094.74			
8265.990	8268.262			0.1			12094.44			
8266.106	8268.378		0.1		0.2b		12094.27			
8266.468	8268.741				0.2b		12093.74			
8266.687	8268.959		0.1		0.2b		12093.42	3A 0	2B 9	P 2
8266.926	8269.199		0.1				12093.07			
8267.138	8269.411		0.1	0.1d			12092.76			
8267.528	8269.801		0.2	0.2			12092.19			
8267.774	8270.047		0.1	0.2			12091.83	3A 0	2B 9	P 1
8267.856	8270.129				0.1		12091.71	V 0	2C 0	Q 4
8268.239	8270.512		0.2	0.3			12091.15			
8268.506	8270.779		1.5	1.1	0.9		12090.76	Z 2	2C 2	R 4
8269.149	8271.422		0.3	0.2d			12089.82			

λ_{air}	λ_{vac}	Ref.	I_1	I_2	I_3	I_4	ν	Upper	Lower	Br.
8269.450	8271.723	G	2.4	1.9	1.2		12089.38	2A 4	2B 4	P 2
8269.956	8272.229		1.5	1.1			12088.64	3E 3	2C 3	P 3
8270.086	8272.359		0.4				12088.45			
8270.435	8272.708			0.2b			12087.94			
8270.572	8272.845		0.3				12087.74			
8270.777	8273.050	G	1.6	1.3	0.9		12087.44	3B 4	2A 2	R 0
8271.078	8273.351		0.2	0.2			12087.00	V 0	2C 0	P 3
8271.146	8273.420		0.2				12086.90			
8271.372	8273.646		0.4	0.3			12086.57			
8271.502	8273.776		0.2				12086.38			
8271.639	8273.913		0.3	0.1			12086.18			
8271.797	8274.070		0.2b	0.1			12085.95	3D 1	2B10	P 2
8272.111	8274.385		0.3	0.3			12085.49			
8272.200	8274.474			0.5			12085.36			
8272.337	8274.611		0.2	0.1			12085.16			
8272.556	8274.830		0.6	0.3			12084.84			
8272.714	8274.988			0.2			12084.61			
8272.755	8275.029		0.4		0.3		12084.55			
8272.830	8275.104			0.4			12084.44			
8272.988	8275.262			0.6			12084.21			
8273.077	8275.351						12084.08			
8273.282	8275.556	G	3.7	3.0	2.3		12083.78	2A 2	2B 1	R 0
8273.617	8275.892		1.0	0.3			12083.29			
8273.761	8276.036		1.0				12083.08			
8273.809	8276.083			0.9	1.4		12083.01	3E 1	2C 1	P 7
8274.008	8276.282		0.4	0.2			12082.72			
8274.241	8276.515		1.5	1.5	2.0		12082.38			
8274.474	8276.748		0.4	0.2			12082.04			
8274.699	8276.974		0.2	0.2			12081.71			
8274.816	8277.090		0.2	0.1			12081.54			
8275.001	8277.275		0.4	0.7	1.3		12081.27			
8275.193	8277.467		0.3	0.6	1.0		12080.99			
8275.419	8277.693		0.3	0.2			12080.66			
8275.631	8277.906				0.1		12080.35			
8275.741	8278.015		0.2				12080.19			
8275.823	8278.098			0.1			12080.07			
8275.974	8278.248		0.1				12079.85			
8276.179	8278.454			0.1d			12079.55	3c 0	2a 2	R 6
8276.638	8278.913		0.3	0.3b			12078.88	3D 3	2B14	P 3
8276.947	8279.222		0.1				12078.43			
8277.097	8279.372			0.2b			12078.21			
8277.447	8279.722		0.1	0.2			12077.70			
8277.742	8280.017		0.2	0.3			12077.27			
8278.084	8280.360			0.2			12076.77			
8278.331	8280.607		0.1	0.3			12076.41			
8278.482	8280.757			0.1			12076.19			
8278.687	8280.963			0.1			12075.89			
8278.907	8281.182		0.3	0.3			12075.57			
8279.387	8281.663		0.1	0.2			12074.87			
8279.606	8281.882		0.2	0.3			12074.55			
8279.970	8282.246		0.2				12074.02	3d 2	2c 4	P 1
8280.676	8282.952		0.2	0.1			12072.99			
8280.964	8283.240		1.7	1.4	0.9		12072.57	3D 2	2C 2	P 4
8281.218	8283.494		0.5	0.3			12072.20	X 2	2C 3	R 1
8281.506	8283.782		0.4	0.3			12071.78			
8282.096	8284.373		0.3	0.3			12070.92	3F 0	2B 9	R 3
8282.439	8284.716		0.1				12070.42			
8282.618	8284.894			0.1			12070.16			
8283.180	8285.457		0.0	0.1			12069.34			
8283.558	8285.835			0.2			12068.79			

λ_{air}	λ_{vac}	Ref.	I_1	I_2	I_3	I_4	ν	Upper	Lower	Br.
8283.874	8286.151		1.5	1.3	1.3		12068.33	3D 1	2C 1	Q 5
8284.340	8286.618			0.1			12067.65			
8284.505	8286.782		0.1	0.1			12067.41			
8284.800	8287.078		1.5	1.3	1.6		12066.98	3E 2	2C 2	P 6
8286.407	8288.685			0.1			12064.64			
8286.696	8288.974		0.3	0.6	0.3		12064.22	3F 2	2B13	Q 3
8287.170	8289.448	G	1.8	1.4	1.0		12063.53	3D 3	2C 3	P 2
8287.170	8289.448	G	1.8	1.4	1.0		12063.53	Z 2	2B12	P 3
8289.327	8291.606			0.2			12060.39			
8289.582	8291.860			0.1			12060.02			
8289.850	8292.128		0.1	0.2			12059.63	3C 2	2A 3	Q 4
8290.111	8292.390		0.8	0.5			12059.25			
8290.413	8292.692			0.1			12058.81			
8290.764	8293.043			0.0d			12058.30	4F 0	2C 3	P 4
8291.218	8293.497			0.0d			12057.64			
8291.459	8293.738			0.2			12057.29	Z 0	2C 0	P 4
8291.575	8293.854		0.2	0.5			12057.12	4E 1	2C 4	P 2
8291.844	8294.123			0.1			12056.73			
8292.057	8294.336		0.0	0.1			12056.42	3F 1	2B11	Q 4
8292.298	8294.577		0.2	0.5	1.1		12056.07	3d 0	2c 2	P 3
8292.298	8294.577		0.2	0.5	1.1		12056.07	3d 1	2c 3	P 3
8292.642	8294.921			0.1			12055.57		-	
8292.903	8295.182		0.0	0.2	0.2b		12055.19			
8293.130	8295.409			0.1			12054.86			
8294.217	8296.497			0.0			12053.28			
8294.472	8296.752		0.1	0.3	0.1		12052.91	3C 2	2A 3	R 2
8294.658	8296.937		0.1	0.2			12052.64			
8294.974	8297.254			0.1			12052.18			
8295.298	8297.578		0.3	0.5			12051.71			
8295.731	8298.011			0.0			12051.08	4F 0	2C 3	Q 2
8295.731	8298.011			0.0			12051.08	3D 0	2B 8	R 3
8296.048	8298.328			0.0			12050.62			
8296.365	8298.645		0.9	0.9	1.3		12050.16			
8296.785	8299.065		0.2	0.2			12049.55			
8296.922	8299.203		0.3	1.1	1.2		12049.35	3D 0	2C 0	Q 6
8297.150	8299.430		0.7	1.1	0.1		12049.02			
8297.453	8299.733			0.1			12048.58			
8297.632	8299.912			0.1			12048.32			
8297.886	8300.167		0.4	0.7	1.2		12047.95	X 2	2B14	R 0
8298.183	8300.464			0.2			12047.52			
8298.369	8300.649		0.2	0.1			12047.25	3d 2	2c 4	R 3
8298.575	8300.856		0.2	0.5	1.4		12046.95	X 2	2B14	R 1
8298.575	8300.856		0.2	0.5	1.4		12046.95	3f 3	3b 0	R 2
8298.858	8301.139			0.1			12046.54			
8299.071	8301.352		1.5	1.0	0.2		12046.23	X 2	2C 3	Q 1
8299.368	8301.649			0.1			12045.80			
8299.561	8301.842			0.2			12045.52			
8299.788	8302.069		0.3	0.6	0.7		12045.19	3D 2	2C 2	Q 5
8299.788	8302.069		0.3	0.6	0.7		12045.19	X 2	2B14	R 2
8300.112	8302.393			0.0			12044.72			
8300.484	8302.765			0.3			12044.18			
8300.780	8303.062		0.5	0.6	0.7		12043.75			
8301.118	8303.400		0.1	0.2	0.1b		12043.26			
8301.407	8303.689		0.3	0.2			12042.84	Z 1	2B10	R 2
8301.780	8304.062		0.0	0.2	0.4		12042.30			
8301.986	8304.268			0.0			12042.00			
8302.242	8304.524		0.1	0.0			12041.63	X 2	2B14	R 3
8302.476	8304.758			0.0			12041.29			
8302.786	8305.068			0.1d			12040.84			
8302.973	8305.255			0.1			12040.57			

λ_{air}	λ_{vac}	Ref.	I_1	I_2	I_3	I_4	ν	Upper	Lower	Br.
8303.276	8305.558		0.1	0.1			12040.13			
8303.504	8305.786		1.5	0.9	1.0		12039.80	2A 4	2B 4	P 3
8303.717	8306.000	G	2.2	1.8	1.3		12039.49	3E 3	2C 3	P 3
8304.138	8306.421			0.2			12038.88			
8304.380	8306.662						12038.53			
8304.469	8306.752		0.1	0.2			12038.40			
8304.800	8307.083			0.1d			12037.92	3d 1	2c 3	Q 4
8305.125	8307.407		0.4	0.7			12037.45	Z 2	2C 2	Q 4
8305.463	8307.745			0.1			12036.96	V 0	2C 0	P 4
8306.118	8308.401				0.1		12036.01			
8306.560	8308.843			0.3			12035.37			
8306.656	8308.940				0.2		12035.23	3c 0	2a 2	R 5
8307.651	8309.934			0.1			12033.79			
8307.782	8310.065		0.1				12033.60			
8307.913	8310.196			0.2			12033.41			
8308.217	8310.500		0.1	0.2b			12032.97			
8308.569	8310.852			0.1			12032.46	3D 3	2B14	P 4
8308.721	8311.004		0.2				12032.24			
8308.859	8311.143			0.5	0.1		12032.04			
8308.983	8311.267			0.4	0.8		12031.86	3d 0	2c 2	Q 4
8309.197	8311.481			0.1d			12031.55			
8309.570	8311.854		0.5	0.5	0.8		12031.01			
8309.701	8311.985		0.1				12030.82	3F 3	2B15	Q 2
8310.088	8312.372			0.2d			12030.26			
8310.537	8312.821			0.2b			12029.61			
8310.924	8313.208		0.2	0.3			12029.05	3D 2	2B12	P 6
8310.924	8313.208		0.2	0.3			12029.05	X 2	2B14	R 1
8311.463	8313.748			0.1			12028.27			
8311.871	8314.155			0.3			12027.68			
8312.230	8314.515			0.1			12027.16	4c 3	2a 8	Q 1
8312.590	8314.874			0.2			12026.64			
8313.799	8316.084		0.2	0.2			12024.89	3e 0	2c 2	P 7
8314.041	8316.326			0.3			12024.54			
8314.297	8316.582			0.2			12024.17	4E 0	2B14	R 0
8314.657	8316.942		0.2	0.2			12023.65			
8314.837	8317.122			0.0			12023.39			
8315.092	8317.378			0.1			12023.02			
8315.404	8317.689		1.5	1.0			12022.57	4E 0	2C 3	Q 1
8315.404	8317.689		1.5	1.0			12022.57	Y 2	2B13	P 4
8315.729	8318.014			0.1			12022.10			
8315.929	8318.215			0.0			12021.81			
8315.999	8318.284		0.1				12021.71			
8316.192	8318.478		0.2	0.1			12021.43			
8316.469	8318.755			0.1			12021.03			
8316.704	8318.990			0.2			12020.69			
8317.112	8319.398			0.3b			12020.10			
8317.451	8319.737	G	1.9	2.0	2.5		12019.61	3b 3	2a 2	P 5
8317.777	8320.063		0.2	0.3			12019.14	3D 0	2B 8	R 2
8317.887	8320.174			0.2			12018.98	4f 2	3b 2	R 4
8318.040	8320.326			0.1			12018.76			
8318.137	8320.423		0.1				12018.62			
8318.275	8320.561		0.3	0.6	1.1		12018.42			
8318.621	8320.907		0.3	0.6	0.9		12017.92	3D 1	2B10	P 3
8318.967	8321.254			0.1d			12017.42			
8319.175	8321.461		0.1	0.1			12017.12			
8319.376	8321.662		0.4	0.8	0.2		12016.83	3E 2	2C 2	P 6
8319.514	8321.801		0.1	0.2			12016.63			
8319.729	8322.015	G	0.9	1.3	1.7		12016.32	3B 4	2A 2	R 3
8320.220	8322.507			0.1			12015.61	3D 0	2B 8	R 0
8320.490	8322.777		0.2	0.2			12015.22			

λ_{air}	λ_{vac}	Ref.	I_1	I_2	I_3	I_4	ν	Upper	Lower	Br.
8320.726	8323.013			0.3	0.1		12014.88	3d 2	2c 4	Q 2
8321.031	8323.318		0.3	0.2			12014.44			
8321.176	8323.463		0.1	0.3	0.2		12014.23			
8321.557	8323.844			0.1d			12013.68			
8321.651	8323.938						12013.54	Z 1	2B10	R 0
8322.042	8324.329		0.5	0.7	0.8		12012.98	X 0	2C 1	R 1
8322.125	8324.412		0.3				12012.86			
8322.347	8324.634			0.1			12012.54	3D 0	2B 8	R 1
8323.469	8325.757			0.1			12010.92	Z 1	2C 1	P 4
8323.795	8326.083		0.5	0.7	0.9		12010.45			
8323.920	8326.208		0.1				12010.27			
8324.031	8326.318			0.2			12010.11			
8324.239	8326.526		0.3	0.8	1.1		12009.81			
8324.502	8326.790		0.1	0.2			12009.43			
8324.897	8327.185		0.0	0.1			12008.86			
8325.327	8327.615			0.2			12008.24			
8325.528	8327.816			0.1			12007.95			
8325.715	8328.003		0.0	0.2			12007.68			
8326.187	8328.475		0.0	0.1			12007.00			
8326.499	8328.787			0.1			12006.55			
8327.192	8329.481		0.3	0.3	0.2b		12005.55	3F 2	2B13	R 1
8327.366	8329.654		0.2				12005.30			
8327.581	8329.869			0.1			12004.99			
8327.706	8329.994		0.2b				12004.81			
8327.948	8330.237		0.2				12004.46			
8328.032	8330.320			0.3d			12004.34	3D 0	2C 0	P 5
8328.365	8330.654		0.2				12003.86			
8328.517	8330.806		0.3	0.2			12003.64			
8328.802	8331.091		0.3	0.1			12003.23			
8328.961	8331.250		0.2	0.1			12003.00			
8329.287	8331.577		0.5				12002.53			
8329.378	8331.667		0.7	1.0	1.2		12002.40			
8329.718	8332.007		0.6	0.3			12001.91			
8329.877	8332.167		0.4	0.1			12001.68			
8329.995	8332.285			0.2			12001.51			
8330.155	8332.445			0.5			12001.28			
8330.231	8332.521						12001.17			
8330.433	8332.722	G	3.4	2.7	2.1		12000.88	3B 4	2A 2	P 1
8330.433	8332.722	G	3.4	2.7	2.1		12000.88	2A 2	2B 1	P 1
8330.787	8333.076		1.0	0.2			12000.37			
8330.891	8333.181		1.0	0.3			12000.22			
8331.106	8333.396		0.4				11999.91			
8331.266	8333.556			0.2	0.1		11999.68			
8331.377	8333.667		0.3				11999.52	3D 3	2B14	P 6
8331.467	8333.757			0.2			11999.39			
8331.780	8334.069		1.0	0.8	0.7		11998.94			
8332.050	8334.340		0.2	0.3			11998.55			
8332.245	8334.535				0.2		11998.27			
8332.377	8334.667		0.3	0.1			11998.08			
8332.523	8334.813		0.2				11997.87			
8332.682	8334.973		0.3	0.6	0.7		11997.64	3D 0	2B 8	P 1
8332.863	8335.153		0.5	0.3			11997.38	Z 2	2B12	P 4
8333.141	8335.431			0.1			11996.98			
8333.439	8335.730			0.1			11996.55			
8333.703	8335.994				0.1		11996.17			
8333.856	8336.147		0.2				11995.95			
8334.099	8336.390		0.2				11995.60			
8334.377	8336.668	G	2.0	1.5	1.9		11995.20	3E 3	2C 3	P 4
8334.579	8336.870		1.9	1.5	1.9		11994.91	3D 3	2C 3	Q 3
8334.780	8337.071			0.9	1.5b		11994.62	3b 2	2a 1	P 9

λ_{air}	λ_{vac}	Ref.	I_1	I_2	I_3	I_4	ν	Upper	Lower	Br.
8335.399	8337.690	G	0.1				11993.73	4E 0	2B14	P 4
8335.698	8337.989		0.1				11993.30			
8335.927	8338.218		0.1				11992.97			
8336.108	8338.399		0.2	0.1			11992.71			
8336.358	8338.649		0.4	0.7	1.2		11992.35	3F 1	2B11	R 2
8336.796	8339.087		0.1	0.2	0.2		11991.72			
8337.081	8339.372		0.6	1.3	1.7		11991.31	4E 0	2B14	Q 4
8337.728	8340.019		0.2	0.2			11990.38	X 2	2B14	P 2
8337.728	8340.019		0.2	0.2			11990.38	3f 3	3b 0	R 1
8338.048	8340.339		1.6	1.4	0.6		11989.92	2K 4	2B 3	R 4
8338.048	8340.339		1.6	1.4	0.6		11989.92	3D 3	2B14	P 5
8338.479	8340.771			0.1			11989.30			
8338.555	8340.847		0.1				11989.19			
8338.743	8341.035		0.4	0.2			11988.92			
8339.035	8341.327			0.1			11988.50			
8339.369	8341.661			0.2b			11988.02			
8339.675	8341.967		0.2	0.7	0.4		11987.58			
8340.093	8342.385			0.2	0.1		11986.98	3c 0	2a 2	R 4
8341.018	8343.311			0.1			11985.65			
8341.324	8343.617		1.0	1.0	0.8		11985.21	X 2	2C 3	R 2
8341.770	8344.062		0.2b	0.3			11984.57	3c 2	2a 4	R 5
8342.083	8344.376			0.1			11984.12			
8342.396	8344.689			0.2			11983.67			
8342.758	8345.051			0.1			11983.15			
8343.079	8345.371			0.3			11982.69	3B 4	2K 2	P 4
8343.357	8345.650		0.7	0.8	0.9		11982.29			
8343.663	8345.957			0.2			11981.85			
8343.991	8346.284	G	1.8	1.5	1.4		11981.38	2K 4	2B 3	R 3
8344.262	8346.556		1.9	1.6	1.4		11980.99	2A 4	2B 4	P 4
8344.645	8346.939		0.1	0.1			11980.44			
8344.889	8347.183		0.1				11980.09			
8345.063	8347.357		0.2	0.1	0.8		11979.84			
8345.314	8347.608		0.2	0.2d	0.1		11979.48			
8345.502	8347.796						11979.21			
8345.774	8348.068		0.6b	1.5	1.4		11978.82			
8345.969	8348.263		0.3				11978.54			
8346.094	8348.388			0.1			11978.36			
8346.331	8348.625		0.4	0.3			11978.02	X 1	2C 2	R 1
8346.666	8348.960			0.2			11977.54			
8346.777	8349.071		0.3				11977.38	4F 0	2C 3	P 5
8346.896	8349.190			0.3			11977.21			
8347.070	8349.364		0.3b	0.2	0.2		11976.96			
8347.251	8349.545			0.3			11976.70			
8347.419	8349.713		0.2b	0.2d			11976.46			
8347.614	8349.908		0.4	0.3			11976.18			
8347.781	8350.075			0.3			11975.94			
8347.906	8350.201		0.2				11975.76			
8348.157	8350.452		1.6	1.3	1.1		11975.40	X 0	2C 1	R 2
8348.325	8350.619		0.4				11975.16			
8348.541	8350.835		0.2	0.2			11974.85			
8348.827	8351.121		1.0	0.8	1.1		11974.44			
8349.252	8351.547						11973.83			
8349.336	8351.630						11973.71	Z 0	2C 0	Q 5
8349.538	8351.833	G	4.0	3.2	2.7		11973.42	2A 2	2B 1	P 2
8349.887	8352.181						11972.92			
8349.998	8352.293		1.5	0.9			11972.76			
8350.166	8352.460		0.5	0.2			11972.52			
8350.298	8352.593			0.3			11972.33			
8350.500	8352.795		0.4	0.2			11972.04			
8350.633	8352.928				0.2		11971.85			

λ_{air}	λ_{vac}	Ref.	I_1	I_2	I_3	I_4	ν	Upper	Lower	Br.
8350.856	8353.151		0.6	0.8	0.5		11971.53			
8351.086	8353.381	G	2.4	2.0	1.9		11971.20	2K 4	2B 3	R 2
8351.463	8353.758		0.4	0.2			11970.66			
8351.603	8353.898		0.3				11970.46			
8351.728	8354.023		0.3	0.3b			11970.28			
8352.028	8354.323		0.2	0.1			11969.85			
8352.251	8354.547		0.2	0.2			11969.53			
8352.565	8354.861		0.2b	0.2	0.1		11969.08			
8353.082	8355.378		0.3	0.6	0.1		11968.34			
8353.222	8355.517		0.2				11968.14			
8353.480	8355.776		0.4	0.8	0.9		11967.77	3D 2	2C 2	R 6
8353.480	8355.776		0.4	0.8	0.9		11967.77	3D 1	2B10	P 4
8353.710	8356.006		0.2	0.1			11967.44			
8354.024	8356.320		0.2	0.2			11966.99			
8354.255	8356.551		0.6	0.8	1.0		11966.66	4E 0	2B14	P 2
8354.255	8356.551		0.6	0.8	1.0		11966.66	Z 1	2B10	P 1
8354.520	8356.816		0.1				11966.28	3F 2	2B13	P 4
8354.694	8356.991		0.2				11966.03			
8355.030	8357.326		0.2	0.2			11965.55	4D 0	2B14	R 5
8355.274	8357.570		0.1	0.7			11965.20			
8355.365	8357.661			0.2d			11965.07	3A 2	2B14	R 3
8355.875	8358.171		0.1	0.2d			11964.34			
8356.154	8358.450			0.1			11963.94			
8356.301	8358.597				0.2		11963.73			
8356.531	8358.828			0.1d			11963.40			
8356.859	8359.156			0.4b	0.3		11962.93			
8356.922	8359.219		0.4				11962.84			
8357.153	8359.450	G	2.0	1.6	1.0		11962.51	X 0	2C 1	Q 1
8357.314	8359.610		1.0	0.8	1.2		11962.28	3D 3	2C 3	R 4
8357.663	8359.960		0.7	0.3			11961.78	X 2	2C 3	Q 2
8357.872	8360.169		0.7	0.4			11961.48			
8358.243	8360.540		0.5	1.1	1.0		11960.95			
8358.341	8360.638				0.2		11960.81			
8358.599	8360.896		0.2	0.2			11960.44			
8358.886	8361.183		0.9	0.9	1.0		11960.03			
8359.033	8361.330		0.2	0.5			11959.82			
8359.270	8361.568				0.1		11959.48			
8359.431	8361.728		0.1	0.2			11959.25			
8359.787	8362.085	G	2.1	1.7	1.2		11958.74	2K 4	2B 3	R 1
8359.787	8362.085	G	2.1	1.7	1.2		11958.74	3E 3	2C 3	P 4
8360.158	8362.456			0.1			11958.21	3F 2	2B13	Q 2
8360.361	8362.658			0.2			11957.92			
8360.584	8362.882		0.2	0.3			11957.60	4E 2	2B19	Q 2
8360.773	8363.071		0.1	0.1			11957.33			
8361.074	8363.372		1.6	1.3	1.0		11956.90	3B 4	2A 2	P 2
8361.563	8363.861		0.1	0.2			11956.20	4E 0	2C 3	Q 3
8361.857	8364.155		0.1				11955.78			
8362.983	8365.282		0.1				11954.17			
8363.284	8365.583			0.2			11953.74	3C 2	2A 3	Q 3
8363.487	8365.786		0.3	0.3			11953.45	4E 0	2C 3	R 3
8363.725	8366.024		0.3	0.7	1.0		11953.11	3b 4	2a 3	R 3
8363.977	8366.275			0.1			11952.75	3C 2	2A 3	Q 3
8364.173	8366.471		0.3	0.6	0.9		11952.47			
8364.712	8367.010		0.1	0.2d			11951.70			
8364.901	8367.200				0.2		11951.43	3F 1	2B11	P 5
8365.006	8367.305		0.1	0.2d			11951.28	3d 1	2c 3	Q 5
8365.363	8367.662		0.7	0.3			11950.77			
8365.692	8367.991			0.2			11950.30			
8365.993	8368.292		1.0	0.9			11949.87	X 1	2C 2	Q 1
8366.196	8368.495		0.2	0.2			11949.58	Z 2	2B12	P 5

λ_{air}	λ_{vac}	Ref.	I_1	I_2	I_3	I_4	ν	Upper	Lower	Br.
8366.336	8368.635		0.2	0.2			11949.38			
8366.602	8368.901		0.5	0.4			11949.00			
8366.931	8369.230	G	3.1	2.6	2.2		11948.53	2A 2	2B 1	P 3
8366.931	8369.230	G	3.1	2.6	2.2		11948.53	4g 0	2a 5	P 1
8367.246	8369.546		0.5	0.5			11948.08	3C 2	2A 3	R 1
8367.393	8369.693			0.3			11947.87			
8367.659	8369.959		0.2	0.1			11947.49	4E 1	2C 4	P 3
8368.002	8370.302		0.8	0.8	1.1		11947.00	Z 2	2C 2	P 4
8368.359	8370.659		0.6	1.0	0.2		11946.49	3B 4	2K 2	R 2
8368.647	8370.947			0.1			11946.08			
8368.822	8371.122			0.2			11945.83			
8369.032	8371.332		0.1	0.2			11945.53			
8369.369	8371.669		0.1	0.1d			11945.05			
8369.740	8372.040		0.3	0.2			11944.52			
8370.076	8372.376	G	2.6	2.1	1.5		11944.04	2K 3	2B 2	R 0
8370.504	8372.804		0.2b	0.2d			11943.43			
8370.805	8373.105		0.2	0.1d	0.1		11943.00			
8371.057	8373.358		0.1	0.1			11942.64			
8371.205	8373.505		0.1				11942.43			
8371.352	8373.652		0.2	0.2			11942.22			
8371.492	8373.793		0.2				11942.02	4F 0	2C 3	P 4
8371.751	8374.052	G	2.3	2.0	1.7		11941.65	2K 3	2B 2	R 1
8371.751	8374.052	G	2.3	2.0	1.7		11941.65	3c 2	2a 4	R 4
8371.751	8374.052	G	2.3	2.0	1.7		11941.65	3b 4	2a 3	R 2
8372.060	8374.361	G	2.8	2.2	1.5		11941.21	2K 4	2B 3	R 0
8372.347	8374.648			0.1			11940.80	4f 3	3b 3	R 3
8372.509	8374.810		0.2b	0.2			11940.57			
8372.838	8375.139		0.1	0.2			11940.10			
8372.971	8375.272		0.1				11939.91			
8373.063	8375.364			0.2			11939.78			
8373.315	8375.616		0.2	0.5	0.8		11939.42			
8373.554	8375.855			0.1			11939.08	4E 0	2C 3	R 2
8373.806	8376.107		0.1	0.2d			11938.72	3d 0	2c 2	Q 5
8374.002	8376.304		0.2				11938.44	X 2	2B14	P 3
8374.241	8376.542	G	1.5	1.2	0.9		11938.10	3B 4	2K 2	P 3
8374.606	8376.907		0.2	0.1			11937.58	3F 0	2B 9	P 5
8375.069	8377.370			0.2			11936.92			
8375.385	8377.686		0.9	1.0	1.0		11936.47			
8375.777	8378.079		1.0	0.5			11935.91	X 0	2C 1	P 1
8376.002	8378.304		0.4	0.2			11935.59			
8376.128	8378.430				0.2b		11935.41	3c 0	2a 2	R 3
8376.262	8378.564						11935.22	4E 0	2C 3	R 1
8376.311	8378.613		1.7	1.2	1.0		11935.15	3D 3	2C 3	P 3
8376.725	8379.027		0.2	0.1b			11934.56	Z 1	2B10	P 2
8377.146	8379.448		0.2	0.9	1.0		11933.96			
8377.602	8379.905				0.2		11933.31			
8377.883	8380.186		0.1				11932.91			
8378.108	8380.410			0.1			11932.59			
8378.529	8380.832		0.1	0.2			11931.99			
8378.803	8381.106			0.2			11931.60			
8379.119	8381.422		0.9	1.4	2.0		11931.15	3D 0	2B 8	P 2
8379.526	8381.829			0.0			11930.57	3D 2	2C 2	Q 6
8380.088	8382.391			0.2			11929.77			
8380.397	8382.700		0.2	0.4	0.2		11929.33			
8380.601	8382.904		0.1	0.2			11929.04			
8380.868	8383.171		0.6	0.7			11928.66	4E 0	2C 3	Q 1
8381.177	8383.480	G	1.9	1.9	2.0		11928.22	3b 4	2a 3	R 1
8381.486	8383.790		0.2	0.2			11927.78			
8381.641	8383.944		0.2	0.5			11927.56			
8381.852	8384.155			0.1			11927.26			

λ_{air}	λ_{vac}	Ref.	I_1	I_2	I_3	I_4	ν	Upper	Lower	Br.
8381.943	8384.247		0.1				11927.13			
8382.140	8384.443		0.1	0.2d			11926.85			
8382.288	8384.591		0.1				11926.64			
8382.456	8384.760		0.2	0.5	0.9		11926.40	3D 1	2C 1	Q 6
8383.103	8385.407			0.2			11925.48			
8383.525	8385.829		0.1				11924.88			
8383.693	8385.997		0.1	0.2			11924.64			
8384.059	8386.363		1.7	1.2	0.8		11924.12			
8384.481	8386.785			0.1			11923.52			
8384.706	8387.010	1			0.3		11923.20			
8385.114	8387.418			0.0			11922.62			
8385.501	8387.805		0.0				11922.07			
8386.290	8388.595						11920.95	6c 0	2a 8	Q 1
8386.290	8388.595						11920.95	3D 1	2C 1	P 5
8386.380	8388.685		0.1	0.2b			11920.82			
8386.479	8388.783		0.0				11920.68			
8386.943	8389.248		0.2	0.3d			11920.02	3d 1	2c 3	P 4
8387.231	8389.536	G	2.2	1.9	1.8		11919.61	2K 3	2B 2	R 2
8387.231	8389.536	G	2.2	1.9	1.8		11919.61	2A 1	2B 0	R 5
8387.682	8389.987		0.2b	0.2	0.2b		11918.97			
8387.801	8390.106						11918.80			
8387.907	8390.212		0.0				11918.65			
8388.357	8390.663			0.2d			11918.01			
8388.428	8390.733			0.5			11917.91			
8388.533	8390.839		0.6	0.5			11917.76			
8389.315	8391.620		0.1	0.2d	0.7		11916.65			
8389.660	8391.965		0.6	0.9	1.2		11916.16	3E 2	2C 2	P 7
8389.927	8392.233		0.2	0.2			11915.78	3D 1	2B10	P 5
8390.188	8392.493		0.5	0.7	0.6		11915.41	4E 0	2B14	Q 3
8390.188	8392.493		0.5	0.7	0.6		11915.41	X 1	2C 2	R 2
8390.667	8392.972		0.1				11914.73			
8390.828	8393.134			0.1d			11914.50			
8391.131	8393.437		0.5	0.8	1.3		11914.07			
8391.441	8393.747		0.2	0.2	0.2		11913.63			
8391.688	8393.994		0.3	0.7	0.8		11913.28	X 0	2C 1	R 3
8391.948	8394.255			0.2			11912.91			
8392.244	8394.551		0.1	0.2	0.2d		11912.49			
8392.463	8394.769		0.2	0.7	0.9		11912.18	2A 4	2B 4	P 5
8392.463	8394.769		0.2	0.7	0.9		11912.18	3F 0	2B 9	Q 4
8393.505	8395.812			0.1			11910.70	3d 0	2c 2	P 4
8393.823	8396.129			0.1			11910.25	4F 0	2C 3	P 3
8394.147	8396.454			0.5			11909.79			
8394.478	8396.785			0.2			11909.32			
8394.760	8397.067		0.2	0.6	0.9		11908.92			
8395.599	8397.906			0.1			11907.73			
8395.944	8398.252			0.2b			11907.24	3c 1	2a 3	R 4
8396.375	8398.682		0.2	0.3			11906.63	4E 0	2C 3	P 2
8396.727	8399.035			0.0			11906.13	3F 1	2B11	Q 3
8397.045	8399.352			0.1			11905.68			
8397.284	8399.592			0.1			11905.34			
8397.538	8399.846		0.2	0.3			11904.98			
8397.694	8400.001		0.2	0.2			11904.76			
8397.849	8400.157			0.3			11904.54			
8397.962	8400.270			0.5			11904.38			
8398.053	8400.361		0.6				11904.25			
8398.279	8400.587	G	3.1	2.7	2.5		11903.93	2K 2	2B 1	P 4
8398.625	8400.933		0.3	0.2	0.2		11903.44	4E 0	2B14	Q 1
8398.738	8401.046		0.3	0.3			11903.28			
8398.935	8401.243		0.1	0.2			11903.00			
8399.062	8401.370			0.1			11902.82			

λ_{air}	λ_{vac}	Ref.	I_1	I_2	I_3	I_4	ν	Upper	Lower	Br.
8399.260	8401.568		0.1	0.2			11902.54			
8399.464	8401.773		0.2	0.2	0.2		11902.25			
8399.754	8402.062	G		1.2	1.3		11901.84	3b 4	2a 3	R 0
8400.142	8402.450		0.3	0.6	1.1		11901.29	3D 0	2C 0	Q 7
8400.467	8402.775			0.1			11900.83			
8400.841	8403.150			0.1			11900.30			
8401.215	8403.524		0.1	0.2			11899.77			
8401.434	8403.743			0.1			11899.46			
8401.752	8404.060		0.9	1.0	0.8		11899.01	X 0	2C 1	Q 2
8402.253	8404.562			0.2			11898.30			
8402.458	8404.767		0.1				11898.01			
8402.691	8405.000		0.3	0.3	0.2		11897.68			
8402.966	8405.276			0.1			11897.29			
8403.221	8405.530		0.2	0.3	0.2b		11896.93			
8403.383	8405.692			0.1			11896.70			
8403.595	8405.904			0.1			11896.40	5c 0	2a 6	R 2
8403.870	8406.180		0.2	0.3			11896.01			
8404.217	8406.526		0.3	0.3			11895.52			
8404.457	8406.766		0.1				11895.18			
8404.697	8407.007		0.2	0.3	0.4		11894.84	U 1	2B11	R 4
8404.874	8407.183		0.1				11894.59	3c 1	2a 3	R 3
8405.149	8407.459		0.3	0.4	0.2		11894.20	3c 2	2a 4	R 3
8405.411	8407.721			0.2			11893.83	Z 0	2B 8	R 4
8405.665	8407.975			0.1d			11893.47			
8405.948	8408.258		0.1				11893.07			
8406.082	8408.392				0.3		11892.88			
8406.266	8408.576				0.4		11892.62	3A 2	2B14	P 6
8406.351	8408.661			0.2			11892.50			
8406.414	8408.724		0.2				11892.41			
8406.683	8408.993	G	2.0	1.7	1.2		11892.03	2K 3	2B 2	P 1
8407.022	8409.333		0.1	0.1	0.4		11891.55	Z 1	2B10	P 3
8407.376	8409.686			0.0			11891.05	Z 0	2B 8	R 3
8407.468	8409.778				0.7		11890.92			
8407.638	8409.948		0.1	0.3			11890.68			
8407.708	8410.019				0.7		11890.58			
8407.949	8410.259						11890.24			
8408.196	8410.507	1	2.2	0.4	0.8		11889.89			
8408.557	8410.868				1.0		11889.38			
8408.677	8410.988				1.0		11889.21			
8408.847	8411.157				0.6		11888.97	3F 2	2B13	P 3
8409.328	8411.639		0.2	0.4			11888.29			
8409.420	8411.731				0.3		11888.16			
8409.646	8411.957		0.4	1.0	0.9		11887.84	3b 3	2a 2	P 6
8409.901	8412.212				0.2		11887.48			
8410.155	8412.467			0.1	0.3		11887.12	3f 3	3b 0	Q 4
8410.318	8412.629			0.1	0.3		11886.89			
8410.580	8412.891			0.2			11886.52	3f 3	3b 0	Q 3
8410.700	8413.011		0.5	0.7	0.3		11886.35			
8410.913	8413.224		0.6	0.7	0.7		11886.05	Z 0	2B 8	R 2
8411.422	8413.734		0.1	0.2			11885.33			
8411.606	8413.918			0.1	0.2b		11885.07			
8411.932	8414.243			0.1	0.2b		11884.61			
8412.293	8414.604			0.2			11884.10	3f 3	3b 0	Q 2
8412.413	8414.725				0.2		11883.93			
8412.533	8414.845		0.2	0.4			11883.76	3c 3	2a 5	R 3
8412.746	8415.057			0.1	0.2		11883.46			
8413.001	8415.312				0.1		11883.10	3f 3	3b 0	Q 5
8413.213	8415.525		0.2	0.2			11882.80			
8413.539	8415.851	G	2.3	1.7	1.0		11882.34	2K 4	2B 3	P 1
8413.673	8415.985						11882.15			

λ_{air}	λ_{vac}	Ref.	I_1	I_2	I_3	I_4	ν	Upper	Lower	Br.
8413.836	8416.148		0.1	0.2	0.1		118 81.92	X 2	2B14	P 4
8414.006	8416.318		0.1		0.1		11881.68			
8414.141	8416.453		0.2	0.2			11881.49			
8414.474	8416.786	G	1.5	1.5	1.6		11881.02	3c 0	2a 2	R 2
8414.806	8417.119		0.1	0.2			11880.55			
8415.118	8417.430			0.1			11880.11			
8415.628	8417.941		0.2	0.2			11879.39	4E 0	2B14	Q 2
8415.628	8417.941		0.2	0.2			11879.39	3C 2	2A 3	Q 2
8416.124	8418.437			0.2d			11878.69	Z 0	2B 8	R 1
8416.259	8418.571		0.1		0.5		11878.50			
8416.422	8418.734		0.7	1.0	1.3		11878.27	2K 3	2B 2	R 3
8416.584	8418.897				0.6		11878.04			
8416.698	8419.011				0.2		11877.88			
8416.868	8419.181			0.2			11877.64			
8417.194	8419.507		0.2	0.2	0.2b		11877.18			
8417.513	8419.826		0.1	0.1			11876.73			
8417.789	8420.102		0.5	1.1	1.2		11876.34	3E 3	2C 3	P 5
8418.179	8420.492		0.3	0.3			11875.79			
8418.434	8420.748			0.6			11875.43			
8418.682	8420.996		0.1	0.2			11875.08	4g 0	2a 5	R 0
8418.938	8421.251		0.3	0.1			11874.72			
8419.044	8421.358				0.3		11874.57	3E 3	2C 3	P 5
8419.129	8421.443		0.1	0.2			11874.45			
8419.377	8421.691			0.3	0.4		11874.10			
8419.569	8421.882			0.1			11873.83			
8419.718	8422.031						11873.62	3D 0	2B 8	P 3
8419.781	8422.095		0.2	0.3	0.2		11873.53			
8419.980	8422.294		1.5	0.9	0.9		11873.25			
8420.370	8422.684		0.2		0.2		11872.70			
8420.640	8422.953			0.2			11872.32			
8420.789	8423.102				0.2d		11872.11			
8420.902	8423.216		0.1				11871.95			
8421.257	8423.571		0.3	0.5	0.9		11871.45			
8421.505	8423.819			0.1			11871.10			
8421.675	8423.990		0.1				11870.86			
8421.782	8424.096			0.2	0.3d		11870.71			
8422.115	8424.429	G	1.9	1.6	1.3		11870.24	3B 4	2K 2	R 1
8422.449	8424.763		0.3	0.3	0.4d		11869.77	X 1	2C 2	Q 2
8422.690	8425.004				0.6d		11869.43			
8422.825	8425.139			0.2			11869.24			
8422.995	8425.310	i	0.1		0.5d		11869.00			
8423.165	8425.480			0.1	0.3d		11868.76			
8423.456	8425.771				0.4		11868.35	3C 2	2A 3	P 4
8423.648	8425.963		1.5	0.9	0.5		11868.08			
8423.911	8426.225				1.0		11867.71			
8424.031	8426.346		0.1	0.2			11867.54	3C 1	2K 2	R 3
8424.116	8426.431				0.9		11867.42			
8424.351	8426.666			0.1			11867.09			
8424.613	8426.928	G	2.5	0.6	0.8		11866.72	3C 2	2A 3	R 0
8424.670	8426.985	i			1.5b		11866.64			
8425.039	8427.354			0.2			11866.12			
8425.295	8427.610			0.0	0.8		11865.76			
8425.572	8427.887		0.2	0.0	0.3		11865.37	V 0	2B 8	R 2
8425.842	8428.157				0.4		11864.99			
8425.927	8428.242			0.1			11864.87			
8426.240	8428.555		0.2	0.4	0.9		11864.43			
8426.275	8428.590				0.5b		11864.38			
8426.502	8428.818			0.2			11864.06			
8426.595	8428.910				0.4		11863.93			
8426.779	8429.095		0.1		0.4		11863.67			

λ_{air}	λ_{vac}	Ref.	I_1	I_2	I_3	I_4	ν	Upper	Lower	Br.
8427.447	8429.763		0.1	0.1	0.3d		11862.73	4E 1	2C 4	P 4
8427.760	8430.075		0.2	0.4d	0.3d		11862.29			
8428.001	8430.317				0.2b		11861.95			
8428.086	8430.402			0.0			11861.83	Z 1	2B10	P 4
8428.364	8430.679				0.2		11861.44			
8428.555	8430.871			0.2			11861.17			
8428.761	8431.078			0.2			11860.88	Z 0	2B 8	R 0
8428.847	8431.163				0.2		11860.76			
8429.010	8431.326			0.4			11860.53			
8429.124	8431.440		0.8	0.9	0.7		11860.37			
8429.181	8431.497			0.6b	0.3		11860.29	3E 0	2B 9	R 4
8429.479	8431.796			0.1			11859.87	4F 0	2B15	R 3
8429.735	8432.052		0.3	0.4	0.7		11859.51			
8430.069	8432.386		0.3	0.3			11859.04	Y 1	2C 1	R 1
8430.069	8432.386		0.3	0.3			11859.04	3d 2	2c 4	P 3
8430.403	8432.720			0.2	0.2		11858.57			
8430.567	8432.883		0.4	0.5	0.7		11858.34	3D 3	2C 3	Q 4
8431.008	8433.324		0.1				11857.72			
8431.918	8434.235			0.3	0.2		11856.44			
8432.025	8434.342		0.4				11856.29			
8432.266	8434.583			0.6	0.9		11855.95			
8432.615	8434.932		0.2	0.4	0.2		11855.46	3A 2	2C 3	R 4
8433.049	8435.366		0.1	0.2	0.7		11854.85			
8433.148	8435.466				0.3		11854.71			
8433.469	8435.786		0.3	0.4	0.7		11854.26			
8433.796	8436.113		0.1				11853.80			
8433.896	8436.213		0.1	0.1			11853.66			
8434.237	8436.555		0.4	0.5	0.6		11853.18	Z 2	2C 2	Q 5
8434.550	8436.868			0.1			11852.74			
8434.806	8437.124			0.2			11852.38			
8435.148	8437.466		0.2	0.3	0.1		11851.90			
8435.596	8437.914			0.2			11851.27	3F 0	2B 9	R 2
8435.682	8438.000		0.1				11851.15			
8435.931	8438.249			0.2			11850.80	4D 0	2B14	R 4
8436.251	8438.569		0.8	0.9	0.9		11850.35	3c 1	2a 3	R 2
8436.607	8438.925			0.1			11849.85	3d 0	2c 2	Q 6
8436.949	8439.267			0.1d			11849.37			
8437.091	8439.410		0.1				11849.17			
8437.326	8439.645		0.3	0.4	0.7		11848.84			
8437.704	8440.022		0.1	0.2d			11848.31			
8438.017	8440.336			0.1			11847.87			
8438.345	8440.663		0.4	0.4			11847.41			
8438.865	8441.183		0.1				11846.68			
8439.470	8441.789		0.1				11845.83			
8439.841	8442.160		0.1				11845.31			
8440.111	8442.431		0.1				11844.93	T 0	2B14	R 2
8440.518	8442.837		0.7	0.8			11844.36	X 2	2C 3	Q 3
8440.518	8442.837		0.7	0.8			11844.36	3c 2	2a 4	R 2
8440.853	8443.172		1.5	1.2	1.4		11843.89	X 0	2C 1	P 2
8440.853	8443.172		1.5	1.2	1.4		11843.89	Z 0	2C 0	P 5
8440.981	8443.300		0.5	0.7	0.5		11843.71			
8441.159	8443.478		0.2	0.3			11843.46			
8441.430	8443.749		0.1	0.1			11843.08			
8441.551	8443.871		0.1		0.2		11842.91			
8441.644	8443.963		0.2	0.1			11842.78			
8442.086	8444.405		0.4	0.2			11842.16			
8442.385	8444.705		0.2				11841.74			
8442.656	8444.976		0.9	0.5			11841.36			
8442.813	8445.133		0.5	0.9	1.2		11841.14			
8443.041	8445.361		0.4	0.3	0.2		11840.82			

λ_{air}	λ_{vac}	Ref.	I_1	I_2	I_3	I_4	ν	Upper	Lower	Br.
8443.255	8445.575		0.5	0.2			11840.52			
8443.583	8445.903	G	3.0	2.4	1.6		11840.06	2K 4	2B 3	P 2
8443.797	8446.117				0.4		11839.76			
8443.868	8446.188			0.1			11839.66			
8443.925	8446.245		0.2				11839.58			
8444.104	8446.424		0.3	0.1			11839.33	X 1	2C 2	P 2
8444.318	8446.638			0.2			11839.03			
8444.389	8446.709		0.4				11838.93			
8444.646	8446.966	G	2.9	2.4	1.8		11838.57	2K 3	2B 2	P 2
8444.646	8446.966	G	2.9	2.4	1.8		11838.57	X 2	2B14	P 5
8445.010	8447.330		0.3				11838.06			
8445.138	8447.458		0.3	0.2	0.2		11837.88			
8445.338	8447.658		0.2				11837.60			
8445.552	8447.872		0.2				11837.30			
8445.766	8448.086		0.2	0.4	1.6		11 37.00	Z 2	2B12	P 6
8446.058	8448.379		0.1				11836.59	3F 1	2B11	R 1
8446.358	8448.679			0.1	0.8		11836.17			
8446.758	8449.079				0.6		11835.61			
8447.386	8449.707		0.2	0.3			11834.73	3c 3	2a 5	R 2
8447.578	8449.900		0.1				11834.46			
8447.735	8450.057	G	2.1				11834.24			
8447.871	8450.192		1.5b	1.5	0.9		11834.05	4f 0	3b 0	R 4
8448.549	8450.871			0.2			11833.10			
8448.835	8451.156			0.9	1.2		11832.70			
8448.878	8451.199		0.6				11832.64	3C 2	2A 3	Q 1
8449.228	8451.549			0.1			11832.15			
8449.335	8451.656		0.1				11832.00			
8449.749	8452.071			0.1			11831.42			
8450.170	8452.492		0.1	0.2			11830.83			
8451.099	8453.421		0.1	0.2			11829.53			
8451.606	8453.929			0.2			11828.82			
8451.942	8454.264		0.6	0.7	1.0		11828.35			
8452.521	8454.844		0.0				11827.54			
8452.750	8455.072		0.0	0.0			11827.22			
8453.064	8455.387		0.1	0.2			11826.78			
8453.343	8455.666		0.1	0.3d			11826.39			
8453.900	8456.223		0.0				11825.61			
8454.415	8456.738		0.2	0.5	0.6		11824.89	3D 0	2B 8	P 4
8454.780	8457.103		0.2	0.3	0.1		11824.38	3c 0	2a 2	R 1
8455.159	8457.482		0.2	0.3	0.2		11823.85			
8455.409	8457.732		0.1				11823.50			
8455.724	8458.047			0.1			11823.06			
8456.432	8458.755			0.3d			11822.07			
8456.718	8459.042		0.1	0.2			11821.67			
8456.933	8459.256						11821.37			
8457.011	8459.335		1.6	1.9	1.9		11821.26	Y 1	2C 1	Q 1
8457.040	8459.364						11821.22			
8457.362	8459.686		0.1	0.1			11820.77			
8457.641	8459.965		0.9	1.2	1.7		11820.38	2K 3	2B 2	R 4
8458.135	8460.459		0.1	0.2			11819.69	4E 0	2C 3	Q 2
8458.535	8460.860			0.1			11819.13			
8458.586	8460.910		0.4				11819.06			
8458.807	8461.132		0.1	0.6	1.2		11818.75	3b 2	2a 1	P10
8458.807	8461.132		0.1	0.6	1.2		11818.75	4f 2	3b 2	R 3
8459.473	8461.797			0.1	0.1		11817.82	4E 0	2C 3	P 2
8459.709	8462.034		0.0	0.1	0.1		11817.49			
8459.845	8462.170				0.1		11817.30			
8460.117	8462.442			0.2			11816.92			
8460.669	8462.993			0.2			11816.15			
8461.027	8463.352			0.1			11815.65			

λ_{air}	λ_{vac}	Ref.	I_1	I_2	I_3	I_4	ν	Upper	Lower	Br.
8461.607	8463.932		0.2	0.4	0.7		11814.84			
8461.800	8464.125		0.1				11814.57			
8461.943	8464.268			0.0			11814.37			
8462.832	8465.157		0.1				11813.13	3F 1	2B11	P 4
8462.832	8465.157		0.1				11813.13	5c 0	2a 6	R 1
8463.290	8465.615		0.2	0.3			11812.49			
8463.584	8465.909	G	1.7	1.4	1.6		11812.08			
8464.042	8466.368		0.1	0.1			11811.44			
8464.358	8466.683			0.7	0.9		11811.00			
8464.444	8466.770		0.6				11810.88			
8464.824	8467.149		0.1	0.1			11810.35			
8465.146	8467.472		1.5	0.9	0.7		11809.90	3D 0	2B 8	P 5
8465.440	8467.766			0.1			11809.49			
8465.877	8468.203		0.1				11808.88			
8466.171	8468.498		1.0	0.9	0.7		11808.47	3B 4	2K 2	R 0
8466.544	8468.870			0.1			11807.95			
8466.809	8469.136		0.1	0.1			11807.58			
8467.146	8469.473		1.6	1.3	1.0		11807.11	X 0	2C 1	Q 3
8467.598	8469.925		0.3	0.3			11806.48	3C 2	2A 3	P 3
8467.598	8469.925		0.3	0.3			11806.48	3A 1	2C 2	R 2
8467.598	8469.925		0.3	0.3			11806.48	3A 2	2C 3	R 2
8467.842	8470.169			0.1			11806.14			
8468.108	8470.434		0.7	0.7			11805.77		2B	P 5
8468.488	8470.815		0.1				11805.24			
8468.631	8470.958		0.1	0.0			11805.04			
8469.047	8471.374			0.1			11804.46			
8469.442	8471.769		0.2	0.2			11803.91			
8469.772	8472.099		0.6	0.5	1.8		11803.45	Y 1	2C 1	R 2
8470.088	8472.415	G	2.0	1.8	1.8		11803.01			
8470.224	8472.551						11802.82			
8470.425	8472.752		1.5	1.3	0.2		11802.54	2K 2	2B 1	P 5
8470.425	8472.752		1.5	1.3	0.2		11802.54	3b 4	2a 3	P 1
8470.425	8472.752		1.5	1.3	0.2		11802.54	3c 1	2a 3	R 1
8470.712	8473.040			0.1			11802.14			
8470.949	8473.276		0.1				11801.81			
8471.451	8473.779		0.1				11801.11			
8471.810	8474.138		0.4	0.6	0.2b		11800.61	Y 1	2C 1	P 1
8472.536	8474.863		0.0	0.2			11799.60			
8472.851	8475.180		0.0	0.1			11799.16			
8473.433	8475.761		0.0				11798.35			
8473.878	8476.207		0.3				11797.73			
8473.929	8476.257			0.7	0.8		11797.66			
8474.223	8476.552			0.2			11797.25			
8474.482	8476.810		0.1	0.2			11796.89			
8474.784	8477.112		0.1	0.3	0.3		11796.47			
8474.985	8477.313			0.1			11796.19			
8475.215	8477.543		0.4	0.8	1.1		11795.87	X 0	2C 1	R 4
8475.653	8477.982		0.1				11795.26			
8475.998	8478.327		0.2	0.2			11794.78	3A 2	2C 3	R 1
8476.350	8478.679			0.0			11794.29			
8476.738	8479.067		0.2	0.5	0.7		11793.75			
8477.133	8479.463		0.2	0.4			11793.20	T 0	2C 3	R 1
8477.385	8479.714			0.1			11792.85			
8477.651	8479.980			0.2			11792.48			
8477.917	8480.246				0.2		11792.11			
8477.989	8480.318		0.6	0.8	0.7		11792.01	3c 2	2a 4	R 1
8478.320	8480.649			0.3	0.1b		11791.55			
8478.715	8481.045	G	2.3	1.8	1.3		11791.00	2K 4	2B 3	P 3
8479.118	8481.448		0.1	0.1			11790.44			
8479.578	8481.908		0.1	0.1			11789.80	3E 2	2B13	Q 6

λ_{air}	λ_{vac}	Ref.	I_1	I_2	I_3	I_4	ν	Upper	Lower	Br.
8479.729	8482.059						11789.59			
8479.859	8482.189		0.1				11789.41			
8480.139	8482.469			0.0			11789.02			
8480.715	8483.045			0.1			11788.22			
8481.405	8483.736			0.4	0.7		11787.26	3F 1	2B11	Q 2
8481.866	8484.196		0.1	0.1			11786.62			
8482.219	8484.549			0.4	0.2		11786.13			
8482.456	8484.787		0.6	0.9	1.0		11785.80			
8482.945	8485.276		0.5	0.7	0.8		11785.12	3c 3	2a 5	R 1
8482.945	8485.276		0.5	0.7	0.8		11785.12	T 0	2C 3	R 2
8483.579	8485.910			0.1			11784.24	3C 1	2K 2	Q 4
8483.860	8486.191			0.1			11783.85	3d 1	2c 3	P 5
8484.126	8486.457		0.1	0.2			11783.48	4E 1	2C 4	P 5
8484.601	8486.933		0.4	0.7	1.0		11782.82	3D 1	2C 1	Q 7
8484.601	8486.933		0.4	0.7	1.0		11782.82	V 0	2C 0	P 5
8484.861	8487.192		0.1	0.1			11782.46			
8485.091	8487.422			0.0			11782.14			
8485.343	8487.675			0.2			11781.79			
8485.473	8487.804		0.1	0.2			11781.61			
8485.639	8487.970			0.2			11781.38			
8485.754	8488.085			0.4	0.2		11781.22			
8486.085	8488.417	G	2.5	2.3	2.7		11780.76	2A 1	2B 0	R 4
8486.431	8488.763		0.0	0.2			11780.28			
8486.546	8488.878			0.3			11780.12			
8486.913	8489.245		0.2d	0.3	0.4		11779.61			
8487.173	8489.505		0.2	0.4	1.1		11779.25			
8487.439	8489.771		0.2	0.3			11778.88			
8487.929	8490.262			0.1			11778.20			
8488.261	8490.593			0.3			11777.74	3C 2	2A 3	P 2
8488.895	8491.228		0.2	0.2			11776.86	U 1	2B11	R 3
8489.234	8491.567		1.6	1.0	0.7		11776.39	Z 0	2B 8	P 2
8489.595	8491.927		0.1	0.1			11775.89	4D 0	2C 3	P 1
8489.710	8492.042		0.1				11775.73	3E 2	2B13	R 3
8489.919	8492.252		1.5	0.9	0.7		11775.44	3D 3	2C 3	P 4
8490.301	8492.634	i		0.1	0.2		11774.91			
8491.001	8493.333			0.4	0.3		11773.94			
8491.253	8493.586		0.1	0.2			11773.59			
8491.534	8493.867		0.0	0.2			11773.20			
8492.501	8494.834			0.1			11771.86			
8492.826	8495.159		0.1	0.4			11771.41			
8493.129	8495.462		0.0	0.2			11770.99			
8493.533	8495.866			0.2			11770.43			
8493.865	8496.198		0.1	0.5			11769.97			
8494.240	8496.574		0.2	0.5	0.7		11769.45	3A 2	2B14	R 1
8494.529	8496.863		0.1	0.3			11769.05			
8494.695	8497.029			0.1			11768.82			
8494.954	8497.288		0.2	0.3			11768.46	T 0	2B14	R 1
8495.359	8497.693		0.4		0.2		11767.90			
8495.604	8497.938			0.1			11767.56			
8495.821	8498.155			0.2			11767.26	4g 0	2a 5	R 1
8496.160	8498.494	G	2.1	1.7	1.4		11766.79	2K 3	2B 2	P 3
8496.586	8498.921		0.2	0.2			11766.20	4c 2	2a 7	R 1
8496.781	8499.116			0.1			11765.93	3c 0	2a 2	R 0
8496.781	8499.116			0.1			11765.93	3d 0	2c 2	Q 7
8497.041	8499.376			0.1			11765.57			
8497.352	8499.686		0.2	0.2			11765.14			
8497.532	8499.867			0.1			11764.89	3A 1	2C 2	R 1
8497.800	8500.134		0.1	0.3	0.2		11764.52	3d 0	2c 2	P 5
8498.038	8500.373			0.2			11764.19			
8498.291	8500.626		0.1	0.0			11763.84			

λ_{air}	λ_{vac}	Ref.	I$_1$	I$_2$	I$_3$	I$_4$	ν	Upper	Lower	Br.
8498.796	8501.131		0.1	0.0			11763.14			
8499.143	8501.478		0.1	0.2			11762.66			
8499.331	8501.666						11762.40			
8499.533	8501.869		0.0	0.2			11762.12			
8500.032	8502.367		0.1				11761.43	3A 2	2B14	P 5
8500.393	8502.729			0.1			11760.93			
8500.776	8503.112		0.2	0.3			11760.40			
8501.550	8503.886		0.1		0.1		11759.33			
8501.825	8504.161		0.1				11758.95			
8501.984	8504.320		0.2	0.2			11758.73	X 1	2B12	R 1
8502.360	8504.696		0.2	0.2			11758.21			
8502.541	8504.877		0.1	0.2			11757.96	X 1	2B12	R 0
8503.011	8505.347		0.1	0.2			11757.31	4f 3	3b 3	R 2
8503.141	8505.477		0.1				11757.13			
8503.322	8505.658			0.2	0.1b		11756.88			
8504.096	8506.432			0.0			11755.81			
8504.334	8506.671		0.3	0.6	1.1		11755.48			
8504.776	8507.112			0.1d			11754.87			
8505.101	8507.438		0.4	0.7	0.8		11754.42	X 2	2C 3	Q 2
8505.101	8507.438		0.4	0.7	0.8		11754.42	X 1	2B12	R 2
8505.101	8507.438		0.4	0.7	0.8		11754.42	X 1	2C 2	Q 3
8505.470	8507.807		0.1	0.2			11753.91			
8506.281	8508.618			0.1			11752.79			
8506.606	8508.944			0.2			11752.34			
8506.780	8509.117		0.3				11752.10			
8507.070	8509.407		0.1				11751.70			
8507.837	8510.175		0.1				11750.64	4D 0	2B14	P 1
8507.837	8510.175		0.1				11750.64	3c 6	2a 8	Q 1
8508.503	8510.841			0.2			11749.72			
8508.844	8511.182		0.6	0.4			11749.25	3c 1	2a 3	R 0
8509.054	8511.392		0.2	0.2			11748.96			
8509.170	8511.507				0.3b		11748.80			
8509.293	8511.631	G	1.9	1.6	1.3		11748.63	3B 4	2K 2	P 1
8509.662	8512.000			0.2			11748.12			
8509.887	8512.225			0.3			11747.81	3C 1	2A 2	R 2
8510.350	8512.688		0.0				11747.17			
8510.792	8513.131		0.2	0.6	1.0		11746.56			
8511.075	8513.413			0.2			11746.17			
8511.271	8513.609		0.1	0.1			11745.90			
8511.466	8513.805		0.2	0.6	0.2		11745.63			
8511.691	8514.029			0.1			11745.32			
8511.901	8514.240		0.5	0.8	0.2		11745.03	Y 1	2C 1	Q 2
8512.285	8514.624		0.4	1.2	2.1		11744.50	3b 3	2a 2	P 7
8512.285	8514.624		0.4	1.2	2.1		11744.50	4D 0	2B14	R 0
8512.611	8514.950	G	1.7	1.4			11744.05	3B 4	2K 2	P 2
8512.959	8515.298				0.2		11743.57			
8513.068	8515.407			0.1			11743.42			
8513.351	8515.690		0.1	0.3			11743.03	4D 0	2C 3	Q 1
8513.575	8515.914			0.7	1.0		11742.72			
8513.923	8516.263		0.1	0.1			11742.24			
8514.250	8516.589			0.1			11741.79			
8514.474	8516.814			0.2			11741.48	3E 3	2B15	Q 6
8514.678	8517.017			0.1			11741.20	5c 0	2a 6	Q 3
8514.990	8517.329			0.1			11740.77			
8515.425	8517.764			0.3			11740.17			
8515.700	8518.040			0.1			11739.79			
8516.005	8518.345		0.2	0.2			11739.37	Z 0	2B 8	P 3
8516.310	8518.649		0.4	0.3			11738.95	5c 0	2a 6	R 0
8516.600	8518.940			0.2			11738.55			
8516.680	8519.020		0.4				11738.44	3c 2	2a 4	R 0

λ_{air}	λ_{vac}	Ref.	I_1	I_2	I_3	I_4	ν	Upper	Lower	Br.
8516.897	8519.237	G	2.5	2.1	1.8		11738.14	2K 4	2B 3	P 4
8517.231	8519.571			0.2			11737.68			
8517.369	8519.709		0.1	0.2			11737.49			
8517.558	8519.898		0.1	0.1			11737.23			
8517.877	8520.217		0.1	0.2d			11736.79	4E 0	2C 3	P 4
8518.029	8520.370			0.2d			11736.58	3F 0	2B 9	Q 3
8518.233	8520.573		0.4	0.6	0.9		11736.30			
8518.399	8520.740			0.2			11736.07			
8518.625	8520.965			0.1			11735.76			
8518.879	8521.219			0.2			11735.41			
8519.162	8521.502		0.2	0.2b			11735.02			
8519.336	8521.677			0.2			11734.78			
8519.583	8521.923		0.2	0.6	0.3b		11734.44			
8519.757	8522.098			0.2			11734.20	4E 0	2C 3	Q 4
8519.982	8522.323			0.6			11733.89			
8520.098	8522.439		0.3				11733.73			
8520.360	8522.701	G	2.8	3.0	3.1		11733.37	3b 0	2a 0	R 3
8520.360	8522.701	G	2.8	3.0	3.1		11733.37	3c 3	2a 5	R 0
8520.730	8523.071	G	2.4				11732.86	3B 4	2A 2	P 3
8520.897	8523.238				0.9		11732.63			
8521.071	8523.412			0.2			11732.39			
8521.376	8523.718	1			1.5b		11731.97			
8521.456	8523.797		1.8	0.3			11731.86	V 0	2B 8	R 1
8521.754	8524.095			0.3	0.7		11731.45			
8521.914	8524.255			0.2	0.7		11731.23			
8522.248	8524.590	G	1.7	1.5	1.6		11730.77	3b 4	2a 3	P 2
8522.597	8524.938			0.2			11730.29			
8522.691	8525.033		0.2				11730.16	4D 0	2B14	R 2
8522.953	8525.295	G	1.8	2.1	2.5		11729.80	3b 0	2a 0	R 4
8523.389	8525.731			0.2	0.3		11729.20			
8523.600	8525.941				0.2		11728.91			
8523.701	8526.043			0.3			11728.77			
8523.876	8526.218		0.9	1.2	1.1		11728.53	3A 0	2C 1	R 2
8524.297	8526.639	G	1.6				11727.95			
8524.741	8527.083		0.1				11727.34			
8524.828	8527.170			0.2	0.2d		11727.22	3D 3	2C 3	Q 5
8525.141	8527.483			0.4			11726.79	3f 3	3b 0	P 3
8525.250	8527.592		0.2				11726.64			
8525.853	8528.195		0.1				11725.81			
8526.071	8528.414			0.2			11725.51			
8526.348	8528.690			0.3	0.4		11725.13			
8526.900	8529.243			0.1d			11724.37			
8527.038	8529.381						11724.18			
8527.264	8529.607				0.2b		11723.87			
8527.344	8529.687			0.2			11723.76			
8527.569	8529.912				0.2		11723.45			
8527.751	8530.094			0.3			11723.20			
8527.824	8530.167		0.2				11723.10			
8528.093	8530.436	G	2.6	2.7	2.6		11722.73	3b 0	2a 0	R 2
8528.537	8530.880		0.5	0.3	0.3		11722.12	4g 0	2a 5	P 2
8528.821	8531.164			0.2	0.3b		11721.73			
8529.061	8531.404		0.2	0.1			11721.40	X 0	2C 1	P 3
8529.235	8531.579		0.4	0.8	1.1		11721.16			
8529.956	8532.299				0.4		11720.17			
8530.065	8532.409				0.2		11720.02			
8530.145	8532.489		0.1	0.1b			11719.91			
8530.698	8533.042		0.2	0.2d			11719.15			
8531.426	8533.770			0.0			11718.15	T 0	2B14	R 0
8531.878	8534.222		0.1				11717.53			
8532.577	8534.921			0.3			11716.57			

λ_{air}	λ_{vac}	Ref.	I_1	I_2	I_3	I_4	ν	Upper	Lower	Br.
8532.752	8535.096		0.3	0.7	1.0		11716.33	3D 2	2C 2	Q 7
8532.752	8535.096		0.3	0.7	1.0		11716.33	T 0	2C 3	Q 1
8533.189	8535.533		0.1	0.1d			11715.73			
8533.596	8535.941		0.5	1.0	1.6		11715.17			
8533.764	8536.109		0.4				11714.94			
8534.106	8536.451		0.2	0.2			11714.47			
8534.354	8536.699		0.2	0.7	1.1		11714.13			
8534.529	8536.874		0.3				11713.89			
8534.937	8537.282		0.1	0.2			11713.33			
8535.039	8537.384				0.2b		11713.19			
8535.352	8537.697			0.2			11712.76	3F 1	2B11	P 3
8535.702	8538.047	G	1.7	2.2	2.8		11712.28	3b 0	2a 0	R 5
8536.103	8538.448		0.1	0.1	0.4		11711.73			
8536.555	8538.900		0.1	0.2			11711.11	X 1	2C 2	P 3
8537.437	8539.783			0.1	0.1		11709.90			
8537.736	8540.081		0.2	0.3	0.3b		11709.49			
8538.144	8540.490		0.2	0.7			11708.93			
8538.502	8540.847			0.1			11708.44			
8538.815	8541.161		0.2	0.8			11708.01			
8539.136	8541.482		0.1				11707.57			
8539.267	8541.613			0.1d			11707.39			
8539.742	8542.088		0.2	0.2			11706.74	X 2	2C 3	Q 4
8540.077	8542.423			0.1			11706.28			
8540.500	8542.847		0.2	0.8	1.1		11705.70	4D 0	2B14	R 1
8540.807	8543.153		0.3	1.0	1.4		11705.28	3c 0	2a 2	Q 1
8540.807	8543.153		0.3	1.0	1.4		11705.28	U 1	2B11	R 2
8541.245	8543.591		0.1	0.2	0.7		11704.68	3E 2	2B13	P 6
8541.412	8543.759			0.1			11704.45			
8541.704	8544.051	G	2.1	1.8	1.5		11704.05	2K 6	2B 5	R 2
8541.704	8544.051	G	2.1	1.8	1.5		11704.05	3c 5	2a 7	Q 1
8542.128	8544.474		0.1	0.2			11703.47			
8542.266	8544.613		0.2		0.1		11703.28	3A 2	2B14	P 4
8542.493	8544.839			0.2			11702.97	3c 0	2a 2	Q 2
8542.653	8545.000		0.8	0.9	1.0		11702.75	Y 1	2C 1	P 2
8542.982	8545.329		0.2	0.2	0.1		11702.30			
8543.128	8545.475				0.1		11702.10			
8543.354	8545.701			0.2			11701.79			
8543.507	8545.854				0.2		11701.58			
8543.726	8546.074		0.2	0.2			11701.28			
8543.924	8546.271			0.2			11701.01			
8544.048	8546.395		0.2				11700.84			
8544.296	8546.643		0.3	0.2b	0.3b		11700.50			
8544.427	8546.775			0.8			11700.32			
8544.508	8546.855		0.3				11700.21			
8544.698	8547.045				0.1		11699.95			
8545.055	8547.403		0.2	0.2b			11699.46	3c 0	2a 2	Q 3
8545.158	8547.505			0.7			11699.32			
8545.508	8547.856		0.6	0.4	0.3		11698.84			
8545.633	8547.980						11698.67			
8545.749	8548.097		0.4				11698.51			
8545.910	8548.258			0.9			11698.29			
8546.246	8548.594	G	3.3	3.2	3.0		11697.83	3b 0	2a 0	R 1
8546.597	8548.945		0.6		0.4		11697.35			
8546.736	8549.084		0.7	0.6			11697.16			
8546.882	8549.230						11696.96			
8546.926	8549.274		0.4				11696.90			
8547.123	8549.471		0.3	0.1			11696.63	3E 0	2B 9	R 3
8547.211	8549.559		0.2				11696.51			
8547.481	8549.829		0.3	0.3	0.1d		11696.14			
8547.759	8550.107		0.2	0.2			11695.76			

λ_{air}	λ_{vac}	Ref.	I_1	I_2	I_3	I_4	ν	Upper	Lower	Br.
8548.102	8550.451		1.8	1.5	1.2		11695.29	2K 6	2B 5	R 1
8548.102	8550.451		1.8	1.5	1.2		11695.29	3c 0	2a 2	Q 4
8548.234	8550.582		0.4				11695.11			
8548.438	8550.787		0.3	0.2	0.1b		11694.83	3c 4	2a 6	Q 2
8548.672	8551.021		0.6				11694.51			
8549.126	8551.474		0.2	0.7	0.1		11693.89			
8549.725	8552.074		0.1	0.1d			11693.07			
8549.901	8552.250		0.1				11692.83			
8550.215	8552.564		0.6	0.9	0.9		11692.40	3c 1	2a 3	Q 1
8550.215	8552.564		0.6	0.9	0.9		11692.40	X 0	2C 1	Q 4
8550.471	8552.820		0.2	0.1			11692.05			
8550.698	8553.047		0.7	0.9	0.9		11691.74	Z 0	2B 8	P 4
8551.020	8553.369		0.2				11691.30			
8552.022	8554.371		0.1				11689.93	3c 0	2a 2	Q 5
8552.212	8554.561		0.7	1.2	1.3		11689.67	3B 4	2A 2	P 4
8552.212	8554.561		0.7	1.2	1.3		11689.67	5c 0	2a 6	Q 1
8552.812	8555.161		0.1	0.2			11688.85			
8553.661	8556.011			0.1			11687.69			
8554.071	8556.421		0.7	1.3	1.6		11687.13	3c 1	2a 3	Q 3
8554.071	8556.421		0.7	1.3	1.6		11687.13	4D 0	2B14	P 2
8554.356	8556.706		0.2				11686.74			
8554.458	8556.809			0.1			11686.60			
8554.744	8557.094		0.3	0.3	0.2		11686.21			
8554.920	8557.270		0.6	0.8			11685.97	3c 4	2a 6	Q 1
8555.169	8557.519			0.1			11685.63			
8555.315	8557.665			0.1			11685.43			
8555.513	8557.863		0.2	0.2			11685.16			
8555.806	8558.156		1.6	1.3	1.4		11684.76	2K 4	2B 3	P 5
8556.091	8558.442		0.3	0.2			11684.37	3A 2	2C 3	Q 2
8556.377	8558.727		0.1	0.2			11683.98	3c 0	2a 2	Q 6
8556.552	8558.903		0.1				11683.74	3c 4	2a 6	Q 3
8556.823	8559.174		0.8	0.9	0.2		11683.37	3c 2	2a 4	Q 1
8557.043	8559.394		0.2	0.2			11683.07	T 0	2C 3	Q 2
8557.043	8559.394		0.2	0.2			11683.07	3c 1	2a 3	Q 4
8557.292	8559.643			0.6			11682.73			
8557.395	8559.746		0.3	0.2			11682.59			
8557.571	8559.922				0.1		11682.35			
8557.746	8560.097			0.1d			11682.11			
8557.849	8560.200		0.1				11681.97			
8558.186	8560.537			0.2			11681.51	3c 2	2a 4	Q 2
8558.186	8560.537			0.2			11681.51	T 0	2B14	P 4
8558.186	8560.537			0.2			11681.51	U 1	2B11	R 1
8558.347	8560.698		0.2				11681.29	3C 0	2A 1	R 4
8558.464	8560.816		0.5	1.4	2.2		11681.13	3b 0	2a 0	R 6
8558.721	8561.072		0.1	0.1			11680.78			
8558.860	8561.211			0.2			11680.59	3f 3	2c 0	P 4
8559.058	8561.409		0.8	0.9			11680.32	3c 3	2a 5	Q 1
8559.314	8561.666			0.1			11679.97			
8559.578	8561.930		0.2	0.2			11679.61			
8559.915	8562.267	G	2.5	1.8	1.2		11679.15	2K 6	2B 5	R 0
8560.223	8562.575		0.2	0.3			11678.73	3c 3	2a 5	Q 2
8560.223	8562.575		0.2	0.3			11678.73	4E 0	2C 3	P 3
8560.223	8562.575		0.2	0.3			11678.73	3c 2	2a 4	Q 3
8560.399	8562.751			0.2			11678.49	X 1	2B12	P 2
8560.670	8563.022		1.5	1.4	1.7		11678.12	3c 1	2a 3	Q 5
8561.052	8563.403		0.1	0.1			11677.60			
8561.323	8563.675		0.2	0.2d			11677.23			
8561.653	8564.005	G	2.3	2.0	1.9		11676.78	2K 3	2B 2	P 4
8561.653	8564.005	G	2.3	2.0	1.9		11676.78	3c 3	2a 5	Q 3
8562.056	8564.408		0.1				11676.23	3A 0	2C 1	R 1

λ_{air}	λ_{vac}	Ref.	I_1	I_2	I_3	I_4	ν	Upper	Lower	Br.
8562.855	8565.208			0.2			11675.14	3E 1	2B11	Q 6
8562.855	8565.208			0.2			11675.14	3c 2	2a 4	Q 4
8563.361	8565.714			0.2			11674.45	4g 0	2a 5	R 2
8563.655	8566.007			0.1			11674.05			
8563.963	8566.316			0.2			11673.63	3c 3	2a 5	Q 4
8564.227	8566.580						11673.27			
8564.308	8566.661			0.1			11673.16	T 0	2C 3	P 1
8564.594	8566.947			0.2			11672.77	3c 1	2a 3	Q 6
8564.594	8566.947			0.2			11672.77	4b 2	2a 6	R 1
8565.108	8567.461			0.1			11672.07	3A 2	2B14	R 0
8565.460	8567.813		0.3	1.1	0.2		11671.59	V 0	2B 8	P 4
8565.739	8568.092			0.0			11671.21	3F 0	2B 9	R 1
8565.981	8568.334			0.2			11670.88	3c 2	2a 4	Q 5
8566.304	8568.657		0.5	0.7			11670.44	3A 2	2C 3	Q 1
8566.935	8569.289			0.1d			11669.58			
8567.302	8569.656		0.2				11669.08	3E 1	2B11	R 3
8567.361	8569.714		1.8	1.4	1.2		11669.00	3A 1	2C 2	Q 1
8568.382	8570.735			0.2			11667.61			
8568.675	8571.029			0.2			11667.21	4F 0	2B15	P 5
8568.962	8571.316			0.2			11666.82			
8569.248	8571.602			0.2			11666.43			
8569.535	8571.889			0.1			11666.04	3E 2	2B13	R 2
8569.829	8572.183		0.1	0.3			11665.64	3A 2	2C 3	Q 3
8570.042	8572.396			0.2			11665.35	4c 2	2a 7	Q 1
8570.505	8572.859		0.2				11664.72			
8570.997	8573.351		0.3	0.9	0.9		11664.05	3C 1	2A 2	R 1
8571.269	8573.623			0.1	0.1		11663.68			
8571.541	8573.895		0.1	0.2			11663.31			
8571.768	8574.123			0.2			11663.00			
8572.055	8574.410			0.7	1.0		11662.61			
8572.327	8574.682		0.2	0.8	1.1		11662.24	3C 1	2A 2	R 3
8572.658	8575.013		0.1	0.1			11661.79			
8572.930	8575.285		0.1	0.1			11661.42			
8573.209	8575.564		0.3	0.5			11661.04	3A 2	2B14	P 3
8574.158	8576.513		0.1	0.2			11659.75	T 0	2B14	P 3
8574.312	8576.668			0.2			11659.54			
8574.459	8576.815		0.2	0.1			11659.34			
8574.570	8576.925		0.4	0.3			11659.19			
8574.915	8577.271	G	2.7	2.7	2.3		11658.72	3b 0	2a 0	R 0
8575.254	8577.609		0.3	0.1			11658.26			
8575.379	8577.734		0.2	0.2			11658.09			
8575.570	8577.926		0.1				11657.83			
8575.916	8578.271			0.1			11657.36			
8576.269	8578.625		0.3	0.6			11656.88	3A 1	2C 2	Q 2
8576.497	8578.853			0.0			11656.57			
8576.791	8579.147		0.1	0.6	0.8		11656.17			
8577.167	8579.523			0.2d			11655.66			
8578.138	8580.494		0.3	0.5	0.8		11654.34			
8578.440	8580.796			0.3			11653.93			
8578.705	8581.061			0.1			11653.57			
8578.984	8581.341			0.2			11653.19			
8579.478	8581.835			0.2			11652.52			
8580.516	8582.873			0.2d			11651.11			
8580.906	8583.264		0.2	0.5	0.9		11650.58			
8581.474	8583.831			0.1d			11649.81	Z 0	2B 8	P 5
8582.026	8584.384			0.2d			11649.06			
8582.335	8584.693		0.2	0.3			11648.64			
8582.483	8584.840			0.5	0.1		11648.44			
8582.822	8585.179		0.1	0.1	0.7		11647.98	T 0	2B14	P 1
8583.065	8585.423			0.1			11647.65			

λ_{air}	λ_{vac}	Ref.	I_1	I_2	I_3	I_4	ν	Upper	Lower	Br.
8583.271	8585.629		0.0	0.2	0.1		11647.37			
8583.441	8585.799		0.1	0.2			11647.14			
8583.684	8586.042			0.2			11646.81	X 0	2B10	R 4
8584.016	8586.374	G	2.6	2.2	2.3		11646.36	2A 1	2B 0	R 3
8584.347	8586.706			0.1			11645.91			
8584.428	8586.787		0.7				11645.80			
8584.591	8586.949		0.3	0.3d	0.2		11645.58	3A 1	2B12	R 2
8584.591	8586.949		0.3	0.3d	0.2		11645.58	3A 1	2C 2	Q 3
8584.922	8587.281		0.2	0.2			11645.13			
8585.261	8587.620	G	2.2	2.1	2.1		11644.67	3b 4	2a 3	P 3
8585.571	8587.930			0.2			11644.25			
8585.741	8588.099			0.1			11644.02	T 0	2C 3	Q 3
8586.250	8588.608			0.1			11643.33			
8586.485	8588.844			0.1			11643.01			
8586.832	8589.191			0.2d			11642.54			
8587.223	8589.582		0.0	0.3			11642.01	3C 1	2K 2	Q 3
8587.430	8589.789			0.0			11641.73			
8587.880	8590.239		0.2	0.7	0.9		11641.12			
8588.212	8590.571				0.1		11640.67			
8588.411	8590.770			0.1			11640.40			
8588.853	8591.213			0.5			11639.80			
8589.185	8591.545			0.2	0.1d		11639.35			
8589.503	8591.862		0.1	0.3d	0.2		11638.92			
8589.901	8592.261		0.2				11638.38			
8590.706	8593.066			0.2d			11637.29			
8591.023	8593.383			0.9			11636.86			
8591.090	8593.450		1.0	1.9	2.4		11636.77	3b 0	2a 0	R 7
8591.171	8593.531			0.9			11636.66	3C 0	2A 1	R 3
8591.393	8593.753		0.4	0.1	0.3		11636.36	3A 2	2C 3	P 1
8591.644	8594.004			0.2			11636.02			
8591.917	8594.277		0.4	0.3	0.9		11635.65			
8592.124	8594.484			0.8			11635.37			
8592.264	8594.624		0.5	0.4			11635.18	Y 1	2C 1	Q 3
8592.389	8594.750		0.3	1.1	1.3		11635.01	3f 3	3b 0	P 5
8593.807	8596.168			0.1			11633.09			
8594.369	8596.730			0.1			11632.33			
8594.702	8597.062		0.8	1.3	1.8		11631.88	2K 4	2B 3	P 6
8595.041	8597.402		0.1	0.1d			11631.42			
8595.359	8597.720		0.6				11630.99	4D 0	2B14	P 6
8595.477	8597.839			0.2	0.1		11630.83			
8595.877	8598.238		0.2b	0.2			11630.29	3A 2	2C 3	Q 4
8596.534	8598.896		0.2d	0.5	0.7		11629.40			
8597.118	8599.480		0.4				11628.61			
8597.251	8599.613		0.2				11628.43			
8597.377	8599.739			0.2d			11628.26			
8598.035	8600.397		0.2	0.1			11627.37	4D 0	2C 3	P 2
8598.442	8600.804			0.3	0.8		11626.82			
8599.263	8601.625			0.1			11625.71			
8599.633	8601.995			0.2			11625.21	4f 2	3b 2	R 2
8600.040	8602.402		0.6	1.2	1.5		11624.66	3B 4	2A 2	P 5
8600.373	8602.735		2.0	1.4	1.1		11624.21	2K 6	2B 5	P 1
8600.705	8603.068			0.1			11623.76			
8601.009	8603.372		0.4	0.9	0.8		11623.35	X 1	2B12	P 3
8601.364	8603.727		1.0	0.6			11622.87	3A 1	2C 2	P 1
8601.949	8604.312		0.2	0.1d			11622.08	3d 0	2c 2	P 6
8602.104	8604.467		0.2				11621.87			
8602.541	8604.904		0.3	0.2			11621.28			
8603.866	8606.230			0.1			11619.49			
8604.162	8606.526		0.1	0.2			11619.09			
8604.770	8607.133			0.0			11618.27			

λ_{air}	λ_{vac}	Ref.	I_1	I_2	I_3	I_4	ν	Upper	Lower	Br.
8605.325	8607.689		0.2d	0.8	0.3		11617.52			
8605.688	8608.052			0.1			11617.03			
8605.762	8608.126				0.7		11616.93			
8605.999	8608.363		0.3	0.6	0.9		11616.61			
8606.385	8608.749			0.2			11616.09	3C 2	2K 4	R 3
8607.200	8609.564			0.2			11614.99	3A 2	2B14	P 2
8608.067	8610.431			0.2			11613.82			
8608.534	8610.898			0.2	0.2		11613.19	4E 1	2B17	Q 4
8609.460	8611.825		0.1	0.2			11611.94	4f 0	3b 0	R 3
8610.209	8612.574				0.1		11610.93	3A 2	2B14	P 1
8610.469	8612.834			0.0			11610.58			
8610.788	8613.153			0.2			11610.15			
8611.211	8613.576			0.2			11609.58			
8611.515	8613.880			0.1			11609.17			
8612.078	8614.444			0.2			11608.41			
8612.375	8614.741		0.4	1.7	0.8		11608.01			
8612.516	8614.882		0.2				11607.82			
8612.657	8615.023			0.1			11607.63	3E 2	2B13	Q 5
8613.095	8615.461		0.2	0.3			11607.04			
8613.421	8615.787		0.2	0.3			11606.60			
8614.000	8616.367			0.2			11605.82			
8614.268	8616.634			0.1			11605.46	X 0	2B10	R 2
8614.564	8616.931			0.2			11605.06	4e 3	3b 3	R 4
8615.790	8618.156		0.2				11603.41			
8616.317	8618.684			0.2b			11602.70	3F 0	2B 9	Q 2
8616.896	8619.263			0.2d			11601.92			
8617.810	8620.177		0.2	0.2			11600.69	T 0	2C 3	P 2
8617.810	8620.177		0.2	0.2			11600.69	3C 2	2K 4	Q 5
8619.065	8621.433			0.1			11599.00			
8619.310	8621.678		0.2	0.2			11598.67			
8619.883	8622.250			0.2d			11597.90	V 0	2B 8	R 0
8619.883	8622.250			0.2d			11597.90	4g 0	2a 5	R 3
8619.883	8622.250			0.2d			11597.90	4F 0	2B15	Q 4
8620.366	8622.734		0.2	0.2	0.5		11597.25	U 1	2B11	P 2
8620.678	8623.046			0.0			11596.83	4D 0	2B14	P 3
8621.072	8623.440		0.5	0.3			11596.30	3C 1	2A 2	R 0
8621.295	8623.663		0.2	0.9	0.8		11596.00			
8623.214	8625.582		0.2	0.2			11593.42	3f 3	3b 0	P 6
8623.504	8625.873			0.1			11593.03			
8623.764	8626.133		0.3	0.9	1.4		11592.68	V 0	2B 8	P 3
8623.943	8626.312			0.2			11592.44			
8624.419	8626.788			0.0d	0.1		11591.80			
8624.828	8627.197			0.1d			11591.25	4f 3	3b 3	R 1
8625.148	8627.517			0.5	1.4		11590.82			
8625.327	8627.696			0.1			11590.58			
8625.528	8627.897			0.2			11590.31	3D 2	2B13	R 6
8626.346	8628.716			0.1d			11589.21			
8626.808	8629.177		0.2	0.5	0.8		11588.59	V 0	2B 8	P 5
8627.388	8629.758				0.1		11587.81			
8627.537	8629.907			0.1			11587.61			
8628.312	8630.682		0.1	0.0			11586.57			
8628.684	8631.054		0.2	0.2			11586.07			
8629.034	8631.404	G	3.8	2.1	1.4		11585.60	2K 6	2B 5	P 2
8629.034	8631.404	G	3.8	2.1	1.4		11585.60	3c 0	2a 2	P 2
8629.340	8631.710		1.5b	1.7	1.2		11585.19			
8629.675	8632.045			0.1			11584.74	3E 3	2B15	R 3
8630.122	8632.492		0.1		0.1		11584.14			
8631.374	8633.744			0.1			11582.46			
8631.776	8634.147			0.2	0.1		11581.92			
8632.946	8635.318		0.3	0.2			11580.35	3c 3	2a 5	P 2

λ_{air}	λ_{vac}	Ref.	I_1	I_2	I_3	I_4	ν	Upper	Lower	Br.
8633.386	8635.757		0.4	1.2	1.7		11579.76	3b 0	2a 0	R 8
8633.684	8636.056		0.2	0.2			11579.36			
8633.923	8636.294			0.1			11579.04			
8634.191	8636.563		0.3	0.6			11578.68	W 1	2C 2	R 1
8634.601	8636.973		0.2				11578.13	3c 1	2a 3	P 2
8635.295	8637.667				0.1		11577.20			
8635.825	8638.197		0.2	0.2			11576.49	3c 2	2a 4	P 2
8636.362	8638.734			0.3			11575.77	X 0	2B10	R 1
8636.362	8638.734			0.3			11575.77	3C 1	2A 2	Q 2
8636.601	8638.973		0.2	0.7	1.1		11575.45			
8637.063	8639.436			0.1			11574.83			
8637.399	8639.772		0.3				11574.38	5c 0	2a 6	P 3
8637.518	8639.891			0.1d			11574.22			
8637.929	8640.302			0.2d			11573.67	X 0	2C 1	P 4
8638.392	8640.764			0.1d			11573.05	3E 2	2B13	P 5
8638.839	8641.212			0.2			11572.45			
8639.175	8641.548			0.1			11572.00			
8639.489	8641.862		0.9	1.1	0.8		11571.58	W 1	2C 2	R 2
8639.489	8641.862		0.9	1.1	0.8		11571.58	3E 1	2B11	R 2
8639.489	8641.862		0.9	1.1	0.8		11571.58	3C 0	2A 1	R 2
8640.460	8642.833			0.1			11570.28	4D 0	2B14	P 5
8641.132	8643.505		0.5	1.1	1.3		11569.38	2K 3	2B 2	P 5
8641.535	8643.909		0.2	0.1			11568.84			
8641.632	8644.006		0.2				11568.71			
8641.894	8644.268			0.2			11568.36			
8642.230	8644.604			0.2			11567.91			
8642.559	8644.933		0.2	0.2			11567.47	Y 1	2C 1	P 3
8643.463	8645.837			0.2			11566.26			
8643.822	8646.196			0.1			11565.78			
8644.188	8646.562			0.3			11565.29			
8645.160	8647.534		0.2	0.2			11563.99			
8646.057	8648.432			0.1			11562.79	3E 0	2B 9	P 5
8646.057	8648.432			0.1			11562.79	X 1	2B12	P 4
8646.229	8648.604		0.1	0.2			11562.56	3A 0	2C 1	Q 2
8646.580	8648.955			0.2			11562.09			
8647.276	8649.651			0.1			11561.16			
8647.358	8649.733		0.1				11561.05			
8647.657	8650.033		0.2				11560.65			
8647.814	8650.190		0.2	0.1			11560.44	3A 0	2C 1	Q 1
8647.979	8650.354		0.4	0.8	0.8		11560.22	3A 0	2C 1	Q 3
8648.203	8650.579		0.2	0.0			11559.92			
8648.510	8650.886		0.2	0.2	0.1		11559.51			
8648.757	8651.133		0.2	0.1			11559.18			
8648.899	8651.275		0.4				11558.99			
8649.041	8651.417		0.7	0.3	0.1		11558.80			
8649.281	8651.656		0.6	0.3			11558.48	3A 1	2C 2	P 2
8649.573	8651.948		0.2	0.2			11558.09			
8649.977	8652.353		0.2				11557.55	U 1	2B11	P 3
8650.748	8653.124		0.3				11556.52			
8651.496	8653.873		0.3		0.1		11555.52			
8651.676	8654.052		0.3	0.2			11555.28	U 1	2B11	P 5
8651.841	8654.217		0.2				11555.06	W 1	2B12	R 4
8652.036	8654.412		1.0	0.2			11554.80	4f 3	3b 3	P 5
8652.238	8654.614		0.1				11554.53	3A 2	2C 3	P 2
8652.387	8654.764		0.8	0.3			11554.33	W 1	2C 2	R 3
8652.492	8654.869		0.3				11554.19			
8652.672	8655.049		1.5	0.9	0.7		11553.95			
8652.837	8655.213		0.4	0.1			11553.73			
8653.076	8655.453		0.7	0.1			11553.41			
8653.226	8655.603		0.1				11553.21			

λ_{air}	λ_{vac}	Ref.	I$_1$	I$_2$	I$_3$	I$_4$	ν	Upper	Lower	Br.
8653.413	8655.790		0.2	0.2			11552.96			
8653.571	8655.948		0.8				11552.75	X 0	2B10	R 0
8653.676	8656.052		1.0	0.9	0.7		11552.61			
8654.005	8656.382		0.9	0.2	0.1		11552.17			
8654.222	8656.599		0.3				11551.88			
8654.365	8656.742		0.3				11551.69			
8654.500	8656.877		0.1				11551.51			
8654.657	8657.034		0.4	0.1			11551.30			
8654.904	8657.282		0.1				11550.97			
8655.264	8657.641		0.2	0.2			11550.49			
8655.841	8658.218		0.1	0.1d			11549.72			
8656.118	8658.496		0.5	0.2d			11549.35			
8656.238	8658.616		0.6				11549.19			
8656.373	8658.751		0.3				11549.01			
8656.516	8658.893		1.5	1.1	0.9		11548.82			
8656.680	8659.058		0.2				11548.60	3C 1	2A 2	Q 1
8657.010	8659.388		0.4	0.2			11548.16			
8657.228	8659.606	G	1.5b	1.9	2.2		11547.87	3b 4	2a 3	P 4
8657.228	8659.606	G	1.5b	1.9	2.2		11547.87	3E 0	2B 9	R 2
8657.498	8659.875		0.1				11547.51			
8657.648	8660.026		1.0	0.3			11547.31			
8658.097	8660.476			0.1d			11546.71			
8658.210	8660.588		0.3				11546.56			
8658.405	8660.783		0.3				11546.30			
8658.555	8660.933		0.1				11546.10			
8658.705	8661.083		0.5	0.2d			11545.90	3A 1	2B12	R 1
8658.840	8661.218		0.6				11545.72	W 2	2B14	R 3
8659.110	8661.488		0.3				11545.36			
8659.267	8661.646		0.1		0.8		11545.15			
8659.440	8661.818		0.2	0.0			11544.92	3E 1	2B11	Q 5
8659.732	8662.111		0.1				11544.53			
8659.987	8662.366		0.4				11544.19			
8660.100	8662.479		0.1				11544.04			
8660.280	8662.659		0.9				11543.80			
8660.700	8663.079		0.2				11543.24			
8660.993	8663.372		0.4	0.2d			11542.85			
8661.173	8663.552		0.1				11542.61			
8661.308	8663.687		0.1	0.2			11542.43			
8661.668	8664.047		0.2	0.2			11541.95	4E 1	2B17	Q 3
8661.893	8664.272		0.3	0.3	0.2		11541.65			
8662.066	8664.445		0.2				11541.42			
8662.171	8664.550		0.3	0.3			11541.28			
8662.404	8664.783			0.2			11540.97			
8662.689	8665.068		0.2	0.2			11540.59			
8663.042	8665.421		1.9	1.5	1.0		11540.12	2K 6	2B 5	P 3
8663.282	8665.661			0.2			11539.80	V 1	2B 9	R 2
8663.440	8665.819		0.2				11539.59			
8663.545	8665.924		1.5	0.5			11539.45			
8663.897	8666.277	G	3.3	3.1	2.7		11538.98	3b 0	2a 0	P 1
8664.235	8666.615		0.6	0.2			11538.53			
8664.378	8666.758		0.6	0.3			11538.34			
8664.573	8666.953		0.3	0.1			11538.08	3E 0	2B 9	Q 5
8664.874	8667.254		0.3				11537.68			
8664.934	8667.314			0.2			11537.60			
8665.137	8667.517		0.3	0.9	1.0		11537.33			
8665.317	8667.697			0.1			11537.09			
8665.580	8667.960		1.0	0.3	0.3		11536.74	4b 1	2a 5	R 2
8665.813	8668.193		0.2				11536.43			
8665.888	8668.268		0.3	0.2	0.1		11536.33			
8665.970	8668.351		0.3				11536.22			

λ_{air}	λ_{vac}	Ref.	I_1	I_2	I_3	I_4	ν	Upper	Lower	Br.
8666.136	8668.516		0.2	0.2d	0.9		11536.00			
8666.414	8668.794		0.2	0.2			11535.63	3C 1	2A 2	Q 5
8666.601	8668.982		0.1				11535.38			
8666.789	8669.170		0.2				11535.13			
8666.887	8669.267		0.2	0.1			11535.00	3C 1	2K 2	R 2
8667.195	8669.575		0.9				11534.59			
8667.518	8669.899		0.9				11534.16			
8667.676	8670.057				0.3		11533.95			
8667.909	8670.290		0.4	0.8	0.1b		11533.64			
8668.240	8670.620		0.1	0.2	0.2		11533.20			
8668.555	8670.936		0.2	0.2			11532.78			
8668.811	8671.192		0.2	0.2			11532.44			
8669.096	8671.478		0.2	0.1			11532.06	3e 3	3b 0	R 4
8669.284	8671.666		0.2	0.2			11531.81	3c 3	2a 5	P 3
8669.593	8671.974		0.9	0.3			11531.40			
8669.675	8672.057			0.9			11531.29			
8669.803	8672.184			0.1			11531.12			
8670.051	8672.433		0.3	0.3			11530.79			
8670.202	8672.583		0.3	0.2			11530.59			
8670.344	8672.726			0.2			11530.40			
8670.472	8672.854		0.6				11530.23			
8670.517	8672.899		1.5				11530.17			
8670.818	8673.200	G	3.6	3.2	2.8		11529.77	2A 1	2B 0	R 2
8670.818	8673.200	G	3.6	3.2	2.8		11529.77	3F 0	2B 9	P 3
8671.142	8673.523		0.5	0.2			11529.34	4D 0	2C 3	Q 3
8671.285	8673.666		0.5	0.3	0.2		11529.15			
8671.495	8673.877		0.3				11528.87			
8671.555	8673.937		0.3b	0.1			11528.79			
8671.811	8674.193		0.2	0.0			11528.45			
8672.067	8674.449		0.2	0.7			11528.11			
8672.255	8674.637		0.1				11527.86			
8672.518	8674.900		0.3	0.3	0.1		11527.51			
8672.729	8675.111		0.3b				11527.23			
8672.827	8675.209		0.2	0.1			11527.10			
8673.000	8675.382		0.2	0.2			11526.87			
8673.383	8675.766		0.4	0.2			11526.36			
8673.572	8675.954		0.2				11526.11			
8673.715	8676.097			0.0			11525.92	U 1	2B11	P 4
8674.430	8676.812		0.9				11524.97			
8674.595	8676.978			0.7			11524.75			
8674.723	8677.106		0.3				11524.58			
8675.197	8677.580			0.0			11523.95			
8675.551	8677.934		0.9	0.7			11523.48	3c 2	2a 4	P 3
8676.116	8678.499		0.1	0.1			11522.73			
8676.477	8678.860		0.9	0.7	0.9		11522.25	X 0	2C 1	Q 5
8676.568	8678.951				0.1		11522.13			
8676.666	8679.049		0.1				11522.00			
8677.840	8680.224		0.1				11520.44	4f 2	2c 2	P 6
8677.961	8680.344		0.2	0.0			11520.28			
8678.074	8680.458		0.3				11520.13			
8678.406	8680.789				0.2		11519.69			
8678.684	8681.068		0.2				11519.32			
8678.918	8681.302		1.5b	0.7	0.8		11519.01	3c 1	2a 3	P 3
8679.438	8681.822		0.3	0.2	0.2		11518.32			
8679.912	8682.297			0.0d			11517.69			
8680.244	8682.628		0.1				11517.25			
8680.448	8682.832		0.6	0.2			11516.98			
8681.073	8683.458		0.2				11516.15			
8681.307	8683.691		0.9	0.7	1.0		11515.84			
8681.669	8684.053		0.3				11515.36			

λ_{air}	λ_{vac}	Ref.	I_1	I_2	I_3	I_4	ν	Upper	Lower	Br.
8681.925	8684.310				0.2		11515.02			
8682.408	8684.792		1.0				11514.38			
8682.928	8685.313		0.4	0.2			11513.69	4g 0	2a 5	P 3
8683.245	8685.630		1.0	0.2			11513.27	V 0	2B 8	P 2
8683.396	8685.781						11513.07			
8684.014	8686.399		0.4	0.2			11512.25	X 1	2C 2	P 4
8684.391	8686.776		0.3	0.2			11511.75			
8684.701	8687.086		0.0				11511.34			
8684.912	8687.297			0.1	0.2b		11511.06			
8685.221	8687.607		0.7	0.7	1.6		11510.65	3b 0	2a 0	R 9
8685.448	8687.833		0.1				11510.35			
8685.644	8688.029		0.5				11510.09			
8686.066	8688.452		0.2				11509.53			
8686.315	8688.701		0.2				11509.20			
8686.542	8688.928		1.7	1.3	0.9		11508.90	W 1	2C 2	Q 1
8686.806	8689.192		0.2				11508.55			
8686.995	8689.381		0.4	0.2			11508.30			
8687.221	8689.607		0.2				11508.00	W 2	2B14	R 2
8687.546	8689.932						11507.57			
8687.712	8690.098		0.2				11507.35	4c 1	2a 6	R 2
8687.938	8690.325		0.2				11507.05			
8688.165	8690.551			0.1			11506.75			
8688.384	8690.770		0.6	0.2	0.1		11506.46			
8688.505	8690.891		0.2				11506.30			
8688.776	8691.163		1.0	0.3	0.2		11505.94			
8688.882	8691.269			1.1	1.5		11505.80			
8688.958	8691.344		1.5	0.4	0.2		11505.70	3C 0	2A 1	R 1
8689.101	8691.488		0.4				11505.51			
8689.275	8691.661		0.1	0.0			11505.28			
8689.592	8691.979		0.1				11504.86			
8689.811	8692.198		0.8	0.5			11504.57	W 1	2C 2	R 4
8689.985	8692.372		0.1				11504.34			
8690.196	8692.583		0.2				11504.06	4E 1	2B17	Q 1
8690.997	8693.384		0.2b				11503.00			
8691.556	8693.944		1.5	0.6			11502.26			
8691.813	8694.201		0.2				11501.92			
8692.040	8694.427		0.1				11501.62			
8692.327	8694.714		0.2				11501.24	3E 2	2B13	P 4
8692.713	8695.100		0.1				11500.73	3E 2	2B13	Q 4
8693.242	8695.629		0.7				11500.03	3A 0	2C 1	P 1
8693.461	8695.849		0.1	0.1			11499.74			
8693.612	8696.000		0.1				11499.54			
8693.839	8696.227		1.7	1.3	1.0		11499.24			
8694.104	8696.492		0.1				11498.89	4E 1	2B17	Q 2
8694.300	8696.688		0.2				11498.63	4E 0	2C 3	P 5
8694.300	8696.688		0.2				11498.63	X 0	2B10	P 1
8694.512	8696.900		0.3	0.1			11498.35			
8694.882	8697.271		0.5	0.1			11497.86	3A 1	2C 2	P 3
8695.041	8697.429						11497.65	4b 0	2a 4	R 4
8695.427	8697.815		0.3				11497.14	4F 0	2B15	R 2
8695.722	8698.110		0.2				11496.75			
8695.919	8698.307		0.2				11496.49			
8696.047	8698.436		0.7	0.2			11496.32			
8696.168	8698.557		0.2				11496.16			
8696.297	8698.685		0.8	0.6	1.0		11495.99			
8696.539	8698.927		0.3				11495.67	W 2	2C 3	R 1
8696.758	8699.147		0.2				11495.38			
8697.522	8699.911		0.3	0.2			11494.37			
8697.712	8700.100		0.2				11494.12			
8698.022	8700.411			0.1			11493.71			

λ_{air}	λ_{vac}	Ref.	I_1	I_2	I_3	I_4	ν	Upper	Lower	Br.
8698.362	8700.751		1.5	1.0	0.8		11493.26			
8698.582	8700.971		0.2				11492.97			
8698.786	8701.176		0.3	0.1			11492.70			
8699.051	8701.440		2.1	1.6	1.0		11492.35			
8699.293	8701.683		0.2				11492.03			
8699.460	8701.849		0.4				11491.81			
8699.611	8702.001		0.2	0.2			11491.61	3C 1	2A 2	Q 4
8700.157	8702.546		0.1	0.0d			11490.89			
8700.316	8702.705		0.4				11490.68			
8701.618	8704.008		0.2				11488.96			
8701.868	8704.258		0.2	0.1			11488.63			
8702.126	8704.516		2.2	1.7	1.5		11488.29	2K 6	2B 5	P 4
8702.315	8704.705		0.3	1.7			11488.04			
8702.520	8704.910			0.0			11487.77			
8702.679	8705.069		0.2				11487.56			
8702.830	8705.220		0.1				11487.36			
8704.065	8706.456		0.2		0.1		11485.73	3e 3	3b 0	R 3
8704.429	8706.819		0.2	0.2			11485.25	3C 2	2K 4	Q 4
8704.694	8707.085		0.2	0.1			11484.90	4f 3	3b 3	Q 4
8705.020	8707.411		0.4				11484.47			
8705.119	8707.509		0.3	0.2			11484.34	3C 2	2K 4	R 2
8705.210	8707.600		0.2				11484.22			
8705.475	8707.866		0.3				11483.87			
8705.627	8708.018		0.6	0.2			11483.67			
8706.097	8708.488		0.2				11483.05			
8706.461	8708.852		0.5	0.2			11482.57	3c 3	2a 5	P 4
8706.688	8709.079		0.9	0.7	0.9		11482.27			
8706.809	8709.201		0.5				11482.11	4b 1	2a 5	R 1
8707.204	8709.595		0.2				11481.59			
8707.333	8709.724			0.2			11481.42			
8707.697	8710.088		2.1	1.6	1.2		11480.94	2K 5	2B 4	R 1
8708.114	8710.505			0.1			11480.39			
8708.228	8710.619		0.2				11480.24			
8708.448	8710.839		0.2	0.1			11479.95			
8708.622	8711.014		0.9				11479.72	3C 1	2K 2	P 3
8708.842	8711.234		2.5	2.0	1.7		11479.43	2K 5	2B 4	R 2
8709.115	8711.507		1.5b	0.7			11479.07			
8709.373	8711.765		0.2				11478.73			
8709.472	8711.864		0.1				11478.60			
8709.745	8712.137		0.1				11478.24			
8710.071	8712.463			0.5			11477.81			
8710.215	8712.608		0.2	0.5			11477.62			
8710.891	8713.283						11476.73	3A 1	2B12	R 0
8711.437	8713.830		0.5				11476.01			
8711.620	8714.012		0.7	0.5			11475.77			
8711.840	8714.232		0.2				11475.48			
8712.136	8714.528		1.5	1.0			11475.09			
8712.447	8714.840		0.3	0.0			11474.68	3F 0	2C 1	R 5
8712.713	8715.106		0.2				11474.33	3E 2	2B13	R 0
8713.055	8715.448		0.4	0.2			11473.88	X 0	2B10	P 2
8714.042	8716.435		0.2				11472.58			
8714.353	8716.747			0.1			11472.17			
8714.574	8716.967		0.7b				11471.88			
8714.764	8717.157		0.6b	0.2			11471.63	3c 2	2a 4	P 4
8715.090	8717.484	G	2.8	2.2	1.5		11471.20	2K 5	2B 4	R 0
8715.090	8717.484	G	2.8	2.2	1.5		11471.20	3D 2	2B13	R 5
8715.090	8717.484	G	2.8	2.2	1.5		11471.20	W 1	2C 2	P 1
8715.356	8717.750		0.7				11470.85			
8715.470	8717.864		0.2	0.1			11470.70	W 2	2B14	R 1
8715.698	8718.092		1.5	0.9			11470.40	W 1	2C 2	Q 2

λ_{air}	λ_{vac}	Ref.	I_1	I_2	I_3	I_4	ν	Upper	Lower	Br.
8715.979	8718.373		0.2				11470.03			
8716.199	8718.593		1.5	1.0	0.8		11469.74			
8716.359	8718.753		0.4				11469.53			
8716.792	8719.186		0.4	0.1d			11468.96	3E 1	2B11	R 1
8717.385	8719.779		0.1	0.0d			11468.18			
8717.682	8720.076		1.0	1.2	1.2		11467.79	2K 5	2B 4	R 3
8717.796	8720.190		0.2				11467.64			
8718.214	8720.608		0.8	0.2			11467.09			
8718.708	8721.103		0.5	0.0			11466.44	W 1	2B12	R 3
8718.822	8721.217						11466.29	4e 2	3b 2	R 4
8719.271	8721.665		0.9	0.2			11465.70	X 1	2C 2	P 2
8719.271	8721.665		0.9	0.2			11465.70	3c 0	2a 2	P 4
8719.423	8721.818			0.5			11465.50			
8719.719	8722.114		0.5				11465.11			
8719.978	8722.373		0.2	0.1			11464.77	3E 3	2B15	R 1
8720.168	8722.563		0.1				11464.52			
8720.358	8722.753		0.5	0.2			11464.27			
8720.533	8722.928		0.7				11464.04			
8720.761	8723.157		0.2	0.1			11463.74			
8721.111	8723.507		0.4	0.2			11463.28			
8721.332	8723.727		0.2				11462.99			
8721.500	8723.895		0.6				11462.77			
8721.842	8724.237		0.2				11462.32			
8722.101	8724.496		0.4	0.2			11461.98			
8722.253	8724.648		0.2				11461.78			
8722.519	8724.915		0.3	0.2			11461.43			
8722.649	8725.044		0.3				11461.26			
8722.854	8725.250		0.1				11460.99			
8723.113	8725.509		0.2	0.2			11460.65			
8723.440	8725.836		0.2	0.2			11460.22			
8723.585	8725.981		0.2	0.2			11460.03			
8723.676	8726.072		1.0	0.2			11459.91			
8723.866	8726.262		0.7	0.3			11459.66			
8724.194	8726.590	G	3.0	3.0	2.5		11459.23	3b 0	2a 0	P 2
8724.506	8726.902		0.5				11458.82			
8724.681	8727.077		0.2	0.2			11458.59			
8725.001	8727.397		1.0b	1.5	2.1		11458.17	2A 1	2B 0	P 7
8725.214	8727.610		0.2				11457.89			
8725.443	8727.839		0.1	0.0			11457.59			
8725.701	8728.098		0.3b	0.2b			11457.25			
8725.945	8728.342		0.2				11456.93	4f 2	3b 2	R 1
8726.204	8728.601		0.9	0.8	0.9		11456.59	3c 1	2a 3	P 4
8726.471	8728.867		0.2				11456.24			
8726.623	8729.020		0.2	0.2			11456.04			
8726.890	8729.286		1.0	0.9	1.0		11455.69			
8727.362	8729.759		0.1				11455.07			
8727.819	8730.216		0.8				11454.47			
8728.124	8730.521		0.2				11454.07	4f 3	3b 3	P 4
8728.231	8730.628		1.0	0.7	0.8		11453.93			
8728.520	8730.917			0.0			11453.55	3A 1	2B12	P 4
8728.894	8731.291		0.2				11453.06			
8729.138	8731.535			0.0			11452.74			
8729.488	8731.886		0.8	0.8	1.6		11452.28	3b 4	2a 3	P 5
8730.151	8732.549		0.2				11451.41			
8730.334	8732.732		0.2	0.0			11451.17			
8730.655	8733.052			0.0			11450.75			
8730.731	8733.129		0.1				11450.65			
8731.028	8733.426		0.9	0.7	0.9		11450.26	3E 2	2B13	P 3
8731.226	8733.624		0.2				11450.00			
8731.471	8733.869		0.3	0.0			11449.68			

λ_{air}	λ_{vac}	Ref.	I_1	I_2	I_3	I_4	ν	Upper	Lower	Br.
8731.753	8734.151		0.2				11449.31			
8732.020	8734.418		0.8	0.2			11448.96			
8732.233	8734.631		1.5	1.3	1.5		11448.68	2K 5	2B 4	R 4
8732.515	8734.914		1.5	0.2			11448.31	W 2	2C 3	R 2
8732.767	8735.165		0.9				11447.98			
8732.927	8735.326		1.5	1.4	1.7		11447.77	2K 3	2B 2	P 6
8732.927	8735.326		1.5	1.4	1.7		11447.77	3E 1	2B11	P 5
8733.286	8735.684		0.5	0.7			11447.30			
8733.377	8735.776		0.1				11447.18			
8733.751	8736.150			0.1			11446.69			
8733.858	8736.257		0.1				11446.55			
8734.133	8736.531		0.8	0.2			11446.19	3A 2	2C 3	P 4
8734.354	8736.753		0.3				11445.90			
8734.583	8736.982		0.4	0.1			11445.60			
8734.881	8737.280		0.1				11445.21			
8735.056	8737.455		0.8				11444.98	3F 1	2C 2	R 5
8735.346	8737.745		0.2	0.2			11444.60			
8735.652	8738.051		0.7	0.2	0.2		11444.20			
8735.842	8738.242		0.2				11443.95			
8736.102	8738.501		2.0	1.6	1.2		11443.61			
8736.362	8738.761		0.1				11443.27	4F 0	2B15	P 4
8736.575	8738.975		0.1				11442.99	3E 1	2B11	Q 4
8737.270	8739.670		0.2				11442.08	X 0	2B10	P 3
8737.270	8739.670		0.2				11442.08	4b 0	2a 4	R 3
8737.492	8739.891		0.6				11441.79	3A 0	2C 1	P 2
8737.850	8740.250		0.0				11441.32			
8738.423	8740.823		1.0	0.6			11440.57	3C 0	2A 1	R 0
8738.805	8741.205		0.2				11440.07			
8738.958	8741.358		0.1				11439.87	W 2	2B14	R 0
8739.134	8741.534		1.0	0.3			11439.64			
8739.332	8741.732		1.9	2.0	2.4		11439.38	3B 1	2A 0	R 3
8739.714	8742.115		0.4				11438.88	3F 0	2C 1	R 4
8740.089	8742.489		1.5b				11438.39	3D 0	2C 3	Q 4
8740.135	8742.535			1.2	1.9		11438.33	3B 1	2A 0	R 4
8740.135	8742.535			1.2	1.9		11438.33	W 2	2C 3	Q 1
8740.517	8742.917		0.3	0.1			11437.83	3e 3	3b 0	R 2
8741.312	8743.712		0.6	0.7	0.9		11436.79	4f 3	3b 3	Q 3
8741.312	8743.712		0.6	0.7	0.9		11436.79	3C 1	2A 2	Q 3
8741.526	8743.926		0.3	0.2			11436.51			
8741.816	8744.217		0.2				11436.13			
8741.954	8744.354		0.1				11435.95			
8742.137	8744.538		0.5	0.2			11435.71			
8742.741	8745.142		1.0	0.8			11434.92			
8742.955	8745.356		0.7	0.2			11434.64			
8743.230	8745.631		0.3	0.1			11434.28	3E 3	2B15	R 2
8743.376	8745.777		0.7				11434.09	4b 2	2a 6	P 3
8743.704	8746.106		0.4				11433.66	Y 1	2B10	R 2
8743.865	8746.266		0.9	0.8			11433.45			
8744.140	8746.542		0.2				11433.09			
8744.309	8746.710		0.3	0.1			11432.87			
8744.446	8746.848		0.2				11432.69			
8744.752	8747.154		0.3d	0.1d			11432.29			
8744.905	8747.307		0.3d				11432.09			
8745.081	8747.483		0.4d				11431.86			
8745.280	8747.682						11431.60			
8745.502	8747.904		0.2				11431.31			
8745.793	8748.194			0.1			11430.93			
8745.961	8748.363		0.2				11430.71	3D 2	2B13	R 4
8746.191	8748.593		0.9	0.3			11430.41			
8746.405	8748.807		0.9	1.2	1.3		11430.13	3b 0	2a 0	R10

λ_{air}	λ_{vac}	Ref.	I_1	I_2	I_3	I_4	ν	Upper	Lower	Br.
8746.566	8748.968		0.9d				11429.92			
8746.680	8749.082		0.4	0.2			11429.77			
8747.040	8749.442	G	3.4	2.8	2.3		11429.30	2A 1	2B 0	R 1
8747.285	8749.687		0.2				11428.98			
8747.430	8749.833		0.2	0.1			11428.79	3F 3	2B16	R 2
8747.691	8750.093		1.8	1.2	1.6		11428.45	3F 1	2B12	Q 6
8747.867	8750.269		0.7	0.3	0.6		11428.22			
8748.073	8750.476			0.1			11427.95			
8748.165	8750.568		0.2				11427.83			
8748.395	8750.797		1.6	1.7	2.0		11427.53	3B 1	2A 0	R 2
8748.395	8750.797		1.6	1.7	2.0		11427.53	4b 1	2a 5	R 0
8748.793	8751.195		0.2	0.0			11427.01			
8749.053	8751.456		0.6	0.6	1.0		11426.67			
8749.421	8751.824		0.1				11426.19			
8750.049	8752.452		0.3	0.0			11425.37			
8750.225	8752.628		0.2	0.0			11425.14			
8750.424	8752.827		0.1				11424.88	3E 2	2B13	Q 3
8750.608	8753.011		1.0	1.2	2.2		11424.64	3B 1	2A 0	R 5
8750.608	8753.011		1.0	1.2	2.2		11424.64	T 0	2C 3	P 4
8750.608	8753.011		1.0	1.2	2.2		11424.64	X 0	2B10	P 4
8750.929	8753.333		0.8	0.2			11424.22			
8751.320	8753.724		0.2d				11423.71			
8751.994	8754.398		0.3				11422.83			
8752.401	8754.804		0.4	0.0			11422.30			
8752.638	8755.042		0.2				11421.99	Y 1	2B10	R 1
8753.128	8755.532		0.5	0.0			11421.35			
8753.282	8755.686		1.5	0.8	0.9		11421.15			
8753.443	8755.847		0.2				11420.94			
8753.634	8756.038		0.9	0.2			11420.69	W 2	2C 3	P 1
8753.903	8756.307		0.1				11420.34			
8754.133	8756.537		0.3				11420.04	3c 2	2a 4	P 5
8754.271	8756.675		1.5	0.6			11419.86			
8754.416	8756.820		0.4				11419.67			
8754.792	8757.196		0.2				11419.18			
8755.352	8757.756		0.2				11418.45	w 0	3b 1	R 0
8755.559	8757.963		0.5				11418.18			
8755.766	8758.170		0.7	0.2			11417.91			
8756.057	8758.462		2.5	1.8	1.1		11417.53	2K 5	2B 4	P 1
8756.325	8758.730		0.2	0.1			11417.18			
8756.609	8759.014		0.5	0.1			11416.81			
8756.816	8759.221		0.2				11416.54	3E 0	2B 9	R 1
8757.422	8759.827		0.1				11415.75			
8757.929	8760.334		0.2				11415.09	3E 3	2B15	Q 3
8758.074	8760.480		0.7	0.2			11414.90			
8758.305	8760.710		0.1				11414.60			
8758.535	8760.940		1.5	0.7	0.8		11414.30			
8758.765	8761.170		0.1	0.1			11414.00			
8759.034	8761.439		0.8	0.6	0.9		11413.65	V 1	2B 9	R 1
8759.318	8761.723		0.1				11413.28	W 2	2B14	P 6
8759.548	8761.953		0.2	0.1			11412.98			
8759.694	8762.099		0.5				11412.79			
8760.039	8762.445		0.1				11412.34			
8760.231	8762.637		0.1				11412.09			
8760.438	8762.844		0.8	0.6			11411.82	3C 1	2K 2	R 1
8760.569	8762.975		0.2				11411.65	Y 1	2B10	R 0
8760.784	8763.190			0.6			11411.37			
8760.860	8763.266		0.1				11411.27	w 0	3b 1	P 1
8761.137	8763.543		0.1	0.0			11410.91	W 1	2B12	R 2
8761.390	8763.796		1.6	1.2	0.9		11410.58	W 1	2C 2	Q 3
8761.651	8764.057				0.2		11410.24	3A 1	2B12	P 3

λ_{air}	λ_{vac}	Ref.	I_1	I_2	I_3	I_4	ν	Upper	Lower	Br.
8762.711	8765.118			0.2			11408.86			
8762.819	8765.225						11408.72			
8762.965	8765.371		0.4	0.2			11408.53			
8763.341	8765.748						11408.04			
8763.679	8766.086		0.1				11407.60			
8763.863	8766.270		0.4				11407.36	3c 0	2a 2	P 5
8764.171	8766.578		0.9	0.2			11406.96	2A 3	2B 3	R 3
8764.378	8766.785		1.0	0.9	1.1		11406.69	3D 2	2B13	R 3
8764.778	8767.185		1.5	0.6	1.4		11406.17	4f 2	3b 2	P 5
8765.162	8767.569		2.5	2.2	2.0		11405.67	2A 3	2B 3	R 2
8765.446	8767.854		0.3	0.1			11405.30			
8765.761	8768.169		0.1	0.1			11404.89	3E 2	2B13	P 2
8766.038	8768.445		0.3				11404.53			
8766.192	8768.599		0.6	0.2			11404.33			
8766.338	8768.745		0.2				11404.14			
8766.745	8769.153			0.2			11403.61	3A 1	2B12	P 1
8766.984	8769.391		0.6d				11403.30	3A 1	2C 2	P 4
8766.984	8769.391		0.6d				11403.30	X 0	2C 1	P 5
8767.176	8769.583			0.2			11403.05	4f 0	3b 0	R 2
8767.422	8769.830			0.9			11402.73			
8767.522	8769.930		2.5	2.6	2.4		11402.60	3B 1	2A 0	R 1
8767.776	8770.183		0.2d	0.0			11402.27	3f 0	2c 3	R 5
8767.976	8770.383		0.1				11402.01			
8768.129	8770.537		0.9	0.2			11401.81	Y 1	2C 1	P 4
8768.437	8770.845		0.2				11401.41			
8768.637	8771.045		0.1				11401.15			
8768.798	8771.207		0.3	0.7			11400.94	3E 3	2B15	R 0
8768.960	8771.368		0.2				11400.73			
8769.475	8771.884		0.2b				11400.06	3f 0	2c 3	R 4
8769.883	8772.291			0.1			11399.53			
8769.975	8772.384		0.3				11399.41			
8770.229	8772.638		1.5	0.8	1.5		11399.08			
8770.514	8772.922		0.7	0.8	1.5		11398.71			
8770.722	8773.130		0.3				11398.44	4F 0	2B15	Q 3
8770.845	8773.254		1.5	0.9			11398.28	W 1	2C 2	P 2
8771.176	8773.584		0.4	0.2d			11397.85	X 0	2B10	P 5
8771.414	8773.823		0.2				11397.54	3A 1	2B12	P 2
8771.668	8774.077		0.5	0.2	0.9		11397.21			
8771.876	8774.285		0.1	0.0			11396.94			
8772.169	8774.578		1.5b	1.4	1.0		11396.56			
8772.561	8774.970		0.3b	0.1			11396.05			
8772.961	8775.371						11395.53	W 2	2B14	P 1
8773.454	8775.864		0.3				11394.89	4e 3	3b 3	R 2
8774.055	8776.464		0.2				11394.11			
8774.301	8776.711		0.2				11393.79	3c 1	2a 3	P 5
8774.848	8777.258		0.4d	0.1d			11393.08			
8775.041	8777.450		0.5d				11392.83			
8775.333	8777.743		2.4	1.8	1.1		11392.45	3E 1	2B11	P 4
8775.610	8778.020		0.5				11392.09	3E 3	2B15	P 5
8776.296	8778.706		0.2	0.1			11391.20			
8776.550	8778.961		1.0	0.7			11390.87			
8776.774	8779.184			0.1			11390.58			
8776.890	8779.300		0.6d				11390.43			
8777.098	8779.508		2.4	2.0	1.5		11390.16	2A 3	2B 3	R 1
8777.398	8779.808			0.1			11389.77			
8777.691	8780.101		1.0				11389.39			
8777.830	8780.240		1.0	0.8			11389.21			
8777.953	8780.364			0.2b			11389.05			
8778.608	8781.019		0.6	0.6			11388.20	3C 2	2K 4	Q 3
8778.970	8781.381		0.3				11387.73			

λ_{air}	λ_{vac}	Ref.	I_1	I_2	I_3	I_4	ν	Upper	Lower	Br.
8779.209	8781.620		0.2				11387.42			
8779.526	8781.937		0.1				11387.01	4b 0	2a 4	R 2
8780.019	8782.430			0.0d			11386.37			
8780.366	8782.777		1.0	1.2	1.5		11385.92			
8780.620	8783.032		0.2				11385.59			
8780.752	8783.163		0.2				11385.42	3D 2	2B13	R 2
8780.883	8783.294		0.8	0.1d			11385.25	W 2	2C 3	Q 2
8781.137	8783.549		0.2				11384.92	3A 0	2C 1	P 3
8781.546	8783.958		0.2	0.0d			11384.39			
8782.240	8784.652		0.2				11383.49			
8782.927	8785.339		0.7	0.9			11382.60	3D 2	2B13	P 1
8783.159	8785.571		0.1				11382.30	W 2	2B14	P 2
8783.390	8785.802		0.3	0.1			11382.00			
8783.776	8786.188		0.2				11381.50			
8783.930	8786.342		0.3				11381.30			
8784.108	8786.520		0.4	0.1			11381.07			
8784.417	8786.829		0.2	0.1			11380.67			
8784.609	8787.022		0.2				11380.42			
8784.795	8787.207		1.7	1.5	1.6		11380.18	3E 3	2B15	P 4
8785.312	8787.725		0.1				11379.51			
8785.474	8787.887		0.2				11379.30			
8785.698	8788.111		0.2				11379.01			
8785.930	8788.342		0.7	0.8	1.0		11378.71			
8786.161	8788.574			0.2			11378.41			
8786.254	8788.667		0.2				11378.29			
8786.509	8788.922		0.4	0.1			11377.96			
8786.756	8789.169		1.5	1.2	1.1		11377.64	3C 0	2A 1	Q 2
8786.980	8789.393		0.3	0.1			11377.35	3F 3	2B16	Q 3
8787.320	8789.733		0.3b				11376.91			
8787.552	8789.965		0.4b	0.1			11376.61	3F 0	2C 1	R 4
8787.706	8790.119		0.3d				11376.41	W 2	2C 3	R 4
8788.038	8790.451		0.2d				11375.98			
8788.316	8790.729		0.3d	0.0			11375.62			
8788.633	8791.047		0.5d	1.0	1.2		11375.21	3C 0	2A 1	Q 3
8788.633	8791.047		0.5d	1.0	1.2		11375.21	3C 0	2A 1	Q 4
8788.633	8791.047		0.5d	1.0	1.2		11375.21	3C 0	2A 1	Q 5
8788.857	8791.271		0.2				11374.92			
8789.066	8791.479		0.5				11374.65	3D 1	2B11	R 6
8789.274	8791.688		0.6	0.1			11374.38			
8789.583	8791.997		3.1	2.5	1.7		11373.98	2K 5	2B 4	P 2
8789.583	8791.997		3.1	2.5	1.7		11373.98	4F 0	2B15	R 1
8789.877	8792.291			0.2			11373.60	4e 2	3b 2	R 3
8790.209	8792.623		2.5	2.6	2.7		11373.17	2A 1	2B 0	P 6
8790.209	8792.623		2.5	2.6	2.7		11373.17	4f 3	2c 3	P 3
8790.727	8793.141		0.1	0.1d			11372.50			
8791.021	8793.435		0.8	0.2			11372.12	3C 2	2K 4	R 1
8791.230	8793.644		0.7				11371.85	3e 3	3b 0	R 1
8791.423	8793.837			1.0			11371.60			
8792.073	8794.487						11370.76			
8792.273	8794.688			0.1			11370.50	3E 1	2B11	Q 3
8792.498	8794.912		0.3	0.2			11370.21			
8792.691	8795.106		0.4				11369.96			
8792.869	8795.283			0.2			11369.73			
8793.008	8795.423		0.2				11369.55			
8793.116	8795.531			0.2			11369.41			
8793.318	8795.732		0.5	0.2			11369.15	4f 2	3b 2	Q 4
8793.689	8796.104			0.2			11368.67			
8793.828	8796.243			1.0			11368.49	3E 2	2B13	Q 2
8794.153	8796.568		0.3	0.3			11368.07			
8794.292	8796.707			0.2			11367.89			

λ_{air}	λ_{vac}	Ref.	I_1	I_2	I_3	I_4	ν	Upper	Lower	Br.
8794.439	8796.854			0.2			11367.70			
8794.571	8796.986		0.8	0.5			11367.53			
8794.919	8797.334	G	3.5	1.5b	3.1		11367.08	3b 0	2a 0	P 3
8795.282	8797.698		0.5	0.3	0.1		11366.61			
8795.422	8797.837		0.5				11366.43			
8795.623	8798.038		0.2	0.1			11366.17	3D 2	2B13	R 1
8795.801	8798.216		0.5	0.3	1.4		11365.94			
8795.987	8798.402		0.3	0.2			11365.70			
8796.343	8798.758		3.1	2.7	1.9		11365.24	2A 3	2B 3	R 0
8796.815	8799.231		2.2	1.8	1.8		11364.63	3B 1	2A 0	R 0
8796.815	8799.231		2.2	1.8	1.8		11364.63	3E 0	2B 9	P 4
8797.179	8799.594		0.2b	0.1d			11364.16	3C 2	2K 4	P 5
8797.179	8799.594		0.2b	0.1d			11364.16	4f 3	3b 3	Q 2
8797.349	8799.765		0.2				11363.94			
8797.519	8799.935			0.2d			11363.72	3E 1	2B11	R 0
8797.666	8800.082						11363.53			
8797.798	8800.214			0.1d			11363.36	Y 1	2B10	P 1
8798.146	8800.562		0.4	1.3	0.1d		11362.91	3C 1	2K 2	Q 2
8798.340	8800.756		0.4				11362.66			
8798.580	8800.996		0.8	0.0			11362.35			
8798.967	8801.383		0.2	0.2			11361.85	W 2	2B14	P 3
8799.293	8801.709			0.0			11361.43			
8799.649	8802.065		0.9	0.3	1.8		11360.97			
8799.997	8802.414		0.5				11360.52	4c 1	2a 6	Q 3
8800.392	8802.809		0.5				11360.01	w 0	3b 1	R 1
8802.035	8804.452		0.1				11357.89			
8802.260	8804.677		0.3				11357.60			
8802.578	8804.995		0.3	0.1			11357.19			
8802.764	8805.181		0.2				11356.95	V 1	2B 9	P 4
8802.996	8805.414		0.5				11356.65			
8803.430	8805.848		0.2				11356.09	3F 1	2C 2	R 4
8803.663	8806.080			0.3			11355.79			
8803.771	8806.189		0.2				11355.65			
8804.081	8806.499		0.8				11355.25			
8804.229	8806.646		0.8	1.0			11355.06			
8804.454	8806.871		0.1				11354.77			
8804.826	8807.244		0.8	1.1	1.1		11354.29	3E 3	2B15	P 3
8805.260	8807.678		0.2				11353.73			
8805.516	8807.934		0.2				11353.40			
8805.694	8808.113		0.3b	0.2b			11353.17			
8806.036	8808.454		0.2	0.1			11352.73			
8806.501	8808.920		0.2	0.1			11352.13			
8806.649	8809.067			1.0			11351.94			
8806.974	8809.393		0.3	0.2			11351.52			
8807.114	8809.532			0.1			11351.34			
8807.308	8809.727		0.2	0.2	0.1		11351.09			
8807.409	8809.827		0.9	0.3			11350.96			
8807.743	8810.161	G	4.0	3.4	2.5		11350.53	2A 1	2B 0	R 0
8808.100	8810.518		0.5				11350.07	3c 0	2a 2	P 6
8808.239	8810.658		0.5	0.2			11349.89			
8808.441	8810.860		0.3	0.1			11349.63			
8808.627	8811.046		1.0	1.3			11349.39			
8808.744	8811.163		0.2				11349.24			
8809.000	8811.419		0.2				11348.91			
8809.473	8811.892		0.2				11348.30			
8809.792	8812.211		0.3				11347.89			
8809.994	8812.413		0.2	0.2b			11347.63			
8810.203	8812.622		0.2				11347.36			
8810.382	8812.801		0.2				11347.13	3D 3	2B15	R 4
8810.700	8813.120		0.2				11346.72			

λ_{air}	λ_{vac}	Ref.	I_1	I_2	I_3	I_4	ν	Upper	Lower	Br.
8810.801	8813.220		0.5	0.9			11346.59			
8811.197	8813.617			0.1			11346.08			
8811.422	8813.842		0.9				11345.79			
8811.966	8814.386		0.2	0.1			11345.09	3E 1	2B11	P 3
8812.292	8814.712		0.4	0.8			11344.67	W 2	2B14	P 4
8812.517	8814.937		0.2	0.2			11344.38	3D 2	2B13	R 0
8813.061	8815.481						11343.68	u 2	2a 5	R 3
8813.830	8816.251				0.2		11342.69	W 2	2B14	P 5
8813.830	8816.251				0.2		11342.69	3E 0	2B 9	Q 4
8814.095	8816.515		0.2	0.1			11342.35			
8814.390	8816.810		0.5	0.8			11341.97			
8814.685	8817.106		0.3				11341.59			
8815.097	8817.518		0.4				11341.06			
8815.299	8817.720		0.2				11340.80			
8815.820	8818.241		0.6	0.2			11340.13			
8816.497	8818.918		0.2	0.2			11339.26			
8816.745	8819.166		0.3				11338.94			
8816.854	8819.275		1.5	1.5	2.2		11338.80	3b 0	2a 0	R11
8818.130	8820.551		0.1	0.1			11337.16			
8818.277	8820.699						11336.97			
8818.534	8820.956			0.0			11336.64			
8818.736	8821.158		0.3				11336.38			
8818.892	8821.314		0.4	0.7			11336.18	W 1	2C 2	Q 4
8818.892	8821.314		0.4	0.7			11336.18	4e 0	3b 0	R 6
8819.460	8821.882						11335.45			
8819.849	8822.271		0.3	0.1			11334.95			
8820.137	8822.559			0.1			11334.58			
8820.635	8823.057		0.2				11333.94			
8821.125	8823.547		0.2				11333.31			
8821.483	8823.906		0.3		0.1		11332.85	Y 1	2B10	P 2
8821.483	8823.906		0.3		0.1		11332.85	4e 1	3b 1	R 4
8821.616	8824.038		0.6	0.6			11332.68	4b 0	2a 4	R 1
8821.616	8824.038		0.6	0.6			11332.68	6c 0	2a 7	Q 1
8822.028	8824.451		0.9	0.9	0.2		11332.15	3E 2	2B13	Q 1
8822.449	8824.871		0.1				11331.61			
8822.900	8825.323		0.1				11331.03			
8823.251	8825.673		0.2				11330.58			
8823.360	8825.782				0.1		11330.44			
8823.866	8826.289		0.1				11329.79			
8824.512	8826.935		0.5	0.6			11328.96	3F 0	2C 1	R 3
8824.801	8827.224		0.2	0.1			11328.59			
8825.042	8827.465		0.3	0.5			11328.28			
8825.268	8827.691		0.3				11327.99			
8825.743	8828.167		0.2				11327.38	4c 0	2a 5	R 4
8826.016	8828.440		0.2				11327.03			
8826.172	8828.595		0.3				11326.83	3A 0	2C 1	P 4
8826.561	8828.985		0.2				11326.33	4e 3	3b 3	Q 4
8827.403	8829.827		0.2	0.1			11325.25			
8827.738	8830.162		0.1				11324.82			
8828.027	8830.451		0.2				11324.45			
8828.268	8830.693		0.3	0.5			11324.14			
8828.931	8831.355			0.5			11323.29	4c 1	2a 6	Q 1
8829.025	8831.449		0.3	0.2			11323.17	4b 1	2a 5	P 1
8829.360	8831.784		0.2				11322.74	W 2	2C 3	P 2
8829.360	8831.784		0.2				11322.74	4g 0	2a 5	P 4
8829.773	8832.198		0.2				11322.21	t 1	2a 4	R 1
8829.976	8832.401		0.7	0.7			11321.95	3E 1	2B11	Q 2
8830.389	8832.814		0.2				11321.42			
8830.631	8833.056		0.2				11321.11			
8831.248	8833.672		0.2d				11320.32			

λ_{air}	λ_{vac}	Ref.	I_1	I_2	I_3	I_4	ν	Upper	Lower	Br.
8831.482	8833.907		0.3	0.1			11320.02			
8831.723	8834.149		2.4	1.8			11319.71	2K 5	2B 4	P 3
8831.880	8834.305		0.2				11319.51			
8832.106	8834.531		0.2d				11319.22			
8832.363	8834.789		0.4	0.1d			11318.89	3E 3	2B15	Q 1
8833.503	8835.928		0.1				11317.43			
8833.628	8836.053		0.3				11317.27	3C 1	2K 2	R 0
8833.807	8836.233		0.2				11317.04			
8834.127	8836.553		0.1				11316.63	W 1	2B12	R 0
8834.424	8836.849		0.2				11316.25			
8834.744	8837.170						11315.84			
8835.400	8837.826		0.2	0.1			11315.00	3D 3	2B15	R 3
8836.477	8838.904				0.1		11313.62			
8836.595	8839.021		0.2				11313.47			
8837.048	8839.474		0.1	0.0d			11312.89			
8837.219	8839.646		0.2				11312.67			
8837.454	8839.880		0.4	0.2			11312.37			
8837.860	8840.287		0.5	0.2	0.2		11311.85			
8838.009	8840.435			0.6			11311.66			
8838.102	8840.529		0.4	0.2			11311.54			
8838.243	8840.670		0.3				11311.36	W 1	2C 2	P 3
8838.243	8840.670		0.3				11311.36	3D 3	2B15	R 1
8838.657	8841.084		0.3	0.2			11310.83	3C 2	2K 4	Q 2
8839.329	8841.756		0.1				11309.97	4D 0	2C 3	P 4
8839.759	8842.186				0.1		11309.42			
8840.072	8842.499						11309.02			
8840.166	8842.593		0.2				11308.90			
8840.267	8842.695		0.3				11308.77			
8840.580	8843.007		0.2d				11308.37			
8840.768	8843.195			0.2	0.1		11308.13	3D 3	2B15	R 0
8840.893	8843.320		0.2d				11307.97			
8841.127	8843.555		2.8	2.5	0.3		11307.67	2A 1	2B 0	P 5
8841.323	8843.750		0.7				11307.42	W 2	2C 3	Q 3
8841.596	8844.024			0.0	0.1		11307.07			
8841.831	8844.259						11306.77			
8842.300	8844.728		0.2	0.1			11306.17			
8842.574	8845.002		0.3				11305.82			
8842.871	8845.299		0.2b				11305.44			
8843.200	8845.628						11305.02			
8843.622	8846.050		0.7	0.5			11304.48	3F 2	2B14	Q 4
8843.958	8846.387		0.4	0.5			11304.05			
8844.240	8846.669		0.3				11303.69	Y 2	2C 3	R 1
8844.240	8846.669		0.3				11303.69	3c 1	2a 3	P 6
8844.827	8847.256		0.2d				11302.94			
8845.132	8847.561						11302.55			
8845.297	8847.725		0.2				11302.34			
8845.469	8847.897		0.2				11302.12			
8845.782	8848.211		0.4b				11301.72	3e 3	3b 0	R 0
8845.970	8848.399		0.7b	0.1			11301.48	4f 2	3b 2	P 4
8846.040	8848.469		0.2				11301.39			
8846.275	8848.704		2.9	2.2			11301.09	2A 3	2B 3	P 1
8846.870	8849.299		0.2				11300.33			
8847.661	8850.090			0.0			11299.32			
8847.982	8850.411		2.0	1.3			11298.91			
8848.326	8850.756		0.1d				11298.47			
8848.538	8850.967		1.0	0.8			11298.20			
8849.211	8851.641			0.1			11297.34			
8849.822	8852.252		0.8	0.2			11296.56	3C 1	2K 2	Q 1
8849.963	8852.393		0.2				11296.38			
8850.112	8852.542		0.7	0.2			11296.19			

λ_{air}	λ_{vac}	Ref.	I_1	I_2	I_3	I_4	ν	Upper	Lower	Br.
8850.332	8852.762		0.1				11295.91			
8850.770	8853.201		0.2				11295.35			
8851.099	8853.530		0.2	0.1			11294.93			
8851.342	8853.773						11294.62			
8851.836	8854.267		0.2b				11293.99	3C 2	2K 4	P 4
8852.103	8854.533		0.4	0.2			11293.65			
8852.228	8854.659						11293.49			
8852.338	8854.768		0.3	0.2			11293.35	3E 0	2B 9	R 0
8852.902	8855.333						11292.63	3E 1	2B11	P 2
8853.012	8855.443		0.1				11292.49			
8853.302	8855.733				0.1		11292.12	Y 2	2B14	R 2
8853.623	8856.054		0.2d				11291.71			
8854.188	8856.619		0.4				11290.99			
8854.949	8857.380		0.1				11290.02			
8855.192	8857.623		0.3				11289.71			
8855.662	8858.094			0.0			11289.11			
8856.031	8858.463		0.3b	0.2			11288.64			
8856.902	8859.334		0.2	0.0			11287.53			
8857.012	8859.444				0.1		11287.39	3F 3	2B16	Q 2
8857.012	8859.444				0.1		11287.39	3E 1	2B11	Q 1
8857.553	8859.986		0.2				11286.70			
8857.726	8860.158		0.1				11286.48			
8858.464	8860.896		0.7	0.2			11285.54	3F 1	2C 2	R 3
8858.464	8860.896		0.7	0.2			11285.54	3C 2	2K 4	R 0
8858.644	8861.077		0.2				11285.31			
8858.817	8861.250		0.8	0.2			11285.09			
8859.225	8861.658				0.1		11284.57	4e 2	3b 2	R 2
8860.128	8862.561		1.5	0.3			11283.42			
8861.518	8863.952		0.1				11281.65			
8862.343	8864.777		0.2d				11280.60			
8862.563	8864.997		0.8	0.2			11280.32			
8862.791	8865.225		0.6	0.2			11280.03	4b 0	2a 4	R 0
8863.027	8865.460		0.2				11279.73			
8863.200	8865.633						11279.51			
8863.797	8866.231		0.3	0.1			11278.75	Y 2	2B14	R 1
8864.268	8866.702		0.3	0.2			11278.15	3D 1	2B11	R 5
8864.268	8866.702		0.3	0.2			11278.15	3C 1	2A 2	P 4
8864.544	8866.978		0.1				11277.80			
8865.070	8867.504		0.2				11277.13	3f 0	2c 3	R 4
8865.518	8867.953		0.1				11276.56			
8865.801	8868.236		0.4	0.2			11276.20			
8866.084	8868.519		0.4				11275.84			
8866.195	8868.629		0.4	0.3			11275.70	3b 5	21 4	R 2
8866.942	8869.376		0.1				11274.75	3D 2	2B13	P 2
8866.942	8869.376		0.1				11274.75	3F 0	2C 1	R 3
8867.177	8869.612		0.9	0.2			11274.45	3C 1	2A 2	P 3
8867.177	8869.612		0.9	0.2			11274.45	4D 0	2B15	R 6
8867.177	8869.612		0.9	0.2			11274.45	4b 1	2a 5	P 2
8867.358	8869.793			0.1			11274.22	3C 1	2A 2	P 3
8867.760	8870.194			0.1			11273.71			
8867.885	8870.320		0.1				11273.55	u 2	2a 5	R 2
8868.043	8870.478		1.0	0.4	0.2		11273.35	3b 5	2a 4	R 1
8868.546	8870.981		0.5	0.1			11272.71			
8868.759	8871.194		0.1				11272.44	Z 2	2B13	R 0
8868.924	8871.359			0.1			11272.23			
8869.073	8871.509						11272.04	V 1	2B 9	P 3
8869.726	8872.162		0.1				11271.21			
8869.970	8872.406		0.2				11270.90			
8870.183	8872.618		0.3	0.2			11270.63	3C 1	2A 2	P 5
8870.183	8872.618		0.3	0.2			11270.63	4f 0	3b 0	P 6

λ_{air}	λ_{vac}	Ref.	I_1	I_2	I_3	I_4	ν	Upper	Lower	Br.
8870.584	8873.020		0.1	0.1			11270.12			
8870.852	8873.288		0.2	0.1			11269.78			
8871.308	8873.744		0.1				11269.20			
8871.623	8874.059		0.5	0.0			11268.80	w 0	3b 1	P 2
8872.009	8874.445		0.9	0.3			11268.31	3F 1	2B12	Q 5
8872.245	8874.682		0.2				11268.01			
8872.458	8874.894		0.6				11267.74			
8872.655	8875.091			0.0			11267.49			
8873.064	8875.501		0.3	0.2			11266.97	3F 1	2C 2	R 3
8873.064	8875.501		0.3	0.2			11266.97	3F 2	2C 3	R 3
8873.064	8875.501		0.3	0.2			11266.97	W 1	2C 2	Q 5
8873.442	8875.879		0.2				11266.49			
8873.852	8876.288				0.1		11265.97	4e 0	3b 0	R 5
8874.096	8876.533		0.2	0.2			11265.66			
8874.340	8876.777						11265.35	Y 2	2B14	R 0
8874.340	8876.777						11265.35	Z 2	2B13	R 1
8874.577	8877.013			0.2			11265.05	3E 0	2B 9	Q 3
8874.844	8877.281		0.2b				11264.71			
8875.018	8877.455			0.2			11264.49			
8875.325	8877.762	G	2.4	1.5b	0.2		11264.10	2A 3	2B 3	P 2
8875.672	8878.109			0.1	0.1		11263.66	Y 2	2C 3	Q 1
8875.672	8878.109			0.1	0.1		11263.66	3C 0	2A 1	P 2
8875.948	8878.385		0.9d		0.8		11263.31	3b 0	2a 0	P 4
8875.948	8878.385		0.9d		0.8		11263.31	4F 0	2B15	P 3
8876.082	8878.519	G	3.7	1.5b	2.4		11263.14	2A 1	2B 0	P 4
8876.160	8878.598		1.5b		0.7		11263.04			
8876.515	8878.952		0.2	0.1			11262.59			
8876.688	8879.126		0.2	0.2	0.1		11262.37			
8876.909	8879.346		0.2	0.1			11262.09			
8877.114	8879.552		0.1				11261.83			
8877.414	8879.851		0.2	0.2d	0.1		11261.45			
8877.713	8880.151		2.2	1.5b	2.1		11261.07	3b 1	2a 1	R 3
8878.178	8880.616		0.2	0.2			11260.48			
8878.604	8881.042		0.3	0.3			11259.94			
8878.817	8881.255		0.2				11259.67	Z 2	2B13	R 2
8879.014	8881.452		0.4	0.0			11259.42	3E 3	2B15	Q 2
8879.866	8882.304		0.5				11258.34			
8880.844	8883.282		0.7				11257.10			
8880.970	8883.409			0.0			11256.94			
8881.199	8883.638		0.9	0.2			11256.65	3C 2	2K 4	Q 1
8881.404	8883.843		0.2				11256.39			
8881.799	8884.237		0.2	0.1			11255.89			
8882.067	8884.506		0.3	0.7	0.4		11255.55	3b 1	2a 1	R 4
8882.241	8884.679		2.6	0.9			11255.33	2K 5	2B 4	P 4
8882.643	8885.082		0.4	0.0			11254.82	Y 2	2C 3	P 1
8882.643	8885.082		0.4	0.0			11254.82	4f 1	3b 1	Q 4
8883.133	8885.572		0.2				11254.20			
8883.504	8885.943						11253.73			
8883.590	8886.029		0.2				11253.62			
8883.796	8886.235			0.1			11253.36			
8884.072	8886.511		0.7d				11253.01	3C 1	2K 2	P 2
8884.340	8886.780		1.0d		0.4		11252.67	3b 1	2a 1	R 2
8884.340	8886.780		1.0d		0.4		11252.67	Y 1	2B10	P 4
8884.538	8886.977	G	2.7	1.5b	0.2		11252.42	2A 1	2B 0	P 1
8884.538	8886.977	G	2.7	1.5b	0.2		11252.42	Z 2	2B13	R 3
8884.980	8887.419		0.3	0.2	0.2		11251.86	W 1	2B12	P 1
8885.114	8887.554		0.4	0.2			11251.69	3b 5	2a 4	R 0
8885.612	8888.051		0.2				11251.06			
8885.936	8888.375		0.2	0.1			11250.65	W 1	2B12	P 6
8886.196	8888.636		2.7	1.5b	0.4		11250.32	3B 1	2A 0	P 1

λ_{air}	λ_{vac}	Ref.	I_1	I_2	I_3	I_4	ν	Upper	Lower	Br.
8886.504	8888.944		0.1	0.1d			11249.93			
8886.836	8889.276		0.1				11249.51	3D 2	2B13	P 3
8887.073	8889.513		0.2	0.0			11249.21	4e 1	3b 1	R 3
8887.673	8890.114		0.3				11248.45			
8887.887	8890.327			0.1			11248.18	W 1	2B12	P 5
8888.053	8890.493		0.2				11247.97			
8888.250	8890.691		0.7				11247.72			
8888.558	8890.999		0.2				11247.33	4c 0	2a 5	R 3
8888.867	8891.307		0.5	0.2	0.1		11246.94			
8889.096	8891.536		0.3	0.1			11246.65	3D 0	2B 9	R 5
8889.444	8891.884		0.2	0.2			11246.21			
8889.926	8892.367		0.1	0.2			11245.60	3D 2	2B13	P 6
8890.266	8892.707		0.2				11245.17			
8891.135	8893.577		0.1				11244.07			
8891.373	8893.814		0.4	0.2			11243.77			
8891.974	8894.415		0.2				11243.01			
8892.132	8894.573		0.2				11242.81			
8892.274	8894.716		0.3	0.2			11242.63			
8892.899	8895.341			0.1			11241.84			
8893.050	8895.491		0.3				11241.65			
8893.160	8895.602			0.2			11241.51			
8893.619	8896.061		2.2	0.5			11240.93	3C 2	2K 4	P 3
8893.817	8896.259		0.1				11240.68			
8894.023	8896.465		0.3	0.2			11240.42	3A 2	2B15	R 4
8894.379	8896.821			0.1			11239.97			
8894.687	8897.130		0.2	0.2			11239.58			
8895.091	8897.533		0.7	0.1			11239.07			
8895.495	8897.937			0.2			11238.56	Z 2	2B13	R 4
8896.001	8898.444				0.1		11237.92	3f 0	2c 3	R 3
8896.247	8898.689		0.3	0.2			11237.61	W 1	2B12	P 2
8896.445	8898.887		0.2	0.3	0.2		11237.36	3b 0	2a 0	R12
8896.761	8899.204	G	3.5	1.5b	0.3		11236.96	2A 1	2B 0	P 3
8896.927	8899.370		0.3				11236.75			
8897.165	8899.608		0.5	0.0			11236.45			
8897.300	8899.743		0.3	0.8	1.9		11236.28	3b 1	2a 1	R 5
8897.300	8899.743		0.3	0.8	1.9		11236.28	3F 2	2B14	R 2
8897.529	8899.972		0.2				11235.99			
8897.712	8900.154			0.0	0.1		11235.76			
8898.068	8900.511			0.4	0.1		11235.31	2A 2	2B 2	R 4
8898.195	8900.638		0.2				11235.15	3E 3	2B15	P 2
8898.456	8900.899		0.2	0.3	0.1		11234.82			
8898.797	8901.240	G	4.4	1.5b	0.5		11234.39	2A 1	2B 0	P 2
8899.177	8901.620		0.2b	0.1			11233.91			
8899.351	8901.794		0.4	0.3			11233.69			
8899.533	8901.977		0.2	0.2			11233.46			
8900.088	8902.531		0.1				11232.76			
8900.452	8902.896		0.2				11232.30	Y 2	2C 3	R 2
8900.872	8903.316		0.2				11231.77			
8901.245	8903.689			0.2			11231.30			
8901.681	8904.125		0.2	0.3			11230.75	3f 1	2c 4	R 3
8902.014	8904.458	G	3.3	1.5b	1.8		11230.33	3b 1	2a 1	R 1
8902.394	8904.838		0.2		0.1		11229.85			
8902.521	8904.965		0.7	0.2			11229.69	Y 2	2B14	P 1
8903.298	8905.742			0.1			11228.71	4b 1	2a 5	P 3
8903.544	8905.988		0.3				11228.40			
8903.837	8906.282		0.5	0.3			11228.03			
8904.020	8906.464		0.2				11227.80			
8904.186	8906.631		0.8	0.2			11227.59			
8904.805	8907.250		0.1d				11226.81			
8905.114	8907.559		0.2d	0.2			11226.42	4f 2	3b 2	Q 2

λ_{air}	λ_{vac}	Ref.	I_1	I_2	I_3	I_4	ν	Upper	Lower	Br.
8905.281	8907.726						11226.21			
8905.455	8907.900		0.2	0.2			11225.99			
8905.709	8908.154		0.5				11225.67			
8905.995	8908.440		0.2				11225.31			
8906.233	8908.678		0.3				11225.01			
8906.455	8908.900			0.1			11224.73			
8906.677	8909.122		0.8				11224.45	3E 0	2B 9	Q 2
8906.828	8909.273		2.4	0.4			11224.26	4f 2	3b 2	P 3
8907.145	8909.591		0.8	0.3	0.2		11223.86	W 1	2B12	P 3
8907.526	8909.972		0.2				11223.38	4f 0	3b 0	R 1
8907.661	8910.107		0.2	0.2			11223.21			
8907.963	8910.408		0.2				11222.83			
8908.288	8910.734		2.6	0.9			11222.42	2A 3	2B 3	P 3
8909.471	8911.917						11220.93	u 2	2a 5	R 1
8909.820	8912.267			0.1			11220.49	3D 2	2B13	P 4
8909.916	8912.362		0.3				11220.37			
8910.591	8913.037		0.8	0.2			11219.52			
8910.877	8913.323			0.2			11219.16	W 1	2B12	P 4
8911.623	8914.070						11218.22			
8911.965	8914.412		0.2d				11217.79			
8912.291	8914.738		0.2				11217.38			
8912.458	8914.904		0.8	0.3			11217.17			
8913.363	8915.811		0.3				11216.03			
8914.548	8916.995		0.6	0.2			11214.54	3D 3	2B15	R 2
8915.716	8918.164		0.2	0.2			11213.07			
8916.345	8918.793				0.1		11212.28			
8916.551	8918.999			0.1			11212.02	3E 0	2B 9	P 2
8916.774	8919.222				0.1		11211.74			
8916.933	8919.381			0.1			11211.54			
8917.538	8919.986		0.4	0.2			11210.78	4e 3	3b 3	Q 2
8917.538	8919.986		0.4	0.2			11210.78	W 2	2C 3	Q 4
8917.880	8920.328		0.2b	0.0			11210.35	3C 0	2A 1	P 3
8918.246	8920.694						11209.89			
8918.580	8921.028		0.1				11209.47			
8918.826	8921.275		0.2				11209.16	3C 2	2K 4	P 2
8919.359	8921.808						11208.49			
8919.471	8921.920		0.2				11208.35	3D 1	2B11	R 4
8919.662	8922.111						11208.11			
8919.845	8922.294		0.1				11207.88	Y 2	2B14	P 2
8920.092	8922.541		0.2				11207.57	4e 3	3b 3	R 0
8920.394	8922.843		0.3				11207.19	3E 0	2B 9	Q 1
8921.373	8923.823						11205.96	4f 2	3b 2	Q 3
8921.970	8924.420						11205.21			
8922.599	8925.049						11204.42			
8923.173	8925.623		0.4	0.3	0.3		11203.70	3b 1	2a 1	R 6
8923.444	8925.894		0.2				11203.36	3D 2	2B13	P 5
8923.611	8926.061		0.1				11203.15			
8923.818	8926.268		2.2	0.3			11202.89	V 0	2B 9	P 2
8925.451	8927.902		0.4	0.2			11200.84	3D 3	2B15	P 3
8926.408	8928.858		0.2				11199.64	3F 0	2C 1	R 2
8926.734	8929.185						11199.23	4e 2	3b 2	R 1
8927.157	8929.608		0.3	0.1			11198.70			
8927.667	8930.118		0.9	0.2			11198.06	3F 1	2B12	R 3
8928.807	8931.259		0.2	0.2			11196.63			
8930.083	8932.535			0.2			11195.03			
8930.538	8932.990			0.2			11194.46			
8930.881	8933.333		2.7	1.5	0.3		11194.03	3b 1	2a 1	R 0
8931.456	8933.908		0.2				11193.31	Y 2	2C 3	Q 2
8932.030	8934.483		0.3				11192.59			
8932.453	8934.906		0.1				11192.06	3F 2	2C 3	R 2

λ_{air}	λ_{vac}	Ref.	I_1	I_2	I_3	I_4	ν	Upper	Lower	Br.
8932.453	8934.906		0.1				11192.06	4e 0	3b 0	R 4
8933.403	8935.856		0.1				11190.87			
8934.353	8936.806						11189.68			
8934.569	8937.022		0.2				11189.41			
8934.712	8937.165						11189.23			
8936.174	8938.627						11187.40			
8937.348	8939.802						11185.93			
8938.099	8940.553		0.2	0.2			11184.99	3F 1	2C 2	R 2
8938.099	8940.553		0.2	0.2			11184.99	4b 1	2c 5	P 4
8938.947	8941.401		0.2	0.1			11183.93			
8939.330	8941.784		0.1				11183.45			
8939.586	8942.040		0.4	0.3			11183.13	3B 4	2K 3	R 3
8940.146	8942.600		0.6	0.3			11182.43	2K 5	2B 4	P 5
8940.322	8942.776		0.4	0.1			11182.21			
8940.513	8942.968		0.1				11181.97			
8940.697	8943.152		0.2	0.1d			11181.74			
8941.025	8943.480		0.1				11181.33			
8941.409	8943.864		0.2	0.2d			11180.85	4f 1	3b 1	Q 3
8941.625	8944.080		0.2				11180.58	u 2	2a 5	R 0
8941.881	8944.336		0.5	0.0d			11180.26			
8942.201	8944.656						11179.86			
8942.457	8944.912		0.1				11179.54			
8944.001	8946.456		0.1				11177.61	3F 1	2C 2	R 2
8944.001	8946.456		0.1				11177.61	3A 0	2B10	P 4
8944.257	8946.712		0.2b				11177.29			
8944.369	8946.825	G					11177.15			
8944.721	8947.177		0.2	0.1			11176.71			
8945.009	8947.465		0.2				11176.35	3F 2	2C 3	R 2
8945.161	8947.617		0.9	0.3			11176.16	4b 0	2a 4	P 1
8945.546	8948.001			0.1	0.1		11175.68	4E 0	2B15	R 2
8945.833	8948.290			0.2	0.2		11175.32			
8946.194	8948.650		2.4	0.9	0.3		11174.87	3B 1	2A 0	P 2
8946.610	8949.066			0.1			11174.35			
8946.994	8949.451		0.3				11173.87			
8947.243	8949.699		0.1				11173.56			
8947.587	8950.043		0.3	0.2			11173.13			
8947.763	8950.220		1.5	0.2			11172.91	t 1	2a 4	R 3
8947.979	8950.436		2.8	0.9	0.2		11172.64	2A 3	2B 3	P 4
8948.636	8951.093		0.1				11171.82	3F 0	2C 1	R 2
8948.636	8951.093		0.1				11171.82	3D 1	2B11	R 3
8949.149	8951.606			0.0			11171.18			
8949.397	8951.854		0.4	0.2			11170.87	3D 1	2B11	R 0
8949.397	8951.854		0.4	0.2			11170.87	4c 0	2a 5	R 2
8950.519	8952.976				0.1		11169.47	Y 2	2B14	P 3
8950.671	8953.129		0.2				11169.28			
8951.016	8953.473		0.1				11168.85			
8951.208	8953.666		0.1				11168.61	3F 2	2B14	Q 3
8951.425	8953.882		0.2				11168.34	W 1	2C 2	P 4
8953.725	8956.184		0.3	0.1			11165.47			
8954.223	8956.681		0.2				11164.85	3F 0	2C 1	Q 5
8954.471	8956.930			0.1			11164.54			
8954.656	8957.114		0.7	0.2			11164.31			
8954.904	8957.363		0.1				11164.00			
8956.653	8959.112		0.2d				11161.82	3D 1	2B11	R 1
8956.653	8959.112		0.2d				11161.82	3C 0	2A 1	P 4
8956.990	8959.449		0.3	0.1			11161.40			
8957.215	8959.674		0.2d	0.1			11161.12			
8957.954	8960.413		0.1				11160.20	3D 3	2B15	P 4
8957.954	8960.413		0.1				11160.20	3D 1	2B11	P 1
8959.591	8962.051		0.5	0.3	1.6		11158.16			

λ_{air}	λ_{vac}	Ref.	I_1	I_2	I_3	I_4	ν	Upper	Lower	Br.
8959.760	8962.220		0.9	0.3			11157.95			
8960.081	8962.541				0.1		11157.55			
8960.258	8962.718		0.2				11157.33			
8960.740	8963.200			0.0			11156.73			
8961.013	8963.473		0.8	0.2			11156.39	3b 1	2a 1	R 7
8961.334	8963.794			0.1			11155.99			
8961.615	8964.076		0.6	0.2			11155.64	3b 5	2a 4	P 1
8962.089	8964.550		0.2				11155.05			
8962.242	8964.702		0.2	0.2			11154.86			
8962.844	8965.305		0.3				11154.11			
8962.973	8965.434						11153.95			
8963.423	8965.884		0.1				11153.39	3F 3	2C 4	R 1
8963.423	8965.884		0.1				11153.39	3f 0	2c 3	R 3
8964.291	8966.752						11152.31	X 2	2B15	P 1
8964.870	8967.331			0.2			11151.59			
8965.288	8967.749						11151.07	4E 0	2B15	R 0
8965.288	8967.749						11151.07	u 1	2a 4	R 3
8966.068	8968.529						11150.10			
8966.333	8968.795			0.2			11149.77			
8966.558	8969.020			0.1			11149.49			
8966.792	8969.253			0.2			11149.20			
8967.138	8969.599		2.7	1.5b	3.0		11148.77	3b 0	2a 0	P 5
8967.612	8970.074			0.2	0.1		11148.18	Y 2	2C 3	P 2
8967.612	8970.074			0.2	0.1		11148.18	3f 0	2c 3	Q 5
8968.384	8970.846			0.1			11147.22			
8969.044	8971.506		0.6	0.3	0.1d		11146.40	2A 2	2B 2	R 3
8969.438	8971.901			0.1			11145.91	4f 0	3b 0	Q 4
8969.438	8971.901			0.1			11145.91	4f 0	3b 0	P 5
8970.895	8973.358		0.3	0.1			11144.10			
8971.040	8973.503						11143.92			
8971.153	8973.616		0.2	0.1			11143.78			
8971.861	8974.324		0.1				11142.90			
8974.482	8976.946		0.2				11139.65	3D 3	2B15	P 6
8975.140	8977.604		0.1				11138.83	4e 3	3b 3	P 6
8975.599	8978.063		0.2				11138.26			
8975.696	8978.160						11138.14			
8976.058	8978.522			0.1			11137.69			
8977.001	8979.466		0.2				11136.52			
8977.267	8979.732		0.3	0.1			11136.19			
8977.364	8979.829						11136.07			
8977.461	8979.925		0.9	0.2			11135.95			
8977.711	8980.175		0.1				11135.64			
8978.356	8980.821		0.2b				11134.84			
8978.533	8980.998			0.1			11134.62			
8978.888	8981.353		2.7	0.6	0.2		11134.18		2B 1	P 2
8979.944	8982.410				0.1		11132.87			
8980.993	8983.459						11131.57			
8981.429	8983.895		0.2	0.2			11131.03			
8982.623	8985.089		0.1				11129.55	4F 0	2B15	Q 2
8983.438	8985.905				0.1		11128.54			
8984.415	8986.882		0.2				11127.33			
8984.714	8987.181		0.4	0.2	0.1		11126.96	3F 2	2C 3	R 1
8984.892	8987.358		0.1				11126.74			
8985.158	8987.625		0.5	0.3	0.4		11126.41	3b 0	2a 0	R13
8985.586	8988.053			0.0			11125.88			
8985.901	8988.368		0.2d	0.1			11125.49			
8986.241	8988.707		0.7	0.3			11125.07			
8986.475	8988.942		0.1				11124.78			
8986.669	8989.136		0.2	0.2			11124.54			
8987.145	8989.612						11123.95			

λ_{air}	λ_{vac}	Ref.	I_1	I_2	I_3	I_4	ν	Upper	Lower	Br.
8987.784	8990.251		0.2	0.2			11123.16			
8988.373	8990.841			0.0			11122.43	3D 3	2B15	P 5
8989.513	8991.981		0.6	0.3			11121.02			
8990.103	8992.571		0.2				11120.29			
8990.491	8992.959						11119.81			
8990.799	8993.267		0.1				11119.43			
8991.033	8993.501		0.6	0.3			11119.14	3F 2	2C 3	R 1
8991.033	8993.501		0.6	0.3			11119.14	4e 0	3b 0	R 3
8991.154	8993.623		0.2				11118.99	4E 0	2B15	Q 4
8991.316	8993.784		0.5	0.3			11118.79	3C 0	2A 1	P 5
8991.316	8993.784		0.5	0.3			11118.79	4e 2	3b 2	R 0
8991.623	8994.092			0.0			11118.41			
8991.923	8994.391		0.4	0.2			11118.04	4b 0	2a 4	P 2
8992.359	8994.828		0.1				11117.50			
8992.723	8995.192				0.1		11117.05			
8994.212	8996.681			0.1			11115.21			
8994.681	8997.151		0.4	0.2			11114.63			
8994.965	8997.434						11114.28			
8995.442	8997.911		0.2				11113.69			
8995.717	8998.187		0.1	0.1			11113.35			
8995.944	8998.414			0.2			11113.07			
8997.223	8999.693						11111.49	w 0	3b 1	P 3
8997.920	9000.390		0.1				11110.63			
8998.284	9000.754			0.0			11110.18	2A 3	2B 3	P 5
8998.519	9000.989		0.3	0.2			11109.89			
8998.794	9001.265		0.4	0.3			11109.55	Y 2	2C 3	Q 3
9000.293	9002.764		0.2				11107.70			
9000.471	9002.942		0.4	0.2			11107.48			
9000.998	9003.469			0.1			11106.83			
9001.655	9004.126						11106.02	t 1	2a 4	P 1
9001.963	9004.434		0.1				11105.64	3A 0	2B10	P 3
9003.268	9005.739		0.4	0.3	0.1		11104.03	2K 5	2B 4	P 6
9003.730	9006.202		0.3	0.3	0.1		11103.46			
9004.606	9007.078		0.2		0.1		11102.38	4f 0	3b 1	Q 2
9004.719	9007.191			0.1			11102.24			
9004.979	9007.451		0.3d	0.2	0.2		11101.92	3F 2	2B14	R 1
9005.247	9007.719			0.0			11101.59			
9005.660	9008.133		0.1				11101.08			
9006.407	9008.879		0.2	0.2	0.3		11100.16	4c 0	2a 5	R 1
9007.632	9010.105		0.2	0.3			11098.65			
9007.835	9010.308		0.2				11098.40			
9008.411	9010.884			0.3			11097.69	4D 0	2B15	R 5
9009.418	9011.891			0.3			11096.45			
9010.230	9012.703			0.4			11095.45			
9011.359	9013.833						11094.06			
9011.562	9014.036		0.2				11093.81	u 1	2a 4	R 2
9011.700	9014.174		0.4				11093.64			
9011.814	9014.288		0.2	0.4			11093.50			
9011.936	9014.410		0.2				11093.35			
9012.423	9014.897		0.2				11092.75			
9012.545	9015.019		0.5	0.5			11092.60	3F 1	2C 2	R 1
9013.268	9015.742						11091.71	4e 3	3b 3	P 4
9013.390	9015.864		0.2	0.2			11091.56			
9014.373	9016.848		0.4	0.4			11090.35	3F 1	2C 2	R 1
9014.373	9016.848		0.4	0.4			11090.35	4e 1	3b 1	R 1
9014.967	9017.441		0.4	0.6	0.1		11089.62	3b 5	2a 4	P 2
9015.568	9018.043						11088.88			
9015.877	9018.352			0.2			11088.50			
9016.007	9018.482		0.1				11088.34			
9016.243	9018.718		2.8	0.9	0.6		11088.05	3B 1	2A 0	P 3

λ_{air}	λ_{vac}	Ref.	I_1	I_2	I_3	I_4	ν	Upper	Lower	Br.
9016.243	9018.718		2.8	0.9	0.6		11088.05	W 2	2C 3	P 4
9016.878	9019.353		0.8	0.5			11087.27	3F 0	2C 1	Q 5
9017.081	9019.556		0.1				11087.02			
9017.374	9019.849		0.1				11086.66			
9017.561	9020.036		2.9	0.8	0.2		11086.43	2A 2	2B 2	R 2
9017.561	9020.036		2.9	0.8	0.2		11086.43	3A 2	2B15	R 3
9018.529	9021.004		0.2	0.3			11085.24			
9018.846	9021.322		0.1				11084.85			
9019.448	9021.924			0.2			11084.11			
9020.978	9023.455		0.6				11082.23			
9021.117	9023.593						11082.06			
9022.004	9024.481		0.2b	0.2	0.1		11080.97			
9022.330	9024.806		3.3	1.5b	0.5		11080.57	3b 1	2a 1	P 1
9023.559	9026.036						11079.06	3F 0	2C 1	R 1
9024.561	9027.039		0.1	0.3			11077.83			
9026.077	9028.555		0.2				11075.97	4e 3	3b 3	P 2
9027.568	9030.047			0.4			11074.14			
9028.359	9030.838		0.1				11073.17	3F 2	2B14	P 4
9029.101	9031.580		0.3	0.6			11072.26	3A 0	2B10	P 2
9030.985	9033.464		0.2				11069.95			
9031.271	9033.750		0.2	0.0d			11069.60	3F 0	2C 1	R 1
9031.557	9034.036		0.1				11069.25	3B 4	2K 3	R 2
9031.728	9034.207		0.2	0.8	0.1		11069.04	4f 0	3b 0	Q 3
9032.462	9034.942			0.3			11068.14			
9032.813	9035.293			0.3			11067.71	3F 1	2C 2	Q 4
9034.013	9036.493			0.3d			11066.24			
9035.074	9037.555			0.0			11064.94			
9035.499	9037.979		2.7	0.6	0.1		11064.42	2K 2	2B 2	R 1
9035.744	9038.224		0.2				11064.12	3D 0	2B 9	R 4
9035.744	9038.224		0.2				11064.12	4b 0	2a 4	P 3
9036.659	9039.139		0.1	0.3			11063.00	3A 0	2B10	P 1
9036.659	9039.139		0.1	0.3			11063.00	3F 2	2B14	Q 2
9037.402	9039.883		0.1				11062.09			
9038.440	9040.921			0.3			11060.82	t 0	2a 3	R 0
9040.173	9042.654			0.3			11058.70			
9041.031	9043.513			0.4			11057.65			
9041.833	9044.314			0.4			11056.67			
9042.937	9045.419		0.3	0.4			11055.32			
9043.403	9045.885		0.2				11054.75			
9043.575	9046.057		3.0	0.7	0.2		11054.54	2K 2	2B 2	R 0
9044.311	9046.794		0.5	0.4			11053.64	t 1	2a 4	P 3
9045.416	9047.899			0.0d			11052.29			
9046.775	9049.258				0.1		11050.63			
9047.135	9049.618		0.1	0.5			11050.19	t 0	2a 3	P 1
9047.135	9049.618		0.1	0.5			11050.19	u 1	2a 4	R 1
9047.839	9050.323			0.2			11049.33	4e 0	3b 0	R 2
9048.101	9050.585		0.2	0.0d			11049.01			
9048.560	9051.043			0.1			11048.45			
9048.912	9051.396		0.2				11048.02			
9049.166	9051.650		0.4	0.9			11047.71		2B 1	P 2
9049.166	9051.650		0.4	0.9			11047.71	3F 2	2C 3	Q 4
9049.878	9052.362		0.2				11046.84			
9050.239	9052.723			0.6			11046.40			
9052.345	9054.830		0.2b	0.5			11043.83			
9053.271	9055.756			0.2			11042.70			
9054.042	9056.527			0.4			11041.76	3F 0	2C 1	Q 4
9055.863	9058.349		0.3	1.0	0.1		11039.54	3B 4	2K 3	P 5
9056.150	9058.636		0.1				11039.19			
9056.667	9059.153		0.1	0.8	0.2		11038.56			
9056.979	9059.464		0.2				11038.18	3D 1	2B11	P 3

λ_{air}	λ_{vac}	Ref.	I_1	I_2	I_3	I_4	ν	Upper	Lower	Br.
9057.799	9060.285			0.2d			11037.18	4E 0	2B15	Q 3
9058.349	9060.835			0.3			11036.51	4f 0	3b 0	P 4
9059.211	9061.697		0.2	0.4			11035.46	3F 2	2C 3	Q 3
9059.605	9062.092		0.1				11034.98			
9060.089	9062.576		0.1				11034.39	4c 0	2a 5	R 0
9060.393	9062.880			0.2			11034.02	3f 0	2c 3	Q 4
9061.214	9063.702		0.8	1.5	0.1d		11033.02	3f 0	2c 3	R 2
9063.514	9066.002		0.2	1.5	0.3		11030.22			
9064.722	9067.211		0.6	1.5			11028.75	2A 4	2B 5	R 2
9065.150	9067.638		0.1				11028.23			
9065.429	9067.918		0.3	1.5	0.1		11027.89	u 2	2a 5	P 2
9066.712	9069.201			0.2d			11026.33	4E 0	2B15	Q 1
9067.172	9069.661			0.0d			11025.77			
9067.584	9070.073		0.7	0.3	0.1		11025.27	t 0	2a 3	R 1
9067.937	9070.426			0.2			11024.84	3F 1	2C 2	Q 4
9067.937	9070.426			0.2			11024.84	2A 4	2B 5	R 3
9068.324	9070.813		0.4	1.5b	0.9		11024.37	3b 0	2a 0	P 6
9068.324	9070.813		0.4	1.5b	0.9		11024.37	3F 1	2B12	R 2
9068.612	9071.101		0.2				11024.02	3F 1	2C 2	Q 3
9069.451	9071.940			0.5			11023.00	3f 0	2c 3	Q 5
9070.060	9072.549		0.4	0.8			11022.26	2A 4	2B 5	R 1
9071.393	9073.883			0.6			11020.64			
9073.130	9075.621			0.3			11018.53	3F 1	2C 2	Q 3
9074.135	9076.626			0.3			11017.31	4e 1	3b 1	R 0
9074.745	9077.235		0.3	0.3			11016.57	3F 2	2C 3	Q 2
9075.461	9077.952			0.5	0.1		11015.70			
9075.799	9078.290			0.1			11015.29			
9076.483	9078.974			0.5			11014.46	4b 0	2a 4	P 4
9076.920	9079.411		0.3	0.3			11013.93			
9078.156	9080.648		0.2				11012.43			
9078.807	9081.299			0.6			11011.64			
9079.178	9081.671			0.5			11011.19			
9079.607	9082.099			0.1			11010.67			
9080.786	9083.279		0.5	1.5			11009.24	X 2	2B15	P 2
9082.577	9085.070			0.1d			11007.07			
9082.965	9085.458		2.7	0.6	0.2		11006.60	3b 0	2a 0	R14
9082.965	9085.458		2.7	0.6	0.2		11006.60	2A 4	2B 5	R 0
9083.658	9086.151						11005.76			
9084.186	9086.680			0.0d	0.1		11005.12	t 0	2a 3	R 2
9084.450	9086.944		0.4	0.5	0.2		11004.80	3b 5	2a 4	P 3
9084.624	9087.117		0.1				11004.59			
9084.863	9087.357		2.9	1.0b	0.4		11004.30	3b 1	2a 1	P 2
9085.135	9087.629		0.1				11003.97	t 0	2a 3	R 3
9085.548	9088.042		0.3	0.5			11003.47	3F 2	2C 3	Q 2
9086.341	9088.835			0.1			11002.51			
9088.398	9090.893		0.8	1.0	0.1		11000.02	2K 2	2B 2	P 1
9089.274	9091.769						10998.96	3D 0	2B 9	R 3
9090.431	9092.926			0.3			10997.56			
9091.373	9093.869			0.2			10996.42	3E 2	2B14	R 4
9092.043	9094.539			0.0d			10995.61			
9092.605	9095.101			0.0d			10994.93			
9092.994	9095.490			0.3			10994.46	3F 0	2C 1	Q 3
9093.432	9095.928			0.3			10993.93			
9093.896	9096.392			0.1			10993.37	3F 2	2B14	P 3
9094.996	9097.492			0.1			10992.04			
9095.435	9097.931		0.2	0.0d			10991.51			
9096.063	9098.560		0.7	1.5b	0.3		10990.75	3B 1	2A 0	P 4
9096.237	9098.734		0.1				10990.54			
9096.932	9099.429			0.3			10989.70	3E 0	2C 1	R 4
9097.098	9099.595			0.0			10989.50	4f 0	3b 0	Q 2

λ_{air}	λ_{vac}	Ref.	I_1	I_2	I_3	I_4	ν	Upper	Lower	Br.
9097.437	9099.934		0.1	0.2			10989.09			
9098.174	9100.672		0.3				10988.20			
9099.110	9101.608		0.2	0.3b			10987.07			
9101.280	9103.778		0.3	0.0d			10984.45	3F 1	2C 2	Q 2
9101.488	9103.986			0.5			10984.20	4e 0	3b 0	R 1
9102.242	9104.740			0.7			10983.29	3D 1	2B11	P 4
9103.012	9105.511			0.4			10982.36			
9104.074	9106.572		0.4	0.2			10981.08			
9104.364	9106.863		0.9	0.9			10980.73	3B 4	2K 3	R 1
9104.712	9107.211			0.1	0.1		10980.31			
9105.176	9107.675		0.5	0.5			10979.75	3F 1	2C 2	Q 2
9105.359	9107.858		0.1				10979.53	3F 1	2B12	P 5
9105.583	9108.082		0.2				10979.26	4D 0	2B15	R 4
9107.026	9109.526		0.4	0.3			10977.52	4d 3	3b 3	R 0
9108.121	9110.621		1.0	0.4			10976.20			
9108.370	9110.870		0.1d	0.0d			10975.90			
9109.980	9112.481			0.2			10973.96			
9110.379	9112.879		0.5	0.3			10973.48			
9111.060	9113.560		0.3	0.7			10972.66			
9111.483	9113.984			0.1d			10972.15			
9112.247	9114.748		0.3	0.3			10971.23			
9113.211	9115.712			0.1	0.1		10970.07	3D 0	2B 9	R 2
9113.892	9116.394			0.1	0.2		10969.25			
9115.338	9117.840			0.1	0.1		10967.51	u 2	2a 5	P 3
9115.695	9118.197		0.8	1.5b	0.3		10967.08	3B 2	2A 1	R 3
9115.695	9118.197		0.8	1.5b	0.3		10967.08	4d 3	3b 3	P 1
9116.543	9119.045		0.2	0.2			10966.06	4e 2	3b 2	P 3
9117.383	9119.885			0.3			10965.05	3F 0	2C 1	Q 2
9117.483	9119.985			0.0			10964.93			
9118.822	9121.324		0.5	1.0	0.2		10963.32	3D 0	2B 9	R 0
9119.437	9121.940		0.2	1.5	0.2		10962.58	3B 2	2A 1	R 4
9119.870	9122.373		0.6	1.5	0.2		10962.06	3B 2	2A 1	R 2
9119.870	9122.373		0.6	1.5	0.2		10962.06	3D 0	2B 9	R 1
9120.244	9122.747				0.1		10961.61			
9120.519	9123.022			0.3	0.1		10961.28			
9120.918	9123.422			0.0d	0.2		10960.80	4e 2	3b 2	P 5
9121.301	9123.805			0.2			10960.34	4e 2	3b 2	P 4
9121.709	9124.212			0.1	0.1		10959.85	4f 0	3b 0	P 3
9121.959	9124.462				0.1		10959.55	4e 2	3b 2	Q 3
9122.150	9124.654			0.2	0.2		10959.32			
9122.375	9124.878				0.1		10959.05			
9122.624	9125.128				0.5		10958.75			
9122.991	9125.495			1.0b	0.1		10958.31			
9123.332	9125.836				0.3		10957.90			
9123.424	9125.928		0.2	0.2			10957.79	3B 4	2K 3	P 4
9123.482	9125.986				0.3		10957.72			
9123.682	9126.186				0.2		10957.48			
9123.998	9126.502				0.1		10957.10			
9124.290	9126.794				0.1		10956.75			
9124.806	9127.310				0.1		10956.13			
9125.056	9127.560				0.2		10955.83			
9126.089	9128.594			0.1			10954.59	3E 0	2C 1	R 3
9126.655	9129.160			0.1			10953.91			
9126.880	9129.385		0.2d				10953.64			
9127.430	9129.935			0.2			10952.98			
9128.605	9131.111		1.5	0.7			10951.57	2A 4	2B 5	P 1
9128.789	9131.294		0.1				10951.35			
9128.964	9131.469		0.2	0.8			10951.14			
9129.089	9131.594		0.2	0.0d			10950.99	3F 0	2C 1	Q 2
9130.039	9132.545						10949.85	3D 1	2B11	P 5

λ_{air}	λ_{vac}	Ref.	I_1	I_2	I_3	I_4	ν	Upper	Lower	Br.
9130.264	9132.770		0.1d	0.1			10949.58			
9130.398	9132.904			0.1	0.1		10949.42			
9130.698	9133.204		3.3	1.5b	0.3		10949.06	2K 2	2B 2	P 2
9132.074	9134.581				0.2		10947.41			
9132.283	9134.789		0.5	0.2			10947.16	3f 0	2c 3	Q 4
9132.541	9135.048			0.1	0.1		10946.85	3D 0	2B 9	P 1
9132.625	9135.131		0.1				10946.75			
9132.842	9135.348		0.3	1.0b	0.3		10946.49	3B 2	2A 1	R 5
9133.276	9135.782		0.1	0.1			10945.97	4e 3	3b 3	R 3
9134.068	9136.575						10945.02			
9134.494	9137.001		0.2	0.2			10944.51			
9135.529	9138.036		0.3	0.5			10943.27	4d 3	3b 3	R 1
9136.113	9138.621			0.1			10942.57			
9136.448	9138.955		1.5b	1.5	0.3		10942.17	3B 2	2A 1	R 1
9136.932	9139.439		0.1				10941.59	3C 2	2K 5	Q 5
9137.166	9139.673		1.5	0.8			10941.31			
9139.789	9142.297			0.1			10938.17			
9140.006	9142.514			0.2			10937.91			
9141.092	9143.601			0.1			10936.61			
9141.945	9144.454		0.2d	0.3			10935.59			
9143.090	9145.600		0.2	0.4			10934.22	3f 0	2c 3	R 1
9143.090	9145.600		0.2	0.4			10934.22	3f 1	2c 4	Q 3
9143.759	9146.269			0.2			10933.42			
9145.474	9147.984						10931.37			
9145.951	9148.461			0.3			10930.80	3F 1	2B12	Q 3
9148.797	9151.308			0.3b			10927.40			
9149.065	9151.576						10927.08			
9150.765	9153.276		0.2d	0.6			10925.05	3f 0	2c 3	Q 3
9151.661	9154.173				0.1		10923.98			
9151.912	9154.424						10923.68			
9152.281	9154.793			0.3			10923.24	4e 2	3b 2	Q 4
9152.767	9155.279		0.2				10922.66			
9153.538	9156.050						10921.74			
9153.789	9156.302			0.3			10921.44	u 2	2a 5	P 4
9153.932	9156.444						10921.27			
9154.695	9157.207		0.2				10920.36			
9154.804	9157.316			1.0	0.2d		10920.23			
9156.128	9158.641			0.1			10918.65	3f 0	2c 3	R 1
9156.548	9159.061		0.2	0.2b			10918.15			
9157.345	9159.858		0.2b	0.6			10917.20			
9157.697	9160.210			0.2			10916.78			
9158.016	9160.529		1.0d	0.5			10916.40	3B 4	2K 3	R 0
9158.167	9160.680			0.3	0.1		10916.22			
9158.519	9161.033		3.5	1.5b	0.9b		10915.80	3b 1	2a 1	P 3
9158.838	9161.352		0.1	0.1			10915.42			
9158.997	9161.511			0.2			10915.23			
9159.383	9161.897		0.4	0.4			10914.77	3F 3	2C 4	P 3
9159.761	9162.275			0.1			10914.32			
9160.004	9162.518			0.2			10914.03			
9160.332	9162.846		0.4	0.0	0.1		10913.64	U 1	2B12	R 4
9160.827	9163.341		1.5b	1.5	0.2		10913.05	2A 4	2B 5	P 2
9161.398	9163.912		0.1				10912.37			
9161.901	9164.416			0.3			10911.77			
9162.363	9164.878		0.2				10911.22			
9162.850	9165.365						10910.64	4e 3	3b 3	R 1
9162.850	9165.365						10910.64	4d 3	3b 3	R 2
9163.640	9166.155		1.0	1.5	0.2		10909.70	3B 2	2A 1	R 0
9164.379	9166.894						10908.82			
9165.328	9167.844		0.3	0.9			10907.69			
9165.791	9168.306			0.1			10907.14			

λ_{air}	λ_{vac}	Ref.	I_1	I_2	I_3	I_4	ν	Upper	Lower	Br.
9167.581	9170.097			0.2			10905.01			
9168.220	9170.736			0.2			10904.25			
9168.481	9170.997			0.0			10903.94			
9168.926	9171.443		0.2	0.0d			10903.41			
9169.397	9171.914		0.2	0.7	0.1		10902.85	3b 5	2a 4	P 4
9169.801	9172.317			0.1			10902.37			
9170.146	9172.662		0.0				10901.96			
9170.836	9173.353		0.6	0.4			10901.14			
9171.206	9173.723		0.1d				10900.70	U 1	2C 2	Q 1
9171.458	9173.975		0.3				10900.40			
9171.711	9174.228		0.4	0.3			10900.10			
9172.073	9174.590		0.9	0.9			10899.67	3B 4	2K 3	P 3
9172.493	9175.010			0.0d			10899.17			
9172.897	9175.415		0.1	0.1			10898.69	3C 2	2K 5	R 3
9172.897	9175.415		0.1	0.1			10898.69	4d 2	3b 2	R 0
9173.360	9175.878		0.2	0.9	0.2		10898.14	4E 0	2B15	Q 2
9173.739	9176.257		0.1				10897.69	U 1	2C 2	R 2
9173.739	9176.257		0.1				10897.69	3f 1	2c 4	Q 2
9174.404	9176.922			0.1			10896.90			
9175.120	9177.638		0.2	0.2			10896.05	3E 0	2C 1	R 2
9175.330	9177.848			0.0			10895.80			
9175.760	9178.278			0.0			10895.29			
9176.088	9178.606			0.2			10894.90	4D 0	2B15	R 3
9176.214	9178.733			0.7			10894.75			
9177.630	9180.148		0.1	0.4			10893.07	3A 2	2B15	R 1
9177.630	9180.148		0.1	0.4			10893.07	4d 2	3b 2	P 1
9177.630	9180.148		0.1	0.4			10893.07	4e 2	3b 2	Q 5
9178.953	9181.472			0.0			10891.50			
9179.298	9181.817		0.8	1.5b	1.0b		10891.09	3b 0	2a 0	P 7
9179.745	9182.264			0.0			10890.56			
9180.579	9183.099		1.5b	1.5	0.2d		10889.57	2K 2	2B 2	P 3
9180.579	9183.099		1.5b	1.5	0.2d		10889.57	U 1	2C 2	R 3
9181.574	9184.094		0.2				10888.39			
9181.684	9184.204			0.2			10888.26	V 1	2C 1	R 2
9182.840	9185.360			0.2	0.1		10886.89			
9183.978	9186.499			0.4			10885.54	3A 2	2B15	R 2
9184.822	9187.343						10884.54			
9185.117	9187.638				0.1		10884.19			
9185.379	9187.900		0.7	1.5b	0.6		10883.88	3B 1	2A 0	P 5
9186.282	9188.803						10882.81			
9187.067	9189.588			0.2			10881.88	3F 3	2C 4	P 4
9187.067	9189.588			0.2			10881.88	3D 0	2B 9	P 2
9187.067	9189.588			0.2			10881.88	3E 3	2B16	Q 6
9187.641	9190.163		0.4	0.8			10881.20	2K 2	2B 2	R 2
9187.641	9190.163		0.4	0.8			10881.20	4c 0	2a 5	P 2
9187.802	9190.323		0.3				10881.01			
9189.736	9192.258		0.2				10878.72			
9189.863	9192.385		0.3	0.9	0.3		10878.57	3b 0	2a 0	R15
9190.276	9192.799		1.5	0.9			10878.08			
9191.164	9193.686			0.0			10877.03	3f 0	2c 3	Q 3
9191.916	9194.438		0.3	0.3			10876.14			
9192.854	9195.377		0.3	0.4			10875.03			
9193.226	9195.749		0.2d	0.3			10874.59	4D 0	2B15	P 1
9193.911	9196.434			0.2			10873.78	3F 1	2B12	R 1
9194.630	9197.153			0.6	0.3		10872.93			
9194.841	9197.364		0.2				10872.68			
9195.289	9197.813		0.1	0.0			10872.15	Y 2	2C 3	P 4
9196.736	9199.260			0.0			10870.44			
9197.751	9200.275			0.4			10869.24			
9198.462	9200.986			0.2	0.1		10868.40			

λ_{air}	λ_{vac}	Ref.	I_1	I_2	I_3	I_4	ν	Upper	Lower	Br.
9199.080	9201.604			1.0	0.1		10867.67	4e 1	3b 1	P 2
9199.596	9202.121		0.8	1.0			10867.06	2A 4	2B 5	P 3
9199.774	9202.299		0.6	0.3			10866.85	3B 4	2K 3	P 2
9200.358	9202.883			0.3			10866.16			
9200.655	9203.179						10865.81			
9201.865	9204.391		0.3	0.9			10864.38	3D 0	2B 9	P 5
9202.077	9204.603		0.5				10864.13			
9203.077	9205.602		0.2	0.2			10862.95	U 1	2C 2	R 4
9203.348	9205.874		0.6	1.0b	0.2		10862.63	2A 2	2B 2	P 6
9203.348	9205.874		0.6	1.0b	0.2		10862.63	3F 2	2C 3	P 3
9203.788	9206.314			0.1			10862.11			
9204.009	9206.535						10861.85	3E 2	2B14	Q 6
9204.458	9206.984				0.1		10861.32	4e 3	3b 3	Q 1
9204.780	9207.306			0.3			10860.94			
9205.840	9208.366			0.1			10859.69			
9206.314	9208.841		0.9	0.9			10859.13	3B 4	2K 3	P 1
9207.230	9209.757			0.4			10858.05	4d 2	3b 2	R 1
9208.120	9210.647		0.3	1.0	0.2		10857.00			
9209.223	9211.750			0.3			10855.70	Z 0	2B 9	R 4
9211.183	9213.711		0.6	1.0			10853.39	2K 2	2B 2	R 3
9211.311	9213.838				0.1		10853.24			
9212.643	9215.172		0.3	1.0			10851.67	2A 2	2B 2	R 1
9214.214	9216.743		0.2d				10849.82			
9214.375	9216.904			0.7			10849.63			
9214.800	9217.329		0.2	0.0d			10849.13			
9215.633	9218.162			0.4			10848.15	3B 3	2K 2	R 4
9217.221	9219.751		0.2	0.8			10846.28	4c 0	2a 5	P 3
9217.221	9219.751		0.2	0.8			10846.28	4d 3	3b 3	P 2
9218.207	9220.737			0.3d			10845.12	3d 3	3b 0	R 0
9219.168	9221.698			0.2			10843.99			
9219.516	9222.047		1.5b	1.0b	0.3		10843.58	2A 2	2B 2	P 4
9219.797	9222.327		0.1				10843.25			
9221.039	9223.569			0.2			10841.79			
9221.600	9224.131			0.1			10841.13	u 0	2a 3	R 2
9223.038	9225.569			0.5			10839.44			
9224.187	9226.718				0.2		10838.09	Z 0	2B 9	R 3
9224.357	9226.888			0.1			10837.89			
9224.493	9227.024				0.1		10837.73			
9224.723	9227.254		0.5	1.0b	0.6		10837.46	2A 2	2B 2	P 5
9225.012	9227.544		0.5	0.2	0.2		10837.12			
9225.412	9227.944		0.2d	0.1	0.1		10836.65	Z 0	2B 9	R 2
9225.412	9227.944		0.2d	0.1	0.1		10836.65	3d 3	3b 0	R 1
9226.068	9228.600		0.2	0.3			10835.88			
9226.758	9229.290			0.8	0.1		10835.07			
9227.371	9229.903		0.2				10834.35			
9228.248	9230.781		0.2				10833.32			
9228.691	9231.224			0.3			10832.80			
9229.126	9231.658		0.3b	0.5			10832.29	3F 1	2B12	P 4
9230.267	9232.800			0.2			10830.95	4E 1	2B18	Q 4
9230.267	9232.800			0.2			10830.95	3A 2	2B15	P 4
9230.608	9233.141		0.2	0.2			10830.55	3f 0	2c 3	Q 2
9231.742	9234.275			0.2			10829.22	U 1	2C 2	Q 2
9231.742	9234.275			0.2			10829.22	4e 0	3b 0	Q 1
9231.742	9234.275			0.2			10829.22	3C 1	2K 3	R 3
9231.742	9234.275			0.2			10829.22	4D 0	2B15	R 1
9233.524	9236.058		0.1	0.4			10827.13	3F 1	2C 2	P 6
9234.010	9236.544			0.3			10826.56			
9234.991	9237.525		0.2	0.2			10825.41	3F 1	2C 2	P 3
9234.991	9237.525		0.2	0.2			10825.41	4e 2	3b 2	Q 6
9235.529	9238.063		0.2				10824.78	X 1	2B13	R 1

λ_{air}	λ_{vac}	Ref.	I_1	I_2	I_3	I_4	ν	Upper	Lower	Br.
9235.529	9238.063		0.2				10824.78	X 1	2B13	R 2
9235.819	9238.353		0.4	0.9			10824.44	3E 0	2C 1	R 1
9236.083	9238.618		0.1				10824.13			.
9236.595	9239.130			0.2			10823.53			
9237.150	9239.685			0.2			10822.88			
9238.379	9240.914		0.5	0.2			10821.44	V 1	2C 1	R 1
9238.379	9240.914		0.5	0.2			10821.44	4d 1	3b 1	P 1
9238.780	9241.316		0.4	0.5			10820.97			
9238.977	9241.512		0.2				10820.74	3D 0	2B 9	P 3
9238.977	9241.512		0.2				10820.74	4d 1	3b 1	R 0
9239.165	9241.700		0.4				10820.52			
9239.361	9241.896			0.9			10820.29			
9239.600	9242.136		0.1				10820.01			
9240.070	9242.605		0.2	0.3			10819.46			
9240.540	9243.075		0.2				10818.91	3F 1	2B12	Q 2
9240.540	9243.075		0.2				10818.91	3F 0	2C 1	P 5
9240.540	9243.075		0.2				10818.91	4c 0	2a 5	P 4
9240.540	9243.075		0.2				10818.91	4d 2	3b 2	R 2
9240.676	9243.212		0.3	0.6			10818.75	3F 1	2C 2	P 3
9241.283	9243.819			0.1			10818.04	X 1	2B13	R 0
9241.881	9244.417		0.3	0.3			10817.34			
9242.838	9245.374			0.1			10816.22	3F 2	2C 3	P 4
9242.838	9245.374			0.1			10816.22	V 0	2B 9	R 2
9243.034	9245.571		0.2				10815.99			
9243.214	9245.750		1.0	1.5b	0.4		10815.78	3b 1	2a 1	P 4
9243.641	9246.178			0.0d			10815.28			
9244.085	9246.622		0.2				10814.76			
9244.761	9247.298		0.2	0.4			10813.97	3f 0	2c 3	Q 2
9245.376	9247.913		0.2d				10813.25	3F 0	2B10	P 5
9245.616	9248.153		0.9	1.5			10812.97	2A 4	2B 5	P 4
9246.411	9248.948		0.1				10812.04	3F 2	2C 3	P 5
9247.309	9249.847		0.3	0.7			10810.99			
9248.592	9251.130			0.3			10809.49	T 0	2B15	P 4
9248.592	9251.130			0.3			10809.49	4e 1	3b 1	P 5
9249.328	9251.866		1.0	0.7			10808.63	3d 3	3b 0	R 2
9250.252	9252.791		0.2	0.3			10807.55			
9251.365	9253.904		0.2	0.3			10806.25			
9251.828	9254.366			0.1			10805.71			
9252.607	9255.146		0.1				10804.80			
9252.829	9255.369		1.5b	1.5b	0.3		10804.54	3B 2	2A 1	P 1
9253.163	9255.703			0.1			10804.15			
9253.489	9256.028		0.2	0.3			10803.77			
9254.226	9256.765		0.2	0.8	0.1		10802.91			
9254.577	9257.116		0.2				10802.50			
9254.757	9257.296		0.9	0.8			10802.29	2A 2	2B 2	R 0
9255.545	9258.085		0.4	0.9	0.2		10801.37	U 1	2B12	R 3
9255.545	9258.085		0.4	0.9	0.2		10801.37	4e 1	3b 1	P 6
9256.025	9258.565						10800.81			
9256.350	9258.891			0.3			10800.43	3F 1	2C 2	P 4
9257.370	9259.911			0.2			10799.24	3A 2	2B15	R 0
9258.074	9260.614			0.0			10798.42			
9258.262	9260.803			0.0			10798.20			
9258.451	9260.991			0.5			10797.98	3F 0	2C 1	P 4
9259.951	9262.492		0.1	0.4			10796.23			
9260.775	9263.316			0.4			10795.27			
9261.530	9264.071			0.1			10794.39			
9262.362	9264.904			0.3			10793.42	3C 2	2K 5	Q 4
9262.980	9265.522						10792.70			
9263.555	9266.097			0.3			10792.03			
9263.735	9266.277						10791.82			

λ_{air}	λ_{vac}	Ref.	I_1	I_2	I_3	I_4	ν	Upper	Lower	Br.
9264.439	9266.982			0.2			10791.00	u 0	2a 3	R 1
9265.470	9268.012		0.2	0.9			10789.80	2K 4	2B 4	R 4
9266.483	9269.026			0.0d			10788.62			
9266.818	9269.361		0.7	1.5b	0.5		10788.23	3b 2	2a 2	R 3
9267.617	9270.160		0.2d	0.5			10787.30	3F 1	2C 2	P 5
9267.617	9270.160		0.2d	0.5			10787.30	3F 0	2C 1	P 3
9268.115	9270.659						10786.72	3D 0	2B 9	P 4
9268.579	9271.123			0.1			10786.18			
9268.932	9271.475		0.2	1.5	0.2		10785.77			
9269.258	9271.802						10785.39			
9269.370	9271.914		0.7	0.2			10785.26			
9269.559	9272.103		0.2	1.0	0.2		10785.04			
9270.642	9273.186		0.2	0.3			10783.78			
9271.442	9273.986		0.2				10782.85			
9271.631	9274.175		0.6	1.5	0.3		10782.63	3b 2	2a 2	R 2
9271.631	9274.175		0.6	1.5	0.3		10782.63	3A 2	2B15	P 3
9271.631	9274.175		0.6	1.5	0.3		10782.63	3F 0	2C 1	P 3
9272.328	9274.872		0.1	0.4			10781.82	3F 1	2C 2	P 4
9272.328	9274.872		0.1	0.4			10781.82	3E 1	2C 2	R 3
9272.328	9274.872		0.1	0.4			10781.82	4d 2	3b 2	R 3
9273.798	9276.343			1.5	0.3		10780.11	3b 2	2a 2	R 4
9274.039	9276.584		0.2				10779.83			
9274.246	9276.791			0.0d			10779.59			
9275.020	9277.565			0.4			10778.69			
9275.502	9278.047		0.1	0.4			10778.13			
9276.036	9278.581			0.3			10777.51	V 0	2B 9	R 3
9277.181	9279.726		0.2	0.8			10776.18	2K 4	2B 4	R 3
9277.964	9280.510			0.3			10775.27			
9278.489	9281.035			0.2			10774.66			
9278.679	9281.225		0.1	0.0			10774.44			
9279.471	9282.017			0.3			10773.52	3F 0	2C 1	P 4
9279.471	9282.017			0.3			10773.52	3d 3	3b 0	P 1
9280.823	9283.370		0.1	0.3			10771.95	4d 1	3b 1	R 1
9281.435	9283.982			0.1			10771.24	T 0	2B15	P 1
9281.435	9283.982			0.1			10771.24	3F 1	2C 2	P 7
9281.435	9283.982			0.1			10771.24	3E 2	2B14	R 2
9282.245	9284.792		0.2	0.4			10770.30			
9283.090	9285.637			0.4			10769.32			
9283.831	9286.379		0.2	1.0b	0.4		10768.46	3B 1	2A 0	P 6
9284.228	9286.776		0.2		0.1		10768.00	3F 0	2C 1	P 5
9285.349	9287.897			0.2d			10766.70			
9287.109	9289.657			0.3			10764.66	3F 0	2C 1	P 6
9287.644	9290.192		0.2d	0.1			10764.04	3E 1	2C 2	R 2
9288.023	9290.572			0.0	0.1		10763.60			
9288.368	9290.917		1.5b	1.5b	0.4		10763.20	3b 2	2a 2	R 1
9288.368	9290.917		1.5b	1.5b	0.4		10763.20	Z 0	2B 9	P 1
9288.869	9291.418			0.1d			10762.62	3F 0	2C 1	P 7
9288.869	9291.418			0.1d			10762.62	4F 0	2B16	Q 4
9289.275	9291.824		0.9	1.5	0.1		10762.15	2K 4	2B 4	R 2
9290.432	9292.981		0.1	0.4			10760.81			
9291.105	9293.654		0.1				10760.03			
9291.519	9294.069			0.5	0.3		10759.55	3C 1	2K 3	Q 4
9291.787	9294.337		0.2				10759.24			
9292.029	9294.579		0.9	0.3b			10758.96			
9292.236	9294.786		0.1				10758.72	4e 0	3b 0	Q 2
9292.452	9295.002		0.2	1.5b	0.4		10758.47	3b 2	2a 2	R 5
9292.452	9295.002		0.2	1.5b	0.4		10758.47	4d 2	3b 2	P 2
9293.808	9296.359			0.2			10756.90			
9294.137	9296.687		0.1				10756.52			
9294.413	9296.964			0.1			10756.20	3E 3	2B16	Q 5

λ_{air}	λ_{vac}	Ref.	I_1	I_2	I_3	I_4	ν	Upper	Lower	Br.
9294.932	9297.482			0.3			10755.60			
9295.597	9298.148		0.1	0.6	0.2		10754.83			
9295.943	9298.494		0.3	0.1			10754.43	4F 0	2B16	R 2
9296.652	9299.203			0.0			10753.61			
9296.868	9299.419		0.1	0.2			10753.36			
9298.026	9300.578			0.0			10752.02			
9298.165	9300.716			0.1			10751.86	4E 1	2B18	Q 3
9299.869	9302.421		0.5	1.0b	0.8		10749.89	2A 4	2B 5	P 5
9299.869	9302.421		0.5	1.0b	0.8		10749.89	u 0	2a 3	R 0
9299.869	9302.421		0.5	1.0b	0.8		10749.89	3b 0	2a 0	P 8
9300.552	9303.104		0.1				10749.10	4D 0	2B15	R 2
9301.279	9303.831		0.4	0.5			10748.26			
9301.747	9304.299		0.2				10747.72			
9302.119	9304.671		0.1d				10747.29			
9302.318	9304.870		0.8	0.9			10747.06	2K 4	2B 4	R 1
9302.742	9305.295			0.0			10746.57	4d 2	3b 2	R 4
9303.097	9305.650			0.5			10746.16			
9304.049	9306.602			0.3			10745.06	3E 3	2B16	R 3
9305.348	9307.902			0.2			10743.56			
9306.007	9308.560			0.6	0.2d		10742.80	3b 0	2a 0	R16
9307.090	9309.643		0.4	0.2			10741.55	t 0	2a 3	P 3
9307.324	9309.877		0.2d	0.4d			10741.28			
9307.514	9310.068		0.1				10741.06			
9308.225	9310.779			0.1			10740.24			
9308.892	9311.446			0.3			10739.47	3E 2	2C 3	R 4
9308.892	9311.446			0.3			10739.47	3E 0	2C 1	Q 7
9309.976	9312.530			0.3			10738.22			
9310.427	9312.981		0.2	0.4			10737.70	3F 1	2B12	P 3
9310.947	9313.502		0.2				10737.10	U 1	2B12	R 2
9310.947	9313.502		0.2				10737.10	3a 1	2c 4	R 3
9312.048	9314.604			0.4			10735.83	4d 0	3b 0	R 0
9312.742	9315.298			0.4			10735.03			
9313.575	9316.131		0.1	0.1			10734.07	3E 0	2C 1	Q 4
9314.660	9317.216		1.5b	1.5	0.2		10732.82	3B 2	2A 1	P 2
9314.903	9317.459		0.2				10732.54			
9315.684	9318.240		0.1				10731.64	4d 0	3b 0	P 1
9317.177	9319.734		1.0	1.5	0.2		10729.92	3b 2	2a 2	R 0
9318.662	9321.219		1.5b	1.0	0.1		10728.21	2K 4	2B 4	R 0
9319.001	9321.558		0.2	0.5			10727.82	3E 0	2C 1	Q 3
9319.175	9321.732		0.2	0.2			10727.62			
9319.696	9322.253		0.4	0.4			10727.02	Z 0	2B 9	P 2
9320.044	9322.601		0.3	0.6			10726.62	4d 1	3b 1	R 2
9320.478	9323.036		0.1				10726.12			
9321.356	9323.914		0.2	0.4			10725.11	4d 3	3b 3	P 3
9321.660	9324.218			0.9	0.2		10724.76	3E 1	2C 2	R 1
9322.799	9325.357						10723.45	3b 2	2a 2	R 6
9324.016	9326.575		1.5b	0.9			10722.05			
9324.877	9327.436		1.0	0.6			10721.06	2A 2	2B 2	P 1
9325.103	9327.662		0.1				10720.80			
9325.634	9328.193						10720.19	4e 0	3b 0	P 3
9326.286	9328.845			0.1			10719.44			
9326.974	9329.533			0.2			10718.65	4D 0	2B15	P 3
9326.974	9329.533			0.2			10718.65	3E 1	2C 2	R 6
9328.271	9330.830			0.2d			10717.16			
9329.272	9331.831			0.5			10716.01	3E 0	2C 1	Q 5
9329.272	9331.831			0.5			10716.01	3d 3	3b 0	R 4
9331.614	9334.175		0.3	1.0	0.2		10713.32	3B 3	2K 2	R 3
9332.206	9334.767		0.1d	0.4	0.1		10712.64			
9332.903	9335.464			0.4			10711.84	3E 0	2C 1	R 7
9332.903	9335.464			0.4			10711.84	3E 0	2C 1	R 6

λ_{air}	λ_{vac}	Ref.	I_1	I_2	I_3	I_4	ν	Upper	Lower	Br.
9332.903	9335.464			0.4			10711.84	3E 2	2B14	Q 5
9333.400	9335.961		0.2	0.1			10711.27	3E 2	2C 3	R 2
9333.714	9336.275			0.4			10710.91	3E 0	2C 1	R 5
9333.714	9336.275			0.4			10710.91	3E 0	2C 1	Q 2
9334.673	9337.234			0.4			10709.81	3E 0	2C 1	Q 6
9335.335	9337.896		0.2	0.3			10709.05	3E 0	2C 1	R 4
9335.823	9338.385		0.1	0.4			10708.49	U 1	2B12	R 0
9337.419	9339.981			0.0d			10706.66	4E 1	2B18	Q 2
9338.125	9340.687		0.2				10705.85	4D 0	2B15	P 2
9338.125	9340.687		0.2				10705.85	3E 0	2C 1	R 3
9338.335	9340.897			0.1			10705.61			
9338.509	9341.072		0.1				10705.41	3C 2	2K 5	P 5
9338.684	9341.246		0.9	1.5b	0.9		10705.21	3b 1	2a 1	P 5
9339.120	9341.682		0.2		0.1		10704.71			
9339.556	9342.119			0.3			10704.21	Z 0	2B 9	P 5
9341.912	9344.476			0.0			10701.51			
9342.794	9345.358			0.3			10700.50	3E 0	2C 1	R 2
9343.982	9346.546		0.1	0.5			10699.14	*		
9344.969	9347.533			0.5			10698.01			
9345.563	9348.127		0.3	0.2d			10697.33	2K 3	2B 3	R 0
9345.563	9348.127		0.3	0.2d			10697.33	4E 1	2B18	Q 1
9345.991	9348.555		1.5b	0.9			10696.84	2A 2	2B 2	P 2
9346.375	9348.940		0.2	0.0			10696.40	2K 3	2B 3	R 1
9347.651	9350.216			0.4			10694.94	4D 0	2B15	P 4
9347.651	9350.216			0.4			10694.94	3f 0	2c 3	P 5
9348.359	9350.924			0.2			10694.13	3E 1	2B12	R 3
9350.781	9353.347			0.1			10691.36	3E 3	2B16	R 2
9350.921	9353.487		0.2	0.6			10691.20	3E 0	2C 1	R 1
9351.770	9354.336		0.2	0.3			10690.23			
9352.330	9354.896		0.2	0.5			10689.59			
9353.100	9355.666			0.3			10688.71			
9353.283	9355.850		0.2				10688.50	3E 2	2C 3	R 1
9353.931	9356.497		0.1				10687.76			
9354.202	9356.769			0.7	0.5		10687.45			
9355.121	9357.688		0.1	0.3			10686.40	Z 0	2B 9	P 3
9355.121	9357.688		0.1	0.3			10686.40	3d 3	3b 0	P 2
9355.121	9357.688		0.1	0.3			10686.40	4d 1	3b 1	R 3
9355.892	9358.459			0.4			10685.52	3C 1	2K 3	P 5
9357.039	9359.606			0.5			10684.21	4d 0	3b 0	R 1
9357.626	9360.193			0.2			10683.54			
9358.432	9360.999		0.1d				10682.62	3e 1	2c 4	R 1
9358.791	9361.359		0.2	0.5			10682.21			
9359.536	9362.104			0.0d			10681.36	V 0	2B 9	R 1
9359.728	9362.296		0.2				10681.14			
9359.877	9362.446			0.1			10680.97			
9360.535	9363.103		0.2	1.5	0.1		10680.22			
9361.446	9364.015			0.4			10679.18	T 0	2B15	P 2
9361.683	9364.252			0.3			10678.91	3E 0	2C 1	Q 1
9363.147	9365.716		0.2	1.5	0.1		10677.24	2K 3	2B 3	R 2
9363.261	9365.830		0.4				10677.11			
9363.586	9366.155		0.2	0.1			10676.74	2A 2	2B 2	P 3
9363.586	9366.155		0.2	0.1			10676.74	u 1	2a 4	P 5
9364.235	9366.804		0.2d	1.5	0.3d		10676.00	3b 2	2a 2	R 7
9366.095	9368.664			0.3			10673.88			
9366.718	9369.288		0.2d	0.9	0.2d		10673.17	3f 0	2c 3	P 4
9366.851	9369.421						10673.02	7c 0	2a 8	R 1
9368.622	9371.193		0.2				10671.00	3C 2	2K 5	Q 3
9368.622	9371.193		0.2				10671.00	U 1	2C 2	Q 4
9368.622	9371.193		0.2				10671.00	4d 1	3b 1	P 2
9368.921	9371.492		1.5b	1.0d			10670.66	2K 4	2B 4	P 1

λ_{air}	λ_{vac}	Ref.	I_1	I_2	I_3	I_4	ν	Upper	Lower	Br.
9368.921	9371.492		1.5b	1.0d			10670.66	3E 1	2C 2	R 7
9370.019	9372.589		0.2d				10669.41			
9370.177	9372.748			0.5			10669.23			
9371.319	9373.890		0.2	0.8			10667.93			
9371.881	9374.452		0.2	0.4			10667.29	4e 0	3b 0	Q 4
9373.023	9375.595		0.2				10665.99	3E 1	2C 2	R 5
9373.129	9375.700			0.7			10665.87			
9375.115	9377.687			0.3			10663.61	3E 2	2B14	R 1
9376.249	9378.822		0.1				10662.32	X 0	2B11	R 4
9376.645	9379.218			0.0			10661.87	3D 2	2B14	R 6
9376.882	9379.455		0.1				10661.60			
9377.384	9379.957			0.4			10661.03	3E 1	2C 2	R 4
9377.771	9380.344		0.2				10660.59			
9378.211	9380.784		0.3	0.3			10660.09	3E 3	2B16	Q 4
9379.619	9382.192		0.4	0.6			10658.49			
9381.247	9383.821			0.3			10656.64			
9382.259	9384.833			0.3d			10655.49			
9382.462	9385.036		0.1				10655.26			
9382.867	9385.441			0.1			10654.80			
9382.955	9385.529			0.8			10654.70			
9383.272	9385.846			0.0			10654.34	X 0	2B11	R 3
9384.012	9386.586		0.2	0.8			10653.50	Z 0	2B 9	P 4
9384.012	9386.586		0.2	0.8			10653.50	3E 1	2C 2	R 3
9385.034	9387.609		0.1	0.8			10652.34	W 1	2B13	R 4
9385.034	9387.609		0.1	0.8			10652.34	4e 0	3b 0	P 5
9385.633	9388.208			0.3			10651.66	4d 1	3b 1	R 4
9385.985	9388.561		0.2				10651.26			
9386.849	9389.425						10650.28			
9387.016	9389.592		0.3	0.4			10650.09	3E 1	2B12	Q 6
9389.538	9392.114			0.1			10647.23			
9389.697	9392.273		0.2				10647.05			
9389.917	9392.494		2.9	1.5b	0.4		10646.80	3B 2	2A 1	P 3
9389.917	9392.494		2.9	1.5b	0.4		10646.80	2K 3	2B 3	P 1
9390.420	9392.996			0.1			10646.23			
9391.020	9393.596		0.2	1.5b	0.4		10645.55	3B 1	2A 0	P 7
9391.770	9394.346		0.2	0.3			10644.70			
9392.820	9395.397		3.0				10643.51	V 0	2B 9	P 5
9393.270	9395.847		0.2	0.1			10643.00			
9393.411	9395.988		0.4	1.5	0.2d		10642.84	3B 3	2A 2	R 2
9393.976	9396.553			0.4			10642.20			
9394.876	9397.454			0.0d			10641.18	3e 0	2c 3	R 5
9395.786	9398.364			0.2			10640.15	2K 3	2B 3	R 3
9397.358	9399.936		0.3	0.6			10638.37	2K 2	2B 2	P 4
9397.888	9400.466		0.1				10637.77			
9398.374	9400.953			0.1			10637.22			
9398.886	9401.465		0.2				10636.64			
9399.293	9401.872			0.2			10636.18			
9399.408	9401.987		0.1				10636.05			
9399.991	9402.570		0.1				10635.39			
9400.415	9402.995			0.3			10634.91			
9400.955	9403.534		0.1	0.0			10634.30	4e 0	3b 0	Q 5
9401.715	9404.294		0.4	0.3			10633.44	3A 2	2B15	P 2
9401.936	9404.515		0.1				10633.19	V 0	2B 9	P 4
9403.041	9405.621		0.1	0.5			10631.94	4d 0	3b 0	R 2
9403.652	9406.232		0.2d				10631.25			
9403.873	9406.453		1.5b	1.0	0.2		10631.00	2K 4	2B 4	P 2
9404.846	9407.426		0.2				10629.90			
9405.536	9408.116		0.2	0.3			10629.12	U 1	2B12	P 2
9405.536	9408.116		0.2	0.3			10629.12	3C 2	2K 5	R 1
9406.058	9408.639		0.1	0.1d			10628.53	4e 0	3b 0	P 6

λ_{air}	λ_{vac}	Ref.	I_1	I_2	I_3	I_4	ν	Upper	Lower	Br.
9406.209	9408.789			0.4			10628.36			
9406.766	9409.347			0.4			10627.73			
9407.793	9410.374		0.2	0.2			10626.57			
9408.652	9411.233			0.5			10625.60	U 1	2C 2	Q 5
9409.148	9411.729		0.4	0.6			10625.04	3E 1	2C 2	R 1
9409.148	9411.729		0.4	0.6			10625.04	4d 2	3b 2	P 3
9409.750	9412.332		0.1				10624.36			
9409.936	9412.518		0.6	1.0			10624.15			
9410.512	9413.094		0.2				10623.50			
9410.848	9413.430			0.1d	0.1		10623.12			
9411.194	9413.776		1.5b	1.5b	0.3		10622.73	3b 2	2a 2	P 1
9412.257	9414.839		0.1	0.3			10621.53	3E 3	2B16	R 1
9412.488	9415.070			0.5	0.1		10621.27			
9413.826	9416.409		0.1	0.0			10619.76			
9415.369	9417.952		0.3				10618.02			
9417.231	9419.815			0.7	0.1		10615.92			
9417.737	9420.320		0.1	0.3			10615.35	3E 0	2C 1	Q 1
9418.535	9421.119				0.1		10614.45			
9418.668	9421.252			0.2			10614.30			
9419.458	9422.042			0.2			10613.41			
9420.009	9422.593			0.2			10612.79	3e 0	2c 3	R 4
9421.269	9423.854			0.5			10611.37	t 0	2a 3	P 4
9421.527	9424.111		0.2	0.5			10611.08			
9422.832	9425.417			0.2			10609.61	U 1	2B12	P 6
9422.832	9425.417			0.2			10609.61	4e 0	3b 0	P 7
9423.987	9426.572			0.2			10608.31	4e 0	3b 0	Q 6
9424.582	9427.168		0.1				10607.64	3E 2	2B14	Q 4
9424.582	9427.168		0.1				10607.64	4d 3	3b 3	P 4
9424.840	9427.425		0.2	0.2			10607.35			
9425.746	9428.332			0.1			10606.33			
9426.573	9429.159		0.1	0.4			10605.40			
9427.284	9429.870			0.2			10604.60			
9428.084	9430.670			0.3d			10603.70	3C 1	2K 3	Q 3
9428.671	9431.257		0.2	0.3			10603.04	3E 1	2B12	R 2
9428.671	9431.257		0.2	0.3			10603.04	3E 1	2C 2	Q 3
9428.671	9431.257		0.2	0.3			10603.04	3a 0	2c 3	R 3
9429.587	9432.174		0.1	0.1			10602.01	3C 2	2K 5	P 4
9429.845	9432.432		0.6	1.5b	1.0		10601.72	3b 0	2a 0	P 9
9430.174	9432.761		0.1	0.0			10601.35	3a 1	2c 4	R 1
9430.325	9432.912		0.1				10601.18	2B19	2A 0	R 1
9430.690	9433.277				0.1		10600.77			
9431.081	9433.669				0.1		10600.33			
9431.580	9434.167		0.2	0.8	0.2		10599.77	3b 0	2a 0	R17
9432.024	9434.612			0.1			10599.27			
9432.229	9434.817		0.1				10599.04			
9432.434	9435.021		1.5b	1.5	0.2		10598.81	3B 3	2A 2	R 1
9432.808	9435.395			0.0			10598.39			
9433.920	9436.508			0.3			10597.14			
9434.775	9437.363		0.7	0.7			10596.18	2K 3	2B 3	P 2
9434.775	9437.363		0.7	0.7			10596.18	3E 1	2C 2	Q 1
9435.710	9438.298		0.1	0.6			10595.13			
9436.342	9438.931		0.2				10594.42			
9436.422	9439.011			0.7	0.2		10594.33			
9437.233	9439.822			0.3			10593.42			
9438.008	9440.597			0.6			10592.55	3b 6	2a 5	R 2
9438.685	9441.275		0.2d				10591.79			
9439.291	9441.881			0.5			10591.11			
9439.915	9442.505		0.2	0.3			10590.41	4e 0	3b 0	P 8
9440.530	9443.120		0.1				10589.72	3E 2	2C 3	R 5
9441.538	9444.128		0.2	0.3			10588.59			

λ_{air}	λ_{vac}	Ref.	I_1	I_2	I_3	I_4	ν	Upper	Lower	Br.
9442.412	9445.002			0.3			10587.61	2K 3	2B 3	R 4
9442.412	9445.002			0.3			10587.61	4d 0	3b 0	R 3
9443.054	9445.645		0.1				10586.89	3e 0	2c 3	R 3
9444.017	9446.608		1.0	1.0	0.1		10585.81	2K 4	2B 4	P 3
9444.017	9446.608		1.0	1.0	0.1		10585.81	X 0	2B11	R 1
9444.883	9447.474		0.3	1.5b	0.4		10584.84	3b 1	2a 1	P 6
9445.963	9448.554		0.4	0.9			10583.63	U 1	2B12	P 5
9446.811	9449.402		0.1				10582.68			
9447.132	9449.724		0.1	0.5			10582.32	U 1	2B12	P 3
9447.641	9450.233		0.1				10581.75			
9447.980	9450.572		0.3	0.2			10581.37			
9448.462	9451.054			0.2			10580.83			
9449.382	9451.974		0.1				10579.80			
9450.258	9452.850		0.3	0.5			10578.82	u 0	2a 3	P 2
9451.204	9453.797		0.2	1.0	0.2		10577.76	3B 3	2A 2	R 5
9451.204	9453.797		0.2	1.0	0.2		10577.76	3E 2	2C 3	Q 2
9451.919	9454.512		0.3	0.8			10576.96	3C 2	2K 5	Q 2
9451.919	9454.512		0.3	0.8			10576.96	4d 0	3b 0	P 2
9452.733	9455.326		0.1				10576.05			
9452.956	9455.549		0.4	0.9			10575.80	3D 2	2B14	R 5
9452.956	9455.549		0.4	0.9			10575.80	3E 3	2B16	Q 3
9453.627	9456.220		0.1				10575.05			
9453.895	9456.488			0.1			10574.75			
9454.324	9456.917			0.3			10574.27			
9455.146	9457.740		0.1				10573.35	3E 1	2B12	Q 5
9455.907	9458.501			0.3d			10572.50	3E 0	2C 1	Q 2
9456.479	9459.073		0.2				10571.86			
9457.105	9459.700			0.5			10571.16			
9458.179	9460.774		0.2	0.8			10569.96	3E 2	2C 3	R 3
9458.179	9460.774		0.2	0.8			10569.96	3E 2	2C 3	Q 3
9458.179	9460.774		0.2	0.8			10569.96	3e 0	2c 4	R 2
9459.092	9461.687			0.3			10568.94			
9460.408	9463.003		0.8	0.9			10567.47	3b 6	2a 5	R 1
9460.676	9463.271		2.8				10567.17	3E 2	2B14	R 0
9461.974	9464.570		0.2	0.3			10565.72	3E 2	2C 3	Q 1
9461.974	9464.570		0.2	0.3			10565.72	3e 0	2c 3	R 2
9463.005	9465.600			0.5			10564.57			
9463.963	9466.559		0.1				10563.50	3E 1	2C 2	Q 4
9464.151	9466.748		0.2	0.2			10563.29			
9464.259	9466.855		0.2d	0.6d			10563.17			
9464.734	9467.330		0.1	0.0			10562.64			
9465.038	9467.635		0.1				10562.30			
9465.289	9467.886		0.1				10562.02			
9465.459	9468.056		0.3	0.6			10561.83			
9466.589	9469.186		0.1				10560.57			
9466.858	9469.455		0.2				10560.27	3E 0	2C 1	P 2
9467.082	9469.679		0.6	1.0			10560.02	2K 6	2B 6	R 2
9468.221	9470.818		0.9	1.0			10558.75	3B 3	2A 2	R 0
9468.221	9470.818		0.9	1.0			10558.75	4d 0	3b 0	R 4
9468.633	9471.231			0.4			10558.29	4F 0	2B16	Q 3
9468.633	9471.231			0.4			10558.29	X 0	2B11	R 0
9469.691	9472.289			0.2			10557.11	7c 0	2a 8	Q 1
9471.082	9473.680			0.2			10555.56			
9471.315	9473.913		0.1				10555.30			
9471.692	9474.291		0.1				10554.88	3B 3	2A 2	R 4
9472.087	9474.685			0.9			10554.44			
9472.356	9474.955			0.8			10554.14	3E 3	2B16	R 0
9472.428	9475.027		0.2				10554.06			
9473.227	9475.826		0.1				10553.17			
9473.694	9476.293		0.6	1.5b			10552.65	3B 2	2A 1	P 4

λ_{air}	λ_{vac}	Ref.	I_1	I_2	I_3	I_4	ν	Upper	Lower	Br.
9473.694	9476.293		0.6	1.5b			10552.65	3E 3	2B16	P 5
9473.694	9476.293		0.6	1.5b			10552.65	3E 2	2C 3	R 2
9474.520	9477.119			0.2			10551.73			
9475.041	9477.640		0.2				10551.15	V 1	2C 1	P 2
9475.041	9477.640		0.2				10551.15	W 1	2B13	P 3
9475.373	9477.972		0.1	0.2			10550.78	3A 1	2B13	P 4
9475.858	9478.457		0.2	0.9			10550.24			
9476.406	9479.005		1.5b	1.5	0.3		10549.63	3b 2	2a 2	P 2
9476.406	9479.005		1.5b	1.5	0.3		10549.63	4d 0	3b 0	R 6
9476.729	9479.329		0.2b				10549.27			
9477.026	9479.625		0.4	0.6			10548.94	2K 6	2B 6	R 1
9478.490	9481.090			0.4			10547.31			
9479.038	9481.639						10546.70	4d 0	3b 0	R 5
9479.317	9481.917		0.1	0.3			10546.39	3C 1	2K 3	P 4
9481.115	9483.716		0.3				10544.39	2K 2	2B 2	P 5
9481.484	9484.085		0.2	0.9			10543.98			
9482.050	9484.651			0.3			10543.35	3E 1	2C 2	Q 1
9482.554	9485.155		0.2	0.5			10542.79	5c 1	2a 8	Q 1
9483.013	9485.614		0.1				10542.28			
9483.786	9486.388		0.2	0.3			10541.42			
9484.398	9487.000		0.2	0.5			10540.74			
9485.604	9488.206			0.5			10539.40			
9486.306	9488.908		0.2				10538.62	3C 2	2K 5	R 0
9486.594	9489.196		0.1				10538.30			
9486.837	9489.440		1.0	1.0	0.2		10538.03	2K 4	2B 4	P 4
9487.701	9490.304		0.1				10537.07	3D 2	2B14	R 4
9488.836	9491.439						10535.81			
9489.521	9492.124			0.1			10535.05			
9490.376	9492.980				0.1		10534.10			
9491.277	9493.881		0.2	0.8			10533.10	3E 0	2C 1	P 3
9491.277	9493.881		0.2	0.8			10533.10	2B19	2A 0	R 2
9492.602	9495.206		1.5b	1.0			10531.63	2K 6	2B 6	R 0
9492.945	9495.549		0.1				10531.25			
9493.251	9495.855			0.2			10530.91	4F 0	2B16	R 1
9494.595	9497.199		0.4	0.7			10529.42	3E 2	2B14	Q 3
9494.595	9497.199		0.4	0.7			10529.42	3E 2	2C 3	R 1
9495.262	9497.867		0.3	0.5			10528.68	2K 3	2B 3	P 3
9495.578	9498.182				0.2d		10528.33			
9496.010	9498.615		0.2				10527.85			
9496.453	9499.058		0.3	0.6			10527.36			
9497.283	9499.888			0.2			10526.44			
9497.526	9500.131			0.2			10526.17	3E 0	2B10	R 2
9498.275	9500.881			0.3			10525.34			
9498.807	9501.413		0.2				10524.75	4F 0	2C 4	Q 4
9499.810	9502.415		0.3	0.4			10523.64	3C 2	2K 5	P 3
9499.810	9502.415		0.3	0.4			10523.64	4d 1	3b 1	P 3
9499.810	9502.415		0.3	0.4			10523.64	3a 0	2c 3	R 2
9501.317	9503.924		0.2	0.6			10521.97	3C 1	2K 3	R 1
9502.220	9504.827			0.2			10520.97			
9503.078	9505.685			0.4			10520.02			
9503.828	9506.435		0.2	0.2			10519.19	W 2	2B15	P 1
9505.382	9507.990				0.2		10517.47			
9505.572	9508.180		0.2				10517.26			
9505.699	9508.306			0.3			10517.12			
9506.639	9509.247		0.2	1.5	0.2		10516.08	3E 3	2B16	Q 2
9506.775	9509.382				0.2		10515.93			
9507.407	9510.015		0.2d				10515.23			
9507.679	9510.287			0.3			10514.93			
9508.140	9510.748		0.2	0.3			10514.42	3E 3	2B16	P 3
9508.140	9510.748		0.2	0.3			10514.42	3E 1	2C 2	Q 2

λ_{air}	λ_{vac}	Ref.	I_1	I_2	I_3	I_4	ν	Upper	Lower	Br.
9508.827	9511.436		0.3	0.4			10513.66	3C 2	2K 5	Q 1
9509.433	9512.042		0.4	0.4			10512.99			
9509.949	9512.557		0.1	0.6			10512.42	3D 3	2B16	R 4
9510.899	9513.508		0.3	0.2			10511.37			
9511.767	9514.377			1.5	0.2		10510.41	3D 2	2B14	R 3
9512.817	9515.427		0.2d	0.6			10509.25			
9513.677	9516.287		0.1				10508.30	3F 0	2B10	P 3
9513.876	9516.486		0.3	1.5	0.2		10508.08	3B 3	2A 2	R 3
9514.764	9517.374		0.3	0.6			10507.10	u 0	2a 3	P 3
9515.615	9518.225		0.3	0.8			10506.16			
9515.896	9518.506		0.3				10505.85	3E 1	2B12	R 1
9516.575	9519.186		0.2				10505.10			
9516.729	9519.340			0.4			10504.93			
9516.892	9519.503		0.2b				10504.75			
9517.137	9519.748		0.2				10504.48	3E 2	2C 3	Q 5
9518.378	9520.989			0.2			10503.11	t 0	2a 3	P 5
9519.475	9522.086			0.4			10501.90			
9520.300	9522.911		0.3	0.5	0.1		10500.99	3E 0	2C 1	P 2
9520.726	9523.338		0.1				10500.52			
9521.560	9524.172			0.3			10499.60			
9522.957	9525.569			0.3			10498.06			
9523.982	9526.595		0.4				10496.93			
9524.073	9526.686			0.4			10496.83			
9525.443	9528.056		0.3	0.9			10495.32	4d 2	3b 2	P 4
9526.242	9528.855		0.3	0.5			10494.44	3E 0	2C 1	P 4
9526.242	9528.855		0.3	0.5			10494.44	3A 1	2B13	P 3
9527.132	9529.745		0.1	0.3			10493.46			
9528.258	9530.871			0.2			10492.22	3E 3	2B16	P 2
9529.520	9532.134		0.4	1.0			10490.83	2K 4	2B 4	P 5
9530.429	9533.043		0.2	0.2			10489.83			
9531.428	9534.043		0.2	0.4			10488.73			
9532.437	9535.052		0.2				10487.62	X 0	2B11	P 2
9532.582	9535.197			0.3			10487.46	3D 2	2C 3	R 1
9533.301	9535.916			0.6	0.4		10486.67			
9533.637	9536.252		0.0				10486.30	V 1	2C 1	P 4
9534.364	9536.979		0.4	0.6			10485.50	3E 1	2C 2	P 2
9534.364	9536.979		0.4	0.6			10485.50	3C 1	2K 3	Q 2
9534.728	9537.343		0.0				10485.10			
9534.974	9537.589			0.3			10484.83			
9536.356	9538.972		0.2				10483.31	W 2	2B15	P 3
9536.356	9538.972		0.2				10483.31	3E 1	2C 2	Q 3
9536.938	9539.554		0.4	0.4			10482.67			
9538.130	9540.746		0.0				10481.36			
9538.367	9540.983			0.1			10481.10	2A 1	2B 1	R 4
9538.367	9540.983			0.1			10481.10	W 1	2B13	R 2
9539.140	9541.757		0.1	0.1			10480.25			
9539.668	9542.285			0.1			10479.67			
9540.806	9543.423			0.2			10478.42			
9541.089	9543.706		0.1				10478.11			
9541.316	9543.933		1.0	0.7			10477.86	2K 6	2B 6	P 1
9542.437	9545.054			0.3			10476.63			
9542.637	9545.255		0.0				10476.41			
9543.648	9546.266		0.2	0.3			10475.30	3C 2	2K 5	P 2
9545.981	9548.599		0.4	0.5d			10472.74	W 2	2B15	P 4
9546.163	9548.782						10472.54	3E 0	2C 1	Q 4
9546.428	9549.046			0.5			10472.25			
9546.938	9549.557		0.0				10471.69	3D 3	2B16	R 2
9546.938	9549.557		0.0				10471.69	3e 0	2c 3	Q 1
9547.859	9550.478		0.2				10470.68			
9548.835	9551.454		1.5	1.5	0.2		10469.61	3B 3	2A 2	P 1

λ_{air}	λ_{vac}	Ref.	I_1	I_2	I_3	I_4	ν	Upper	Lower	Br.
9548.835	9551.454		1.5	1.5	0.2		10469.61	3e 1	2c 4	Q 1
9549.264	9551.883			0.5			10469.14	3A 1	2B13	P 1
9549.346	9551.965		0.2				10469.05			
9550.230	9552.850		0.1d				10468.08	3A 1	2B13	P 2
9550.230	9552.850		0.1d				10468.08	3D 3	2B16	R 1
9550.796	9553.416		0.3	0.5			10467.46	W 1	2B13	P 6
9552.010	9554.630			0.3d			10466.13	3B 3	2K 2	P 4
9552.941	9555.561				0.1		10465.11			
9553.361	9555.981		0.1d				10464.65			
9553.653	9556.273		1.5b	1.0b	0.5		10464.33	3b 2	2a 2	P 3
9554.082	9556.703		0.3				10463.86	V 0	2B 9	P 2
9554.876	9557.497			0.8	0.1		10462.99			
9555.123	9557.744		0.1				10462.72	3D 2	2B14	R 1
9556.237	9558.859			0.4			10461.50	3D 3	2B16	R 0
9556.237	9558.859			0.4			10461.50	3E 1	2B12	Q 4
9557.059	9559.681		0.1	0.5			10460.60	3E 3	2C 4	R 5
9557.608	9560.229			0.3			10460.00			
9558.640	9561.262		0.2	0.4			10458.87			
9559.545	9562.167		0.5	0.5			10457.88			
9559.682	9562.304		0.1d				10457.73	3E 1	2C 2	P 3
9560.469	9563.091		0.1d				10456.87			
9560.734	9563.356		0.2	0.8			10456.58			
9561.557	9564.179		0.3	1.5b	0.7		10455.68	3b 1	2a 1	P 7
9562.471	9565.094		0.2	0.9			10454.68	3B 4	2A 3	R 3
9562.471	9565.094		0.2	0.9			10454.68	3a 0	2c 3	R 1
9563.450	9566.073		0.2d	0.4			10453.61	3E 2	2C 3	Q 1
9564.667	9567.291		0.1d				10452.28	3E 3	2C 4	R 4
9566.772	9569.396		0.3	0.5			10449.98	3b 0	2a 0	R18
9566.772	9569.396		0.3	0.5			10449.98	3E 2	2C 3	P 2
9567.129	9569.753		0.1				10449.59			
9567.367	9569.991		0.4	1.5b	0.3		10449.33	3B 2	2A 1	P 5
9569.034	9571.659		0.2	1.0b	0.5		10447.51	3b 0	2a 0	P10
9569.034	9571.659		0.2	1.0b	0.5		10447.51	u 0	2a 3	P 4
9570.170	9572.795		0.4	0.3			10446.27			
9570.573	9573.198		0.3				10445.83			
9571.187	9573.812		0.4	1.0			10445.16	2K 4	2B 4	P 6
9572.067	9574.692		0.5	0.6			10444.20	2K 3	2B 3	P 4
9572.067	9574.692		0.5	0.6			10444.20	3E 0	2C 1	P 5
9572.351	9574.976		0.2				10443.89			
9574.487	9577.113		1.5b	1.5			10441.56	2K 6	2B 6	P 2
9574.560	9577.186				0.1		10441.48	3C 1	2K 3	P 3
9575.147	9577.773		0.4	0.9			10440.84	3e 0	2c 3	Q 2
9576.037	9578.663			0.2			10439.87	X 0	2B11	P 4
9576.651	9579.278				0.1		10439.20			
9576.899	9579.526		0.0				10438.93	3E 0	2B10	P 5
9577.073	9579.700			0.3			10438.74			
9578.431	9581.059			0.2			10437.26	3D 2	2B14	R 0
9579.992	9582.619		0.4	0.5			10435.56	3D 2	2C 3	Q 1
9580.634	9583.262			0.2			10434.86	3E 1	2C 2	P 2
9581.718	9584.346		0.1	0.8			10433.68	3E 3	2C 4	R 3
9582.820	9585.448		0.2b	0.8			10432.48	X 0	2B11	P 5
9582.820	9585.448		0.2b	0.8			10432.48	3F 2	2B15	Q 4
9583.987	9586.615			0.2			10431.21			
9584.694	9587.323		0.3				10430.44	3D 2	2C 3	R 2
9585.071	9587.700		0.2	0.8			10430.03	3B 3	2K 2	R 2
9585.071	9587.700		0.2	0.8			10430.03	3a 1	2c 4	P 1
9585.714	9588.344		0.2				10429.33			
9586.211	9588.840		0.3	1.0			10428.79	3E 2	2B14	Q 1
9586.744	9589.373		0.9	1.5	0.1		10428.21	3B 3	2A 2	P 2
9587.654	9590.284			0.1			10427.22			

λ air	λ vac	Ref.	I₁	I₂	I₃	I₄	ν	Upper	Lower	Br.
9588.105	9590.734			0.3			10426.73			
9588.325	9590.955				0.1		10426.49	2B19	2A 0	R 3
9589.024	9591.654			0.2			10425.73			
9589.484	9592.114				0.1		10425.23	W 1	2B13	R 1
9589.484	9592.114				0.1		10425.23	3C 1	2K 3	R 0
9589.668	9592.298			0.2			10425.03			
9592.567	9595.198			0.5			10421.88			
9593.819	9596.450			0.3			10420.52			
9595.062	9597.693			0.1			10419.17			
9596.176	9598.808		0.4	0.9			10417.96			
9596.471	9599.103		0.2	0.4			10417.64	3D 1	2C 2	R 1
9596.932	9599.564		0.2				10417.14			
9598.572	9601.204			0.4			10415.36	4D 0	2B16	R 6
9599.410	9602.043		1.0	1.5b	0.2		10414.45	3B 3	2K 2	P 3
9599.410	9602.043		1.0	1.5b	0.2		10414.45	3E 0	2B10	Q 5
9599.604	9602.237		0.1				10414.24	3e 1	2c 4	Q 3
9600.028	9602.661		0.2				10413.78	3E 2	2C 3	Q 2
9601.162	9603.795		0.1	0.5			10412.55			
9603.302	9605.936		0.4	0.5			10410.23			
9603.809	9606.443		0.4	0.4			10409.68	3e 0	2c 3	Q 3
9604.898	9607.532			0.2			10408.50			
9606.255	9608.889		0.2	0.5			10407.03	3C 1	2K 3	Q 1
9607.464	9610.099			0.2			10405.72	3A 2	2B16	R 4
9608.618	9611.254		0.3d	0.5			10404.47	3E 0	2C 1	P 3
9609.939	9612.575		0.3				10403.04	3D 1	2C 2	Q 1
9610.142	9612.778		0.2	0.6			10402.82			
9610.918	9613.554		0.1				10401.98	Y 2	2B15	R 1
9610.918	9613.554		0.1				10401.98	X 0	2B11	P 6
9611.787	9614.423		0.4	0.3			10401.04	2B 1		R 0
9611.787	9614.423		0.4	0.3			10401.04	W 2	2B15	P 2
9612.702	9615.338		0.2	0.4			10400.05			
9613.219	9615.856		0.7	1.0			10399.49	2K 6	2B 6	P 3
9613.219	9615.856		0.7	1.0			10399.49	3D 3	2B16	P 2
9615.651	9618.289		0.2	0.8			10396.86			
9617.057	9619.695		0.1				10395.34	3E 1	2B12	Q 3
9617.224	9619.861		0.8	0.9			10395.16			
9618.797	9621.435			0.1			10393.46			
9619.454	9622.092				0.1		10392.75	Y 2	2B15	R 0
9619.454	9622.092				0.1		10392.75	3b 6	2a 5	P 2
9619.843	9622.481		0.2	0.9			10392.33			
9620.898	9623.537			0.2			10391.19			
9622.528	9625.167			0.2			10389.43			
9624.418	9627.057		0.3	0.3			10387.39	3E 0	2B10	R 1
9625.363	9628.003		0.4				10386.37	Z 2	2C 3	R 1
9625.493	9628.133			0.7			10386.23			
9625.947	9628.587				0.1		10385.74			
9626.225	9628.865		0.4	0.7			10385.44			
9628.153	9630.794			0.2			10383.36			
9628.941	9631.582			0.3			10382.51	3E 3	2C 4	R 1
9629.396	9632.037			0.2	0.1		10382.02	4F 0	2C 4	Q 3
9629.924	9632.566				0.1		10381.45	4d 1	3b 1	P 4
9630.184	9632.826		0.2	15.b			10381.17	3e 0	2c 3	Q 4
9630.258	9632.900				0.3		10381.09			
9630.964	9633.605				0.1		10380.33			
9631.279	9633.921			0.3			10379.99			
9633.785	9636.427		0.2	0.2			10377.29	W 1	2B13	R 0
9633.785	9636.427		0.2	0.2			10377.29	3D 2	2C 3	Q 2
9635.428	9638.071		0.2	0.8			10375.52	3B 4	2A 3	R 2
9635.428	9638.071		0.2	0.8			10375.52	3C 1	2K 3	P 2
9635.428	9638.071		0.2	0.8			10375.52	3f 2	3b 0	R 5

λ_{air}	λ_{vac}	Ref.	I_1	I_2	I_3	I_4	ν	Upper	Lower	Br.
9636.794	9639.437			0.2			10374.05			
9636.998	9639.641		0.2b				10373.83			
9637.221	9639.864			0.4			10373.59	Y 1	2B11	P 1
9639.405	9642.048			0.1			10371.24			
9640.399	9643.043		0.2	0.2			10370.17	3E 1	2B12	P 3
9640.399	9643.043		0.2	0.2			10370.17	3D 0	2C 1	R 1
9642.789	9645.434		0.7	1.0b			10367.60	3b 2	2a 2	P 4
9642.873	9645.517				0.2		10367.51	3e 0	2c 3	P 2
9643.375	9646.020		0.4	0.4			10366.97			
9644.017	9646.662			0.8			10366.28			
9645.822	9648.468			0.2			10364.34	Z 2	2B14	R 2
9646.502	9649.147		0.4	0.4			10363.61	Z 2	2C 3	Q 1
9647.367	9650.013		0.3	0.5			10362.68			
9648.345	9650.991				0.1		10361.63	Z 2	2B14	R 1
9648.485	9651.131		0.4	0.4			10361.48			
9649.071	9651.718		0.4	0.6			10360.85	3D 3	2B16	P 3
9649.071	9651.718		0.4	0.6			10360.85	3E 1	2C 2	P 4
9649.621	9652.267		0.0				10360.26			
9650.925	9653.572			0.1			10358.86			
9651.148	9653.795		0.0				10358.62	3E 3	2C 4	Q 1
9651.148	9653.795		0.0				10358.62	3e 0	2c 3	Q 5
9651.698	9654.345		0.3				10358.03			
9652.108	9654.755			0.3			10357.59			
9654.373	9657.021		0.3	0.4			10355.16	3D 1	2C 2	R 2
9654.457	9657.105				0.1		10355.07	3D 0	2C 1	P 1
9655.893	9658.542			0.9			10353.53	3E 1	2B12	Q 2
9655.977	9658.625		0.4				10353.44	3E 1	2C 2	P 3
9657.003	9659.652		0.1				10352.34	3E 3	2C 4	Q 2
9657.199	9659.848		0.8	1.0			10352.13	2K 6	2B 6	P 4
9657.582	9660.230				0.2		10351.72			
9657.768	9660.417		0.2	1.5			10351.52			
9657.852	9660.501				0.1		10351.43			
9658.011	9660.660				0.6		10351.26			
9658.356	9661.005		0.3		0.1		10350.89	3E 2	2C 3	P 4
9658.636	9661.285				0.1		10350.59	3E 2	2C 3	Q 4
9660.335	9662.984		0.3	0.5			10348.77	3D 0	2C 1	Q 1
9660.923	9663.572			0.3			10348.14	W 1	2B13	P 5
9662.379	9665.029		0.4	0.8			10346.58	3B 3	2K 2	R 1
9662.379	9665.029		0.4	0.8			10346.58	Y 1	2B11	P 2
9663.304	9665.954		0.1				10345.59			
9663.491	9666.141		0.8	0.8			10345.39	Z 2	2B14	R 4
9663.631	9666.281		0.1				10345.24	3E 0	2B10	P 4
9665.285	9667.935			0.4			10343.47	2K 3	2B 3	P 5
9665.453	9668.104		0.1d	0.4			10343.29	3E 2	2C 3	P 2
9666.238	9668.889			0.2			10342.45			
9666.958	9669.609			0.2			10341.68	3a 1	2c 4	P 3
9667.958	9670.609						10340.61			
9669.379	9672.031		0.1d	1.5b	0.2		10339.09	2A 1	2B 1	R 3
9671.278	9673.930			0.5			10337.06	V 1	2B10	P 4
9671.437	9674.090		0.1				10336.89			
9672.045	9674.698		0.4	0.8			10336.24			
9672.916	9675.568		0.2				10335.31			
9673.487	9676.140		0.3	1.5			10334.70			
9675.481	9678.134		0.3	0.8			10332.57			
9676.708	9679.361		0.2	0.2			10331.26	3E 0	2B11	R 6
9677.401	9680.055		0.2				10330.52			
9678.028	9680.683		0.1d				10329.85			
9678.394	9681.048		0.2	0.6			10329.46			
9678.919	9681.573		0.1d				10328.90			
9679.153	9681.807		0.6	1.0b			10328.65	2A 0	2B 0	R 8

λ_{air}	λ_{vac}	Ref.	I_1	I_2	I_3	I_4	ν	Upper	Lower	Br.
9679.153	9681.807		0.6	1.0b			10328.65	X 2	2B16	R 0
9680.418	9683.073			0.2			10327.30	X 2	2B16	R 1
9681.843	9684.498			0.5			10325.78	3D 3	2B16	P 4
9682.753	9685.408			0.2			10324.81	3D 1	2B12	P 2
9683.963	9686.618			0.1			10323.52	3E 0	2B10	Q 4
9684.113	9686.769		0.1				10323.36	X 2	2B16	R 3
9684.610	9687.266			0.1			10322.83			
9685.304	9687.960			0.1	0.1		10322.09			
9686.449	9689.106			0.3			10320.87	X 2	2B16	R 2
9687.378	9690.035		0.3	0.3			10319.88	3D 2	2C 3	P 2
9688.374	9691.031						10318.82	3E 3	2C 4	Q 3
9688.561	9691.218		0.2	1.0b	0.3		10318.62	3b 1	2a 1	P 8
9689.021	9691.679		0.3				10318.13	3D 0	2C 1	R 2
9691.670	9694.328			0.2			10315.31			
9693.108	9695.766		0.9	1.5			10313.78	3b 6	2a 5	P 3
9693.691	9696.349		0.1d				10313.16			
9694.001	9696.659		0.7	1.5			10312.83	3B 4	2A 3	R 1
9694.659	9697.318		0.1	0.3			10312.13			
9695.533	9698.192		1.0	1.0			10311.20			
9695.731	9698.390		0.1				10310.99	Y 2	2B15	R 2
9696.784	9699.443		0.2	1.5			10309.87	3b 3	2a 3	R 3
9698.374	9701.033		0.1				10308.18	W 1	2B13	P 3
9698.374	9701.033		0.1				10308.18	Y 1	2B11	P 3
9698.854	9701.513		0.3	0.8			10307.67	W 1	2B13	P 2
9699.503	9702.163		0.2	1.0			10306.98	3b 3	2a 3	R 2
9700.708	9703.368			0.2			10305.70	3D 1	2B12	R 5
9701.470	9704.131		0.4	0.3			10304.89	Z 2	2C 3	R 2
9702.110	9704.771			0.2			10304.21			
9703.127	9705.788		0.4	0.9			10303.13	2K 5	2B 5	R 2
9704.041	9706.702		0.4	0.7			10302.16	2K 5	2B 5	R 1
9704.041	9706.702		0.4	0.7			10302.16	3E 0	2C 1	P 4
9704.748	9707.409			0.2			10301.41			
9705.558	9708.219		0.2	0.4			10300.55	3d 3	3b 0	P 5
9706.378	9709.040		0.2	0.7			10299.68			
9706.821	9709.483			0.7			10299.21	3b 3	2a 3	R 4
9708.234	9710.897		0.2d	0.3			10297.71			
9710.714	9713.378			0.3			10295.08	2K 5	2B 5	R 3
9712.148	9714.812		0.2	0.6			10293.56			
9712.894	9715.558		0.3	0.2			10292.77			
9713.403	9716.067			0.3			10292.23	3D 1	2C 2	P 2
9714.074	9716.738						10291.52	Z 1	2C 2	R 1
9714.074	9716.738						10291.52	Y 2	2B15	P 3
9714.074	9716.738						10291.52	3f 2	3b 0	R 4
9714.366	9717.030		0.8	0.9			10291.21	2K 5	2B 5	R 0
9714.895	9717.559						10290.65		2B 1	P 2
9715.206	9717.871		0.6	1.5			10290.32	3b 3	2a 3	R 1
9715.206	9717.871		0.6	1.5			10290.32	3F 2	2B15	Q 3
9716.604	9719.269			0.2			10288.84			
9717.256	9719.921		0.2	1.5			10288.15	3b 0	2a 0	P11
9718.229	9720.894		0.3	0.5			10287.12			
9719.230	9721.895			0.3			10286.06	3e 0	2c 3	P 3
9720.903	9723.569			0.3			10284.29	4E 0	2B16	Q 4
9720.903	9723.569			0.3			10284.29	3D 1	2C 2	R 3
9722.255	9724.921		0.1d				10282.86	3D 3	2B16	P 5
9722.387	9725.053		0.3	0.4			10282.72			
9723.928	9726.595		0.2				10281.09			
9724.505	9727.172			0.4			10280.48	3D 2	2C 3	R 4
9724.505	9727.172			0.4			10280.48	2K 5	2B 5	R 4
9725.187	9727.854		0.3	0.3			10279.76	3B 3	2K 2	R 0
9725.187	9727.854		0.3	0.3			10279.76	3D 3	2B16	P 6

λ_{air}	λ_{vac}	Ref.	I$_1$	I$_2$	I$_3$	I$_4$	ν	Upper	Lower	Br.
9728.205	9730.873		0.3				10276.57	3a 0	2c 3	P 1
9728.594	9731.261		0.3	0.6			10276.16	Z 2	2C 3	Q 2
9729.540	9732.208			1.0			10275.16	3b 3	2a 3	R 5
9730.781	9733.449		0.2	0.4			10273.85			
9731.311	9733.980						10273.29			
9731.700	9734.368		0.5	0.8			10272.88	3E 2	2C 3	P 3
9732.609	9735.278		0.2	0.3			10271.92			
9732.950	9735.620		0.1				10271.56			
9733.301	9735.970						10271.19			
9733.708	9736.378		0.5	0.6			10270.76			
9734.647	9737.316			0.4			10269.77			
9734.827	9737.496		0.1				10269.58			
9735.339	9738.009						10269.04	3E 0	2B10	R 0
9735.746	9738.416		0.5	0.4			10268.61			
9738.003	9740.674		0.4	0.5			10266.23			
9739.151	9741.822			0.4			10265.02	3E 1	2C 2	P 4
9740.224	9742.895		0.2	0.4			10263.89			
9741.903	9744.575		0.5	0.6			10262.12	3B 4	2A 3	R 0
9742.768	9745.439		0.6	0.5			10261.21	4F 0	2C 4	Q 2
9743.698	9746.370		0.7	1.5b			10260.23	3b 2	2a 2	P 5
9743.698	9746.370		0.7	1.5b			10260.23	2B19	2A 0	R 4
9744.002	9746.674		0.4				10259.91	3b 3	2a 3	R 0
9744.002	9746.674		0.4				10259.91	4f 2	3b 3	R 4
9744.724	9747.396			0.6			10259.15			
9745.123	9747.795		0.3				10258.73	4D 0	2B16	R 5
9747.384	9750.057			0.5			10256.35	3D 0	2C 1	R 3
9747.498	9750.171		0.3				10256.23	3D 0	2C 1	Q 2
9749.285	9751.959			0.3			10254.35			
9750.036	9752.710			0.2			10253.56	3E 0	2B10	P 3
9751.168	9753.842			0.2			10252.37			
9751.891	9754.565			0.3			10251.61			
9751.996	9754.670		0.3				10251.50	Z 1	2C 2	Q 1
9752.966	9755.641			0.1			10250.48	V 1	2B10	P 3
9754.251	9756.926			0.2			10249.13	Z 1	2C 2	R 2
9754.384	9757.059		0.2				10248.99	3E 3	2C 4	Q 4
9754.965	9757.640		0.3	0.1			10248.38	4d 2	3b 2	P 6
9755.384	9758.059		0.3	0.2			10247.94	Z 2	2C 3	P 2
9756.516	9759.192		0.2	0.3			10246.75	4d 1	3b 1	P 5
9757.212	9759.887		0.5	0.3			10246.02	3A 2	2B16	R 3
9757.373	9760.049			0.1			10245.85			
9759.688	9762.365			0.4			10243.42	3E 0	2B10	Q 3
9760.908	9763.585						10242.14			
9761.518	9764.195		0.3	1.5b			10241.50			
9763.386	9766.064		0.2	0.4			10239.54			
9763.482	9766.159						10239.44			
9763.939	9766.617		0.1				10238.96			
9764.130	9766.808		0.8	0.5			10238.76	2K 5	2B 5	P 1
9764.483	9767.161		0.1				10238.39			
9764.731	9767.409		0.4	0.5			10238.13			
9765.542	9768.220		0.2	1.0b			10237.28			
9766.858	9769.537			0.2			10235.90	Z 0	2C 1	R 1
9767.879	9770.558			0.8			10234.83			
9770.180	9772.859		0.3	0.6			10232.42	Z 1	2C 2	P 1
9770.753	9773.432		0.1	0.3			10231.82			
9771.870	9774.550		0.2	0.2			10230.65			
9772.711	9775.391		0.2	0.6			10229.77	2K 3	2B 3	P 6
9773.246	9775.926			0.3			10229.21			
9774.077	9776.757		0.1d	0.2			10228.34			
9775.606	9778.287			0.2			10226.74	Y 2	2B15	P 2
9777.232	9779.913		0.2b	0.2			10225.04			

λ_{air}	λ_{vac}	Ref.	I_1	I_2	I_3	I_4	ν	Upper	Lower	Br.
9777.844	9780.525			0.2			10224.40			
9778.054	9780.735						10224.18			
9778.571	9781.252		0.5	1.0			10223.64			
9779.479	9782.161			0.1			10222.69			
9780.608	9783.290		0.1d				10221.51			
9781.450	9784.133		0.4	0.3			10220.63			
9782.724	9785.406			0.2			10219.30			
9783.403	9786.086			0.1			10218.59			
9784.016	9786.699						10217.95			
9784.571	9787.254		0.5	1.5			10217.37	2A 1	2B 1	R 2
9784.571	9787.254		0.5	1.5			10217.37	3B 3	2K 2	P 1
9785.999	9788.682						10215.88			
9786.487	9789.171		0.6	0.7r			10215.37	3B 3	2K 2	P 2
9786.918	9789.602			0.3			10214.92			
9787.014	9789.698		0.6				10214.82			
9787.790	9790.474		0.1				10214.01			
9788.816	9791.500			0.3			10212.94			
9790.225	9792.909		0.1	0.1			10211.47			
9790.858	9793.542			0.4			10210.81	3D 1	2B12	R 0
9790.963	9793.648		0.2				10210.70			
9791.816	9794.501		0.2				10209.81		2a 1	P 3
9792.392	9795.077		1.0	1.0b			10209.21	3B 3	2A 2	P 3
9794.042	9796.728		0.1d				10207.49			
9794.752	9797.438		0.2				10206.75			
9795.894	9798.580			0.5			10205.56	3e 0	2c 3	P 4
9795.894	9798.580			0.5			10205.56	3f 2	3b 0	R 3
9797.152	9799.838			0.2b			10204.25			
9798.333	9801.020			0.1			10203.02	3E 0	2B10	Q 2
9798.333	9801.020			0.1			10203.02	3d 0	2c 3	P 1
9799.457	9802.144		0.1				10201.85	2A 3	2B 4	R 3
9800.321	9803.009			0.3			10200.95	3F 2	2B15	P 4
9801.599	9804.287		0.1	0.1			10199.62			
9802.358	9805.046			0.3			10198.83	3E 0	2C 1	P 5
9802.608	9805.296		0.2				10198.57	3D 3	2C 4	Q 2
9803.079	9805.767						10198.08	3D 1	2B12	R 1
9803.483	9806.171		1.0	1.0			10197.66	2K 5	2B 5	P 2
9803.483	9806.171		1.0	1.0			10197.66	4E 0	2B16	Q 3
9804.464	9807.152		0.2	0.6			10196.64	3D 1	2B12	R 3
9804.464	9807.152		0.2	0.6			10196.64	2A 3	2B 4	R 2
9804.791	9807.479		0.1				10196.30			
9805.993	9808.682			0.2b			10195.05	3E 2	2C 3	P 4
9805.993	9808.682			0.2b			10195.05	3a 0	2c 3	P 3
9807.041	9809.730		0.3	0.3			10193.96	3D 1	2C 2	Q 3
9807.041	9809.730		0.3	0.3			10193.96	Z 0	2C 1	Q 1
9807.609	9810.298			0.5			10193.37	3D 0	2C 1	R 4
9807.696	9810.385						10193.28	3A 0	2B11	P 4
9808.129	9810.818		0.2b				10192.83			
9808.446	9811.136			1.0v			10192.50			
9809.139	9811.829		0.2	0.6			10191.78			
9809.274	9811.964						10191.64			
9809.688	9812.378		0.8	0.9			10191.21	3B 4	2A 3	P 1
9810.265	9812.955						10190.61	3B 4	2A 3	P 4
9810.660	9813.350		0.3	0.9			10190.20	3E 0	2B10	P 2
9811.353	9814.043		0.1	0.5			10189.48			
9811.690	9814.381		0.3	0.4			10189.13			
9812.490	9815.180			0.3			10188.30			
9813.424	9816.115			0.2			10187.33			
9814.272	9816.963		0.1	0.1			10186.45			
9814.898	9817.589		0.1	0.7			10185.80			
9815.707	9818.399		0.2				10184.96	Z 0	2C 1	R 2

λ_{air}	λ_{vac}	Ref.	I_1	I_2	I_3	I_4	ν	Upper	Lower	Br.
9816.103	9818.794			0.8			10184.55			
9817.192	9819.884			0.3			10183.42	4E 0	2B16	Q 1
9817.346	9820.038		0.1				10183.26	Z 1	2C 2	Q 2
9817.847	9820.539						10182.74			
9818.204	9820.896		0.8	0.6			10182.37			
9819.487	9822.179			0.1			10181.04	V 1	2B10	P 2
9821.484	9824.177			0.2			10178.97			
9823.443	9826.136			0.4			10176.94	3D 1	2C 2	R 4
9824.070	9826.764		0.2	0.5			10176.29	3D 2	2C 3	Q 4
9825.045	9827.739		0.2				10175.28	3d 1	2c 4	P 1
9825.354	9828.048						10174.96			
9825.663	9828.358		0.4	1.5b			10174.64			
9826.262	9828.957		0.2				10174.02			
9826.639	9829.333		0.1				10173.63	3E 1	2C 2	P 5
9826.890	9829.584		0.4				10173.37	3A 2	2B16	P 6
9827.160	9829.855		0.6	1.0b			10173.09	3B 3	2A 2	P 4
9827.160	9829.855		0.6	1.0b			10173.09	3b 1	2a 1	P 9
9827.160	9829.855		0.6	1.0b			10173.09	3C 2	2A 4	R 3
9827.460	9830.155		0.1d				10172.78			
9827.914	9830.609		0.2	0.3			10172.31			
9828.658	9831.353						10171.54			
9828.948	9831.643		0.8	1.5			10171.24	3B 4	2A 3	P 3
9828.948	9831.643		0.8	1.5			10171.24	Z 0	2C 1	P 1
9829.692	9832.387		0.1				10170.47			
9830.881	9833.577		0.1				10169.24			
9831.616	9834.312		0.2	0.5			10168.48			
9832.795	9835.492		0.3	0.3			10167.26			
9833.637	9836.333			0.6			10166.39			
9834.111	9836.807		0.4	0.3			10165.90			
9835.843	9838.540			0.1			10164.11	Z 2	2C 3	Q 3
9835.911	9838.608		0.2				10164.04			
9836.472	9839.169			0.2			10163.46			
9839.066	9841.764		0.6	0.9			10160.78			
9840.955	9843.653		0.6	1.5			10158.83	3b 3	2a 3	P 1
9840.955	9843.653		0.6	1.5			10158.83		2B 1	R 0
9843.862	9846.561			0.3			10155.83			
9843.968	9846.668		0.2				10155.72	4E 0	2B16	Q 2
9847.275	9849.975			0.2			10152.31	2A 3	2B 4	R 0
9847.673	9850.373			0.2			10151.90			
9849.050	9851.751			0.5			10150.48	3D 1	2C 2	P 3
9849.681	9852.382		0.2	0.4			10149.83	3D 0	2C 1	Q 3
9850.651	9853.353			0.2			10148.83			
9851.418	9854.120		0.1	0.4			10148.04			
9852.428	9855.129		0.5	0.8			10147.00	2K 5	2B 5	P 3
9852.943	9855.644			0.3			10146.47			
9853.564	9856.266		0.1				10145.83			
9854.205	9856.907		0.3	0.5			10145.17			
9854.846	9857.548			0.4			10144.51	3a 0	2c 3	P 5
9855.497	9858.200		0.4	0.6			10143.84	4D 0	2B16	R 4
9856.148	9858.851		0.2	1.5b			10143.17	3b 2	2a 2	P 6
9858.996	9861.699		0.2	0.3			10140.24			
9860.017	9862.721			0.4b			10139.19			
9861.038	9863.742		0.0				10138.14			
9861.992	9864.696			0.3			10137.16			
9862.731	9865.435		0.3	0.7			10136.40			
9863.383	9866.088		0.3	0.3			10135.73			
9864.035	9866.740			0.2			10135.06			
9865.184	9867.889		0.1				10133.88			
9866.498	9869.203			0.2			10132.53	X 2	2C 4	Q 1
9866.498	9869.203			0.2			10132.53	3E 1	2C 2	P 6

λ_{air}	λ_{vac}	Ref.	I_1	I_2	I_3	I_4	ν	Upper	Lower	Br.
9868.280	9870.986		0.2b	0.5			10130.70			
9869.060	9871.766		0.1b				10129.90			
9869.995	9872.701		0.3	0.7			10128.94	3e 0	2c 3	P 5
9872.403	9875.109			0.2			10126.47	3d 0	2c 3	P 2
9874.440	9877.148		0.2	1.5			10124.38	3b 0	2a 0	P12
9874.440	9877.148		0.2	1.5			10124.38	3E 2	2B15	R 4
9875.143	9877.850			0.3			10123.66			
9875.533	9878.241		0.1				10123.26	3D 0	2B10	R 5
9876.304	9879.012		0.3	0.4			10122.47	Z 0	2C 1	Q 2
9878.948	9881.657			0.3			10119.76	4f 3	3b 4	R 2
9881.663	9884.373		0.2				10116.98			
9882.191	9884.900		0.4	1.0r			10116.44	3B 3	2A 2	P 5
9882.191	9884.900		0.4	1.0r			10116.44	3F 0	2B11	Q 5
9882.679	9885.389		0.3				10115.94			
9885.767	9888.478			0.2			10112.78	2A 1	2B 1	R 1
9886.813	9889.524			0.3			10111.71			
9889.073	9891.784			0.2			10109.40			
9889.669	9892.381		0.7	0.5			10108.79	4E 0	2C 4	Q 1
9889.669	9892.381		0.7	0.5			10108.79	4F 0	2C 4	P 5
9890.139	9892.850			0.4			10108.31			
9890.354	9893.066		0.1				10108.09			
9892.096	9894.808			0.2			10106.31			
9893.672	9896.385			0.2			10104.70			
9895.288	9898.001		0.2	0.7			10103.05			
9895.856	9898.569			0.5			10102.47			
9896.179	9898.893		0.1				10102.14			
9896.679	9899.392		0.4	0.8			10101.63	2A 0	2B 0	P12
9897.620	9900.333			0.2			10100.67	3C 1	2A 3	R 3
9898.629	9901.343		0.5				10099.64			
9898.805	9901.519			1.5b			10099.46	3D 3	2C 4	Q 3
9899.031	9901.745		0.1				10099.23			
9899.599	9902.314		0.2	1.0			10098.65			
9900.139	9902.853		0.1				10098.10			
9900.776	9903.490		0.4	1.5			10097.45	2A 0	2B 0	R 7
9900.776	9903.490		0.4	1.5			10097.45	3E 0	2C 1	P 6
9902.208	9904.923			0.2			10095.99	Z 0	2C 1	R 3
9902.983	9905.698			0.1			10095.20			
9904.111	9906.826		0.2d	0.1			10094.05			
9904.690	9907.405			0.2			10093.46			
9904.984	9907.700		0.1				10093.16			
9905.897	9908.613		0.2	0.7			10092.23	2A 1	2B 1	P 6
9906.427	9909.143			0.6			10091.69			
9907.350	9910.066			0.2			10090.75			
9908.666	9911.382		0.2				10089.41	2A 3	2B 4	P 1
9909.196	9911.913		0.5	1.0			10088.87	3b 3	2a 3	P 2
9909.196	9911.913		0.5	1.0			10088.87		2B 1	R 0
9910.699	9913.416		0.6	1.0			10087.34	2K 5	2B 5	P 4
9911.505	9914.222		0.1d				10086.52			
9911.908	9914.625			0.2			10086.11			
9913.048	9915.766		0.0				10084.95			
9914.916	9917.634			0.2			10083.05			
9915.683	9918.401			0.2			10082.27	3F 2	2B15	Q 2
9916.391	9919.110			0.2			10081.55			
9919.264	9921.983		0.4	0.6			10078.63	3D 0	2C 1	P 3
9919.264	9921.983		0.4	0.6			10078.63	4g 0	2a 6	P 1
9920.130	9922.850			0.1			10077.75			
9920.789	9923.509		0.1	0.3			10077.08			
9922.385	9925.105			0.2b			10075.46			
9922.532	9925.253		0.2	0.3			10075.31	Z 0	2C 1	P 2
9923.389	9926.110			0.3			10074.44			

λ_{air}	λ_{vac}	Ref.	I_1	I_2	I_3	I_4	ν	Upper	Lower	Br.
9924.217	9926.938			0.1			10073.60	2B19	2A 0	R 5
9924.926	9927.647			0.1			10072.88	3A 0	2B11	P 1
9926.454	9929.175			0.1d			10071.33			
9927.705	9930.427		0.3	0.3			10070.06	4D 2	2C 3	P 4
9928.396	9931.118			0.1			10069.36	3F 0	2B11	R 3
9929.017	9931.739			0.1			10068.73			
9929.835	9932.558			0.2			10067.90			
9930.703	9933.426			0.6			10067.02			
9931.473	9934.196			0.4			10066.24	3D 1	2C 2	Q 4
9933.190	9935.913			0.2			10064.50	4c 2	2a 8	Q 1
9934.029	9936.753		0.2	1.0			10063.65	3C 2	2A 4	Q 4
9934.957	9937.681			0.2			10062.71	4F 0	2C 4	P 4
9934.957	9937.681			0.2			10062.71	4E 1	2B19	Q 4
9935.786	9938.510			0.0			10061.87	3E 0	2B11	R 5
9936.912	9939.636			0.2			10060.73	4E 0	2C 4	Q 3
9937.791	9940.516		0.2	0.2			10059.84			
9938.433	9941.158			0.2			10059.19			
9938.937	9941.662			0.2b			10058.68			
9939.026	9941.751		0.6	0.3			10058.59		2B 1	R 0
9939.718	9942.443			0.2			10057.89	4E 0	2C 4	R 3
9939.718	9942.443			0.2			10057.89	X 2	2C 4	Q 2
9940.449	9943.175		0.2	0.2			10057.15	3f 3	3b 1	R 3
9941.062	9943.788			0.2			10056.53			
9941.626	9944.351		0.2	1.0			10055.96			
9942.506	9945.231			0.2			10055.07	4D 0	2B16	R 3
9942.506	9945.231			0.2			10055.07	2A 3	2B 4	P 2
9942.585	9945.311		0.3				10054.99			
9946.789	9949.516		0.1	0.2			10050.74			
9947.551	9950.278			0.2			10049.97	3A 2	2B16	R 1
9949.274	9952.001			0.1			10048.23			
9951.710	9954.438		0.1	0.2			10045.77			
9952.344	9955.073			0.2			10045.13	3F 0	2B11	P 6
9952.344	9955.073			0.2			10045.13	3D 2	2C 3	Q 5
9953.365	9956.094		0.2	0.2			10044.10	3D 0	2B10	R 4
9954.376	9957.105			0.2			10043.08	3f 2	3b 0	R 1
9955.337	9958.067			0.1d			10042.11	3d 0	2c 3	Q 3
9956.596	9959.326			0.3			10040.84			
9959.572	9962.302			0.1			10037.84			
9960.217	9962.948			0.0			10037.19			
9961.279	9964.010			0.3			10036.12			
9962.153	9964.884		0.2	0.3			10035.24	4E 0	2C 4	R 2
9962.153	9964.884		0.2	0.3			10035.24	3D 0	2C 1	Q 4
9963.969	9966.701			0.1			10033.41			
9964.734	9967.466			0.2			10032.64			
9965.191	9967.923		0.2				10032.18	2A 1	2B 1	R 0
9965.400	9968.132			0.3			10031.97			
9966.194	9968.927			0.2			10031.17	4D 0	2B16	P 1
9966.999	9969.732			0.3			10030.36	3C 2	2A 4	R 2
9968.073	9970.805			0.2			10029.28			
9969.246	9971.979			0.2b			10028.10			
9969.942	9972.675			0.2			10027.40			
9971.115	9973.849			0.4			10026.22			
9971.632	9974.366			0.2			10025.70			
9972.040	9974.774			0.1			10025.29	4D 0	2B16	R 0
9972.776	9975.510		0.2b	1.0			10024.55	Z 1	2C 2	P 3
9973.721	9976.456			0.0			10023.60			
9974.398	9977.132			0.0			10022.92	3f 2	3b 0	Q 6
9975.174	9977.909		0.8	0.6			10022.14	Z 2	2C 3	Q 4
9975.632	9978.367			0.2			10021.68	4E 0	2C 4	R 1
9976.160	9978.895			0.2			10021.15			

λ_{air}	λ_{vac}	Ref.	I_1	I_2	I_3	I_4	ν	Upper	Lower	Br.
9977.155	9979.891		0.1	0.4			10020.15	2K 5	2B 5	P 5
9980.024	9982.760		0.2	1.5b			10017.27	3b 2	2a 2	P 7
9980.024	9982.760		0.2	1.5b			10017.27	2A 3	2B 4	P 3
9980.591	9983.328		0.2	0.3			10016.70	2A 1	2B 1	P 5
9980.591	9983.328		0.2	0.3			10016.70	Z 0	2C 1	Q 3
9983.671	9986.408			0.2			10013.61	Z 1	2B12	P 1
9988.409	9991.148			0.1			10008.86			
9990.056	9992.795			0.3			10007.21			
9990.546	9993.284		0.6	1.5b			10006.72	3b 3	2a 3	P 3
9991.384	9994.123			0.1			10005.88	4D 0	2B16	R 2
9991.973	9994.713			0.0			10005.29			
9992.523	9995.262			0.2			10004.74			
9992.663	9995.402		0.1				10004.60			
9993.212	9995.952			0.2			10004.05			
9993.841	9996.581			0.2			10003.42			
9994.621	9997.361			0.7			10002.64	2A 2	2B 3	R 4
9994.621	9997.361			0.7			10002.64	4g 0	2a 6	R 0
9995.230	9997.970			0.1			10002.03	3E 2	2B15	Q 6
9995.230	9997.970			0.1			10002.03	3E 2	2B15	R 3
9995.230	9997.970			0.1			10002.03	3f 2	3b 0	Q 5
9995.800	9998.540			0.1			10001.46	3d 3	3b 0	P 7
9996.779	9999.520		0.1	0.1d			10000.48			
9997.839	10000.580			0.1			9999.42			
9998.269	10001.010			0.2			9998.99	3E 0	2C 1	P 7
9998.939	10001.680		0.2	0.3			9998.32			

λ_{air}	λ_{vac}	Ref.	I_1	I_2	I_3	I_4	ν	Upper	Lower	Br.
10001.43	10004.17			0.1			9995.83	V 0	2C 1	Q 3
10001.43	10004.17			0.1			9995.83	4d 0	3b 0	P 6
10002.15	10004.89			0.2			9995.11			
10002.92	10005.66			0.2			9994.34			
10003.77	10006.51		0.2	0.3			9993.49	3f 3	3b 1	R 2
10004.69	10007.44			0.2			9992.57	3E 1	2C 2	P 7
10004.69	10007.44			0.2			9992.57	3B 3	2K 3	R 5
10004.69	10007.44			0.2			9992.57	3C 1	2A 3	Q 4
10005.41	10008.16			0.2			9991.85			
10005.51	10008.26			0.2			9991.75			
10006.26	10009.01			0.1			9991.00			
10007.27	10010.01			0.2			9990.00			
10007.80	10010.54			0.1			9989.47			
10008.48	10011.22			0.3			9988.79			
10009.38	10012.12		0.1	0.2			9987.89			
10010.46	10013.21		0.3	0.5			9986.81			
10011.55	10014.29			0.1			9985.73	4D 0	2B16	R 1
10012.22	10014.96		0.1				9985.06	Z 0	2C 1	R 4
10013.10	10015.84		0.3	0.3			9984.18	4E 1	2B19	Q 3
10014.31	10017.06			0.1d			9982.97			
10017.44	10020.19			1.0			9979.85	Z 1	2B12	P 2
10018.28	10021.02			0.4			9979.02	3D 3	2C 4	Q 4
10018.28	10021.02			0.4			9979.02	3f 2	3b 0	Q 4
10018.63	10021.38			0.4			9978.67			
10018.75	10021.50		0.3	0.3			9978.55		2B 1	P 2
10020.57	10023.31			0.2			9976.74	3D 0	2B10	R 3
10021.17	10023.92			0.1			9976.14			
10022.01	10024.76		0.1	0.4			9975.30	V 0	2C 1	Q 2
10023.15	10025.90			0.3			9974.17			
10023.92	10026.67			0.1			9973.40			
10024.22	10026.97		0.1				9973.10			
10024.91	10027.66		0.3	0.3			9972.42		2B	P 1
10024.91	10027.66		0.3	0.3			9972.42	2A 3	2B 4	P 4
10025.78	10028.53			0.2			9971.55	3c 2	2a 5	R 4
10025.78	10028.53			0.2			9971.55	3d 0	2c 3	P 3
10027.91	10030.66			0.2			9969.43			
10029.70	10032.44			0.2			9967.66			
10033.75	10036.50			0.4			9963.63	3C 1	2A 3	R 2
10034.04	10036.79		0.3	0.6			9963.34	2A 1	2B 1	P 4
10034.04	10036.79		0.3	0.6			9963.34	4D 0	2B16	P 2
10035.18	10037.93			0.2			9962.21	4e 3	3b 4	R 4
10036.33	10039.08			0.1			9961.07			
10039.54	10042.29			0.0			9957.89	3d 0	2c 3	Q 4
10040.46	10043.22		0.2	1.0b			9956.97			
10041.13	10043.88			0.1			9956.31	3f 2	3b 0	Q 3
10041.70	10044.46			0.2			9955.74			
10042.76	10045.52			0.3			9954.69			
10043.24	10045.99			0.3			9954.22			
10044.03	10046.79			0.1			9953.43			
10044.92	10047.68			0.2			9952.55	3A 2	2B16	R 0
10045.52	10048.27			0.1			9951.96	3D 1	2C 2	P 4
10047.67	10050.42			0.2			9949.83			
10048.30	10051.06			0.1			9949.20	4f 3	3b 4	R 1
10048.99	10051.75			0.5			9948.52	X 2	2C 4	Q 3
10048.99	10051.75			0.5			9948.52	2K 5	2B 5	P 6
10049.12	10051.88			0.6v			9948.39			
10049.37	10052.12		0.3b				9948.15	3D 0	2B10	R 2
10049.49	10052.24			0.3b			9948.03			
10052.00	10054.76			0.3			9945.54			
10052.09	10054.85			0.3			9945.45	3C 2	2A 4	Q 3

λ_{air}	λ_{vac}	Ref.	I_1	I_2	I_3	I_4	ν	Upper	Lower	Br.
10055.81	10058.57			0.2			9941.77			
10056.41	10059.17			0.1			9941.18	T 0	2B16	P 3
10056.98	10059.73			0.8			9940.62	3d 1	2c 4	P 3
10057.06	10059.82		0.2	0.8			9940.54			
10057.93	10060.69		0.2	0.3			9939.68			
10059.21	10061.97			0.1			9938.41	4E 1	2B19	Q 2
10059.21	10061.97			0.1			9938.41	4F 0	2B17	Q 4
10059.98	10062.74			0.3			9937.65			
10060.95	10063.70			0.1			9936.70	T 0	2B16	P 2
10060.95	10063.70			0.1			9936.70	Z 1	2B12	P 3
10061.55	10064.31		0.2d	0.2			9936.10	2A 1	2B 1	P 1
10061.98	10064.74			0.2			9935.68	3f 2	3b 0	Q 2
10062.70	10065.46			0.3			9934.97			
10063.76	10066.52			0.0			9933.92	4F 0	2B17	R 2
10064.39	10067.15			0.1			9933.30			
10065.90	10068.66			0.2			9931.81			
10066.89	10069.65			0.1			9930.83	V 0	2C 1	Q 1
10066.89	10069.65			0.1			9930.83	3f 3	3b 1	R 1
10067.70	10070.46			0.3			9930.03	2A 1	2B 1	P 3
10067.70	10070.46			0.3			9930.03	X 1	2B14	R 2
10068.34	10071.10			0.3			9929.40	3F 1	2B13	P 4
10069.16	10071.92		0.2	0.8			9928.59			
10069.97	10072.74			0.1			9927.79	T 0	2B16	P 1
10070.53	10073.29			0.1			9927.24			
10071.05	10073.81			0.2			9926.73			
10072.13	10074.89			0.4			9925.67			
10073.30	10076.06		0.3	0.3			9924.51			
10074.14	10076.90			0.3			9923.69			
10074.48	10077.24			0.3			9923.35			
10074.94	10077.70			0.2			9922.90			
10075.47	10078.23			0.2			9922.38			
10075.96	10078.72		0.3	0.4			9921.89	2A 1	2B 1	P 2
10076.70	10079.47			0.3			9921.16	X 1	2B14	R 1
10077.62	10080.38			0.3			9920.26			
10078.16	10080.92			0.3			9919.73	3D 0	2C 1	Q 5
10078.61	10081.38			0.1			9919.28	4E 1	2B19	Q 1
10080.49	10083.26			0.2d			9917.43	3D 0	2B10	P 1
10082.37	10085.14			0.2			9915.58			
10083.07	10085.83			0.1			9914.90			
10083.99	10086.76			0.2			9913.99	4E 0	2C 4	P 2
10083.99	10086.76			0.2			9913.99	3C 2	2A 4	R 1
10084.82	10087.58			1.0			9913.18	3b 3	2a 3	P 4
10085.56	10088.32			0.2			9912.45	3c 2	2a 5	R 3
10086.26	10089.03			0.1			9911.76	4D 0	2B16	P 6
10087.07	10089.84			0.2			9910.96	X 1	2B14	R 0
10087.89	10090.65			0.1			9910.16			
10088.37	10091.13			0.2			9909.69	X 1	2C 3	Q 1
10088.84	10091.60			0.1			9909.23			
10089.34	10092.11			0.3			9908.73			
10089.84	10092.61			0.5			9908.24	2A 2	2B 3	R 3
10089.84	10092.61			0.5			9908.24	3F 0	2B11	Q 4
10090.50	10093.26			0.1			9907.60			
10091.26	10094.03			0.0			9906.85			
10092.24	10095.00			0.2			9905.89			
10094.41	10097.18			0.1			9903.76			
10095.48	10098.25			0.2			9902.71			
10096.55	10099.32			0.1			9901.66			
10097.58	10100.35			0.2			9900.65			
10098.33	10101.10			0.2			9899.91			
10099.02	10101.79			0.3			9899.24			

λ_{air}	λ_{vac}	Ref.	I_1	I_2	I_3	I_4	ν	Upper	Lower	Br.
10100.73	10103.50			0.1			9897.56	4g 0	2a 6	R 1
10102.14	10104.91			0.2			9896.18	Z 1	2B12	P 4
10102.90	10105.66			0.2			9895.44	3C 0	2K 2	R 4
10103.53	10106.30			0.1			9894.82			
10104.19	10106.96			0.2			9894.17			
10104.84	10107.61			0.2			9893.54			
10105.45	10108.22			0.5			9892.94			
10105.98	10108.75			0.1			9892.42			
10106.53	10109.30		0.4	0.5			9891.88			
10107.26	10110.03			0.2			9891.17			
10108.05	10110.82			0.4			9890.39			
10108.68	10111.45			0.1d			9889.78	4d 0	3b 0	P 7
10109.31	10112.08			0.2			9889.16			
10110.52	10113.29			0.1b			9887.98			
10111.15	10113.92			0.4			9887.36			
10111.97	10114.74			0.3			9886.56	U 1	2B13	R 3
10111.97	10114.74			0.3			9886.56	3c 0	2a 3	R 6
10112.45	10115.22						9886.09			
10113.04	10115.82			0.1d			9885.51			
10114.66	10117.43						9883.93	2A 4	2B 6	R 3
10115.21	10117.99		0.3	1.5b			9883.39	2A 0	2B 0	R 6
10116.18	10118.95			0.8			9882.45			
10117.31	10120.08		0.3	0.9			9881.34			
10117.94	10120.71			0.3			9880.73			
10118.62	10121.40			0.2			9880.06			
10119.43	10122.21			0.2			9879.27			
10120.44	10123.21			0.2			9878.29	3d 0	2c 3	Q 5
10121.48	10124.25			0.3			9877.27			
10122.38	10125.16			0.1			9876.39			
10123.02	10125.79			0.2			9875.77	2A 4	2B 6	R 1
10123.59	10126.37			0.3			9875.21	Z 0	2C 1	Q 4
10123.59	10126.37			0.3			9875.21	3C 1	2A 3	Q 3
10124.21	10126.98			0.1			9874.61			
10124.78	10127.56			0.2			9874.05			
10125.49	10128.26			0.4			9873.36			
10126.26	10129.03			0.2			9872.61	3C 2	2A 4	P 4
10126.89	10129.67			0.2			9871.99			
10127.70	10130.48			0.3			9871.20			
10128.84	10131.62			0.2			9870.09	X 1	2C 3	R 2
10129.76	10132.53		0.2d	1.5b			9869.20			
10130.59	10133.37			0.2			9868.39			
10132.72	10135.50			0.2			9866.31	3E 1	2B13	Q 6
10133.71	10136.49			0.2d			9865.35	3D 1	2C 2	Q 6
10133.71	10136.49			0.2d			9865.35	3f 3	3b 1	Q 5
10135.16	10137.94			0.3			9863.94			
10137.55	10140.33			0.2d			9861.61			
10138.69	10141.47			0.2			9860.50	3D 0	2B10	P 2
10139.57	10142.35			0.2			9859.65			
10140.17	10142.94			0.2			9859.07	2A 4	2B 6	R 0
10142.68	10145.46			0.3b			9856.63	3C 2	2A 4	Q 2
10143.34	10146.12			0.2			9855.98	3E 0	2B11	R 4
10143.92	10146.70			0.1			9855.42			
10144.56	10147.34			0.3			9854.80	3f 3	3b 1	Q 4
10145.34	10148.12			0.4			9854.04	3C 1	2A 3	R 1
10145.95	10148.73			0.1			9853.45	3c 0	2a 5	R 2
10146.70	10149.48			0.2			9852.72	Z 2	2C 3	P 5
10147.81	10150.59			0.3			9851.64			
10148.67	10151.45			0.1			9850.81			
10149.50	10152.28			0.2			9850.00			
10150.01	10152.79			0.3			9849.51			

λ_{air}	λ_{vac}	Ref.	I_1	I_2	I_3	I_4	ν	Upper	Lower	Br.
10150.45	10153.23		0.2d				9849.08			
10150.65	10153.43			0.1			9848.89			
10151.26	10154.05			0.1			9848.29	3F 0	2B11	P 5
10152.69	10155.47			0.2			9846.91			
10153.18	10155.97			0.1			9846.43			
10153.69	10156.47			0.2			9845.94	Z 1	2C 2	P 4
10154.36	10157.14			0.2			9845.29	4f 2	3b 3	R 2
10155.12	10157.90			0.1			9844.55			
10155.62	10158.40		0.3	0.9			9844.07	2A 2	2B 3	R 2
10155.62	10158.40		0.3	0.9			9844.07	3E 2	2B15	Q 5
10155.62	10158.40		0.3	0.9			9844.07	3b 4	2a 4	R 3
10156.46	10159.25			0.1d			9843.25	3F 0	2B11	R 2
10157.85	10160.63			0.2			9841.91	X 0	2C 2	R 1
10157.85	10160.63			0.2			9841.91	3e 2	3b 0	R 5
10157.85	10160.63			0.2			9841.91	3f 3	3b 1	Q 3
10158.61	10161.39			0.3			9841.17			
10159.78	10162.56			0.3			9840.04			
10163.38	10166.17			0.3			9836.55	3d 0	2c 3	P 4
10165.22	10168.01		0.2d				9834.77			
10168.09	10170.88			0.3			9831.99			
10169.30	10172.09			0.2			9830.82	3f 3	3b 1	Q 2
10170.51	10173.30			0.2			9829.65	4D 0	2C 4	Q 1
10171.50	10174.29			0.2			9828.70			
10172.49	10175.28			0.2			9827.74	X 2	2C 4	Q 4
10173.06	10175.85			0.3			9827.19			
10173.86	10176.65			0.7			9826.42			
10174.63	10177.42			0.2			9825.67	3C 2	2A 4	R 0
10175.71	10178.50			0.2			9824.63	3c 0	2a 3	R 5
10176.36	10179.15			0.1			9824.00	3b 4	2a 4	R 2
10176.78	10179.57			0.1			9823.60			
10177.19	10179.98			0.3			9823.20	3B 3	2K 3	R 4
10177.79	10180.58			0.5			9822.62	3F 1	2B13	P 3
10178.76	10181.55			0.2			9821.69	3c 3	2a 6	Q 3
10178.86	10181.65			0.3			9821.59			
10179.88	10182.67			0.2b			9820.61			
10180.60	10183.39			0.1			9819.91			
10181.35	10184.14		0.2	0.5			9819.19	2K 2	2B 3	R 1
10182.42	10185.21			0.2d			9818.16			
10183.52	10186.31			0.2b			9817.10			
10184.87	10187.66			0.3			9815.80			
10185.73	10188.52			0.3			9814.97	3c 3	2a 6	Q 2
10186.25	10189.04			0.6			9814.47	V 0	2C 1	Q 4
10186.99	10189.78			0.2			9813.75			
10187.21	10190.00		0.1				9813.54			
10187.68	10190.47			0.2d			9813.09			
10188.38	10191.18			0.2			9812.41	V 0	2C 1	P 2
10189.06	10191.85		0.2	0.6			9811.76	X 0	2C 2	R 2
10189.66	10192.45			0.1			9811.18	4f 3	3b 4	P 4
10190.11	10192.90			0.4			9810.75	3c 1	2a 4	R 4
10190.11	10192.90			0.4			9810.75	4g 0	2a 6	R 2
10190.11	10192.90			0.4			9810.75	3c 3	2a 6	Q 1
10190.96	10193.75			0.2			9809.93			
10191.88	10194.68		0.2	1.5b			9809.04	3b 3	2a 3	P 5
10192.63	10195.43			0.1			9808.32			
10193.12	10195.91		0.3	0.5			9807.85	2K 2	2B 3	R 0
10193.60	10196.39			0.5			9807.39	U 1	2B13	R 2
10194.44	10197.23			0.2			9806.58			
10195.35	10198.14			0.2			9805.71	3D 0	2C 1	Q 6
10195.80	10198.60		0.1				9805.27	2A 4	2B 6	P 1
10196.06	10198.86			0.9			9805.02			

λ_{air}	λ_{vac}	Ref.	I_1	I_2	I_3	I_4	ν	Upper	Lower	Br.
10196.79	10199.59			0.2			9804.32	3b 4	2a 4	R 1
10197.64	10200.44			0.9			9803.50	4d 0	3b 0	P 8
10198.40	10201.20			0.2			9802.77	T 0	2C 4	Q 1
10198.87	10201.67			0.3			9802.32	3D 2	2B15	R 6
10199.45	10202.25			0.1			9801.76	4f 0	3b 1	R 4
10200.78	10203.57		0.2	0.7			9800.49			
10201.54	10204.33			0.2			9799.76			
10202.16	10204.96			0.1			9799.16	3D 0	2B10	P 3
10203.25	10206.05		0.1	0.3d			9798.11	3C 2	2A 4	Q 1
10203.25	10206.05		0.1	0.3d			9798.11	3C 2	2A 4	P 3
10203.25	10206.05		0.1	0.3d			9798.11	V 0	2C 1	P 3
10204.54	10207.34		0.3	0.3			9796.87			
10205.98	10208.78			0.2			9795.49	3c 2	2a 5	R 1
10207.02	10209.82			0.2			9794.49	V 0	2B10	R 2
10207.55	10210.34			0.2			9793.99	X 1	2C 3	P 2
10208.17	10210.97			0.2			9793.39	4f 2	3b 3	P 6
10208.98	10211.78			0.3			9792.61	3e 2	3b 0	R 4
10209.41	10212.21		0.1				9792.20			
10209.89	10212.69			0.3			9791.74	3C 1	2A 3	Q 2
10210.64	10213.44		0.2	0.3			9791.02	X 0	2C 2	Q 1
10211.81	10214.61			0.2d			9789.90	3E 2	2B15	R 2
10212.58	10215.38			0.1d			9789.16			
10213.95	10216.75			0.3			9787.85			
10214.46	10217.26			0.2			9787.36			
10215.19	10217.99			0.9			9786.66	3E 2	2B15	R 1
10216.12	10218.92			0.1			9785.77	3c 1	2a 4	R 3
10217.11	10219.91			0.2			9784.82			
10218.68	10221.48			0.2			9783.32	4E 0	2C 4	P 3
10218.68	10221.48			0.2			9783.32	Z 0	2C 1	P 4
10219.57	10222.37			0.1			9782.47			
10220.35	10223.15			0.3d			9781.72			
10221.24	10224.04			0.1			9780.87			
10221.88	10224.68			0.3			9780.26	3A 2	2C 4	Q 2
10222.50	10225.30			0.4			9779.66	3C 1	2A 3	P 4
10223.19	10225.99			0.4			9779.00	T 0	2C 4	Q 2
10225.02	10227.82			0.2			9777.25			
10225.57	10228.37			0.2			9776.73			
10226.39	10229.20			0.3			9775.94			
10227.44	10230.24			0.2			9774.94	3b 4	2a 4	R 0
10228.20	10231.01			0.3			9774.21			
10230.36	10233.16			0.3			9772.15			
10231.31	10234.12			0.1			9771.24	3C 1	2A 3	R 0
10232.17	10234.97		0.2d				9770.42	3F 3	2B18	Q 3
10232.35	10235.15			0.2			9770.25	3A 2	2C 4	Q 3
10233.09	10235.90			0.6			9769.54			
10233.64	10236.44		0.3	0.4			9769.02	2A 4	2B 6	P 2
10233.64	10236.44		0.3	0.4			9769.02	U 1	2B13	R 0
10234.50	10237.30			0.2			9768.20			
10235.13	10237.94			0.2			9767.59			
10235.63	10238.43			0.1			9767.12	3D 0	2B10	P 4
10236.35	10239.16			0.4			9766.43	X 0	2C 2	R 3
10236.95	10239.75			0.4			9765.86			
10237.98	10240.78			0.2			9764.88	X 0	2C 2	P 1
10238.67	10241.47			0.2			9764.22			
10241.19	10244.00			0.2			9761.81			
10241.71	10244.52			0.1			9761.32	3c 0	2a 3	R 4
10243.26	10246.07			0.2			9759.84			
10245.25	10248.05			0.2			9757.95			
10246.35	10249.16			0.4			9756.90	4e 3	3b 4	R 2
10246.54	10249.35		0.2b				9756.72	3A 2	2C 4	Q 1

λ_{air}	λ_{vac}	Ref.	I_1	I_2	I_3	I_4	ν	Upper	Lower	Br.
10247.37	10250.18			0.2			9755.93	V 0	2B10	R 3
10248.62	10251.43		0.2	0.3			9754.74	2K 2	2B 3	P 1
10249.42	10252.23			0.4			9753.98			
10250.40	10253.21		0.1	0.3			9753.04			
10251.02	10253.83			0.1			9752.45			
10251.77	10254.58			0.5			9751.74	3A 2	2C 4	Q 5
10252.65	10255.46			0.2			9750.90	3A 2	2C 4	Q 4
10253.34	10256.15			0.6			9750.25			
10253.68	10256.49			0.9			9749.92			
10253.88	10256.69						9749.73			
10256.29	10259.10			0.2			9747.44			
10256.70	10259.51			0.4			9747.05			
10257.30	10260.11		0.1	0.2			9746.48	3D 0	2C 1	P 5
10257.86	10260.67			0.1			9745.95			
10258.31	10261.13			0.2			9745.52			
10259.17	10261.98			0.3			9744.71			
10259.77	10262.58			0.2			9744.14			
10260.58	10263.39			0.1			9743.37	4g 0	2a 6	R 3
10261.37	10264.18			1.5b			9742.62			
10262.34	10265.15			0.2d			9741.70	4F 0	2B17	Q 3
10262.57	10265.38			0.2			9741.48			
10264.30	10267.11			0.2			9739.84			
10265.02	10267.84			0.2			9739.15	3C 1	2A 3	Q 1
10265.02	10267.84			0.2			9739.15	3c 2	2a 5	R 0
10265.02	10267.84			0.2			9739.15	3E 3	2B17	P 5
10266.25	10269.06			0.1d			9737.99	4f 3	3b 4	P 3
10267.12	10269.93			0.1			9737.16			
10269.30	10272.12			0.1			9735.09	3F 0	2B11	Q 3
10269.91	10272.72			0.3			9734.52			
10270.95	10273.77			0.3			9733.53	X 0	2C 2	Q 2
10270.95	10273.77			0.3			9733.53	Z 0	2B10	P 1
10272.36	10275.18			0.2			9732.19	6c 0	2a 8	Q 1
10272.36	10275.18			0.2			9732.19	3c 1	2a 4	R 2
10272.80	10275.61			0.1			9731.78			
10273.58	10276.39			0.2			9731.04			
10274.63	10277.45			0.2			9730.04			
10275.45	10278.26			0.2			9729.27	X 1	2C 3	Q 3
10276.54	10279.35			0.2			9728.24			
10277.32	10280.13			0.2			9727.50			
10278.08	10280.89			0.9			9726.78			
10278.35	10281.17		0.1				9726.52	2A 4	2B 6	P 3
10278.97	10281.78			0.2			9725.94			
10279.73	10282.54			0.8			9725.22			
10280.24	10283.06			0.2			9724.73			
10280.93	10283.75			0.2			9724.08			
10282.94	10285.76			0.2			9722.18			
10283.32	10286.14			0.1			9721.82			
10283.86	10286.68			0.4			9721.31	3E 3	2B17	P 4
10285.62	10288.44			0.1			9719.65			
10286.22	10289.04		0.1				9719.08			
10286.64	10289.46			0.2			9718.68			
10287.62	10290.44			0.2			9717.76	3c 3	2a 6	P 2
10289.31	10292.13			0.4			9716.16	4D 0	2C 4	Q 2
10289.97	10292.79			0.2			9715.54			
10290.60	10293.42			0.2			9714.94			
10291.28	10294.10			0.2			9714.30			
10292.19	10295.01			0.2			9713.44			
10292.85	10295.67		0.1	0.3			9712.82	3C 1	2A 3	P 3
10293.41	10296.23			0.3			9712.29			
10294.09	10296.91			0.1			9711.65	Z 0	2C 1	Q 5

λ_{air}	λ_{vac}	Ref.	I_1	I_2	I_3	I_4	ν	Upper	Lower	Br.
10294.64	10297.46			0.5			9711.13			
10295.58	10298.41			0.2			9710.24			
10296.56	10299.38		0.1b	0.9			9709.32			
10297.30	10300.12			0.1			9708.62	4F 0	2B17	R 1
10298.42	10301.24			0.1			9707.57	3D 2	2B15	R 5
10299.35	10302.17		0.4	0.6			9706.69	2K 2	2B 3	P 2
10299.35	10302.17		0.4	0.6			9706.69	4e 2	3b 3	R 4
10299.96	10302.78			0.3			9706.12			
10300.87	10303.69			0.2			9705.26	Z 0	2B10	P 2
10301.62	10304.44			0.2			9704.55			
10302.18	10305.01			0.2			9704.02	3d 0	2c 3	P 5
10303.04	10305.87		0.1	0.6			9703.21			
10303.63	10306.45			0.2			9702.66			
10304.15	10306.97		0.1				9702.17	4E 0	2C 4	P 4
10305.19	10308.01			0.2			9701.19			
10307.82	10310.65		0.1				9698.71			
10308.83	10311.66		0.3	1.0			9697.76	3E 3	2B17	P 3
10308.83	10311.66		0.3	1.0			9697.76	3c 0	2a 3	R 3
10309.57	10312.39			0.1			9697.07			
10310.15	10312.98		0.1	0.2			9696.52			
10310.68	10313.51			0.2			9696.02	3C 0	2K 2	R 3
10311.62	10314.45			0.9			9695.14	3E 0	2B11	R 3
10311.62	10314.45			0.9			9695.14	3b 3	2a 3	P 6
10313.36	10316.19			0.5			9693.50			
10313.94	10316.77			0.1			9692.96			
10314.56	10317.38			0.2			9692.38	3D 1	2C 2	Q 7
10315.38	10318.20		0.2d	0.3			9691.61			
10316.24	10319.07			0.1			9690.80			
10317.13	10319.96			0.2			9689.96			
10318.04	10320.87			0.3			9689.11			
10319.36	10322.19		0.2	1.5r			9687.87	2A 0	2B 0	R 5
10319.36	10322.19		0.2	1.5r			9687.87	Y 1	2C 2	R 1
10321.86	10324.69			0.4			9685.52			
10323.67	10326.50		0.3	1.5b			9683.82	2A 0	2B 0	P10
10323.67	10326.50		0.3	1.5b			9683.82	U 1	2B13	P 5
10324.59	10327.42			0.3			9682.96	4e 3	3b 4	Q 4
10325.81	10328.64			0.2			9681.82	3C 1	2A 3	P 2
10327.32	10330.15			0.3			9680.40	X 0	2C 2	P 2
10327.96	10330.79			0.2			9679.80	2B18	2A 0	P 2
10328.67	10331.50			0.3			9679.14			
10329.03	10331.86			0.2			9678.80	3D 0	2C 1	Q 7
10329.03	10331.86			0.2			9678.80	3c 1	2a 4	R 1
10329.03	10331.86			0.2			9678.80	3b 4	2a 4	P 1
10329.89	10332.72			0.3			9677.99			
10330.48	10333.31			0.4			9677.44			
10331.16	10333.99			0.4			9676.80	2A 4	2B 6	P 4
10331.16	10333.99			0.4			9676.80	3c 3	2a 6	P 3
10332.09	10334.92		0.2	0.1			9675.93	3F 3	2B18	P 4
10332.09	10334.92		0.2	0.1			9675.93	3E 2	2B15	P 3
10332.72	10335.55			0.5			9675.34			
10333.21	10336.05			0.3			9674.88	3B 3	2K 3	R 3
10333.54	10336.38		0.2	0.2			9674.57			
10334.11	10336.94		0.3	0.4			9674.04	3F 3	2B18	Q 2
10335.55	10338.39			0.2			9672.69	3E 1	2B13	Q 5
10335.55	10338.39			0.2			9672.69	4b 1	2a 6	R 2
10336.26	10339.09			0.2			9672.03			
10338.79	10341.63			0.2			9669.66			
10339.45	10342.29			0.1			9669.04			
10340.22	10343.06			0.2			9668.32			
10340.98	10343.82			0.2			9667.61			

λ_{air}	λ_{vac}	Ref.	I_1	I_2	I_3	I_4	ν	Upper	Lower	Br.
10341.62	10344.45			0.2			9667.02	U 1	2B13	P 3
10342.95	10345.79			0.3d			9665.77			
10343.77	10346.60			0.2			9665.01	Z 0	2B10	P 3
10343.77	10346.60			0.2			9665.01	3D 2	2B15	R 4
10344.55	10347.38			0.2			9664.28	3F 0	2B11	P 4
10345.29	10348.12		0.2b	0.3			9663.59	3e 2	3b 0	R 2
10345.93	10348.76			0.2			9662.99			
10346.44	10349.28			0.2			9662.51			
10346.93	10349.77			0.2			9662.05			
10347.60	10350.43			0.3			9661.43	4f 2	3b 3	P 5
10348.48	10351.31			0.2			9660.61			
10349.53	10352.36			0.2			9659.63			
10350.36	10353.20			0.2			9658.85	4g 0	2a 6	P 3
10351.61	10354.44			0.2			9657.69	3A 1	2B14	P 4
10352.27	10355.11			0.2			9657.07			
10354.28	10357.11			0.1			9655.20	W 1	2B14	R 3
10355.16	10357.99			0.4			9654.38			
10355.82	10358.66			0.3			9653.76			
10357.14	10359.98			0.2			9652.53			
10357.57	10360.41			0.2d			9652.13	3F 0	2B11	R 1
10357.57	10360.41			0.2d			9652.13	V 0	2B10	R 1
10358.32	10361.16			0.4			9651.43	2K 2	2B 3	P 3
10358.87	10361.71			0.3			9650.92	X 0	2C 2	Q 3
10358.87	10361.71			0.3			9650.92	3E 2	2B15	Q 3
10359.54	10362.38			0.1			9650.29			
10360.04	10362.88		0.2	0.3			9649.83	Y 1	2C 2	Q 1
10360.50	10363.34			0.2			9649.40			
10361.74	10364.58			0.2			9648.24	3f 3	3b 1	P 4
10362.77	10365.61			0.1			9647.29			
10363.49	10366.33			0.4			9646.62			
10364.84	10367.68			0.2			9645.36			
10365.66	10368.50			0.5			9644.60	2A 2	2B 3	P 6
10366.49	10369.33			0.0			9643.82			
10368.16	10371.00		0.1				9642.27	U 1	2B13	P 4
10369.07	10371.92			0.2			9641.42			
10369.81	10372.65		0.3	0.4			9640.74			
10370.66	10373.50			0.2			9639.95	Y 1	2C 2	R 2
10371.29	10374.13			0.3			9639.36			
10372.07	10374.91			0.1			9638.64	2K 2	2B 3	R 2
10372.70	10375.54			0.3			9638.05			
10373.43	10376.27		0.1	0.7			9637.37	3B 4	2K 4	P 5
10374.28	10377.13		0.1	0.7			9636.58			
10374.77	10377.61			0.4			9636.13			
10375.97	10378.82			0.2			9635.01			
10376.35	10379.19			0.2			9634.66			
10377.01	10379.85			0.1			9634.05	Z 0	2B10	P 4
10377.01	10379.85			0.1			9634.05	3c 0	2a 3	R 2
10377.01	10379.85			0.1			9634.05	4D 0	2C 4	Q 3
10377.88	10380.72			0.4			9633.24			
10378.90	10381.75			0.5			9632.29	3D 2	2B15	R 3
10380.50	10383.34			0.3			9630.81	X 0	2C 2	R 4
10381.14	10383.99			0.2			9630.21	4E 0	2C 4	P 5
10381.97	10384.82			0.2			9629.44	Y 1	2C 2	P 1
10382.65	10385.50		0.2	0.3			9628.81	3A 1	2C 3	Q 1
10383.33	10386.18			0.1			9628.18			
10383.85	10386.70		0.1	0.3			9627.70			
10384.74	10387.58		0.2				9626.88			
10385.59	10388.43			0.2			9626.09			
10386.34	10389.19			0.1			9625.39			
10386.94	10389.78			0.2			9624.84			

λ_{air}	λ_{vac}	Ref.	I_1	I_2	I_3	I_4	ν	Upper	Lower	Br.
10387.35	10390.19		0.1				9624.46			
10387.64	10390.48			0.1			9624.19			
10388.16	10391.00			0.2			9623.71			
10389.04	10391.89			0.3b			9622.89	3A 1	2C 3	Q 2
10390.26	10393.11			0.8			9621.76	2K 4	2B 5	R 4
10390.70	10393.55			0.0			9621.35			
10391.15	10393.99			0.1			9620.94			
10391.81	10394.65			0.3			9620.33	3A 1	2C 3	Q 3
10392.29	10395.14			0.3			9619.88			
10392.88	10395.72			0.3			9619.34	2A 4	2B 6	P 5
10393.90	10396.75			0.2			9618.39			
10394.65	10397.50			0.3			9617.70			
10395.24	10398.09			0.2			9617.15	3f 3	3b 1	P 5
10395.92	10398.77			0.4			9616.52	3B 4	2K 4	P 4
10397.25	10400.10			0.3			9615.29	2K 2	2B 3	R 3
10397.25	10400.10			0.3			9615.29	3B 4	2K 4	P 1
10398.17	10401.02			0.1			9614.44			
10398.83	10401.68		0.2b	1.0			9613.83	V 0	2B10	P 4
10399.84	10402.69			0.5			9612.90	3b 4	2a 4	P 2
10400.51	10403.36			0.1d			9612.28	4E 0	2B17	R 2
10400.51	10403.36			0.1d			9612.28	4b 1	2a 6	R 1
10401.25	10404.10		0.1	0.4			9611.60	2A 2	2B 3	P 5
10401.93	10404.78		0.3	0.7			9610.97	2A 2	2B 3	P 4
10405.58	10408.43			0.2			9607.60			
10406.45	10409.30		0.1	0.2			9606.79	4F 0	2B17	P 3
10406.94	10409.79			0.3			9606.34	2A 2	2B 3	R 1
10407.42	10410.27			0.3			9605.90			
10407.66	10410.51		0.2	0.5			9605.68	3B 4	2K 4	P 3
10408.93	10411.79		0.3	0.9			9604.50	3B 4	2K 4	P 2
10410.08	10412.94			0.5			9603.44	2K 4	2B 5	R 3
10411.56	10414.41			0.1			9602.08	3D 2	2B15	P 1
10411.56	10414.41			0.1			9602.08	4e 2	3b 3	R 3
10413.34	10416.19			0.3			9600.44			
10413.91	10416.76			0.6			9599.91			
10414.57	10417.43			0.1			9599.30	3A 1	2B14	P 3
10415.46	10418.32			0.2d			9598.48			
10416.14	10418.99			0.3			9597.86	3F 2	2B16	Q 4
10417.18	10420.03		0.2	1.0			9596.90			
10417.98	10420.83			0.2			9596.16	3E 0	2B11	P 6
10417.98	10420.83			0.2			9596.16	3c 1	2a 4	Q 5
10418.75	10421.61			0.9			9595.45			
10419.75	10422.61			0.2			9594.53	3F 0	2B11	Q 2
10420.72	10423.57			0.1			9593.64	3f 3	3b 1	P 6
10421.24	10424.09			0.1			9593.16			
10421.80	10424.66			0.2			9592.64			
10422.87	10425.72			0.1			9591.66			
10423.42	10426.28			0.1			9591.15	3E 0	2B11	Q 6
10423.98	10426.83		0.2d	0.2			9590.64	3A 1	2C 3	P 1
10425.07	10427.93			0.3			9589.63	3C 0	2K 2	Q 4
10425.90	10428.76			0.4			9588.87			
10426.54	10429.40			0.2			9588.28			
10427.64	10430.50			0.2			9587.27			
10428.74	10431.60			0.1			9586.26	Z 0	2C 1	P 5
10428.74	10431.60			0.1			9586.26	W 1	2B14	R 2
10429.26	10432.12		0.3	0.6			9585.78	2K 4	2B 5	R 2
10429.26	10432.12		0.3	0.6			9585.78	3D 2	2B15	R 1
10430.12	10432.98			0.2			9584.99			
10430.90	10433.76			0.3			9584.27	3e 2	3b 0	R 1
10431.51	10434.37			0.1			9583.71			
10432.17	10435.03			0.1			9583.11			

λ_{air}	λ_{vac}	Ref.	I_1	I_2	I_3	I_4	ν	Upper	Lower	Br.
10432.98	10435.84		0.2	0.2			9582.36			
10433.64	10436.50			0.1			9581.76			
10434.24	10437.09			0.1			9581.21	3A 2	2B17	R 4
10434.24	10437.09			0.1			9581.21	4c 1	2a 7	Q 1
10434.87	10437.73			0.2			9580.63	Z 0	2B10	P 5
10435.97	10438.83			0.2			9579.62	Y 1	2C 2	Q 2
10439.51	10442.37			0.2			9576.37			
10440.93	10443.79			0.1d			9575.07			
10442.01	10444.87			0.7			9574.08			
10443.85	10446.71			1.5b			9572.39			
10444.61	10447.48			0.2			9571.69			
10445.60	10448.46			0.1			9570.79			
10446.14	10449.00			0.1			9570.29			
10446.69	10449.55			0.3			9569.79			
10447.32	10450.18			0.2			9569.21			
10447.75	10450.61			0.2			9568.82			
10448.37	10451.23			0.3			9568.25	2K 4	2B 5	R 1
10449.57	10452.43			0.2			9567.15	4f 0	3b 1	R 3
10450.90	10453.77			0.2b			9565.93	3D 1	2B13	R 6
10450.90	10453.77			0.2b			9565.93	3A 1	2B14	P 1
10450.90	10453.77			0.2b			9565.93	4F 0	2B17	Q 2
10451.94	10454.80			0.2			9564.98	3A 0	2C 2	R 2
10453.54	10456.40			0.2b			9563.52			
10454.87	10457.73		0.1	0.5			9562.30			
10455.91	10458.77		0.1d				9561.35			
10456.74	10459.61			0.1			9560.59			
10458.42	10461.29			0.1			9559.05	4D 0	2C 4	Q 4
10458.42	10461.29			0.1			9559.05	3E 1	2B13	Q 4
10459.13	10461.99			0.3			9558.41			
10459.66	10462.53			0.2			9557.92			
10461.61	10464.48			0.2			9556.14			
10462.25	10465.11		0.2	0.3			9555.56	2A 2	2B 3	R 0
10463.00	10465.87			0.1			9554.87			
10463.41	10466.27			0.2			9554.50	4b 1	2a 6	R 0
10464.16	10467.03			0.3			9553.81			
10464.85	10467.72			0.2			9553.18	3C 0	2A 2	R 2
10465.35	10468.22			0.3			9552.72	3B 3	2K 3	R 2
10466.03	10468.90			0.2			9552.10	3E 2	2B15	Q 1
10466.55	10469.42			0.1			9551.63			
10467.62	10470.49			0.2d			9550.65			
10468.23	10471.09			0.2			9550.10			
10469.63	10472.50		0.1	0.3d			9548.82			
10470.09	10472.96			1.5			9548.40	3C 1	2K 4	Q 5
10470.28	10473.15		0.3				9548.23	X 0	2C 2	Q 4
10470.28	10473.15		0.3				9548.23	2K 4	2B 5	R 0
10473.29	10476.16		0.2	1.5b			9545.48	X 0	2B12	P 1
10474.32	10477.19		0.1	0.1d			9544.54	3D 3	2B17	P 3
10475.14	10478.01			0.2			9543.80			
10476.45	10479.32			0.1			9542.60	4f 2	3b 3	P 4
10477.63	10480.50			0.2			9541.53			
10478.06	10480.93			1.0			9541.14			
10479.01	10481.88			0.4			9540.27	3E 0	2B11	R 2
10480.13	10483.00			0.2			9539.25	Y 1	2C 2	P 2
10480.74	10483.61		0.2	0.1			9538.70			
10481.13	10484.00			0.3			9538.34			
10481.87	10484.74			0.2			9537.67			
10482.76	10485.63			0.3			9536.86			
10483.65	10486.52			0.2			9536.05			
10483.96	10486.83		0.1	0.5			9535.77	3b 4	2a 4	P 3
10484.63	10487.50			0.3			9535.16	3C 1	2K 4	R 3

λ_{air}	λ_{vac}	Ref.	I_1	I_2	I_3	I_4	ν	Upper	Lower	Br.
10485.32	10488.19			0.2			9534.53			
10486.21	10489.08			0.1			9533.72			
10487.04	10489.91			0.1d			9532.97			
10487.60	10490.47			0.3			9532.46			
10488.14	10491.01			0.1			9531.97	V 1	2B11	R 2
10488.84	10491.72			0.2			9531.33	3B 3	2K 3	P 5
10489.31	10492.19			0.3			9530.90			
10490.14	10493.01			0.1			9530.15			
10490.90	10493.77			0.2			9529.46			
10491.68	10494.56			0.3			9528.75	3E 2	2B15	P 2
10492.24	10495.12			0.2			9528.24	3F 0	2B11	P 3
10492.68	10495.56		0.2d				9527.84			
10492.91	10495.79			0.3			9527.63	4b 0	2a 5	R 4
10493.76	10496.64			0.2			9526.86			
10494.45	10497.32		0.1	0.3			9526.24	W 1	2C 3	R 2
10495.04	10497.92			0.2			9525.70			
10495.58	10498.46			0.1			9525.21	V 0	2C 1	P 5
10496.76	10499.64		0.3	0.4			9524.14			
10497.28	10500.15			0.1			9523.67			
10497.95	10500.83			0.2			9523.06			
10498.85	10501.73		0.2				9522.24			
10499.07	10501.95			0.3			9522.04	V 1	2B14	R 1
10501.62	10504.50			0.1d			9519.73	V 0	2B10	P 5
10501.62	10504.50			0.1d			9519.73	X 0	2B12	P 2
10503.13	10506.01			0.3d			9518.36	V 0	2B10	P 3
10503.13	10506.01			0.3d			9518.36	2B18	2A 0	P 3
10504.02	10506.89			0.2			9517.56			
10504.70	10507.58		0.2	0.4			9516.94			
10505.89	10508.77			0.2			9515.86			
10506.52	10509.40			0.4			9515.29			
10507.27	10510.15			0.1			9514.61			
10507.77	10510.65			0.2			9514.16			
10508.39	10511.27			0.2			9513.60			
10508.96	10511.84			0.2			9513.08	3A 1	2C 3	P 2
10509.57	10512.45			0.3			9512.53	3F 2	2B16	R 2
10510.22	10513.10		0.5	1.5b			9511.94	2A 0	2B 0	R 4
10511.31	10514.19			0.2			9510.96			
10512.13	10515.01			0.2			9510.21	X 1	2C 3	P 4
10512.13	10515.01			0.2			9510.21	3c 0	2a 3	R 0
10513.11	10515.99			0.3			9509.33	3D 2	2B15	R 2
10514.04	10516.92		0.2	0.3			9508.49			
10514.69	10517.57		0.2d	0.6			9507.90			
10515.47	10518.35			0.2			9507.19			
10516.09	10518.97		0.1				9506.63			
10516.44	10519.32			0.1d			9506.32			
10517.31	10520.19		0.2	1.0			9505.53	2A 0	2B 0	P 9
10517.31	10520.19		0.2	1.0			9505.53	3A 0	2C 2	R 1
10518.64	10521.52		0.2b	0.3			9504.33	4e 2	3b 3	R 2
10519.89	10522.77			0.1d			9503.20			
10520.70	10523.58			0.5			9502.47			
10521.17	10524.06			0.2			9502.04	W 1	2C 3	R 4
10521.90	10524.79			0.1			9501.38	3D 3	2B17	P 4
10522.68	10525.56			0.2			9500.68			
10523.76	10526.65			0.1			9499.70			
10525.22	10528.10			0.1			9498.39			
10525.93	10528.81			0.3			9497.75	3C 2	2K 6	R 3
10526.87	10529.75		0.2	0.3			9496.90			
10527.87	10530.75			0.2			9496.00			
10528.70	10531.58			0.1			9495.25			
10529.32	10532.20			0.3			9494.69			

λ_{air}	λ_{vac}	Ref.	I_1	I_2	I_3	I_4	ν	Upper	Lower	Br.
10529.95	10532.83			0.1			9494.12			
10530.37	10533.26		0.1				9493.74			
10530.98	10533.87			0.2			9493.19			
10531.54	10534.42		0.1	0.3			9492.69	3E 2	2B15	Q 2
10531.54	10534.42		0.1	0.3			9492.69	Z 2	2B15	R 0
10532.63	10535.52		0.2	0.3			9491.70	2K 4	2B 5	P 1
10533.05	10535.93			0.2d			9491.33	3c 0	2a 3	Q 6
10533.66	10536.54			0.3			9490.78	3C 0	2A 2	R 3
10534.20	10537.09			0.2			9490.29	4D 0	2C 4	Q 5
10534.73	10537.62			0.1			9489.81			
10535.35	10538.24			0.2			9489.25			
10536.06	10538.95			0.2			9488.61	u 2	2a 6	R 3
10536.73	10539.62			0.2			9488.01			
10537.31	10540.20			0.3			9487.49			
10538.37	10541.26			0.2b			9486.53	X 2	2B17	P 1
10538.37	10541.26			0.2b			9486.53	X 0	2B12	P 3
10539.49	10542.37			0.2			9485.53			
10539.95	10542.84			0.2			9485.11			
10540.41	10543.30		0.2	0.5			9484.70	Z 2	2B15	R 1
10540.41	10543.30		0.2	0.5			9484.70	2K 3	2B 4	R 1
10540.85	10543.74		0.4	0.7			9484.30	2K 3	2B 4	R 0
10542.08	10544.96			0.2			9483.20	4E 0	2B17	R 0
10544.32	10547.21			0.2			9481.18	W 2	2C 4	Q 2
10545.26	10548.14			0.1			9480.34	3E 1	2B13	Q 3
10546.10	10548.99			0.2			9479.58	3c 0	2a 3	Q 5
10546.78	10549.67			0.3			9478.97	Y 1	2C 2	Q 3
10546.78	10549.67			0.3			9478.97	Y 1	2B12	R 2
10547.68	10550.57			0.2			9478.16	Z 2	2B15	R 3
10548.45	10551.34			0.1d			9477.47	3D 3	2B17	P 6
10549.13	10552.02			0.4			9476.86			
10549.67	10552.56			0.1			9476.37			
10550.26	10553.15		0.1	0.2			9475.84	2A 2	2B 3	P 1
10551.28	10554.17			0.2			9474.93			
10552.13	10555.03			0.2			9474.16			
10553.08	10555.97			0.2			9473.31	Z 2	2B15	R 4
10553.08	10555.97			0.2			9473.31	3E 0	2B11	P 5
10554.45	10557.34			0.6			9472.08			
10555.59	10558.48			0.2			9471.06			
10556.37	10559.26			0.0			9470.36	V 1	2B14	R 0
10556.81	10559.71		0.1				9469.96	3c 0	2a 3	Q 4
10557.59	10560.49			0.2			9469.26	3D 3	2B17	P 5
10558.12	10561.01		0.2				9468.79			
10558.26	10561.16			0.3			9468.66	W 1	2C 3	Q 1
10558.26	10561.16			0.3			9468.66	Y 1	2B12	R 1
10558.71	10561.60		0.4				9468.26			
10558.84	10561.74			0.9			9468.14	2K 3	2B 4	R 2
10560.00	10562.90			0.3d			9467.10			
10561.04	10563.93			0.1			9466.17			
10562.06	10564.95			0.2			9465.26			
10563.34	10566.23			0.4			9464.11			
10564.17	10567.06			0.1			9463.37			
10564.77	10567.66			0.3			9462.83			
10565.53	10568.42			0.2			9462.15	3c 0	2a 3	Q 3
10566.31	10569.20			0.3			9461.45			
10567.68	10570.58			0.2			9460.22	X 0	2B12	P 5
10567.68	10570.58			0.2			9460.22	3c 1	2a 4	P 2
10567.85	10570.75		0.1				9460.07	4E 0	2B17	Q 4
10568.26	10571.16			0.2b			9459.70			
10569.94	10572.84			0.1			9458.20			
10570.73	10573.63			0.4			9457.49	3D 2	2B15	P 6

λ_{air}	λ_{vac}	Ref.	I_1	I_2	I_3	I_4	ν	Upper	Lower	Br.
10570.73	10573.63			0.4			9457.49	Y 1	2B12	R 0
10571.02	10573.92		0.2				9457.23			
10571.21	10574.11			0.5			9457.06	3B 3	2K 3	R 1
10571.64	10574.54			0.4			9456.68			
10572.49	10575.39			0.3			9455.92	3c 0	2a 3	Q 2
10573.22	10576.11		0.1	0.1			9455.27	3D 2	2B15	P 4
10573.85	10576.75		0.3	0.6			9454.70	2A 2	2B 3	P 2
10573.85	10576.75		0.3	0.6			9454.70	2K 4	2B 5	P 2
10574.60	10577.50			0.3			9454.03	3C 0	2A 2	R 1
10575.50	10578.39			0.1			9453.23	4b 1	2a 6	P 1
10576.20	10579.10			0.2d			9452.60	4f 2	3b 3	P 3
10577.07	10579.97			0.2d			9451.82	3c 0	2a 3	Q 1
10578.10	10581.00			0.3			9450.90	3b 4	2a 4	P 4
10578.50	10581.39			0.3			9450.55			
10578.96	10581.85			0.2			9450.14	3F 2	2B16	Q 3
10579.41	10582.31			0.1			9449.73	3a 2	3b 0	P 3
10581.32	10584.22		0.1	0.3			9448.03	2K 6	2B 7	R 2
10584.16	10587.06			0.0			9445.49	X 2	2B17	P 2
10585.16	10588.06			0.5			9444.60	3F 1	2C 3	R 5
10586.54	10589.44		0.2	0.2			9443.37			
10587.18	10590.08			1.5b			9442.80			
10588.20	10591.10			0.9			9441.89	V 0	2B10	P 2
10589.05	10591.95			0.1d			9441.13	3B 3	2K 3	P 4
10589.97	10592.87			0.2			9440.31			
10590.78	10593.68		0.1	0.3			9439.59	3D 2	2B15	P 5
10591.29	10594.20			0.7			9439.13	W 1	2C 3	P 1
10591.95	10594.85			0.1			9438.55	2A 2	2B 3	P 3
10592.50	10595.40			0.2			9438.06			
10593.21	10596.12			0.1			9437.42			
10593.82	10596.72			0.4			9436.88			
10594.56	10597.46			0.2			9436.22	W 1	2C 3	Q 2
10594.56	10597.46			0.2			9436.22	4e 1	3b 2	R 4
10595.75	10598.65		0.5				9435.16			
10595.92	10598.82			0.7			9435.01	2K 3	2B 4	P 1
10595.92	10598.82			0.7			9435.01	2K 3	2B 4	R 3
10595.92	10598.82			0.7			9435.01	3C 0	2K 2	Q 3
10595.92	10598.82			0.7			9435.01	2K 6	2B 7	R 1
10597.60	10600.51		0.2d	0.4			9433.51			
10598.23	10601.14			0.5			9432.95			
10598.51	10601.42			0.5			9432.70			
10599.23	10602.14			0.1			9432.06			
10599.71	10602.61			0.2			9431.64			
10600.45	10603.35		0.2				9430.98			
10600.58	10603.49			1.0b			9430.86			
10601.68	10604.59			0.1d			9429.88	3A 2	2B17	R 3
10602.91	10605.81			0.4			9428.79			
10603.43	10606.33			0.5			9428.33			
10605.07	10607.97			0.2b			9426.87			
10606.14	10609.04		0.1				9425.92			
10607.53	10610.44			0.2			9424.68			
10608.21	10611.12			0.2			9424.08	3E 1	2B13	Q 2
10608.79	10611.70			0.5			9423.56			
10611.99	10614.90			0.2			9420.72	Y 1	2C 2	P 3
10612.80	10615.71			0.3			9420.00			
10613.48	10616.39		0.2b	0.5			9419.40			
10614.11	10617.02			0.2			9418.84			
10614.76	10617.67			0.1			9418.26	3C 1	2K 4	Q 4
10615.40	10618.30			0.2			9417.70			
10616.00	10618.91			0.1			9417.16			
10617.13	10620.04		0.1	0.2			9416.16	2K 6	2B 7	R 0

λ_{air}	λ_{vac}	Ref.	I_1	I_2	I_3	I_4	ν	Upper	Lower	Br.
10618.10	10621.01			0.1			9415.30			
10619.89	10622.80			0.2			9413.71			
10620.59	10623.50		0.2	0.3			9413.09	2K 4	2B 5	P 3
10620.59	10623.50		0.2	0.3			9413.09	4e 2	3b 3	R 1
10620.59	10623.50		0.2	0.3			9413.09	4F 0	2B18	R 3
10621.50	10624.41			0.1			9412.29			
10622.16	10625.07			0.2d			9411.70	W 2	2C 4	Q 3
10623.02	10625.93		0.3	0.4			9410.94	3e 2	3b 0	Q 1
10623.94	10626.85			0.3			9410.13	Y 1	2B12	P 1
10623.94	10626.85			0.3			9410.13	3c 1	2a 4	P 3
10623.94	10626.85			0.3			9410.13	4b 1	2a 6	P 2
10624.73	10627.64			0.3			9409.43	u 2	2a 6	R 2
10625.56	10628.47			0.1			9408.69	X 0	2C 2	P 4
10625.95	10628.86			0.2			9408.35			
10626.42	10629.33			0.1			9407.93			
10627.04	10629.95			0.1			9407.38			
10627.65	10630.56			0.3			9406.84			
10628.17	10631.08			0.3			9406.38	3D 1	2B13	R 5
10628.72	10631.64			0.2			9405.89	2K 2	2B 3	P 4
10629.41	10632.33			0.2			9405.28			
10630.01	10632.92			0.2			9404.75			
10630.85	10633.76		0.2	0.3			9404.01	3A 0	2B12	R 2
10630.85	10633.76		0.2	0.3			9404.01	3a 2	3b 0	P 4
10631.64	10634.55		0.3	0.9			9403.31	X 2	2B17	P 3
10632.36	10635.28			0.2			9402.67			
10633.11	10636.02			0.3			9402.01			
10634.31	10637.22			0.2d			9400.95	3A 1	2C 3	P 4
10635.40	10638.31			0.1			9399.99			
10636.07	10638.99			0.3			9399.39			
10636.69	10639.60			0.3			9398.85	3D 2	2B15	P 2
10637.58	10640.49			0.2			9398.06			
10639.27	10642.18			0.2			9396.57	4b 0	2a 5	R 2
10641.62	10644.54			0.2			9394.49	V 1	2B11	R 1
10642.34	10645.25			0.2			9393.86			
10643.17	10646.09		0.1	0.5			9393.12	X 0	2C 2	Q 5
10644.21	10647.12		0.1	0.5			9392.21	4c 0	2a 6	R 3
10644.99	10647.90			0.2			9391.52			
10645.76	10648.67		0.1	0.5			9390.84			
10646.51	10649.42			0.2			9390.18	3E 1	2B13	Q 1
10647.16	10650.08			0.1			9389.60			
10647.70	10650.61			0.3			9389.13			
10648.57	10651.49			0.2			9388.36			
10649.38	10652.29		0.2	1.0			9387.65	2K 3	2B 4	R 4
10649.38	10652.29		0.2	1.0			9387.65	3B 3	2K 3	R 0
10650.01	10652.93		0.7	1.0			9387.09	2K 3	2B 4	P 2
10650.01	10652.93		0.7	1.0			9387.09	3e 2	3b 0	Q 2
10650.01	10652.93		0.7	1.0			9387.09	3F 1	2B14	R 3
10650.84	10653.76			0.2			9386.36			
10651.95	10654.87		0.2	0.3			9385.38	W 1	2C 3	Q 3
10654.18	10657.10			0.1b			9383.42	Z 2	2B15	R 2
10655.19	10658.11			0.6			9382.53	3B 3	2A 0	R 1
10655.64	10658.56			1.0			9382.13	3F 2	2B16	R 1
10657.13	10660.05			0.2			9380.82	4E 0	2B17	Q 3
10659.06	10661.98			0.2			9379.12			
10659.96	10662.88		0.2	0.7			9378.33	Y 1	2B12	P 2
10660.61	10663.53			0.1			9377.76			
10660.98	10663.90			0.5			9377.43			
10661.53	10664.45			0.1			9376.95	X 0	2B12	P 6
10662.14	10665.06			0.2			9376.41	3C 0	2K 2	P 4
10662.60	10665.52		0.2d	0.4			9376.01	3B 3	2K 3	P 3

λ_{air}	λ_{vac}	Ref.	I_1	I_2	I_3	I_4	ν	Upper	Lower	Br.
10663.40	10666.33			0.1			9375.30	3C 0	2A 2	R 0
10663.92	10666.84			0.2			9374.85			
10664.96	10667.88			0.2			9373.93	Z 2	2B15	P 3
10665.46	10668.39		0.1	0.1			9373.49	4b 1	2a 6	P 3
10665.91	10668.83			0.1			9373.10			
10667.99	10670.91			0.2			9371.27	3F 2	2C 4	R 3
10669.42	10672.35		0.2	0.6			9370.01	2K 4	2B 5	P 4
10669.42	10672.35		0.2	0.6			9370.01	3b 4	2a 4	P 5
10670.05	10672.97			0.0			9369.46			
10670.55	10673.47			0.2			9369.02			
10671.12	10674.04			0.2			9368.52			
10671.65	10674.57			0.2			9368.06			
10672.89	10675.81			0.1			9366.97			
10673.59	10676.52			1.5b			9366.35	3F 2	2B16	P 4
10674.76	10677.68			0.3			9365.33			
10675.26	10678.18			0.1			9364.89			
10676.19	10679.12			0.4			9364.07			
10677.41	10680.34			0.2d			9363.00			
10678.01	10680.93			0.3			9362.48			
10678.46	10681.39		0.1	0.2			9362.08	3a 2	3b 0	P 5
10678.95	10681.88			0.4			9361.65			
10679.57	10682.49			0.1			9361.11	4E 0	2B17	Q 1
10680.20	10683.12			0.5			9360.56	3E 0	2B11	P 4
10680.63	10683.56			0.3			9360.18	3c 1	2a 4	P 4
10681.29	10684.22			0.3			9359.60			
10681.80	10684.73			0.9			9359.15	3C 0	2A 2	Q 2
10682.31	10685.23			0.5			9358.71			
10682.91	10685.84			0.1			9358.18			
10683.44	10686.36			0.3			9357.72	4f 1	3b 2	Q 4
10683.93	10686.85			0.2			9357.29	3C 2	2K 6	R 2
10684.68	10687.61		0.5	1.5b			9356.63	2A 0	2B 0	R 3
10685.69	10688.61			0.2d			9355.75	3C 0	2A 2	Q 5
10687.00	10689.93			0.2			9354.60	3e 2	3b 0	Q 3
10688.38	10691.31			0.2			9353.39			
10688.87	10691.80			0.3			9352.96	W 1	2C 3	P 2
10688.87	10691.80			0.3			9352.96	V 1	2B11	P 4
10689.78	10692.71			0.3			9352.17			
10690.27	10693.20			0.2			9351.74			
10691.23	10694.16			0.2			9350.90	X 2	2B17	P 4
10691.23	10694.16			0.2			9350.90	u 2	2a 6	P 1
10692.41	10695.34			0.2			9349.87	Y 2	2C 4	Q 1
10692.89	10695.82			0.1			9349.45	4f 0	3b 1	R 2
10693.61	10696.54		0.5	1.5b			9348.82	2A 0	2B 0	P 8
10694.26	10697.19			0.3			9348.25			
10694.66	10697.59			0.2			9347.90			
10699.67	10702.60			0.3			9343.52			
10699.93	10702.86		0.1				9343.30			
10700.43	10703.36			0.2			9342.86			
10701.01	10703.94			0.3			9342.35			
10701.63	10704.56			0.2			9341.81			
10702.54	10705.47			0.3			9341.02	3F 1	2C 3	R 4
10702.54	10705.47			0.3			9341.02	3e 2	3b 0	P 2
10703.33	10706.26			0.2			9340.33	3C 0	2K 2	R 2
10703.33	10706.26			0.2			9340.33	4e 1	3b 2	R 3
10704.71	10707.65			0.2			9339.12	3E 0	2B11	Q 4
10704.71	10707.65			0.2			9339.12	3F 2	2B16	Q 2
10705.59	10708.52			0.3			9338.36			
10705.95	10708.89		0.2	0.3			9338.04	3B 3	2K 3	P 2
10706.55	10709.48			0.1d			9337.52			
10707.44	10710.38			0.3			9336.74			

λ_{air}	λ_{vac}	Ref.	I_1	I_2	I_3	I_4	ν	Upper	Lower	Br.
10708.18	10711.11			0.2			9336.10	4b 0	2a 5	R 1
10708.66	10711.59		0.1	0.6			9335.68	4e 0	3b 1	R 6
10709.75	10712.68			0.2			9334.73	4E 0	2B17	Q 2
10710.84	10713.77			0.1			9333.78	4d 3	3b 4	R 0
10712.24	10715.17			0.2			9332.56	Y 1	2B12	P 3
10712.82	10715.76			0.3			9332.05	W 2	2C 4	Q 4
10713.72	10716.65			0.2			9331.27			
10714.86	10717.79		0.3	0.7			9330.28			
10715.67	10718.61		0.1	0.4			9329.57	Z 2	2B15	P 4
10715.67	10718.61		0.1	0.4			9329.57	2K 6	2B 7	P 2
10715.67	10718.61		0.1	0.4			9329.57	4e 2	3b 3	R 0
10716.42	10719.35			0.1			9328.92	3A 0	2C 2	P 1
10716.86	10719.79		0.1	0.3			9328.54	2K 4	2B 5	P 5
10716.86	10719.79		0.1	0.3			9328.54	2B18	2A 0	P 4
10717.64	10720.57		0.3	0.4			9327.86	3B 3	2K 3	P 1
10718.13	10721.07			0.2			9327.43			
10719.00	10721.94			0.2b			9326.67			
10719.27	10722.20		0.1				9326.44	Z 2	2B15	P 2
10719.75	10722.69			0.2			9326.02	3C 0	2A 2	Q 1
10720.94	10723.87			0.3			9324.99	3D 1	2B13	R 4
10720.94	10723.87			0.3			9324.99	4d 3	3b 4	P 1
10722.71	10725.64		0.3	0.7			9323.45	2K 3	2B 4	P 3
10723.04	10725.98						9323.16			
10723.33	10726.26			0.2			9322.91			
10724.09	10727.02			0.1			9322.25			
10724.89	10727.83			0.1			9321.55	3A 2	2B17	R 2
10725.58	10728.52			0.3			9320.95	W 1	2C 3	Q 4
10726.00	10728.93			0.3			9320.59	3e 2	3b 0	P 3
10726.56	10729.50			0.2d			9320.10			
10727.45	10730.39			0.4			9319.33			
10728.14	10731.08			0.1			9318.73	2K 2	2B 3	P 5
10728.97	10731.91			0.1			9318.01	X 2	2B17	P 5
10729.54	10732.48			0.2			9317.51			
10730.72	10733.66			0.2			9316.49	3e 2	3b 0	Q 4
10731.98	10734.92			0.3			9315.39			
10732.58	10735.52			0.2			9314.87			
10733.86	10736.80			0.3			9313.76			
10734.20	10737.14			0.3			9313.47			
10734.80	10737.74			0.4			9312.95			
10737.0	10739.9		1.5				9311.04	4e 3	3b 4	R 3
10737.71	10740.65		0.2	0.9			9310.42			
10738.64	10741.58			0.1			9309.62	3C 1	2K 4	Q 3
10739.95	10742.89			0.2			9308.48			
10740.65	10743.60			0.1			9307.87	u 2	2a 6	R 0
10741.7	10744.6		1.2				9306.96	4c 0	2a 6	R 2
10743.87	10746.82			1.0			9305.08			
10744.57	10747.51			0.2			9304.48			
10745.91	10748.85			0.2			9303.32			
10746.40	10749.35			0.2			9302.89			
10747.22	10750.17			0.1			9302.18			
10747.84	10750.78			0.3			9301.65	4d 3	3b 4	R 1
10748.80	10751.74		0.2	0.5			9300.82			
10749.41	10752.35			0.5			9300.29	T 0	2B17	R 2
10750.02	10752.97			0.1			9299.76			
10750.67	10753.61			0.2			9299.20			
10751.43	10754.38			0.1			9298.54			
10752.10	10755.05			0.6			9297.96			
10752.60	10755.54			0.2			9297.53			
10753.14	10756.09			0.2			9297.06	3e 2	3b 0	P 4
10753.95	10756.90		0.1	0.3			9296.36	3C 0	2A 2	Q 4

λ_{air}	λ_{vac}	Ref.	I_1	I_2	I_3	I_4	ν	Upper	Lower	Br.
10757.01	10759.95			0.2			9293.72			
10757.79	10760.74			0.2			9293.04			
10758.43	10761.38		0.1	0.2			9292.49			
10759.17	10762.12			0.4			9291.85			
10759.37	10762.32		0.2	.			9291.68			
10759.77	10762.72			0.3			9291.33			
10760.48	10763.43			0.1			9290.72	2K 6	2B 7	P 3
10761.75	10764.70			0.6			9289.62	2K 4	2B 5	P 6
10761.75	10764.70			0.6			9289.62	Y 2	2C 4	Q 2
10762.68	10765.63			0.2			9288.82			
10763.62	10766.57			0.3			9288.01	3F 2	2C 4	R 2
10763.62	10766.57			0.3			9288.01	3c 0	2a 3	P 3
10764.88	10767.83			0.2			9286.92	Y 1	2B12	P 4
10765.88	10768.83			0.2			9286.06	Z 2	2B15	P 5
10765.88	10768.83			0.2			9286.06	3A 0	2B12	R 1
10766.85	10769.80			0.7			9285.22	3b 5	2a 5	R 2
10767.31	10770.26		0.1				9284.83			
10768.07	10771.02			0.1			9284.17			
10769.04	10771.99		0.1	0.7			9283.34	3C 1	2K 4	P· 5
10769.70	10772.65			0.1			9282.77			
10770.40	10773.35			0.2			9282.16			
10771.46	10774.41			0.2			9281.25	3D 1	2B13	R 3
10772.16	10775.11			0.3			9280.65	4b 0	2a 5	R 0
10772.95	10775.90			0.2			9279.97			
10773.40	10776.35			0.3			9279.58	3F 1	2B14	Q 4
10774.28	10777.23			0.1			9278.82			
10774.98	10777.93			0.6			9278.22	3A 0	2C 2	P 2
10774.98	10777.93			0.6			9278.22	3C 1	2K 4	R 1
10775.93	10778.88			0.2			9277.40			
10776.64	10779.59			0.5			9276.79	3b 5	2a 5	R 1
10777.73	10780.68			0.2			9275.85			
10778.42	10781.37			0.2			9275.26			
10779.14	10782.09			0.2			9274.64	3E 0	2B11	R 0
10779.14	10782.09			0.2			9274.64	3F 2	2B16	P 3
10780.06	10783.01			0.1			9273.85	3F 0	2C 2	R 4
10780.06	10783.01			0.1			9273.85	3E 0	2B11	P 3
10780.60	10783.55			0.2			9273.38			
10781.17	10784.12			0.3			9272.89	4d 3	3b 4	R 2
10781.17	10784.12			0.3			9272.89	3e 2	3b 0	Q 5
10782.79	10785.74			0.2			9271.50	3D 1	2B13	R 0
10782.79	10785.74			0.2			9271.50	4f 1	3b 2	Q 3
10782.79	10785.74			0.2			9271.50	W 2	2C 4	Q 5
10783.23	10786.18			0.2			9271.12			
10783.71	10786.66			0.2			9270.71	3D 0	2B11	R 6
10783.71	10786.66			0.2			9270.71	V 1	2B11	P 3
10784.93	10787.88			0.6r			9269.66	3C 0	2K 2	P 3
10784.93	10787.88			0.6r			9269.66	3C 2	2K 6	Q 3
10786.18	10789.14			0.2			9268.58	4e 3	3b 4	R 1
10786.85	10789.80			0.5			9268.01			
10787.65	10790.61			0.1			9267.32			
10788.95	10791.91			0.2			9266.20	V 1	2B11	R 0
10788.95	10791.91			0.2			9266.20	3e 2	3b 0	P 5
10789.87	10792.83			0.2			9265.41	Z 2	2B15	P 6
10790.69	10793.65			0.2			9264.71	3D 1	2B13	R 2
10791.31	10794.26			0.2			9264.18	3D 1	2B13	R 1
10792.20	10795.16			0.2			9263.41	3E 0	2B11	Q 3
10792.86	10795.81		0.2	0.9			9262.85			
10793.38	10796.34			0.6			9262.40	3D 1	2B13	P 1
10794.02	10796.98			0.3			9261.85			
10794.90	10797.85			0.2			9261.10			

λ_{air}	λ_{vac}	Ref.	I_1	I_2	I_3	I_4	ν	Upper	Lower	Br.
10795.98	10798.94		0.1	0.2			9260.17			
10796.73	10799.68			0.2			9259.53			
10797.44	10800.40			0.2			9258.92			
10798.15	10801.11			0.3			9258.31			
10798.62	10801.57			0.3			9257.91			
10799.25	10802.20			0.1			9257.37			
10800.24	10803.20			0.4			9256.52			
10800.77	10803.73		0.2	0.3			9256.06			
10801.30	10804.26			0.2			9255.61			
10801.81	10804.77			0.1			9255.17			
10802.54	10805.50			0.2			9254.55			
10802.93	10805.89			0.3			9254.21			
10803.52	10806.48		0.1	0.2			9253.71			
10804.18	10807.14			0.2			9253.14			
10805.32	10808.28			0.3			9252.17	3b 5	2a 5	R 0
10806.09	10809.05			0.2			9251.51			
10806.86	10809.82			0.3			9250.85			
10807.60	10810.57			0.2			9250.21			
10808.32	10811.28			0.2			9249.60	4e 1	3b 2	R 2
10809.10	10812.06		0.1	0.4			9248.93			
10809.58	10812.54			0.4			9248.52	4e 0	3b 1	R 5
10811.03	10813.99			0.2			9247.28			
10811.70	10814.66			0.3			9246.71			
10812.99	10815.96		0.1				9245.60			
10813.64	10816.60			0.2d			9245.05	3C 0	2A 2	P 2
10814.03	10817.00		0.1				9244.71			
10814.81	10817.77		0.4	1.0			9244.05	2K 3	2B 4	P 4
10815.70	10818.66			0.3			9243.29			
10816.60	10819.56			0.1			9242.52			
10818.39	10821.35			0.3			9240.99	3A 2	2B17	P 5
10818.39	10821.35			0.3			9240.99	3C 2	2K 6	R 1
10818.91	10821.88			0.1			9240.54	3c 0	2a 3	P 4
10818.91	10821.88			0.1			9240.54	4e 2	3b 3	P 6
10819.43	10822.39			0.6			9240.10			
10821.41	10824.37			0.2d			9238.41	3A 0	2C 2	P 3
10822.16	10825.12		0.2	1.5b			9237.77	3b 0	2a 1	R 4
10823.28	10826.25		0.2d	0.2d			9236.81	Y 1	2C 2	P 4
10824.03	10827.00		0.2	1.5b			9236.17	3b 0	2a 1	R 5
10824.96	10827.92			0.1			9235.38			
10826.31	10829.27			0.2d			9234.23			
10827.08	10830.05			1.0			9233.57			
10827.37	10830.34		0.2				9233.32			
10828.11	10831.08			0.3d			9232.69			
10829.10	10832.07		0.9				9231.85			
10829.61	10832.58		0.2	0.2d			9231.41			
10830.31	10833.27		1.5b				9230.82			
10830.71	10833.68			0.2d			9230.47			
10831.34	10834.31		0.2				9229.94	4c 0	2a 6	R 1
10831.95	10834.92		0.1	0.4			9229.42	3C 0	2A 2	Q 3
10832.27	10835.23		0.1	0.3			9229.15	3e 2	3b 0	P 6
10833.03	10836.00		0.5	1.5b			9228.50	3b 0	2a 1	R 3
10833.75	10836.71			0.3			9227.89	3A 2	2B17	R 1
10834.96	10837.92			0.1			9226.86	3e 2	3b 0	Q 6
10835.81	10838.78			0.1d			9226.13	T 0	2B17	R 1
10837.27	10840.24			0.2b			9224.89			
10838.44	10841.41			1.5			9223.89	3b 0	2a 1	R 6
10839.35	10842.32		1.0b	1.5b			9223.12	2A 0	2B 0	R 2
10839.35	10842.32		1.0b	1.5b			9223.12	3C 1	2K 4	Q 2
10840.39	10843.36			0.2d			9222.23			
10841.24	10844.21			0.2d			9221.51			

λ_{air}	λ_{vac}	Ref.	I_1	I_2	I_3	I_4	ν	Upper	Lower	Br.
10842.03	10845.00			0.2d			9220.84			
10842.92	10845.89			0.5			9220.08			
10843.85	10846.82			0.3			9219.29	4e 3	3b 4	Q 1
10844.73	10847.70			0.0			9218.54			
10845.12	10848.09		0.1				9218.21			
10845.85	10848.82			0.1			9217.59			
10847.02	10849.99			0.1			9216.60	3E 0	2B11	Q 2
10847.50	10850.47			0.2			9216.19	4e 2	3b 3	P 5
10848.32	10851.29		0.3	1.5b			9215.49	3B 0	2A 0	P 1
10848.32	10851.29		0.3	1.5b			9215.49	2A 1	2B 2	R 4
10849.05	10852.02		0.3	1.5b			9214.87	2A 0	2B 0	P 7
10850.57	10853.54			0.3			9213.58			
10852.20	10855.17			0.2			9212.20	3A 0	2B12	P 4
10853.49	10856.47			0.2			9211.10			
10854.11	10857.08			0.2			9210.58			
10854.66	10857.63		0.2	0.1d			9210.11			
10855.11	10858.08			0.1			9209.73			
10855.58	10858.55		0.2	0.3			9209.33			
10855.92	10858.89		0.4	1.5b			9209.04			
10857.0	10860.0		8.5				9208.13	3b 0	2a 1	R 2
10857.0	10860.0		8.5				9208.13	4d 3	3b 4	P 2
10857.0	10860.0		8.5				9208.13	4D 0	2B17	P 1
10857.91	10860.89			0.1			9207.35			
10858.37	10861.35			0.2			9206.96			
10859.31	10862.28			0.2			9206.17			
10860.19	10863.17			0.2			9205.42	3F 2	2C 4	R 1
10860.34	10863.32		0.2				9205.29			
10860.60	10863.58			0.2			9205.07	3C 1	2K 4	P 4
10861.11	10864.09			0.2			9204.64			
10861.85	10864.83			0.2			9204.01	3E 0	2B11	P 2
10862.57	10865.55			0.2			9203.40			
10863.13	10866.10			0.2			9202.93			
10863.85	10866.82			0.2			9202.32	3C 0	2K 2	R 1
10863.85	10866.82			0.2			9202.32	3D 1	2B13	P 2
10865.10	10868.08		0.2	1.5b			9201.26	3b 0	2a 1	R 7
10865.10	10868.08		0.2	1.5b			9201.26	4e 2	3b 3	P 4
10865.94	10868.92			0.2b			9200.55			
10867.14	10870.12			0.1			9199.53	3F 1	2B14	R 2
10867.88	10870.85			0.1			9198.91			
10868.76	10871.74			0.2			9198.16	4e 2	3b 3	P 2
10869.52	10872.50			0.9			9197.52			
10870.35	10873.32			0.6			9196.82	3A 0	2B12	R 0
10870.35	10873.32			0.6			9196.82	3c 0	2a 3	P 5
10871.37	10874.35			0.2			9195.95	Y 2	2C 4	Q 3
10872.21	10875.19			0.2			9195.24	V 1	2B11	P 2
10872.21	10875.19			0.2			9195.24	4e 2	3b 3	P 3
10873.07	10876.04			0.7			9194.52			
10873.48	10876.46			0.2			9194.17			
10874.83	10877.81			0.2			9193.03			
10875.83	10878.81			0.2b			9192.18			
10878.45	10881.43			0.2			9189.97	3C 1	2K 4	R 0
10879.60	10882.58			0.2			9189.00			
10880.92	10883.90		0.1	0.6			9187.88	3E 0	2B11	Q 1
10881.8	10884.8		0.8				9187.14			
10882.79	10885.78			0.1			9186.30			
10884.51	10887.49			0.1			9184.85	4f 1	3b 2	Q 2
10885.35	10888.34			0.2			9184.14	3F 1	2B14	P 5
10885.86	10888.85			0.2			9183.71	3C 2	2K 6	Q 2
10886.43	10889.41			0.2			9183.23			
10887.05	10890.03			0.2			9182.71			

λ_{air}	λ_{vac}	Ref.	I_1	I_2	I_3	I_4	ν	Upper	Lower	Br.
10887.56	10890.54			0.1			9182.28	3F 0	2C 2	R 3
10888.18	10891.16			0.2			9181.76			
10888.99	10891.98			0.1b			9181.07			
10890.77	10893.76			0.3d			9179.57	4b 0	2a 5	P 1
10891.54	10894.53			0.1			9178.92			
10892.38	10895.36			0.3			9178.22	3e 2	3b 0	Q 7
10893.36	10896.34			0.2			9177.39	T 0	2B17	R 0
10894.10	10897.08		0.8	1.5b			9176.77	3b 0	2a 1	R 1
10894.10	10897.08		0.8	1.5b			9176.77	3B 2	2K 2	R 4
10895.15	10898.14			0.2b			9175.88			
10896.95	10899.93			0.3			9174.37			
10898.50	10901.49			0.2			9173.06			
10899.93	10902.91			0.5			9171.86	3B 4	2K 5	R 3
10899.93	10902.91			0.5			9171.86	3A 2	2B17	P 4
10900.99	10903.97			0.1			9170.97			
10902.03	10905.02			0.4			9170.09			
10902.50	10905.48			0.2			9169.70	u 1	2a 5	R 3
10903.60	10906.59			0.2d	0.9		9168.77	3b 0	2a 1	R 8
10904.61	10907.60			0.2			9167.92			
10905.42	10908.41			0.4			9167.24			
10906.1	10909.1		1.0				9166.67	4e 1	3b 2	R 1
10906.87	10909.86			0.5			9166.02	W 1	2C 3	P 4
10906.87	10909.86			0.5			9166.02	4c 0	2a 6	Q 3
10908.06	10911.05			0.2			9165.02			
10908.66	10911.65			0.1			9164.52	3e 3	3b 1	Q 1
10909.62	10912.61			0.2			9163.71	u 2	2a 6	P 2
10909.62	10912.61			0.2			9163.71	4f 0	3b 1	R 1
10910.56	10913.55			0.2d			9162.92	3C 1	2K 4	Q 1
10911.67	10914.66			0.8			9161.99	3A 0	2C 2	P 4
10911.67	10914.66			0.8			9161.99	4c 0	2a 6	R 0
10912.56	10915.55			0.2			9161.24			
10913.22	10916.21			0.2			9160.69			
10914.35	10917.34			0.2d			9159.74	4e 0	3b 1	R 4
10914.54	10917.53		0.1				9159.58	2K 5	2B 6	R 2
10915.15	10918.14		0.4	0.9			9159.07	3b 5	2a 5	P 1
10915.15	10918.14		0.4	0.9			9159.07	4e 3	3b 4	Q 3
10915.15	10918.14		0.4	0.9			9159.07			
10915.84	10918.83		0.1				9158.49	Z 1	2B13	R 2
10916.23	10919.22			0.2			9158.16			
10917.26	10920.25			0.2			9157.30	3D 0	2B11	R 5
10917.26	10920.25			0.2			9157.30	3c 0	2a 3	P 6
10919.02	10922.01		0.3				9155.82	2K 5	2B 6	R 1
10919.08	10922.07			1.5			9155.77	3B 1	2A 1	R 5
10919.57	10922.56			0.6			9155.36			
10920.66	10923.65			0.6			9154.45	2K 5	2B 6	R 3
10922.77	10925.76			0.1			9152.68			
10923.5	10926.5		1.8				9152.07	3C 2	2K 6	R 0
10924.36	10927.35			0.2			9151.35			
10925.48	10928.47			0.2b			9150.41	T 0	2B17	P 4
10925.48	10928.47			0.2b			9150.41	3A 0	2B12	P 3
10926.46	10929.45			0.6			9149.59	2K 3	2B 4	P 5
10927.62	10930.61			0.8			9148.62	3B 1	2A 1	R 4
10929.18	10932.18			0.2			9147.31	3C 1	2K 4	P 3
10929.18	10932.18			0.2			9147.31	3D 1	2B13	P 3
10929.98	10932.98			0.2			9146.64	3e 3	3b 1	Q 2
10930.71	10933.71			0.2			9146.03	3C 0	2K 2	Q 2
10931.45	10934.45			0.2			9145.41			
10932.55	10935.55		0.2	0.8			9144.49	2K 5	2B 6	R 4
10932.55	10935.55		0.2	0.8			9144.49	3F 0	2B12	Q 5
10933.49	10936.48		0.5	0.6			9143.71	2K 5	2B 6	R 0

λ_air	λ_vac	Ref.	I₁	I₂	I₃	I₄	ν	Upper	Lower	Br.
10933.49	10936.48	0.5	0.6				9143.71	3F 1	2C 3	R 2
10934.56	10937.56		0.2				9142.81	3E 2	2B16	Q 6
10935.35	10938.35		0.1				9142.15	4D 0	2B17	P 2
10936.51	10939.51		0.1b				9141.18	4F 0	2B18	R 2
10937.22	10940.21		0.4				9140.59			
10938.07	10941.06	0.5b					9139.88	3F 2	2C 4	Q 3
10938.15	10941.15		1.0b				9139.81	3D 1	2B13	P 6
10938.15	10941.15		1.0b				9139.81	3F 1	2C 3	R 2
10939.68	10942.68		0.1				9138.53	Z 1	2B13	R 1
10940.80	10943.79		0.2				9137.60	3E 3	2B18	R 3
10940.80	10943.79		0.2				9137.60	4c 0	2a 6	Q 2
10941.9	10944.9	2.3					9136.68			
10944.63	10947.63	0.5	1.5				9134.40	3b 0	2a 1	R 0
10944.63	10947.63	0.5	1.5				9134.40	4c 0	2a 6	Q 2
10945.46	10948.45		0.4				9133.71			
10946.63	10949.63	0.1	1.5				9132.73	3B 1	2A 1	R 3
10947.85	10950.85		0.2				9131.71			
10948.66	10951.65		0.2				9131.04	3A 2	2B17	R 0
10949.53	10952.53		0.4				9130.31			
10950.25	10953.25		0.1				9129.71			
10950.72	10953.72		0.4				9129.32			
10951.28	10954.28		0.2				9128.85			
10953.41	10956.41		0.6				9127.08	4b 0	2a 5	P 2
10953.81	10956.81		0.6				9126.74	3b 0	2a 1	R 9
10954.8	10957.8	1.3					9125.92	3A 2	2B17	P 3
10954.8	10957.8	1.3					9125.92	3C 2	2K 6	Q 1
10955.42	10958.42		0.3				9125.40			
10956.61	10959.61		0.1				9124.41	T 0	2B17	P 3
10956.97	10959.97	0.1	0.8				9124.11			
10957.8	10960.8	1.5					9123.42			
10958.68	10961.68		0.1				9122.69	3C 2	2K 6	P 3
10958.68	10961.68		0.1				9122.69	3f 0	2c 4	R 3
10959.61	10962.62		0.2				9121.91	3e 3	3b 1	Q 3
10960.52	10963.52		0.2				9121.16			
10961.49	10964.49		0.2				9120.35			
10961.95	10964.95		0.4				9119.97	Z 1	2B13	R 0
10961.95	10964.95		0.4				9119.97	3F 1	2B14	Q 3
10962.99	10965.99		0.3				9119.10			
10963.80	10966.80		0.1				9118.43			
10964.83	10967.83		0.1				9117.57	3A 0	2B12	P 2
10964.83	10967.83		0.1				9117.57	U 1	2B14	R 4
10965.31	10968.32		0.6				9117.17			
10966.58	10969.58		0.1				9116.12	T 0	2B17	P 2
10967.65	10970.65		0.0d				9115.23			
10968.49	10971.49		0.6				9114.53			
10969.64	10972.65		0.1				9113.57	3C 1	2K 4	P 2
10970.55	10973.55	0.1	0.3				9112.82	3F 2	2C 4	Q 2
10970.55	10973.55	0.1	0.3				9112.82	4f 0	3b 1	Q 4
10970.55	10973.55	0.1	0.3				9112.82	3d 2	3b 0	R 0
10971.26	10974.26	0.9	1.5b				9112.23	2A 0	2B 0	R 1
10971.26	10974.26	0.9	1.5b				9112.23	u 2	2a 6	P 3
10972.17	10975.18		0.2				9111.47			
10972.74	10975.74		0.2				9111.00			
10973.41	10976.42		0.9				9110.44	3B 1	2A 1	R 2
10974.45	10977.45		0.0				9109.58	3A 0	2B12	P 1
10974.45	10977.45		0.0				9109.58	4d 2	3b 3	R 0
10975.36	10978.37		0.1d				9108.82	4D 0	2B17	P 6
10976.36	10979.37		0.2				9107.99			
10977.11	10980.12		0.1				9107.37			
10977.71	10980.72		0.1				9106.87	4d 2	3b 3	P 1

λ_{air}	λ_{vac}	Ref.	I_1	I_2	I_3	I_4	ν	Upper	Lower	Br.
10978.43	10981.43			0.2			9106.28			
10979.26	10982.26		0.2	1.0			9105.59	T 0	2B17	P 1
10980.15	10983.16		0.2				9104.85			
10980.84	10983.85		0.4	1.5b			9104.28	2A 0	2B 0	P 6
10982.08	10985.09			0.2			9103.25	u 1	2a 5	R 2
10982.51	10985.52			0.4			9102.89			
10983.58	10986.58			0.1			9102.01	W 2	2B17	R 4
10983.58	10986.58			0.1			9102.01	2B18	2A 0	P 5
10984.54	10987.55			0.2			9101.21			
10987.05	10990.06			0.3			9099.13	3C 0	2A 2	P 5
10987.05	10990.06			0.3			9099.13	3b 5	2a 5	P 2
10987.70	10990.71			0.2			9098.59			
10988.54	10991.55			0.1			9097.90			
10989.03	10992.04			0.1			9097.49			
10989.54	10992.55			0.2			9097.07			
10990.22	10993.23			0.2			9096.51	3C 0	2K 2	R 0
10990.91	10993.92			0.2			9095.94			
10991.55	10994.56			0.1			9095.41			
10993.32	10996.33			0.1			9093.94	3d 2	3b 0	R 1
10994.56	10997.57			0.1			9092.92			
10995.15	10998.16		0.4	0.3			9092.43	2K 5	2B 6	P 1
10996.08	10999.09			0.1			9091.66			
10996.82	10999.83			0.2			9091.05	3e 3	3b 1	Q 4
10997.40	11000.41			0.2			9090.57	4d 3	3b 4	P 3
10997.40	11000.41			0.2			9090.57	4e 1	3b 2	R 0
10998.22	11001.23			0.4			9089.89			
10999.02	11002.03			0.2			9089.23	3e 3	3b 1	P 2
11000.67	11003.68			0.1			9087.87			
11001.33	11004.34			0.2			9087.32			
11001.76	11004.77			0.1			9086.97			
11002.42	11005.43			0.1			9086.42			
11002.98	11005.99			0.1d			9085.96			
11003.46	11006.48		0.1d				9085.56			
11003.74	11006.75			0.4			9085.33			
11004.33	11007.35			0.2			9084.84			
11004.95	11007.97			0.2			9084.33			
11005.42	11008.44			0.2d			9083.94			
11006.20	11009.21			0.1d			9083.30	3C 0	2A 2	P 4
11006.67	11009.69			0.2d			9082.91	4b 0	2a 5	P 3
11007.18	11010.20			0.1			9082.49			
11007.81	11010.83			0.1			9081.97	3C 2	2K 6	P 2
11009.44	11012.45			0.2			9080.63			
11010.09	11013.11			0.2			9080.09			
11010.69	11013.70			0.2			9079.60			
11011.39	11014.40			0.3			9079.02	Z 1	2B13	P 1
11012.18	11015.19			0.1			9078.37	3D 1	2B13	P 5
11012.37	11015.39		0.1				9078.21	3E 1	2B18	R 2
11012.67	11015.69			0.1d			9077.96			
11013.16	11016.18		0.1				9077.56			
11014.08	11017.10		0.4	1.5b			9076.80	3B 1	2A 1	R 1
11015.33	11018.35			0.7			9075.77	3b 0	2a 1	R10
11016.00	11019.02			0.1			9075.22			
11017.00	11020.01			0.1			9074.40	4g 1	2a 8	R 1
11017.00	11020.01			0.1			9074.40	4e 0	3b 1	R 3
11017.97	11020.98			0.3			9073.60	3C 0	2K 2	Q 1
11018.66	11021.68			0.3			9073.03	3B 4	2K 5	R 2
11020.5	11023.5		3.1				9071.51	4d 2	3b 3	R 1
11021.51	11024.53		0.3	1.5b			9070.68	2A 0	2B 1	R 8
11022.58	11025.60			0.3			9069.80	3A 2	2B17	P 2
11023.20	11026.22			0.1			9069.29			

λ_{air}	λ_{vac}	Ref.	I_1	I_2	I_3	I_4	ν	Upper	Lower	Br.
11023.68	11026.70			0.1			9068.90	3A 2	2B17	P 1
11024.66	11027.68			0.2			9068.09			
11025.26	11028.28		0.2	1.5			9067.60	3B 0	2A 0	P 3
11025.26	11028.28		0.2	1.5			9067.60	2A 1	2B 2	R 3
11026.17	11029.19		0.1				9066.85			
11026.61	11029.63			0.1d			9066.49			
11027.39	11030.41			0.7			9065.85			
11027.76	11030.78			0.3			9065.54			
11028.26	11031.28			0.1			9065.13			
11029.00	11032.02			0.2			9064.52	3C 0	2A 2	P 3
11029.83	11032.85			0.1			9063.84	3e 3	3b 1	P 3
11030.52	11033.55			0.1			9063.27	2B17	2A 0	R 0
11031.16	11034.18			0.2			9062.75			
11032.36	11035.38			0.3			9061.76	4D 0	2B17	P 3
11033.03	11036.05			0.2			9061.21	3d 2	3b 0	R 2
11033.59	11036.61			0.2			9060.75	3F 1	2C 3	R 1
11034.41	11037.43			0.1			9060.08			
11034.99	11038.01			0.4			9059.60	3D 0	2B11	R 4
11036.19	11039.21			0.3			9058.62			
11036.82	11039.84			0.3			9058.10			
11037.4	11040.4		1.7				9057.62			
11038.66	11041.68			0.5			9056.59	3B 3	2A 3	R 4
11041.00	11044.02			0.1			9054.67			
11042.28	11045.31		0.5	0.7			9053.62	3E 3	2B18	Q 4
11042.28	11045.31		0.5	0.7			9053.62	u 1	2a 5	R 1
11042.28	11045.31		0.5	0.7			9053.62	2K 5	2B 6	P 2
11043.90	11046.93			0.2			9052.29			
11044.67	11047.70			0.1			9051.66			
11045.61	11048.64			0.1			9050.89			
11046.33	11049.36			0.3			9050.30	Z 1	2B13	P 2
11046.33	11049.36			0.3			9050.30	3F 1	2C 3	R 1
11046.94	11049.97			0.2			9049.80	4D 0	2B17	P 5
11046.94	11049.97			0.2			9049.80	3d 2	3b 0	P 1
11047.89	11050.92			0.2			9049.02			
11048.72	11051.75			0.5			9048.34			
11049.53	11052.56		0.2	0.5			9047.68			
11050.39	11053.41			0.2			9046.98	3F 3	2B19	R 2
11051.61	11054.63			0.1d			9045.98			
11053.90	11056.93			0.2			9044.10	X 1	2B15	R 1
11054.38	11057.41			0.2			9043.71			
11055.11	11058.14		0.1	0.9			9043.11	2K 3	2B 4	P 6
11055.97	11059.00			0.1			9042.41			
11056.59	11059.62			0.3			9041.90			
11057.51	11060.54			0.3			9041.15			
11058.39	11061.42			0.1d			9040.43	2B17	2A 0	R 1
11058.64	11061.67			0.1			9040.23			
11059.62	11062.64			0.3d			9039.43	3e 3	3b 1	P 4
11062.43	11065.46			0.1			9037.13			
11064.05	11067.08			0.4			9035.81	3F 1	2B14	R 1
11064.05	11067.08			0.4			9035.81	4D 0	2B17	P 4
11064.05	11067.08			0.4			9035.81	3F 0	2C 2	R 2
11064.95	11067.98			0.2			9035.07			
11065.44	11068.47		0.2	0.6			9034.67	3B 1	2A 1	R 0
11066.30	11069.33			0.1			9033.97			
11066.78	11069.81			0.1			9033.58			
11067.32	11070.35			0.3			9033.14			
11068.49	11071.52			0.2			9032.18	3C 0	2K 2	P 2
11069.25	11072.28			0.2			9031.56			
11070.04	11073.07			0.1			9030.92			
11070.59	11073.62			0.2d			9030.47			

λ_{air}	λ_{vac}	Ref.	I_1	I_2	I_3	I_4	ν	Upper	Lower	Br.
11070.85	11073.88			0.2			9030.26			
11071.59	11074.63			0.1			9029.65			
11072.59	11075.62			0.2			9028.84	2A 3	2B 5	R 3
11073.32	11076.36			0.5			9028.24			
11074.13	11077.17			0.1			9027.58			
11074.61	11077.64			0.1			9027.19			
11075.48	11078.52		0.1b	1.0			9026.48	3B 2	2K 2	R 3
11076.28	11079.31			0.3			9025.83			
11077.63	11080.66		1.5b	1.5b			9024.73	2A 0	2B 0	R 0
11078.86	11081.89			0.6			9023.73	3b 5	2a 5	P 3
11078.86	11081.89			0.6			9023.73	4f 0	3b 1	Q 3
11079.63	11082.67			0.3			9023.10			
11080.10	11083.13			0.2			9022.72			
11081.14	11084.18			0.1d			9021.87			
11082.35	11085.38			0.1			9020.89	Z 1	2B13	P 3
11083.12	11086.15		0.3	1.0			9020.26	2A 3	2B 5	R 2
11084.47	11087.51			0.1			9019.16	3d 2	3b 0	R 3
11085.23	11088.27			0.2			9018.54			
11086.00	11089.03		0.8	1.5b			9017.92	3b 0	2a 1	P 1
11086.00	11089.03		0.8	1.5b			9017.92	4c 0	2a 6	P 2
11086.00	11089.03		0.8	1.5b			9017.92	2A 0	2B 0	P 5
11087.14	11090.18			0.1			9016.99			
11087.91	11090.95		0.1	0.8			9016.36			
11088.96	11092.00			0.2			9015.51			
11090.63	11093.67			0.1b			9014.15			
11091.51	11094.54			0.3			9013.44			
11093.83	11096.87			0.1d			9011.55			
11094.76	11097.79			0.1			9010.80	W 2	2B17	R 3
11094.76	11097.79			0.1			9010.80	4d 2	3b 3	R 3
11096.21	11099.25			0.1			9009.62	4e 1	3b 2	Q 1
11096.8	11099.8		1.0				9009.14			
11097.50	11100.54			0.5			9008.57			
11099.7	11102.7		1.0				9006.79	u 1	2a 5	R 0
11100.22	11103.26		0.2	0.4			9006.36	2K 5	2B 6	P 3
11101.52	11104.56			0.1			9005.31			
11102.0	11105.0		1.4				9004.92	3E 2	2B16	Q 5
11103.91	11106.95			0.1			9003.37			
11105.7	11108.7		1.0				9001.92	3e 3	3b 1	P 5
11106.45	11109.49			0.3			9001.31			
11107.02	11110.06			0.1			9000.85			
11107.67	11110.72			0.2			9000.32			
11108.45	11111.49		0.3	0.6			8999.69	2A 3	2B 5	R 1
11109.91	11112.95			0.3			8998.51			
11110.66	11113.70			0.2			8997.90			
11111.40	11114.45			0.3			8997.30			
11111.9	11114.9		1.4				8996.90	3D 0	2B11	R 3
11112.75	11115.79			0.2			8996.21	4e 0	3b 1	R 2
11114.01	11117.05			0.1			8995.19			
11114.44	11117.49			0.1			8994.84			
11115.00	11118.04			0.2			8994.39	3F 1	2B14	Q 2
11115.68	11118.72		0.2	0.4			8993.84	Z 1	2B13	P 4
11116.74	11119.78			0.2			8992.98			
11117.42	11120.46			0.2			8992.43	3E 3	2B18	R 1
11118.76	11121.80			0.2			8991.35	4c 0	2a 6	P 3
11119.78	11122.83			0.2			8990.52	U 1	2B14	R 3
11120.45	11123.50			0.3			8989.98			
11123.93	11126.97			0.1			8987.17	4d 2	3b 3	R 4
11125.66	11128.71			0.3			8985.77			
11126.34	11129.39			0.3b			8985.22			
11127.75	11130.80			0.1			8984.08			

λ_{air}	λ_{vac}	Ref.	I_1	I_2	I_3	I_4	ν	Upper	Lower	Br.
11128.46	11131.51			0.2			8983.51			
11129.65	11132.70			0.3			8982.55			
11131.11	11134.16			0.3d			8981.37			
11132.31	11135.36			0.1d			8980.40			
11133.22	11136.27			0.1			8979.67			
11133.89	11136.94			0.2			8979.13	2B17	2A 0	P 1
11134.67	11137.72			0.2			8978.50	3B 4	2K 5	P 5
11135.47	11138.52			0.2			8977.85	4d 2	3b 3	P 2
11136.50	11139.55			0.2			8977.02			
11137.68	11140.73			0.2			8976.07			
11138.32	11141.37			0.2			8975.56			
11140.13	11143.18			0.7			8974.10			
11140.75	11143.80			0.0			8973.60			
11141.83	11144.88			0.1			8972.73			
11142.45	11145.50		0.5	0.7			8972.23	2A 3	2B 5	R 0
11143.75	11146.81			0.1			8971.18	3d 2	3b 0	R 4
11145.88	11148.93			0.1			8969.47			
11146.72	11149.78			0.2			8968.79	3E 3	2B18	Q 3
11146.72	11149.78			0.2			8968.79	3e 3	3b 1	P 6
11150.4	11153.5		1.4				8965.83	4d 3	3b 4	P 4
11150.4	11153.5		1.4				8965.83	3F 2	2C 4	P 3
11153.03	11156.08			0.1			8963.72	W 2	2B17	R 2
11153.90	11156.95			0.1			8963.02			
11155.36	11158.41			0.2d			8961.85	3D 0	2B11	R 2
11155.87	11158.92			0.2d			8961.44			
11157.36	11160.42		0.1	0.1			8960.24			
11158.83	11161.89			0.2			8959.06			
11159.84	11162.89			0.3			8958.25			
11160.91	11163.97			0.3			8957.39	4e 1	3b 2	Q 2
11161.38	11164.44			0.2			8957.01			
11162.12	11165.18		0.9	1.5b			8956.42	2A 0	2B 0	P 4
11163.23	11166.28			0.1d			8955.53			
11164.31	11167.37			0.1			8954.66	3F 2	2C 4	P 4
11165.52	11168.58			0.2d			8953.69	3d 2	3b 0	P 2
11166.36	11169.42			0.4			8953.02			
11166.86	11169.91			0.1			8952.62			
11168.08	11171.14			0.1			8951.64	4F 0	2B18	Q 3
11168.67	11171.72		0.2	0.6			8951.17	2K 5	2B 6	P 4
11170.52	11173.58			0.2d			8949.68	4e 1	3b 2	P 2
11171.34	11174.40			0.3			8949.03	X 1	2B15	R 2
11171.82	11174.88			0.4			8948.64			
11173.56	11176.62			0.2			8947.25			
11174.56	11177.62			0.8			8946.45	3B 3	2A 3	R 3
11175.33	11178.39			0.1			8945.83			
11175.93	11178.99			0.1			8945.35			
11176.78	11179.84		0.5	1.5b			8944.67	3b 0	2a 1	P 2
11176.78	11179.84		0.5	1.5b			8944.67	3D 0	2B11	R 0
11177.58	11180.64			0.1			8944.03			
11178.18	11181.24			0.7			8943.55	3B 2	2A 2	R 2
11178.18	11181.24			0.7			8943.55	3F 2	2C 4	P 5
11179.1	11182.2		3.5				8942.82	3D 0	2B11	R 1
11179.1	11182.2		3.5				8942.82	3D 2	2B16	R 6
11179.58	11182.64						8942.43			
11181.85	11184.91		0.6	1.5b			8940.62	2A 1	2B 2	R 2
11182.88	11185.95			0.1			8939.79			
11184.32	11187.38			0.3d			8938.64	3E 3	2B18	P 4
11185.07	11188.14			0.1			8938.04	3E 1	2B14	Q 6
11185.07	11188.14			0.1			8938.04	4e 1	3b 2	P 6
11185.66	11188.72			0.1			8937.57	3f 0	2c 4	R 1
11186.36	11189.42			0.2			8937.01			

λ_{air}	λ_{vac}	Ref.	I_1	I_2	I_3	I_4	ν	Upper	Lower	Br.
11186.94	11190.00			0.1			8936.55			
11187.61	11190.68			0.2			8936.01	4f 0	3b 1	Q 2
11188.43	11191.49			0.1			8935.36			
11189.12	11192.18			0.2d			8934.81			
11189.99	11193.06			0.4			8934.11	3B 4	2K 5	R 0
11190.67	11193.73			0.4			8933.57	4e 1	3b 2	P 3
11191.53	11194.60			0.2			8932.88	3b 5	2a 5	P 4
11192.13	11195.20			0.1			8932.40			
11192.64	11195.70			0.1			8932.00			
11193.19	11196.25			0.2			8931.56			
11194.50	11197.57			0.1			8930.51			
11195.10	11198.17			0.2			8930.03			
11195.98	11199.05			0.1d			8929.33			
11198.00	11201.07			0.2			8927.72	4e 1	3b 2	P 5
11198.47	11201.53			0.0			8927.35	3F 0	2B12	Q 4
11199.24	11202.31			0.8			8926.73	3B 2	2A 2	R 5
11199.24	11202.31			0.8			8926.73	3F 1	2B14	P 3
11200.65	11203.72			0.2			8925.61	4e 1	3b 2	P 4
11201.41	11204.48			0.0			8925.00	3B 4	2K 5	P 4
11202.04	11205.11		0.2	0.8			8924.50	3B 1	2A 1	P 1
11202.04	11205.11		0.2	0.8			8924.50	4e 0	3b 1	R 1
11202.8	11205.9		4.5				8923.90			
11203.45	11206.52			0.1			8923.38			
11204.60	11207.67		1.0	1.5			8922.46	2A 0	2B 0	P 1
11207.13	11210.20			0.2			8920.45			
11207.86	11210.93		0.8	1.5b			8919.87	2A 0	2B 0	P 3
11209.03	11212.09			0.2d			8918.94			
11210.24	11213.31			0.1			8917.97	3f 0	2c 4	R 2
11210.99	11214.06			0.2d			8917.38			
11212.28	11215.35			0.2d			8916.35			
11214.55	11217.62			0.1			8914.55			
11215.04	11218.11			0.2			8914.16			
11216.12	11219.19			0.1			8913.30			
11217.20	11220.27			0.2			8912.44	U 1	2B14	R 2
11218.13	11221.20			0.2			8911.70			
11218.51	11221.58			0.2			8911.40			
11219.10	11222.17			0.2			8910.93			
11219.43	11222.50		0.4	0.3			8910.67	2A 3	2B 5	P 1
11220.27	11223.34			0.0d			8910.00			
11221.25	11224.33			0.2			8909.22			
11222.07	11225.15		1.5b	1.5b			8908.57	2A 0	2B 0	P 2
11223.37	11226.44		0.2d	0.4			8907.54	3E 3	2B18	P 3
11224.32	11227.39			0.3			8906.79			
11225.19	11228.26			0.1			8906.10	3D 3	2B18	R 4
11228.0	11231.1		1.2				8903.87	3E 3	2B18	Q 2
11230.75	11233.82			0.1d			8901.69	4F 0	2B18	R 1
11232.93	11236.01			0.2			8899.96			
11233.95	11237.03			0.1			8899.15	W 2	2B17	R 0
11234.66	11237.74			1.5b			8898.59	3b 1	2a 2	R 4
11235.90	11238.97			0.2b			8897.61	U 1	2B14	P 6
11235.90	11238.97			0.2b			8897.61	4d 1	3b 2	P 1
11237.36	11240.44			0.3			8896.45	4e 1	3b 2	Q 4
11238.38	11241.46			0.1d			8895.64			
11239.57	11242.65			1.5b			8894.70	3b 1	2a 2	R 5
11240.67	11243.75			0.2			8893.83			
11241.42	11244.50		0.1	0.6			8893.24			
11242.49	11245.57		0.1				8892.39			
11242.76	11245.84			0.1			8892.18			
11243.48	11246.56		0.2	1.5b			8891.61	3b 1	2a 2	R 3
11243.48	11246.56		0.2	1.5b			8891.61	W 2	2B17	P 6

λ_{air}	λ_{vac}	Ref.	I_1	I_2	I_3	I_4	ν	Upper	Lower	Br.
11244.82	11247.90		0.3	1.0			8890.55	3B 2	2A 2	R 1
11245.63	11248.71			0.1			8889.91			
11246.24	11249.32			0.3			8889.43	2K 5	2B 6	P 5
11246.86	11249.94			0.2			8888.94			
11247.56	11250.64			0.3			8888.38	3B 4	2K 5	P 3
11248.48	11251.56			0.2			8887.66			
11249.16	11252.24			0.2d			8887.12	W 1	2B15	R 4
11250.02	11253.10			0.3			8886.44			
11252.11	11255.19			0.1d			8884.79			
11252.92	11256.00			0.1			8884.15			
11253.39	11256.47			0.5			8883.78	3B 2	2A 2	R 4
11254.8	11257.9		1.0				8882.67			
11255.38	11258.46			0.2			8882.21			
11257.67	11260.75			0.9			8880.40	U 1	2B14	R 1
11257.67	11260.75			0.9			8880.40	3b 1	2a 2	R 6
11257.67	11260.75			0.9			8880.40	4e 1	3b 2	Q 5
11259.09	11262.17			0.1			8879.28	3E 3	2B18	P 2
11259.89	11262.97		0.5	0.6			8878.65	2A 3	2B 5	P 2
11260.47	11263.56			0.2			8878.19			
11263.59	11266.68			0.1			8875.73	2B17	2A 0	R 3
11264.89	11267.97			0.1			8874.71			
11265.52	11268.61			0.2			8874.21	4e 1	3b 2	Q 6
11265.99	11269.08		0.2	1.0b			8873.84	3b 1	2a 2	R 2
11265.99	11269.08		0.2	1.0b			8873.84	2B17	2A 0	P 2
11267.10	11270.18			0.1			8872.97			
11268.05	11271.14		0.1	0.2			8872.22	3B 4	2K 5	P 1
11269.04	11272.13			0.2			8871.44			
11269.55	11272.64			0.1			8871.04			
11270.15	11273.23			0.2			8870.57	3C 0	2K 3	R 4
11270.15	11273.23			0.2			8870.57	3B 4	2K 5	P 2
11271.00	11274.08			0.2d			8869.90			
11271.84	11274.92			0.1			8869.24			
11272.82	11275.90			0.2			8868.47	3D 2	2B16	R 5
11274.4	11277.5		0.8				8867.22	3D 3	2B18	R 3
11276.12	11279.21			0.1			8865.87			
11277.0	11280.1		2.4				8865.18			
11277.70	11280.79			0.1			8864.63			
11278.08	11281.17		0.1				8864.33			
11278.63	11281.72			0.1			8863.90			
11279.24	11282.33			0.3			8863.42			
11280.03	11283.12			0.1			8862.80			
11280.78	11283.87		0.5	1.5b			8862.21	3b 0	2a 1	P 3
11280.78	11283.87		0.5	1.5b			8862.21	U 1	2B14	R 0
11282.13	11285.22			0.3			8861.15			
11283.11	11286.20			0.3			8860.38	U 1	2C 3	Q 1
11283.87	11286.96			0.5			8859.78			
11284.89	11287.98			0.4			8858.98	3B 3	2A 3	R 2
11285.43	11288.52			0.1			8858.56			
11285.81	11288.90			0.1			8858.26	3D 3	2B18	R 2
11286.45	11289.54			0.5			8857.76	3B 1	2A 1	P 2
11286.60	11289.69		0.2d				8857.64			
11287.49	11290.58						8856.94	3F 1	2B15	Q 6
11288.71	11291.81			1.0b			8855.98			
11289.44	11292.53			0.2			8855.41			
11290.46	11293.55			0.4			8854.61			
11292.09	11295.18			0.2			8853.33	W 2	2B17	P 1
11292.9	11296.0		1.2				8852.70	4d 2	3b 3	P 3
11293.71	11296.81			0.2			8852.06			
11294.30	11297.39			0.2			8851.60	Z 0	2B11	R 4
11294.85	11297.94			0.1			8851.17			

λ_{air}	λ_{vac}	Ref.	I_1	I_2	I_3	I_4	ν	Upper	Lower	Br.
11295.61	11298.71			0.1			8850.57			
11297.16	11300.25			0.2d			8849.36			
11298.32	11301.41			0.1d			8848.45			
11300.19	11303.28			0.1			8846.99	3E 3	2B18	Q 1
11302.46	11305.55		0.7	1.5b			8845.21	3b 1	2a 2	R 1
11302.46	11305.55		0.7	1.5b			8845.21	3B 2	2A 2	R 0
11303.37	11306.46			0.3			8844.50	2A 3	2B 5	P 3
11304.66	11307.75			0.1			8843.49			
11305.41	11308.51			0.8			8842.90	2A 1	2B 2	P 6
11306.75	11309.85			0.0			8841.85			
11311.4	11314.5		1.3				8838.22	3D 3	2B18	R 1
11311.4	11314.5		1.3				8838.22	W 2	2B17	P 2
11313.04	11316.13			0.1			8836.94	Z 0	2B11	R 3
11314.11	11317.21		0.1				8836.10			
11314.38	11317.48			0.1d			8835.89	2B18	2A 0	P 6
11315.33	11318.43			0.2d			8835.15			
11315.9	11319.0		1.5				8834.70			
11317.24	11320.34			0.1			8833.66			
11317.96	11321.05		0.2	0.5			8833.10	2A 1	2B 2	R 1
11318.77	11321.87			0.1			8832.46			
11319.40	11322.50			0.2			8831.97			
11321.25	11324.35			0.1			8830.53	3D 2	2B16	R 4
11321.25	11324.35			0.1			8830.53	3d 2	3b 0	P 3
11322.67	11325.77			0.1			8829.42			
11323.38	11326.48			0.1			8828.87	Z 0	2B11	R 2
11324.13	11327.23			0.1			8828.28			
11324.99	11328.09			0.5			8827.61	2A 0	2B 1	R 7
11325.81	11328.92			0.1			8826.97	W 2	2B17	P 3
11326.33	11329.43		0.1d				8826.57			
11328.51	11331.61			0.1			8824.87	3E 0	2C 2	R 4
11329.38	11332.48			0.3			8824.19	2K 5	2B 6	P 6
11330.42	11333.53			0.1			8823.38			
11331.27	11334.37			0.2d			8822.72			
11332.39	11335.49			0.4			8821.85	W 2	2B17	P 5
11333.13	11336.24			0.5			8821.27	3B 2	2A 2	R 3
11333.97	11337.07			0.1			8820.62			
11334.83	11337.93			0.1			8819.95			
11336.91	11340.02			0.1			8818.33			
11338.20	11341.30			0.2			8817.33			
11341.64	11344.75			0.1			8814.65			
11342.26	11345.37			0.1			8814.17			
11342.88	11345.99			0.1			8813.69	W 2	2B17	P 4
11347.22	11350.33			0.2			8810.32			
11348.04	11351.15			0.1			8809.68			
11348.97	11352.08			0.2			8808.96	Z 0	2B11	R 1
11350.7	11353.8		1.8				8807.62	3E 0	2C 2	R 3
11353.12	11356.23		0.4	0.8			8805.74	3b 1	2a 2	R 0
11354.73	11357.84			0.5			8804.49	2A 3	2B 5	P 4
11354.73	11357.84			0.5			8804.49	U 1	2B14	P 2
11355.84	11358.95			0.1			8803.63			
11357.21	11360.32			0.4			8802.57	2A 2	2B 4	R 4
11357.88	11360.99			0.3			8802.05			
11358.00	11361.11		0.1d				8801.96			
11362.2	11365.3		0.7				8798.70	3A 2	2B18	R 4
11364.3	11367.4		0.7				8797.08			
11366.2	11369.3		0.7				8795.61	U 1	2C 3	Q 2
11367.63	11370.74		0.1	0.1			8794.50			
11369.18	11372.29			0.2			8793.30			
11370.23	11373.34			0.2			8792.49	3D 2	2B16	R 3
11371.4	11374.5		1.7				8791.58			

λ_{air}	λ_{vac}	Ref.	I_1	I_2	I_3	I_4	ν	Upper	Lower	Br.
11372.78	11375.89		0.1				8790.52			
11374.52	11377.64		0.2d	0.5			8789.17	3B 3	2A 3	R 1
11376.27	11379.39			0.1			8787.82	U 1	2B14	P 5
11378.7	11381.8		1.1				8785.94	3A 1	2B15	P 4
11378.7	11381.8		1.1				8785.94	3D 3	2B18	P 2
11383.6	11386.7		1.5				8782.16	3d 3	3b 1	R 0
11384.56	11387.68			0.6			8781.42	3B 1	2A 1	P 3
11385.63	11388.74			0.1			8780.60			
11386.53	11389.65			0.1			8779.90			
11387.13	11390.25			0.3			8779.44	3F 1	2C 3	P 4
11387.96	11391.08			0.1			8778.80	X 0	2B13	R 4
11388.75	11391.87			0.5			8778.19	3b 1	2a 2	R 9
11389.5	11392.6		3.3				8777.61	3E 1	2B14	Q 5
11389.5	11392.6		3.3				8777.61	W 1	2B15	R 3
11390.49	11393.61			0.1d			8776.85	3d 3	3b 1	R 1
11392.26	11395.37			0.2			8775.49	2A 4	2B 7	R 3
11393.68	11396.80			0.2			8774.39			
11395.02	11398.14			0.1d			8773.36	4F 0	2B18	Q 2
11395.85	11398.97			0.3			8772.72	2A 4	2B 7	R 2
11397.74	11400.86		0.2d	0.8			8771.27	3b 0	2a 1	P 4
11397.74	11400.86		0.2d	0.8			8771.27	U 1	2B14	P 3
11398.63	11401.75			0.1			8770.58	3C 0	2K 3	Q 5
11401.01	11404.13			0.3			8768.75			
11403.1	11406.2		1.8				8767.14	3B 2	2K 2	P 4
11404.35	11407.48						8766.18	3D 2	2B16	R 2
11406.2	11409.3		0.7				8764.76	X 0	2B13	R 3
11408.92	11412.05			0.1			8762.67			
11410.32	11413.44			0.2			8761.60	2A 4	2B 7	R 1
11412.50	11415.63			0.1d			8759.92	3F 0	2B12	Q 3
11413.44	11416.57			0.1			8759.20	3D 2	2B16	P 1
11414.11	11417.23			0.4			8758.69	2A 1	2B 2	P 5
11415.27	11418.39			0.1			8757.80			
11415.92	11419.04			0.2			8757.30			
11416.54	11419.67						8756.82			
11419.74	11422.87			0.1d			8754.37	2A 3	2B 5	P 5
11420.69	11423.82		0.1				8753.64	3D 3	2B18	P 3
11421.07	11424.20			0.2			8753.35			
11421.61	11424.73		0.2	0.4			8752.94	3B 2	2A 2	P 1
11421.61	11424.73		0.2	0.4			8752.94	4d 1	3b 2	P 2
11422.40	11425.53			0.2			8752.33			
11423.3	11426.4		3.6				8751.64			
11424.50	11427.63		0.4	0.5			8750.72	2A 1	2B 2	R 0
11426.3	11429.4		3.4				8749.34	3E 2	2B16	Q 2
11426.3	11429.4		3.4				8749.34	U 1	2B14	P 4
11429.0	11432.1		1.8				8747.28			
11431.18	11434.31			0.2d			8745.61			
11432.2	11435.3		1.0				8744.83			
11433.00	11436.13		0.1				8744.22	Z 0	2B11	P 1
11433.79	11436.92		0.2	0.3			8743.61	2A 4	2B 7	R 0
11435.4	11438.5		1.0				8742.38	3D 2	2B16	R 1
11436.21	11439.34			0.1			8741.76			
11438.35	11441.48			0.2			8740.13			
11440.1	11443.2		0.8				8738.79			
11441.83	11444.96			0.3			8737.47			
11444.15	11447.28			0.1			8735.70	4d 2	3b 3	P 4
11445.00	11448.13			0.2			8735.05			
11446.3	11449.4		1.6				8734.06	3E 0	2C 2	R 2
11447.13	11450.27		0.1	0.2			8733.42	3B 3	2A 3	R 0
11449.18	11452.31			0.1d			8731.86			
11450.69	11453.82			0.2			8730.71	3B 2	2K 2	R 2

λ_air	λ_vac	Ref.	I₁	I₂	I₃	I₄	ν	Upper	Lower	Br.
11451.61	11454.74			0.1			8730.01			
11452.8	11455.9		1.2				8729.10			
11453.78	11456.92			0.2			8728.35	3b 6	2a 6	R 2
11455.07	11458.21			0.1			8727.37	2B17	2A 0	R 4
11456.37	11459.51			0.1			8726.38	3D 3	2B18	P 6
11457.24	11460.37			0.3			8725.72	3b 1	2a 2	R10
11457.24	11460.37			0.3			8725.72	2B17	2A 0	P 3
11458.76	11461.90			0.1			8724.56	V 1	2C.2	R 2
11459.64	11462.78			0.1			8723.89			
11460.52	11463.66			0.3			8723.22	u 0	2a 4	R 2
11461.3	11464.4		1.7				8722.63	3d 3	3b 1	R 3
11462.06	11465.19			0.1			8722.05			
11462.67	11465.81			0.3			8721.58	X 0	2B13	R 2
11463.55	11466.69			0.1			8720.91	3E 2	2C 4	R 5
11463.55	11466.69			0.1			8720.91	3A 1	2B15	P 3
11464.4	11467.5		0.7				8720.27	3D 3	2B18	P 4
11464.4	11467.5		0.7				8720.27	Z 0	2B11	P 2
11467.50	11470.64			0.1d			8717.91	3D 2	2B16	R 0
11469.3	11472.4		0.5				8716.54			
11470.93	11474.07			0.1d			8715.30	3f 0	2c 4	Q 2
11472.25	11475.39			0.3d			8714.30	3B 2	2A 2	P 2
11472.25	11475.39			0.3d			8714.30	3d 3	3b 1	P 1
11475.15	11478.29			0.2			8712.10			
11475.74	11478.88		0.1				8711.65	U 1	2C 3	Q 3
11476.28	11479.42			0.1			8711.24			
11479.4	11482.5		0.7				8708.87	3E 2	2B16	Q 1
11482.15	11485.29			0.1			8706.79			
11482.91	11486.05			0.4			8706.21	3B 2	2K 2	P 3
11484.0	11487.1		2.6				8705.38	4e 0	3b 1	Q 2
11485.46	11488.60			0.2			8704.28			
11487.08	11490.22			0.2			8703.05	2A 2	2B 4	R 3
11488.04	11491.19			0.2			8702.32			
11489.06	11492.21			0.1			8701.55			
11489.73	11492.88			0.3			8701.04	3B 1	2A 1	P 4
11489.73	11492.88			0.3			8701.04	3E 2	2C 4	R 4
11489.94	11493.09		0.1				8700.88			
11492.23	11495.38			0.2d			8699.15			
11493.54	11496.68			0.2			8698.16			
11494.07	11497.21		0.3	0.7			8697.76	2A 1	2B 2	P 4
11494.07	11497.21		0.3	0.7			8697.76	3b 6	2a 6	R 1
11495.81	11498.96			0.1			8696.44			
11497.12	11500.27		0.4	0.8			8695.45	3b 1	2a 2	P 1
11498.1	11501.2		9.9				8694.71	3d 2	3b 0	P 4
11499.28	11502.42			0.1d			8693.82			
11503.11	11506.26			0.2			8690.92	2A 4	2B 7	P 1
11503.11	11506.26			0.2			8690.92	3D 3	2B18	P 5
11504.1	11507.2		1.2				8690.17			
11505.69	11508.84			0.1			8688.97	3A 1	2B15	P 1
11506.67	11509.82			0.1			8688.23	X 0	2B13	R 1
11507.50	11510.65			0.2			8687.61			
11508.57	11511.72			0.2			8686.80			
11509.24	11512.39			0.2			8686.29			
11510.13	11513.28			0.2			8685.62			
11511.03	11514.18			0.1			8684.94	Z 0	2B11	P 3
11511.71	11514.86			0.2			8684.43			
11514.2	11517.4		1.0				8682.55	3d 3	3b 1	R 4
11516.0	11519.2		0.8				8681.19			
11517.13	11520.29			0.1			8680.34	4E 0	2B18	P 4
11519.96	11523.11			0.1			8678.21			
11520.23	11523.38			0.2			8678.01	4E 0	2B18	Q 4

λ_{air}	λ_{vac}	Ref.	I_1	I_2	I_3	I_4	ν	Upper	Lower	Br.
11521.37	11524.52			0.1			8677.15			
11522.72	11525.88			0.1d			8676.13	4e 0	3b 1	P 3
11523.88	11527.03			0.2			8675.26			
11526.53	11529.69			0.1			8673.26	4d 0	3b 1	R 0
11527.36	11530.51			0.8			8672.64	3b 0	2a 1	P 5
11527.36	11530.51			0.8			8672.64	3b 6	2a 6	R 0
11528.37	11531.52			0.1			8671.88	4d 0	3b 1	P 1
11534.6	11537.8		0.5				8667.20	u 0	2a 4	R 1
11534.6	11537.8		0.5				8667.20	3E 1	2B14	Q 4
11537.78	11540.93			0.3			8664.81			
11539.27	11542.43			0.1			8663.69			
11540.29	11543.45			0.2			8662.92	4e 0	3b 1	Q 3
11540.43	11543.58			0.2			8662.82			
11541.9	11545.1		2.3				8661.71	3C 1	2K 5	R 2
11544.24	11547.40			0.2			8659.96	3B 3	2A 3	P 1
11545.17	11548.33			0.1			8659.26	X 0	2B13	R 0
11546.57	11549.73		0.1d	0.3			8658.21	2A 1	2B 2	P 3
11548.21	11551.37		0.2	0.2			8656.98	2A 4	2B 7	P 2
11549.05	11552.21						8656.35			
11549.29	11552.45		0.3	0.2			8656.17	2A 1	2B 2	P 1
11549.29	11552.45		0.3	0.2			8656.17	U 1	2C 3	Q 4
11550.05	11553.21		0.1				8655.60	3D 2	2B16	P 2
11551.57	11554.74		0.1				8654.46			
11553.2	11556.4		1.1				8653.24	3E 0	2C 2	R 1
11555.8	11559.0		1.0				8651.29	4e 0	3b 1	P 4
11555.8	11559.0		1.0				8651.29	3E 0	2C 2	R 6
11555.8	11559.0		1.0				8651.29	V 1	2C 2	R 1
11559.4	11562.6		1.8				8648.60	3E 0	2C 2	Q 6
11559.4	11562.6		1.8				8648.60	3E 0	2C 2	Q 7
11559.4	11562.6		1.8				8648.60	3E 2	2C 4	R 2
11559.4	11562.6		1.8				8648.60	Z 0	2B11	P 4
11560.79	11563.96			0.3			8647.56	3B 3	2A 3	P 3
11561.6	11564.8		3.1				8646.95			
11563.91	11567.07		0.4	0.4			8645.23	2A 1	2B 2	P 2
11563.91	11567.07		0.4	0.4			8645.23	4F 0	2B19	R 3
11564.56	11567.73						8644.74	W 1	2B15	R 1
11565.04	11568.21			0.2			8644.38	3B 3	2A 3	P 2
11571.3	11574.5		1.3				8639.71	3d 3	3b 1	R 5
11571.3	11574.5		1.3				8639.71	3A 2	2B18	R 3
11572.33	11575.49			0.2			8638.94			
11572.63	11575.80		0.1				8638.71			
11573.16	11576.33			0.2			8638.32	3B 2	2K 2	R 1
11574.34	11577.50			0.1			8637.44			
11575.5	11578.7		2.6				8636.57	3D 2	2B16	P 3
11577.66	11580.83		0.1d	0.3			8634.96	2A 2	2B 4	R 2
11577.66	11580.83		0.1d	0.3			8634.96	4e 0	3b 1	P 5
11577.66	11580.83		0.1d	0.3			8634.96	4e 0	3b 1	P 8
11578.4	11581.6		3.4				8634.41	4e 0	3b 1	Q 4
11581.2	11584.4		1.3				8632.32	3d 3	3b 1	P 2
11583.5	11586.7		1.1				8630.61	4e 0	3b 1	P 7
11583.5	11586.7		1.1				8630.61	4E 0	2B18	P 2
11590.40	11593.57		0.2	0.5			8625.47	3b 1	2a 2	P 2
11590.40	11593.57		0.2	0.5			8625.47	U 1	2C 3	Q 5
11591.1	11594.3		9.1				8624.95	4d 0	3b 1	R 1
11591.1	11594.3		9.1				8624.95	4d 2	3b 3	P 5
11591.1	11594.3		9.1				8624.95	3C 1	2K 5	P 5
11600.9	11604.1		1.9				8617.66	4f 2	3b 4	R 4
11600.9	11604.1		1.9				8617.66	2A 4	2B 7	P 3
11601.93	11605.10			0.2			8616.90	4e 0	3b 1	Q 5
11603.38	11606.56			0.1b			8615.82	3E 2	2C 4	R 1

λ_{air}	λ_{vac}	Ref.	I_1	I_2	I_3	I_4	ν	Upper	Lower	Br.
11604.50	11607.67			0.2			8614.99	3B 1	2A 1	P 5
11604.50	11607.67			0.2			8614.99	Z 0	2B11	P 5
11604.50	11607.67			0.2			8614.99	4d 1	3b 2	P 3
11605.55	11608.73			0.1			8614.21			
11608.15	11611.33			0.1			8612.28			
11612.6	11615.8		1.4				8608.98	4e 0	3b 1	Q 6
11614.65	11617.83			0.1			8607.46	2K 2	2B 4	R 1
11616.40	11619.58			0.1			8606.16			
11621.37	11624.55			0.4			8602.48	2A 0	2B 1	R 6
11625.75	11628.93			0.1			8599.24			
11626.55	11629.73			0.1			8598.65			
11629.2	11632.4		1.4				8596.69	W 1	2B15	R 0
11630.27	11633.45			0.2			8595.90			
11631.72	11634.90			0.2			8594.83	2K 2	2B 4	R 0
11633.5	11636.7		1.5				8593.51			
11635.8	11639.0		1.5				8591.81	3A 1	2B15	P 2
11635.8	11639.0		1.5				8591.81	Z 0	2B11	P 6
11636.60	11639.79			0.1			8591.22			
11637.50	11640.68			0.2			8590.56	4E 0	2B18	Q 3
11637.50	11640.68			0.2			8590.56	3B 3	2K 4	R 5
11639.6	11642.8		1.2				8589.01	3E 0	2C 2	Q 4
11641.3	11644.5		0.8				8587.76			
11643.1	11646.3		0.9				8586.43	3E 0	2C 2	Q 5
11644.8	11648.0		0.9				8585.17			
11646.1	11649.3		1.4				8584.22	3E 1	2B14	Q 3
11648.9	11652.1		1.1				8582.15	3E 0	2C 2	R 5
11648.9	11652.1		1.1				8582.15	3E 1	2C 3	Q 2
11650.3	11653.5		1.2				8581.12			
11652.3	11655.5		0.9				8579.65			
11653.25	11656.44			0.1			8578.95			
11654.3	11657.5		0.5				8578.18	4d 0	3b 1	R 2
11654.3	11657.5		0.5				8578.18	3E 1	2C 3	Q 3
11656.2	11659.4		0.8				8576.78			
11660.2	11663.4		0.6				8573.84			
11662.15	11665.34			0.2			8572.40	2A 4	2B 7	P 4
11663.35	11666.54			0.1			8571.52	3E 0	2C 2	Q 3
11664.27	11667.47			0.1			8570.84	3E 0	2B12	P 6
11667.2	11670.4		1.6				8568.69	3f 1	3b 0	R 6
11668.70	11671.89			0.2			8567.59			
11669.33	11672.52			0.3			8567.13	3b 0	2a 1	P 6
11674.5	11677.7		0.8				8563.33	3C 0	2K 3	Q 4
11674.5	11677.7		0.8				8563.33	V 1	2B12	R 2
11677.8	11681.0		1.2				8560.91			
11678.58	11681.78			0.2b			8560.34			
11679.3	11682.5		1.3				8559.81	3f 0	2c 4	P 3
11681.4	11684.6		1.5				8558.28			
11684.2	11687.4		2.3				8556.22	3E 1	2C 3	Q 1
11686.57	11689.77			0.3			8554.49	3b 2	2a 3	R 4
11686.57	11689.77			0.3			8554.49	4E 0	2B18	Q 1
11692.02	11695.22			0.6			8550.50	3b 2	2a 3	R 3
11695.59	11698.79			0.4			8547.89	3b 2	2a 3	R 5
11695.59	11698.79			0.4			8547.89	3E 1	2C 3	Q 4
11697.69	11700.89			0.6			8546.36	3b 1	2a 2	P 3
11702.2	11705.4		0.8				8543.06	4d 0	3b 1	R 3
11702.2	11705.4		0.8				8543.06	2K 2	2B 4	P 1
11704.4	11707.6		0.9				8541.46	4E 0	2B18	Q 2
11708.3	11711.5		1.6				8538.61	2B17	2A 0	P 4
11712.27	11715.47			0.3			8535.72	3b 2	2a 3	R 2
11712.27	11715.47			0.3			8535.72	3C 1	2K 5	R 1
11713.24	11716.45			0.1			8535.01	3E 0	2C 2	R 2

λ_{air}	λ_{vac}	Ref.	I_1	I_2	I_3	I_4	ν	Upper	Lower	Br.
11717.25	11720.46			0.1			8532.09			
11718.87	11722.08			0.3			8530.91	3b 2	2a 3	R 6
11718.87	11722.08			0.3			8530.91	2B17	2A 0	R 5
11719.4	11722.6		3.1				8530.53	3A 2	2B18	R 2
11726.79	11730.00			0.1			8525.15			
11729.4	11732.6		1.0				8523.25	4d 0	3b 1	P 2
11731.6	11734.8		1.1				8521.65	4d 2	3b 3	P 6
11734.66	11737.87			0.1d			8519.43	3E 0	2C 2	R 1
11738.4	11741.6		0.7				8516.72			
11741.4	11744.6		0.9				8514.54	3C 1	2K 5	P 4
11746.71	11749.93			0.1			8510.69			
11747.47	11750.69			0.4			8510.14	3b 2	2a 3	R 1
11748.91	11752.12			0.1d			8509.10			
11753.4	11756.6		1.3				8505.85			
11756.12	11759.34			0.2			8503.88	3b 2	2a 3	R 7
11760.17	11763.39			0.2			8500.95	3B 2	2A 2	P 3
11760.17	11763.39			0.2			8500.95	3B 2	2K 2	P 1
11764.77	11767.99			0.1			8497.63	2K 2	2B 4	P 2
11770.0	11773.2		1.3				8493.85			
11771.60	11774.82			0.1			8492.70			
11774.1	11777.3		0.3				8490.89			
11779.0	11782.2		0.1				8487.36	3a 0	2c 4	R 3
11781.2	11784.4		0.6				8485.78	4d 1	3b 2	P 4
11781.2	11784.4		0.6				8485.78	3E 1	2B14	Q 1
11781.2	11784.4		0.6				8485.78	2K 4	2B 6	R 4
11785.1	11788.3		0.7				8482.97			
11787.8	11791.0		1.0				8481.03	3B 3	2K 4	R 4
11787.8	11791.0		1.0				8481.03	4F 0	2B19	P 5
11791.2	11794.4		0.7				8478.58			
11794.8	11798.0		1.1				8475.99			
11797.87	11801.10			0.2			8473.79	3b 2	2a 3	R 0
11797.87	11801.10			0.2			8473.79	3B 2	2A 2	P 4
11801.2	11804.4		0.4				8471.40			
11804.1	11807.3		0.3				8469.31			
11807.37	11810.60			0.1d			8466.97			
11809.6	11812.8		0.7				8465.37			
11811.2	11814.4		0.8				8464.22			
11813.3	11816.5		0.8				8462.72	3A 2	2B18	P 5
11813.3	11816.5		0.8				8462.72	3C 0	2K 3	R 2
11813.3	11816.5		0.8				8462.72	2K 4	2B 6	R 3
11818.77	11822.01			0.3			8458.80	3b 1	2a 2	P 4
11823.26	11826.50			0.2			8455.59	3b 0	2a 1	P 7
11829.47	11832.71			0.2			8451.15	2A 0	2B 1	P10
11834.3	11837.5		1.9				8447.70	4D 0	2B18	R 3
11836.0	11839.2		2.0				8446.49	3B 4	2A 4	R 3
11836.0	11839.2		2.0				8446.49	2K 2	2B 4	P 3
11840.9	11844.1		1.4				8442.99	3C 1	2K 5	R 0
11842.64	11845.89			0.1			8441.75	2K 4	2B 6	R 2
11842.64	11845.89			0.1			8441.75	3B 2	2K 3	R 5
11844.4	11847.6		0.4				8440.50			
11847.5	11850.7		0.5				8438.29			
11849.9	11853.1		1.1				8436.58			
11851.7	11854.9		1.1				8435.30			
11855.5	11858.7		1.2				8432.60			
11857.76	11861.00			0.1			8430.99	V 1	2B12	R 1
11859.71	11862.96			0.2			8429.60	3B 2	2A 2	P 5
11859.71	11862.96			0.2			8429.60	2K 2	2B 4	R 2
11859.71	11862.96			0.2			8429.60	3C 1	2K 5	P 3
11865.1	11868.3		0.7				8425.77			
11867.0	11870.2		1.2				8424.42			

λ_air	λ_vac	Ref.	I₁	I₂	I₃	I₄	ν	Upper	Lower	Br.
11869.2	11872.4		1.1				8422.86	2K 4	2B 6	R 1
11871.99	11875.24			0.1			8420.88	3A 2	2B18	R 1
11882.2	11885.5		1.1				8413.65			
11884.4	11887.7		1.5				8412.09	4f 2	3b 4	R 3
11886.6	11889.9		3.0				8410.53	3d 3	3b 1	P 4
11886.6	11889.9		3.0				8410.53	2A 2	2B 4	P 4
11886.6	11889.9		3.0				8410.53	4D 0	2B18	R 0
11886.6	11889.9		3.0				8410.53	2K 2	2B 4	R 3
11891.0	11894.3		3.0				8407.42	3d 2	3b 0	P 6
11891.0	11894.3		3.0				8407.42	3F 1	2B15	Q 4
11894.9	11898.2		1.0				8404.66			
11900.2	11903.5		1.6				8400.92	3E 1	2C 3	P 2
11900.2	11903.5		1.6				8400.92	2K 4	2B 6	R 0
11902.0	11905.3		1.9				8399.65			
11904.2	11907.5		1.8				8398.10	u 0	2a 4	P 3
11905.9	11909.2		2.4				8396.90	2A 0	2B 1	R 5
11905.9	11909.2		2.4				8396.90	3C 0	2K 3	Q 3
11905.9	11909.2		2.4				8396.90	3E 0	2C 2	P 2
11911.5	11914.8		0.6				8392.95	2A 2	2B 4	R 1
11911.5	11914.8		0.6				8392.95	4D 0	2B18	R 2
11918.7	11922.0		1.0				8387.88	3E 0	2C 2	P 3
11918.7	11922.0		1.0				8387.88	4F 0	2B19	Q 4
11924.7	11928.0		0.9				8383.66	3D 1	2C 3	P 1
11926.5	11929.8		1.3				8382.40			
11928.4	11931.7		1.4				8381.06	3B 3	2K 4	R 3
11931.0	11934.3		1.0				8379.23	3C 1	2K 5	P 2
11931.0	11934.3		1.0				8379.23	3E 0	2B12	P 4
11932.4	11935.7		1.2				8378.25			
11934.4	11937.7		0.9				8376.85	4d 0	3b 1	P 3
11936.8	11940.1		0.6				8375.16			
11939.7	11943.0		1.2				8373.13	4F 0	2B19	R 2
11944.68	11947.94			0.2			8369.64	3b 2	2a 3	P 1
11953.27	11956.54			0.3			8363.62	3b 1	2a 2	P 5
11959.0	11962.3		0.9				8359.62			
11962.8	11966.1		0.8				8356.96	3a 0	2c 4	Q 1
11962.8	11966.1		0.8				8356.96	3e 0	2c 4	Q 1
11962.8	11966.1		0.8				8356.96	3f 2	3b 1	R 5
11962.8	11966.1		0.8				8356.96	4D 0	2B18	R 1
11966.3	11969.6		0.5				8354.52	2K 6	2B 8	R 1
11970.4	11973.7		1.0				8351.65	3A 0	2B13	R 1
11970.4	11973.7		1.0				8351.65	3B 4	2A 4	R 2
11972.2	11975.5		1.5				8350.40	4f 3	3b 5	Q 3
11972.2	11975.5		1.5				8350.40	u 0	2a 4	P 4
11976.2	11979.5		0.5				8347.61	5c 0	2a 8	Q 1
11978.5	11981.8		1.4				8346.01	2K 4	2B 6	P 1
11979.8	11983.1		1.8				8345.10	2A 4	2C 0	R 1
11984.3	11987.6		2.0				8341.97	2A 2	2B 4	R 0
11988.6	11991.9		1.4				8338.98	3a 0	2c 4	R 1
11988.6	11991.9		1.4				8338.98	3b 0	2a 1	P 8
11991.5	11994.8		1.0				8336.96	4g 0	2a 7	P 1
11991.5	11994.8		1.0				8336.96	3A 2	2B18	R 0
11992.8	11996.1		1.0				8336.05	3A 2	2B18	P 3
11992.8	11996.1		1.0				8336.05	3E 0	2C 2	P 2
11996.2	11999.5		2.0				8333.69	2K 6	2B 8	R 0
11999.0	12002.3		1.0				8331.75			
12002.9	12006.2		0.7				8329.04	3E 0	2C 2	P 4
12002.9	12006.2		0.7				8329.04	3E 1	2C 3	P 3
12008.1	12011.4		0.7				8325.43	3f 1	3b 0	R 4
12017.8	12021.1		0.9				8318.71	2B17	2A 0	P 5
12019.9	12023.2		1.1				8317.26			

λ_{air}	λ_{vac}	Ref.	I_1	I_2	I_3	I_4	ν	Upper	Lower	Br.
12024.3	12027.6		1.1				8314.22			
12028.1	12031.4		0.6				8311.59	3C 0	2K 3	R 1
12030.7	12034.0		3.3				8309.79	3A 0	2B13	P 4
12030.7	12034.0		3.3				8309.79	2K 4	2B 6	P 2
12037.2	12040.5		1.5				8305.31	2K 3	2B 5	R 0
12037.2	12040.5		1.5				8305.31	2K 3	2B 5	R 1
12037.2	12040.5		1.5				8305.31	V 1	2B12	R 0
12040.96	12044.26			0.2			8302.71	3b 2	2a 3	P 2
12047.4	12050.7		0.7				8298.28	2A 4	2C 0	Q 1
12047.4	12050.7		0.7				8298.28	3E 0	2B12	P 3
12051.9	12055.2		0.7				8295.18	3e 0	2c 4	Q 3
12055.1	12058.4		0.5				8292.97			
12056.7	12060.0		0.6				8291.87	2K 3	2B 5	R 2
12058.3	12061.6		0.6				8290.77	3B 3	2K 4	R 2
12060.1	12063.4		0.6				8289.54	2A 4	2C 0	R 2
12062.3	12065.6		0.4				8288.02	4e 3	3b 5	R 2
12068.4	12071.7		0.5				8283.84	2K 6	2B 8	P 1
12070.5	12073.8		1.0				8282.39	3d 3	3b 1	P 5
12073.7	12077.0		1.1				8280.20			
12076.2	12079.5		1.5				8278.49	3B 4	2A 4	R 1
12078.0	12081.3		0.7				8277.25	3A 2	2B18	P 2
12081.96	12085.26			0.1			8274.54			
12082.57	12085.88			0.1			8274.12	2A 4	2C 0	P 1
12085.0	12088.3		1.2				8272.46	2K 4	2B 6	P 3
12087.5	12090.8		0.9				8270.75			
12090.1	12093.4		0.5				8268.97	3C 0	2K 3	Q 2
12093.4	12096.7		1.1				8266.71			
12097.9	12101.2		1.7				8263.64	2A 2	2B 4	P 1
12100.4	12103.7		4.9				8261.93	3b 1	2a 2	P 6
12100.4	12103.7		4.9				8261.93	3A 2	2B18	P 1
12100.4	12103.7		4.9				8261.93	2K 3	2B 5	R 3
12106.3	12109.6		0.9				8257.90	3f 2	3b 1	R 4
12107.5	12110.8		0.9				8257.08	3A 0	2B13	R 0
12107.5	12110.8		0.9				8257.08	4f 3	3b 5	Q 2
12108.8	12112.1		1.1				8256.20	4d 1	3b 2	P 6
12111.9	12115.2		0.5				8254.08			
12116.4	12119.7		0.9				8251.02	3E 1	2C 3	P 4
12120.3	12123.6		1.5				8248.36	3E 0	2C 2	P 3
12120.3	12123.6		1.5				8248.36	2K 6	2B 8	P 2
12122.3	12125.6		1.2				8247.00	3D 1	2C 3	P 2
12123.6	12126.9		1.2				8246.12	2A 2	2B 4	P 2
12126.9	12130.2		1.1				8243.87			
12128.8	12132.1		0.7				8242.58	3F 1	2B15	Q 3
12132.5	12135.8		0.6				8240.07			
12135.5	12138.8		0.6				8238.03			
12137.5	12140.8		0.7				8236.67	3F 0	2B13	P 6
12139.0	12142.3		1.0				8235.66	3a 0	2c 4	Q 2
12139.0	12142.3		1.0				8235.66	3E 0	2B12	P 2
12139.0	12142.3		1.0				8235.66	3A 0	2B13	P 3
12142.1	12145.4		1.7				8233.55	4d 0	3b 1	P 4
12142.1	12145.4		1.7				8233.55	4F 0	2B19	P 4
12142.1	12145.4		1.7				8233.55	2K 4	2B 6	P 4
12142.1	12145.4		1.7				8233.55	2A 2	2B 4	P 3
12150.11	12153.44			0.1			8228.12	3f 3	3b 2	R 4
12152.41	12155.73			0.2			8226.57	3b 2	2a 3	P 3
12157.0	12160.3		0.6				8223.46	2A 4	2C 0	Q 2
12160.6	12163.9		1.0				8221.03	3B 4	2A 4	R 0
12163.7	12167.0		0.9				8218.93	2K 3	2B 5	R 4
12166.3	12169.6		2.4				8217.18	3b 0	2a 1	P 9
12173.67	12177.00			0.1			8212.20	2A 0	2B 1	R 4

λ_{air}	λ_{vac}	Ref.	I_1	I_2	I_3	I_4	ν	Upper	Lower	Br.
12179.3	12182.6		0.7				8208.41			
12180.66	12183.99			0.1			8207.49	4f 2	3b 4	R 2
12185.0	12188.3		1.1				8204.57	3C 0	2K 3	R 0
12185.0	12188.3		1.1				8204.57	2K 2	2B 4	P 4
12186.9	12190.2		2.1				8203.29	3b 3	2a 4	R 4
12186.9	12190.2		2.1				8203.29	3f 1	3b 0	R 3
12191.3	12194.6		9.9				8200.33	3b 3	2a 4	R 3
12195.8	12199.1		0.5				8197.30	2K 4	2B 6	P 5
12197.7	12201.0		0.4				8196.02			
12202.29	12205.63			0.0			8192.94	3b 3	2a 4	R 5
12203.0	12206.3		5.2				8192.46			
12208.9	12212.2		4.0				8188.51	3b 3	2a 4	R 2
12208.9	12212.2		4.0				8188.51	3A 0	2B13	P 2
12214.2	12217.5		0.7				8184.95	3C 0	2K 3	Q 1
12216.6	12219.9		0.7				8183.34	3D 0	2C 2	P 1
12216.6	12219.9		0.7				8183.34	4F 0	2B19	Q 3
12219.8	12223.1		0.6				8181.20			
12223.3	12226.6		1.3				8178.86			
12225.6	12228.9		1.3				8177.32	3D 0	2C 2	Q 1
12227.1	12230.4		1.5				8176.32			
12228.7	12232.0		1.5				8175.25	3A 0	2B13	P 1
12230.8	12234.1		2.0				8173.84	2A 4	2C 0	P 2
12230.8	12234.1		2.0				8173.84	3E 1	2C 3	P 5
12232.4	12235.7		2.9				8172.77	2K 6	2B 8	P 4
12232.4	12235.7		2.9				8172.77	3c 3	2a 7	R 1
12238.6	12241.9		2.6				8168.63			
12242.4	12245.7		3.5				8166.10	3b 3	2a 4	R 1
12242.4	12245.7		3.5				8166.10	2K 4	2B 6	P 6
12248.6	12252.0		1.6				8161.96			
12251.2	12254.6		1.2				8160.23	3f 2	3b 1	R 3
12257.6	12261.0		2.1				8155.97	3E 2	2B17	P 5
12261.9	12265.3		7.5				8153.11	3b 1	2a 2	P 7
12261.9	12265.3		7.5				8153.11	3d 3	3b 1	P 6
12261.9	12265.3		7.5				8153.11	3B 3	2K 4	R 0
12261.9	12265.3		7.5				8153.11	3B 2	2K 3	R 4
12269.3	12272.7		1.6				8148.19	3f 3	3b 2	R 3
12270.5	12273.9		1.7				8147.40			
12274.7	12278.1		3.0				8144.61			
12279.7	12283.1		6.6				8141.29	3b 2	2a 3	P 4
12286.2	12289.6		0.7				8136.99			
12289.3	12292.7		1.1				8134.93			
12292.6	12296.0		1.6				8132.75	3b 3	2a 4	R 0
12297.2	12300.6		1.0				8129.71	X 1	2B16	P 2
12298.8	12302.2		1.5				8128.65	3B 3	2K 4	P 5
12301.4	12304.8		0.6				8126.93			
12304.2	12307.6		0.7				8125.08	2K 2	2B 4	P 5
12305.6	12309.0		0.9				8124.16	4F 0	2B19	R 1
12309.2	12312.6		1.2				8121.78	W 2	2B18	R 1
12312.4	12315.8		1.2				8119.67			
12314.3	12317.7		0.8				8118.42			
12317.2	12320.6		0.4				8116.51			
12319.6	12323.0		0.8				8114.93			
12323.1	12326.5		1.5				8112.62			
12326.1	12329.5		1.0				8110.65	2A 4	2C 0	Q 3
12329.0	12332.4		0.4				8108.74			
12331.7	12335.1		0.7				8106.96			
12333.4	12336.8		0.7				8105.85	W 2	2B18	R 0
12335.1	12338.5		0.8				8104.73			
12336.8	12340.2		1.0				8103.61	3C 0	2A 3	R 4
12336.8	12340.2		1.0				8103.61	4d 0	3b 1	P 5

λ_{air}	λ_{vac}	Ref.	I_1	I_2	I_3	I_4	ν	Upper	Lower	Br.
12339.0	12342.4		1.1				8102.17			
12341.1	12344.5		1.2				8100.79	3B 3	2K 4	P 4
12343.6	12347.0		1.1				8099.15			
12346.6	12350.0		0.8				8097.18			
12351.4	12354.8		1.1				8094.03	X 1	2B16	P 3
12351.4	12354.8		1.1				8094.03	3b 0	2a 1	P10
12354.3	12357.7		2.4				8092.13	3C 1	2A 4	R 3
12354.3	12357.7		2.4				8092.13	3f 1	3b 0	Q 7
12356.5	12359.9		5.3				8090.69	2A 0	2B 1	P 8
12356.5	12359.9		5.3				8090.69	3D 0	2C 2	Q 2
12362.1	12365.5		1.4				8087.03	4e 3	3b 5	R 0
12366.7	12370.1		1.3				8084.02	3B 3	2K 4	P 1
12366.7	12370.1		1.3				8084.02	3f 1	3b 0	R 2
12369.9	12373.3		2.0				8081.93	3B 3	2K 4	P 3
12376.2	12379.6		0.8				8077.81			
12379.0	12382.4		0.8				8075.99	3B 3	2K 4	P 2
12379.0	12382.4		0.8				8075.99	3f 3	3b 2	R 2
12382.9	12386.3		1.1				8073.44			
12386.8	12390.?		0.8				8070.90			
12389.8	12393.2		0.4				8068.95	3c 3	2a 7	Q 1
12393.1	12396.5		0.6				8066.80	3D 0	2C 2	P 2
12393.1	12396.5		0.6				8066.80	3f 2	3b 1	R 2
12395.7	12399.1		1.0				8065.11	Z 0	2C 2	R 1
12398.5	12401.9		1.1				8063.29	4e 2	3b 4	R 4
12399.9	12403.3		1.0				8062.38			
12403.5	12406.9		0.4				8060.04	X 1	2B16	P 4
12403.5	12406.9		0.4				8060.04	3f 2	3b 1	Q 7
12405.6	12409.0		0.4				8058.67			
12407.5	12410.9		1.0				8057.44	2B17	2A 0	P 6
12407.5	12410.9		1.0				8057.44	3c 2	2a 6	R 3
12412.9	12416.3		0.8				8053.93			
12415.2	12418.6		1.5				8052.44	2B16	2A 0	P 2
12419.9	12423.3						8049.39	2A 0	2B 1	R 3
12419.9	12423.3						8049.39	3b 2	2a 3	P 5
12419.9	12423.3						8049.39	4F 0	2B19	P 3
12424.2	12427.6		1.3				8046.61	2K 5	2B 7	R 2
12424.2	12427.6		1.3				8046.61	3f 1	3b 0	Q 6
12427.1	12430.5		0.9				8044.73	2K 5	2B 7	R 3
12431.1	12434.5		1.0				8042.14	2K 5	2B 7	R 1
12434.7	12438.1		2.5				8039.81	3b 1	2a 2	P 8
12434.7	12438.1		2.5				8039.81	2K 5	2B 7	R 4
12442.8	12446.2		4.8				8034.58	3b 3	2a 4	P 1
12449.0	12452.4		0.8				8030.58	3A 2	2B19	R 4
12450.5	12453.9		1.4				8029.61	4f 2	3b 4	R 1
12453.2	12456.6		1.5				8027.87	2K 5	2B 7	R 0
12459.1	12462.5		1.4				8024.07	3f 2	3b 1	Q 6
12464.0	12467.4		1.4				8020.91	Z 0	2C 2	R 2
12464.0	12467.4		1.4				8020.91	Z 0	2C 2	Q 1
12464.0	12467.4		1.4				8020.91	3d 3	3b 1	P 7
12466.5	12469.9		0.8				8019.30	3E 2	2B17	P 3
12468.9	12472.3		0.8				8017.76			
12471.9	12475.3		0.9				8015.83			
12475.2	12478.6		1.0				8013.71			
12476.6	12480.0		1.2				8012.81	3F 1	2B15	Q 2
12480.3	12483.7		1.4				8010.44			
12482.7	12486.1		1.7				8008.90			
12486.4	12489.8		1.4				8006.52	3f 3	3b 2	R 1
12489.8	12493.2		1.5				8004.34	4F 0	2B19	Q 2
12492.2	12495.6		0.8				8002.81			
12497.1	12500.5		0.7				7999.67	V 0	2C 2	R 2

λ_{air}	λ_{vac}	Ref.	I_1	I_2	I_3	I_4	ν	Upper	Lower	Br.
12500.0	12503.4		1.2				7997.81	3f 1	3b 0	Q 5
12502.0	12505.4		0.8				7996.53			
12503.5	12506.9		0.5				7995.57	4d 0	3b 1	P 6
12506.4	12509.8		1.4				7993.72	3D 0	2C 2	Q 3
12506.4	12509.8		1.4				7993.72	3D 0	2B12	R 2
12507.6	12511.0		1.4				7992.95			
12509.0	12512.4		1.4				7992.06			
12510.9	12514.3		1.5				7990.84			
12513.7	12517.1		1.5				7989.06	3B 2	2K 3	R 3
12513.7	12517.1		1.5				7989.06	3c 2	2a 6	R 2
12520.2	12523.6		1.6				7984.91	3D 0	2B12	R 0
12520.2	12523.6		1.6				7984.91	3f 2	3b 1	Q 5
12523.0	12526.4		3.7				7983.12	2A 1	2B 3	R 4
12523.0	12526.4		3.7				7983.12	3f 2	3b 1	R 1
12526.6	12530.0		1.3				7980.83	4c 1	2a 8	Q 1
12528.6	12532.0		1.3				7979.55	3D 0	2B12	R 1
12530.5	12533.9		1.5				7978.34	2K 5	2B 7	P 1
12534.8	12538.2		1.6				7975.61	3F 0	2B13	P 5
12534.8	12538.2		1.6				7975.61	3f 1	3b 0	R 1
12537.7	12541.1		1.5				7973.76			
12540.1	12543.5		2.6				7972.24			
12542.0	12545.4		3.7				7971.03	3b 3	2a 4	P 2
12549.2	12552.6		1.8				7966.46	4e 2	3b 4	R 3
12561.8	12565.2		1.3				7958.47			
12574.3	12577.7		2.0				7950.55	3b 2	2a 3	P 6
12574.3	12577.7		2.0				7950.55	3f 1	3b 0	Q 4
12574.3	12577.7		2.0				7950.55	Z 0	2C 2	R 3
12579.3	12582.7		1.7				7947.39	3f 2	3b 1	Q 4
12583.3	12586.7		1.3				7944.87	2A 0	2B 1	P 7
12586.4	12589.8		0.7				7942.91			
12588.5	12591.9		0.9				7941.59	3C 1	2A 4	R 2
12588.5	12591.9		0.9				7941.59	2K 5	2B 7	P 2
12598.6	12602.0		1.8				7935.22			
12618.6	12622.1		1.5				7922.64	3b 1	2a 2	P 9
12618.6	12622.1		1.5				7922.64	3e 1	3b 0	R 4
12622.6	12626.1		0.5				7920.13			
12625.6	12629.1		0.6				7918.25	V 0	2C 2	R 1
12629.3	12632.8		0.6				7915.93			
12637.8	12641.3		6.1				7910.61	2A 0	2B 1	R 2
12637.8	12641.3		6.1				7910.61	4d 0	3b 1	P 7
12637.8	12641.3		6.1				7910.61	3f 2	3b 1	Q 3
12644.8	12648.3		0.5				7906.23	3f 1	3b 0	Q 3
12646.5	12650.0		0.5				7905.16	2B16	2A 0	P 3
12646.5	12650.0		0.5				7905.16	4f 0	3b 2	R 4
12653.5	12657.0		0.7				7900.79			
12657.8	12661.3		3.9				7898.11	3b 3	2a 4	P 3
12657.8	12661.3		3.9				7898.11	2K 5	2B 7	P 3
12662.3	12665.8		0.6				7895.30			
12666.3	12669.8		0.6				7892.81			
12667.8	12671.3		0.6				7891.87	3D 0	2C 2	Q 4
12672.1	12675.6		0.4				7889.19			
12673.3	12676.8		0.5				7888.45	2A 3	2B 6	R 3
12673.3	12676.8		0.5				7888.45	V 0	2C 2	R 3
12677.7	12681.2		0.6				7885.71	X 0	2B14	R 4
12682.3	12685.8		0.7				7882.85	3f 2	3b 1	Q 2
12687.4	12690.9		0.6				7879.68	Y 2	2B18	P 1
12698.9	12702.4		0.2				7872.54			
12700.2	12703.7		0.3				7871.74	3A 2	2B19	R 3
12702.2	12705.7		0.3				7870.50	Y 2	2B18	P 2
12704.6	12708.1		0.3				7869.01			

λ_{air}	λ_{vac}	Ref.	I_1	I_2	I_3	I_4	ν	Upper	Lower	Br.
12706.3	12709.8		0.3				7867.96			
12707.7	12711.2		0.6				7867.09			
12709.3	12712.8		0.5				7866.10	3c 2	2a 6	R 0
12709.3	12712.8		0.5				7866.10	4e 2	3b 4	R 2
12709.3	12712.8		0.5				7866.10	3f 1	3b 0	Q 2
12712.1	12715.6		0.6				7864.37			
12715.9	12719.4		0.8				7862.02	2A 4	2C 0	P 4
12715.9	12719.4		0.8				7862.02	3b 4	2a 5	R 3
12718.1	12721.6		0.8				7860.66	Z 0	2C 2	Q 3
12729.9	12733.4		0.5				7853.37	2A 3	2B 6	R 1
12729.9	12733.4		0.5				7853.37	3B 2	2K 3	R 2
12731.5	12735.0		0.5				7852.39	3e 1	3b 0	R 3
12735.3	12738.8		0.5				7850.04			
12738.1	12741.6		0.7				7848.32	4d 3	3b 5	P 1
12738.1	12741.6		0.7				7848.32	4d 0	3b 1	P 8
12741.5	12745.0		1.0				7846.22	2K 5	2B 7	P 4
12744.4	12747.9		2.6				7844.44	3b 2	2a 3	P 7
12744.4	12747.9		2.6				7844.44	3B 2	2K 3	P 5
12744.4	12747.9		2.6				7844.44	Y 2	2B18	P 3
12750.3	12753.8		0.6				7840.81	2A 0	2B 2	R 8
12750.3	12753.8		0.6				7840.81	3c 1	2a 5	R 4
12753.4	12756.9		0.4				7838.90			
12755.4	12758.9		0.5				7837.67			
12757.9	12761.4		0.6				7836.14			
12764.3	12767.8		0.9				7832.21	3b 4	2a 5	R 2
12769.6	12773.1		1.0				7828.96	2A 1	2B 3	R 3
12775.3	12778.8		0.6				7825.46	2A 3	2B 6	R 0
12775.3	12778.8		0.6				7825.46	X 0	2B14	R 2
12779.7	12783.2		4.1				7822.77	2A 0	2B 1	P 6
12779.7	12783.2		4.1				7822.77	3f 1	3b 0	P 8
12783.7	12787.2		0.4				7820.32	Z 0	2C 2	R 4
12784.8	12788.3		0.5				7819.65	3C 1	2A 4	R 1
12788.0	12791.5		0.9				7817.69			
12791.2	12794.7		1.9				7815.74	Y 2	2B18	P 4
12798.4	12801.9		0.7				7811.34	4b 2	3e 0	R 1
12805.0	12808.5		1.7				7807.31	3b 4	2a 5	R 1
12818.4	12821.9						7799.15			
12824.3	12827.8		2.3				7795.56	2A 0	2B 1	R 1
12830.6	12834.1		0.5				7791.74	4e 3	3b 5	R 1
12833.1	12836.6		1.1				7790.22	3D 0	2C 2	Q 5
12833.1	12836.6		1.1				7790.22	2K 5	2B 7	P 5
12840.8	12844.3		0.4				7785.55			
12842.7	12846.2		0.5				7784.40	X 0	2B14	R 1
12844.0	12847.5		0.4				7783.61			
12849.0	12852.5		0.4				7780.58	3F 0	2B13	P 4
12855.0	12858.5		0.6				7776.95			
12858.3	12861.8		0.6				7774.95	3b 4	2a 5	R 0
12861.1	12864.6		0.7				7773.26	3e 1	3b 0	R 2
12863.8	12867.3		0.8				7771.63	4e 2	3b 4	R 1
12871.6	12875.1		0.4				7766.92	X 0	2B14	R 3
12886.6	12890.1		0.4				7757.88			
12892.7	12896.2		0.4				7754.21			
12896.7	12900.2		0.4				7751.80	3D 0	2C 2	P 4
12896.7	12900.2		0.4				7751.80	2B17	2A 0	P 7
12896.7	12900.2		0.4				7751.80	X 0	2B14	R 0
12900.0	12903.5		0.6				7749.82	X 0	2C 3	Q 1
12901.5	12905.0		0.7				7748.92	3B 2	2K 3	R 1
12903.3	12906.8		0.8				7747.84			
12904.9	12908.4		0.7				7746.88			
12912.7	12916.2		0.4				7742.20	3B 2	2K 3	P 4

λ_{air}	λ_{vac}	Ref.	I_1	I_2	I_3	I_4	ν	Upper	Lower	Br.
12913.7	12917.2		0.5				7741.60	3c 1	2a 5	R 2
12914.9	12918.4		0.4				7740.88			
12918.2	12921.7		0.5				7738.90	4d 3	3b 5	P 2
12918.2	12921.7		0.5				7738.90	3f 2	3b 1	P 7
12923.2	12926.7		0.3				7735.91	3b 2	2a 3	P 8
12923.2	12926.7		0.3				7735.91	3f 1	3b 0	P 5
12926.3	12929.8		1.5				7734.05	2A 3	2B 6	P 2
12926.3	12929.8		1.5				7734.05	3f 2	3b 1	P 3
12931.9	12935.4		0.3				7730.70	Z 0	2C 2	Q 4
12933.6	12937.1		0.3				7729.69	3C 1	2A 4	R 0
12939.1	12942.6		7.2				7726.40	2A 0	2B 1	P 5
12939.1	12942.6		7.2				7726.40	3b 3	2a 4	P 5
12945.8	12949.3		0.6				7722.40	3f 2	3b 1	P 4
12945.8	12949.3		0.6				7722.40	3c 2	2a 6	P 2
12947.5	12951.0		0.7				7721.39	2B16	2A 0	P 4
12947.5	12951.0		0.7				7721.39	3f 2	3b 1	P 5
12951.6	12955.1		0.5				7718.94			
12954.3	12957.8		0.5				7717.33	3f 1	3b 0	P 4
12956.4	12959.9		0.5				7716.08			
12959.7	12963.2		0.3				7714.12	3f 1	3b 0	P 3
12962.4	12965.9		0.5				7712.51			
12965.2	12968.7		0.9				7710.85	4b 1	2a 7	P 1
12973.8	12977.3		4.9				7705.73	2A 0	2B 1	R 0
12977.4	12980.9		0.4				7703.60	2A 3	2B 6	P 3
12979.2	12982.7		0.4				7702.53			
12981.4	12985.0		0.3				7701.22			
12984.2	12987.8		0.6				7699.56	X 0	2C 3	Q 2
12987.5	12991.1		0.5				7697.61	2A 1	2B 3	R 2
12990.9	12994.5		0.5				7695.59			
13003.2	13006.8		0.5				7688.31	3e 2	3b 1	R 3
13005.4	13009.0		0.4				7687.01	3c 2	2a 6	P 3
13007.5	13011.1		0.5				7685.77	4e 2	3b 4	R 0
13011.4	13015.0		0.5				7683.47	3e 1	3b 0	R 1
13014.0	13017.6		1.1				7681.93	3b 4	2a 5	P 1
13014.0	13017.6		1.1				7681.93	3c 1	2a 5	R 1
13021.0	13024.6		0.6				7677.80	3b 1	2a 2	P11
13023.5	13027.1		0.8				7676.33			
13027.9	13031.5		0.6				7673.74	3B 2	2K 3	R 0
13030.1	13033.7		0.6				7672.44			
13057.5	13061.1						7656.34	2A 0	2B 1	P 4
13061.1	13064.7		0.4				7654.23			
13063.8	13067.4		0.4				7652.65	2A 4	2C 0	P 5
13065.4	13069.0		0.7				7651.71			
13067.0	13070.6		0.5				7650.77	V 0	2C 2	P 3
13070.3	13073.9		0.6				7648.84	V 0	2B12	P 4
13070.3	13073.9		0.6				7648.84	V 0	2C 2	P 2
13073.3	13076.9		0.2				7647.09			
13077.5	13081.1		0.6				7644.63	3C 0	2A 3	R 1
13083.9	13087.5		0.6				7640.89	3A 2	2B19	R 1
13086.9	13090.5		0.6				7639.14	3F 0	2B13	P 3
13086.9	13090.5		0.6				7639.14	4d 3	3b 5	P 3
13095.1	13098.7		0.5				7634.36	2A 2	2B 5	R 4
13097.2	13100.8		0.4				7633.13			
13102.9	13106.5		1.2				7629.81	4e 2	3b 4	Q 1
13108.1	13111.7		0.6				7626.78	X 0	2C 3	Q 3
13112.5	13116.1		2.1				7624.23	2A 1	2B 3	P 6
13112.5	13116.1		2.1				7624.23	2K 5	2C 0	R 1
13112.5	13116.1		2.1				7624.23	3B 2	2K 3	P 2
13112.5	13116.1		2.1				7624.23	3b 2	2a 3	P 9
13117.1	13120.7		0.9				7621.55	3b 4	2a 5	P 2

λ_{air}	λ_{vac}	Ref.	I_1	I_2	I_3	I_4	ν	Upper	Lower	Br.
13121.7	13125.3		1.5				7618.88	3F 0	2B14	Q 6
13126.0	13129.6		0.8				7616.38	3c 1	2a 5	Q 4
13133.4	13137.0		3.9				7612.09	2A 0	2B 1	P 3
13133.4	13137.0		3.9				7612.09	2A 4	2B 8	P 1
13133.4	13137.0		3.9				7612.09	3B 2	2K 3	P 1
13137.5	13141.1		0.6				7609.72	Y 1	2C 3	Q 1
13137.5	13141.1		0.6				7609.72	3e 2	3b 1	R 2
13139.6	13143.2		0.8				7608.50			
13142.8	13146.4		1.5				7606.65	2A 0	2B 1	P 1
13145.1	13148.7		1.9				7605.32			
13149.4	13153.0		0.7				7602.83	3e 3	3b 2	R 4
13152.2	13155.8		0.5				7601.21			
13157.3	13160.9		0.7				7598.27	T 0	2B19	P 4
13157.3	13160.9		0.7				7598.27	V 0	2C 2	P 4
13162.6	13166.2		•				7595.21	2A 0	2B 1	P 2
13167.5	13171.1		0.2				7592.38			
13171.7	13175.3		0.2				7589.96			
13173.2	13176.8		0.4				7589.09	3D 0	2C 2	Q 7
13176.6	13180.2		1.5				7587.14	2A 1	2B 3	R 1
13176.6	13180.2		1.5				7587.14	3e 1	3b 0	R 0
13176.6	13180.2		1.5				7587.14	2A 0	2B 2	R 7
13185.1	13188.7		0.3				7582.24	Z 0	2B 2	Q 5
13185.1	13188.7		0.3				7582.24	2K 5	2C 0	Q 1
13185.1	13188.7		0.3				7582.24	V 0	2B12	P 5
13187.0	13190.6		0.4				7581.15	3c 1	2a 5	Q 2
13190.4	13194.0		0.2				7579.20	3A 0	2B14	R 2
13193.0	13196.6		0.4				7577.70			
13195.0	13198.6		0.5				7576.56	3E 0	2B13	Q 5
13195.0	13198.6		0.5				7576.56	3F 1	2C 4	R 5
13196.7	13200.3		0.5				7575.58	2A 4	2B 8	P 2
13199.9	13203.5		0.4				7573.74			
13203.5	13207.1		0.6				7571.68	3c 1	2a 5	Q 1
13205.0	13208.6		0.7				7570.82	w 0	3b 3	P 2
13206.2	13209.8		0.6				7570.13			
13209.9	13213.5		0.3				7568.01	3A 2	2B19	R 0
13211.6	13215.2		0.3				7567.04	T 0	2B19	P 3
13214.6	13218.2		0.3				7565.32			
13216.8	13220.4		0.5				7564.06	2K 5	2C 0	R 2
13218.0	13221.6		0.5				7563.37	V 0	2B12	P 3
13220.3	13223.9		0.5				7562.06			
13222.9	13226.5		0.7				7560.57	2K 5	2C 0	P 1
13225.5	13229.1		0.5				7559.08			
13230.5	13234.1		0.4				7556.23	T 0	2B19	P 2
13234.9	13238.5		2.2				7553.71	3b 4	2a 5	P 3
13234.9	13238.5		2.2				7553.71	4e 2	3b 4	Q 3
13239.4	13243.0		0.3				7551.15			
13241.5	13245.1		0.5				7549.95	3C 0	2A 3	R 0
13249.1	13252.7		0.5				7545.62	Y 1	2C 3	Q 2
13253.5	13257.1		0.5				7543.11			
13256.0	13259.6		0.6				7541.69	2A 4	2B 8	P 3
13257.8	13261.4		0.8				7540.67			
13260.8	13264.4		0.5				7538.96	2B18	2A 1	R 0
13262.8	13266.4		0.4				7537.82			
13264.6	13268.2		0.2				7536.80			
13267.3	13270.9						7535.27			
13271.3	13274.9		1.2				7533.00	2A 1	2B 3	P 5
13271.3	13274.9		1.2				7533.00	X 0	2C 3	Q 4
13271.3	13274.9		1.2				7533.00	3e 3	3b 2	R 3
13279.8	13283.4		2.5				7528.18	3C 0	2K 4	R 4
13279.8	13283.4		2.5				7528.18	3F 0	2B14	R 4

λ_air	λ_vac	Ref.	I₁	I₂	I₃	I₄	ν	Upper	Lower	Br.
13291.4	13295.0		0.5				7521.60			
13293.1	13296.7		0.4				7520.64	T 0	2B19	P 1
13293.1	13296.7		0.4				7520.64	4e 2	3b 4	Q 4
13298.3	13301.9		0.2				7517.70			
13300.0	13303.6		0.4				7516.74			
13302.8	13306.4		0.7				7515.16			
13305.4	13309.0		0.4				7513.69			
13323.1	13326.7		2.1				7503.71	2K 5	2C 0	Q 2
13323.1	13326.7		2.1				7503.71	2A 1	2B 3	R 0
13323.1	13326.7		2.1				7503.71	3E 2	2B18	P 6
13329.1	13332.7		0.5				7500.33	3b 2	2a 3	P10
13335.8	13339.4		0.6				7496.56	2A 4	2B 8	P 4
13338.3	13341.9		0.9				7495.16	2B16	2A 0	P 5
13343.6	13347.2		0.5				7492.18			
13346.9	13350.5		0.4				7490.33			
13352.2	13355.9		0.7				7487.35	3B 4	2K 6	P 3
13352.2	13355.9		0.7				7487.35	4c 0	2a 7	R 1
13352.2	13355.9		0.7				7487.35	V 0	2B12	P 2
13359.2	13362.9		0.3				7483.43	3B 4	2K 6	P 1
13362.8	13366.5		0.6				7481.42	3b 4	2a 5	P 4
13364.2	13367.9		1.1				7480.63			
13368.2	13371.9		0.6				7478.39			
13371.2	13374.9		0.2				7476.72	3B 4	2K 6	P 2
13371.2	13374.9		0.2				7476.72	2B15	2A 0	R 1
13375.1	13378.8		0.3				7474.54			
13376.5	13380.2		0.3				7473.75			
13378.2	13381.9		0.3				7472.80	4e 0	3b 2	R 6
13381.7	13385.4		0.3				7470.85	3B 3	2K 5	P 5
13392.5	13396.2		3.4				7464.82	2A 1	2B 3	P 4
13392.5	13396.2		3.4				7464.82	4d 2	3b 4	R 0
13392.5	13396.2		3.4				7464.82	2K 5	2C 0	R 3
13401.2	13404.9		0.3				7459.98			
13403.6	13407.3		0.5				7458.64	2A 2	2B 5	R 2
13403.6	13407.3		0.5				7458.64	2K 5	2C 0	P 2
13403.6	13407.3		0.5				7458.64	2A 4	2B 8	P 5
13412.7	13416.4		0.3				7453.58	Y 1	2C 3	Q 3
13419.1	13422.8		0.2				7450.03			
13422.4	13426.1		0.3				7448.20	3A 0	2B14	R 1
13422.4	13426.1		0.3				7448.20	2B18	2A 1	R 2
13429.1	13432.8		0.3				7444.48	3e 1	3b 0	Q 1
13430.7	13434.4		0.3				7443.59			
13434.6	13438.3		0.3				7441.43			
13437.8	13441.5		0.3				7439.66			
13440.7	13444.4		0.3				7438.05			
13442.3	13446.0		0.3				7437.17	3e 2	3b 1	R 0
13445.0	13448.7		0.3				7435.68			
13453.0	13456.7		0.3				7431.25	4f 0	3b 2	R 2
13453.0	13456.7		0.3				7431.25	w 0	3b 3	P 3
13455.3	13459.0		0.4				7429.98	4d 2	3b 4	R 1
13458.9	13462.6		1.3				7428.00	2K 2	2B 5	R 1
13458.9	13462.6		1.3				7428.00	3e 1	3b 0	P 2
13462.6	13466.3		0.2				7425.96			
13467.7	13471.4		0.9				7423.14			
13470.3	13474.0		0.9				7421.71	3b 5	2a 6	R 2
13473.0	13476.7		1.3				7420.22	2A 1	2B 3	P 3
13473.0	13476.7		1.3				7420.22	3e 1	3b 0	P 3
13476.7	13480.4		0.3				7418.19	3e 1	3b 0	Q 2
13483.2	13486.9		0.6				7414.61	2K 2	2B 5	R 0
13485.9	13489.6		0.5				7413.13			
13490.3	13494.0		0.8				7410.71	2A 1	2B 3	P 1

λ_{air}	λ_{vac}	Ref.	I_1	I_2	I_3	I_4	ν	Upper	Lower	Br.
13494.5	13498.2		0.3				7408.40	3B 3	2K 5	P 4
13496.8	13500.5		0.8				7407.14	3b 5	2a 6	R 1
13496.8	13500.5		0.8				7407.14	3e 1	3b 0	P 4
13505.0	13508.7		3.6				7402.64	2A 1	2B 3	P 2
13525.4	13529.1		0.6				7391.48			
13530.1	13533.8		0.3				7388.91	3D 2	2B18	R 6
13533.0	13536.7		0.3				7387.32	3e 3	3b 2	R 1
13535.1	13538.8		0.3				7386.18	2K 5	2C 0	Q 3
13535.1	13538.8		0.3				7386.18	3e 1	3b 0	Q 3
13537.1	13540.8		0.3				7385.09	3B 2	2A 3	R 4
13537.1	13540.8		0.3				7385.09	3e 1	3b 0	P 5
13539.3	13543.0		0.7				7383.89			
13542.4	13546.1		0.2				7382.20	2K 4	2B 7	R 4
13546.5	13550.2		0.2				7379.96	3b 5	2a 6	R 0
13550.0	13553.7		0.9				7378.06	3E 2	2B18	P 5
13556.8	13560.5		0.3				7374.36			
13559.6	13563.3		0.2				7372.83	3E 0	2B13	Q 3
13559.6	13563.3		0.2				7372.83	3c 0	2a 4	Q 4
13567.6	13571.3		0.3				7368.49	3C 0	2K 4	Q 5
13569.8	13573.5		0.3				7367.29			
13571.0	13574.7		0.4				7366.64	4e 0	3b 2	R 5
13572.4	13576.1		0.3				7365.88			
13575.5	13579.2		0.4				7364.20	3B 3	2K 5	P 3
13575.5	13579.2		0.4				7364.20	3C 0	2K 4	R 3
13575.5	13579.2		0.4				7364.20	2K 2	2B 5	P 1
13582.5	13586.2		0.3				7360.40			
13584.2	13587.9		0.3				7359.48	3e 1	3b 0	P 6
13589.4	13593.1		0.2				7356.67	3a 1	3b 0	P 2
13591.4	13595.1		0.3				7355.58			
13592.9	13596.6		0.3				7354.77	2K 4	2B 7	R 3
13596.0	13599.7		1.2				7353.09	2A 0	2B 2	R 6
13596.0	13599.7		1.2				7353.09	3c 0	2a 4	Q 3
13601.2	13604.9		0.5				7350.28	3A 0	2B14	R 0
13601.2	13604.9		0.5				7350.28	2B18	2A 1	R 3
13601.2	13604.9		0.5				7350.28	3e 1	3b 0	Q 4
13602.2	13605.9		0.5				7349.74	3F 0	2B14	Q 5
13607.1	13610.8		0.2				7347.10	3F 1	2C 4	R 3
13609.2	13612.9		0.1				7345.96	3F 0	2C 3	R 5
13611.3	13615.0						7344.83			
13616.1	13619.8		1.2				7342.24	3B 3	2K 5	P 1
13616.1	13619.8		1.2				7342.24	3B 3	2K 5	P 2
13633.3	13637.0		0.3				7332.98			
13637.7	13641.4		0.3				7330.61	2K 5	2C 0	R 4
13637.7	13641.4		0.3				7330.61	3e 1	3b 0	P 7
13641.0	13644.7		0.3				7328.84	2K 4	2B 7	R 2
13641.0	13644.7		0.3				7328.84	3c 0	2a 4	Q 1
13643.8	13647.5		0.2				7327.33			
13645.7	13649.4		0.3				7326.31			
13647.5	13651.2		0.3				7325.35			
13649.0	13652.7		0.3				7324.54			
13653.6	13657.3		0.1				7322.07	2K 2	2B 5	P 2
13657.2	13660.9		0.3				7320.14			
13659.2	13662.9		0.3				7319.07	3E 0	2B13	Q 2
13659.2	13662.9		0.3				7319.07	2B15	2A 0	R 2
13661.9	13665.6		0.3				7317.63	2K 6	2B 9	R 2
13663.8	13667.5		0.2				7316.61	2K 5	2C 0	P 3
13667.3	13671.0		0.3				7314.73	3e 1	3b 0	Q 5
13667.3	13671.0		0.3				7314.73	u 1	2a 6	R 3
13669.4	13673.1		0.3				7313.61	3a 1	3b 0	P 3
13674.6	13678.3		0.3				7310.83	2B15	2A 0	P 1

λ_{air}	λ_{vac}	Ref.	I_1	I_2	I_3	I_4	ν	Upper	Lower	Br.
13674.6	13678.3		0.3				7310.83	3e 3	3b 2	R 0
13678.1	13681.8		0.4				7308.96	4b 0	2a 6	P 1
13681.5	13685.2		0.1				7307.14	2K 4	2B 7	R 1
13683.9	13687.6		0.1				7305.86			
13686.6	13690.3		0.1				7304.44	2K 6	2B 9	R 1
13688.9	13692.6		0.2				7303.19			
13692.1	13695.8		0.2				7301.49	2A 2	2B 5	P 6
13701.4	13705.1		0.1				7296.53	3A 0	2C 3	P 1
13703.6	13707.3		0.2				7295.36	3E 2	2B18	P 4
13707.6	13711.3		0.2				7293.23			
13710.3	13714.0		0.2				7291.79	3E 0	2B13	Q 1
13715.5	13719.2		0.6				7289.03	3b 5	2a 6	P 1
13725.2	13729.0		0.3				7283.88	3F 0	2B14	R 3
13725.2	13729.0		0.3				7283.88	2K 4	2B 7	R 0
13736.5	13740.3		0.2				7277.88	3D 2	2B18	R 5
13741.9	13745.7		0.4				7275.03	3a 1	3b 0	P 4
13751.9	13755.7		0.2				7269.73	3C 1	2K 6	R 2
13759.5	13763.3		0.3				7265.72			
13764.3	13768.1		0.2				7263.19	4b 0	2a 6	P 2
13764.3	13768.1		0.2				7263.19	4e 0	3b 2	R 4
13767.0	13770.8		0.2				7261.76			
13773.5	13777.3		0.2				7258.33	3B 2	2A 3	R 3
13775.4	13779.2		0.2				7257.33			
13783.3	13787.1		0.9				7253.17	2K 2	2B 5	R 2
13788.2	13792.0		0.2				7250.60			
13789.7	13793.5		0.2				7249.81			
13799.1	13802.9		0.7				7244.87	2A 0	2B 2	P10
13803.6	13807.4		0.2				7242.51	2A 2	2B 5	P 4
13803.6	13807.4		0.2				7242.51	2B16	2A 0	P 6
13803.6	13807.4		0.2				7242.51	3a 1	3b 0	P 5
13808.2	13812.0		0.2				7240.09	3e 3	3b 2	Q 1
13809.4	13813.2		0.3				7239.46	3F 1	2C 4	R 2
13809.4	13813.2		0.3				7239.46	4f 0	3b 2	R 1
13809.4	13813.2		0.3				7239.46	u 1	2a 6	R 2
13811.4	13815.2		0.1				7238.42			
13813.0	13816.8		0.2				7237.58	2K 2	2B 5	R 3
13813.8	13817.6		0.1				7237.16			
13815.3	13819.1		0.2				7236.37			
13817.3	13821.1		0.5				7235.33	3b 5	2a 6	P 2
13820.3	13824.1		0.1				7233.76	3A 0	2C 3	P 2
13823.4	13827.2		0.2				7232.13	2K 6	2B 9	P 1
13823.4	13827.2		0.2				7232.13	2K 5	2C 0	Q 4
13824.8	13828.6		0.3				7231.40	2K 4	2B 7	P 1
13826.9	13830.7		0.2				7230.30			
13828.2	13832.0		0.1				7229.62	3E 2	2B18	P 3
13829.3	13833.1		0.2				7229.05	3e 3	3b 2	Q 2
13830.7	13834.5		0.2				7228.32			
13831.9	13835.7		0.1				7227.69	4b 0	2a 6	P 3
13833.5	13837.3		0.2				7226.85			
13835.9	13839.7		0.1				7225.60			
13839.7	13843.5		0.2				7223.62	3D 2	2B18	R 4
13839.7	13843.5		0.2				7223.62	4e 1	3b 3	Q 1
13841.4	13845.2		0.2				7222.73	3C 0	2K 4	Q 4
13845.0	13848.8		0.2				7220.85	3c 0	2a 4	P 2
13852.4	13856.2		0.3				7216.99	3F 0	2C 3	R 4
13852.4	13856.2		0.3				7216.99	2B15	2A 0	R 3
13854.3	13858.1		0.5				7216.00	2A 2	2B 5	R 1
13854.3	13858.1		0.5				7216.00	2B18	2A 1	R 4
13856.8	13860.6		0.4				7214.70			
13858.6	13862.4		0.6				7213.76	3a 1	3b 0	P 6

λ_{air}	λ_{vac}	Ref.	I_1	I_2	I_3	I_4	ν	Upper	Lower	Br.
13862.1	13865.9		0.3				7211.94	2B18	2A 1	P 3
13867.0	13870.8		0.2				7209.39			.
13870.6	13874.4		0.2				7207.52	2B15	2A 0	P 2
13872.6	13876.4		0.2				7206.48			
13882.7	13886.5		0.2				7201.24	3d 1	3b 0	R 0
13884.6	13888.4		0.2				7200.26	3C 0	2K 4	R 2
13886.9	13890.7		0.1				7199.06	2K 4	2B 7	P 2
13886.9	13890.7		0.1				7199.06	2K 6	2B 9	P 2
13890.7	13894.5		0.4				7197.09			
13904.2	13908.0		0.2				7190.11	3a 1	3b 0	P 7
13913.0	13916.8		0.3				7185.56	3D 2	2B18	R 3
13917.5	13921.3		0.2				7183.23	u 1	2a 6	R 1
13923.4	13927.2		0.2				7180.19			
13932.2	13936.0		0.2				7175.66	3d 1	3b 0	R 1
13939.4	13943.2		0.8				7171.95	3E 2	2B18	P 2
13939.4	13943.2		0.8				7171.95	3F 1	2C 4	Q 4
13947.6	13951.4		0.6				7167.73	3b 5	2a 6	P 3
13951.1	13954.9		0.3				7165.93	4e 0	3b 2	R 3
13957.9	13961.7		1.0				7162.44	2A 2	2B 5	R 0
13957.9	13961.7		1.0				7162.44	2B15	2A 0	P 3
13957.9	13961.7		1.0				7162.44	2K 4	2B 7	P 3
13957.9	13961.7		1.0				7162.44	2K 6	2B 9	P 3
13962.2	13966.0		0.2				7160.24	3A 0	2C 3	P 4
13962.2	13966.0		0.2				7160.24	3B 2	2A 3	R 2
13962.2	13966.0		0.2				7160.24	2K 3	2B 6	R 1
13966.5	13970.3		0.2				7158.03	2K 3	2B 6	R 0
13987.5	13991.3		0.4				7147.29	2K 3	2B 6	R 2
13987.5	13991.3		0.4				7147.29	4e 1	3b 3	Q 6
13994.4	13998.2		0.3				7143.76	3c 0	2a 4	P 4
14003.9	14007.7		0.2				7138.92	2A 0	2B 2	R 5
14003.9	14007.7		0.2				7138.92	3d 1	3b 0	R 2
14006.7	14010.5		0.1				7137.49	2K 5	2C 0	P 4
14006.7	14010.5		0.1				7137.49	3F 0	2C 4	R 1
14006.7	14010.5		0.1				7137.49	4e 1	3b 3	Q 4
14011.5	14015.3		0.1				7135.04	3B 1	2A 2	R 5
14011.5	14015.3		0.1				7135.04	u 1	2a 6	R 0
14011.5	14015.3		0.1				7135.04	2K 6	2B 9	P 4
14011.5	14015.3		0.1				7135.04	4e 1	3b 3	Q 5
14016.6	14020.4		0.4				7132.45			
14019.2	14023.0		0.4				7131.13	3F 0	2B14	Q 4
14019.2	14023.0		0.4				7131.13	2K 4	2B 7	P 4
14028.9	14032.7		0.5				7126.19			
14050.7	14054.5		0.4				7115.14			
14054.7	14058.5		0.1				7113.11	3D 2	2B18	R 1
14056.8	14060.6		0.2				7112.05	3C 0	2K 4	P 5
14060.4	14064.2		0.3				7110.23	2K 3	2B 6	P 1
14065.9	14069.7		0.4				7107.45			
14071.0	14074.8		0.2				7104.87	4d 1	3b 3	R 0
14073.8	14077.6		0.2				7103.46	3D 2	2B18	R 0
14073.8	14077.6		0.2				7103.46	3F 1	2C 4	Q 3
14077.7	14081.5		0.4				7101.49	3C 0	2K 4	Q 3
14079.6	14083.4		0.3				7100.53			
14081.7	14085.5		0.2				7099.47			
14083.6	14087.5		0.3				7098.52	2K 4	2B 7	P 5
14083.6	14087.5		0.3				7098.52	3d 1	3b 0	R 3
14097.9	14101.8		0.4				7091.32	3B 1	2A 2	R 2
14100.6	14104.5		0.5				7089.96			
14104.5	14108.4		0.3				7088.00			
14108.6	14112.5		0.4				7085.94			
14110.1	14114.0		0.6				7085.19	2A 2	2B 5	P 1

λ_{air}	λ_{vac}	Ref.	I_1	I_2	I_3	I_4	ν	Upper	Lower	Br.
14113.6	14117.5		0.3				7083.43	2K 3	2B 6	R 4
14122.8	14126.7		0.4				7078.81	4e 0	3b 2	R 2
14127.7	14131.6		0.2				7076.36			
14130.1	14134.0		0.2				7075.16	2B15	2A 0	R 4
14134.8	14138.7		0.5				7072.80			
14136.3	14140.2		0.3				7072.05			
14142.9	14146.8		0.1				7068.75	2A 2	2B 5	P 2
14142.9	14146.8		0.1				7068.75	3B 1	2A 2	R 4
14142.9	14146.8		0.1				7068.75	3C 0	2K 4	R 1
14156.9	14160.8		0.5				7061.76	4d 1	3b 3	R 1
14159.9	14163.8		1.0				7060.27	2A 2	2B 5	P 3
14162.6	14166.5		0.5				7058.92			
14167.0	14170.9		0.4				7056.73	3B 0	2A 1	R 1
14167.0	14170.9		0.4				7056.73	3d 1	3b 0	R 4
14167.0	14170.9		0.4				7056.73	3C 1	2K 6	R 0
14168.7	14172.6		0.4				7055.88			
14171.7	14175.6		0.6				7054.39			
14176.2	14180.1		0.5				7052.15			
14178.7	14182.6		0.3				7050.91	3F 0	2B14	R 2
14180.8	14184.7		0.2				7049.86	3d 2	3b 1	R 0
14187.0	14190.9		0.3				7046.78	3F 1	2C 4	Q 2
14191.3	14195.2		0.3				7044.65	2K 5	2C 0	Q 5
14194.4	14198.3		0.5				7043.11	3d 1	3b 0	P 2
14196.8	14200.7		0.7				7041.92	2A 0	2B 2	P 9
14199.6	14203.5		0.3				7040.53			
14202.7	14206.6		0.4				7038.99	2B18	2A 1	P 4
14205.6	14209.5		0.5				7037.55	2K 2	2B 5	P 4
14209.7	14213.6		0.6				7035.52			
14210.3	14214.2		0.5				7035.23	3d 2	3b 1	R 1
14212.9	14216.8		0.5				7033.94			
14214.8	14218.7		0.3				7033.00			
14218.7	14222.6		0.5				7031.07			
14222.2	14226.1		0.7				7029.34	3C 1	2K 6	P 3
14222.2	14226.1		0.7				7029.34	4d 1	3b 3	R 2
14229.1	14233.0		0.4				7025.93	3B 1	2A 2	R 1
14233.7	14237.6		0.6				7023.66			
14237.4	14241.3		0.4				7021.83			
14239.6	14243.5		0.4				7020.75			
14241.3	14245.2		0.4				7019.91	3B 2	2A 3	R 0
14246.9	14250.8		0.5				7017.15	W 1	2B17	P 6
14250.9	14254.8		0.4				7015.18			
14252.7	14256.6		0.5				7014.30	3d 1	3b 0	R 5
14255.6	14259.5		0.4				7012.87	X 0	2B15	R 4
14255.6	14259.5		0.4				7012.87	2B18	2A 1	R 5
14257.7	14261.6		0.2				7011.84			
14260.9	14264.8		0.5				7010.26	3C 0	2K 4	P 4
14260.9	14264.8		0.5				7010.26	2K 3	2B 6	P 3
14265.4	14269.3		0.5				7008.05	3d 2	3b 1	R 2
14268.0	14271.9		0.6				7006.78	3C 0	2K 4	Q 2
14268.0	14271.9		0.6				7006.78	4d 1	3b 3	R 3
14273.8	14277.7		0.2				7003.93			
14279.0	14282.9		0.7				7001.38			
14281.2	14285.1		0.4				7000.30	4e 0	3b 2	R 1
14282.7	14286.6		0.3				6999.56			
14284.2	14288.1		0.2				6998.83			
14288.7	14292.6		0.4				6996.62			
14291.7	14295.6		0.7				6995.16	4d 1	3b 3	R 4
14298.9	14302.8		0.4				6991.63			
14300.9	14304.8		0.4				6990.66	3F 0	2C 3	R 2
14302.2	14306.1		0.4				6990.02	X 0	2B15	R 3

λ_{air}	λ_{vac}	Ref.	I_1	I_2	I_3	I_4	ν	Upper	Lower	Br.
14308.0	14311.9		0.5				6987.19	3B 1	2A 2	R 3
14308.0	14311.9		0.5				6987.19	3C 1	2K 6	P 2
14313.8	14317.7		0.5				6984.36			
14317.3	14321.2		0.3				6982.65			
14335.0	14338.9		0.6				6974.03	3d 2	3b 1	R 3
14338.9	14342.8		0.7				6972.13	3F 0	2C 3	R 2
14338.9	14342.8		0.7				6972.13	3d 1	3b 0	R 6
14341.1	14345.0		0.5				6971.06	2K 5	2B 8	R 3
14344.0	14347.9		0.4				6969.65	3B 1	2A 2	R 0
14344.0	14347.9		0.4				6969.65	3C 0	2K 4	R 0
14349.6	14353.5		0.5				6966.93			
14351.6	14355.5		0.5				6965.96	2K 5	2B 8	R 2
14351.6	14355.5		0.5				6965.96	2K 5	2B 8	R 4
14355.8	14359.7		0.5				6963.92			
14359.6	14363.5		0.8				6962.08	2K 5	2B 8	R 1
14359.6	14363.5		0.8				6962.08	2K 2	2B 5	P 5
14364.6	14368.5		0.4				6959.66			
14370.3	14374.2		0.6				6956.90			
14373.3	14377.2		1.3				6955.44	3b 6	2a 7	R 1
14375.5	14379.4		0.8				6954.38			
14384.8	14388.7		0.5				6949.88			
14387.0	14390.9		0.6				6948.82	3F 0	2B14	Q 3
14387.0	14390.9		0.6				6948.82	2B16	2A 0	P 7
14392.0	14395.9		3.0				6946.41	2A 0	2B 2	R 4
14392.0	14395.9		3.0				6946.41	U 1	2C 4	Q 1
14397.1	14401.0		0.4				6943.95	2K 5	2B 8	R 0
14399.7	14403.6		0.4				6942.69			
14403.8	14407.7		0.4				6940.72	3C 0	2K 4	Q 1
14403.8	14407.7		0.4				6940.72	2K 3	2B 6	P 4
14407.6	14411.5		0.4				6938.88	3d 2	3b 1	R 4
14413.0	14416.9		0.4				6936.28	3C 0	2K 4	P 3
14418.3	14422.2		0.3				6933.74			
14421.5	14425.4		0.9				6932.20	W 1	2B17	P 5
14421.5	14425.4		0.9				6932.20	3d 1	3b 0	R 7
14427.1	14431.0		0.3				6929.51	t 0	2a 5	P 1
14428.8	14432.7		0.5				6928.69			
14429.5	14433.4		0.6				6928.35			
14430.9	14434.8		0.4				6927.68			
14433.2	14437.1		0.4				6926.58			
14439.3	14443.2		0.5				6923.65	2K 5	2C 0	P 5
14441.7	14445.6		0.4				6922.50			
14447.4	14451.3		0.5				6919.77			
14455.1	14459.1		0.4				6916.08			
14458.8	14462.8		0.4				6914.31	3E 2	2B19	R 4
14460.0	14464.0		0.2				6913.74			
14462.8	14466.8		0.3				6912.40	3d 1	3b 0	P 3
14472.2	14476.2		0.2				6907.91	X 0	2B15	R 1
14477.1	14481.1		0.6				6905.57			
14483.6	14487.6		0.4				6902.47			
14498.2	14502.2		0.8				6895.52			
14500.2	14504.2		0.5				6894.57	3b 0	2a 2	R 5
14504.6	14508.6		0.5				6892.48	U 1	2C 4	Q 2
14504.6	14508.6		0.5				6892.48	3C 0	2K 4	P 2
14504.6	14508.6		0.5				6892.48	W 1	2B17	P 4
14516.6	14520.6		0.5				6886.78			
14520.4	14524.4		0.5				6884.98			
14524.4	14528.4		0.4				6883.08			
14527.0	14531.0		0.3				6881.85	2B15	2A 0	P 4
14528.7	14532.7		0.4				6881.05			
14529.8	14533.8		0.4				6880.53	3b 0	2a 2	R 4

λ_{air}	λ_{vac}	Ref.	I_1	I_2	I_3	I_4	ν	Upper	Lower	Br.
14531.7	14535.7		0.4				6879.63	3B 1	2K 2	R 2
14531.7	14535.7		0.4				6879.63	X 0	2B15	R 0
14537.5	14541.5		0.2				6876.88	W 1	2B17	P 3
14539.3	14543.3		0.5				6876.03	3F 0	2C 3	R 1
14541.0	14545.0		0.4				6875.23			
14543.1	14547.1		0.7				6874.23	3F 0	2C 3	Q 4
14545.3	14549.3		0.3				6873.19	3B 1	2A 2	P 1
14549.2	14553.2		0.3				6871.35	W 1	2B17	P 1
14551.2	14555.2		0.5				6870.41			
14554.6	14558.6		0.3				6868.80	2B15	2A 0	R 5
14556.1	14560.1		0.6				6868.09	W 1	2B17	P 2
14561.1	14565.1		0.3				6865.74			
14569.8	14573.8		2.0				6861.64	2A 0	2B 2	P 8
14574.6	14578.6		0.6				6859.38	3b 0	2a 2	R 3
14576.9	14580.9		0.4				6858.29	3F 0	2C 3	R 1
14579.7	14583.7		0.5				6856.98	2K 3	2B 6	P 5
14583.2	14587.2		0.4				6855.33	3d 3	3b 2	R 0
14589.4	14593.4		0.5				6852.42	3d 3	3b 2	R 0
14593.5	14597.5		0.5				6850.49	3F 0	2B14	R 1
14597.6	14601.6		0.4				6848.57			
14606.3	14610.3		0.4				6844.49	X 0	2B15	R 2
14609.7	14613.7		0.5				6842.90	4d 1	3b 3	P 3
14612.4	14616.4		0.5				6841.63			
14619.0	14623.0		0.3				6838.54	3B 1	2A 2	P 2
14619.0	14623.0		0.3				6838.54	3C 0	2K 2	R 4
14622.6	14626.6		0.4				6836.86	3d 3	3b 2	R 2
14627.2	14631.2		0.4				6834.71			
14630.6	14634.6		0.4				6833.12	2B18	2A 1	P 5
14633.7	14637.7		0.3				6831.67			
14636.1	14640.1		0.2				6830.55			
14638.2	14642.2		0.3				6829.57	3b 0	2a 2	R 2
14640.9	14644.9		0.6				6828.31			
14643.4	14647.4		0.5				6827.15			
14646.9	14650.9		0.5				6825.52			
14651.6	14655.6		1.3				6823.33			
14668.2	14672.2		0.5				6815.61	U 1	2C 4	Q 3
14671.2	14675.2		0.4				6814.21	2K 4	2C 0	P 1
14671.2	14675.2		0.4				6814.21	3d 3	3b 2	R 3
14674.1	14678.1		0.4				6812.87			
14676.6	14680.6		0.3				6811.71			
14682.4	14686.4		0.3				6809.01			
14685.1	14689.1		0.5				6807.76			
14688.3	14692.3		0.5				6806.28			
14692.2	14696.2		0.5				6804.47			
14695.7	14699.7		0.6				6802.85	3D 1	2B16	R 5
14698.7	14702.2		1.1				6801.70	3F 0	2B14	Q 2
14702.4	14706.4		0.5				6799.75			
14712.0	14716.0		0.4				6795.32			
14716.9	14720.9		0.6				6793.05			
14719.5	14723.5		0.9				6791.85	3b 0	2a 2	R 1
14722.1	14726.1		0.8				6790.65			
14724.6	14728.6		0.6				6789.50			
14726.7	14730.7		0.5				6788.53			
14732.9	14736.9		1.0				6785.68			
14739.2	14743.2		2.0				6782.77	2A 1	2B 4	R 4
14749.8	14753.8		1.5				6777.90	2A 0	2B 2	R 3
14749.8	14753.8		1.5				6777.90	U 1	2C 4	Q 4
14756.1	14760.1		0.5				6775.01			
14758.9	14762.9		0.7				6773.72	3d 1	3b 0	P 4
14761.4	14765.4		0.7				6772.57	3B 1	2K 2	R 1

λ_{air}	λ_{vac}	Ref.	I_1	I_2	I_3	I_4	ν	Upper	Lower	Br.
14770.6	14774.6		0.5				6768.36			
14786.7	14790.7		0.5				6760.99			
14795.4	14799.4		0.3				6757.01	U 1	2C 4	Q 5
14798.6	14802.6		1.3				6755.55			
14804.7	14808.7		0.3				6752.77			
14809.4	14813.4		0.4				6750.62			
14813.7	14817.7		0.2				6748.66			
14817.2	14821.2		0.2				6747.07			
14820.8	14824.9		0.2				6745.43	3b 0	2a 2	R 0
14820.8	14824.9		0.2				6745.43	4d 0	3b 2	R 0
14825.4	14829.5		0.2				6743.34	3F 0	2C 3	Q 2
14832.6	14836.7		0.4				6740.06	2A 3	2B 7	R 1
14851.6	14855.7		2.0				6731.44			
14903.1	14907.2		0.5				6708.18	2A 3	2B 7	R 0
14903.1	14907.2		0.5				6708.18	3C 0	2K 5	Q 5
14910.2	14914.3		1.1				6704.99	2A 0	2B 2	P 7
14914.8	14918.9		0.2				6702.92			
14919.4	14923.5		0.4				6700.85	4d 0	3b 2	R 1
14922.0	14926.1		0.5				6699.68			
14941.2	14945.3		0.9				6691.07	3B 1	2K 2	R 0
14950.3	14954.4		0.6				6687.00	3b 1	2a 3	R 6
14957.8	14961.9		0.9				6683.65	3b 1	2a 3	R 5
14962.2	14966.3		0.3				6681.68	3E 0	2B14	Q 5
14964.2	14968.3		0.2				6680.79			
14965.9	14970.0		0.5				6680.03			
14969.6	14973.7		0.5				6678.38	3B 3	2A 4	P 4
14981.9	14986.0		0.2				6672.90	3b 1	2a 3	R 4
14981.9	14986.0		0.2				6672.90	4d 1	3b 3	P 2
14993.7	14997.8		0.2				6667.65	3D 1	2B16	R 3
15001.9	15006.0		0.2				6664.00			
15012.6	15016.7		0.3				6659.25			
15016.6	15020.7		0.2				6657.48			
15018.2	15022.3		0.2				6656.77			
15025.5	15029.6		1.0				6653.53	3b 1	2a 3	R 3
15025.5	15029.6		1.0				6653.53	2B17	2A 1	P 1
15027.5	15031.6		0.5				6652.65			
15039.0	15043.1		0.3				6647.56	2A 4	2B 9	R 3
15039.0	15043.1		0.3				6647.56	4d 0	3b 2	R 5
15039.0	15043.1		0.3				6647.56	3C 0	2K 5	R 3
15046.5	15050.6		0.3				6644.25	3D 1	2B16	R 0
15046.5	15050.6		0.3				6644.25	3D 1	2B16	R 2
15049.7	15053.8		0.3				6642.84	2A 4	2B 9	R 2
15049.7	15053.8		0.3				6642.84	2A 0	2B 3	R 8
15055.5	15059.6		0.5				6640.28	3D 1	2B16	R 1
15057.9	15062.0		0.4				6639.22	3B 3	2A 4	P 3
15057.9	15062.0		0.4				6639.22	3D 1	2B16	P 1
15062.4	15066.5		0.3				6637.23			
15070.6	15074.7		5.0				6633.62	2A 0	2B 2	R 2
15070.6	15074.7		5.0				6633.62	4d 0	3b 2	R 3
15086.6	15090.7		0.4				6626.59	3b 1	2a 3	R 2
15086.6	15090.7		0.4				6626.59	3B 3	2A 4	P 1
15092.4	15096.5		1.1				6624.04	2A 1	2B 4	R 3
15095.1	15099.2		0.5				6622.86	2A 3	2B 7	P 2
15095.1	15099.2		0.5				6622.86	3B 3	2A 4	P 2
15101.6	15105.7		0.3				6620.01	4d 1	3b 3	P 5
15108.6	15112.7		0.4				6616.94			
15118.4	15122.5		0.3				6612.65			
15122.0	15126.1		0.2				6611.08			
15127.4	15131.5		0.5				6608.72			
15129.1	15133.2		0.5				6607.97	2A 4	2B 9	R 0

λ_{air}	λ_{vac}	Ref.	I_1	I_2	I_3	I_4	ν	Upper	Lower	Br.
15134.5	15138.6		0.3				6605.61			
15136.6	15140.7		0.3				6604.70			
15143.0	15147.1		0.5				6601.91			
15149.1	15153.2		0.3				6599.25			
15151.6	15155.7		0.6				6598.16	2K 6	2C 1	Q 2
15151.6	15155.7		0.6				6598.16	3A 0	2B15	R 2
15151.6	15155.7		0.6				6598.16	3D 2	2B19	R 6
15165.9	15170.0		1.1				6591.94	3b 1	2a 3	R 1
15165.9	15170.0		1.1				6591.94	X 1	2B18	R 2
15169.6	15173.7		0.2				6590.33	3B 2	2K 4	R 2
15178.2	15182.3		0.3				6586.60	3E 0	2C 3	Q 5
15178.2	15182.3		0.3				6586.60	U 1	2B17	R 4
15182.3	15186.4		0.2				6584.82			
15188.2	15192.4		0.3				6582.26	2K 4	2C 0	P 3
15188.2	15192.4		0.3				6582.26	3D 1	2B16	P 2
15188.2	15192.4		0.3				6582.26	3E 0	2C 3	R 5
15201.3	15205.5		0.4				6576.59	V 1	2B14	P 4
15201.3	15205.5		0.4				6576.59	X 1	2B18	R 0
15207.5	15211.7		3.8				6573.91	2A 0	2B 2	P 6
15207.5	15211.7		3.8				6573.91	3E 0	2C 3	Q 4
15211.1	15215.3		0.4				6572.35	3A 0	2B15	R 1
15211.1	15215.3		0.4				6572.35	X 1	2B18	R 1
15233.7	15237.9		0.6				6562.60	3E 0	2B14	Q 4
15248.2	15252.4		0.5				6556.36	2B17	2A 1	P 2
15254.1	15258.3		0.5				6553.82			
15256.9	15261.1		0.2				6552.62			
15259.9	15264.1		0.2				6551.33			
15264.5	15268.7		0.4				6549.36	3E 0	2C 3	R 4
15264.5	15268.7		0.4				6549.36	3b 1	2a 3	R 0
15266.7	15270.9		0.3				6548.41			
15270.3	15274.5		0.7				6546.87	3E 0	2C 3	Q 3
15288.8	15293.0		0.2				6538.95	V 1	2C 3	Q 3
15293.3	15297.5		0.2				6537.02			
15296.4	15300.6		0.2				6535.70	3E 2	2B19	R 1
15303.6	15307.8		0.2				6532.63	3C 0	2K 5	Q 4
15303.6	15307.8		0.2				6532.63	3D 1	2B16	P 3
15310.7	15314.9		0.1				6529.60	4d 1	3b 3	P 6
15317.5	15321.7		0.3				6526.70	2A 4	2B 9	P 2
15327.5	15331.7		0.4				6522.44			
15330.5	15334.7		0.3				6521.16	3D 2	2B19	R 5
15333.8	15338.0		0.3				6519.76			
15338.5	15342.7		0.2				6517.76			
15343.0	15347.2		1.5				6515.85	2A 0	2B 2	R 1
15355.1	15359.3		0.6				6510.72	3E 0	2C 3	Q 2
15357.4	15361.6		0.4				6509.74			
15360.9	15365.1		0.7				6508.26	V 1	2C 3	Q 2
15363.7	15367.9		0.5				6507.07			
15369.0	15373.2		0.3				6504.83	3B 2	2K 4	R 1
15371.9	15376.1		0.2				6503.60			
15376.1	15380.3		0.2				6501.82			
15380.0	15384.2		0.6				6500.17	3D 1	2B16	P 4
15384.6	15388.8		1.0				6498.23	2A 2	2B 6	R 4
15387.1	15391.3		0.6				6497.18	2K 6	2C 1	Q 3
15399.7	15403.9		0.3				6491.86	3D 1	2B16	P 6
15402.2	15406.4		0.2				6490.81			
15407.0	15411.2		2.3				6488.78	2A 1	2B 4	R 2
15407.0	15411.2		2.3				6488.78	2A 4	2B 9	P 3
15417.4	15421.6		0.2				6484.41	3A 1	2B18	P 3
15417.4	15421.6		0.2				6484.41	V 1	2B14	P 3
15421.2	15425.4		0.5				6482.81			

λ_{air}	λ_{vac}	Ref.	I_1	I_2	I_3	I_4	ν	Upper	Lower	Br.
15425.8	15430.0		0.4				6480.88			
15429.9	15434.1		0.7				6479.15	3E 0	2C 3	R 1
15434.8	15439.0		0.3				6477.10	3A 0	2B15	R 0
15434.8	15439.0		0.3				6477.10	3E 0	2B14	Q 3
15436.9	15441.1		0.3				6476.21			
15440.3	15444.5		0.4				6474.79	3D 1	2B16	P 5
15446.6	15450.8		0.4				6472.15			
15451.4	15455.6		0.4				6470.14	u 0	2a 5	P 2
15455.2	15459.4		1.8				6468.55	2A 0	2B 2	P 5
15455.2	15459.4		1.8				6468.55	3f 3	3b 3	R 4
15460.6	15464.8		0.6				6466.29	3C 0	2K 5	R 2
15462.3	15466.5		0.8				6465.58	3b 2	2a 4	R 5
15463.8	15468.0		0.7				6464.95			
15468.8	15473.0		0.6				6462.86	3E 2	2B19	R 0
15471.2	15475.4		0.4				6461.86	3A 0	2B15	P 3
15478.4	15482.6		0.6				6458.85	2A 4	2B 9	P 4
15478.4	15482.6		0.6				6458.85	V 1	2C 3	Q 1
15482.7	15486.9		0.5				6457.06	3b 2	2a 4	R 4
15486.4	15490.6		0.7				6455.51	3D 2	2B19	R 4
15486.4	15490.6		0.7				6455.51	U 1	2B17	R 3
15504.0	15508.2		0.5				6448.19			
15515.3	15519.5		0.5				6443.49			
15518.3	15522.5		0.7				6442.24	3b 1	2a 3	P 1
15518.3	15522.5		0.7				6442.24	3B 2	2K 4	P 5
15521.3	15525.5		1.3				6441.00	3b 2	2a 4	R 3
15530.1	15534.3		2.1				6437.35	2A 1	2B 4	P 6
15530.1	15534.3		2.1				6437.35	3B 2	2K 4	R 0
15561.7	15566.0		2.6				6424.28	2A 0	2B 2	R 0
15561.7	15566.0		2.6				6424.28	3E 0	2B14	Q 2
15574.9	15579.2		0.2				6418.83	2K 4	2C 0	P 4
15574.9	15579.2		0.2				6418.83	2B17	2A 1	P 3
15577.6	15581.9		0.5				6417.72	3D 2	2B19	R 3
15577.6	15581.9		0.5				6417.72	3b 2	2a 4	R 2
15579.6	15583.9		0.6				6416.90	u 0	2a 5	P 3
15585.3	15589.6		0.2				6414.55			
15588.2	15592.5		0.5				6413.36			
15601.2	15605.5		0.4				6408.01			
15609.8	15614.1		0.5				6404.48	2B15	2A 0	P 6
15618.9	15623.2		0.4				6400.75	3B 2	2K 4	P 4
15618.9	15623.2		0.4				6400.75	V 1	2B14	P 2
15622.8	15627.1		0.4				6399.15			
15626.3	15630.6		0.5				6397.72			
15628.8	15633.1		0.5				6396.70			
15632.4	15636.7		0.4				6395.22	3A 0	2B15	P 1
15640.0	15644.3		0.6				6392.12			
15644.2	15648.5		6.1				6390.40	2A 0	2B 2	P 4
15647.9	15652.2		0.6				6388.89	2A 2	2B 6	R 3
15656.2	15660.5		0.9				6385.50	3b 2	2a 4	R 1
15656.2	15660.5		0.9				6385.50	3C 0	2K 5	Q 3
15656.2	15660.5		0.9				6385.50	3D 2	2B19	R 2
15662.6	15666.9		0.5				6382.89			
15670.5	15674.8		0.3				6379.67			
15673.1	15677.4		0.4				6378.62	2A 0	2B 3	R 7
15673.1	15677.4		0.4				6378.62	3b 1	2a 3	P 2
15678.4	15682.7		0.5				6376.46	2A 1	2B 4	R 1
15678.4	15682.7		0.5				6376.46	3f 3	3b 3	R 3
15681.4	15685.7		0.9				6375.24			
15686.2	15690.5		0.3				6373.29	3B 2	2K 4	P 3
15688.0	15692.3		0.4				6372.56			
15700.3	15704.6		0.6				6367.57	3B 2	2K 4	P 1

λ_{air}	λ_{vac}	Ref.	I_1	I_2	I_3	I_4	ν	Upper	Lower	Br.
15700.3	15704.6		0.6				6367.57	U 1	2B17	R 2
15702.6	15706.9		0.7				6366.63	4f 2	3b 5	P 3
15705.5	15709.8		0.5				6365.46			
15707.6	15711.9		0.5				6364.61			
15711.7	15716.0		0.9				6362.95	3B 2	2K 4	P 2
15718.8	15723.1		0.4				6360.07			
15721.9	15726.2		0.4				6358.82	3A 1	2B18	R 1
15724.4	15728.7		0.3				6357.81	3d 1	3b 0	P 7
15736.5	15740.8		0.7				6352.92	3D 2	2B19	P 1
15749.8	15754.1		0.2				6347.55	3b 2	2a 4	R 0
15762.0	15766.3		0.4				6342.64			
15767.2	15771.5		2.4				6340.55	2A 0	2B 2	P 3
15773.0	15777.3		0.5				6338.22	2A 1	2B 4	P 5
15773.0	15777.3		0.5				6338.22	U 1	2B17	R 1
15787.1	15791.4		0.4				6332.56	3D 2	2B19	R 0
15804.8	15809.1		1.2				6325.46	2A 0	2B 2	P 1
15804.8	15809.1		1.2				6325.46	3C 0	2K 5	R 1
15812.5	15816.8		0.5				6322.38	2A 3	2C 0	R 1
15812.5	15816.8		0.5				6322.38	U 1	2B17	R 0
15822.0	15826.3		5.3				6318.59	3C 0	2K 5	P 4
15822.0	15826.3		5.3				6318.59	2A 0	2B 2	P 2
15826.9	15831.2		0.2				6316.63			
15828.8	15833.1		0.4				6315.87			
15831.8	15836.1		0.3				6314.68	2A 2	2B 6	R 2
15834.1	15838.4		0.7				6313.76			
15838.4	15842.7		0.5				6312.04	3A 0	2B15	P 2
15845.0	15849.3		0.3				6309.42			
15848.8	15853.1		0.9				6307.90	3b 1	2a 3	P 3
15851.1	15855.4		0.3				6306.99			
15854.3	15858.6		0.4				6305.71	2K 4	2B 8	R 4
15862.5	15866.8		0.2				6302.45			
15864.6	15868.9		0.3				6301.62	3E 0	2C 3	P 2
15875.2	15879.5		0.3				6297.41	Z 1	2C 4	Q 1
15879.0	15883.3		0.1				6295.91	3A 1	2B18	R 0
15879.0	15883.3		0.1				6295.91	2K 6	2B10	R 2
15879.0	15883.3		0.1				6295.91	3f 3	3b 3	R 2
15890.5	15894.8		0.1				6291.35	2A 1	2B 4	R 0
15893.4	15897.7		1.3				6290.20			
15933.2	15937.6		0.2				6274.49	3A 1	2B18	P 3
15933.2	15937.6		0.2				6274.49	3D 2	2B19	P 2
15933.2	15937.6		0.2				6274.49	2K 6	2B10	R 1
15939.4	15943.8		0.4				6272.05	3C 0	2K 5	Q 2
15947.2	15951.6		0.1				6268.98			
15950.8	15955.2		0.2				6267.57	2K 2	2B 6	R 0
15953.9	15958.3		0.1				6266.35			
15958.5	15962.9		3.5				6264.54	2A 1	2B 4	P 4
15958.5	15962.9		3.5				6264.54	2A 3	2C 0	Q 1
15966.5	15970.9		0.2				6261.40			
15969.5	15973.9		0.7				6260.23	3D 2	2B19	P 3
15979.7	15984.1		0.2				6256.23	2K 6	2B10	R 0
15984.8	15989.2		0.2				6254.23	3D 2	2B19	P 5
15987.7	15992.1		0.5				6253.10	3D 2	2B19	P 6
15995.1	15999.5		0.2				6250.21			
15998.5	16002.9		0.1				6248.88	2K 4	2B 8	R 2
15998.5	16002.9		0.1				6248.88	2B17	2A 1	P 4
16002.5	16006.9		0.2				6247.32			
16004.7	16009.1		0.3				6246.46			
16007.0	16011.4		0.9				6245.56	3b 2	2a 4	P 1
16007.0	16011.4		0.9				6245.56	3D 2	2B19	P 4
16007.0	16011.4		0.9				6245.56	Z 1	2C 4	Q 2

λ_{air}	λ_{vac}	Ref.	I_1	I_2	I_3	I_4	ν	Upper	Lower	Br.
16028.9	16033.3		0.4				6237.03			
16033.4	16037.8		0.2				6235.28	3A 1	2B18	P 2
16039.8	16044.2		1.2				6232.79	3b 1	2a 3	P 4
16039.8	16044.2		1.2				6232.79	2A 3	2C 0	P 1
16039.8	16044.2		1.2				6232.79	3b 3	2a 5	R 4
16042.4	16046.8		0.4				6231.78	2K 4	2C 0	P 5
16046.4	16050.8		0.3				6230.23			
16052.7	16057.1		0.3				6227.78	2K 4	2B 8	R 1
16055.3	16059.7		0.2				6226.77	3d 1	3b 0	P 8
16057.7	16062.1		0.2				6225.84			
16059.1	16063.5		0.3				6225.30			
16067.3	16071.7		0.2				6222.12	3C 0	2K 5	R 0
16067.3	16071.7		0.2				6222.12	3E 0	2C 3	P 3
16070.3	16074.7		0.3				6220.96	3C 0	2K 5	P 3
16070.3	16074.7		0.3				6220.96	3f 3	3b 3	R 1
16075.5	16079.9		0.7				6218.95	3b 3	2a 5	R 3
16078.4	16082.8		0.2				6217.83	2K 2	2B 6	P 1
16080.2	16084.6		0.1				6217.13	3A 1	2B18	P 1
16080.2	16084.6		0.1				6217.13	3f 2	3b 2	Q 7
16083.0	16087.4		0.2				6216.05			
16087.0	16091.4		1.3				6214.50	2A 1	2B 4	P 3
16117.7	16122.1		0.3				6202.66			
16120.0	16124.4		0.2				6201.78	2A 3	2C 0	Q 2
16120.0	16124.4		0.2				6201.78	2K 4	2B 8	R 0
16129.3	16133.7		0.9				6198.20	2A 1	2B 4	P 1
16129.3	16133.7		0.9				6198.20	2A 3	2C 0	R 3
16129.3	16133.7		0.9				6198.20	3b 3	2a 5	R 2
16131.4	16135.8		0.4				6197.40	3C 0	2K 5	Q 1
16132.7	16137.1		0.2				6196.90			
16134.9	16139.3		0.1				6196.05			
16138.0	16142.4		0.3				6194.86	2A 1	2B 4	P 2
16143.8	16148.2		0.3				6192.64			
16168.3	16172.7		0.7				6183.25	3c 2	2a 7	R 1
16177.9	16182.3		0.2				6179.58			
16184.9	16189.3		0.6				6176.91	2A 2	2B 6	P 6
16184.9	16189.3		0.6				6176.91	2K 2	2B 6	P 2
16191.2	16195.6		0.2				6174.51			
16206.5	16210.9		0.1				6168.68	3b 3	2a 5	R 1
16210.6	16215.0		0.4				6167.12	Z 1	2C 4	Q 3
16211.9	16216.3		0.2				6166.62			
16222.4	16226.8		0.2				6162.63	3B 3	2K 6	R 2
16228.2	16232.6		0.1				6160.43			
16233.3	16237.7		0.1				6158.49	3C 0	2K 5	P 2
16233.3	16237.7		0.1				6158.49	3f 2	3b 2	Q 6
16241.4	16245.8		0.4				6155.42			
16250.2	16254.6		0.3				6152.09	3b 1	2a 3	P 5
16272.3	16276.7		0.2				6143.73	3E 0	2C 3	P 4
16297.6	16302.1		0.2				6134.20	2A 0	2B 3	R 6
16299.5	16304.0		0.3				6133.48	3b 3	2a 5	R 0
16299.5	16304.0		0.3				6133.48	2K 2	2B 6	P 3
16305.8	16310.3		0.2				6131.11			
16308.1	16312.6		0.3				6130.25			
16324.0	16328.5		0.2				6124.28	2A 2	2B 6	P 5
16343.3	16347.8		0.4				6117.04	3b 2	2a 4	P 3
16347.1	16351.6		0.3				6115.62			
16348.7	16353.2		0.3				6115.02			
16350.3	16354.8		0.3				6114.43			
16366.1	16370.6		0.7				6108.52	3C 0	2A 4	R 4
16366.1	16370.6		0.7				6108.52	4e 2	3b 5	P 3
16366.1	16370.6		0.7				6108.52	2K 2	2B 6	R 2

λ_{air}	λ_{vac}	Ref.	I_1	I_2	I_3	I_4	ν	Upper	Lower	Br.
16373.0	16377.5		0.1				6105.95	2A 2	2B 6	P 4
16397.2	16401.7		0.3				6096.94	2K 2	2B 6	R 3
16399.9	16404.4		0.1				6095.93			
16406.9	16411.4		0.3				6093.33			
16409.3	16413.8		0.1				6092.44			
16425.8	16430.3		0.1				6086.32			
16432.9	16437.4		0.8				6083.69			
16438.0	16442.5		0.1				6081.80	3B 3	2K 6	R 1
16446.6	16451.1		0.1				6078.62			
16463.2	16467.7		0.1				6072.49			
16467.60	16472.10		3.13				6070.87			
16471.55	16476.05		3.57				6069.42	2A 2	2B 6	R 1
16471.55	16476.05		3.57				6069.42	3E 0	2C 3	P 5
16473.15	16477.65		3.27				6068.83	3b 1	2a 3	P 6
16475.60	16480.10		2.9				6067.92	2A 0	2B 3	P10
16476.55	16481.05		3.39				6067.57			
16479.80	16484.30		3.39				6066.38			
16484.90	16489.40		3.27				6064.50			
16490.84	16495.35		2.7				6062.31			
16499.84	16504.35		3.13				6059.01	3f 2	3b 2	R 1
16499.84	16504.35		3.13				6059.01	w 0	3b 4	P 1
16519.59	16524.10		3.0				6051.77			
16524.09	16528.60		3.3r				6050.12	3f 2	3b 2	Q 4
16529.80	16534.32		3.13				6048.03			
16531.05	16535.57		3.32				6047.57			
16532.10	16536.62		3.38				6047.18			
16535.40	16539.92		3.32				6045.98	3f 1	3b 1	Q 6
16538.00	16542.52		3.4b				6045.03	3b 2	2a 4	P 4
16538.00	16542.52		3.4b				6045.03	2K 3	2B 7	R 1
16552.20	16556.72		2.7b				6039.84	2B19	2K 2	P 3
16556.24	16560.76		3.46				6038.37	3b 3	2a 5	P 1
16560.90	16565.42		3.46				6036.67			
16563.41	16567.93		2.7				6035.76	2K 3	2B 7	R 2
16599.47	16604.00		3.1				6022.65			
16604.65	16609.19		3.36				6020.76	2B19	2K 2	P 4
16607.41	16611.95		2.7				6019.76	3B 3	2K 6	R 0
16609.46	16614.00		3.36				6019.02			
16611.42	16615.96		3.54				6018.31			
16615.09	16619.63		3.21				6016.98			
16616.05	16620.59		3.45				6016.63			
16620.35	16624.89		3.79				6015.08	2A 2	2B 6	R 0
16623.61	16628.15		3.9r				6013.90	2A 3	2C 0	P 3
16633.15	16637.69		3.25				6010.45			
16637.68	16642.22		3.25				6008.81			
16641.38	16645.93		3.24				6007.47	3b 4	2a 6	R 3
16642.99	16647.54		3.25				6006.89			
16647.61	16652.16		2.9				6005.23			
16653.91	16658.46		2.7				6002.96	3f 2	3b 2	Q 3
16654.85	16659.40		2.9				6002.62	3f 3	3b 3	P 3
16654.85	16659.40		2.9				6002.62	3f 3	3b 3	P 6
16658.74	16663.29		2.7				6001.22			
16673.67	16678.22		3.4v				5995.84	2K 3	2B 7	P 1
16682.13	16686.69		2.7v				5992.80	2A 4	2C 1	Q 1
16684.51	16689.07		3.23				5991.95	3f 3	3b 3	P 4
16688.46	16693.02		3.1				5990.53	2A 4	2C 1	R 2
16688.46	16693.02		3.1				5990.53	3f 3	3b 3	P 5
16696.08	16700.64		3.31				5987.79	X 0	2B16	P 1
16696.08	16700.64		3.31				5987.79	4d 2	3b 5	P 1
16713.98	16718.55		3.24				5981.38			
16716.63	16721.20		3.48				5980.43	3b 3	2a 5	P 2

λ_{air}	λ_{vac}	Ref.	I_1	I_2	I_3	I_4	ν	Upper	Lower	Br.
16716.63	16721.20		3.48				5980.43	3f 1	3b 1	Q 5
16720.12	16724.69		3.0				5979.18	2K 3	2B 7	R 4
16722.21	16726.78		2.9				5978.44			
16740.45	16745.02		3.0b				5971.92			
16743.38	16747.95		3.0				5970.88	X 0	2B16	P 2
16748.01	16752.58		3.36				5969.23	3b 4	2a 6	R 2
16748.01	16752.58		3.36				5969.23	3e 1	3b 1	R 5
16753.01	16757.59		3.62				5967.45	3b 2	2a 4	P 5
16772.92	16777.50		3.40				5960.36			
16774.66	16779.24		3.19				5959.75			
16776.94	16781.52		3.84				5958.94			
16779.65	16784.23		1.9				5957.97	X 0	2B16	P 3
16779.65	16784.23		1.9				5957.97	X 0	2B16	P 5
16782.84	16787.42		3.56				5956.84	X 0	2B16	P 4
16786.05	16790.64		3.28				5955.70	4f 1	3b 4	Q 4
16788.83	16793.42		3.0				5954.71	2K 3	2B 7	P 2
16809.01	16813.60		2.9b				5947.57			
16813.50	16818.09		2.4b				5945.98	X 0	2B16	P 6
16818.13	16822.72		3.41				5944.34	3f 0	3b 0	Q 6
16819.79	16824.38		3.27				5943.76			
16833.31	16837.91		3.3r				5938.98	2A 2	2B 6	P 1
16836.20	16840.80		3.59				5937.96	3b 4	2a 6	R 1
16843.31	16847.91		3.55				5935.45			
16847.84	16852.44		3.25				5933.86			
16850.78	16855.38		3.55				5932.82	w 0	3b 4	P 2
16852.58	16857.18		3.25				5932.19			
16853.36	16857.96		3.25				5931.92			
16864.81	16869.42		3.71				5927.86			
16873.01	16877.62		4.06				5925.01	2A 2	2B 6	P 2
16873.01	16877.62		4.06				5925.01	2A 4	2C 1	Q 2
16887.23	16891.84		3.61				5920.02	2A 2	2B 6	P 3
16892.37	16896.98		3.7v				5918.22	3f 1	3b 1	Q 4
16896.56	16901.18		3.60				5916.75	3a 1	3b 1	R 6
16898.20	16902.82		3.76				5916.17	3b 3	2a 5	P 3
16898.20	16902.82		3.76				5916.17	2K 5	2B 9	R 2
16901.04	16905.66		3.2v				5915.18			
16904.41	16909.03		2.7				5914.00			
16907.18	16911.80		3.19				5913.03	2A 3	2B 8	R 5
16907.18	16911.80		3.19				5913.03	2A 0	2B 3	R 5
16909.20	16913.82		3.19				5912.33	2B19	2K 2	P 1
16915.69	16920.31		3.0				5910.06	2B19	2K 2	P 2
16921.09	16925.71		3.0				5908.17			
16929.67	16934.29		3.11				5905.18	2A 4	2C 1	R 3
16933.55	16938.18		3.69				5903.82			
16936.87	16941.50		3.0v				5902.67	3b 4	2a 6	R 0
16940.46	16945.09		3.97				5901.41	2K 2	2B 6	P 4
16940.46	16945.09		3.97				5901.41	2K 3	2B 7	P 3
16943.53	16948.16		3.19				5900.35			
16944.49	16949.12		3.11				5900.01			
16946.62	16951.25		3.36				5899.27			
16952.69	16957.32		3.0				5897.16			
16962.96	16967.59		3.2v				5893.59	2B16	2A 1	R 1
16965.39	16970.02		2.7				5892.74			
16966.69	16971.32		3.31				5892.29	2K 5	2B 9	R 0
16969.33	16973.97		2.7				5891.37			
16974.89	16979.53		3.19				5889.44	3e 1	3b 1	R 4
16981.16	16985.80		3.04				5887.27	4c 0	2a 8	R 1
16981.16	16985.80		3.04				5887.27	4f 0	3b 3	R 3
16986.41	16991.05		3.3v				5885.45			
17004.17	17008.81		3.85				5879.31			

λ_{air}	λ_{vac}	Ref.	I_1	I_2	I_3	I_4	ν	Upper	Lower	Br.
17008.24	17012.89		3.0				5877.90	3c 1	2a 6	R 2
17014.36	17019.01		3.0				5875.78			
17027.76	17032.41		3.0				5871.16	4d 2	3b 5	P 2
17027.76	17032.41		3.0				5871.16	3f 0	3b 0	Q 5
17029.34	17033.99		3.0				5870.62			
17037.89	17042.54		3.35				5867.67	3f 1	3b 1	P 8
17042.57	17047.23		3.29				5866.06	4f 1	3b 4	Q 3
17048.15	17052.81		3.41				5864.14	3f 2	3b 2	P 6
17050.41	17055.07		3.14				5863.36	3e 2	3b 2	R 4
17051.22	17055.88		3.43				5863.08			
17073.92	17078.58		4.11				5855.29			
17079.44	17084.11		2.7				5853.39	2A 0	2B 3	P 9
17080.77	17085.44		3.1				5852.94			
17083.45	17088.12		2.7				5852.02			
17098.70	17103.37		3.25				5846.80			
17100.16	17104.83		3.34				5846.30	3b 3	2a 5	P 4
17103.93	17108.60		3.89				5845.01			
17105.49	17110.16		3.0				5844.48			
17111.44	17116.11		3.66				5842.45			
17115.52	17120.19		3.33				5841.06	2B16	2A 1	R 2
17117.64	17122.32		3.30				5840.33			
17120.16	17124.84		3.14				5839.47	3f 2	3b 2	P 5
17121.26	17125.94		3.14				5839.10			
17123.70	17128.38		3.34				5838.26	3E 0	2B15	P 5
17125.88	17130.56		2.9				5837.52			
17136.61	17141.29		3.79				5833.87			
17138.46	17143.14		3.52				5833.24			
17139.95	17144.63		2.7				5832.73			
17141.66	17146.34		2.9				5832.15	2K 2	2B 6	P 5
17149.86	17154.54		3.34				5829.36	3a 1	3b 1	R 5
17155.53	17160.22		4.16				5827.43			
17158.44	17163.13		3.88				5826.44			
17160.02	17164.71		3.22				5825.91	3f 2	3b 2	P 3
17160.02	17164.71		3.22				5825.91	3f 2	3b 2	P 4
17162.11	17166.80		3.0				5825.20			
17164.41	17169.10		2.7				5824.42			
17174.31	17179.00		2.9				5821.06			
17175.95	17180.64		3.60				5820.50			
17178.21	17182.90		3.60				5819.74	3e 0	3b 0	R 4
17184.02	17188.71		3.4r				5817.77			
17186.08	17190.77		3.54				5817.08			
17188.33	17193.02		3.14				5816.31	3D 1	2B17	P 1
17189.72	17194.42		3.24				5815.84			
17191.18	17195.88		2.9				5815.35	V 0	2C 3	Q 3
17198.62	17203.32		3.73				5812.83	3b 4	2a 6	P 1
17198.62	17203.32		3.73				5812.83	3f 1	3b 1	P 7
17209.66	17214.36		3.14				5809.10			
17218.57	17223.27		2.7				5806.10			
17220.27	17224.97		3.0				5805.53			
17221.51	17226.21		3.0				5805.11			
17233.20	17237.91		2.93				5801.17			
17234.61	17239.32		3.32				5800.69			
17235.99	17240.70		3.8v				5800.23	3f 0	3b 0	Q 4
17240.73	17245.44		2.9r				5798.63			
17244.33	17249.04		3.04				5797.42			
17249.47	17254.18		4.15				5795.70			
17256.27	17260.98		2.7				5793.41			
17259.28	17263.99		3.31				5792.40			
17261.92	17266.63		3.0				5791.52			
17263.97	17268.68		2.7				5790.83			

λ_{air}	λ_{vac}	Ref.	I_1	I_2	I_3	I_4	ν	Upper	Lower	Br.
17269.92	17274.64		3.2r				5788.83			
17273.48	17278.20		2.9				5787.64			
17275.84	17280.56		3.39				5786.85			
17279.25	17283.97		4.13				5785.71			
17281.56	17286.28		3.69				5784.93			
17284.22	17288.94		3.51				5784.04			
17286.29	17291.01		2.9				5783.35			
17287.27	17291.99		3.74				5783.02			
17288.24	17292.96		3.45				5782.70			
17289.49	17294.21		3.19				5782.28			
17292.55	17297.27		4.00				5781.26			
17294.96	17299.68		3.84				5780.45			
17296.35	17301.07		4.11				5779.99	3e 2	3b 2	R 3
17299.81	17304.54		3.57				5778.83			
17300.51	17305.24		3.19				5778.60			
17305.09	17309.82		3.96				5777.07			
17307.06	17311.79		3.11				5776.41			
17309.96	17314.69		3.47				5775.44	V 0	2C 3	Q 2
17312.12	17316.85		3.0r				5774.72			
17318.65	17323.38		2.7				5772.55			
17321.73	17326.46		3.62				5771.52			
17322.58	17327.31		3.19				5771.24			
17324.92	17329.65		3.36				5770.46			
17334.18	17338.91		3.49				5767.38	4d 2	3b 5	P 3
17336.42	17341.15		3.0				5766.63	4f 1	3b 4	Q 2
17348.38	17353.12		3.75				5762.65	3D 1	2B17	P 2
17348.38	17353.12		3.75				5762.65	3f 1	3b 1	P 6
17353.03	17357.77		3.28				5761.11	3e 3	3b 3	R 3
17356.34	17361.08		3.46				5760.01			
17361.15	17365.89		3.11				5758.41	3b 4	2a 6	P 2
17368.76	17373.50		3.36				5755.89			
17371.85	17376.59		3.1b				5754.87	3f 0	3b 0	P 7
17375.73	17380.48		2.9r				5753.58			
17385.86	17390.61		3.0				5750.23	3c 1	2a 6	R 0
17388.16	17392.91		3.11				5749.47			
17392.24	17396.99		3.36				5748.12			
17396.49	17401.24		3.30				5746.72	3C 0	2A 4	R 2
17399.17	17403.92		3.0				5745.83			
17402.33	17407.08		3.96				5744.79	2B16	2A 1	R 3
17402.33	17407.08		3.96				5744.79	3a 1	3b 1	R 4
17404.53	17409.28		3.68				5744.06			
17409.13	17413.88		3.72				5742.55	3e 0	3b 0	R 3
17413.07	17417.83		3.0				5741.24			
17415.33	17420.09		2.4				5740.50			
17417.78	17422.54		3.60				5739.69			
17427.75	17432.51		3.58				5736.41			
17429.19	17433.95		3.7r				5735.93			
17436.03	17440.79		3.18				5733.69	3f 0	3b 0	Q 3
17439.83	17444.59		3.26				5732.44			
17444.39	17449.15		3.3r				5730.94			
17447.42	17452.19		3.54				5729.94			
17457.34	17462.11		3.0				5726.68			
17459.67	17464.44		3.19				5725.92			
17461.86	17466.63		3.36				5725.20			
17463.06	17467.83		3.9r				5724.81			
17466.04	17470.81		3.3b				5723.83	4b 0	2a 7	R 1
17476.73	17481.50		3.49				5720.33			
17479.16	17483.93		3.47				5719.54	3e 1	3b 1	R 2
17480.83	17485.60		3.77				5718.99	V 0	2C 3	Q 1
17483.83	17488.61		4.06				5718.01	3c 1	2a 6	Q 2

λ_air	λ_vac	Ref.	I₁	I₂	I₃	I₄	ν	Upper	Lower	Br.
17483.83	17488.61		4.06				5718.01	3f 1	3b 1	P 5
17487.95	17492.73		3.79				5716.66	3D 1	2B17	P 3
17490.86	17495.64		3.90				5715.71			
17494.31	17499.09		3.85				5714.58			
17496.02	17500.80		3.7r				5714.02	t 1	2a 7	R 1
17496.02	17500.80		3.7r				5714.02	2A 0	2B 3	R 4
17498.82	17503.60		2.7				5713.11			
17501.38	17506.16		3.69				5712.27			
17503.06	17507.84		3.75				5711.73	3E 0	2B15	P 4
17516.60	17521.38		4.21				5707.31			
17519.83	17524.62		2.7b				5706.26	4b 2	3e 1	Q 2
17536.97	17541.76		4.31				5700.68			
17538.59	17543.38		3.81				5700.16			
17541.85	17546.64		3.51				5699.10	3b 4	2a 6	P 3
17544.71	17549.50		3.11				5698.17			
17547.97	17552.76		4.51				5697.11			
17551.14	17555.93		3.36				5696.08	4b 2	3e 1	R 1
17556.88	17561.67		3.36				5694.22	3c 0	2a 5	R 4
17559.57	17564.37		2.9				5693.34			
17561.21	17566.01		3.45				5692.81			
17564.44	17569.24		3.55				5691.77	3e 2	3b 2	R 2
17566.35	17571.15		3.24				5691.15			
17570.13	17574.93		3.81				5689.92	3E 0	2B15	Q 4
17570.13	17574.93		3.81				5689.92	3f 0	3b 0	P 6
17571.06	17575.86		4.20				5689.62	3D 1	2B17	P 6
17572.06	17576.86		4.13				5689.30			
17577.97	17582.77		3.0				5687.39			
17579.51	17584.31		3.86				5686.89			
17580.64	17585.44		4.48				5686.52	3e 3	3b 3	R 2
17585.01	17589.81		3.65				5685.11	3f 1	3b 1	P 4
17588.14	17592.94		2.9				5684.10	2A 4	2C 1	Q 4
17591.53	17596.33		3.41				5683.00	2A 3	2B 8	R 2
17593.18	17597.98		3.4r				5682.47			
17596.15	17600.96		2.83				5681.51			
17601.71	17606.52		2.3				5679.71			
17609.02	17613.83		3.1				5677.36	W 1	2B19	R 4
17616.76	17621.57		3.1				5674.86	3D 1	2B17	P 4
17622.20	17627.01		3.2v				5673.11	3f 0	3b 0	Q 2
17639.38	17644.20		2.6				5667.58			
17646.77	17651.59		3.72				5665.21			
17647.98	17652.80		3.27				5664.82	3b 5	2a 7	R 1
17650.67	17655.49		3.65				5663.96	3a 1	3b 1	R 3
17654.44	17659.26		3.27				5662.75	3e 0	3b 0	R 2
17656.17	17660.99		3.49				5662.20	2A 0	2B 3	P 8
17656.17	17660.99		3.49				5662.20	3D 1	2B17	P 5
17663.45	17668.27		2.6				5659.86			
17669.33	17674.16		2.6				5657.98			
17678.74	17683.57		3.01				5654.97	V 0	2C 3	Q 4
17680.25	17685.08		3.25				5654.48			
17692.67	17697.50		2.9				5650.52			
17694.36	17699.19		2.9				5649.98			
17695.19	17700.02		2.9				5649.71			
17695.78	17700.61		2.9				5649.52			
17703.29	17708.12		3.42				5647.13			
17708.75	17713.59		3.01				5645.38			
17718.63	17723.47		2.3				5642.24			
17722.67	17727.51		3.40				5640.95			
17732.59	17737.43		3.25				5637.80			
17735.89	17740.73		2.6				5636.75			
17744.50	17749.35		3.67				5634.01			

λ_{air}	λ_{vac}	Ref.	I_1	I_2	I_3	I_4	ν	Upper	Lower	Br.
17757.09	17761.94		2.9				5630.02			
17760.96	17765.81		3.23				5628.79			
17764.29	17769.14		3.36				5627.73	3f 0	3b 0	P 5
17770.26	17775.11		3.35				5625.84	2A 3	2B 8	R 0
17770.26	17775.11		3.35				5625.84	2A 4	2B10	R 3
17778.49	17783.35		3.0b				5623.24			
17779.72	17784.58		3.0r				5622.85			
17783.31	17788.17		3.0				5621.71			
17787.25	17792.11		2.8				5620.47	2A 4	2B10	R 2
17790.38	17795.24		2.6				5619.48			
17802.82	17807.68		3.40				5615.55			
17805.46	17810.32		3.82				5614.72	2A 1	2B 5	R 4
17807.33	17812.19		2.6				5614.13			
17810.55	17815.41		3.1r				5613.12			
17812.96	17817.82		2.9				5612.36			
17814.15	17819.02		2.9				5611.98			
17816.82	17821.69		3.40				5611.14			
17819.63	17824.50		3.16				5610.26	3C 0	2A 4	R 1
17821.34	17826.21		3.56				5609.72			
17823.90	17828.77		2.7				5608.91	3E 0	2B15	P 3
17824.88	17829.75		2.7				5608.60			
17828.02	17832.89		2.8r				5607.62	3c 0	2a 5	R 3
17831.54	17836.41		2.8				5606.51			
17837.18	17842.05		2.7				5604.74			
17843.04	17847.91		3.34				5602.90			
17847.13	17852.00		3.0				5601.61	2A 4	2B10	R 1
17847.13	17852.00		3.0				5601.61	3e 3	3b 3	R 1
17850.94	17855.82		3.13				5600.41	3e 2	3b 2	R 1
17854.86	17859.74		3.91				5599.19	3E 0	2B15	Q 3
17862.04	17866.92		3.56				5596.94			
17872.41	17877.29		2.6				5593.69			
17877.92	17882.80		3.52				5591.97			
17881.73	17886.61		3.1v				5590.77			
17886.44	17891.32		2.6				5589.30			
17890.96	17895.85		4.22				5587.89	3a 1	3b 1	R 2
17904.17	17909.06		3.2v				5583.77	2A 4	2B10	R 0
17914.99	17919.88		3.24				5580.39			
17924.13	17929.02		4.35				5577.55			
17941.00	17945.90		3.83				5572.30	3f 0	3b 0	P 4
17945.99	17950.89		3.44				5570.75			
17953.20	17958.10		3.29				5568.52			
17956.37	17961.27		3.17				5567.54	4b 0	2a 7	P 1
17964.55	17969.46		3.23				5565.00			
17970.38	17975.29		3.03				5563.19			
17972.97	17977.88		3.83				5562.39	W 1	2B19	R 3
17982.54	17987.45		2.8				5559.43			
17987.94	17992.85		2.6				5557.76			
17997.64	18002.55		2.7				5554.77			
18007.02	18011.94		3.0b				5551.87			
18012.95	18017.87		3.97				5550.05			
18022.54	18027.46		2.7				5547.09	3b 5	2a 7	P 1
18027.83	18032.75		2.94				5545.47			
18034.66	18039.58		4.17				5543.37			
18041.25	18046.18		3.37				5541.34	2A 3	2B 8	P 2
18046.33	18051.26		3.33				5539.78	2A 0	2B 3	R 3
18046.33	18051.26		3.33				5539.78	4e 1	3b 4	Q 2
18051.40	18056.33		3.23				5538.22			
18055.67	18060.60		2.7b				5536.91			
18059.00	18063.93		3.39				5535.89			
18070.97	18075.90		3.36				5532.23			

λ_{air}	λ_{vac}	Ref.	I_1	I_2	I_3	I_4	ν	Upper	Lower	Br.
18073.70	18078.64		4.43				5531.39	2A 4	2B10	P 1
18073.70	18078.64		4.43				5531.39	3f 0	3b 0	P 3
18082.56	18087.50		3.35				5528.68			
18091.28	18096.22		2.7				5526.02			
18093.39	18098.33		3.73				5525.37	3c 0	2a 5	R 2
18098.05	18102.99		3.79				5523.95			
18102.73	18107.67		3.14				5522.52	3e 3	3b 3	R 0
18105.02	18109.96		3.14				5521.82	3D 0	2B15	R 5
18110.00	18114.95		3.86				5520.30			
18111.64	18116.59		3.02				5519.80	2A 3	2B 8	P 3
18119.50	18124.45		3.96				5517.41	3a 1	3b 1	R 1
18122.39	18127.34		3.24				5516.53			
18123.35	18128.30		3.77				5516.24			
18125.15	18130.10		4.05				5515.69			
18126.76	18131.71		3.36				5515.20	4e 1	3b 4	Q 3
18128.11	18133.06		3.0				5514.79			
18130.30	18135.25		3.2b				5514.12			
18143.35	18148.30		3.75				5510.16	3E 0	2B15	Q 1
18143.35	18148.30		3.75				5510.16	3e 2	3b 2	R 0
18155.37	18160.33		2.8				5506.51			
18161.52	18166.48		2.9b				5504.64	2A 4	2B10	P 2
18163.64	18168.60		2.8				5504.00			
18165.95	18170.91		2.8				5503.30			
18174.69	18179.65		2.94				5500.66			
18176.93	18181.89		2.7				5499.98			
18180.26	18185.22		2.5				5498.97			
18186.41	18191.38		2.5b				5497.11			
18189.68	18194.65		3.09				5496.12	2A 0	2B 3	P 7
18203.80	18208.77		4.1v				5491.86			
18214.51	18219.48		3.28				5488.63	2A 3	2B 8	P 4
18226.31	18231.29		3.86				5485.08			
18243.10	18248.08		2.8v				5480.03	W 1	2B19	R 2
18254.08	18259.06		3.36				5476.73	3F 0	2C 4	R 5
18256.10	18261.09		3.7r				5476.12			
18263.75	18268.74		4.1v				5473.83			
18266.30	18271.29		3.96				5473.07			
18269.36	18274.35		3.5b				5472.15	2A 0	2B 4	R 8
18271.02	18276.01		3.58				5471.65	3e 0	3b 0	R 0
18277.97	18282.96		3.01				5469.57			
18280.09	18285.08		3.7r				5468.94			
18283.36	18288.35		4.2r				5467.96	2A 4	2B10	P 3
18286.40	18291.39		3.21				5467.05			
18287.75	18292.74		3.58				5466.65			
18290.36	18295.35		2.9				5465.87			
18293.86	18298.86		3.3v				5464.82			
18295.16	18300.16		3.32				5464.43			
18297.76	18302.76		3.58				5463.66			
18299.85	18304.85		3.32				5463.03	2A 3	2B 8	P 5
18302.26	18307.26		3.53				5462.31			
18304.21	18309.21		3.95				5461.73			
18307.91	18312.91		4.09				5460.63	4d 1	3b 4	R 0
18311.16	18316.16		3.9r				5459.66	3a 0	3b 0	R 3
18322.16	18327.16		3.31				5456.38			
18324.01	18329.01		2.9r				5455.83	2B18	2A 2	R 1
18328.28	18333.29		3.7v				5454.56			
18330.65	18335.66		3.21				5453.85	3e 3	3b 3	Q 1
18332.30	18337.31		4.21				5453.36	3a 1	3b 1	R 0
18332.30	18337.31		4.21				5453.36	4f 0	3b 3	R 1
18338.38	18343.39		3.2b				5451.56	2A 1	2B 5	R 3
18341.63	18346.64		3.0				5450.59	3c 0	2a 5	R 1

λ_{air}	λ_{vac}	Ref.	I_1	I_2	I_3	I_4	ν	Upper	Lower	Br.
18344.07	18349.08		3.0r				5449.86			
18346.93	18351.94		3.01				5449.02			
18348.62	18353.63		3.72				5448.51	3e 3	3b 3	Q 2
18353.97	18358.98		3.21				5446.93			
18361.36	18366.37		3.47				5444.73			
18368.98	18374.00		3.11				5442.47	3E 0	2B15	Q 2
18372.64	18377.66		3.33				5441.39	3e 3	3b 3	Q 3
18374.17	18379.19		3.33				5440.94			
18376.66	18381.68		3.0				5440.20			
18379.98	18385.00		3.0r				5439.22	2A 4	2B10	P 4
18383.18	18388.20		2.9				5438.27			
18386.29	18391.31		3.29				5437.35			
18394.25	18399.27		3.76				5435.00	3e 3	3b 3	Q 4
18395.58	18400.60		3.15				5434.61			
18401.27	18406.29		3.01				5432.93			
18402.71	18407.74		3.31				5432.50			
18408.67	18413.70		3.11				5430.74			
18412.65	18417.68		2.6				5429.57	2B18	2A 2	R 2
18413.97	18419.00		3.43				5429.18			
18416.12	18421.15		3.21				5428.54			
18421.88	18426.91		3.43				5426.85	3e 2	3b 2	Q 1
18425.30	18430.33		2.9				5425.84			
18427.36	18432.39		3.67				5425.23			
18429.60	18434.63		3.55				5424.57			
18431.62	18436.65		2.8				5423.98			
18440.71	18445.75		2.8				5421.30			
18446.35	18451.39		2.6				5419.65	4d 1	3b 4	R 1
18449.35	18454.39		3.3b				5418.76			
18452.43	18457.47		3.11				5417.86			
18455.01	18460.05		2.8				5417.10			
18458.94	18463.98		2.9				5415.95			
18460.31	18465.35		4.01				5415.55	3e 2	3b 2	Q 2
18463.97	18469.01		3.2b				5414.48			
18468.24	18473.28		3.7r				5413.22			
18472.34	18477.38		3.8v				5412.02			
18476.99	18482.04		3.35				5410.66	3D 0	2B15	R 4
18481.52	18486.57		3.52				5409.33			
18488.87	18493.92		2.8				5407.18			
18492.75	18497.80		2.3				5406.05			
18496.05	18501.10		2.6v				5405.08			
18498.19	18503.24		3.12				5404.46			
18503.41	18508.46		3.40				5402.93	3c 0	2a 5	Q 4
18508.09	18513.14		3.58				5401.57	3e 2	3b 2	Q 3
18513.82	18518.88		3.68				5399.89			
18514.95	18520.01		3.53				5399.57			
18517.92	18522.98		3.61				5398.70			
18525.80	18530.86		3.2v				5396.40			
18528.32	18533.38		3.26				5395.67			
18529.33	18534.39		3.13				5395.38	2B18	2A 2	P 1
18532.59	18537.65		3.56				5394.43	2A 2	2B 7	R 4
18532.59	18537.65		3.56				5394.43	W 1	2B19	R 1
18534.58	18539.64		3.66				5393.85			
18539.58	18544.64		3.40				5392.39			
18542.24	18547.30		3.9r				5391.62	2A 0	2B 3	R 2
18544.90	18549.96		3.20				5390.85	4d 1	3b 4	R 2
18544.90	18549.96		3.20				5390.85	3e 3	3b 3	P 2
18547.51	18552.57		3.01				5390.09			
18549.17	18554.23		3.69				5389.61	2A 4	2B10	P 5
18552.19	18557.26		3.36				5388.73			
18553.17	18558.24		3.16				5388.44			

λ_{air}	λ_{vac}	Ref.	I_1	I_2	I_3	I_4	ν	Upper	Lower	Br.
18554.40	18559.47		2.6				5388.08			
18557.90	18562.97		3.96				5387.07	3e 2	3b 2	Q 4
18564.48	18569.55		2.8				5385.16	3e 1	3b 1	Q 1
18566.32	18571.39		4.0r				5384.63			
18569.10	18574.17		3.56				5383.82	3e 3	3b 3	P 3
18571.48	18576.55		3.61				5383.13	3c 0	2a 5	R 0
18571.48	18576.55		3.61				5383.13	3e 3	3b 3	P 4
18573.72	18578.79		4.47				5382.48	3a 0	3b 0	R 2
18576.75	18581.82		3.44				5381.60			
18578.19	18583.26		3.21				5381.19			
18581.99	18587.06		3.50				5380.09			
18583.28	18588.35		3.15				5379.71			
18588.57	18593.65		3.42				5378.18			
18592.47	18597.55		3.4b				5377.05	3e 3	3b 3	P 6
18595.01	18600.09		3.98				5376.32	3e 1	3b 1	P 3
18598.36	18603.44		2.9				5375.35	3e 3	3b 3	P 5
18600.91	18605.99		4.60				5374.61	3e 1	3b 1	P 2
18600.91	18605.99		4.60				5374.61	3e 1	3b 1	P 4
18603.82	18608.90		3.5r				5373.77	3e 2	3b 2	Q 5
18613.50	18618.58		2.9				5370.98			
18614.11	18619.19		3.24				5370.80			
18615.28	18620.36		3.4b				5370.47			
18617.79	18622.87		3.56				5369.74	3e 2	3b 2	P 2
18618.69	18623.77		3.68				5369.48			
18624.07	18629.16		3.89				5367.93	3e 2	3b 2	P 4
18625.63	18630.72		3.87				5367.48	3e 1	3b 1	P 5
18626.47	18631.56		3.65				5367.24	3e 2	3b 2	P 3
18626.47	18631.56		3.65				5367.24	3e 2	3b 2	P 5
18629.35	18634.44		3.0				5366.41	W 1	2B19	R 0
18633.76	18638.85		3.42				5365.14	3e 2	3b 2	P 6
18635.62	18640.71		4.14				5364.60	3e 1	3b 1	Q 2
18637.72	18642.81		3.1b				5364.00			
18641.48	18646.57		3.7r				5362.92	3e 2	3b 2	Q 6
18641.48	18646.57		3.7r				5362.92	3e 2	3b 2	P 7
18653.47	18658.56		3.35				5359.47			
18657.26	18662.35		3.96				5358.38	3e 1	3b 1	P 6
18660.46	18665.56		3.4r				5357.46			
18665.50	18670.60		3.85				5356.01	2A 0	2B 3	P 6
18667.85	18672.95		3.09				5355.34	3b 6	2a 8	R 1
18669.27	18674.37		2.6				5354.93	3e 2	3b 2	Q 7
18671.01	18676.11		2.8				5354.43			
18681.20	18686.30		3.42				5351.51			
18685.95	18691.05		3.77				5350.15	3e 1	3b 1	P 7
18692.12	18697.22		2.9b				5348.39			
18693.47	18698.57		3.51				5348.00	3c 0	2a 5	Q 2
18697.67	18702.78		3.29				5346.80			
18701.54	18706.65		3.72				5345.69	3a 1	3b 1	P 1
18704.65	18709.76		3.68				5344.80	2B18	2A 2	P 2
18708.10	18713.21		3.76				5343.82	4f 0	3b 3	Q 3
18716.97	18722.08		3.67				5341.29	3e 1	3b 1	Q 3
18723.74	18728.85		2.9v				5339.36			
18726.33	18731.44		2.8				5338.62			
18728.49	18733.60		2.9				5338.00	3F 0	2C 4	R 4
18730.42	18735.53		2.6				5337.45			
18751.15	18756.27		5.04				5331.55	3e 0	3b 0	Q 1
18751.15	18756.27		5.04				5331.55	3c 0	2a 5	Q 1
18798.88	18804.01		4.08				5318.01	3e 1	3b 1	Q 4
18801.03	18806.16		3.73				5317.41			
18807.73	18812.87		3.31				5315.51			
18809.60	18814.74		3.96				5314.98	3a 0	3b 0	R 1

λ_{air}	λ_{vac}	Ref.	I_1	I_2	I_3	I_4	ν	Upper	Lower	Br.
18817.18	18822.32		4.26				5312.84	3e 0	3b 0	P 2
18817.18	18822.32		4.26				5312.84	2A 1	2B 5	R 2
18818.74	18823.88		3.69				5312.40	3e 0	3b 0	P 3
18827.87	18833.01		3.2v				5309.83			
18853.20	18858.35		3.0b				5302.69			
18856.97	18862.12		3.12				5301.63			
18862.74	18867.89		4.16				5300.01	3e 0	3b 0	Q 2
18865.77	18870.92		3.4r				5299.16			
18874.86	18880.01		4.1v				5296.61	3e 1	3b 1	Q 5
18874.86	18880.01		4.1v				5296.61	3e 0	3b 0	P 4
18884.83	18889.99		3.1v				5293.81	3F 1	2B19	R 3
18887.15	18892.31		3.1r				5293.16			
18891.76	18896.92		2.8				5291.87			
18894.42	18899.58		2.8				5291.12			
18895.24	18900.40		2.3				5290.89			
18902.45	18907.61		2.8				5288.88			
18906.90	18912.06		2.3				5287.63			
18912.84	18918.00		2.8				5285.97			
18919.54	18924.71		2.6				5284.10			
18920.81	18925.98		2.6				5283.74			
18922.23	18927.40		2.6				5283.35			
18924.66	18929.83		3.66				5282.67	2A 1	2B 5	P 6
18930.57	18935.74		2.9				5281.02	2A 2	2B 7	R 3
18937.68	18942.85		3.80				5279.04	3e 1	3b 1	Q 6
18948.86	18954.03		3.46				5275.92	3e 0	3b 0	P 5
18957.05	18962.23		2.8				5273.64			
18967.46	18972.64		3.4r				5270.75	2A 0	2B 3	R 1
18984.73	18989.91		4.53				5265.95	3D 0	2B15	R 0
18984.73	18989.91		4.53				5265.95	3e 0	3b 0	Q 3
18987.32	18992.50		3.12				5265.24	3D 0	2B15	R 1
19006.39	19011.58		3.2r				5259.95			
19013.53	19018.72		3.06				5257.98			
19019.23	19024.42		3.78				5256.40	3e 0	3b 0	P 6
19021.61	19026.80		3.37				5255.74			
19023.40	19028.59		2.8				5255.25			
19025.31	19030.50		3.0				5254.72			
19027.61	19032.81		2.8				5254.08			
19033.42	19038.62		2.4				5252.48			
19034.29	19039.49		2.6v				5252.24			
19039.82	19045.02		2.8				5250.72	3a 0	3b 0	R 0
19041.38	19046.58		3.25				5250.29	3D 0	2B15	P 1
19059.06	19064.26		3.0				5245.42			
19067.23	19072.44		3.48				5243.17	2A 0	2B 3	P 5
19069.15	19074.36		3.29				5242.64	3a 1	3b 1	P 4
19072.04	19077.25		2.9				5241.85			
19077.09	19082.30		3.33				5240.46			
19086.82	19092.03		2.3				5237.79	4f 0	3b 3	Q 2
19098.21	19103.42		4.11				5234.66	3e 0	3b 0	Q 4
19110.26	19115.48		3.07				5231.36			
19119.35	19124.57		3.52				5228.88			
19136.09	19141.31		2.8				5224.30	3a 1	3b 1	P 5
19144.20	19149.43		3.0r				5222.09			
19152.26	19157.49		3.1r				5219.89			
19165.56	19170.79		2.7b				5216.27			
19174.99	19180.23		3.17				5213.70	3a 1	3b 1	P 6
19179.50	19184.74		3.07				5212.48			
19181.12	19186.36		3.37				5212.04			
19186.44	19191.68		3.11				5210.59	3a 1	3b 1	P 7
19200.83	19206.07		3.59				5206.69	3e 0	3b 0	Q 5
19207.45	19212.69		3.25				5204.89	2B18	2A 2	R 3

λ_{air}	λ_{vac}	Ref.	I_1	I_2	I_3	I_4	ν	Upper	Lower	Br.
19213.68	19218.93		2.4				5203.20	3D 0	2B15	P 6
19215.85	19221.10		3.66				5202.62	2A 2	2B 7	R 2
19234.70	19239.95		3.87				5197.52	2A 1	2B 5	R 1
19234.70	19239.95		3.87				5197.52	3c 0	2a 5	P 3
19239.19	19244.44		2.8				5196.31			
19284.08	19289.35		3.15				5184.21			
19287.66	19292.93		3.90				5183.25	3e 0	3b 0	Q 6
19302.35	19307.62		2.9				5179.30			
19306.45	19311.72		3.68				5178.20	2A 0	2B 3	R 0
19312.14	19317.41		3.2v				5176.68	2A 1	2B 5	P 5
19315.81	19321.08		2.9v				5175.69			
19321.37	19326.65		2.7				5174.20			
19324.00	19329.28		3.18				5173.50	3c 0	2a 5	P 4
19326.47	19331.75		3.49				5172.84			
19344.35	19349.63		3.02				5168.06	2K 2	2B 7	R 1
19350.74	19356.02		2.9				5166.35			
19354.86	19360.14		3.30				5165.25	X 0	2B17	P 1
19356.88	19362.16		3.30				5164.71			
19358.15	19363.44		3.30				5164.37			
19363.32	19368.61		2.4				5162.99	3e 0	3b 0	Q 7
19364.57	19369.86		3.02				5162.66			
19367.31	19372.60		3.08				5161.93			
19374.30	19379.59		2.4				5160.07			
19380.02	19385.31		4.05				5158.55	2A 0	2B 3	P 4
19381.99	19387.28		2.9				5158.02			
19386.22	19391.51		2.7				5156.90			
19391.53	19396.82		2.7				5155.48			
19393.97	19399.27		2.7				5154.83	3D 0	2B15	P 3
19397.92	19403.22		2.9r				5153.78			
19405.23	19410.53		3.31				5151.84	2K 2	2B 7	R 0
19407.81	19413.11		3.59				5151.16	2B15	2A 1	R 1
19411.51	19416.81		3.2b				5150.18	X 0	2B17	P 2
19422.02	19427.32		3.18				5147.39			
19435.70	19441.01		2.9v				5143.77	X 0	2B17	P 5
19435.70	19441.01		2.9v				5143.77	X 0	2B17	P 6
19440.52	19445.83		2.9				5142.49			
19445.61	19450.92		3.6v				5141.14	X 0	2B17	P 3
19450.67	19455.98		3.2b				5139.81	3D 0	2B15	P 5
19453.62	19458.93		3.87				5139.03	3d 1	3b 1	R 0
19455.51	19460.82		3.38				5138.53			
19464.13	19469.44		3.62				5136.25			
19467.27	19472.58		3.82				5135.43			
19492.77	19498.09		3.31				5128.71			
19500.22	19505.54		2.9r				5126.75			
19510.25	19515.58		3.40				5124.11			
19515.98	19521.31		3.7v				5122.61	3d 2	3b 2	R 0
19530.83	19536.16		3.90				5118.71			
19540.08	19545.41		3.7r				5116.29	3d 1	3b 1	R 1
19546.72	19552.06		3.9v				5114.55			
19557.11	19562.45		2.4				5111.83			
19560.36	19565.70		3.52				5110.99	2A 1	2B 5	R 0
19561.88	19567.22		3.41				5110.59	3d 2	3b 2	R 1
19574.36	19579.70		4.27				5107.33			
19576.69	19582.03		3.75				5106.72			
19578.68	19584.03		3.17				5106.20			
19587.19	19592.54		3.11				5103.98	2K 2	2B 7	P 1
19590.87	19596.22		3.7v				5103.02	2A 0	2B 3	P 3
19593.01	19598.36		2.8				5102.47			
19613.81	19619.16		3.94				5097.06	2A 1	2B 5	P 4
19621.08	19626.44		3.00				5095.17			

λ_{air}	λ_{vac}	Ref.	I_1	I_2	I_3	I_4	ν	Upper	Lower	Br.
19624.04	19629.40		2.83				5094.40	3F 1	2B19	R 2
19635.34	19640.70		3.8r				5091.47	3d 0	3b 0	R 1
19635.34	19640.70		3.8r				5091.47	3d 1	3b 1	P 1
19640.84	19646.20		3.56				5090.04			
19642.49	19647.85		3.99				5089.62	3d 2	3b 2	R 2
19646.99	19652.35		2.96				5088.45			
19654.25	19659.62		2.96				5086.57			
19656.60	19661.97		4.3r				5085.96	3d 1	3b 1	R 2
19661.27	19666.64		3.48				5084.75			
19662.35	19667.72		3.06				5084.47			
19673.59	19678.96		3.09				5081.57			
19675.96	19681.33		3.31				5080.96	2A 0	2B 3	P 1
19678.19	19683.56		2.4				5080.38			
19691.10	19696.48		3.99				5077.05	2A 0	2B 3	P 2
19722.46	19727.84		2.7				5068.98			
19724.01	19729.39		2.7				5068.58	3F 0	2C 4	R 2
19725.43	19730.82		3.10				5068.21			
19726.26	19731.65		2.8				5068.00	2B15	2A 1	R 0
19731.86	19737.25		3.15				5066.56	3d 3	3b 3	R 1
19733.35	19738.74		3.40				5066.18	3d 2	3b 2	P 1
19733.35	19738.74		3.40				5066.18	3d 3	3b 3	R 0
19734.82	19740.21		3.64				5065.80	3d 2	3b 2	R 3
19734.82	19740.21		3.64				5065.80	2B18	2A 2	P 3
19737.02	19742.41		3.41				5065.24	2K 2	2B 7	P 2
19751.02	19756.41		2.4				5061.65	t 0	2a 6	R 0
19759.97	19765.36		2.4				5059.36			
19763.26	19768.66		2.4				5058.51	t 0	2a 6	R 3
19769.45	19774.85		3.69				5056.93	3d 3	3b 3	R 2
19772.02	19777.42		3.20				5056.27			
19773.81	19779.21		4.48				5055.81	3d 0	3b 0	R 2
19779.54	19784.94		3.81				5054.35	3d 1	3b 1	R 3
19785.44	19790.84		3.08				5052.84			
19790.96	19796.36		3.64				5051.43	3a 0	3b 0	P 3
19793.02	19798.42		2.8				5050.91			
19810.85	19816.26		3.0r				5046.36			
19817.81	19823.22		2.8b				5044.59			
19821.32	19826.73		2.8				5043.70			
19824.71	19830.12		2.92				5042.83	3d 3	3b 3	R 3
19827.00	19832.41		4.08				5042.25	3d 2	3b 2	R 4
19827.00	19832.41		4.08				5042.25	2A 1	2B 5	P 3
19838.02	19843.44		3.02				5039.45			
19839.75	19845.17		2.92				5039.01			
19843.01	19848.43		2.8				5038.18			
19863.25	19868.67		2.4				5033.05	t 0	2a 6	R 2
19866.95	19872.37		2.4				5032.11			
19867.89	19873.31		2.7				5031.87			
19884.34	19889.77		3.6v				5027.71	3d 3	3b 3	R 4
19893.25	19898.68		3.17				5025.46	2A 2	2B 7	P 5
19898.46	19903.89		4.24				5024.14	3d 1	3b 1	R 4
19898.46	19903.89		4.24				5024.14	2K 2	2B 7	P 3
19912.42	19917.86		3.6r				5020.62	3d 2	3b 2	R 5
19912.42	19917.86		3.6r				5020.62	2A 1	2B 5	P 1
19920.11	19925.55		3.94				5018.68	3d 0	3b 0	R 3
19922.73	19928.17		2.9				5018.02			
19924.93	19930.37		3.75				5017.47	2A 1	2B 5	P 2
19930.05	19935.49		4.13				5016.18	3a 0	3b 0	P 4
19939.14	19944.58		3.2b				5013.89	3d 3	3b 3	R 5
19944.48	19949.92		2.8b				5012.55			
19945.40	19950.84		2.9b				5012.32			
19954.47	19959.92		2.6				5010.04			

λ_{air}	λ_{vac}	Ref.	I_1	I_2	I_3	I_4	ν	Upper	Lower	Br.
19957.34	19962.79		3.13				5009.32	Y 1	2B17	P 2
19959.95	19965.40		3.04				5008.66			
19962.62	19968.07		2.8				5008.00			
19969.76	19975.21		2.8				5006.21			
19972.66	19978.11		2.9b				5005.48			
19980.48	19985.93		3.4b				5003.52	3d 3	3b 3	P 1
19983.18	19988.64		2.9				5002.84			
19985.49	19990.95		4.04				5002.26	3d 2	3b 2	R 6
19985.49	19990.95		4.04				5002.26	2A 2	2B 7	P 4
20004.54	20010.00		3.05				4997.50	2K 2	2B 7	R 2
20007.73	20013.19		3.61				4996.70	3d 1	3b 1	R 5
20017.20	20022.66		2.9				4994.34			
20022.03	20027.50		3.5v				4993.13	2B18	2A 2	R 5
20026.03	20031.50		2.63				4992.14	3a 0	3b 0	P 5
20029.13	20034.60		2.77				4991.36			
20032.05	20037.52		2.6b				4990.64			
20037.46	20042.93		3.38				4989.29	3d 1	3b 1	P 2
20038.11	20043.58		3.38				4989.13	3d 2	3b 2	R 7
20046.73	20052.20		2.63				4986.98	Y 1	2B17	P 3
20060.09	20065.57		4.31				4983.66	3d 0	3b 0	R 4
20066.29	20071.77		3.18				4982.12	3d 2	3b 2	P 2
20067.47	20072.95		2.8b				4981.83	V 0	2B15	P 4
20075.32	20080.80		3.73				4979.88			
20078.24	20083.72		3.4r				4979.16			
20084.02	20089.50		2.77				4977.72			
20093.39	20098.88		3.1v				4975.40			
20103.97	20109.46		3.88				4972.78	3d 1	3b 1	R 6
20111.71	20117.20		2.3				4970.87			
20113.51	20119.00		2.3				4970.43			
20120.46	20125.95		3.02				4968.71			
20122.59	20128.08		3.1r				4968.18			
20124.56	20130.05		2.3				4967.70			
20127.44	20132.93		2.3b				4966.99			
20133.19	20138.69		3.0v				4965.57			
20135.66	20141.16		2.8				4964.96	2K 3	2B 8	R 1
20143.11	20148.61		3.19				4963.12			
20146.32	20151.82		2.6				4962.33			
20154.56	20160.06		2.6				4960.30			
20154.81	20160.31		2.8				4960.24	Y 1	2B17	P 4
20156.42	20161.92		2.6				4959.85	2B18	2A 2	P 4
20165.79	20171.30		2.93				4957.54			
20168.50	20174.01		2.9				4956.87			
20169.41	20174.92		2.8				4956.65	4d 0	3b 3	R 0
20173.69	20179.20		2.3				4955.60	2A 2	2B 7	R 1
20178.12	20183.63		2.88				4954.51	3d 1	3b 1	R 7
20178.12	20183.63		2.88				4954.51	2K 3	2B 8	R 2
20184.22	20189.73		3.3v				4953.01	2a 8	3b 0	R 0
20184.22	20189.73		3.3v				4953.01	3d 1	3b 1	R 7
20187.88	20193.39		3.69				4952.12	3d 0	3b 0	R 5
20192.69	20198.20		2.43				4950.94			
20199.59	20205.10		2.43				4949.25			
20202.97	20208.49		2.3				4948.42	2A 0	2B 4	R 6
20210.80	20216.32		2.6				4946.50			
20218.40	20223.92		2.6				4944.64	3F 0	2C 4	R 1
20226.62	20232.14		3.34				4942.63	3d 1	3b 1	R 8
20232.25	20237.77		2.6				4941.26			
20234.60	20240.12		2.6b				4940.68			
20237.91	20243.43		2.3				4939.87			
20242.37	20247.90		2.3				4938.78			
20246.27	20251.80		2.57				4937.83	2K 3	2B 8	R 3

λ_{air}	λ_{vac}	Ref.	I_1	I_2	I_3	I_4	ν	Upper	Lower	Br.
20251.07	20256.60		2.57				4936.66			
20254.45	20259.98		2.6				4935.84			
20259.24	20264.77		3.16				4934.67	3d 3	3b 3	P 2
20271.17	20276.70		2.51				4931.77			
20287.96	20293.50		2.2				4927.69			
20297.26	20302.80		2.2				4925.43			
20300.48	20306.02		4.01				4924.65	3d 0	3b 0	R 6
20305.24	20310.78		3.2r				4923.49			
20308.54	20314.08		2.3				4922.69			
20313.16	20318.70		2.68				4921.57			
20315.75	20321.30		2.8				4920.95			
20320.75	20326.30		2.51				4919.73			
20329.85	20335.40		2.43				4917.53	t 0	2a 6	P 2
20331.78	20337.33		2.8				4917.07			
20336.92	20342.47		2.3				4915.82	2K 3	2B 8	P 1
20339.83	20345.38		2.3				4915.12			
20341.39	20346.94		3.08				4914.74	4d 0	3b 3	R 1
20344.15	20349.70		3.16				4914.08			
20349.04	20354.60		2.68				4912.89			
20356.90	20362.46		2.8b				4911.00			
20363.44	20369.00		2.57				4909.42	2B15	2A 1	R 3
20368.87	20374.43		2.6b				4908.11	3F 1	2B19	R 1
20380.55	20386.11		2.6				4905.30			
20386.58	20392.15		2.8				4903.85	2K 3	2B 8	R 4
20389.33	20394.90		2.57				4903.19			
20393.53	20399.10		3.31				4902.18			
20396.63	20402.20		3.29				4901.43	3d 0	3b 0	R 7
20404.03	20409.60		2.50				4899.66	2A 2	2B 7	R 0
20414.23	20419.80		2.73				4897.21			
20419.13	20424.70		2.7v				4896.03			
20432.32	20437.90		2.87				4892.87			
20434.42	20440.00		2.77				4892.37			
20438.52	20444.10		3.10				4891.39			
20461.11	20466.70		2.61				4885.99			
20466.91	20472.50		2.30				4884.60			
20475.71	20481.30		3.54				4882.50	3d 0	3b 0	R 8
20482.01	20487.60		3.23				4881.00			
20497.50	20503.10		3.17				4877.31	3d 2	3b 2	P 3
20501.10	20506.70		2.82				4876.46			
20510.20	20515.80		2.24				4874.29	V 0	2B15	P 3
20513.60	20519.20		2.98				4873.48	2K 3	2B 8	P 2
20519.60	20525.20		2.3b				4872.06			
20524.30	20529.90		2.89				4870.94			
20530.90	20536.50		2.3b				4869.38			
20537.79	20543.40		3.16				4867.74	3d 0	3b 0	R 9
20537.79	20543.40		3.16				4867.74	3d 1	3b 1	P 3
20548.09	20553.70		2.42				4865.30			
20561.89	20567.50		2.73				4862.04			
20583.98	20589.60		3.23				4856.82	3d 0	3b 0	R10
20592.18	20597.80		2.40				4854.89			
20601.38	20607.00		2.46				4852.72	3b 6	3e 0	Q 3
20606.67	20612.30		2.80				4851.47			
20611.87	20617.50		3.12				4850.25	3b 6	3e 0	Q 1
20615.57	20621.20		2.98				4849.38	3b 6	3e 0	Q 2
20619.27	20624.90		2.48				4848.51	3d 3	3b 3	P 3
20631.67	20637.30		3.1v				4845.60			
20636.57	20642.20		2.69				4844.44			
20644.86	20650.50		2.74				4842.50			
20647.26	20652.90		2.59				4841.93			
20653.36	20659.00		1.9				4840.51			

λ_{air}	λ_{vac}	Ref.	I_1	I_2	I_3	I_4	ν	Upper	Lower	Br.
20659.76	20665.40		2.57				4839.01			
20665.76	20671.40		2.44				4837.60			
20671.26	20676.90		2.38				4836.31	3b 6	3e 0	R 1
20674.26	20679.90		2.28				4835.61			
20678.26	20683.90		2.44				4834.68			
20681.35	20687.00		2.44				4833.95			
20691.35	20697.00		2.8r				4831.62			
20697.35	20703.00		2.46				4830.22			
20703.25	20708.90		2.20				4828.84			
20709.95	20715.60		2.6v				4827.28			
20715.85	20721.50		2.6v				4825.91	2K 3	2B 8	P 3
20721.94	20727.60		2.83				4824.49	2A 2	2B 7	P 1
20729.04	20734.70		2.55				4822.83			
20736.24	20741.90		3.3v				4821.16			
20740.94	20746.60		2.14				4820.07			
20744.64	20750.30		2.54				4819.21	3C 0	2K 6	P 3
20747.94	20753.60		2.19				4818.44			
20754.53	20760.20		2.04				4816.91			
20765.53	20771.20		2.69				4814.36			
20771.33	20777.00		2.90				4813.01	2A 2	2B 7	P 2
20778.43	20784.10		2.12				4811.37	2A 2	2B 7	P 3
20781.93	20787.60		2.12				4810.56			
20787.93	20793.60		2.68				4809.17			
20794.72	20800.40		2.10				4807.60			
20799.62	20805.30		2.12				4806.47	3b 6	3e 0	R 2
20807.72	20813.40		2.88				4804.60			
20815.32	20821.00		2.14				4802.84			
20831.31	20837.00		2.57				4799.16			
20834.51	20840.20		2.52				4798.42			
20840.51	20846.20		2.60				4797.04	2K 2	2B 7	P 4
20851.41	20857.10		2.36				4794.53			
20859.01	20864.70		2.16				4792.78			
20865.50	20871.20		2.42				4791.29			
20869.10	20874.80		2.86				4790.47			
20872.70	20878.40		2.47				4789.64			
20876.40	20882.10		2.6b				4788.79			
20896.00	20901.70		2.22				4784.30	2B15	2A 1	R 4
20901.39	20907.10		2.20				4783.06			
20908.29	20914.00		2.26				4781.49			
20917.19	20922.90		2.50				4779.45			
20925.19	20930.90		2.87				4777.63			
20936.29	20942.00		3.53				4775.09			
20941.88	20947.60		2.55				4773.82			
20953.98	20959.70		3.25				4771.06			
20960.58	20966.30		2.5b				4769.56			
20966.08	20971.80		2.05				4768.31			
20977.37	20983.10		2.73				4765.74	3C 0	2K 6	P 2
20977.37	20983.10		2.73				4765.74	3d 2	3b 2	P 4
20985.57	20991.30		2.65				4763.88			
20989.87	20995.60		2.25				4762.90			
21001.57	21007.30		2.99				4760.25	2K 3	2B 8	P 4
21007.77	21013.50		2.2b				4758.85			
21011.96	21017.70		2.11				4757.89			
21016.46	21022.20		2.17				4756.88			
21025.06	21030.80		2.15				4754.93			
21029.86	21035.60		2.83				4753.85	3d 3	3b 3	P 4
21036.96	21042.70		2.37				4752.24			
21043.56	21049.30		2.15				4750.75	2K 2	2C 0	R 1
21057.45	21063.20		2.15				4747.62			
21077.55	21083.30		2.57				4743.09			

λ_{air}	λ_{vac}	Ref.	I_1	I_2	I_3	I_4	ν	Upper	Lower	Br.
21083.05	21088.80		2.45				4741.85			
21088.14	21093.90		2.75				4740.71	3d 1	3b 1	P 4
21097.64	21103.40		2.5r				4738.57			
21104.84	21110.60		2.15				4736.96			
21111.74	21117.50		2.40				4735.41	3F 0	2C 4	P 4
21120.53	21126.30		2.15				4733.44			
21127.63	21133.40		2.25				4731.85	3f 2	3b 3	R 5
21139.83	21145.60		2.82				4729.12			
21174.12	21179.90		2.05				4721.46			
21182.92	21188.70		2.9r				4719.50	2A 0	2B 4	R 5
21191.82	21197.60		2.57				4717.52			
21196.41	21202.20		3.03				4716.49			
21223.31	21229.10		2.66				4710.52			
21226.61	21232.40		2.30				4709.78			
21233.10	21238.90		2.06				4708.34			
21244.00	21249.80		2.48				4705.93	3F 0	2C 4	P 3
21267.79	21273.60		2.78				4700.66			
21295.69	21301.50		2.42				4694.51	4d 2	3b 6	P 1
21302.39	21308.20		2.20				4693.03	2A 0	2B 4	P 9
21305.68	21311.50		2.68				4692.30			
21317.98	21323.80		2.76				4689.60	3d 0	3b 0	P 4
21336.58	21342.40		2.46				4685.51			
21354.07	21359.90		2.26				4681.67	V 0	2B15	P 2
21365.97	21371.80		2.54				4679.06			
21399.06	21404.90		3.56				4671.83			
21402.96	21408.80		2.30				4670.98			
21409.86	21415.70		2.16				4669.47			
21425.35	21431.20		2.56				4666.09			
21436.65	21442.50		2.84				4663.64			
21449.15	21455.00		3.15				4660.92			
21451.94	21457.80		2.69				4660.31			
21457.14	21463.00		2.46				4659.18			
21470.74	21476.60		3.38				4656.23			
21475.84	21481.70		2.26				4655.13			
21482.94	21488.80		3.6v				4653.59			
21491.03	21496.90		2.06				4651.83	2A 3	2B 9	R 3
21495.43	21501.30		2.06				4650.88			
21499.93	21505.80		2.85				4649.91			
21513.13	21519.00		2.95				4647.06			
21523.23	21529.10		2.73				4644.88			
21528.42	21534.30		2.20				4643.75			
21533.82	21539.70		2.16				4642.59			
21543.92	21549.80		3.01				4640.41			
21547.12	21553.00		2.54				4639.73			
21555.42	21561.30		2.16				4637.94			
21563.71	21569.60		2.36				4636.15			
21575.61	21581.50		3.2r				4633.60	2A 3	2B 9	R 2
21588.21	21594.10		3.09				4630.89			
21612.30	21618.20		2.36				4625.73	3E 1	2B19	P 5
21643.39	21649.30		2.38				4619.09			
21654.99	21660.90		3.5b				4616.61			
21661.89	21667.80		2.66				4615.14			
21664.49	21670.40		2.90				4614.59	3d 1	3b 1	P 5
21670.69	21676.60		3.13				4613.27			
21684.98	21690.90		3.06				4610.23			
21698.58	21704.50		2.06				4607.34			
21707.18	21713.10		2.8v				4605.51			
21712.47	21718.40		2.87				4604.39			
21735.17	21741.10		3.06				4599.58	2B15	2A 1	R 5
21762.46	21768.40		2.32				4593.81			

λ_{air}	λ_{vac}	Ref.	I_1	I_2	I_3	I_4	ν	Upper	Lower	Br.
21768.16	21774.10		2.16				4592.61			
21779.06	21785.00		2.59				4590.31			
21792.15	21798.10		2.68				4587.56	3b 6	3f 0	R 2
21808.95	21814.90		2.54				4584.02			
21819.64	21825.60		2.39				4581.78			
21824.44	21830.40		2.4r				4580.77	4d 2	3b 6	P 2
21841.94	21847.90		2.06				4577.10			
21846.04	21852.00		2.06				4576.24	3b 1	2a 4	R 4
21863.53	21869.50		3.10				4572.58			
21878.03	21884.00		2.5b				4569.55			
21891.03	21897.00		2.62				4566.84			
21910.12	21916.10		2.68				4562.86			
21915.92	21921.90		2.36				4561.65	3d 3	3b 3	P 6
21936.32	21942.30		2.06				4557.41			
21948.61	21954.60		2.50				4554.85			
21966.70	21972.70		2.16				4551.10			
21972.90	21978.90		2.06				4549.82			
22001.10	22007.10		2.6b				4543.99	3d 2	3b 2	P 6
22020.39	22026.40		2.64				4540.01			
22025.19	22031.20		3.08				4539.02			
22033.59	22039.60		2.32				4537.29			
22038.19	22044.20		2.06				4536.34			
22047.08	22053.10		2.62				4534.51			
22060.18	22066.20		2.36				4531.82			
22061.88	22067.90		2.36				4531.47			
22066.28	22072.30		2.59				4530.57			
22071.18	22077.20		2.32				4529.56			
22076.77	22082.80		2.30				4528.41			
22085.97	22092.00		2.06				4526.53			
22092.87	22098.90		2.30				4525.11			
22108.77	22114.80		2.26				4521.86			
22116.76	22122.80		2.4r				4520.22			
22123.06	22129.10		2.36				4518.94	2A 3	2B 9	P 1
22130.56	22136.60		2.44				4517.41	2A 3	2B 9	P 5
22140.76	22146.80		2.73				4515.33			
22147.46	22153.50		2.88				4513.96	2A 0	2B 4	R 4
22147.46	22153.50		2.88				4513.96	2K 2	2C 0	R 2
22156.95	22163.00		2.38				4512.02			
22159.35	22165.40		2.42				4511.54			
22164.45	22170.50		2.06				4510.50			
22167.65	22173.70		2.06				4509.85			
22177.65	22183.70		2.26				4507.81			
22196.14	22202.20		2.06				4504.06			
22204.44	22210.50		2.36				4502.38			
22252.73	22258.80		2.9v				4492.61	2A 0	2B 4	P 8
22255.53	22261.60		3.41				4492.04	2A 3	2B 9	P 2
22264.02	22270.10		2.06				4490.33			
22268.12	22274.20		2.67				4489.50			
22276.72	22282.80		2.41				4487.77			
22317.91	22324.00		3.42				4479.48			
22321.91	22328.00		2.83				4478.68	2A 1	2B 6	R 4
22362.60	22368.70		2.95				4470.53			
22369.10	22375.20		2.33				4469.23			
22384.59	22390.70		2.52				4466.14	3E 1	2B19	P 3
22401.49	22407.60		2.26				4462.77			
22416.48	22422.60		2.01				4459.79	4d 2	3b 6	P 3
22421.28	22427.40		2.48				4458.83			
22433.88	22440.00		2.78				4456.33			
22451.97	22458.10		2.08				4452.74			
22478.17	22484.30		2.55				4447.55			

λ_{air}	λ_{vac}	Ref.	I_1	I_2	I_3	I_4	ν	Upper	Lower	Br.
22480.86	22487.00		3.31				4447.01			
22492.76	22498.90		2.86				4444.66			
22501.86	22508.00		2.33				4442.86			
22573.04	22579.20		2.43				4428.85			
22639.12	22645.30		2.72				4415.93			
22685.31	22691.50		2.52				4406.94	3D 0	2B16	P 1
22685.31	22691.50		2.52				4406.94	2K 2	2C 0	R 3
22696.91	22703.10		2.10				4404.68			
22702.20	22708.40		2.46				4403.66			
22750.39	22756.60		3.52				4394.33	3E 1	2B19	P 2
22761.59	22767.80		2.48				4392.17			
22830.87	22837.10		2.51				4378.84			
22843.57	22849.80		2.44				4376.41			
22849.26	22855.50		2.11				4375.31			
22855.86	22862.10		2.11				4374.05			
22868.16	22874.40		2.79				4371.70			
22873.96	22880.20		2.47				4370.59			
22878.46	22884.70		2.46				4369.73	3f 2	3b 3	R 2
22954.74	22961.00		2.12				4355.21	2B19	2A 3	P 1
22959.93	22966.20		2.80				4354.22			
23005.82	23012.10		2.12				4345.54			
23016.72	23023.00		2.12				4343.48	3D 0	2B16	P 6
23022.92	23029.20		2.37				4342.31			
23030.22	23036.50		2.88				4340.94			
23058.01	23064.30		2.37				4335.70			
23063.91	23070.20		2.61				4334.60	2A 0	2B 4	R 3
23083.90	23090.20		2.33				4330.84			
23093.60	23099.90		2.0				4329.02			
23120.79	23127.10		3.04				4323.93			
23155.08	23161.40		2.45				4317.53	2A 0	2B 4	P 7
23169.58	23175.90		2.86				4314.83	3D 0	2B16	P 3
23186.37	23192.70		2.31				4311.70			
23190.07	23196.90		2.43				4311.01	2A 1	2B 6	R 3
23225.26	23231.60		2.56				4304.48			
23265.05	23271.40		2.14				4297.12			
23270.55	23276.90		2.7r				4296.10			
23377.82	23384.20		2.35				4276.39			
23397.52	23403.90		2.15				4272.79	2B19	2A 3	P 3
23397.52	23403.90		2.15				4272.79	3f 2	3b 3	R 1
23470.30	23476.70		2.35				4259.54			
23492.29	23498.70		3.05				4255.55			
23497.69	23504.10		2.42				4254.58			
23511.48	23517.90		3.4r				4252.08			
23521.28	23527.70		2.18				4250.31			
23550.97	23557.40		2.56				4244.95			
23602.06	23608.50		2.40				4235.76			
23637.25	23643.70		2.81				4229.46	2B19	2A 3	P 4
23650.65	23657.10		2.29				4227.06			
23654.34	23660.80		2.55				4226.40			
23683.34	23689.80		2.53				4221.23			
23693.83	23700.30		2.75				4219.36			
23745.22	23751.70		2.75				4210.22			
23771.71	23778.20		2.56				4205.53	2A 2	2B 8	R 3
23827.80	23834.30		2.31				4195.63	3b 4	2a 7	R 1
23839.49	23846.00		2.35				4193.58			
23844.99	23851.50		2.64				4192.61			
23862.39	23868.90		3.13				4189.55			
23902.38	23908.90		3.06				4182.54	2A 0	2B 4	R 2
23914.77	23921.30		2.80				4180.37	4e 1	3b 5	R 0
23924.97	23931.50		2.90				4178.59	3d 0	3b 0	P 8

λ_{air}	λ_{vac}	Ref.	I_1	I_2	I_3	I_4	ν	Upper	Lower	Br.
23958.06	23964.60		2.37				4172.82			
23978.46	23985.00		3.11				4169.27	2A 0	2B 4	P 6
23981.16	23987.70		2.87				4168.80	2A 1	2B 6	R 2
23999.85	24006.40		2.64				4165.56			
24041.74	24048.30		2.96				4158.30	2A 1	2B 6	P 6
24082.23	24088.80		2.40				4151.31			
24159.31	24165.90		2.83				4138.06			
24228.09	24234.70		2.66				4126.31			
24257.48	24264.10		2.55				4121.32	2A 2	2B 8	R 2
24263.08	24269.70		2.43				4120.36			
24269.68	24276.30		2.43				4119.24			
24306.27	24312.90		2.53				4113.04			
24317.76	24324.40		2.43				4111.10			
24324.66	24331.30		2.71				4109.93			
24375.55	24382.20		2.42				4101.35			
24396.74	24403.40		2.41				4097.79			
24405.24	24411.90		2.85				4096.36			
24441.83	24448.50		3.40				4090.23			
24455.43	24462.10		2.64				4087.96			
24473.62	24480.30		3.06				4084.92			
24516.51	24523.20		2.51				4077.77			
24521.51	24528.20		2.71				4076.94			
24532.61	24539.30		2.75				4075.10			
24564.50	24571.20		2.60				4069.81	3b 4	2a 7	P 1
24564.50	24571.20		2.60				4069.81	3c 1	2a 7	R 1
24629.48	24636.20		2.59				4059.07	2A 0	2B 4	R 1
24637.58	24644.30		3.06				4057.73			
24662.07	24668.80		2.41				4053.70	3d 0	3b 0	P 9
24678.47	24685.20		2.63				4051.01	2A 1	2B 6	R 1
24689.46	24696.20		2.63				4049.21	2A 0	2B 4	P 5
24700.26	24707.00		2.93				4047.44			
24709.36	24716.10		2.56				4045.95	2A 1	2B 6	P 5
24709.36	24716.10		2.56				4045.95	3f 1	3b 2	P 8
24805.53	24812.30		2.63				4030.26	2C 4	2K 2	Q 3
24812.03	24818.80		2.31				4029.20			
24833.62	24840.40		2.91				4025.70			
24845.62	24852.40		3.56				4023.76	2K 2	2B 8	P 1
24869.41	24876.20		2.55				4019.91			
24894.81	24901.60		2.78				4015.81			
24898.41	24905.20		2.77				4015.23			
24913.70	24920.50		2.99				4012.76			
24932.10	24938.90		2.60				4009.80			
24942.59	24949.40		2.40				4008.11	3e 2	3b 3	R 3
24966.89	24973.70		2.25				4004.21			
24972.09	24978.90		2.19				4003.38			
24986.18	24993.00		3.04				4001.12			
25000.08	25006.90		2.33				3998.90			
25011.98	25018.80		2.38				3996.99			
25020.47	25027.30		2.18				3995.64			
25036.17	25043.00		2.89				3993.13	3f 1	3b 2	R 1
25063.76	25070.60		2.78				3988.74			
25076.56	25083.40		2.64				3986.70			
25079.86	25086.70		2.87				3986.18			
25094.05	25100.90		2.35				3983.92	2K 2	2B 8	P 2
25125.14	25132.00		2.83				3978.99			
25162.83	25169.70		2.60				3973.03			
25179.03	25185.90		3.56				3970.48			
25193.73	25200.60		2.89				3968.16			
25212.62	25219.50		2.96				3965.19	2A 0	2B 4	R 0
25223.12	25230.00		3.0v				3963.54	2A 1	2B 6	R 0

λ_{air}	λ_{vac}	Ref.	I_1	I_2	I_3	I_4	ν	Upper	Lower	Br.
25230.42	25237.30		3.0b				3962.39			
25239.91	25246.80		3.17				3960.90	2A 1	2B 6	P 4
25255.51	25262.40		3.45				3958.45	2A 0	2B 4	P 4
25270.01	25276.90		2.7r				3956.18			
25287.70	25294.60		2.6b				3953.41			
25300.70	25307.60		2.59				3951.38			
25312.89	25319.80		2.15				3949.48			
25322.29	25329.20		2.7r				3948.01	3a 1	3b 2	R 5
25336.59	25343.50		3.5b				3945.78			
25343.39	25350.30		2.25				3944.73			
25349.68	25356.60		2.51				3943.75			
25375.58	25382.50		2.15				3939.72			
25381.67	25388.60		2.35				3938.78			
25406.47	25413.40		2.34				3934.93			
25430.66	25437.60		2.79				3931.19			
25445.56	25452.50		2.82				3928.89			
25457.65	25464.60		2.90				3927.02	2A 2	2B 8	P 4
25457.65	25464.60		2.90				3927.02	2A 3	2C 1	P 1
25461.15	25468.10		2.44				3926.48			
25474.55	25481.50		3.23				3924.42			
25484.05	25491.00		2.38				3922.95	2K 5	2B11	R 4
25505.24	25512.20		2.68				3919.69			
25519.84	25526.80		2.30				3917.45			
25546.53	25553.50		3.82				3913.36			
25557.63	25564.60		2.34				3911.66	3e 2	3b 3	R 2
25575.62	25582.60		2.26				3908.91	2K 5	2B11	R 2
25585.22	25592.20		2.70				3907.44			
25608.91	25615.90		2.87				3903.83			
25623.91	25630.90		2.89				3901.54	2A 1	2B 6	P 3
25648.50	25655.50		2.81				3897.80	2A 0	2B 4	P 3
25661.20	25668.20		2.48				3895.87			
25671.40	25678.40		2.30				3894.32			
25694.29	25701.30		2.54				3890.85			
25713.28	25720.30		2.60				3887.98			
25765.87	25772.90		2.54				3880.04			
25781.67	25788.70		2.14				3877.67			
25784.76	25791.80		2.60				3877.20			
25793.96	25801.00		2.72				3875.82			
25805.36	25812.40		2.52				3874.11	2A 1	2B 6	P 1
25805.36	25812.40		2.52				3874.11	2K 5	2B11	R 0
25809.96	25817.00		3.19				3873.42	2A 1	2B 6	P 2
25825.75	25832.80		3.32				3871.05			
25837.15	25844.20		3.24				3869.34	2A 0	2B 4	P 1
25840.95	25848.00		2.40				3868.77			
25846.15	25853.20		3.24				3867.99	2A 0	2B 4	P 2
25854.55	25861.60		2.38				3866.74			
25885.44	25892.50		3.30				3862.12			
25923.93	25931.00		2.34				3856.39			
25948.02	25955.10		2.84				3852.81			
25979.11	25986.20		2.50				3848.20	3a 1	3b 2	R 4
25984.61	25991.70		2.24				3847.38			
26008.20	26015.30		2.14				3843.89			
26011.50	26018.60		2.14				3843.40			
26019.90	26027.00		2.44				3842.16			
26036.30	26043.40		2.34				3839.74			
26043.29	26050.40		2.24				3838.71			
26059.49	26066.60		2.34				3836.33	3f 1	3b 2	P 5
26068.89	26076.00		2.34				3834.94			
26081.78	26088.90		2.26				3833.05			
26089.28	26096.40		2.69				3831.95			

λ_{air}	λ_{vac}	Ref.	I_1	I_2	I_3	I_4	ν	Upper	Lower	Br.
26096.58	26103.70		2.20				3830.87			
26109.08	26116.20		2.50				3829.04			
26113.08	26120.20		2.64				3828.45			
26133.67	26140.80		2.4b				3825.44			
26235.44	26242.60		3.72				3810.60	3e 3	3b 4	Q 2
26242.34	26249.50		2.90				3809.60			
26251.64	26258.80		4.13				3808.25			
26276.43	26283.60		2.14				3804.65			
26294.93	26302.10		2.80				3801.98			
26299.42	26306.60		2.0				3801.33			
26313.62	26320.80		2.25				3799.28			
26321.22	26328.40		2.8b				3798.18			
26326.92	26334.10		3.08				3797.36			
26348.61	26355.80		3.31				3794.23			
26389.20	26396.40		2.71				3788.40	3f 1	3b 2	P 4
26441.89	26449.10		2.3b				3780.85			
26456.28	26463.50		2.66				3778.79			
26461.88	26469.10		2.32				3777.99			
26478.68	26485.90		2.73				3775.59			
26488.47	26495.70		2.33				3774.20			
26503.97	26511.20		3.13				3771.99	2C 4	2K 2	Q 2
26522.86	26530.10		2.33				3769.30			
26539.06	26546.30		2.84				3767.00			
26543.26	26550.50		2.18				3766.41			
26573.75	26581.00		2.12				3762.09			
26622.64	26629.90		2.93				3755.18	3a 1	3b 2	R 3
26650.03	26657.30		2.59				3751.32			
26665.43	26672.70		2.63				3749.15			
26703.41	26710.70		2.39				3743.82			
26751.90	26759.20		3.54				3737.03	4b 2	3e 2	Q 2
26780.09	26787.40		2.82				3733.10			
26787.59	26794.90		2.40				3732.05	4b 0	3e 0	Q 5
26790.19	26797.50		2.20				3731.69	2A 2	2B 8	P 2
26835.88	26843.20		2.31				3725.34	3a 2	3b 3	P 3
26840.38	26847.70		2.4				3724.71	2A 3	2C 1	P 3
26848.18	26855.50		2.32				3723.63	2C 4	2K 2	Q 1
26855.17	26862.50		2.52				3722.66			
26878.97	26886.30		3.10				3719.37			
26882.67	26890.00		3.03				3718.85			
26904.16	26911.50		2.34				3715.88	3a 2	3b 3	P 4
26909.86	26917.20		2.74				3715.10			
26916.46	26923.80		2.24				3714.19			
26970.44	26977.80		3.19				3706.75			
27057.32	27064.70		3.26				3694.85	4b 0	3e 0	Q 4
27064.22	27071.60		2.99				3693.91			
27107.20	27114.60		2.32				3688.05			
27128.30	27135.70		2.48				3685.18			
27189.48	27196.90		2.61				3676.89	4b 1	3e 1	Q 3
27195.08	27202.50		2.51				3676.13			
27200.18	27207.60		2.45				3675.44			
27240.77	27248.20		2.9v				3669.97	3a 1	3b 2	R 2
27270.46	27277.90		2.58				3665.97			
27293.35	27300.80		2.75				3662.90	4b 0	3e 0	R 4
27296.85	27304.30		3.49				3662.43			
27299.15	27306.60		2.89				3662.12			
27343.34	27350.80		2.90				3656.20	4b 0	3e 0	Q 3
27358.94	27366.40		3.46				3654.12			
27411.22	27418.70		2.41				3647.15			
27421.22	27428.70		3.18				3645.82	2B18	2A 3	R 1
27421.22	27428.70		3.18				3645.82	2B18	2A 3	R 2

λ_{air}	λ_{vac}	Ref.	I$_1$	I$_2$	I$_3$	I$_4$	ν	Upper	Lower	Br.
27421.22	27428.70		3.18				3645.82	4b 1	3e 1	Q 2
27442.21	27449.70		2.98				3643.03			
27450.41	27457.90		2.91				3641.94	4b 1	3e 1	R 2
27497.30	27504.80		3.64				3635.73	4b 1	3e 1	R 1
27513.29	27520.80		3.20				3633.62			
27518.09	27525.60		3.30				3632.98	4b 0	3e 0	R 3
27600.87	27608.40		2.73				3622.09			
27607.47	27615.00		2.58				3621.22			
27618.87	27626.40		3.04				3619.73	3f 0	3b 1	Q 2
27631.96	27639.50		3.38				3618.01	4b 0	3e 0	Q 2
27691.85	27699.40		2.93				3610.19	4b 0	3e 0	R 2
27730.14	27737.70		3.37				3605.20	4b 0	3e 0	R 1
27754.43	27762.00		3.37				3602.05			
27883.59	27891.20		2.68				3585.36	4b 0	3e 0	Q 1
27912.59	27920.20		2.48				3581.64			
27974.87	27982.50		2.49				3573.66	3b 5	3a 0	P 2
28125.33	28133.00		2.72				3554.54			
28178.11	28185.80		3.57				3547.89			
28273.49	28281.20		4.06				3535.92			
28290.98	28298.70		3.35				3533.73			
28333.27	28341.00		2.7b				3528.46			
28348.67	28356.40		2.79				3526.54	3a 1	3b 2	R 0
28608.90	28616.70		2.79				3494.46			
28667.68	28675.50		3.14				3487.30			